Clocking in Modern VLSI Systems

## Series on Integrated Circuits and Systems

Series Editor: Anantha Chandrakasan
Massachusetts Institute of Technology
Cambridge, Massachusetts

---

For other titles published in this series, go to
http://www.springer.com/Series/7236

Thucydides Xanthopoulos
Editor

# Clocking in Modern VLSI Systems

Springer

*Editor*
Thucydides Xanthopoulos
Cavium Networks
Marlboro, MA 01752
USA

ISSN 1558-9412
ISBN 978-1-4614-2441-3 e-ISBN 978-1-4419-0261-0
DOI 10.1007/978-1-4419-0261-0
Springer Dordrecht Heidelberg London New York



Printed on acid-free paper

Springer is part of Springer Science+Business Media (www.springer.com)

Στη Μαργαρίτα, το Νικόλα και τη Μαρία-Ελένη που μας έπιασε στα μισά

:-)

To Margarita, Nicholas and Maria-Helen who joined us half-way through this book

:-)

# Preface

... εἰωθότες οἱ ἄνθρωποι οὗ μὲν ἐπιθυμοῦσιν ἐλπίδι ἀπερισκέπτῳ διδόναι, ὃ δὲ μὴ προσίενται λογισμῷ αὐτοκράτορι διωθεῖσθαι.

THUCYDIDIS HISTORIAE IV:108
C. Hude ed., Teubner, Lipsiae MCMXIII

Ὁι ἄνθρωποι, ἄλλωστε, συνειθίζουν νὰ ἐμπιστεύωνται εἰς τὴν ἀπερίσκεπτον ἐλπίδα ἐκεῖνο ποὺ ἐπιθυμοῦν καὶ ν' ἀποκρούουν δι' αὐθαιρέτου συλλογισμοῦ ἐκεῖνο ποὺ ἀποστέργουν.

ΘΟΥΚΥΔΙΔΟΥ ΙΣΤΟΡΙΑΙ Δ:108
Κατά Μετάφρασιν Ἐλευθερίου Βενιζέλου
Δ. Κακλαμάνος Εκδ.
Σμυρνιωτάκης, Αθήνα

It being the fashion of men, what they wish to be true to admit even upon an ungrounded hope, and what they wish not, with a magistral kind of arguing to reject.

Thucydides (the Peloponnesian War Part I), IV:108
Thomas Hobbes Trans., Sir W. Molesworth ed.
In *The English Works of Thomas Hobbes of Malmesbury, Vol. VIII*

I have been introduced to clock design very early in my professional career when I was tapped right out of school to design and implement the clock generation and distribution of the Alpha 21364 microprocessor. Traditionally, Alpha processors exhibited highly innovative clocking systems, always worthy of ISSCC/JSSC publications and for a while Alpha processors were leading the industry in terms of clock performance. I had huge shoes to fill. Obviously, I was overwhelmed, confused and highly confident that I would drag the entire project down. When a few years later

Carl Harris asked me to do a book on clocking for the Springer Integrated Circuits and Systems Series, I readily agreed with the hope that I could save young and aspiring clock designers substantial time and frustration by providing leads and maybe answers to the questions that I had when I was embarking on the Alpha clock design quest. As my choice of opening quotation would suggest, clock design can be a minefield of misconceptions based on little more than a reluctance to apply Kirchhoff's laws, basic constituent relationships, and a little bit of common sense.

In addition to my personal design experience, the choice of material for this book has been heavily informed by my long tenure in the International Solid-State Circuits Conference (ISSCC) program committee. The subjects covered reflect to a large extent the collective interests and foci of both industry and academia with respect to clocking based on ISSCC submissions. The only exception is that there is no coverage of phase locked loop design since there are a number of recent texts available on this subject matter.

It is my hope that this book will help engineers and students interested in clock design obtain the appropriate mental models and design viewpoints, capture design trends that have appeared over the last few years, and provide a comprehensive list of references for further study. I am indebted to my co-authors for providing precise, structured and complete coverage in their respective chapters in addition to maintaining a viewpoint that is very up to date and highly reflective of current trends in the industry. I hope that the reader will not find "ungrounded hopes" and "magistral arguings" in this book.

Carl Harris and Katelyn Stanne of Springer deserve special thanks for helping me throughout the preparation of the manuscript. I wish to acknowledge a number of colleagues at Cavium Networks for their helpful and stimulating discussions and excellent feedback: Scott Meninger, Ethan Crain, David Lin, and Suresh Balasubramanian. I would like to thank my bosses at Cavium Networks Anil Jain and Syed Ali for building a great semiconductor company from the ground up and an excellent working environment that fosters creativity and innovation in addition to maintaining a sharp focus on product development and company value. I would especially like to thank Anil Jain for entrusting me with the Alpha clocking project while being my boss at Compaq Computer which helped me acquire the background and skills necessary to produce this book. Above all, I would like to thank my wife Margarita not only for putting up with my constant working on this book but also for typesetting the entire manuscript in LaTeX and retouching figures as needed. I could not have done this without her.

Boston, Massachusetts *T.X.*
December 2008

# Contents

# List of Figures

# List of Tables

# List of Contributors

**Georgios Konstadinidis**
Sun Microsystems

**Scott Meninger**
Cavium Networks

**Simon Tam**
Intel Corporation (SC12-408)

**James Tschanz**
Intel Corporation

**Nestoras Tzartzanis**
Fujitsu Laboratories of America

**James Warnock**
IBM T.J. Watson Research Center

**Thucydides Xanthopoulos**
Cavium Networks

# 1

# Introduction and Overview

Thucydides Xanthopoulos

Cavium Networks

Clock frequency is a major attribute of any microprocessor design. Early on, during product definition, it constitutes a major business or marketing decision and it is usually the result of a trade-off among customer needs, competitive landscape, and time-to-market. As soon as the frequency target is handed down the food chain to silicon implementation, it will affect all project design aspects from the day that the project is kicked off until it tapes out (and in most cases well beyond this point too). It is not surprising therefore that the job of generating, distributing, and analyzing the clocks in complex chips is considered to be an important and visible assignment. Clock design has traditionally been an area of innovation and has been in the spotlight in technical conferences and journals.

Why is clock frequency such an important microprocessor aspect? For a number of applications it is only loosely correlated with performance with other design aspects such as memory system, parallelism, and hardware acceleration being equally or even more effective. Nevertheless, it is a single number that is widely understood by both technical and nontechnical audiences and in certain situations has strong correlation with single-thread performance.

Clock frequency, although very important, is only one aspect of clock design. Other aspects include power dissipation, efficient clock signal distribution in large and complex chips, coping with variation and uncertainty, managing multiple clock domains in the context of highly integrated system-on-a-chip (SoC) designs, and multicore integration, providing good voltage/frequency scalability to support a wide product roadmap, tuning capabilities for yield enhancement and postsilicon optimization, and sophisticated active power management features.

The purpose of this book is to introduce a designer to important aspects of state-of-the-art clock design by exposing methodology steps and analytical modelling techniques, providing design examples and case studies and enumerating a long list of references for further study.

T. Xanthopoulos (ed.), *Clocking in Modern VLSI Systems*, Integrated Circuits and Systems,
DOI 10.1007/978-1-4419-0261-0_1, © Springer Science+Business Media, LLC 2009

## 1.1 The Clock Design Problem

The plots of Figs. 1.1–1.3 provide good insight into the problem faced by clock designers today. Moore's plot (Fig. 1.1) states that integration keeps on increasing at historical rates. Transistor density is spearheaded by large multicore server processor chips with large caches, integrated memory controllers and multiple high-speed *I/O* links. Although the datapoints in Fig. 1.1 only extend to 2008, in ISSCC 2009 a server processor [1] has been introduced that contains 2.3B transistors and exceeds all previously reported chips in transistor count. For the clock designer, increasing integration means more far reaching and complex clock distribution, more clock domains, and more testability and yield enhancement features.

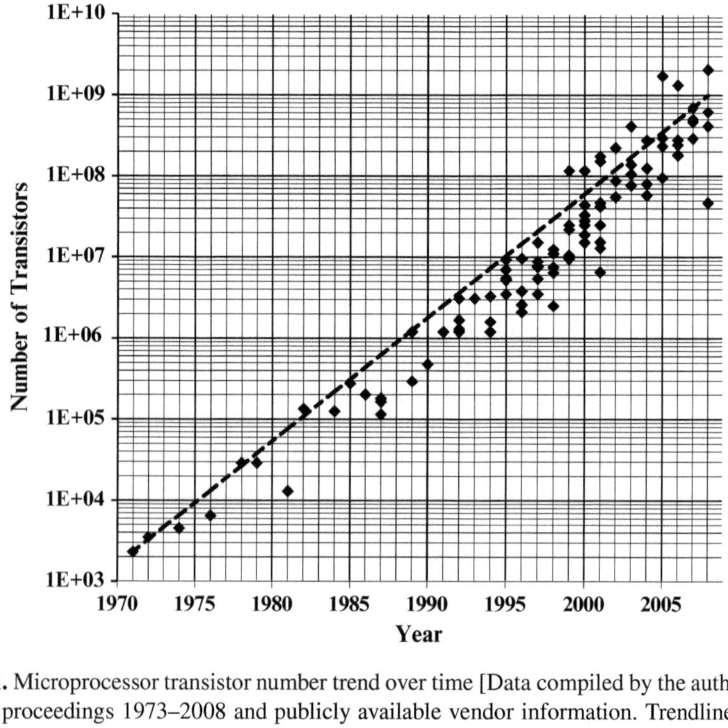


**Fig. 1.1.** Microprocessor transistor number trend over time [Data compiled by the author from ISSCC proceedings 1973–2008 and publicly available vendor information. Trendline is "visually" fitted and signifies a rate of doubling every 2 years]

On the other hand, frequency and power seem to be levelling off (Figs. 1.2 and 1.3). Microprocessors in excess of 3–4GHz are rarely reported these days. Frequency stopped scaling due to excessive power dissipation penalties. Instead, the industry chose to keep on scaling performance through multicore integration. On the power side, the industry seems to converge to a fixed power envelope for both server (130W) and desktop (60–70W) processors. As far as clock design is concerned, this

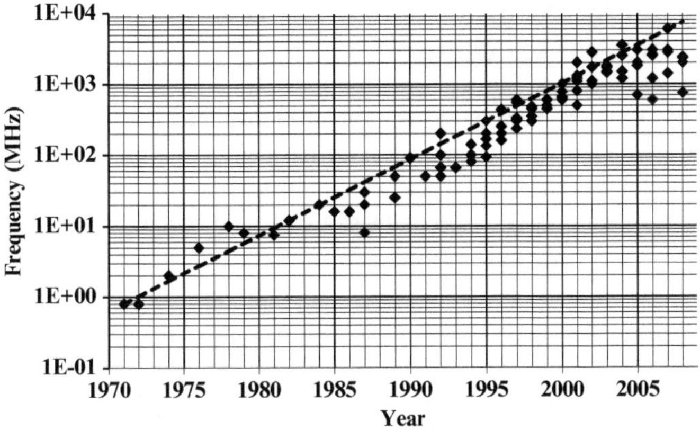


**Fig. 1.2.** Microprocessor frequency trend over time [Data compiled by the author from ISSCC proceedings 1973–2008 and publicly available vendor information. Trendline is "visually" fitted and signifies a rate of doubling every 3 years]

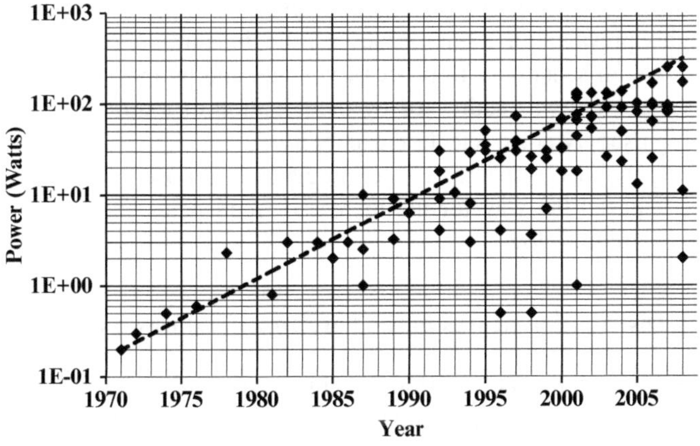


**Fig. 1.3.** Microprocessor power trend over time [Data compiled by the author from ISSCC proceedings 1973–2008 and publicly available vendor information. Trendline is "visually" fitted and signifies a rate of doubling every 3.5 years]

translates to increased number of features and complexity to support fine-grain clock and power gating, variable voltages, and frequencies in addition to other sophisticated active power management related attributes.

In addition to the above evolving specifications, one must add increased variation in device and interconnect characteristics present in advanced process nodes and increased voltage and temperature variation due to higher integration and more complex interactions. We end up with a multidimensional design problem requiring substantial resources for each product generation.

## 1.2 Some Subjective Milestones in the History of Microprocessor Clocking

In the last 15–20 years there have been major innovations in microprocessor clocking that have resulted in large increases in frequency as well as substantial clock design methodology changes. This section lists some of those important milestones so that we can observe the industry trends and speculate on potential future directions.

### 1.2.1 Integrating the PLL

An integrated PLL for a microprocessor application was first reported in 1992 [2]. The motivation for an integrated PLL at the time was the desire to keep the processor and the external bus in phase so as to minimize timing constraints and reach the maximum possible system frequency given the synchronous nature of the system. Additional goals were to clock the processor at even higher frequencies ($2x$) than the bus in addition to running the VCO at twice the processor frequency with a factor of two postscaling for duty cycle fidelity.

Some of the original problems that the designers faced were the digital noise which is highly prevalent on a microprocessor die, in addition to low quality passive devices and overall sensitivity to process, voltage, and temperature variations. The supply noise problem was addressed by the adoption of a differential CML-based VCO with high supply noise rejection capability.

Integrating the PLL was a major step in general purpose microprocessor clocking because it paved the way for the increased frequencies and complex clocking schemes that were to follow.

### 1.2.2 Clock Distribution Moves to the Forefront: The Dawn of the GHz Race

The original DEC Alpha microprocessor [3] moved clock design to the forefront. It was operating at 200MHz, approximately a factor of 2 faster than other processors at the time (Fig. 1.2). The DEC Alpha design introduced the low-skew and high-power grid clock distribution coupled with detailed RC-based skew simulation and construction of skew contour maps that are taken into account during timing closure: The Alpha pipeline was based on level-sensitive latches. Race-through in this context is a major functional concern. The radial profile of clock skew from the center of the chip to the periphery was taken into account while floorplanning the pipelines in order to guarantee that the skew would improve the functional race margins.

The design also featured important contributions from a process and physical design perspective. The process technology featured a thick low resistance metal 3 layer used exclusively for power, clock, and a handful of critical signals. On-chip decoupling capacitors built with thin oxide devices were also used in close proximity to the clock drivers in order to address $Ldi/dt$ concerns arising from driving a highly capacitive final clock node with a very fast edge rate. A 10:1 ratio of decoupling capacitance to switching capacitance was maintained throughout the design. Many of these contributions are still in use today in current processor designs.

Arguably, the DEC Alpha designs started the GHz race among microprocessor vendors that culminated in the deep pipelines and multiGHz designs of present high performance chips.

### 1.2.3 Delay Lock Techniques

Simple first-order mechanisms to achieve phase lock [4] have been used extensively in processor designs in the last 10–15 years in order to simplify the clock distribution problem: A large distribution throughout a big die is broken into pieces tailored for each chip partition, and each partial distribution is phase locked using delay locked loops. There are multiple such examples discussed in Chap. 2. Such a partitioning helps with design time, power dissipation, testability, and manufacturing yield. It is definitely possible to achieve the same goal with higher order systems (i.e. distributed PLLs) and this has been demonstrated in the literature [5, 6]. Yet, nothing beats the simplicity of a digital delay line controlled by a basic finite state machine. More details on DLL design will be presented in Chap. 6.

### 1.2.4 Exploiting Inductance for Oscillation and Distribution

The notion of being able to return energy back to the clock generator has been rather intriguing and holds a lot of promise. Resonant clock drivers have been originally introduced as an off-chip solution in the context of powering adiabatic circuits [7, 8]. In the past several years, there has been renewed interest in this technique due to the fact that digital frequencies have become consistent with resonant frequencies of fully integrated passive devices. The strong motivation is the potential of saving substantial clock power by using LC resonance for clock pulse generation. LC techniques have been recently augmented with transmission-line-based techniques (traveling waves [9], standing waves [10] and salphasic [11]) that address both clock generation and low (or highly predictable) skew distribution. Commercial applications of these techniques are already prevalent. Chapter 4 explores this subject in more detail.

### 1.2.5 Variable Frequency (and Voltage)

A server processor design [12] introduced the idea of a constant power envelope and variable voltage/frequency. This active power management scheme uses an integrated ammeter that monitors incoming current, a clocking scheme that can generate variable frequencies with fast adjustment time and an on-chip micro-controller that monitors power/ temperature and controls frequency and core supply voltage through an external regulator. This technique is discussed in more detail in Chap. 2.

Variable frequency clocking methods are rapidly becoming mainstream [13, 14] as part of sophisticated power management methodologies designed to control thermal design power in large multicore chips. All digital methods ensure repeatability in a production environment.

### 1.2.6 Frequency Increase (or Supply Lowering) Through Resiliency

Commercial designs have substantial frequency margins to address issues such as lack of total coverage in production tests, unanticipated corner cases and noise patterns, noncompliant system specifications and device aging. This margin can be substantially reduced or even eliminated if the underlying hardware has error detection and correction capabilities. The margin can then be used as performance benefit by increasing frequency, power benefit by dropping voltage below the specified $V_{\mathrm{MIN}}$, or even yield enhancement by populating existing frequency bins with parts that under different scenarios would not make it.

The Razor technique [15] addresses the frequency/voltage margin issue by instituting the capability of timing error detection and correction in the processor pipeline. Performance can be maximized (or power minimized) by increasing the frequency (or lowering the supply) up to the point where the overhead of error correction will start exceeding the performance or power benefit. The Razor technique is addressed in more detail in Chaps. 3 and 7.

It is not easy to predict the future but given the current industry trends one can conclude that the clocking system will be part of an increasingly sophisticated active power management scheme: Highly sophisticated firmware threads implementing complex control algorithms will be running in parallel with the application. They will be receiving input such as on-chip and system temperature, current measurements, error rates, moving averages of architectural events, and cues from the application and they will control supply voltage, clock frequencies, and higher level architectural events such as clock and power gating, instruction issue rate, and pipeline stalls. To some extent, this is already happening.

## 1.3 Overview of this Book

Chapter 2 introduces the fundamental setup and hold constraints, and defines basic clock attributes such as skew, jitter, latency, and duty cycle distortion. It introduces basic clock distribution methods such as balanced tree, central spines, and grids and examines them from a performance and power perspective. Numerous case studies from commercial microprocessors are presented and a number of advanced topics such as global and local skew compensation, on-die attribute measurements, various techniques for locating critical paths and synchronization methods are discussed.

Chapter 3 constitutes a detailed discussion of clocked elements (level-sensitive latches and flip-flops) from the viewpoint of latency, hold time requirements, power dissipation, and testability. The focus is primarily on state-of-the-art designs with advanced topics such as process variation and reliability addressed in detail.

As mentioned in Sect. 1.2.4, exploiting inductance for clock generation and distribution makes perfect sense: Inductance can produce oscillations with lower energy since an LC-based system inherently recycles energy between the capacitive and inductive elements. Moreover, oscillator phase noise is less because all active-device-related noise sources do not exist. Chapter 4 presents detailed background information on integrated inductors and transmission lines. Furthermore, it discusses examples of LC-based oscillators and transmission-line-based clock generation and distribution schemes.

Jitter analysis is very important in clock system design. If not properly managed, jitter can be the limiting factor in both core and *I/O* clocking. Chapter 5.1 defines all jitter types relevant to clock design and establishes their relationship to phase noise using Parceval's theorem. Furthermore, it enumerates all noise sources inside a clock generator and clearly shows with numerical examples how a PLL transfers jitter from input to output. Based on this analysis, it establishes the importance of reference clock phase noise and jitter regarding the quality of the multiplied output clock. The domain seamlessly moves from frequency to time using mathematical "filter" functions to transform phase jitter to period jitter which is more relevant for core clock generation. Since jitter has a random component, which is theoretically unbounded, the chapter establishes an MTBF-based statistical analysis for estimating the effect of jitter on critical paths. A serial link discussion is also presented, which shows how reference clock jitter can be removed from the total link budget under certain conditions.

Chapter 6 is an attempt at textbook-like coverage of digital delay locked loops that are used extensively in clocking systems. The chapter is design-oriented and contains detailed discussions and analyses on all DLL components and presents a number of relevant applications. It also contains a detailed analysis of metastability in the context of phase detection and a simple analytical model for supply-induced jitter on long delay lines and/or clock buffers.

Advanced process nodes exhibit large variation and uncertainty in device and interconnect parameters. Chapter 7 presents methods of addressing this issue on the design front, manufacturing, and postsilicon tuning front and also by using resilient methods involving hardware timing error detection and correction.

Finally, Chap. 8 addresses process, voltage, and temperature variation issues from a physical design perspective. Clock skew components in the context of setup and hold constraints are redefined with a statistical approach and all variation sources are taken into account. Sources of transistor and interconnect variation are enumerated, quantified, and explained. Methods of accounting for voltage and temperature variations are discussed. In the end, important physical design guidelines are presented to minimize uncertainty and variation in clock-related circuits.

## References


[1] S. Rusu, S. Tam, H. Muljono, J. Stinson, D. Ayers, J. Chang, R. Varada, M. Ratta, and S. Kottapalli, "A 45nm 8-core enterprise Xeon® processor," in *Digest of Technical Papers IEEE International Solid-State Circuits Conference (ISSCC 2009)*, 2009, pp. 56–57.

[2] I. A. Young, J. K. Greason, and K. L. Wong, "A PLL clock generator with 5 to 10 MHz of lock range for microprocessors," *IEEE Journal of Solid-State Circuits*, vol. 27, no. 11, pp. 1599–1607, Nov. 1992.

[3] D. Dobberpuhl, R. Witek, R. Allmon, R. Anglin, D. Bertucci, S. Britton, L. Chao, R. Conrad, D. Dever, B. Gieseke, S. Hassoun, G. Hoeppner, K. Kuchler, M. Ladd, B. Leary, L. Madden, E. McLellan, D. Meyer, J. Montanaro, D. Priore, V. Rajagopalan, S. Samudrala, and S. Santhanam, "A 200-MHz 64-b dual-issue CMOS microprocessor," *IEEE Journal of Solid-State Circuits*, vol. 27, no. 11, pp. 1555–1567, Nov. 1992.

[4] M. Johnson and E. Hudson, "A variable delay line PLL for CPU-coprocessor synchronization," *IEEE Journal of Solid-State Circuits*, vol. 23, no. 5, pp. 1218–1223, 1988.

[5] G. A. Pratt and J. Nguyen, "Distributed synchronous clocking," in *Proceedings of Sixteenth Conference on Advanced Research in VLSI*, 27–29 March 1995, pp. 316–330.

[6] V. Gutnik and A. Chandrakasan, "Active GHz clock network using distributed PLLs," *IEEE Journal of Solid-State Circuits*, vol. 35, no. 11, pp. 1553–1560, Nov. 2000.

[7] C. L. Seitz, A. H. Frey, S. Mattison, S. D. Rabin, D. A. Speck, and J. L. A. van de Snepscheut, "Hot-clock NMOS," in *Proceedings of Chapel Hill Conference VLSI*, 1985, pp. 1–17.

[8] R. Feynman, T. Hey, and R. Allen, *Feynman Lectures on Computation*. Westview Press, Boulder, CO, 2000.

[9] J. Wood, S. Lipa, P. Franzon, and M. Steer, "Multi-gigahertz low-power low-skew rotary clock scheme," in *Digest of Technical Papers IEEE International Solid-State Circuits Conference (ISSCC 2001)*, 2001, pp. 400–401, 470.

[10] F. O'Mahony, C. Yue, M. Horowitz, and S. Wong, "A 10-GHz global clock distribution using coupled standing-wave oscillators," *IEEE Journal of Solid-State Circuits*, vol. 38, no. 11, pp. 1813–1820, Nov. 2003.

[11] V. L. Chi, "Salphasic distribution of clock signals for synchronous systems," *IEEE Transactions on Computers*, vol. 43, no. 5, pp. 597–602, May 1994.

[12] S. Naffziger, B. Stackhouse, and T. Grutkowski, "The implementation of a 2-core multi-threaded Itanium family processor," in *Digest of Technical Papers IEEE International Solid-State Circuits Conference (ISSCC 2005)*, 2005, pp. 182–183, 592.

[13] R. Kumar and G. Hinton, "A family of 45nm IA processors," in *Digest of Technical Papers IEEE International Solid-State Circuits Conference (ISSCC 2009)*, 2009, pp. 58–59.

[14] A. Allen, J. Desai, F. Verdico, F. Anderson, D. Mulvihill, and D. Krueger, "Dynamic frequency-switching clock system on a quad-core Itanium® processor," in *Digest of Technical Papers IEEE International Solid-State Circuits Conference (ISSCC 2009)*, 2009, pp. 62–63.

[15] D. Ernst, N. S. Kim, S. Das, S. Pant, R. Rao, T. Pham, C. Ziesler, D. Blaauw, T. Austin, K. Flautner, and T. Mudge, "Razor: a low-power pipeline based on circuit-level timing speculation," in *Proceedings of 36th Annual IEEE/ACM International Symposium on MICRO-36 Microarchitecture*, 2003, pp. 7–18.

# 2

# Modern Clock Distribution Systems

Simon Tam

Intel Corporation (SC12-408)

## 2.1 Introduction

Modern clock distribution design continues to face challenges in spite of significant advances in the last decade. We can distinguish three primary challenges. The first is the need to support higher clock frequencies based on the strong correlation between frequency and chip performance. Figure 2.1 shows processor clock frequency trend suggesting a continuous exponential increase in clock frequency with variable rates. Second, process technology scaling allows higher level of integration and larger die size leading to higher clock loading and larger distances the clock network needs to traverse. The final challenge is that technology scaling leads to an increase in on-die variations that may degrade clock performance if not properly addressed.

In order to address these design challenges successfully, it is necessary to understand the fundamental clocking requirements, key design parameters that affect clock performance, different clock distribution topologies and their trade-offs, and design techniques needed to overcome certain limitations. In this chapter, the following topics are presented:

- Definitions and Design Requirements
- Clock Distribution Topologies
- Microprocessor Clock Distributions
- Clock Design for Test and Manufacturing
- Elements of Clock Distribution Circuits
- Clock DFX (Design-for-Test and Design-for-Manufacturing) Techniques
- Multiclock Domain Distributions
- Future Directions

T. Xanthopoulos (ed.), *Clocking in Modern VLSI Systems*, Integrated Circuits and Systems, DOI 10.1007/978-1-4419-0261-0_2, © Springer Science+Business Media, LLC 2009

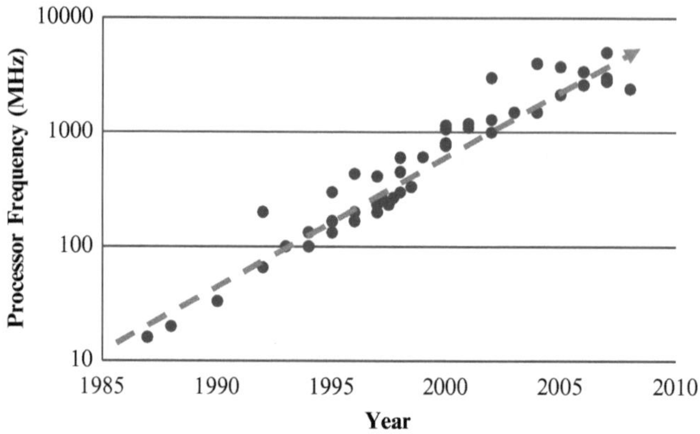


**Fig. 2.1.** Processor clock frequency trend [1–30]

## 2.2 Definitions and Design Requirements

Synchronous circuits may be simplified to have two timing limitations: setup (MAX delay) and hold (MIN delay). Setup specifies whether the digital signal from one stage of the sequential structure has sufficient time to travel to and "set-up" before being captured by the next stage of the sequential structure. Hold specifies whether the digital signal from the current state within a sequential structure is immune from contamination by a signal from a future state due to a fast path. Figure 2.2 shows a typical synchronous sequential structure bounded by two flip-flops with a logic circuit that exhibits a circuit delay of value $T_{\mathrm{d}}$. The sequential elements are clocked by a source clock Ck1 and a destination clock Ck2.

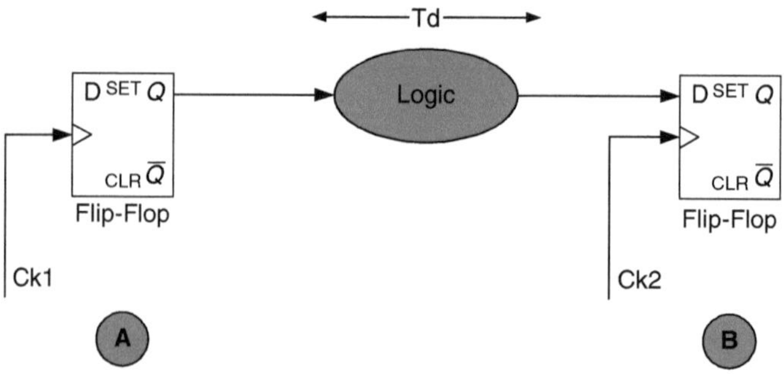


**Fig. 2.2.** Sequential structure bounded by flip-flops

Clocks Ck1 and Ck2 can be spatially far apart on die as shown in Fig.2.3. In this illustration, clocks Ck1 and Ck2 have their root at a common point (Clock Gen.) and are routed through the on-die clock distribution before arriving at their respective destinations. Locations A and B constitute the source and destination of the sequential path. The transit time (clock latency[1]) of CK1 and Ck2, their latency difference, their variations, and the design structure to minimize the above are the main topics of discussion in subsequent sections. As will be shown later, the timing uncertainty and the timing differences of Ck1 and Ck2 will play a fundamental role in determining whether the setup and the hold constraints can be robustly met.

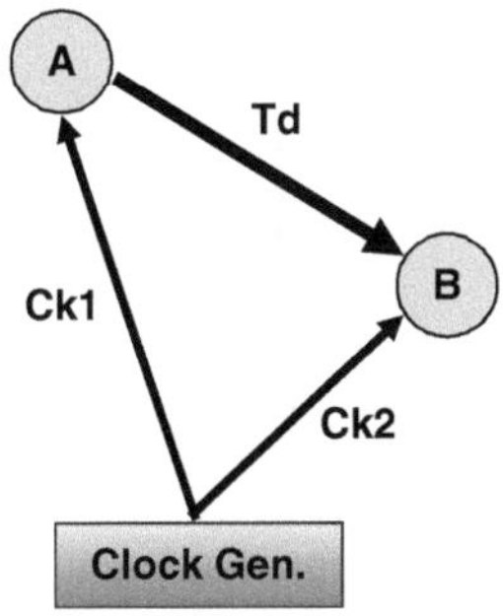


**Fig. 2.3.** Sequential path showing explicit clock distribution

### 2.2.1 Setup and Hold Timing Constraints

This section will present a brief outline of the formulation and the key parameters affecting the setup and the hold constraints.

The setup constraint specifies how data from the source sequential stage at cycle $N$ can be captured reliably at the destination sequential stage at cycle $N+1$. This situation is illustrated in Fig. 2.4 in which the source clock Ck1 is shown to lag behind the receive clock Ck2 due to clock uncertainty. The constraint for the source data to be received reliably by the receiver is defined in inequality 2.1, where $T_{\text{d-slow}}$ is the slowest (maximum) data path delay, $T_{\text{su}}$ is the setup time for the receiver flip-flop, $T_{\text{per}}$ is the clock period, $T_{\text{Ck1}}$ and $T_{\text{Ck2}}$ are the arrival times for clocks Ck1 and Ck2 (at cycle $N$) respectively.

$$T_{\text{per}} \geq T_{\text{d-slow}} + T_{\text{su}} + |T_{\text{Ck1}} - T_{\text{Ck2}}| \,. \tag{2.1}$$

In the setup constraint situation, the available time for data propagation is reduced by the clock uncertainty defined as the absolute difference of the clock arrival times. This uncertainty $|T_{\text{Ck1}} - T_{\text{Ck2}}|$ can originate from various sources and their classification will be discussed in subsequent sections. In order to accommodate the clock uncertainty and meet the inequality in (2.1), either clock period must be extended or path delay must be reduced. In either case, power and operating frequency may be affected.

The hold constraint is shown in Fig. 2.5. This case specifies the situation where the data propagation delay is fast, and clock uncertainty makes the problem even

[1] The latency is referenced to the root of the distribution.

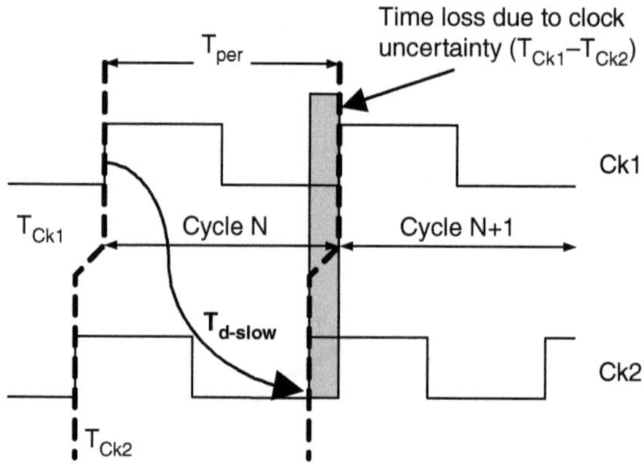


**Fig. 2.4.** Timing diagram for the setup constraint

worse and the data intended to be captured at cycle $N+1$ is erroneously captured at cycle $N$, corrupting the receiver state. In order to ensure that the hold constraint is not violated, the design has to guarantee that the minimum data propagation delay is sufficiently long to satisfy inequality (2.2):

$$T_{\mathrm{d-fast}} \geq T_{\mathrm{hold}} + |T_{\mathrm{Ck1}} - T_{\mathrm{Ck2}}|, \tag{2.2}$$

where $T_{\mathrm{hold}}$ is the hold time requirement for the receive flip-flop.

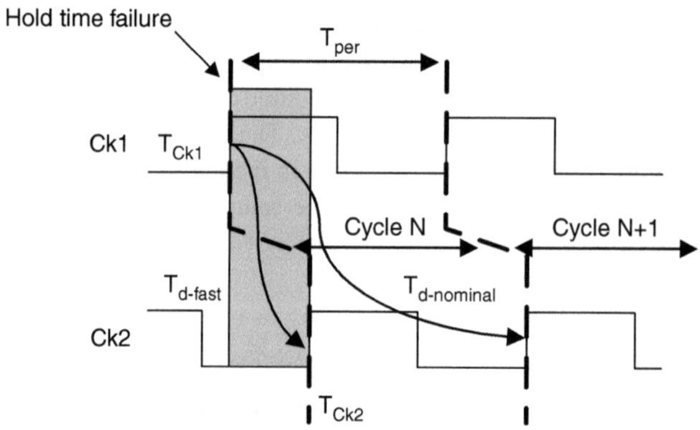


**Fig. 2.5.** Timing diagram for the hold constraint

In the discussion above, the following relationship is expected to hold (inequality 2.3):

$$T_{\mathrm{d-fast}} < T_{\mathrm{d-nominal}} < T_{\mathrm{d-slow}}. \tag{2.3}$$

Meeting the hold constraint in (2.2) with large clock uncertainty could result in setup violation due to (2.3) since the slowest manifestation of the same path could violate the delay requirement in (2.1). Such two-sided constraints are not uncommon in modern design if the clock uncertainty is high.

Central to the discussion above is the clock uncertainty defined by the absolute difference of delays $T_{Ck1}$ and $T_{Ck2}$. A typical clock distribution structure (Fig. 2.6) relies on buffer[2] stages to amplify the clock from the clock generator to the respective receivers (shown as the sequential elements FF in Fig. 2.6). In general, when measuring the clock arrival time at the end points of a clock distribution, the clock latencies with respect to the distribution common point ($T_{DELAY}$) will exhibit a statistical distribution as shown in Fig. 2.6. This statistical distribution is attributed to various static or dynamic sources. For example, design mismatches and on-die process variations will result in static delay mismatches. Clock generator (e.g. PLL) jitter or dynamic voltage variations can introduce dynamic clock uncertainties. Minimizing $T_{DELAY}$ will also minimize clock uncertainty and improve setup and hold margins.

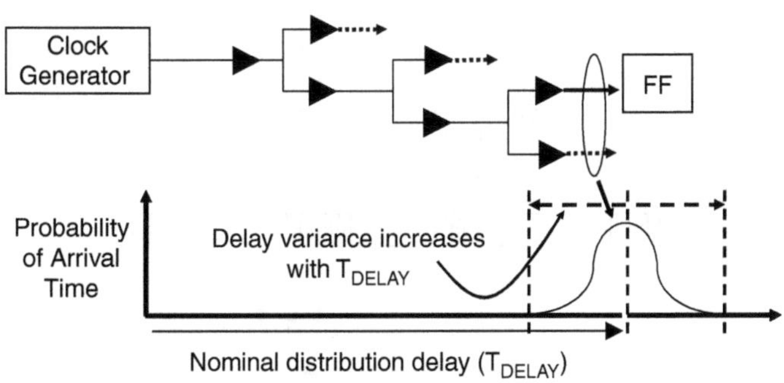


**Fig. 2.6.** Statistical nature of clock arrival times

### 2.2.2 Clock Attributes

We use this term to denote clock characteristics that affect the timing constraints described in Sect. 2.2.1. The key attributes are:

1. Clock uncertainties (skew and jitter)
2. Clock distribution latency
3. Clock duty cycle

The first and the last play an explicit role in the timing constraints of a synchronous design. The clock latency by itself does not affect the sequential timing constraints but plays a critical role in determining the other two.

[2] In this context, a buffer stage could be any gate that exhibits gain.

**Static and Dynamic Clock Uncertainties**

Clock uncertainties can be classified as static or dynamic. Static uncertainty does not vary or varies very slowly with time. Process variation induced clock uncertainty is such an example. On the other hand, dynamic uncertainty varies with time. Dynamic power supply induced delay variation is an example of a dynamic uncertainty.

In Fig. 2.7, the clock attributes $T_{\text{skew}}$ and $T_{\text{jitter}}$ are defined on clock waves Ck1 and Ck2. Taking the wave Ck1 as an example, when one of the clock edges is repeatedly sampled with an ideal reference, a timing histogram will result. A timing histogram exists for every clock edge and is characterized by a mean value and a peak-to-peak range (Fig. 2.7). The difference between the mean of two corresponding clock edges (example: between edge A and edge B) is defined as skew ($T_{\text{skew}}$) and is treated as a static uncertainty. The peak-to-peak range of a single edge is specified as the jitter ($T_{\text{jitter}}$) and its character is dynamic.

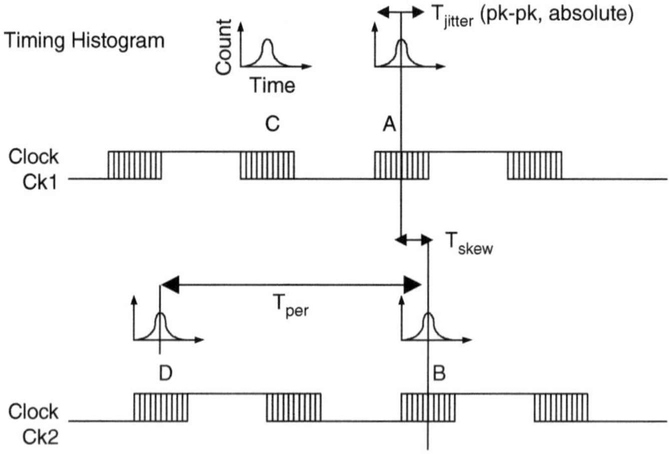


**Fig. 2.7.** Clock skew and jitter definitions

Table 2.1 highlights the sources of the static and dynamic clock uncertainties. Sources of static clock uncertainties are:

1. Intentional or unintentional design mismatches
2. On-die process variations
3. Loading variations (mismatch) at the intermediate or final stage of the clock distribution

Design mismatches arise because of a number of factors. For example, a nonbalanced clock distribution may be necessary due to floorplan constraints. A poorly chosen distribution topology could lead to structural design mismatches. In other situations, the clock arrival times at certain receivers are intentionally skewed to facilitate time borrowing across sequential boundaries due to nonuniform data path lengths. On-die device mismatch due to on-die process variations is a significant factor. Additionally, nonuniform clock loading is common in highly integrated designs. All skew sources mentioned above remain constant over time (except through the slow process of transistor aging) and are treated as static. Figure 2.8 shows an empirical breakdown of skew contributors.

**Table 2.1.** Sources of static and dynamic clock uncertainties

| Clock uncertainties | Sources |
|---|---|
| Static (skew) | Intentional or unintentional design mismatches<br>On-die process variations<br>Final or intermediate loading variations |
| Dynamic (jitter) | Voltage droop and dynamic voltage variations<br>Temperature gradient due to activity variations<br>Clock generator jitter |

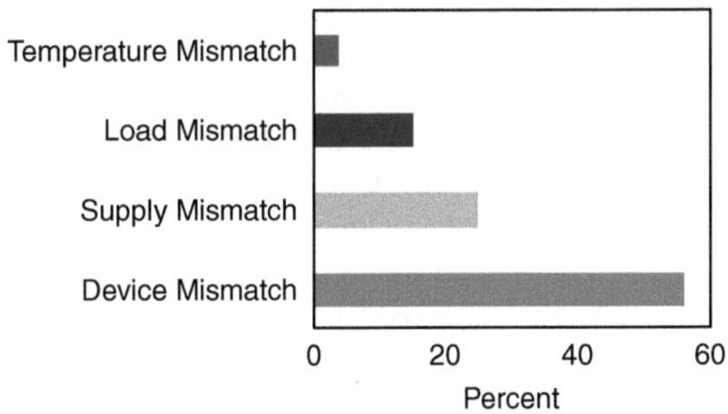


**Fig. 2.8.** Factors affecting clock skew. Among all the sources, device mismatch is the dominant contributor. Reproduced with permission from [31], ©1998 IEEE

Inherently static clock uncertainties can be corrected either by careful pre-silicon analysis and design or post-silicon adaptive compensation. Accurate pre-silicon analysis can be time consuming and iterative. On the other hand, post-silicon adaptive compensation is flexible and significantly more suited to high volume manufacturing.

Figure 2.9 shows clock skew as a percentage of cycle time vs. processor frequency for a number of recent designs. On average, the trend suggests that the skew as a fraction of the clock cycle time stays at about 4.5–5%. The ability of the trend to continue is attributed to the adoption of clock distribution topologies that are skew tolerant, more robust design flow, and more importantly the incorporation of robust post-silicon compensation techniques.

Clock uncertainties caused by voltage variation, temperature variation, and clock generator jitter are dynamic in nature. We use the term jitter to encompass all

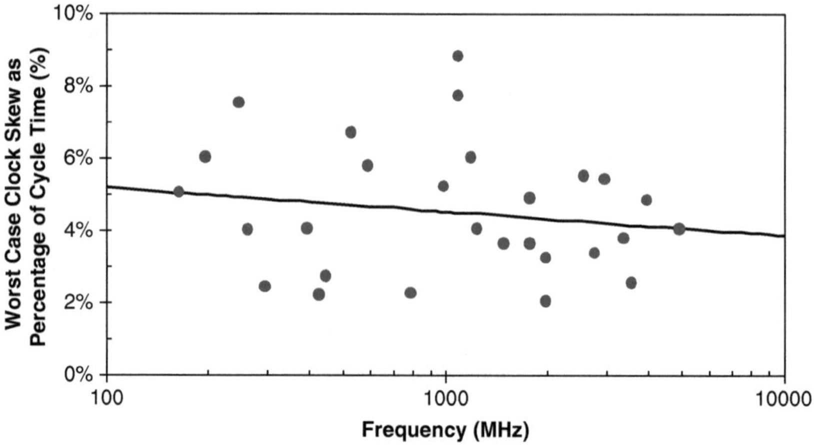


**Fig. 2.9.** Clock skew as percentage of cycle time vs. processor frequency [8, 15–17, 19–21, 24–26, 28–30, 32–46]

dynamic uncertainties. Voltage variation is the dominant source and it can be due to local switching events affecting specific areas of the clock distribution in a nonuniform fashion. A mathematical model of supply-induced jitter based on additive sinusoidal supply noise is developed in Section 6.8. Global voltage droop and clock generator jitter are common to the entire distribution and contribute to the setup constraint by modulating the cycle time. Clock generator jitter is addressed in detail in Chapter 5. Temperature variation has a long time constant and its impact is usually minor as seen in Fig. 2.8. Figure 2.10 shows the trend of peak-to-peak clock jitter as a fraction of cycle time. The average reduction of effective clock cycle time due to jitter is about 5.5% and minimizing this is critical for performance reasons.

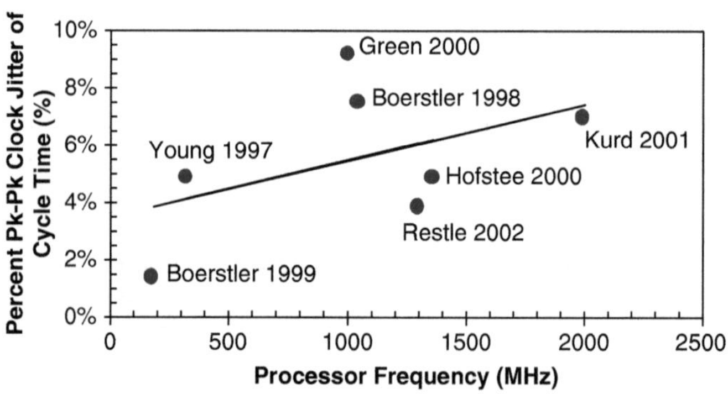


**Fig. 2.10.** Pk–pk clock jitter as a fraction of clock cycle time vs. processor frequency [6, 17, 18, 36, 47–49]

We now present a simple mathematical skew and jitter model based on clock distribution latency (Fig.2.11). The figure shows a source clock path with $M$ buffer stages and a receiver clock path with $N$ buffer stages resulting in delays of $T_{\mathrm{Ck1}}$ and $T_{\mathrm{Ck2}}$ respectively. The point-of-divergence (POD) delay is defined as the sum of the source clock delay and the receiver clock delay measured from a common origin. In Fig.2.11, the POD delay equals the sum of $T_{\mathrm{Ck1}}$ and $T_{\mathrm{Ck2}}$. Assuming $T_i$ is the actual delay at buffer stage $i$ and $\tau$ is the average delay per stage, the source clock and the receiver clock delays are:

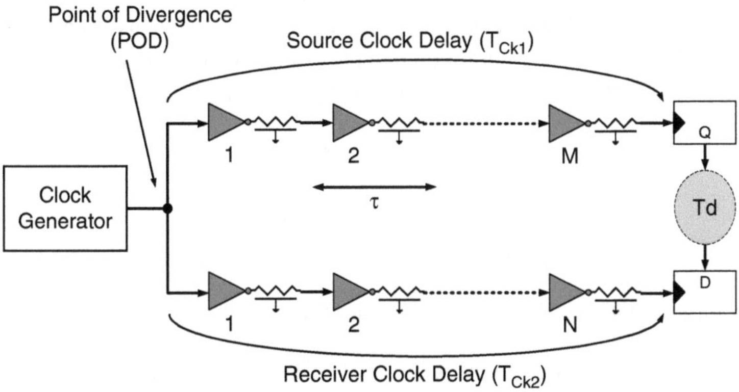


**Fig. 2.11.** Sample clock distribution for skew and jitter model

$$T_{\mathrm{Ck1}} = \sum_{1}^{M} T_i \cong M\tau, \tag{2.4}$$

$$T_{\mathrm{Ck2}} = \sum_{1}^{N} T_i \cong N\tau. \tag{2.5}$$

The average delay per stage is determined by the drive current of the driver stage ($I_{\mathrm{d}}$), the output loading capacitance ($C_{\mathrm{l}}$), and the output voltage swing ($V_{\mathrm{CC}}$):

$$\tau = \frac{C_{\mathrm{l}} V_{\mathrm{CC}}}{I_{\mathrm{d}}}. \tag{2.6}$$

Using a simple linearized model, the change in delay per stage ($\Delta\tau$) can be formulated as:

$$\Delta\tau = \frac{\partial\tau}{\partial V_{\mathrm{CC}}}\Delta V_{\mathrm{CC}} + \frac{\partial\tau}{\partial C_{\mathrm{l}}}\Delta C_{\mathrm{l}} + \frac{\partial\tau}{\partial I_{\mathrm{d}}}\Delta I_{\mathrm{d}}. \tag{2.7}$$

Evaluating the partial derivatives yields:

$$\Delta\tau = \frac{C_{\mathrm{l}}}{I_{\mathrm{d}}}\Delta V_{\mathrm{CC}} + \frac{V_{\mathrm{cc}}}{I_{\mathrm{d}}}\Delta C_{\mathrm{l}} - \frac{C_l V_{\mathrm{cc}}}{I_{\mathrm{d}}^2}\Delta I_{\mathrm{d}}. \tag{2.8}$$

Finally, substituting the expression for $\tau$ (2.6) results in the following expression:

$$\Delta\tau = \tau\left(\frac{\Delta V_{CC}}{V_{CC}} + \frac{\Delta C_l}{C_l} - \frac{\Delta I_d}{I_d}\right). \tag{2.9}$$

Equation 2.9 states that the delay variation at each stage can be approximated to be proportional to the stage delay. Additionally, we make the assumption that the delay per stage is a random variable, normally distributed with the following standard deviation:

$$\sigma(\tau) \cong \alpha\tau, \tag{2.10}$$

where $\alpha$ is the proportionality constant predicted by (2.9). We assume that $\alpha$ is on the order of 5%. Under the assumption that each stage delay is independent and identically distributed, the standard deviations of $T_{Ck1}$, $T_{Ck2}$, and $|T_{Ck1} - T_{Ck2}|$ become:

$$\sigma(T_{Ck1}) \cong \sqrt{M}\alpha\tau, \tag{2.11}$$

$$\sigma(T_{Ck2}) \cong \sqrt{N}\alpha\tau, \tag{2.12}$$

$$\sigma(|T_{Ck1} - T_{Ck2}|) \cong (\sqrt{M+N})\alpha\tau. \tag{2.13}$$

The standard deviations of skew and jitter are therefore[3]:

$$\sigma\left[T_{skew}(\mathrm{Ck1},\mathrm{Ck2})\right] = \left(\sqrt{M+N}\right)\alpha_{skew}\tau, \tag{2.14}$$

$$\sigma\left[T_{jitter}(\mathrm{Ck1})\right] = \left(\sqrt{M}\right)\alpha_{jitter}\tau, \tag{2.15}$$

$$\sigma\left[T_{jitter}(\mathrm{Ck2})\right] = \left(\sqrt{N}\right)\alpha_{jitter}\tau. \tag{2.16}$$

where $\alpha_{skew}$ and $\alpha_{jitter}$ represent the variation coefficients for static and dynamic clock uncertainties, respectively. Typical clock distribution will have $M = N$ and the formulation will be reduced to:

$$\sigma\left[T_{skew}(\mathrm{Ck1},\mathrm{Ck2})\right] = \left(\sqrt{2M}\right)\alpha_{skew}\tau, \tag{2.17}$$

$$\sigma\left[T_{jitter}(\mathrm{Ck1})\right] = \left(\sqrt{M}\right)\alpha_{jitter}\tau, \tag{2.18}$$

$$\sigma\left[T_{jitter}(\mathrm{Ck2})\right] = \left(\sqrt{M}\right)\alpha_{jitter}\tau. \tag{2.19}$$

Equations 2.17–2.19 show that the skew and jitter variations will grow as the square-root of the number of distribution buffering stages and linearly with the nominal delay per stage. This formulation can be generalized for any pair of clocks that share a common point of clock divergence. The sum of the variation coefficients for modern process technology and design is between 5 and 10%.

[3] $T_{skew}$(Ck1, Ck2) means the skew between clocks Ck1 and Ck2 and $T_{jitter}$(Ck#) is the jitter of Ck#.

**Distribution Delay**

Equations 2.17–2.19 suggest that the clock distribution delay (latency) is a key component in determining the overall clock uncertainties. In order to handle the final clock loading and to traverse the distances needed to reach the loads, a clock network has to rely on a series of clock buffers for gain and signal propagation. In a typical processor, the number of clock buffer[4] stages may exceed 20 resulting in clock latency that can approach 1 ns. Minimizing clock distribution latency is a primary design objective irrespective of the distribution topology.

**Duty Cycle**

Duty cycle (Fig. 2.12) is the relative percentage of the clock high phase time vs. low phase time. Except for special clocking applications such as pulse generators and clocks for dynamic and memory circuits, a 50% clock duty cycle is considered optimal. This is particularly important for a latch-based designs and memory circuits where any offset between the high phase and low phase can lead to phase paths that are more difficult to meet timing constraints. In a phase path, time lost due to duty cycle distortion will subtract directly from the total available phase time. Cycle-based sequential designs using edge-triggered flip-flops are more immune to clock duty cycle distortion. In a clock distribution, duty cycle distortion is introduced when there is asymmetry between rising and falling edge delays.

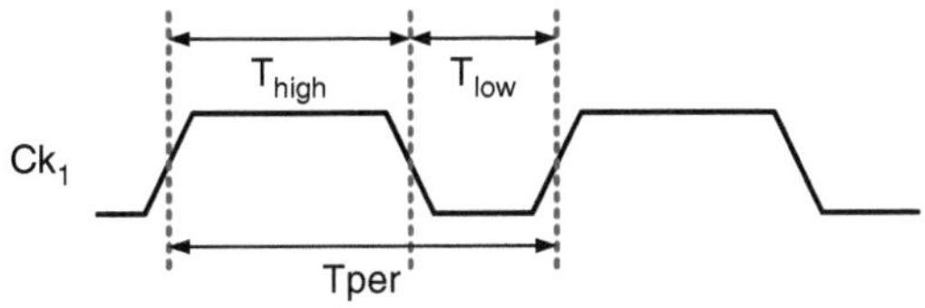


**Fig. 2.12.** Clock duty cycle

## 2.2.3 Clock Distribution Power

Power dissipation attributed to the clock distribution has emerged as a critical constraint in multi-GHz multicore processors with large on-die caches. In this section, we will develop a model for clock distribution power. Typically, the total end-of-distribution loading due to sequential elements strongly determines the overall clock network power. Let us consider the switching power of an unconditional clock at the final distribution stage $M$:

$$Pd_{\mathrm{CK\_}M} = C_{\mathrm{L\_}M} V_{\mathrm{CC}}^2 f, \tag{2.20}$$

[4] In this context, a clock buffer stage represents one unit-inverter stage.

where $C_{L\_M}$ is the stage load that encompasses both gates and interconnect, $V_{CC}$ is the power supply voltage, and $f$ is the clock frequency. Total power will include components for short circuit and leakage, but the dynamic component dominates. If the fan out per stage is $k$, the clock dynamic power consumed at stage $M-1$ is:

$$Pd_{Ck\_M-1} = C_{L\_M-1} V_{CC}^2 f = \frac{C_{L\_M}}{k} V_{CC}^2 f. \quad (2.21)$$

Assuming that the fan out is constant across all M stages, the total clock distribution dynamic power is:

$$Pd_{Ck\_Total} = \sum_{i=1}^{M} Pd_{Ck\_i} = \sum_{i=1}^{M} C_{L\_i} V_{CC}^2 f, \quad (2.22)$$

$$Pd_{Ck\_Total} = V_{CC}^2 f C_{L\_M} \left[ \frac{1-\left(\frac{1}{k}\right)^M}{1-\frac{1}{k}} \right]. \quad (2.23)$$

Let us define the clock load multiplier as the ratio of the total clock distribution load capacitance to the end-of-distribution load capacitance:

$$\text{Clock load multiplier} = \frac{\sum_{i=1}^{M} C_{L\_i}}{C_{L\_M}} = \left[ \frac{1-\left(\frac{1}{k}\right)^M}{1-\frac{1}{k}} \right]. \quad (2.24)$$

Figure 2.13 shows the clock load multiplier vs. the number of distribution stages. Decreasing the stage fan out will lead to higher total network capacitance that approaches 1.5 at a stage fan out of 3. Note that this parameter is not very sensitive to the number of buffer stages in the clock distribution network. Figure 2.14 shows the

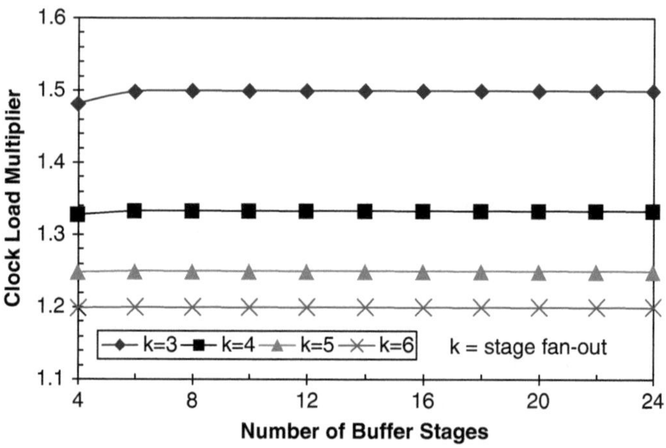


**Fig. 2.13.** Clock loading multiplier of a clock distribution

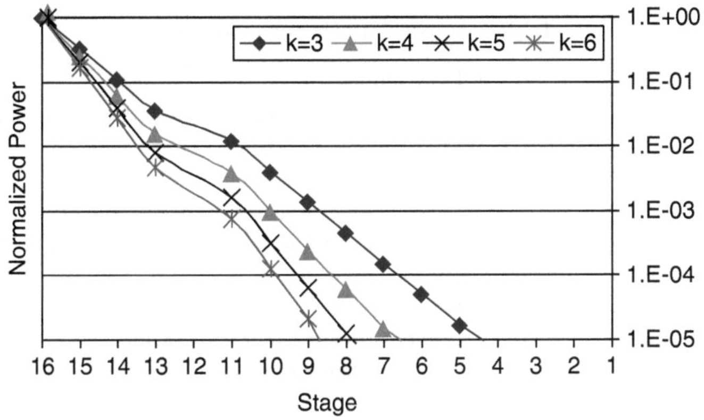


**Fig. 2.14.** Normalized clock stage power vs. stage number

power dissipation at each of the stage normalized to the power dissipation at the last stage. It can be seen that power dissipation of the clock distribution is dominated by the final end-of-distribution loading and that the last couple of stages in the distribution will account for more than 90% of the total clock power. An implication of this analysis is that the manner in which clock is distributed at the final stages of the distribution will ultimately determine the overall clock power and that the distribution topology upstream will not have a strong impact. In a typical processor, clock distribution (excluding the last stage) should not exceed 10% of total chip power and have a design goal of 5–8%.

## 2.3 Clock Distribution Topologies

In this section, various clock distribution topologies are described. Table 2.2 lists distribution topologies encountered in modern processors. While discussing these topologies, we will focus on the same attributes described in Sect.2.2.2. In addition, ease of implementation will be considered. Ease of implementation is subjective and depends heavily on historical design styles and prior art.

### 2.3.1 Unconstrained Tree

An unconstrained tree style clock distribution is illustrated in Fig.2.15. It is commonly used in automatic synthesis flows and usually placed with little or no restriction on the number of buffer stages and explicit matching between interconnect delays and the buffer delays. The network design is accomplished with a cost function that minimizes the delay differences across all clock branches. Figure 2.15

**Table 2.2.** Clock distribution topologies

| Style | Description |
|---|---|
| Unconstrained tree | Automated buffer placements with unconstrained trees |
| Balanced tree | Multiple levels of balanced tree segments<br>H-tree is most common |
| Central spine | Central clock driver |
| Spines with matched branches | Multiple central structures with length (or delay) matched branches |
| Grid | Interconnected (shorted) clock structure |
| Hybrid distribution | Combination of multiple techniques<br>Common theme is tree + grid or spine + grid |

shows an unconstrained clock distribution tree with $K$ branches. A cost function ($\vartheta$) that minimizes the delay differences can be used in the construction of the network:

$$\vartheta = \sum_{i=1}^{K} \left(T_{\mathrm{Ck}i} - T_{\mathrm{Ck\_Average}}\right)^2, \tag{2.25}$$

$$T_{\mathrm{Ck\_Average}} = \frac{1}{N}\sum_{i=1}^{K} T_{\mathrm{Ck}i}. \tag{2.26}$$

In the primitive form and specifically without explicit structural matching, a clock network with dissimilar buffer and interconnect delay composition may result in radically different branches that will exhibit significant mis-tracking across process, voltage, and temperature variations. More sophisticated optimizing algorithms can be incorporated to improve PVT tracking. Due to this limitation, this style of clock distribution is usually restricted to small functional blocks within a larger design.

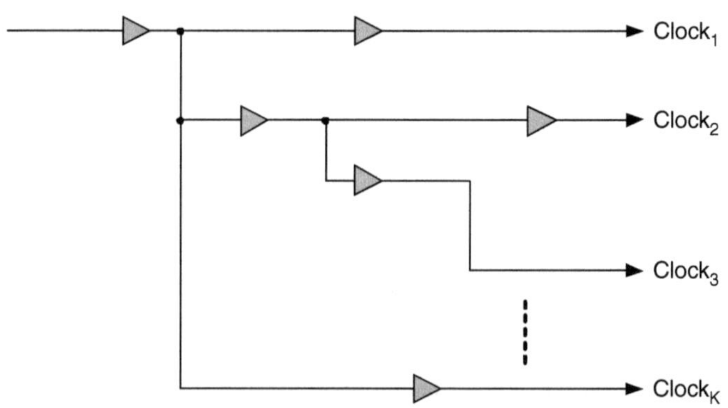


**Fig. 2.15.** Unconstrained tree clock network

### 2.3.2 Balanced Tree

Figure 2.16 shows a balanced H-tree clock topology. Due to the structural symmetry, a balanced tree exhibits identical nominal delay and identical buffer and interconnect segments from the root of the distribution to all branches. If the matching is adhered to, structural skew can be zero. With identical buffer and interconnect segments, an idealized balanced tree clock distribution will exhibit good tracking across PVT compared to the unconstrained network described earlier.

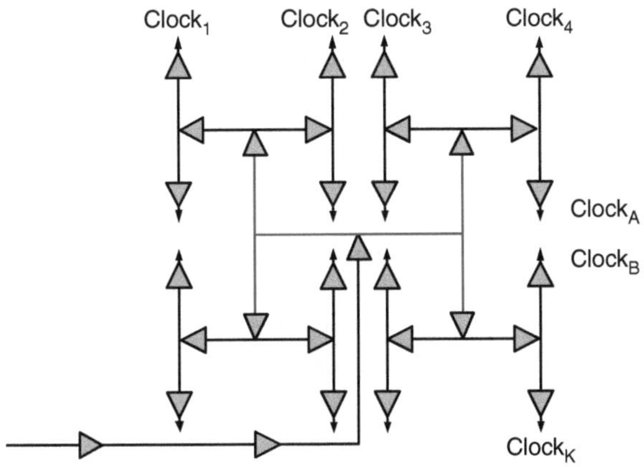


**Fig. 2.16.** Balanced H-tree clock network

Figure 2.17 shows alternative balanced tree topologies: the X-tree and a tapered H-tree. The X-tree incorporates nonrectilinear clock trunks in the physical implementation but exhibits the same properties as the H-tree. The trunk widths in a tapered H-tree increase geometrically toward the root of the distribution to maintain impedance matching at the T-junctions. One important characteristic of the aforementioned tree structures is that by continuing to expand the buffer hierarchy, balanced trees are capable of delivering the clock to all part of the silicon die. Typically, the clocks at the end-of-distribution branches will serve a small local region. The size and number of the local regions will determine the depth of the tree. A larger number of regions requires a tree with more depth.

Full balanced tree topologies are designed to span the entire die in both the horizontal and vertical dimensions. They are capable of delivering the clock to all regions of the die. A binary tree on the other hand (Fig. 2.18) is intended to deliver the clock in a balanced manner in either the vertical or horizontal dimension.

All branches of a binary tree exhibit identical buffer-interconnect segments, zero structural skew, and similar PVT tracking. In contrast to the H-tree, the buffers in a binary tree can be designed to co-locate in close proximity along a centralized stripe. The closer physical proximity of the buffers in a binary tree can result in

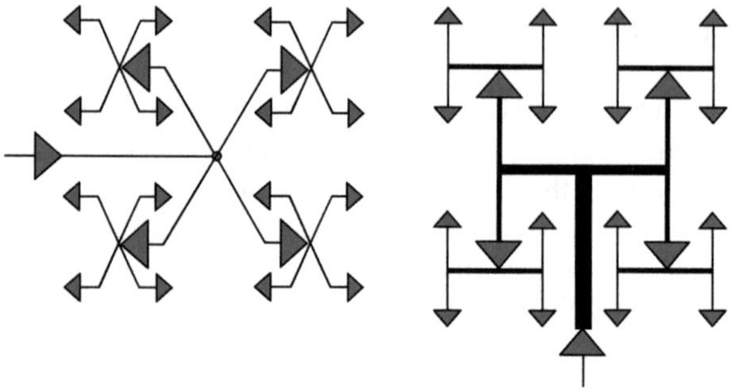


**Fig. 2.17.** Variations on the balanced tree topology

reduced sensitivity to on-die variation. Moreover, physical placement of the clock buffers in close proximity will minimize floorplan disruptions. On the other hand, the idealized buffer placements associated with an H-tree may be difficult to achieve. Due to these reasons, binary trees are often the preferred structure over an idealized H-tree. Figure 2.19 shows a binary tree distribution with intermediate shorting. The benefits of shorting will be discussed in a later section.

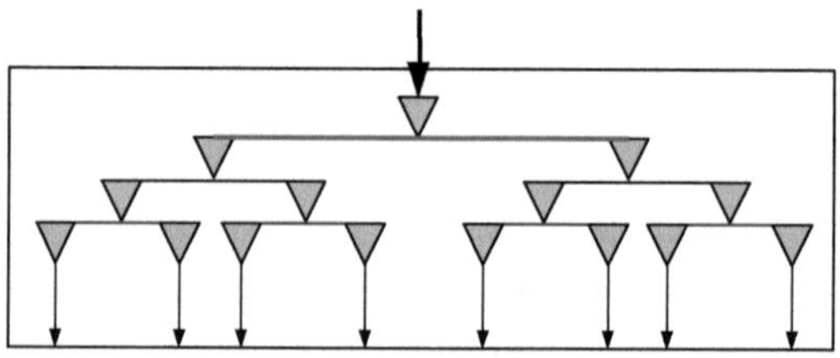


**Fig. 2.18.** Binary tree clock distribution

A balanced tree will exhibit nonzero clock uncertainty among branches due to nonzero POD delays (2.17–2.19). Let us consider branches $Clock_4$ and $Clock_A$ in Fig. 2.16. The point-of-divergence is two buffer-interconnect segments apart. In contrast, branches $Clock_A$ and $Clock_B$ are six buffer-interconnect segments apart resulting in higher POD delay and higher skew uncertainty. Therefore, among pairs of equivalent branches in a balanced tree, nonuniform skew uncertainty will result and will depend on point-of-divergence delay. In summary, a balanced tree is capable of delivering the clock from the root to all regions of the die with good structural

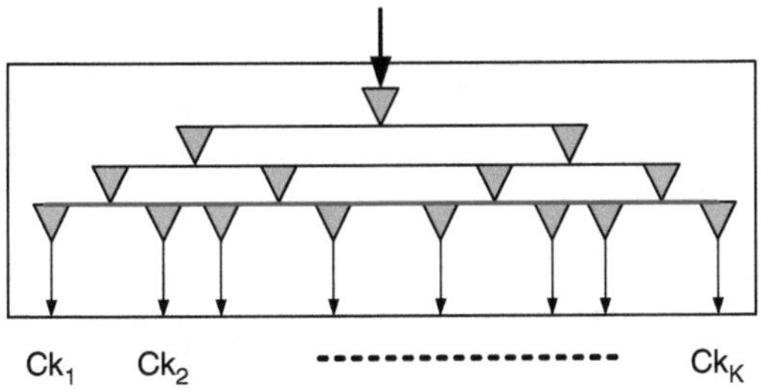


**Fig. 2.19.** Binary tree clock distribution with intermediate shorting

matching, efficient clock power and low structural latency. On the other hand, it will exhibit nonuniform POD-induced clock skew uncertainty.

Before proceeding it should be noted that a nonsymmetric tree can also be used in this context. A nonsymmetric tree usually maintains the same number of buffer stages but will not have delay matching on a per stage basis. Delay adjustment for overall branch equalization is done in a fashion similar to the unconstrained tree. Intense computational effort usually is needed for this design and its application is less common.

### 2.3.3 Central Spine

A central spine clock distribution is a specific implementation of a binary tree. Figure 2.20 shows an idealized central spine implementation with the final branches serving all parts of the die. The binary tree is shown to have embedded shorting at all distribution levels and unconstrained routing to the local loads at the final branches. In this configuration, the clock can be transported in a balanced fashion across one dimension of the die with low structural skew. The unconstrained branches are simple to implement although there will be residual skew due to asymmetry (Fig. 2.20).

### 2.3.4 Spines with Matched Branches

An extension of the central spine structure can be realized by replacing the unconstrained end-of-distribution branches with delay matched routes as shown in Fig. 2.21. In this implementation, the longest branch determines the delay from the output of the central spine to the end loads. Serpentine routes are added to the shorter branches for delay matching. Figure 2.21 shows a structure with three central spines. Multiple central spines are needed when the routing distance of the local branches is increased. Dividing the chip into several sectors served by multiple spines is a practical topology to ensure small local branch delays.

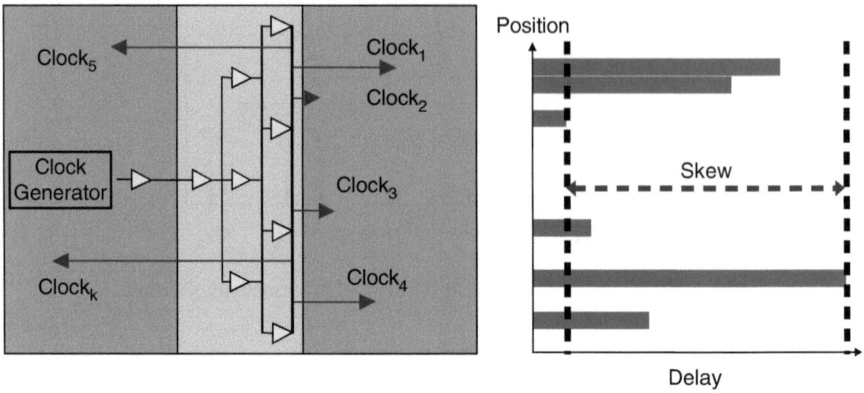


**Fig. 2.20.** Central clock spine distribution

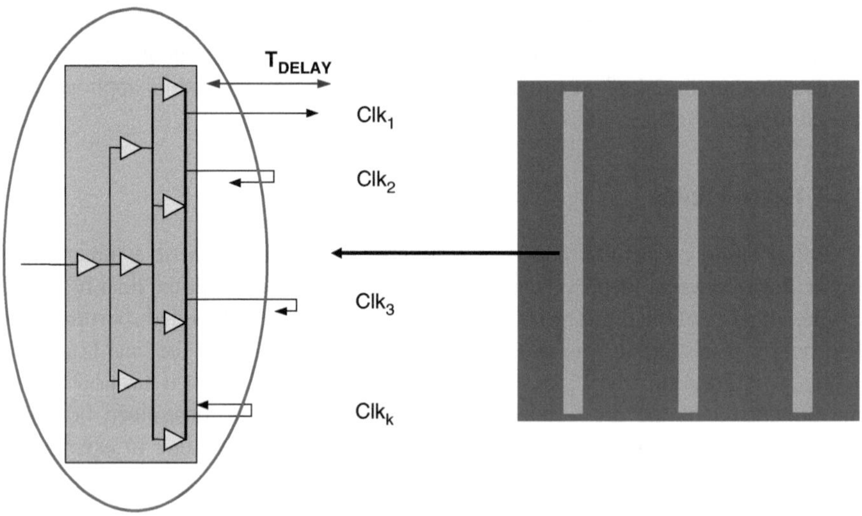


**Fig. 2.21.** Multiple clock spines with matched branches

### 2.3.5 Grid

The tree style distributions described in previous sections rely on individual branches to deliver the clock to the local (end-of-distribution) points of clock consumption (e.g. local flip-flops). A processor will have a large number of these local points and will require a large number of branches and therefore a deep distribution tree. A deep distribution tree will exhibit large POD delays and degraded clock performance. Subdividing the die into a smaller number of clock regions and applying a grid to serve each region can be a superior solution.

Figure 2.22 schematically shows a 2-dimensional grid serving one of these clock regions. This clock grid resembles a mesh with fully connected clock tracks in both dimensions and grid drivers located on all four sides. Local loads within a region can be directly connected to the grid. The grid effectively shorts the output of all drivers and helps minimize delay mismatches. Figure 2.22 shows an idealized delay profile of a 2-dimensional grid assuming uniform loading. The shorted grid node helps balance the load nonuniformities and results in a more gradual delay profile across the region. Additionally, since the grid drivers are shorted, the POD delay to all the loads within a region is limited to the interconnect delay of the grid which is typically small and results in lower clock skew uncertainty across the region. Grid drivers may also be placed on two sides leading to a structure and delay profile shown in Fig. 2.23. Critical design parameters for the grid are grid driver locations and pitch in addition to the grid metal pitches and dimensions.

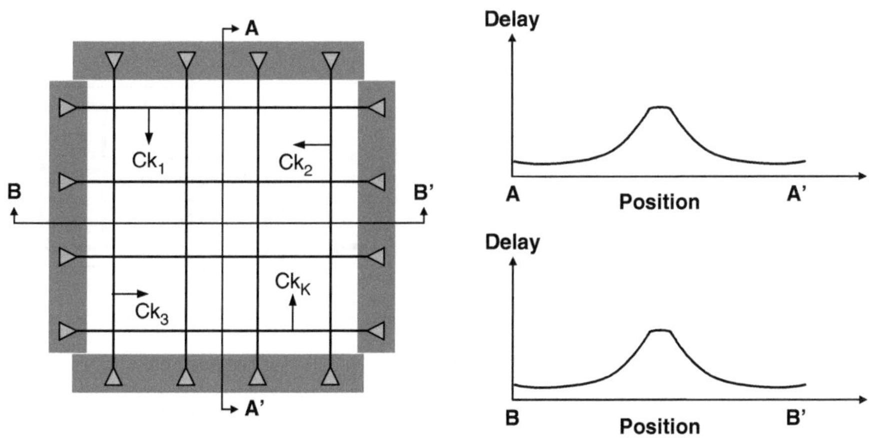


**Fig. 2.22.** Clock grid with 2-dimensional clock drivers

The recombinant tile structure is an enhancement over the conventional grid structure and incorporates the properties of a balanced tree [29]. Figure 2.24 shows the evolution of the recombinant tile structure from a regular H-tree segment to a tile template and to the final tile assembly. A typical implementation will have uniform interconnect pitches in both the $x$ and $y$ dimensions.[5] The pitches are determined by the intrinsic interconnect segment delays and the edge rate requirements. The uniformity of the segment pitches allows all intermediate buffers to be placed in predetermined locations.

The following analysis highlights the benefits of shorting intermediate stages in reducing clock uncertainties. Figure 2.25 shows two clock branches $Ck_A$ and $Ck_B$ exhibiting input skew $skew_{IN}$. Let us assume that the two branches are identical

[5] Note that the $x$ and $y$ pitches do not need to be equal.

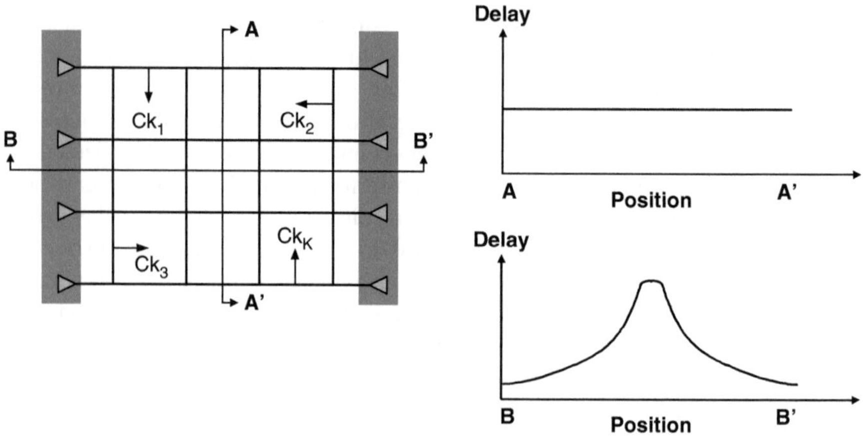


**Fig. 2.23.** Clock grid with 1-dimensional drivers

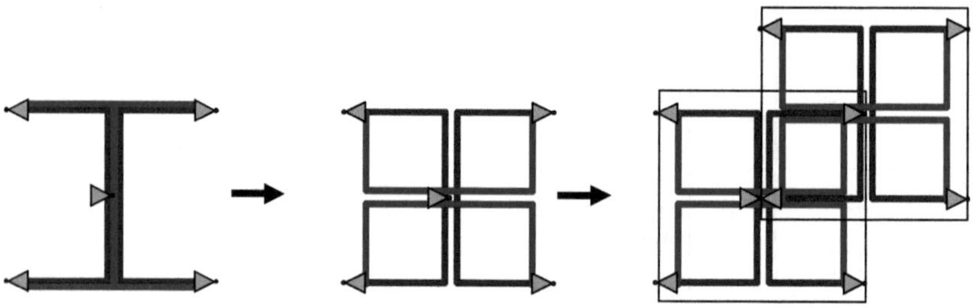


**Fig. 2.24.** Recombinant tile clock structure. Reproduced with permission from [29], ©2003 IEEE

(A-X-Y and B-U-V) and that $R_{\text{SHORT}}$ in Fig.2.25 is an open circuit ($R_{\text{SHORT}} = \infty$). In this case, the skew between $\text{Ck}_\text{U}$ and $\text{Ck}_\text{Y}$ will be the same as the skew between $\text{Ck}_\text{A}$ and $\text{Ck}_\text{B}$. With an ideal short circuit ($R_{\text{SHORT}} = 0\Omega$), the skew between $\text{Ck}_\text{U}$ and $\text{Ck}_\text{Y}$ will be zero. Hence, with a non zero and finite $R_{\text{SHORT}}$, the output skew between $\text{Ck}_\text{U}$ and $\text{Ck}_\text{Y}$ will be proportional to the incoming skew:

$$\text{skew}\,(\text{Ck}_\text{U}, \text{Ck}_\text{Y}) = \gamma \text{skew}(\text{Ck}_\text{A}, \text{Ck}_\text{B}), \tag{2.27}$$

where $\gamma$ is a skew averaging coefficient ($\gamma \geq 0$). The averaging coefficient is dependent upon the local POD induced delay mismatch and will scale with the spatial separation between the output nodes. For example, when the shorting resistor $R_{\text{SHORT}}$ in Fig.2.25 exhibits near zero resistance, the skew between $\text{Ck}_\text{Y}$ and $\text{Ck}_\text{U}$ will be close to zero suggesting that $\gamma$ is near 0. As $R_{\text{SHORT}}$ approaches an open, $\gamma$ will be equal to or larger than 1. For example, in a 90nm technology, a typical $R_{\text{SHORT}}$ will

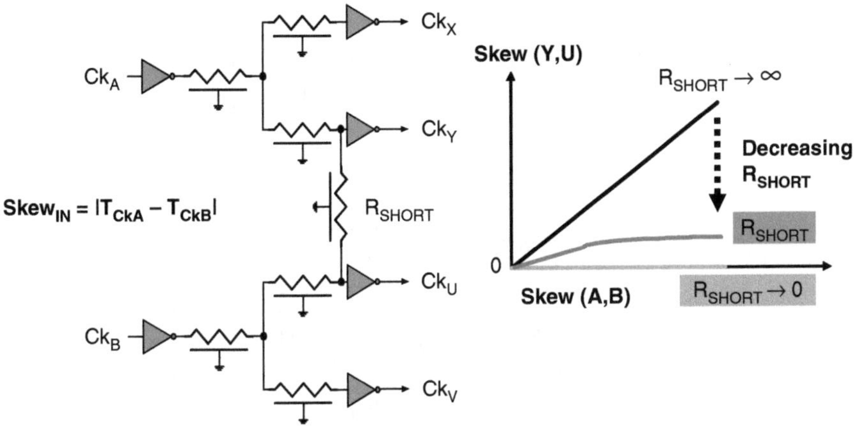


**Fig. 2.25.** Effect of shorting on clock skew [50]

result in $\gamma$ in the range of 0.2–0.8 for a few hundred microns of spatial separation [29]. When this structure is cascaded in a clock distribution consisting of $S$ cascaded stages ($S > 20$), the skew at the output of stage $S$ will be (for small $\gamma$) [50]:

$$\text{skew}_S \approx \gamma^2(\text{skew}_{IN}) + \gamma(\text{skew}_{IN}). \qquad (2.28)$$

Equation 2.28 states that the skew at the output of the distribution will stay relatively independent of the number of stages and therefore less sensitive to die size. The recombinant tile structure can be easily scaled to accommodate new designs. Comparing the recombinant tile of Fig. 2.24 to the grid structures of Figs. 2.22 and 2.23, reveals that they share similar skew benefits due to averaging. However, the grid structures require a pregrid clock distribution network to drive the grid or collection of grids. A hybrid clock distribution topology can meet this need.

### 2.3.6 Hybrid Distribution

A hybrid clock distribution incorporates a combination of earlier described topologies. Common configurations are spines-grid distribution or tree-grid distribution. Figure 2.26 shows the topology of a tree-grid distribution. It employs a multilevel H-tree driving a common grid. Specifically, the multilevel H-tree delivers the clock from the clock generator (PLL in Fig. 2.26) to various regions of the die. Regional buffers (labeled as level 4 buffers in Fig. 2.26) residing at the end of the multilevel H-tree drive a common grid that includes all local loads. As an alternative, there can be multiple regional grids each served by a branch of the pregrid H-tree. Partitioning the design enables intentional skew rebalancing across the regions.

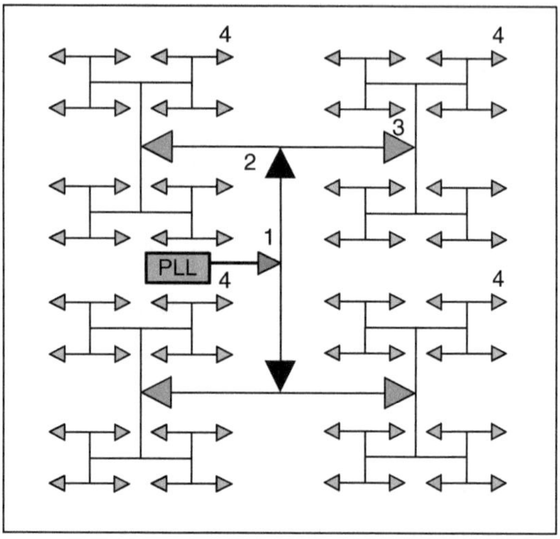


**Fig. 2.26.** Hybrid clock distribution consisting of balanced H-Tree and Grid

## 2.4 Microprocessor Clock Distributions

Due to the design complexity and the significant mis-tracking to process, voltage and temperature, a fully unconstrained clock distribution network is rarely (if ever) applied to a processor design.

The closest example is a hybrid combination of symmetric and asymmetric clock trees. Figure 2.27 shows an example of a processor clock distribution with a first level H-tree connected to multiple secondary trees that are asymmetric but delay balanced.

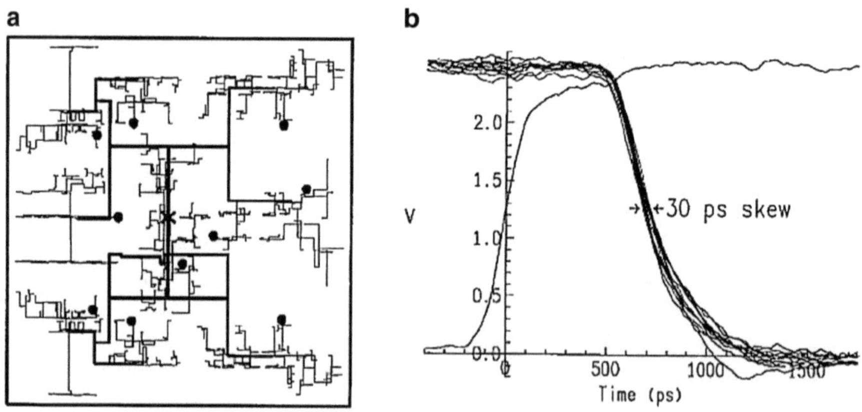


**Fig. 2.27.** Asymmetric clock tree distribution network based on delay matching. Reproduced with permission from [51], ©1998 IEEE

The construction of the trees was based on a custom methodology that matches the wire delays by tuning the metal widths, spacings, and lengths. Specifically, the first level symmetric H-tree (thick lines in Fig. 2.27a) routes the global clock from the center of the die to 9 sector buffers. The 9 sector buffers rely on multiple delay-matched secondary trees (light lines in Fig. 2.27a) to distribute the clock to 580 global clock receivers. Figure 2.27b shows the measured skew.

Figure 2.28 shows another example of a hybrid multilevel clock tree design [52]. The cache area (un-core) of the processor is partitioned into 13 regional clock zones served by the secondary clock trees. Postlayout extraction-based simulation models were used to perform tree optimization and delay matching. Additional examples of a hybrid multilevel clock tree for processors are found in [10, 12, 13, 43].

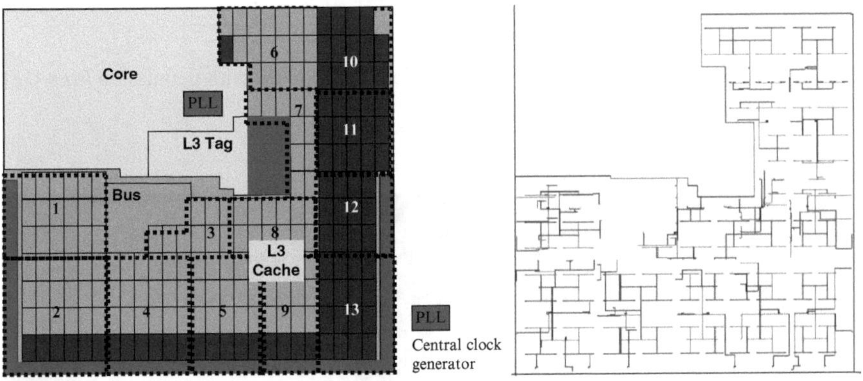


**Fig. 2.28.** Asymmetric clock tree distribution with multiple regions. Reproduced with permission from [52], ©2005 IEEE

Hybrid clock distributions that consist of multilevel symmetric trees and grids have been applied to a number of processors [17–20]. Figure 2.29 shows an example [18]. In this implementation, the clock from the clock generator is distributed from a central clock buffer through two levels of buffering and three levels of delay tuned H-trees before reaching a main grid covering most of the chip, and two smaller grids covering two units that require delayed clocks.

Figure 2.30 shows another example of a hybrid tree-grid design in [17]. A multi-level tree structure delivers the clock to 64 sector buffers driving a common grid via multiple second level tuned trees.

Figure 2.31 shows the floorplan and the clock skew profile of three generations of Alpha®[6] processors. These designs followed a common strategy of having one or more centralized clock spines to drive a common grid. The first generation design relied on a single spine to support the entire die whereas the third generation design

[6] Other names and brands may be claimed as property of others.

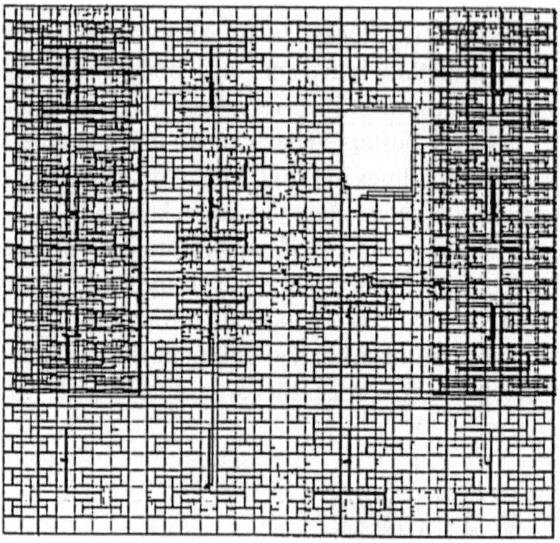

Fig. 2.29. Multilevel symmetric H-Tree distribution. Reproduced with permission from [18], ©2000 IEEE

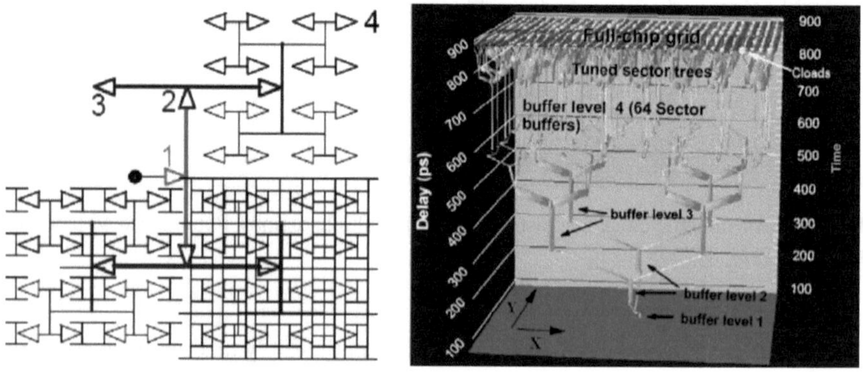


Fig. 2.30. Delay characteristics of a multilevel tree-grid distribution. Reproduced with permission from [17], ©2002 IEEE

utilized 16 central spines organized in a 2-dimensional fashion to drive the common grid. By partitioning the die into smaller regions. the third generation design reduced the clock skew across the grid.

Figure 2.32 shows the application of recombinant tiles to a multi-GHz IA processor fabricated in 90nm [29].[7] The buffers needed for the recombinant tiles are embedded in eight central clock stripes. The recombinant tile distribution consists of

[7] Other names and brands may be claimed as property of others.

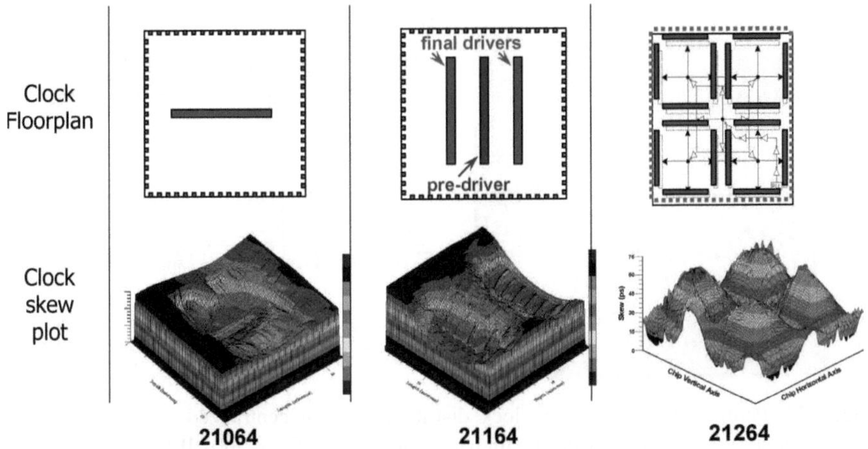


**Fig. 2.31.** Centralized clock drivers with grids on three generations of the Alpha® microprocessor. Reproduced with permission from [53], ©1998 IEEE

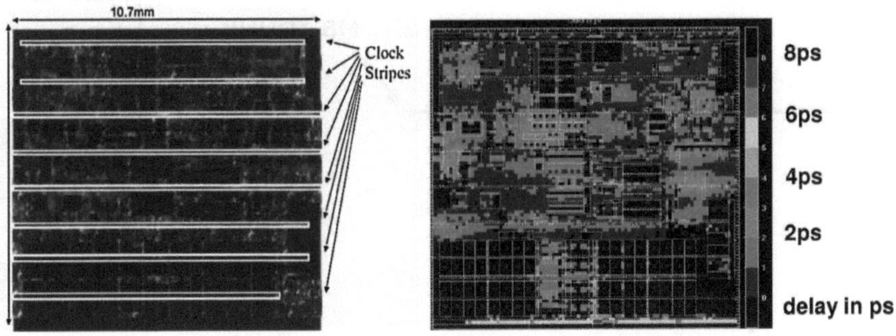


**Fig. 2.32.** Recombinant clock tiles on a 90nm processor. Reproduced with permission from [50], ©2003 IEEE

27 inversion stages and a total of 1,474 grid drivers (each driver is an inverter). An automated grid driver sizing flow was used to minimize grid driver oversizing for power efficiency. The simulated delay profile is shown in Fig. 2.32. A global skew of less than 10ps was achieved with this design.

An example of centralized spines with delay-matched branches is the clock distribution of the 180nm Pentium® 4 processor [8][8]. Binary distribution trees embedded in three central clock spines drive the local loads with delay-matched branches. The binary trees embedded in middle spine buffer the clock from the central PLL and deliver it in a balanced fashion to the other spines. The final clock drivers use matched

[8] Other names and brands may be claimed as property of others.

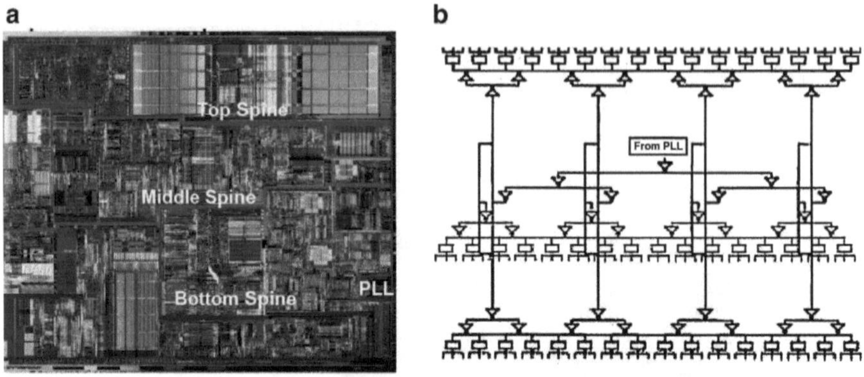


**Fig. 2.33.** Pentium® 4 processor clock distribution using centralized spines with delay matched final branches. Reproduced with permission from [49], ©2001 IEEE

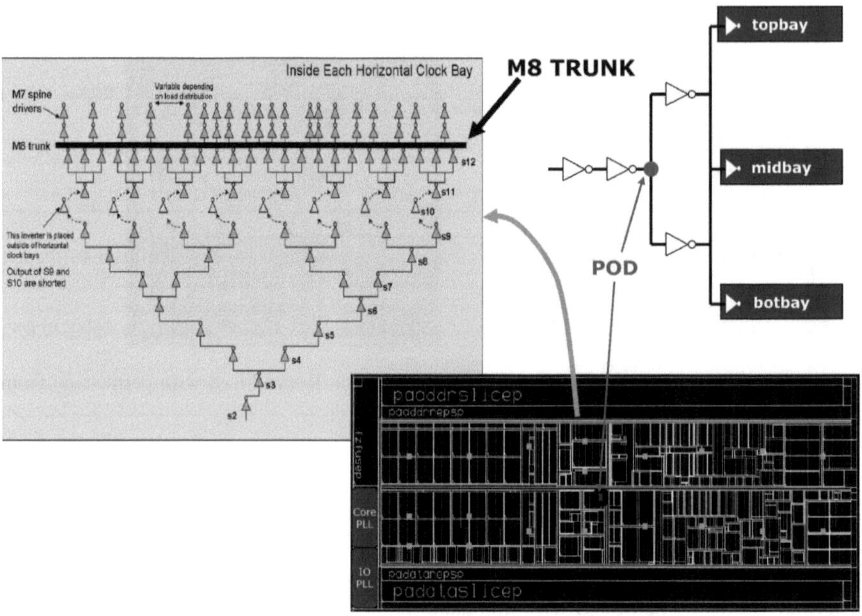


**Fig. 2.34.** Clock distribution of a low power IA processor consisting of binary trees embedded in the centralized spines. Reproduced with permission from [55], ©2008 IEEE

branches to support the local loads. The final drivers incorporate delay tuning capability to further optimize the skew via post-silicon compensation. Without engaging compensation, the measured skew is $\pm 32$ps.

The low power IA processor in [54] is another example of the application of the clock distribution topology consisting of centralized spines and delay-matched branches. The global binary tree with limited clock recombination is embedded in the spines. The spine output drivers are shorted with a high layer metal (M8) to equalize the driver delays. The highly selective application of clock recombination and other power saving schemes enables this design to achieve total clock power dissipation that is less than 10% of the total processor power.

The 65nm dual-core Xeon® processor employed a hybrid spine-grid clock distribution topology to support the multicore and uncore clock domains [45].[9] Figure 2.35 shows the multiple clock domains and the distribution design of this processor. The processor has two cores operating at the high frequency MCLK. The uncore is supported by the SCLK at half the MCLK frequency and the ZCLK dedicated for the I/O circuits at 4 times the system clock frequency. Binary trees embedded in the horizontal and vertical clock spines in the uncore distribute the clocks to the SCLK and ZCLK grids. The core employs the recombinant tile clock distribution similar to that described in [29]. Operational flexibility is achieved by keeping core and uncore clock regions independent.

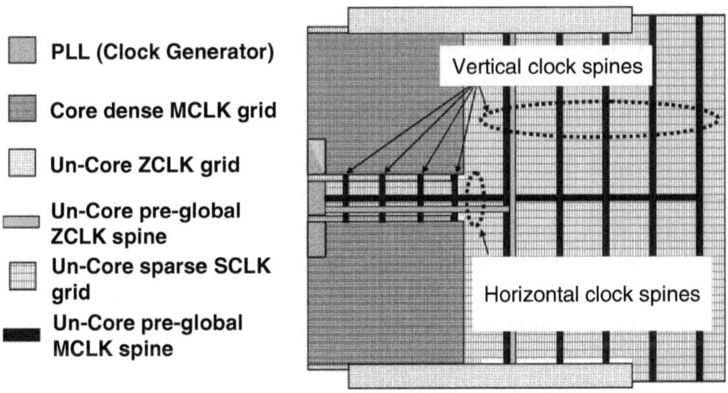


**Fig. 2.35.** Hybrid spine-grid clock distribution in a dual core processor. Reproduced with permission from [45], ©2006 IEEE

Figure 2.36 shows the details of the uncore grid implementation and the local clocking in [45]. The preglobal clock and the final grid clock driver are embedded in the vertical clock spines. A common SCLK grid covers the entire uncore area to serve the local logic units. Local clocking consists of two buffer stages featuring clock gating and delay tuning. The local clock buffers are placed inside the local logic unit with direct connection to the overlying global grid. Support for multiple local clock flavors is achieved with a family of local clock buffer macros.

[9] Other names and brands may be claimed as property of others.

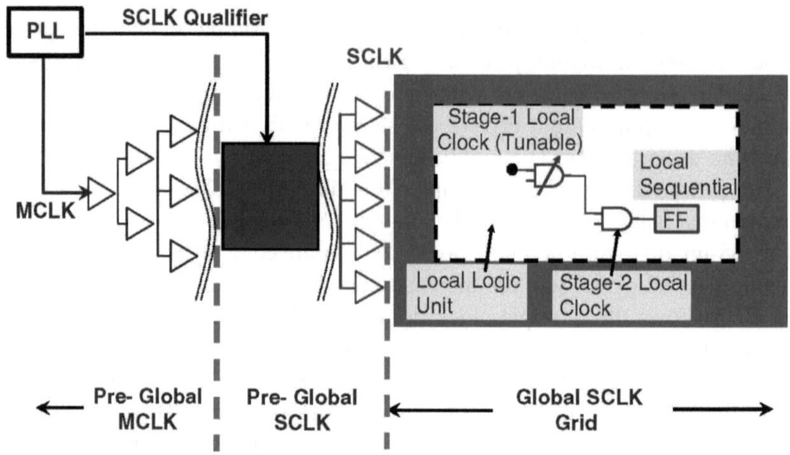


**Fig. 2.36.** Local clock distribution of the hybrid spine-grid clock distribution. Reproduced with permission from [45], ©2006 IEEE

## 2.5 Clock Design for Test and Manufacturing

### 2.5.1 Global and Local Clock Compensations

It should be obvious from the earlier discussions that the primary objective of the aforementioned clock distribution topologies is to deliver the clock to all corners of the die with low skew. For example, the application of averaging in the recombinant tile clock distribution will result in a scalable network that exhibits very low skew. However, implementation of the recombinant tile network will encounter floorplan constraints leading to nonideal driver placements and loss of performance. Floorplan constrains will also affect other topologies such as the H-tree distribution. Moreover, in many situations, intentional clock skew between specific regions of the global clock network is needed to rebalance the path timings. Therefore, a clock distribution network with adaptive delay compensation[10] is superior to the conventional design that has fixed delays, even if the adaptive design may exhibit higher initial skew. Additionally, adaptive delay compensation will adjust to skew caused by process variations and will overcome difficulties related to the construction of a pre-silicon design model and design flow that accurately and exhaustively accounts for all process effects (i.e. SOI dynamic switching effects). As an example, the evolution of the processor clock distribution designs in [17–19] eventually incorporated adaptive clock compensation in the latest implementation [20]. By allowing for slightly high initial skew, the physical design resource needs (e.g. metal tracks, floorplan

[10] The terms "adaptive delay compensation," "active deskew," and "skew rebalancing" are used interchangeably in this discussion.

restrictions, pre-silicon analysis, etc.) for a clock network with adaptive compensation are expected to be lower. In the following sections, we discuss adaptive global and local clock compensation architectures.

### 2.5.2 Global Clock Compensation Architecture

Figure 2.37 shows a dual-zone adaptive deskew clock distribution architecture implemented in a 450MHz microprocessor [32]. The global clock distribution of this processor is partitioned into two planes supported by the "left" and the "right" clock spines. Binary clock trees embedded in the spines deliver the global clock from the clock generator to the spine drivers. Two digital delay lines with 17 delay adjustment steps and approximately 12ps average delay step size reside near the root of their respective clock spines. A phase detector (PD in Fig. 2.37) strategically placed in the microprocessor core compares the phase difference of the clocks between the left and the right planes. A digital control logic unit (Control FSM and Delay SR) interprets the phase detector output and makes adjustments to the delay lines. The construction of the delay line is shown in Fig. 2.38. It consists of two cascaded inverters with switchable load capacitors at each stage. The delay shift register (Delay SR) generates a thermometer-coded delay adjustment and sequentially enables the load capacitors between the two stages. A 60ps skew was reported with adaptive deskewing disabled and 15ps with the mechanism engaged.

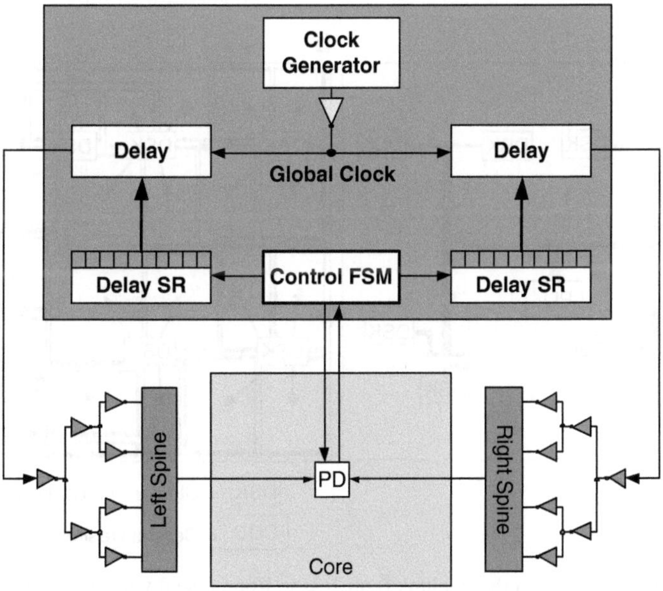


**Fig. 2.37.** Dual-zone deskew architecture. Reproduced with permission from [32], ©1998 IEEE

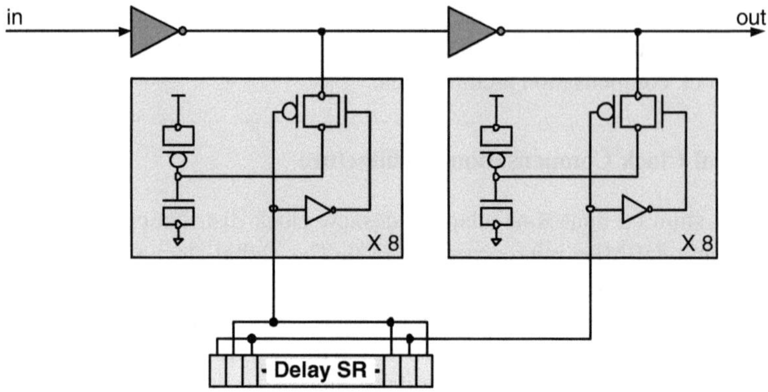


**Fig. 2.38.** Deskew delay line structure. Reproduced with permission from [32], ©1998 IEEE

The previous implementation [32] embodied only two independent deskew regions. Figure 2.39 shows a deskew architecture in the Itanium® processor[11] that supports 30 independent deskew regions [36, 56]. An H-tree distributes the global clock from the central PLL to eight clusters of deskew buffers (DSK in Fig. 2.39) serving 30 independent deskew regions. Each DSK cluster may contain up to four independent deskew buffers.

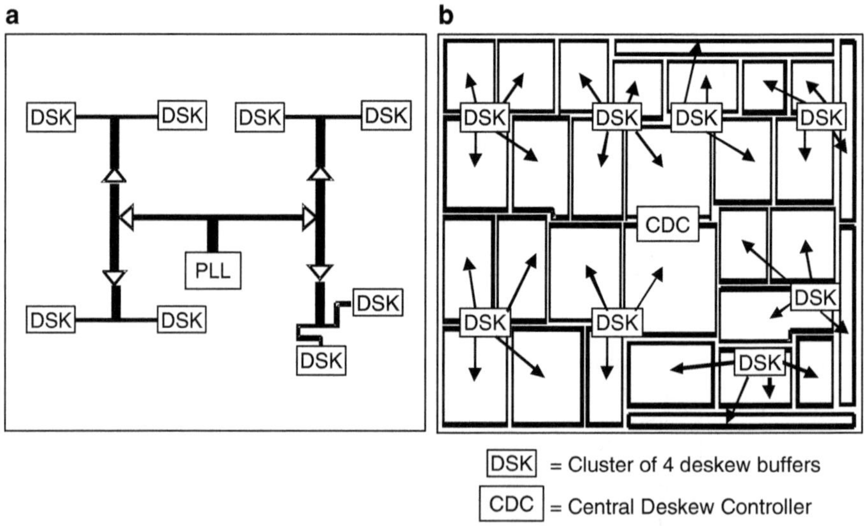


**Fig. 2.39.** Deskew zones in the itanium® processor. Reproduced with permission from [56], ©2000 IEEE

[11] Other names and brands may be claimed as property of others.

Figure 2.40 shows the detailed clock distribution architecture of [36] that encompasses the H-tree global distribution, the grid structure for the regional distribution, and the local buffers for the local distribution. In this design, the grid drivers (RCD in Fig. 2.40) are located at the top and bottom of grid. In addition to the global clock (main clock in Fig. 2.40), a dedicated and tightly matched reference clock is routed from the central clock generator to the eight DSK clusters (reference clock in Fig. 2.40). The purpose of the reference clock is to act as the reference to deskew the global clock.

Figure 2.41 shows the details of the deskew buffer architecture and the delay circuit design. The delay circuit design is similar to the previous design consisting of two inverter stages with switchable capacitor loads using a 20b thermometer code with a tri-state controllable output stage. In this implementation, the total skew is 28ps with deskew turned on. The skew increases by a factor of 4 with the deskew mechanism disabled. Additional details of this active deskew architecture are discussed in Sect. 7.3.2 in the context of addressing variations in the clock network.

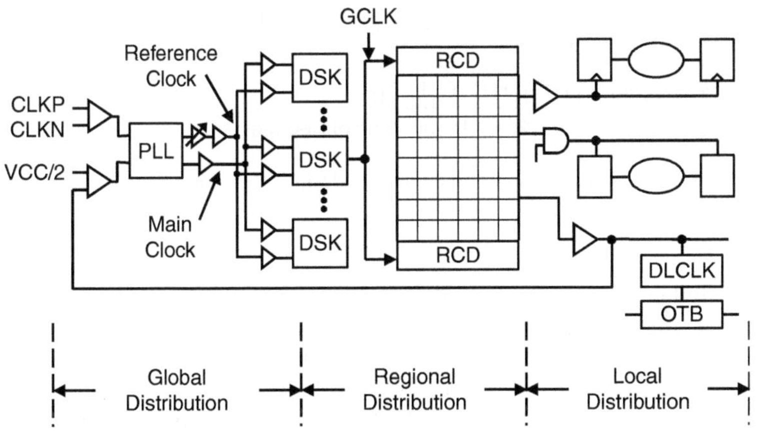


**Fig. 2.40.** Clocking architecture of the first itanium® processor. Reproduced with permission from [56], ©2000 IEEE

The 180nm Pentium® 4 processor[12] clock distribution described in Sect. 2.4 employs a hierarchical deskew architecture consisting of 47 adjustable clock zones, and 3 levels of deskew hierarchy with 46 phase detectors [49]. The left panel in Fig. 2.42 shows the deskew system architecture whereas the right panel shows the details of the deskew hierarchy. Phase detectors are placed in between the deskew hierarchies. For example, phase detectors between the single primary reference and the secondary references will detect the delay differences between the reference clock zone and the secondary zones. The clock latencies to the secondary zones are adjusted via corresponding deskew buffers (DB2–DB47). Application of hierarchical

[12] Other names and brands may be claimed as property of others.

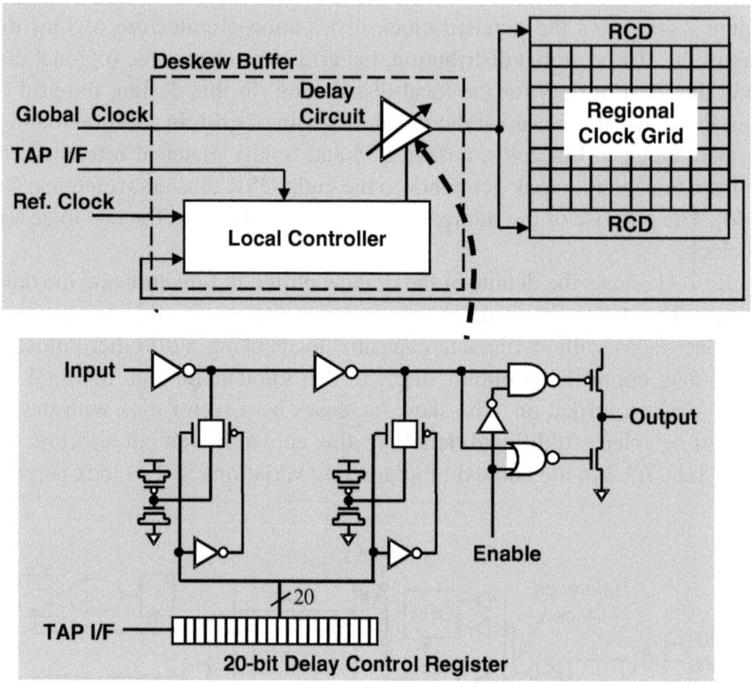


**Fig. 2.41.** Deskew controller and deskew buffer design. Reproduced with permission from [56], ©2000 IEEE

deskew eliminated the need for a tightly matched reference clock as in [56]. The deskew hierarchy depth and the phase detector quantization error should be kept low to ensure that there is no excessive accumulation of residual mismatches between the primary reference and the final clock zones. In this implementation, the depth of the deskew hierarchy is 3. Figure 2.43 shows the skew profile before and after

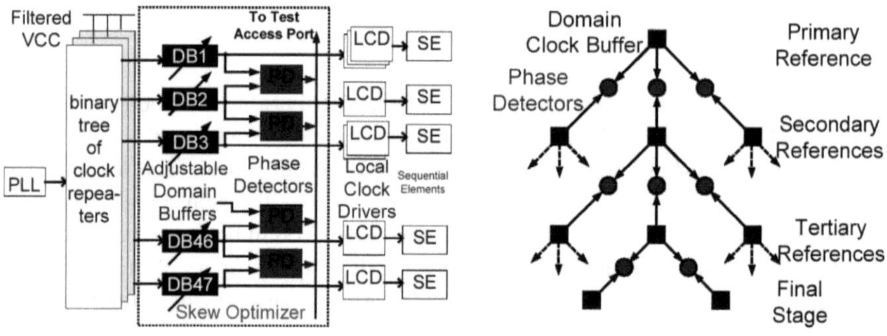


**Fig. 2.42.** Pentium® 4 processor deskew architecture. Reproduced with permission from [49], ©2001 IEEE

deskew. With deskew enabled, the in-die skew was adjusted to ±8ps and limited by the resolution of the adjustable delay elements. The preadjustment skew was at about 64ps. A hierarchical deskew architecture with deeper hierarchy was used in the 90nm dual-core Itanium® processor [43][13] and is shown in Fig. 2.44.

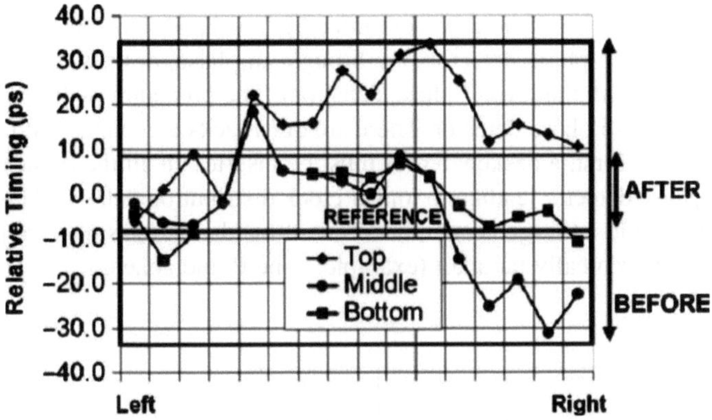


**Fig. 2.43.** Before and after skew profile of the Pentium® 4 processor. Reproduced with permission from [49], ©2001 IEEE

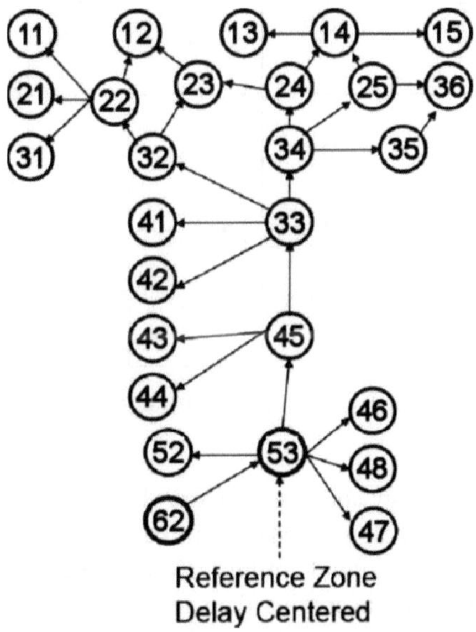


**Fig. 2.44.** Hierarchical deskew architecture of a dual-core processor. Reproduced with permission from [43], ©2005 IEEE

[13] Other names and brands may be claimed as property of others.

The hierarchical deskew scheme can be generalized to cover any neighboring clock zones in an H-tree style implementation (Fig. 2.45) or a mesh style implementation (Fig. 2.46) [57]. In the H-tree topology shown in Fig. 2.45, deskew is accomplished hierarchically. For example, the level-1 phase detector placed at the boundary between zone D and zone H will determine which of these two clocks is early and the control logic associated with this phase detector will adjust the clock delays to reduce the skew to within some predetermined guard-band. Once the zones associated with the level-1 phase detectors are brought in phase, deskew will continue in zones associated with the level-2 phase detectors. The procedure will continue until it reaches the level-4 phase detector. There are drawbacks associated with the H-tree deskew topology. First, a deskew buffer must be inserted in all the branches associated with a phase detector leading to longer clock distribution delay and higher jitter. Second, there could be large accumulated guard-bands in zones that are physically adjacent but hierarchically far apart (example: zone B and zone C).

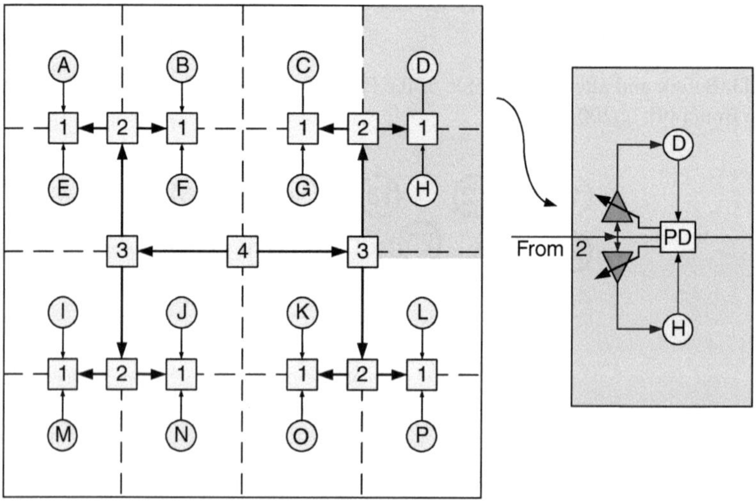


**Fig. 2.45.** H-tree deskew topology (PD=Phase detector) [57]

The mesh deskew topology addresses the drawbacks of the H-tree deskew topology. In the mesh deskew topology, deskew will be performed on all adjacent zones via averaging. For example, region C in Fig. 2.46 is deskewed with respect to zones B, D, & G simultaneously by averaging their respective phase detector outputs. To ensure stability, a mesh deskew algorithm has been proposed that takes into account potential conflicts and the impact due to guard-band accumulations [57].

Table 2.5.2 summarizes clock distribution characteristics of various commercial processors. The prevalence of adaptive skew compensation techniques is evident.

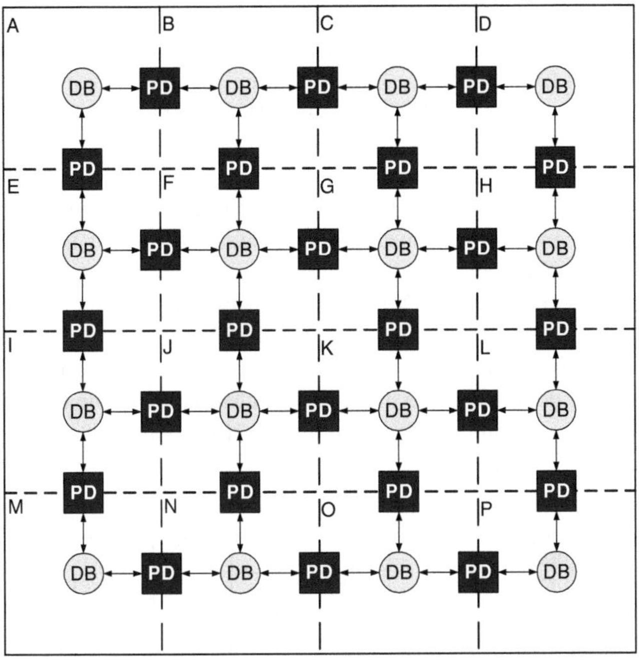


**Fig. 2.46.** Mesh deskew topology (PD = Phase detector, DB = Deskew buffer) [57]

**Table 2.3.** Clock distribution characteristics of commercial processors

| Name | Ref. | Frequency (MHz) | skew (ps) | Technology (nm) | Clock Dist. style | Deskew |
|---|---|---|---|---|---|---|
| Merom | [30] | 3,000 | 18 | 65 | Tree/Grid | Yes |
| Power6® | [20] | 5,000 | 8 | 65 | Sym. H-Tree/Grid | Yes |
| Quad-Core Opteron™ | [16] | 2,800 | 12 | 65 | Tree/Grid | |
| Xeon® processor | [45] | 3,400 | 11 | 65 | Tree/Grid | Yes |
| Itanium® 2 processor | [43] | >2,000 | 10 | 90 | Asymmetric tree | Yes |
| Power5® | [19] | >1,500 | 27 | 130 | Sym. H-Tree/Grid | No |
| Pentium® 4 processor | [29] | 3,600 | 7 | 90 | Recombinant tile | Yes |
| Itanium® 2 processor | [42] | 1,500 | 24 | 130 | Asymmetric tree | Yes |
| Power4® | [17] | >1,000 | 25 | 180 | Tree/Grid | No |
| Itanium® 2 processor | [35] | 1,000 | 52 | 180 | Asymmetric tree | No |
| Pentium® 4 processor | [8] | >2,000 | 16 | 180 | Spine/Grid | Yes |
| Itanium® processor | [36] | 800 | 28 | 180 | H-Tree/Grid | Yes |

Note: Other names and brands may be claimed as property of others

### 2.5.3 Local Clock Compensation Architecture

The concept of global clock compensation outlined in the last section and aimed at optimizing clock skews (intentional or unintentional) among global clock zones can be easily extended to the local level to address path specific clock timings. Specifically, locating critical paths (LCP) buffers with adjustable delays can replace regular local buffers with fixed delays [7, 30, 45]. Figure 2.47 shows an implementation example [30]. The local clock buffers are replaced by LCP buffers to create the LCP domains. They are controlled by the LCP control chain and the chain setting is usually determined post-silicon. Since each LCP buffer is targeted at a limited fan out, the number of LCP domains could be too many to manage. To overcome this difficulty, a processor implementation will cluster the LCP domains resulting in hundreds of LCP zones. The LCP technique is highly effective in resolving clock timing marginalities post-silicon. For example, a MAX timing path between "A" and "B" in Fig. 2.47 can be compensated by intentionally delaying the LCP buffer at the receiver (B). On the contrary, a MIN path marginality between "A" and "B" can be resolved by delaying the LCP buffer at domain A. The highly distributed nature of the LCP methodology permits fine-grain post-silicon clock timing adjustments in contemporary processor implementations.

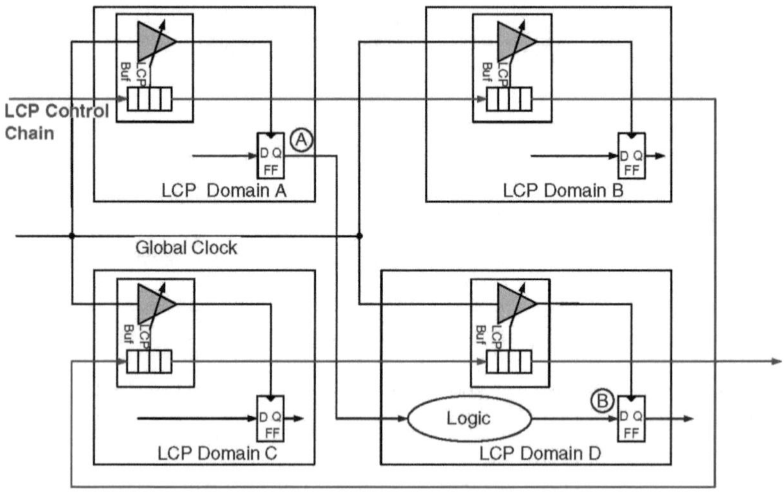


**Fig. 2.47.** Local Clock Compensation [30]

## 2.6 Elements of Clock Distribution Circuits

### 2.6.1 Clock Duty Cycle

Maintaining the clock duty cycle as close to 50% as possible is important and particularly critical for high performance processors. Specifically, high performance processors will have many phase paths in which any duty cycle distortion will unnecessarily penalize the design. Let us assume a clock period of $T_{per}$ with a duty cycle error of Q%. The corresponding high and low phase times become:

$$T_{\mathrm{phH}} = T_{\mathrm{per}} \times \frac{(50+Q)}{100}, \tag{2.29}$$

$$T_{\mathrm{phL}} = T_{\mathrm{per}} \times \frac{(50-Q)}{100}. \tag{2.30}$$

Let $F_{50\%}$ be the maximum operating frequency achievable with a 50% duty cycle. Then, the corresponding high and low phase paths will correspond to effective frequencies $F_{\mathrm{maxH}}$ and $F_{\mathrm{maxL}}$:

$$F_{\mathrm{max\,H}} = F_{50\%} \times \frac{100}{(100+2Q)}, \tag{2.31}$$

$$F_{\mathrm{max\,L}} = F_{50\%} \times \frac{100}{(100-2Q)}. \tag{2.32}$$

Figure 2.48 shows the effective $F_{\mathrm{max}}$ increase necessary for a phase path to the meet timing due to duty cycle distortion. The effective maximum frequency increase is approximately twice that of the percentage duty cycle offset from 50%.

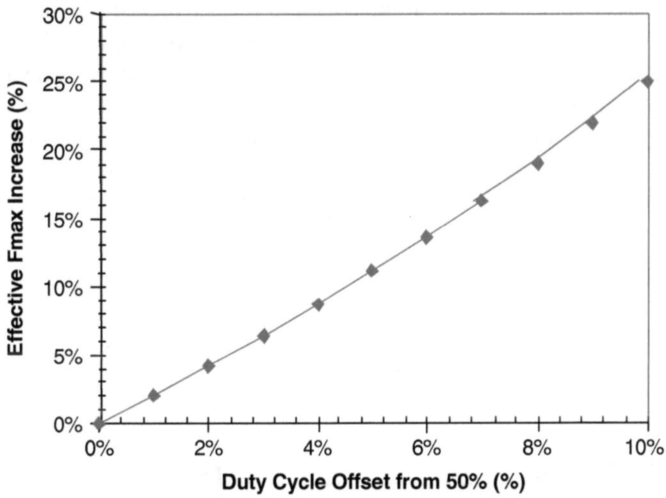


**Fig. 2.48.** $F_{\mathrm{max}}$ shift caused by duty cycle distortion in a phase-path dominated design

Clock distribution induced duty cycle error is mainly attributed to asymmetry in the clock distribution repeaters. Figure 2.49 compares a buffer-based (two inverters) clock distribution design vs. an inverter-based clock distribution. In a buffer-based design, an incoming clock edge undergoes asymmetric rise and fall edges: Two fall edges experience gate loading only whereas two rise edges experience interconnect loading. In contrast, in the inverter-based implementation, both positive and negative edges experience similar loads. By having loading symmetry between rise and fall edges, an inverter-based clock distribution network is more robust in maintaining duty cycle fidelity.

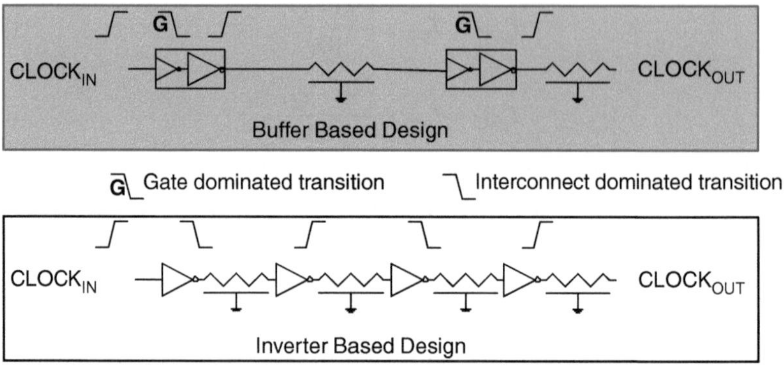


**Fig. 2.49.** Duty cycle distortion due to asymmetric edge propagation between a buffer-based clock distribution and an inverter-based clock distribution

In applications such as double data rate circuits where attaining a near perfect 50% duty cycle is critical, a differential clock distribution is the preferred solution. Sending the clock differentially improves noise immunity and duty cycle fidelity. Moreover, cross coupled stages can be added to further reduce duty cycle error by introducing constraints between true and complement clock signals. Finally, active duty cycle correction can be added to dynamically adjust the duty cycle across process, voltage, and temperature variations. Figure 2.50 shows an active duty cycle correction system with the corrector and detector circuits shown in Fig.2.51. In a differential clocking environment, the duty cycle can be easily determined by sensing the clock common mode and making the appropriate adjustments.

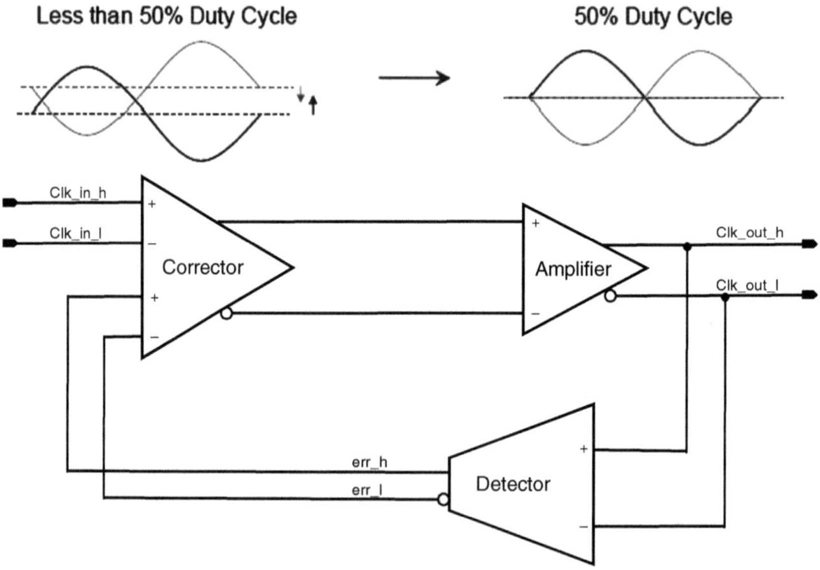


**Fig. 2.50.** Duty cycle corrector. Reproduced with permission from [13], ©2008 IEEE

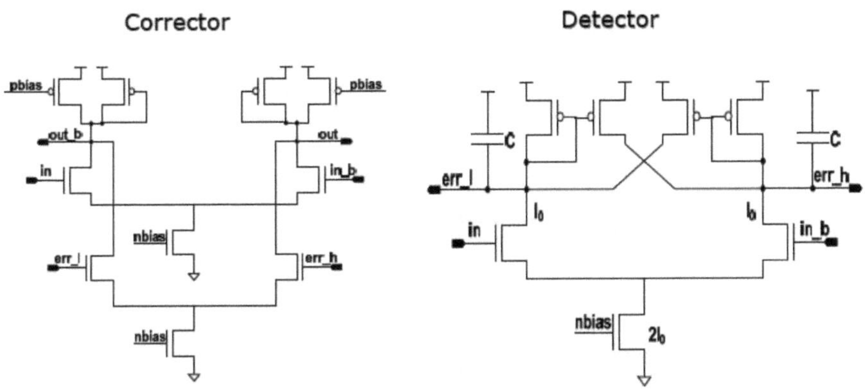


**Fig. 2.51.** Duty cycle corrector circuits. Reproduced with permission from [13], ©2008 IEEE

### 2.6.2 Power Supply

In (2.9), it was noted that the clock distribution delay variation is proportional to the distribution latency of which the delay sensitivity to power supply variation is a key factor. Power supply droop due to $\mathrm{d}i/\mathrm{d}t$ and power supply noise due to switching are two significant sources.

Power supply decoupling capacitors are used extensively to reduce the power supply droop and noise effects. Wong et al. [58] studied improving the processor timing margin by enhancing its immunity to power supply noise by compensating the clock delay against the data delay. In addition to on-die decoupling capacitors, on-die power supply filters can reduce the impact of power supply noise. Figure 2.52 shows two possible implementations [16, 49]. The implementation in [49] demonstrated a 5× reduction in power supply noise with filtering (Fig. 2.53). It should be noted that the circuits shown in Fig. 2.52, if not designed with care, can experience excessive DC power supply drop leading to delay degradations.

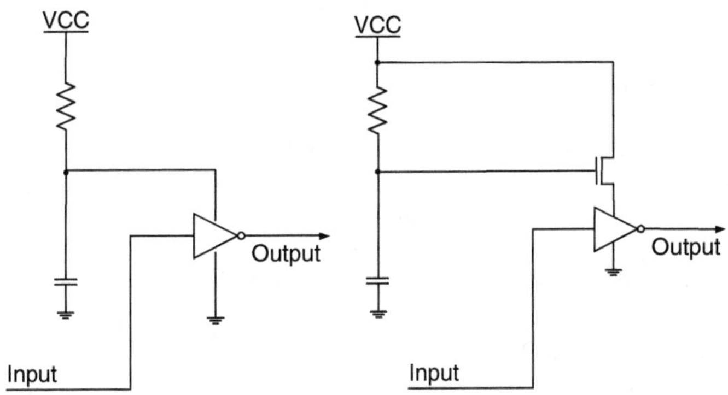


**Fig. 2.52.** Clock buffer design with power-supply filters. Reproduced with permission in a form similar to that in [49] and [16], ©2001, 2007 IEEE

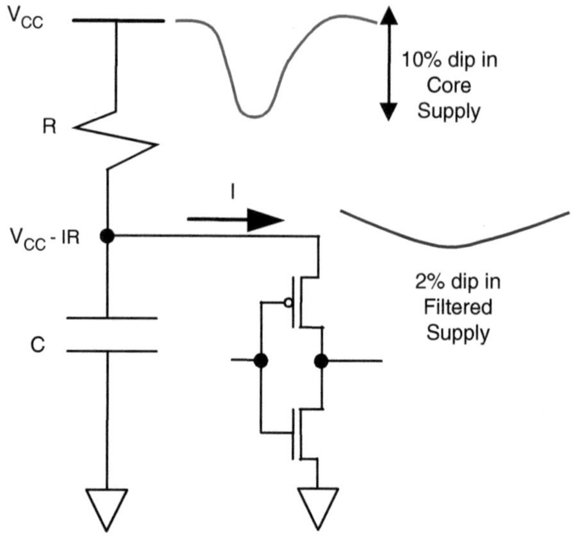


**Fig. 2.53.** On-die clock tree filter circuit. Reproduced with permission from [49], ©2001 IEEE

The reader is encouraged to review Sect. 6.8 for an analytical model of supply noise induced jitter as a function of noise frequency and amplitude and clock network latency.

## 2.7 Clock DFX Techniques

Clock DFX refers to the design-for-test or design-for-manufacturing features. DFX capabilities are critical for the design of the clock distribution network. For example, methods to measure on-die clock skew and jitter are needed for post-silicon validation of the design. Additionally, the ability to manipulate the clock edges at some specific clock cycle is a critical timing debug feature.

### 2.7.1 Optical Probing

Due to the increased number of metal layers and flip-chip packaging, optical probing from the back-side of the die is a standard technique to probe die internal nodes [59]. This technique is quite suitable for on-die clock measurement due to the periodic nature of the clock. The back-side optical probing technique monitors the infrared photons that are emitted by switching transistors. Figures 2.54 and 2.55 illustrate the emission mechanism and measurement methodology [60]. The photon emission intensity is proportional to the transistor switching current and is at a maximum during the switching transient. Therefore, the clock transition edge can be determined by correlating the photon emission peak with a time base. By monitoring the clock

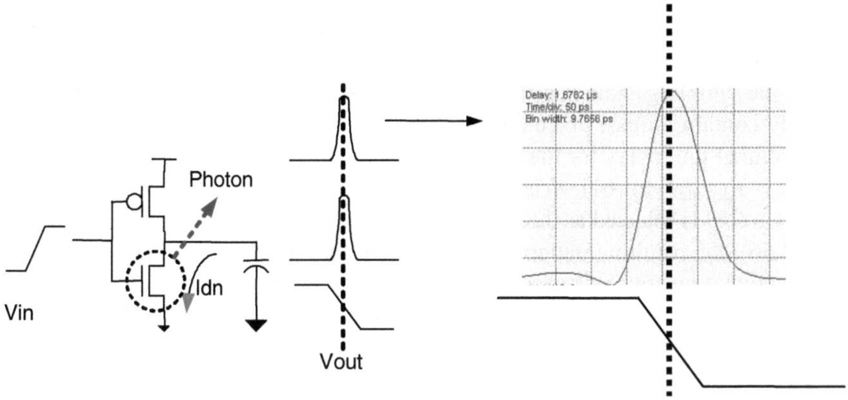


**Fig. 2.54.** Back-side optical probing technique (**a**). Reproduced with permission from [60], ©2004 IEEE

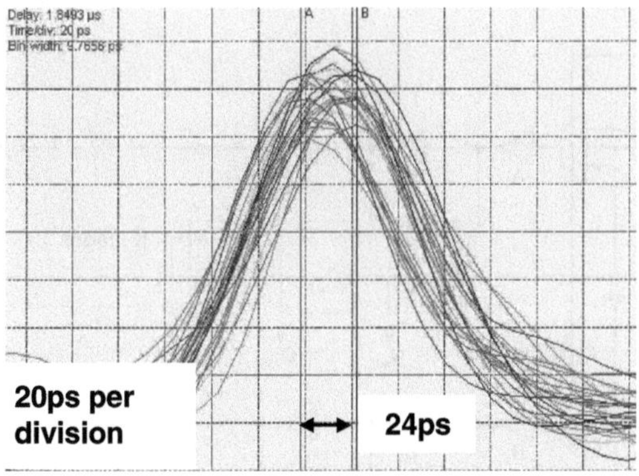


**Fig. 2.55.** Back-side optical probing technique (**b**). Reproduced with permission from [60]. ©2004 IEEE

drivers in various die locations, this technique provides static clock skew measurement capability. Limitations of this technique are instrumentation complexity, timing resolution and its inability to measure dynamic clock uncertainties (jitter) due to the intrinsic averaging effect of the measurement.

### 2.7.2 On-Die Measurement

On-die skew and jitter measurement schemes are more suitable for high volume data collection when compared to the optical back-side probing technique outlined in the last section. Figure 2.56 shows a circuit capable of measuring both skew and jitter [61]. The circuit contains an inverter based delay line (inverters $A_1$ to $A_{K+1}$) and a chain of flip-flops configured to sample the taps along the delay line. The inverters are designed to be nominally identical and with low fan out. The length of the delay line and the sampling chain $K$ is designed to be sufficiently long so that the total delay will fully contain at least one clock period at the lowest frequency of interest. If $\tau$ is the nominal unit delay for the inverter and $T_{\text{per\_max}}$ is the maximum clock period, then $K\tau > T_{\text{per\_max}}$. In [61], $K$ is set to 128 and the inverter has a nominal delay of 8ps. When $Ck_A$ is selected as the input to the delay line (via the MUX in "$\text{Ck}_{\text{Sample1}}$" in Fig. 2.56), the resulting pattern from the sampling flip-flops is shown in Fig. 2.57 with the clock transition edges denoted by two consecutive 0s or two consecutive 1s. If the number of inverter stages between the clock transition edges is $N_x$, the phase time $T_{\text{ph}}$ is bounded by:

$$N_x\tau \leq T_{\text{ph}} \leq (N_x + 1)\tau. \tag{2.33}$$

If there is no jitter on the incoming clock, the stage count $N_x$ and the pattern will be the same in every cycle. With nonzero clock jitter, the sample pattern will not be identical in every cycle and the sampled locations of the clock edges will vary.

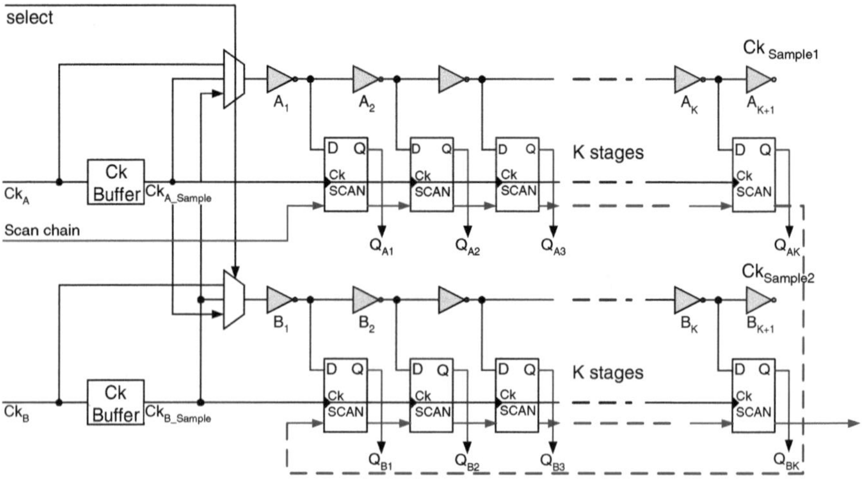


**Fig. 2.56.** Skew and jitter measurement circuit. Reproduced with permission from [61], ©2004 IEEE

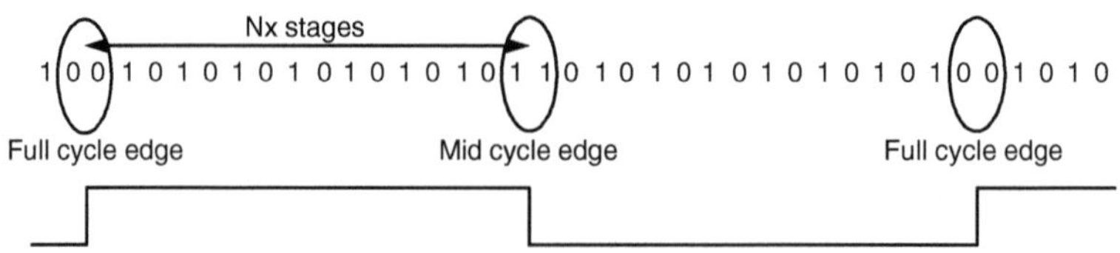


**Fig. 2.57.** Sampled delay pattern of the skew and jitter measurement circuit

To measure the clock skew between clock $Ck_A$ and $Ck_B$, the coupled configuration shown in Fig. 2.56 is used. Clock $Ck_{B\_Sample}$ is multiplexed to the delay chain in circuit "$Ck_{Sample1}$" and $Ck_{A\_Sample}$ is multiplexed to the delay chain in circuit "$Ck_{Sample2}$." If there is no skew between $Ck_A$ and $Ck_B$, the captured patterns from the two circuits will be identical. With finite skew, the difference in the position of the sampled clock edges is the measured skew.

Shortcomings of the aforementioned technique are the limited measurement resolution due to the discrete unit delay of the inverter and the delay variation among the inverters in the delay line [62, 63]. A vernier delay line (VDL) as shown in Fig. 2.58 can be used to improve the measurement resolution. With $\tau_A \neq \tau_B$ (Fig. 2.58), the timing resolution becomes $\delta = |\tau_A - \tau_B|$. Although this scheme can in principle achieve higher resolution, delay variation among the stages will impose a limitation on the results.

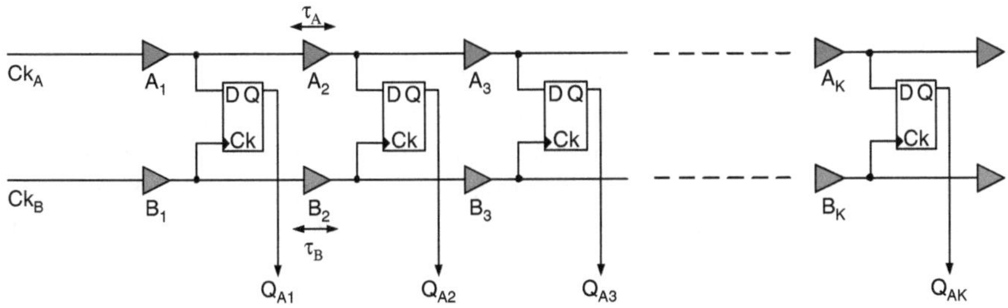


**Fig. 2.58.** Vernier delay line

Figure 2.59 shows a timing diagram of the Vernier delay line with $Ck_A$ and $Ck_B$ skewed by $3\delta$ where $\delta$ is the delay difference between $\tau_A$ and $\tau_B$. The sampled

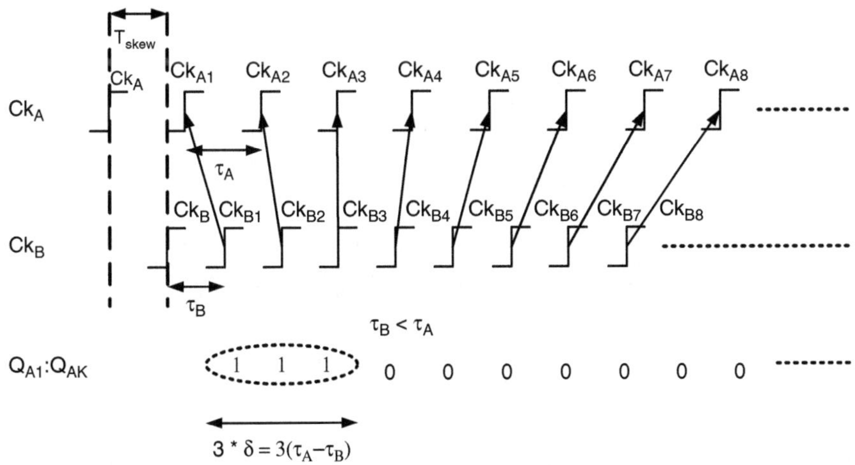


**Fig. 2.59.** Vernier delay line timing example

pattern $Q_{A1}$ to $Q_{AK}$ will show three consecutive ones representing the skew between the two clocks. In general, if the skew between the clock inputs to the VDL is $T_{skew}$ and the stage delays are $\tau_A$ and $\tau_B$ respectively, the number of ones ($N_y$) in the observed pattern will be:

$$N_y = \left\lfloor \frac{T_{skew}}{\delta} \right\rfloor, \tag{2.34}$$

$$\delta = |\tau_A - \tau_B| \tag{2.35}$$

The limitation of the VDL technique is attributed to the stage-to-stage delay variations making measurement resolutions in the sub-picosecond range difficult to achieve without incorporating additional mismatch compensation schemes.

### 2.7.3 Locating Critical Path

In addition to being an important post-silicon (local) clock compensation technique, the locating critical path (LCP) scheme is also a valuable post-silicon timing and clock debug tool [7, 30]. By intentionally skewing the clock arrival times between local clock zones, physical locations of marginal timing paths can be identified for further analysis. When coupled with the on-die clock shrink method (Sect. 2.7.4), the LCP technique has been shown to be a critical post-silicon clock and timing debug tool.

### 2.7.4 On-Die-Clock Shrink

In conjunction with the LCP technique, the on-die clock shrink (ODCS) method [56] is an effective clock edge manipulation technique that has been used extensively in post-silicon timing debug. The main concept behind ODCS is the capability to manipulate specific clock edges for phase or cycle expansion or contraction. Figure 2.60

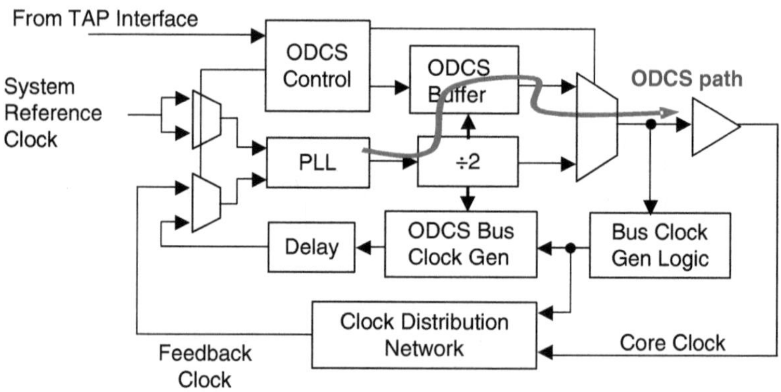


**Fig. 2.60.** On-die-clock shrink architecture. Reproduced with permission from [56] ©2000 IEEE

shows the clock generation architecture capable of supporting this feature. An ODCS buffer is inserted in series with the core clock generation path. The ODCS buffer is designed to be able to stretch or shrink the phase or cycle time.

Figure 2.61 shows an example of the ODCS operating principle. The input to the ODCS buffer is assumed to have equal high and low phase times. The initial rise and fall default setting are equal (10 arbitrary units in Fig. 2.61). With this setting, the default ODCS buffer output will also have equal high and low phase times. The settings in cycles $N+1$, $N+2$, and $N+3$ shown in Fig. 2.61 will shrink the low phase at cycle $N+1$ and then restore the clock waveform to the default.[14] Figure 2.62 summarizes the overall capabilities of the ODCS technique. Implementation of ODCS requires the incorporation of ODCS rise and fall registers and a state machine to cycle through these registers in a deterministic manner.

In an actual experiment to locate the source of the timing failure during timing debug, the default operating frequency of the device under test is first relaxed and the ODCS phase or period shrinkage is applied systematically across a specific range of tester cycles to locate the actual cycle of failure and the failure timing margin. In this manner, the failure pattern can be correlated to a specific test cycle to identify root cause.

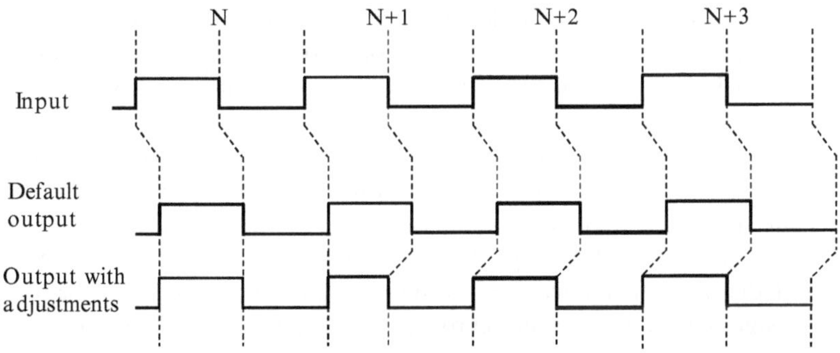


| State | Rise Setting (arbitrary time unit) | Fall Setting (arbitrary time unit) |
|---|---|---|
| Current | 10U | 10U |
| N+1 | 10U | 8U |
| N+2 | 8U | 8U |
| N+3 | 8U | 8U |

**Fig. 2.61.** ODCS clock waveform. Reproduced with permission from [56] ©2000 IEEE

[14] In this implementation, the $N+1$ "Rise" setting actually affects the rising edge at cycle $N+2$ since this setting is applied after the fall edge of cycle $N+1$. The $N+1$ "Fall" setting affects the falling edge at cycle $N+1$ because it is also applied after the rising edge of cycle $N+1$.

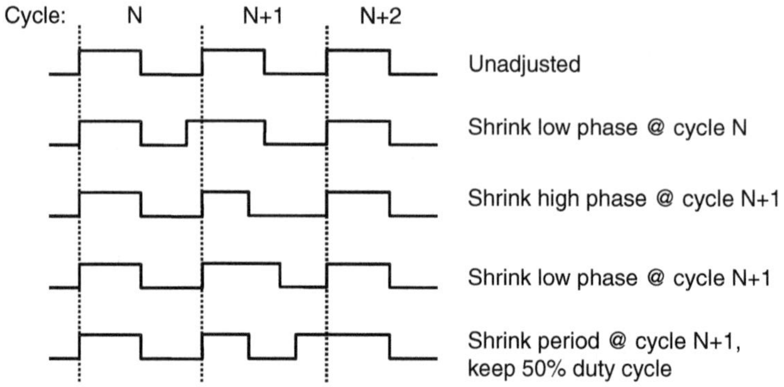


**Fig. 2.62.** ODCS capabilities. Reproduced with permission from [56] ©2000 IEEE

## 2.8 Multiclock Domain Distributions

As digital designs move towards multicores and systems-on-chip (SOC) architectures, multiple clock domains will become a prevalent design style and the clock distribution schemes will need to be enhanced to fulfill this need. A multiclock domain design typically embodies multiple islands operating synchronously and served by independent clocks within the domain and with dedicated interfaces to manage the inter-domain communications. In a multicore processor, each of the cores can be a separate clock domain. Similarly, different functional units in an SoC can be on separate clock domains.

In multicore processors and SoC designs, it is advantageous to implement multiple clock domains because this scheme provides functional flexibility for each of the domains to operate at the optimal frequency and to minimize the complexity and power associated with distributing a low skew clock to the entire die. The multidomain clock distribution architectures for multicore processors and SoCs belong to a class of designs called globally asynchronous and locally synchronous (GALS) [64]. Table 2.4 summarizes synchronization categories within the GALS class.

**Table 2.4.** Clock synchronization categories

| Type | Characteristics of distribution |
|---|---|
| Synchronous | Single distribution point-of-divergence (POD) with known static delay offsets among all the branches and single operating frequency.<br>Encompasses all of the clock distribution styles described in Sects. 2.3 and 2.4 |
| Mesochronous | Single distribution POD but with nonconstant delay offset among the branches |
| Plesiochronous | Multiple distribution PODs but with nominally identical frequency among all the domains |
| Heterochronous | Multiple distribution PODs with nominally different operating frequencies among the domains |

Figure 2.63 is a generic illustration of the GALS design style. Multiple clock domains are embedded in a single silicon die. The chip may receive multiple copies of the system clock and multiple PLLs are used to generate the clocks for each of the synchronous units. Clocks Ck1, Ck2, ..., Ck4 may have different phases or frequencies. The clock distribution within Each synchronous unit will rely on any of the topologies described in Sect. 2.3 to achieve low skew and fully synchronous operation.

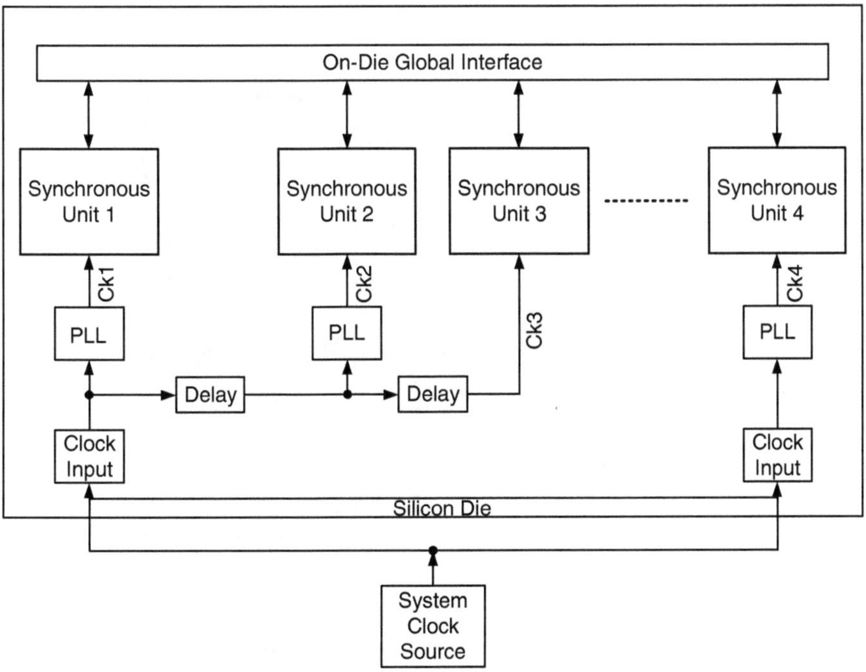


**Fig. 2.63.** Globally synchronous and locally synchronous architecture. Reproduced with permission from [64], ©2007 IEEE

Since each of the synchronous units can operate at different frequency and phase from the rest, the on-die global interfaces will be responsible for the data transfer among the domains. Many modern multicore processors and SoCs only adopt the loosely synchronous styles of Table 2.4 to avoid the significant complexity associated with truly asynchronous design.

### 2.8.1 Multicore Processor Clock Distribution

A plesiochronous clock distribution involves multiple distribution PODs but with a nominally identical operating frequency. The 65 nm dual core processor described in [45] is an example. Figure 2.64 illustrates this scheme which consists of two domains for the two cores and the uncore and *I/O* domain with the interface operating at the same frequency. The system clock input to the three clock generators (PLLs) is routed in the package. There are three independent distribution PODs for the cores and the uncore. A recombinant style distribution for low skew is used in the core whereas the uncore employs a hybrid spine-grid structure for lower power and simpler implementation. Data communication between the cores and the uncore is accomplished via the pipelined deskew logic (PDSL) interfaces operating at the same frequency. Introducing independent clock domains for the cores permitted a modular design.

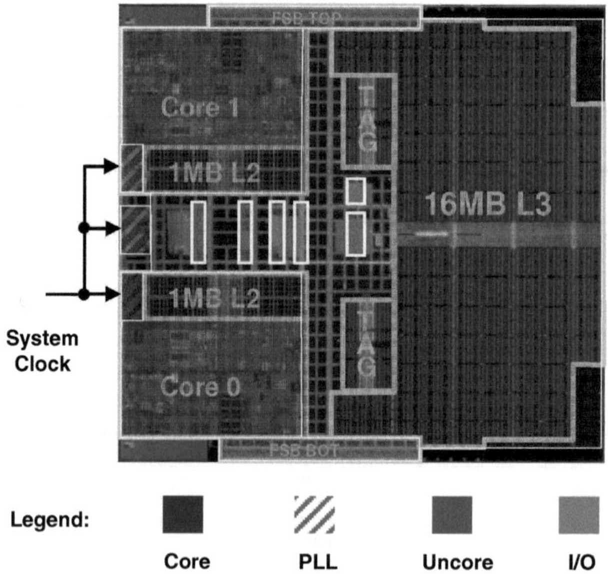


**Fig. 2.64.** Multidomain clocking in a dual-core processor. Reproduced with permission from [45], ©2006 IEEE

A mesochronous clock distribution has a single distribution POD while exhibiting nonconstant delays to the branches. The 80-tile processor design in [65, 66] is one example of a mesochronous clock distribution. This processor has a single PLL and therefore a single distribution POD (Fig. 2.65). Distributing the global clock to the respective tiles is achieved using chains of clock buffers embedded within each tile. The daisy-chaining of the global clock buffers suggests nonequal clock latencies to each processor tile. The systematic clock skews across the processor tiles help spread the clock switching power over an entire cycle. Interprocessor-tile communication is achieved with a skew insensitive FIFO interface [66]. Within each processor tile, a balanced H-tree is responsible for the distribution and all circuits within each processor tile operate synchronously. Figure 2.65 (left) shows details of the distribution scheme. Figure 2.65 (right) shows the relative arrival times of the clocks at the processor tiles. In a processor design having a large number of processor cores operating at the same frequency, this clock distribution style has the potential of reducing the impact of simultaneous switching on the on-chip power delivery network.

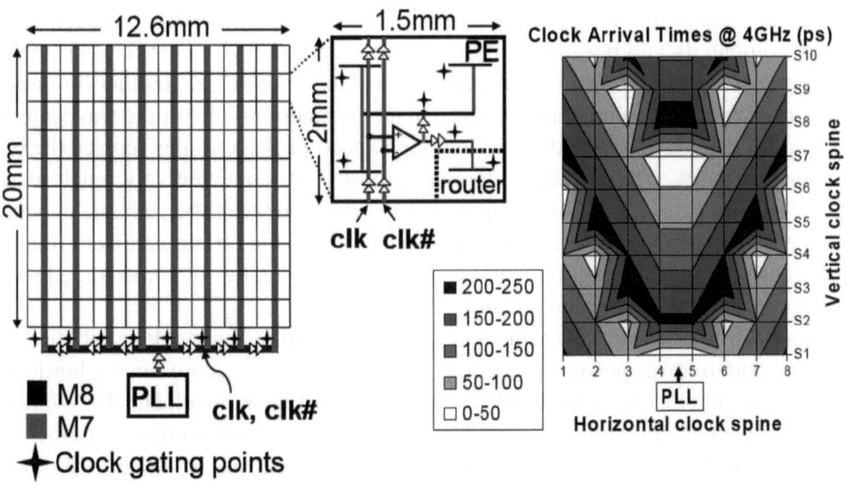


**Fig. 2.65.** Clock distribution of an 80-tile processor design. Reproduced with permission from [66], ©2008 IEEE

A heterochronous clock distribution has multiple distribution PODs and can support nominally different operating frequencies across the domains [12, 13, 16, 67]. The 90 nm 2-core Itanium® processor [12] and the 65 nm quad-core Itanium® processor [13] have adaptive core frequency switching implemented with multiple PLLs. Low failure probability FIFOs are placed at the interface of the corresponding clock domains. Similarly, [16] and [67] have independent PLLs for the cores, uncore, and the *I/O* that are capable of operating at different frequencies. Clock domain crossing is accomplished with low-latency FIFO schemes.

Despite the potential design benefits of the loosely synchronous methods described above, the requirement of a deterministic domain crossing will impose higher design complexity at the interfaces. FIFO-like structures are used at the cost of additional transfer latency in order to compensate for the higher crossdomain clock uncertainties. The PDSL interface in [45] incurs two additional cycles of latency when compared to a single-cycle transfer in a fully synchronous design. Similarly, [66] incorporates a 4-deep FIFO to ensure robust crossdomain transfer. Better partitioning that reduces the crossdomain traffic will reduce the performance penalty due to the increased synchronization latency.

## 2.9 Future Directions

Alternative clock distribution methods have been proposed in the literature. Among them are optical clock distribution [68, 69], package level clock distribution [70], clock network with distributed PLLs [71] and resonant clocking [72, 73].

The main advantage of an optical clock distribution scheme vs. a conventional one is the potential for higher performance and short-term jitter reduction. This is due to the ability of an optical distribution network to have fewer repeaters and be first-order immune to on-chip power supply noise. A significant disadvantage is the design complexity and performance limitations (e.g. power & electrical robustness) of the signal conversion from optical to electrical and vice versa. Another limitation is due to the process complexity involved in the formation of the on-die waveguides.

In a package level clock distribution, high speed package routes replace the on-die clock network. The package interconnect will have comparatively much lower *RC* delay. The main limitation is due to the interface between the package network and the on-die receivers. Additional skew and jitter at the on-die receivers (due to buffer design and ESD requirements) and testing complexities in prepackaged parts are the most important implementation impediments.

Application of on-die resonant clocking is a very promising clock distribution alternative due to lower power dissipation and jitter reduction. A significant hurdle of on-die resonant clocking is in the physical limitation of the network and the naturally narrow frequency operating window. Numerous enhancements to resonant clocking have been proposed and some of these techniques can become practical. A resonant clock distribution network was applied to a 90nm processor by adding a new top layer metal for the *LC* tank inductors and capacitors [74]. This design demonstrated basic functionality and power savings at frequencies above 3GHz. A detailed discussion of resonant clocking techniques is presented in Chap. 4.

All of the alternatives above try to address limitations of electrical clock interconnect such as power and noise. With the strong emergence of multidomain clock distribution in multicore processors and SOCs, constraints due to the very large on-die electrical interconnect are significantly alleviated. Since the industry is moving away from a single large distribution network with long latency and large number of clock buffers, on-die electrical clock distribution will continue to be the predominant clocking technology in the near future.

## 2.10 Conclusion

In this chapter, the requirements for synchronous clock distributions have been described. Numerous clock distribution topologies have been described with spine-grid and tree-grid topologies emerging as the dominant ones. On-die deskew has became common practice in order to contain skew without excessive design complexity and use of interconnect resources. Various clock debug features have been summarized that encompass circuits for skew and jitter measurement, design for test and critical path location. Additionally, the chapter has also outlined the basics of multidomain clock distribution topologies that are becoming the industry norm. Finally, the chapter concludes by mentioning three alternative clock distribution techniques that may demonstrate practical design benefits in the future.

## References


[1] J. Schutz, "A CMOS 100MHz microprocessor," in *Digest of Technical Papers IEEE International Solid-State Circuits Conference (ISSCC 1991)*, 1991, pp. 90–91, 294.

[2] J. Schutz, "A 3.3V 0.6μm BiCMOS superscalar microprocessor," in *Digest of Technical Papers IEEE International Solid-State Circuits Conference (ISSCC 1994)*, 1994, pp. 202–203.

[3] R. Colwell and R. Steck, "A 0.6μm BiCMOS processor with dynamic execution," in *Digest of Technical Papers IEEE International Solid-State Circuits Conference (ISSCC 1995)*, 1995, pp. 176–177, 361.

[4] M. Choudhury and J. Miller, "A 300 MHz CMOS microprocessor with multimedia technology," in *Digest of Technical Papers IEEE International Solid-State Circuits Conference (ISSCC 1997)*, 1997, pp. 170–171, 450.

[5] J. Schutz and R. Wallace, "A 450MHz IA32 P6 family microprocessor," in *Digest of Technical Papers IEEE International Solid-State Circuits Conference (ISSCC 1998)*, 1998, pp. 236–237.

[6] P. Green, "A 1GHz IA-32 architecture microprocessor implemented on 0.18μ technology with aluminum interconnect," in *Digest of Technical Papers IEEE International Solid-State Circuits Conference (ISSCC 2000)*, 2000, pp. 98–99, 449.

[7] J. Schutz and C. Webb, "A scalable x86 CPU design for 90nm process," in *Digest of Technical Papers IEEE International Solid-State Circuits Conference (ISSCC 2004)*, 2004, pp. 62–63, 513.

[8] N. Kurd, J. Barkatullah, R. Dizon, T. Fletcher, and P. Madland, "Multi-GHz clocking scheme for Intel® Pentium® 4 microprocessor," in *Digest of Technical Papers IEEE International Solid-State Circuits Conference (ISSCC 2001)*, 2001, pp. 404–405.

[9] G. Singer and S. Rusu, "The first IA-64 microprocessor: a design for highly-parallel execution," in *Digest of Technical Papers IEEE International Solid-State Circuits Conference (ISSCC 2000)*, 2000, pp. 422–423.

[10] S. Naffziger and G. Hammond, "The implementation of the next-generation 64 b Itanium™ microprocessor," in *Digest of Technical Papers IEEE International Solid-State Circuits Conference (ISSCC 2002)*, vol. 1, 2002, pp. 344–345, 472.

[11] J. Stinson and S. Rusu, "A 1.5GHz third generation Itanium® processor," in *Digest of Technical Papers IEEE International Solid-State Circuits Conference (ISSCC 2003)*, 2003, pp. 252–253, 492.

[12] S. Naffziger, B. Stackhouse, and T. Grutkowski, "The implementation of a 2-core multi-threaded Itanium family processor," in *Digest of Technical*

*Papers IEEE International Solid-State Circuits Conference (ISSCC 2005)*, 2005, pp. 182–183, 592.

[13] B. Stackhouse, B. Cherkauer, M. Gowan, P. Gronowski, and C. Lyles, "A 65nm 2-billion-transistor quad-core Itanium® processor," in *Digest of Technical Papers IEEE International Solid-State Circuits Conference (ISSCC 2008)*, 2008, pp. 92–93, 598.

[14] S. Rusu, S. Tam, H. Muljono, D. Ayers, and J. Chang, "A dual-core multi-threaded Xeon® processor with 16MB L3 cache," in *Digest of Technical Papers IEEE International Solid-State Circuits Conference (ISSCC 2006)*, 2006, pp. 315–324.

[15] M. Golden, S. Arekapudi, G. Dabney, M. Haertel, S. Hale, L. Herlinger, Y. Kim, K. McGrath, V. Palisetti, and M. Singh, "A 2.6GHz dual-core 64bx86 microprocessor with DDR2 memory support," in *Digest of Technical Papers IEEE International Solid-State Circuits Conference (ISSCC 2006)*, 2006, pp. 325–332.

[16] J. Dorsey, S. Searles, M. Ciraula, S. Johnson, N. Bujanos, D. Wu, M. Braganza, S. Meyers, E. Fang, and R. Kumar, "An integrated quad-core Opteron processor," in *Digest of Technical Papers IEEE International Solid-State Circuits Conference (ISSCC 2007)*, 2007, pp. 102–103.

[17] P. Restle, C. Carter, J. Eckhardt, B. Krauter, B. McCredie, K. Jenkins, A. Weger, and A. Mule, "The clock distribution of the Power4 microprocessor," in *Digest of Technical Papers IEEE International Solid-State Circuits Conference (ISSCC 2002)*, vol. 1, 3–7 Feb. 2002, pp. 144–145.

[18] P. Hofstee, N. Aoki, D. Boerstler, P. Coulman, S. Dhong, B. Flachs, N. Kojima, O. Kwon, K. Lee, D. Meltzer, K. Nowka, J. Park, J. Peter, S. Posluszny, M. Shapiro, J. Silberman, O. Takahashi, and B. Weinberger, "A 1GHz single-issue 64b PowerPC processor," in *Digest of Technical Papers IEEE International Solid-State Circuits Conference (ISSCC 2000)*, 2000, pp. 92–93.

[19] J. Clabes, J. Friedrich, M. Sweet, J. Dilullo, S. Chu, D. Plass, J. Dawson, P. Muench, L. Powell, M. Floyd, B. Sinharoy, M. Lee, M. Goulet, J. Wagoner, N. Schwartz, S. Runyon, G. Gorman, P. Restle, R. Kalla, J. McGill, and S. Dodson, "Design and implementation of the POWER5$^{TM}$ microprocessor," in *Digest of Technical Papers IEEE International Solid-State Circuits Conference (ISSCC 2004)*, 2004, pp. 56–57.

[20] J. Friedrich, B. McCredie, N. James, B. Huott, B. Curran, E. Fluhr, G. Mittal, E. Chan, Y. Chan, D. Plass, S. Chu, H. Le, L. Clark, J. Ripley, S. Taylor, J. Dilullo, and M. Lanzerotti, "Design of the POWER6 microprocessor," in *Digest of Technical Papers IEEE International Solid-State Circuits Conference (ISSCC 2007)*, 2007, pp. 96–97.

[21] C. Webb, C. Anderson, L. Sigal, K. Shepard, J. Liptay, J. Warnock, B. Curran, B. Krumm, M. Mayo, P. Camporese, E. Schwarz, M. Farrell, P. Restle, I. Averill, R.M., T. Slegel, W. Houtt, Y. Chan, B. Wile, T. Nguyen, P. Emma, D. Beece, C.-T. Chuang, and C. Price, "A 400-MHz S/390 microprocessor," *IEEE Journal of Solid-State Circuits*, vol. 32, no. 11, pp. 1665–1675, Nov. 1997.

[22] G. Northrop, R. Averill, K. Barkley, S. Carey, Y. Chan, Y. Chan, M. Check, D. Hoffman, W. Huott, B. Krumm, C. Krygowski, J. Liptay, M. Mayo, T. McNamara, T. McPherson, E. Schwarz, L. Siegel, C. Webb, D. Webber, and P. Williams, "600MHz G5 S/390 microprocessor," in *Digest of Technical Papers IEEE International Solid-State Circuits Conference (ISSCC 1999)*, 1999, pp. 88–89.

[23] T. McPherson, R. Averill, D. Balazich, K. Barkley, S. Carey, Y. Chan, Y. Chan, R. Crea, A. Dansky, R. Dwyer, A. Haen, D. Hoffman, A. Jatkowski, M. Mayo, D. Merrill, T. McNamara, G. Northrop, J. Rawlins, L. Sigal, T. Slegel, D. Webber, P. Williams, and F. Yee, "760 MHz G6 S/390 microprocessor exploiting multiple Vt and copper interconnects," in *Digest of Technical Papers IEEE International Solid-State Circuits Conference (ISSCC 2000)*, 2000, pp. 96–97.

[24] D. Dobberpuhl, R. Witek, R. Allmon, R. Anglin, D. Bertucci, S. Britton, L. Chao, R. Conrad, D. Dever, B. Gieseke, S. Hassoun, G. Hoeppner, K. Kuchler, M. Ladd, B. Leary, L. Madden, E. McLellan, D. Meyer, J. Montanaro, D. Priore, V. Rajagopalan, S. Samudrala, and S. Santhanam, "A 200-MHz 64-b dual-issue CMOS microprocessor," *IEEE J. Solid-State Circuits*, vol. 27, no. 11, pp. 1555–1567, Nov. 1992.

[25] B. Benschneider, A. Black, W. Bowhill, S. Britton, D. Dever, D. Donchin, R. Dupcak, R. Fromm, M. Gowan, P. Gronowski, M. Kantrowitz, M. Lamere, S. Mehta, J. Meyer, R. Mueller, A. Olesin, R. Preston, D. Priore, S. Santhanam, M. Smith, and G. Wolrich, "A 300-mhz 64-b quad-issue CMOS RISC microprocessor," *IEEE J. Solid-State Circuits*, vol. 30, no. 11, pp. 1203–1214, Nov. 1995.

[26] P. Gronowski, W. Bowhill, D. Donchin, R. Blake-Campos, D. Carlson, E. Equi, B. Loughlin, S. Mehta, R. Mueller, A. Olesin, D. Noorlag, and R. Preston, "A 433-MHz 64-b quad-issue RISC microprocessor," *IEEE Journal of Solid-State Circuits*, vol. 31, no. 11, pp. 1687–1696, Nov. 1996.

[27] D. Bailey and B. Benschneider, "Clocking design and analysis for a 600-mhz Alpha microprocessor," *IEEE J. Solid-State Circuits*, vol. 33, no. 11, pp. 1627–1633, Nov. 1998.

[28] T. Xanthopoulos, D. Bailey, A. Gangwar, M. Gowan, A. Jain, and B. Prewitt, "The design and analysis of the clock distribution network for a 1.2 GHz Alpha microprocessor," in *Digest of Technical Papers IEEE International Solid-State Circuits Conference (ISSCC 2001)*, 2001, pp. 402–403.

[29] N. Bindal, T. Kelly, N. Velastegui, and K. Wong, "Scalable sub-10ps skew global clock distribution for a 90nm multi-GHz IA microprocessor," in *Digest of Technical Papers IEEE International Solid-State Circuits Conference (ISSCC 2003)*, 2003, pp. 346–347, 498.

[30] N. Sakran, M. Yuffe, M. Mehalel, J. Doweck, E. Knoll, and A. Kovacs, "The implementation of the 65nm dual-core 64b Merom processor," in *Digest of Technical Papers IEEE International Solid-State Circuits Conference (ISSCC 2007)*, 2007, pp. 106–107, 590.

[31] G. Geannopoulos and X. Dai, "An adaptive digital deskewing circuit for clock distribution networks," in *Visuals Supplement IEEE International Solid-State Circuits Conference (ISSCC 1998)*, 1998, pp. 326–327, 468–469.
[32] G. Geannopoulos and X. Dai, "An adaptive digital deskewing circuit for clock distribution networks," in *Digest of Technical Papers IEEE International Solid-State Circuits Conference (ISSCC 1998)*, 1998, pp. 400–401.
[33] A. Chamas, A. Dalal, P. deDood, P. Ferolito, B. Frederick, O. Geva, D. Greenhill, H. Hingarh, J. Kaku, L. Kohn, L. Lev, M. Levitt, R. Melanson, S. Mitra, R. Sundar, M. Tamjidi, P. Wang, D. Wendell, R. Yu, and G. Zyner, "A 64 b microprocessor with multimedia support," in *Digest of Technical Papers IEEE International Solid-State Circuits Conference (ISSCC 1995)*, 1995, pp. 178–179.
[34] G. Konstadinidis, K. Normoyle, S. Wong, S. Bhutani, H. Stuimer, T. Johnson, A. Smith, D. Cheung, F. Romano, S. Yu, S.-H. Oh, V. Melamed, S. Narayanan, D. Bunsey, C. Khieu, K. Wu, R. Schmitt, A. Dumlao, M. Sutera, J. Chau, K. Lin, and W. Coates, "Implementation of a third-generation 1.1-GHz 64-bit microprocessor," *IEEE Journal of Solid-State Circuits*, vol. 37, no. 11, pp. 1461–1469, 2002.
[35] F. Anderson, J. Wells, and E. Berta, "The core clock system on the next generation Itanium$^{TM}$ microprocessor," in *Digest of Technical Papers IEEE International Solid-State Circuits Conference (ISSCC 2002)*, vol. 1, 2002, pp. 146–147, 453.
[36] S. Rusu and S. Tam, "Clock generation and distribution for the first IA-64 microprocessor," in *Digest of Technical Papers IEEE International Solid-State Circuits Conference (ISSCC 2000)*, 2000, pp. 176–177.
[37] M. Golden, S. Hesley, A. Scherer, M. Crowley, S. Johnson, S. Meier, D. Meyer, J. Moench, S. Oberman, H. Partovi, F. Weber, S. White, T. Wood, and J. Yong, "A seventh-generation x86 microprocessor," *IEEE Journal of Solid-State Circuits*, vol. 34, no. 11, pp. 1466–1477, 1999.
[38] D. Draper, M. Crowley, J. Holst, G. Favor, A. Schoy, J. Trull, A. Ben-Meir, R. Khanna, D. Wendell, R. Krishna, J. Nolan, D. Mallick, H. Partovi, M. Roberts, M. Johnson, and T. Lee, "Circuit techniques in a 266-MHz MMX-enabled processor," *IEEE Journal of Solid-State Circuits*, vol. 32, no. 11, pp. 1650–1664, 1997.
[39] G. Gerosa, M. Alexander, J. Alvarez, C. Croxton, M. D'Addeo, A. Kennedy, C. Nicoletta, J. Nissen, R. Philip, P. Reed, H. Sanchez, S. Taylor, and B. Burgess, "A 250-MHz 5-w PowerPC microprocessor with on-chip L2 cache controller," *IEEE Journal of Solid-State Circuits*, vol. 32, no. 11, pp. 1635–1649, 1997.
[40] E. Cohen, J. Ballard, J. Blomgren, C. Brashears, V. Moldenhauer, and J. Pattin, "A 533 MHz BiCMOS superscalar microprocessor," in *Digest of Technical Papers IEEE International Solid-State Circuits Conference (ISSCC 1997)*, 1997, pp. 164–165, 448.

[41] E. Fayneh and E. Knoll, "Clock generation and distribution for intel Banias mobile microprocessor," in *Proceedings of Digest of Technical Papers VLSI Circuits 2003 Symposium*, 2003, pp. 17–20.

[42] S. Tam, U. Desai, and R. Limaye, "Clock generation and distribution for the third generation Itanium® processor," in *Proceedings of Digest of Technical Papers VLSI Circuits 2003 Symposium*, 2003, pp. 9–12.

[43] P. Mahoney, E. Fetzer, B. Doyle, and S. Naffziger, "Clock distribution on a dual-core, multi-threaded Itanium®-family processor," in *Digest of Technical Papers IEEE International Solid-State Circuits Conference (ISSCC 2005)*, 2005, pp. 292–293, 599.

[44] D. Pham, S. Asano, M. Bolliger, M. Day, H. Hofstee, C. Johns, J. Kahle, A. Kameyama, J. Keaty, Y. Masubuchi, M. Riley, D. Shippy, D. Stasiak, M. Suzuoki, M. Wang, J. Warnock, S. Weitzel, D. Wendel, T. Yamazaki, and K. Yazawa, "The design and implementation of a first-generation CELL processor," in *Digest of Technical Papers IEEE International Solid-State Circuits Conference (ISSCC 2005)*, 2005, pp. 184–185,592 Vol. 1.

[45] S. Tam, J. Leung, R. Limaye, S. Choy, S. Vora, and M. Adachi, "Clock generation and distribution of a dual-core Xeon® processor with 16MB L3 cache," in *Digest of Technical Papers IEEE International Solid-State Circuits Conference (ISSCC 2006)*, 2006, pp. 1512–1521.

[46] J. Hart, K. Lee, D. Chen, L. Cheng, C. Chou, A. Dixit, D. Greenley, G. Gruber, K. Ho, J. Hsu, N. Malur, and J. Wu, "Implementation of a fourth-generation 1.8-GHz dual-core SPARC V9 microprocessor," *IEEE Journal of Solid-State Circuits*, vol. 41, no. 1, pp. 210–217, 2006.

[47] D. Boerstler, "A low-jitter PLL clock generator for microprocessors with lock range of 340–612 MHz," *IEEE J. Solid-State Circuits*, vol. 34, no. 4, pp. 513–519, April 1999.

[48] D. Boerstler and K. Jenkins, "A phase-locked loop clock generator for a 1 GHz microprocessor," in *Proceedings of Digest of Technical Papers VLSI Circuits 1998 Symposium*, 11–13 June 1998, pp. 212–213.

[49] N. Kurd, J. Barkarullah, R. Dizon, T. Fletcher, and P. Madland, "A multigigahertz clocking scheme for the Pentium® 4 microprocessor," *IEEE Journal of Solid-State Circuits*, vol. 36, no. 11, pp. 1647–1653, Nov. 2001.

[50] N. Bindal, T. Kelly, N. Velastegui, and K. Wong, "Scalable sub-10ps skew global clock distribution for a 90nm multi-GHz IA microprocessor," in *Visuals Supplement IEEE International Solid-State Circuits Conference (ISSCC 2003)*, 2003, pp. 281–284.

[51] P. Restle, K. Jenkins, A. Deutsch, and P. Cook, "Measurement and modeling of on-chip transmission line effects in a 400MHz microprocessor," *IEEE Journal of Solid-State Circuits*, vol. 33, no. 4, pp. 662–665, April 1998.

[52] J. Chang, S. Rusu, J. Shoemaker, S. Tam, M. Huang, M. Haque, S. Chiu, K. Truong, M. Karim, G. Leong, K. Desai, R. Goe, and S. Kulkarni, "A 130-nm triple-$V_t$ 9-MB third-level on-die cache for the 1.7-GHz Itanium® 2 processor," *IEEE Journal of Solid-State Circuits*, vol. 40, no. 1, pp. 195–203, Jan. 2005.

[53] P. Gronowski, W. Bowhill, R. Preston, M. Gowan, and R. Allmon, "High-performance microprocessor design," *IEEE Journal of Solid-State Circuits*, vol. 33, no. 5, pp. 676–686, May 1998.
[54] G. Gerosa, S. Curtis, M. D'Addeo, B. Jiang, B. Kuttanna, F. Merchant, B. Patel, M. Taufique, and H. Samarchi, "A sub-1w to 2w low-power IA processor for mobile internet devices and ultra-mobile PCs in 45nm Hi-K metal gate CMOS," in *Digest of Technical Papers IEEE International Solid-State Circuits Conference (ISSCC 2008)*, 2008, pp. 256–257, 611.
[55] G. Gerosa, S. Curtis, M. D'Addeo, B. Jiang, B. Kuttanna, F. Merchant, B. Patel, M. Taufique, and H. Samarchi, "A sub-1w to 2w low-power IA processor for mobile internet devices and ultra-mobile PCs in 45nm Hi-K metal gate CMOS," in *Visuals Supplement IEEE International Solid-State Circuits Conference (ISSCC 2008)*, 2008.
[56] S. Tam, S. Rusu, U. Nagarji Desai, R. Kim, J. Zhang, and I. Young, "Clock generation and distribution for the first IA-64 microprocessor," *IEEE Journal of Solid-State Circuits*, vol. 35, no. 11, pp. 1545–1552, Nov. 2000.
[57] C. Dike, N. Kurd, P. Patra, and J. Barkatullah, "A design for digital, dynamic clock deskew," in *Proceedings of Digest of Technical Papers VLSI Circuits 2003 Symposium*, 12–14 June 2003, pp. 21–24.
[58] K. Wong, T. Rahal-Arabi, M. Ma, and G. Taylor, "Enhancing microprocessor immunity to power supply noise with clock-data compensation," *IEEE Journal of Solid-State Circuits*, vol. 41, no. 4, pp. 749–758, April 2006.
[59] S. Rusu, S. Seidel, G. Woods, D. Grannes, H. Muljono, J. Rowlette, and K. Petrosky, "Backside infrared probing for static voltage drop and dynamic timing measurements," in *Digest of Technical Papers IEEE International Solid-State Circuits Conference (ISSCC 2001)*, 2001, pp. 276–277, 454.
[60] S. Tam, R. Limaye, and U. Desai, "Clock generation and distribution for the 130nm Itanium® 2 processor with 6-MB on-die L3 cache," *IEEE Journal of Solid-State Circuits*, vol. 39, no. 4, pp. 636–642, April 2004.
[61] P. Restle, R. Franch, N. James, W. Huott, T. Skergan, S. Wilson, N. Schwartz, and J. Clabes, "Timing uncertainty measurements on the Power5 microprocessor," in *Digest of Technical Papers IEEE International Solid-State Circuits Conference (ISSCC 2004)*, 2004, pp. 354–355.
[62] P. Dudek, S. Szczepanski, and J. Hatfield, "A high-resolution CMOS time-to-digital converter utilizing a Vernier delay line," *IEEE J. Solid-State Circuits*, vol. 35, no. 2, pp. 240–247, Feb. 2000.
[63] T. Hashimoto, H. Yamazaki, A. Muramatsu, T. Sato, and A. Inoue, "Time-to-digital converter with vernier delay mismatch compensation for high resolution on-die clock jitter measurement," in *Proceedings of IEEE Symposium on VLSI Circuits*, 18–20 June 2008, pp. 166–167.
[64] P. Teehan, M. Greenstreet, and G. Lemieux, "A survey and taxonomy of GALS design styles," *IEEE Design & Test of Computers*, vol. 24, no. 5, pp. 418–428, Sept.–Oct. 2007.
[65] S. Vangal, J. Howard, G. Ruhl, S. Dighe, H. Wilson, J. Tschanz, D. Finan, P. Iyer, A. Singh, T. Jacob, S. Jain, S. Venkataraman, Y. Hoskote, and N. Borkar,

"An 80-tile 1.28TFLOPS network-on-chip in 65nm CMOS," in *Digest of Technical Papers IEEE International Solid-State Circuits Conference (ISSCC 2007)*, 2007, pp. 98–99, 589.

[66] S. Vangal, J. Howard, G. Ruhl, S. Dighe, H. Wilson, J. Tschanz, D. Finan, A. Singh, T. Jacob, S. Jain, V. Erraguntla, C. Roberts, Y. Hoskote, N. Borkar, and S. Borkar, "An 80-tile sub-100-w TeraFLOPS processor in 65-nm CMOS," *IEEE Journal of Solid-State Circuits*, vol. 43, no. 1, pp. 29–41, Jan. 2008.

[67] N. Kurd, J. Douglas, P. Mosalikanti, and R. Kumar, "Next generation Intel® micro-architecture (Nehalem) clocking architecture," in *Proceedings of IEEE Symposium on VLSI Circuits*, 18–20 June 2008, pp. 62–63.

[68] A. Mule, S. Schultz, T. Gaylord, and J. Meindl, "An optical clock distribution network for gigascale integration," in *Proceedings of IEEE 2000 International Interconnect Technology Conference*, 5–7 June 2000, pp. 6–8.

[69] A. Mule, E. Glytsis, T. Gaylord, and J. Meindl, "Electrical and optical clock distribution networks for gigascale microprocessors," *IEEE Transactions of VLSI Systems*, vol. 10, no. 5, pp. 582–594, Oct. 2002.

[70] Q. Zhu and S. Tam, "Package clock distribution design optimization for high-speed and low-power VLSIs," *IEEE Transactions on Components, Packaging, and Manufacturing Technology, Part B: Advanced Packaging*, vol. 20, no. 1, pp. 56–63, Feb. 1997.

[71] V. Gutnik and A. Chandrakasan, "Active GHz clock network using distributed PLLs," *IEEE Journal of Solid-State Circuits*, vol. 35, no. 11, pp. 1553–1560, Nov. 2000.

[72] F. O'Mahony, C. Yue, M. Horowitz, and S. Wong, "A 10-GHz global clock distribution using coupled standing-wave oscillators," *IEEE Journal of Solid-State Circuits*, vol. 38, no. 11, pp. 1813–1820, Nov. 2003.

[73] S. Chan, P. Restle, K. Shepard, N. James, and R. Franch, "A 4.6ghz resonant global clock distribution network," in *Digest of Technical Papers IEEE International Solid-State Circuits Conference (ISSCC 2004)*, 2004, pp. 342–343, Vol. 1.

[74] S. Chan, P. Restle, T. Bucelot, S. Weitzel, J. Keaty, J. Liberty, B. Flachs, R. Volant, P. Kapusta, and J. Zimmerman, "A resonant global clock distribution for the cell broadband-engine™processor," in *Digest of Technical Papers IEEE International Solid-State Circuits Conference (ISSCC 2008)*, 2008, pp. 512–513.

# 3

# Clocked Elements

James Warnock

IBM T.J. Watson Research Center

## 3.1 Introduction

One of the most critical design considerations during the planning of any large VLSI structure is the definition and implementation of the clocked storage elements (CSEs) and the circuitry which drives the local clocks to these elements [1–4]. The nature of the solutions adopted will affect almost every aspect of the design, including its manufacturability, testability, reliability, power consumption, and operating frequency, while the complexity and style of latches and flip-flops employed will affect almost every design automation tool, from high level logic simulation methodology and logic synthesis engines, to circuit tools for detailed device tuning, timing, and testability analyses. A modern microprocessor chip may contain from 0.75 to 1.5M latches and flip-flops [5, 6], and clocked elements may account for 30–40% of the total chip AC power dissipation [7, 8]. Furthermore, the delay overhead or latency of these elements is typically in the range of 2–3 FO4 for modern high-speed designs [5, 9] which may account for 10–25% of the design cycle time for designs spanning the range from 10 FO4 (performance-only optimization) up to about 30 FO4, typically the upper end of the range for power performance optimized designs [10]. Thus the CSE definition is of fundamental importance to any VLSI project; the correct selection, optimization, and implementation will be a basic part of the global design strategy. The goal of this chapter is first to provide a high level overview of the design space of these elements covering the basic design metrics, issues, and trade offs, and second to look at several families of CSEs. This will be followed by a more detailed discussion on aspects of test and testability, design robustness against variability, reliability and soft error rate (SER) considerations.

T. Xanthopoulos (ed.), *Clocking in Modern VLSI Systems*, Integrated Circuits and Systems, DOI 10.1007/978-1-4419-0261-0_3, © Springer Science+Business Media, LLC 2009 

## 3.2 CSE Design Issues

In this section some of the basic CSE design issues are discussed, including latency, hold time, power, and scan testing. This foundation will serve as a useful reference when describing the design space covered by the various CSE families.

### 3.2.1 Latency

One of the primary considerations of any digital design element is its delay, ideally characterized and compared at a constant, fixed power dissipation. For CSEs, the definition of latency is a little more complicated, due to the interaction of the data with the clock edge. If the data arrives at the CSE well before the activating clock edge, the time delay between when the data arrives at the input (labeled for reference as "$d$") to when it is propagated to the output (labeled for reference as "$q$"), will be very long since the data has to wait for the arrival of the clock before it may propagate to the output of the CSE. The total time from $d$ to $q$ will thus decrease linearly as d arrives later (as the wait time is reduced). However, a point will be reached where, if the data arrives too late, the time from clock arrival to the output transition ($q$) starts to increase, as the point is approached where the data arrives too late to be captured or propagated to $q$. Therefore, it can be seen that, when measuring the latency of the CSE there will generally be some optimum specification for the data arrival time relative to the clock edge, which produces the minimum overall delay [11], and this optimum point may be used for design comparisons. Since the clock arrival time typically marks the cycle timing boundary, the total $d-q$ delay is often broken into two segments describing the behavior on either side of the boundary; the setup time (amount of time that the data needs to arrive in advance of the clock), and clock-to-$q$ delay (time measured from arrival of the clock and the appearance of the new data at "$q$").

Having described the conditions under which CSE latencies may be compared, it should be noted that the shape of the latency curves described above is also an important design consideration. This effect is illustrated in Fig. 3.1, for two hypothetical CSEs, plotting the time from $d$ to $q$ vs. the $d$ arrival time relative to the clock. It can be seen from this example that "design A" has a lower latency than "design B," since it offers the minimum latency (d to q time). However, this design (A) exhibits a "hard" cycle boundary behavior. Data cannot arrive much later than at the optimal setup time point before the element will fail, and getting the data to the latch early will provide no benefit to the timing of the downstream logic on the next cycle. Design B exhibits a "soft" cycle boundary. The data arrival time can be targeted to fall in the middle of the "window" where the latency is minimal. Provided that the data arrives somewhere in the window, earlier arriving data will show up at q earlier (since the $d-q$ latency is about constant), to the benefit of the downstream logic on the next cycle. Later arriving data will not cause a failure until the data falls outside this "transparent" window, or until the timing impact on the downstream logic causes a failure.

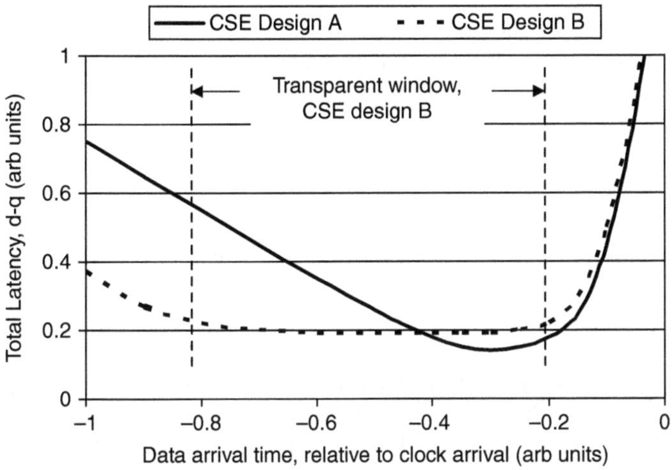


**Fig. 3.1.** Latency ($d-q$ time) vs. data arrival time for two hypothetical CSE designs

This concept of "soft" vs "hard" cycle boundaries is very important for CSE design, and particularly as technology variability increases, a simple latency metric is insufficient to characterize a particular CSE style. In many situations, with all other things equal, a design like "B" may be preferred over "A," and may actually give better results in hardware, even though the absolute latency is worse.

### 3.2.2 Hold Time

Just as the data arriving at the CSE must arrive early enough to get properly written, the data must also hold its value for some amount of time (relative to the clock), if the proper value is to be written. Hold time errors may result, for example, from a situation where new data arrives very quickly at the beginning of the cycle, overwriting the data that was supposed to be captured and latched at the end of the previous cycle.

Hold time specifications are not typically considered when comparing power performance characteristics of CSEs. However, long hold times may drive the addition of a substantial number of delay-padding elements to eliminate any possible early transition at the CSE input. Thus indirectly, long hold times may cost both area and power. In addition, the amount of effort devoted to detection, analysis, and the fixing of hold time misses reduces resources available for other forms of design optimization. Finally, design solutions with longer hold times are inherently riskier, and will be more prone to unexpected hardware issues. In many designs there is a natural trade off between the hold time and the size of the transparent window at the cycle boundary. As the window is widened (making the timing on cycle-limiting speed paths less sensitive to clock skew and process variation), the hold time also increases (making the design more prone to clock-skew-induced or variation-induced functional failure through hold time misses).

### 3.2.3 Power

Power is often considered in conjunction with latency, since the two metrics typically have an inverse dependency on each other. There have been many comparative power performance studies of various latch and flip-flop families [3, 9, 11–14], but the analysis is generally not as straight-forward as it may seem at first. In the same way that the action of the clock signal complicates the latency considerations, it also complicates power comparisons as well. In a given design, the total AC power dissipation is the sum of the power dissipated by clock nets, and other nodes regularly charged and discharged by the clock, and the nodes that switch whenever the input (and output) data change. Thus, when considering a particular latch or flip-flop solution for a specific design application, it is important to characterize the design with the correct weighting between clock switch events and data switch events.

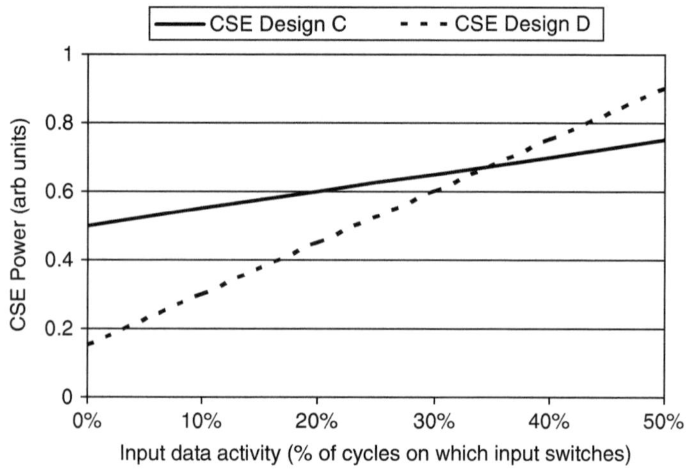


**Fig. 3.2.** Power vs. input data switching factor for two hypothetical CSE designs

This is illustrated in Fig. 3.2, for two hypothetical CSE designs "C" and "D." Design "C" has a relatively high capacitance on clock nets and/or nodes which charge and discharge every cycle independent of the input data, but uses this dynamic action to limit the amount of capacitance switched by changing data inputs. This sort of behavior is typical of many dynamic CSE designs. Design "D" has much less clock capacitance, but has more capacitance switched when the data changes state. Power comparisons at high data switching factors, for example at 50%, would tend to favor the dynamic design "C," while comparisons at lower switching factors, for example 10–20%, would tend to favor design "D." Complicating the situation further is the fact that in modern microprocessors, the local clocks are gated off whenever it is known enough ahead of time that a group of CSEs does not need to be updated on a given cycle. Thus for power comparisons, it is necessary to understand the average data switching factor on cycles when the clock is active. Depending on the type of CSE, it may also be necessary to consider the power dissipated by data switching activity when the clock is gated off as well.

At this point, it is worth mentioning that the ideal situation, with regard to power consumption, would be to activate the clocks to a given CSE only when the new input data is different from the old data, and there have been several schemes proposed along these lines [12, 15]. However, the power, area, and delay overhead associated with determining whether or not to fire the clocks may be significant, and these techniques have never been widely deployed.

### 3.2.4 Scan Design for CSEs

Given the complexity of modern VLSI systems, the ability to directly observe the contents of critical system storage elements during design test and debug, as well as the ability to apply arbitrary initialization patterns to these elements, is another important design consideration. Typically this is done by linking the CSEs together in a serial fashion, using a secondary input port as shown schematically in Fig. 3.3. Level sensitive scan design (LSSD) [16] is one methodology which accomplishes this goal, although there are other more general ways of providing scan capability in a design.

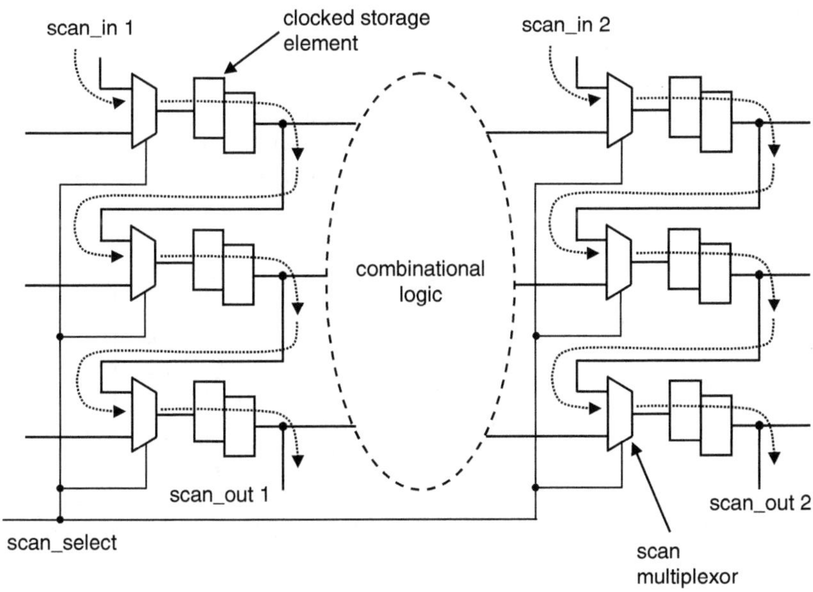


**Fig. 3.3.** Basic scan design. Scan data flow is indicated by the *dotted lines*

The key features of a scannable design include the scan multiplexor, to allow choosing between scan and normal functional data to load into the CSE, and the scan chain itself which needs to be configured in the style of a shift register to allow movement of the data in and out of the CSEs in an orderly fashion. In practice, the goal is usually to minimize the overhead of the extra logic needed for scan and test functionality, without sacrificing the desired test features; for example the multiplexor in Fig. 3.3 is often integrated directly into the CSE itself. Some of the detailed requirements, and ideal test features are described later in this chapter. In the CSE discussion that follows, for each design style, the implications and overhead associated with scan functionality will be described. In modern VLSI designs, it should be assumed that a significant fraction of CSEs will need to be scannable, and therefore this is a significant design consideration.

## 3.3 Static Latch Designs

In this section, examples of several basic styles and classes of static latches will be presented and discussed, including a discussion of the local clock generation aspects, comparative merits and general features of each of the families in light of some of the design considerations described earlier.

### 3.3.1 Master–Slave Latches

For general logic design, the master–slave latch (MSL) is probably the most widely used CSE, with many topological variants and clocking styles described over the years. A typical transmission-gate master–slave latch [17] is shown in Fig. 3.4, using the style and naming conventions of reference [5]. The two clocks, "dclk" (or "data capture clock") and "lclk" (or "launch clock") are roughly 180° out of phase. At the beginning of a cycle (cycle boundary), dclk falls while lclk rises, simultaneously blocking any writes to the master latch while writing the slave latch and propagating

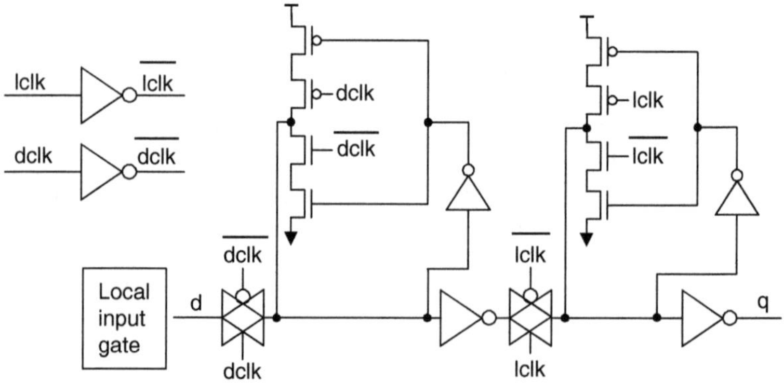


**Fig. 3.4.** Master–slave latch. Reproduced with permission from [5], ©2006 IEEE

new data to the latch output $q$. At the middle of the cycle, dclk rises while lclk falls, allowing the master latch to accept new data until the end of the cycle, while blocking any update of the data in the slave latch so that the data is held until the beginning of the next cycle.

In high-speed designs, the dclk and lclk waveforms are designed to overlap (both clocks high) by some amount at the cycle boundary in order to allow late-arriving data to propagate through the MSL with minimal delay by avoiding interference from the clock edges. This technique not only minimizes latency [5], but also provides for soft cycle boundaries; if the data is timed to arrive in the middle of the window when both clocks are high, then later-arriving data can still propagate through, stealing time from the next cycle. Also, the MSL becomes tolerant to some amount of local clock skew depending on the width of this window, and to the extent that the data arrives in the middle of the window, away from the clock edges. The trade off though is that as the falling dclk is delayed to improve latency and widen the window of transparency, the hold time increases, creating race hazards, and requiring delay padding on short latch-to-latch paths. This trade off is a fundamental consideration for MSL operation, and is shown in Fig.3.5 [5]. Here the delay unit "FO4" refers to a technology independent delay metric, the delay of an inverter with a fan out of 4 [18]. It should be noted here that since the latency of the MSL is highly dependent on the relative timing of the master and slave clocks, any power performance comparisons including MSL elements (often cited as a standard benchmark in comparisons), must provide detailed information on the clocking assumptions made, since otherwise any latency quoted is not very meaningful.

Delaying dclk with respect to lclk also improves the situation at the middle of the cycle. Here, the desire is for the two clocks to be well un-overlapped, with dclk rising well after lclk falls in order to avoid potential race-through of the wrong data at mid-cycle.

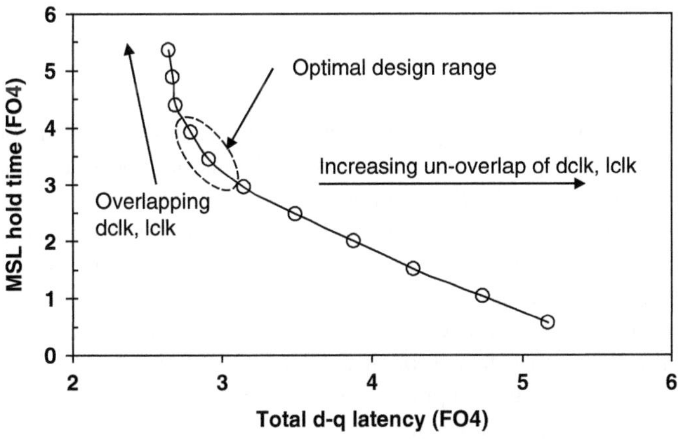


**Fig. 3.5.** MSL hold time vs. total MSL latency as the cycle boundary overlap of dclk and lclk is varied. Reproduced with permission from [5], ©2006 IEEE

It should be noted here that there are many other MSL circuit topologies, some of which are inherently more robust against race or min-path issues [19]. However, these will generally be higher-latency solutions, and by their very nature will not possess the soft cycle boundary properties which provide protection against delay variability and clock skew. For designs with long cycle times (for example, cycle times larger than 40–50 FO4), it may be more important to have a simple, robust solution, than to get the ultimate latency, or optimal power performance. It is expected though, that modern high-speed microprocessors will continue to need the more difficult design solutions, especially as technology variability (and the benefit of soft cycle boundaries) continues to increase. Increasing sophistication of design analysis tools and techniques will help designers to deal with these more complicated solutions.

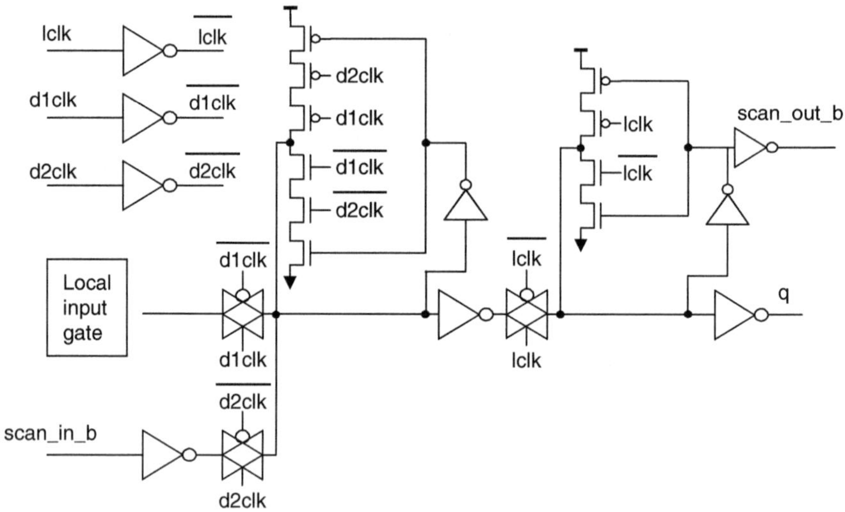


**Fig. 3.6.** Scannable MSL. Reproduced with permission from [5], ©2006 IEEE

The MSL shown in Fig. 3.4 has no provision for scan testing. One of the most common ways of adding this functionality is to add another port to the master latch, controlled by its own clock. This second port may be controlled by a symmetric twin of the first dclk, as shown in Fig. 3.6. In this case the MSL design can support scan shifts and test sequences at frequencies up to the functional design frequency, although other considerations (such as power, and noise) may often impose practical limits on the scan shift frequency. A set of clock waveforms is shown in Fig. 3.7, illustrating the shifting of data through the scan chains, and functional operation. This type of design solution supports a seamless transition from scan shifting to functional operation, and vice versa.

Other scannable MSL topologies have also been used. One alternate version uses a multiplexor on the slave latch feedback path [20] instead of directly at the input to

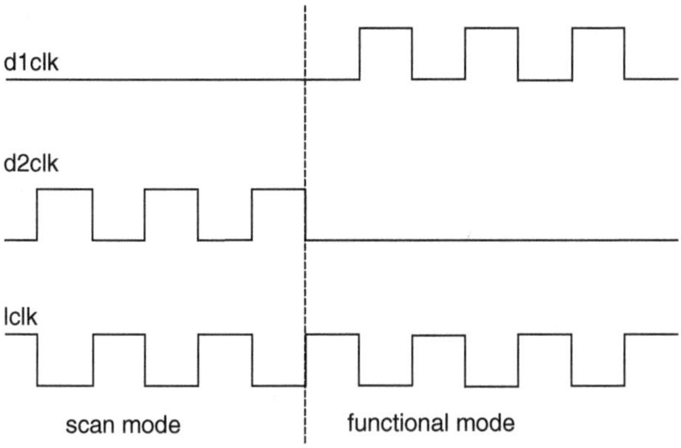


**Fig. 3.7.** Sample set of clock waveforms for scan shifting and for functional operation, for the MSL of Fig. 3.6

the latch node, as shown in Fig. 3.8. In LSSD methodologies, the second port to the slave latch is generally a low-speed scan clock with its own dedicated distribution network. Even with a low-speed scan clock, it is still possible to do full-frequency AC testing on the part. After scanning the desired data in to all scannable latches and appropriately initializing any nonscan elements, the system can be started synchronously in functional mode at speed, with the first at-speed lclk launching out either the data from the last scan clock, or the functional data waiting at the data port input, depending on the type of test desired. Such LSSD schemes may provide very flexible and robust scan design solutions [21], even if the rate of data shifting through the scan strings is limited by the speed of the secondary LSSD clock distributions.

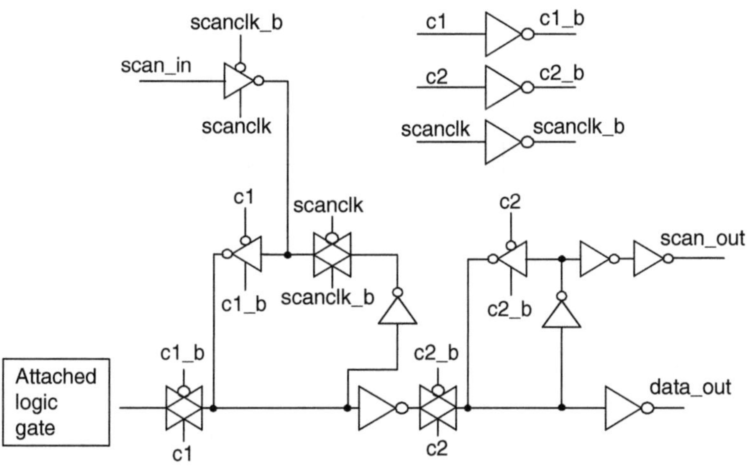


**Fig. 3.8.** MSL with scan MUX in feedback path. Reproduced with permission from [20], ©2002 IBM

Many VLSI designs have been implemented with MSLs as the main "workhorse" latch, often with sophisticated clock edge controls [22]. However, the variation-tolerant soft cycle boundary concept can only be taken so far because of the hold time increase associated with the transparent overlapped clocks at the cycle boundary. In addition, designers have looked for lower power solutions, or higher speed solutions for specific critical applications. These considerations have led to some of the styles described in the following sections.

### 3.3.2 Two-Phase Level-Sensitive Latches

It has long been recognized that the use of level sensitive latches, as shown in Fig. 3.9, will improve the clock-skew and delay-variability tolerance of the design, while simultaneously offering more flexibility for balancing logic delays across multiple cycles. This feature, or technique has been referred to as "cycle stealing" [23] since it allows circuits on either side of the latch to "steal" time from the previous or next cycle. In comparison with the MSL style described above, this technique can be conceptualized as the result of splitting apart some, or all of the master and slave latches, with logic interspersed between master and slave, as well as between the slave and master. Any path through the logic will pass alternately through level sensitive latches clocked with opposite phases. It can be shown that, with about one-half of a cycle of logic between each latch, the data arrival can be timed such that data arrives at each latch in the middle of its active clock window. Therefore, this scheme can be made robust to clock skew and local delay variability. It can also be made very robust against hold time issues, by shrinking the active phase of each clock, with no latency penalty as long as the data is timed to propagate through the latches near the middle of each active clock phase.

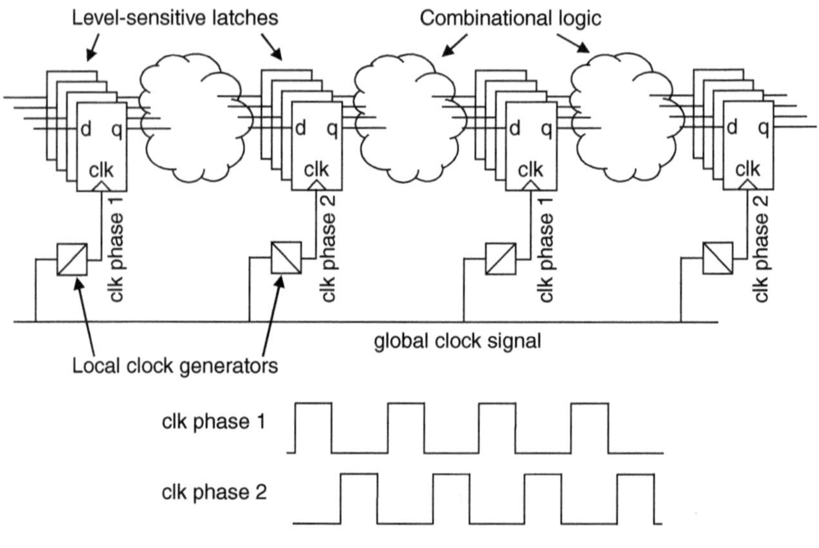


**Fig. 3.9.** Basic two-phase level-sensitive latch scheme, with sample local clock waveforms

This scheme would seem to offer the most tolerance to process variability and clock skew, and in fact these techniques have been used in some large microprocessor designs [20, 24]. From a power perspective, it would seem that the power dissipated should be comparable to, or somewhat less than that of a design using an MSL methodology, since the extra flexibility offered might help to cut down the overall number of latches needed.

However, there are some negative aspects of this design style which must be weighed in the balance. The timing analysis for split-latch designs can be very complicated, with long multicycle (or multihalf-cycle) critical paths, timing loops, and a general inability to define any clear cycle boundaries [25–27]. In fact, the more optimal the design (in terms of lining up data arrival at the latch with the midpoint of the appropriate clock phase), the more complicated the timing analysis becomes. In addition, the split master and slave latches cannot generally share any of the overhead for test control and clock gating in the generation of the local clocks, and will usually need separate global clock taps. The split latches themselves may impose a certain overhead on the design, especially if small devices are used in the latches in order to save clock power. Each latch may become a "bottleneck," requiring stages of gain afterwards to drive the surrounding logic, or, to avoid this situation, clocked devices may have to be increased in size relative to that of an MSL design.

Finally, there are some test issues that need to be considered. Each scannable split-latch requires an additional scan-latch in order to build a robust scan chain, as shown in the example in Fig. 3.10 [20]. Also, AC test sequences need to consider the

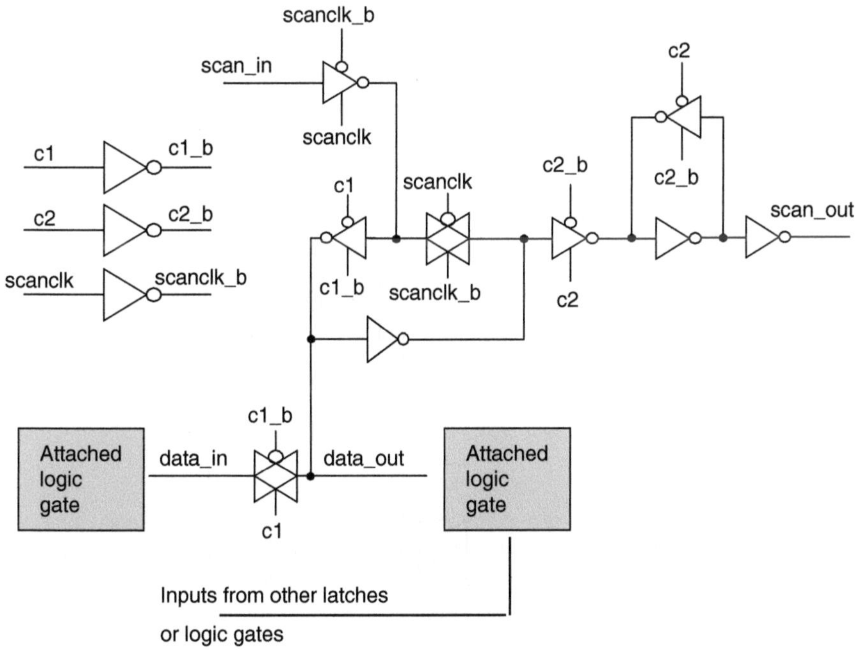


**Fig. 3.10.** Scannable level-sensitive latch. Reproduced with permission from [20], ©2002 IBM

implications of having the scan data propagating into the downstream logic as soon as the clock fires. This means that the initial timing of signals launched from the level-sensitive latch during test is likely to be different from that during functional operation, when transitions may flush through later in the cycle. Therefore AC test sequences may become more complicated, with several functional clock cycles required.

The above design and test issues probably explain why this design style has never gained widespread acceptance for large, high-speed microprocessor designs, even as design for variability tolerance becomes an increasingly significant consideration.

### 3.3.3 Pulsed-Clock Static Level-Sensitive Latches

The use of static level-sensitive latches with single-phase pulsed-clocks appears to be gaining favor for use in large microprocessor designs [5, 8, 22], even as device variability and presumably the hazards associated with hold time issues have been increasing. It is worthwhile to study this particular CSE implementation in order to understand the benefits and hazards associated with it.

An example of a very simple static pulsed level-sensitive latch is shown in Fig. 3.11 [5]. This style has minimal overhead, whether measured in terms of latency, power dissipation, or area. The latch itself is very simple, but the clock design is more delicate. A locally-generated, self-timed clock pulse is generated from either a global or local clock at the cycle boundary, and is used to write the latch (or, more generally, a group of latches). The width of the pulse must be wide enough to reliably write the latch across the full process/voltage/temperature (PVT) space, including all margins for local variability, noise, and any modeling uncertainties. However, the wider the pulse, the longer the hold time and the more difficult it becomes to protect against min-path or hold time failures. Therefore, careful design of the pulse-generating circuitry is of utmost importance, and the pulse width is a critical design parameter.

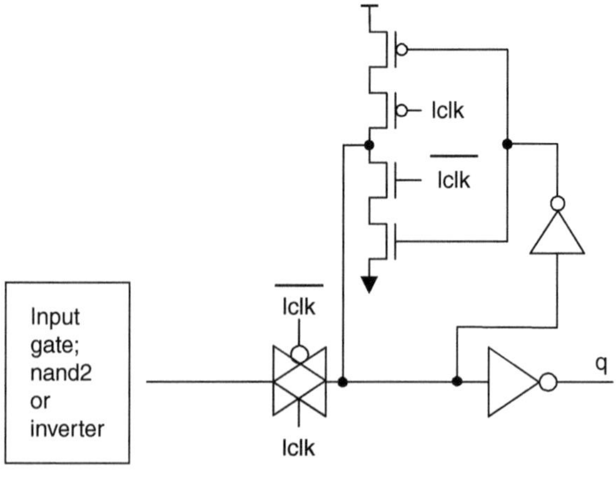


**Fig. 3.11.** Simple nonscan pulsed-clock latch. Reproduced with permission from [5], ©2006 IEEE

Despite the issues associated with pulsed-clock operation of these latches, there are several significant advantages with this design style. Reduced power consumption is one important benefit. In contrast to the conventional MSL solution, or the 2-phase level-sensitive scheme, only one active clock is needed, cutting clock power by about a factor of 2. In fact, this may be the most important benefit of pulsed clocking. One power-saving strategy along these lines is to use regular MSLs, but force the master clock to stay high all the time, while pulsing the slave clock [5, 22, 28]. This approach delivers significant power savings while preserving a fall-back strategy to conventional two-phase clocking in case problems are seen in the hardware. However, this approach does not provide the benefits of reduced area and latency offered by dedicated pulsed-clock latch solutions.

Another important feature of the pulsed latch is its ability to provide a soft cycle boundary, similar to that of the MSLs with overlapped clocks. Unlike the MSL situation though, the amount of cycle boundary transparency cannot be tuned below a certain value (set by the minimum pulse width needed). Thus, even as the benefit of such a pulsed-clock soft cycle boundary will increase as the technology variability increases, this benefit will be offset by the difficulty of simultaneously ensuring writeability and maintaining margin against data races. The design cycle time figures prominently in this trade off. The tolerable upper limit for pulse width would be expected to scale with the overall logic depth, with a practical maximum pulse width of probably about 1/4 to 1/3 of the design cycle time. With wider pulses, the difficulty of padding all the potential short data paths, while avoiding any impact to the cycle-limiting paths, is likely to become unmanageable. At a constant design cycle time (as measured in FO4) increasing technology variability may push pulse-widths up to their practical limit, necessitating larger devices (with faster write-speed but more power dissipation) or other design restrictions for continued use of pulsed-clock latches.

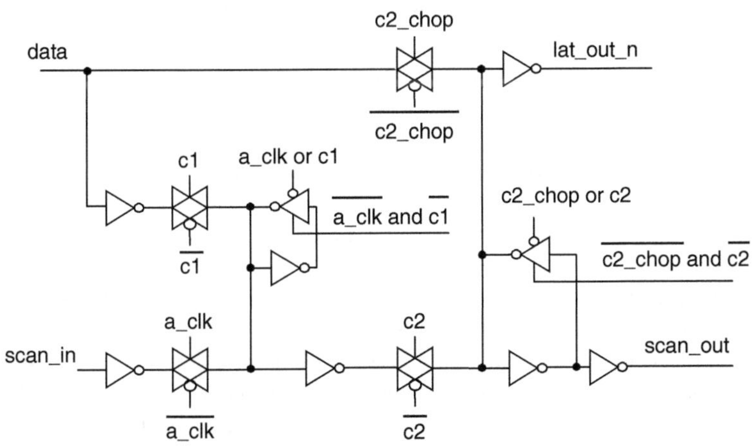


**Fig. 3.12.** Scannable pulsed-clock (c2_chop) latch with built-in MSL-mode fallback (using c1 and c2 clocks). Reproduced with permission from [28], ©2007 IBM

There are some special test issues for pulsed-latch clocking. In general, it would not be desirable to have to pad the scan path, so usually an additional scan latch is needed. This makes the serial shift operation along the scan chain look like a series of MSLs, whose clocks may be unoverlapped as desired to avoid any possibility of race issues during scan. Thus most of the area advantage of the pulsed-clock solution is likely to be lost in situations where a scannable design is needed. An example of a scannable pulsed latch is shown in Fig. 3.12 [28]. This particular design achieves a low latency typical of dedicated pulsed-clock latches, but still maintains the ability to revert back to an MSL mode of operation if problems develop, although at a reduced performance level. In this sense, the extra scan latch added not only converts the scan chain into a series of MSLs, but also can provide a similar function on the data path as well, if conditions warrant.

## 3.4 Flip-Flop Designs

Flip-flops are inherently edge-triggered designs, as opposed to the latches described in the previous section, which are inherently level-sensitive structures. There are a large variety of flip-flop designs which have been employed over the years, and it would be impossible to cover all of them here. This section will describe some of the more common styles that have been used, comparing some of the design issues and the merits of each.

### 3.4.1 Sense-Amp Style Flip-Flop

The proto typical sense-amp flip-flop (SAFF) is shown in Fig. 3.13 [29–31]. This type of design consists of a sense-amplifier coupled with a slave latch to hold the output data when the sense-amp is being reset. This style is more naturally suited for use with dynamic logic on the input side, where data values will hold until they are reset, and where this data reset will occur at about the same time that the flip-flop itself is being reset. Furthermore, with dual-rail dynamic logic, if both true and complement inputs are precharged low, then the activating transition may arrive somewhat later than the clock edge, effectively borrowing time from the next cycle. Since there are internal nodes floating when both inputs are low, and the clock is high, the amount of time borrowing may be limited, depending on the process, noise, and environmental conditions. Also, with dual-rail logic the extra overhead for deriving the other data phase may not be an issue.

Given the generally high power dissipation of dynamic circuit styles, especially dual-rail, sensitivity to noise and process variations, and the general difficulties associated with using such circuits on a large scale in a modern microprocessor, it is unlikely that the industry will see a widespread resurgence in the use of dynamic circuits for general purpose processors. Therefore, this type of design should really be considered in the context of the static circuits which would usually surround it. In a static design, it can be seen that the critical path through the flip-flop traverses 4 stages of logic (the local inverter to generate the complement data, the pull down of one side and subsequent pull-up of the output, and then the pull down of the other side), and so it would not be expected to have a very low latency. Also, power dissipation is likely to be an issue unless very small (and therefore slow, and more variable) devices are used, since one side or the other will need to be charged and discharged in each cycle that the clock is active, independent of the data input pattern. Furthermore, with static input signals, this design does not provide for soft cycle boundaries; whatever data is present when the clock edge arrives is written into the latch. The corollary though is that the hold time for this type of CSE is generally small.

Finally, it can be seen that if the input data were to change before the clock were to fall, internal nodes would be left floating, potentially resulting in a failure to retain the proper state at the output. This particular issue is relatively easy to remedy, by the addition of a single additional small NMOS transistor between nodes $A$ and $A'$ in Fig. 3.13, with gate tied to the supply [30]. This fix will, however, have some impact on the power/performance characteristics. The additional transistor not only adds device and interconnect capacitance to both sides of the sense amplifier (one side of which must be charged and discharged every cycle), but also partially discharges the nonactive side of the sense-amp on every cycle. In addition, there will be a slight degradation of the differential resolution due to the weak coupling between the opposite arms of the sense-amp structure, leading to a corresponding increase in setup time.

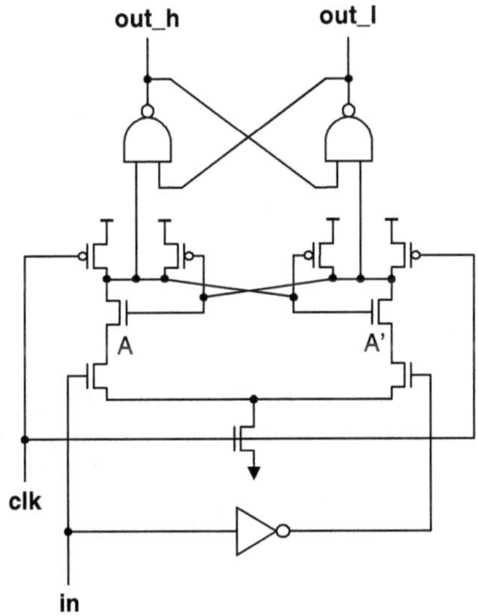


**Fig. 3.13.** Sense-Amp flip-flop. Reproduced with permission from [31], ©2006 IEEE

For improved testability, modifications have been described to make these designs scannable [32], an example of which is shown in Fig. 3.14. This technique can be optimized for AC test, since the paths from input to output are similar for both the functional data input and the scan input. The scan data input is implemented

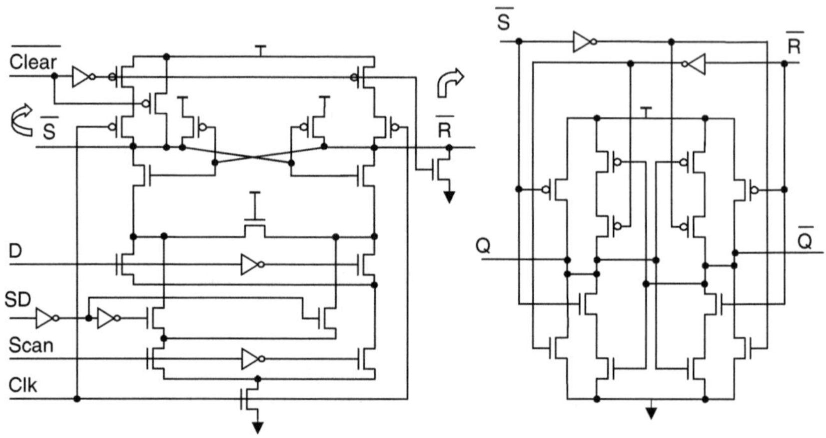


**Fig. 3.14.** Improved scannable sense-amp flip-flop, with asynchronous reset. Sense-amp "pulse-generator" on the *left* drives the latchings stage on the *right*. Reproduced with permission from [32], ©2000 IEEE

as a second port in the first "pulse-generator" stage of the flip-flop However, this will add capacitance to nodes which need to be precharged every cycle, thereby increasing power dissipation. Other methods [13] may avoid this power cost, but may be less well suited for AC test. In addition, there have been many modifications proposed to this basic design including techniques for "conditional precharge" to try to avoid charging and discharging of internal nodes when input data is not changing from cycle to cycle [33, 34] as well as other variations aimed at improving the power performance characteristics [32, 35, 36]. The resulting designs tend to be more complicated, and may still suffer from one or several of the drawbacks mentioned above. Therefore it seems likely that these SAFF will remain confined to only special-purpose applications in the future.

### 3.4.2 Hybrid Latch Flip-Flop

The goal of the hybrid latch flip-flop (HLFF) [37] was to try to combine some of the best features of flip-flops, including edge-triggered sensing, low latency, and low input clock load, with some of the best features of latches, including a soft cycle boundary. The result, shown in Fig. 3.15 is a flip-flop-style front end, activated by what is effectively a locally generated clock pulse, with a static capture latch back-end.

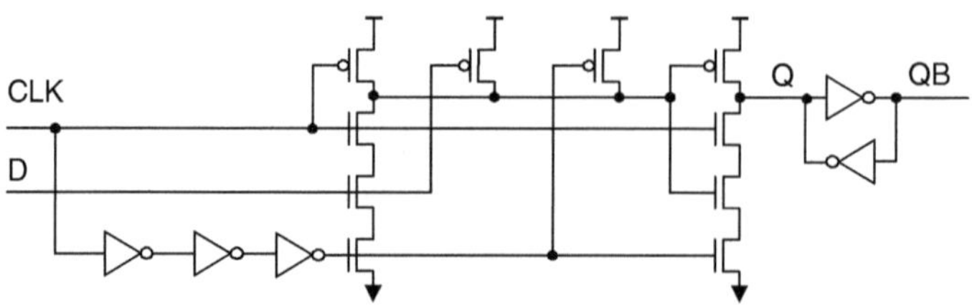


**Fig. 3.15.** Hybrid latch flip-flop. Reproduced with permission from [37], © 1996 IEEE

With only two gate delays from input to output, low latency can be achieved. Also, the local delay chain sets up a transparent window at the cycle boundary, allowing some amount of clock skew/delay variability tolerance. This particular design suffers from the fact that the static capture latch must be small enough to be quickly overwritten in either direction, meaning that the output will be very sensitive to any noise when only the capture latch holds the state of the output. In realistic applications, an additional local output gate would probably be needed. Also, this type of design will consume significant power, due to the precharge/discharge which occurs every cycle whenever the input data is high, and also the switching in the local delay chain inverters. Finally, the output is subject to glitching when the input data is high, since the output stage will begin to discharge when the clock edge is received, pulling back high only after the first stage output transitions low. Various improvements on this design have been reported [38, 39], but in general microprocessor designers have not found this type of flip-flop to offer any compelling benefits, at least in its original form.

### 3.4.3 Semi-Dynamic Flip-Flop

The term semi-dynamic flip-flop (SDFF) [40] was coined to refer to a design style which includes a dynamic front end followed by a static latch [5, 28, 41, 42]. In some ways, this is very similar to the HLFF approach discussed above, but now expanded to provide a means of incorporating a stage of high-speed dynamic logic at each cycle boundary. This technique still provides an easy interface to surrounding static logic, accepting normal static inputs, and providing static outputs. A typical design is shown in Fig. 3.16. These designs rely on some form of a pulsed clock to limit the hold time at the input dynamic stage. This may be accomplished by providing an explicit pulsed clock [28, 40, 41], or by ANDing in a delayed complement clock either in the dynamic pull down tree [40], or somewhere in the cone of logic for all the data inputs [5]. The dynamic stage is followed by a static set–reset latch (SRL), which holds the output while the dynamic stage is being precharged for the next cycle. For situations when the dynamic stage pull down path is cut off while the clock input is still high, a full keeper is usually used on the dynamic node in order to ensure that the dynamic node is held solidly low after being discharged. This keeper can be gated with the clock input to avoid drive fights during the precharge operation.

The advantage of this design is that it provides a way to incorporate a stage of dynamic logic into an otherwise fully-static design. It can also be extended to add additional dynamic logic stages after the first stage. In this case, only the first stage, with static inputs, needs a footer device and a pulsed evaluation clock. The SRL is moved downstream to the final dynamic stage, providing dynamic-to-static signal conversion. This technique can be applied at both clock edges [43], so that in such a two-phase system, pulsed clocks are no longer necessary.

The SDFF is a very useful construct for extremely critical paths in a design, where a wide OR, or wide multiplexor is required at the cycle boundary. However, there are some design costs associated with this solution. There is a relatively large capacitance which may be charged and discharged every cycle, even when there is

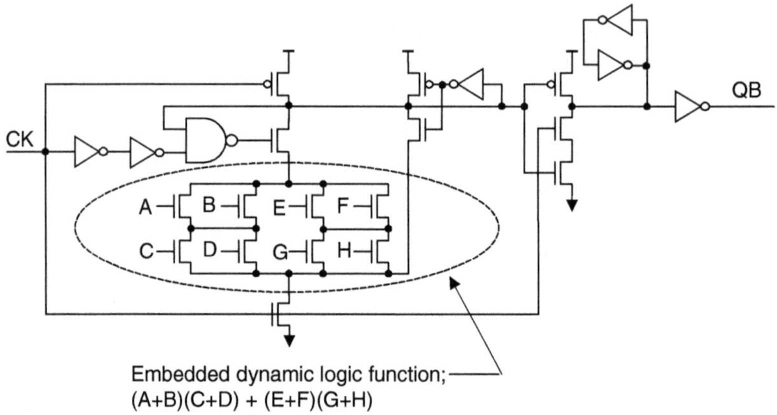


**Fig. 3.16.** Semi-dynamic flip-flop with embedded dynamic logic. Reproduced with permission from [40], ©1999 IEEE

no switching activity present on the data inputs. Thus, power consumption tends to be high for these designs, especially when used in contexts which do not see particularly high data switching factors, or when efficient clock gating is difficult. Power consumption may also be increased by glitching at the output when holding a 1. This glitch is similar in nature to that observed in the HLFF, occurring since the path to pull the output down (via the reset action of the static latch) is usually faster than the path to force the output high (via the dynamic pull down). The flip side of this glitch though, is that the latch will generally write a 0 faster than a 1, and downstream static logic can be skewed to take advantage of this fact (and also absorb the glitch).

The cycle boundary also is not easily softened to absorb timing variability, in contrast to the HLFF, due to the dynamic input stage. In principle, late rising input transitions may arrive at the dynamic gate after the clock rises, provided that they still have enough time to discharge the dynamic node, but falling transitions must meet a strict setup criterion. Similarly, the hold times for these designs tend to be relatively long, and are also asymmetric due to the action of the dynamic gate. Inputs only need stay high long enough to be able to discharge the dynamic node, but inputs initially low must stay low until near the end of the clock pulse. Just as for the pulsed level-sensitive latches, the clock pulse width must be wide enough to reliably write the latch across the process and application space of interest, but not so wide that the length of the hold time becomes an issue. For the SDFF, it is necessary to consider the range of dynamic input gates used in order to determine the minimum pulse width required.

For test, there are several strategies for making the design scannable [5, 28, 41]. One method involves adding a port into the static latch, and also adding in an additional scan latch, as was required for the static pulsed designs [28, 41]. An example of this approach is shown in Fig.3.17. With this scheme it should be noted that for AC test, the launch of scan data from the static latch bypasses the input dynamic stage, and so if the test sequence involves a launch from the scan clock, the timing characteristics may differ from that in functional operation. It is possible to more closely match the functional path with the scan path by using an additional input into the front-end dynamic stage [5]. In this case, all the pull down paths through the functional data ports need to be disabled to avoid interference with the scan path.

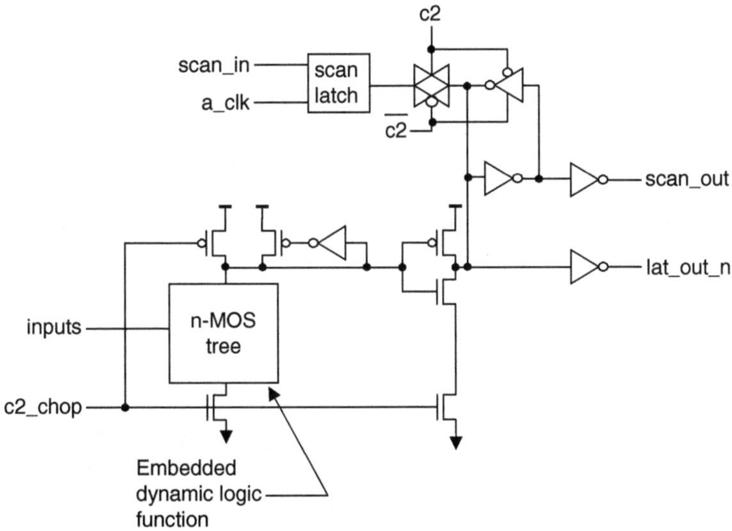


**Fig. 3.17.** Scannable semi-dynamic flip-flop. "c2_chop" is a pulsed clock input. Reproduced with permission from [28], ©2007 IBM

## 3.5 Test and Debug Considerations

As microprocessor frequency scaling has run up against severe thermal, physical and electrical constraints, the industry has turned to multicore architectures in order to allow continued performance gains from generation to generation [44]. As a result, microprocessor chips are now being fabricated with over 2 billion transistors [45], and with density scaling still expected to continue for at least several more generations in the future [46], observability and testability, including both DC and AC test coverage, have become of paramount importance. For this reason, large microprocessor design projects have by and large adopted scan design methodologies [5, 28, 41, 47, 48] and it is expected that scan design and test/debug methodologies will continue to need careful consideration in future designs. It will be important to minimize the overhead of these scan methodologies, but it will not be tolerable to make significant sacrifices to the testability of the design.

For DC test coverage, it may not be important exactly how the scan data is written into the latch, since the timing of the launch into the downstream logic is unimportant. Also, in the downstream logic, it is unimportant how each node is switched; the activating transition is unimportant, as long as a pattern can be found which tests the node in each state. However, for high-speed AC test, there are a number of key features which are desirable for a robust and flexible test methodology. The highest test coverage is obtained if each CSE is designed with an extra scan-only storage element, as shown in Fig. 3.18a, used only for test purposes, which can store an independent data value to be used to launch a transition into the downstream logic [49]. In this way, each scannable latch can be independently configured to launch an arbitrary transition into the downstream logic. Referring to the initial input vector as "V1" and the transition vector as "V2," it can be seen in this case that "V1" and "V2" are independent and unconstrained. This enhanced scan design provides the maximum test coverage, but the overhead is generally quite high. Although this technique can be confined to be used only on critical CSEs [50], and selective deployment has been reported in critical areas of some microprocessors [41], the cost is too high for techniques such as these to be used commonly in the industry.

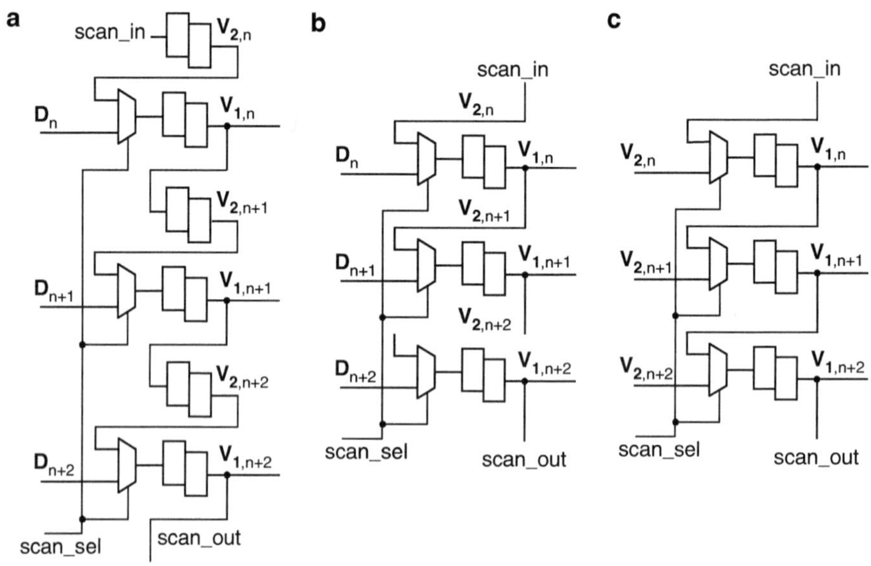


**Fig. 3.18.** Scan test configurations (**a**) enhanced scan, (**b**) skewed load, (**c**) broadside test

At the other extreme, the simplest AC test sequence relies on loading scan data into all scannable latches, then switching to functional mode to launch and capture data for test. In this case, V1 is unconstrained, but V2 is the next state determined from combinational logic response to the V1 vector, i.e., the functional data at the inputs of the CSEs given the V1 values at their outputs, as shown in Fig. 3.18c. This has been called a "broadside test" [51] sequence, and places minimal constraints on the design of the scannable CSEs. The speed at which the scan data can be written into the CSEs is unimportant, as are the details of the switch from scan mode to functional mode. The downside of this method is that AC test coverage can be low, with few, or no options for improvement through DFT-related design changes. In addition, analysis and debug are more complicated, since V2 depends on the response of all the upstream combinational logic and the V1 state of all CSEs in the cone of logic, increasing simulation time for test analysis and making it harder to change V2 in a systematic way.

A compromise between the two above approaches is the concept of a "skewed load" sequence [52]. In this type of sequence, V1 is again scanned into the CSEs, but now V2 comes from the upstream data in the scan string as shown in Fig. 3.18b. During the high-speed test sequence, data must be loaded into each CSE from the scan port, and launched at speed into the downstream logic. On the next clock cycle, data is captured in each CSE through the functional data port. Sample clock waveforms for such a sequence are shown in Fig. 3.19a for the scannable MSL shown in Fig. 3.6, contrasting with those for the broadside load sequence (Fig. 3.19b). To make this sequence work properly, all CSEs must be designed such that the clock-Q delay from the scan port is a close match to that from the data port, otherwise the AC characteristics of the test will not match that of the functional operation. Also, the CSE clock control system has to be capable of switching from "scan mode" to "functional mode" within a single cycle, in order to launch scan data, and then to capture data from the functional port. This will generally require an accurate pipelining and distribution scheme for at least this one high-speed global test control signal.

One advantage of the skewed load test sequence is that it provides enhanced AC test coverage compared to the broadside test, although there are still coverage limitations which may arise from the adjacency of V1 and V2 in the scan string. Furthermore, the test patterns applied are very flexible, easy to modify in a systematic

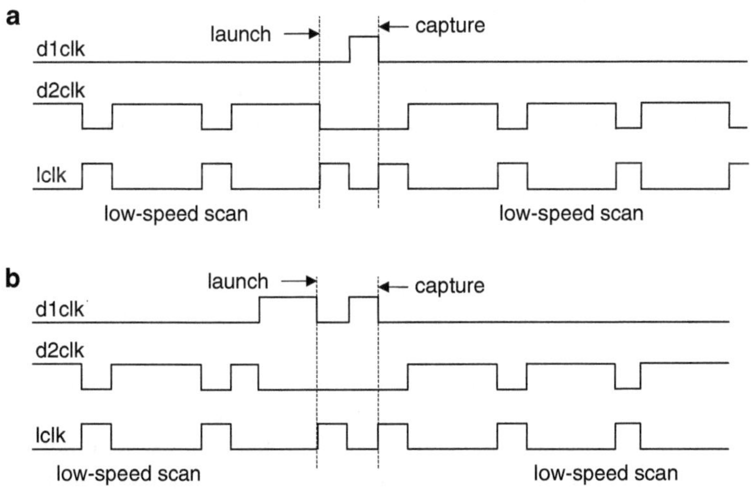


**Fig. 3.19.** Sample clock waveforms for AC test using the MSL from Fig. 3.6. (**a**) skewed load test. Note that data from scan port (d2clk) is launched. (**b**) Broadside test. Note that data from the functional port (d1clk) is launched

fashion, and failure analysis needs only to consider a single cycle's worth of logic instead of two as in the broadside case.

Finally, even though much of the discussion above has focused on single-cycle test sequences, which offer an appealing simplicity from a test point of view, various other CSE design factors may result in the requirement for longer at speed test sequences. Inclusion of nonscan elements (which may help to reduce chip area, latency, and power dissipation) will force the use of longer sequences to test all paths in and out of these components. Also, the use of soft cycle boundaries may also affect AC test operation, since certain critical paths may effectively become more than one cycle long.

## 3.6 CSE Design for Variability

Design for variability has become an increasingly important consideration as the relative level of parameter and device variability has been increasing recently and is expected to continue to increase rapidly with future technology scaling [53]. The importance of this fact is magnified by the expected continuation of device density scaling; this increased variability is manifested in ever larger collections of devices on a single chip. There are many ways in which variability may degrade the quality of a design, and this topic is addressed in greater detail in Chap. 7 of this book. The discussion here will focus on two particular aspects of variability-induced degradation which are relevant to CSE design, namely operating frequency degradation, which involves the cycle limiting paths in the design, and functional failures, which may result from racing paths with insufficient margin, or other design vulnerabilities.

### 3.6.1 Variability-Induced Frequency Degradation

The key aspect of cycle limiting paths in any given design is that there is usually a relatively large number of logic gates involved, including not only the logic between the launching and capturing CSEs, but also the CSEs themselves, and the circuitry driving the local clock signals to these CSEs. Thus local, uncorrelated, random fluctuations in device parameters are less likely to have a large impact on the timings of these paths, since these will average over the large numbers of devices involved. Problems are more likely to arise from more global, systematic variations which apply to all the circuits in a given region of the design. Some examples here might include PFET to NFET strength ratios, wire speed to device speed ratios, design pattern-factor-induced effects, chip mean device threshold voltage variability, etc. In principle, given an accurate knowledge of all the parameter distributions, and a complete statistical timing methodology [54], it would be possible to predict more accurately which speed paths are most likely to limit the design, and in fact it is expected that this sort of analysis will become more prevalent in the future. However, the underlying variabilities and uncertainties will not go away, even as the ability to accommodate them improves, and so it is important to understand how specific CSE design techniques might improve the situation.

The concept of soft cycle boundaries has already been discussed at some length, and this is an important method of providing some protection against variability or unknown factors. The 2-phase level-sensitive latch methodology probably provides the most benefit in this regard, but in light of the issues described earlier, it is unlikely that the usage of this style will become widespread as a solution for improved variability tolerance. Rather, it is expected that efforts to soften the cycle boundaries are more likely to be concentrated on MSFF and pulsed level-sensitive designs. Furthermore, given that an MSFF with overlapped clocks and pulsed level-sensitive latches have similar hold time issues for a given transparent window size, but unequal power and latency characteristics [5], it seems that pulsed clocking techniques would be the preferred method of providing a soft cycle boundary to protect against variability-induced frequency degradation. However, the local clock edges for the MSFF solution can be tuned to optimize the trade off between hold time and transparent window size without the minimum pulse-width constraints of pulsed-clock designs, and widespread use of pulsed-clock latches will make the design more prone to variability-induced functional failure, as described in the next section.

In addition to the use of soft cycle boundaries, many chips are being designed with programmable clock edges which can also be used to mitigate the effects of variability and uncertainty [28]. Delaying specific clock edges can mitigate the longest timing paths found in the hardware. It is expected that techniques such as this will become more prevalent in the future, as variability continues to increase, and as the improvement in technology decreases from generation to generation. One might expect to see more automated techniques used to optimize clock control settings on a chip-specific basis, or the use of adaptive or autonomous techniques, moving from the more global adaptive deskew techniques in use today [55] (also described in Chaps. 2 and 7), to more fine-grained adjustments in the future.

### 3.6.2 Variability-Induced Functional Failures

In contrast to variability-induced frequency degradation, a race path or other functional hazard may involve only a small number of devices, so that local random variations may play a big role in determining the margins required to protect against such failures. For specific circuit configurations within certain CSE topologies, it will be necessary to carry out a complete statistical analysis at the desired conditions for use, and at various process corners, to make sure that the design is robust enough against any probable statistical variation [6]. However, in the general case, it will be impossible to analyze all race conditions between all CSEs to that level of detail without a global statistical timing methodology. Also, since the amount of variability will depend on the detailed circuit parameters (for example channel widths involved, device types, the types and amounts of parasitic resistance and capacitances, etc.) it will become extremely difficult to guardband every potential race path against the worst-case combination of statistical fluctuations. Thus it is likely that advanced CSE designs and the need to protect against statistical-variability-induced failures will drive the need for an increasingly sophisticated true statistical timing methodology.

In fact, the need for such a methodology is arguably even greater here than for the operating frequency analysis discussed in the preceding section.

Although there are many sources of variability, any one of which may lead to failure, it is expected that future designs will continue to push towards lower supply voltages, while device dimensions continue to shrink. This means that factors leading to threshold voltage variability are likely to play an increasingly significant role [56]. For this reason, statistical analyses of race conditions will be needed at the minimum operating voltage, and this low-voltage corner will often impose more stringent restrictions on the design than will the high-voltage corner, especially if the write of the latch must overcome a "weak" keeper device. Recent effort has been reported on measurement of hold time variability, using structures designed to look at race conditions under realistic design conditions, showing directly, in a hardware-based measurement, the increase in margins needed as the technology is scaled to finer dimensions [57], and/or as the operating voltage is lowered [58].

Pulsed static, or pulsed semi-dynamic designs may be particularly susceptible to fluctuation-induced writeability failures. For this reason, designers have added local pulse-width controls to the pulse generators used for these latches [6, 28], for test purposes and also as an emergency option in case of unexpected hardware problems. Such an example is shown in Fig. 3.20. Another technique mentioned earlier in the context of power-reduction features is to configure the clock drivers for MSFF latches such that they can provide either regular clocks, or a pulsed clock to the slave with the master clock held at a constant high value [5, 22, 28]. This scheme maintains a fail-safe option in case of either writeability issues or data races. In principle, with techniques like this it should be possible to cut down on the margins normally needed for guaranteeing race-free operation; registers containing latches which are seen to fail in the hardware with pulsed-mode clocking could be configured to run in MSFF mode.

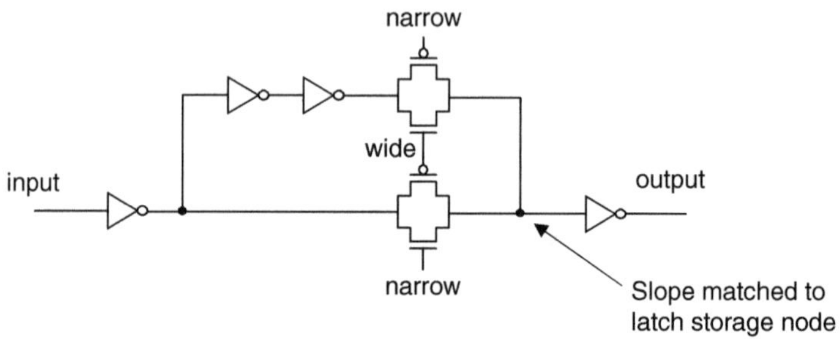


**Fig. 3.20.** Programmable delay line defining the trailing edge of a local clock pulse. Transmission gate structure is designed to match that in the latch. Reproduced with permission from [6], ©2008 IEEE

Techniques like these may be used not only to guard against unexpected problems, but may also help with test and debug, and one may imagine that in the future, designs may be automatically configured with the optimal pulse-width settings for simultaneously maximizing cycle stealing across soft cycle boundaries while maintaining a safe margin against race failures.

## 3.7 Reliability Issues

While issues related to process variability or random parameter fluctuations can, in principle, be screened or protected against by proper testing of the part, reliability issues can be much more difficult to protect against. It will be no surprise to the reader that adequate reliability of CSEs is becoming increasingly difficult to guarantee, as the number of CSEs on a single chip continues to increase and as the technology feature size continues to shrink. This section will examine two types of reliability issues, soft error upsets which can disturb the data in the CSEs, and wear out mechanisms which may cause failure after a prolonged period of use.

### 3.7.1 Soft Error Rate Considerations

Several factors combine to make SER robustness an increasingly important design concern in future systems. Although many designers may consider SER to be important only for situations where extremely high reliability is required, or a problem only for large, high-density SRAMs, recent work has shown that this assumption is no longer true, and as a result, soft errors in CSEs have been getting an increasing amount of attention in recent years.

As CMOS technology has continued to scale well into the sub-100nm dimensions, the combination of feature size reduction, and especially power supply voltage reduction has resulted in a steadily increasing susceptibility of individual latches and flip-flops to soft error upset [59–61], and by some accounts the SER rate in a typical latch or flip-flop has already eclipsed the SRAM SER rate in unprotected rate-per-bit comparisons [62]. Furthermore, error correction codes may be used to help improve the SER in SRAMs, with a relatively low overhead, but no such low-overhead solution is currently available for the collection of CSEs in a typical microprocessor. The voltage factor is especially important, and as systems become ever more adept at lowering chip voltage whenever possible to cut back on power dissipation, there will be a price to be paid in terms of SER. At the same time, the microprocessor industry has seen a shift away from steady clock frequency increases for system performance improvements, towards increased parallelism, increased functionality, and increased integration levels. Therefore, it is expected that the number of CSEs per chip will continue to grow rapidly. At the same time, the SER per element is also growing, leading to a looming “SER crisis” if no action is taken.

To improve the SER situation, in addition to specific CSE design techniques, there are various system, error checking/detection, and technology options which have been discussed in the literature [41, 60, 63–66]. Improvements in processes and materials may also help to reduce contaminants and/or provide increased immunity. However, this section will focus on local CSE design options, and methods therein to improve SER robustness. Perhaps the first approach to try might be to target improvements in SER by careful layout optimization. However, studies of the SER as a function of the CSE layout details have not shown any significant dependencies [62], and there appears to be little success reported in this regard. Another possibility is to improve the SER robustness of individual latches or flip-flops by simply adding additional capacitance to sensitive nodes [67], or by selectively increasing the size of devices holding sensitive nodes [68]. These techniques have been shown to help, and overall area cost may not be too excessive in some situations [69]. However, these techniques are more useful for specific design areas where high reliability is needed, and operating frequency/power is not a big issue. These hardening techniques are not very suitable for general use in microprocessor design. Moreover, these SER mitigation strategies will not scale very well to future technologies.

Future strategies for improving the soft error characteristics of CSEs are more likely to involve topological modifications to the circuits themselves. It is expected that the industry is likely to make more use of circuit related hardening techniques such as the Dual Interlocked storage Cell (DICE) latch [70]. The DICE latch uses redundant and interconnected storage nodes such that the cell cannot be flipped by the upset of a single device. An example [71] is shown in Fig.3.21. This design is resistant to single device upsets, although it can be seen that certain nodes may be left floating for a short period of time after a strike occurs. Depending on the operating voltage, experimental data on this cell showed a factor of 30× to 100× improvement in SER robustness with the observed errors likely caused by multiple upsets. Since the failure mechanism for this type of CSE involves charge collection on multiple nodes, care should be taken during layout to separate critical devices in order to reduce the probability of an error caused in this fashion [71].

DICE latches have been employed in a recent high-end microprocessor design [6], illustrating the combination of pulsed-clock techniques for power reduction and soft cycle boundaries, scan capability for test and debug, and special design tech-

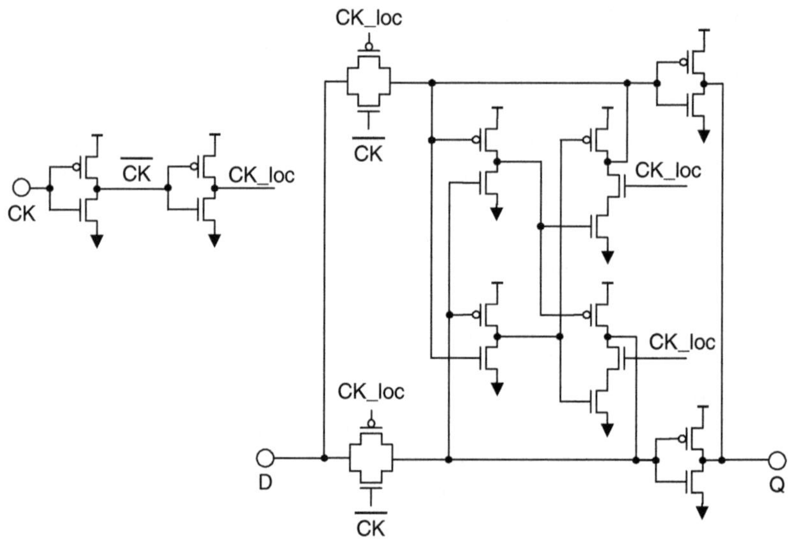


**Fig. 3.21.** DICE latch topology. Reproduced with permission in a form similar to that in [71], ©2004 IEEE

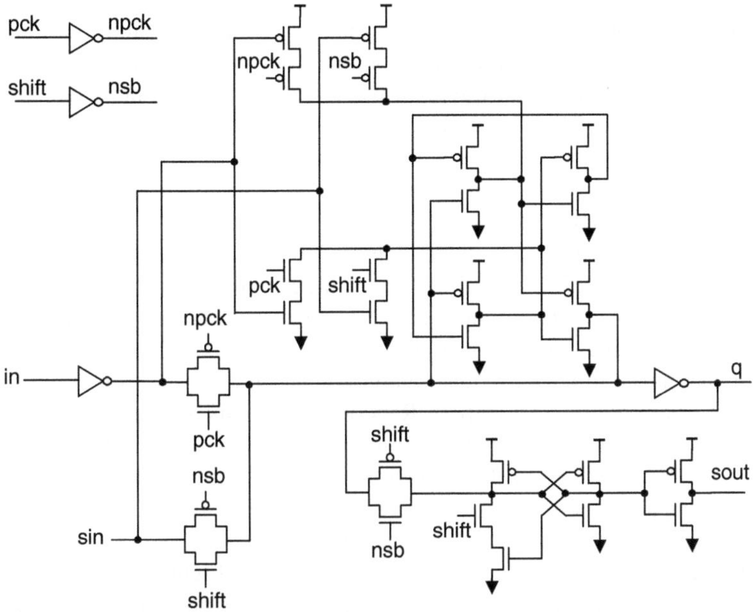


**Fig. 3.22.** Scannable pulsed-clock DICE latch. Reproduced with permission in a form similar to that in [6], ©2008 IEEE

niques for SER robustness. This scannable, pulsed-clock DICE latch is shown in Fig. 3.22. The DICE method is expensive; an overall latch area increase of about 35% and power increase of about 25% has been reported [6]. However, there will be many situations where error checking may not be practical, or the overhead may be too high (much higher than that of switching to DICE elements), and it is expected that techniques such as these will become more widespread in the future as designers are forced to turn their attention to this problem.

### 3.7.2 End of Life Considerations for CSE Design

There are a number of wear out mechanisms which affect the operation of silicon CMOS circuits, some of which may result in relatively sudden catastrophic failures, while others result in a more gradual parametric device degradation over time, leading eventually to failure. The former type of problem is difficult to deal with at the single element level; usually higher level checking or redundancy schemes are needed, which are beyond the scope of this chapter. However, the second type of problem is more amenable to local circuit solutions, and some of these solutions will be described in this section.

Two major phenomena which may lead to gradual circuit degradation are hot carrier injection (HCI) and negative bias temperature instability (NBTI) [72]. HCI has long been recognized as a potential problem for CMOS circuits [73, 74], but as modern CMOS technology has moved to lower voltages, degradation due to HCI has tended to diminish in terms of its importance. This is especially true for digital CMOS circuits where the current flow through the devices is very transient in nature, and devices are not biased in a way that would subject them to hot carrier stress for long periods of time. Thus for CSE designs with reasonable signal slews, HCI is unlikely to cause significant degradation, although with the advent of new materials this an area that will bear watching for the future [75]. Both HCI and NBTI are addressed in more detail in Sect. 8.3.5

For CSE designs in today's technologies, NBTI is a more important concern. NBTI is specific to PFETs, increasing the magnitude of the threshold voltage, and decreasing carrier mobility over time, depending on the stress conditions. The customary manner of treating this issue is to design test sequences such that all parts which pass are guaranteed to have adequate margin under all operating conditions against the impact of any future degradation. Random collections of parts may be stressed over periods of time to assess their reliability, given a particular set of screening tests at time zero. However, given the intrinsic statistical nature of the phenomenon involved, the shrinking of the device geometries with the resultant growing variability, and the increasing numbers of devices integrated on a single chip, it may be necessary to ask whether the margins needed during test will remain within reasonable bounds in the future, and, given the power/performance cost of maintaining such margins, whether there are CSE design techniques which could be used to reduce this overhead.

Most of the work on this subject is focused on the issue of frequency degradation, or the slow down of components which occurs over time. One method that has been studied is the so called Razor technique, originally proposed [76] to allow aggressive dynamic voltage scaling, but also applicable to wearout-induced circuit degradation as well. A Razor MSL is shown in Fig. 3.23. The clock to the shadow latch is delayed enough to guarantee that the incoming data is successfully latched, even when the main latch fails to capture the correct data. In the event that the two latches contain different data, an error signal is registered, and it is possible for the

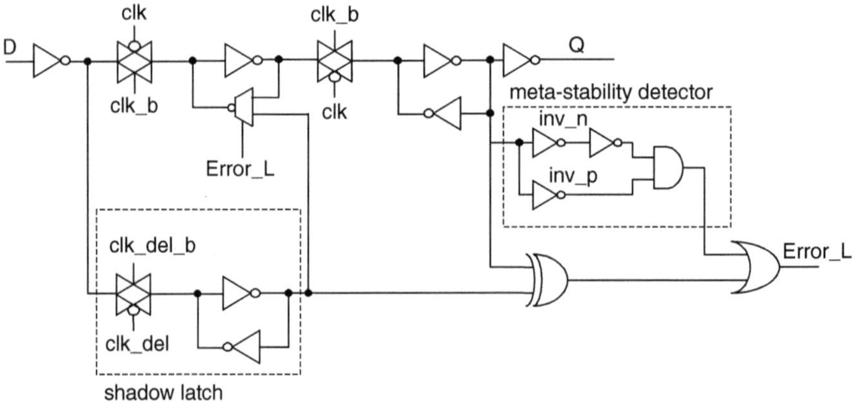


**Fig. 3.23.** Razor master–slave latch. Reproduced with permission from [76], ©2003 IEEE

system to recover by later swapping in the correct data from the shadow latch (at the cost of a few cycles, depending on the exact recovery scheme used). An error monitor could take appropriate action to maintain a reasonably low error rate, for example by adjusting the processor voltage or frequency, thereby avoiding excessive performance loss. The optimal error rate could then represent a trade off between the performance degradation caused by the overhead of error correction and the benefits of a higher frequency or lower operating voltage. The Razor technique applies considerable overhead to a typical MSL (not to mention the recovery logic overhead) but would not be needed on all CSEs. The Razor MSFF is itself vulnerable to hold time issues, and in fact the delayed clock to the shadow latch significantly increases the overall hold time for the CSE. Another issue of such a scheme would be that either the power or performance of the system could change over time, as adjustments were made for NBTI degradation (or simply in response to environmental changes), perhaps requiring re-instatement of some of the guardbands that were to be avoided in the first place. Sections 7.4.1–7.4.3 describe the Razor methodology in more detail from the viewpoint of addressing process variation through resiliency.

Other research has focused on trying to predict impending errors, and take action before the error actually occurs [78]. In this case, a transition detector watches for transitions which are arriving very late at the capturing CSE, and triggers if these transitions fall within the defined danger window. Chip monitoring hardware or software may then take action before any errors occur. This eliminates the logic complexity and overhead associated with the error recovery mechanisms. Such transition detectors may also be integrated into pulsed latches and used as error detectors [77, 79] in a similar fashion to the original Razor design, but with less overhead inside the CSE. An example of such a scheme is shown in Fig. 3.24. A more detailed description of these techniques appears in Sect. 7.4. One drawback of these transition detector schemes is that, in order to reliably signal the presence of an error (or an impending error), such circuits will need enough built-in margin to work reliably; it has to be guaranteed that the transition is always detected before an error actually occurs. This built-in margin will tend to lower the achievable operating frequency.

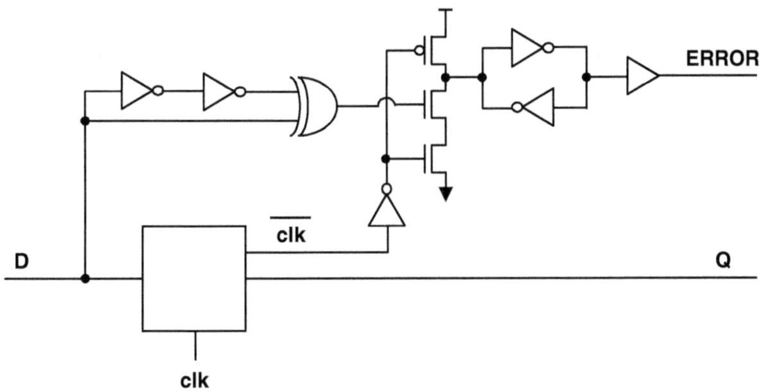


**Fig. 3.24.** Transition detection scheme. Reproduced with permission from [77], ©2008 IEEE

As a "last resort," redundant logic techniques have also been studied, where the CSE has the ability to automatically swap out a whole block of logic on sensing an impending fail [80], replacing it with an equivalent set of logic gates. Of course, the overhead here is extremely high. Regardless of whether or not techniques like these ever become adopted in a widespread fashion by the industry, it is likely that future microprocessors will require more advanced techniques to guard against wear out-induced reliability failures, either locally at the CSE-level, as described above, or by using more global monitoring and checking algorithms.

Finally, throughout all of the above, it has been assumed that race conditions, pulse-width and/or latch writeability margins can be ensured through a combination of design margin and test conditions, without too much overhead. While this latter assumption may still hold true for some time in the future, usage of large numbers of pulsed-clock components in future technologies are likely to drive the need for more advanced testing techniques, including both race path and pulse-width stressing using some of the special clock pulse width/and/or clock edge control features described earlier. In addition, that fact that race path or hold time failures generally involve a small number of logic gates means that they will tend to be more sensitive to variabilities inherent in the various device degradation mechanisms. As this variability increases, it may no longer be possible to ensure adequate margin through specific test voltage and temperature conditions alone.

## 3.8 Conclusion

In light of the ongoing power crisis in modern microprocessors, tomorrow's high-performance processors are likely to continue the push towards aggressive use of pulsed-static latches, which require only a single clock and provide for a soft cycle boundary. Improved analysis tools will be needed to guarantee robust operation of a large collection of such circuits across the full PVT space, and especially to be able to handle the ever-increasing impact of random local fluctuations. To ensure the highest quality, reliability, and system performance, future designs will use an increasingly sophisticated collection of special features for test, debug, and chip optimization. Finally, SER-related reliability will become a key issue for CSE design in the future. Tomorrow's latch and flip-flop designers will need to consider not only the usual power/performance aspects of their solutions, but will also need to design for enhanced testability, robustness against statistical variations, and high levels of reliability, and SER immunity.

## Acknowledgements

The author would like to acknowledge the feedback and comments from Leon Sigal and Thucydides Xanthopoulos.

## References


[1] S. Unger and C.-J. Tan, "Clocking schemes for high-speed digital systems," *IEEE Trans. Comput.*, vol. 35, no. 10, pp. 880–895, Oct. 1986.

[2] K. Wagner, "Clock system design," *IEEE Des. Test Comput.*, vol. 5, no. 5, pp. 9–27, Oct. 1988.

[3] V. Oklobdzija, V. Stojanovic, D. Markovic, and N. Nedovic, *Digital System Clocking*. Wiley-IEEE Press, New York, 2003.

[4] V. Oklobdzija, "Clocking and clocked storage elements in a multi-gigahertz environment," *IBM J. Res. Dev.*, vol. 47, no. 5/6, pp. 567–583, September/November 2003.

[5] J. Warnock, D. Wendel, T. Aipperspach, E. Behnen, R. Cordes, S. Dhong, K. Hirairi, H. Murakami, S. Onishi, D. Pham, J. Pille, S. Posluszny, O. Takahashi, and H. Wen, "Circuit design techniques for a first-generation Cell broadband engine processor," *IEEE J. Solid-State Circuits*, vol. 41, no. 8, pp. 1692–1706, Aug. 2006.

[6] D. Krueger, E. Francom, and J. Langsdorf, "Circuit design for voltage scaling and SER immunity on a quad-core Itanium® processor," in *Digest of Technical Papers IEEE International Solid-State Circuits Conference (ISSCC 2008)*, 2008, pp. 94–95.

[7] J. Friedrich, B. McCredie, N. James, B. Huott, B. Curran, E. Fluhr, G. Mittal, E. Chan, Y. Chan, D. Plass, S. Chu, H. Le, L. Clark, J. Ripley, S. Taylor, J. Dilullo, and M. Lanzerotti, "Design of the POWER6 microprocessor," in *Digest of Technical Papers IEEE International Solid-State Circuits Conference (ISSCC 2007)*, 2007, pp. 96–97.

[8] S. Naffziger, G. Colon-Bonet, T. Fischer, R. Riedlinger, T. Sullivan, and T. Grutkowski, "The implementation of the Itanium 2 microprocessor," *IEEE J. Solid-State Circuits*, vol. 37, no. 11, pp. 1448–1460, Nov. 2002.

[9] C. Giacomotto, N. Nedovic, and V. Oklobdzija, "The effect of the system specification on the optimal selection of clocked storage elements," *IEEE J. Solid-State Circuits*, vol. 42, no. 6, pp. 1392–1404, June 2007.

[10] V. Zyuban, D. Brooks, V. Srinivasan, M. Gschwind, P. Bose, P. Strenski, and P. Emma, "Integrated analysis of power and performance for pipelined microprocessors," *IEEE Trans. Comput.*, vol. 53, no. 8, pp. 1004–1016, Aug. 2004.

[11] V. Stojanovic and V. Oklobdzija, "Comparative analysis of master-slave latches and flip-flops for high-performance and low-power systems," *IEEE J. Solid-State Circuits*, vol. 34, no. 4, pp. 536–548, April 1999.

[12] D. Markovic, B. Nikolic, and R. Brodersen, "Analysis and design of low-energy flip-flops," in *Proceedings of the Low Power Electronics and Design, International Symposium*, 6–7 Aug. 2001, pp. 52–55.

[13] V. Zyuban, "Optimization of scannable latches for low energy," *IEEE Trans. VLSI Syst.*, vol. 11, no. 5, pp. 778–788, Oct. 2003.

[14] J. Tschanz, S. Narendra, Z. Chen, S. Borkar, M. Sachdev, and V. De, "Comparative delay and energy of single edge-triggered and dual edge-triggered pulsed flip-flops for high-performance microprocessors," in *Proceedings of the Low*

*Power Electronics and Design, International Symposium*, 6–7 Aug. 2001, pp. 147–152.

[15] M. Hamada, H. Hara, T. Fujita, C. K. Teh, T. Shimazawa, N. Kawabe, T. Kitahara, Y. Kikuchi, T. Nishikawa, M. Takahashi, and Y. Oowaki, "A conditional clocking flip-flop for low power H.264/MPEG-4 audio/visual codec LSI," in *Proceedings of the IEEE Custom Integrated Circuits Conference (CICC 2005)*, 18–21 Sept. 2005, pp. 527–530.

[16] S. DasGupta, E. Eichelberger, and T. Williams, "LSI chip design for testability," in *Digest of Technical Papers IEEE International Solid-State Circuits Conference (ISSCC 1978)*, 1978, pp. 216–217.

[17] G. Gerosa, S. Gary, C. Dietz, D. Pham, K. Hoover, J. Alvarez, H. Sanchez, P. Ippolito, T. Ngo, S. Litch, J. Eno, J. Golab, N. Vanderschaaf, and J. Kahle, "A 2.2 W, 80 MHz superscalar RISC microprocessor," *IEEE J. Solid-State Circuits*, vol. 29, no. 12, pp. 1440–1454, Dec. 1994.

[18] R. Ho, K. Mai, and M. Horowitz, "The future of wires," *Proceedings of the IEEE*, vol. 89, no. 4, pp. 490–504, April 2001.

[19] Y. Suzuki, K. Odagawa, and T. Abe, "Clocked CMOS calculator circuitry," *IEEE J. Solid-State Circuits*, vol. 8, no. 6, pp. 462–469, Dec 1973.

[20] J. Warnock, J. Keaty, J. Petrovick, J. Clabes, C. Kircher, B. Krauter, P. Restle, B. Zoric, and C. Anderson, "The circuit and physical design of the POWER4 microprocessor," *IBM J. Res. Dev.*, vol. 46, no. 1, pp. 27–51, January 2002.

[21] D. Lackey, "Efficient latch and clock structures for system-on-chip test flexibility," in *Proceedings of the IEEE International Test Conference ITC '06*, Oct. 2006, pp. 1–7.

[22] B. Stolt, Y. Mittlefehldt, S. Dubey, G. Mittal, M. Lee, J. Friedrich, and E. Fluhr, "Design and implementation of the POWER6 microprocessor," *IEEE J. Solid-State Circuits*, vol. 43, no. 1, pp. 21–28, Jan. 2008.

[23] I. Lin, J. Ludwig, and K. Eng, "Analyzing cycle stealing on synchronous circuits with level-sensitive latches," in *Proceedings of the ACM/IEEE Design Automation Conference*, 8–12 June 1992, pp. 393–398.

[24] E. Shriver, D. Hall, N. Nassif, N. Rahman, N. Rethman, G. Watt, and J. Farrell, "Timing verification of the 21264: A 600 MHz full-custom microprocessor," in *Proceedings of the International Conference on Computer Design: VLSI in Computers and Processors ICCD '98*, 5–7 Oct. 1998, pp. 96–103.

[25] K. Sakallah, T. Mudge, and O. Olukotun, "$checkT_c$ and $minT_c$: Timing verification and optimal clocking of synchronous digital circuits," in *Proceedings of the IEEE International Conference on Computer-Aided Design ICCAD-90. Digest of Technical Papers*, 11–15 Nov. 1990, pp. 552–555.

[26] T. Szymanski and N. Shenoy, "Verifying clock schedules," in *Proceedings of the IEEE/ACM International Conference on Computer-Aided Design ICCAD-92. Digest of Technical Papers*, 8–12 Nov. 1992, pp. 124–131.

[27] T. Burks, K. Sakallah, and T. Mudge, "Identification of critical paths in circuits with level-sensitive latches," in *Proceedings of the IEEE/ACM International Conference on Computer-Aided Design ICCAD-92. Digest of Technical Papers*, 8–12 Nov. 1992, pp. 137–141.

[28] B. Curran, "Power-constrained high-frequency circuits for the IBM POWER6 microprocessor," *IBM J. Res. Dev.*, vol. 51, no. 6, pp. 715–731, November 2007.

[29] M. Matsui, H. Hara, Y. Uetani, L.-S. Kim, T. Nagamatsu, Y. Watanabe, A. Chiba, K. Matsuda, and T. Sakurai, "A 200 MHz 13 $mm^2$-2D DCT macrocell using sense-amplifying pipeline flip-flop scheme," *IEEE J. Solid-State Circuits*, vol. 29, no. 12, pp. 1482–1490, Dec. 1994.

[30] J. Montanaro, R. Witek, K. Anne, A. Black, E. Cooper, D. Dobberpuhl, P. Donahue, J. Eno, W. Hoeppner, D. Kruckemyer, T. Lee, P. Lin, L. Madden, D. Murray, M. Pearce, S. Santhanam, K. Snyder, R. Stehpany, and S. Thierauf, "A 160-MHz, 32-b, 0.5-W CMOS RISC microprocessor," *IEEE J. Solid-State Circuits*, vol. 31, no. 11, pp. 1703–1714, Nov. 1996.

[31] P. Gronowski, W. Bowhill, R. Preston, M. Gowan, and R. Allmon, "High-performance microprocessor design," *IEEE J. Solid-State Circuits*, vol. 33, no. 5, pp. 676–686, May 1998.

[32] B. Nikolic, V. Oklobdzija, V. Stojanovic, W. Jia, J. K.-S. Chiu, and M. Ming-Tak Leung, "Improved sense-amplifier-based flip-flop: design and measurements," *IEEE J. Solid-State Circuits*, vol. 35, no. 6, pp. 876–884, June 2000.

[33] Y. Zhang, H. Yang, and H. Wang, "Low clock-swing conditional-precharge flip-flop for more than 30% power reduction," *Electron. Lett.*, vol. 36, no. 9, pp. 785–786, 2000.

[34] T. Darwish and M. Bayoumi, "Reducing the switching activity of modified SAFF flip-flop for low power applications," in *Proceedings of the 14th International Conference on 2002 – ICM Microelectronics*, 11–13 Dec. 2002, pp. 96–99.

[35] J.-C. Kim, Y.-C. Jang, and H.-J. Park, "CMOS sense amplifier-based flip-flop with two $N-C^2MOS$ output latches," *Electron. Lett.*, vol. 36, no. 6, pp. 498–500, 16 March 2000.

[36] A. Strollo, D. De Caro, E. Napoli, and N. Petra, "A novel high-speed sense-amplifier-based flip-flop," *IEEE Trans. VLSI Syst.*, vol. 13, no. 11, pp. 1266–1274, Nov. 2005.

[37] H. Partovi, R. Burd, U. Salim, F. Weber, L. DiGregorio, and D. Draper, "Flow-through latch and edge-triggered flip-flop hybrid elements," in *Digest of Technical Papers IEEE International Solid-State Circuits Conference (ISSCC 1996)*, 1996, pp. 138–139.

[38] N. Nedovic and V. Oklobdzija, "Dynamic flip-flop with improved power," in *Proceedings of the 26th European ESSCIRC Solid-State Circuits Conference '00*, 19–21 Sept. 2000, pp. 376–379.

[39] N. Nedovic and V. Oklobdzija, "Hybrid latch flip-flop with improved power efficiency," in *Proceedings of the 13th Symposium on Integrated Circuits and Systems Design*, 18–24 Sept. 2000, pp. 211–215.

[40] F. Klass, C. Amir, A. Das, K. Aingaran, C. Truong, R. Wang, A. Mehta, R. Heald, and G. Yee, "A new family of semidynamic and dynamic flip-flops with embedded logic for high-performance processors," *IEEE J. Solid-State Circuits*, vol. 34, no. 5, pp. 712–716, May 1999.

[41] C. Webb, C. Anderson, L. Sigal, K. Shepard, J. Liptay, J. Warnock, B. Curran, B. Krumm, M. Mayo, P. Camporese, E. Schwarz, M. Farrell, P. Restle, I. Averill, R.M., T. Slegel, W. Houtt, Y. Chan, B. Wile, T. Nguyen, P. Emma, D. Beece, C.-T. Chuang, and C. Price, "A 400-MHz S/390 microprocessor," *IEEE J. Solid-State Circuits*, vol. 32, no. 11, pp. 1665–1675, Nov. 1997.

[42] R. Heald, K. Aingaran, C. Amir, M. Ang, M. Boland, P. Dixit, G. Gouldsberry, D. Greenley, J. Grinberg, J. Hart, T. Horel, W.-J. Hsu, J. Kaku, C. Kim, S. Kim, F. Klass, H. Kwan, G. Lauterbach, R. Lo, H. McIntyre, A. Mehta, D. Murata, S. Nguyen, Y.-P. Pai, S. Patel, K. Shin, K. Tam, S. Vishwanthaiah, J. Wu, G. Yee, and E. You, "A third-generation SPARC V9 64-b microprocessor," *IEEE J. Solid-State Circuits*, vol. 35, no. 11, pp. 1526–1538, Nov. 2000.

[43] W. Belluomini, D. Jamsek, A. Martin, C. McDowell, R. Montoye, T. Nguyen, H. Ngo, J. Sawada, I. Vo, and R. Datta, "An 8 GHz floating-point multiply," in *Digest of Technical Papers IEEE International Solid-State Circuits Conference (ISSCC 2005)*, 2005, pp. 374–375, 604.

[44] J. Parkhurst, J. Darringer, and B. Grundmann, "From single core to multi-core: Preparing for a new exponential," in *Proceedings of the IEEE/ACM International Conference on Computer-Aided Design ICCAD '06*, 2006, pp. 67–72.

[45] B. Stackhouse, B. Cherkauer, M. Gowan, P. Gronowski, and C. Lyles, "A 65nm 2-billion-transistor quad-core Itanium® processor," in *Digest of Technical Papers IEEE International Solid-State Circuits Conference (ISSCC 2008)*, 2008, pp. 92–93, 598.

[46] C.-G. Hwang, "New paradigms in the silicon industry," in *Proceedings of the International Electron Devices Meeting IEDM '06*, 11–13 Dec. 2006, pp. 1–8.

[47] D. Josephson, S. Poehlman, V. Govan, and C. Mumford, "Test methodology for the McKinley processor," in *Proceedings of the International Test Conference*, 30 Oct.–1 Nov. 2001, pp. 578–585.

[48] R. Molyneaux, T. Ziaja, H. Kim, S. Aryani, S. Hwang, and A. Hsieh, "Design for testability features of the SUN Microsystems Niagara2 CMP/CMT SPARC chip," in *Proceedings of the IEEE International Test Conference ITC 2007*, 21–26 Oct. 2007, pp. 1–8.

[49] S. DasGupta, R. Walther, T. Williams, and E. Eichelberger, "An enhancement to LSSD and some applications of LSSD in reliability, availability, and serviceability," in *Proceedings of the Twenty-Fifth International Symposium on Fault-Tolerant Computing, ' Highlights from Twenty-Five Years '*, 1995, p. 289.

[50] K.-T. Cheng, S. Devadas, and K. Keutzer, "A partial enhanced-scan approach to robust delay-fault test generation for sequential circuits," in *Proceedings of the International Test Conference*, 26–30 Oct 1991, p. 403.

[51] J. Savir and S. Patil, "On broad-side delay test," *IEEE Trans. VLSI Syst.*, vol. 2, no. 3, pp. 368–372, Sept. 1994.

[52] J. Savir and S. Patil, "Scan-based transition test," *IEEE Trans. Comput.-Aided Design Integr. Circuits Syst.*, vol. 12, no. 8, pp. 1232–1241, Aug. 1993.

[53] A. Asenov, "Simulation of statistical variability in nano MOSFETs," in *Proceedings of the IEEE Symposium on VLSI Technology*, 12–14 June 2007, pp. 86–87.

[54] C. Visweswariah, "Death, taxes and failing chips," in *Proceedings of the Design Automation Conference*, 2–6 June 2003, pp. 343–347.

[55] E. Fetzer, "Using adaptive circuits to mitigate process variations in a microprocessor design," *IEEE Design & Test Comput.*, vol. 23, no. 6, pp. 476–483, June 2006.

[56] G. Roy, A. Brown, F. Adamu-Lema, S. Roy, and A. Asenov, "Simulation study of individual and combined sources of intrinsic parameter fluctuations in conventional Nano-MOSFETs," *IEEE Trans. Electron. Devices*, vol. 53, no. 12, pp. 3063–3070, Dec. 2006.

[57] G. Neuberger, F. Kastensmidt, R. Reis, G. Wirth, R. Brederlow, and C. Pacha, "Statistical analysis of systematic and random variability of flip-flop race immunity in 130nm and 90nm CMOS technologies," in *Proceedings of the IFIP International Conference on Very Large Scale Integration VLSI - SoC 2007*, 15–17 Oct. 2007, pp. 78–83.

[58] F. Klass, A. Jain, G. Hess, and B. Park, "An all-digital on-chip process-control monitor for process-variability measurements," in *Digest of Technical Papers IEEE International Solid-State Circuits Conference (ISSCC 2008)*, 3–7 Feb. 2008, pp. 408–409, 623.

[59] T. Karnik and P. Hazucha, "Characterization of soft errors caused by single event upsets in CMOS processes," *IEEE Trans. Dependable Secure Comput.*, vol. 1, no. 2, pp. 128–143, April–June 2004.

[60] R. Baumann, "Radiation-induced soft errors in advanced semiconductor technologies," *IEEE Trans. Device Mater. Rel.*, vol. 5, no. 3, pp. 305–316, Sept. 2005.

[61] H. Fukui, M. Hamaguchi, H. Yoshimura, H. Oyamatsu, F. Matsuoka, T. Noguchi, T. Hirao, H. Abe, S. Onoda, T. Yamakawa, T. Wakasa, and T. Kamiya, "Comprehensive study on layout dependence of soft errors in CMOS latch circuits and its scaling trend for 65 nm technology node and beyond," in *Proceedings of the Digest of Technical Papers VLSI Technology 2005 Symposium*, 14–16 June 2005, pp. 222–223.

[62] T. Heijmen, P. Roche, G. Gasiot, K. Forbes, and D. Giot, "A comprehensive study on the soft-error rate of flip-flops from 90-nm production libraries," *IEEE Trans. Device Mater. Rel.*, vol. 7, no. 1, pp. 84–96, March 2007.

[63] F. Wang and V. D. Agrawal, "Single event upset: An embedded tutorial," in *Proceedings of the 21st International Conference on VLSI Design VLSID 2208*, 4–8 Jan. 2008, pp. 429–434.

[64] M. Nicolaidis, "Design for soft error mitigation," *IEEE Trans. Device Mater. Rel.*, vol. 5, no. 3, pp. 405–418, Sept. 2005.

[65] P. Meaney, S. Swaney, P. Sanda, and L. Spainhower, "IBM z990 soft error detection and recovery," *IEEE Trans. Device Mater. Rel.*, vol. 5, no. 3, pp. 419–427, Sept. 2005.

[66] S. Mitra, M. Zhang, N. Seifert, T. Mak, and K. S. Kim, "Built-in soft error resilience for robust system design," in *Proceedings of the IEEE International Conference on Integrated Circuit Design and Technology ICICDT '07*, May 30 2007–June 1 2007, pp. 1–6.

[67] Y. Dhillon, A. Diril, A. Chatterjee, and A. Singh, "Analysis and optimization of nanometer CMOS circuits for soft-error tolerance," *IEEE Trans. VLSI Syst.*, vol. 14, no. 5, pp. 514–524, May 2006.

[68] Q. Zhou and K. Mohanram, "Gate sizing to radiation harden combinational logic," *IEEE Trans. Comput.-Aided Design Integr. Circuits Syst.*, vol. 25, no. 1, pp. 155–166, Jan. 2006.

[69] A. KleinOsowski, E. Cannon, M. Gordon, D. Heidel, P. Oldiges, C. Plettner, K. Rodbell, R. Rose, and H. Tang, "Latch design techniques for mitigating single event upsets in 65 nm SOI device technology," *IEEE Trans. Nucl. Sci.*, vol. 54, no. 6, pp. 2021–2027, Dec. 2007.

[70] T. Calin, M. Nicolaidis, and R. Velazco, "Upset hardened memory design for submicron CMOS technology," *IEEE Trans. Nucl. Sci.*, vol. 43, no. 6, pp. 2874–2878, Dec. 1996.

[71] P. Hazucha, T. Karnik, S. Walstra, B. Bloechel, J. Tschanz, J. Maiz, K. Soumyanath, G. Dermer, S. Narendra, V. De, and S. Borkar, "Measurements and analysis of SER-tolerant latch in a 90-nm dual-Vt CMOS process," *IEEE J. Solid-State Circuits*, vol. 39, no. 9, pp. 1536–1543, Sept. 2004.

[72] D. Schroder and J. Babcock, "Negative bias temperature instability: Road to cross in deep submicron silicon semiconductor manufacturing," *J. Appl. Phys.*, vol. 94, p. 1, 2003.

[73] T. Ning, P. Cook, R. Dennard, C. Osburn, S. Schuster, and H. Yu, "1 $\mu$m MOSFET VLSI technology: Part IV – hot-electron design constraints," *IEEE Trans. Electron. Devices*, vol. 26, no. 4, pp. 346–353, Apr 1979.

[74] C. Hu, S. C. Tam, F.-C. Hsu, P.-K. Ko, T.-Y. Chan, and K. Terrill, "Hot-electron-induced MOSFET degradation – model, monitor, and improvement," *IEEE J. Solid-State Circuits*, vol. 20, no. 1, pp. 295–305, Feb 1985.

[75] J. McPherson, "Reliability trends with advanced CMOS scaling and the implications for design," in *Proceedings of the IEEE Custom Integrated Circuits Conference CICC '07*, 16–19 Sept. 2007, pp. 405–412.

[76] D. Ernst, N. S. Kim, S. Das, S. Pant, R. Rao, T. Pham, C. Ziesler, D. Blaauw, T. Austin, K. Flautner, and T. Mudge, "Razor: a low-power pipeline based on circuit-level timing speculation," in *Proceedings of the 36th Annual IEEE/ACM International Symposium on MICRO-36 Microarchitecture*, 2003, pp. 7–18.

[77] K. Bowman, J. Tschanz, N. S. Kim, J. Lee, C. Wilkerson, S.-L. Lu, T. Karnik, and V. De, "Energy-efficient and metastability-immune timing-error detection and instruction-replay-based recovery circuits for dynamic-variation tolerance," in *Digest of Technical Papers IEEE International Solid-State Circuits Conference (ISSCC 2008)*, 3–7 Feb. 2008, pp. 402–403, 623.

[78] M. Agarwal, B. Paul, M. Zhang, and S. Mitra, "Circuit failure prediction and its application to transistor aging," in *Proceedings of the 25th IEEE VLSI Test Symposium*, 6–10 May 2007, pp. 277–286.

[79] D. Blaauw, S. Kalaiselvan, K. Lai, W.-H. Ma, S. Pant, C. Tokunaga, S. Das, and D. Bull, "Razor II: In situ error detection and correction for PVT and SER tolerance," in *Digest of Technical Papers IEEE International Solid-State Circuits Conference (ISSCC 2008)*, 2008, pp. 400–401, 622.

[80] T. Nakura, K. Nose, and M. Mizuno, "Fine-grain redundant logic using defect-prediction flip-flops," in *Digest of Technical Papers IEEE International Solid-State Circuits Conference (ISSCC 2007)*, 2007, pp. 402–403, 611.

# 4

# Exploiting Inductance

Nestoras Tzartzanis

Fujitsu Laboratories of America

In this chapter, the benefits of using inductance for generating and distributing clocks are explored. Starting with the implementation and modeling of spiral inductors and transmission lines the focus, subsequently, shifts into LC and transmission-line oscillators for generating two or more phases with low phase noise. Finally, resonant clock distribution methods including rotary traveling-wave oscillator arrays, standing wave oscillators and grids, and inductor-based clock grids are presented.

## 4.1 Introduction

Although inductive components have been used as off-chip components for several decades, their first practical implementation on silicon was reported in early nineties in silicon bipolar and BiCMOS processes [1, 2]. There are two reasons for this delay. First, the low Q obtained for inductors in early silicon processes with only two or three metal layers compared to off-chip inductors was prohibitive for their use in any practical system. Second, the low-frequency requirements for the inductors made system-level integration in a single die very difficult. Since inductor size is inversely proportional to the square of the frequency, the inductors required at the system level were too large to accommodate for on-chip implementations.

Following the evolution of CMOS processes, new applications and new challenges emerged. CMOS has been the technology of choice for large scale applications such as microprocessors in which large integration is the primary goal. However, the rapid increase of CMOS transition (cut-off) frequency ($f_T$) coupled with its high energy efficiency and high integration scale compared to competing technologies (e.g., SiGe, GaAs) makes CMOS the primary choice for high frequency wireless, wireline, and optical communication applications. Utilizing on-chip inductors to build oscillators became attractive for communication applications as their operating frequency increased to GHz range. Furthermore, one of the drawbacks of CMOS technology is its large process variation which translates into uneven phase generation for ring oscillators and clock skew for clock distribution networks. Oscillator phase deviation can be improved by using LC-based oscillators instead

T. Xanthopoulos (ed.), *Clocking in Modern VLSI Systems*, Integrated Circuits and Systems, DOI 10.1007/978-1-4419-0261-0_4, © Springer Science+Business Media, LLC 2009 

of ring oscillators. Also, LC-based oscillators have lower phase noise compared to ring oscillators since they exhibit better supply noise rejection. The clock skew issue gets worse for increasing clock load since it is proportional to the number of buffer stages required to drive the load [3]. To make things even worse, clock skew takes a larger fraction of the clock cycle as frequency increases, even if the clock distribution network and load remain the same. Clock skew can be addressed by replacing conventional buffer-based clock networks with transmission-line-based and LC-based clock distribution.

Spiral inductors and transmission lines are the two dominant structures for forming on-chip inductive components. In this chapter, both spiral inductors and transmission lines are described including implementation and modeling methods. Subsequently, LC-based Voltage-Controlled Oscillators (VCO) for generating two or more phases as well as transmission-line based distributed VCOs are described. In Sect. 4.4, clock distribution methods based on resonance are presented. Finally, Sect. 4.5 summarizes the chapter.

## 4.2 Monolithic Inductance

### 4.2.1 Spiral Inductors

The implementation of spiral inductors depends mostly on the target application and the process parameters. For instance, the inductor quality ($Q$) factor is a key parameter for spiral inductors designed for LC VCOs. Maximizing the $Q$ of the inductor around the VCO operating frequency results in decreasing the VCO phase noise. Furthermore, to meet the VCO frequency range, it is necessary to accurately predict the inductance $L$ of the spiral inductor. For other applications (e.g., shunt and series peaking [4], etc.), inductor area is more important than its $Q$ or inductance accuracy. Finally, for practical verification purposes, an inductor model should be readily available. Inductor models vary depending on the application. Narrowband models are necessary for LC VCOs since they operate around a well-determined frequency whereas broadband models are required for other applications. In this chapter, we focus on inductor implementation and modeling for LC VCOs.

Spiral inductors can be implemented in various shapes. Figure 4.1 shows examples of 2-turn square, octagonal, and symmetrical octagonal spiral inductors. Other common shapes include hexagonal and circular spiral inductors. Typically, high-$Q$ spiral inductors are implemented using low sheet resistance, high-level metal layers. It is also possible to strap together multiple metal layers in order to reduce the inductor resistance. Using high-level metals results in low inductor capacitance and substrate losses as we see next.

The inductance $L$ of spiral inductors can be estimated using Greenhouse's method [5]. However, simpler analytical models are also available [4, 6, 7] based on the inductor geometry (shape, number of turns, radius, metal width, metal spacing, and metal thickness). The inductance model presented in [6] is based on inductor segment decomposition. The length, $l$, of square spiral inductors is given by:

$$l = (4n+1)\,d_{\text{in}} + (4N_{\text{i}}+1)\,N_{\text{i}}\,(w+s)\,, \tag{4.1}$$

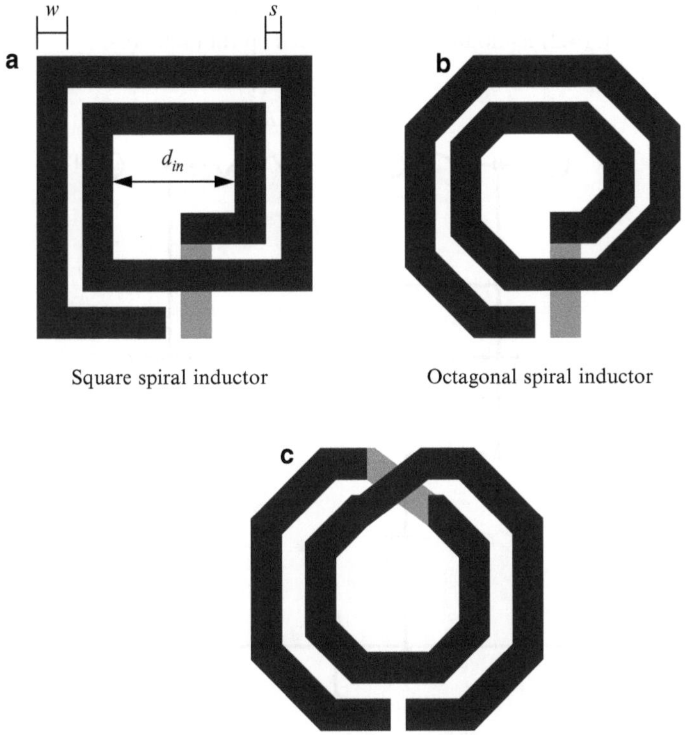


Octagonal symmetrical spiral inductor

**Fig. 4.1.** Common 2-turn spiral inductor shapes

where $n$ is the number of turns, $N_{\mathrm{i}}$ is the integer part of $n$, $d_{\mathrm{in}}$ is the inner diameter of the inductor, $w$ is the metal width, and $s$ is the metal spacing.

The inductance $L$ is estimated taking into consideration the self inductance of the spiral inductor, the negative mutual inductance between segments, and the positive mutual inductance between segments. The total inductance of the square spiral inductance is given by [6]:

$$L=\frac{\mu_0}{2\pi}l\left[\begin{array}{l}\ln\frac{l}{n(w+t)}-0.2-0.47n\\ +(n+1)\left(\ln\left(\sqrt{1+\left(\frac{l}{4nd_{\mathrm{av}}}\right)^2}+\frac{l}{4nd_{\mathrm{av}}}\right)-\sqrt{1+\left(\frac{4nd_{\mathrm{av}}}{l}\right)^2}+\frac{4nd_{\mathrm{av}}}{l}\right)\end{array}\right], \tag{4.2}$$

where $\mu_0$ is the permeability of vacuum, $l$ and $n$ are the inductor length and number of turns, $w$ and $t$ are the metal width and thickness, and $d_{\mathrm{av}}$ is the average segment distance for the positive mutual inductance given by (4.3).

$$d_{\mathrm{av}}=(w+s)\,\frac{(3n-2N_{\mathrm{i}}-1)\,(N_{\mathrm{i}}+1)}{3\,(2n-N_{\mathrm{i}}-1)}. \tag{4.3}$$

Inductance analytical models can be derived for different shapes similar to square inductors. In [6], (4.2) is modified for octagonal spiral inductors.

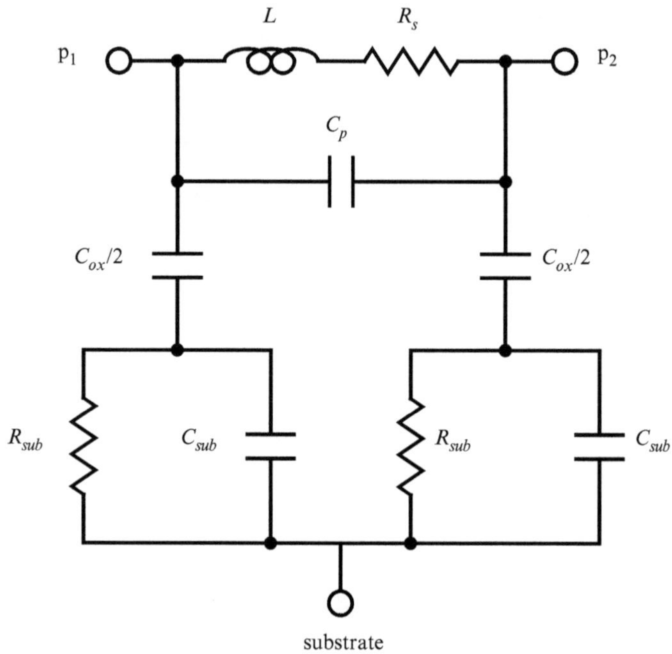


**Fig. 4.2.** Spiral inductor narrowband lumped model

Deriving lumped element models for spiral inductors is of utmost importance. First, they facilitate circuit simulations (especially in time domain). Second, they provide a well-defined description of spiral inductors that can be used to improve their design. Figure 4.2 shows a widely used lumped element model for spiral inductors [4, 8, 9], which is more appropriate for narrowband applications. A broadband inductor model has also been proposed [10], but such models are beyond the scope of this chapter. The narrowband model has two signal ports $p_1$ and $p_2$ and the substrate port which typically connects to ground. It includes the inductance $L$ connected to a series resistance $R_s$, a shunt capacitance $C_p$ and the inductor to substrate capacitance $C_{ox}$. Finally, $R_{sub}$ and $C_{sub}$ model substrate resistance and capacitance respectively.

The series resistance $R_s$ models losses due to the metal resistivity and the skin effect. The latter term is used to describe that at high frequency, the current tends to flow at the surface of a conductor causing its resistance to increase. Therefore, $R_s$ is frequency dependent and is given by [4, 9]:

$$R_s = \frac{\rho l}{w\delta\left(1 - e^{-t/\delta}\right)}, \tag{4.4}$$

where $\rho$ is the metal resistivity at DC, $l$ is the inductor length, $w$ is the metal width, $t$ is the metal thickness and $\delta$ is the metal skin depth obtained by (4.5):

$$\delta = \sqrt{\frac{2\rho}{\omega\mu_0}}. \tag{4.5}$$

Other losses not modeled with (4.4) and (4.5) include proximity effect losses and eddy currents losses to the substrate. The proximity effect term describes the phenomenon of the currents of two parallel conductors runing along the side of their surface that is close to each other [4]. This effect is caused by the magnetic field between the two conductors. Substrate losses due to eddy currents depend on the substrate doping and can be neglected for lightly doped substrates.

Capacitance $C_p$ models the overlap between the metal layer used for the inductor and the underpass metal to connect to the other port:

$$C_p = nw^2 \frac{\varepsilon_{oxM}}{t_{oxM}}, \tag{4.6}$$

where $n$ is the number of turns, $w$ is the metal width, $\varepsilon_{oxM}$ and $t_{oxM}$ are the oxide permittivity and the oxide thickness between the two metal layers respectively.

$C_{ox}$ is the capacitance between the inductor metal layers and the substrate modeled with:

$$C_{ox} = wl \frac{\varepsilon_{ox}}{t_{ox}}, \tag{4.7}$$

where $\varepsilon_{ox}$ and $t_{ox}$ are the oxide permittivity and the oxide thickness between the inductor and the substrate, respectively.

Resistance $R_{sub}$ and capacitance $C_{sub}$ model losses to the substrate and are fitting parameters based on measurements. $R_{sub}$ models loss due to current capacitively coupled to substrate and is given by [4, 9]:

$$R_{sub} = \frac{2}{lwG_{sub}}, \tag{4.8}$$

where $G_{sub}$ is the substrate conductance per unit area. $C_{sub}$ is given by [4, 9]:

$$C_{sub} = \frac{lwC_{subu}}{2}, \tag{4.9}$$

where $C_{subu}$ is the substrate capacitance per unit area.

The analytical expressions presented in this section can be used as first-order approximations to develop inductor models. There are effects that cannot be accounted for by using the analytical equations. For instance dummy metal fills are inserted in modern chemical mechanical polishing (CMP) processes which impact at least the inductor to substrate capacitance. More accurate inductor models are based on hardware characterization and parameter fitting for the components shown in Fig. 4.2. Generally, this task should be performed for each inductor used in a circuit. However, it is possible to develop scalable inductor models for varying inductor radius if the other parameters such as shape, number of turns, width, and space are kept constant. Although this restriction imposes limitations to the designers, it can speed up the inductor modeling process. Only a few inductors with different radius must be characterized whereas a full scalable model can be developed through extrapolation from the measured data. In addition, several EM analysis tools [11, 12] are available as an option between simple analytical models and hardware characterization. There is a trade-off between their accuracy and the inductor analysis time and they are more effective when their process settings are calibrated based on hardware measurements to produce accurate models for fast or modest analysis time.

Besides the inductance, the $Q$ factor is the other key parameter for inductors used for clock generation. Analytical models for inductor $Q$ factor based on the inductor lumped model have been developed [9, 13, 14]. However, the $Q$ factor which is a function of frequency can be extracted through hardware measurements for better accuracy. In general the $Q$ of an inductor can be improved by reducing losses. Certain processes provide options to boost the $Q$ of inductors. For example, low-resistivity thick high-level metal layers can be available for inductor implementation. If such option is not available, multiple metal layers can be strapped together to form the inductor. Another practical issue that can impact the inductor $Q$ factor is the necessity of dummy metal fills in modern processes which increase inductor coupling to the substrate. If it is not possible to exclude the dummy metal fills below the inductor, they can be added manually. This serves two purposes. First, only the minimum amount of dummy metal fills is inserted to satisfy process requirements. Second, it resolves the issue of random dummy metal fills which depend on the location of the inductor in the die. Another technique to improve inductor $Q$ is the insertion of patterned ground shields [14] using a low metal layer. Shielding the inductor from the substrate reduces inductor coupling to the substrate as well as noise that reaches the inductor from the substrate. The drawback is the increase of the inductor capacitance since the shield is closer to the inductor than the substrate.

### 4.2.2 Transmission Lines

As the integrated circuit operating frequency increases, long interconnects behave as distributed instead of lumped elements. This happens when the length of the interconnect approaches the wavelength of the signal highest frequency. In this mode of operation, long interconnects behave as transmission lines and they cannot be modeled using simple RC-based models. Figure 4.3 shows a lumped model for such interconnect. Components $L$, $R$, $C$, and $G$ represent inductance, resistance, capacitance, and conductance per unit length respectively. The inductance $L$ and capacitance $C$ model the magnetic and electric energy-stored components, respectively. The resistance and conductance model energy losses in the transmission line. Specifically, $R$ models losses due to the series resistance of the transmission line including the skin effect, whereas $G$ models losses due to the dielectric conductance.

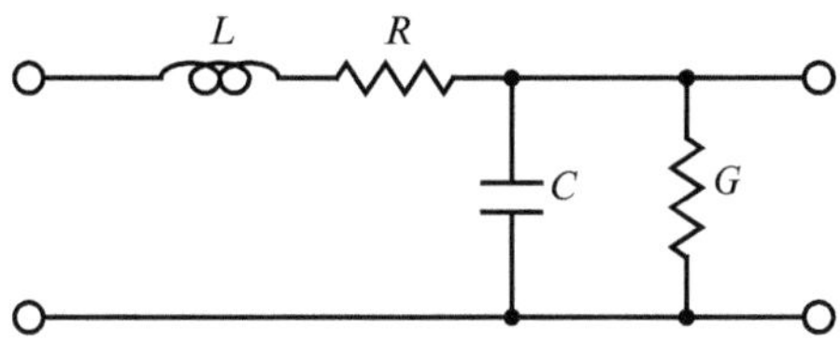


**Fig. 4.3.** Transmission line lumped model

The characteristic impedance $Z_0$ of a transmission line determines the voltage to current ratio of a signal propagating along its length and is given by:

$$Z_0 = \sqrt{\frac{R + j\omega L}{G + j\omega C}}. \tag{4.10}$$

If the losses (i.e., $R$ and $G$) are negligible or if RC=GL, then (4.10) simplifies to:

$$Z_0 = \sqrt{\frac{L}{C}}. \tag{4.11}$$

Another important parameter of a transmission line is its propagation constant, $\gamma$, that determines the signal attenuation through the line. The voltage $V$ at any point in the transmission line that is in distance $z$ from its beginning, is given by:

$$V(z) = V_0 e^{-\gamma z}, \tag{4.12}$$

where $V_0$ is the voltage at $z = 0$.

The propagation constant, $\gamma$, is given by:

$$\gamma = \sqrt{(R + j\omega L)(G + j\omega C)}, \tag{4.13}$$

According to (4.13), $\gamma$ can be generalized into the following complex form [4]:

$$\gamma = \alpha + j\beta, \tag{4.14}$$

where $\alpha$ determines the attenuation in the line due to the distance increase and $\beta$ has only phase contribution.

If the losses are negligible, (4.13) reduces to:

$$\gamma = j\omega\sqrt{LC}. \tag{4.15}$$

As expected, there is no attenuation to signal amplitude, only its phase changes proportionally to its frequency. As shown in [4], the delay of a lossless line is a constant independent from frequency, but proportional to $\sqrt{LC}$. At the same time, the bandwidth in the lossless transmission line is infinite because there is no signal attenuation. As long as $R$ and $G$ can be neglected, increasing $L$ and $C$ causes the transmission line delay to increase, but has no effect on its bandwidth.

If RC=GL, then (4.13) simplifies to:

$$\gamma = \sqrt{RG} + j\omega\sqrt{LC}. \tag{4.16}$$

In this case, the transmission line has losses proportional to $\sqrt{RG}$, and therefore limited bandwidth. However, its delay is still independent from the frequency and its characteristic impedance is still given by (4.11) similar to the lossless line. Therefore, the transmission line is distortionless which improves the transmission of broadband signals. This property of the transmission line was first recognized by Heaviside [15] and Pupin [16] and led to the improvement of the telephony industry through the periodic insertion of coils so that the RC and GL products match. The same approach was also used recently [17, 18] to implement distortionless differential transmission lines by adding periodic resistances to increase their shunt conductance $G$ in order for RC and GL products to match. Using distortionless transmission lines is beneficial for transmitting broadband signals because it reduces their intersymbol interference (ISI) since their loss and group delay are frequency independent. However, such an improvement is not necessary for transmitting clock signals because their frequency is constant. Although not directly related to the potential distortionless nature of transmission lines, [19] has demonstrated that on-chip interconnect transmission-line effects can be used to correct duty cycle deviations as well as sensitivity to PVT variations.

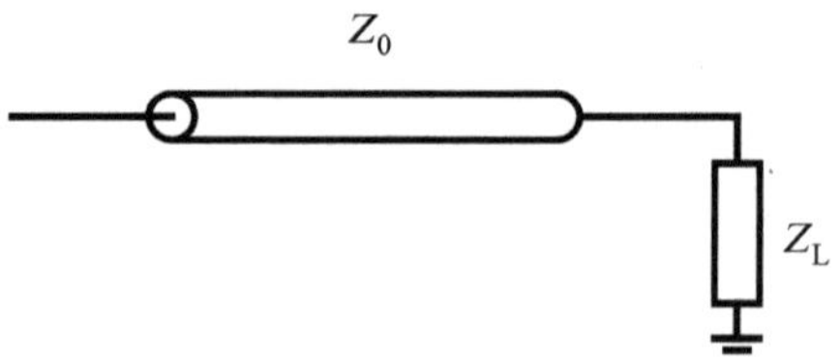


**Fig. 4.4.** Finite-length transmission line with termination

The transmission line characteristics presented above apply to transmission lines with infinite length. In practice, transmission lines have a finite length and are used to transmit a signal between two points. In order to ensure that transmission-line characteristics resemble those of an infinite line, a termination impedance must be inserted at the receiving end with impedance $Z_L$, equal to the characteristic impedance $Z_0$ of the line (Fig.4.4). If the termination impedance does not match the line characteristic impedance, there is a reflection propagating in the opposite direction of the transmitted signal. Let $\Gamma_L$ denote the reflection coefficient defined as the ratio of reflected to incident voltage and current at the receiving end of the line. Then $\Gamma_L$ is given by:

$$\Gamma_L = \frac{Z_L - Z_0}{Z_L + Z_0}. \tag{4.17}$$

The reflection coefficient absolute value can vary between 0, when $Z_L$ perfectly matches $Z_0$, and $\pm 1$, when termination is an open or short. As discussed in Sects. 4.3 and 4.4, certain circuit topologies for clock generation and distribution are based on circular transmission lines eliminating the need for impedance termination.

Similar to spiral inductors, transmission lines are implemented in silicon using low-resistive, thick, high-level metals. In general there are two major forms of transmission lines: microstrips and coplanar waveguides. Microstrips are implemented as metal lines on the top of substrate or a ground shield using a low-level metal layer (Fig. 4.5). Coplanar waveguides include side ground shields for ground current returns and can also run either on top of the substrate or a ground shield (Fig. 4.6). Coplanar waveguides are better shielded from surrounding circuits and other interconnects at the expense of area overhead for the side ground shields. Since different width and spacing settings can give the same characteristic impedance $Z_0$, there are several different parameters that need to be considered for their implementation:

- Low-resistivity substrate with epi layer have higher losses compared to high-resistivity substrate. Therefore, bottom ground metal shield may be used to reduce losses in the former case.
- Using ground metal shield reduces losses to the substrate as well as substrate noise, but at the same time it increases the transmission-line capacitance since

a

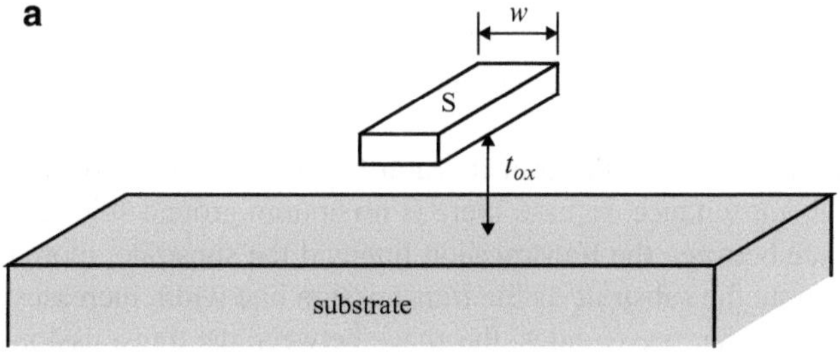


Microstrip transmission line

b

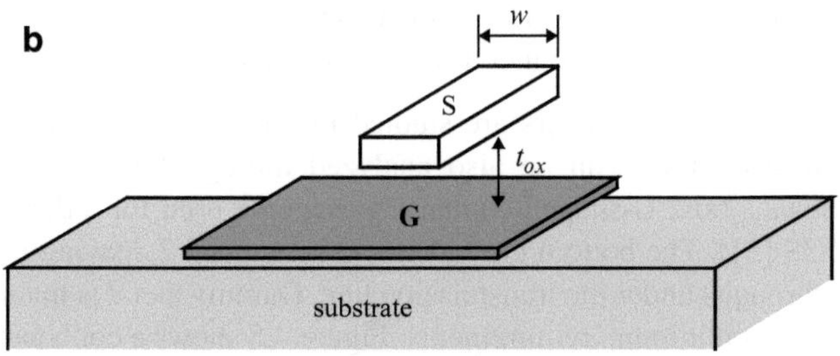


Microstrip transmission line with metal shield

**Fig. 4.5.** Microstrip transmission line with and without ground shield

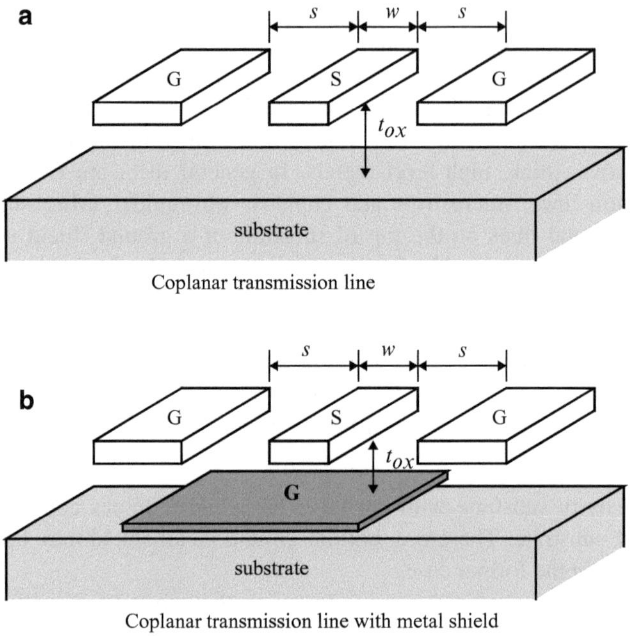


**Fig. 4.6.** Coplanar transmission line with and without ground shield

$t_{ox}$ is reduced. Another issue with bottom ground shield is that transmission line current will tend to run along its bottom side due to the proximity and skin effects.

- Increasing the width of the transmission line reduces resistive losses but it increases the capacitance. In case there is no bottom ground shield, depending on the distance between the transmission line and the substrate, more electric field lines penetrate the substrate as the transmission line width increases.
- In case of coplanar waveguides, the space between the transmission line and the side ground lines affects both the inductance and capacitance.
- Dummy metal insertion required by certain processes can change both the characteristic impedance and the losses of a transmission line. The dummy metal effect can be reduced if these metals are placed manually.

Trade-offs of these parameters are studied in [20] through hardware measurements. Transmission lines can be also analyzed through EM software [11, 12]. Figure 4.7 shows a 60Ω U-shaped coplanar waveguide used for a distributed VCO in 90nm CMOS [21]. The bottom ground shield is in metal 2, leaving metal 1 available for feed-throughs under the transmission line. Dummy metal is manually placed to meet the process minimum requirements. Figure 4.8 shows a coplanar differential waveguide. The space between the two differential signals is twice the space between each line and the side ground line.

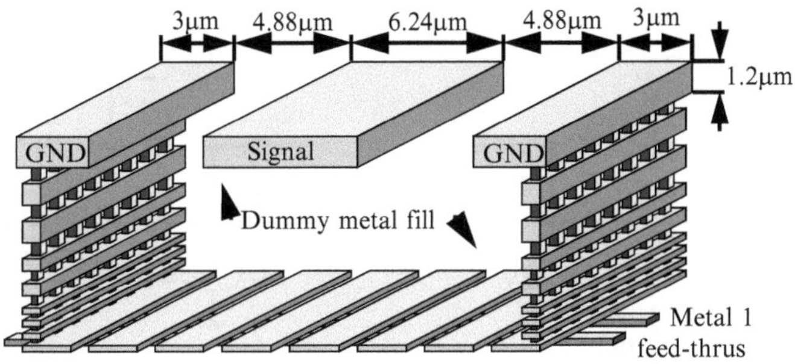


**Fig. 4.7.** U-shaped coplanar waveguide. Reproduced with permission from [21], © 2006 IEEE

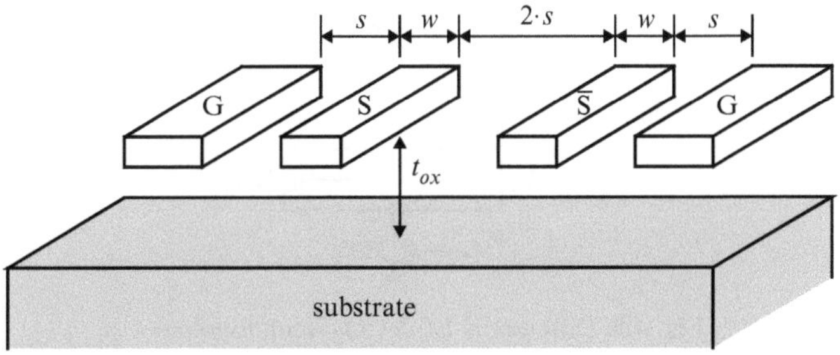


**Fig. 4.8.** Coplanar differential transmission line without ground shield

## 4.3 Inductor-Based Clock Generation

### 4.3.1 Differential LC VCO

As was noted before, LC VCO implementation was one of the first applications of on-chip spiral inductors [2]. In general, LC oscillators consist of two main components: the LC tank and the negative resistance. The latter is an active circuit necessary to start and maintain the oscillation [22]. Figure 4.9 shows commonly used differential LC VCO circuit topologies in CMOS. The first two circuits are similar with the main difference being the bias configuration (top biased vs. bottom biased). They both rely on cross-coupled NFETs $MN_0$ and $MN_1$ to form the negative resistance. The third circuit uses cross-coupled NFETs and PFETs for negative resistance with the advantage of requiring a single inductor for both LC tanks. The other two topologies require two inductors or one center-tapped inductor. In any case, the physical placement and wiring is greatly simplified when the VCO topology requires a single inductor. Another advantage for the third VCO topology is that the clock phases do not exceed the supply voltage. In contrast, it is possible especially for the clock phases of the first topology to surpass the supply voltage, which can cause process

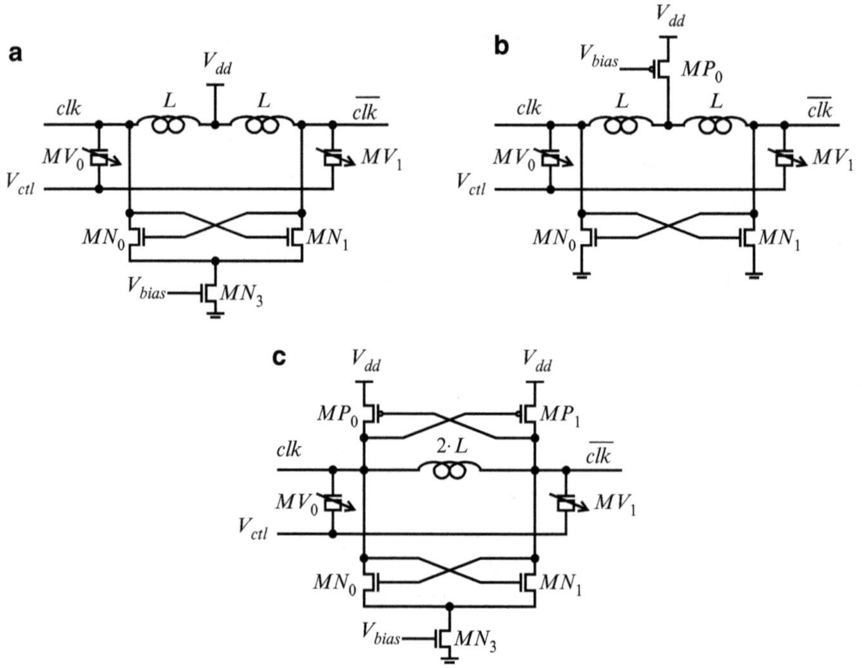


**Fig. 4.9.** Differential LC VCO circuit topologies

break down assuming the VCO supply voltage is the same as the process maximum allowed voltage. On the other hand, the common-mode voltage of the first VCO is equal to the supply voltage allowing for more voltage headroom.

In LC VCOs, the frequency is controlled by varying the voltage across the varactors. There are two types of varactors: PN junctions and MOSFET [23, 24]. $MV_0$ and $MV_1$ represent MOSFET varactors in Fig. 4.9 while $V_{\mathrm{ctl}}$ denotes an analog control voltage. The frequency of the VCO is given by:

$$f = \frac{1}{2\pi\sqrt{LC}}, \tag{4.18}$$

where $L$ and $C$ denote the VCO inductance and capacitance respectively. The capacitance consists of a fixed part (transistor gate, drain, and parasitics) and a varying part (varactors) with the latter determining the frequency range. Another method to adjust the VCO frequency is to use magnetic tuning through transformers [25] which effectively changes the inductance factor in (4.18).

Besides the VCO frequency and range, its phase noise is the most important characteristic, especially for communication applications in which low clock jitter is necessary in order to meet system jitter specifications for transmitting and receiving data. Starting with the generic Leeson model for oscillator phase noise [26], several theories have been developed recently specifically for LC oscillators [27–30]. Leeson's equation for oscillator phase noise due to thermal noise is:

$$L(\omega_m) = F\frac{1}{V^2}\frac{kT}{C}\frac{\omega_0}{Q}\frac{1}{\omega_m^2}, \tag{4.19}$$

where $\omega_0$ and $\omega_m$ are the center and offset frequency respectively, $kT/C$ is the thermal noise density, $V$ is the voltage amplitude, $Q$ is the resonator quality factor, and $F$ is a circuit-dependent noise factor. Extending Leeson's linear equation in order to more accurately model the inherently nonlinear behavior of the LC oscillators, it has been shown that for the VCO topology of Fig. 4.9a the noise factor $F$ is given by [30, 31]:

$$F = 1 + \frac{4\gamma IR}{\pi V} + \gamma\frac{4}{9}g_{\mathrm{mtail}}R, \tag{4.20}$$

where $\gamma$ is the FET channel noise coefficient, $I$ is the tail current, $V$ is the voltage amplitude, $R$ is the tank resistance, and $g_{\mathrm{mtail}}$ is the transconductance of the current-source FET. Based on the phase noise equations above, it appears that increasing the VCO voltage amplitude reduces phase noise since it improves signal to noise ratio. However, in practice there is a limit on how large the voltage amplitude can be before in fact the phase noise starts increasing. As the voltage amplitude increases, the differential pair FETs start operating in the linear region for part of the cycle time. Depending on the circuit topology and process parameters, it is possible for the differential pair FETs to completely turn off. Furthermore, the current-source FET may also go out of saturation into the linear region of operation. In fact, there are two modes of operation for LC VCOs [31–33]: current-limited and voltage-limited mode. In current-limited mode operation, the clock voltage amplitude is determined by the tail current and the equivalent tank resistance. In voltage-limited mode operation, the amplitude approaches or exceeds the supply voltage for the VCOs shown in Fig. 4.9c and Fig. 4.9a, respectively.

One of the main contributors to phase noise is the tail current source noise. This noise is both high frequency and low frequency. The former is mainly due to the second harmonic of the VCO frequency since even harmonics flow into the common-mode path. This noise can be filtered by placing a large capacitor parallel to the current source to short the noise frequencies around the second harmonic and an inductor between the current source and the tail sized properly to provide high impedance at the second harmonic [31]. Low-frequency noise in the tail current source can also be suppressed using inductive degeneration and capacitive filtering which require off chip inductor and capacitor, respectively [34].

To ensure that the VCO frequency range has enough margin to meet its specification despite process, voltage, and temperature variations, extra varactor banks are typically included. The problem is exacerbated in applications that require operation in multiple frequencies [35]. The VCO gain is proportional to the varactor size. Furthermore, the control voltage usually varies around half of the supply voltage where the VCO gain is close to its maximum. Any VCO noise (control voltage as well as supply and substrate noise) gets converted into phase noise in proportion with its gain [36]. Since the gain of the VCO is minimized when the control voltage is set to ground or to the supply voltage, phase noise can be reduced by operating the VCO at a lower gain. This can be accomplished by splitting the varactor banks into digitally-controlled and analog-controlled [36] with the majority of them being controlled digitally. VCOs with hybrid analog/digital control complicate the design at the system-level (e.g., PLL, CDR, etc.) since they require both coarse and fine tuning. There is also the extreme case of completely digitally-controlled oscillators (DCO) [37].

### 4.3.2 Quadrature LC VCO

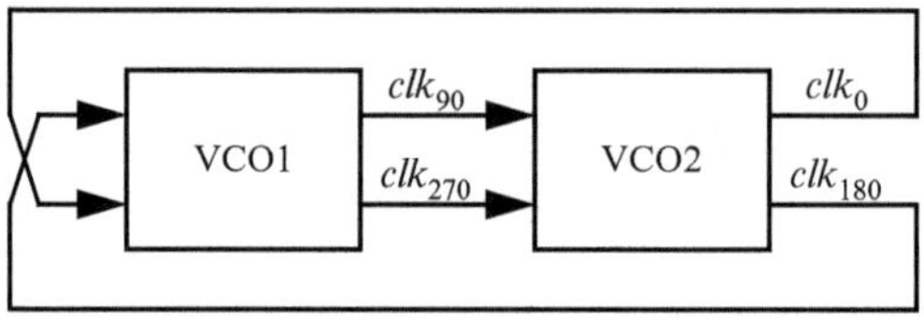


**Fig. 4.10.** Quadrature LC VCO block diagram

Several communication applications require four clock phases with 90° phase difference between consecutive phases. Such applications include half-rate CDR architectures [38, 39] and wireless transceivers [40]. To meet strict phase deviation and phase noise requirements imposed by these applications at multigigahertz frequency range, using LC VCOs is the primary choice especially in CMOS. A quadrature LC VCO can be formed by coupling two differential LC VCOs (Fig.4.10). Starting with a typical differential LC VCO (Fig.4.9a), coupling can be introduced either in parallel or in series with the cross-coupled pair transistors. Figure 4.11a shows a parallel-coupled quadrature LC VCO [41]. Varactors are omitted for clarity. $MN_4$, $MN_5$, $MN_6$, and $MN_7$ are the coupling transistors connected in parallel with the stage internal cross-coupled switch transistors $MN_0$, $MN_1$, $MN_2$, and $MN_3$ respectively. Assuming that coupling and switch transistors have the same length, the coupling strength, $\alpha$, is defined as the ratio between the width of the coupling transistors to the width of the switch transistors [42]. Coupling strength $\alpha$ has a value between 0 and 1. It has been shown [42] that in case of parallel coupling both the phase noise and phase deviation depend on $\alpha$ in an inverse fashion. As $\alpha$ increases, phase deviation decreases whereas phase noise increases. Figure 4.11b shows an alternative parallel-coupled quadrature VCO in which the coupling parallel path is biased using a different tail-current transistor. In this case, $\alpha$ is determined by the bias current ratio between the coupling path and the switch path.

Figure 4.12 shows two series-coupled quadrature LC VCOs [42, 43]. In both cases, the coupling transistors in each stage are connected in series with the cross-coupled transistors. In Fig.4.12a, the coupling transistors $MN_4$–$MN_7$ are on the

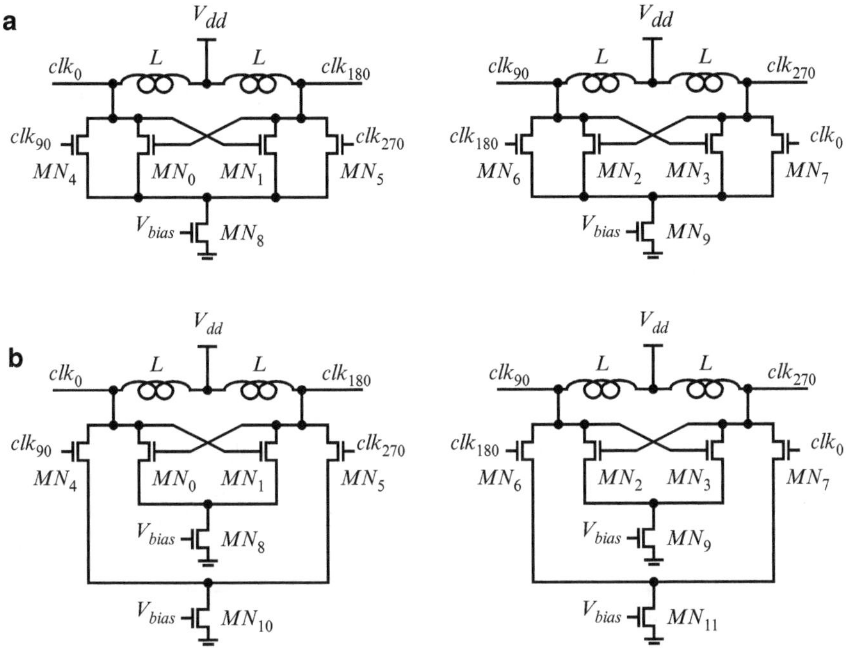


**Fig. 4.11.** Parallel-coupled quadrature LC VCOs

top whereas in Fig.4.12b they are on the bottom. As has been reported [42, 43], the phase deviation in series-coupled quadrature LC VCOs is independent of the coupling strength for all reasonable values (e.g., between 1/3 and 1/2). Therefore, the coupling strength can be optimized for phase noise without affecting phase deviation between the four clock phases. In studies comparing parallel- and series-coupled quadrature VCOs, it has been shown [42, 43] that series-coupled VCOs exhibit better phase noise characteristics compared to equivalent parallel-coupled VCOs. Phase noise and oscillation stability of parallel-coupled quadrature VCOs can be improved by shifting the coupling phases by 90° (Fig.4.13) [44–46]. The outputs of the phase shifters that are used to couple the VCO stages are in phase with their respective cross-coupled phases that drive the other branch of the parallel structure. Phase shifter circuits can be based on a differentiator or an integrator [45] or a passive RC-CR filter [46]. As shown in [46], the accuracy of the phase shift is not essential since same results can be obtained with smaller than 90° phase shift.

Other quadrature LC VCO architectures have been proposed that require two inductors instead of four [47, 48]. In the first design, the coupling is introduced using PFETs. The second design introduces both pull-down and pull-up parallel coupling networks using NFETs and PFETs, respectively.

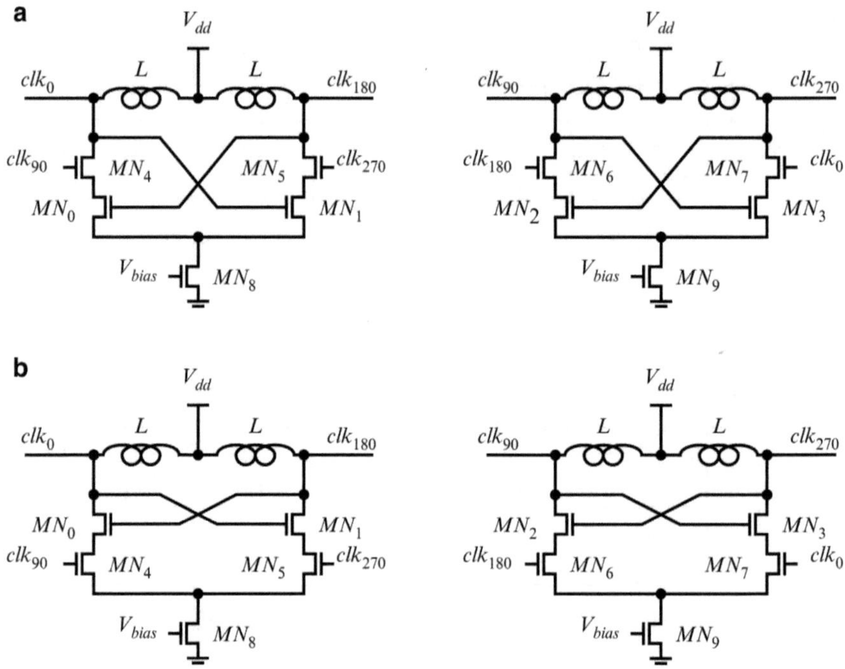


**Fig. 4.12.** Series-coupled quadrature LC VCOs

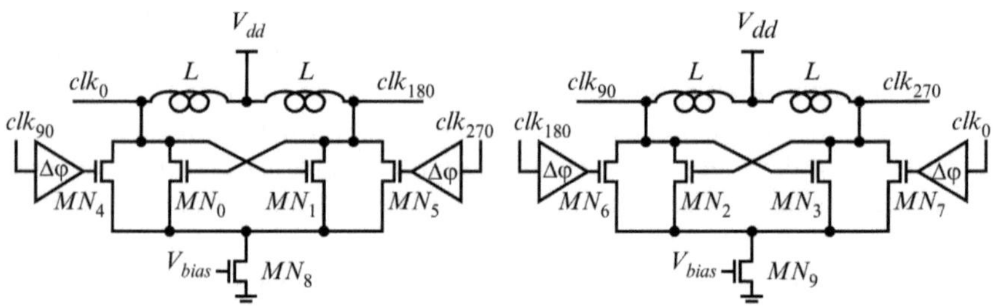


**Fig. 4.13.** Parallel-coupled quadrature LC VCO with 90° phase shift in the coupling phase

### 4.3.3 Distributed VCO

Distributed amplifiers are used in applications with high bandwidth requirements. Despite the physical challenges associated with distributed amplifier implementation, they can achieve high bandwidth because parasitic capacitance (including transistor gate and drain) is absorbed in the transmission lines. Distributed oscillators can be implemented based on distributed amplifiers by feeding the output of the amplifier back to its input [49, 50]. A 3-stage distributed oscillator is shown in Fig. 4.14. Com-

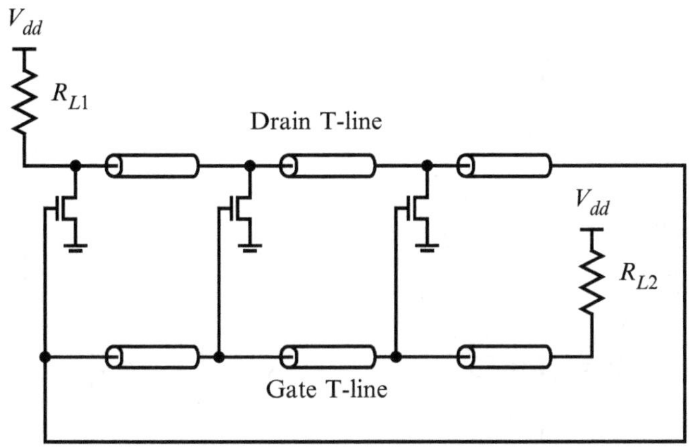


**Fig. 4.14.** Distributed oscillator

pared to lumped LC oscillators, distributed oscillators can be operated at frequencies close to the $f_T$ of the process. As reported in [50], using conventional varactors for frequency tuning can result in loading the transmission line excessively which would cause the oscillator operating frequency to be reduced. An alternative frequency tuning method is proposed in [50]. Specifically, the oscillator frequency is tuned by effectively changing the length of the transmission line. This can be achieved by shortening or lengthening the path of the current either in the drain or the gate transmission line. A variation of the distributed oscillator shown in Fig. 4.14 has also been implemented using discrete components [51–53]. In this design, the drain line connects to the gate line at the drain line load.

There are several issues associated with the implementation of conventional distributed oscillators. First, the voltage amplitude across both the drain and gate lines varies depending on the distance to the respective load resistors. Furthermore, the termination resistors are sources of thermal noise that affect the oscillator phase noise. Finally, it is difficult to implement differential distributed oscillators since they require two gate and two drain lines.

### 4.3.4 Poly-Phase Circularly Distributed VCO

Communication applications such as quarter-rate and/or oversampled CDRs require eight [54] or more [18] clock phases equally-spaced in time. Distributed oscillators offer the ability to generate multiple clock phases. However, as was discussed before conventional distributed oscillators have several drawbacks that make their use difficult. Another type of distributed oscillators can be formed using circular transmission line [21, 54] as shown in Fig. 4.15 which are based on rotary traveling waves [55, 56]. The distributed oscillator shown in Fig. 4.15 produces eight clock phases with 45° phase difference between any two consecutive phases. The transmission line is folded so that clock phases are delivered as four differential pairs ($\text{clk}_i$ and

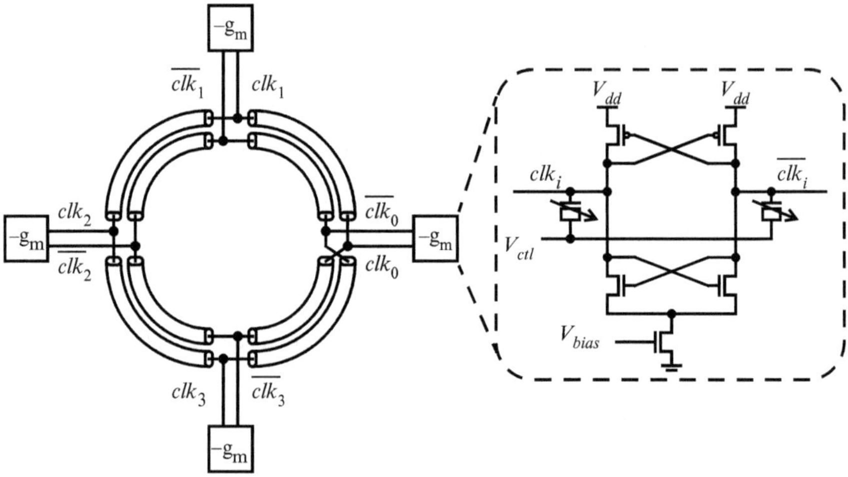


**Fig. 4.15.** Circular distributed oscillator. Reproduced with permission from [21], ©2006 IEEE

$\overline{\text{clk}_i}$, $i = 1,..., 3$). Four negative-$g_\text{m}$ circuits connected to the differential pairs start and sustain the oscillation. The negative-$g_\text{m}$ circuits can be cross-coupled NFET/PFET pairs. Varactors are included so that the oscillator frequency can be adjusted through a control voltage, $V_\text{ctl}$. The circular distributed oscillator overcomes all issues related to the amplifier-based distributed oscillator. All clock phases have the same amplitude regardless of their position. The circular transmission line has virtually infinite length, and therefore eliminates the need for termination resistors. Finally, the clock phases are inherently delivered as differential pairs.

A distributed oscillator is useful only if the direction of the clock phases is controllable. Circular distributed oscillators are symmetric resulting in uncertainty for the clock direction. The clock phases can propagate either clockwise or counter-clockwise. After power up $\text{clk}_i$ can either lead or trail $\text{clk}_{i-1}$ depending on several physical design factors such as the location of the pads, the supply distribution network, and the physical arrangement of the transmission lines. In [21], a method to control the direction of the clock phases has been demonstrated on a 24-phase closed-loop distributed VCO. The method is outlined in Fig. 4.16. In each stage two NFETs are added that conditionally pull down $\text{clk}_i$ and $\overline{\text{clk}_i}$. When on, these NFETs add positive resistance that cancels out the negative-$g_\text{m}$ circuits. Their gates are controlled by signal ccw/cw starting with $\text{clk}_0$ and ending with $\overline{\text{clk}_3}$ or the other way around, following the VCO loop. Pulling up the control signal stops the VCO from oscillating and resets the clock phases to a DC voltage determined by the strength of the FETs in the negative-$g_\text{m}$ circuit and the positive-resistance NFETs. After all phases are reset, the control signal is pulled down either from the ccw side ($\text{clk}_0$ leads) or from the cw side ($\overline{\text{clk}_3}$ leads), releasing the clock phases in a predetermined order.

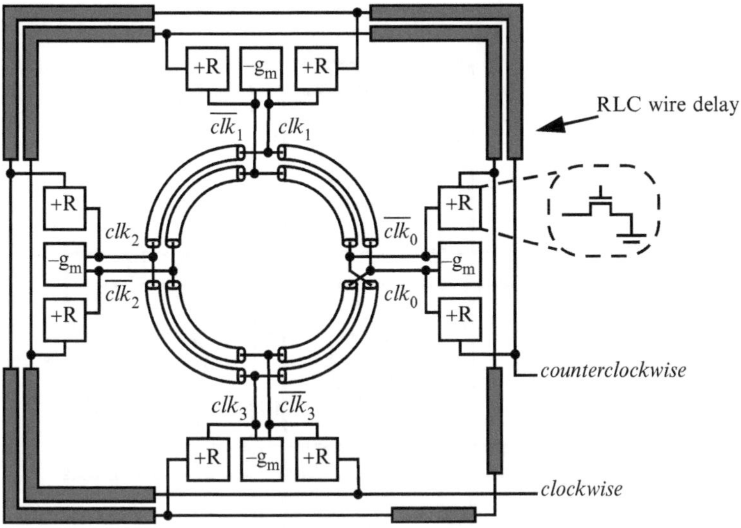


**Fig. 4.16.** Circular distributed oscillator with clock direction control. Reproduced with permission from [21], ©2006 IEEE

## 4.4 Clock Distribution Using Inductance

### 4.4.1 Rotary Traveling-Wave Oscillator Arrays

Rotary traveling-wave oscillator arrays are formed using multiple rotary traveling-wave oscillators [55, 56] as shown in Fig.4.17. Assuming all loops are symmetrically matched, each loop has a traveling wave. The cross-coupled inverters are used to start and sustain the operation due to losses in the transmission-line loops. They also serve as latches that amplify the differential signal and generate square waveform clocks instead of sinusoidal. The individual loops are synchronized at their junctions through a phase-locking phenomenon. In any four port junction there are two pulses that are arriving simultaneously and are propagated to the output ports if the impedances are matched. The clock frequency is determined by the characteristics of a single loop and is given by [56]:

$$f = \frac{1}{2\sqrt{L_{\mathrm{lp}}C_{\mathrm{lp}}}}, \tag{4.21}$$

where $L_{\mathrm{lp}}$ and $C_{\mathrm{lp}}$ are the total loop inductance and capacitance, respectively. The factor of 2 is inserted because the phase wave has to go around the loop twice per clock cycle. As is the case with other oscillators, the frequency can be tuned by placing varactor banks along the transmission line.

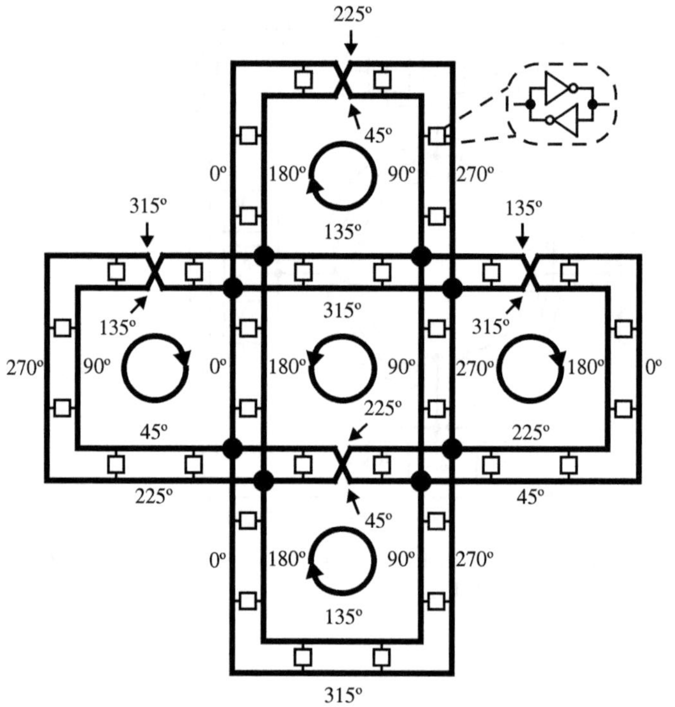


**Fig. 4.17.** Rotary traveling-wave arrays. Reproduced with permission from [55], ©2001 IEEE

As shown in [56], rotary traveling-wave oscillator arrays have low skew even in the presence of mismatches between the various loops. Furthermore, they operate adiabatically [57] when the losses in the transmission line and the gate resistance of the cross-coupled inverters are small. Finally, as suggested by Fig. 4.17, rotary traveling-wave oscillator arrays are expandable since more loops can be connected together to form larger arrays. Rotary traveling-wave clocks operate well with digital logic that requires two phases. In addition to conventional logic, a new two-phase adiabatic logic style operating with such a clock has been demonstrated [58]. However, careful consideration must be given at the system level since the phase of the clocks depends on their location within each loop. For instance, moving data globally may require retiming to a different clock domain. Another issue is coupling between the clock transmission lines and signal interconnects that need to be routed underneath the transmission lines.

### 4.4.2 Standing Wave Oscillator and Grid

Another type of oscillators based on transmission line signal propagation is standing wave oscillators [59]. The oscillating signals are formed in the transmission line by the incident and the reflected waves that travel in opposite directions. A standing

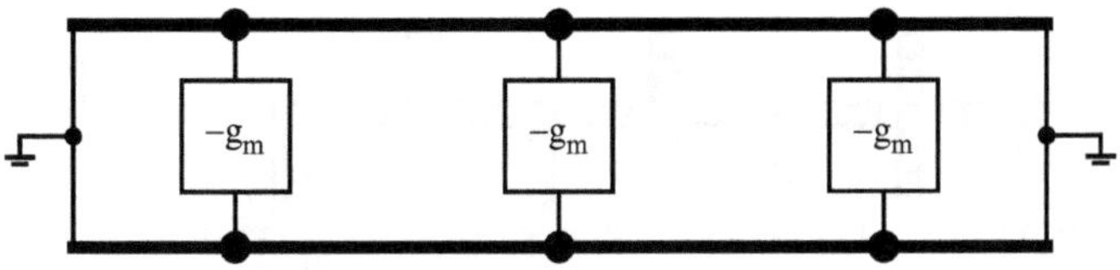


**Fig. 4.18.** Standing wave oscillator

wave oscillator (Fig.4.18) consists of a differential transmission line and several negative-$g_m$ cells distributed along the line. The transmission lines are shorted to ground at both ends creating a half-wave ($\lambda/2$) resonator, where $\lambda$ is the wavelength. The negative-$g_m$ cells used in [59, 60] contain cross-coupled NFETs to compensate for gain losses in the transmission lines and diode-connected PFETs for loads to set the common mode voltage. Unlike other LC-based oscillators, standing-wave oscillators may require clock injection in order to oscillate. Using diode connected loads simplifies clock injection as shown in [60]. Compared to rotary traveling-wave oscillators, standing wave oscillator clocks have the same phase irrespective of their position along the line. However, clock amplitude is position-dependent and varies sinusoidally along the transmission line. Specifically, clock amplitude is high in the middle of the transmission line and low at the two ends. Therefore, clock phases close to the transmission line ends require amplification to be usable. As shown in [60], a clock buffer consisting of a limiting amplifier and a sine-to-square converter is required especially for digital applications. The standing wave oscillator oscillates at a frequency $f$ if the following equations are satisfied [59]:

$$g < \frac{1}{n}\frac{R_{tl}C_{tl}}{L_{tl}} \tag{4.22}$$

and

$$l = \frac{1}{2f\sqrt{L_{tl}C_{tl}}}, \tag{4.23}$$

where $R_{tl}$, $C_{tl}$, and $L_{tl}$ are the resistance, capacitance, and inductance of the distributed transmission line, $l$ is the length of the transmission line, $n$ is the number of negative-$g_m$ cells along the transmission line, and $g$ is the transconductance of each one of the negative-$g_m$ cells. A more generic analytical model for the operation of standing wave oscillators is derived in [60].

Multiple standing wave oscillators can be coupled into clock grids as shown in Fig.4.19 [59, 60]. A single standing wave oscillator is diagrammed in the top right corner of the grid. The ends of standing wave oscillators are folded outside the grid since the low clock amplitude in these segments makes the clocks unusable. Each oscillator has five negative-$g_m$ cells. Although not shown in Fig.4.19, each oscillator can contain varactors in order to adjust its resonance frequency so that it matches the injected clock frequency. The clock injection point is also indicated in Fig.4.19. Similar to rotary traveling-wave oscillators, standing wave oscillators exhibit a phase averaging effect when connected in grids resulting in overall low-jitter and low-skew clock distribution.

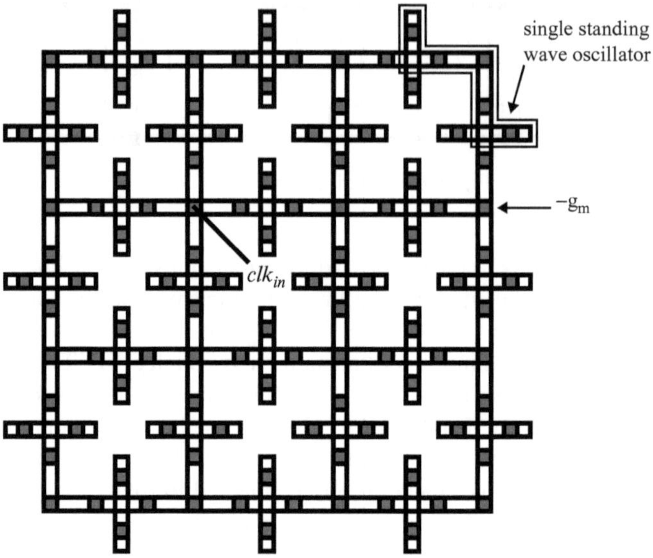


**Fig. 4.19.** Coupled standing wave oscillators used for clock distribution network. Reproduced with permission from [60], ©2003 IEEE

One of the drawbacks of the standing wave oscillator shown in Fig. 4.18 is the low voltage amplitude for a rather large portion of the transmission line length. To overcome this issue, spiral inductors can be placed at the two ends of the differential transmission line (Fig. 4.20a) [61, 62]. The resulting oscillator can have the same frequency while exhibiting a small voltage variation along its transmission line. Furthermore, its resonance frequency does not depend on the transmission line length as is the case for the half-wave oscillator. For instance, the transmission line length is 3.6mm for a 20GHz oscillator when fixed to $\lambda/2$ [61]. Since the transmission line is relatively short, only two negative-$g_m$ cells are placed at both ends of the line. In [61], the negative-$g_m$ cell consists of cross-coupled NFETs and PFETs and a single inductor is placed between the differential signals at each end of the line (i.e., a total of two inductors). In [62], the negative-$g_m$ cell consists of cross-coupled NFETs and two inductors are placed at each end connected to the supply voltage. The oscillator frequency is given by [62]:

$$f = \frac{1}{2\pi}\frac{Z_0}{L_{ld}}\tan(\pi - \beta l), \tag{4.24}$$

where $Z_0$ is the transmission line characteristic impedance, $L_{ld}$ is the inductance of the load, $\beta$ is the transmission line phase constant, and $l$ is the transmission line length.

In [62], multiple standing wave oscillators with inductive loads are coupled forming a clock distribution grid (Fig. 4.20b). After power up, all the oscillators are synchronized due to the inductive coupling between adjacent oscillators. The clock

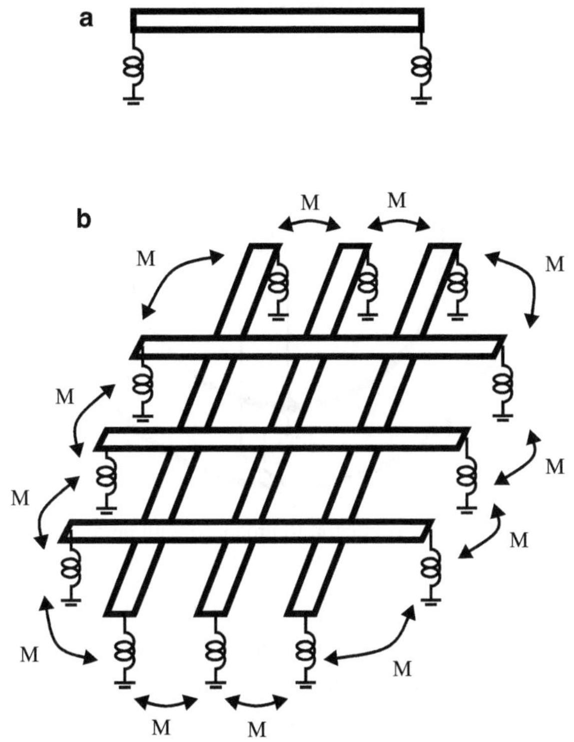


**Fig. 4.20.** (**a**) Standing-wave oscillator with inductive loads and (**b**) Multiple oscillators magnetically coupled. Reproduced with permission from [62], ©2007 IEEE

frequency is also given by (4.24). However, in this case mutual inductance must be taken into account, and thus $L_{\rm ld}$ equals $(1+2k)L$ where $L$ is the oscillator load inductance and $k = M/L$ with $M$ being the mutual inductance.

Another standing wave oscillator topology is presented in [63]. In this oscillator, the differential transmission line is only terminated in one end, whereas a negative-$g_{\rm m}$ cell is placed in the other end resulting in a $\lambda/4$ standing wave oscillator. The voltage amplitude is position-dependent having a maximum value at the end where the negative-$g_{\rm m}$ cell is located and being reduced sinusoidally along the transmission line until it becomes 0 at the other end where the short is located. To remedy the voltage amplitude reduction along the transmission line, a tapered line is used instead of a uniform line resulting in lower phase noise.

In [64], a circular standing wave oscillator is introduced (Fig. 4.21). In this standing wave oscillator the lines are not shorted, but they are circularly connected. Two negative-$g_{\rm m}$ cells are placed in two opposite sides. The circular standing wave oscillator requires that voltage $V(\phi)$ for any angle $\phi$ be equal to $V(\phi + 2\pi)$. Therefore, the length $l$ of the circumference is an integer multiple $n$ of the wavelength $\lambda$ of the mode. To suppress the even mode, the differential nodes from the two negative-$g_{\rm m}$

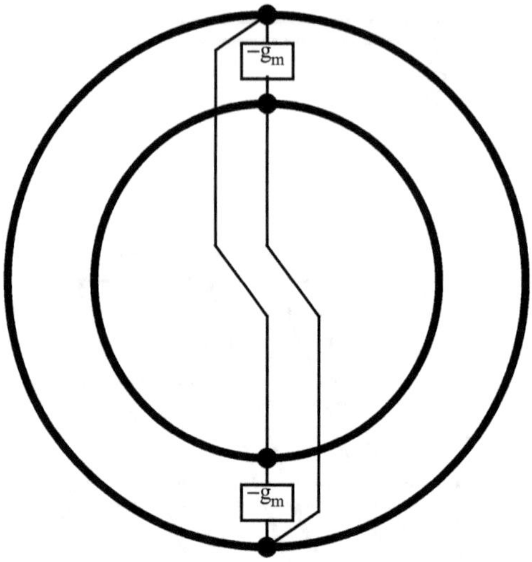


**Fig. 4.21.** Circular standing wave oscillator. Reproduced with permission from [64], ©2004 IEEE

cells are connected as shown in Fig. 4.21. To be able to make these connections, the circular differential transmission line is formed so that the two negative-$g_m$ cells are physically placed close to each other. The voltage amplitude is also position-dependent with the negative-$g_m$ cell nodes having the highest amplitude.

### 4.4.3 Inductor-Based Resonant Global Clock Distribution

Resonant global clock distribution was primarily used as a means to significantly reduce power dissipation through adiabatic charging and energy recovery [57]. The resonantly generated clocks power both the clock network as well as circuit nodes. A differential LC clock generator and driver [65] is used to produce two non-overlapping clock phases. The active components of the clock generator are integrated in the same chip with the rest of the circuit. Furthermore, in [66] the active components are distributed along the clock grid forming a distributed clock network. However, the two inductors are placed off chip [66–68]: First, the $Q$ of on-chip inductors was too small for these CMOS processes. Second, the inductors would be too large to integrate since their size is inversely proportional to the operating frequency. As CMOS processes advanced, both the inductor $Q$ increased and clock frequency reached the GHz-range, allowing for on-chip inductor integration. In [69], an adiabatic logic family is demonstrated to operate from a fully-integrated resonant clock driver.

Other LC-based clocking methods target power reduction only for the clock network (not data nodes) as well as low clock jitter and skew. Figure 4.22 illustrates one of those methods [70] which is based on conventional clock distribution. Specifically, the clock network is divided into sectors with a clock buffer chain driving each sector. One such sector is shown in Fig. 4.22. The clock sector consists of a global H-tree with the ends of the tree driving a global grid. Local clock buffers connect to the clock grid and deliver the clock to local blocks. To enhance clock distribution, spiral inductors are attached to certain nodes in the H-tree. Decoupling capacitors are placed in series to the inductors so that one port of the inductors connects to the clock tree whereas the other port connects to the decoupling capacitors. Figure 4.23 shows a simplified lumped circuit for a clock sector [70]. Capacitors $C_{\mathrm{clk}}$ and $C_{\mathrm{dec}}$ indicate the clock and decoupling capacitances respectively, $L$ indicates the spiral inductance, and $R_{\mathrm{clk}}$ and $R_{\mathrm{ind}}$ represent loads in the clock network and the inductor parasitic resistance respectively. The decoupling capacitance must be an order of magnitude larger than the clock capacitance so that the corresponding pole introduced does not

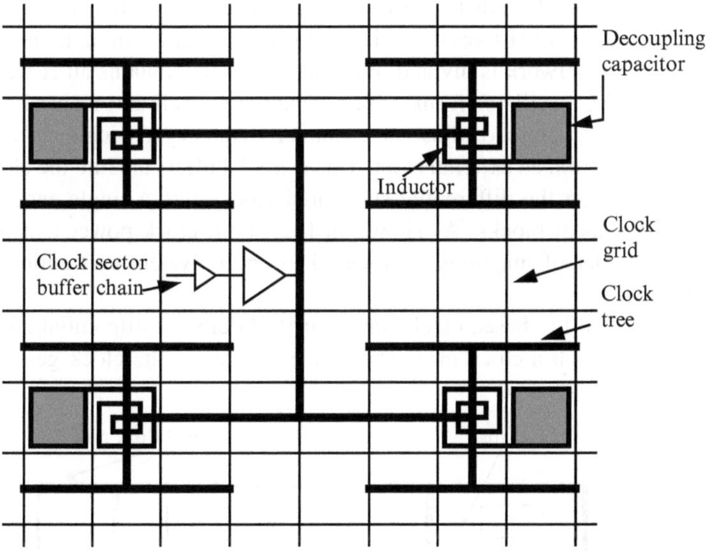


**Fig. 4.22.** Global clock distribution with resonant load. Reproduced with permission from [70], ©2005 IEEE

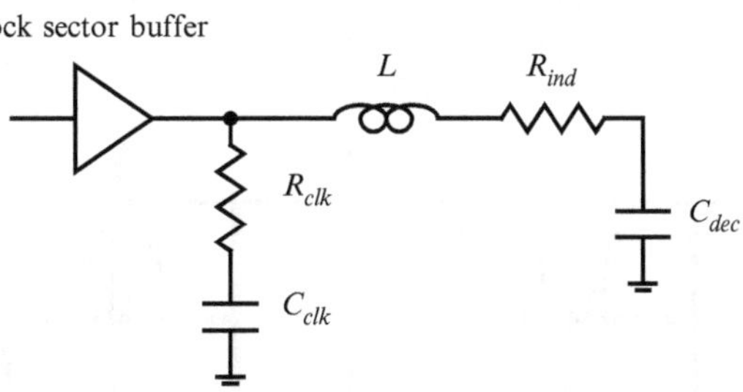


**Fig. 4.23.** Lumped circuit model of the clock sector. Reproduced with permission from [70], ©2005 IEEE

interfere with the resonant frequency of this circuit. When the clock frequency is approximately equal to the resonant frequency (i.e., $1/\left(2\pi\sqrt{LC_{\text{clk}}}\right)$), the inductive reactance of the spiral inductor cancels the capacitive reactance of the clock load resulting in significantly lower power dissipation compared to an equivalent conventional clock network [70]. In addition to the energy being recovered and re-used through the LC tank, the sector clock buffers do not need to be as strong as in the conventional case. The reduction of clock latency in the distribution network combined with the resonance causes both clock skew and jitter to be reduced as well. To further demonstrate the viability of this method, the conventional clock distribution of a commercial microprocessor was modified to implement this clocking method [71]. Specifically, a thick top metal layer was added to accommodate the spiral inductors without affecting the existing microprocessor design.

Another LC-based clock distribution (Fig. 4.24) uses differential clocks [72]. The clock network consists of H-trees that drive a global clock grid similar to the previous method. The clock network is divided into tiles. Each tile contains an H-tree driven by a differential LC-oscillator formed by a negative-$g_m$ cell and a spiral inductor. Each tile contains varactors to adjust its resonance frequency. A single buffer is used for clock injection from an external clock. Local clock buffers tap into the clock grid. These buffers convert the differential sinusoidal clocks into a single-ended square clock that drives local blocks. As shown in [72], both clock power and jitter are reduced by an order of magnitude compared to an equivalent conventional clock distribution network.

In [73], a central LC-based clock generator [65] delivers differential clocks that are distributed through a clock grid. The active devices of the clock generator are

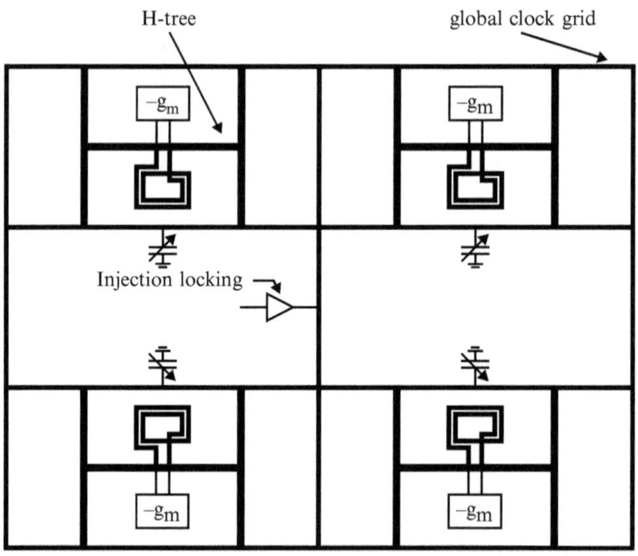


**Fig. 4.24.** Distributed differential global clock network. Reproduced with permission from [72], ©2006 IEEE

distributed in the clock grid similar to [66]. Latches that can operate from sinusoidal clocks are used as clocked storage elements similar to [68]. In [74], a hybrid injection-locked oscillator and LC-based buffer is used to drive the clock load. This technique can perform an interesting trade-off between input clock and clock buffer jitter.

## 4.5 Conclusion

In this chapter both the implementation and use of monolithic inductance were discussed. Most common monolithic inductors are spiral inductors and transmission lines. In addition to the implementation of spiral inductors and transmission lines, their accurate modeling is of utmost importance to circuit designers. A narrowband model for spiral inductors was presented. Modeling of transmission lines is simpler than spiral inductors. Their characteristics (i.e., inductance, capacitance, resistance, and conductance) are proportional to their length and can be modeled with simpler extraction methods.

The second part of the chapter focused on clock generation. LC VCOs are commonly used for high-speed wireless, wireline, and optical communications due to their low phase noise, small phase deviation, and low-power dissipation compared to ring oscillators. LC VCOs can efficiently generate up to four phases. Although in theory, more LC tanks can be coupled to generate more than four phases, this is not practically possible due to the physical arrangement of the LC tanks, coupling connections between the LC tanks, and potentially asymmetric clock phase routing to their destination. For all these reasons, distributed oscillators become the primary choice for generating more than four phases.

The third part of the chapter covered resonant-based methods to improve global clock distribution. Chapter 2 has demonstrated that conventional global clock distribution methods are more and more challenged to meet clock skew and jitter requirements for increasing clock load and clock frequency. Moreover, a large portion of total power dissipation is typically attributed to clock networks. Resonant-based methods can remedy these problems. Transmission-line-based methods such as rotary traveling-wave oscillator arrays and standing wave grids were presented. Furthermore, transmission lines can be efficient for transmitting high-frequency clocks without buffering as shown in [75]. The section concludes with LC-based global clock grids which are extensions of conventional global grids with the addition of spiral inductors.

## Acknowledgments

I would like to thank Fujitsu Laboratories of America for providing me the time to complete this manuscript and W. Walker for discussions about spiral inductor and transmission line modeling. Last but not least, I would like to thank my wife, Malina, for support.

## References


[1] N. Nguyen and R. Meyer, "Si IC-compatible inductors and LC passive filters," *IEEE J. Solid-State Circuits*, vol. 25, no. 4, pp. 1028–1031, 1990.

[2] N. Nguyen and R. Meyer, "A 1.8-GHz monolithic LC voltage-controlled oscillator," *IEEE J. Solid-State Circuits*, vol. 27, no. 3, pp. 444–450, 1992.

[3] P. Restle, T. McNamara, D. Webber, P. Camporese, K. Eng, K. Jenkins, D. Allen, M. Rohn, M. Quaranta, D. Boerstler, C. Alpert, C. Carter, R. Bailey, J. Petrovick, B. Krauter, and B. McCredie, "A clock distribution network for microprocessors," *IEEE J. Solid-State Circuits*, vol. 36, no. 5, pp. 792–799, 2001.

[4] T. Lee, *The design of CMOS radio-frequency integrated circuits, 2nd Edition*. Cambridge, United Kingdom: Cambridge University Press, 2004.

[5] H. Greenhouse, "Design of planar rectangular microelectronic inductors," *IEEE Trans Parts, Hybrids, and Packaging*, vol. 10, no. 2, pp. 101–109, June 1974.

[6] S. Jenei, B. Nauwelaers, and S. Decoutere, "Physics-based closed-form inductance expression for compact modeling of integrated spiral inductors," *IEEE J. Solid-State Circuits*, vol. 37, no. 1, pp. 77–80, Jan. 2002.

[7] S. Mohan, M. del Mar Hershenson, S. Boyd, and T. Lee, "Simple accurate expressions for planar spiral inductances," *IEEE J. Solid-State Circuits*, vol. 34, no. 10, pp. 1419–1424, Oct. 1999.

[8] K. Ashby, I. Koullias, W. Finley, J. Bastek, and S. Moinian, "High Q inductors for wireless applications in a complementary silicon bipolar process," *IEEE J. Solid-State Circuits*, vol. 31, no. 1, pp. 4–9, Jan. 1996.

[9] C. Yue, C. Ryu, J. Lau, T. Lee, and S. Wong, "A physical model for planar spiral inductors on silicon," in *Proceedings of International Electron Devices Meeting*, 8–11 Dec. 1996, pp. 155–158.

[10] Y. Cao, R. Groves, X. Huang, N. Zamdmer, J.-O. Plouchart, R. Wachnik, T.-J. King, and C. Hu, "Frequency-independent equivalent-circuit model for on-chip spiral inductors," *IEEE J. Solid-State Circuits*, vol. 38, no. 3, pp. 419–426, March 2003.

[11] "Sonnet user's guide," Sonnet High Frequency Electromagnetic Software.

[12] A. Niknejad, "ASITIC." [Online]. Available: http://rfic.eecs.berkley.edu/niknejad/asitic.html

[13] J. Long and M. Copeland, "The modeling, characterization, and design of monolithic inductors for silicon RF IC's," *IEEE J. Solid-State Circuits*, vol. 32, no. 3, pp. 357–369, March 1997.

[14] C. Yue and S. Wong, "On-chip spiral inductors with patterned ground shields for Si-based RF ICs," *IEEE J. Solid-State Circuits*, vol. 33, no. 5, pp. 743–752, May 1998.

[15] O. Heaviside, "Electromagnetic induction and its propagation," *The Electrician*, vol. 19, no. 40, pp. 79–81, 1887.

[16] M. Pupin, "Art of reducing attenuation of electrical apparatus therefore," U.S. Patent no. 652,230, June 1900.

[17] H. Chen, R. Shi, C.-K. Cheng, and D. Harris, "Surfliner: a distortionless electrical signaling scheme for speed of light on-chip communications," in *Proceedings of the IEEE International Conference on Computer Design: VLSI in Computers and Processors ICCD 2005*, 2–5 Oct. 2005, pp. 497–502.

[18] N. Nedovic, N. Tzartzanis, H. Tamura, F. Rotella, M. Wiklund, Y. Mizutani, Y. Okaniwa, T. Kuroda, J. Ogawa, and W. Walker, "A 40-to-44 Gb/s 3x oversampling CMOS CDR/1:16 DEMUX," *IEEE J. Solid-State Circuits*, vol. 42, no. 12, pp. 2726–2735, Dec. 2007.

[19] M. Thomson, P. Restle, and N. James, "A 5GHz duty-cycle correcting clock distribution network for the POWER6 microprocessor," in *Digest of Technical Papers IEEE International Solid-State Circuits Conference (ISSCC 2006)*, 2006, pp. 384–385.

[20] B. Kleveland, C. Diaz, D. Vook, L. Madden, T. Lee, and S. Wong, "Exploiting CMOS reverse interconnect scaling in multigigahertz amplifier and oscillator design," *IEEE J. Solid-State Circuits*, vol. 36, no. 10, pp. 1480–1488, Oct. 2001.

[21] N. Tzartzanis and W. Walker, "A reversible poly-phase distributed VCO," in *Digest of Technical Papers IEEE International Solid-State Circuits Conference (ISSCC 2006)*, 2006, pp. 596–597.

[22] N. Nguyen and R. Meyer, "Start-up and frequency stability in high-frequency oscillators," *IEEE J. Solid-State Circuits*, vol. 27, no. 5, pp. 810–820, May 1992.

[23] P. Andreani and S. Mattisson, "A 1.8-GHz CMOS VCO tuned by an accumulation-mode MOS varactor," in *Proceedings of the ISCAS 2000 Geneva Circuits and Systems the 2000 IEEE International Symposium*, vol. 1, 28–31 May 2000, pp. 315–318.

[24] A.-S. Porret, T. Melly, C. Enz, and E. Vittoz, "Design of high-Q varactors for low-power wireless applications using a standard CMOS process," *IEEE J. Solid-State Circuits*, vol. 35, no. 3, pp. 337–345, March 2000.

[25] G. Cusmai, M. Repossi, G. Albasini, A. Mazzanti, and F. Svelto, "A magnetically tuned quadrature oscillator," *IEEE J. Solid-State Circuits*, vol. 42, no. 12, pp. 2870–2877, Dec. 2007.

[26] D. Leeson, "A simple model of feedback oscillator noise spectrum," *Proc. IEEE*, vol. 54, no. 2, pp. 329–330, Feb. 1966.

[27] A. Hajimiri and T. Lee, "A general theory of phase noise in electrical oscillators," *IEEE J. Solid-State Circuits*, vol. 33, no. 2, pp. 179–194, Feb. 1998.

[28] T. Lee and A. Hajimiri, "Oscillator phase noise: A tutorial," *IEEE J. Solid-State Circuits*, vol. 35, no. 3, pp. 326–336, March 2000.

[29] A. Demir, A. Mehrotra, and J. Roychowdhury, "Phase noise in oscillators: A unifying theory and numerical methods for characterization," *IEEE Trans. Circuits Syst. I*, vol. 47, no. 5, pp. 655–674, May 2000.

[30] J. Rael and A. Abidi, "Physical processes of phase noise in differential LC oscillators," in *Proc. IEEE Custom Integrated Circuits Conference (CICC 2000)*, 21–24 May 2000, pp. 569–572.

[31] E. Hegazi, H. Sjoland, and A. Abidi, "A filtering technique to lower LC oscillator phase noise," *IEEE J. Solid-State Circuits*, vol. 36, no. 12, pp. 1921–1930, Dec. 2001.
[32] A. Hajimiri and T. Lee, "Design issues in CMOS differential LC oscillators," *IEEE J. Solid-State Circuits*, vol. 34, no. 5, pp. 717–724, May 1999.
[33] D. Ham and A. Hajimiri, "Concepts and methods in optimization of integrated LC VCOs," *IEEE J. Solid-State Circuits*, vol. 36, no. 6, pp. 896–909, June 2001.
[34] P. Andreani and H. Sjoland, "Tail current noise suppression in RF CMOS VCOs," *IEEE J. Solid-State Circuits*, vol. 37, no. 3, pp. 342–348, March 2002.
[35] H. Nosaka, E. Sano, K. Ishii, M. Ida, K. Kurishima, S. Yamahata, T. Shibata, H. Fukuyama, M. Yoneyama, T. Enoki, and M. Muraguchi, "A 39-to-45-Gbit/s multi-data-rate clock and data recovery circuit with a robust lock detector," *IEEE J. Solid-State Circuits*, vol. 39, no. 8, pp. 1361–1365, Aug. 2004.
[36] T.-H. Lin and W. Kaiser, "A 900-MHz 2.5-mA CMOS frequency synthesizer with an automatic SC tuning loop," *IEEE J. Solid-State Circuits*, vol. 36, no. 3, pp. 424–431, March 2001.
[37] R. Staszewski, D. Leipold, C.-M. Hung, and P. Balsara, "A first digitally-controlled oscillator in a deep-submicron CMOS process for multi-GHz wireless applications," in *Proceedings of the IEEE Radio Frequency Integrated Circuits (RFIC) Symposium*, 8–10 June 2003, pp. 81–84.
[38] J. Hauenschild, C. Dorschky, T. Winkler von Mohrenfels, and R. Seitz, "A plastic packaged 10 Gb/s BiCMOS clock and data recovering 1:4-demultiplexer with external VCO," *IEEE J. Solid-State Circuits*, vol. 31, no. 12, pp. 2056–2059, Dec. 1996.
[39] M. Rau, T. Oberst, R. Lares, A. Rothermel, R. Schweer, and N. Menoux, "Clock/data recovery PLL using half-frequency clock," *IEEE J. Solid-State Circuits*, vol. 32, no. 7, pp. 1156–1159, July 1997.
[40] A. Rofougaran, G. Chang, J. Rael, J.-C. Chang, M. Rofougaran, P. Chang, M. Djafari, M.-K. Ku, E. Roth, A. Abidi, and H. Samueli, "A single-chip 900-MHz spread-spectrum wireless transceiver in 1-mm CMOS – Part I: Architecture and transmitter design," *IEEE J. Solid-State Circuits*, vol. 33, no. 4, pp. 515–534, April 1998.
[41] A. Rofougaran, J. Rael, M. Rofougaran, and A. Abidi, "A 900 MHz CMOS LC-oscillator with quadrature outputs," in *Digest of Technical Papers IEEE International Solid-State Circuits Conference (ISSCC 1996)*, 1996, pp. 392–393.
[42] P. Andreani, A. Bonfanti, L. Romano, and C. Samori, "Analysis and design of a 1.8-GHz CMOS LC quadrature VCO," *IEEE J. Solid-State Circuits*, vol. 37, no. 12, pp. 1737–1747, Dec. 2002.
[43] P. Andreani and X. Wang, "On the phase-noise and phase-error performances of multiphase LC CMOS VCOs," *IEEE J. Solid-State Circuits*, vol. 39, no. 11, pp. 1883–1893, Nov. 2004.
[44] P. Vancorenland and M. Steyaert, "A 1.57-GHz fully integrated very low-phase-noise quadrature VCO," *IEEE J. Solid-State Circuits*, vol. 37, no. 5, pp. 653–656, 2002.

[45] J. van der Tang, P. van de Ven, D. Kasperkovitz, and A. van Roermund, "Analysis and design of an optimally coupled 5-GHz quadrature LC oscillator," *IEEE J. Solid-State Circuits*, vol. 37, no. 5, pp. 657–661, 2002.

[46] A. Mirzaei, M. Heidari, R. Bagheri, S. Chehrazi, and A. Abidi, "The quadrature LC oscillator: A complete portrait based on injection locking," *IEEE J. Solid-State Circuits*, vol. 42, no. 9, pp. 1916–1932, 2007.

[47] C.-Y. Wu and H.-S. Kao, "A 1.8 GHz CMOS quadrature voltage-controlled oscillator (VCO) using the constant-current LC ring oscillator structure," in *Proceedings of the IEEE International Symposium on Circuits and Systems ISCAS '98*, vol. 4, pp. 378–381, 1998.

[48] M. Tiebout, "Low-power low-phase-noise differentially tuned quadrature VCO design in standard CMOS," *IEEE J. Solid-State Circuits*, vol. 36, no. 7, pp. 1018–1024, 2001.

[49] B. Kleveland, C. Diaz, D. Vock, L. Madden, T. Lee, and S. Wong, "Monolithic CMOS distributed amplifier and oscillator," in *Digest of Technical Papers IEEE International Solid-State Circuits Conference (ISSCC 1999)*, 1999, pp. 70–71.

[50] H. Wu and A. Hajimiri, "Silicon-based distributed voltage-controlled oscillators," *IEEE J. Solid-State Circuits*, vol. 36, no. 3, pp. 493–502, 2001.

[51] Z. Skvor, S. Saunders, and C. Aitchison, "Novel decade electronically tunable microwave oscillator based on the distributed amplifier," *Electron. Lett.*, vol. 28, no. 17, pp. 1647–1648, 1992.

[52] L. Divina and Z. Skvor, "Experimental verification of a distributed amplifier oscillator," in *Proceedings of the 25th European Microwave Conference*, vol. 2, 1995, pp. 1163–1167.

[53] L. Divina and Z. Skvor, "The distributed oscillator at 4 GHz," *IEEE Trans. Microw. Theory Tech.*, vol. 46, no. 12, pp. 2240–2243, 1998.

[54] J. Lee and B. Razavi, "A 40-Gb/s clock and data recovery circuit in 0.18-$\mu$m CMOS technology," *IEEE J. Solid-State Circuits*, vol. 38, no. 12, pp. 2181–2190, 2003.

[55] J. Wood, S. Lipa, P. Franzon, and M. Steer, "Multi-gigahertz low-power low-skew rotary clock scheme," in *Digest of Technical Papers IEEE International Solid-State Circuits Conference (ISSCC 2001)*, 2001, pp. 400–401, 470.

[56] J. Wood, T. Edwards, and S. Lipa, "Rotary traveling-wave oscillator arrays: A new clock technology," *IEEE J. Solid-State Circuits*, vol. 36, no. 11, pp. 1654–1665, 2001.

[57] W. Athas, L. Svensson, J. Koller, N. Tzartzanis, and E. Ying-Chin Chou, "Low-power digital systems based on adiabatic-switching principles," *IEEE Trans. VLSI Syst.*, vol. 2, no. 4, pp. 398–407, Dec. 1994.

[58] J. Wood, T. Edwards, and C. Ziesler, "A 3.5GHz rotary-traveling-wave-oscillator clocked dynamic logic family in 0.25μm CMOS," in *Digest of Technical Papers IEEE International Solid-State Circuits Conference (ISSCC 2006)*, 2006, pp. 390–391.

[59] F. O'Mahony, C. Yue, M. Horowitz, and S. Wong, "A 10-GHz global clock distribution using coupled standing-wave oscillators," in *Digest of Technical*

*Papers IEEE International Solid-State Circuits Conference (ISSCC 2003)*, 2003, pp. 428–429.

[60] F. O'Mahony, C. Yue, M. Horowitz, and S. Wong, "A 10-GHz global clock distribution using coupled standing-wave oscillators," *IEEE J. Solid-State Circuits*, vol. 38, no. 11, pp. 1813–1820, Nov. 2003.

[61] M. Sasaki, M. Shiozaki, A. Mori, A. Iwata, and H. Ikeda, "17GHz fine grid clock distribution with uniform-amplitude standing-wave oscillator," in *Proceedings of Digest of Technical Papers VLSI Circuits 2006 Symposium*, 2006, pp. 124–125.

[62] M. Sasaki, M. Shiozaki, A. Mori, A. Iwata, and H. Ikeda, "12GHz low-area-overhead standing-wave clock distribution with inductively-loaded and coupled technique," in *Digest of Technical Papers IEEE International Solid-State Circuits Conference (ISSCC 2007)*, 2007, pp. 180–181.

[63] W. Andress and D. Ham, "Standing wave oscillators utilizing wave-adaptive tapered transmission lines," *IEEE J. Solid-State Circuits*, vol. 40, no. 3, pp. 638–651, Mar. 2005.

[64] D. Ham and W. Andress, "A circular standing wave oscillator," in *Digest of Technical Papers IEEE International Solid-State Circuits Conference (ISSCC 2004)*, 2004, pp. 380–381.

[65] W. Athas, L. Svensson, and N. Tzartzanis, "A resonant signal driver for two-phase, almost-non-overlapping clocks," in *Proceedings of the IEEE International Symposium on Circuits and Systems ISCAS '96, 'Connecting the World'*, vol. 4, 12–15 May 1996, pp. 129–132.

[66] W. Athas, N. Tzartzanis, L. Svensson, and L. Peterson, "A low-power microprocessor based on resonant energy," *IEEE J. Solid-State Circuits*, vol. 32, no. 11, pp. 1693–1701, Nov. 1997.

[67] N. Tzartzanis and W. Athas, "Clock-powered logic for a 50 MHz low-power RISC datapath," in *Digest of Technical Papers IEEE International Solid-State Circuits Conference (ISSCC 1997)*, 1997, pp. 338–339, 482.

[68] W. Athas, N. Tzartzanis, W. Mao, L. Peterson, R. Lal, K. Chong, J.-S. Moon, L. Svensson, and M. Bolotski, "The design and implementation of a low-power clock-powered microprocessor," *IEEE J. Solid-State Circuits*, vol. 35, no. 11, pp. 1561–1570, Nov. 2000.

[69] V. S. Sathe, J.-Y. Chueh, and M. C. Papaefthymiou, "Energy-efficient GHz-class charge-recovery logic," *IEEE J. Solid-State Circuits*, vol. 42, no. 1, pp. 38–47, Jan. 2007.

[70] S. Chan, K. Shepard, and P. Restle, "Uniform-phase uniform-amplitude resonant-load global clock distributions," *IEEE J. Solid-State Circuits*, vol. 40, no. 1, pp. 102–109, Jan. 2005.

[71] S. Chan, P. Restle, T. Bucelot, S. Weitzel, J. Keaty, J. Liberty, B. Flachs, R. Volant, P. Kapusta, and J. Zimmerman, "A resonant global clock distribution for the cell broadband-engine™processor," in *Digest of Technical Papers IEEE International Solid-State Circuits Conference (ISSCC 2008)*, 2008, pp. 512–513.

[72] S. Chan, K. Shepard, and P. Restle, "Distributed differential oscillators for global clock networks," *IEEE J. Solid-State Circuits*, vol. 41, no. 9, pp. 2083–2094, Sept. 2006.
[73] V. Sathe, J. Kao, and M. Papaefthymiou, "Resonant-clock latch-based design," *IEEE J. Solid-State Circuits*, vol. 43, no. 4, pp. 864–873, April 2008.
[74] L. M. Lee and C.-K. K. Yang, "An adaptive low-jitter LC-based clock distribution," in *Digest of Technical Papers IEEE International Solid-State Circuits Conference (ISSCC 2007)*, 2007, pp. 182–183.
[75] F. O'Mahony, M. Mansuri, B. Casper, J. Jaussi, and R. Mooney, "A low-jitter PLL and repeaterless clock distribution network for a 20Gb/s link," in *Proceedings of the Digest of Technical Papers VLSI Circuits 2006 Symposium*, 2006, pp. 29–30.

# 5

# Phase Noise and Jitter

Scott Meninger

Cavium Networks

## 5.1 Introduction

In this chapter we examine variations that occur in the edge locations of the clock signal in a synchronous system. These edge variations are referred to in the time domain as *jitter* and in the frequency domain as *phase noise*. We also describe the various mechanisms that can cause these non-idealities and present techniques to analyze their individual contributions to total jitter. We begin by defining precisely what we mean by jitter and relate the various types of jitter to one another. Next, we explore the relationship between the time domain representation of timing error as jitter and the frequency domain representation of timing error as phase noise. The fundamental relationship between phase jitter (also denoted as absolute jitter) and phase noise will provide a useful basis for analysis of all types of jitter via simple frequency domain filter functions.

Building on this frequency domain foundation, we explore the jitter behavior of the phase locked loop (PLL) system that is most often used to generate on-chip clock signals. We utilize a control system block diagram model [1] that allows for simple analysis to determine how the PLL dynamics filter the noise sources internal to the PLL (*intrinsic* noise) as well as how the PLL reacts to noise sources external to the PLL (*extrinsic* noise).

We then use this model to examine jitter performance and reference clock specification for the PLLs used by two different types of systems. The first is a microprocessor core clock PLL, where the key requirement is that the minimum clock period produced by the PLL does not violate a minimum cycle time requirement. The second system we examine is a serial link that has the clock embedded within the data stream and uses PLLs on both ends of the link to recover the clock local to their respective systems.

The overall goal for this chapter is to give the reader an understanding of the relationship between phase noise and jitter in PLL based systems and to see that by using phase noise as a basis for calculation, system jitter analysis can be performed using simple frequency domain filtering techniques.

T. Xanthopoulos (ed.), *Clocking in Modern VLSI Systems*, Integrated Circuits and Systems, DOI 10.1007/978-1-4419-0261-0_5, © Springer Science+Business Media, LLC 2009

## 5.2 Timing Error in the Time Domain: Jitter

The first task we face is to define what we mean by jitter. To do so, we use terms consistent with industry standards [2] and notation proposed by a high speed serial data (PCI Express) jitter subcommittee [3, 4].

Jitter can be caused by many different processes. Duty cycle distortion of a waveform can be thought of as a source of jitter, if both edges of the clock are used in the system. Power supply noise may induce jitter in a PLL either directly, by moving the edge crossing of the PLL's output signal by changing the delay through a buffer, or indirectly, by changing the oscillator control voltage inside the PLL and thereby changing the PLL output frequency. We will focus our discussion in this chapter on jitter mechanisms that are caused by noise processes within circuit elements that make up the PLL.

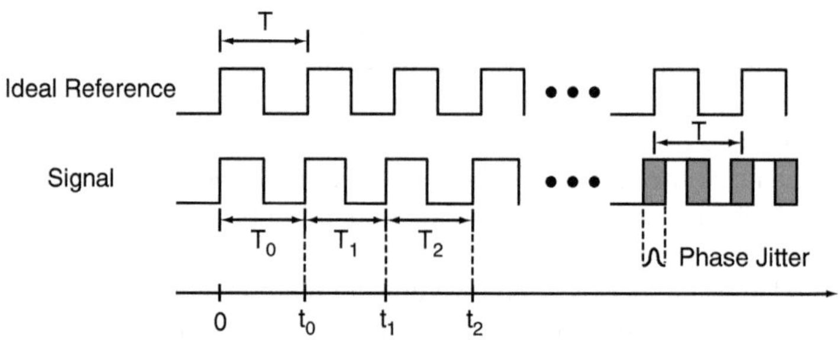


$$\text{Phase jitter} = \Phi[n] = t_n - nT$$

$$\text{Period jitter} = \Phi'[n] = (t_n - t_{n-1}) - T = T_n - T = \Phi[n] - \Phi[n-1]$$

$$\text{Cycle to cycle jitter} = \Phi''[n] = (t_n - t_{n-1}) - (t_{n-1} - t_{n-2}) = T_n - T_{n-1} = \Phi'[n] - \Phi'[n-1]$$

$$n = 0, 1, 2, \ldots.$$

**Fig. 5.1.** Jitter definitions

Figure 5.1 visually presents the definitions that we will use for the three different types of jitter commonly discussed in the literature [4]. These are phase jitter, period jitter, and cycle-to-cycle (c2c) jitter. In all cases, we use $n$ in the standard way, namely as an index to indicate the edge sample ($t_n$) or period sample ($T_n$) under discussion. In general, jitter is described by statistical properties (rms or peak-to-peak values), which indicate that a large number of samples have been aggregated in the analysis. We will discuss the nature of jitter distributions in Sect. 5.6.

### 5.2.1 Phase Jitter

*Phase* jitter, or *absolute* jitter (sometimes also referred to as time interval error (TIE)) is defined as the difference between the edges of the clock signal being measured and the ideal locations where the edges would occur in the absence of jitter. In Fig. 5.1, we set up an ideal reference that allows us to determine the absolute jitter as the difference in edge locations between our signal and the ideal reference. We can express this phase jitter as:

$$\Phi[n] = t_n - nT \tag{5.1}$$

where $t_n$ is the $n^{\text{th}}$ signal edge and $nT$ is the $n^{\text{th}}$ period of the ideal reference. It is apparent that $nT$ corresponds to the ideal $n^{\text{th}}$ edge location, so Eq. (5.1) is consistent with our definition.

Phase jitter is the jitter type of interest in systems where absolute time difference between clock and signal matters. For example, in a serial data link, where the clock is recovered from the data, it is important to know the difference between the data edge and the recovered clock edge since data is being captured by the recovered clock. Phase jitter is, therefore, the jitter metric appropriate for serial data applications.

### 5.2.2 Period Jitter

*Period* jitter is defined as the difference between a given period of the signal, and the average period of the signal. In a synchronous system with no frequency offset, such as a PLL, the average period will be the same as the ideal period, $T$. The expression for period jitter is shown in Fig. 5.1 to be:

$$\Phi'[n] = (t_n - t_{n-1}) - T = T_n - T \tag{5.2}$$

which can also be expressed as:

$$\Phi'[n] = \Phi[n] - \Phi[n-1] \tag{5.3}$$

Equation (5.3) explains why the calculus "tick" notation is used to denote period jitter; namely, period jitter is simply the first difference of phase jitter. The importance of this relationship will become apparent in the analysis portion of this chapter, where the difference function will be utilized in the frequency domain as a filter function.

Period jitter is defined as the first difference of phase jitter, so we can say that phase jitter is a more fundamental jitter type. We also note that because period jitter can be derived from phase jitter through a difference operation, period jitter is a more high frequency phenomenon than phase jitter.[1]

[1] Finally, note that we lose some information in going from phase jitter to period jitter. That is to say that while we can construct a period jitter sequence from a phase jitter sequence, we cannot do the opposite because the starting phase is unknown. This is equivalent to saying that while we can differentiate a function without loss of information, we cannot integrate absolutely without knowing the initial condition (or offset).

Period jitter is the type of jitter of concern within a digital system where the minimum (or maximum) time period is of importance. For example, a microprocessor has a minimum cycle time that is allowed before its critical path undergoes a timing violation. In this case, knowledge of period jitter is critical since it will determine how much of the overall system timing budget will be occupied by the clock source.

### 5.2.3 Cycle-to-Cycle Jitter

*Cycle-to-cycle (c2c)* jitter is defined as the difference between successive periods of a signal. It is expressed as:

$$\Phi''[n] = T_n - T_{n-1} \tag{5.4}$$

which can also be written as:

$$\Phi''[n] = \Phi'[n] - \Phi'[n-1] \tag{5.5}$$

Just as period jitter is the first difference of phase jitter, so c2c jitter is the first difference of period jitter, or the second difference of phase jitter. Cycle-to-cycle jitter is, therefore, a very high frequency phenomenon. Once again, some information is lost in going from period jitter to c2c jitter, as the difference function removes any offset information.

## 5.3 Timing Error in the Frequency Domain: Phase Noise

The physical mechanisms that cause random jitter are noise processes, and these are best described and analyzed in the frequency domain. The frequency domain equivalent of jitter is phase noise.

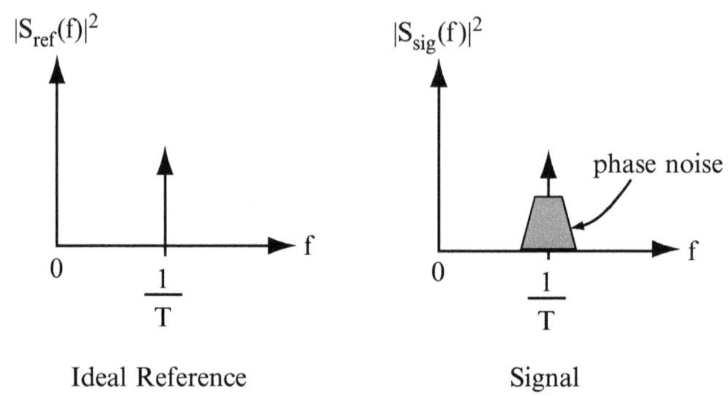


**Fig. 5.2.** Phase noise

Figure 5.2 depicts the spectra associated with the waveforms of Fig. 5.1. $|S_{\text{ref}}(f)|^2$ is the spectrum of the ideal reference, while $|S_{\text{sig}}(f)|^2$ is the spectrum of the signal[2].

[2] $S_{\text{ref}}(f)$ and $S_{\text{sig}}(f)$ are the power spectral densities of the time-domain signals $s_{\text{ref}}(t)$ and $s_{\text{sig}}(t)$, respectively. [26]

The ideal reference appears as an impulse at its frequency, $1/T$.[3] The signal spectrum, by contrast, has some degree of spreading over frequency. Phase noise is the term used to represent this spectral spreading. On average, the signal frequency is equal to $1/T$.[4] However, the instantaneous deviations from ideal behavior that appear as jitter in the time domain translate to phase noise in the frequency domain. The units of phase noise are dBc/Hz, and it represents the amount of noise power present in a 1 Hz bandwidth of spectrum.

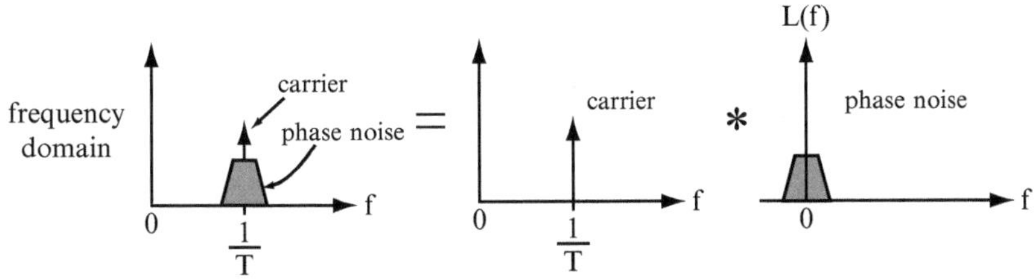


**Fig. 5.3.** Phase noise decomposition into carrier and L(f)

Since the system operates around a fundamental frequency $1/T$, it is possible to separate the noise spectrum from the carrier, as depicted in Fig.5.3. The phase noise spectrum is denoted as $L(f)$, which is centered around DC. Because $L(f)$ is symmetric, it is general practice to use a single-sided version of $L(f)$, wherein $L(f)$ is plotted only for positive frequencies.

### 5.3.1 Relationship Between Phase Noise and Jitter

In [5], arguments are made using Parceval's theorem to show the equivalence between the energy in the time domain jitter and frequency domain phase noise representations. We will simply present the result as:

$$\sigma_{\Phi}^2 = \left(\frac{T}{2\pi}\right)^2 \int_{-\infty}^{\infty} L(f)\mathrm{d}f \tag{5.6}$$

Equation (5.6) is a powerful relationship. It states that the variance of the phase jitter of a signal is the integral over frequency of its phase noise spectrum, scaled by a factor that simply relates how much time corresponds to $2\pi$ radians of phase. Under the assumption that the signal jitter statistics can be described by a normal distribution, which is a good assumption under most PLL analysis cases[5], we can

[3] We have not plotted the impulses appearing at odd harmonics of the fundamental frequency caused by the square-wave nature of the ideal reference for simplicity.

[4] In the case of a PLL, which is the focus of our discussion, it is a fundamental property that the average output frequency is equal to the ideal output frequency.

[5] It is a property of systems with large numbers of independent noise processes that the overall noise distribution of the system will appear normal as described by the Central Limit Theorem [6].

directly calculate the rms and peak-to-peak values of phase jitter if we know the phase noise spectrum, $L(f)$. The rms value of the phase noise is just the square root of the variance given by Eq. (5.6). The peak-to-peak value is determined by how many standard deviations are required for a particular application. As we will show in Sect. 5.4.11, period jitter and cycle-to-cycle jitter may also be determined from $L(f)$ by applying the appropriate filter functions.

## 5.4 Frequency Domain Modeling of PLLs

Now that we have established the relationship between phase noise and phase jitter, we look at the phase noise properties of PLLs.

### 5.4.1 PLL Phase Noise

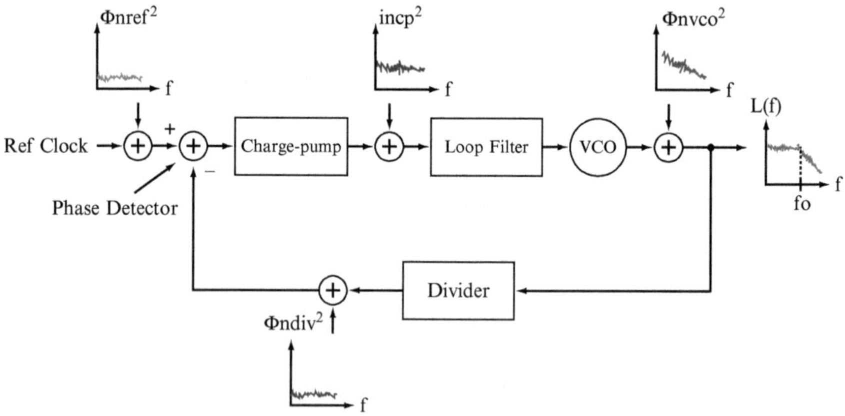


**Fig. 5.4.** PLL block diagram with noise sources included

Figure 5.4 presents a block diagram of a typical charge-pump type PLL along with the noise sources associated with each of the circuit elements that comprise the PLL. These noise sources are all contributors to the overall phase noise output of the PLL. $\Phi n_{\mathrm{ref}}^2$ represents reference clock phase noise, $in_{\mathrm{cp}}^2$ is a combination of phase detector, charge-pump and loop filter noise current, $\Phi n_{\mathrm{vco}}^2$ is the output referred VCO phase noise, and $\Phi n_{\mathrm{div}}^2$ is the feedback divider phase noise. The PLL closed loop response appears as a low-pass filter to charge-pump, loop filter, feedback divider, phase detector, and reference clock phase noise, and as a high-pass filter to VCO phase noise [1]. The overall PLL phase noise, $L(f)$, is the sum of all of the individual filtered components and has a low-pass nature, as depicted in Fig.5.4.

We divide the noise sources into two categories, *intrinsic* noise and *extrinsic* noise [7]. Intrinsic noise is associated with the components necessary to build the

PLL, namely the charge-pump, loop filter, VCO, feedback divider, and phase detector. Extrinsic noise is the noise that is processed by the PLL and is present either at the PLL input (i.e., the reference clock) or is due to undesired non-ideal system characteristics (such as poor supply noise rejection or duty cycle distortion). We will deal solely with reference clock noise as an extrinsic noise source in the analyses presented in this chapter. It is useful to go through the different noise sources in turn to develop a feel for how they impact the PLL output phase noise.

### 5.4.2 PLL Intrinsic Noise: VCO

VCO phase noise analysis is a rich field of research, with many different papers written on how to analyze the contributions of individual transistors to VCO output phase noise. Several interesting methods for VCO phase noise analysis are presented in [8–10], and [11] and [12] are journal paper collections with more references.

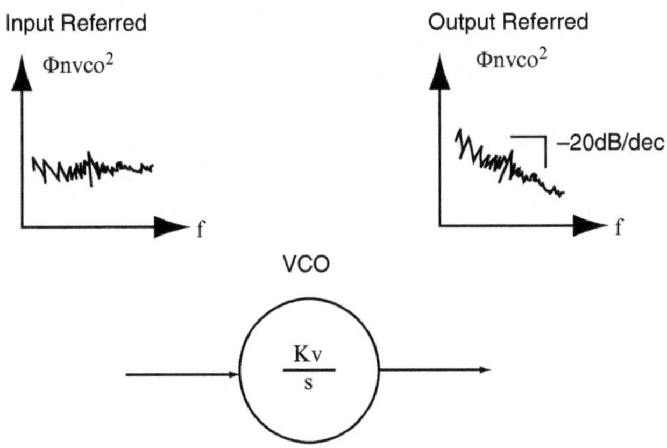


**Fig. 5.5.** VCO noise modeling

VCO noise behavior can be summarized by a few key characteristics. First, the VCO takes as its input a voltage[6] and outputs a specific frequency. The gain through the VCO will be denoted by $K_v$ and has units of Hz/V or (Radians/s)/V. From the perspective of a PLL, which locks based on *phase*, the VCO transfer function from input to output looks like an ideal integrator, since phase is the integration of frequency. Second, while the VCO phase noise profile changes slope across frequency, it is generally a good assumption that in the PLL region of interest, namely around the PLL closed loop bandwidth, the VCO phase noise rolls off at −20 dB/dec.

[6] There are also current controlled oscillators (CCOs) and digitally controlled oscillators (DCOs), which have similar relationships between the control node or bus and VCO output phase. We will use VCOs in our analysis as they are still the most common oscillator type in use.

Since the VCO is modeled as an integrator, and in the frequency range of interest its phase noise rolls off at −20 dB/dec, its phase noise can be input referred as a white noise source that is then integrated, as depicted in Fig. 5.5. This is a useful point to keep in mind when analyzing noise on the VCO control voltage input due to other sources, such as, for example, the loop filter. If the additional noise is white in the frequency range of interest, it is indistinguishable from the input referred VCO noise.

### 5.4.3 PLL Intrinsic Noise: Feedback Divider

As Fig. 5.4 shows, the feedback divider is not easily distinguishable from the reference clock input in the block diagram. Fortunately, divider phase noise is generally very low, well below other system components that contribute phase noise in the same frequency band [12]. The nature of the divider phase noise will depend on its design. For example, a synchronous counter will typically exhibit less phase noise than an asynchronous cascade of divide-by stages because the synchronous design will have its output edge set directly by the VCO, whereas the asynchronous design will have its output edge set by a cascade of edges, each of which may contribute some jitter.

The challenge in modern PLLs, which operate in the multi-GHz range, is that it is difficult to create high speed synchronous dividers that do not consume significant power while operating. Therefore, lower power asynchronous dividers are very popular in high speed PLLs [12]. As phase noise requirements become more stringent, designers are developing techniques to re-synchronize the divider output to the VCO, effectively resetting its phase noise/jitter, and thereby minimizing its impact [13, 14].

In our analysis, we will assume that divider phase noise has a white profile and is below the reference clock phase noise magnitude. In such a case, we can approximate the divider phase noise as being zero.

### 5.4.4 PLL Intrinsic Noise: Phase Detector

Phase detectors (PD) are difficult to analyze for several reasons. First, they can be very non-linear in nature[7] [14]. The non-linearity is present, for example, in tri-state phase/frequency detectors (PFDs)[8] like the one depicted in Fig. 5.6. Tri-state PFDs are of popular use in the charge-pump PLL, which is among the most popular PLL architectures in use. The PFD in Fig. 5.6 compares the reference clock (REF) phase to the feedback divider (DIV) phase and creates output pulses UP or DOWN that are used to steer positive (UP) or negative (DOWN) current onto a capacitor. The more phase difference between the REF and DIV signals, the longer the current sources are on, and the more resultant positive charge $Q_u$ or negative charge $Q_d$ is integrated onto the capacitor.

[7] Which, of course, makes their inclusion in a linear model difficult.

[8] Tri-state PFDs offer the benefit that, in addition to phase detection, frequency detection is obtained "for free" meaning that harmonic false locking problems associated with pure phase detectors are eliminated [12].

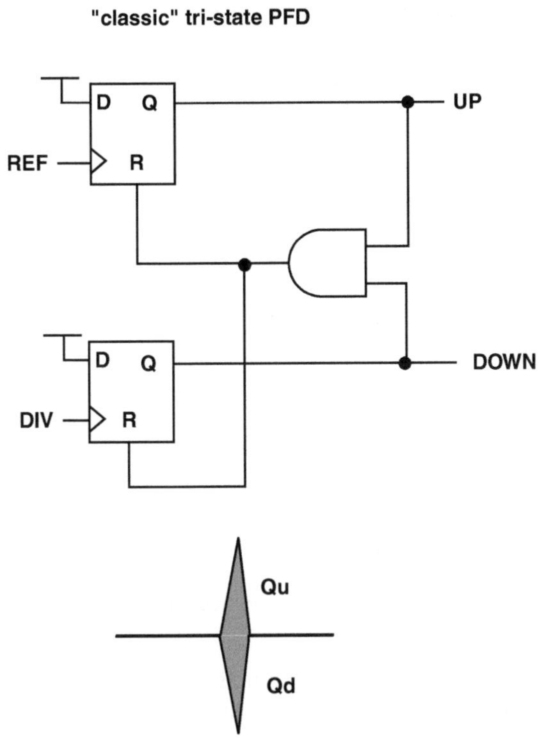


**Fig. 5.6.** Tri-state phase-frequency detector (PFD)

Under locked conditions, this PFD operates with very small phase differences between REF and DIV and, therefore, must produce either very short duration output pulses, or no pulses at all. Any time digital circuitry is asked to operate with narrow pulse inputs and produce narrow pulse outputs, there will be a nonlinearity present in the transfer from input pulse width to output pulse width. The simplest way to think of this is that finite edge rates in any real system will impact the pulse duration, and matching input and output edge rates perfectly is not practical. In Fig. 5.6, we show one possible steady-state locked condition, where both flip-flops fire for a short duration and produce charge packets that cancel. It is apparent that this is a situation ripe with non-linear possibilities, since either REF or DIV may occur first, and any edge-rate differences or delay differences in the reset path will allow the ordering and length of pulses to swap as the PLL feedback action seeks to correct any instantaneous errors.

There are linearization techniques to improve performance that mainly center around making the tri-state PFD operate with forced finite output pulse durations [15]. One possible configuration of this linearization technique is shown in Fig. 5.7 and results in the input phases being skewed apart enough so that the PFD operates linearly with respect to the input phase error. In this case, the cell that delays the reset signal to the REF flip-flop causes the $Q_u$ charge packet to occur later in time than the

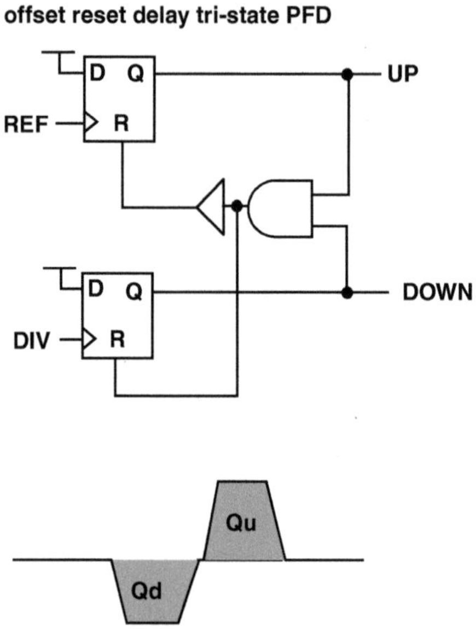


**Fig. 5.7.** Modified tri-state phase-frequency detector (PFD)

$Q_d$ packet. This creates a constant (within noise limitations) positive charge packet $Q_u$ that acts as a reference charge. REF is constrained to always occur after DIV, and no short pulses are generated. The system, therefore, behaves much more linearly under locked conditions than the PFD of Fig.5.6. The tradeoff is that, by having the PFD on longer, any noise associated with charge packet generation has a larger impact on overall PLL noise. In the case of the tri-state PFD, it is the charge-pump, which the PFD controls, that is the noise source of interest. This will be made clear in the next section.[9]

Generally speaking, the PFD itself is not considered as a noise source in phase noise analysis. This is because the phase comparison circuitry is synchronous with respect to the reference clock and feedback divider output. Any nonlinear behavior is best explored with behavioral simulators such as MATLAB or Cppsim [16]. It is mainly the way in which the PFD controls the charge-pump outputs that PLL noise is affected. Therefore, it is not uncommon to lump the charge-pump and PFD together for analysis.

### 5.4.5 PLL Intrinsic Noise: Charge Pump

The charge pump is used as a transducer that takes the phase difference information encoded in the PFD output pulse width and translates it into a correction signal

[9] It is also interesting to note that the tri-state PFD really has more than three states since there are two bits controlling the output. The "tri-state" name really comes into play when the charge-pump is considered. When we do this, we see that it is possible to have either a net negative charge, a net positive charge, or no charge integrated onto the loop filter (hence three states).

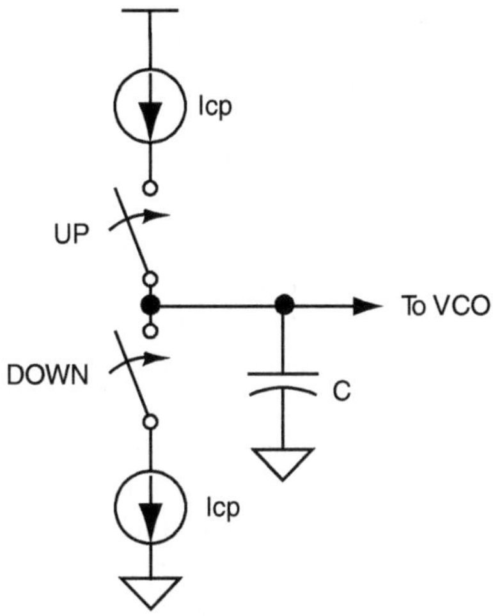


**Fig. 5.8.** Charge-pump architecture

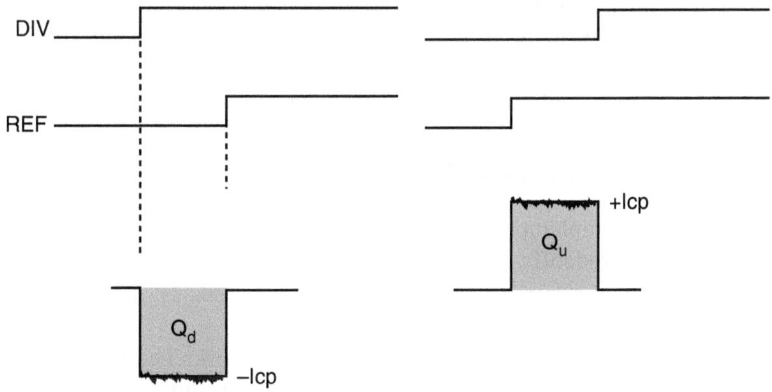


**Fig. 5.9.** Charge-pump operation

by integrating charge onto the loop filter capacitance. Figure 5.8 depicts a typical charge-pump used by a PLL. The UP and DOWN signals produced by the PFD close switches that cause either a positive or negative current to be routed to a loop filter, which in this case is represented by a simple capacitor.

Figure 5.9 shows the charge-pump of Fig.5.8 in operation when controlled by a tri-state PFD under conditions where the PLL is not yet locked, and the charge-pump is on. The phase detector output turns on either a positive or negative current source that is integrated onto a loop filter capacitor. The corresponding correction charge, $Q_u$ or $Q_d$, moves the VCO control voltage either down or up, depending on whether the divider phase leads or lags the reference clock. Thermal noise generated within the current sources supplying the charge appear in $Q_u$ or $Q_d$ and are depicted in Fig.5.9 as the variations in the magnitude of $+Icp$ and $-Icp$. The noise current results in a noise voltage, which then produces output phase noise by modulating the VCO. It is apparent from the figure that the longer the charge-pump currents are on in steady-state, the more noise they contribute.

An offset tri-state PFD and charge-pump combination, such as the one shown in Fig.5.7, produces a steady-state ripple on the control voltage due to the $Q_d$ and $Q_u$ charge packets continuously lowering and then raising the VCO control voltage back to its starting point when the PLL is locked. This periodic modulation of the VCO control voltage occurs at the reference frequency, and appears in the VCO output spectrum as a spur. This spur contains noise energy and, therefore, adds to PLL output jitter.

Sample-and-hold type loop filters can be used to dramatically reduce the spur by adding a sampling network to a tri-state PFD and charge-pump based PLL [14, 17]. The idea is to construct a circuit that samples the loop filter voltage after the up and down pulses have completed. Under a locked condition, the net excursion of the voltage due to up and down currents will be zero except for any charge-pump noise. The sampled voltage is, therefore, theoretically spur-free when presented to the VCO control input. Practical limitations in implementing the sample and hold circuitry will most likely result in reference spur suppression on the order of 20–40 dB, which is still a significant improvement.

Charge-pump noise itself is a device current noise. In MOS designs, it will, therefore, have a $1/f$ component and a thermal noise component.

### 5.4.6 PLL Intrinsic Noise: Loop Filter

Loop filters may be implemented in many different ways. Active or passive loop filters have different tradeoffs, as do split loop filters employed for faster acquisition of lock. The reader is referred to [11] and [12] for details on various loop filter structures.

The simplest loop filter is a totally passive RC network. The resistor elements will generate thermal noise that is filtered by the capacitor(s). Figure 5.10 depicts such a filter for use in a charge-pump PLL. The VCO is included in the figure for context.

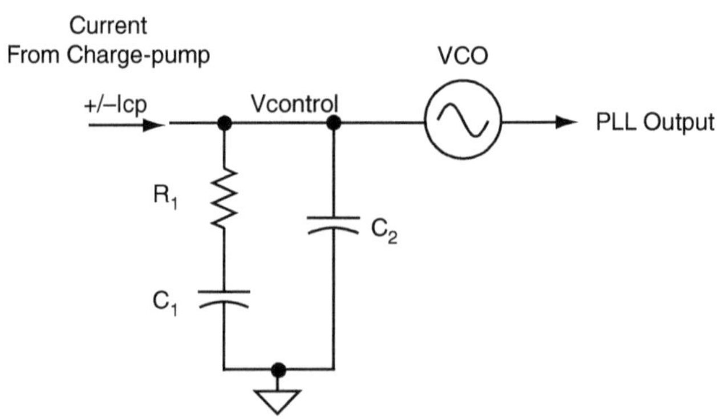


**Fig. 5.10.** Typical charge-pump PLL passive RC filter

Capacitor $C_1$ is large and acts as an integrator for the charge-pump current. $R_1$ adds a zero to the closed loop PLL response for stability reasons, and $C_2$ is a small capacitor that acts to add another pole to filter the control voltage ripple introduced by adding $R_1$. $R_1$ noise will appear directly at the VCO input and, therefore, should be considered when designing the PLL. Recall from the discussion of VCO noise modeling that the input referred phase noise profile of the VCO in the frequency range of interest appears white. The loop filter thermal noise power spectral density (PSD) will also be a white voltage source of magnitude $4kTR^2$ and is filtered by a single pole low-pass filter due to $R_1$ and the series combination of $C_1$ and $C_2$. It is, therefore, necessary to do a quick check of the resistor thermal noise and compare it to the VCO input referred noise and make sure that the overall phase noise is acceptable. Note that as the VCO is tuned to different voltages, its control voltage must change accordingly. Therefore, if the current sources internal to the charge-pump have finite output resistance, any change in control voltage could modulate the value of $+Icp$ or $-Icp$ and, thereby, change PLL behavior. In cases where exact control of $Icp$ is important, an active loop filter may be employed. In an active loop filter implementation, a feedback op-amp is used to keep the charge-pump output node at some desired voltage while the loop filter output changes to the value that is required by the VCO. A downside to such an architecture is that the op-amp becomes another noise generator appearing at the VCO input in the frequency range where the VCO input referred noise should be dominant. See [14] for an example of an active loop filter implementation along with the relevant analysis.

In the analyses presented in this chapter, we will combine the loop filter noise with the charge-pump noise for simplicity, since they can be both approximated by white noise sources over a large frequency range, and they add in that range.

### 5.4.7 PLL Extrinsic Noise: Reference Clock

We finally come to the first extrinsic noise source of interest, namely reference clock phase noise. Reference clock phase noise is a key design parameter in PLL based systems, yet is often not known a-priori by the PLL designer, since the reference clock is picked by the system designer (the PLL end user). As we will show in Sect. 5.4.10, the PLL filters the reference clock phase noise as it proceeds to the PLL output. Therefore, the PLL filters the reference clock jitter, potentially reducing it significantly.[10]

The key concept to keep in mind is that filter analysis is much more easily performed in the frequency domain, so it is necessary that we know the frequency response of the PLL as well as the phase noise profile of the reference clock to perform this kind of analysis. Simply specifying an rms or peak-to-peak jitter number is not enough, since it does not tell us anything about the shape of the noise content of the reference clock, and, therefore, we can not predict precisely how the PLL will filter the noise.

---

[10] One PLL application is to be a so-called "clean-up" or "jitter filter" PLL, where a jittery system clock is passed through a PLL solely for the purpose of jitter reduction.

Reference clock phase noise can have different properties, depending on what the clock source is. For example, a crystal oscillator has a complex low frequency (sub kHz) response that levels out at higher frequencies to a white noise floor profile over the frequency range of interest to the PLL designer. The reference clock phase noise of a crystal can, therefore, often be approximated as white for analysis purposes.

Alternatively, the reference clock may be supplied by a PLL instead of a crystal oscillator. In this case, the PLL generating the reference clock will have its own intrinsic noise and filter response[11]. Knowledge of the frequency response and intrinsic phase noise of the clock generator PLL is, therefore, necessary for such a configuration.

### 5.4.8 PLL Extrinsic Noise: Supply Noise

We mention here that supply noise also impacts overall PLL performance. In particular, supply noise coupling to the VCO control voltage node will directly modulate the VCO output phase. Because this noise appears inside the PLL loop, it has the same impact as an intrinsic noise source and, therefore, must be minimized. The transfer of supply noise to the VCO input is highly dependent on VCO and loop filter architecture, so we will ignore supply noise in our analysis, assuming that the design has been constructed to minimize its impact.

### 5.4.9 PLL Extrinsic Noise: Buffer Delay and Noise

Supply noise outside the loop also impacts overall PLL noise performance. For example, the PLL output may be buffered before it is distributed throughout the chip to its various loads. Supply noise will impact the switching point of these buffers and modulate the buffer delay, adding to jitter. One can argue that this noise is outside the context of the PLL and, therefore, a different problem, but for the system clock designer, it is a phenomenon that has to be considered. In general, this type of added noise scales with the total delay through the buffers. Simply put, the more delay you create in initially buffering the clock, the more buffer jitter you will have. This statement has been quantified to first order in Eq. (2.9).

In our analysis, we focus on the PLL itself and, therefore, don't account for buffer phase noise/jitter. In a well designed system, buffer induced jitter will be very small compared to PLL jitter, especially in the context of phase jitter[12]. Section 6.8, presents an analytical model for clock buffer supply noise-induced jitter. The modeling focuses on period jitter ($\Phi'[n]$) validating the claim that the effect on phase jitter is limited.

---

[11] The clock generator PLL will also require its own reference, meaning there is probably a crystal oscillator somewhere in the system.

[12] This is because buffer jitter only impacts the clock edge on which the noise event occurs, whereas intrinsic noise sources in a PLL affect not only the current edge, but all edges afterward, since the VCO integrates noise at its input.

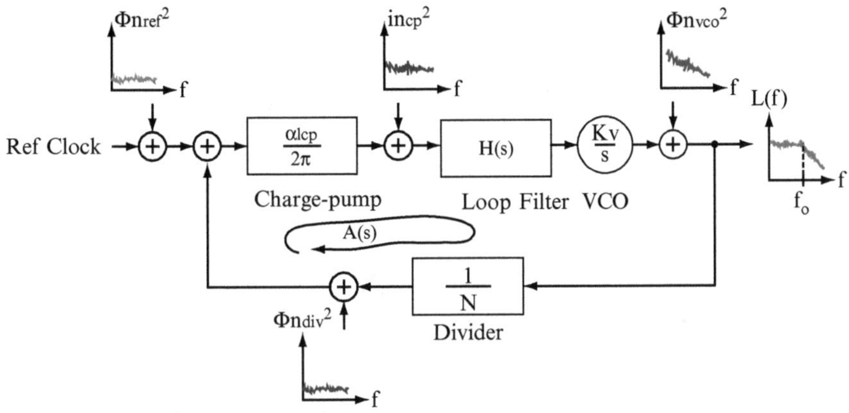


**Fig. 5.11.** PLL noise analysis model proposed in [1]

### 5.4.10 PLL Phase Noise Filtering

We now explore how the PLL filters noise. We use as our basis a simplified version of the noise model proposed in [1], which is depicted in Fig.5.11. The individual circuit blocks are replaced by their frequency domain models. The feedback divider is represented by its phase gain, which is $1/N$, where $N$ is the PLL multiplier. The charge-pump and PFD are combined into a single block and comprise a gain term $(\alpha Icp)/(2\pi)$, where $\alpha$ is the charge-pump phase gain ($\alpha = 1$ for a tri-state PFD [1]), and $I_{cp}$ is the charge-pump current magnitude. The loop filter is represented by its Laplace transform, $H(s)$, and its noise has been lumped together with the charge-pump noise. The VCO is depicted as an integrator with gain $Kv$, and its phase noise is output referred.

In [1], a closed loop function $G(s)$ is constructed from the loop gain, $A(s)$, where:

$$A(s) = \frac{\alpha I_{cp} K_v H(s)}{2\pi N s} \tag{5.7}$$

and

$$G(s) = \frac{A(s)}{1+A(s)} \tag{5.8}$$

$G(s)$ has a low-pass nature with a bandwidth of $f_o$ Hz and represents the closed loop dynamics of the PLL. Also, in [1], the block diagram is manipulated such that the transfer functions from each noise source to the PLL output are calculated in terms of $G(s)$ and the individual gain terms comprising $A(s)$. This is depicted in Fig.5.12[13]. We use the notation that the overall phase noise profile $L(f)$ is composed of the sum of the intrinsic phase noise $L_{\text{int}}(f)$ and the extrinsic phase noise $L_{\text{ext}}(f)$.

---

[13] As is noted in Fig.5.12, the transfer functions must be squared when processing the noise PSDs.

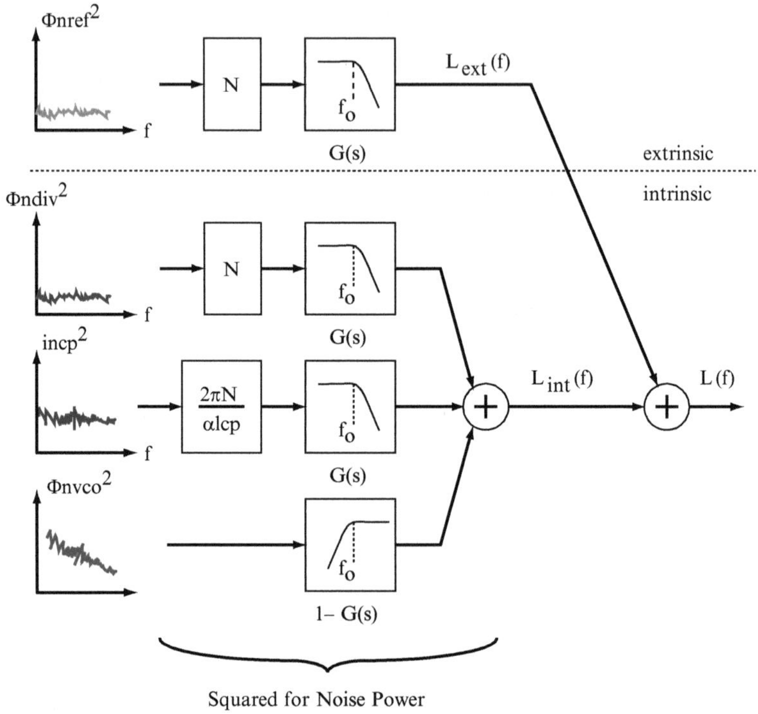


**Fig. 5.12.** PLL noise analysis model proposed in [1]

There are two key observations that we make from Fig. 5.12. First, all intrinsic phase noise sources are low-pass filtered by the PLL with the exception of VCO phase noise, which is high-pass filtered. Second, the extrinsic reference clock phase noise is low-pass filtered by the PLL with a noise gain of $N^2$. The phase noise transfer for the reference clock noise power is

$$L_{\mathrm{ext}}(f) = \Phi n_{\mathrm{ref}}^2(f)\left(N^2|G(f)|^2\right) \tag{5.9}$$

where we have substituted $s = j2\pi f$ in $G(s)$. We can also calculate the intrinsic phase noise as

$$L_{\mathrm{int}}(f) = N^2|G(f)|^2\Phi n_{\mathrm{div}}^2(f) + \left(\frac{2\pi N}{\alpha Icp}\right)^2|G(f)|^2 in_{\mathrm{cp}}^2(f) + |1-G(f)|^2\Phi n_{\mathrm{vco}}^2(f) \tag{5.10}$$

The total phase noise is just the linear sum of the noise powers[14] and is expressed as

$$L(f) = L_{\mathrm{int}}(f) + L_{\mathrm{ext}}(f) \tag{5.11}$$

[14] The linear sum assumes that there is no correlation between the various noise sources, which is reasonable.

Once we have phase noise, we can convert to rms jitter using Eqs. (5.6) and (5.11):

$$\sigma_{\Phi}^2 = \left(\frac{T}{2\pi}\right)^2 \int_{-\infty}^{\infty} (L_{\text{int}}(f) + L_{\text{ext}}(f))\,\mathrm{d}f \tag{5.12}$$

where, to be consistent with earlier discussions, $T$ is the PLL output period. Since integration is a linear operation, we can separate the components in Eq. (5.12) and express the total jitter variance as

$$\sigma_{\Phi}^2 = \sigma_{\Phi,\text{int}}^2 + \sigma_{\Phi,\text{ext}}^2 \tag{5.13}$$

The rms PLL phase jitter is just the square root of the variance[15]:

$$\sigma_{\Phi} = \sqrt{\sigma_{\Phi,\text{int}}^2 + \sigma_{\Phi,\text{ext}}^2} \tag{5.14}$$

Using Eq. (5.14), the designer can calculate the rms phase jitter resulting from his or her particular PLL design from the intrinsic and extrinsic phase noise sources. While intrinsic noise is within the control of the PLL designer, extrinsic reference clock induced phase noise is determined by the system designer's choice of reference clock. It is, therefore, necessary to properly specify the limitations on the reference clock so that total PLL output jitter is within the requirements of the desired application.

**Some Intuition on Reference Clock Phase Noise (or Jitter) Filtering**

As mentioned above, it is often necessary for the PLL designer to specify how much extrinsic noise is allowed at the PLL input for the PLL output to meet a required specification. Historically, this has been done by specifying a jitter requirement for the reference clock. The preceding analysis in this chapter has shown that specifying a jitter number alone does not really address the problem. The reason is that jitter, which is the result of noise processes, is filtered by the PLL according to the PLL closed loop response and reference clock phase noise profile. Jitter is the integration of this noise over frequency. So, specifying a scalar jitter quantity as an input specification does not allow us to determine how the PLL will filter the reference clock phase noise. For example, if the reference clock jitter has a phase noise profile that is entirely within the PLL bandwidth, the PLL does not filter it at all and merely passes the reference clock jitter through. If, on the other hand, the reference clock jitter has a phase noise profile with frequency components outside the PLL bandwidth, then the high frequency components are filtered by the PLL, and the reference clock contribution to total PLL jitter is reduced. Finally, there is the case where the reference clock is generated by a PLL with a bandwidth very similar to the PLL under

[15] Note that the intrinsic noise itself may be expanded and that $\sigma_{\Phi,\text{int}} = \sqrt{\sigma_{\text{vco}}^2 + \sigma_{\text{div}}^2 + \sigma_{\text{cp}}^2}$.

analysis. If both PLLs have peaking in their respective $G(f)$'s, then reference clock jitter (the output from the first PLL) may actually be *amplified* by the second PLL![16]

The PLL output jitter variance component due to the reference clock phase noise can be found using Eqs. (5.6) and (5.9) and is expressed as:

$$\sigma^2_{\Phi,\mathrm{ext}} = \left(\frac{T}{2\pi}\right)^2 \int_{-\infty}^{\infty} N^2 \Phi n^2_{\mathrm{ref}}(f)|G(f)|^2 \mathrm{d}f \tag{5.15}$$

Equation (5.15) states that the reference clock phase noise is multiplied by $N^2$ and filtered as it passes through the PLL. The reason for the $N^2$ factor has to do with the fact that the reference clock phase noise is referred to the reference clock period, $T_{\mathrm{ref}}$. The PLL output phase, however, is referred to the PLL output period, $T$, which is $N$ times smaller than $T_{\mathrm{ref}}$. Any noise variations in the reference clock edge at the PLL input that get through the PLL and pass to the PLL output occupy the same amount of time duration, but $N$ times more phase at the output than at the input. Noise calculations are done in noise power, so the multiplicative factor for noise analysis becomes $N^2$.

We can substitute $T = T_{\mathrm{ref}}/N$ in Eq. (5.15) to get

$$\sigma^2_{\Phi,\mathrm{ext}} = \left(\frac{T_{\mathrm{ref}}}{2\pi}\right)^2 \int_{-\infty}^{\infty} \Phi n^2_{\mathrm{ref}}(f)|G(f)|^2 \mathrm{d}f \tag{5.16}$$

What Eq. (5.16) shows is that it is possible to perform analysis on the extrinsic reference clock phase noise contribution to jitter at the PLL *output* using the reference clock phase noise profile present at the PLL *input* without explicitly taking into account the $N^2$ factor, as long as we remember to scale the phase noise at the PLL output by $T_{ref}$ rather than $T$. From this viewpoint, the PLL is truly just a phase noise filter, $G(f)$, from input to output. Figure 5.13 depicts the simplification in the model, where the $N^2$ factor has been be factored out.

### 5.4.11 Phase Noise to Period Jitter and Phase Noise to C2C Jitter

In Sect. 5.2, we described phase jitter as being the most fundamental jitter type, with period and c2c jitter defined to be the first and second difference of phase jitter, respectively. Using this information, we can derive a frequency domain expression for period and c2c jitter based on phase jitter.

A discrete time difference function $y[n] = x[n] - x[n-1]$ is described in the frequency domain by its $z$ transform:

$$Y(z) = X(z)(1 - z^{-1}) \tag{5.17}$$

[16] This is why it is a good practice to make sure that, when cascading PLLs, the bandwidths of the PLLs in question are "sufficiently different", where "sufficiently different" means that the two PLL bandwidths are far enough apart that any peaking does not get amplified when their closed loop responses are multiplied.

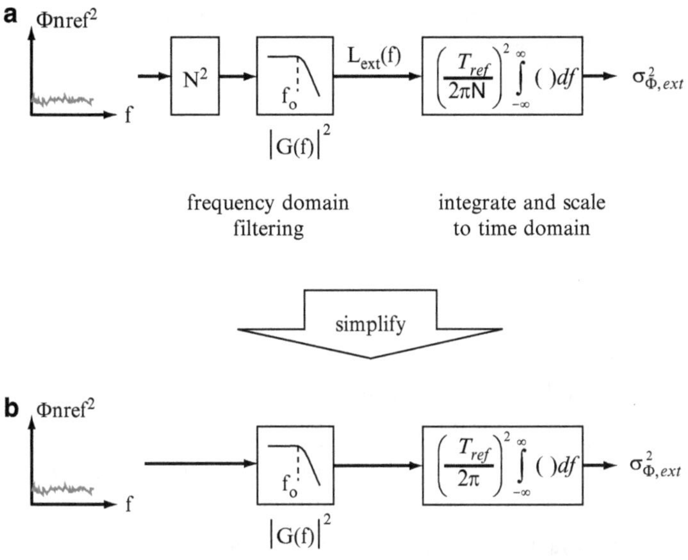


**Fig. 5.13.** Reference clock phase noise to jitter modeling

The transfer function from $x$ to $y$, which we will call $D(z)$, is:

$$D(z) = \frac{Y(z)}{X(z)} = 1 - z^{-1} \tag{5.18}$$

Making the substitution $z = e^{j2\pi fT}$, where $T$ is the discrete time sample period, we arrive at:

$$D(f) = 1 - e^{-j2\pi fT} \tag{5.19}$$

The log magnitude of Eq. (5.19) is plotted in Fig.5.14 assuming $T = 1\,\text{ns}$ to be consistent with examples that will be presented later on. $D(f)$ is periodic in frequency with nulls appearing at multiples of $1/T$. It is the "filter function" we apply to the PLL phase noise output, $L(f)$, in order to determine period jitter from phase noise. For c2c jitter, we apply $D(f)^2$ as the filter function since we know it is the second difference of phase jitter. Note that the nature of $D(f)$ is high-pass, as we expected from discussions in Sect. 5.2. We only show one "period" of $D(f)$ (up to the first null) for a reason that is fundamental to PLLs. $D(f)$ is periodic at multiples of the PLL output frequency, $1/T$. For reasons of stability [15], PLLs are designed with bandwidths significantly lower than their output frequency. A common rule of thumb is that the PLL bandwidth be set to at least $10\times$ lower than the reference clock frequency, which is $N$ times lower than the output frequency. This implies that the PLL bandwidth is at least $10N$ times lower than $1/T$. Since the PLL is low-pass in

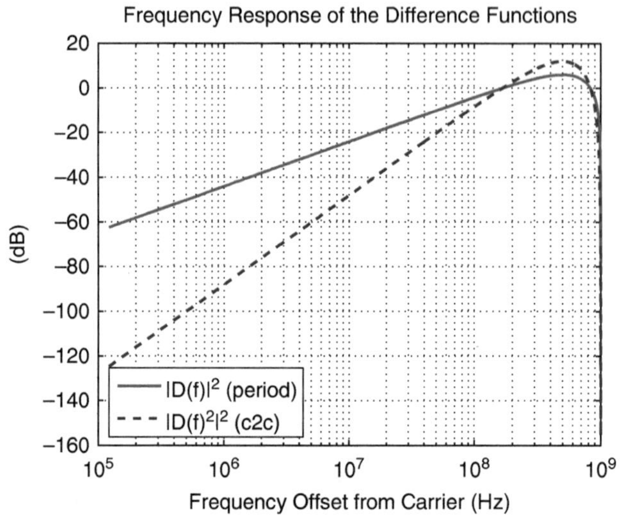


**Fig. 5.14.** Phase to period and C2C jitter difference functions

nature, it greatly attenuates the impact of $D(f)$ and/or $D(f)^2$ at high frequencies where they have significant magnitude.[17]

To determine period jitter from phase noise, we use Eq. (5.20):

$$\sigma^2_{\Phi'} = \left(\frac{T}{2\pi}\right)^2 \int_{-\infty}^{\infty} L(f)|D(f)|^2 \mathrm{d}f \tag{5.20}$$

To determine c2c jitter from phase noise, we use Eq. (5.21):

$$\sigma^2_{\Phi''} = \left(\frac{T}{2\pi}\right)^2 \int_{-\infty}^{\infty} L(f)|D(f)|^4 \mathrm{d}f \tag{5.21}$$

To determine the rms jitter contribution of any individual phase noise source to either period or c2c jitter, we simply substitute the appropriate transfer function for $G(f)$ (as depicted in Fig.5.11) and the individual phase noise profile for $L(f)$ in Eqs. (5.20) and (5.21), respectively. We will give examples in Sects. 5.5 and 5.7 regarding calculating the reference clock contribution to PLL output jitter.

[17] Here, we make the assumption that the PLL is of sufficiently high order that it rolls off fast enough to attenuate $D(f)$ or $D(f)^2$ significantly at high frequencies, which is reasonable because PLLs are typically designed to be 2nd order closed loop systems, which will roll off at the same rate that the $D(f)^2$ function rises. Unavoidable high frequency parasitic poles that are well outside the PLL closed loop bandwidth (typically 4× higher or more in frequency) will further attenuate any impact of the $D(f)$ and $D(f)^2$ filter functions.

### 5.4.12 Phase, Period, and C2C Jitter Examples

Before moving on to practical examples of jitter analysis, we will first use the information obtained thus far to calculate the various jitter types from a phase noise profile for an example PLL. In order to verify the frequency domain calculations, we will also explicitly measure phase, period, and c2c jitter from the time domain edge crossings of the PLL output and use the jitter definitions presented in Sect. 5.2.

To carry out the PLL simulations presented in this chapter, the CppSim behavioral modeling tool was used [1, 16]. CppSim is a constant time step C++ based simulator that is particularly useful for PLL analysis. CppSim's constant time step approach makes performing frequency domain analysis on the time domain waveforms simple via the use of FFTs.

#### Phase Jitter

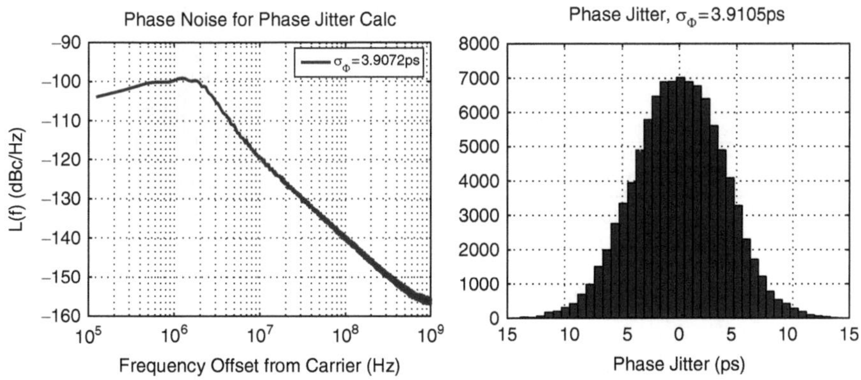


**Fig. 5.15.** Reference clock to phase jitter model

Figure 5.15 presents the PLL simulation results. The left plot is the PLL output $L(f)$, which includes contributions from all of the intrinsic and extrinsic phase noise sources in the system. Also shown on the plot is the calculated value of rms phase jitter using Eq. (5.6) and the simulated profile for $L(f)$. The PLL output frequency was set to 1 GHz, therefore $T = 1$ ns. The right plot shows a histogram of time domain phase jitter calculated according to the definition in Eq. (5.1). To perform this calculation, an ideal non-jittery timing reference at 1 GHz was synchronized to a PLL output edge once the PLL locked, and then the difference in edge locations between the ideal reference and the PLL output were computed over one million cycles. This large sample was taken so that the rms value calculated from the distribution would be based on a significant number of samples. Also, to a first order, we can think of this process as capturing noise frequency information down to about 1 KHz, since we have one million periods of 1 ns average length, which corresponds to 1 ms of time information, or down to 1 KHz of frequency information. The FFT in the plot does not go down to the KHz range in frequency purely for computational reasons in generating such a large FFT. This is acceptable because the phase noise components in the very low frequency range do not contribute significantly to total PLL noise power.

The frequency domain, phase noise based, rms jitter calculation is 3.9 ps. The direct measurement from the time domain simulation using the time domain definition for phase jitter is also 3.9 ps, showing excellent agreement between the two methods. This result validates our earlier discussions and Eq. (5.6). Note also that the shape of the time domain jitter measurement looks Gaussian, in agreement with our earlier assertion.

## Period Jitter

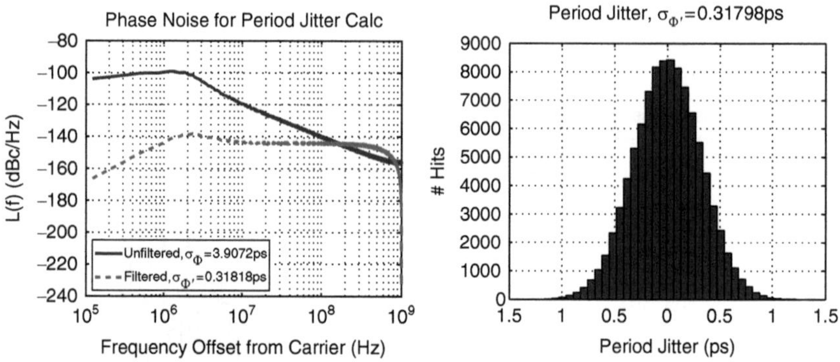


**Fig. 5.16.** Reference clock to period jitter model

Figure 5.16 shows the result of applying $D(f)$ from Fig. 5.14 to the PLL output $L(f)$ of Fig. 5.15. The unfiltered $L(f)$ is also shown in the plot as a reference for comparison. We see that the high-pass nature of $D(f)$ removes a significant amount of low frequency noise in the period jitter calculation. The low-pass nature of $G(f)$ attenuates the high frequency component of $D(f)$[18]. The right plot shows a direct measurement of period jitter as the difference between each period and the average period (1 ns).

We once again have excellent agreement between the rms jitter value based on the frequency domain analysis and the time domain measurement. This experiment validates Eqs. (5.19) and (5.20).

## C2C Jitter

Finally, for completeness, we show the results in Fig. 5.17 for the c2c jitter calculation based on both frequency and time domain analysis. They both agree very

[18] Also, as mentioned previously, in a real PLL, there will be additional high frequency poles in $G(f)$ in the many MHz to GHz range that will further attenuate $D(f)$. These poles were left out in the simulation for simplicity.

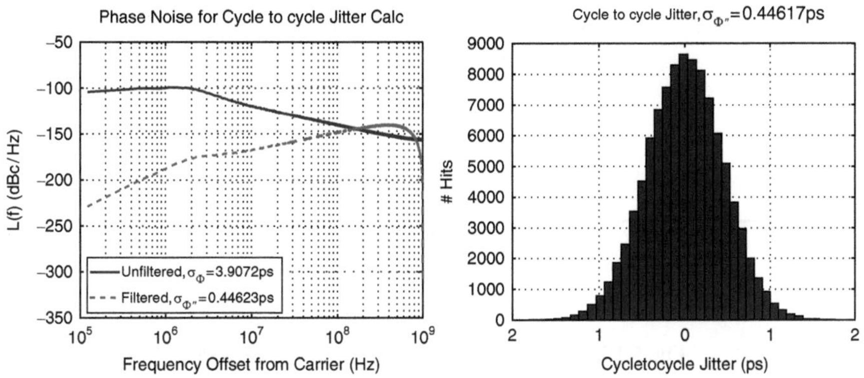


**Fig. 5.17.** Reference clock to C2C jitter model

well, which is to be expected since the phase and period jitter calculations were in agreement.

Having verified through simulation that our analysis methodology and jitter definitions are in agreement, we proceed with two examples of reference clock phase noise impact on system design.

## 5.5 Reference Clock Jitter Transfer Example: Microprocessor

In this section, we analyze the contribution of a reference clock to the output period jitter of a microprocessor core clock generator PLL. Since the microprocessor operates in its own time domain, we are primarily concerned with period jitter because it impacts the maximum operating frequency. Typically, the clock has been routed throughout the chip in a well controlled fashion (see Chap. 2) with minimum skew. The primary constraint for this exercise is to ensure that the clock period generated at the PLL output does not shrink substantially from the nominal value so that there will be a critical path violation and a catastrophic failure.

### 5.5.1 A Proposed Core Clock Methodology Using Mean Time Between Failures (MTBF)

Using all of the information presented thus far, we employ the model shown in Fig.5.18 to calculate the impact of reference clock phase noise on the PLL output jitter. The reference clock phase noise, $\Phi n_{\mathrm{ref}}^2$, passes through the PLL filter function, $G(f)$, and then is processed by the period jitter difference function, $D(f)$, before it is integrated and scaled to calculate the output rms extrinsic jitter component[19].

[19] Note that while the scale factor in the integral is $T_{\mathrm{ref}}$, due to the simplification proposed in Sect. 5.4.10, $D(f)$ is still periodic at the PLL output frequency, $1/T$.

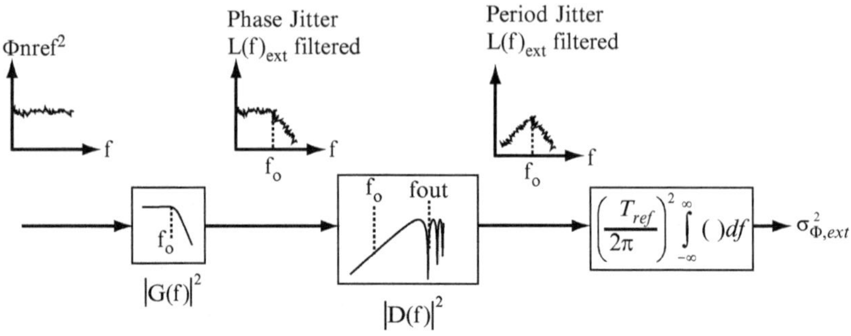


**Fig. 5.18.** Reference clock jitter to period jitter model

The output rms jitter due to reference clock, $\sigma_{\Phi,\text{ext}}$, is summed with the PLL intrinsic rms jitter according to Eq. (5.14) to determine total system jitter. The question then becomes: "Given an rms period jitter value, how many standard deviations should be used to determine a peak-to-peak value?" We propose that the metric to be employed to specify the number of standard deviations used to calculate peak-to-peak reference clock jitter for a microprocessor should be Mean Time Between Failures (MTBF).

MTBF in our case means how long the microprocessor can run continuously without exhibiting any timing failures, which is to say that the microprocessor does not experience any periods smaller than a critical value. If the MTBF for a 1 GHz processor is 1 year, we calculate the number of clock periods in 1 year as:

$$\text{num\_per} = (1\,\text{year})\left(365.25\frac{\text{days}}{\text{year}}\right)\left(86400\frac{\text{s}}{\text{day}}\right)\left(\frac{1\text{ period}}{1E-9\text{ s}}\right) = 3.16\text{E}+16\text{ periods} \tag{5.22}$$

We can think of $1/\text{num\_per} = 3.2\text{E}-17$ as equivalent to a bit error rate (BER), which we will use in Sect. 5.7 as the key component in determining how many $\sigma$ to use for calculating peak-to-peak PLL jitter for a serial link. Simply stated, we are allowing 1 period out of $3.16\text{E}+16$ periods to be too small. This corresponds to an error rate of $3.2\text{E}-17$, which is also equivalent to a probability measure.

We utilize the fact that the jitter distribution is Gaussian (Normal) to determine how many $\sigma$ are required to achieve a probability of error of $1/\text{num\_per}$ using an approximation to the *error function*, $Q(\sigma)$. For $\sigma > 3$, the error function can be approximated by [5]:

$$Q(\sigma) = \frac{1}{\sigma\sqrt{2\pi}}e^{\frac{-\sigma^2}{2}} \tag{5.23}$$

and the number of $\sigma$ for a given peak-to-peak error of $1/\mathit{num\_per}$ can be read from the plot of $Q(\sigma)$ vs. $\sigma$.

Figure 5.19 shows the result of sweeping $\sigma$ in $Q(\sigma)$. We can find the point on the chart that corresponds to the desired BER and determine how many $\sigma$ should be used to determine the peak-to-peak value. In our example, we require a peak-to-peak value of $16.7\sigma$, which corresponds to $\pm 8.35\sigma$ about the mean of zero.

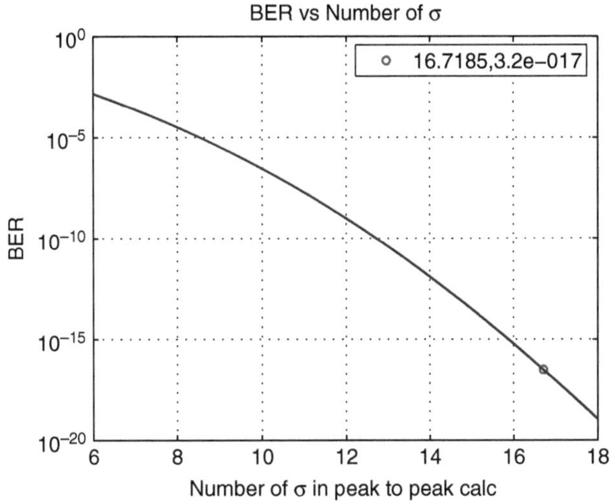


**Fig. 5.19.** Number of $\sigma$ in peak-to-peak calculation for normal distribution

Jitter statistics reported on data sheets and in the literature typically provide rms and peak-to-peak values, where the peak-to-peak value is only seven to ten times the rms value. Part of the reason peak-to-peak values reported in data sheets vary in the number of standard deviations, $\sigma$, for which they are calculated, may be due to limits in the number of samples taken. This is an area that is often not well defined or explained in the average data sheet. We will proceed using the rms jitter value and MTBF calculation with the understanding that it may be on the conservative side.

In our example, we require $\pm 8.4\sigma$ to calculate peak-to-peak jitter. The problem is reduced to calculating rms period jitter at the PLL output. Figure 5.20 presents the filter functions that we will use in this example. The PLL closed loop transfer function, $G(f)$, is plotted along with the period jitter difference function, $D(f)$, and the cascaded transfer function, $G(f)D(f)$, which filters the reference clock phase noise. The PLL is modeled as a type II[20], second order system with a 10 MHz bandwidth and approximately 1.8 dB of peaking. $D(f)$ is the same as depicted in Fig. 5.14 for a 1 GHz PLL output clock.

Figure 5.20 shows that the shape of the filter cascade, $G(f)D(f)$, follows $D(f)$ at low frequencies and then flattens out when the falling slope of $G(f)$ equalizes the rising slope of $D(f)$. The cascade function then decreases at very high frequencies where $D(f)$ rolls off again.

To calculate the reference clock induced output jitter, we process the reference clock phase noise, $\Phi n_{ref}^2$, with the $G(f)D(f)$ cascade, as depicted in the top plot of Fig. 5.21. $\Phi n_{ref}^2$ has been set to have $1/f$ and $1/f^2$ regions at low frequencies, and a white thermal noise floor of $-120$ dBc/Hz at higher frequencies. This is consistent

[20] This means that there are two integrators inside the PLL loop. For example, the VCO is an ideal integrator, and the charge-pump approximates an ideal integrator very well.

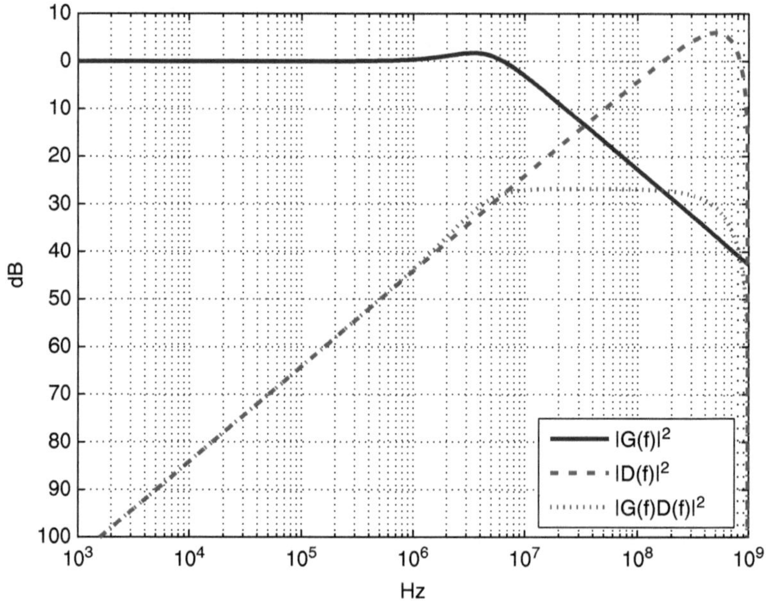


**Fig. 5.20.** Example transfer functions for period jitter calculation

with a crystal oscillator phase noise profile. Real crystals also have $1/f^3$ and $1/f^4$ regions at lower and lower frequencies, which we have left out for simplicity. Real crystals also typically do not report phase noise at frequencies above a few hundred MHz because of limitations in measurement equipment. We have extended the thermal noise floor up to 1 GHz for this analysis to demonstrate the impact of $D(f)$.

The filtered[21] $\Phi n^2_{ref}$ is then integrated and scaled in accordance with the model shown in Fig.5.18. The bottom plot presents the result of the integration. The period jitter value increases as the noise components are added up over frequency. We observe that the noise increases most at high frequencies where $\Phi n^2_{ref}$ flattens out. This is intuitive because Fig.5.21 has a logarithmic x-axis, so the flat region at very high frequencies contains more noise energy than the low frequency range.

Integration of the phase noise profile results in an rms period jitter value of $\sigma_{\Phi',\mathrm{ext}}$ =240 fs. Earlier,we calculated that we required approximately $17\sigma$ to determine the peak-to-peak value for a 1-year MTBF. The total result is $\sigma_{\Phi',\mathrm{pp,ext}}$ =4 ps, where $\sigma_{\Phi',\mathrm{pp,ext}}$ is the peak-to-peak period jitter at the PLL output due to the reference clock. This number is RSS (root sum of squares) summed with the PLL intrinsic period jitter to determine overall period jitter:

[21] We are being a little loose with the term "filter" in our discussions. $G(f)$ truly does filter the reference clock phase noise in the traditional sense. $D(f)$, by contrast, is a mathematical by-product of the definition of period jitter. So it is not a true, physical filter. It does have a frequency response that has an impact equivalent to a filter, so we use the term "filter" to describe the cascade of $G(f)$ and $D(f)$.

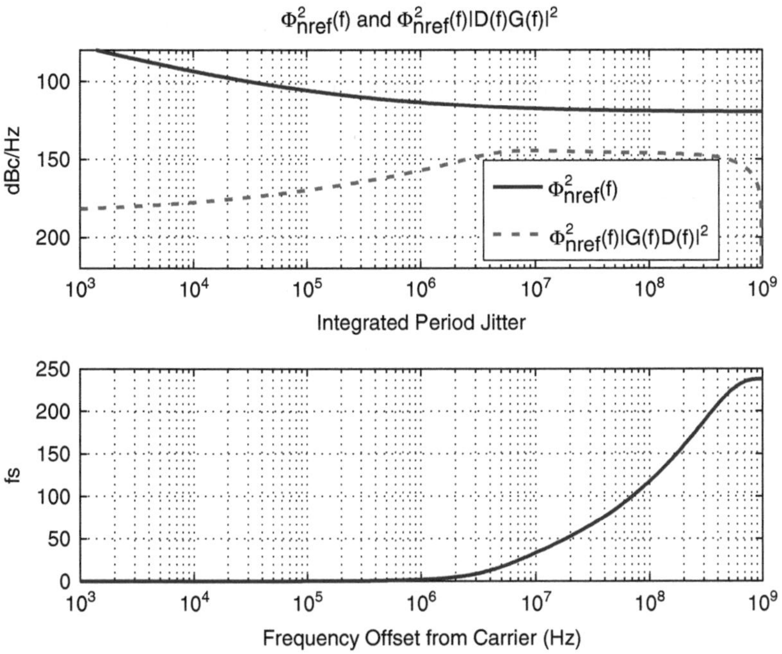


**Fig. 5.21.** Example reference clock $\Phi n_{ref}^2$ and filtered $\Phi n_{ref}^2$

$$\sqrt{\sigma^2_{\Phi',\text{pp,ext}} + \sigma^2_{\Phi',\text{pp,int}}} \leq t_{budget} \tag{5.24}$$

The total peak-to-peak period jitter number is then divided by two, and compared with the critical path budget, to ensure that timing margins are not violated. The reason for the division by two is that, under the assumption that the jitter is Gaussian with zero mean, half of the period jitter results in longer periods, which will not violate a critical path, while the other half of the period jitter is associated with shorter periods that will potentially violate the critical path specification.

For situations where the designer is attempting to define a reference clock phase noise profile, the analysis can be performed with a family of different phase noise responses. By taking the profile that results in the maximum PLL output period jitter, the designer can define the reference clock phase noise mask. Such a group of simulations is not trivial, so some assumptions need to be made. One reasonable assumption is that a crystal oscillator supplying the reference clock will generally have a broadband phase noise profile that is lower than the low frequency noise limit if the reference clock is supplied by a PLL.[22] Additionally, if a PLL is used to supply the reference clock, worst case output jitter will result if the bandwidth of the two

[22] A PLL that will generally have a crystal reference supplied to its reference clock input.

PLLs align, an effect that is amplified if both PLLs have peaking in their respective $G(f)$'s.[23]

As a final demonstration, we examine the impact of a high frequency pole in the PLL closed loop transfer function on total period jitter.

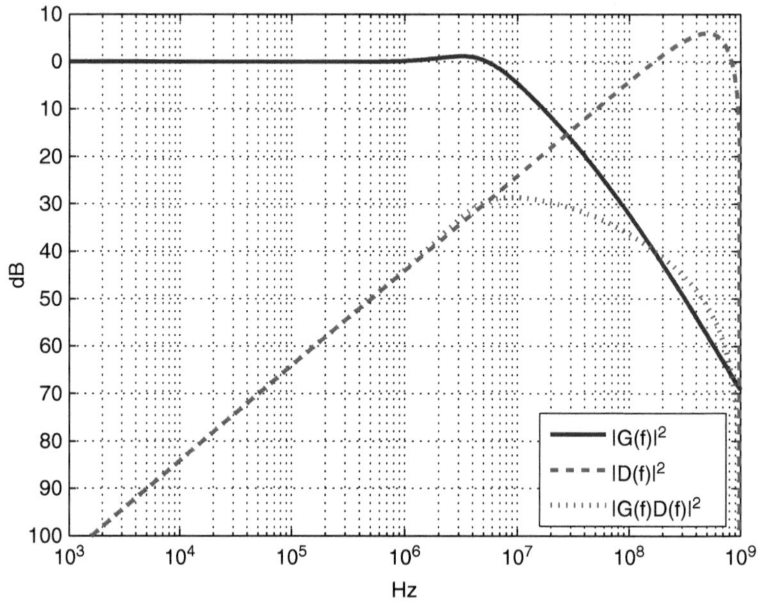


**Fig. 5.22.** Example transfer functions for period jitter calculation with added pole

Figures 5.22 and 5.23 present the results of the period jitter analysis with a 50 MHz pole added to the PLL closed loop transfer function. A high frequency pole is often added by the PLL designer to suppress high frequency noise caused by the phase comparison process occurring at the PFD. In some cases, several poles at high frequency are added to obtain additional suppression. As the figures show, the added pole helps greatly to reduce the high frequency phase noise present in the period jitter profile. The rms jitter value is reduced from 240 fs to 77 fs, and the peak-to-peak value from 4 ps to 1.3 ps. Clearly, high frequency poles in $G(f)$, whether explicit or parasitic, help reduce total period jitter.

## 5.6 Non-Random Jitter Distributions

Thus far the assumption of zero mean, Gaussian, random jitter distributions has been utilized when performing analysis. In practical systems, there will be non-random

[23] This is a phenomenon known as jitter peaking, which is of particular concern in long haul data links [5].

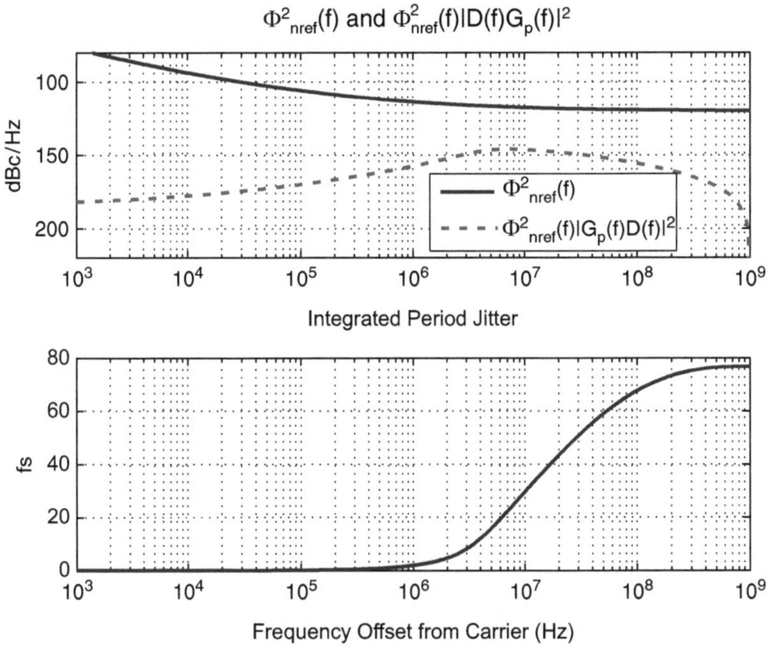


**Fig. 5.23.** Example reference clock $\Phi^2_{nref}$ and filtered $\Phi^2_{nref}$ with added pole

(deterministic) jitter sources that impact overall PLL jitter. Deterministic jitter is sometimes referred to as *static* jitter because unlike random jitter, which is defined by an rms value and whose peak-to-peak value depends on the application requirements, deterministic jitter can be defined by a single peak-to-peak value and manifests itself as a non-random modulation of clock phase. In this section, we present some sources of deterministic jitter and methods to include them in analysis.

### 5.6.1 Reference Spurs in PLLs

The feedback action inside a PLL compares the divider output to the reference input via a PFD. The output of the PFD is then filtered and used to control the VCO. Ideally, the only error components present in the signal at the VCO input relate to the phase error of the PLL or the noise sources inherent to the circuitry that comprises the PLL. In most practical PLLs, however, there is an additional error component due either to non-idealities in the PFD, or caused by design techniques aimed at improving the linearity of the PFD.

In Sect. 5.4.5, we discussed how an explicit offset is sometimes added in the reset path of a tri-state PFD to improve the PFD/charge-pump combination's linearity. We also discussed how the periodic modulation of the VCO control voltage introduced by the offset tri-state PFD appears in the PLL output spectrum as a spur. In the time domain, the reference spur appears as a deterministic jitter component.

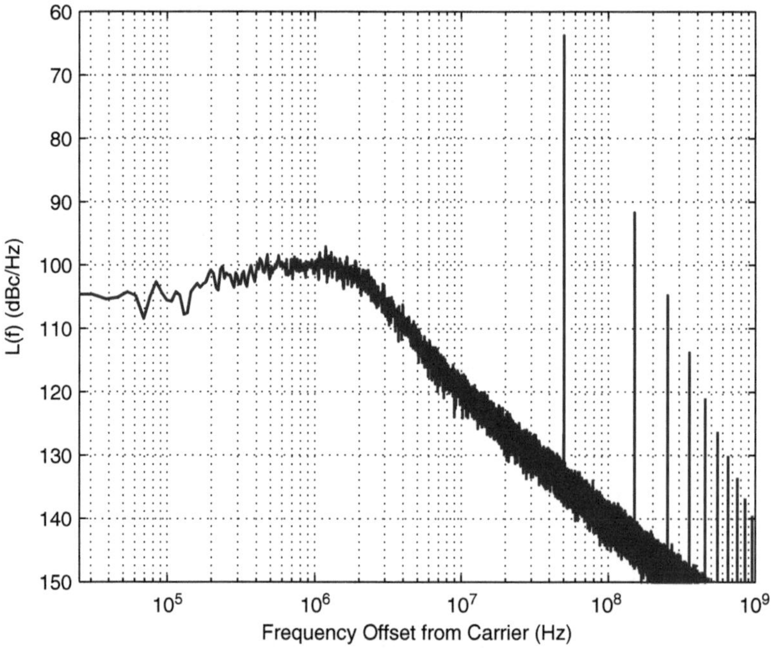


**Fig. 5.24.** PLL phase noise spectrum showing reference spurs

Figure 5.24 shows the phase noise profile of a PLL that exhibits reference spurs. For the PLL simulated, the reference clock frequency was set to 50 MHz and an offset tri-state PFD was used. For this PLL, the reference clock spur is significant, and we observe spurs at odd harmonics of the reference clock. The spurs are the result of the explicit offset introduced between the reference and divider flip-flop resets to improve PFD linearity. For a mathematical method to calculate the spur contribution to output jitter, see [18].

In the time domain, these spurs are observed as a deterministic component in the PLL output jitter, as shown in Fig. 5.25. We observe that the Gaussian jitter profile we expect from random noise sources is still present, but now there are two Gaussian profiles that overlap and whose peaks are separated by approximately 40 ps. The total jitter profile can, therefore, be thought of as a Gaussian jitter distribution convolved with two deterministic jitter impulses appearing 40 ps apart. For this example, the deterministic jitter component is expressed as a peak-to-peak value of approximately 40 ps.

Several techniques exist to reduce the reference spur. The simplest, and most commonly used technique, is to add high frequency poles to the PLL filter function, $G(f)$. Since the reference spur and its harmonics occur outside the PLL bandwidth, high order poles will suppress their impact while not significantly changing the PLL closed loop response. A second technique is to use a sample-and-hold loop filter [14, 17]. A sample-and-hold loop filter works by waiting until the PFD has completed its UP and DOWN pulses, and then sampling the loop filter cap. In this

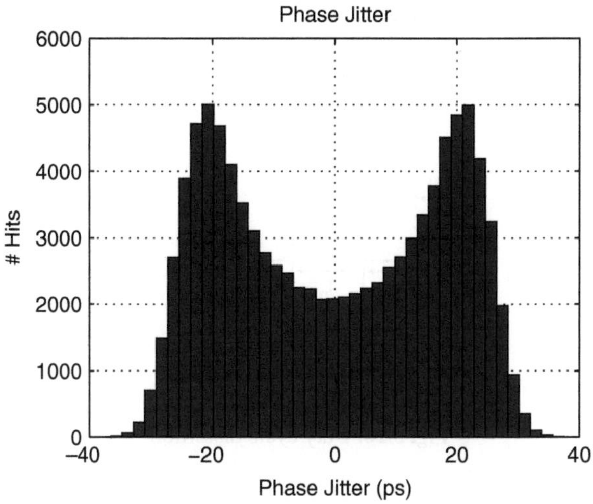


**Fig. 5.25.** Deterministic jitter histogram due to reference spurs

way, any periodic modulation of the VCO control voltage due to PFD operation is minimized.[24] Finally, the PFD itself can be architected to minimize the reference spur. In a tri-state PFD, this can be accomplished by minimizing the on-time of the UP and DOWN pulses when the PLL is in lock. The tradeoff with this approach is that any non-linearities present in the PFD and charge-pump combination will be exposed, as discussed in Sect. 5.4.4.

### 5.6.2 Duty Cycle Distortion (DCD)

In double data rate (DDR) systems, *both* edges of the clock are used as timing references. The clock frequency is, therefore, half the data rate, which eases bandwidth requirements for the clock distribution network, and, potentially, for a number of sub-systems that run at the clock rate.

The effective clock rate in a DDR system, which corresponds to the data rate, is, therefore, twice the clock frequency. However, if the clock signal exhibits any error in its duty cycle, the instantaneous data rate of the system varies from the nominal value according to the amount of duty cycle distortion present in the clock. Figure 5.26 depicts the impact of DCD in a DDR system.

$CLK0$ represents an ideal DDR clock. In this case, the time from rising edge to falling edge, $t_{p0}$, equals the time from falling edge to rising edge, $t_{n0}$, and the data rate, $1/T_d$, is equal to twice the clock frequency, $1/T$. To summarize:

$$t_{n0} = t_{p0} = T_d = \frac{1}{2}T \tag{5.25}$$

[24] Ideally, the periodic modulation of the VCO control voltage would be eliminated, but practical sample-and-hold loop filters will have some reference clock feed-through, and so a residual reference spur will remain.

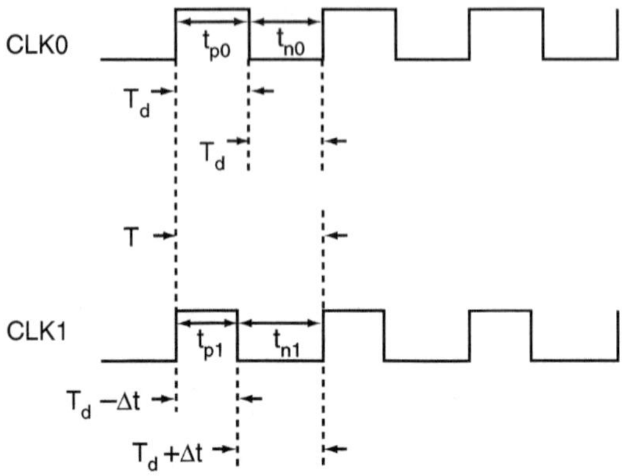


**Fig. 5.26.** Duty cycle distortion

*CLK*1, by contrast, exhibits DCD. While the frequency of *CLK*1 is equal to $1/T$ and is, therefore, the same as that of *CLK*0, the time from rising edge to falling edge, $t_{p1}$, does not equal the time from falling edge to rising edge, $t_{n1}$, and the data rate changes from bit to bit. For $t_{p1}$, the bit period is

$$t_{p1} = T_d - \Delta t \tag{5.26}$$

while for $t_{n1}$ the bit period is

$$t_{n1} = T_d + \Delta t \tag{5.27}$$

The average bit period can be expressed as

$$< bitperiod >= \frac{t_{p1} + t_{n1}}{2} = T_d \tag{5.28}$$

The average bit period for the DCD case is, therefore, the same as that for the ideal case, but the instantaneous bit periods exhibit a deterministic jitter component with a peak-to-peak value $2\Delta t$.

### 5.6.3 Power Supply Noise

Power supply noise is another potential source of deterministic jitter. As mentioned in Sect. 5.4.8, power supply noise can couple directly into the VCO control voltage node and modulate the PLL output frequency. In this case, the impact of the supply noise lingers in the system because of the integration of phase that takes place in the VCO. Additionally, supply noise can modulate the delay through a buffer chain used to distribute the clock across chip (Sect. 6.8). In both the cases, if the supply noise is caused by a periodic signal (such as the clock itself, or sub-harmonics of the clock introduced by clock dividers or a piece of software running a loop on a microprocessor), the noise will be exhibited as a periodic modulation in the clock output phase, which appears as a deterministic jitter component.

### 5.6.4 Inter-Symbol Interference (ISI)

Yet another source of deterministic jitter in a communication system is inter-symbol interference (ISI). ISI is a phenomenon of particular concern in a serial data link, where the finite bandwidth of the channel filters the data being sent through it. To illustrate the impact of ISI, a CppSim behavioral simulation of a 2.5 Gb/s random data signal passing through a channel modeled as a simple low-pass filter with a $-3$ dB bandwidth of 1 GHz was performed. The results are presented in Figs. 5.27 and 5.28.

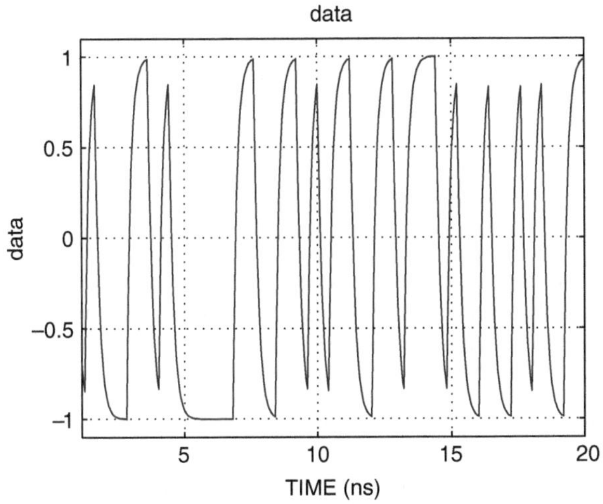


**Fig. 5.27.** Random data signal filtering by a channel

In Fig. 5.27, the filtered data signal is plotted. Note that sequences of multiple bits of the same polarity (sequences of multiple ones or multiple zeroes) have a larger final magnitude than one-zero-one or zero-one-zero sequences. This is because the finite bandwidth of the channel filters the high frequency data patterns more than the low frequency patterns.

An eye diagram of the filtered data signal is plotted in Fig. 5.28. An eye diagram is created by chopping the data signal in fixed increments of time, and then overlaying the sections. In Fig. 5.28, we plot the eye for two data periods. We observe that the finite bandwidth of the simple RC filter model for the channel results in incomplete settling for some bits and complete settling for others. Incomplete settling corresponds to the high frequency data sequences, while complete settling corresponds to the lower frequency data sequences.

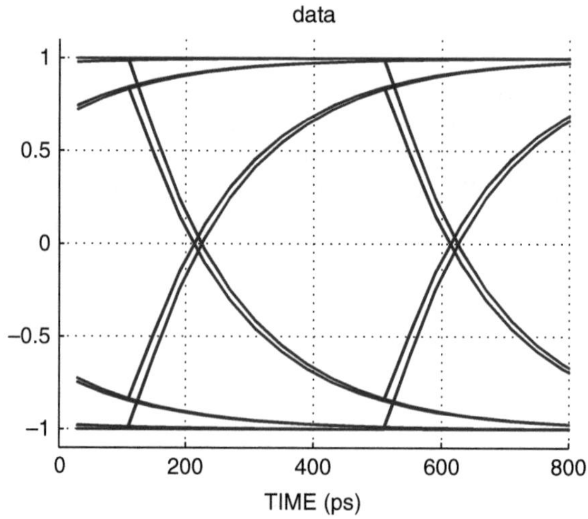


**Fig. 5.28.** Eye diagram of filtered data signal

Incomplete bit settling appears as a fixed component to the data jitter at the end of the channel, as observed in the two zero crossings of the data in the eye diagram. ISI can, therefore, be modeled as a deterministic jitter component.[25]

### 5.6.5 Including Deterministic Jitter in Analysis

The examples of non-random jitter mentioned above all have different origins and may introduce subtly different deterministic jitter components in the clock signal, but they can all be analyzed in a similar fashion.

As mentioned in Sect. 5.6.1, deterministic jitter can be modeled as impulses that occur at a finite peak-to-peak distance apart [19, 20]. These impulses are convolved with the random jitter distribution to calculate the overall jitter profile. For example, if the ISI induced deterministic jitter depicted in Fig. 5.28 has a peak-to-peak value of 10 ps and system random jitter has a calculated peak-to-peak value of 50 ps, we calculate that the total peak-to-peak system jitter is 60 ps. If there are additional uncorrelated deterministic jitter sources, we add them linearly; that is, we add their peak-to-peak values. If there are additional random jitter sources, we add them in an RSS sense, under the assumption that they are statistically independent.

[25] In addition to ISI, cross-talk can introduce jitter components in a serial link. Cross-talk describes coupling between lanes within a link, or across links, in a serial data system. Coupling of an aggressor signal to the data signal can move the data signal's edge. Cross-talk is very specific to the exact board layout of a system and is, therefore, beyond the scope of this text and assumed to be minimal in our analyses.

## 5.7 Reference Clock Jitter Transfer Example: Serial Link

In this section, we present an analysis of the reference clock contribution to the output phase jitter of a PLL used in a serial link. The jitter analysis for a common reference clock architecture follows the example set forth in [4] and illustrates an important point about serial links, which is that if the reference clock is common to both ends of the link, the reference clock phase noise requirements may be greatly relaxed.

### 5.7.1 Serial Link Budgeting

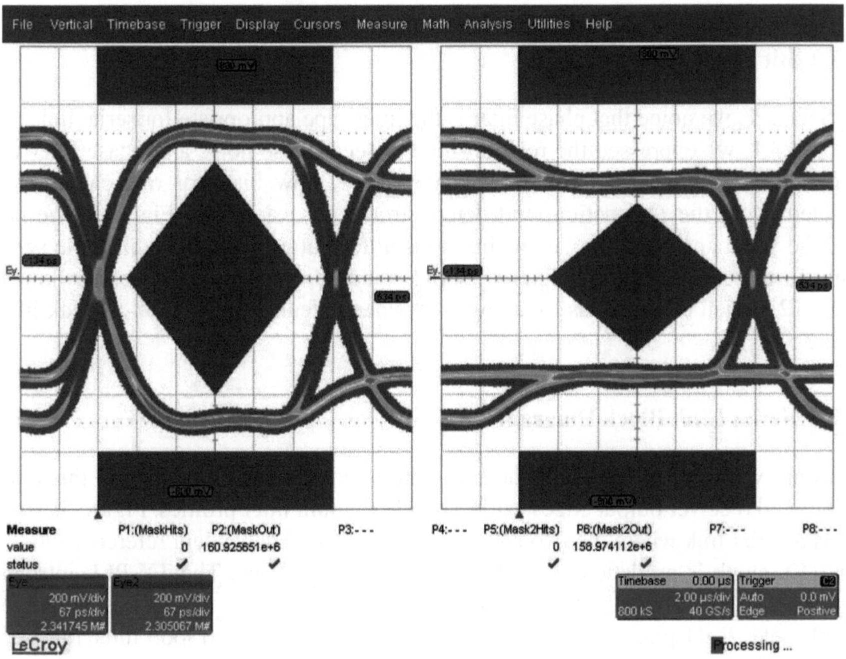


**Fig. 5.29.** Eye diagram

Serial link budgets are expressed in terms of an eye diagram. Figure 5.29 depicts the measured eye diagram of a 2.5 Gb/s transmitter (TX) with the PCI Express transmitter mask superimposed [21]. The eye mask has both amplitude requirements (the transmitter output must fall within the eye limits) and jitter requirements (the transmitter total output jitter must be less than the mask limits). We will focus on the jitter mask limit. In the case of the PCI Express specification, the transmitter applies some pre-filtering to overcome the finite bandwidth of the data channel. This technique, known as de-emphasis, results in two separate eye specifications. The first eye is for "transition" bits, which are bits that correspond to the data transitioning from zero to one to "don't care" or from one to zero to "don't care". The second eye is for "non-transition" bits, which are patterns of successive ones or zeros (one-one-"don't care", zero-zero-"don't care"). De-emphasis as a technique transmits larger amplitudes for transition bits and lower amplitudes for non-transition bits. One can think of this as placing emphasis on edge transitions in the data, which corresponds to a high-pass filter action. This high-pass filter action counters the low-pass filter nature of the channel.

The eye limits for jitter are expressed in UI, which represent a percentage of a "unit interval". A unit interval is simply the bit time. For a 2.5 Gb/s serial data link, 1 UI = 400 ps. The PCI Express mask depicted in Fig.5.29 requires that the total transmitter output signal not touch a mask that is 750 mUI wide. This means that total transmitter output jitter must be less than 1 UI–750 mUI = 250 mUI, which corresponds to 100 ps.

### 5.7.2 Bit Error Rate

In Sect. 5.2, we noted that phase jitter is the jitter type appropriate for serial links. In Sect. 5.3.1, we expressed the relationship between phase noise and phase jitter using Eq. (5.6). We also discussed, in Sect. 5.5.1, how bit error rate can be calculated using the $Q$ function, and demonstrated an example in Fig.5.19. We can use the same approach for calculating transmitter output jitter in an example serial link. In the case of the PCI Express serial standard, the link bit error rate limit is $1\mathrm{E}-12$, which corresponds to taking $14\sigma$ when calculating the peak-to-peak jitter value.[26]

### 5.7.3 Serial Link Block Diagram

For our example, we will make the assumption that the PLLs that control the transmitter and receiver output edges exhibit purely random jitter profiles. Figure 5.30 depicts a serial link where the two ends of the link have independent reference clocks. The TX clock is produced by a PLL internal to the TX IC. The TX PLL intrinsic jitter, $\sigma^2_{\Phi,int1}$ is summed with the extrinsic jitter introduced by the TX PLL reference clock, ref1. Ref1 phase noise is filtered by the TX PLL closed loop filter function, $G1(f)$. Given a phase noise profile for ref1, and given $G1(f)$, the total phase jitter at the output of the TX PLL can be calculated using the filter analysis techniques previously described in this chapter. The rms extrinsic jitter is then RSS summed with the intrinsic jitter as:

$$\sigma_{\Phi,TX} = \sqrt{\sigma^2_{\Phi,int1} + \sigma^2_{\Phi,ext1}} \tag{5.29}$$

where $\sigma^2_{\Phi,ext1}$ is the extrinsic jitter variance and the peak-to-peak output jitter for a BER $= 1\mathrm{E}-12$ is $14\sigma_{\Phi,TX}$.

While our goal is to calculate the total TX output jitter, and we have just enumerated a method to do so, it is worth considering the receiver (RX) jitter as well. We see

[26] Actually, $14.07\sigma$, but who's counting...

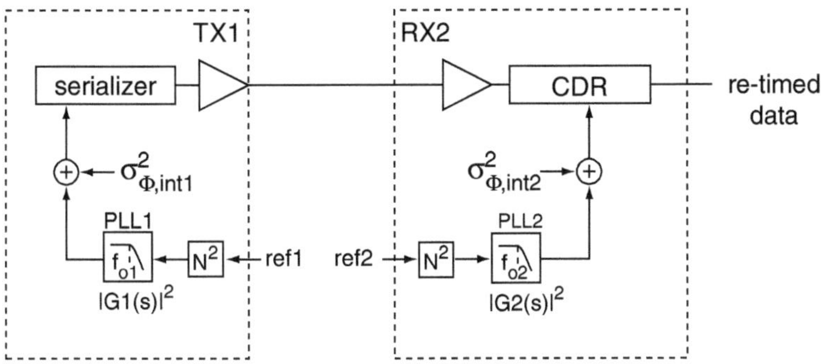


**Fig. 5.30.** Serial link with separate TX and RX reference clocks

that the receiver consists of a PLL and a clock and data recovery (CDR) block. This is a typical serial link design, where a single PLL is used to create a timing reference for the RX lane and then a CDR in the form of a DLL is used to recover the clock [5, 22]. This architecture allows a single RX PLL to be shared among several serial links, reducing overall power and area compared to an approach that uses one PLL per RX lane.

The RX jitter can be calculated in exactly the same manner as was the TX jitter. $G2(f)$, along with the phase noise profile of ref2 and $\sigma_{\Phi,int2}$, are used to calculate a total jitter for the RX link. If the sum of total TX and RX jitter components, along with any timing margin requirements in the RX circuitry and any distortion contributed by the channel (ISI, cross-talk, etc.) is less than 1 UI, the link operates successfully. For an accurate calculation, and under the assumption that the PLL jitters are totally random, we add the rms RX jitter and rms TX jitter in an RSS sense to get a total rms value for the link, multiply this value by 14 to get a peak-to-peak value, and then add this total to any deterministic components present elsewhere in the system.

In the architecture of Fig. 5.30, the fact that the reference clocks at either end of the link are independent means that their jitter is uncorrelated and so each reference clock phase noise adds in an RSS sense to total link jitter. A different approach, that can relax reference clock phase noise requirements, is depicted in Fig. 5.31.

In Fig. 5.31, the reference clock is common to both the TX and RX PLLs, and so this link structure is called a "common clock" architecture [4]. On the basis of early discussions in this chapter, we know that any reference clock phase noise appearing at frequencies inside the PLL bandwidth are not filtered by the PLL. This means that the jitter transfer from reference clock to PLL output for in-band noise is unity, which has an interesting consequence in the link architecture presented in Fig. 5.31. Since both PLL's are provided with the same reference clock, they both see the same in-band phase noise. To the extent that the PLL filter functions $G1(f)$ and $G2(f)$ match, both PLLs will track the in-band reference clock noise in the same way, and the net effect will be that the in-band portion of the reference clock phase noise will

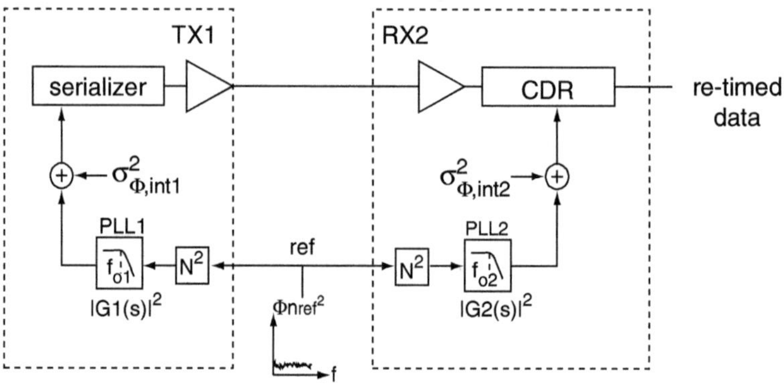


**Fig. 5.31.** Serial link with common reference clock

largely cancel and not contribute to the overall jitter budget! This behavior would, therefore, result in a relaxed specification for the reference clock phase noise.[27]

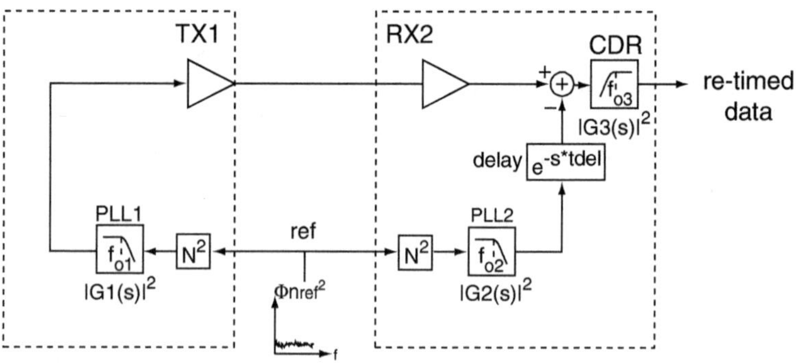


**Fig. 5.32.** Serial link model with common reference clock and skew

Figure 5.32 presents a modified noise model for the common clock serial link. The intrinsic jitter of each PLL has been removed, since we are now primarily interested in seeing how much reference clock phase noise impacts the jitter budget of the common reference clock link.[28] Additionally, we have introduced a delay term,

[27] This also explains why serial links that use spread spectrum clocking (SSC) to reduce electromagnetic interference (EMI) by modulating the reference clock frequency with a slow signal often require a "common clock" architecture. Since the reference clock spreading is in-band, if a common reference clock is used, both PLLs track the modulation identically, and there is minimal impact to system jitter performance.

[28] The intrinsic TX and RX jitter would, of course, be included in a total link jitter calculation.

$e^{-st_{del}}$, into the path for the RX PLL, where $t_{del}$ represents any skew between the distribution of the reference clock to PLL1 (TX) and PLL2 (RX). A skew amounts to a phase difference, which impacts the degree to which the in-band reference clock induced jitter tracked by PLL1 matches that of PLL2, and therefore impacts the degree to which the in-band reference clock jitter cancels across the link.

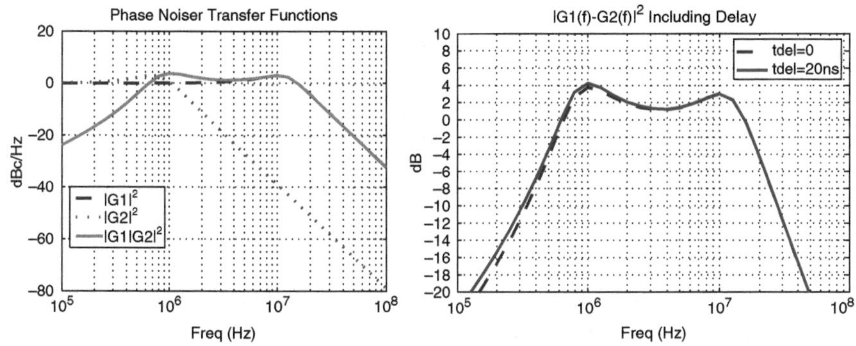


**Fig. 5.33.** Cancelation of reference clock noise in common clock architecture 1

Figure 5.33 shows results of jitter filter calculation for the model presented in Fig.5.32. $G1(f)$ has been set up as a second order PLL with a bandwidth of 22 MHz, while $G2(f)$ has been set up as a second order PLL with a bandwidth of 1.5 MHz. These values represent the extremes of the PLL bandwidth limits specified in the PCI Express 1.1 specification [21]. The PLL filter functions $G1(f)$ and $G2(f)$ are plotted, along with the difference between them, $|G1(f) - G2(f)|$. The skew between the two reference clock paths has been set to zero in the left plot. The right plot is a zoom-in of the difference function for two different values of skew. There are three points to note. First, low frequency noise will be tracked identically by both PLLs, and, therefore, cancels. Second, at higher frequencies, the difference function increases until it follows the shape of the higher of the two individual filter functions where the PLLs do not track one another. The net result is a band-pass filter function. Third, there is more suppression at low frequencies for a low value of skew than for a large value of skew. However, the impact of skew is relatively small, because, for in-band frequencies where the two individual PLL filter functions track one another, the skew value is a relatively small portion of the period of the frequencies in question. For example, the in-band cancelation has a low frequency $-3$ dB bandwidth of approximately 500 KHz, which corresponds to a noise period of $2\mu s$. A 20 ns skew, therefore, represents 1% of the noise period at this frequency, which is not enough to result in significant cancelation error.

To calculate the reference clock impact on the jitter budget, we follow the same procedure as for an individual PLL, with the difference that we process the reference clock phase noise profile with the difference function, $G1(f) - G2(f)$, rather than a single PLL filter function, $G(f)$.

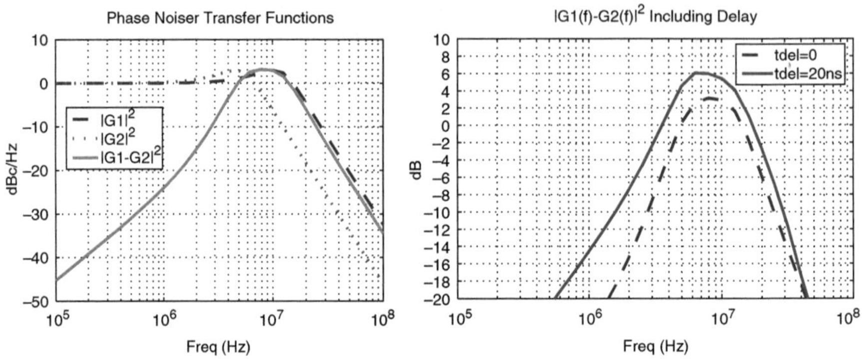


**Fig. 5.34.** Cancellation of reference clock noise in common clock architecture 2

The questions arises, "What happens if the individual PLL filter functions are closer together?" We examine the impact of such a situation in Fig. 5.34. PLL1 bandwidth is kept at 22 MHz, while PLL2 bandwidth is changed to 10 MHz. In this case, there is more low frequency reference clock phase noise cancellation, since PLL2 bandwidth has been increased. We also see that the peak value of the difference function is higher (6 dB) for the 20 ns skew case than for the previous example (4 dB), but this is counteracted by the fact that the bandwidth of the band-pass difference function has decreased. Whether this is an overall advantage or not depends on the shape of the reference clock phase noise profile, and whether or not it has any peaking (if it is provided by a PLL), and where that peaking occurs relative to the difference function peaking. We also see that since, with more closely matched bandwidths, while the difference filter function suppresses noise up to a much higher frequency, it varies more with changes in the reference clock distribution skew. The difference function $-3$ dB bandwidth is now between 3 MHz and 5 MHz, and so a 20 ns skew results in a 6% to 10% error in cancelation, an amount which is significant.

For the zero skew case, there appears to be a clear advantage of having the PLL bandwidths more closely matched, as there is a narrower band-pass bandwidth in addition to a lower overall difference filter function peak. The skew case performing worse makes sense, since any skew will produce phase differences between the two paths that grow worse with increasing noise frequency.

Overall, use of a common clock architecture can potentially reduce the contribution of reference clock phase noise in a serial link. The exact extent to which the common clock architecture is advantageous depends on the difference filter function, $G1(f) - G2(f)$, the total skew in the reference clock distribution, and the shape of the reference clock phase noise profile.

## 5.8 Delay Locked Loops (DLLs) and Jitter

In this chapter, we make note of the fact that there are other systems in VLSI circuits that process timing signals and generate, and/or filter, jitter. One such system is the delay locked loop (DLL). Since DLLs are discussed in detail in Chap. 6, we will only mention one point about the jitter behavior of DLLs in this section.

There is no VCO in a DLL, but rather a delay line is used to align the phase of a reference with the output signal. As discussed earlier in this chapter, a delay element will exhibit jitter caused by supply noise only during the edges when the noise event occurs. Another way of stating this is that there is no memory in a delay element. This behavior is in contrast to a VCO, where noise introduced to the VCO control voltage node will remain in the system up to the point at which the closed loop can respond [9].[29] DLLs, therefore, have inherent noise advantages over PLLs [22]. The tradeoff, of course, is that frequency multiplication with DLLs is more difficult. Multiplying DLLs (MDLLs) attempt a compromise by having the equivalent of a gated ring VCO, where the phase of the oscillator is "resynchronized" to a reference every M cycles, where M is the multiplication factor [23, 24]. In this way, the memory impact associated with a VCO is reduced by the phase reset action. The most significant challenge with an MDLL is that the oscillator synchronization operation results in a potentially significant deterministic jitter component at the reference frequency, which must then be addressed by additional design techniques [25].

## 5.9 Conclusion

In this chapter, we have defined three primary types of jitter and related these jitter types to one another through simple discrete time filter functions. We have presented the relationship between phase noise and jitter to establish a frequency domain based approach for PLL jitter filtering analysis. Through the use of the behavioral model proposed in [1] and frequency domain filter analysis techniques, we have examined how the various intrinsic and extrinsic noise sources present in a PLL are filtered by the PLL closed loop transfer function, $G(f)$. In cases where it is desired to calculate period jitter or cycle-to-cycle jitter, simple discrete time filter functions may be included in the analysis. Examples for a core clock microprocessor period jitter calculation and a serial-link PLL phase jitter calculation were used to demonstrate the effectiveness of frequency domain filter based analysis in determining overall system jitter budgeting.

## Acknowledgements

The author would like to thank Duke Xanthopoulos for providing the opportunity to contribute this chapter as well as thoughtful feedback. He would also like to thank his wife Kim for her encouragement and support throughout the writing process.

[29] In an open loop VCO, the impact of noise on the VCO control node stays around forever, since there is no feedback action to suppress it [9].

## References


[1] M. Perrott, M. Trott, and C. Sodini, A modeling approach for $\Sigma - \Delta$ fractional-N frequency synthesizers allowing straightforward noise analysis. *IEEE J. Solid-State Circuits*, 37(8), 1028–1038, 2002.
[2] JEDEC Standard JESD65B, 2003.
[3] PCI Express Jitter Modeling, Revision 1.0RD, July 2004.
[4] PCI Express Jitter and BER, Revision 1.0, February 2005.
[5] B. Razavi, *Design of Integrated Circuits for Optical Communications*. McGraw-Hill, New York, 2002.
[6] D. P. Bertsekas and J. N. Tsitsiklis, *Introduction to Probability*. Athena Scientific, Belmont, MA, 2002.
[7] S. Meninger and M. Perrott, Bandwidth extension of low noise fractional-N synthesizers. In: *Proc. Digest of Papers Radio Frequency integrated Circuits (RFIC) Symposium 2005 IEEE*, 12–14 June 2005, pp. 211–214.
[8] T. Weigandt, B. Kim, and P. Gray, Analysis of timing jitter in CMOS ring oscillators. In: *Proc. IEEE International Symposium on Circuits and Systems ISCAS '94*, 4, 27–30, 1994.
[9] J. McNeill, Jitter in ring oscillators. *IEEE J. Solid-State Circuits*, 32(6), 870–879, 1997.
[10] A. Hajimiri and T. Lee, A general theory of phase noise in electrical oscillators. *IEEE J. Solid-State Circuits*, 33(2), 179–194, 1998.
[11] B. Razavi, (ed.) *Monolithic Phase-Locked Loops and Clock Recovery Circuits: Theory and Design*. Wiley-IEEE Press, New York, 1996.
[12] B. Razavi, (ed.) *Phase-Locking in High-Performance Systems: From Devices to Architectures*. Wiley-IEEE Press, New York, 2003.
[13] H.-M. Chien, T.-H. Lin, B. Ibrahim, L. Zhang, M. Rofougaran, A. Rofougaran, and W. Kaiser, A 4 GHz fractional-N synthesizer for IEEE 802.11a. In: *Proc. Digest of Technical Papers VLSI Circuits 2004 Symposium*, 17–19 June 2004, 46–49.
[14] S. Meninger, Low phase noise, high bandwidth frequency synthesis techniques. Ph.D. dissertation, Massachusetts Institute of Technology, 2005.
[15] W. F. Egan, *Frequency Synthesis by Phase Lock*, 2nd edn. Wiley-Interscience, New York, 1999.
[16] M. Perrott, CppSim behavioral simulation package. [Online]. Available: http://www.cppsim.com
[17] M. Cassia, P. Shah, and E. Bruun, Analytical model and behavioral simulation approach for a $\Sigma\Delta$ fractional-N synthesizer employing a sample-hold element. *IEEE Trans. Circuits Syst. II*, 50(11), 850–859, 2003.
[18] Integrated phase noise. Silicon Laboratories AN256, 2006.
[19] Measuring Jitter in Digital Systems. Agilent Technologies Application Note 1448-1.
[20] M. Li, J. Wilstrup, R. Jessen, and D. Petrich, New jitter decomposition method and its applications. Wavecrest Corporation, MN, 1999.
[21] PCI Express Base Specification Revision 1.1, March 2005.

[22] S. Sidiropoulos and M. Horowitz, A semidigital dual delay-locked loop. *IEEE J. Solid-State Circuits*, 32(11), 1683–1692, 1997.
[23] R. Farjad-Rad, W. Dally, H.-T. Ng, R. Senthinathan, M.-J. Lee, R. Rathi, and J. Poulton, A low-power multiplying DLL for low-jitter multigigahertz clock generation in highly integrated digital chips. *IEEE J. Solid-State Circuits*, 37(12), 1804–1812, 2002.
[24] S. Ye, L. Jansson, and I. Galton, A multiple-crystal interface PLL with VCO realignment to reduce phase noise. *IEEE J. Solid-State Circuits*, 37(12), 1795–1803, 2002.
[25] B. Helal, M. Straayer, G.-Y. Wei, and M. Perrott, A highly digital MDLL-based clock multiplier that leverages a self-scrambling time-to-digital converter to achieve subpicosecond jitter performance. *IEEE J. Solid-State Circuits*, 43(4), 855–863, 2008.
[26] B. Razavi, RF Microelectronics. Prentice Hall, New Jersy, 1998.

# 6

# Digital Delay Lock Techniques

Thucydides Xanthopoulos

Cavium Networks

## 6.1 Introduction

Digital delay locked loops are highly prevalent in integrated systems. They are essentially delay lines under feedback control that can generate derived clocks based on an input reference. Applications include clock distribution, I/O interfaces, clock generation, and frequency multiplication. Digital delay locked loops also have time-to-digital conversion properties and can be used in monitoring and sensing applications.

While DLLs can be designed with digital-only methods, their design involves direct manipulation of clock signals. Therefore, additional techniques are involved as opposed to standard custom digital datapath design. This chapter presents an identification of all essential digital delay locked loop components and addresses relevant design aspects for each part. It concludes with global design issues and an overview of advanced applications.

## 6.2 What Constitutes a Digital Delay Locked Loop?

The digital delay locked loop (DLL henceforth) is a simple closed loop system that is capable of generating a clock signal that has a precise phase relationship with an input reference clock. Because of feedback, this phase relationship tracks across input frequencies, process, voltage, and temperature. The accuracy of the phase relationship between input and output clocks depends on DLL design parameters, process mismatch characteristics, and on deterministic noise sources such as independent supply noise and forms of coupling.

Digital DLLs can easily be unconditionally stable and are analyzed in the time domain. Because of their all digital nature, they can be ported across process nodes, can be simulated using fast digital simulators, and can be easily monitored and characterized in silicon.

Figure 6.1 shows a simplified DLL block diagram, which identifies the three main system components: The phase detector, the control block, and the delay line.

T. Xanthopoulos (ed.), *Clocking in Modern VLSI Systems*, Integrated Circuits and Systems,
DOI 10.1007/978-1-4419-0261-0_6, 

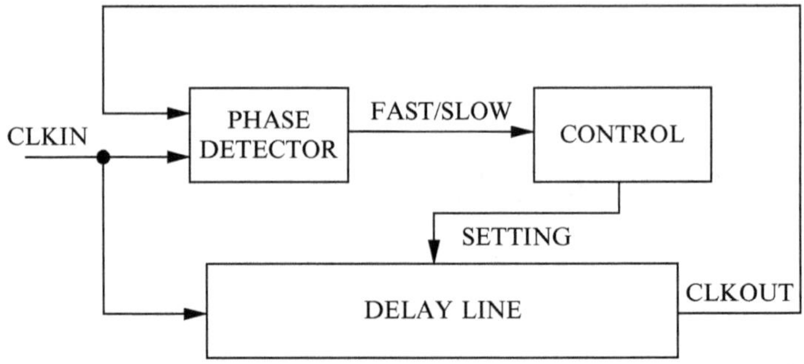


**Fig. 6.1.** Generalized DLL block diagram

The type of phase detector that will concern us in this chapter has a single bit output that only changes on the positive edge of the reference clock. It is commonly referred to as "bang–bang" in the literature. Such a phase detector can be thought of as a system that has the transfer function depicted in Fig.6.2. It compares the phase difference between two clocks and outputs a single logic value indicating which clock is ahead in time. It can be thought of as a flip-flop that has the reference clock (CLKIN) as its clock input and the controlled clock (CLKOUT) as the data input. When the flop evaluates to a logic 1, it means the controlled clock is faster than the reference. On the other hand, when the controlled clock is slower than the reference, the flop will evaluate to logic 0. This behavior can be guaranteed as long as the data input is faster than the clock input at least by the flop setup time ($T_s$ in Fig.6.2) or slower at least by the flop hold time ($T_h$ in Fig.6.2). If the timing between the two clocks falls within the gray area of Fig.6.2, the phase detector behavior is not defined and the output can be either a logic 0, a logic 1 or potentially a metastable value. The width of the gray rectangle ($T_s + T_h$) is the primary figure of merit of phase detectors and one of the main DLL design parameters. It is called the phase detector *sampling window* $d_{sw}$ or *dead zone*. It affects the phase locking accuracy of the entire system in addition to other important specifications. Phase detectors will be discussed in Sect. 6.4.

The most design intensive component of the DLL is the Digitally Controlled Delay Line (DCDL). A sample transfer function is shown in Fig.6.3. A DCDL is a combinational circuit that delays its input by an open loop value that typically has a monotonic relationship with the digital setting input. Such delay value is not precisely defined and is subject to process, voltage, and temperature conditions. A DCDL is primarily characterized by three design parameters: its minimum delay $D_{min}$ (delay value at setting 0), its maximum delay $D_{max}$ (delay value at the maximum setting $N-1$), and its resolution $d_r$ (incremental delay per setting). The dynamic range is defined as $D_{max} - D_{min}$ and is directly related to the capability of the overall DLL to track significant PVT variations or work with an extended range of input clock frequencies. The resolution $d_r$ affects the DLL accuracy along with $d_{sw}$. DCDL design will be discussed in Sect. 6.5.

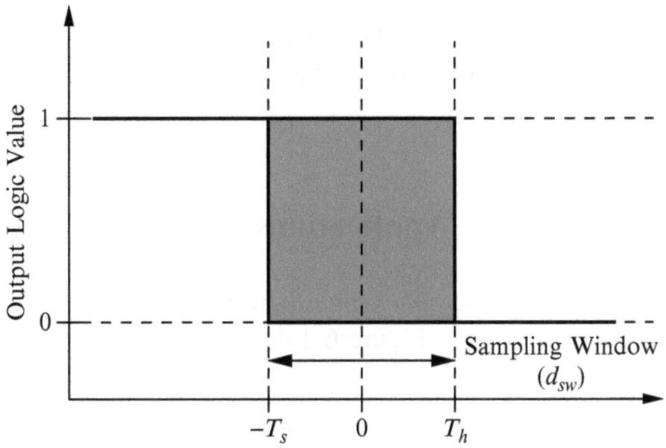


**Fig. 6.2.** Phase detector transfer function

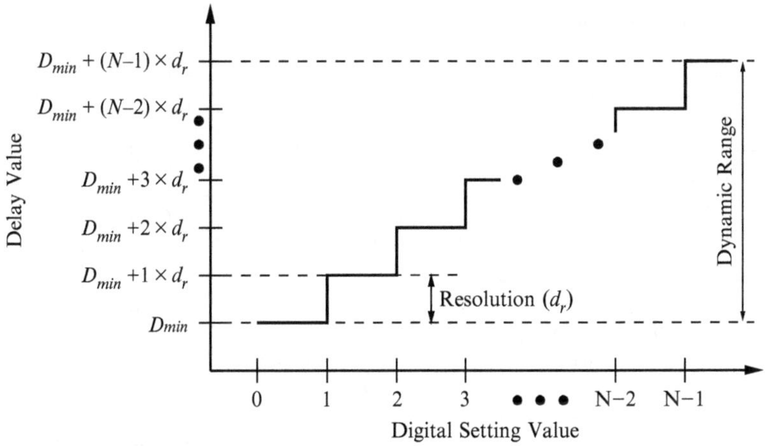


**Fig. 6.3.** Digitally controlled delay line transfer function

The final building block of Fig.6.1 is the control module. The control block increases and decreases the DCDL settings based on the output of the phase detector. In its simplest form, it is an up/down counter controlled by the phase detector. In its most general form, it is a finite state machine (FSM) that controls the DCDL settings based on the output of the phase detector and internal state. The inclusion of additional state information can support more complex behavior and extend the DLL capabilities. Control structures will be addressed in Sect. 6.6.

The combination of a phase detector, a delay line, and a control block produces a simple and useful feedback system that can find multiple applications in modern systems-on-a-chip. The DLL of Fig.6.1 will adjust its delay line until CLKIN and

CLKOUT are matched in phase. At this point, the DLL has *locked*, and the delay through its DCDL is one CLKIN period (or potentially an integral multiple of input clock periods).

## 6.3 An Overview of DLL Applications

The average modern microprocessor contains multiple digital delay locked loops embedded in various subsystems. Figure 6.4 demonstrates different uses of a basic DLL structure.

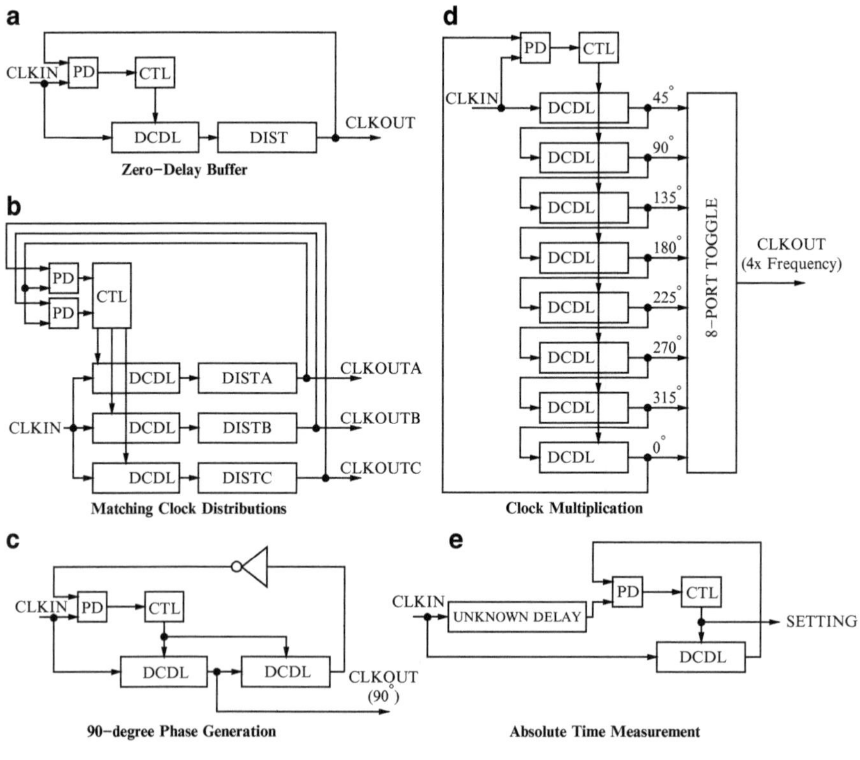


**Fig. 6.4.** DLL applications

The vast majority of DLL applications are related to clocking. Figure 6.4a demonstrates *zero-delay buffering*. Such a topology is well suited for synchronous I/O interfaces (e.g., PCI/PCI-X). A common clock (CLKIN) is being distributed to multiple bus end points. Each end point is buffering and distributing it to a number of flip-flops. A DLL in the loop ensures that the buffered clock version (CLKOUT) is phase-locked to the master interface clock (CLKIN).

Figure 6.4b shows an application where three separate clock domains are synchronized. Each domain has a separate clock distribution (DISTA, DISTB, DISTC) involving multiple buffer stages. Open loop matching is infeasible in the presence of PVT and random variations. A DLL-controlled delay line at the root of each distribution can guarantee phase matching among all three clocks. The DLL controller is more complex than the one shown in Fig.6.1, and it uses information from two separate phase detectors. First, CLKOUTA and CLKOUTB are phase locked. As soon as this happens, the controller locks CLKOUTC to CLKOUTA. This scheme is extensible to multiple clocks.

Quadrature clock generation is another application suitable for a DLL. Such a topology is shown in Fig.6.4c. When this DLL locks, the total delay through both DCDLs and the feedback path inverter is one half period of CLKIN (180°). Therefore, the delay through one DCDL is virtually a quarter period (90°) and CLKOUT is a quadrature clock.

The previous technique can be extended and used for constant factor clock multiplication. In Fig.6.4d, the delay through all eight DCDLs is one CLKIN period. The delay through a single DCDL is 45°. The outputs of the eight DCDLs are equally spaced phases spanning the entire CLKIN period. The toggle element can be thought of as a toggle flop with eight independent clock ports. Every positive edge of each of the eight phases can toggle the flop, thus, producing a CLKOUT that has a frequency equal to four times that of the input.

The final example (Fig.6.4e) is a deviation from strictly clocking applications. A DLL can be used for absolute measurements of unknown delays (time-to-digital conversion). First, a 2-point calibration is necessary. The DLL is placed in calibration mode (unknown delay is bypassed) and an input clock of a known period $T_0$ is fed into the CLKIN input. The DLL locks and the setting is recorded ($s_0$). The input clock period is changed to $T_1$, the DLL is allowed to lock and the setting is recorded again ($s_1$). We now have a system of two equations with two unknowns:

$$D_{\mathrm{min}} + s_0 \cdot d_{\mathrm{r}} = T_0, \tag{6.1}$$

$$D_{\mathrm{min}} + s_1 \cdot d_{\mathrm{r}} = T_1, \tag{6.2}$$

where $D_{\mathrm{min}}$ is the DCDL delay at the minimum delay setting and $d_{\mathrm{r}}$ is the DCDL resolution. We can solve the above system and obtain values for $D_{\mathrm{min}}$ and $d_{\mathrm{r}}$. The DLL now is placed out of calibration mode, and the unknown delay is multiplexed into the system. The DLL locks and the setting ($s_u$) is recorded. The absolute delay is, therefore, $D_{\mathrm{min}} + s_u \cdot d_{\mathrm{r}}$.

## 6.4 Phase Detectors

In a digital delay locked loop, the output of the phase detector is typically processed by a digital circuit such as an up/down counter or in the most general case an FSM controller. The most common phase detector for such an application is a specially-designed flip-flop that has a single bit output indicating leading or lagging feedback clock as described in Sect. 6.2 and shown in Fig.6.2. Traditionally, such a structure with a single-bit digital output is called a "bang–bang" phase detector. The main design goals for a flip-flop used as a phase detector are:

1. It must be a fully static design containing cross-coupled nodes exhibiting exponential voltage development with time (Sect. 6.4.1) to minimize time spent in a potential metastable state and prevent system failure.
2. It must have a small sampling window $d_{\mathrm{sw}}$, which is the sum of the underlying flop setup and hold times ($T_{\mathrm{s}} + T_{\mathrm{h}}$) to guarantee good phase matching between feedback clock and reference clock.
3. The setup and hold times must be well-balanced to avoid deterministic bias during phase detection and result in a significant systematic phase error between feedback clock and reference clock. A differential design can be desirable, but it is not a requirement.
4. The open loop gain of the cross coupled gates must be high to guarantee quick exit from a potential metastable condition (Sect. 6.4.1).
5. The capacitance of the cross-coupled nodes must be kept to a minimum given the other constraints in order to guarantee quick exit from metastability (Sect. 6.4.1). This will also help minimize setup requirements but may hurt hold time.
6. Unlike regular flop designs, a short clock-to-q delay is not a critical design requirement, since there is typically a full reference clock cycle available until the phase detector output needs to be setup and processed by the FSM controller. This path is unlikely to be critical. Moreover, unlike a proportional phase detector, balancing clock-to-q delays for a 0-to-1 vs. a 1-to-0 transition is not necessary. There is no phase information encoded in this delay for bang–bang operation.

The edge-triggered fully static flop designs of Chap. 3 such as the master-slave latch (flop) of Fig.3.4 or the sense-amp flip-flop of Fig.3.13 can be used as a bang–bang phase detector assuming that the design is tuned to meet the design goals outlined above. One design goal which won't be met is the balancing of the setup and hold times (and tracking across PVT) due to the assymetry in the clock and data path in any regular flip-flop. One way around this problem is illustrated in Fig.6.5 [1].

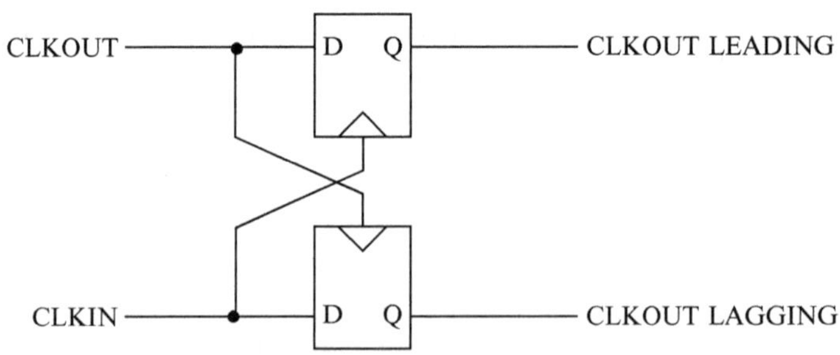


**Fig. 6.5.** Symmetric phase detector out of asymmetric flops

In a single-flop phase detector, the feedback clock leading decision depends on the setup time and the feedback clock lagging decision depends on the hold time. Asymmetry between these two properties can introduce a systematic phase error. In the coupled flops of Fig. 6.5, both decisions depend on the setup time of the flop and assuming identical flops, the systematic phase error is removed. The coupled flops of Fig. 6.5 are also a virtual ternary phase detector where a 11 or 00 state can be interpreted as a NOP (no operation) where no adjustment to the delay line takes place.

A flip-flop design that addresses all design goals outlined above is certainly possible. Figure 6.6 shows a non-traditional flop design used as a phase detector in [2]. CLKIN constitutes the clock input and CLKOUT is the data input. This edge-triggered flop is composed of three separate RS latches. Latch (A1,A2) is the master latch, latch (C1,C2) is the slave latch, and latch (B1,B2) is an auxiliary latch whose additional state is necessary for correct operation. Setup and hold timing behavior is entirely set by the master latch. The auxiliary latch is not in the signal path during sampling and the slave latch only affects clock-Q delay which is not relevant for bang–bang phase detector operation. Gates D1 and D2 are only present for load balancing along the CLKIN and CLKOUT paths.

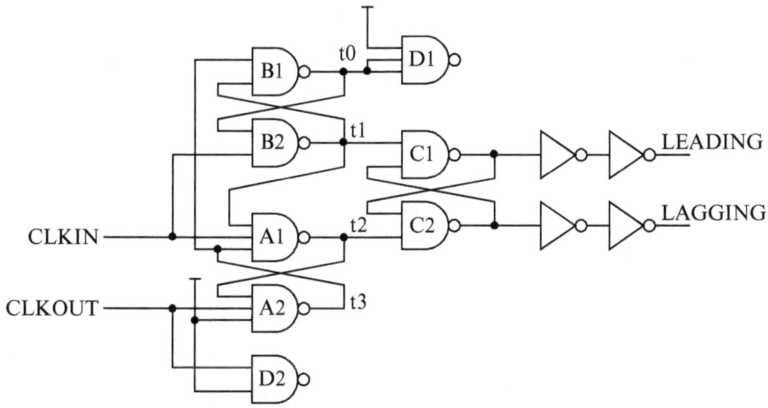


**Fig. 6.6.** Bang–bang phase detector. Reproduced with permission in a form similar to that in [2], ©1988 IEEE

Table 6.1 shows a truth table for the internal nodes $t0$–$t3$ of Fig. 6.6. We observe that when CLKIN $= 0$, the inputs $t1$ and $t2$ of the slave latch are at logic 1 which means that the phase detector outputs are holding. On the positive edge of CLKIN, the internal nodes will assume a value conditional on the state of CLKOUT around the sampling edge, and the slave latch will be updated accordingly. The full operation of this structure is illustrated in the state sequence Tables 6.2 and 6.3 which show the sequence of internal node and output values for CLKOUT lagging and leading, respectively.

**Table 6.1.** Truth table for bang–bang phase detector internal nodes

| CLKIN | CLKOUT | $t0$ | $t1$ | $t2$ | $t3$ |
|---|---|---|---|---|---|
| 0 | 0 | 0 | 1 | 1 | 1 |
| 0 | 1 | 1 | 1 | 1 | 0 |
| 1 | 0 | Hold | Hold | $\overline{t1}$ | 1 |
| 1 | 1 | $\overline{t3}$ | $t3$ | Hold | Hold |

**Table 6.2.** State sequence for CLKOUT lagging CLKIN

| CLKIN | CLKOUT | $t0$ | $t1$ | $t2$ | $t3$ | *Leading* | *Lagging* |
|---|---|---|---|---|---|---|---|
| 0 | 0 | 0 | 1 | 1 | 1 | Hold | Hold |
| 1 | 0 | 0 | 1 | 0 | 1 | 0 | 1 |
| 1 | 1 | 0 | 1 | 0 | 1 | 0 | 1 |
| 0 | 1 | 1 | 1 | 1 | 0 | 0 (Hold) | 1 (Hold) |
| 0 | 0 | 0 | 1 | 1 | 1 | 0 (Hold) | 1 (Hold) |

**Table 6.3.** State sequence for CLKOUT leading CLKIN

| CLKIN | CLKOUT | $t0$ | $t1$ | $t2$ | $t3$ | *Leading* | *Lagging* |
|---|---|---|---|---|---|---|---|
| 0 | 0 | 0 | 1 | 1 | 1 | Hold | Hold |
| 0 | 1 | 1 | 1 | 1 | 0 | Hold | Hold |
| 1 | 1 | 1 | 0 | 1 | 0 | 1 | 0 |
| 1 | 0 | 1 | 0 | 1 | 1 | 1 | 0 |
| 0 | 0 | 0 | 1 | 1 | 1 | 1 (Hold) | 0 (Hold) |

The sampling window of this phase detector is set entirely by the master latch (A1,A2). The setup and hold times of interest occur when the CLKIN and CLKOUT positive edges are virtually coincident. The setup time is set by the delay difference of NAND gate A2 on a 1-0 transition on node $t3$ and the delay of NAND gate A1 on a 1-0 transition on node $t2$:

$$T_{\mathrm{s}} = t_{d(\mathrm{A2})} - t_{d(\mathrm{A1})}. \tag{6.3}$$

Similarly, the hold time is:

$$T_{\mathrm{h}} = t_{d(\mathrm{A1})} - t_{d(\mathrm{A2})}. \tag{6.4}$$

Even though the setup and hold time of this phase detector can be minimized by removing all static bias from physical design, the true parameters should be determined statistically through Monte Carlo simulations. A positive setup and hold time must be determined that will guarantee correct decision with arbitrarily high probability in the presence of random process variations.

The phase detector of Fig. 6.6 can satisfy all design requirements including being fully static and having nominally small and equal setup and hold times. Moreover, it is a true single phase design which requires no inversion on the clock or data inputs, thus, minimizing setup and hold requirements. The author has been hard pressed to find a better overall design with good portability across process nodes.

### 6.4.1 Metastability

A bang–bang phase detector will produce the correct lead/lag decision if the controlled clock input falls outside its sampling window ($d_{\mathrm{sw}}$). A DLL in the locked state though will cause frequent controlled clock edges to fall within $d_{\mathrm{sw}}$. A naive approach to this issue would be to assume that in this situation a wrong decision coupled with a small DCDL resolution $d_{\mathrm{r}}$ will not cause a DLL failure but rather a small increase in the DLL phase error which can be accommodated by the application.

A fourteenth century French philosopher Jean Buridan postulated that a donkey located at equal distances between two bales of hay should theoretically starve to death because it will be equally attracted to both [3, 4]. A more precise articulation of this principle in the words of Lamport [4] is as follows:

> *Buridan's Principle:* A discrete decision based upon an input having a continuous range of values cannot be made within a bounded length of time.

The continuous variable in the Buridan example is the position of the donkey along the axis connecting the two bales as a function of time and initial position. The discrete decision is which bale of hay to consume. In the case of the phase detector, the continuous variable is the voltage at the cross-coupled nodes of the flip-flop state element as a function of time and initial voltage. In this case, the initial voltage refers to the voltage established at the cross-coupled nodes right after sampling the flop data input. The discrete decision is whether a particular node will converge to logic 0 or logic 1.

This is of course the well-known synchronization failure problem [3]. No one can build a phase detector that can guarantee a valid logic output in bounded time if the controlled clock falls within the $d_{\mathrm{sw}}$ established by the sampling clock. In this case, the output can be at an undefined logic level (metastable state) for an arbitrarily long time, and eventually this undefined level may be interpreted by two separate logic receivers as different logic values. This can cause catastrophic DLL failure because it can drive the control automaton into a wrong or undefined state. Metastability in static flops has been extensively studied and observed in practice [5].

Although it is not possible to build a phase detector which will never experience metastability for unbounded time, it is possible to design a phase detector that minimizes the probability of system failure. We will study this problem by deriving an expression for the voltage in the cross coupled gates inside the phase detector as a function of time and initial conditions. Then, we will assign a probability measure to two events: Entering a metastable condition at time $t = 0$ and still being in a metastable condition at time $t = t_d$. Finally, we will derive an expression for the mean-time-between-failures (MTBF) which provides an indication of how often we can expect catastrophic synchronization failures in a DLL.

We develop a first order metastability model based on the analysis of Veendrick [6]. We begin by assuming a linear voltage transfer function for a CMOS inverter (for normalized supply voltage):

$$V_o = -A(V_i - V_{sw}) + 0.5, \tag{6.5}$$

where $V_i$ and $V_o$ are the voltages at the input and output of the inverter, respectively, $V_{sw}$ is the switching threshold (defined as the value of $V_i$ for which $V_o = 0.5$) and $A$ is a large positive number denoting the inverter gain. There is an underlying assumption that $V_o$ is further processed by a nonlinear limiter which clamps $V_o$ to 1 for all $V_o > 1$ and also clamps $V_o$ to 0 for all $V_o < 0$. All voltages are normalized $(V_i, V_o, V_{sw})$ and assume values between 0 and 1. Figure 6.7 plots Eq. (6.5) for $A = 10$, $V_{sw} = 0.5$. Figure 6.8 plots Eq. (6.5) for various $A$ and $V_{sw}$ parameter values.

Cross-coupling two inverters results in three equilibrium positions (Fig.6.9): Two stable positions (a small voltage perturbation will result in the system canceling it out and remaining in the same state) that signify the two memory states, and an unstable equilibrium position (a small voltage perturbation will result in the system leaving this state and assuming one of the two stable states) signifying the metastable state. The metastable voltage $V_m$ that the two system nodes converge is a function of the inverter gain $A$ and switching threshold $V_{sw}$ and can be computed easily by setting $V_i = V_o = V_m$ in Eq. (6.5) (Fig.6.10):

$$V_m = \frac{2AV_{sw} + 1}{2(A+1)}. \tag{6.6}$$

We introduce transient behavior in the cross-coupled inverter model by adding an *RC* output stage to the ideal gain elements along the lines of [6] as shown in Fig.6.11. We omit the limiters and make the assumption that this model is only valid for inverter input voltages that fall in the linear region of the transfer function $V_{sw} - 0.5/A \leq V_i \leq V_{sw} + 0.5/A$ (Fig.6.7).

We assume that the sampling switch is ideal, and that clocking this latch consists of establishing the following initial conditions at $t = 0$ (please note the change of variable names $V_1(t)$ and $V_2(t)$ in Fig.6.11):

$$V_2(0) = V_0, \tag{6.7}$$

$$V_1(0) = -A(V_0 - V_{sw}) + 0.5, \tag{6.8}$$

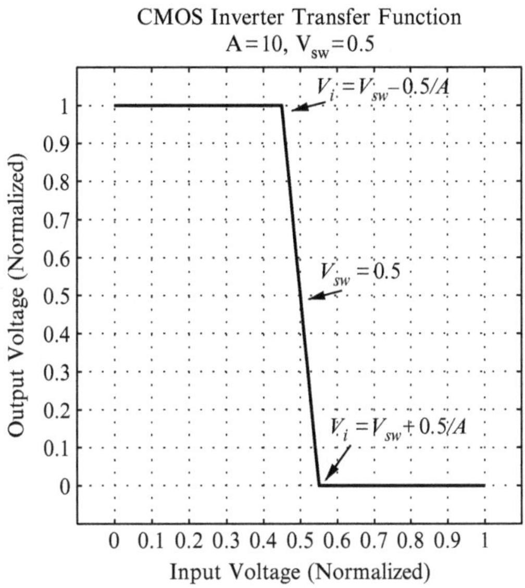


**Fig. 6.7.** CMOS inverter voltage transfer function

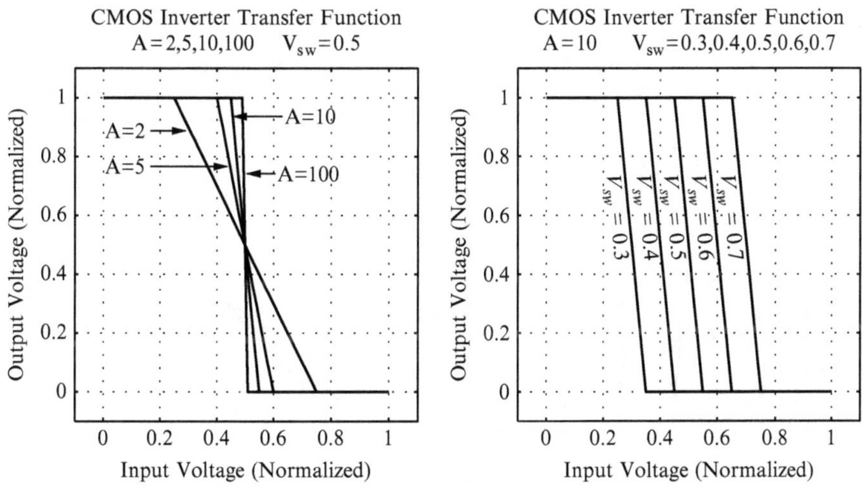


**Fig. 6.8.** CMOS inverter voltage transfer function parametrization

We want to derive an analytical expression for $V_1(t)$. We begin by enforcing KCL at $V_1(t)$ and $V_2(t)$, respectively:

$$\frac{\mathrm{d}V_1(t)}{\mathrm{d}t} + \frac{1}{RC}V_1(t) + \frac{A}{RC}V_2(t) - \frac{AV_{\mathrm{sw}} + 0.5}{RC} = 0, \tag{6.9}$$

$$\frac{\mathrm{d}V_2(t)}{\mathrm{d}t} + \frac{1}{RC}V_2(t) + \frac{A}{RC}V_1(t) - \frac{AV_{\mathrm{sw}} + 0.5}{RC} = 0. \tag{6.10}$$

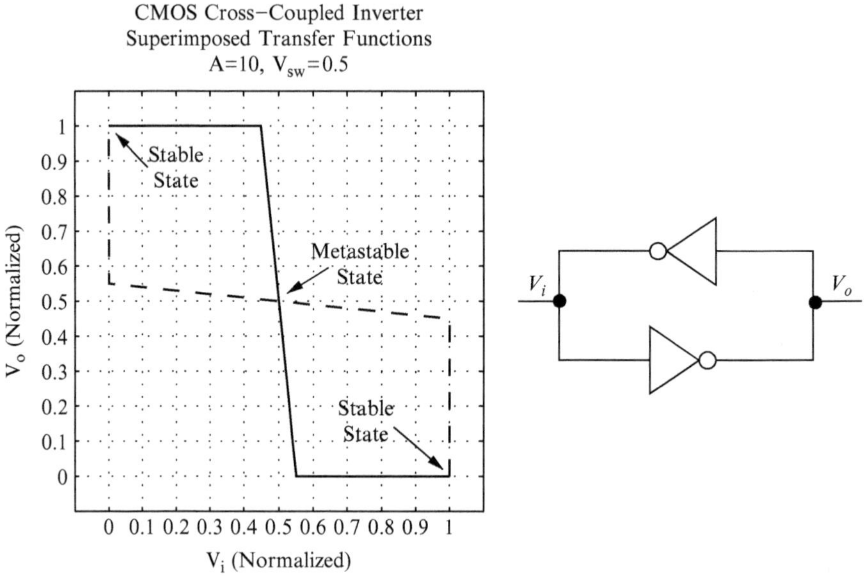


**Fig. 6.9.** Metastable state in CMOS cross-coupled inverters

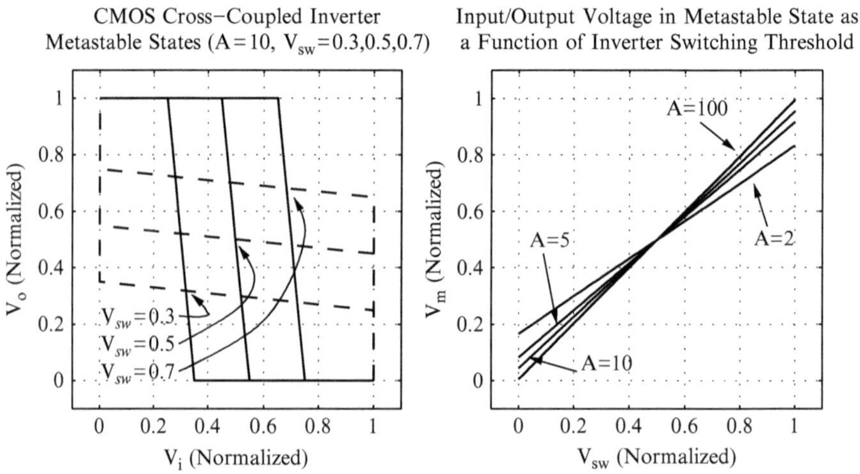


**Fig. 6.10.** Voltage in metastable state

In order to simplify the analysis, we introduce two new functions:

$$V_d(t) = V_1(t) - V_2(t),$$
$$V_s(t) = V_1(t) + V_2(t).$$

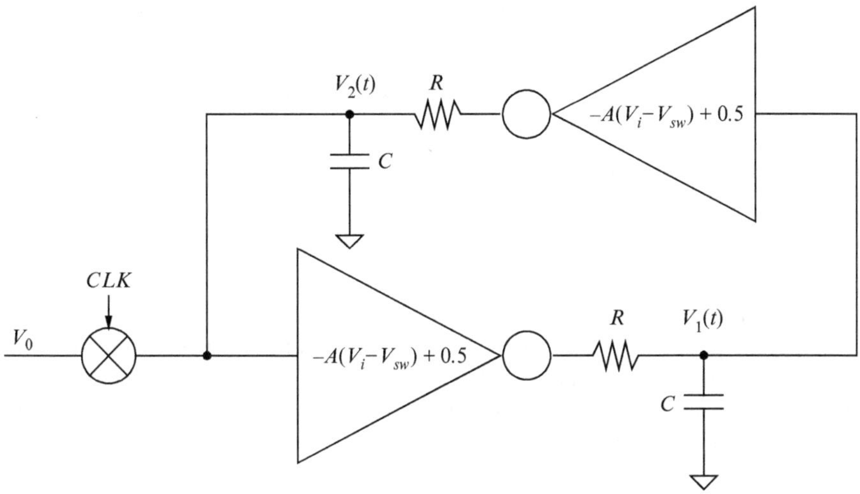


**Fig. 6.11.** Small signal model of cross-coupled inverters. Reproduced with permission in a form similar to that in [6], ©1980 IEEE

Subtracting (6.10) from (6.9) yields:

$$\frac{\mathrm{d}V_d(t)}{\mathrm{d}t} - \frac{A-1}{RC}V_d(t) = 0. \tag{6.11}$$

Adding (6.10) to (6.9) yields:

$$\frac{\mathrm{d}V_s(t)}{\mathrm{d}t} + \frac{A+1}{RC}V_s(t) - \frac{2AV_{\mathrm{sw}}+1}{RC} = 0. \tag{6.12}$$

Initial conditions for $V_{\mathrm{d}}(t)$ and $V_{\mathrm{s}}(t)$ have been established during sampling:

$$V_{\mathrm{d}}(0) = V_1(0) - V_2(0),$$

$$V_{\mathrm{s}}(0) = V_1(0) + V_2(0).$$

The solution of (6.11) is

$$V_d(t) = [V_1(0) - V_2(0)]\mathrm{e}^{\frac{A-1}{RC}t}. \tag{6.13}$$

The solution of (6.12) is

$$V_s(t) = \left(V_1(0) + V_2(0) - \frac{2AV_{\mathrm{sw}}+1}{A+1}\right)\mathrm{e}^{-\frac{A+1}{RC}t} + \frac{2AV_{\mathrm{sw}}+1}{A+1}. \tag{6.14}$$

Expression (6.14) can be simplified by substituting the expression for the metastable voltage (6.6):

$$V_s(t) = [V_1(0) + V_2(0) - 2V_\mathrm{m}]\,\mathrm{e}^{-\frac{A+1}{RC}t} + V_\mathrm{m}. \tag{6.15}$$

Reverting back to the original variable $V_1(t)$, we have

$$V_1(t) = \left(\frac{V_1(0)+V_2(0)}{2} - V_\mathrm{m}\right)\mathrm{e}^{-\frac{A+1}{RC}t} + \frac{V_1(0)-V_2(0)}{2}\mathrm{e}^{\frac{A-1}{RC}t} + V_\mathrm{m}. \tag{6.16}$$

Equation (6.16) can be simplified and become more meaningful if we make the following observation: The decaying exponential quickly vanishes with increasing $t$ and the expression is dominated by the increasing exponential. We can approximate the expression above by only keeping the positive exponential with its coefficient adjusted for the initial condition:

$$\boxed{V_1(t) = (V_1(0) - V_\mathrm{m})\mathrm{e}^{\frac{A-1}{RC}t} + V_\mathrm{m}.} \tag{6.17}$$

Equation (6.17) is fundamental in the description of cross-coupled circuits and states that a perturbation from the unstable equilibrium position $V_\mathrm{m}$ results in a time-exponential path toward a stable state. We note that Eq. (6.17) has a consistent solution for $V_1(t=0)$ and also that for $V_1(0) = V_\mathrm{m}$, we have $V_1(t) = V_\mathrm{m}$ for all $t$ which is another way of saying that a metastable state may persist indefinitely. We also note that Eq. (6.17) is identical to Eq. (1-18) in [3] although the derivation is very different. Mead and Conway derive the cross-coupled inverter node equation starting from basic circuit relationships in a depletion-loaded NMOS cross-coupled pair.

It is important to understand that Eq. (6.17) is only valid for node voltages in close proximity to the metastable voltage $V_\mathrm{m}$. Both initial conditions $V_1(0)$ and $V_2(0)$ must lie in the shaded regions of Fig. 6.12 and be related through the gain curve. Unless this is true, the assumptions made during the derivation of (6.17) (both inverters are in their linear region) are no longer operative. We can still make the assertion that Eq. (6.17) fully describes the metastable state because reaching the voltage limit of the above equation signifies the exit from metastability: At that point, the node voltage has reached a value outside the linear gain limits and any receiver with a similar gain curve will interpret it as a well-defined logic state. Figure 6.13 plots (6.17) for various initial conditions of $V_1(0) - V_\mathrm{m}$ and exhibits the exponential nature of the trajectory leading the node voltage out of the metastable state.

Metastability is a stochastic phenomenon and must be studied with probabilistic tools. Equation (6.17) states that there exists an initial condition $V_1(0) = V_\mathrm{m}$ which will cause the phase detector to lie in the metastable state indefinitely. At the same time though, if $V_1(t)$ is modeled as a continuous random variable, then the probability of $V_1(t)$ assuming a discrete value $V_\mathrm{m}$ vanishes. It is not possible to design a perfect phase detector, one that is guaranteed not to lie in the metastable state indefinitely.

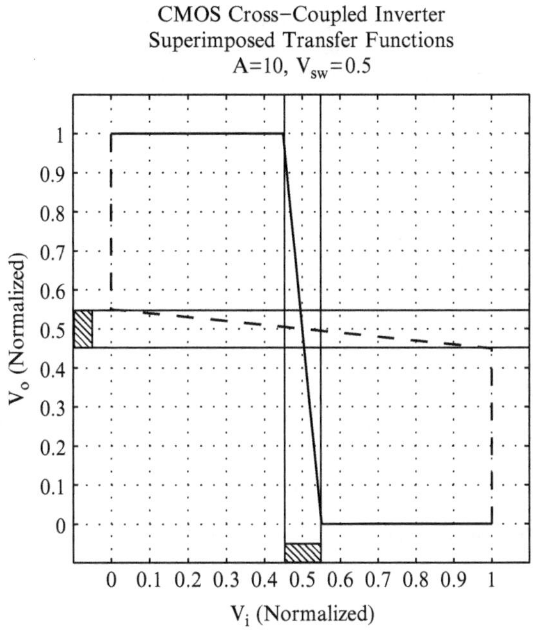


**Fig. 6.12.** Valid node voltage region for Eq. (6.17)

It is possible though to design a phase detector that has an arbitrarily small conditional probability of being in the metastable state at a given time $t_{\mathrm{d}}$ assuming it was in the metastable state at time $t = 0$.

Before proceeding to define the problem in stochastic terms, let us simplify Eq. (6.17) by referring all voltages to $V_{\mathrm{m}}$ instead of ground [6]:

$$V(t) = V_1 \mathrm{e}^{\frac{t}{\tau_m}}, \tag{6.18}$$

$$V(t) = V_1(t) - V_{\mathrm{m}}, \tag{6.19}$$

$$V_1 = V_1(0) - V_{\mathrm{m}}, \tag{6.20}$$

$$\tau_m = \frac{RC}{A-1}. \tag{6.21}$$

Furthermore, we need to state two assumptions:

- We consider a phase detector to be metastable when its output voltage $V_1(t)$ is within $\Delta V$ of $V_{\mathrm{m}}$:

$$|V(t)| < \Delta V, \tag{6.22}$$

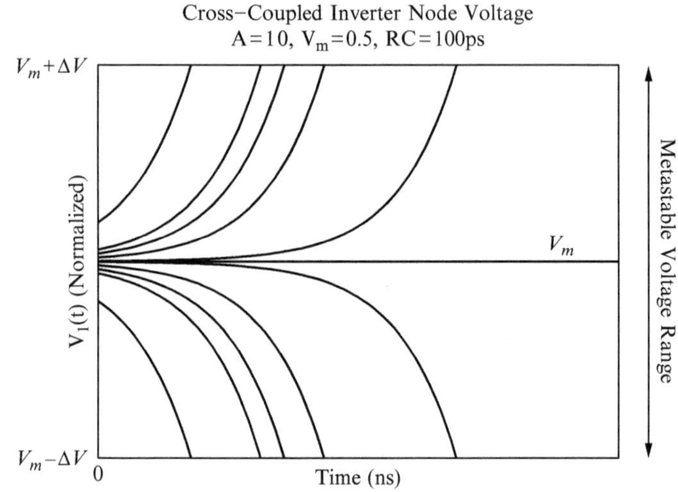


**Fig. 6.13.** Exponential trajectory towards a stable state

where $\Delta V$ is a phase detector/synchronizer parameter and is the maximum voltage deviation from $V_m$ which will still guarantee that both cross-coupled nodes are in the linear gain region of the two cross-coupled CMOS gates.

- The initial condition $V_1$ is a random variable uniformly distributed between $-\Delta V$ and $\Delta V$. By definition then, the system is metastable at $t = 0$.

We wish to quantify the probability that the phase detector is still metastable at $t = t_d$ given that it is metastable at $t = 0$:

$$\Pr(|V(t_d)| < \Delta V) = \Pr\left(\left|V_1 \mathrm{e}^{\frac{t_d}{\tau_m}}\right| < \Delta V\right). \tag{6.23}$$

Multiplying both sides of the event inequality in (6.23) with $\mathrm{e}^{-\frac{t_d}{\tau_m}}$ yields:

$$\Pr(|V_1| < \Delta V \mathrm{e}^{-\frac{t_d}{\tau_m}}) = \mathrm{e}^{-\frac{t_d}{\tau_m}}. \tag{6.24}$$

The last equation stems from the fact that $V_1$ is uniformly distributed between $-\Delta V$ and $\Delta V$. We can arrive at the same result using a different approach [3]: Let us assume that the exit from the metastable state can be modeled as the first arrival of a Poisson process with rate $\lambda$. Modeling it as a Poisson arrival implies that the probability of leaving the metastable state within a very small time interval $\Delta t$ is proportional to the duration of the interval and equal to $\lambda \Delta t$. This makes physical sense since in general, Poisson processes are heavily used to model simple stochastic phenomena in continuous time. The probability density function of the time until the first Poisson arrival ($l_1$) is given by the following Eq. ([7]):

$$f_{l_1}(t) = \lambda e^{-\lambda t}. \tag{6.25}$$

The probability that the phase detector is still metastable at time $t_d$ is given by the following expression:

$$\Pr(l_1 > t_d) = \int_{t_d}^{\infty} \lambda e^{-\lambda t} dt = e^{-\lambda t_d}. \tag{6.26}$$

For $\lambda = 1/\tau_m$, expression (6.26) is in agreement with (6.24).

Equation (6.24) describes a conditional probability where the conditioning event is $|V_1| < \Delta V$ (phase detector is metastable at time $t = 0$). We will now quantify the probability of the conditioning event and determine the overall unconditional probability of phase detector failure. Let us assume that in the DLL locked state, the feedback clock edges are uniformly distributed between $-d_r$ and $d_r$ with respect to the sampling reference clock edge (where $d_r$ is the DCDL resolution). This is a reasonable assumption to make assuming that the digital control module will be designed with a small $\pm 1$ LSB limit cycle and various noise processes will add uncertainty to the feedback clock edges. We further assume a feedback clock slew rate of $L$ in V/s. Figure 6.14 shows graphically that such a distribution of feedback edges will cause a uniform distribution of sampled voltages with a range equal to $2Ld_r$. The probability of the conditioning event is, therefore, $\Delta V/(Ld_r)$, and the overall phase detector unconditional failure probability at $t_d$ is,

$$\boxed{\Pr(\text{Metastable at } t_d) = \frac{\Delta V}{Ld_r} e^{-\frac{t_d}{\tau_m}}.} \tag{6.27}$$

So far, we have made enough assumptions and simplified our model sufficiently to be able to derive a simple formula for $\Delta V$, the sampled voltage deviation from $V_m$ which will place a cross-coupled structure in a metastable state. Without loss of generality, we can simplify the analysis by assuming $V_m = V_{sw} = 0.5$ in Eq. (6.5) (all voltages are normalized with respect to the nominal supply voltage). The range of input voltages which will cause a cross-coupled structure to become metastable should be determined by requiring that the result of applying the inverter transfer function (6.5) to such input voltage should yield a voltage which is still in the linear range of the same transfer function (shown in Fig. 6.7). If this is not the case, then a logic gate with similar gain will interpret the cross-coupled output as a discrete logic value. Moreover, the second cross-coupled gate will also interpret such an output as

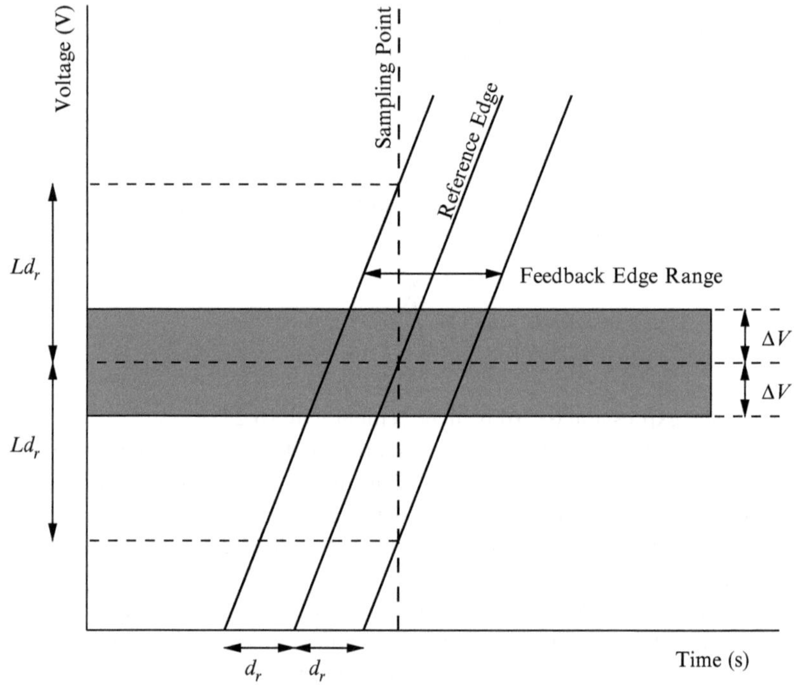


**Fig. 6.14.** Determining the probability of entering metastability

a discrete logic value, and the latch will converge to a non-metastable state as soon as the sample switch opens up. It is straightforward to calculate that the above condition will be satisfied for:

$$\Delta V = \frac{1}{2A^2}. \tag{6.28}$$

Let us now briefly focus on $t_d$, the allowed metastability resolution time. For a digital DLL with a bang–bang phase detector, $t_d$ is at least equal to the reference clock period $T$. It is also common to add additional synchronization stages and effectively extend the resolution time to $nT$. Updating Eq. (6.27) with our latest observations and also expanding the expression for $\tau_m$ we have,

$$\boxed{\mathrm{Pr(PD\ Failure)} = \frac{1}{2nA^2Ld_\mathrm{r}}\mathrm{e}^{-\frac{A-1}{r_oC}nT}}, \tag{6.29}$$

where

$d_\mathrm{r}$ DCDL resolution

$L$ Feedback clock edge rate

$r_\mathrm{o}$ Small signal output resistance of gate around $V_\mathrm{m}$.
Should be small signal saturation resistance of pullup device in parallel with pulldown device

*A* Small signal gate gain. Should be $g_m r_o$ where $g_m$ is pullup transconductance in saturation in parallel with pulldown transconductance.
*C* Cross-coupled node capacitance
*T* Reference clock period
*n* Number of synchronization stages including the phase detector

An additional scaling factor of $1/n$ has been included to account for the fact that a digital controller typically implements a $1/n$ correction issue rate in order to minimize limit cycle amplitude and guarantee stability (Sect. 6.6). In such a case, the controller only looks at the output of the phase detector once every $n$ cycles and the probability of entering a metastable state must be scaled accordingly. Equation (6.29) quantifies the risk of DLL failure for $T$ seconds of operation. An alternative way of expressing this risk is the mean-time-between-failures (MTBF):

$$\text{MTBF} = T\Pr(\text{PD Failure}). \tag{6.30}$$

The MTBF figure of merit is typically expressed in years and signifies the average duration of error-free operation. In digital system applications, an MTBF in excess of 10–100 years is typically desired.

### An Example of Phase Detector Failure Calculation

It is important to realize that Eq. (6.29) is based on a number of assumptions and approximations such as the inverter piecewise linear transfer function of Eq. (6.5). Nevertheless, it captures the right dependencies on design parameters and can be used as a phase detector design guideline. The MTBF value predicted by (6.30) should have substantial margin of a few orders of magnitude to guarantee correct operation even in the presence of modeling inaccuracies that can affect the final result in an exponential nature.

Although it is certainly possible to assign circuit parameters to all contributing factors of Eq. (6.29) [6], it is probably much easier to estimate MTBF using spice small signal analysis around $V_m$. Figure 6.15 shows the setup of a spice AC analysis for a 45 nm cross-coupled inverter using 45 nm bulk spice predictive models [8–10]. A self-biased inverter is used to compute the metastable voltage ($V_{bias}$ in Fig. 6.15) which is then used as the bias point through an ideal buffer for a cross-coupled pair of identical sizing. An AC source is cascaded in series with the bias point to calculate gain, output resistance, and node capacitance.

For this example, Table 6.4 shows the MTBF calculation under certain assumptions for $L$, $n$, $T$, and $d_r$. The MTBF calculation is very sensitive to the metastability time constant ($\tau_m$), and the parameters that affect it because of the exponential dependence. As an example, if the node capacitance gets doubled from the value in Table 6.4 due to the necessary addition of a receiving gate, MTBF becomes $4.7624 \times 10^9$ years. If it triples due to poor phase detector design, MTBF will become 24.58 years.

**Table 6.4.** Example MTBF calculation for the circuit of Fig. 6.15

| Parameter | Value | Units |
|---|---|---|
| Clock period ($T$) | $1 \times 10^{-9}$ | s |
| Slew rate ($L$) | $1 \times 10^{10}$ | V/s |
| DLL subsampling factor ($n$) | 1 | – |
| DLL resolution ($d_r$) | $50 \times 10^{-12}$ | s |
| Gain ($A$) | 4.5456 | – |
| Output resistance ($r_o$) | $9.3079 \times 10^{3}$ | $\Omega$ |
| Node capacitance ($C$) | $3.3271 \times 10^{-15}$ | F |
| Probability of entering metastability: $\frac{1}{2nA^2Ld_r}$ | 0.0484 | – |
| Metastability time constant: $\tau_m = \frac{r_oC}{A-1}$ | $8.7343 \times 10^{-12}$ | s |
| Conditional probability of being metastable at $nT$: $e^{-\frac{nT}{\tau_m}}$ | $1.8928 \times 10^{-50}$ | – |
| Unconditional synchronization failure probability: $\frac{1}{2nA^2Ld_r}e^{-\frac{nT}{\tau_m}}$ | $9.16 \times 10^{-52}$ | – |
| MTBF | $3.4615 \times 10^{34}$ | years |

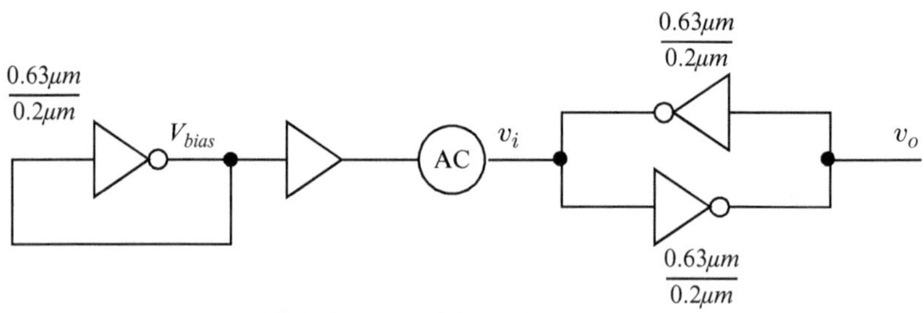

All devices are minimum length (45nm)

**Fig. 6.15.** Determining MTBF with a spice simulation

## 6.5 DCDL Design

There is a very broad range of delay line design options limited only by designer imagination. This section provides a structured overview of various DCDL design approaches. The structure based on DCDL characteristics adopted for this section does not follow a widely accepted classification in the field but is merely done for ease of presentation. Moreover, this section is by no means an all-inclusive exposition of all possible design options. It can be considered a stratified sampling of the design space that presents various alternatives based on the main DCDL characteristics of Sect. 6.2 ($D_{min}, D_{max}, d_r$). Additional characteristics (such as input capacitance, linearity, and potential for synchronous vs. asynchronous setting change) will also be identified.

For ease of presentation, we divide DCDLs into two main categories based on $d_r$: Gate-delay DCDLs (coarse) and Subgate-delay DCDLs (fine).

### 6.5.1 Gate-Delay DCDLs

Gate-delay DCDLs can be constructed by cascading standard CMOS logic gates to form a delay chain with intermediate outputs routed to a high-fanin multiplexing structure. Alternatively, the chain length can be modulated using low-fanin distributed multiplexing structures as part of the delay cell. They exhibit relatively small $D_{min}$, arbitrarily high $D_{max}$ but relatively coarse $d_r$ by design.

A very popular coarse delay line is shown in Fig.6.16 [11, 12]. This particular implementation has four hierarchical delay stages between input $A$ and output $Y$. It is controlled by a bidirectional shift register with one-hot encoding ($Q[3:0]$). Resolution $d_r$ is $2T_G$, where $T_G$ is the average CMOS gate delay. NAND gates labeled with the letter B form a distributed multiplexer that controls the entry point of the clock ($A$ input) in this cascaded structure. One important observation is that this design presents substantial load to the clock especially for a large number of delay stages since all delay element inputs are shorted. Clock buffering may be necessary which will increase $D_{min}$ to a value larger than $2T_G$. An arbitrarily high number of stages can be cascaded to increase dynamic range provided that input $A$ is sufficiently buffered. $D_{max}$ is $2NT_G$, where $N$ is the number of delay stages.

The folded design of Fig.6.17 eliminates the heavily loaded $A$ input by modulating the delay line length in a telescopic fashion. This design is also controlled by a bidirectional one-hot shift register. For $Q[3:0] = 0001$ the signal path is Cell $0.A +$ Cell $0.B$. For $Q[3:0] = 0010$ the path is Cell $0.C +$ Cell $1.A +$ Cell $1.B +$ Cell $0.B$. The length modulation of this structure can be compared to the sliding action of a trombone. The wrap-around connection between $IN1$ and $OUT1$ of Delay Cell 3 is not part of the delayed signal path but is necessary to establish the propagating condition for the B NAND gate associated with the delay cell that constitutes the final telescope link (CTL $= 1$). If the delay cell had non inverting forward and reverse paths, then input $IN1$ of Delay Cell 3 could have been hard-wired to ground. The wrap-around connection makes it toggle between odd and even settings to ensure that the signal will propagate for all setting values. The characteristics of the telescopic line are identical to the previous one with the sole exception that $D_{min}$ can be $2T_G$ since there is no need for input buffering.

A straightforward coarse DCDL implementation is shown in Fig.6.18. Two cascaded inverters constitute the delay element and a binary multiplexer selects the appropriate output tap. $D_{min}$ is equal to $2T_G + \log_2 N \times T_G$ and $d_r$ is still $2T_G$. This delay line can be controlled by a binary counter ($Q[1:0]$) requiring $\log_2 N$ storage elements as opposed to the $N$ storage elements of the previous two designs. The control structure of this DCDL scales better with $N$ as opposed to the NAND-based ones. $D_{min}$ also increases though with increasing $N$ and for certain applications this may be a concern.

DCDLs with a single $T_G$ resolution are also possible. Figure 6.19 illustrates one based on a differential delay element. The relative sizing of the cross-coupled devices

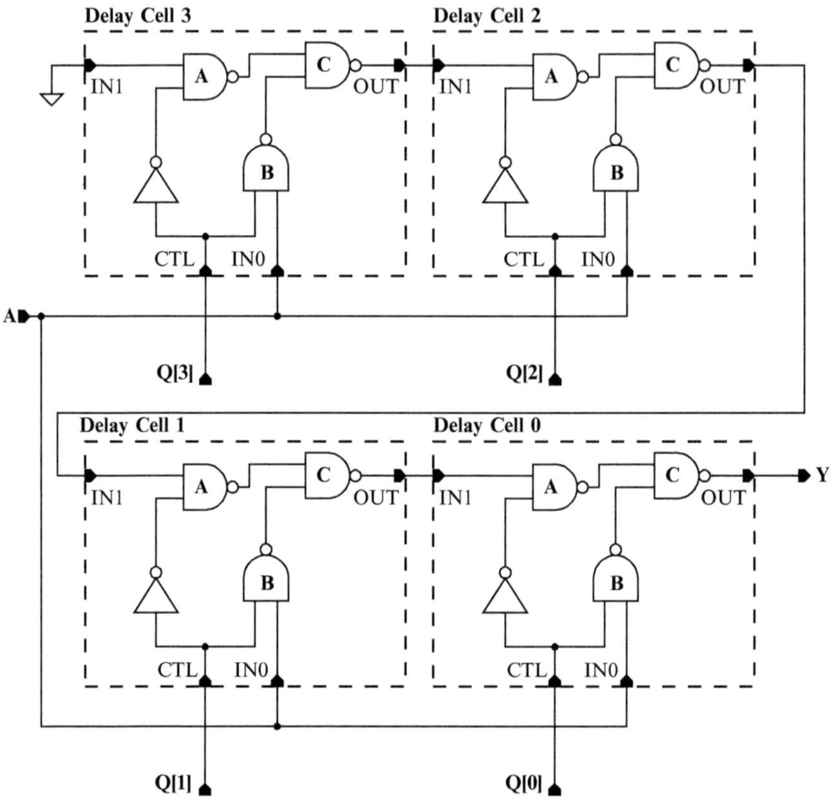


**Fig. 6.16.** NAND-based registered-controlled 4-stage delay line

vs. the forward devices can be used to fine tune $d_r$ as long as writing the cross-coupled pair can be safely guaranteed across all PVT conditions. Figure 6.20 shows a single-ended single $T_G$ DCDL with a conditionally inverting output stage. The output XNOR gate must be designed with equal delays from both input polarities to the output to ensure linearity.

Design options are certainly not limited to the five examples shown above. Not only different organizations are possible but also different logic families (i.e., dynamic or low swing current model logic) can be used as the base for the delay cell and the multiplexing structure. The characteristics of the examples presented so far are summarized in Table 6.5. Further assumptions are that the single-ended-to-differential converter of Fig.6.19 and the XNOR of Fig.6.20 can be implemented with two gates and, therefore, add $2T_G$ to $D_{min}$ and $D_{max}$ of the corresponding DCDLs.

Physical design is very important in precisely controlling DCDL specifications. All delay cells should be identical, and metal capacitance should be thoroughly characterized since it will probably be responsible for a substantial percentage of the cell delay. Post-layout simulation-based characterization across all PVT corners should

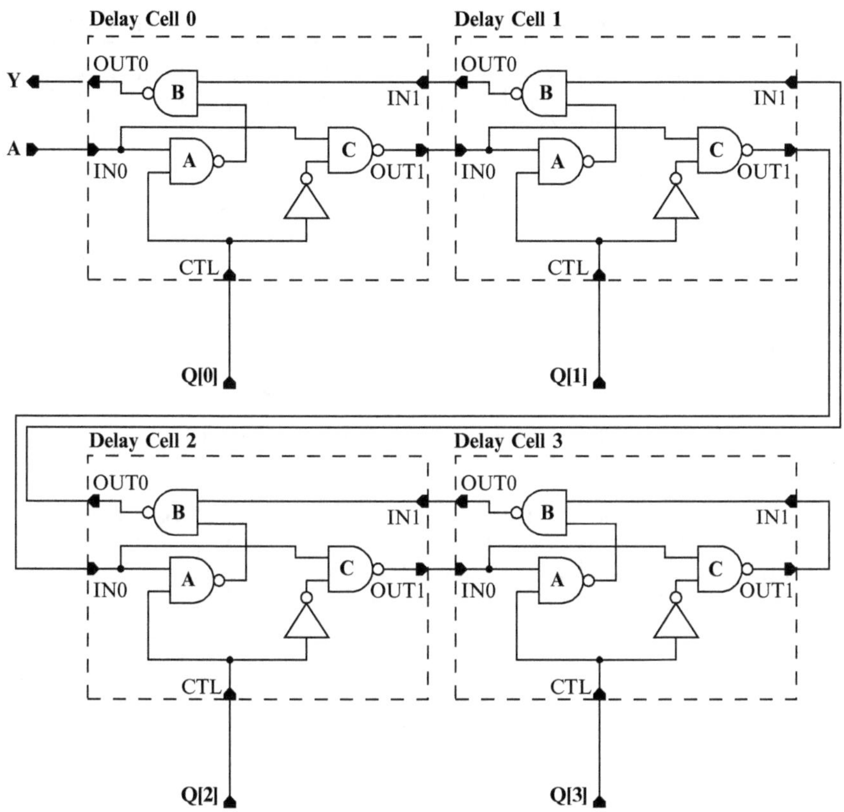


**Fig. 6.17.** NAND-based telescopic 4-stage delay line

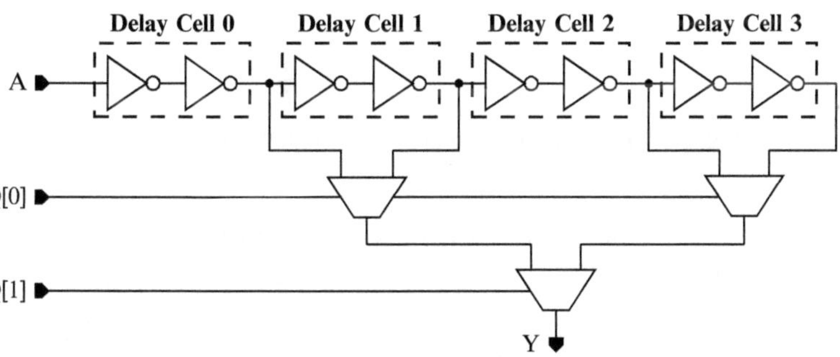


**Fig. 6.18.** Inverter-based logarithmic 4-stage delay line

be done in order to establish the ranges for $D_{\min}$, $D_{\max}$, and $d_r$ and ensure that the application requirements are met. In most cases, coarse DCDLs will be operating on clock signals, where duty cycle fidelity is important. If there are significant

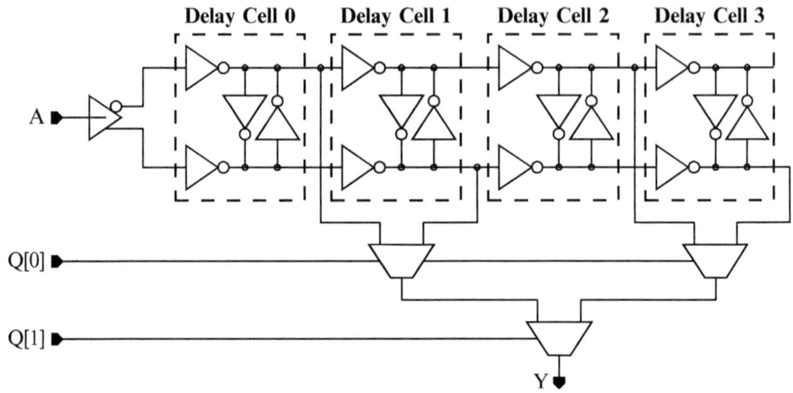


**Fig. 6.19.** Inverter-based differential 4-stage delay line

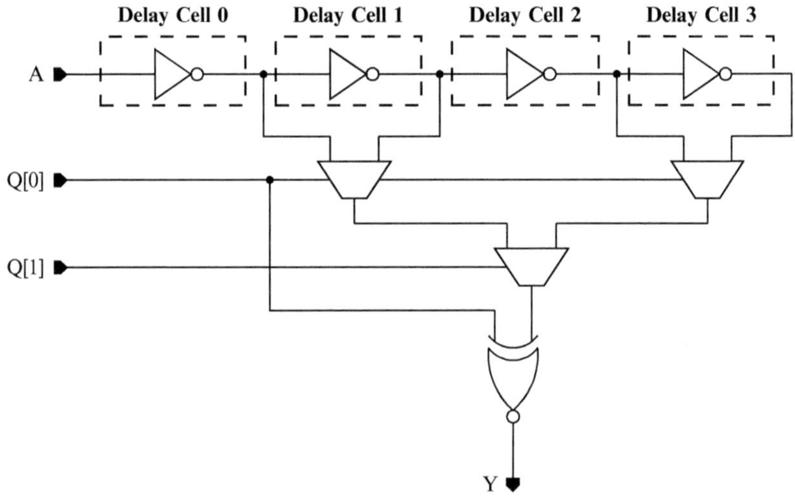


**Fig. 6.20.** Inverter-based conditional-output 4-stage delay line

**Table 6.5.** Characteristics of coarse DCDLs

| | $D_{\min}$ | $D_{\max}$ | $d_r$ | Dynamic range | Linearity | CTL flops |
|---|---|---|---|---|---|---|
| NAND (Fig. 6.16) | $> 2T_G$ | $> 2NT_G$ | $2T_G$ | $2(N-1)T_G$ | Good | $N$ |
| Tel. (Fig. 6.17) | $2T_G$ | $2NT_G$ | $2T_G$ | $2(N-1)T_G$ | Good | $N$ |
| Log. (Fig. 6.18) | $T_G(2+\log_2 N)$ | $T_G(2N+\log_2 N)$ | $2T_G$ | $2(N-1)T_G$ | Good | $\log_2 N$ |
| Diff. (Fig. 6.19) | $T_G(3+\log_2 N)$ | $T_G(2+N+\log_2 N)$ | $T_G$ | $(N-1)T_G$ | Good | $\log_2 N$ |
| Cond. (Fig. 6.20) | $T_G(3+\log_2 N)$ | $T_G(2+N+\log_2 N)$ | $T_G$ | $(N-1)T_G$ | Good | $\log_2 N$ |

imbalances between rise vs. fall delay in the delay cell and the multiplexing structure, severe duty cycle distortion may occur at the output signal especially for large delay settings. In extreme cases of high frequency clocks and long DCDLs, the square wave may completely disappear and a DC value may be observed at the output. NMOS vs. PMOS ratios for equal rise and fall times should be used and if possible inverting logic gates should be instantiated in cascaded pairs with similar fanouts to ensure equal treatment of positive vs. negative edges and good duty cycle PVT tracking.

Under certain circumstances, the NAND-based DCDLs of Figs. 6.16 and 6.17 can have a power advantage. Their switching activity is setting dependent and will exhibit reduced switching power at lower settings. A complete power comparison of the DCDLs presented so far is process and application dependent and will provide little if any additional insight.

### Synchronous vs. Asynchronous Operation in Coarse DCDLs

DCDL outputs are typically clock signals and as such must always be well behaved and not undergo spurious transitions (glitches) that may cause erroneous circuit operation. In order to demonstrate an important differentiation among DCDLs, we will adopt the ad hoc definition that a DCDL capable of asynchronous operation is a DCDL that can undergo a valid setting change ($\pm 1$) at any time without the possibility of a glitch at the output. On the other hand, a DCDL capable of synchronous only operation is a DCDL that requires that valid setting changes be timed synchronously with respect to the input clock in order to guarantee glitch-free output behavior.

DCDLs are essentially combinational (stateless) circuits with clock being simply another input. As a result, spurious transitions at the output are certainly possible unless all inputs are guaranteed to transition at the same time and all inputs have equal delay paths to the output. Neither is true in the DCDL case. Clock is not guaranteed to transition at the same time as the control settings and clock has multiple paths to the DCDL output through the delay element chain. Under certain circumstances though, and using simple analysis, we can convince ourselves that a certain class of DCDLs can have a glitch-free output. Before proceeding with the analysis, we make two important assumptions: We are considering DCDLs where all the control setting inputs have equal delay paths to the output, and we only consider setting changes that increment or decrement the current setting by one position.

A simplified delay line that is consistent with these assumptions is shown in Fig.6.21. It contains a single delay element of delay $d_r$ and a 2-input multiplexer. For simplicity, we assume that the multiplexer input-to-output and select-to-output delays are both equal to $t_m$. The timing diagram on the right shows the conditions that can generate a spurious output transition. The select line needs to transition while the clock edge goes through the delay element. Furthermore, $d_r$ needs to be large with respect to $t_m$ for such a glitch to be formed. In reality, if $d_r$ is reasonably close to $t_m$ the glitch will never form because of low pass filtering. The right design approach is to set up the conditions of Fig.6.21 and simulate across all PVT conditions to ensure that the glitch won't form.

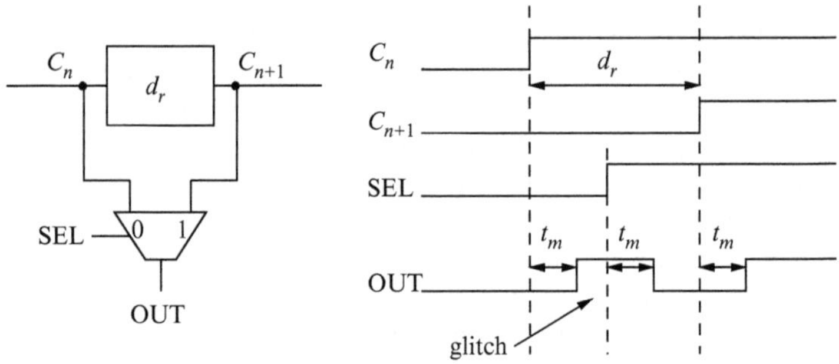


**Fig. 6.21.** DCDL spurious output transition

Interestingly enough, none of the DCDLs in Figs. 6.16 through 6.20 are glitch-free. Their setting vectors do not have equal delays to the output, and therefore the simple analysis of Fig. 6.21 does not apply. They have a lot more exposure to spurious transitions, and their analysis is rather ad-hoc and can be quite cumbersome. It is safe to create a design restriction stating that *only coarse DCDLs with equal setting-to-output delays should be considered when glitch free operation is necessary*. Distributed and logarithmic multiplexer structures are not allowed.

The simple DCDL of Fig. 6.18 can be a candidate for glitch-free operation if the output logarithmic multiplexer is replaced by a one-hot NAND-based (sum-of-products) multiplexing structure where each control line is at the same logic depth. Another potential solution is to pad the most significant bit with additional delay and the problem transforms to heterogeneous structure delay tracking across PVT and can be rather process-specific.

The coarse DCDLs of Figs. 6.16 through 6.20 are typically used in applications where the clock is not utilized while settings transition. Alternatively, they can be used in a synchronous fashion. The control inputs must be timed in a way that guarantees that when they change, the entire delay chain is at a constant value. This can be accomplished by latching the settings with an appropriate strobe which can either be the input clock or a delayed version of it (perhaps a particular delay chain tap) that meets this timing restriction. Extensive validation is required.

They can also be used asynchronously by adding significant hardware resources and control complexity. Such an example is shown in Fig. 6.22. We have a pair of identical delay lines. Only one is on-line at any single time. When there is a settings change, it is performed on the off-line DCDL. When the change has stabilized and all spurious transitions are gone, the output multiplexer is flipped and the new settings change shows up at the output. At this point, the analysis of Fig. 6.21 applies. In addition to consuming large hardware resources, this structure has a very subtle issue. It is based on the assumption that if we have two instantiations of a delay line on silicon and one has a setting equal to $n$ and the other has a setting equal to $n+1$, then the one with the $n+1$ setting will have a longer delay. This may not be true in deep submicron processes due to large random $V_t$ variation. Monte Carlo analysis can show a substantial probability of such a case. If this happens, then the composite delay line is non-monotonic. This will affect performance, and under certain circumstances, it may confuse the control automaton and cause catastrophic failure. A better approach is shown in Fig. 6.23 where a single dual output delay line is used. It uses less hardware and does not have potential monotonicity issues. It has the same control complexity though.

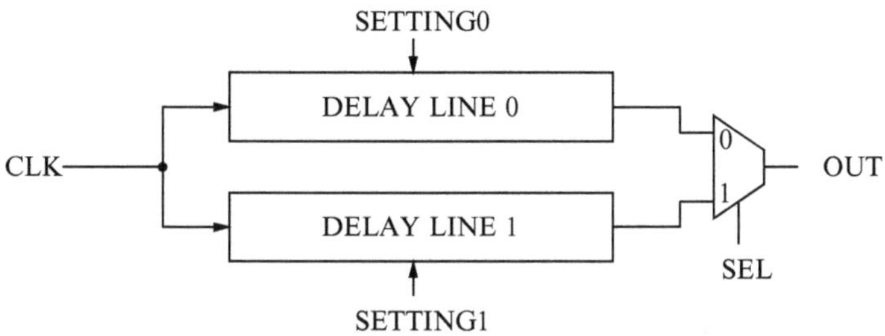


**Fig. 6.22.** Duplicating DCDLs for glitch suppression

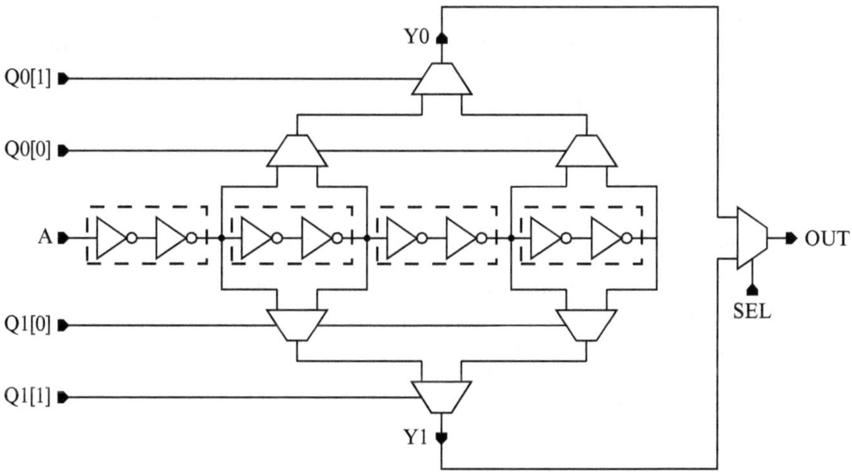


**Fig. 6.23.** Dual output DCDL for glitch suppression

### 6.5.2 Subgate-Delay DCDLs

Phase lock accuracy requirements can be tighter than a gate delay. In such applications, a subgate-delay (fine) DCDL must be employed.

A class of fine DCDLs relies on variable RC delays for delay generation. Two examples of fine DCDL stages are shown in Fig. 6.24. The delay stage on the left is based on variable cell driving resistance. Turning on more device branches

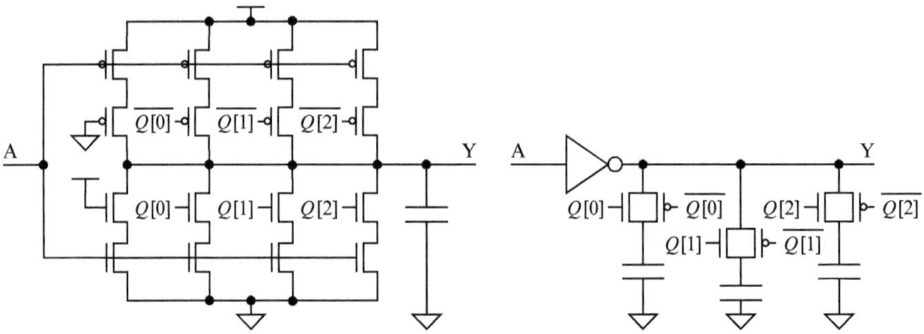


**Fig. 6.24.** RC-Based fine DCDL stages

through control vector $Q[2:0]$ reduces the overall RC delay through the stage. Transistors in the variable branches are sized for linear control vector encodings. Control bits $Q[2:0]$ are thermometer encoded and have a fixed switching order (i.e., $000 \rightarrow 001 \rightarrow 011 \rightarrow 111$). Logarithmic encoding is not possible due to the non-linearity of parallel resistance addition. Table 6.6 shows a normalized example of branch sizing for linearity. The example should be considered as a starting point only and process specific fine tuning will be required. The number of variable branches is limited by the extreme sizing that will be required to maintain linearity for an increasing number of control bits. The dynamic range can be extended by cascading multiple such stages at the expense of increasing $D_{\min}$.

**Table 6.6.** Variable resistance DCDL branch sizing (thermometer-encoded control)

| | Normalized width | Equivalent resistance | Capacitance | RC delay |
|---|---|---|---|---|
| Always-on branch (AO) | 1 | 1 | 4 | 4 |
| Branch 0 + AO | 1/3 + 1 | 3/4 | 4 | 3 |
| Branch 1 + 0 + AO | 2/3 + 1/3 + 1 | 1/2 | 4 | 2 |
| Branch 2 + 1 + 0 + AO | 2 + 2/3 + 1/3 + 1 | 1/4 | 4 | 1 |

The stage on the right is based on variable output capacitance. Turning on more capacitive branches increases the RC delay through the cell. Both linear (thermometer) and logarithmic (binary) control encoding are possible due to the linearity of parallel capacitance addition. Output capacitors can be metal-based or device-based depending on accuracy and linearity requirements. Per-stage dynamic range is limited by the maximum edge rate tolerated by the design. The dynamic range can be extended by cascading multiple stages as in the previous case.

An alternative method of constructing fine delay lines is shown in Fig. 6.25. It is based on the delay difference between the top vs. the bottom delay path. Control encoding is thermometer-based. This organization can achieve PVT-independent lin-

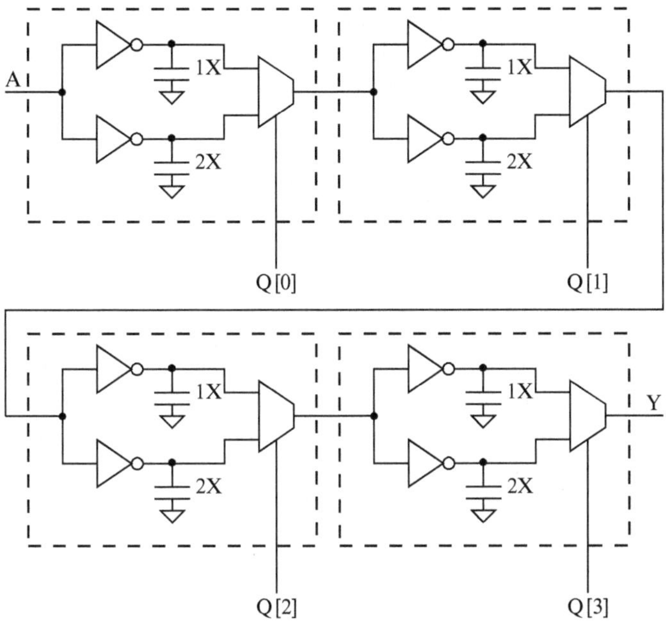


**Fig. 6.25.** Fine DCDL based on delay differences

earity subject only to random process variation. The main drawback is that both linearity and dynamic range are achieved through large $D_{\mathrm{min}}$ increases which can be a problem for certain applications.

An additional method of constructing subgate delays based on phase interpolation will be discussed in Sect. 6.5.3.

The delay lines of Figs. 6.24 and 6.25 can have resolution on the order of a few picoseconds but a very limited dynamic range. They can be operated asynchronously without the possibility of a glitch forming at the output. The DCDLs of Fig.6.24 vary resistance and capacitance. Settings changes are not considered combinational logic switching events. On the other hand, the multiplexer-based DCDL of Fig.6.25 can be analyzed using the method of Fig.6.21 with a much smaller $d_{\mathrm{r}}$.

### 6.5.3 Resolution vs. Dynamic Range in DCDLs

In typical applications, both fine resolution and large dynamic range are highly desirable. High resolution directly affects phase match accuracy. A large dynamic range ensures that the DLL won't be the limiting factor in selecting input clock frequencies or $V_{\mathrm{MIN}}$ for the overall design.

From the previous analysis on DCDLs, it would seem that designers must perform a tradeoff between resolution and range. This is rarely true in real applications since both are essential. Instead of using a single delay line that compromises $d_{\mathrm{r}}$ in favor of dynamic range for a given number of settings, multiple delay lines can be employed with different properties so that the overall design target is met [13]. Typically, a coarse high dynamic range DCDL is cascaded with a fine low dynamic range subgate DCDL to achieve both specifications. This design decision involves additional complexity in the control automaton which must first lock the coarse line to an appropriate setting (which tracks input clock frequency and static process and voltage variations) and then engage the fine delay line which tracks dynamic voltage and temperature variations. Depending on control implementation, the coarse DCDL can readily be of the synchronous type (as described in Sect. 6.5.1) if it is locked only once and settings never change while in mission mode. In this case, the designer must ensure that the fine delay line has enough dynamic range on either side of the lock point to guarantee phase lock maintenance in the presence of voltage and temperature ramps spanning the entire product voltage-temperature range in the worst possible scenario. If this is not possible, the control algorithm must ensure repeated engagement of the coarse DCDL to compensate for large V-T ramps without the possibility of a large phase change affecting system functionality. Such an example is a DDR memory incoming strobe phase lock system which engages the coarse DCDL only when there is no memory transaction present on the bus and, therefore, no one is looking at the data strobe. While memory reads or writes are in progress, only the fine DCDL is engaged which should guarantee DLL-induced deterministic jitter on the order of a few picoseconds and no spurious clock transitions. An additional concern with cascaded DCDLs is increased $D_{\mathrm{min}}$ which can be shown to be prohibitive for certain applications since it directly affects the overall minimum supply voltage allowed ($V_{\mathrm{MIN}}$).

There exists a powerful method of achieving both low $d_{\mathrm{r}}$ (high resolution) and arbitrarily high dynamic range with low $D_{\mathrm{min}}$ and without cascading DCDLs. It is based on the "ping-pong" DCDL arrangement of Figs. 6.22 and 6.23 with a phase interpolator as the output stage instead of the 2-to-1 multiplexer. A phase interpolator receives two input clocks with a phase difference on the order of 1-2 gate delays and a digital setting value. For a setting equal to zero, the interpolator output simply delays the early input phase by a fixed amount. For a setting equal to the maximum possible value, the interpolator output delays the late input phase by the same amount. For any intermediate setting, the interpolator produces an output clock with a phase proportional to the applied setting as measured using the output at setting 0 as a reference.

A sample schematic is shown in Fig. 6.26. The two separate branches are implemented with tri-state inverters controlled by complementary settings. Typically, the output is restored with an output inverter (not shown) which also makes the overall structure non-inverting. For a first order analysis, we will use the methodology and terminology established in [14] for current-mode reduced-swing structures. The time delay between the two interpolated phases is $\Delta t$, and it will be used as a parameter throughout the analysis. We make the following assumptions:

1. The overall output resistance of a branch when fully enabled is $R$ and has no voltage dependence.

2. The interpolation setting can be represented by a continuous variable $w$ where $0 \leq w \leq 1$. When $w = 1$, the early phase is fully enabled, and the late phase does not affect the output. When $w = 0$, the early phase is disabled and only the late phase controls the output. For any other value, both the early and the late phase affect the output delay with weights equal to $w$ and $1 - w$, respectively.

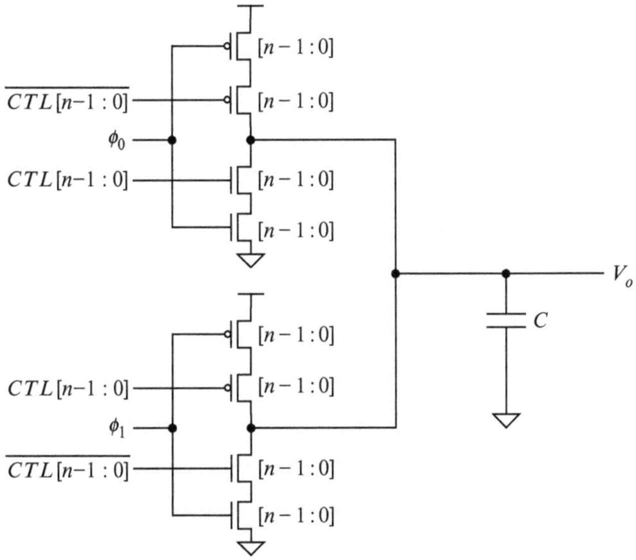


**Fig. 6.26.** Full swing phase interpolator ($\log_2 n$-bit control)

The interpolator time constant $RC$ (where $R$ is the on-resistance of a fully enabled branch and $C$ is the output capacitance in Fig.6.26) is a fundamental property which will affect transfer function linearity as will be shown. Without loss of generality, we assume that the interpolator is mixing two falling edges ($\phi_0$ and $\phi_1$) so that the output waveform is rising. Phase $\phi_0$ occurs at time 0 and $\phi_1$ occurs at time $\Delta t$. For $0 \leq t < \Delta t$, the equivalent circuit is shown in Fig.6.27a. Writing KCL at $V_o$ yields equation:

$$\frac{dV_o(t)}{dt} + \frac{1}{RC}V_o(t) - \frac{wV_{DD}}{RC} = 0, \tag{6.31}$$

which has the following solution assuming $V_o(0) = 0$ :

$$V_o(t) = wV_{DD}(1 - e^{-\frac{t}{RC}}). \tag{6.32}$$

For $t \geq \Delta t$, both branches pull high and resistors $R/w$ and $R/(1-w)$ are connected in parallel. The equivalent circuit is shown in Fig. 6.27b. The governing equation is:

$$\frac{\mathrm{d}V_o(t)}{\mathrm{d}t} + \frac{1}{RC}V_o(t) - \frac{V_{\mathrm{DD}}}{RC} = 0, \tag{6.33}$$

with the following well-known solution (assuming zero initial conditions):

$$V_o(t) = V_{\mathrm{DD}}(1 - \mathrm{e}^{-\frac{t}{RC}}). \tag{6.34}$$

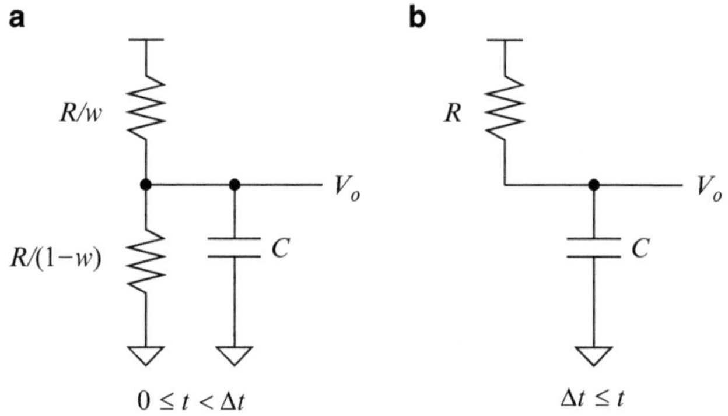


**Fig. 6.27.** Phase interpolator equivalent circuits

The overall solution is a composite waveform that consists of Eq. (6.32) for $0 \leq t < \Delta t$ and Eq. (6.34) for $t \geq \Delta t$ time-shifted by $\Delta t$ and adjusted by the initial condition established by Eq. (6.32) at $t = \Delta t$:

$$V_o(t) = wV_{\mathrm{DD}}(1 - \mathrm{e}^{-\frac{t}{RC}})u(t)u(\Delta t - t) + V_{\mathrm{DD}}[1 - (1 - w + w\mathrm{e}^{-\frac{\Delta t}{RC}})\mathrm{e}^{-\frac{t-\Delta t}{RC}}]u(t - \Delta t), \tag{6.35}$$

where $u(t)$ is the unit step function.

Figure 6.28 plots Eq. (6.35) for various $\Delta t$ using the interpolator weight $w$ as a parameter. It is obvious from the $V_{\mathrm{DD}}/2$ crossing point that linearity with respect to $w$ is a strong function of $\Delta t$. The larger $\Delta t$ is with respect to the interpolator time constant $RC$, the larger the deviation from linearity with respect to interpolation weight $w$. This is more clearly shown in Fig. 6.29 that shows interpolator delay transfer function with respect to interpolation weight $w$ using $\Delta t$ (delay between two input phases) as a parameter. The top graph shows absolute inteprolator delay as a function of $w$ adjusted so that the delay for $w = 1$ (early phase only affecting the output) is zero. The bottom graph shows the same data but normalized to the same $0-1$ range. Linearity is quite good for $\Delta t \ll RC$ but deteriorates quickly for increased spacing between input phases $\phi_0$ and $\phi_1$. The interpolator time constant $RC$ is a fundamental property that has a strong effect on its linearity. Larger time constants are

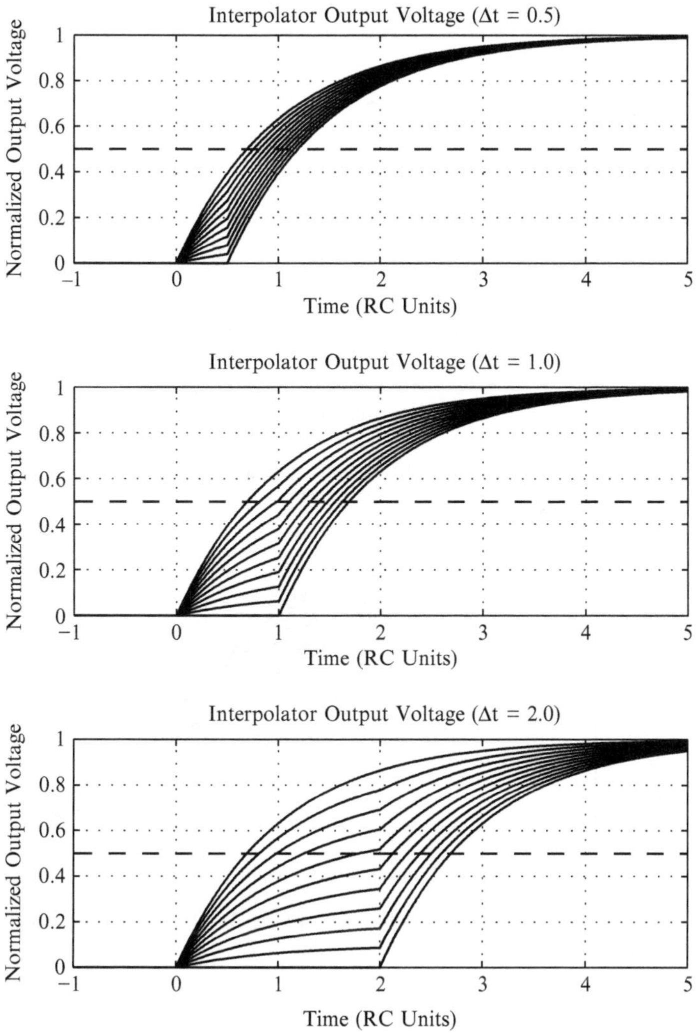


**Fig. 6.28.** Phase interpolator normalized voltage output for $\Delta t = 0.5, 1, 2$ ($w$ is decreasing from 1 to 0 in steps of 0.1 from left to right in all three plots)

desired if linearity is important. An effect similar to an increased interpolator *RC* can be achieved by slowing the edge rates of input phases $\phi_0$ and $\phi_1$.

An interpolator based DCDL can achieve both high resolution and high dynamic range at the expense of increased control complexity. Such a DCDL can easily be designed to be free from spurious output transitions. When a coarse setting changes that affects one input of the interpolator, the interpolator weight should be such that only the other non-changing input affects the output.

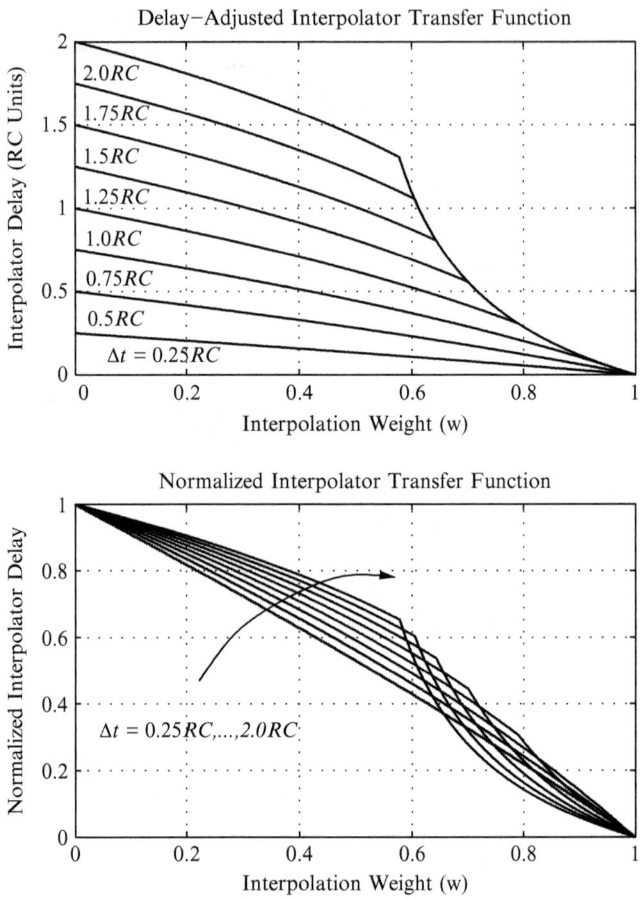


**Fig. 6.29.** Phase interpolator transfer function for varying $\Delta t$. Top graph reproduced in a form similar to that in [14], ©1998 Stefanos Sidiropoulos

## 6.6 Control

There is one major classification that needs to be made regarding digital DLL control: first order and higher order. A first order digital DLL contains a single DCDL state element (counter or shift register) that maintains the delay line setting and gets updated through some mechanism based on the phase difference between reference and controlled clock. In contrast, a second order DLL will have a second counter which maintains some representation of the *frequency* or rate of phase change between the two clocks. The DCDL is then adjusted by combining both such state variables. All DLL examples in this chapter assume that the reference clock at the phase detector and the source clock of the DCDL are the same. Therefore, there is no possibility of a frequency difference between the two inputs of the phase detector. An application where a second order control loop would make sense is clock-data recovery, which is briefly described in Sect. 6.9.4. The rest of this section is only concerned with first order control and identical frequencies of reference and feedback clocks.

A typical DLL control block contains a state structure that holds the DCDL settings and a mechanism to update the settings based on input from the phase detector. The state element can be a binary up/down counter if the DCDL has binary encoding. Different DCDL setting encodings can be easily handled with the addition of combinational decoding logic past the main counter. The designer should always keep in mind that post-flop combinational logic can generate spurious transitions and create more conditions for glitches at the output of the DCDL. For one-hot DCDL configurations, it is more common to use a bidirectional shift register which requires no decoding and will produce glitch free control outputs.

The simplest control arrangement is shown in Fig.6.30a. It involves a single up/down counter (or bidirectional shift register) under the direct control of the phase detector. When the phase detector evaluates to a 1 (controlled clock faster with respect to the reference), the counter is incremented and the DCDL has a larger (slower) setting. When the PD evaluates to a 0, the counter is decremented and the DCDL speeds up. A more general control arrangement is shown in Fig.6.30b. In this case, a Finite State Machine (FSM) is inserted between the phase detector output and the counter increment/decrement. The FSM will make setting change decisions based not only on the phase detector output but also based on internal state. In this fashion, more complex control algorithms can be implemented which can have a strong impact on DLL capabilities such as sensitivity to initial conditions, dynamic range, and stability.

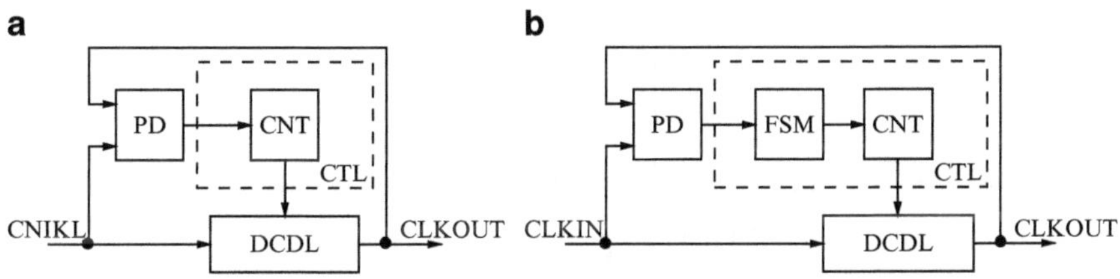


**Fig. 6.30.** DLL control options

### 6.6.1 Sensitivity to Initial Phase

Even though a DLL may contain a DCDL with a large dynamic range, a simplistic control design may result in an inability to achieve phase lock. An example is shown in Fig.6.31. Let us assume that the DLL configuration is the one in Fig.6.30a with $D_{\min} < 0.5T$ where $T$ is the CLKIN period. The waveforms right after system reset (zero DCDL setting) are shown on the top part of Fig.6.31. The stateless control design of Fig.6.30a will attempt to push the DCDL toward the left and match edge 1 of CLKIN with edge 1 of CLKOUT. If the settings counter wraps around, a very large delay will be added to the DCDL and the DLL will either lock to an undesirable setting (not the minimum delay setting that will achieve phase lock) or will degenerate to an oscillation between $D_{\min}$ and $D_{\max}$ without ever achieving lock. If the settings counter is designed not to wrap-around but saturate at the minimum setting, the DCDL will stay fixed at $D_{\min}$ and again phase lock will never be achieved.

If on the other hand the FSM of Fig. 6.31 (*bottom*) is inserted between the phase detector and the settings counter, lock can be easily achieved. The FSM will keep on incrementing the DCDL settings until a 1-to-0 transition is observed at the phase detector independent of the initial state of the phase detector output. Such a transition indicates a lock condition and the state machine will be bouncing back and forth between states INC and DEC. Edge 1 of CLKOUT will be aligned to edge 2 of CLKIN.

A word of caution is absolutely essential when a state machine similar to the one depicted in Fig. 6.31 (*bottom*) is used. This particular FSM is designed to disregard a 0-to-1 transition from the phase detector and essentially lock the system when a 1-to-0 transition is observed. In reality, a 1-to-0 transition can easily be observed right after a 0-to-1 when the controlled clock (CLKOUT) phase traverses a negative edge of the reference clock (CLKIN). When the positive edge of CLKIN is aligned to the negative edge of CLKOUT the phase detector can undergo multiple transitions due to internal factors such as a large dead zone where its output is undefined or external factors such as supply noise and reference clock jitter. If state DEC is entered (Fig. 6.31) while we have positive to negative edge alignment, the system will never

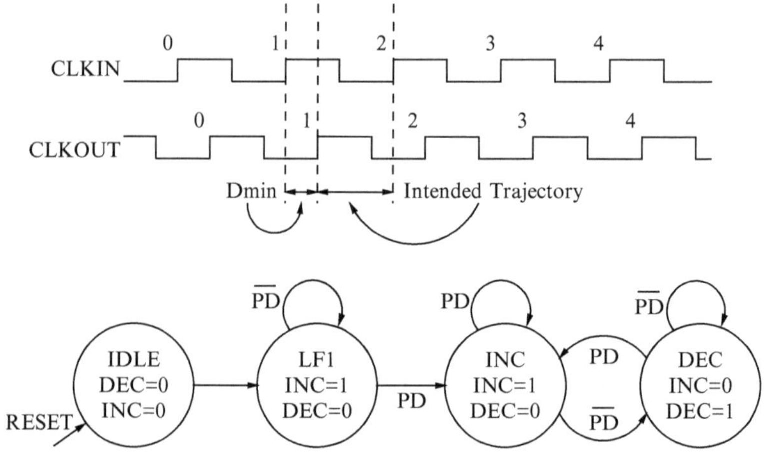


RESET: State Machine Reset
PD: Phase Detector Output (1 = CLKOUT leads, 0 = CLKOUT lags)
INC: Increment DCDL Settings Counter (Active High)
DEC: Decrement DCDL Settings Counter (Active High)

**Fig. 6.31.** DLL FSM example for initial condition flexibility achieve lock but will likely be pushed back to its reset state. State DEC must only be entered when a positive CLKOUT edge has just started to lag a positive CLKIN edge. Otherwise, the phase detector decisions won't be interpreted correctly and the DCDL will be moved to the wrong direction.

There are multiple ways that this issue can be addressed. Digital low pass filtering of the phase detector output can be employed to remove spurious transitions. This method can reduce the probability of a false lock, but it is hard to show that it can totally eliminate it. A more robust method is to insert an intermediate state between LF1 and INC in which the controller will reside for $n$ clock cycles. In this state, the DCDL setting is incremented $n$ times independent of the phase detector output. In this way, we can guarantee that when we have negative edge traversal, the DCDL will be incremented enough to push CLKOUT out of the phase detector sampling window or dead zone ($d_{\mathrm{sw}}$) given worst case supply noise and jitter conditions. The number $n$ should be large enough to guarantee $n \cdot d_{\mathrm{r}} \gg d_{\mathrm{sw}}$ and at the same time small enough to guarantee $n \cdot d_{\mathrm{r}} \ll 0.5T$ (where $T$ is the reference clock period) so that the next positive CLKIN edge is not traversed and the phase detector output does not change interpretation. This should be a rather loose constraint for most applications.

### 6.6.2 Dynamic Range Increase

It is possible to increase the dynamic range of a DLL with the addition of little extra hardware and incremental complexity in the control automaton. This method assumes that the DCDL output can be conditionally inverted (i.e., XOR output stage) and that the settings counter or bidirectional shift register can provide a carry out (CO) output to the FSM indicating that its maximum value has been reached. The concept is very simple. If the DLL reaches the maximum value of the settings counter without achieving lock (the phase detector output has not undergone a 1-to-0 transition), then the DCDL output is inverted, the counter is cleared and the state machine goes back to the initial state and attempts to lock the inverted DCDL. The inversion has the effect of adding 180° of phase and virtually doubles the DLL dynamic range. This is shown conceptually in Fig. 6.32 (*top*). An example FSM for this configuration is shown at the bottom of the figure.

### 6.6.3 Stability and Bandwidth

At the beginning of this chapter, a claim was made that a digital DLL can be unconditionally stable. In this section, we qualify this claim. The stability of feedback systems is typically analyzed in the frequency domain. A frequency domain model of a digital DLL is non-trivial and not particularly insightful due to the nonlinearity of a bang–bang phase detector and the sampled nature of the system. Instead, we choose to revisit the analysis of a linear (analog) delay locked loop, establish stability conditions and apply the intuition developed to the non-linear digital counterpart. We follow the methodology and notation of Maneatis in [15]. A very similar approach is also presented by Yang in [16].

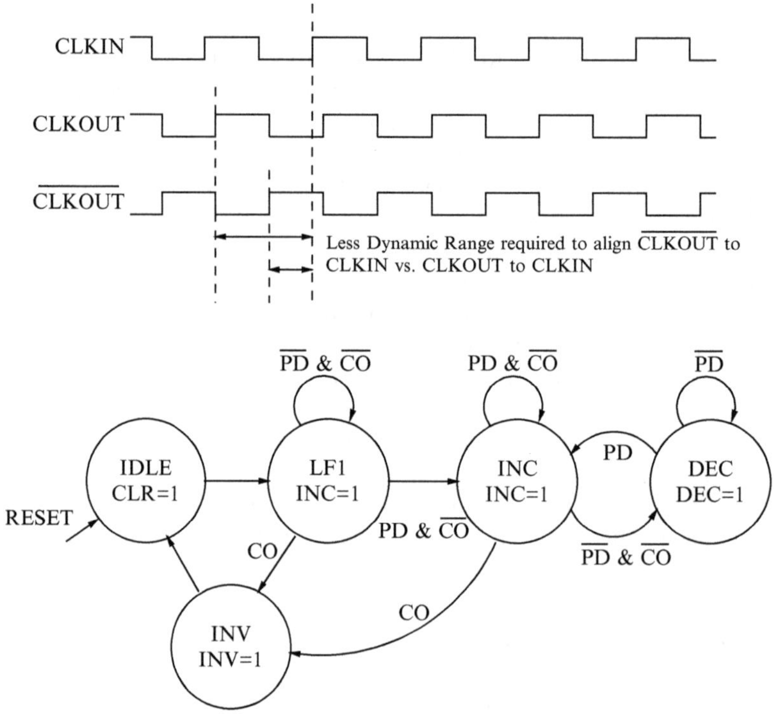


RESET: State Machine Reset
PD: Phase Detector Output (1 = CLKOUT leads, 0 = CLKOUT lags)
INC: Increment DCDL Settings Counter (Active High)
DEC: Decrement DCDL Settings Counter (Active High)
CLR: Clear DCDL Settings Counter (Active High)
INV: Invert DCDL Ouput (Active High, Sticky [sets RS Flop])
CO: DCDL Settings Counter Carry Out (Active High)

**Fig. 6.32.** Doubling DLL dynamic range using conditional inversion

Figure 6.33a shows a simple charge-pump based analog DLL. The phase detector is linear rather than bang–bang (Sect. 6.4). Its output is a pulse whose width is proportional to the difference between the CLKIN and CLKOUT phases. The proportional pulse controls a charge pump which integrates a constant current on capacitor *C*. The amount and sign of the integrated current (charge) depend on the amount and sign of the phase difference between the clocks. The capacitor voltage controls an analog delay line which in turn will adjust until CLKOUT matches CLKIN. There are many similarities between this analog DLL and the digital counterpart of Fig. 6.1. The settings counter/accumulator inside the control block performs the same action as the capacitor and essentially integrates the phase detector output. The main difference is that the digital DLL is a non-linear system because the correction in response to the phase error is not proportional but instead has a constant slew rate.

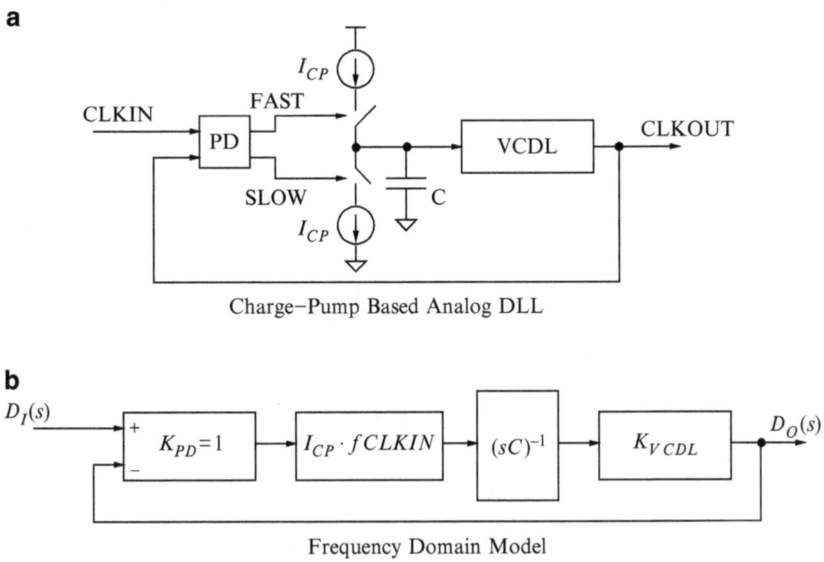


**Fig. 6.33.** Analog DLL frequency domain model

An additional important difference is that a digital DLL may have more delay around the loop due to inherent clock delays in the FSM controller and settings counter. This can be very important for stability and will be addressed in more detail later on.

A frequency domain model of the analog DLL is shown in Fig. 6.33b. The model terminal variables are $d_\mathrm{I}(t)$ (input delay), which is defined as the delay of input CLKIN with respect to an arbitrary point. The output delay of CLKOUT $d_\mathrm{O}(t)$ is also defined with respect to the same arbitrary point. The corresponding Laplace transforms are $D_\mathrm{I}(s)$ and $D_\mathrm{O}(s)$. The frequency domain representation of the phase detector is a simple subtractor with a constant gain of 1 since it generates a pulse with a width equal to the delay difference between CLKIN and CLKOUT. The units at the output of the phase detector are seconds (delay). The charge pump model is also a constant gain since it multiplies the proportional error pulse with a constant current after normalizing it to the CLKIN period ($I_\mathrm{CP} \cdot f_\mathrm{CLKIN}$, where $f_\mathrm{CLKIN}$ is the CLKIN frequency). The units at the output of the charge pump are Amperes (current). The charge pump capacitor $C$ acts as an integrator with a transfer function $(sC)^{-1}$. The units at the integrator output are Volts. Finally, the Voltage-Controlled Delay Line (VCDL) is also modeled with a constant gain ($K_\mathrm{VCDL}$ s/V). The output variable $D_\mathrm{O}(s)$ has units of seconds (delay). Writing the loop equation yields:

$$D_\mathrm{O}(s) = [D_\mathrm{I}(s) - D_\mathrm{O}(s)] \cdot I_\mathrm{CP} \cdot f_\mathrm{CLKIN} \cdot \frac{K_\mathrm{VCDL}}{sC}. \tag{6.36}$$

Further manipulation yields the following transfer function [15]:

$$\frac{D_\mathrm{O}(s)}{D_\mathrm{I}(s)} = \frac{1}{1+\frac{s}{\omega_N}}, \tag{6.37}$$

where

$$\omega_N = \frac{I_{\mathrm{CP}} \cdot K_{\mathrm{VCDL}} \cdot f_{\mathrm{CLKIN}}}{C}. \tag{6.38}$$

The transfer function of Eq. (6.37) is of first order with a single pole at $\omega_N$ (loop bandwidth) and is unconditionally stable with a phase margin of 90°. This is true as long as the delay around the loop is virtually instantaneous and the phase margin is not reduced considerably. Even an analog DLL though is a sampled system and there is loop delay due to the sampled nature of the phase detector: A phase detector measurement won't be taken until the next positive clock edge, thus, reducing the phase margin at unity gain. Such margin reduction though is negligible if $\omega_N \ll 2\pi f_{\mathrm{CLKIN}}$. For all practical purposes, the loop delay can be considered virtually instantaneous if such delay is much lower than the DLL response time. Maneatis [15] sets the unconditional stability criterion at $\omega_N = 2\pi f_{\mathrm{CLKIN}}/10$.

Multiple authors [17, 18] have pointed out that the simple frequency domain model of Fig.6.33 does not apply for the class of delay locked loops where the same reference clock is driving both the phase detector and the DCDL. Figure 6.33 models the delay line as a 2-terminal block (control input and clock output) and does not include a feed-forward path from the input reference clock to the output through the delay line. As a result, such a model cannot accurately model jitter transfer from input to output clock. The stability analysis changes due to the introduction of a zero in the first order model. However, unconditional or conditional stability can still be shown with root locus methods depending on model choice [18].

The main difficulty in applying a similar analysis to the digital DLL lies in the nonlinear transfer function of the phase detector (Fig.6.2). The linear frequency domain analysis is not applicable and the steady state response of such a DLL will be oscillatory (limit cycle). Prior art [19, 20] has linearized the phase detector response around the dead zone, which is a valid approach for high CLKIN frequencies, high DCDL resolutions, and multiple averaged measurements since the expected phase error will be rather small and comparable to $d_{\mathrm{sw}}$. For lower frequency applications though, the phase detector will mostly operate in its nonlinear regime.

Before proceeding, let us first define what we mean by stating that a non-linear DLL is stable. We know for a fact that a DLL with a non-linear (bang–bang) phase detector will exhibit limit cycle behavior due to the binary nature of the phase detector and its inability to encode a zero phase error. For the purposes of this discussion, let us define DLL stability to mean that the expected limit cycle should have by design the minimum possible amplitude of $\pm d_{\mathrm{r}}$. Frequency domain transformation is not really necessary to assess the stability of such a simple system. The time-domain behavior of the digital DLL (Fig.6.1) in the locked state can be summarized by the following difference equations:

$$e[n] = \mathrm{sgn}(d_{\mathrm{O}}[n] - d_{\mathrm{I}}[n]), \tag{6.39}$$

$$d_{\mathrm{O}}[n+1] = d_{\mathrm{O}}[n] + e[n]d_{\mathrm{r}}, \tag{6.40}$$

where $e[n]$ is the error computed by the phase detector ($e[n] \in \{-1, 1\}$), $d_I[n]$ is the input delay at discrete time $n$ ($n$th CLKIN positive edge), $d_O[n]$ is the output delay at discrete time $n$ and $d_r$ is the DCDL resolution. We establish a time-domain stability criterion as follows: *The correction step in Eq. (6.40) at discrete time n should not produce an error ($d_O[n+1] - d_I[n+1]$) at $n+1$ of the same sign and greater magnitude than ($d_O[n] - d_I[n]$) at time n.* This criterion is easily satisfied if the correction term $e[n]d_r$ in (6.40) does not result in crossing over two phase boundaries which will produce a phase error of the same sign ($mod T_{CLKIN}$) and potentially larger magnitude. If $d_r < 0.5T_{CLKIN}$, this should never happen and this constitutes a constraint that is always met for all practical purposes. A DLL with such a large $d_r$ would be highly impractical.

We now focus on the loop delay argument. Eqs. (6.39) and (6.40) do not capture the effect of excessive loop delay and the possibility of reduced "phase margin". A digital DLL has the potential of introducing substantial loop delay through its control mechanism and may turn negative feedback into positive. The reasons for the increased delay can be multiple:

- Additional flip-flop synchronization stages past the phase detector to ensure low probability of metastability.
- Low pass filtering of the phase detector output to remove spurious transitions and potentially emulate "ternary" error detection (fast, slow, NOP).
- Deserialization latency in a CDR (clock-data recovery) loop where phase detection occurs in the divided clock domain on the multiple deserialized bits.
- Pipeline stages in the FSM controller and settings counter.

Let us introduce two more general descriptive DLL design parameters (in addition to $d_r$, $d_{sw}$, $D_{min}$ and $D_{max}$) to help with the analysis of this issue:

$N_d$ indicates the number of CLKIN delay cycles from the input of the phase detector to the setting input of the DCDL. This number is an integer with a minimum value of 1 (delay introduced by the bang–bang phase detector).

$N_{bw}$ is the inverse of the rate at which the FSM controller issues corrections to the DCDL (once every $N_{bw}$th cycle). One can think as $1/N_{bw}$ as the "bandwidth" of this nonlinear system. The minimum value for $N_{bw}$ is 1 (controller issues corrections on every cycle) and the maximum value can be arbitrarily high. $1/N_{bw}$ is also referred to as the DLL sampling rate because it signifies the normalized frequency of looking at the phase detector output.

We introduce a second stability criterion which states that a digital DLL is stable if $N_d < N_{bw}$. This is the equivalent of stating that a *a digital delay locked loop is stable if it won't issue a DCDL correction before the outcome of the previous correction is fully evaluated by its control mechanism.*

We demonstrate this criterion with a cycle-accurate behavioral model of a DLL with the characteristics of Table 6.7 running at 250 MHz.

The controller is equivalent to the FSM of Fig.6.31 with the addition of wait states between the states that update the counter to implement $N_{bw} \neq 1$. The modeled

**Table 6.7.** DLL behavioral model design parameters

| | | | |
|---|---|---|---|
| $D_{\mathrm{min}}$ | 1.525 ns | $D_{\mathrm{max}}$ | 14.275 ns |
| $d_{\mathrm{r}}$ | 0.050 ns | $d_{\mathrm{sw}}$ | 0.0 ns |
| $N_{\mathrm{d}}$ | parameter | $N_{\mathrm{bw}}$ | 4 |

DLL has a built-in $N_{\mathrm{d}}$ of 3 (phase detector plus 2 stages in the FSM/counter). We, therefore, make $N_{\mathrm{bw}}$ equal to 4 to guarantee stability when $N_{\mathrm{d}}$ is minimum. Fig.6.34 shows simulation results (DLL phase error over time) for various $N_{\mathrm{d}}$ values. Case $N_{\mathrm{d}} = 3$ satisfies our stability criterion since the phase error changes sign every $N_{\mathrm{bw}}$ cycles. All other cases fail the criterion and the phase error is amplified. Loop delay causes the DLL to issue correction steps in the wrong direction. The number of wrong correction steps is $\lfloor N_{\mathrm{d}}/N_{\mathrm{bw}} \rfloor$ and the resulting phase error is bounded by $\pm(\lfloor N_{\mathrm{d}}/N_{\mathrm{bw}} \rfloor + 1)d_{\mathrm{r}}$ as opposed to $\pm d_{\mathrm{r}}$ in the stable case.

A digital DLL can be stabilized given a value for $N_{\mathrm{d}}$ by making $N_{\mathrm{bw}} > N_{\mathrm{d}}$. Such an example is shown in Fig.6.35 where a DLL with $N_{\mathrm{d}} = 15$ is stabilized by increasing $N_{\mathrm{bw}}$ from 4 to 16. An expected increase in lock acquisition time is observed due to the reduced rate of issued DCDL corrections. Lock acquisition will be discussed in more detail in Sect. 6.6.4.

Increasing $N_{\mathrm{bw}}$ is analogous to reducing loop bandwidth $\omega_N$ in the analog domain. Higher $N_{\mathrm{bw}}$ increases loop response time and makes it more difficult to track a changing input. Spread spectrum clocking is one important application where tracking a changing input is necessary. In such a case, $N_{\mathrm{bw}}$ should be treated as an important design parameter and the designer must ensure that a changing frequency input won't generate a noticeable increase in phase error. A convenient way of describing the tracking ability of a nonlinear digital DLL is its delay slew rate $r$ defined as follows:

$$r = \frac{d_{\mathrm{r}}}{N_{\mathrm{bw}} \cdot T_{\mathrm{CLKIN}}}. \tag{6.41}$$

The slew rate is a unitless quantity that indicates the amount of delay correction that the DLL can produce per unit time. A simple way to determine whether a DLL can track a given spread spectrum (SS) clock is to calculate an equivalent slew rate $r_{\mathrm{s}}$ for the spread clock and compare it with the DLL rate $r$. SS clocks are typically defined with a spread factor $A_{\mathrm{s}}$ as a percentage which indicates modulation amplitude and the modulation frequency $f_{\mathrm{s}}$ on top of the clock period $T_{\mathrm{CLKIN}}$. The equivalent slew rate of a spread clock indicates the period change per unit time and is given by the following formula:

$$r_{\mathrm{s}} = 4A_{\mathrm{s}}T_{\mathrm{CLKIN}}f_{\mathrm{s}}. \tag{6.42}$$

As an example, a 250 MHz SS clock with a 10% spread factor and a 500 KHz spread frequency will have an $r_{\mathrm{s}}$ of $0.8 \times 10^{-3}$. A DLL with $d_{\mathrm{r}}$ equal to 0.050 ns and a $N_{\mathrm{bw}}$ equal to 4 will have an $r$ of $3.125 \times 10^{-3}$. Since $r > r_{\mathrm{s}}$, we determine that the DLL can track the SS clock without change to its theoretical minimum phase error of $\pm d_{\mathrm{r}}$. Phase error tracking is demonstrated in Fig.6.36 using the same behavioral

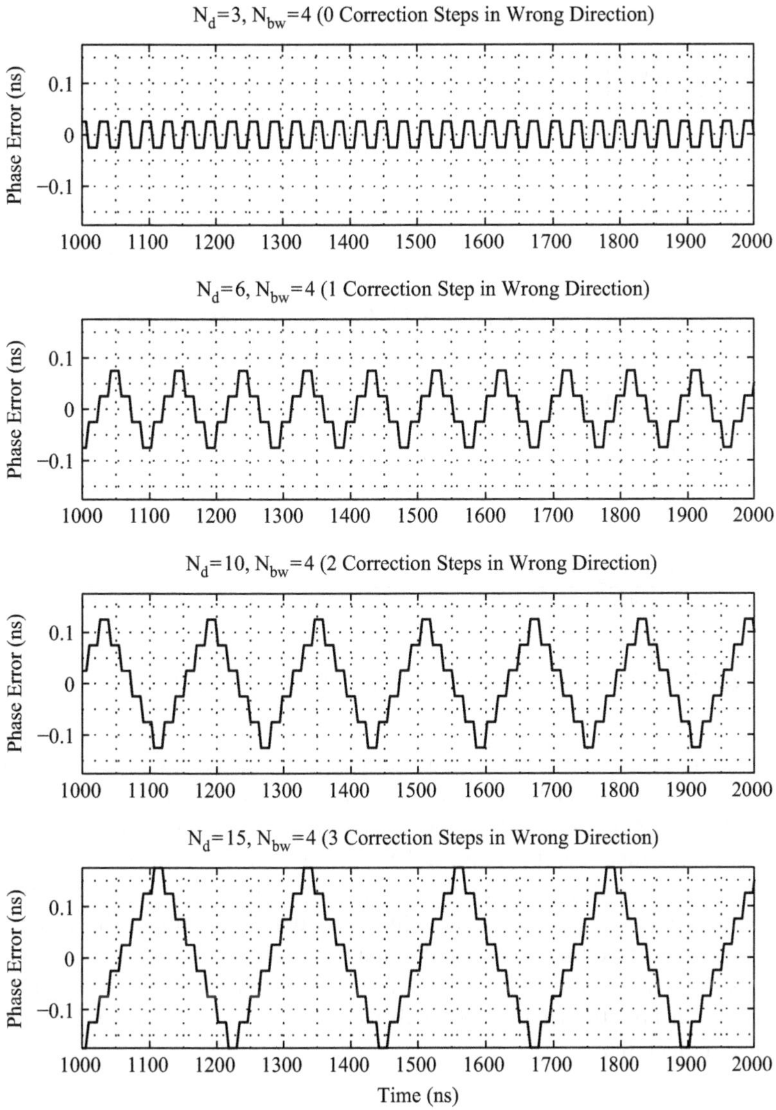


**Fig. 6.34.** DLL limit cycles as a function of $N_d$ in the locked state

DLL model as past examples in this section. Clock is running at 250 MHz with 10%, 500 KHz spread, $N_d$ is fixed at 3 and $N_{bw}$ is parameterized to demonstrate the slew rate tracking criterion. For $N_{bw} = 4$, $r > r_s$ and phase error is close to the $\pm d_r$ structural value and spread spectrum modulation is tracked. For $N_{bw} = 16$, $r$ is slightly less than $r_s$ and phase error starts to get amplified. Finally, at $N_{bw} = 32$, the DLL has very small bandwidth and is incapable of tracking the SS input. Phase error

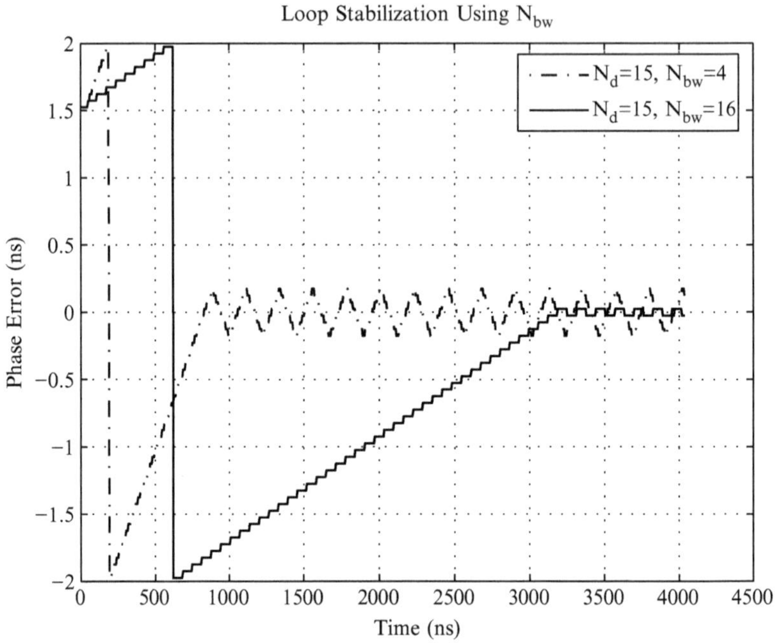


**Fig. 6.35.** DLL loop stabilization using $N_{bw}$ (DLL is running at $T_{CLKIN} = 4$ ns)

is very similar to the SS period variation, and there is little difference between this DLL and an open loop delay line matched to the unmodulated input clock period.

It is worth mentioning that this tracking analysis is only relevant if the DCDL is matched to some function of $T_{CLKIN}$. If the application involves delay-only matching (Fig. 6.4b) then the phase error should not be affected by input clock spreading. All DLL outputs will vary in the same fashion and delay variation will be the same among all outputs resulting in zero additional phase errors.

This section has established two stability criteria and a tracking criterion for digital DLLs. It must be stressed that this analysis applies only to first order digital DLLs as defined in Sect. 6.6.

### 6.6.4 Lock Acquisition

The lock acquisition procedure is the most open-ended controller aspect from a design perspective. Multiple approaches are possible. A simple controller like the one depicted in Fig. 6.31 will produce lock acquisition profiles similar to the ones shown in Fig. 6.35, where lock is achieved with a constant slew rate equal to $d_r/(T_{CLKIN}N_{bw})$. Lock acquisition duration is, therefore, a strong function of $N_{bw}$ selection.

Faster lock acquisition methods are possible such as variable slew rate (higher $r$ during acquisition for lock time reduction and lower $r$ during lock maintenance for stability), binary search [21] and open loop synchronous mirror delay locking

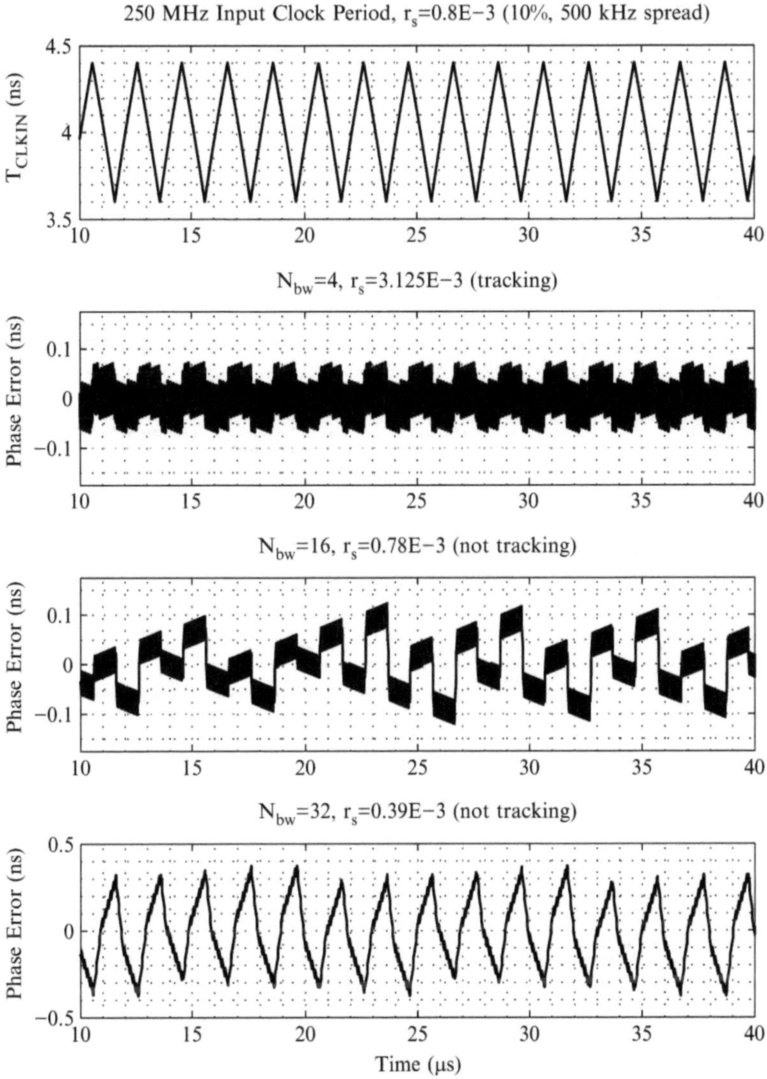


**Fig. 6.36.** DLL phase error tracking of SS clock as a function of $N_{bw}$

[22, 23]. The binary search method computes every bit of a logarithmic delay setting from MSB to LSB in a successive approximation fashion in $\log N$ steps where $N$ is the total number of DCDL settings. The synchronous mirror delay method measures the difference between the unknown delay and a full clock period in a single step. This difference is then installed in the DCDL and the system can either revert to regular feedback DLL mode with standard control or remain open loop until the next time a measurement is made. This method is by far the fastest and can achieve delay lock within a few cycles independent of delay size. All of the above lock acquisition

methods are illustrated in Fig. 6.37 using our standard DLL behavioral model ($N_{bw} = 16$) for $T_{CLKIN} = 250$ MHz. The reduction in lock time is rather obvious going from top to bottom.

Fast lock acquisition can be essential for applications that involve clock gating and suspension modes such as memories and low-power/portable systems. Ability to relock quickly as opposed to saving the previous setting before entering the low power state and reinstating it is much more desirable due to voltage/temperature tracking and is mandatory for dynamic voltage/frequency applications.

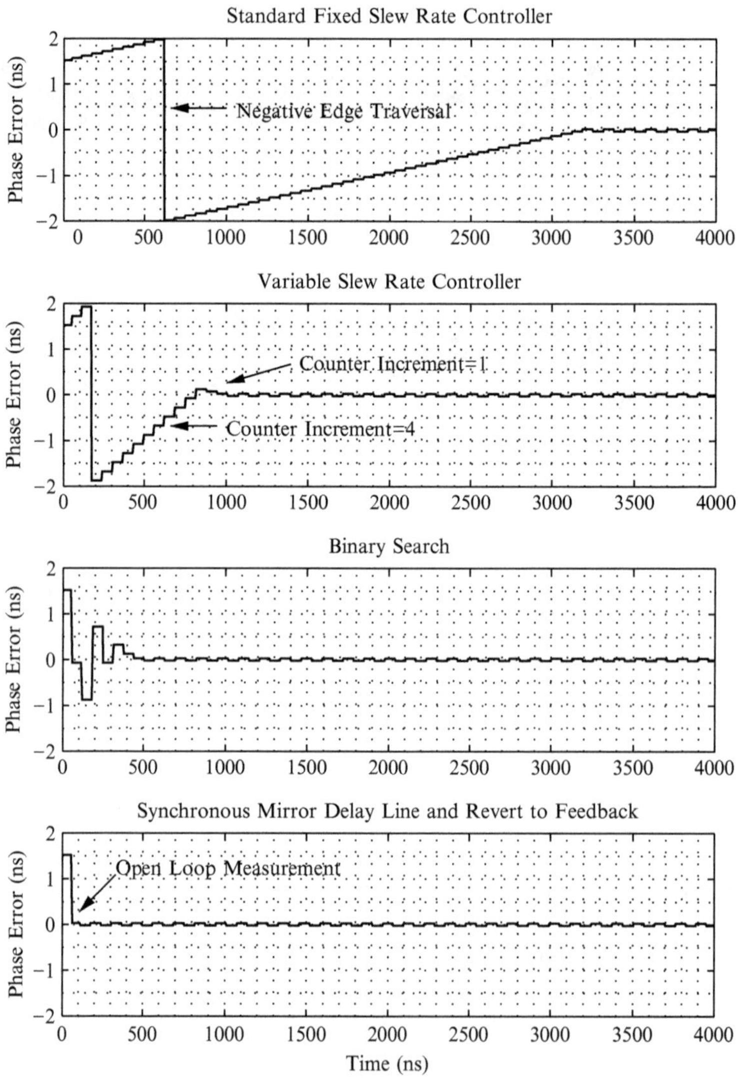


**Fig. 6.37.** DLL lock acquisition profiles ($T_{CLKIN} = 250$ MHz, $N_{bw} = 16$)

## 6.7 Putting it All Together

Sections 6.4 through 6.6 have described in considerable detail all DLL system components. In this section, we briefly review important design parameters and outline a basic design flow.

A digital DLL can be generally described by the following design parameters:

$d_{sw}$ Phase detector sampling window
$d_r$ Delay resolution
$D_{min}$ Minimum DCDL delay
$D_{max}$ Maximum DCDL delay
$N_d$ Control mechanism delay
$N_{bw}$ Inverse of DCDL correction issue rate

Phase detector design/selection should always minimize $d_{sw}$. There is really no tradeoff here. Typically, a designer develops a good phase detector with a small sampling window, and then this can be reused for the vast majority of DLL instances on the chip. Next, an appropriate $d_r$ should be selected based on application requirements and desired DCDL complexity, area and power. $D_{min}$ and $D_{max}$ will be set by the entire range of input clock frequencies that need to be tracked and also by the entire process, voltage, and temperature operating space. Under no circumstances should a DLL prove to be the limiting factor in a chip design in terms of $(F_{MAX}, V_{MAX})$ or $(F_{MIN}, V_{MIN})$. Substantial margin should always be provided to allow for small changes in production specifications. The controller should be designed to perform successful lock acquisition for each PVT and for each input clock frequency. $N_d$ should be minimized to the extent possible. In the lock state, the controller should maintain lock with the minimum theoretical phase error of $\pm d_r$ (wherever possible). $N_{bw}$ should be greater than $N_d$ but small enough to allow for input frequency variation tracking if the application demands it.

Other design aspects such as DCDL output glitching, controller robustness around non-locking edge traversal, etc., should be thoroughly investigated and analyzed because they can prove catastrophic.

## 6.8 Noise Considerations

Phase noise and jitter transfer have been traditionally analyzed in the frequency domain (Chap. 5.1). In this section, we focus on the time domain because of the difficulties introduced by the non-linear nature of the system. For a comprehensive analog DLL noise analysis in the frequency domain, the reader is referred to [17, 18]. Unlike a PLL which can filter reference clock jitter as described in Chap. 5.1, a digital DLL will add jitter to the reference clock through three primary mechanisms:

The first mechanism is the structural limit cycle which in a well designed DLL will add $\pm d_r$ of phase jitter to the output clock. From a frequency domain perspective, this noise will appear around the $f_{CLKIN}/N_{bw}$ frequency band. The second mechanism is the equivalent of jitter peaking in the analog domain: The digital DLL can amplify PLL jitter because of delayed response. This stems from the fact that the DLL is not capable of determining whether a disturbance in the reference vs. output clock must be tracked (i.e., spread spectrum or DCDL environmental change) or not tracked and potentially canceled out (reference clock jitter). Since all DLLs are designed to correct phase error, a jitter disturbance in the reference clock can be tracked before it shows up at the output of the delay line in a cycle later, and this can cause jitter amplification in the next cycle. In general though, this jitter amplification will be limited and on the order of $\pm d_r$. Furthermore, it does not accumulate as in the PLL case (Chap. 5.1) and it will add a small cycle time penalty. It is very simple to alleviate both the structural and the jitter amplification components either by using a ternary phase detector with a NOP state [24] or by opening the loop and tracking the phase error with an auxiliary mechanism [25]. The final and most potent mechanism is supply noise induced jitter. Supply noise is particularly dangerous for DLLs because we have clock propagation through a long delay line which typically has significant delay sensitivity to power supply changes. The remainder of this section frames the problem and presents a simple analytical model.

For the analysis in this section, we will adopt the terminology of Sect. 5.2. We are trying to compute the maximum period jitter generated by a fixed delay $D$ subject to known sinusoidal supply noise, as a function of noise amplitude, noise frequency, delay amount $D$, and clock period $T$. Before proceeding, we define the following symbols:

| | |
|---|---|
| $V(t)$ | Power supply voltage as a function of time |
| $V_{DD}$ | Nominal power supply voltage value |
| $A_n$ | Supply voltage noise amplitude (normalized to $V_{DD}$) |
| $f_n$ | Supply voltage noise frequency |
| $\phi$ | Supply voltage noise phase at $t = 0$ |
| $D$ | DLL delay (also referred to as insertion delay) |
| $T$ | Clock period |
| $t_i$ | Time of $i$th clock positive edge at the input of insertion delay $D$ with respect to an arbitrary reference |

We define the $k$th order period jitter introduced by a DLL with a constant insertion delay $D$ as:

$$\Phi'_k = \max_i |t_{i+k} - t_i - kT| . \tag{6.43}$$

Equation (6.43) is interpreted as the maximum amount that the sum of $k$ consecutive cycles of a given clock of period $T$ can differ from the nominal value. We have introduced an additional subscript $k$ (order) with respect to the jitter definitions of Sect. 5.2 to allow one more parameter in this analysis and have the capability to address multi-cycle paths and long term jitter requirements imposed by certain applications. The subscript $k$ is only relevant for the relative jitter definitions of Sect. 5.2 such as period jitter ($\Phi'[n]$) and cycle-to-cycle jitter ($\Phi''[n]$). The relationship between Eq. (6.43) and the period jitter definition of Sect. 5.2 is

$$\Phi'_1 = \max_n \left|\Phi'[n]\right| . \tag{6.44}$$

In order to derive an analytical expression for $\Phi_k'$ which will help us gain some insight regarding supply noise induced jitter, we start by assuming sinusoidal supply noise:

$$V(t) = V_{\text{DD}}[1 + A_n \sin(2\pi f_n t + \phi)]. \tag{6.45}$$

Period jitter is generated by modulating the insertion delay $D$ through supply noise. We make a second assumption, that the actual value $D_m(t)$ of the insertion delay $D$ at time t (where t is the time that an ideal positive edge arrives at the input of the delay $D$) is given by the following formula:

$$D_m(t) = D \times \left(2 - \frac{\overline{V}(t,D)}{V_{\text{DD}}}\right), \tag{6.46}$$

where $\overline{V}(t,D)$ is simply the forward moving average of $V(t)$ over a period of time equal to insertion delay $D$:

$$\overline{V}(t,D) = \frac{1}{D} \times \int_t^{t+D} V(\tau)d\tau. \tag{6.47}$$

Equation (6.46) simply states that the DLL insertion delay is linearly dependent on its supply voltage averaged over its flight time (i.e., an $x\%$ increase in average supply voltage results in an $x\%$ decrease in delay). This is an acceptable approximation for values of $\overline{V}(t,D)$ reasonably close to nominal $V_{\text{DD}}$. Obviously, this is not true for much larger or smaller values of $\overline{V}(t,D)$ and the axes crossing points implied by Eq. (6.46) are meaningless. There is an additional approximation in Eq. (6.47) since the moving average is computed over a constant time interval $D$ whereas in reality the time interval should vary due to supply noise modulation. This approximation should introduce little error yet make an analytical approach tractable.

Now that we have explicitly defined $D_m(t)$, we can write an expression for $t_i$ (time of $i$th positive edge at the output of delay $D$):

$$t_i = iT + D_m(iT). \tag{6.48}$$

Substituting (6.48) in (6.43) yields

$$\Phi_k' = \max_i |D_m((i+k)T) - D_m(iT)|. \tag{6.49}$$

After substituting (6.46) in (6.49) and with minimal manipulation we have

$$\Phi_k' = \frac{D}{V_{\text{DD}}} \times \max_i \left|\overline{V}(iT,D) - \overline{V}((i+k)T,D)\right|. \tag{6.50}$$

Since we have an expression for supply voltage, we can obtain a closed form expression for the moving average by substituting Eq. (6.45) in (6.47).

$$\overline{V}(t,D) = V_{\text{DD}} + \frac{A_n V_{\text{DD}}}{D} \int_t^{t+D} \sin(2\pi f_n \tau + \phi)d\tau. \tag{6.51}$$

Performing the simple integration yields

$$\overline{V}(t,D) = V_{\mathrm{DD}} + \frac{A_n V_{\mathrm{DD}}}{2\pi f_n D}[\cos(2\pi f_n t + \phi) - \cos(2\pi f_n (t+D) + \phi)]. \tag{6.52}$$

Substituting (6.52) in (6.50) yields

$$\Phi'_k = \frac{A_n}{2\pi f_n} \max_i \left| \begin{array}{l} \cos(2\pi f_n iT + \phi) - \cos(2\pi f_n (iT + D) + \phi) - \\ \cos(2\pi f_n (i+k)T + \phi) + \cos(2\pi f_n ((i+k)T + D) + \phi) \end{array} \right| . \tag{6.53}$$

We now recall the following well-known trigonometric identity:

$$\cos u + \cos v = 2\cos\left(\frac{u+v}{2}\right)\cos\left(\frac{u-v}{2}\right). \tag{6.54}$$

Applying (6.54) on (6.53) yields

$$\Phi'_k = \frac{A_n}{\pi f_n} \max_i |\cos(\pi f_n (2i+k)T + \phi)[\cos(\pi f_n (kT + D)) - \cos(\pi f_n (kT - D))]| . \tag{6.55}$$

Equation (6.55) is expressed in a very convenient form which will enable us to dispense with the maximization operation by inspection: Only the first cosine term ($\cos(\pi f_n (2i+k)T + \phi)$) is a function of $i$, and its maximum value for every possible $i$ is 1. Therefore, Eq. (6.55) becomes

$$\Phi'_k = \frac{A_n}{\pi f_n} |\cos(\pi f_n (kT + D)) - \cos(\pi f_n (kT - D))| . \tag{6.56}$$

Application of one more trigonometric identity results in the following expression which is more desirable since it separates the effects of variables $T$ and $D$:

$$\boxed{\Phi'_k = \frac{2A_n}{\pi f_n} |\sin(\pi f_n kT)\sin(\pi f_n D)| .} \tag{6.57}$$

Equation (6.57) constitutes a closed form analytical expression of worst case period jitter as a function of supply noise frequency and amplitude ($f_n$,$A_n$), clock period ($T$), and DLL insertion delay ($D$). Parameter $k$ denotes whether we are interested in single period jitter ($k = 1$) or longer term (multi-cycle) jitter ($k > 1$). In order to gain some insight into this expression and determine whether it makes physical sense, let us focus on the two plots of Fig. 6.38. Both diagrams plot Eq. (6.57) under certain conditions. The top graph shows $\Phi'_1$ as a function of noise frequency $f_n$ for 2 different insertion delays. We make the following observations: First, we note that for all insertion delay values we have zero jitter at $f_n = 0$ (trivial) and at noise frequencies that are integral multiples of the clock frequency (500 MHz and 1 GHz). This is entirely expected because when this occurs, consecutive clock edges will be delay-modulated identically going through any insertion delay $D$ resulting in zero jitter. Consecutive clock edges will experience the same noise. A second observation is that we have zero jitter at noise frequencies that are integral multiples of the inverse insertion delay (333 MHz and 667 MHz in the top graph for the case where $D = 3$ ns). This is also expected because if the integration interval in the moving average Eq. (6.51) is an integral multiple of the period of the sinusoid that is being integrated, then the integration result is zero. Integrating a sinusoid over an integral multiple of periods always results in zero. This is also clearly observed on the bottom graph of Fig. 6.38 where $\Phi_1'$ is plotted as a function of insertion delay for 3 different noise frequencies. All zeroes on the graph are at insertion delays equal to an integral multiple of the noise period. A final observation is that with increasing noise frequency, period jitter tends to decrease due to the $1/f_n$ term in Eq. (6.57). This happens because as $f_n$ increases, the amount of noise averaging in delay line $D$ also increases and the supply moving average will approach the nominal value.

Figure 6.39 shows the jitter dependence on noise frequency and insertion delay simultaneously in 2 dimensions for 1 GHz and 500 MHz clocks respectively. Such contour plots for each relevant clock frequency are very easy to construct using Eq. (6.57) and can provide a broad perspective early in the design phase in order to help determine important implementation aspects such as:

1. Identification of worst case period jitter and particularly undesired noise frequencies.
2. Derivation of detailed specifications of supply voltage frequency and amplitude if external to the IC.
3. Identification of need for internal voltage regulation to minimize $A_n$ if external supply specifications turn out to be too stringent.

Application of Eq. (6.57) is not restricted to delay locked loops but can be applied to any open-loop clock buffer in order to determine supply induced period jitter. DLLs are very convenient because insertion delay $D$ is well defined and tracked with feedback. In an open-loop buffer, $D$ will vary with PVT and can only be estimated to a limited degree of accuracy. Yet, the analysis above reveals an interesting aspect of supply induced jitter: In an ideal world, it can be canceled out by selecting an appropriate clock buffer delay $D$ as Figs. 6.38 and 6.39 indicate. This is not really possible because the supply noise frequency is never known a priori and typically does not consist of a single sinusoid. A supply voltage waveform will typically contain multiple natural frequencies introduced by the package, board, and external decoupling capacitance network. In addition, it will contain particular solutions related to the system clock frequency. Natural frequencies are unknown at IC design time, since they are system dependent. Particular solutions are also hard to estimate especially in programmable ICs. One could imagine the possibility of constructing a feedback system which measures supply noise frequency and then adjusts clock buffer delay accordingly to minimize jitter. Feedback must be close to instantaneous in order to avoid the worst case, and this makes a digital controller very hard to implement. Assuming that supply noise frequency can actually be measured, such a system could potentially track an undesired natural frequency that should not have been there in the first place. Controlling noise amplitude, undesired natural frequencies and clock

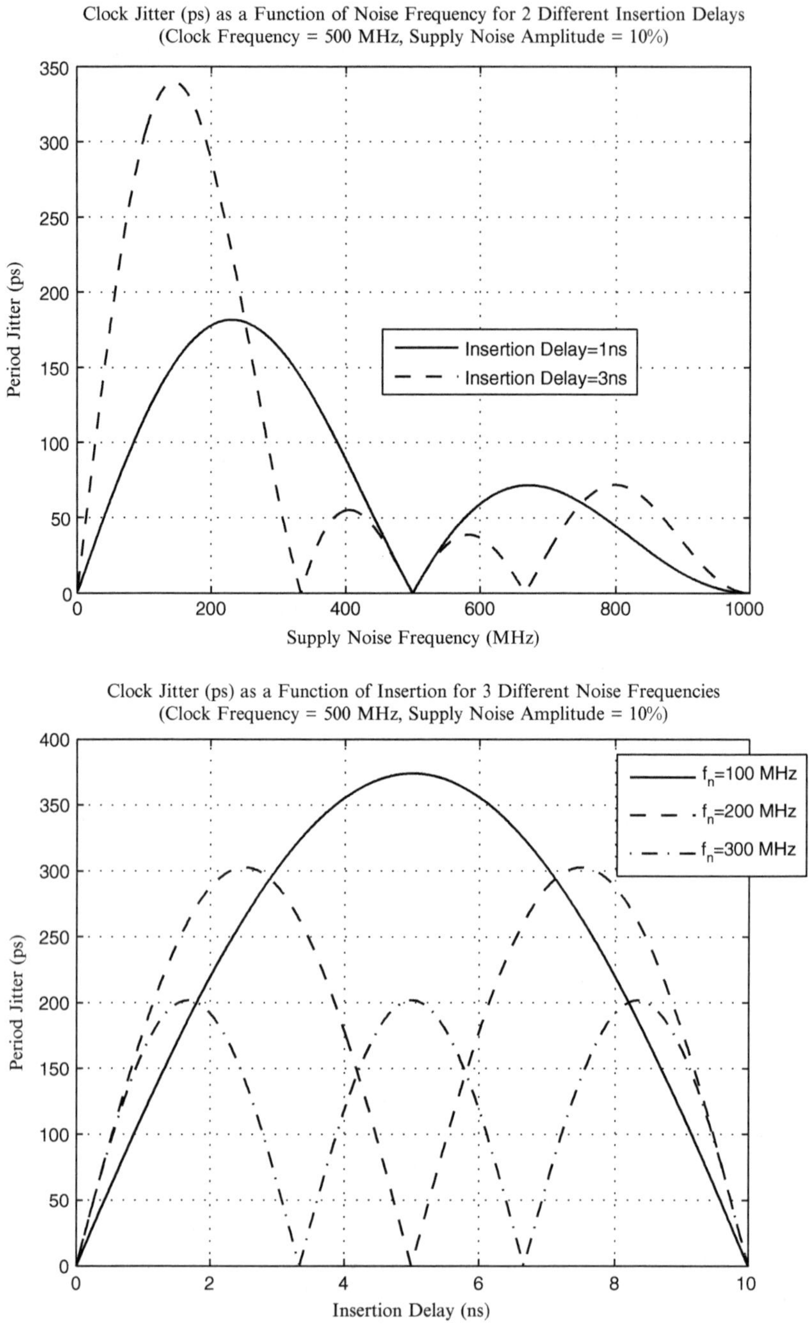


**Fig. 6.38.** First order period jitter ($\Phi_1'$) as a function of supply noise frequency and insertion delay

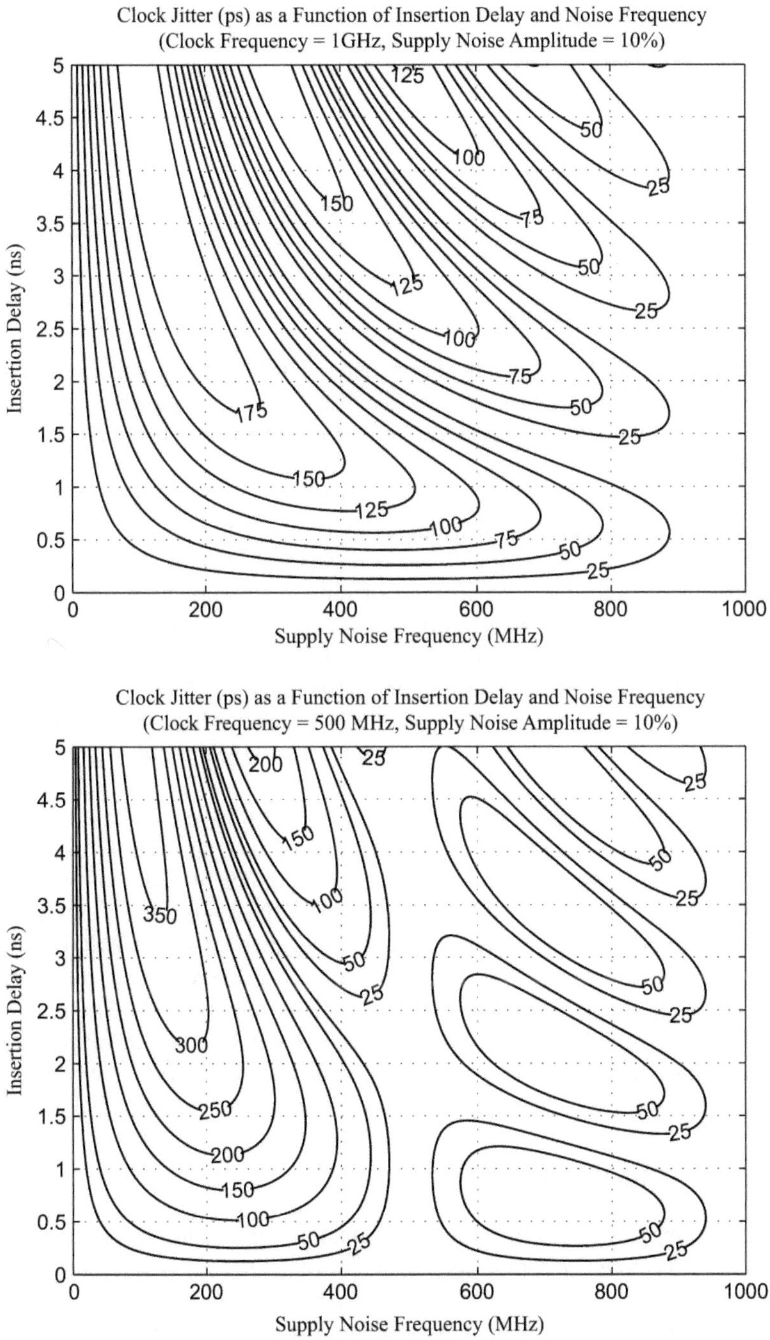


**Fig. 6.39.** First order period jitter ($\Phi_1'$) as a function of insertion delay and supply noise frequency in 2 dimensions

buffer delays are the only effective methods for reducing supply induced jitter that the author is aware of.

## 6.9 Advanced Applications

In this section, we review a number of non-standard DLL applications and demonstrate the versatility of delay lock techniques.

### 6.9.1 Duty Cycle Correction

DLLs are heavily used in I/O applications such as DRAM interfaces. In such applications, 50% duty cycle is very important because the interfaces are typically double-data-rate (DDR) and both clock phases are used for signaling. Duty cycle distortion is a source of deterministic jitter and must be reduced.

It is fairly common for DLLs to include a separate stage of duty cycle correction as an output stage [26] or even distributed duty cycle control at various points along the signal path [27]. These are typically analog methods that involve current-steering based structures. An alternative method of duty cycle correction implemented in DRAMs [28, 29] is shown in Fig. 6.40. It uses two independent DLLs. The first one (PD1, DCDL1, CTL1) is the main loop and is responsible for locking the output to the desired CLKIN phase. The second one (PD2, DCDL2, CTL2) has an inverting delay line and locks the positive edge of CLK2 (which has been generated by a negative edge of CLKIN) to the positive edge of CLK1 (which has been generated by a positive edge of CLKIN).

When the second loop locks, CLK1 and CLK2 will have the phase relationship shown in Fig. 6.40. A phase interpolator with equal weights (similar to the one in Fig. 6.26) is used as the final clock output stage. The positive edge of output clock CLKOUT will be generated by the positive edge of CLK1 since both CLK1 and CLK2 have phase locked positive edges. The negative edge of CLKOUT will be generated by the average of the CLK1 and CLK2 negative edges, resulting in duty cycle correction. Figure 6.40 assumes for simplicity that the phase interpolator is a zero-delay, thus non-causal circuit. In reality, there will be delay involved but it has no effect in the overall correction capability of this method. The amount of duty cycle correction possible depends on the *RC* of the interpolator as described in Sect. 6.5.3.

### 6.9.2 Clock Multiplication

In Sect. 6.3, the basic idea of DLL-based clock multiplication has been presented (Fig. 6.4d). This method of phase-locking a multi-tap DCDL to a reference clock period and then using the intermediate phases to generate a frequency multiple through a clock-multiplying structure such as a multi-port toggle flop has an obvious frequency limitation: it is limited by the minimum delay $D_{\min}$ through each DCDL segment. As an example, the highest output frequency that can be generated by the structure of Fig. 6.4d is subject to the following constraint:

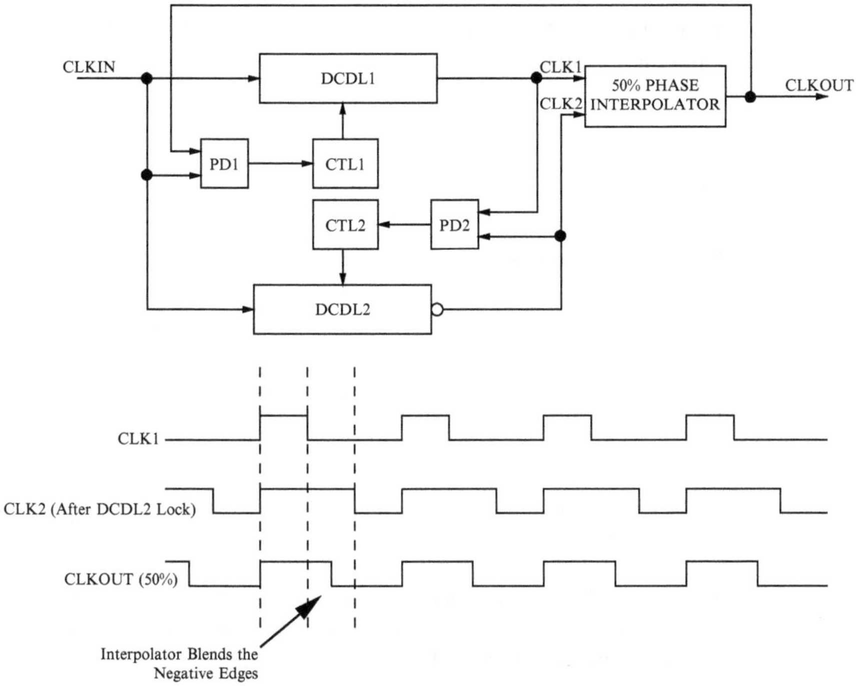


**Fig. 6.40.** Duty cycle correction in a DLL environment

$$f_{\text{out}} \leq \frac{1}{2D_{\text{min}}}, \tag{6.58}$$

where $D_{\text{min}}$ is the minimum possible delay through each of the 8 DCDL segments.

Reference [30] proposes an interesting way to remove this upper bound. The key observation is that the sum of the $N$ DCDLs (where $N/2$ is the intended clock multipler) can be phase-locked to multiple reference clock periods ($M$) instead of 1 and the resulting phases can be re-sorted before driving the final clock multipler stage. For this method to work, $M$ must not divide $N$ exactly but have a non-zero modulus. Figure 6.41 illustrates this concept. We wish to perform clock multiplication by 4 ($N = 8$), but we want to generate an output clock frequency that is faster than the constraint in (6.58) would allow for. Instead of phase-locking all 8 DCDL segments to one reference clock period $T_{\text{ref}}$, we choose to phase-lock them to $3T_{\text{ref}}$ ($M = 3$) which evidently allows for a much larger $D_{\text{min}}$ per DCDL segment. Phases are sorted modulus $T_{\text{ref}}$ before being routed to the final clock multiplication stage. The constraint in (6.58) now becomes

$$f_{\text{out}} \leq \frac{3}{2D_{\text{min}}}, \tag{6.59}$$

and the upper bound for the final output frequency has become more loose by a factor of 3. In [30], a 40 GHz output clock is generated by a 4.4445 GHz reference ($M = 2$,

$N = 9$) by employing an LC-based clock multiplier which can produce a full 360° oscillation per stimulating phase.

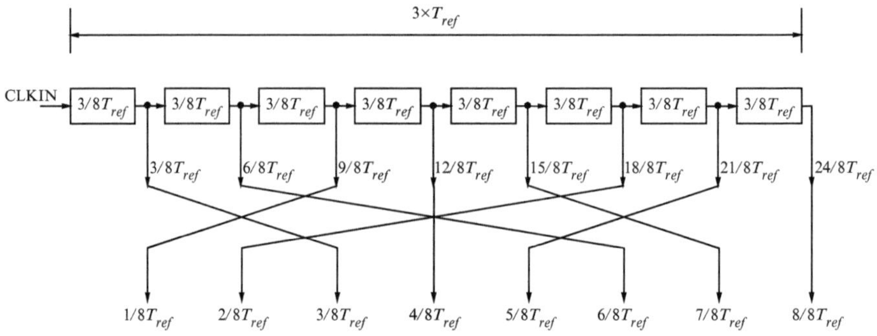


**Fig. 6.41.** Multiperiod DCDL locking for high output frequency clock multiplication

Clock multiplying DLLs with alternative organizations have been proposed in the past years [31, 32] and are actively pursued as a credible alternative to PLLs for clock generation due to superior jitter performance.

### 6.9.3 Infinite Dynamic Range

Lock acquisition and phase capture can be limited in standard DLLs with a reduced DCDL dynamic range as explained in Sects. 6.5 and 6.6. One way to address this issue is to extend the delay line dynamic range through additional DCDL stages or conditional inversion (Sect. 6.6.2) coupled with increased complexity in the FSM controller. An alternative method is to employ a dual DLL architecture where the core loop generates phases that span the input clock period range, and the peripheral loop blends a selected pair of consecutive phases to match the reference clock [1, 33]. This architecture is shown in Fig.6.42. The core DLL phase-locks at 180° and generates 6 phases that are 30° apart. The block labeled "Phase Selection" contains two $3 \times 1$ multiplexers that select a pair of phases $\phi$ and $\psi$ to be interpolated. Before reaching the interpolator, these phases are conditionally inverted inside the "Selective Phase Inversion" block so that the cascade of these three structures can generate all phases at 30° increments that span the entire 0–360° phase interval. Finally, conditionally inverted phases $\phi'$ and $\psi'$ are interpolated. Although the phase interpolator structure in [33] is CML-based, the analysis is very similar to the full swing interpolator of Sect. 6.5.3. The output of the interpolator $\Theta$ is phase-locked to the reference clock ("ref CLK" in Fig.6.42). The feedback of the peripheral DLL closes with a bang–bang phase detector and an FSM that controls the multiplexers, conditional inversion block and phase interpolator. The infinite dynamic range stems from the fact that the FSM controller can keep on selecting phase pairs to be interpolated around the entire 360° phase interval and eventually the selected pair will encompass the phase of the reference clock. Interpolation will then guarantee phase lock within the accuracy of the interpolator. In this architecture, the input core DLL clock ("in CLK") and the reference clock ("ref CLK") can be the same clock at 0-phase or identical frequency clocks at any phase relationship. They can even be plesiochronous clocks (of similar but not equal frequency), and this case will be addressed in Sect. 6.9.4.

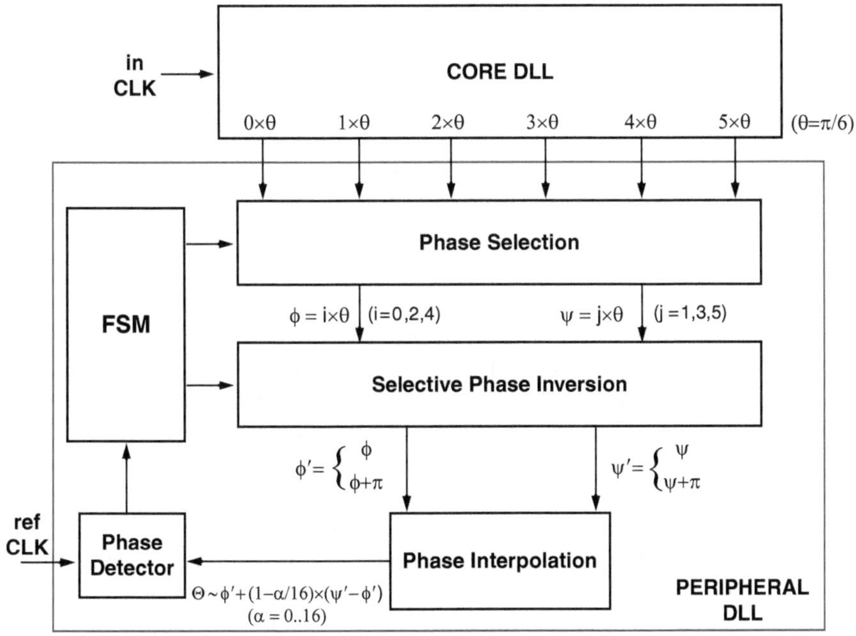


**Fig. 6.42.** Dual DLL architecture for virtually infinite phase capture range. Reproduced with permission from [33], ©1997 IEEE

DLLs with quadrature phase mixing had been proposed before [26] and they also exhibit infinite dynamic range. The main difference between the dual DLL architecture and quadrature mixing is that the phase interpolation is not limited to 90° clocks. Interpolation can blend phases at smaller intervals (30° in this case) which will have fewer slew rate limitations and better noise performance [33].

### 6.9.4 Clock-Data Recovery

The full power of the method of Fig. 6.42 cannot be fully appreciated unless we consider the case where "in CLK" and "ref CLK" are plesiochronous and vary slightly in frequency. This architecture can achieve lock in the plesiochronous case because the FSM controller can change the weight of the phase mixing in each cycle in such a way so as to generate a clock that is effectively faster or slower than "in CLK".

When the interpolator reaches the end of its dynamic range, then the phase selection multiplexer chooses a different pair of phases for mixing that are either leading or lagging the previous pair depending on whether "ref CLK" is faster or slower than "in CLK", respectively. This property makes this method ideally suited for clock-data recovery (CDR) applications where a sampling clock must be generated at the receiver and used to sample the incoming data. The generated clock must match the frequency of the received data using multiple free-running locally-generated phases.

This concept is illustrated in Fig. 6.43. This structure forms the basis of a large class of clock-data recovery architectures implemented as part of various high-speed signaling standards such as PCI Express, XAUI, and FBD (Fully-Buffered-DIMM) [34]. The core DLL of Fig. 6.42 has been replaced with a multi-phase PLL that generates a local high frequency clock which can be plesiochronous to the incoming data. In order to collect enough information for clock recovery, a CDR needs some form of data oversampling. In this case, oversampling is accomplished by generating both an in-phase clock (CLKI) and a quadrature clock (CLKQ). High-speed incoming data are sampled with both clocks using two separate slicers (I-SLICER, Q-SLICER). The data are then deserialized typically by a factor of 10 and retimed in a divided clock domain which is usually a divided-down version of CLKQ. The block labeled "EDGE DETECTION/PHASE DETECTION" examines both I and Q deserialized data and determines whether the current sampling clocks CLKI and CLKQ are fast or slow with respect to the incoming data. This decision can be made deterministically due to oversampling. This decision is then forwarded to the CDR FSM which implements a control algorithm and can change the settings of the phase selectors and interpolators to advance or delay the sampling clocks CLKI and CLKQ. The FSM algorithm will determine important capabilities such as phase capture profile, frequency difference between local PLL and incoming data that can be safely absorbed and limit cycle behavior which will cause deterministic jitter in the system. The output of the Q-DESERIALIZER represents parallel data sampled safely with maximum margin in the middle of the eye diagram and is forwarded to the rest of the system for higher layer processing.

All blocks of Fig. 6.43 with the sole exception of the multi-phase PLL constitute essentially a DLL. The role of the delay line is fulfilled by the combination of phase selectors and interpolators that can effectively delay or advance an input clock and the phase detection is carried out by the slicers, deserializers, and decision block. In a typical SERDES application [34], this DLL structure is present on all receiver bits (*lanes* in SERDES terminology) whereas the multi-phase PLL is amortized over a number of receive and transmit lanes. In this application, it is desirable to implement a second order control structure in order to track more effectively the frequency differences between the incoming data and the locally generated clock. Second order control design has not been addressed in this chapter and is beyond our scope.

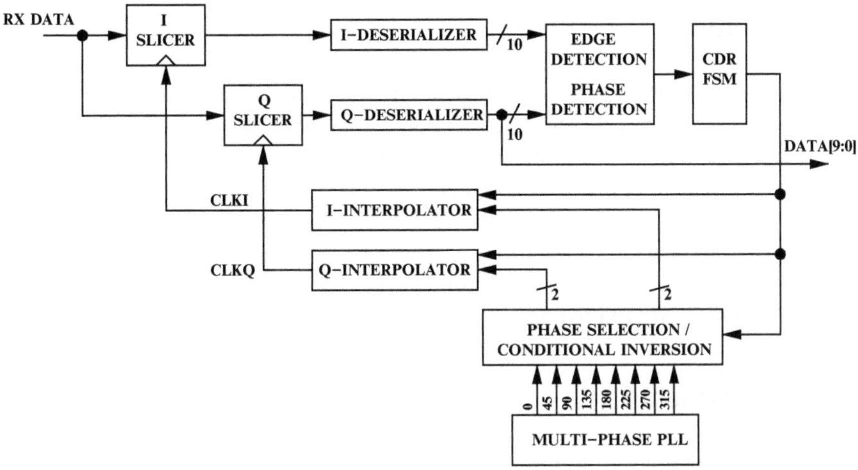


**Fig. 6.43.** DLL-based clock-data recovery loop

### 6.9.5 On-Chip Temperature Sensing

An appropriately calibrated and accurately measured delay line can be used as a temperature sensor since temperature affects digital gate delay. It has been shown [35] that when a delay line is trimmed to a fixed delay at a certain temperature in a production environment across multiple chips, it can act as an accurate temperature sensor. Trimming to a fixed delay at production has the effect of suppressing sensitivity to process and voltage. A trimmed delay line will vary only due to temperature across all fused chips and can have an accuracy of $\pm 2°C$ across the 0–100°C temperature range [35].

Two DLLs constitute this system: A reference DLL is phase locked to a known reference clock which is distributed to the sensor. Tapping at intermediate delay line elements can generate fixed reference delays that are fractions of the original reference clock period. A secondary DLL phase locks the sensing delay line to a fixed delay generated by tapping an intermediate delay output of the reference DLL. This step must be performed at a controlled junction temperature on all chips during manufacturing. The setting in the locked state is then fused and this step performs the trimming (normalization) of the sensing delay line.

During measurement, the temperature can vary and the trimmed sensing line will deviate from its fixed delay value. The secondary DLL then treats it as an unknown delay and measures it using intermediate delay outputs of the reference DLL as the measuring delay line. The digital setting output at the locked state will be an accurate representation of temperature. This dual DLL scheme can provide good accuracy and a high measurement bandwidth (5 ksamples/s) and is appropriate for microprocessor thermal monitoring.

## 6.10 Conclusion

This chapter has presented an overview of the basic digital DLL structure and its components in addition to a survey of a range of applications. DLLs are an essential component of modern VLSI systems. Increased clocking system complexity, process variation, and wider supply voltage ranges will make their use even more prevalent than it is today. Furthermore, their time-to-digital conversion capabilities make them an attractive candidate for measuring analog on-chip quantities (i.e., voltage drop, temperature, process variation) either for valuable data collection or system normalization to dynamically remove variation.

## Acknowledgments

A preliminary version of the power supply noise analytical model in Sect. 6.8 was developed jointly with John Eble when we were colleagues at the Alpha Development Group, Compaq Computer.

## References

[1] W. Dally and J. Poulton, *Digital Systems Engineering*. Cambridge University Press, New York, NY, 1998.

[2] M. Johnson and E. Hudson, A variable delay line PLL for CPU-coprocessor synchronization. *IEEE J. Solid-State Circuits*, 23(5), 1218–1223, 1988.

[3] C. Mead and L. Conway, *Introduction to VLSI Systems*. Addison Wesley, Reading, MA, 1980.

[4] L. Lamport, Buridan's principle, October 1984. [Online]. Available: http://research. microsoft.com/users/lamport/pubs/buridan.pdf

[5] T. Chaney and C. Molnar, Anomalous behavior of synchronizer and arbiter circuits. *IEEE Trans. Comput.*, C-22(4), 421–422, 1973.

[6] H. Veendrick, The behaviour of flip-flops used as synchronizers and prediction of their failure rate. *IEEE J. Solid-State Circuits*, 15(2), 169–176, 1980.

[7] A. Drake, *Fundamentals of Applied Probability Theory*. McGraw-Hill, New York, 1967.

[8] Nanoscale Integration and Modeling Group, Arizona State University, Predictive technology model. [Online]. Available: http://www.eas.asu.edu/ptm.

[9] Y. Cao, T. Sato, M. Orshansky, D. Sylvester, and C. Hu, New paradigm of predictive MOSFET and interconnect modeling for early circuit simulation. In: *Proc. CICC Custom Integrated Circuits Conference*, 2000, pp. 201–204.

[10] W. Zhao and Y. Cao, New generation of predictive technology model for sub-45 nm early design exploration. *IEEE Trans. Electron Devices*, 53(11), 2816–2823, 2006.

[11] A. Hatakeyama, H. Mochizuki, T. Aikawa, M. Takita, Y. Ishii, H. Tsuboi, S. Fujioka, S. Yamaguchi, M. Koga, Y. Serizawa, K. Nishimura, K. Kawabata, Y. Okajima, M. Kawano, H. Kojima, K. Mizutani, T. Anezaki, M. Hasegawa,

and M. Taguchi, A 256-Mb SDRAM using a register-controlled digital DLL. *IEEE J. Solid-State Circuits*, 32(11), 1728–1734, 1997.

[12] F. Lin, J. Miller, A. Schoenfeld, M. Ma, and R. Baker, A register-controlled symmetrical DLL for double-data-rate DRAM. *IEEE J. Solid-State Circuits*, 34(4), 565–568, 1999.

[13] T. Xanthopoulos, D. Bailey, A. Gangwar, M. Gowan, A. Jain, and B. Prewitt, The design and analysis of the clock distribution network for a 1.2 GHz Alpha microprocessor. In: *Digest of Technical Papers IEEE International Solid-State Circuits Conference (ISSCC 2001)*, 2001, pp. 402–403.

[14] S. Sidiropoulos, High performance inter-chip signalling, Ph.D. dissertation, Stanford University, 1998.

[15] J. Maneatis, Design of high-speed CMOS PLLs and DLLs. In: A. Chandrakasan, W. Bowhill, and F. Fox (eds.) *Design of High Performance Microprocessor Circuits*, IEEE Press, New York, 2001, pp. 235–260.

[16] C.-K. K. Yang, Delay-locked loops – an overview. In: B. Razavi, (ed.) *Phase-Locking in High-Peformance Systems*, Wiley-IEEE Press, New York, NY, 2003, pp. 13–22.

[17] M.-J. Lee, W. Dally, T. Greer, H.-T. Ng, R. Farjad-Rad, J. Poulton, and R. Senthinathan, Jitter transfer characteristics of delay-locked loops – theories and design techniques. *IEEE J. Solid-State Circuits*, 38(4), 614–621, 2003.

[18] J. Burnham, G.-Y. Wei, C.-K. K. Yang, and H. Hindi, A comprehensive phase-transfer model for delay-locked loops. In: *Proc. IEEE Custom Integrated Circuits Conference CICC '07*, 2007, pp. 627–630.

[19] J. Lee, K. Kundert, and B. Razavi, Analysis and modeling of bang-bang clock and data recovery circuits. *IEEE J. Solid-State Circuits*, 39(9), 1571–1580, 2004.

[20] J. Sonntag and J. Stonick, A digital clock and data recovery architecture for multi-gigabit/s binary links. *IEEE J. Solid-State Circuits*, 41(8), 1867–1875, 2006.

[21] G.-K. Dehng, J.-M. Hsu, C.-Y. Yang, and S.-I. Liu, Clock-deskew buffer using a SAR-controlled delay-locked loop. *IEEE J. Solid-State Circuits*, 35(8), 1128–1136, 2000.

[22] T. Saeki, Y. Nakaoka, M. Fujita, A. Tanaka, K. Nagata, K. Sakakibara, T. Matano, Y. Hoshino, K. Miyano, S. Isa, S. Nakazawa, E. Kakehashi, J. Drynan, M. Komuro, T. Fukase, H. Iwasaki, M. Takenaka, J. Sekine, M. Igeta, N. Nakanishi, T. Itani, I. Yoshida, K. Yoshino, S. Hashimoto, T. Yoshii, M. Ichinose, T. Imura, M. Uziie, S. Kikuchi, K. Koyama, Y. Fukuzo, and T. Okuda, A 2.5-ns clock access, 250-MHz, 256-Mb SDRAM with synchronous mirror delay. *IEEE J. Solid-State Circuits*, 31(11), 1656–1668, 1996.

[23] K. Sung and L.-S. Kim, A high-resolution synchronous mirror delay using successive approximation register. *IEEE J. Solid-State Circuits*, 39(11), 1997–2004, 2004.

[24] A. Efendovich, Y. Afek, C. Sella, and Z. Bikowsky, Multifrequency zero-jitter delay-locked loop. *IEEE J. Solid-State Circuits*, 29(1), 67–70, 1994.

[25] B. Mesgarzadeh, Low-power low-jitter clock generation and distribution, Ph.D. dissertation, Linkoping University, 2008.

[26] T. Lee, K. Donnelly, J. Ho, J. Zerbe, M. Johnson, and T. Ishikawa, A 2.5 V CMOS delay-locked loop for 18 Mbit, 500 megabyte/s DRAM. *IEEE J. Solid-State Circuits*, 29(12), 1491–1496, 1994.

[27] B. Garlepp, K. Donnelly, J. Kim, P. Chau, J. Zerbe, C. Huang, C. Tran, C. Portmann, D. Stark, Y.-F. Chan, T. Lee, and M. Horowitz, A portable digital DLL for high-speed CMOS interface circuits. *IEEE J. Solid-State Circuits*, 34(5), 632–644, 1999.

[28] J.-B. Lee, K.-H. Kim, C. Yoo, S. Lee, O.-G. Na, C.-Y. Lee, H.-Y. Song, J.-S. Lee, Z.-H. Lee, K.-W. Yeom, H.-J. Chung, I.-W. Seo, M.-S. Chae, Y.-H. Choi, and S.-I. Cho, Digitally-controlled DLL and I/O circuits for 500 Mb/s/pin x16 DDR SDRAM. In: *Digest of Technical Papers IEEE International Solid-State Circuits Conference (ISSCC 2001)*, 2001, pp. 68–69, 431.

[29] W.-J. Yun, H. W. Lee, D. Shin, S. D. Kang, J. Y. Yang, H. O. Lee, D. U. Lee, S. Sim, Y. J. Kim, W. J. Choi, K. S. Song, S. H. Shin, H. H. Choi, H. W. Moon, S. W. Kwack, J. W. Lee, Y. K. Choi, N. K. Park, K. W. Kim, Y. J. Choi, J.-H. Ahn, and Y. S. Yang, A 0.1-to-1.5GHz 4.2mW all-digital DLL with dual duty-cycle correction circuit and update gear circuit for DRAM in 66nm CMOS technology. In: *Digest of Technical Papers IEEE International Solid-State Circuits Conference (ISSCC 2008)*, 2008, pp. 282–283, 613.

[30] C.-N. Chuang and S. luan Liu, A 40GHz DLL-based clock generator in 90nm CMOS technology. In: *Digest of Technical Papers IEEE International Solid-State Circuits Conference (ISSCC 2007)*, 2007, pp. 178–179, 595.

[31] R. Farjad-Rad, W. Dally, H.-T. Ng, R. Senthinathan, M.-J. Lee, R. Rathi, and J. Poulton, A low-power multiplying DLL for low-jitter multigigahertz clock generation in highly integrated digital chips. *IEEE J. Solid-State Circuits*, 37(12), 1804–1812, 2002.

[32] B. Helal, M. Straayer, G.-Y. Wei, and M. Perrott, A highly digital MDLL-based clock multiplier that leverages a self-scrambling time-to-digital converter to achieve subpicosecond jitter performance. *IEEE J. Solid-State Circuits*, 43(4), 855–863, 2008.

[33] S. Sidiropoulos and M. Horowitz, A semidigital dual delay-locked loop. *IEEE J. Solid-State Circuits*, 32(11), 1683–1692, 1997.

[34] D. Stauffer et al., *High Speed Serdes Devices and Applications*. Springer Science and Business Media, New York, 2008.

[35] K. Woo, S. Meninger, T. Xanthopoulos, E. Crain, and D. Ham, Dual-DLL-based CMOS all-digital temperature sensor for microprocessor thermal monitoring. In: *Digest of Technical Papers IEEE International Solid-State Circuits Conference (ISSCC 2009)*, 2009, pp. 68–69.

# 7

# Clocking and Variation

James Tschanz

Intel Corporation

## 7.1 Introduction

As process technology continues to scale, increasing numbers of transistors are integrated onto a single processor die, providing higher levels of performance and additional features (Fig. 1.1). However, the unwanted side-effect of this increased integration is the worsening of variations of all types: static process variations resulting from reduced device dimensions and dynamic voltage and temperature variations. At the same time, transistor degradation and early-life failure are always concerns, but becoming more critical as the number of devices increases.

The clock network on a microprocessor plays a special role in how the processor tolerates variations of all types. On one hand, the clock is very sensitive to variations, and any fluctuations in the clock due to variations can directly impact the frequency of the processor. Thus, it is absolutely critical to design the clock network in such a way that the impact of variations is reduced. On the other hand, variations also present an opportunity in clock network design. Because of the global nature of the clock, it can be used as a "knob" for tolerating variations or reducing their impacts. In this chapter, we will discuss both challenges and opportunities that variations present to clock network design.

## 7.2 Variation Reduction Through Design

The first step in reducing the impacts of variations on the clock network is to design the clocking system in such a way that it is tolerant of variations. This entails properly selecting the clock network topology, driving the clock with a high-quality clock source, and designing and sizing the clock wires and buffers to reduce skew and jitter. These techniques and tradeoffs have already been discussed in Chaps. 2 and 5 and will also be addressed from a physical design perspective in Chap. 8. However, variation-tolerant techniques can also be employed in the datapath to reduce the impacts of variations in both the clock network and the datapath. In this

T. Xanthopoulos (ed.), *Clocking in Modern VLSI Systems*, Integrated Circuits and Systems, 
DOI 10.1007/978-1-4419-0261-0_7, 

chapter, we describe two techniques which employ latches and time-borrowing flip-flops as a method of hiding some of the impact of clock uncertainty and random device variation.

### 7.2.1 Skew and Jitter-Tolerant Design

One method for reducing the impact of clock skew and jitter on the maximum frequency ($F_{MAX}$) of a microprocessor is to employ two-phase latches rather than flip-flops in the critical path [1–3]. This technique has also been discussed in Sect. 3.3.2. Alternately, for speed-critical dynamic circuits, skew-tolerant domino gates can be used with overlapping clocks such that the clock edge never gates the latest-arriving data input [4]. Using latches and dynamic gates in this way provides several important benefits which translate to higher achievable clock frequency. First, this scheme allows time-borrowing: additional slack in one pipeline stage can be taken advantage of in the next pipeline stage so that ideally all stages become equally critical. This is an important feature for designs in which it is not naturally easy to divide the logic into equal-delay stages: rather than one stage becoming the limiter, the logic can be averaged over multiple stages. Secondly, when the data arrive during the transparency window of a latch or dynamic gate, fluctuations in clock edge timing due to clock skew or jitter do not affect the path delay. Therefore, if the transparency window is sufficient, clock skew and jitter do not impact the cycle time of the processor. Of course, proper selection of the transparency window is necessary in order to balance the amount of time-borrowing desired, the variation-tolerance needed, and the power and min-delay overhead for this technique.

Because many designs are flip-flop based and it is sometimes difficult to break up circuits into two phases for a latch-based design, time-borrowing flip-flops are also used [1, 3]. Time-borrowing flip-flops (Fig.7.1) include a transparency window after the rising clock edge such that data arriving in this window are able to propagate through the flip-flop, similar to a latch. Time-borrowing flip-flops require additional clock buffers inside the sequential in order to generate the transparency window; therefore, they have increased power consumption in comparison to a standard non-time-borrowing design; however, this transparency window allows the usage of time-borrowing and tolerance to clock skew and jitter. Additional detailed discussion on time-borrowing flops has been presented in Sect. 3.3.1. Two-phase latches, skew-tolerant domino, and time-borrowing flip-flops are all used commonly in microprocessor circuits to reduce the impacts of clock skew and jitter.

### 7.2.2 Time Borrowing for Datapath Variation Reduction

The previous section described the usage of transparent latches and time-borrowing flip-flops in reducing the effects of clock skew and jitter, providing tolerance to variations in the clocking network of a microprocessor. In this section, we describe how time-borrowing flip-flops may also be used to provide tolerance to datapath variations. While this technique may be applied for any type of datapath block, we focus in this section on long-distance on-chip interconnects, since the differences in

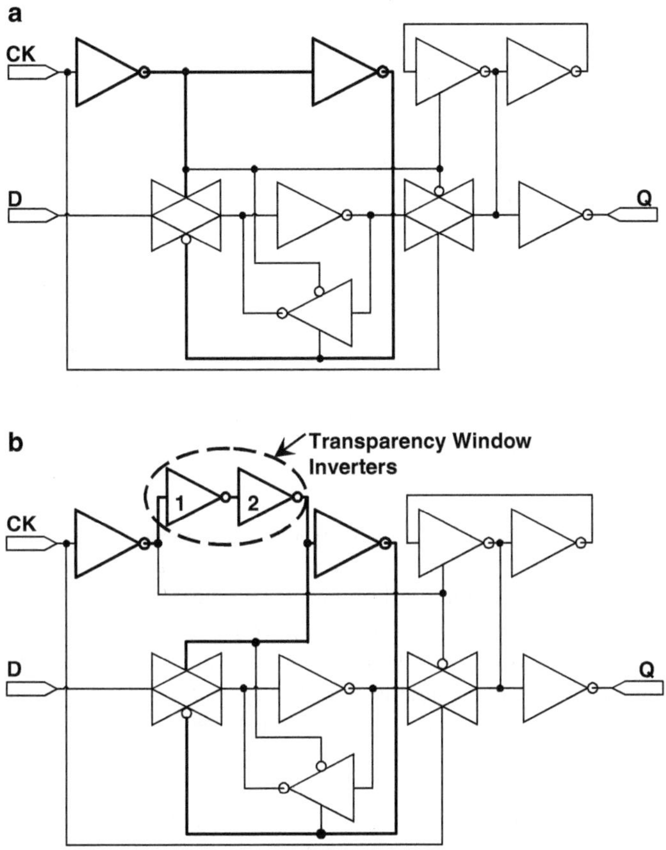


**Fig. 7.1.** (**a**) Standard master-slave flip-flop. (**b**) Creation of a time-borrowing flip-flop by inserting transparency window inverters

max-delay and min-delay for these interconnects are typically much smaller than those in logic blocks, allowing a larger transparency window in the flip-flop without causing min-delay failures. In microprocessor designs, many signals are routed over long repeater-based interconnects, requiring that these signals are propagated across multiple clock cycles. By replacing the flip-flops in these interconnects with time-borrowing flip-flops, delay variations across multiple segments can be averaged, mitigating the impact of systematic and random within-die (WID) variations on mean $F_{\mathrm{MAX}}$. Moreover, the absorption of clock skew and jitter by the transparency windows provides additional slack in the nominal cycle time in comparison with a multi-cycle interconnect with conventional master-slave flip-flops.

Figure 7.2 shows a schematic for a multi-cycle, buffered interconnect using standard non-time-borrowing (NTB) and time-borrowing (TB) flip-flops [5]. The N-cycle interconnect consists of N repeater-based interconnect segments separated by flip-flops, where $t_{\mathrm{DATA}(i)}$ denotes the delay of the $i^{\mathrm{th}}$ interconnect segment, and

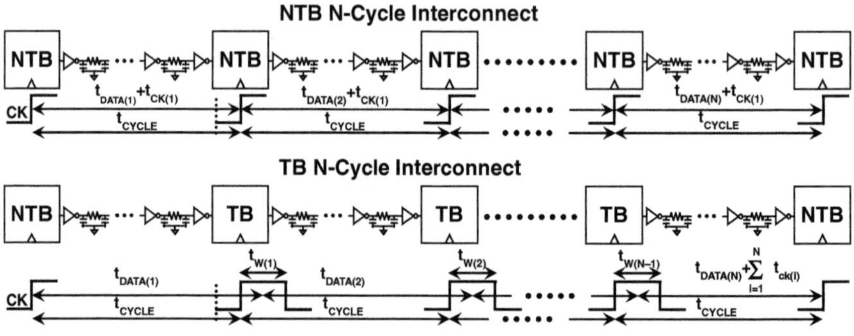


**Fig. 7.2.** NTB and TB N-cycle interconnects, where $t_{\mathrm{CYCLE}}$ is the cycle time, $t_{\mathrm{DATA}(i)}$ is the $i^{\mathrm{th}}$ interconnect segment delay, $t_{W(i)}$ is $i^{\mathrm{th}}$ transparency window, $t_{\mathrm{CK}(1)}$ is single-cycle clock skew and jitter, and $t_{\mathrm{CK}(i)}$ is additional long-term jitter component for $i^{\mathrm{th}}$ cycle ($i > 1$). Reproduced with permission from [5], ©2006 IEEE

$t_{\mathrm{CK}(1)}$ represents the single-cycle clock skew and jitter. As shown in Fig. 7.2, an NTB flip-flop enforces a hard clock edge between the interconnect segments, where data are allowed to pass through the flip-flop only when the data arrive before the setup time. For an NTB interconnect segment, $t_{\mathrm{DATA}(i)}$ equals the sum of the flip-flop clock-to-output delay, the delay through the repeater segments and the protection inverter, and the setup time for the receiving flip-flop. The cycle time of one NTB interconnect segment is the sum of $t_{\mathrm{DATA}(i)}$ and $t_{\mathrm{CK}(1)}$. For an N-cycle interconnect, the slowest interconnect segment limits the cycle time, resulting in more severe degradation in mean $F_{\mathrm{MAX}}$ as N increases.

The TB multi-cycle interconnect is constructed by replacing the flip-flops of the intermediate repeater segments with time-borrowing flip-flops. If the bus segments can be optimized such that the data nominally arrive within the time-borrowing window of the flip-flops, variations in datapath delay arising from process, voltage, or temperature variations along the interconnect can be averaged across multiple segments, and their impact on $F_{\mathrm{MAX}}$ will be reduced. As shown in Fig. 7.2, the cycle time for a TB multi-cycle interconnect, assuming all signals arrive within the time-borrowing window, is based on the average of the segment delays, while total clock skew and jitter (consisting of single-cycle skew and jitter $t_{\mathrm{CK}(1)}$ plus the additional long-term jitter component $t_{\mathrm{CK}(i)}$ in each additional cycle) can be amortized over multiple cycles. Figure 7.3a shows an example nominal timing diagram for a TB 2-cycle interconnect with zero clock skew and jitter. Ideal data delay averaging occurs in a multi-cycle interconnect where $t_{\mathrm{CYCLE}}$ equals the average of the interconnect segment delays, which is the best-case scenario for delay variation tolerance. The amount of delay variation averaging which occurs, then, depends on the delay of the two interconnect segments as well as the size of the time-borrowing window chosen. If $\delta_{\mathrm{DATA}(i)}$ denotes the delay deviation from nominal conditions for the $i^{\mathrm{th}}$ interconnect segment, the "delay variation mismatch" between the first and second interconnect segment is $\delta_{\mathrm{DATA}(1)} - \delta_{\mathrm{DATA}(2)}$. For ideal data delay averaging, the

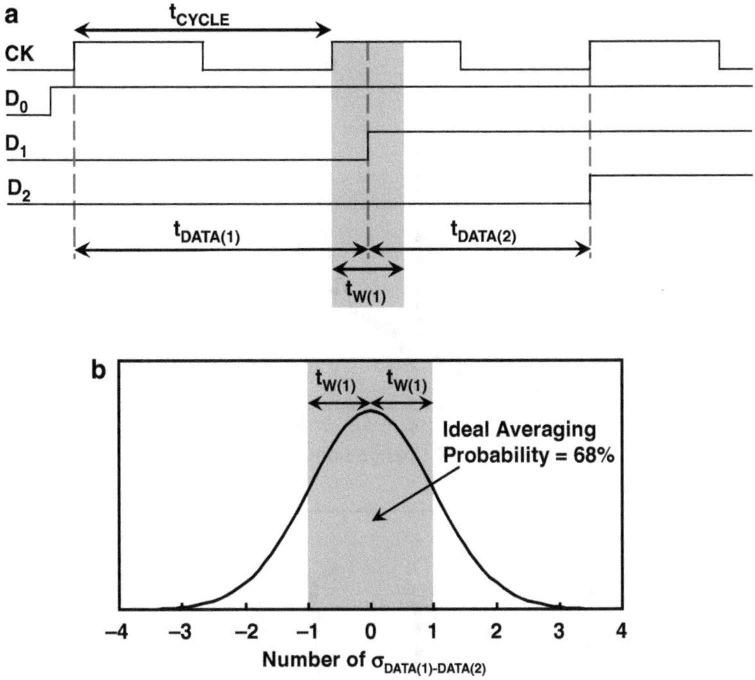


**Fig. 7.3.** (**a**) Nominal timing for a TB 2-cycle interconnect with zero clock skew and jitter, where $D_0$, $D_1$, and $D_2$ are input data for driving, intermediate, and receiving flip-flops, respectively. (**b**) Probability density versus number of standard deviations in $\delta_{\mathrm{DATA}(1)} - \delta_{\mathrm{DATA}(2)}$ ($\sigma_{\mathrm{DATA}(1)-\mathrm{DATA}(2)}$), where $\delta_{\mathrm{DATA}(i)}$ is the 0-mean delay deviation for the $i^{\mathrm{th}}$ interconnect segment. Reproduced with permission from [5], ©2006 IEEE

time-borrowing window must be at least as large as the difference in segment delay variation mismatches, $\delta_{\mathrm{DATA}(1)} - \delta_{\mathrm{DATA}(2)}$. As shown in Fig. 7.3b, the time-borrowing window size ($T_{W(1)}$) can be expressed in terms of the standard deviation of the difference between the two segment delays: $\sigma_{\mathrm{DATA}(1)-\mathrm{DATA}(2)}$. Choosing the TB window as $1\sigma$ ensures that 68% of the time, ideal delay averaging occurs. Choosing a larger TB window provides increased variation tolerance for the TB interconnect, at the cost of increased flip-flop clocking energy.

Simulation results for this technique are described in Fig. 7.4. The design of an NTB interconnect segment in M4 is optimized to provide the minimum energy for a nominal cycle time of 250 ps in a 65 nm CMOS technology [6] with a $V_{\mathrm{DD}}$ of 1.2V and a temperature of 110C, resulting in 6 repeaters, a protection inverter, and a flop-to-flop interconnect segment length of 1,500 $\mu$m. Starting with the optimal NTB interconnect segment, NTB flip-flops are replaced with TB flip-flops for the intermediate segments of a multi-cycle TB interconnect. Since clock skew and jitter are absorbed by the transparency window, additional slack is available, and TB flip-flop and repeater transistors are downsized until the nominal segment delay

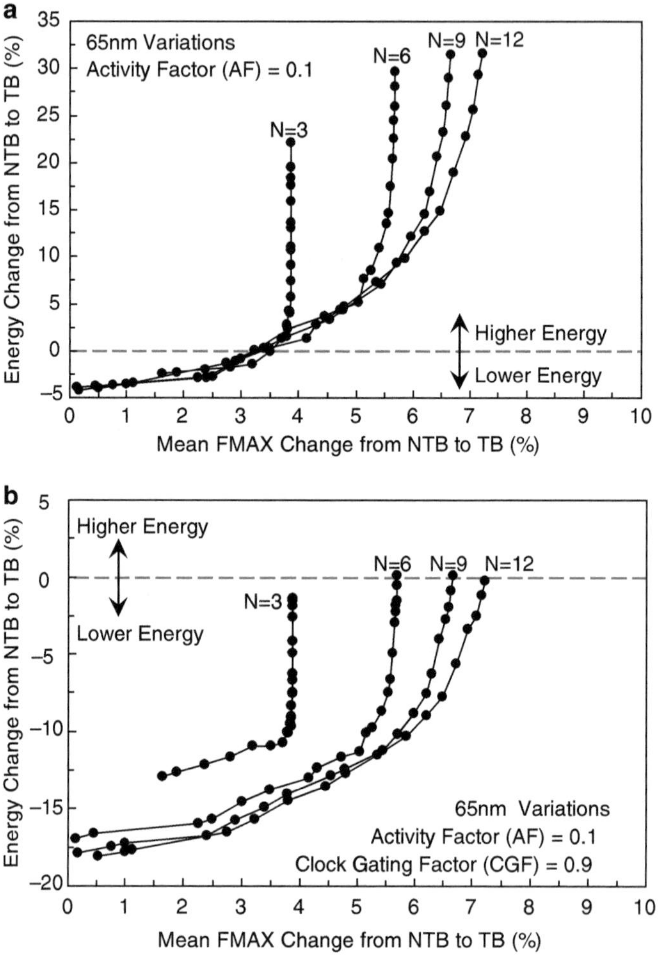


**Fig. 7.4.** Active (**a**) and average (**b**) energies per cycle vs. mean $F_{MAX}$ change from NTB to TB N-cycle interconnect. Reproduced with permission from [5], ©2006 IEEE

equals 250ps. Monte-Carlo simulations are performed to determine the $F_{MAX}$ distribution for these interconnect designs, including WID transistor, interconnect, and voltage variations, and the effects of transparency windows, clock skew, and jitter. In Fig. 7.4, mean $F_{MAX}$, active energy, and average energy of multi-cycle NTB and TB interconnects are compared for a range of cycles, representing present (N=3&6) and future (N=9&12) interconnect designs. The mean $F_{MAX}$ of TB interconnects is 3.5% higher than NTB interconnects at equal active energy. Additional mean $F_{MAX}$ gains can be obtained by expanding the transparency window at the expense of higher clocking energy until averaging of delay variations approaches the ideal case. Since the impact of additional clocking energy in TB flip-flops is mitigated by clock gating, TB interconnects enable 4–6% mean $F_{MAX}$ gain with a corresponding 10% average

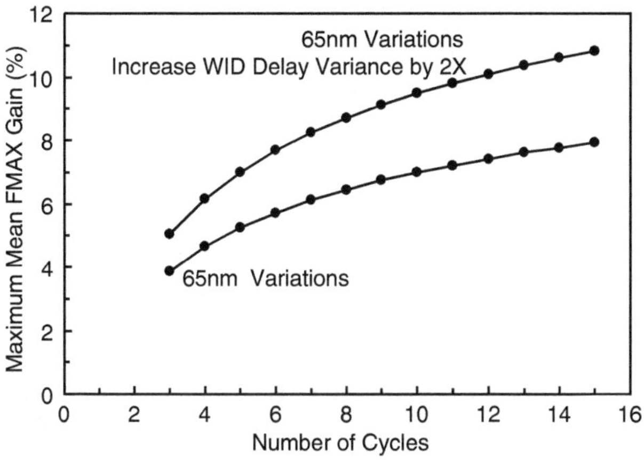


**Fig. 7.5.** Maximum mean $F_{MAX}$ gain from NTB to TB N-cycle interconnect vs. N. Reproduced with permission from [5], ©2006 IEEE

energy savings as shown in Fig. 7.4b. As process technology scales, WID variations become worse, especially for larger dice and longer interconnects. Figure 7.5 shows that if WID delay variance increases by 2X, the maximum mean $F_{MAX}$ gain rises to 5–10%. Thus, the usage of time-borrowing flip-flops can absorb this increased WID variation in the datapath, and at the same time reduce the impact of clock skew and jitter.

## 7.3 Variation Reduction Through Tuning

The previous section described techniques that can be employed in the design phase for reducing the impacts of clock uncertainty and process variations on the performance of a microprocessor. While these techniques ideally can hide some of the effects of clock uncertainty, they do not eliminate clock skew and jitter, and cannot adjust to changing dynamic conditions such as supply voltage and temperature fluctuations. To further improve processor performance and energy-efficiency, tuning can be applied post-fabrication, on a per-die or per-lot basis, to compensate variations. This tuning may be performed only on the clock network, to reduce clock skew and jitter, or may be applied to the entire processor to dynamically adapt to voltage and temperature variations.

In this chapter, we describe three types of post-silicon tuning. The first – manufacturing techniques – encompasses all static techniques which can be applied after fabrication to compensate the effects of process variations. These techniques are applied once, and settings do not change over the lifetime of the part. In contrast to this, the other two techniques described in this chapter are dynamic in nature. Active clock deskew is used to dynamically adjust the timing of the clock network to reduce clock skew as the environmental conditions change, and dynamic frequency adjusts the frequency of the entire clock network as a method of responding to dynamic voltage and temperature changes. Both of these techniques allow the processor to respond to operational conditions rather than assuming a worst-case environment and, therefore, improve performance at lower power consumption.

### 7.3.1 Manufacturing Techniques

Variations that are static in nature can be compensated through a one-time tuning process which is done after fabrication, during the die testing phase. This is typically done for critical analog circuits which are often more sensitive to process variations than the digital circuits in the microprocessor core. Examples of this type of post-silicon tuning include calibration of thermal sensors and I/O receivers, and setting of ideal PLL VCO frequency to minimize PLL random jitter. Once the ideal settings have been determined, they must be permanently programmed into the chip such that they become the default settings each time the microprocessor is powered up. This is typically accomplished through the use of on-chip nonvolatile storage elements such as fuses.

Another way in which post-silicon tuning is used to compensate variations is in the specification of the supply voltage of each processor. Many modern microprocessors include a "voltage identification" (VID) code which allows the processor to specify its supply voltage to the voltage regulator in the system [7]. This VID bus is used, for example, to allow the processor to reduce its supply voltage and frequency during low-power states and then transition back to maximum supply voltage during periods of active computation. However, this feature can also be used to compensate variations by assigning a maximum supply voltage to each processor at test time [8, 9]. Because of process and manufacturing variations, fabricated dice exhibit a range of maximum operating frequencies and standby leakage powers. Those dice that are slow but low-leakage can operate at a higher voltage without violating the maximum power specification, while dice that are fast but high-leakage must operate at a lower voltage. Each die, therefore, has an optimum maximum supply voltage assigned to it that maximizes the processor frequency while ensuring that the power limit is met. This optimum VID setting is fused at test time and becomes the default supply voltage for the part. Modern voltage regulators allow very precise setting of processor supply voltage at a resolution of several millivolts, enabling improvement in microprocessor speed binning and maximizing the number of dice which meet both minimum frequency and maximum power requirements.

### 7.3.2 Active Clock Deskew

High-performance microprocessors often include clock-skew adjustment circuits that can be tuned at test time to reduce skew or improve processor $F_{MAX}$ by optimizing clock timing. However, adjusting such clock skew buffers on a part-by-part basis is extremely expensive, and such a scheme does not account for dynamic variations such as temperature which can cause clock skew to change. To further improve clock timing, active clock deskewing can be used. Figure 7.6 shows the clocking topology for the IA64 microprocessor in a 0.18 $\mu$m CMOS technology, employing

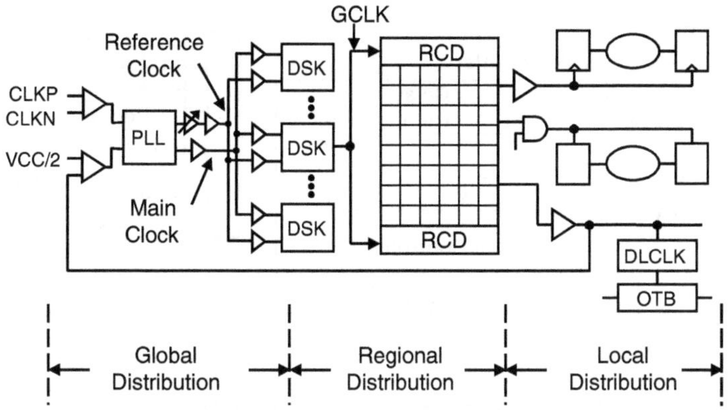


**Fig. 7.6.** Clock distribution topology for a 0.18 $\mu$m, IA-64 microprocessor. Reproduced with permission from [10], ©2000 IEEE

clock deskew buffers (DSKs) to dynamically reduce skew between different regional clock grids [10]. The DSK, shown in Fig. 7.7a, contains a local controller to continuously compare the timing of the regional grid clock to the incoming reference clock, and a variable delay circuit [11] (Fig. 7.7b) to adjust clock delay until skew is minimized. Timing comparison is accomplished through the use of a phase detector in the local controller. This scheme compensates any load variations or WID variations in the core clock distribution. Although the skew of the reference clock, which is distributed to all clock regions on the die, is still included in the final clock skew, the reduced span of the reference clock and the matched load results in significantly less overall skew. Figure 7.8 shows the measured skew. The total skew is 28ps with deskewing and is four times larger with the deskew mechanism disabled. The active deskew technique of the 0.18 $\mu$m IA64 processor (in addition to other processors) has also been discussed in Sect. 2.5.2.

The 90nm Itanium® design [12] also includes active deskew circuitry to deliver a low-skew clock without requiring the use of a high-power, dense clock grid. The clocking architecture for the "Montecito" processor (Fig. 7.9) includes regional active deskew (RAD) circuits to actively manage the clock skew between adjacent clock domains [13]. This system adjusts the delay of tunable second-level clock buffers (SLCBs) as shown in Fig. 7.10. Phase comparators are used to compare the clock outputs of two SLCBs, and the delay lines in the tuning buffers are either incremented or decremented until the two clocks are lined up. The complete RAD system is comprised of 30 phase comparators and 26 second-level clock buffers in each core connected to a tree-like structure. In this structure, all of the clock zones in the core are deskewed to a central anchor zone. The hierarchical nature of the system introduces a finite amount of skew but the RAD system manages this skew, to approximately 10ps across each core. Additional details of this hierarchical deskew algorithm are presented in Sect. 2.5.2.

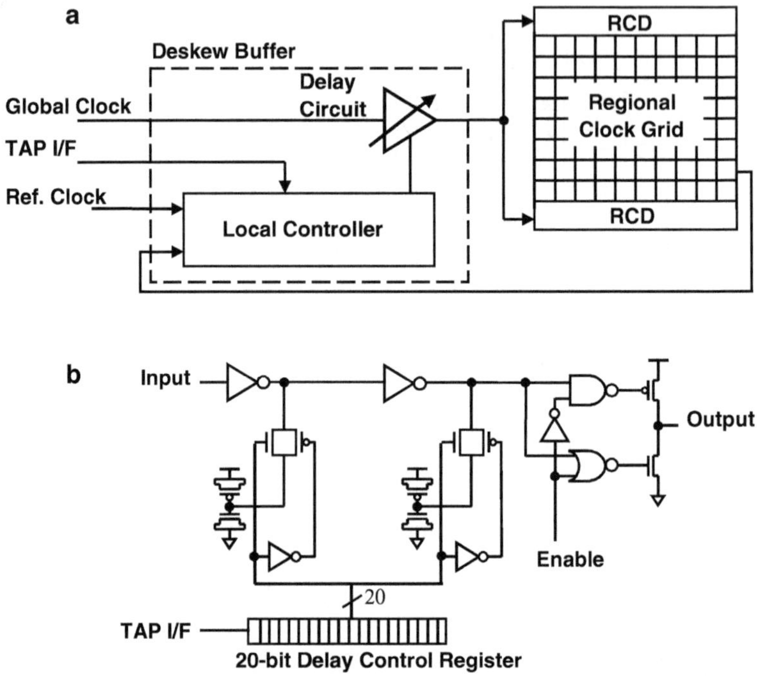


**Fig. 7.7.** (**a**) Deskew buffer (DSK) architecture. (**b**) DSK variable delay circuit. Reproduced with permission from [10], ©2000 IEEE

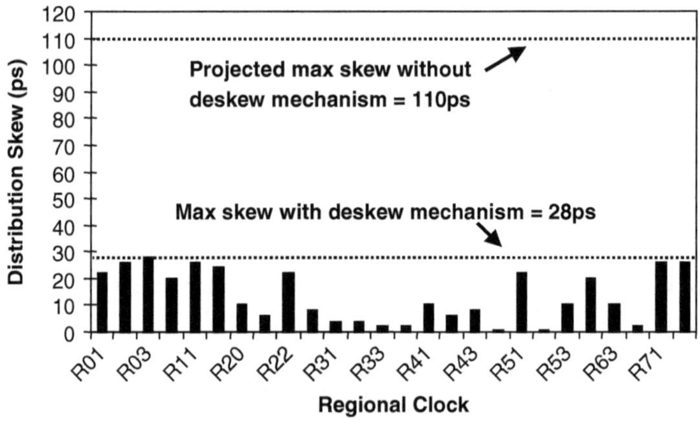


**Fig. 7.8.** Experimental skew measurements. Reproduced with permission from [10], ©2000 IEEE

### 7.3.3 Dynamic Frequency

Dynamic variations such as voltage and temperature fluctuations are typically handled by adding a margin to the clock frequency such that when these variations occur, the computation in the datapath is guaranteed to be correct. This "worst-case" design ensures correct results but requires an overhead in terms of power consumption, or a performance reduction. Because most of the time the nominal operating conditions of a microprocessor are not the worst-case conditions that have been designed for, performance or power are left on the table. To reduce this overhead of worst-case design, it is possible to sense and respond to these dynamic variations by changing clock frequency.

The Intel 90nm Itanium processor design [12–14] contains distributed digital frequency dividers (DFDs) for dynamic clock frequency change coupled with clock vernier devices (CVDs) for local delay fine-tuning (Fig. 7.9). The structure of the DFD is shown in Fig. 7.11: the DFD produces an output clock frequency based on the incoming PLL clock frequency and the current divisor value held in the phase compensator state machine (PCSM). For a divisor value of 0, the DFD clock runs at the frequency of the PLL. For all other divisors, the phase mux synthesizes a clock frequency by selecting the appropriate edges from the 64 phases.

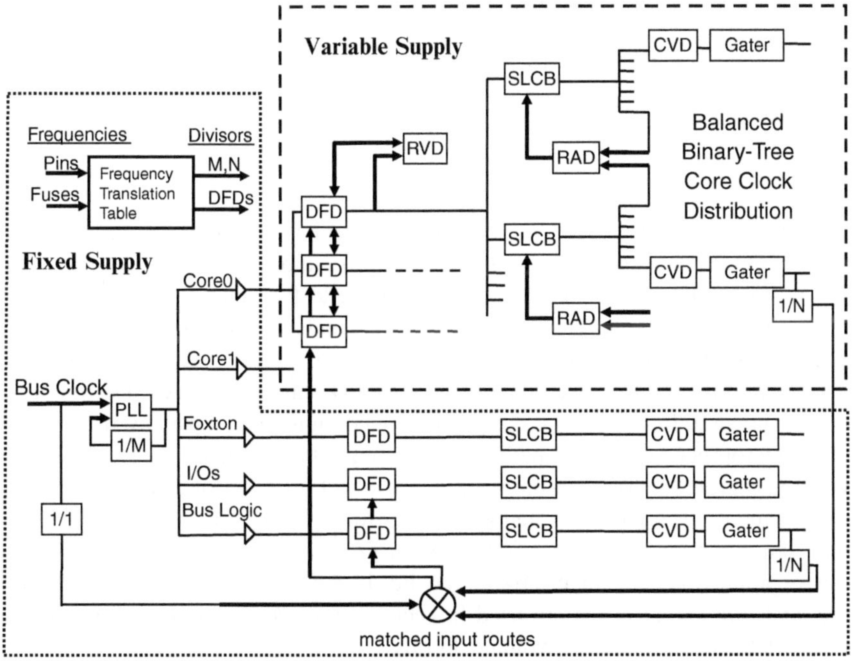


**Fig. 7.9.** Clock system architecture for the 90 nm Itanium® processor (Montecito). Reproduced with permission from [13], ©2006 IEEE

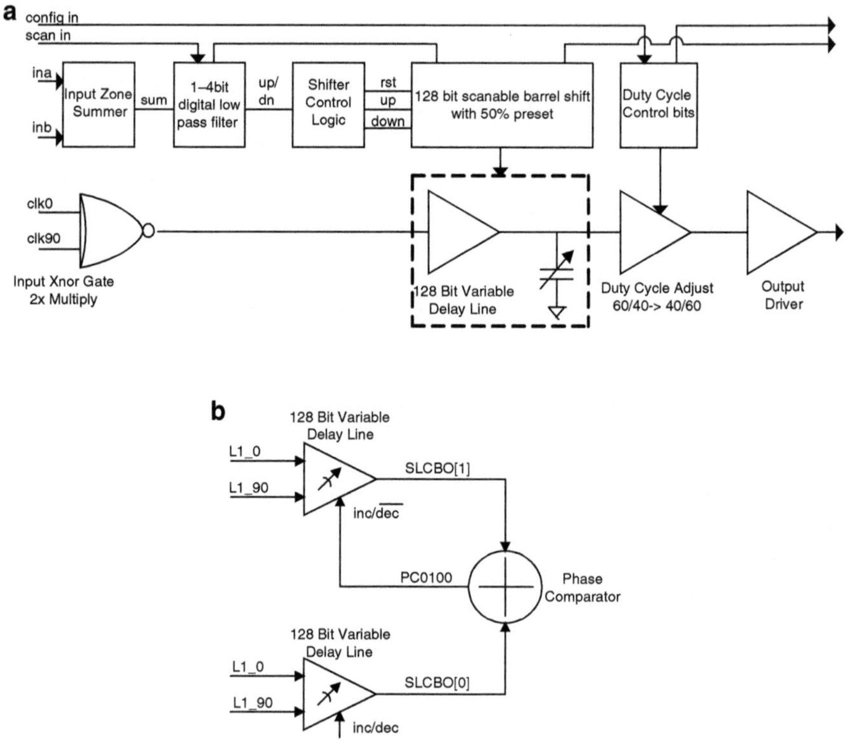


**Fig. 7.10.** (**a**) Second-level clock buffer (SLCB) architecture. (**b**) RAD SLCB/comparator connections. Reproduced with permission from [13], ©2006 IEEE

Variable frequency mode (VFM) allows the clock of the processor to track local supply voltage changes, reducing the guardband required for supply voltage noise. This is accomplished through the use of voltage-to-frequency converters (VFCs) which allow single-cycle response to droop transients and provide frequency steps of 1.5%. The VFC consists of voltage-locked loops in which a regional voltage detector (RVD) is used to sense voltage and set DFD output frequency. Design of the RVD is shown in Fig. 7.12: four delay lines are configured on a part-by-part basis to match microprocessor critical paths in delay and voltage sensitivity. The RVD outputs are generated by two multiplexing phase detector latches and two configurable delay elements. These outputs denote whether timing margin is available (DOWN), ideal timing margins are present (HOLD), or greater margins are needed (UP). To track regional voltage changes around the die, 12 RVDs are distributed around each core, and three DFD zones are used in each core to reduce frequency adjustment latency in the presence of fast voltage droops.

Measured results of the dynamic frequency system response to a 150 mV first-droop in voltage is shown in Fig. 7.13 [13]. The average pre-droop frequency over 32 cycles is 2,272 MHz, and the frequency averaged over the 16 cycles during the event

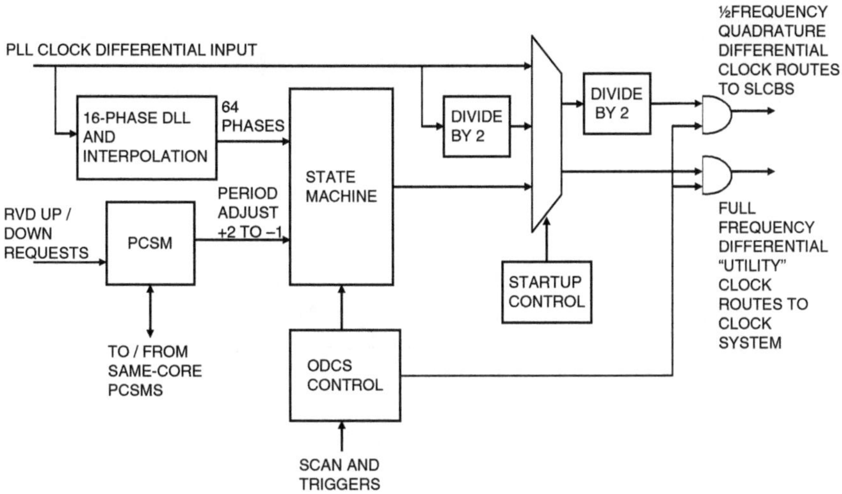


**Fig. 7.11.** Dynamic frequency divider (DFD) and phase compensator state machine (PCSM) block diagram. Reproduced with permission from [13], ©2006 IEEE

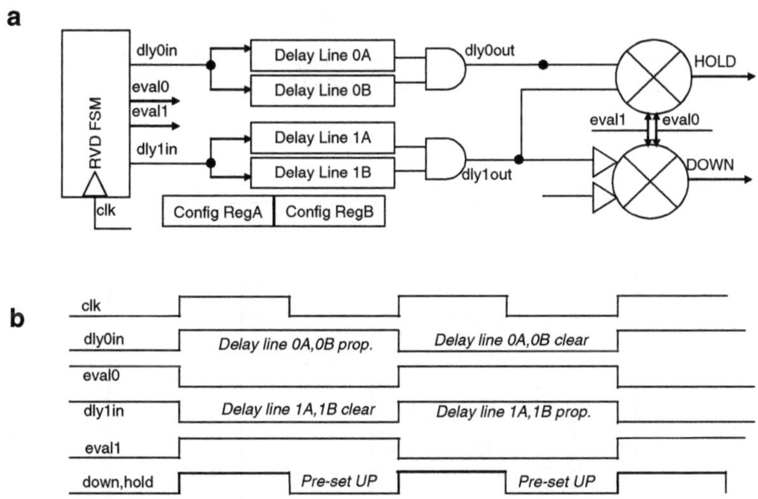


**Fig. 7.12.** Regional voltage detector (RVD) details. (**a**) RVD architecture block diagram. (**b**) RVD internal timing. Reproduced with permission from [13], ©2006 IEEE

is measured at 2,087 MHz. The result is that code can execute at a higher average frequency using VFM than could be achieved at fixed frequency mode where this voltage droop needs to be margined. Comparison of VFM and FFM for different test cases is shown in Fig. 7.14: the DTB and FPU cases are focused on critical path tests,

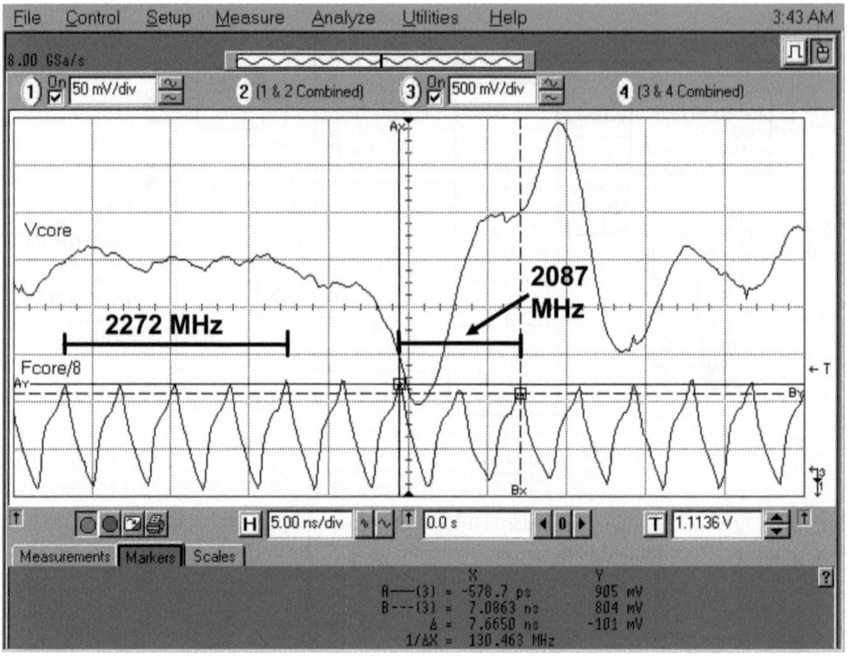


**Fig. 7.13.** Oscilloscope trace of clock system VFM response to 150 mV droop at 2.27 GHz, 1.2 V. Reproduced with permission from [13], ©2006 IEEE

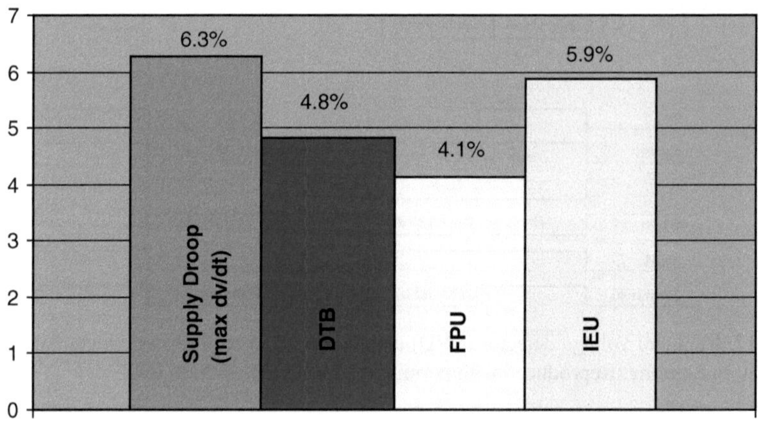


**Fig. 7.14.** Measured variable-frequency mode (VFM) (%) Frequency increase over fixed-frequency mode for multiple test cases. Reproduced with permission from [13], ©2006 IEEE

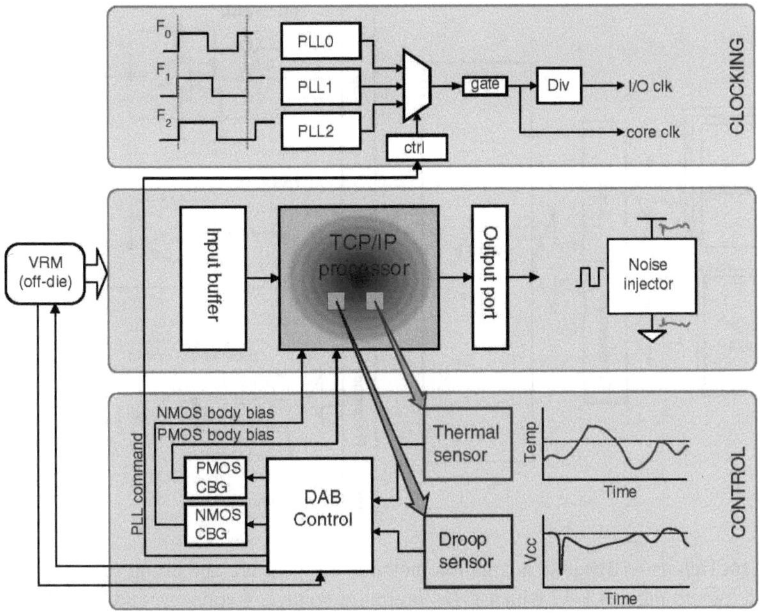


**Fig. 7.15.** Overview of 90 nm TCP/IP processor test chip with dynamic voltage, frequency, and body bias. Reproduced with permission from [15], ©2007 IEEE

while the supply droop case is the maximum $dv/dt$ test. Here, maximum frequency improvement of 4.1% to 6.3% is shown by enabling the variable frequency mode.

The 90nm TCP/IP processor test chip in [15] uses dynamic frequency in a similar fashion in order to reduce the guardbands applied for voltage droop, temperature, and aging. In this case, however, rather than using dynamic voltage-to-frequency converters, analog sensors are used to continuously sense voltage and frequency, and this information is sent to a dynamic adaptive bias (DAB) controller which adjusts frequency, voltage, and body bias (Fig.7.15). Details of the DAB control algorithm are shown in Fig.7.16: multiple temperature and voltage sensors are used to accurately measure the range of voltage and temperature as code is executed on the TCP/IP processor core. These voltage and temperature values are used as an index into a lookup table of die characterization information which is pre-loaded at test time. This lookup table specifies the optimum voltage, body bias, and frequency to be applied for each combination of temperature and voltage measurements. If the DAB control determines that a frequency change is needed, the dynamic clocking block (Fig.7.17) changes the core frequency within a single core-clock cycle, allowing fast response to first-droop changes in voltage. This fast frequency change is accomplished through the use of multiple PLLs, each locked to a different frequency. When the frequency needs to be lowered due to a voltage droop event or temperature change, the slower PLL is selected by the clock select mux, ensuring a transition to the slower clock frequency without any short cycles or glitches. Once the slow PLL is

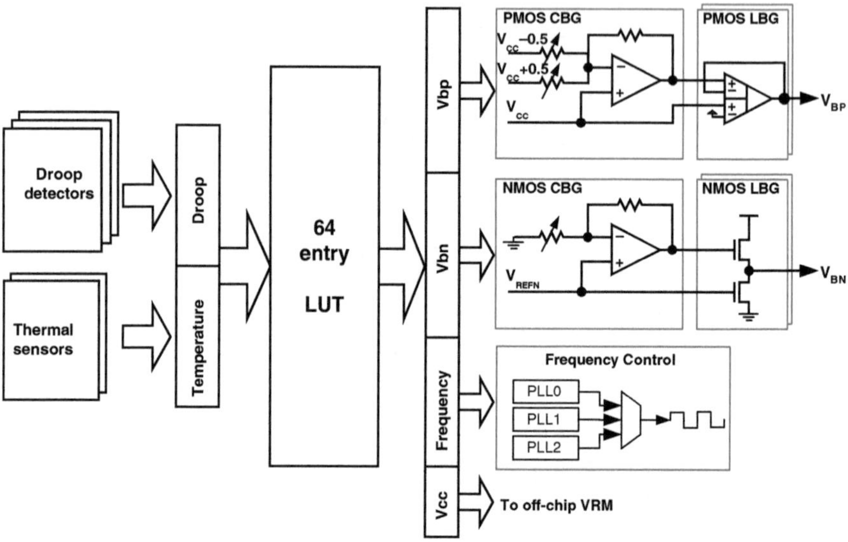


**Fig. 7.16.** Details of dynamic adaptation method. Temperature and droop sensors are used to index a lookup table (LUT) which gives optimum voltage, frequency, and body bias for the given conditions. Reproduced with permission from [15], ©2007 IEEE

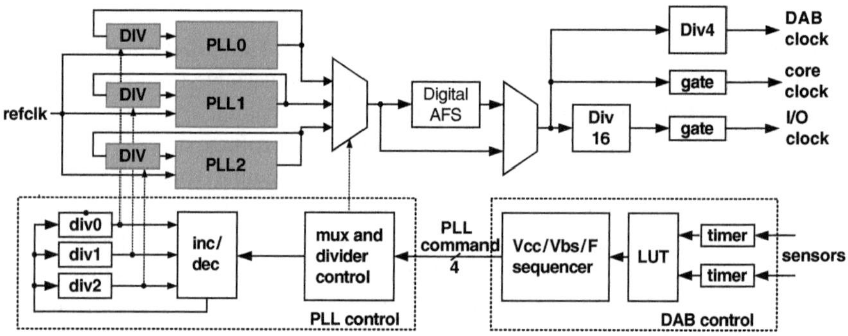


**Fig. 7.17.** Dynamic clocking system details. Multiple PLLs are used to enable fast frequency changes, allowing response to first-droop in voltage. PLL frequency can be changed when not selected, allowing wide frequency range with fine resolution. Reproduced with permission from [15], ©2007 IEEE

selected, the faster PLL can then be re-locked to a new frequency, ensuring frequency change is possible across a wide frequency range with fine (<3%) resolution.

Results from this test chip for supply voltage droop are shown in Fig. 7.18. In this case, a large 20% voltage droop is detected by the voltage sensor, and the DAB control reduces the clock frequency. As the voltage rises at the end of the droop event, this change is again detected and the clock frequency is raised to a new value which

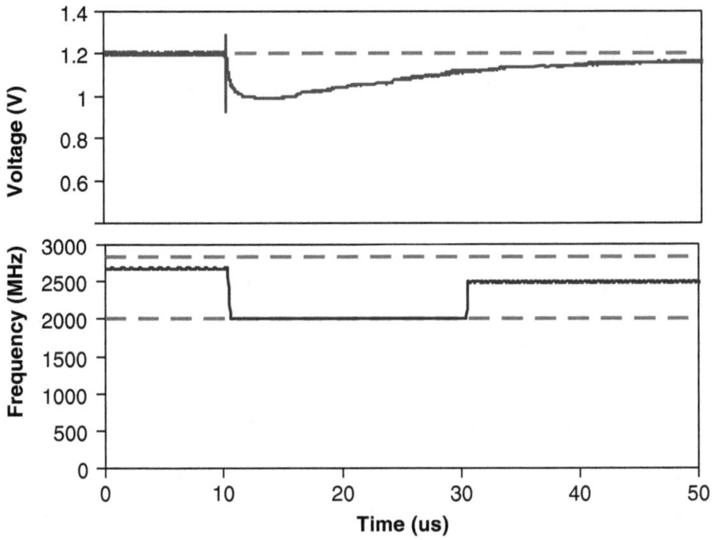


**Fig. 7.18.** Measured dynamic frequency response to a voltage droop. Maximum frequency is 32% higher for the dynamic processor than would be required for a worst-case design with fixed frequency. Reproduced with permission from [15], ©2007 IEEE

accounts for the IR drop based on the increased current load. Without these dynamic frequency features, the clock frequency would have to be set to the frequency supported at the minimum voltage level, representing a large guardband in frequency for these infrequent voltage droops. By dynamically sensing voltage droop and responding, the average frequency can be increased. Both of these design examples demonstrate the large improvement in performance and energy-efficiency that can be obtained by using the clock as part of dynamic variation response system, rather than simply designing for the worst-case conditions.

## 7.4 Variation Reduction Through Resiliency

The previous sections have focused on techniques for tolerating or reducing the impact of variations such that the variations do not produce an error in computation. A combination of variation-tolerant design techniques, coupled with dynamic variation detection and response can be used to reduce the guardbands that must be applied to the clock frequency of microprocessors. However, guardbands must still be applied for variations that cannot be predicted, sensed, or which are too fast to respond to. Examples of these types of variations include very fast $1^{st}$-order voltage droop, delay variation from crosstalk and noise, and within-die differences in transistor degradation. Reducing these guardbands is the focus of this section.

In contrast to variation tolerance which guarantees that errors in the circuit never occur, resilient designs operate correctly even in the presence of errors. Resiliency implies that any errors which result from variations are sensed and corrected so that the final result is error-free, although there may be errors at intermediate stages in the computation. As variations increase while performance targets and power requirements become even more restrictive, resiliency will become much more important in designs.

Resilient design itself is an extremely broad topic – resiliency can be implemented at the circuit level, in the microarchitecture, in the system, or even in software. In this chapter, we focus on circuit-level resiliency as this is most relevant to clock network design. A complete implementation of a resilient microprocessor will combine these resiliency techniques at the circuit level with error correction at the microarchitectural level and dynamic response at the system level.

### 7.4.1 Timing Error Detection – Error Detection Sequentials

Dynamic in situ detection and correction of speed path failures were first proposed as part of the Razor project [16, 17]. In the Razor technique, each delay-critical flip-flop in the datapath is replaced by a Razor flip-flop (Fig. 7.19), which includes a shadow latch controlled by a delayed clock. This technique works on the premise that dynamic variations such as voltage, temperature, noise, or data-dependent delay variation will cause the delay of a critical path to increase. If the frequency of the processor is set by the typical condition, this delay increase will cause the data at the receiving flip-flop to arrive after the sampling clock edge, causing an error in the datapath. However, the data are sampled correctly by the delayed clock of the shadow latch, and by comparing these late-sampled data with the data captured by the datapath flip-flop, the error can be detected. Under nominal conditions, data are sampled correctly by both the datapath flip-flop and the shadow latch, and no error is signaled. When a dynamic variation causes the critical path delay to exceed the clock period, the error signal is asserted and the error in the pipeline must be corrected before program execution can continue. In the original Razor design, the correct data always exist in the shadow latch, even if the datapath flip-flop fails. Thus, the correct data can be injected back into the datapath through the multiplexer shown in Fig. 7.19. The Razor technique along with other similar resiliency techniques has also been discussed in Sect. 3.7.2 from the viewpoint of wearout-induced circuit degradation.

One concern for the Razor-style error-detection sequential (EDS) is in the handling of the metastability condition. Because the processor frequency can now be set by the nominal operating conditions and not by the worst-case conditions, no additional setup time margin is necessary in the clock frequency to allow for dynamic variations. Because of this, the data input to the flip-flop may arrive at the same time as the clock edge, leading to a metastable output. The probability of this occurring is very small; however, it is many times larger than the probability of metastability in the original design with guardbands. If the flip-flop datapath output or error output becomes metastable, there is the possibility of an undetected error: the datapath value is sampled incorrectly, but the error signal is not asserted. Obviously an undetected error cannot be allowed, and the Razor flip-flop includes a metastability detector to

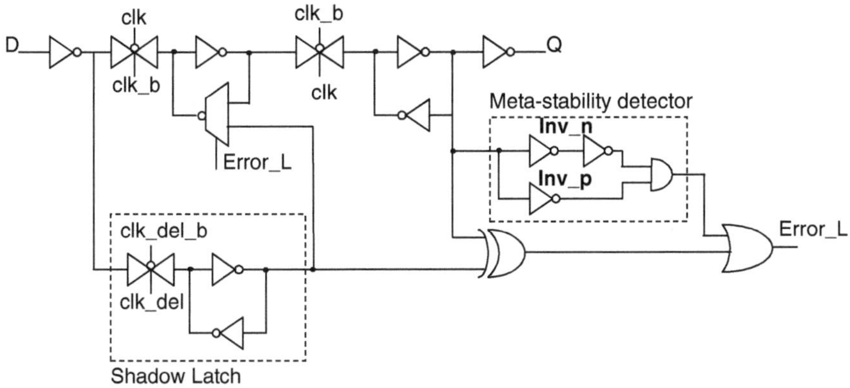


**Fig. 7.19.** Razor error-detection flip-flop. It consists of a datapath flip-flop, a shadow latch for sampling late-arriving data, and a metastability detector. Reproduced with permission from [16], ©2003 IEEE

signal this condition and initiate error recovery. This metastability detector, while effective, contributes in large part to the area overhead in the error-detection flip-flop, and to the overall power overhead for the error detection system, and thus detracts from the energy-efficiency gains possible through resiliency.

Recently, several techniques were independently developed to address the overhead of the original Razor (Razor-I) approach. The versions shown in [18–20] attempt to reduce the clock power overhead and eliminate the metastability concerns by replacing the pipeline flip-flop with a latch. This approach is motivated by the observation that because the critical paths must be designed such that the data never arrive in the nominal condition in the high phase of the clock, the master stage of the datapath flip-flop is unnecessary. Thus this flip-flop can be replaced by a latch which operates essentially as a pulsed latch, where the length of the high phase of the clock determines the transparency window of the latch as well as specifying the min-delay requirement which must be met by each logic path. Because late-arriving data are not gated by the closing of the master stage of the flip-flop, there is no metastability concern around the rising edge of clock. Late-arriving data are still sampled correctly in the current pipeline stage; however, this may cause a critical path and error in the subsequent pipeline stage and, therefore, this late-arriving data must be signaled as an error.

The transition detection with time-borrowing (TDTB) sequential [18] (Fig. 7.20a) uses a pulse generator to generate a pulse each time data transitions. As shown in the example timing diagram (Fig. 7.21), if this data transition occurs during the high clock phase (which defines the error-checking window), an error signal is generated. Because the datapath flip-flop has been replaced by a latch, performance and power consumption are improved over the Razor design. A similar idea is used in the double-sampling with time-borrowing (DSTB) technique [18] (Fig. 7.20b), which is similar to TDTB except that a shadow flip-flop replaces the transition

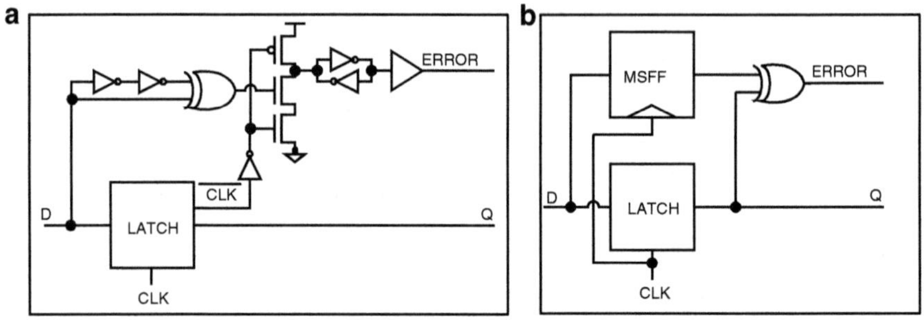


c

| | RFF | TDTB | DSTB |
|---|---|---|---|
| Datapath Metastability | Yes | No | No |
| Clock Energy Relative to MSFF | 36% (T)<br>22% (L) | –9%(T)<br>–13%(L) | 34% (T)<br>14% (L) |
| Clock Energy Relative to Latch | 145% (T)<br>94% (L) | 64% (T)<br>38% (L) | 143% (T)<br>81% (L) |
| D-to-Q Delay Relative to MSFF | 12% | –14% | –12% |
| D-to-Q Delay Relative to Latch | 62% | 23% | 27% |
| Design Complexity | Low | High | Low |
| SER Tolerant Datapath Sequential | Yes | Partial | Yes |
| Ability to Turn Off Error Detection & Operate at Low $F_{CLK}$ with 50% Duty Cycle | Yes | No | No |

**Fig. 7.20.** Error-detection sequential circuits. (**a**) Transition-detection with time-borrowing (TDTB). (**b**) Double-sampling with time-borrowing (DSTB). (**c**) Comparison of Razor-I flip-flop (RFF), TDTB, and DSTB in key circuit metrics. Reproduced with permission from [18], ©2008 IEEE

detector. The clock energy overhead for this EDS is slightly lower than for the Razor design because the larger datapath flip-flop has been replaced by a latch, while the minimum-sized shadow latch has been replaced by a flip-flop. The Razor flip-flop (RFF), TDTB, and DSTB error-detection sequentials are compared in terms of key circuit metrics in Fig. 7.20c.

A later version of the Razor design, denoted Razor-II [19], also replaces the datapath flip-flop with a latch to eliminate metastability concerns. This design (Fig. 7.22) also utilizes a datapath latch for improved performance and handling of metastability. A transition detector generates a pulse whenever the output of the latch switches, and this pulse is qualified with a detection clock to determine whether the output has transitioned because of data which arrived after the clock edge. This timing is demonstrated in Fig. 7.23. Data which arrive late cause a transition detector pulse after the detection clock, and an error signal is generated and propagated down the pipeline. Note that in this version of the Razor flip-flop, the correct data are no longer present in the sequential. Therefore, it cannot be injected into the pipeline as in Razor-I, and the offending instruction must be replayed by the microarchitecture.

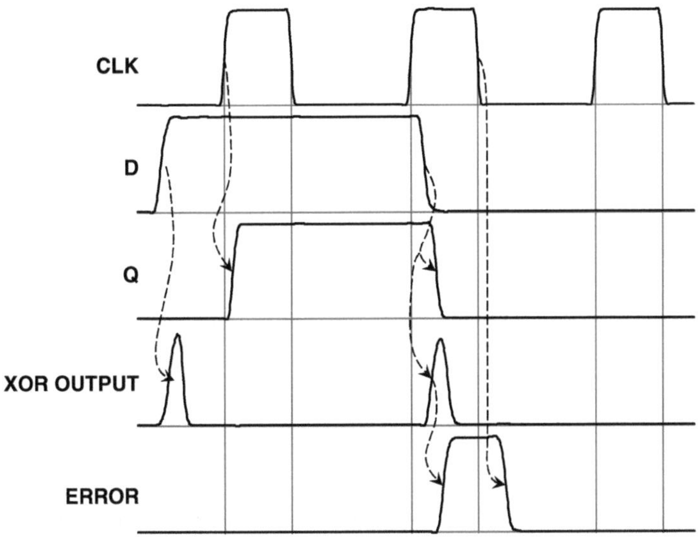


**Fig. 7.21.** Timing diagram for the transition-detection with time-borrowing (TDTB) technique. Data transitions which occur in the high phase of the clock (which sets the error-checking window for the design) generate an error signal. Reproduced with permission from [20], ©2009 IEEE

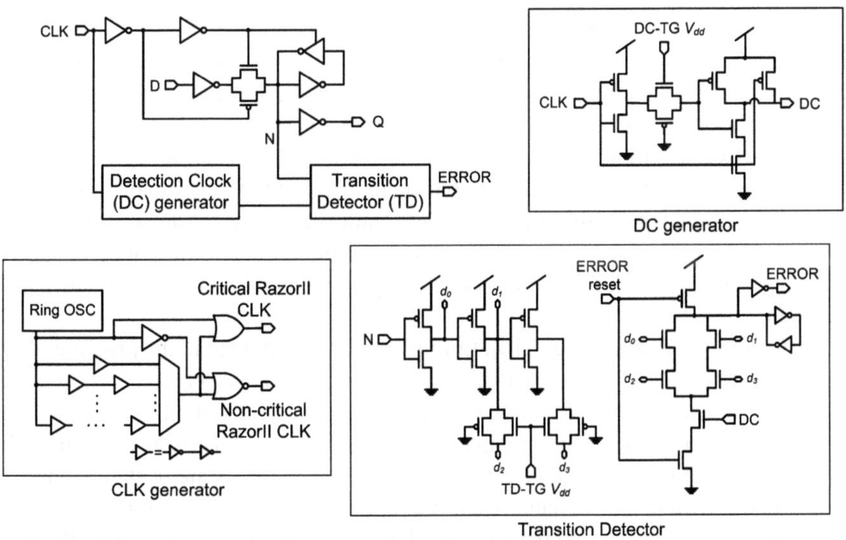


**Fig. 7.22.** Details of the Razor-II sequential design. It includes the latch, clock generator, transition detector (TD), and detection clock (DC) generator. Reproduced with permission from [19], ©2008 IEEE

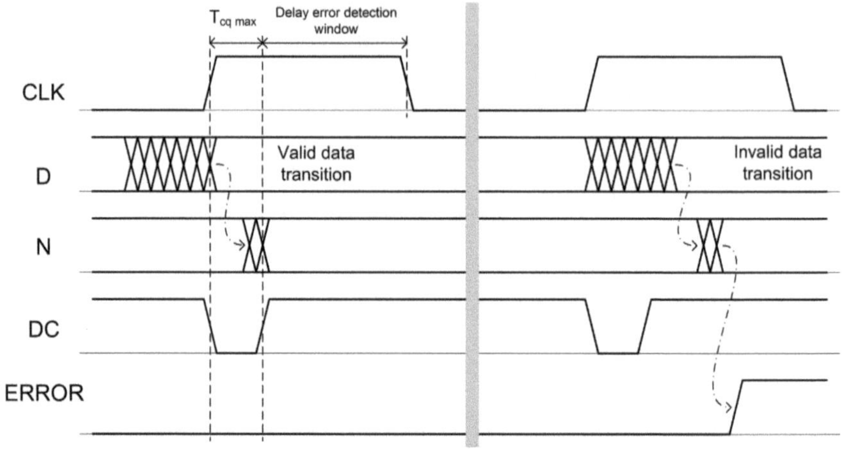


**Fig. 7.23.** Timing diagram for the Razor-II design. Transitions on latch output node "N" which occur after the allowed region (specified by the detection clock DC) are signaled as errors. Reproduced with permission from [19], ©2008 IEEE

### 7.4.2 Timing Error Correction and Recovery

Once timing errors have been detected in a circuit, there must be a method for responding to these errors, negating the offending instructions, and re-executing to get the correct result. As in timing error detection itself, this can be done at different points in the design hierarchy – at the circuit level itself, in the microarchitecture, or with software-level checkpointing. In responding to and correcting errors, there is a trade-off between complexity and performance overhead: correcting errors at the circuit level requires complex hardware, but incurs very small performance overhead, while correcting errors at the software level requires minimal hardware support, but may incur a large performance penalty. Thus, the optimal error rate, where energy-efficiency is maximized, is different depending on the error-recovery method which is chosen.

In the original Razor design [16], timing errors are corrected in the pipeline itself, allowing execution to continue with a one-cycle penalty. This technique relies on the fact that the shadow latch has always captured the correct output, even when the datapath flip-flop is in error. When an error occurs, the offending instruction must be nullified as it passes down the pipeline. This ensures that the state of the machine is not corrupted with incorrect data. Next, the pipeline must be stalled for one cycle as the correct value is forwarded from the shadow latch into the main pipeline. Because the execution continues with the stage and following the stage that caused the error, forward progress of the instruction is guaranteed.

In the design in [18], the error signal initiates the error recovery process in the microarchitecture. The error is treated in a similar way to a branch misprediction. As

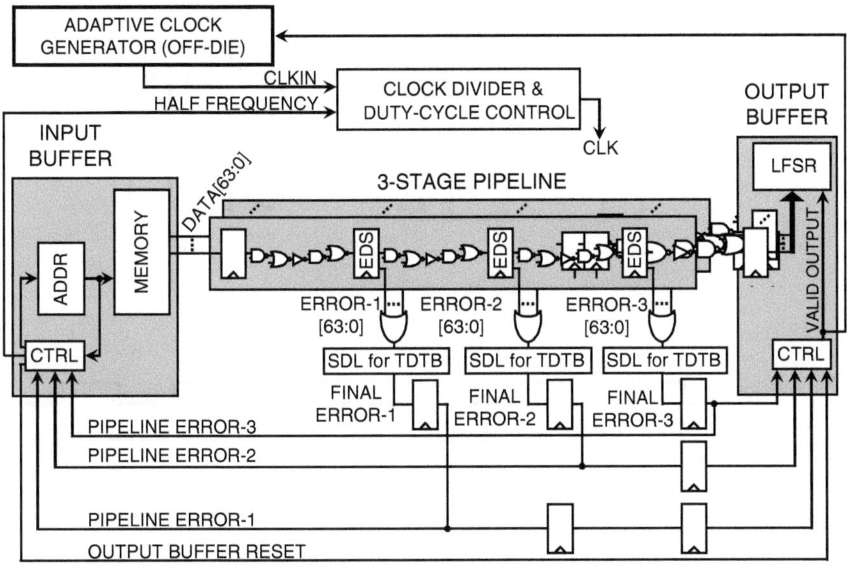


**Fig. 7.24.** Instruction-replay error-recovery design for one error-detection sequential (EDS). Set-dominant latch (SDL) is only used for TDTB. Reproduced with permission from [18], ©2008 IEEE

shown in Fig. 7.24, all of the error signals in a particular pipeline stage are combined using an OR tree to generate a single error signal for that stage. That error signal is then sent down the pipeline, along with the data, to the output buffer to invalidate the data. This is necessary to prevent the failing instruction from corrupting the state of the machine. At the same time, the error signal propagates to the instruction issue stage (input buffer) to initiate the instruction replay sequence. Similar to a branch mispredict, the program counter is loaded with the address of the failing instruction, and the instruction is repeated. If the error was due to a transient environmental condition, the instruction may execute error-free when it is re-executed, and the processor execution continues as usual until the point of the next error. If, however, the error is due to a condition which is set up by the instruction (or group of instructions) itself, it will fail again on re-execution. For this reason, the chip also contains a clock divider circuit (Fig. 7.25) which reduces the clock frequency by half when replaying the instruction. Re-executing the instruction at half frequency guarantees that the instruction will pass the failing pipeline stage and guarantees forward progress for the workload.

The Razor-II design, which was incorporated in a 64-bit, 7-stage Alpha processor design in 0.13 $\mu$m CMOS [19], recovers from errors in a similar way. The last stage of the pipeline is designed to be non-timing critical so that it is known before instruction commit whether an error occurred for that instruction or not (Fig. 7.26). In case

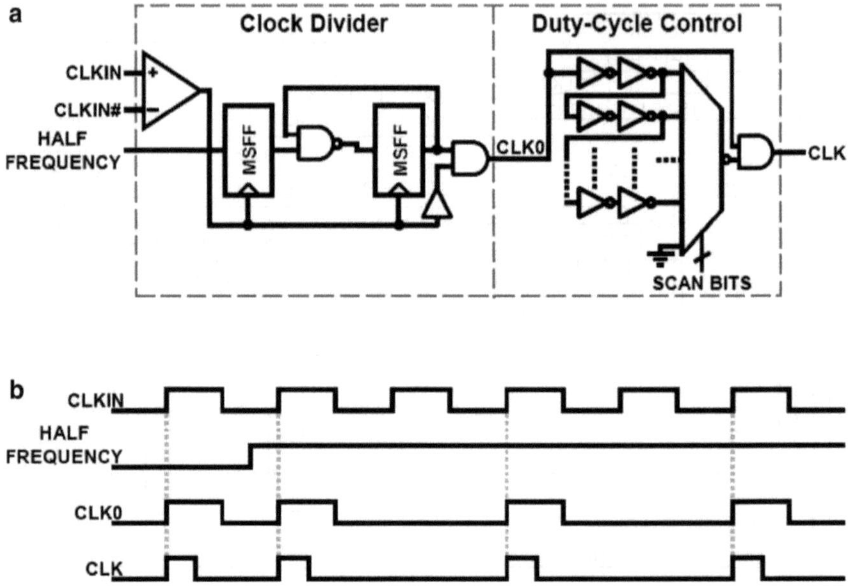


**Fig. 7.25.** Details of the test chip clock generation. (**a**) Clock generator schematic, including the clock divider which reduces clock frequency by half during error replay, and duty cycle control which is necessary for tuning the error-detection window. (**b**) Timing diagram, showing clock frequency reduction during error replay, while high phase of clock is fixed. Reproduced with permission from [18], ©2008 IEEE

of an error, the pipeline is flushed and the offending instruction is re-executed. The error controller can also reduce the clock frequency by half for 8 cycles to ensure that repeatedly-failing instructions are able to execute correctly.

### 7.4.3 Results: Guardband Reduction Through Resiliency

To get the maximum benefit from resiliency, error detection and correction techniques must be combined with adaptive voltage and/or frequency at the system level to respond to slow-changing variations. For example, a temperature increase may persist for milliseconds or longer; hence, the performance of the system would suffer if this results in a large increase in errors that must be corrected. Instead, the error rate is continuously monitored by an error control unit which dynamically changes the frequency or voltage of the processor such that the error rate is regulated at a sufficiently low value. As described before, the ideal error rate for energy-efficiency depends on the type of circuit that is implemented and the energy and performance overhead in doing recovery.

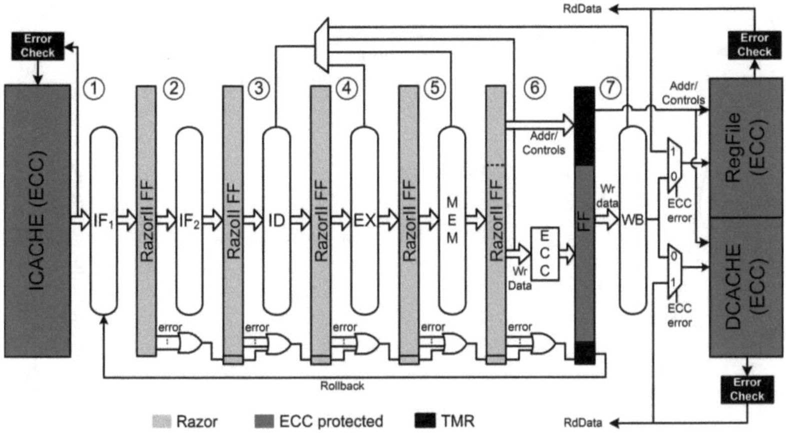


**Fig. 7.26.** Overview of 0.13 $\mu$m, 64-bit, 7-stage alpha processor design incorporating Razor-II error-detection sequentials and instruction replay. Reproduced with permission from [19], ©2008 IEEE

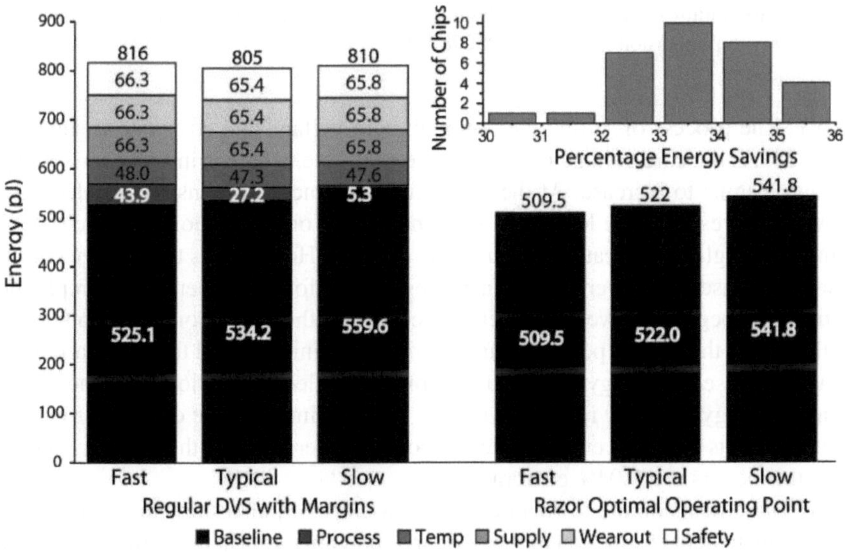


**Fig. 7.27.** Measured Razor-II energy consumption and distribution of energy savings. Reproduced with permission from [19], ©2008 IEEE

Measured energy dissipation results for three Razor-II chips, operating at 0.04% error rate, are shown in Fig. 7.27 [19]. Compared to the energy consumption when voltage is set such that all 31 fabricated dice operate correctly at 85C with 10% margin for wearout, supply fluctuation, and safety, energy dissipation gains are 33.1% to

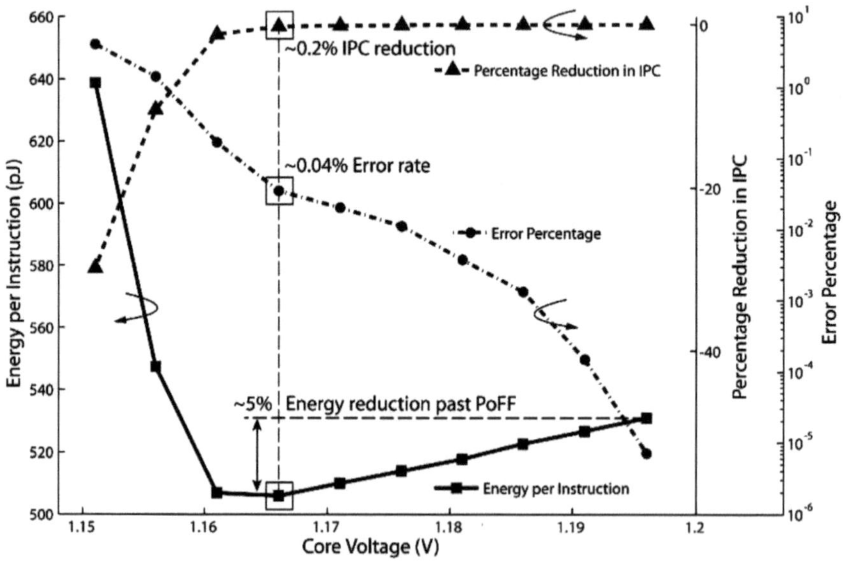


**Fig. 7.28.** Measured energy per instruction and error rate for the Razor-II processor. PoFF (point of first failure) represents the voltage at which the first error in the circuit occurs. Reproduced with permission from [19], ©2008 IEEE

37.5%. The process of setting the optimum core voltage and error rate is shown in Fig. 7.28. As the core voltage is reduced, the error rate, which is initially zero at high voltage, begins to increase. At the same time, the energy per instruction decreases as well as a result of the lower voltage. The impact on instructions per cycle (IPC) is initially negligible because the error rate is small. However, as the voltage continues to decrease and the error rate increases, the performance penalty for replaying instructions begins to have an impact on the IPC. Further reduction in the voltage actually causes the energy per instruction to reach a minimum and then increase – this occurs because the energy overhead of replaying failed instructions becomes larger than the energy saved by reducing voltage. This minimum in the energy per instruction represents the most energy-efficient point of operation for this system – in this case, this occurs at a 0.04% error rate.

Measurements on the resilient circuits test chip [18] also show the performance and energy benefits from reducing guardbands and operating at the optimal error rate. In Fig. 7.29, throughput (TP) and error rate are measured for the TDTB EDS circuit versus $F_{CLK}$, and a range of $V_{CC}$ droop magnitudes and durations are inserted based on data from a recent microprocessor along with assumptions on $V_{CC}$ droop-inducing events. The worst-case $V_{CC}$ droop magnitude is 10% and the worst-case temperature is 110C. Nominal $V_{CC}$ is 1.2V and the nominal operating temperature is assumed to be 60C. In Fig. 7.29a, throughput increases linearly as $F_{CLK}$ increases with no errors. Once errors occur, instructions per cycle (IPC) are reduced as a function of error rate and recovery time. Since $V_{CC}$ droop events are assumed to be infrequent, throughput

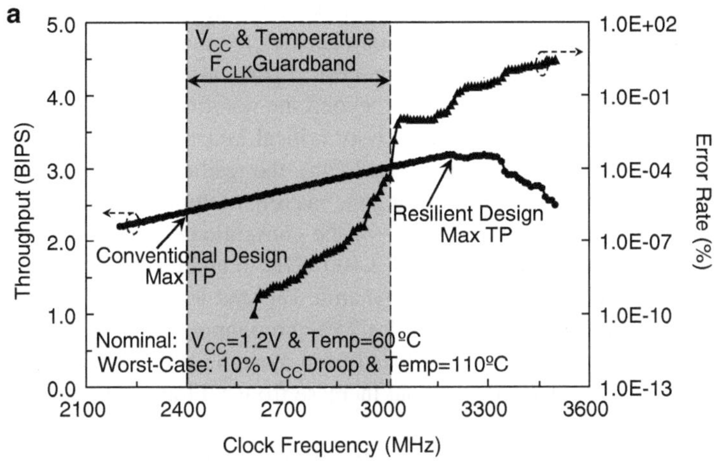


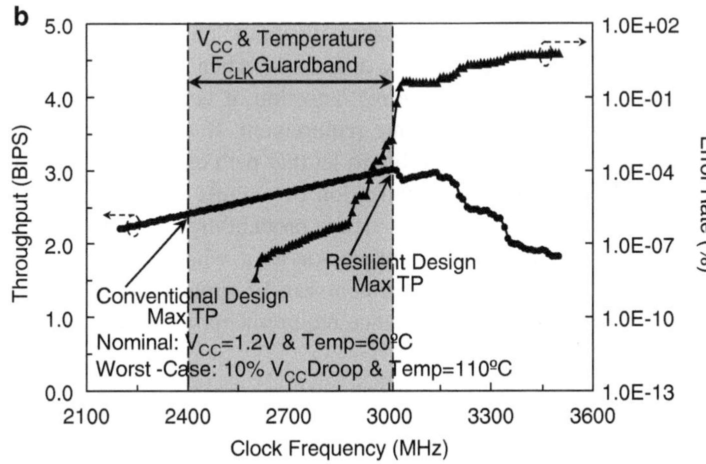


**Fig. 7.29.** Measured throughput and error rate versus clock frequency for resilient design with TDTB EDS circuits for two different path activation examples. (**a**) Critical paths less-frequently activated than non-critical paths. (**b**) All paths have equal probability of activation. Reproduced with permission from [18], ©2008 IEEE

gains continue as $F_{CLK}$ increases into the $V_{CC}$ and temperature guardband region. When $F_{CLK}$ reaches 3,020 MHz, the first path failure occurs under nominal conditions, resulting in a sharp error rate increase. If the most-critical paths on the die are infrequently activated, further throughput gains are achieved at higher $F_{CLK}$, as the frequency is increased past the point of the first critical path failure. The maximum throughput of 3.17 billion instructions per second (BIPS) corresponds to a 3,200 MHz $F_{CLK}$. Increasing $F_{CLK}$ further leads to a larger error rate, where IPC reduction outweighs $F_{CLK}$ gains.

One interesting aspect of built-in error detection sequentials is that the gains achievable depend on the path activation statistics for the critical paths. The previous example in Fig. 7.29a assumed that critical paths are activated less-frequently than non-critical paths, allowing gains beyond the point of first failure. In another scenario, it is assumed that all paths, both critical and non-critical, are activated with equal probability. In this case (Fig. 7.29b), the resilient design cannot exploit the path-activation probabilities, limiting the maximum throughput to 3.01 BIPS. In Fig. 7.29a,b, the maximum throughput for the conventional design with MSFFs is 2.4 BIPS corresponding to an $F_{MAX}$ of 2,400 MHz to guarantee correct functionality within the presence of worst-case dynamic $V_{CC}$ and temperature variations. On the other hand, a resilient design enables 25% throughput gain over a conventional design by eliminating the $F_{CLK}$ guardband from dynamic $V_{CC}$ and temperature variations and an additional 7% throughput increase from exploiting the path-activation probabilities.

These examples demonstrate that by employing error-detection sequentials within the datapath, the processor can reduce or eliminate guardbands that are applied for dynamic variations, critical path activations, and aging. They also point to the necessity of optimizing circuits differently when error detection and correction are used. For example, with error detection, it is no longer necessary that every critical path meets the cycle time requirement. If a critical path is very infrequently activated, it may be desirable to let that path cause an error and execute instead in several clock cycles. Optimization of circuits with error detection and correction needs to consider path activation probabilities to balance the performance and energy-efficiency of the processor as a whole. As demonstrated by these designs, the gains for this investment can be quite large and will become more important as power and performance requirements continue to become more stringent.

## 7.5 Conclusion

As noted in the introduction to this chapter, the clocking network is both a challenge and an opportunity as part of a variation-tolerant processor design. Variation-tolerant design techniques must first be used on both the clock and the datapath to limit the effects of clock uncertainty such as clock skew and jitter and data uncertainty caused by both static and dynamic variations. On top of these design techniques, dynamic tuning techniques can be used after fabrication to allow the clock to dynamically adjust to the current environmental conditions. By employing detectors or sensors for voltage droop, temperature, and/or transistor degradation, the clock frequency can be continually adjusted, allowing the processor to adapt to the actual operating conditions rather than being limited by the constraints imposed by a worst-case design.

Finally, further guardband reduction and energy-efficiency is made possible by dynamically detecting and responding to timing errors in critical paths of the processor pipeline. A combination of error detection techniques at the circuit level and error-recovery methods at the microarchitectural level ensures error-free computation while allowing errors to be generated and corrected in individual pipeline stages. When combined with dynamic frequency and/or voltage capability, this allows a processor to adapt to changing environmental conditions, data workloads, and device degradation without requiring sensors or safety guardbands. In the future, such error detection capability can also be used to improve reliability of the system and aid in test and debug.

## Acknowledgments

The author wishes to acknowledge Keith Bowman (Intel) and David Blaauw (University of Michigan) for the help in the preparation of this manuscript.

## References

[1] V. Oklobdzija (ed.), *The Computer Engineering Handbook*. CRC Press, Boca Raton, 2002.

[2] H. Partovi, R. Burd, U. Salim, F. Weber, L. DiGregorio, and D. Draper, Flow-through latch and edge-triggered flip-flop hybrid elements. In: *Digest of Technical Papers IEEE International Solid-State Circuits Conference (ISSCC 1996)*, 1996, pp. 138–139.

[3] V. Oklobdzija, Clocking and clocked storage elements in a multi-gigahertz environment. *IBM J. Res. Dev.*, 47(5/6), 567–583, 2003.

[4] D. Harris and M. Horowitz, Skew-tolerant domino circuits. *IEEE J. Solid-State Circuits*, 32(11), 1702–1711, 1997.

[5] K. Bowman, J. Tschanz, M. Khellah, M. Ghoneima, Y. Ismail, and V. De, Time-borrowing multi-cycle on-chip interconnects for delay variation tolerance. In: *Proc. International Symposium on ISLPED'06 Low Power Electronics and Design*, 4–6 October 2006, pp. 79–84.

[6] P. Bai, C. Auth, S. Balakrishnan, M. Bost, R. Brain, V. Chikarmane, R. Heussner, M. Hussein, J. Hwang, D. Ingerly, R. James, J. Jeong, C. Kenyon, E. Lee, S.-H. Lee, N. Lindert, M. Liu, Z. Ma, T. Marieb, A. Murthy, R. Nagisetty, S. Natarajan, J. Neirynck, A. Ott, C. Parker, J. Sebastian, R. Shaheed, S. Sivakumar, J. Steigerwald, S. Tyagi, C. Weber, B. Woolery, A. Yeoh, K. Zhang, and M. Bohr, A 65nm logic technology featuring 35nm gate lengths, enhanced channel strain, 8 Cu interconnect layers, low-k ILD and 0.57 $\mu m^2$ SRAM cell. In *Proc. IEDM Technical Digest Electron Devices Meeting IEEE International*, 13–15 December 2004, pp. 657–660.

[7] A. Sarangi, G. F. Taylor, R. J. Parker, E. P. Osburn, and P. J. Ott, Power level management in an IA32 microprocessor. In: *Proc. Electrical Performance of Electronic Packaging*, 21–23 October 2002, pp. 235–238.

[8] J. W. Tschanz, S. Narendra, R. Nair, and V. De, Effectiveness of adaptive supply voltage and body bias for reducing impact of parameter variations in low power and high performance microprocessors. *IEEE J. Solid-State Circuits*, 38(5), 826–829, 2003.

[9] T. Chen and S. Naffziger, Comparison of adaptive body bias (ABB) and adaptive supply voltage (ASV) for improving delay and leakage under the presence of process variation. *IEEE Trans. VLSI Syst.*, 11(5), 888–899, 2003.
[10] S. Tam, S. Rusu, U. Nagarji Desai, R. Kim, J. Zhang, and I. Young, Clock generation and distribution for the first IA-64 microprocessor. *IEEE J. Solid-State Circuits*, 35(11), 1545–1552, 2000.
[11] G. Geannopoulos and X. Dai, An adaptive digital deskewing circuit for clock distribution networks. In: *Digest of Technical Papers IEEE International Solid-State Circuits Conference (ISSCC 1998)*, 1998, pp. 400–401.
[12] S. Naffziger, B. Stackhouse, and T. Grutkowski, The implementation of a 2-core multi-threaded Itanium®-family processor. In: *Digest of Technical Papers IEEE International Solid-State Circuits Conference (ISSCC 2005)*, 2005, pp. 182–183, 592.
[13] T. Fischer, J. Desai, B. Doyle, S. Naffziger, and B. Patella, A 90-nm variable frequency clock system for a power-managed Itanium architecture processor. *IEEE J. Solid-State Circuits*, 41(1), 218–228, 2006.
[14] P. Mahoney, E. Fetzer, B. Doyle, and S. Naffziger, Clock distribution on a dual-core, multi-threaded Itanium®-family processor. In: *Digest of Technical Papers IEEE International Solid-State Circuits Conference (ISSCC 2005)*, February 2005, pp. 292–293, 599.
[15] J. Tschanz, N. S. Kim, S. Dighe, J. Howard, G. Ruhl, S. Vanga, S. Narendra, Y. Hoskote, H. Wilson, C. Lam, M. Shuman, C. Tokunaga, D. Somasekhar, S. Tang, D. Finan, T. Karnik, N. Borkar, N. Kurd, and V. De, Adaptive frequency and biasing techniques for tolerance to dynamic temperature-voltage variations and aging. In: *Digest of Technical Papers IEEE International Solid-State Circuits Conference (ISSCC 2007)*, 11–15 February 2007, pp. 292–293, 604.
[16] D. Ernst, N. S. Kim, S. Das, S. Pant, R. Rao, T. Pham, C. Ziesler, D. Blaauw, T. Austin, K. Flautner, and T. Mudge, Razor: a low-power pipeline based on circuit-level timing speculation. In: *Proc. 36th Annual IEEE/ACM International Symposium on MICRO-36 Microarchitecture*, 2003, pp. 7–18.
[17] S. Das, D. Roberts, S. Lee, S. Pant, D. Blaauw, T. Austin, K. Flautner, and T. Mudge, A self-tuning dvs processor using delay-error detection and correction. *IEEE J. Solid-State Circuits*, 41(4), 792–804, 2006.
[18] K. Bowman, J. Tschanz, N. S. Kim, J. Lee, C. Wilkerson, S.-L. Lu, T. Karnik, and V. De, Energy-efficient and metastability-immune timing-error detection and instruction-replay-based recovery circuits for dynamic-variation tolerance. In: *Digest of Technical Papers IEEE International Solid-State Circuits Conference (ISSCC 2008)*, 3–7 February 2008, pp. 402–403, 623.
[19] D. Blaauw, S. Kalaiselvan, K. Lai, W.-H. Ma, S. Pant, C. Tokunaga, S. Das, and D. Bull, Razor II: In situ error detection and correction for PVT and SER tolerance. In: *Digest of Technical Papers IEEE International Solid-State Circuits Conference (ISSCC 2008)*, 2008, pp. 400–401, 622.
[20] K. A. Bowman, J. W. Tschanz, N. S. Kim, J. C. Lee, C. B. Wilkerson, S.-L. L. Lu, T. Karnik, and V. K. De, Energy-efficient and metastability-immune resilient circuits for dynamic variation tolerance. *IEEE J. Solid-State Circuits*, 44(1), 49–63, 2009.

# 8

# Physical Design Considerations

Georgios Konstadinidis

Sun Microsystems

## 8.1 Introduction

Optimization of the structural skew alone, through balanced H or grid clock design and load balancing, is not adequate. Process, voltage, and temperature (PVT) variations dominate in most cases the total clock skew. While active deskew circuits [1–8] (also thoroughly described in Chaps. 2 and 7, and the use of asynchronous FIFOs in clock domain crossings [9] can reduce the effect of skew, we still need to reduce the variation in the clock network to minimize the overall complexity and design effort. Higher skew would require larger number and extended range in the deskewing circuits. Increased clock skew would mean larger hold time violations that would require additional delay elements inserted in the critical path. This will increase area and power. This chapter focuses on physical design considerations to help minimize overall skew and to avoid overdesign. At first, we provide an overview of various skew components and explain their dependency on the process, voltage, and temperature variations. We describe the main sources of transistor variation including lithographic, layout, proximity, and strain related effects. Similarly, interconnects suffer from lithographic, process, and pattern density effects that add to the variability. We provide recommendations on how to optimize the layout to minimize both the transistor and interconnect variations, and provide answers to fundamental clock designer questions:

- How should I calculate the total process variation along a chain of clock buffers? Should I just add the variations of the individual stages or should I use a Root Mean Square approach? (It turns out none of the above approaches is correct if used in isolation.)
- What is the best approach in dealing with the voltage variation? Do all clock buffers see the same voltage variation?
- How does the temperature variation affect clock skew, and are there ways to compensate for this?
- What is the impact of inductance?

T. Xanthopoulos (ed.), *Clocking in Modern VLSI Systems*, Integrated Circuits and Systems, 275
DOI 10.1007/978-1-4419-0261-0_8, © Springer Science+Business Media, LLC 2009

Good understanding of the behavior of correlated vs. noncorrelated parameters will help us define the correct methodology for the delay variation estimation using well-established statistical techniques.

## 8.2 Clock Skew Components

Clock skew is the difference in the clock arrival time at any source-destination pair of clocked elements. The setup time skew refers to the clock arrival difference (modulo the nominal clock period) at two subsequent clock cycles, while hold time skew applies to the same clock edge [10, 11]. Clock frequency reduction can fix setup related failures. However, failures due to hold time skew are independent of the clock frequency and lead to functional failure and yield loss.

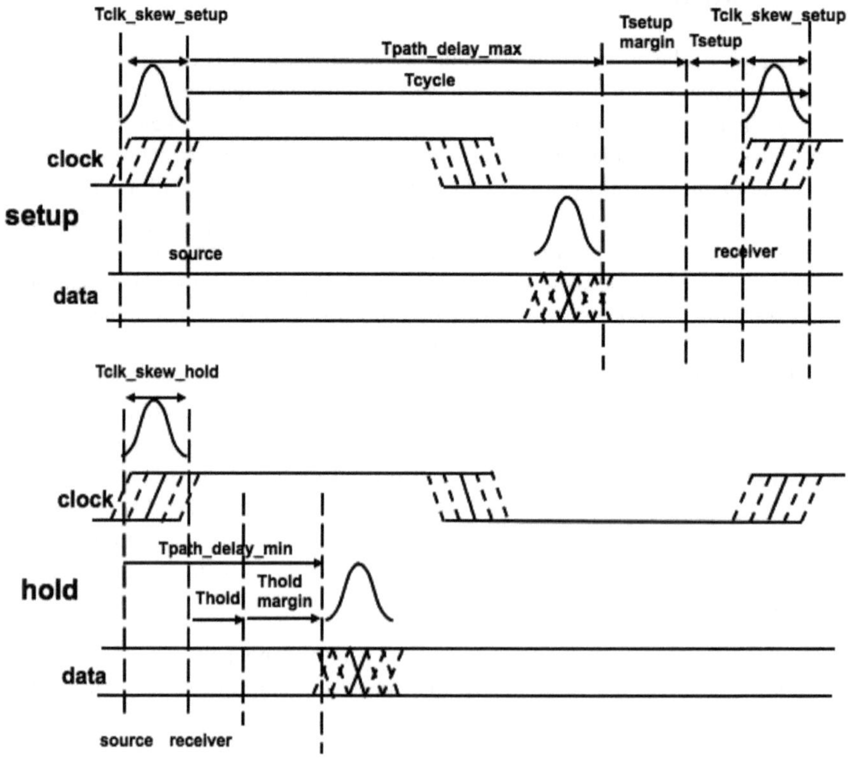


**Fig. 8.1.** Setup and hold time skew definitions considering variation

Figure 8.1 provides the definition of the setup time and hold time skew. The data need to arrive a certain "setup" time earlier than the clock edge. Process, voltage, and temperature variations may cause the late clock arrival at the source flop and perhaps early arrival at the destination flop. This reduces the usable time for logic operations within a clock cycle. In addition to that, the transistor variation from die to die and within the same die can create extra variation in the datapath. The setup check calculation needs to account for the late arrival of the data and potential early arrival of the clock at the receiver flop as shown in Fig. 8.1. The setup margin is defined as:

$$T_{\text{setup\_margin}} = T_{\text{cycle}} - T_{\text{Path\_delay\_max}} - T_{\text{Clk\_skew\_setup}} - T_{\text{setup}}, \tag{8.1}$$

where $T_{\text{cycle}}$ is operation cycle time, $T_{\text{Path\_delay\_max}}$ is the maximum path delay under worst case input conditions in the combinational logic and source flop that result in the largest path delay, $T_{\text{Clk\_skew\_max}}$ is setup time clock skew, and $T_{\text{setup}}$ is the setup time of the receiver clocked element. As we will see in the following paragraphs, there is a difference in the clock skews for setup and hold time analysis. Correct functionality is guaranteed for $T_{\text{setup\_margin}} \geq 0$. Silicon setup failures are fixable by increasing $T_{\text{cycle}}$, i.e., by reducing the operational frequency.

Achieving zero skew is not always the target. In many cases, that involve unbalanced timing paths, the designers add intentional skew in the path in order to "steal" time from the previous or the next cycle that has more margin. However, any deviation of the skew from their intended targets in a nonzero skew clock tree will degrade the performance of the design in the same manner as it does for a zero-skew clock tree. This means that in all cases we need to minimize variation. In the subsequent sections, we assume that the load is balanced through appropriate buffer sizing and interconnect clock network optimization and that we need to deal only with the process, voltage, and temperature variations.

In the hold time case, the data need to be stable for a certain time after the clock edge to guarantee correct capture. The clock at the source flop may arrive early while it can be late for the receiving flop. The datapath delay will also have a delay distribution due to PVT variations. The hold time analysis needs to account for the fast datapath delays. The hold time margin defined as:

$$T_{\text{Hold\_margin}} = T_{\text{Path\_delay\_min}} - T_{\text{Clk\_skew\_hold}} - T_{\text{hold}}, \tag{8.2}$$

where $T_{\text{Path\_delay\_min}}$ is the minimum possible path delay considering the worst case inputs for the combinational logic and the source flops, $T_{\text{Clk\_skew\_hold}}$ is the clock skew for the hold time analysis, and $T_{\text{hold}}$ the receiving flop hold time. $T_{\text{Hold\_margin}}$ should be greater or equal to zero for correct functionality. Since the hold time check refers to the same clock edge, hold time failures cannot be corrected by reducing the frequency of operation. Due to the larger risk for yield loss, it is typical that the designers add extra margin in the hold time analysis. In order to maximize yield especially in ASIC designs, the timing closure methodology calls for simulations covering the worst case process and environmental variations. While this approach increases yield, it frequently leads to overdesign, and it restricts chip performance due to the use of worst case margins.

Statistical clock skew analysis including both device and interconnect variation is essential in order to calculate more accurately the impact of the numerous variation components and to reduce necessary margins. The chip-mean distribution shown in Fig. 8.2 [12] represents the die-to-die and wafer-to-wafer distribution or so called interdie distribution. For a given point on the chip mean distribution, e.g., corresponding to a fast part, there will be a distribution of transistor and interconnect parameters across the chip as well. These intradie or across-chip variations (ACV) have both systematic and random components.

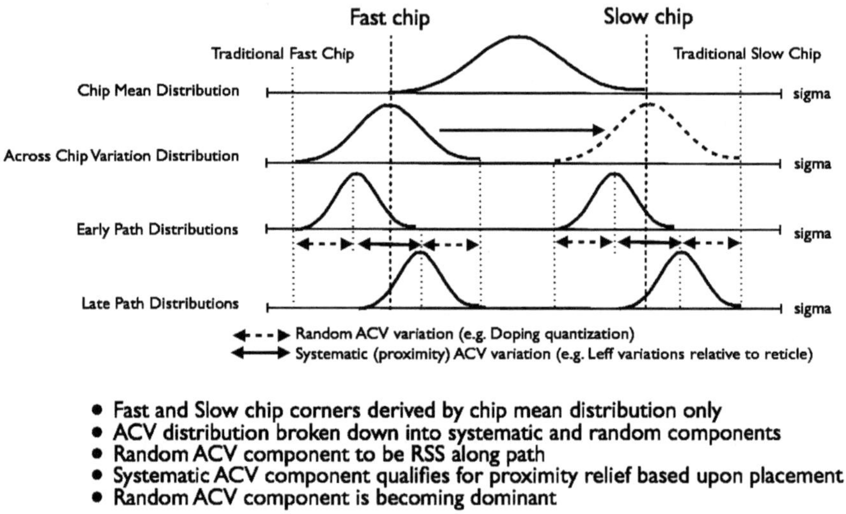


**Fig. 8.2.** Die-to-die and within-die random and systematic variation distributions (courtesy of N. Rohrer and P. Habitz). Reproduced with permission from [12], ©2006 IEEE

The systematic component is influenced by layout topology [13–15] and global scale process steps like chemical mechanical polishing (CMP). The use of localized SiGe in the PMOS devices and dual stress liners [16–28] to improve mobility add a strong systematic layout dependent component. The transistor performance will vary depending on the gate-to-gate spacing, the gate to diffusion edge distance and proximity to other devices. Scattering of moderate-energy well ion implant species at the edge of the photoresist layer used for well formation has been shown to cause a systematic elevation of threshold voltage [29–32]. Random dopant fluctuations become a substantial part of the threshold voltage variation due to the small gate area as result of aggressive process scaling. Line-edge roughness [33–35], gate oxide variation, interface charge nonuniformities [29, 30, 33–43], pocket implant variations [43] and rapid thermal annealing process steps [42] affect the transistor performance. Some of the effects are spatially correlated. Devices in close proximity and of similar layout style behave similarly to each other, while the behavior becomes less and less correlated as the distance between the devices increases. Figure 8.3 gives an overview, basic classification, and examples of the correlated vs. noncorrelated parameters and their spatial dependencies [44]. The predominant components will be analyzed in more detail in the next sections.

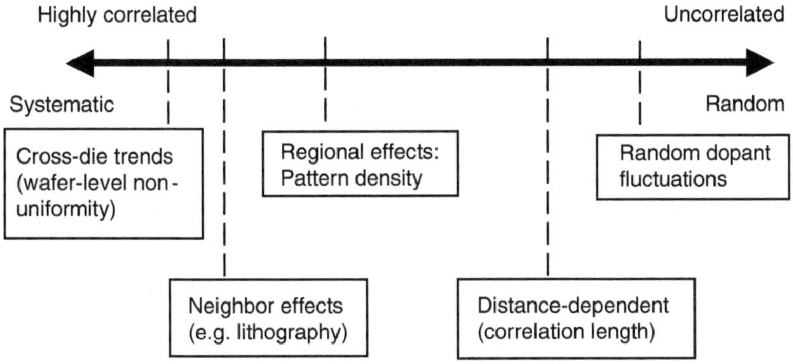


**Fig. 8.3.** Correlated vs. noncorrelated parameters and their spatial dependencies. Reproduced with permission from [44], ©2008 IEEE

In addition to process induced variations, the transistor drive current variation has a dynamic component that depends on supply voltage variations. The impact on the clock buffer behavior will be described in more detail in Sect. 8.4. The wire delay depends not only on systematic layout related and CMP effects, but it also has an additional component that depends on noise due to the activity of the adjacent nets. This component depends on the workload and is technically deterministic, but due to the computational intractability, it is usually modeled as random. Clock jitter is another substantial component of clock skew. This variation is not correlated to the rest of the skew components and can be treated as random. Chapter 5.1 provides a very detailed PLL jitter analysis.

The best way to cope with the various sources of variations and their effect on clock skew is to use well established statistical methodology techniques and not follow a worst case analysis that would lead to overdesign. A statistical model for clock skew based on Monte Carlo analysis is described in [45]. It describes a skew budgeting methodology incorporating both clock and datapath delay variations. It is shown that considering the variations in the datapath, as well as in the skew calculation, is essential to avoid over budgeting for global skew. The analysis in [45] shows that the jitter induced by power supply noise on the clock buffers is the dominant source of clock skew in H-tree networks. It is also shown that the number of paths between clocked elements has an important influence on the most appropriate clock skew budget used in the design phase.

The approach proposed [46] uses joint probability density functions (JPDFs) that preserve the correlation between minimum and maximum delays. The proposed method computes the skew distribution for the entire clock tree as well as the skew distribution of all subtrees simultaneously and, therefore, allows the designer to identify which portions of the clock tree are most prone to process variations.

An analysis of the impact of die-to-die and within-die parameter fluctuations on the maximum clock frequency is provided in [47]. The maximum operation frequency depends on both the clock skew and the datapath variation. It is shown that the within-die fluctuations primarily affect the mean frequency of operation, while die-to-die fluctuations determine the major component of the maximum frequency variance.

Since the Monte Carlo analysis is very time consuming, especially if applied to the whole clock network, the method proposed in [48] includes a three-phase approach to minimize the overall clock analysis effort. The first phase consists of a pruning of noncritical clock network portions, followed by a standard skew analysis with full extraction. Statistical analysis is applied as a third step only to the most critical portions of the clock network.

Despite the fact that general purpose or clock specific statistical timing simulators [49–51] can be used to analyze the clock network and datapath delay sensitivity under process variations, it is important that the designer acquires very good insight on what to anticipate. This basic understanding of the various interactions and process variation sources will be very useful in making the appropriate tradeoffs to achieve a robust design. In the following, we will describe how the correlated and noncorrelated parameters should be treated in the clock skew and datapath analysis. Furthermore, in subsequent sections we will provide more details on the various sources of variations that can affect a digital network delay and identify ways to minimize their impact.

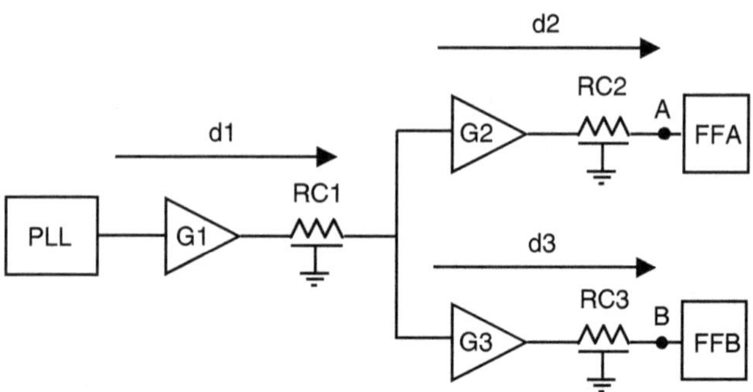


**Fig. 8.4.** Simplified clock network

Figure 8.4 illustrates a subset of a clock network to help define the various skew components and show the impact of systematic and correlated variations as opposed to the random ones.

### 8.2.1 Setup Time Skew

Using the sample clock network shown in Fig. 8.4 as an example, the skew between points A and B for the setup time case can be expressed as

$$T_{\text{Clk\_skew\_setup}} = t_{\text{A}} - t_{\text{B}}, \tag{8.3}$$

where $t_{\text{A}}$ and $t_{\text{B}}$ are the arrival times at points A and B, respectively:

$$t_{\text{A}} = d_1 + d_2, \tag{8.4}$$

$$t_{\text{B}} = d_1 + d_3, \tag{8.5}$$

where $d_i$ represents the combined gate $G$ and $RC$ delay of stage $i$:

$$d_i = G_i + RC_i,\ i = 1,2,3. \tag{8.6}$$

These stage delays have a systematic and random component and the stage delay can be expressed in the simplified form:

$$d_i = d_{i\text{\_sys}} + d_{i\text{\_random}},\ i = 1,2,3. \tag{8.7}$$

The systematic component will depend on the layout and load conditions and will remain constant over time for a given chip. The random components may vary over time even for the same chip. Such random variations include voltage, noise, and temperature variations. These effects will be examined in more detail in the next sections. There will also be random variations that are constant over time but still not predictable. One such example is random dopant fluctuations. These variations may not be Gaussian. Proper analysis of the composite variation will require Monte Carlo analysis. Using basic statistical analysis, the skew can be expressed as:

$$\begin{aligned} T_{\text{Clk\_skew\_setup}} &= t_{\text{A}} - t_{\text{B}} = E[t_{\text{A}}] - E[t_{\text{B}}] \\ &\pm M\sqrt{\sigma_{t_{\text{A}}}^2 + \sigma_{t_{\text{B}}}^2 - 2\rho\sigma_{t_{\text{A}}}\sigma_{t_{\text{B}}} + \sigma_{\text{jitter}}^2} \end{aligned}, \tag{8.8}$$

where $E[t_{\text{A}}]$ and $E[t_{\text{B}}]$ are the mean values of the path delays, $\sigma_{t_{\text{A}}}$ and $\sigma_{t_{\text{b}}}$ are the corresponding standard deviations, and $\rho$ is the relevant correlation coefficient. $M$ is the standard deviation multiplication factor selecting the degree of confidence. While (8.8) is valid for even non-Gaussian distributions, the interpretation of the factor $M$ will be different, depending on the actual distribution. For Gaussian distribution, $M = 3$ will correspond to 99.8% confidence level. The exact $M$ number will depend on the number of critical timing paths only (not all paths). Increased number of critical paths will require larger values of $M$ to guarantee a certain final yield. The correlation coefficient $\rho$ is 0 for statistically independent variables and 1 for completely dependent ones. In general, partially dependent variables will have values of $\rho$ between 0 and 1. The delays are partially dependent on each other, since the gate delay will depend on the input slew rate that depends on the delay of the previous stage. On the other hand, the delay of the previous stage depends on the load that varies when the input capacitance for the following stage is varying due to gate oxide or channel length variations. By substituting (8.4)–(8.7) in (8.8), the skew can be expressed as:

$$T_{\text{Clk\_skew\_setup}} = (E[d_1] + E[d_2]) - (E[d_1] + E[d_3]) \pm M\sqrt{\sum_{i=1}^{3} \sigma^2_{d_{i\_\text{random}}} + \sum_{i=2}^{3} \sigma^2_{d_{i\_\text{sys}}} - 2\sum_{i,j=2:i<j}^{3} \rho_{i,j}\sigma_{d_{i\_\text{sys}}}\sigma_{d_{j\_\text{sys}}} + \sigma^2_{\text{jitter}}}. \tag{8.9}$$

As seen in (8.9), the mean value of the common path $E[d_1]$ does not contribute to the clock skew, and this is a very important property we need to take into account. By removing the common $E[d_1]$ delay, (8.9) is simplified to:

$$T_{\text{Clk\_skew\_setup}} = (E[d_2] - E[d_3]) \pm M\sqrt{\sum_{i=1}^{3} \sigma^2_{d_{i\_\text{random}}} + \sum_{i=2}^{3} \sigma^2_{d_{i\_\text{sys}}} - 2\sum_{i,j=2:i<j}^{3} \rho_{i,j}\sigma_{d_{i\_\text{sys}}}\sigma_{d_{j\_\text{sys}}} + \sigma^2_{\text{jitter}}}. \tag{8.10}$$

In the case of gates in close proximity, the systematic variation that mainly depends on the layout style will be in the same direction, indicating a correlation coefficient value closer to 1. In the special case of gates in close proximity with same layout and load conditions ($\rho_{2,3} = 1$ and $\sigma_{d_{2\_\text{sys}}} = \sigma_{d_{3\_\text{sys}}}$), the systematic variation terms in (8.10) will null each other out resulting in reduced local clock skew:

$$T_{\text{Clk\_skew\_setup}}(\text{local}) = (E[d_2] - E[d_3]) \pm M\sqrt{\sum_{i=1}^{3} \sigma^2_{d_{i\_\text{random}}} + \sigma^2_{\text{jitter}}}. \tag{8.11}$$

Equation 8.11 can be applied for skew estimation for gates up to a couple of hundred microns apart. In the other extreme case for gates across the chip, the covariance term will be zero since the gate delays will be completely independent from each other. The skew across chip will be:

$$T_{\text{Clk\_skew\_setup}}(AC) = (E[d_2] - E[d_3]) \pm M\sqrt{\sum_{i=1}^{3} \sigma^2_{d_{i\_\text{random}}} + \sum_{i=2}^{3} \sigma^2_{d_{i\_\text{sys}}} + \sigma^2_{\text{jitter}}}. \tag{8.12}$$

This will be obviously the worst case skew. For intermediate distances where $0 < \rho_{2,3} < 1$, the systematic variation will be partially reduced providing skew values that are within the range defined by (8.11) and (8.12). The root mean square (RMS) term in the (8.10)–(8.12) indicates an averaging effect of the variation components. In the simplified case where all delay components and variations are equal, the skew will be proportional to the stage delay variation times the square root of the number of stages $N$, rather than $N$ itself. Consequently, simply adding the variation per stage would have resulted in very pessimistic skew estimation.

While the above skew estimation procedure has been applied to a very small network for simplicity, it can easily be extended to large clock tree networks. The total skew will be proportional to the total number of stages in a clock path, and hence proportional to the latency from the PLL output along the clock distribution and grid down to the final clock buffer. This latency is on the order of 700ps to 1ns and the skew is typically on the order of 10% of the cycle time. Typical clock skews values are in the range of 50–70ps when including PVT variations [45].

The final product frequency distribution is affected by skew, setup times, and path delays. Path delays needs to be analyzed in a fashion similar to clock branch delays. The fact that the systematic uncorrelated components are added directly to the cumulative delay causes larger datapath delays overall and will adversely affect frequency distribution. Random variations provide an averaging effect that can reduce total path delay. Statistical timing analysis of both skew and datapath delays is essential to minimize pessimism and prevent overdesign.

### 8.2.2 Hold Time Skew

The big difference between the setup and hold time case is that the hold time skew refers to the same clock edge. In this case, the PLL jitter and the common path delays will be removed from the skew calculation. This is valid for both random and systematic components since they will have just one value at a given time, and can be removed from the skew as common path delay components. In the case of setup time skew, these components may vary from cycle-to-cycle and, therefore, cannot be removed as common path delays. In the hold time case and for the example of Fig. 8.4, components $d_{1\_random}$, $d_{1\_sys}$, and $\sigma^2_{jitter}$ are common and, therefore, removed:

$$T_{Clk\_skew\_hold} = (E[d_2] - E[d_3]) \pm M\sqrt{\sum_{i=2}^{3}\sigma^2_{d_{i\_random}} + \sum_{i=2}^{3}\sigma^2_{d_{i\_sys}} - 2\sum_{i,j=2:i<j}^{3}\rho_{i,j}\sigma_{d_{i\_sys}}\sigma_{d_{j\_sys}}}. \tag{8.13}$$

### 8.2.3 Half-Cycle Setup Skew

In this case, the source uses the rising clock edge and the receiver uses the falling edge. The skew in a half-cycle path will be the same as for the single-cycle skew with the addition of the PLL duty cycle variation in the RSS component.

### 8.2.4 Multiple-Cycle Setup Skew

In the multiple-cycle paths case, the skew will be similar to the single-cycle one. The main difference is the potential extra voltage and temperature variation that may occur during the longer multiple-cycle path.

### 8.2.5 Grid or H-Tree?

Additional averaging effect will result by using a grid rather than a clock network as long the resistance of the wires connecting the clock buffer stages is smaller than the output impedance of the drivers. In the opposite case, the resistive shielding will keep the buffers in isolation and there will be no skew averaging effect. In such a case, the grid will just increase the total power due to the added capacitive load, without offering any skew reduction benefit. The other issue with the use of a grid is that timing tools have a difficult time analyzing this type of networks. The use of H-tree is also necessary in the case of clock gating, since the clock buffers need to be turned off individually.

## 8.3 Transistor Variation

This section provides a description of the various sources of variation in the transistor that have an impact on the clock skew. The transistor channel length variation due to lithographic and several layout dependent effects is a major contributor to the clock delay variation and will be described in Sect. 8.3.1. Random dopant fluctuations affect smaller transistors and they will be discussed in Sect. 8.3.2. Layout dependent effects gained significant importance after the 65nm node and have major impact on transistor performance. Advanced technology nodes use strain in the channel region to improve mobility and the layout style has a major impact on the modulation of this effect. Section 8.3.3 provides a brief overview of these effects and potential remedies.

### 8.3.1 Channel Length Variation

The delay of a clock driver in a first order approximation can be expressed as:

$$\text{gate\_delay} = \frac{C_{\text{load}} \times V_{\text{DD}}}{I_{\text{drain}}}, \tag{8.14}$$

where $C_{\text{LOAD}}$ is the capacitive load, and $V_{\text{DD}}$ is the supply voltage. The drive current is inversely proportional to the channel length and since the gate capacitance is proportional to the transistor width $W$ and channel length $L$, the delay is proportional to $L^2$:

$$C_{\text{load}} \propto W \times L, \tag{8.15}$$

$$I_{\text{drain}} \propto \frac{W}{L}, \tag{8.16}$$

$$\text{gate\_delay} \propto L^2 \times V_{\text{DD}}. \tag{8.17}$$

This means that small variations in the channel length will have a quadratic impact on the gate delay. Short channel effects amplify further this effect by changing the behavior of threshold voltage vs. channel length. Figure 8.5 shows the threshold voltage variation vs. channel length in the linear and saturation regime [26]. At longer channel lengths, the transistor threshold voltage and consequently the drive current variation is less sensitive to channel length variations than in the case of shorter channel

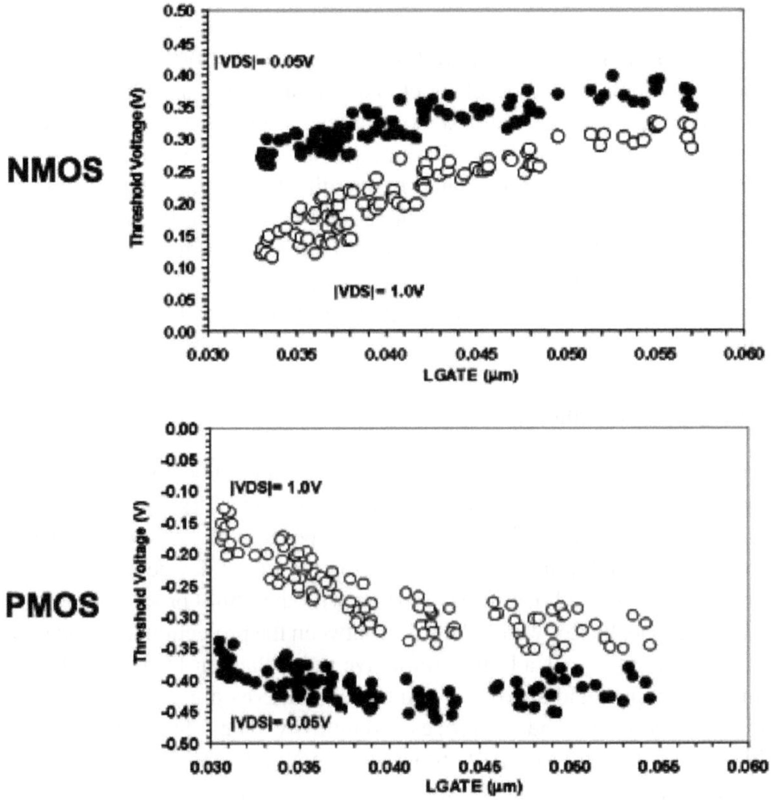


**Fig. 8.5.** Threshold voltage variation vs. channel length at VDS=0.05V, 1.0V. Reproduced with permission from [26], ©2007 IEEE

lengths. Although it is tempting to use longer channel lengths across the board to reduce variation, this would increase path delay, hold time requirements, and power. Ideally, both the datapath and the clock network should use similar channel length devices so that the minimum delay and the hold time requirement track each other in the presence of variations. This is also important when designing a chip that needs to be portable across process nodes. The datapaths that use long channel devices will not speed up in the new node as much as the paths with short channel devices. While using long channel length in short datapaths will not be a problem, substantial redesign will probably be required if the datapath uses short channel devices while the circuitry that affects hold time uses longer channels. In this case, the hold time will not decrease as much as the short datapath will speed up and hold time violations may be created.

The channel length variation has a systematic and a random component with different values for the die-to-die and within-die cases that depend on the layout style. Local variations are usually small and more systematic and correlated (local photolithographic effects), while the across-chip variations are larger and more random in nature. At the local level, variatons result in limited clock skew, and this is especially important in the case of hold time violations that typically happen locally (back-to-back flops). This type of clock skew relaxation can eliminate thousands of false hold time violations and reduce overdesign. On the other hand, systematic variation across chip may be of different sign, depending on location. Systematic variations at point A may be correlated in one direction, e.g., resulting in increased clock branch delay around point A, while around point B on the other side of chip, the delays may be correlated in the opposite direction, resulting in delay reduction in another clock branch. The consequence will be larger across chip skew due to channel length variations. The following sections provide some background regarding root causes.

## Photolithography Challenges

Since the printing features are smaller than the wavelength being used, special optical proximity corrections (OPC) are employed to help print transistor gates [13, 14]. Critical layers in 90nm and 65nm process are exposed with ArF excimer laser light (193nm wavelength). Figure 8.6 [13] shows the minimum device size trend and the exposure wavelength used over time. Starting with the 45nm process node, we need to use immersion technology where the gap between the projection lens and the wafer is filled with water that has a higher refractive index than air [14]. In addition, dual patterning techniques and sacrificial print assist patterns are employed to improve printability at the cost of added process complexity. The 22nm process node will more likely need to use Extreme Ultraviolet Lithography (EUVL). Nanoimprint or Electron Beam Direct Write are the future candidates, but they need further development prior to full deployment in production [13, 14].

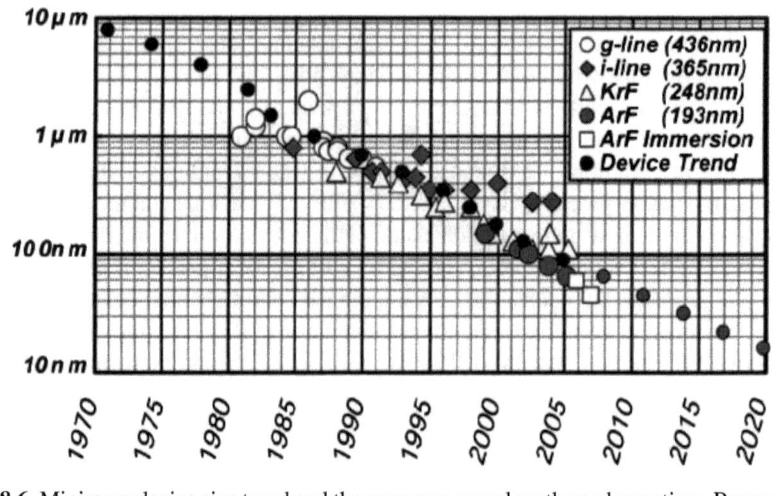


**Fig. 8.6.** Minimum device size trend and the exposure wavelength used over time. Reproduced with permission from [13], ©2006 IEEE

Apart from the global channel length variation that is more random in nature, there are two very important local components: poly flaring and line edge roughness (LER). Poly flaring is a systematic component that depends on the layout style and can lead to channel nonuniformity and increased leakage. LER is the result of polysilicon grain boundary nonuniformity and is more random in nature. Both of these components are examined in more detail in the following sections.

### Poly Flaring and Poly Pullback

The actual structure on silicon is very different from what is drawn in the layout database. Poly flaring as shown in Fig. 8.7 [38] affects the actual channel length at the transistor boundary. The channel width locally is larger causing reduction in drive current. Increase of the distance of the poly jog to active will reduce this effect but it comes at a layout area increase penalty. In addition, the poly extension over active needs to be large enough otherwise the poly pull back during processing will result in a locally very short channel that will cause higher leakage current. This effect will be even worse, if inadequate poly extension is in close proximity to active jogs. The active jog will not be printed as a 90° sharp angle but as a curved edge reducing the poly extension margin and causing higher channel leakage.

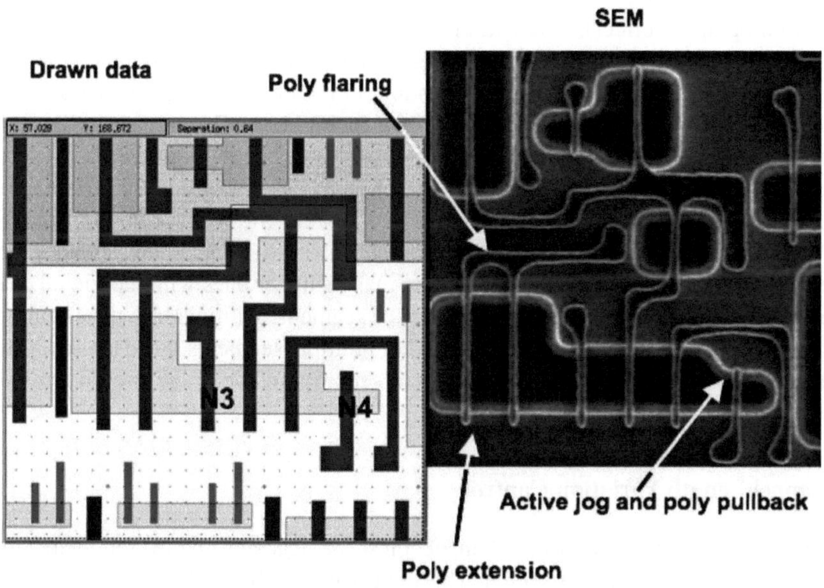


**Fig. 8.7.** "What you draw is not necessarily what you get on silicon" (courtesy of J. Rosal). Reproduced with permission from [38], ©2006 IEEE

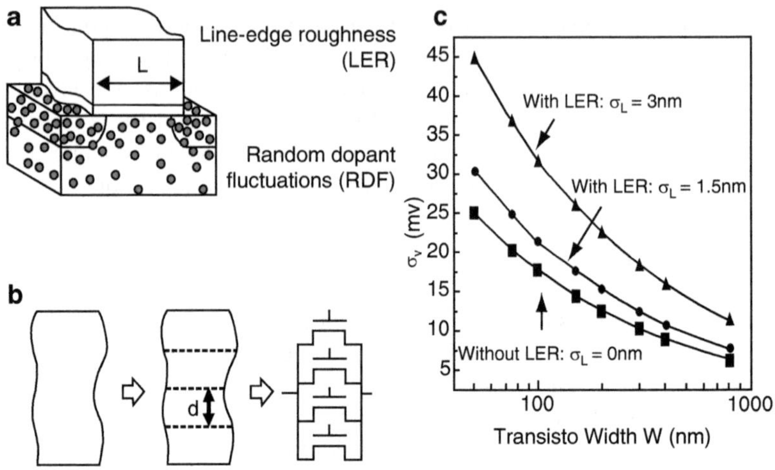


**Fig. 8.8.** Threshold voltage variation vs. line-edge roughness and transistor width. Reproduced with permission from [35], ©2008 ACM/IEEE

## Line Edge Roughness

The poly edge is not a straight line any more exhibiting small curvatures at the edge that are caused by the nonuniformity of the polysilicon grain boundary and photolithographic effects. The end effect is that the channel length is not uniform across the transistor width as shown in Fig. 8.8a, b [35]. This effect is called line edge roughness (LER) and is substantially random in nature. LER impacts the threshold voltage along the transistor width as shown in Fig. 8.8c. The impact is larger for smaller devices and of course it gets worse with larger LER. Figure 8.9 [33] shows actual data from various advanced lithography processes reported by different laboratories indicating that LER does not scale with line-width according to the ITRS roadmap requirements. Circuit simulators that intend to capture the layout dependent effects typically split the transistors in many smaller ones with individual channel lengths per segment. The large number of segments will increase the simulation time. In order to address this issue, instead of multiple segments one can use a single device with the average channel length that is representative of the overall transistor behavior [35].

## Channel Length Variation Control

The channel length has a profound impact on speed, leakage, and overall yield. Good control of this variation is essential. This control cannot be achieved by photolithography alone and requires further assistance by design. In the 90 and 65 nm process nodes, some of the design for manufacturability (DFM) rules used to minimize these effects were just guidelines. In 45 nm and beyond, these rules became required design rules. This affects layout density. Some of the rules require a constant poly pitch and

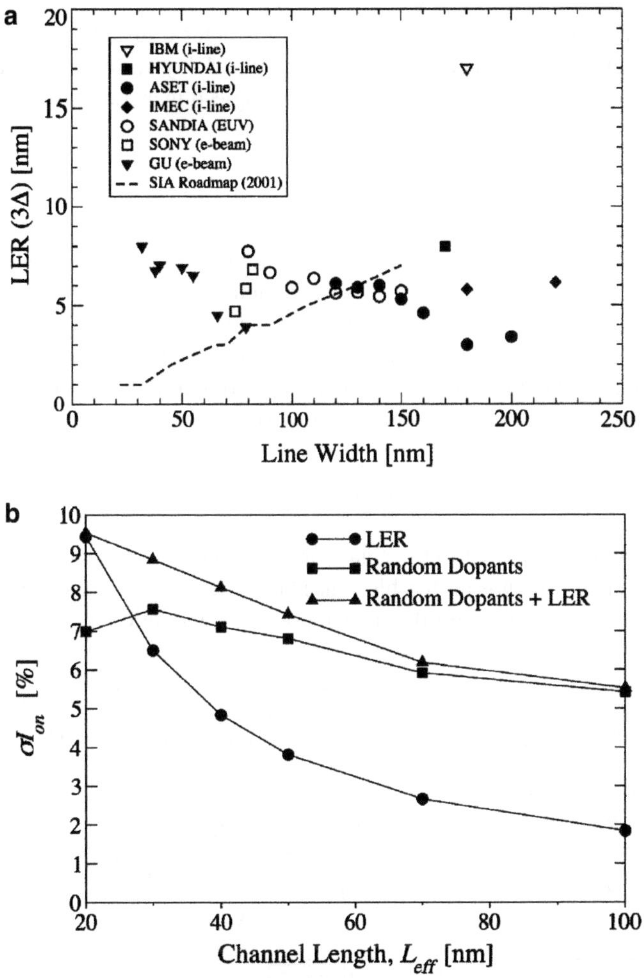


**Fig. 8.9.** (**a**) Line-edge roughness reported by various labs and ITRS (**b**) Threshold voltage variation vs. channel length, random dopant fluctuation and LER. Reproduced with permission from [33], ©2003 IEEE

same poly orientation (horizontal or vertical) across the chip, adding dummy poly gates [52] next to active gates, and strict poly extension requirements. Jogs of active regions in shared transistors are not recommended since they can cause active flaring and poly pullback. For the same reason, the distance between active edge and poly jog also needs to be increased.

Similar layout style (matching layout) can reduce the systematic variation. Poly shields [52] will help maintain uniformity in the channel length of the external transistor "fingers". Using multi-finger devices for clock drivers is better than single-finger devices, since the intermediate fingers act as a poly shield for the neighbors.

In addition, the random variation within a multi-finger transistor provides an averaging effect on the drive current that minimizes the delay variation in a multi-finger device compared with the delay variation of a single-finger transistor with the same total equivalent width.

### 8.3.2 Dopant Fluctuation

Narrow width transistors have literally a few atoms in the channel and their fluctuation affects the transistor threshold voltage and consequently the drive strength. This threshold variation is one of the main components of variation in transistors that have the same layout and are close to each other, and otherwise are expected to have similar behavior. The threshold variation is proportional to the inverse of the square root of the gate area $W \times L$ as shown in Fig. 8.10 [39]:

$$V_{\text{threshold\_shift}} \propto \frac{1}{\sqrt{W \times L}}. \tag{8.18}$$

This means that devices with larger gate area will be less impacted by the dopant fluctuations. Therefore, it is preferable to use larger devices in the clock buffers rather than smaller ones. It is a trade-off between power consumption and skew reduction.

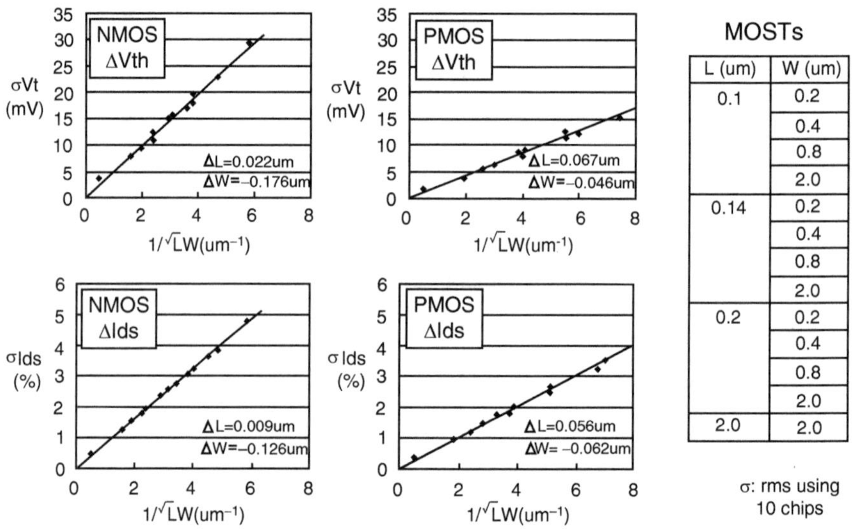


**Fig. 8.10.** Threshold variation due to dopant fluctuations vs. transistor width and channel length. Reproduced with permission from [39], ©2006 IEEE

The dopant fluctuation issue becomes more significant at thinner gate oxides. The increase in the threshold voltage fluctuations amounts to more than 50% in MOSFET with oxide thickness less than 1.5 nm. The poly-Si depletion and the discrete random dopants in the poly-Si gate add an extra 50% in devices with ultrathin oxides [41].

A semianalytical method is proposed in [40] to predict the mean, standard deviation, and probability distribution function of delay in logic circuits considering the random threshold voltage variations, and it was applied to statistical characterization of flip-flops and logic gates.

While this variation is already large and very undesirable, effects like higher local annealing temperature, non uniformities in the gate oxide thickness, and surface states make the problem even worse and require process changes to reduce this variation. Higher local annealing temperature expands junction diffusion both laterally and vertically, resulting in lower $V_T$ due to a combination of short-channel effects and compensation of halo channel doping (for halo/pocket-implanted FETs) [42, 43]. Gate oxide scaling required for better control in the channel area reaches the limits. The gate oxide contains only 4–5 atomic layers. Nonuniformities in the gate oxide thickness and surface states create extra dopant fluctuations [34] and introduce high gate leakage and reliability risks associated with gate oxide integrity. High-K dielectric gates [24, 26] offer a good solution to this problem. This type of gates can achieve the same electrical thickness and control as the traditional gate oxide, but with higher physical gate oxide thickness that reduces gate leakage exponentially [26].

While in the past most of the variability and yield loss came from random defects, in modern technologies yield and performance are substantially impacted by physical layout style and device interaction. Apart from the channel length variation due to lithographic effects as described in previous sections, there are additional effects related to the well proximity effect (WPE), [29–32], and strain effects used to improve the transistor mobility [16–28].

### 8.3.3 Well Proximity Effect

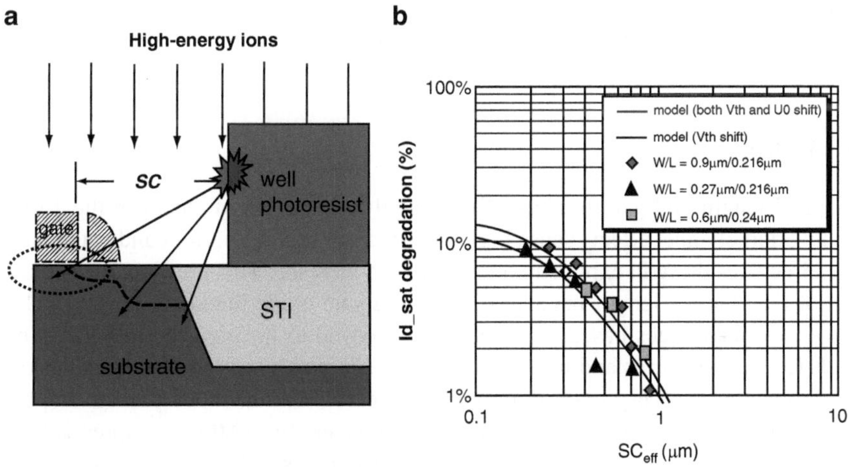


**Fig. 8.11.** (**a**) Ion scattering at the photoresist edge is the cause of the well proximity effect (WPE) (**b**) Drive current degradation due to WPE vs. gate to well edge distance (SC). Reproduced with permission from [29], ©2006 IEEE

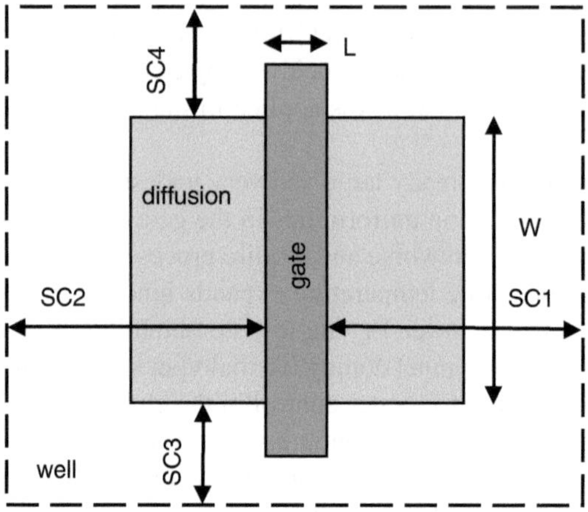


**Fig. 8.12.** Well to gate distance (SC) definitions [53]. WPE reduces as SC [1–4] increase

Lateral scattering of the ion implants at the photoresist edge when forming the MOSFET wells as shown in Fig. 8.11 can land into the channel area, increasing the threshold voltage of the transistor and, therefore, degrading the transistor performance [29]. The amount of dopants collected in the channel area depends on the distance SC of the poly over active to the well edge as shown in Fig. 8.12. From the physical design perspective, what can be done is to increase the spacing of the well to the active poly. This can usually be done for critical devices but not across the whole chip. Possible process improvement steps are described in [30]. Narrow distances on the order of 100nm can cause a 10% degradation in the transistor drive current as shown in Fig. 8.11 [29]. Similar magnitude of degradation is observed in current mirror devices [31] and in circuit delay [32]. The WPE effects are included in the BSIM 4.6.2 spice model [53]. State-of-the-art layout extractors identify the SC gate distances to the well edge, and these are interpreted as layout input parameters to the spice model.

### 8.3.4 Strain

Channel length and gate oxide scaling started running out of steam in the 130nm node. Short channel effects cause high source–drain leakage current and gate tunneling effects resulting in exponential gate leakage increase. This prevents further scaling of the channel length beyond $\approx$30nm and gate oxide thickness beyond 1.2nm [16]. New techniques based on strain are employed to increase the mobility while high-K gate dielectrics and new gate materials are now used to reduce gate leakage while maintaining very good drain current control. The methods used to improve the mobility incorporate highly tensile capping layers for NMOS, compressive capping layers for PMOS, stress memorization techniques, SiGe induced strain in the PMOS devices, and shallow trench mechanical stress (STI) [16–28]. All these techniques introduce systematic layout dependent variations affecting both the clock and the datapath delays. The drive strength of a transistor is affected by the poly pitch, the distance of poly gate to STI, and the surrounding devices. Good understanding of these layout and circuit performance interactions is an essential tool to help minimize clock skew and datapath delay variations.

### Stress Memorization and Tensile Stress Liner

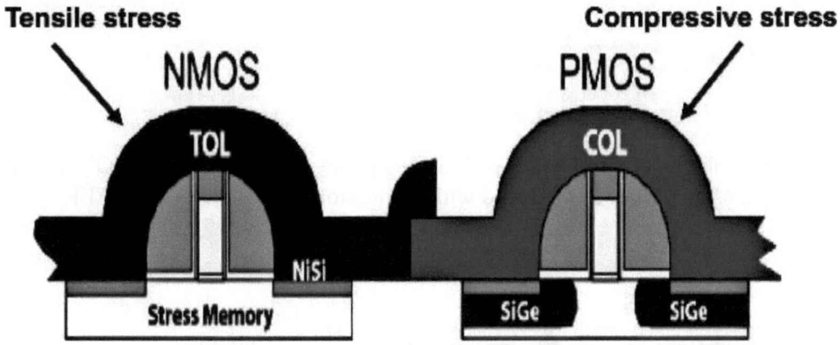


**Fig. 8.13.** Stress induced mobility enhancement techniques: stress memorization and tensile liner for NMOS, SiGe, and compressive stress liner for PMOS. Reproduced with permission from [22], ©2007 IEEE

The stress memorization technique is applied to the NMOS devices during the source–drain spike anneal [19–22]. This thermal treatment creates stress in the NMOS channel area due to the difference in the thermal coefficients of the Si and the nitride sacrificial capping layer used over the poly gate. This stress is tensile in the direction of the current flow and compressive in the vertical direction [21]. Both of these types of stress enhance the NMOS mobility. Further enhancement is achieved by adding a SiN tensile layer on top of the gate (Fig. 8.13). The stress depends on the intrinsic stress of the material used and its thickness, the poly spacing, the gate height, and the contact to gate spacing. Increased poly spacing will allow for more volume of capping layer to create even more stress [16].

### SiGe and Compressive Stress Liner

The PMOS transistor prefers compressive stress parallel to the current flow and tensile stress in the direction perpendicular to the current flow [20]. The compressive cap layer is not as effective as the SiGe [16–19] that is epitaxially grown in the source/drain areas of the PMOS devices as shown in Figs. 8.13 and 8.14. The SiGe has larger lattice constant than Si and introduces compressive stress in the channel. The improvement in whole mobility can be as high as 50% with a 17% Ge composition [17].

The poly pitch optimal distance is not only defined based on the lithographic requirements, but it is also based on the stress inducing capability. Higher gate-to-gate distance will help create more stress due to the higher area of the stress liner

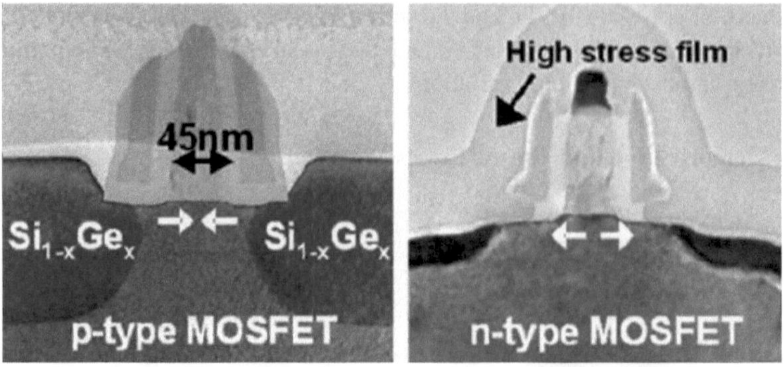


**Fig. 8.14.** TEM photograph of PMOS with SiGe compressive stress, and NMOS with tensile liner stress in a 45 nm Node. Reproduced with permission from [18], ©2004 IEEE

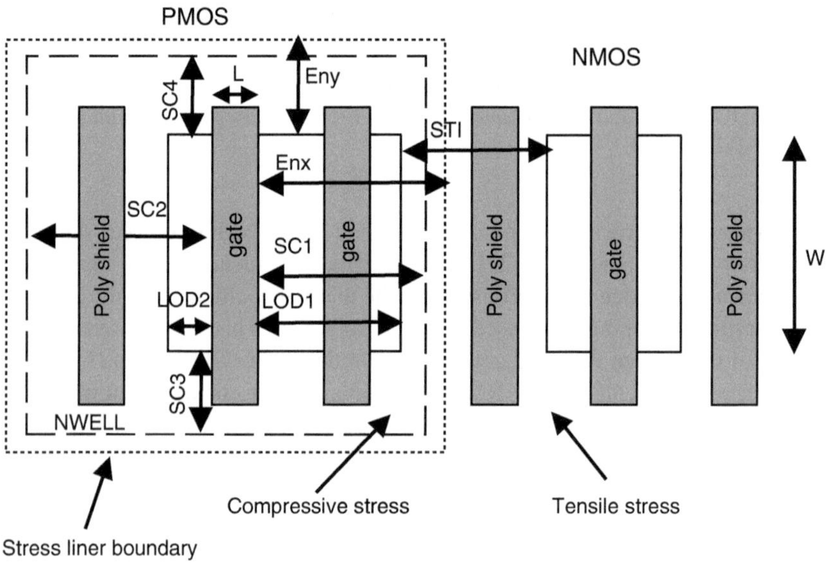


**Fig. 8.15.** Definition of various gate to diffusion and well distances that impact stress magnitude and consequently transistor performance

and the SiGe volume. Similar arguments apply to the compressive stress liner case. Increased stress liner size in the current flow direction *Enx* as defined in Fig.8.15 will help increase the compressive stress in the PMOS device. On the other hand the *Eny* size does not have a predominant effect on PMOS performance [20].

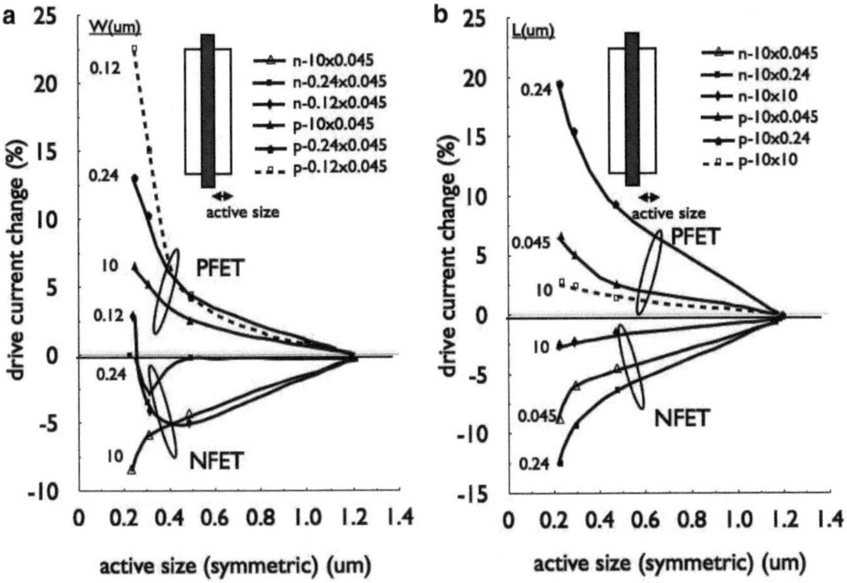


**Fig. 8.16.** Impact of STI stress effect on transistor drive current vs. active size. Reproduced with permission from [23], ©2003 IEEE

### Shallow Trench Isolation

Shallow trench isolation (STI) used to isolate the diffusion areas induces stress between the neighboring devices. This stress is transferred along the device channel and enhances the PMOS mobility, but unfortunately degrades the NMOS mobility. The STI stress depends on the location of the channel within the diffusion region and the STI width defined as STI in Fig.8.15. The STI stress can affect the cell delay by 10–20% depending on the STI-to-poly gate distance [23]. Figure 8.16 shows the NMOS and PMOS current change vs. active area size for various width and channel length combinations. Smaller active size reduces the compressive stress and deteriorates the PMOS device, but it helps the NMOS device that does not fare well under compressive stress. The NMOS is less sensitive to transistor width reduction, but the opposite is true for PMOS. Smaller PMOS width will have more compressive stress and higher drive current. Longer channel devices are relatively insensitive to STI stress. NMOS degrades for intermediate channel lengths and becomes insensitive for small channel lengths. On the contrary, PMOS devices see substantial drive current boost due to STI stress as the channel length becomes shorter. All these layout dependent effects will impact the systematic delay component in the skew calculation as expressed in (8.10).

### New Materials

Further enhancement in the transistor performance is achieved by using high dielectric constant gate interface material and metal gates instead of polysilicon. High-K gate dielectrics are used to reduce gate leakage but at the same time offer very good control of the channel field due to the higher gate dielectric material that forms a better capacitor under the gate without requiring further reduction of the gate physical thickness. This improvement combined with metal gates that prevent polysilicon depletion and the strain techniques described above are the predominant performance boosters for the 45, 32, and 22 nm nodes [25, 26]. New, high mobility, narrow bandgap materials like InxGa1-xAs are strong contenders to replace strained-Si channels in technologies beyond the 22 nm node [28].

### Guidelines

As explained earlier, the performance improvements based on strain depend strongly on the layout style being used. Matching device layout style will help reduce the mismatch in the clock buffers and minimize skew. Optimized layout to maximize the drive strength will help reduce the buffer size and reduce power. This optimized layout comes typically at the cost of extra cell area and larger parasitics that increase the cell delay and power. Obviously there is an optimal trade-off point beyond which there is substantial area and power penalty without a corresponding speed improvement. State-of-the-art layout extractors take into account all these effects, and spice models like BSIM 4.6.2 [53] incorporate these interactions and model the effect on device performance.

While the layout for the clock headers can be matched and minimize the impact of these systematic layout effects, it is obvious that the generic logic gates can be affected substantially and can introduce systematic delay variations in the datapath delays. Since the next generation technologies are running out of steam in terms of electrical performance improvement, a new layout methodology is needed to achieve the best performance and minimum possible variation. The library cell parasitic extraction and electrical characterization need to take into account the fact that the cell will not be in isolation. The impact of the neighboring devices that can create extra stress and affect the performance of the device being analyzed must be included. Dummy devices need to be placed around the actual cell to be characterized, and the analysis must be performed on a netlist extracted from the complete layout including the dummy devices.

## 8.3.5 Long Term Effects on Variation

### NBTI

The negative bias temperature instability effect (NBTI) [54, 55], impacts the PMOS devices and reduces their current driving capability over time. This is due to threshold voltage increase caused by trapped carriers in the gate oxide especially in the presence of hydrogen. Worst-case bias condition is when the gate of the PMOS transistor is turned on and the supply voltage and temperature are high. There is partial recovery under AC conditions [56–58]. The NBTI shift in the early phases of the stress has an exponential behavior, but after a couple of minutes of stress it follows a power law behavior expressed as

$$V_{\text{Threshold\_NBTI\_shift}} \propto \text{Time}^{\frac{1}{6}}. \tag{8.19}$$

Short testing times may indicate very strong dependency on NBTI due to the initial exponential increase. This early stress recovers over time, but the long term effect will not. NBTI stress will be worse under DC conditions. Lower gate-source voltage will induce less stress and increased switching activity will lead to lower overall stress due to averaging effect of the bias conditions. Lower temperature of operation reduces the NBTI effect in an exponential fashion.

This long-term effect can have major impact on the circuit behavior [59], and proper analysis and margining are required to guarantee correct functionality until the end of the product lifetime [60]. It is typical to add an extra 5% frequency margin to cover for NBTI degradation. While the NBTI impact on the threshold voltage is in the order of 50mV in 65nm for normal SiON oxides as shown in Fig. 8.17, it reduces substantially in the case of high-K dielectric gates [26].

The NBTI degradation will lead to increased latency of the clock network and will increase the rise time at the final clock buffer stage. The amount of skew is proportional to the overall clock network latency and, therefore, NBTI will have an impact on the final skew. However, under symmetrical stress in the clock network the impact on delay will be similar on all clock network branches and, therefore, the impact on skew (difference in clock arrival times) will be minimal. This may not

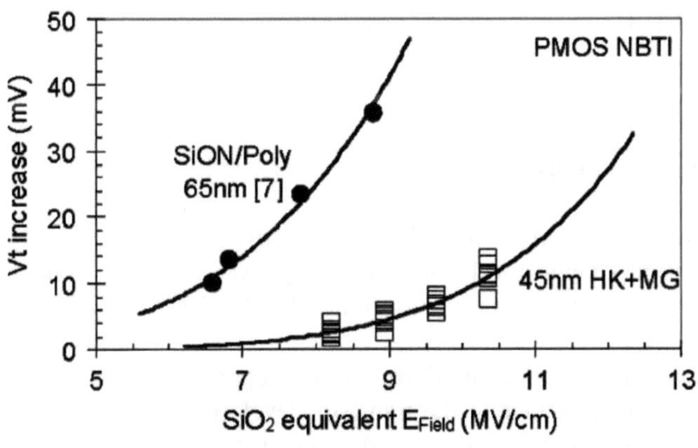


**Fig. 8.17.** NBTI shift dependency on gate material. High K dielectric reduces NBTI shift substantially at equivalent SiON electric field across the gate. Reproduced with permission from [26], ©2007 IEEE

be the case when extensive clock gating is used. Long term clock gating will affect clock buffers differently, depending on their bias conditions. Every other inverter stage will see a different NBTI stress and this can cause duty cycle shift that reduces margin, and it can lead to timing failures in half-cycle paths.

The NBTI effect represents a systematic delay component and will also have major impact on datapath delay. While the pattern in the clock network is highly predictable if no clock gating is involved, in the case of combinational logic the bias condition patterns will depend on the workload and different workloads will create different degrees of NBTI degradation.

#### Hot Carrier Injection

At high electric fields in the channel, the carriers may gain sufficient energy and momentum to enter into the gate oxide layer. This effect called hot carrier injection (*HCI*) or channel hot carrier (*CHC*) effect [61, 62] is predominant on NMOS transistors. Electrons are accelerated much more than holes that have higher effective mass. Moreover, the Si–$SiO_2$ interface barrier is higher for holes. The stored carriers in the oxide increase over time the transistor threshold voltage and reduce the drive current. The degradation will be proportional to the time there is current flow through the channel of the transistor and, therefore, crowbar current, and slow rise times can increase this effect. In the case of clock buffers and repeaters stages in general, one should restrict the output slew rate to minimize HCI. It is also best to use buffer stages built by two inverters rather than single inverter stages since high slew rate at the inverter input will cause higher crowbar current. In the case of two stages that typically use a fanout of 4, the first stage will experience crowbar current but 4 times smaller than in the original single inverter design due to the smaller size of the first stage. The gain from the first stage will help recover the slew rate at the input of the second stage and, therefore, reduce the crowbar current of the output stage. This will also help minimize the electromigration effects at the output of the large driver.

## 8.4 Voltage Variation

The transistor current drive strength depends strongly on voltage. Typically, we see a 10% speed change for 100mV of voltage variation. In 45nm and beyond, this sensitivity is even larger approaching the 15–30% range at lower supply voltages.

The clock and gate switching during normal operation create voltage transients that affect the actual voltage at the transistor level and consequently affect speed. A dynamic IR analysis will be required to identify the exact voltage variation at the various points of the power grid. The voltage variation in the buffers along the main clock tree depends on the chip switching activity. The largest voltage droop will occur when the final clock headers switch, since the power consumed in this case is a substantial part of the total chip power (typically on the order of 30% as

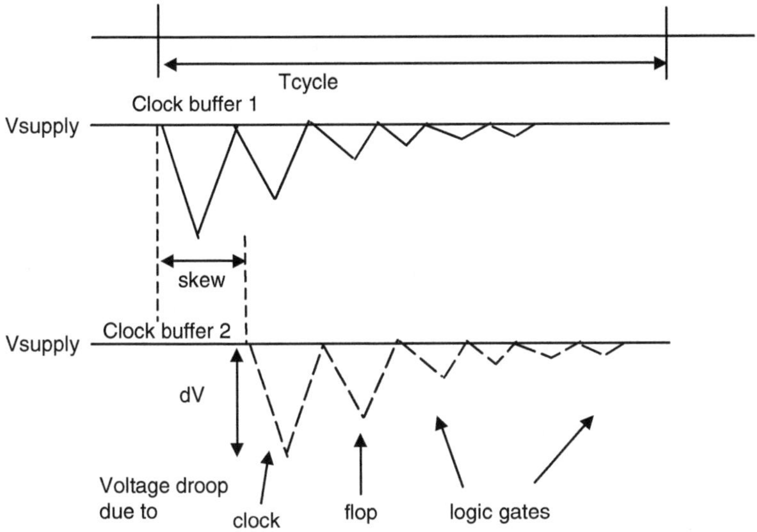


**Fig. 8.18.** Dynamic voltage drop vs. time, clock, flop, and combinational gate activity

mentioned in Sect. 2.2.3). There will be an instantaneous current and voltage surge at the time the final clock buffers switch as shown in Fig. 8.18. After a certain time, the flops will switch creating a second current surge. Subsequent activity will depend on the logic propagation and will gradually reduce as we reach the next clock edge. The voltage spikes at the main clock buffers can be minimized by appropriate use of decoupling capacitors in close proximity. These decoupling capacitors need to be placed close by so that the RC time constant including the power grid connection resistance is at least 5 times smaller than the rise or falling transition time of the voltage spike. Otherwise, the decoupling capacitors will not be able to reduce these fast voltage transients. Differential structures may also be used for the main clock trunk to minimize voltage variation, but this comes at the expense of increased total power since the bias current source used will consume constant power [63].

In the ideal case where the clock skew is zero, the voltage difference at the final buffer stages would be zero as well. However if the skew is larger than the clock fall/rise time, then the two buffers may see a different voltage variation as shown in Fig. 8.18. The voltage variation to be considered for the final stages is much larger than the variation observed at the main clock tree buffers, since the power surge at the time of final buffer switching is very large. In order to minimize the impact of the voltage variation on the PLL jitter, typically the PLL uses a separate supply than the main supply of the chip. Detailed analysis of the PLL jitter is provided in Chap. 5.1 and an analytical model of supply-induced period jitter for long buffer chains is described in Sect. 6.8.

Apart from the high-speed voltage variation, there is a longer-term voltage variation due to package inductance. Typically, the resonance frequency of the package inductance and chip capacitance (large CPU) is on the order of 50–100MHz. The noise period spans several CPU cycles for a GHz clock design. Package decoupling capacitors are used to reduce the noise amplitude, but they cannot filter out the high frequency spikes on the chip. Even in the case of long term $Ldi/dt$ variations, one should calculate the corresponding voltage variation within one cycle and include it in the setup skew calculation. This is not necessary for the hold time skew since it refers to the same clock edge and the $Ldi/dt$ variation for this short time will be negligible. Special attention should also be paid to multi-core designs with power switches used to power down parts of the design. While this separation reduces overall power, it also reduces the total amount of available decoupling connected to the power supply grid leading to larger $Ldi/dt$ noise [64].

## 8.5 Temperature Variation

Both transistor and interconnect performance are affected by temperature. Chip temperature depends on workload conditions. The thermal time constant is on the order of a minute, so it does not track rapid workload variations. However, there is typically a spatial temperature gradient across the chip. Floating point units for example will be much hotter than the L2 cache. Figure 8.19 shows a typical thermal map of a multi-core chip under worst case load conditions [65]. The thermal gradient is 40°C across the chip.

In worst case hold time analysis, one should assume that the source flops are at the lower temperature (i.e., fast clock-to-q delay) and the receiving flops are at the higher temperature and lower voltage (i.e., higher hold time). Low temperature will also reduce the wire resistance which helps achieve higher frequency of operation but may cause hold time violations if this effect is not accounted properly in the hold time analysis.

The transistor drive current is proportional to the carrier mobility and (to first order) to the difference of $V_{DD} - V_{threshold}$. The threshold voltage on the other hand is reduced with temperature, so there should be a voltage point for each process where the rate of mobility decrease vs. temperature is compensated by the drive current due to the threshold voltage reduction. At this particular supply voltage, the device delay becomes independent of temperature. This particular aspect of transistor drive current vs. temperature was used in [66] to develop a self-adjusting clock architecture that minimizes the clock skew dependency on temperature variations.

It is important to note that apart from the impact on delay variation there is a positive aspect of the temperature spatial variation. Various failure mechanisms such as gate oxide integrity, electromigration, and NBTI have a strong dependency on temperature. Taking advantage of a spatial thermal map under worst case load conditions as shown in Fig. 8.19 can help reduce overdesign since the reliability risk for the areas with lower temperature will be smaller.

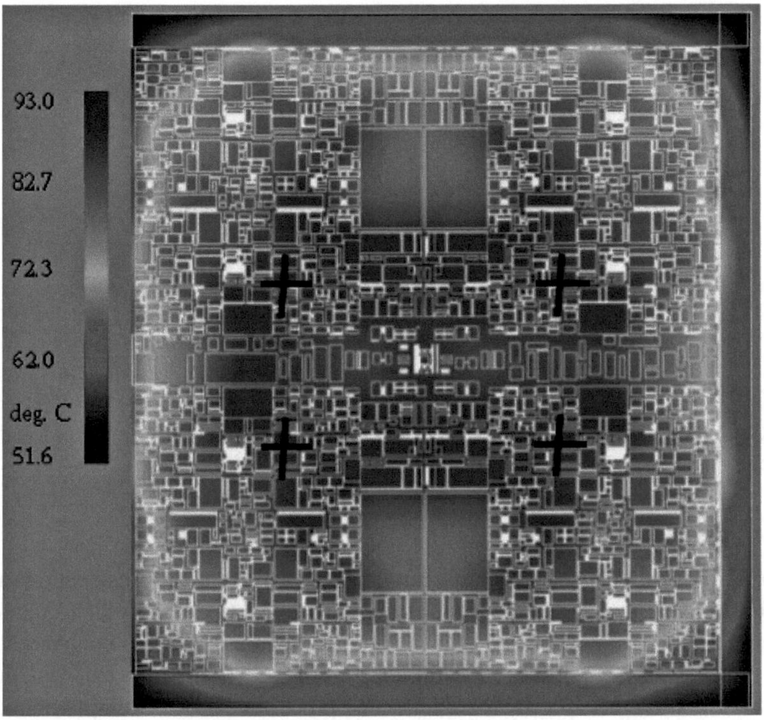


**Fig. 8.19.** Thermal map of a multi-core microprocessor under worst case workload conditions. the crosses represent the thermal sensor locations. Reproduced with permission from [65], ©2008 IEEE

## 8.6 Interconnect Variation

Aggressive scaling and continuously higher demand for more dense routing require substantial improvements in the interconnect stack material and physical construction. However, this comes also with the cost of more challenging reliability requirements, and substantial variations in the wire characteristics affecting the wire delay, the clock skew, and consequently the product performance. In this section, we will focus on the nature of these variations and interconnect design requirements for a robust and predictable clock network.

Similarly to the transistor variation, the interconnect variations have systematic and random components. However, in the interconnect case the systematic variations dominate at least for wider interconnects. Appropriate extraction and modeling are required to allow for a predictable clock skew analysis. Critical systematic variations include nonlinear resistance (NLR) effect, selective process bias (SPB) effect, and thickness variations due to etch and chemical mechanical polishing (CMP). As shown in Fig. 8.20 [15], the sheet resistance varies as a nonlinear function of line

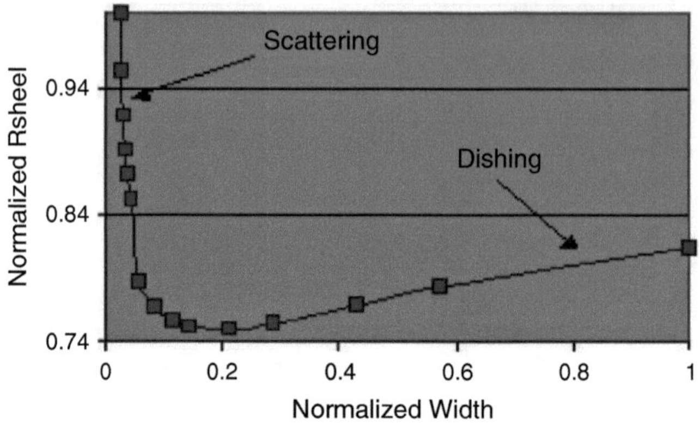


**Fig. 8.20.** Sheet resistance nonlinear behavior vs. interconnect width. Reproduced with permission from [15], ©2004 IEEE

width. Two phenomena cause significant and systematic changes in sheet resistance as a function of line width for copper technologies. The first one is due to scattering of electrons along the sidewall of the conductor and along the grain boundaries. The second includes systematic changes in copper cross-sectional area as a function of line width that are induced by the damascene process flow. An example of such an effect is CMP-induced copper dishing that would increase resistance. For a given pitch, the width and spacing on silicon will be different from the drawn features. Figure 8.21 shows the difference in the width on silicon vs. the normalized drawn width and spacing [15]. Depending on the width and spacing, the CMP will impact the thickness of the wire. In the end, we have a dependency of $R_{\text{sheet}}$ on wire width and spacing as shown in Fig. 8.22 and not only on width as would have been the expectation in the past. Similarly, the coupling capacitance will depend on both the width and spacing and not spacing alone, since the sidewall capacitance depends on the wire thickness that now depends on both wire width and spacing. In addition, $R_{\text{sheet}}$ will have a nonlinear dependency on width and thickness, due to the high resistance sidewalls that reduce the effective wire width and thickness. These sidewalls are required to contain the copper and prevent its diffusion to the surrounding materials. The use of ultra low dielectric constant materials ($K = 2.4$) to reduce the overall capacitance has a strong side effect on the mechanical stability [67]. These materials are "porous", not extremely stable and a challenge for packaging. Special thin Nitride layers used as intermediate layers between the metals increase the mechanical stability but also increase the effective dielectric constant and, therefore, increase the final wire capacitance value. Figure 8.23 shows the combined effect of SPB on wire RC vs. width and spacing [68].

Increase of the aspect ratio (metal thickness over width) to reduce the sheet resistance has the side effect of increased sidewall capacitance and sidewall contours as shown in Fig. 8.24 [67] and creates additional variation in the interconnect per-

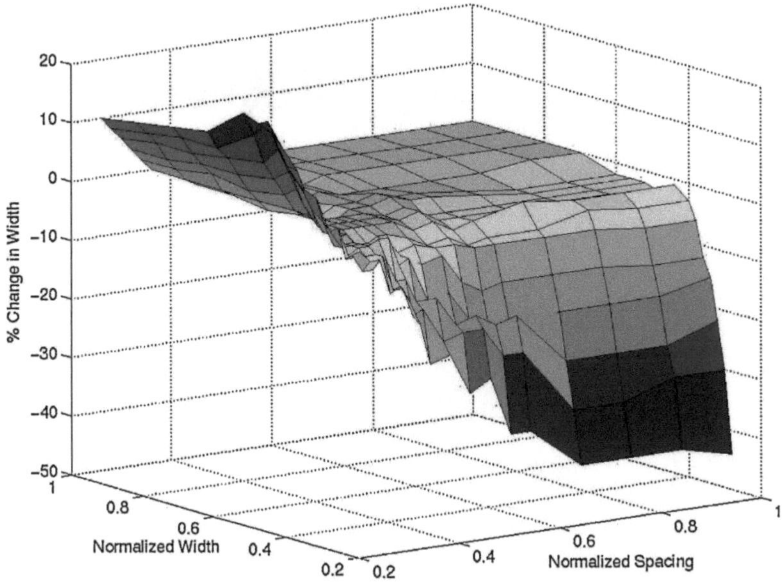


**Fig. 8.21.** Impact of SPB on interconnect width on silicon vs. normalized drawn width and spacing. Reproduced with permission from [15], ©2004 IEEE

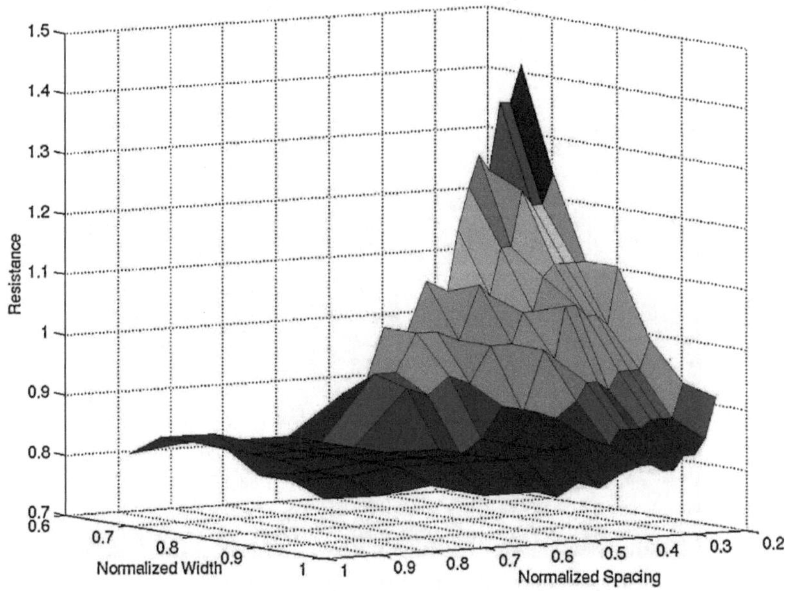


**Fig. 8.22.** Interconnect sheet resistance dependency on width and spacing – not only width – due to SPB, CMP effects. Reproduced with permission from [15], ©2004 IEEE

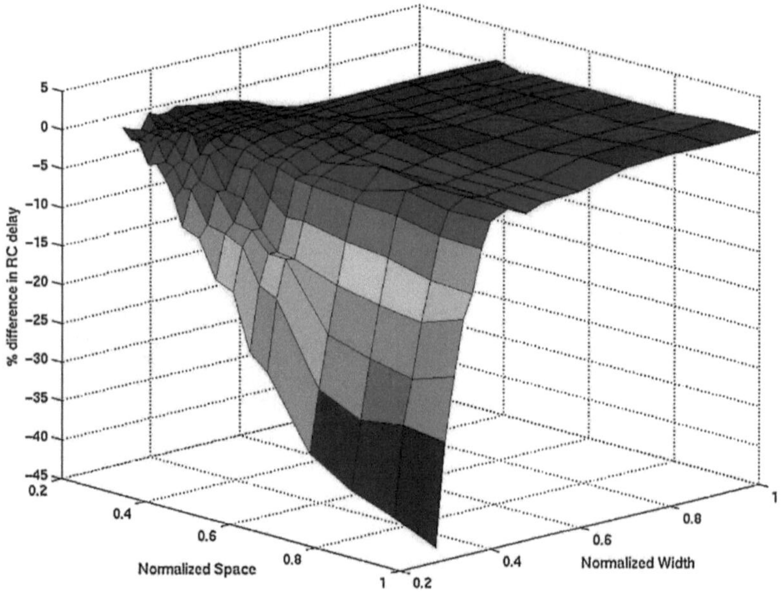


**Fig. 8.23.** SPB impact on RC delay vs. interconnect width and spacing (courtesy of C. Bittlestone and U. Narasimha). Reproduced with permission from [68], ©2005 IEEE

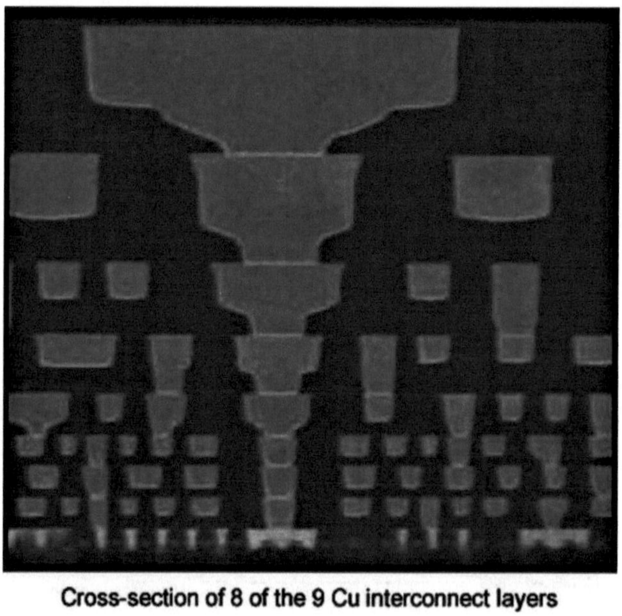


**Fig. 8.24.** Cross section of interconnect stack of a 45 nm process indicating the contours that affect in a nonuniform manner the sidewall capacitance and sheet resistance. Reproduced with permission from [67], ©2006 IEEE

formance. The aspect ratio is typically kept between 1.5 and 2: Lower for the lower level metals where capacitance is more important, and higher for longer upper level interconnects that need to have both low resistance and low capacitance. Due to the higher aspect ratio, the thickness and the sidewall capacitance will be larger, and therefore upper level interconnects will require higher spacing and pitch. This hierarchical stack provides high routing density and low capacitance for local interconnects, and lower routing density but higher performance for upper level interconnects. Typically, the pitch increase in the transition from one routing layer to the next should be accompanied by a similar speed improvement to make the transition worthwhile.

Some examples of the impact of selective process biasing (SPB) on the circuit performance are provided in [69, 70]. Ring oscillator measurements for different interconnect layout patterns show that the stage delay can vary up to 10% [69]. The analysis in [71] shows that the interconnect variations can affect the clock skew by 25%.

The impact of size effects and copper interconnect process variations on the maximum critical path delay of single and multiple core microprocessors is described in [70]. It includes the CMP and edge line roughness effects that will become more important for very narrow interconnects in the future. It looks like that the trend of using multicores instead of monolithic single cores helps contain the issue of variation and speed restrictions posed by long interconnects, since multiple core implementations require typically shorter interconnects. The problem of creating a robust clock automatic routing in the presence of interconnect variations is addressed in [72]. The algorithm calculates the delays and the variations at the various sink points, but instead of using worst case corner approach, it uses a principal component analysis and statistical centering approach to define the optimal clock network implementation considering all sinks simultaneously.

State-of-the-art parasitic extractors account for all these effects, but it is highly desirable for the designer to know in advance what to expect from each interconnect configuration and how to address these issues ahead of time in order to minimize variation. Since these interconnect variations are more systematic, use of similar routing patterns will help minimize variation. Otherwise, one needs to wait for the full extraction and full post layout timing and clock analysis to see the exact result. Isolated lines are impacted more by the CMP and dishing effects and several design rules impose a certain density range per metal layer to minimize this variability and improve yield. In order to achieve this specific metal density, we may need to add metal fill that can be floating or grounded to reduce noise. In order to reduce the coupling capacitance and help speed and power, it is preferable to maximize the spacing of the metal fill to the signal lines. The floating capacitance is typically extracted as grounded and this can affect the final capacitance by a large amount as shown in [15] if the spacing is not adequate.

Additional variation is introduced by capacitive coupling, and this depends on the workload and the resulting signal activity. In setup time analysis in most cases, designers use a worst-case Miller factor of 2 while in hold time analysis, a Miller factor of 0. This can be quite pessimistic and lead to overdesign. Special timing analysis based on timing windows [73] can help reduce this pessimism by calculating the effective Miller factor based on the actual switching activity of the gates in the critical path. Shielding of the clock lines is an effective way to provide a controlled environment for the clock network for both the inductance and noise immunity at minimal expense in routing resources and power for the main clock trunk.

The workload will introduce temperature variations as shown in Fig. 8.19. The wire resistance may vary by 30% going from room temperature to the typical operating condition of 105°C. The temperature gradient will cause additional delay differences that affect both the clock and the datapaths. However, there is a positive aspect in all this. Reduced temperature will help reduce the electromigration risk. A worst case thermal map can be combined with knowledge of the nodal switching activity to develop a more realistic estimate of electromigration risk and help avoid overdesign. In the clock case, we may only need the temperature gradient and not the activity, since in most cases the clock is active most of the time. Still, if we know how much time the clock is gated or is in a low frequency mode (idle mode), we can help prevent overdesign.

In high-speed designs, one should consider the inductive effects as well. There are clock distribution schemes that take advantage of inductive effects to reduce clock power and minimize skew [8, 74–77]. These are described in more detail in Chap. 4. Shielding of the clock lines with $V_{DD}$ and $V_{SS}$ provides both good return paths to control inductance but also shields the clock network from the noise and delay variation due to capacitive coupling from neighboring switching nets. Reducing the variation will improve overall skew and is a strongly recommended practice for high-speed clock networks. The shielding costs a bit of extra power due to the increased capacitive loading. This additional power consumption is not large for the main clock trunk and the shielding is required to minimize noise induced clock skew. In the final stages, the power consumption is higher due to the large fanout, and one could argue that shielding may not be needed since all signals should be stable before the next clock edge. This may be true when there is no cycle stealing involved and negative edge timing is not used architecturally. Yet, there is another reason to maintain shielding across all stages. As has been demonstrated before, the interconnect resistance and capacitance variation can be reduced when using similar dense layout for all wires, and therefore having a well controlled environment helps minimize skew induced by the interconnect performance variation.

In a nutshell, the clock lines need to be designed wide enough to minimize RC variations have good shielding to minimize variation due to noise and inductance, and use a similar routing patterns to minimize the SPB variations. In all cases, full extraction considering all these effects and post layout timing analysis is absolutely required to sign-off the clock design.

## 8.7 Conclusion: Clock Design and Analysis Guidelines: Putting All Together

As a conclusion, in this section, we collect a summary of all analysis and design guidelines presented earlier on.

### 8.7.1 Clock Analysis

- Need to use statistical analysis and not rely on worst case process corners that are very pessimistic and lead to overdesign.
- Need to separate the systematic and correlated variation components.
- Systematic/ correlated components include strain and WPE-induced layout dependent effects, portion of the channel length variation, and CMP, SPB effects for the interconnects. All systematic effects should be modeled with accurate post layout extraction.
- Random components include random dopant fluctuation, line edge roughness (LER) effects, voltage and temperature variation, and PLL jitter.
- Use thermal maps to identify worst case temperature variations.
- Use dynamic IR simulations to identify the local and global IR drop for the clock buffers.

### 8.7.2 Minimizing Variation

- Maximize $W \times L$ to minimize random dopant fluctuations. However, short $L$ is need for speed and power reduction. Therefore, larger $W$ devices at minimum $L$ are appropriate for clock buffers to minimize dopant fluctuations and power dissipation.
- Use same poly orientation and poly shields to minimize poly CD variation.
- Use the maximum allowed poly pitch in technologies using strain to maximize the drive strength through strain.
- Use the same layout style to guarantee the same effect on all transistors.
- Enlarge the spacing of well to active to reduce the well proximity effect that increases transistor $V_T$ and reduces drive strength.
- Use multi-finger devices to create an averaging effect and improve overall strength in processes using strain due to the larger strain in the internal devices.
- Avoid active jogs and increase poly to active spacing to minimize poly flaring and leakage.
- Reduce total clock network latency. Skew will be one way or another proportional to the total number of buffers.
- Reduce voltage variation by using large numbers of decoupling capacitors between supply and ground in close proximity to the clock buffers. The RC time constant of the grid resistance and the decoupling capacitance needs to be at least 5 times smaller than the clock rise/fall times.
- Use clock line shielding for improved noise immunity and good inductive return path to reduce overall delay variation due to noise.

- Minimize the total number of clock buffers used and use large devices of similar total equivalent strength. Variation is proportional to the number of stages and the number of critical paths and clock sink pairs.
- Higher supply voltage will help minimize the impact of $V_T$ variation and partially compensate the impact of NBTI.
- Use H-trees for distribution. Make sure to drive back-to-back flops with the same clock branch. The delay of common branches does not affect the hold time skew.
- For interconnects, use similar layout style with shielding to minimize SPB impact. If possible, use width and spacing combinations that minimize the SPB impact. Use extraction tools that account for CMP and SPB effects.
- Use metal fill to meet density requirements and extract it correctly. Extract grounded metal fill as grounded, floating as floating. If possible, use larger spacing to active signals and clocks to just meet density requirements but minimize the impact on delay due to noise and reduce power consumption by reducing the overall capacitive load.
- Shorting clock buffer outputs can provide some averaging effect, but it increases power and complicates timing analysis. Most timing tools have issues with parallel clock drivers. In order to facilitate clock gating, a tree configuration must be used instead of a shorted-output topology to be able to gate individual clock buffers.
- In the case of clock gating, add enough margin to account for the extra asymmetric NBTI shift. This impacts more the half-cycle paths that depend on both the rise and fall clock transitions.
- Use the same minimum channel length in all clock drivers. Different channel lengths have different sensitivity to CD variations and can impact product speed.
- Use wider metals for clock lines to minimize the RC impact, minimize the nonlinear sheet resistance effect, and reduce the electromigration risk.
- In physical composition, use the same footprint (abstract) for all clock buffers. It will be easier to do post layout buffer fine-tuning in place, without affecting the floorplan.

## Acknowledgments

I would like to acknowledge my wife Anna for her huge patience and support, and Thucydides Xanthopoulos, Antonietta Oliva, Aaron Barker, and David Greenhill for providing thorough reviews of this manuscript.

## References

[1] S. Tam, S. Rusu, U. Nagarji Desai, R. Kim, J. Zhang, and I. Young, "Clock generation and distribution for the first IA-64 microprocessor," *IEEE J. Solid-State Circuits*, vol. 35, no. 11, pp. 1545–1552, Nov. 2000.

[2] T. Xanthopoulos, D. W. Bailey, A. K. Gangwar, M. K. Gowan, A. K. Jain, and B. K. Prewitt, "The design and analysis of the clock distribution network for a 1.2GHz Alpha microprocessor," in *Digest of Technical Papers IEEE International Solid-State Circuits Conference (ISSCC 2001)*, 2001, pp. 402–403.

[3] P. J. Restle, T. G. McNamara, D. A. Webber, P. J. Camporese, K. F. Eng, K. A. Jenkins, D. H. Allen, M. J. Rohn, M. P. Quaranta, D. W. Boerstler, C. J. Alpert, C. A. Carter, R. N. Bailey, J. G. Petrovick, B. L. Krauter, and B. D. McCredie, "A clock distribution network for microprocessors," *IEEE J. Solid-State Circuits*, vol. 36, no. 5, pp. 792–799, May 2001.

[4] P. J. Restle, C. A. Carter, J. P. Eckhardt, B. L. Krauter, B. D. McCredie, K. A. Jenkins, A. J. Weger, and A. V. Mule, "The clock distribution of the Power4 microprocessor," in *Digest of Technical Papers IEEE International Solid-State Circuits Conference (ISSCC 2002)*, vol. 1, 2002, pp. 144–145.

[5] N. Bindal, T. Kelly, N. Velastegui, and K. L. Wong, "Scalable sub-10ps skew global clock distribution for a 90nm multi-GHz IA microprocessor," in *Digest of Technical Papers IEEE International Solid-State Circuits Conference (ISSCC 2003)*, 2003, pp. 346–347,498.

[6] S. Tam, R. D. Limaye, and U. N. Desai, "Clock generation and distribution for the 130-nm Itanium® 2 processor with 6-MB on-die L3 cache," *IEEE J. Solid-State Circuits*, vol. 39, no. 4, pp. 636–642, April 2004.

[7] P. Mahoney, E. Fetzer, B. Doyle, and S. Naffziger, "Clock distribution on a dual-core, multi-threaded Itanium®-family processor," in *Digest of Technical Papers IEEE International Solid-State Circuits Conference (ISSCC 2005)*, 2005, pp. 292–293,599.

[8] M. G. R. Thomson, P. J. Restle, and N. K. James, "A 5GHz duty-cycle correcting clock distribution network for the POWER6 microprocessor," in *Digest of Technical Papers IEEE International Solid-State Circuits Conference (ISSCC 2006)*, 2006, pp. 1522–1529.

[9] G. K. Konstadinidis, K. Normoyle, S. Wong, S. Bhutani, H. Stuimer, T. Johnson, A. Smith, D. Y. Cheung, F. Romano, S. Yu, S.-H. Oh, V. Melamed, S. Narayanan, D. Bunsey, C. Khieu, K. J. Wu, R. Schmitt, A. Dumlao, M. Sutera, J. Chau, K. J. Lin, and W. S. Coates, "Implementation of a third-generation 1.1-GHz 64-bit microprocessor," *IEEE J. Solid-State Circuits*, vol. 37, no. 11, pp. 1461–1469, Nov. 2002.

[10] N. H. E. Weste and D. Harris, *CMOS VLSI Design: A Circuits and System Perspective (third edition)*. Addison Wesley, Reading, MA, 2005.

[11] A. Chandrakasan, W. Bowhill, and F. Fox, *Design of High Performance Microprocessor Circuits*. IEEE Press, 2001.

[12] N. J. Rohrer, "Introduction to statistical variation and techniques for design optimization," ISSCC 2006 Tutorial.

[13] K. Ushida, "Future lithography challenges," in *Proceedings of the IEEE International Symposium on Semiconductor Manufacturing ISSM 2006*, 25–27 Sept. 2006, pp. lv–lx.

[14] S. Sivakumar, "Lithography challenges for 32nm technologies and beyond," in *Proceedings of the International Electron Devices Meeting IEDM '06*, 11–13 Dec. 2006, pp. 1–4.

[15] N. S. Nagaraj, T. Bonifield, A. Singh, R. Griesmer, and P. Balsara, "Interconnect modeling for copper/low-k technologies," in *Proceedings of the 17th International Conference on VLSI Design*, 2004, pp. 425–427.

[16] S. Tyagi, C. Auth, P. Bai, G. Curello, H. Deshpande, S. Gannavaram, O. Golonzka, R. Heussner, R. James, C. Kenyon, S. H. Lee, N. Lindert, M. Liu, R. Nagisetty, S. Natarajan, C. Parker, J. Sebastian, B. Sell, S. Sivakumar, A. S. Amour, and K. Tone, "An advanced low power, high performance, strained channel 65nm technology," in *Proceedings of the IEDM Technical Digest Electron Devices Meeting IEEE International*, 5–7 Dec. 2005, pp. 245–247.

[17] T. Ghani, M. Armstrong, C. Auth, M. Bost, P. Charvat, G. Glass, T. Hoffmann, K. Johnson, C. Kenyon, J. Klaus, B. McIntyre, K. Mistry, A. Murthy, J. Sandford, M. Silberstein, S. Sivakumar, P. Smith, K. Zawadzki, S. Thompson, and M. Bohr, "A 90nm high volume manufacturing logic technology featuring novel 45nm gate length strained silicon CMOS transistors," in *Proceedings of the IEDM '03 Technical Digest Electron Devices Meeting IEEE International*, 8–10 Dec. 2003, pp. 11.6.1–11.6.3.

[18] S. E. Thompson, M. Armstrong, C. Auth, S. Cea, R. Chau, G. Glass, T. Hoffman, J. Klaus, Z. Ma, B. Mcintyre, A. Murthy, B. Obradovic, L. Shifren, S. Sivakumar, S. Tyagi, T. Ghani, K. Mistry, M. Bohr, and Y. El-Mansy, "A logic nanotechnology featuring strained-silicon," *IEEE Electron Device Lett.*, vol. 25, no. 4, pp. 191–193, April 2004.

[19] M. Horstmann, A. Wei, T. Kammler, J. Hntschel, H. Bierstedt, T. Feudel, K. Frohberg, M. Gerhardt, A. Hellmich, K. Hempel, J. Hohage, P. Javorka, J. Klais, G. Koerner, M. Lenski, A. Neu, R. Otterbach, P. Press, C. Reichel, M. Trentsch, B. Trui, H. Salz, M. Schaller, H. J. Engelmann, O. Herzog, H. Ruelke, P. Hubler, R. Stephan, D. Greenlaw, M. Raab, and N. Kepler, "Integration and optimization of embedded-sige, compressive and tensile stressed liner films, and stress memorization in advanced SOI CMOS technologies," in *Proceedings of IEDM Technical Digest Electron Devices Meeting IEEE International*, 5–7 Dec. 2005, pp. 233–236.

[20] P. Grudowski, V. Adams, X.-Z. Bo, K. Loiko, S. Filipiak, J. Hackenberg, M. Jahanbani, M. Azrak, S. Goktepeli, M. Shroff, W.-J. Liang, S. J. Lian, V. Kolagunta, N. Cave, C.-H. Wu, M. Foisy, H. C. Tuan, and J. Cheek, "1-D and 2-D geometry effects in uniaxially-strained dual etch stop layer stressor integrations," in *Proceedings of the Digest of Technical Papers VLSI Technology 2006 Symposium*, 2006, pp. 62–63.

[21] C. Ortolland, P. Morin, C. Chaton, E. Mastromatteo, C. Populaire, S. Orain, F. Leverd, P. Stolk, F. Boeuf, and F. Arnaud, "Stress memorization technique (SMT) optimization for 45nm CMOS," in *Proceedings of Digest of Technical Papers VLSI Technology 2006 Symposium*, 2006, pp. 78–79.

[22] M. Wiatr, T. Feudel, A. Wei, A. Mowry, R. Boschke, P. Javorka, A. Gehring, T. Kammler, M. Lenski, K. Frohberg, R. Richter, M. Horstmann, and

D. Greenlaw, "Review on process-induced strain techniques for advanced logic technologies," in *Proceedings of the 15th International Conference on Advanced Thermal Processing of Semiconductors RTP 2007*, 2–5 Oct. 2007, pp. 19–29.

[23] V. Chan, R. Rengarajan, N. Rovedo, W. Jin, T. Hook, P. Nguyen, J. Chen, E. Nowak, X.-D. Chen, D. Lea, A. Chakravarti, V. Ku, S. Yang, A. Steegen, C. Baiocco, P. Shafer, H. Ng, S.-F. Huang, and C. Wann, "High speed 45nm gate length CMOSFETs integrated into a 90nm bulk technology incorporating strain engineering," in *Proceedings of the IEDM '03 Technical Digest Electron Devices Meeting IEEE International*, 8–10 Dec. 2003, pp. 3.8.1–3.8.4.

[24] C. Auth, A. Cappellani, J. S. Chun, A. Dalis, A. Davis, T. Ghani, G. Glass, T. Glassman, M. Harper, M. Hattendorf, P. Hentges, S. Jaloviar, S. Joshi, J. Klaus, K. Kuhn, D. Lavric, M. Lu, H. Mariappan, K. Mistry, B. Norris, N. Rahhal-orabi, P. Ranade, J. Sandford, L. Shifren, V. Souw, K. Tone, F. Tambwe, A. Thompson, D. Towner, T. Troeger, P. Vandervoorn, C. Wallace, J. Wiedemer, and C. Wiegand, "45nm high-k + metal gate strain-enhanced transistors," in *Proceedings of the Symposium on VLSI Technology*, 17–19 June 2008, pp. 128–129.

[25] P. Ranade, T. Ghani, K. Kuhn, K. Mistry, S. Pae, L. Shifren, M. Stettler, K. Tone, S. Tyagi, and M. Bohr, "High performance 35nm LATE CMOS transistors featuring NiSi metal gate (FUSI), uniaxial strained silicon channels and 1.2nm gate oxide," in *Proceedings of the IEDM Technical Digest Electron Devices Meeting IEEE International*, 5–7 Dec. 2005, p. 4.

[26] K. Mistry, C. Allen, C. Auth, B. Beattie, D. Bergstrom, M. Bost, M. Brazier, M. Buehler, A. Cappellani, R. Chau, C. H. Choi, G. Ding, K. Fischer, T. Ghani, R. Grover, W. Han, D. Hanken, M. Hattendorf, J. He, J. Hicks, R. Huessner, D. Ingerly, P. Jain, R. James, L. Jong, S. Joshi, C. Kenyon, K. Kuhn, K. Lee, H. Liu, J. Maiz, B. McIntyre, P. Moon, J. Neirynck, S. Pae, C. Parker, D. Parsons, C. Prasad, L. Pipes, M. Prince, P. Ranade, T. Reynolds, J. Sandford, L. Shifren, J. Sebastian, J. Seiple, D. Simon, S. Sivakumar, P. Smith, C. Thomas, T. Troeger, P. Vandervoorn, S. Williams, and K. Zawadzki, "A 45nm logic technology with high-k+metal gate transistors, strained silicon, 9 Cu interconnect layers, 193nm dry patterning, and 100% Pb-free packaging," in *Proceedings of the IEEE International Electron Devices Meeting IEDM 2007*, 10–12 Dec. 2007, pp. 247–250.

[27] S. Narasimha, K. Onishi, H. M. Nayfeh, A. Waite, M. Weybright, J. Johnson, C. Fonseca, D. Corliss, C. Robinson, M. Crouse, D. Yang, C. H. J. Wu, A. Gabor, T. Adam, I. Ahsan, M. Belyansky, L. Black, S. Butt, J. Cheng, A. Chou, G. Costrini, C. Dimitrakopoulos, A. Domenicucci, P. Fisher, A. Frye, S. Gates, S. Greco, S. Grunow, M. Hargrove, J. Holt, S. J. Jeng, M. Kelling, B. Kim, W. Landers, G. Larosa, D. Lea, M. H. Lee, X. Liu, N. Lustig, A. McKnight, L. Nicholson, D. Nielsen, K. Nummy, V. Ontalus, C. Ouyang, X. Ouyang, C. Prindle, R. Pal, W. Rausch, D. Restaino, C. Sheraw, J. Sim, A. Simon, T. Standaert, C. Y. Sung, K. Tabakman, C. Tian, R. Van Den Nieuwenhuizen, H. Van Meer, A. Vayshenker, D. Wehella-Gamage,

J. Werking, R. C. Wong, J. Yu, S. Wu, R. Augur, D. Brown, X. Chen, D. Edelstein, A. Grill, M. Khare, Y. Li, S. Luning, J. Norum, S. Sankaran, D. Schepis, R. Wachnik, R. Wise, C. Warm, T. Ivers, and P. Agnello, "High performance 45-nm SOI technology with enhanced strain, porous low-k BEOL, and immersion lithography," in *Proceedings of the International Electron Devices Meeting IEDM '06*, 11–13 Dec. 2006, pp. 1–4.

[28] S. Suthram, Y. Sun, P. Majhi, I. Ok, H. Kim, H. R. Harris, N. Goel, S. Parthasarathy, A. Koehler, T. Acosta, T. Nishida, H.-H. Tseng, W. Tsai, J. Lee, R. Jammy, and S. E. Thompson, "Strain additivity in III-V channels for CMOSFETs beyond 22nm technology node," in *Proceedings of the Symposium on VLSI Technology*, 17–19 June 2008, pp. 182–183.

[29] Y. M. Sheu, K. W. Su, S. Tian, S. J. Yang, C. C. Wang, M. J. Chen, and S. Liu, "Modeling the well-edge proximity effect in highly scaled MOSFETs," *IEEE Trans. Electron Devices*, vol. 53, no. 11, pp. 2792–2798, Nov. 2006.

[30] I. Polishchuk, N. Mathur, C. Sandstrom, P. Manos, and O. Pohland, "CMOS Vt-control improvement through implant lateral scatter elimination," in *Proceedings of the IEEE International Symposium on Semiconductor Manufacturing ISSM 2005*, 13–15 Sept. 2005, pp. 193–196.

[31] P. G. Drennan, M. L. Kniffin, and D. R. Locascio, "Implications of proximity effects for analog design," in *Proceedings of the IEEE Custom Integrated Circuits Conference 2006*, 10–13 Sept. 2006, pp. 169–176.

[32] T. Kanamoto, Y. Ogasahara, K. Natsume, K. Yamaguchi, H. Amishiro, and T. Watanabe, "Impact of well edge proximity effect on timing," in *Proceedings of the ESSCIRC 33rd European Solid State Circuits Conference*, 11–13 Sept. 2007, pp. 115–118.

[33] A. Asenov, S. Kaya, and A. R. Brown, "Intrinsic parameter fluctuations in decananometer MOSFETs introduced by gate line edge roughness," *IEEE Trans. Electron Devices*, vol. 50, no. 5, pp. 1254–1260, May 2003.

[34] A. Asenov, S. Kaya, and J. H. Davies, "Intrinsic threshold voltage fluctuations in decanano MOSFETs due to local oxide thickness variations," *IEEE Trans. Electron Devices*, vol. 49, no. 1, pp. 112–119, Jan. 2002.

[35] Y. Ye, F. Liu, S. Nassif, and Y. Cao, "Statistical modeling and simulation of threshold variation under dopant fluctuations and line-edge roughness," in *Proceedings of the 45th ACM/IEEE Design Automation Conference DAC 2008*, 8–13 June 2008, pp. 900–905.

[36] S. Borkar, "Designing reliable systems from unreliable components: the challenges of transistor variability and degradation," *IEEE Micro*, vol. 25, no. 6, pp. 10–16, Nov.–Dec. 2005.

[37] T. Karnik, V. De, and S. Borkar, "Statistical design for variation tolerance: key to continued Moore's law," in *Proceedings of the International Conference on Integrated Circuit Design and Technology ICICDT '04*, 2004, pp. 175–176.

[38] J. Rosal, "Multi-core SoC test and yield enhancement – Challenges and solutions," ISSCC Microprocessor Forum, February 2006.

[39] H. Masuda, S. Okawa, and M. Aoki, "Approach for physical design in sub-100nm era," in *Proceedings of the IEEE International Symposium on Circuits and Systems ISCAS 2005*, 23–26 May 2005, pp. 5934–5937.

[40] H. Mahmoodi, S. Mukhopadhyay, and K. Roy, "Estimation of delay variations due to random-dopant fluctuations in nanoscale CMOS circuits," *IEEE J. Solid-State Circuits*, vol. 40, no. 9, pp. 1787–1796, Sept. 2005.

[41] A. Asenov, G. Slavcheva, A. R. Brown, J. H. Davies, and S. Saini, "Increase in the random dopant induced threshold fluctuations and lowering in sub-100nm MOSFETs due to quantum effects: a 3-d density-gradient simulation study," *IEEE Trans. Electron Devices*, vol. 48, no. 4, pp. 722–729, April 2001.

[42] I. Ahsan, O. Glushchenkov, R. Logan, E. J. Nowak, H. Kimura, G. Berg, J. Herman, E. Maciejewski, A. Chan, A. Azuma, S. Deshpande, B. Dirahoui, G. Freeman, A. Gabor, M. Gribelyuk, S. Huang, M. Kumar, K. Miyamoto, D. Mocuta, A. Mahorowala, E. Leobandung, and H. Utomo, "RTA-driven intra-die variations in stage delay, and parametric sensitivities for 65nm technology," in *Proceedings of the Digest of Technical Papers VLSI Technology 2006 Symposium*, 2006, pp. 170–171.

[43] T. Tanaka, T. Usuki, T. Futatsugi, Y. Momiyama, and T. Sugii, "$v_{th}$ fluctuation induced by statistical variation of pocket dopant profile," in *Proceedings of the IEDM Technical Digest Electron Devices Meeting International*, 10–13 Dec. 2000, pp. 271–274.

[44] D. S. Boning, K. Balakrishnan, H. Cai, N. Drego, A. Farahanchi, K. M. Gettings, L. Daihyun, A. Somani, H. Taylor, D. Truque, and X. Xiaolin, "Variation," *IEEE Trans. Semicond. Manuf.*, vol. 21, no. 1, pp. 63–71, Feb. 2008.

[45] D. Harris and S. Naffziger, "Statistical clock skew modeling with data delay variations," *IEEE Trans. VLSI Syst.*, vol. 9, no. 6, pp. 888–898, Dec. 2001.

[46] A. Agarwal, V. Zolotov, and D. T. Blaauw, "Statistical clock skew analysis considering intradie-process variations," *IEEE Trans. Comput.-Aided Design Integr. Circuits Syst.*, vol. 23, no. 8, pp. 1231–1242, Aug. 2004.

[47] K. A. Bowman, S. G. Duvall, and J. D. Meindl, "Impact of die-to-die and within-die parameter fluctuations on the maximum clock frequency distribution for gigascale integration," *IEEE J. Solid-State Circuits*, vol. 37, no. 2, pp. 183–190, Feb. 2002.

[48] E. Malavasi, S. Zanella, M. Cao, J. Uschersohn, M. Misheloff, and C. Guardiani, "Impact analysis of process variability on clock skew," in *Proceedings of International Symposium on Quality Electronic Design*, 18–21 March 2002, pp. 129–132.

[49] J. A. G. Jess, K. Kalafala, S. R. Naidu, R. H. J. M. Otten, and C. Visweswariah, "Statistical timing for parametric yield prediction of digital integrated circuits," *IEEE Trans. Comput.-Aided Design Integr. Circuits Syst.*, vol. 25, no. 11, pp. 2376–2392, Nov. 2006.

[50] Y. Zhan, A. J. Strojwas, X. Li, L. T. Pileggi, D. Newmark, and M. Sharma, "Correlation-aware statistical timing analysis with non-Gaussian delay distributions," in *Proceedings of the 42nd Design Automation Conference*, 13–17 June 2005, pp. 77–82.

[51] Z. Feng and P. Li, "A methodology for timing model characterization for statistical static timing analysis," in *Proceedings of the IEEE/ACM International Conference on Computer-Aided Design ICCAD 2007*, 4–8 Nov. 2007, pp. 725–729.

[52] A. B. Kahng, "Key directions and a roadmap for electrical design for manufacturability," in *Proceedings of the ESSCIRC 33rd European Solid State Circuits Conference*, 11–13 Sept. 2007, pp. 83–88.

[53] W. Yang, M. V. Dunga, X. Xi, J. He, W. Liu, M. C. Kanyu, X. Jin, J. J. Ou, M. Chan, A. M. Niknejad, and C. Hu, "BSIM4.6.2 MOSFET model – User's Manual," Department of Electrical Engineering and Computer Sciences, http://www-device.eecs.berkley.edu/ bsim3/BSIM4/BSIM462/ doc/BSIM462_Manual.pdf.

[54] S. Tsujikawa and J. Yugami, "Evidence for bulk trap generation during NBTI phenomenon in pMOSFETs with ultrathin SiON gate dielectrics," *IEEE Trans. Electron Devices*, vol. 53, no. 1, pp. 51–55, Jan. 2006.

[55] A. E. Islam, H. Kufluoglu, D. Varghese, S. Mahapatra, and M. A. Alam, "Recent issues in negative-bias temperature instability: Initial degradation, field dependence of interface trap generation, hole trapping effects, and relaxation," *IEEE Trans. Electron Devices*, vol. 54, no. 9, pp. 2143–2154, Sept. 2007.

[56] H. Kufluoglu and M. A. Alam, "A generalized reaction-diffusion model with explicit H {H}_{2} dynamics for negative-bias temperature-instability (NBTI) degradation," *IEEE Trans. Electron Devices*, vol. 54, no. 5, pp. 1101–1107, May 2007.

[57] S. Mahapatra, P. Bharath Kumar, T. R. Dalei, D. Sana, and M. A. Alam, "Mechanism of negative bias temperature instability in CMOS devices: degradation, recovery and impact of nitrogen," in *Proceedings of IEDM Technical Digest Electron Devices Meeting IEEE International*, 13–15 Dec. 2004, pp. 105–108.

[58] C. R. Parthasarathy, M. Denais, V. Huard, G. Ribes, E. Vincent, and A. Bravaix, "New insights into recovery characteristics during PMOS NBTI and CHC degradation," *IEEE Trans. Device Mater. Rel.*, vol. 7, no. 1, pp. 130–137, March 2007.

[59] W. Wang, S. Yang, S. Bhardwaj, R. Vattikonda, S. Vrudhula, F. Liu, and Y. Cao, "The impact of NBTI on the performance of combinational and sequential circuits," in *Proceedings of the 44th ACM/IEEE Design Automation Conference DAC '07*, 4–8 June 2007, pp. 364–369.

[60] A. Haggag, M. Lemanski, G. Anderson, P. Abramowitz, and M. Moosa, "Realistic projections of product fmax shift and statistics due to HCI and NBTI," in *Proceedings of the Reliability Physics Symposium 45th Annual. IEEE International*, 15–19 April 2007, pp. 93–96.

[61] Y. Taur and T. H. Ning, *Fundamentals of Modern VLSI Devices.* Cambridge University Press, Cambridge, 1998.

[62] K. Bernstein, K. Carrig, C. M. Durham, and P. R. Hansen, *High Speed CMOS Design Styles.* Kluwer, Dordecht, 1998.

[63] Q. K. Zhu and M. Zhang, "Low-voltage swing clock distribution schemes," in *Proceedings of the IEEE International Symposium on Circuits and Systems ISCAS 2001*, vol. 4, 6–9 May 2001, pp. 418–421.

[64] N. James, P. Restle, J. Friedrich, B. Huott, and B. McCredie, "Comparison of split-versus connected-core supplies in the POWER6 microprocessor," in *Digest of Technical Papers IEEE International Solid-State Circuits Conference (ISSCC 2007)*, 11–15 Feb. 2007, pp. 298–299,604.

[65] G. Konstadinidis, M. Tremblay, S. Chaudhry, M. Rashid, P. Lai, Y. Otaguro, Y. Orginos, S. Parampalli, M. Steigerwald, S. Gundala, R. Pyapali, L. Rarick, I. Elkin, Y. Ge, and I. Parulkar, "Architecture and physical implementation of a third generation 65 nm, 16 core, 32 thread chip-multithreading SPARC processor," *IEEE J. Solid-State Circuits*, vol. 44, no. 1, pp. 7–17, Jan. 2009.

[66] J. Long, J. C. Ku, S. O. Memik, and Y. Ismail, "A self-adjusting clock tree architecture to cope with temperature variations," in *Proceedings of the IEEE/ACM International Conference on Computer-Aided Design ICCAD 2007*, 4–8 Nov. 2007, pp. 75–82.

[67] S. Sankaran, S. Arai, R. Augur, M. Beck, G. Biery, T. Bolom, G. Bonilla, O. Bravo, K. Chanda, M. Chae, F. Chen, L. Clevenger, S. Cohen, A. Cowley, P. Davis, J. Demarest, J. Doyle, C. Dimitrakopoulos, L. Economikos, D. Edelstein, M. Farooq, R. Filippi, J. Fitzsimmons, N. Fuller, S. M. Gates, S. E. Greco, A. Grill, S. Grunow, R. Hannon, K. Ida, D. Jung, E. Kaltalioglu, M. Kelling, T. Ko, K. Kumar, C. Labelle, H. Landis, M. W. Lane, W. Landers, M. Lee, W. Li, E. Liniger, X. Liu, J. R. Lloyd, W. Liu, N. Lustig, K. Malone, S. Marokkey, G. Matusiewicz, P. S. McLaughlin, P. V. McLaughlin, S. Mehta, I. Melville, K. Miyata, B. Moon, S. Nitta, D. Nguyen, L. Nicholson, D. Nielsen, P. Ong, K. Patel, V. Patel, W. Park, J. Pellerin, S. Ponoth, K. Petrarca, D. Rath, D. Restaino, S. Rhee, E. T. Ryan, H. Shoba, A. Simon, E. Simonyi, T. M. Shaw, T. Spooner, T. Standaert, J. Sucharitaves, C. Tian, H. Wendt, J. Werking, J. Widodo, L. Wiggins, R. Wisnieff, and T. Ivers, "A 45 nm CMOS node Cu/Low-k/ Ultra Low-k PECVD SiCOH (k=2.4) BEOL technology," in *Proceedings of the International Electron Devices Meeting IEDM '06*, 11–13 Dec. 2006, pp. 1–4.

[68] C. Bittlestone, "Nanometer design effects and modeling," ISSCC Microprocessor Design Forum, February 2005.

[69] M. Kulkarni, N. S. Nagaraj, A. Marshall, and V. Le, "Impact of selective process bias (SPB) of interconnects on circuit delay," in *Proceedings of the IEEE Dallas/CAS Workshop (DCAS-04) Implementation of High Performance Circuits*, 27 Sept. 2004, pp. 155–158.

[70] G. Lopez, R. Murali, R. Sarvari, K. Bowman, J. Davis, and J. Meindl, "The impact of size effects and copper interconnect process variations on the maximum critical path delay of single and multi-core microprocessors," in *Proc. International Interconnect Technology Conference IEEE 2007*, 4–6 June 2007, pp. 40–42.

[71] Y. Liu, S. R. Nassif, L. T. Pileggi, and A. J. Strojwas, "Impact of interconnect variations on the clock skew of a gigahertz microprocessor," in *Proceedings of the 37th Design Automation Conference*, June 5–9, 2000, pp. 168–171.

[72] U. Padmanabhan, J. M. Wang, and J. Hu, "Robust clock tree routing in the presence of process variations," *IEEE Trans. Comput.-Aided Design Integr. Circuits Syst.*, vol. 27, no. 8, pp. 1385–1397, Aug. 2008.

[73] R. Y. Chen, P. Yip, G. Konstadinidis, A. Demas, F. Klass, R. Mains, M. Schmitt, and D. Bistry, "Timing window applications in UltraSPARC-IIIi$^{TM}$ microprocessor design," in *Proceedings of the IEEE International Conference on Computer Design: VLSI in Computers and Processors*, 16–18 Sept. 2002, pp. 158–163.

[74] F. O'Mahony, C. P. Yue, M. Horowitz, and S. S. Wong, "10GHz clock distribution using coupled standing-wave oscillators," in *Digest of Technical Papers IEEE International Solid-State Circuits Conference (ISSCC 2003)*, 2003, pp. 428–429,504.

[75] M. Shen, L.-R. Zheng, E. Tjukanoff, J. Isoaho, and H. Tenhunen, "Concurrent chip package design for global clock distribution network using standing wave approach," in *Proceedings of the Sixth International Symposium on Quality of Electronic Design ISQED 2005*, 21–23 March 2005, pp. 573–578.

[76] S. Chan, P. Restle, T. Bucelot, S. Weitzel, J. Keaty, J. Liberty, B. Flachs, R. Volant, P. Kapusta, and J. Zimmerman, "A resonant global clock distribution for the cell broadband-engine$^{TM}$processor," in *Digest of Technical Papers IEEE International Solid-State Circuits Conference (ISSCC 2008)*, 2008, pp. 512–513.

[77] M. Sasaki, "A 9.5GHz 6ps-skew space-filling-curve clock distribution with 1.8V full-swing standing-wave oscillators," in *Digest of Technical Papers IEEE International Solid-State Circuits Conference (ISSCC 2008)*, 2008, pp. 518–519, 633.

# Index

MIX
Papier aus verantwortungsvollen Quellen
Paper from responsible sources
FSC® C105338

If you have any concerns about our products,
you can contact us on
**ProductSafety@springernature.com**

In case Publisher is established outside the EU,
the EU authorized representative is:
**Springer Nature Customer Service Center GmbH**
**Europaplatz 3, 69115 Heidelberg, Germany**

Printed by Libri Plureos GmbH
in Hamburg, Germany

Christian Laugier, Roland Siegwart (Eds.)

# Field and Service Robotics

## Results of the 6th International Conference

Springer

**Professor Bruno Siciliano,** Dipartimento di Informatica e Sistemistica, Università di Napoli Federico II, Via Claudio 21, 80125 Napoli, Italy, E-mail: siciliano@unina.it

**Professor Oussama Khatib,** Robotics Laboratory, Department of Computer Science, Stanford University, Stanford, CA 94305-9010, USA, E-mail: khatib@cs.stanford.edu

**Professor Frans Groen,** Department of Computer Science, Universiteit van Amsterdam, Kruislaan 403, 1098 SJ Amsterdam, The Netherlands, E-mail: groen@science.uva.nl

## Editors

Christian Laugier
INRIA Grenoble - Rhône-Alpes
Inovallée
655 Avenue de l'Europe
Montbonnot
38334 Saint Ismier cedex
France
Email: Christian.Laugier@inria.fr

Roland Siegwart
Autonomous Systems Lab
Institute of Robotics and Intelligent Systems
ETH Zentrum, CLA E 31
Tannenstrasse 3
CH-8092 Zürich
Switzerland
Email: roland.siegwart@epfl.ch;
rsiegwart@ethz.ch

ISBN 978-3-642-09468-2 e-ISBN 978-3-540-75404-6

DOI 10.1007/978-3-540-75404-6

Springer Tracts in Advanced Robotics ISSN 1610-7438



Printed in acid-free paper

5 4 3 2 1 0

springer.com

STAR (Springer Tracts in Advanced Robotics) has been promoted under the auspices of EURON (European Robotics Research Network)

# Foreword

At the dawn of the new millennium, robotics is undergoing a major transformation in scope and dimension. From a largely dominant industrial focus, robotics is rapidly expanding into the challenges of unstructured environments. Interacting with, assisting, serving, and exploring with humans, the emerging robots will increasingly touch people and their lives.

The goal of the new series of *Springer Tracts in Advanced Robotics (STAR)* is to bring, in a timely fashion, the latest advances and developments in robotics on the basis of their significance and quality. It is our hope that the wider dissemination of research developments will stimulate more exchanges and collaborations among the research community and contribute to further advancement of this rapidly growing field.

Since its inception in 1996, FSR, the *International Conference on Field and Service Robotics* has published archival volumes of high reference value. Starting its edition in 2003, FSR has found its natural home within STAR, together with other thematic symposia devoted to excellence in robotics research

The Sixth edition of *Field and Service Robotics* edited by Christian Laugier and Roland Siegwart offers in its twelve-chapter volume a collection of a broad range of topics spanning underwater robots and systems, autonomous navigation for unmanned aerial vehicles, simultaneous localization and mapping, climbing robots, and sensor fusion among others. The contents of these contributions represent a cross-section of the current state of robotics research from one particular aspect: field and service applications, and how they reflect on the theoretical basis of subsequent developments. Pursuing technologies aimed at realizing robots operating in complex and dynamic environments, as well as robots working closely with humans, is the big challenge running throughout this focused collection.

Rich by topics and authoritative contributors, FSR culminates with this unique reference on the current developments and new directions in field and service robotics. A fine addition to the series!

Naples, Italy
January 2008

Bruno Siciliano
STAR Editor

# Preface

FSR is the leading single track conference in Field and Service Robotics. Its goal is to report and encourage the development and experimental evaluation of field and service robots, and to generate a vibrant exchange and discussion in the community. Field robots are non-factory robots, typically mobile, that operate in complex, and dynamic environments: on the ground (of Earth or planets), under the ground, underwater, in the air or in space. Service robots are those that work closely with humans to help them with their lives.

The first FSR meeting was held in Canberra, Australia, in 1997. Since then the meeting has been held every two years in the pattern Asia, America, Europe. This book presents the results of the sixth edition of Field and Service Robotics, FSR07, held in Chamonix, France, on 9th - 12th July 2007. The conference provided a forum for researchers, professionals and robot manufacturers to exchange latest technical developments and experiences. This book offers a collection of a broad range of topics including: Underwater Robots and Systems, Autonomous Navigation for Unmanned Aerial Vehicles, Simultaneous Localization and Mapping, Climbing Robotics, Sensor Fusion and much more.

This year we had 94 submissions from which we accepted 42 for oral and 16 for poster presentations.

The chairs were

### Chair

| | |
|---|---|
| Christian Laugier | INRIA |

### Co-chairs

| | |
|---|---|
| Roland Siegwart | ETH |

### The papers were thoroughly reviewed by the program committee

| | |
|---|---|
| Asama Hajime | Univ. of Tokyo, Japan |
| Aycard Olivier | IMAG, France |
| Burgard Wolfram | Univ. of Freiburg, Germany |
| Chatila Raja | AAS, France |
| Corke Peter | SIRO, Australia |
| Hollerbach John | Univ. of Utah, USA |

| | |
|---|---|
| Kelly Alonzo | CMU, USA |
| Khatib Oussama | Stanford Univ., USA |
| Laugier Christian | INRIA, France |
| Leonard John | MIT, USA |
| Prassler Erwin | Univ. of Ulm, Germany |
| Roberts Jonathan | CSIRO, Australia |
| Rus Daniela | MIT, USA |
| Salichs Miguel Angel | Univ. Carlos III, Spain |
| Siegwart Roland | ETH Switzerland |
| Singh Sanjiv | CMU, USA |
| Sukkarieh Salah | Univ. of Sydney, Australia |
| Thrun Sebastian | Stanford Univ., USA |
| Tomatis Nicola | Bluebotics, Switzerland |
| Wettergreen David | CMU, USA |
| Yoshida Kazuya | Tohoku University, Japan |
| Yuta Shin'ichi | Univ. of Tsukuba, Japan |
| Zelinsky Alex | CSIRO ICT Centre, Australia |

The conference has been supported and organized by INRIA Grenoble, ETH Zurich and the International Foundation of Robotics Research (IFRR) and organized with a strong support of Elisabeth Beaujard, Martine Consigney, Danièle Herzog and Agostino Martinelli who put together a strong technical and social agenda.

# Contents

## Part I: Mapping



## Part II: Vision for Navigation

# Part I

# Mapping

# Outdoor Radar Mapping Using Measurement Likelihood Estimation

John Mullane, Martin D. Adams, and Wijerupage S. Wijesoma

Nanyang Technological University
School of Electrical and Electronic Engineering
50 Nanyang Avenue, Singapore 639798
john.mullane@pmail.ntu.edu.sg, eadams@ntu.edu.sg, eswwijesoma@ntu.edu.sg

**Summary.** This paper fuses target detection and occupancy mapping theory to develop an improved method for outdoor mapping with a radar sensor. It is shown that the occupancy mapping problem is directly coupled with the signal detection processing which occurs in a range sensor, and that the required measurement likelihoods are those commonly encountered in both the target detection and data association hypotheses decisions. The classical binary Bayes filter approach generally treats these measurement likelihoods as fully known deterministic values. From examination of radar detection theory it is shown these likelihoods are only deterministic under a number of unrealistic assumptions which make it impractical for a real radar system used in a outdoor environment. An algorithm is therefore presented which jointly estimates the measurement likelihoods of each target in the environment and uses a particle filter to propagate their corresponding occupancy estimates. The ideas presented in this paper are demonstrated in the field robotics domain using a millimeter wave radar sensor, and comparisons with laser based maps as well as previous radar models are shown.

## 1 Introduction

Autonomous outdoor navigation is still a very active topic of research due to the presence of unstructured objects and rough terrain in realistic situations. For range/bearing sensors commonly used in robot navigation, such as the polaroid sonar or SICK laser, the target detection algorithm is performed internally resulting in a single $(r, \theta)$ output to the *first* signal considered detected. No other information is returned about the world along the bearing angle $\theta$, however it is typical in the case of most sensor models to assume empty space up to range $r$. This signal may or may not correspond to a target, depending on the environmental properties. These ambiguities are typically resolved in the data association stage by applying a threshold to some statistical distance metric based on the covariance of the predicted and observed feature locations.

However, sensor noise in range/bearing measuring sensors is in fact 3 dimensional as an added uncertainty exists in the detection process itself. This observation is not new and is readily considered in such mapping techniques as occupancy grids [1]. For most occupancy grid maps, the occupancy is distributed in a Gaussian manner as a function of the returned range resulting in *discrete*

C. Laugier and R. Siegwart (Eds.): Field and Service Robotics, STAR 42, pp. 3–12, 2008.
springerlink.com 

observations of occupancy in each cell. The discrete Bayes filter is then used as a solution, which is possible as it uses a completely known occupancy measurement model to update the posterior of the occupancy random variable. However, this is not the case for sensors such as the frequency modulated continuous wave (FMCW) radar[1] where the occupancy measurement models can be estimated directly. FMCW radar sensors are typically applied to outdoor sensing applications as they can operate in hazardous environments where other sensors will fail. This is due to the radar's ability to penetrate dust, fog, and rain [2].

## 2 Related Work

Signal processing problems are not new to the field of autonomous mapping but are generally treated in a simplified manner. In the underwater domain, sonars also return a power versus range vector which is difficult to interpret. In his thesis [3], S. Williams outlined a simple target detection technique for autonomous navigation in a coral reef environment. A constant noise power threshold is used and the maximum signal to noise ratio is chosen as the point target. Clearly this method of extraction results in a large loss of information, which is not desirable for the construction of well defined maps. S. Majumder attempts to overcome this loss by fitting a sum of Gaussian probability density function to the raw sensor data [4], however this represents a likelihood distribution in range of a *single point target* which is misleading as the data can contain multiple targets, leading to the association of non-corresponding points. In field robotics, standard noise power thresholding[2] was again used by S. Clark [5] using an FMCW radar. The range and bearing covariances were then modeled and propagated through an extended Kalman Filter framework to perform autonomous map building (and localisation).

This paper further explores the role of signal detection algorithms in autonomous mapping. Section 3 outlines the general occupancy mapping problem, showing how the occupancy variable becomes directly observable when target detection theory is considered. The problems with a discrete Bayesian filter solution are also discussed. Section 3.1 then outlines the estimation of these likelihoods for the radar sensor. Section 4 then proposes a method of estimating the posterior of each occupancy random variable, using a particle filter to propagate samples from both the prior and measurement likelihoods. Section 5 presents some results of the proposed method using real radar data collected from outdoor field experiments, comparisons are made to occupancy maps generated by SICK laser range finders and previously proposed radar mapping algorithms.

[1] Due to the modulating techniques, a fast fourier transform can be used to return a power value at discrete range increments. Range resolution, beamwidth, and maximum range are dependant on the particular sensor.

[2] Fixed threshold detection is indeed the optimal detector in the case of spatially uncorrelated and homogenous noise distributions of known moments.

## 3 Grid Mapping

Occupancy grid mapping is generally solved by assuming each grid cell to be independent so that the occupancy variable in each cell can independently estimated [1]. Update rules are well established for both probabilistic ($\Theta = \{O, E\}$) [6] and evidential ($\Theta = \{O, E, U\}$)) frameworks [7],

$$\log \frac{P(O|z^t)}{1 - P(O|z^t)} = \log \frac{P(O|z_t)}{1 - P(O|z_t)} + \log \frac{1 - P(O)}{P(O)} + \log \frac{P(O|z^{t-1})}{1 - P(O|z^{t-1})} \tag{1}$$

$$m(X_3) = \frac{\sum_{X_1 \cap X_2 = X_3} m_k(X_1) m_m(X_2)}{1 - \sum_{X_1 \cap X_2 = \emptyset} m_k(X_1) m_m(X_2)} \tag{2}$$

with $O$,$E$ and $U$ denoting occupied, empty and unknown estimates respectively. Also, in eqn 2, $m_k(\cdot)$ and $m_m(\cdot)$ represent mass functions respectively containing the sensor and prior evidences in support of each hypothesis, $\{X_1, X_2, X_3\} \subset \Theta$. These approach requires inverse models, $P(O|z_t)$, which lack rigorous mathematical foundation and may result in inconsistent maps [6]. The measurement likelihood, $P(z_t|O)$, is also used [8] along with the common Bayesian update rule,

$$P(O|z^t) = \gamma P(z_t|O) P(O|z^{t-1}) \tag{3}$$

where $\gamma$ is the normalizing constant. The most common model uses the detected range, $d$, and a Gaussian spread functions with the sensor range covariance, $\sigma^2$. In the case of a 1D the measurement likelihood is written as,

$$p(z_t|O) = \frac{1}{\sqrt{2\pi\sigma^2}} e^{\frac{(z_t - d)^2}{2\sigma^2}}. \tag{4}$$

A key observation is that in this model, the measurement $z_t$ is a range reading. The range at which a sensor reports the presence of a landmark can be used in the filtering of its *location* estimate and does not provide a measurement of the occupancy random variable. To correctly formulate the occupancy estimation problem, the measurement $z_t$ should be redefined as $z_t \in \{Detection, No\, Detection\}$. A simple expansion of eqn 3 shows how the occupancy measurement likelihoods can be obtained when the signal processing stage is considered. Consider the probability of occupancy given a history of measurements,

$$P(O|z^t). \tag{5}$$

The measurement history $z^t$ can now be considered as a sequence of detections, $D$ or non-detections, $\bar{D}$. We can then expand about both measurement possibilities to get,

$$P(O|z_t = D, z^{t-1}) \propto P(z_t = D|O) P(O|z^{t-1}) \tag{6}$$

$$P(O|z_t = \bar{D}, z^{t-1}) \propto P(z_t = \bar{D}|O) P(O|z^{t-1}) \tag{7}$$

These equations calculate, in closed form, the statistically correct posterior of the occupancy random variable, where the measurement likelihoods $P(z_t = D|O)$, $P(z_t = D|E)$, $P(z_t = \bar{D}|O)$ and $P(z_t = \bar{D}|E)$ are common in target detection algorithms. In real radar systems, these measurement likelihoods have to be estimated online therefore stochastic filtering is required. The following section outlines the measurement likelihood estimation process for a radar sensor.

### 3.1 Measurement Likelihood Estimation

This section uses radar target detection theory to estimate the measurement likelihoods, $p(z_t = D|O)$ and $p(z_t = D|E)$. The measurement data is assumed to consist of a vector of K consecutive (in time or range) independent signal intensity measurements, $x$, for each bearing angle, such as that shown in figure 1. A priori signal distribution assumptions are made on both $p(x|\boldsymbol{\Omega}_O, O)$ and $p(x|\boldsymbol{\Omega}_E, E)$, whereas the distribution moments $\boldsymbol{\Omega}_O$ and $\boldsymbol{\Omega}_E$ are unknown and must be estimated.

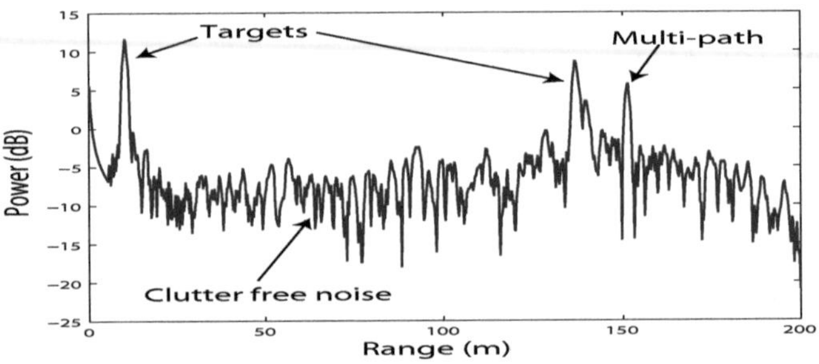


**Fig. 1.** A sample spectrum from a frequency modulated continuous wave radar containing K signal intensities in K range bins. Each measurement is assumed to be a single uncorrelated sample from one of the signal distributions shown in figure.

*Estimating the False Alarm Likelihood*

Assuming $p(x|\boldsymbol{\Omega}_E, E)$ to be exponentially distributed, constant false alarm rate (CFAR) detectors maintain the pre-defined $p(z_t = D|E)$ if the K consecutive intensity measurements are independent samples from identically and independently distributed $p(x|\boldsymbol{\Omega}_E, E)$ and more importantly, *given that the distribution assumption on $p(x|\boldsymbol{\Omega}_E, E)$, is valid.* The density of equation 19 will thus become deterministic.

*Estimating the Detection Likelihood*

Signal detection literature generally considers the probability of detection for targets of known SNR, whereas this work requires $p(z_t = D|O)$ to be estimated with an *unknown* SNR. Therefore an estimate of the targets mean SNR, $\bar{S}$ is required. There are numerous adaptive methods of generating local estimates of the unknown noise distribution parameters [9]. Thus, given an estimate of the local noise intensity from the CFAR detector, $\{x|E\}$ $\bar{S}$ can be estimated as,

$$\bar{S} = \frac{\{x|O\} - \{x|E\}}{\{x|E\}}. \tag{8}$$

An ordered-statistics approach [10], which is adopted in this work, has been shown to be most robust in situations of high clutter and multi-target situations, as is commonly encountered in a field robotics environment. Closed form solutions then exist for $p(z_t = D|O)$,

$$p(z_t = D|O) = \left(1 + \frac{T\{x|E\}}{1+\bar{S}}\right)^{-N} \tag{9}$$

with,

$$p(x|\boldsymbol{\Omega}_E = \mu, E) = \frac{1}{\mu}\exp(-x/\mu)$$
$$p(x|\boldsymbol{\Omega}_O = \{\mu, s\}, O) = \frac{1}{\mu}\exp\left((-x/\mu) + s)\right)I_0\left(2\sqrt{\frac{sx}{\mu}}\right)$$
$$p(s|\bar{S}) = \frac{2s}{\bar{S}}\exp(-s^2/\bar{S}).$$

By approximating each estimate as a Gaussian random variable, we can generate a Gaussian distributed estimate of $p(z_t = D|O)$. These likelihoods are then used during the occupancy measurement update as seen in equation 18 in the following section.

## 4 The Occupancy Filter

This section outlines the formulation for the filtering of an occupancy random variable using the measurement likelihood estimates. The objective of this filter is to propagate the posterior density of the occupancy variable, $O$. As this estimation is now modeled by a density function rather than a discrete random variable, the goal is to propagate the density[3],

$$p(O|z^t). \tag{10}$$

We can expand eqn 10 as before to get,

$$p(O|z^t) = \gamma p(z_t|O)p(O|z^{t-1}). \tag{11}$$

A static process model is assumed such that,

$$p(O_t|O_{t-1}) = p(O|z^{t-1}). \tag{12}$$

[3] As opposed to a deterministic value in the binary Bayes filter.

This assumption is inline with standard autonomous mapping algorithms where a static state is assumed, thus no change occurs to its occupancy estimate during the time update. As the true value of the occupancy state, $O$, is Boolean we must have the constraint,

$$\int p(O|z^t) + \int p(E|z^t) = 1. \tag{13}$$

Thus, given two *distributions* $p(z_t = D|O)$ and $p(z_t = D|E)$ from the measurement likelihood estimator, as well as the two priors $p(O|z^{t-1})$ and $p(E|z^{t-1})$ we can proceed to estimate the posterior, $p(O|z^t)$. The algorithm proceeds as,

1. Draw $N_r$ samples, $O^r$, from one of the prior existence densities,

$$O^r \sim p(O|z^{t-1}), \quad r \in \{1 \dots N_r\} \tag{14}$$

and,

$$E^r = 1 - O^r \tag{15}$$

with,

$$p(O|z^0) = U(0,1). \tag{16}$$

2. Uniformly weight the sets of particles,

$$w_M^r = \frac{1}{2N_r}, \forall r \in \{1 \dots N_r\}, M \in \{O, E\}. \tag{17}$$

Thus the weights for the combined set of particles $O^r$ and $E^r$ are normalized. Therefore $p(O|z^t)$ is approximated by the set $\{(O^r, w_O^r\}$, $p(E|z^t)$ is approximated by the set $\{E^r, w_E^r\}$ $\forall r = \{1 \dots N_r\}$, and the discrete form of equation 13 is satisfied.

3. Draw $N_l$ samples from the measurement likelihood distributions i.e the detection and false alarm densities. The general assumption that $p(z_t = D|O) = 1 - p(z_t = D|E)$ [1] is not valid and independent samples must be drawn.

$$z_O^l \sim p(z_t = D|O) \tag{18}$$

$$z_E^l \sim p(z_t = D|E), \quad l \in \{1 \dots N_l\}. \tag{19}$$

4. Weight the measurement likelihood particles according to,

$$\tilde{v}_O^l = \frac{1}{N_l} p(z_t = D|O) \tag{20}$$

$$\tilde{v}_E^l = \frac{1}{N_l} p(z_t = D|E) \tag{21}$$

and normalize,

$$v_O^l = \frac{\tilde{v}_O^l}{\sum_{j=1}^{N_l} \tilde{v}_O^j} \tag{22}$$

$$v_E^l = \frac{\tilde{v}_E^j}{\sum_{j=1}^{N_l} \tilde{v}_E^j} \tag{23}$$

to generate the two sets of particles and weights $\{z^l_O, v^l_O\}$ and $\{z^l_E, v^l_E\}$ $\forall l = \{1 \ldots N_l\}$ which then approximate both measurement likelihoods.

5. Convolve both sets of measurement likelihood and prior particles through the deterministic Bayesian update rule in equations 6-7 to generate a superset of posterior particles $\{\Theta^k, \Psi^k\}$ with $k \in \{1 \ldots N_r N_l\}$,

$$\Theta^k = \frac{z^l_O O^r}{z^l_O O^r + z^l_E E^r} \tag{24}$$

$$\Psi^k = \frac{v^l_O w^r_O}{v^l_O w^r_O + v^l_E w^r_E} \tag{25}$$

where $r \in \{1 \ldots N_r\}$ and $l \in \{1 \ldots N_l\}$ as before.

6. Resample the superset $\{\Theta_k, \Psi_k\}$, reducing the sample number back down to $N_r$ which, due to the static process model assumption, then give a particle approximation of the prior density, $p(O|z^t)$, at time t+1.

By virtue of the Bayesian normalizing factor, the posterior occupancy particles are guaranteed to exist only within the bounded range [0,1].

## 5 Experiments

The first example propagates an occupancy variable in each range bin of the radar (K = 800) at a single bearing angle. A sample spectrum from this data was

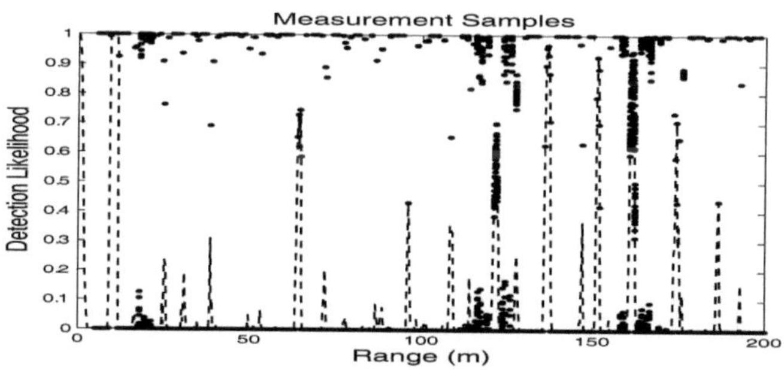


**Fig. 2.** The measurement samples based on the estimated probability of detection and the output hypothesis of the detector. Samples represent particle approximations of $p(z_t = D|O)$ or $p(z_t = D|E)$, in the case of $H_1$ and $H_0$ hypotheses respectively. Larger spreads of particles indicate increased uncertainty in that cell. The dashed plot represents the mean value of $P_d$ in each range bin, independent of the detection hypothesis, thus accommodating for potential missed detections.

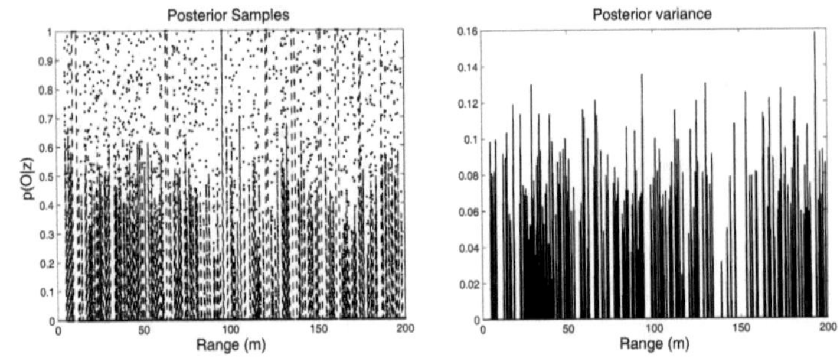


**Fig. 3.** The left hand figure shows the posterior existence samples after a single filter iteration, with the dashed line representing the mean occupancy in each bin. The right hand figure shows the corresponding variance of the estimates in each range bin.

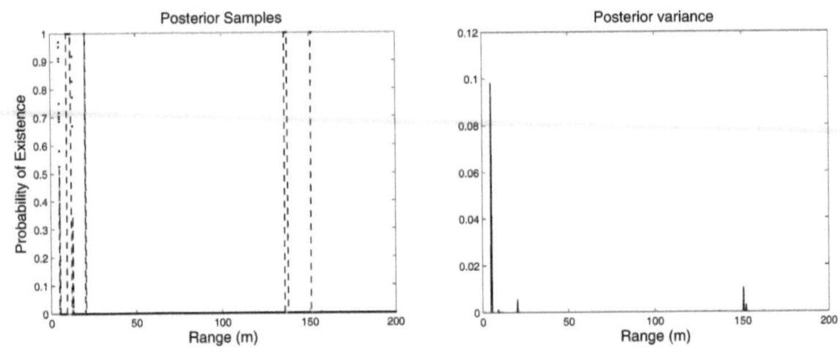


**Fig. 4.** The left hand figure shows the posterior occupancy samples, after five filter iterations, with the dashed line again representing the mean existence in each bin. Updated variances are shown in the right hand figure.

shown previously in figure 1, with three target signals present. Mixing these samples with the prior samples and resampling according to equations 18 - 25, we get the posterior density samples in each range bin as seen in figure 3. The figure also shows the corresponding variance estimated for the occupancy variable. Figure 4 shows the same corresponding plots after 5 iterations, along with update variance in each range bin. It can be seen that the variance remains practically zero for most cells however due to high variance measurement likelihood estimates, particulary at 5m and 150m, there remains some uncertainty in the occupancy estimate. Previous models using deterministic updates fail to account for this uncertainty. A two dimensional map is shown in figure 5 where multiple radar scans are fused in a carpark environment. For comparison purposes an occupancy map built by a laser is included. The result shows promising map building capability using this radar model.

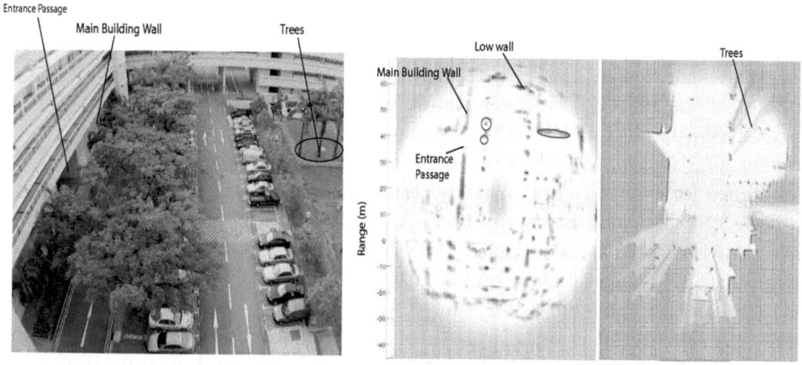


**Fig. 5.** The left hand figure shows an overview of testing ground (Cars were absent at time of experiment). The center figure shows the occupancy map using the radar sensor and the proposed algorithm. The right most figure shows the corresponding map constructed using laser data and an arbitrary sensor model. Highlighted are some landmarks which are detected by the radar but missed by the laser. This is mainly due to the radars increased beamwidth.

## 6 Conclusion

This paper discussed the role played by target detection algorithms in autonomous outdoor mapping. It was shown that the measurement likelihoods used to propagate the occupancy estimate are assumed known in previous mapping sensor models. Once this assumption is relaxed, the measurement likelihoods become stochastic estimates based on the signal detection algorithm and the measurement intensity information. A particle filter approach to the occupancy estimation problem was then proposed. This concept was demonstrated for an FMCW radar sensor which is typically used in an outdoor environment. For most practical scenarios $P_d \gg P_{fa}$, thus Bayesian based mapping method will assign almost unity occupancy to any detection. However, when the assumption on the noise distribution is violated and $P_{fa}$ deviates from its theoretical value, an evidential method should be applied as there would be large ambiguity as to the true false alarm probability. This would further improve the mapping accuracy. Further work also would integrate three dimensional uncertainties (in Euclidean location as well as occupancy) into data association decisions and SLAM algorithms.

## Acknowledgements

This work was funded under the second author's AcRF Grant, RG 10/01, Singapore.

## References


1. Moravec, H., Elfes, A.E.: High resolution maps from wide angle sonar. In: Proceedings of the 1985 IEEE International Conference on Robotics and Automation, March 1985, pp. 116–121 (1985)
2. Brooker, G., Bishop, M., Scheding, S.: Millimetre waves for robotics. In: Australian Conference for Robotics and Automation, Sidney, Australia (November 2001)
3. Williams, S.: Efficient Solutions to Autonomous Mapping and Navigation Problems. PhD thesis, Australian Centre for Field Robotics, University of Sydney (2001)
4. Majumder, S.: Sensor Fusion and Feature Based Navigation for Subsea Robots. PhD thesis, The University of Sydney (August 2001)
5. Clark, S., Durrant-Whyte, H.F.: Autonomous land navigation using millimeter wave radar. In: International Conference on Robotics and Automation (ICRA), Leuven, Belgium, May 1998, pp. 3697–3702. IEEE, Los Alamitos (1998)
6. Thrun, S.: Autonomous robots, vol. 15, pp. 111–127 (2003)
7. Pagac, D., Nebot, E., Durrant-Whyte, H.F.: An evidential approach for map building for autonomous vehicles. IEEE Transactions on Robotics and Automation 14(4) (August 1998)
8. Konolige, K.: Improved occupancy grids for map building. Auton. Robots 4(4), 351–367 (1997)
9. Rohling, H.: Some radar topics: Waveform design, range (cfar) and target recognition. In: NATO: Advances in Sensing with Security Applications, Il Ciocco, Italy (July 2005)
10. Rohling, H., Mende, R.: OS CFAR performance in a 77 GHz radar sensor for car applications. In: CIE International Conference of Radar, October 1996, pp. 109–114 (1996)

# Recovering the Shape of Objects in 3D Point Clouds with Partial Occlusions

Rudolph Triebel[1,2] and Wolfram Burgard[1]

[1] Department of Computer Science, University of Freiburg, Georges-Köhler-Allee 79, 79108 Freiburg, Germany
{triebel,burgard}@informatik.uni-freiburg.de

[2] Autonomous Systems Lab, ETH Zurich, Tannenstraße 3, 8092 Zurich, Switzerland
rudolph.triebel@mavt.ethz.ch

**Summary.** In this paper we present an approach to label data points in 3d range scans and to use these labels to learn prototypical representations of objects. Our approach uses associative Markov networks (AMNs) to calculate the labels and a clustering operation to determine the prototypes of homogeneously labeled regions. These prototypes are then used to replace the original regions. In this way, we obtain more accurate models and additionally are able to recover the structure of partially occluded objects. Our approach has been implemented and evaluated on 3d data of a building acquired with a mobile robot. The experimental results demonstrate that our algorithm can robustly identify objects with the same shape and can use the prototypes of these objects for highly accurate mesh completion in case of occlusions.

## 1 Introduction

Recently, the problem of acquiring three-dimensional models using mobile robots has become quite attractive and a variety of robot systems have been developed that are able to acquire three-dimensional data using laser range scanners [17, 10, 6, 21, 7, 13, 19]. Most of these approaches deal with the problem of how to improve the localization of the robot or how to reduce the huge amount of data by piecewise linear approximations. In this paper we consider the problem of classifying data points in three-dimensional range scans into known classes. The general motivation behind this is to achieve the ability to learn maps that are annotated with symbolic labels. In the past, it has been shown that such information can be utilized to find compact representations of the maps [19], but also to improve the process of joining partial maps into one big map, usually called *map registration* by only considering associations between points belonging to corresponding objects [18, 13]. However, a problem arises when these annotated objects are only seen from a partial view or are occluded by other objects. Even if the annotations or *labels* are correct, corresponding objects can not be found robustly because the entire shape of the objects is not available.

In this paper, we present a supervised learning approach to learn the shape of objects in data points obtained from 3d range scans. Our algorithm applies associative Markov networks (AMNs) [15] to robustly determine segmentations of the data points into different classes. We then perform a clustering operation of the individual segments and

C. Laugier and R. Siegwart (Eds.): Field and Service Robotics, STAR 42, pp. 13–22, 2008.
springerlink.com 

calculate a prototype for each segment. In a final step, we replace the individual segments by their prototypes. As a result, we obtain accurate models and even are able to complete partially scanned objects, which frequently appear in the case of occlusions.

The problem of extracting features from range data has been studied intensively in the field of mobile robotics. For example, Buschka and Saffiotti [4] describe a virtual sensor that is able to identify rooms from range data. Additionally, Simmons and Koenig [9] use a pre-programmed routine to detect doorways from range data. Althaus and Christensen [1] use line features to detect corridors and doorways. Also several authors focused on the problem of extracting planar structures from range scans using the expectation maximization algorithm (EM) [10, 12, 19]. Furthermore, there has been work on employing features extracted from three-dimensional range scans to improve the scan alignment process [18, 13]. The approaches described above either operate on two-dimensional scans, consider single features such as planarity, or apply pre-programmed routines to identify the features in range scans.

In the context of learning annotated 3D maps from point cloud data several authors use the so-called *spin images* [8, 5] to recognize objects. Vandapel *et al.* [20] extract saliency features and apply EM to learn a Gaussian Mixture Model classifier. Another popular object description technique are *shape distributions* [14]. In contrast to these approaches, our algorithm classifies the data by also taking into account the potential labels of neighboring data points. This is modeled in a mathematical framework known as Markov random fields and improves the segmentation by eliminating false classifications. The recent work by Anguelov *et al.* [2] applies a similar approach to label data points. However, this technique does not cluster the segments and also cannot complete objects which have been scanned partially only. Früh and Zakhor [6] generate large-scale 3d-models of urban scenes and apply linear interpolation to deal with partial occlusions. In contrast to this method, our approach extracts object prototypes from range data and uses these prototypes to more accurately recover the structure of the scene in the case of occlusions.

This paper is organized as follows. In Section 2, we introduce our scan-point classification technique and an efficient approach to the learning problem. Then, Section 3 describes our algorithm to handle occlusions in the data sets. Finally, Section 4 presents experimental results illustrating the usefulness of our approach.

## 2 Point Classification Using Markov Networks

The first part of our object recognition algorithm consists of a classification for a set of given data points based on a parameter set that was learned from a hand-labeled training data set. For this classification we use an *associative Markov network* (AMN) [15] which is an instance of a *relational Markov network* (RMN). RMNs and AMNs utilize undirected graphical models to represent the conditional probability $P(\mathbf{y} \mid \mathbf{x})$ where $\mathbf{y}$ is the set of labels and $\mathbf{x}$ the set of features associated to each data point. This is done by defining functions that assign a positive value to each clique in the graph and its associated labeling. These functions are called *clique potentials* and reflect how well a given labeling fits to a specified clique of graph nodes. In AMNs, the maximum size of a clique is 2, so that only *node potentials* $\varphi(\mathbf{x}_i, y_i)$ and *edge potentials* $\psi(\mathbf{x}_{ij}, y_i, y_j)$ exist.

Here, we introduce the feature vector $\mathbf{x}_{ij}$ of features extracted from the edge between nodes $i$ and $j$. Likewise, $y_i$ denotes the label of node $i$. Note that the edge feature vector $\mathbf{x}_{ij}$ does not necessarily have the same size as the node features $\mathbf{x}_i$.

Assuming that the network consists of $N$ nodes and a set $\mathcal{E} = \{(i, j)|i, j \in \{1, \ldots, N\}, i \neq j\}$ of edges, the overall conditional probability represented by the AMN can then be formulated as

$$P(\mathbf{y} \mid \mathbf{x}) = \frac{1}{Z} \prod_{i=1}^{N} \varphi(\mathbf{x}_i, y_i) \prod_{(ij)\in\mathcal{E}} \psi(\mathbf{x}_{ij}, y_i, y_j). \tag{1}$$

where $Z = \sum_{\mathbf{y}'} \prod_{i=1}^{N} \varphi(\mathbf{x}_i, y_i') \prod_{(ij)\in\mathcal{E}} \psi(\mathbf{x}_{ij}, y_i', y_j')$. Usually, $Z$ is denoted as the *partition function* and represents the sum of potentials for all possible labelings $\mathbf{y}'$.

It remains to describe the potentials $\varphi$ and $\psi$. As mentioned above, the potentials reflect how well the features fit to the labels. One simple way to define the potentials is the *log-linear* model [15]. In this model, a weight vector $\mathbf{w}^k$ is introduced for each class label $k = 1, \ldots, K$. The node potential $\varphi$ is then defined so that $\log \varphi(\mathbf{x}_i, y_i) = \mathbf{w}_n^k \cdot \mathbf{x}_i$ where $k = y_i$. Accordingly, the edge potentials are defined as $\log \psi(\mathbf{x}_{ij}, y_i, y_j) = \mathbf{w}_e^{k,l} \cdot \mathbf{x}_i$ where $k = y_i$ and $l = y_j$. Here we distinguish between *weights* $\mathbf{w}_n$ for the nodes and weights $\mathbf{w}_e$ for the edges. Also, we define the weights depending on the class labels $k$ and $l$. This means that the potential of node $i$ depends on the label $k = y_i$ that is assigned to it. Similarly, the edge potential $\psi$ depends on the labels $k$ and $l$ that are assigned to the nodes connected by the edge. The idea here is that of a *relational* modeling where the dependencies of the labels are expressed on a higher level, namely the class level. For example, we can model the fact that classes $A$ and $B$ are more strongly related to each other than, say, classes $A$ and $C$. As a result, the weighting of neighboring points with labels $A$ and $B$ is higher than of points labeled $A$ and $C$.

To summarize, we define the node and edge potentials as:

$$\varphi(\mathbf{x}_i, y_i) = \exp(\mathbf{w}_n^k \cdot \mathbf{x}_i) \tag{2}$$

$$\psi(\mathbf{x}_{ij}, y_i, y_j) = \exp(\mathbf{w}_e^{k,l} \cdot \mathbf{x}_{ij}) \tag{3}$$

## 2.1 Learning and Inference

The scan point segmentation using AMNs is formulated as a supervised learning task. This means that in the learning phase, we are given a set of labels $\hat{\mathbf{y}}$ which were assigned by hand to the training data. Now we learn a set of weights $\mathbf{w}$ so that $P_{\mathbf{w}}(\hat{\mathbf{y}} \mid \mathbf{x})$ is maximized. In the inference phase, these weights are used to find a set of labels $\mathbf{y}$ for the test data set so that $P_{\mathbf{w}}(\mathbf{y} \mid \mathbf{x})$ is maximized. As can be seen from Equation 1, both steps include the computation of the partition function $Z$. In all but the simplest cases, the calculation of $Z$ is intractable. For the inference task we can exploit the fact that $Z$ does not depend on the particular labels $\mathbf{y}$. Therefore the maximization of $P_{\mathbf{w}}(\mathbf{y} \mid \mathbf{x})$ is equivalent to maximizing $ZP_{\mathbf{w}}(\mathbf{y} \mid \mathbf{x})$. This means that $Z$ can simply be neglected in the inference step.

However, in the learning task we need to maximize $P_{\mathbf{w}}(\hat{\mathbf{y}} \mid \mathbf{x})$ over all $\mathbf{w}$, which means to calculate $Z$ for each $\mathbf{w}$. To overcome this problem, Taskar *et al.* [16] propose a different way to learn the weights $\mathbf{w}$. Instead of $P_{\mathbf{w}}(\hat{\mathbf{y}} \mid \mathbf{x})$, they maximize the *margin* between the optimal labeling $\hat{\mathbf{y}}$ and any other labeling $\mathbf{y}$ defined by

$$\log P_{\mathbf{w}}(\hat{\mathbf{y}} \mid \mathbf{x}) - \log P_{\mathbf{w}}(\mathbf{y} \mid \mathbf{x}). \tag{4}$$

This way, the term $Z_{\mathbf{w}}(\mathbf{x})$ cancels out and the maximization can be done efficiently. This method is referred to as *maximum margin* optimization [16]. The details of this formulation are omitted here for the sake of brevity. We only note that the problem is reduced to a quadratic program (QP) of the form:

$$\min \ \frac{1}{2}||\mathbf{w}||^2 + c\xi \tag{5}$$

such that

$$\mathbf{w}\mathbf{X}\hat{\mathbf{y}} + \xi - \sum_{i=1}^{N} \alpha_i \geq N; \quad \mathbf{w}_e \geq 0; \quad \alpha_i - \sum_{ij,ji \in E} \alpha_{ij}^k - \mathbf{w}_n^k \cdot \mathbf{x}_i \geq -\hat{y}_i^k \ \ \forall i,k;$$
$$\alpha_{ij}^k + \alpha_{ji}^k - \mathbf{w}_e^k \cdot \mathbf{x}_{ij} \geq 0 \ \ \forall ij \in E, k; \quad \alpha_{ij}^k, \alpha_{ji}^k \geq 0 \ \ \forall ij \in E, k$$

Here, the variables that are solved for in the QP are the weights $\mathbf{w} = (\mathbf{w}_n, \mathbf{w}_e)$, a slack variable $\xi$ and additional variables $\alpha_i$, $\alpha_{ij}$ and $\alpha_{ji}$. We refer to Taskar *et al.* [15] for details.

In the inference task we want to find labels $\mathbf{y}$ that maximize $\log P_{\mathbf{w}}(\mathbf{y} \mid \mathbf{x})$. As mentioned above, $Z$ does not depend on $\mathbf{y}$ so that the maximization can be done without considering $Z$. This leads to a linear program of the form

$$\operatorname{argmax} \ \sum_{i=1}^{N} \sum_{k=1}^{K} (\mathbf{w}_n^k \cdot \mathbf{x}_i) y_i^k + \sum_{ij \in E} \sum_{k=1}^{K} (\mathbf{w}_e^k \cdot \mathbf{x}_{ij}) y_{ij}^k \tag{6}$$

such that

$$\textbf{a)} \ y_i^k \geq 0, \ \ \forall i,k; \quad \textbf{b)} \ \sum_{k=1}^{K} y_i = 1, \ \ \forall i \quad \textbf{c)} \ y_{ij}^k \leq y_i^k, \ \ y_{ij}^k \leq y_j^k, \ \ \forall ij \in E, k$$

Here, we introduced variables $y_{ij}^k$ representing the labels of two points connected by an edge [15].

## 2.2 Efficient Variant of AMN Learning

Unfortunately, the learning task described in the previous section is computationally expensive with respect to run-time and memory requirements. For each scan point there is one variable and one constraint in the QP. Furthermore, we have two variables and two constraints per edge. This results in a large computational effort (Anguelov *et al.* [2] report one hour run time for about 30,000 scan points). However, in typical data sets, a huge part of the data is redundant and the data can be substantially reduced by downsampling. In our experiments we never found evidence that this reduces the overall performance. The run time dropped from about 20 minutes down to less than a minute while the detection remained at 92%. However, it is not clear how many samples are necessary to obtain good detection rates, because this depends on the data set. A scene with many small objects should not be down-sampled as much as a scene with only few, big objects. Therefore we reduce the data *adaptively*.

The idea of our adaptive data reduction technique is to obtain as much information as necessary from the training data so that still a good recognition rate can be achieved. As this depends on the data set, we need an adaptive data structure. One popular way to adaptively store geometrical data are *kd*-trees. This data structure follows a coarse-to-fine approach: the higher the level in the tree the higher the data abstraction. With *kd*-trees we can reduce the data set by considering only scan points in the tree that are stored in leaf nodes up to a given maximum depth $d_{max}$. All points in deeper branches are then merged into a new leaf node of depth $d_{max}$. The data point in this new leaf node is calculated as the mean of all points from the corresponding subtree. Apart from the reduction in the data complexity, this approach makes the sampling less dependent on the data density. The remaining question is how to select $d_{max}$. As for the uniform down-sampling, this is dependent on the data set. In our current system, we therefore modify the reduction algorithm so that it is parameter-free. Instead of pruning at a fixed level, we merge all points in a subtree whenever all of its labels are equal. Accordingly, for large homogeneous areas, where all points have the same label, we obtain a higher level of abstraction than in heterogeneous areas.

# 3 Occlusion Handling

In most visual recognition tasks we encounter the problem of *occlusion*: objects are partially hidden by others and therefore can not be recognized robustly. Our approach shows a method to overcome this problem using the semantic knowledge acquired in the learning phase. The idea is to compare objects of the same class to each other and to infer the shape of a single object based on the shapes of the other objects in the class. Hereby we assume that the objects in a class are similar to each other. Based on this assumption, we first group all points $\mathcal{P} = \{\mathbf{p}_1, \ldots, \mathbf{p}_n\}$ that were labeled with the same label in the inference step. Next we cluster all obtained groups according to features defined over the point clouds. Then we match the point clouds for each cluster to each other to obtain the prototypical objects. The meshes calculated for the prototypes are then used to replace the original point clouds.

## 3.1 Clustering

First, we cluster $\mathcal{P}$ into contiguous subsets which we call *entities*. The clustering is done using a region-growing algorithm in 3D space where the neighboring relations can be obtained efficiently using a *kd*-tree. Then, we cluster the entities into subclasses. The idea here is that usually a class consists of different *kinds* of objects, e.g., in a class "window" we can find single- and double-size windows as well as windows of different shapes. Of course, the distinction of these subclasses could be done when labeling the training data already. In this case the AMN would automatically yield appropriate labels for the individual subclasses. In our current system we decided to separate the division into sub-classes from the labeling process, because it turned out to be difficult to define features for the different entities on the point level. In our implementation we use three entity features based on the oriented bounding box $B$ of the entity and its point cloud $P$. The features are the volume of $B$, the quotient of the second-longest and the longest edge of $B$ and the *radius* of $P$, defined by the maximal distance of a point in $P$ and the centroid of $P$. Again, for the clustering we apply region-growing, in this case in 3D feature space.

### 3.2 Entity Matching

In the next step, we match the entities that belong to the same subclass to each other. This is the step in which information about the shape of one entity is used to complete the shape of another entity in the same subclass. This assumes that all entities in a subclass have the same shape and that a good matching between entities can be found. The matching is done using the Iterative Closest Point algorithm (ICP) [3]. In our current implementation, we select one entity as a reference frame and match the other entities to the selected one. One could also think of connecting all entities into a clique and match all entities to each other. Then, the matching errors can be reduced by performing a global optimization of the entity poses. In our experiments, we obtained good results with the one-reference-frame technique. After matching the single entities we obtain a merged point cloud as the *prototype* of the subclass.

### 3.3 Mesh Generation

For a better visualization, we generate triangulated meshes from the point cloud resulting from the previous step. To this end, we insert all points into a 3D grid. The size of the grid is defined by the oriented bounding box of the point cloud. For each cell $c$ in the grid we store the expected number of points that fall into $c$, where the probability of falling into $c$ is modeled as an isotropic Gaussian whose mean is the center of $c$. Then we apply the marching cubes contouring algorithm [11] to find the contour that separates occupied cells from free cells. As a result, we obtain a triangular mesh that approximates the volume represented by the point cloud. Finally, we re-project the obtained triangle mesh to all original positions of the singular entities.

## 4 Experiments

We implemented the described algorithm and tested it on a real data set. The data was collected with a mobile outdoor robot that has a SICK laser range finder and a pan/tilt unit mounted on top for 3D data acquisition. We scanned a building with windows of different kinds and sizes (see 1(a)). In a first step, we divided the data into walls by using a plane extraction algorithm (see Figure 1(b)). Then we extracted all points that had a distance of at most $0.5m$ from the planes.

The goal was to classify the scan points into the classes "window", "wall" and "gutter". Accordingly, we labeled the training data set manually as shown in Figure 1(c).

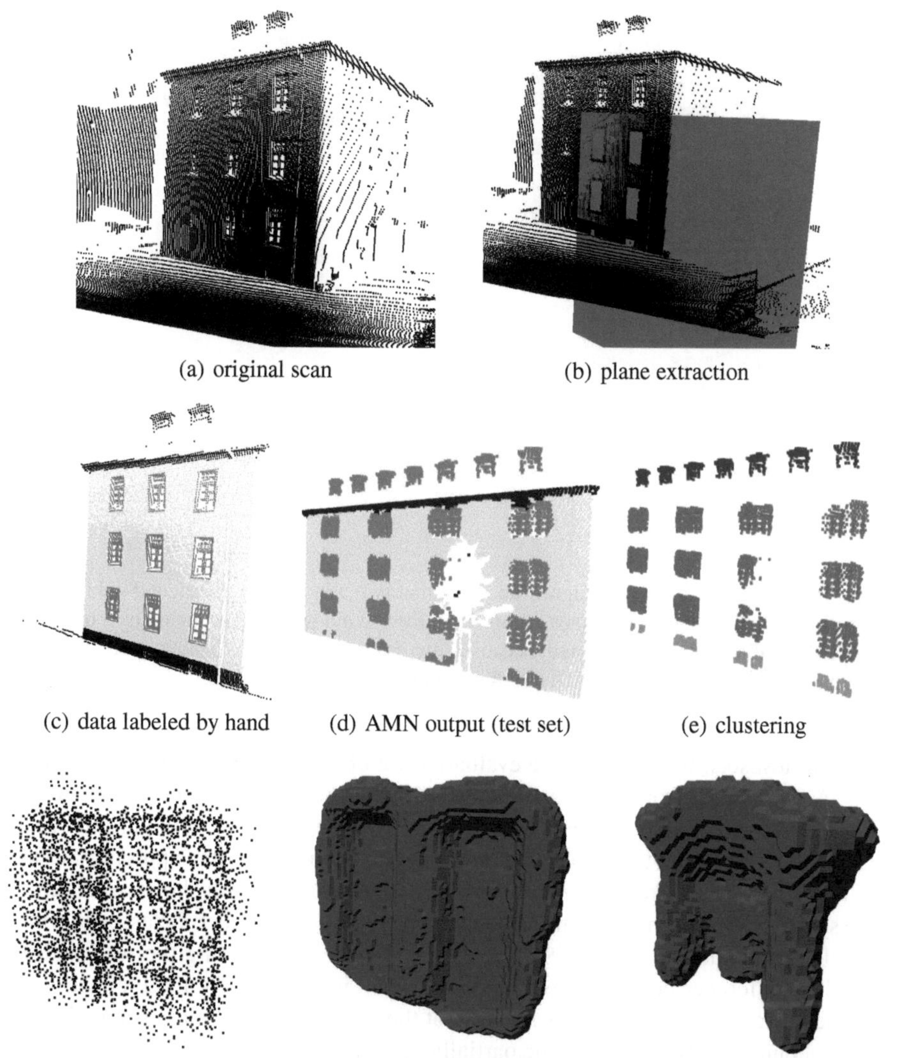

(a) original scan (b) plane extraction

(c) data labeled by hand (d) AMN output (test set) (e) clustering

(f) entity matching (g) mesh generation I (h) mesh generation II

**Fig. 1.** The individual steps of our occlusion handling algorithm. From top left to bottom right: a) original 3D scan (no occlusions) b) plane extraction (only one plane is shown), c) hand-labeling of the training data, d) labeling of a test data set obtained with the AMN approach; note that some windows and the wall are occluded, e) class-wise sub-clustering, here for the window class, f) scan matching of all subclusters, here the big windows, g) mesh generation from prototype shown in f), h) mesh for the roof windows prototype.

It consists of one wall with only single-size windows. The original size of the training data set was 36191 data points, while after adaptive reduction we obtained a reduced set of 3944 data points. Figure 1(d) shows the result of the AMN based classification on

**Fig. 2.** Result obtained with our algorithm. Note that two windows in the second column have been restored. In the original data (see Figure 1(d)) these windows were occluded by a tree. Also note that the wall could not be restored, because only one wall object was encountered in the data set so that no prototype containing data in the occluded areas was obtained.

one of our test sets. In a quantitative evaluation we obtained 93.8% correctly classified labels. As we can see from the figure, there are gaps in the data caused by the occlusions of a tree in front of the building. This results in windows that are only partially seen. Figures 1(d)-1(h) show the remaining steps of our algorithm. After applying the last step, namely the back-projection into the scene we obtain the mesh shown in Figure 2. In the scene, all objects have been replaced by the prototypes of the subclasses in which they fall. Note that this holds for all objects in the scene, including the wall and the gutter. The difference compared to the window class is only that for these classes only one object occurs in the data. This means that the prototype of the class is equal to the object encountered. However, for the partially occluded objects, our algorithm was able to recover the full structure.

## 5 Conclusion

In this paper we presented an approach to segment three-dimensional range data and to use the resulting segments for augmenting the original data. Our approach uses associative Markov networks to robustly extract regions based on an initial labeling obtained with simple geometric features. To efficiently carry out the learning phase, we use an adaptive technique to prune the *kd*-tree. We then cluster the segments and calculate a prototype for each segment. These prototypes are then used to replace the original segments. The advantage of this approach is two-fold. First, it allows to increase the accuracy of the individual regions, and second it allows to complete partially seen objects by the prototypes.

Our approach has been implemented and tested on data acquired with an outdoor-robot equipped with a laser range finder mounted on a pan/tilt unit. In complex data sets containing outer walls of buildings, our approach has successfully been applied to the task of finding a segmentation into walls, windows, and gutters even in the case of partial occlusions.

## Acknowledgements

This work has partly been supported by the German Research Foundation (DFG) within the Research Training Group 1103 and under contract number SFB/TR-8.

## References

1. Althaus, P., Christensen, H.I.: Behaviour coordination in structured environments. Advanced Robotics 17(7), 657–674 (2003)
2. Anguelov, D., Taskar, B., Chatalbashev, V., Koller, D., Gupta, D., Heitz, G., Ng, A.: Discriminative Learning of Markov Random Fields for Segmentation of 3D Range Data. In: Conference on Computer Vision and Pattern Recognition (CVPR) (2005)
3. Besl, P.J., McKay, N.D.: A method for registration of 3-d shapes. IEEE Transactions on Pattern Analysis and Machine Intelligence 14, 239–256 (1992)
4. Buschka, P., Saffiotti, A.: A virtual sensor for room detection. In: Proc. of Intern. Conference on Intelligent Robots and Systems (IROS) (2002)
5. Frome, A., Huber, D., Kolluri, R., Bulow, T., Malik, J.: Recognizing objects in range data using regional point descriptors. In: Proceedings of the European Conference on Computer Vision (ECCV) (2004)
6. Früh, C., Zakhor, A.: 3d model generation for cities using aerial photographs and ground level laser scans. In: Computer Vision and Pattern Recognition Conference (2001)
7. Georgiev, A., Allen, P.K.: Localization methods for a mobile robot in urban environments. IEEE Transactions on Robotics and Automation 20(5), 851–864 (2004)
8. Johnson, A., Hebert, M.: Using spin images for efficient object recognition in cluttered 3d scenes. IEEE Transactions on Pattern Analysis and Machine Intelligence 21(5), 433–449 (1999)
9. Koenig, S., Simmons, R.: Xavier: A robot navigation architecture based on partially observable markov decision process models. In: Kortenkamp, D., Bonasso, R.P., Murphy, R. (eds.) Artificial Intelligence and Mobile Robots, MIT Press, Cambridge (1998)
10. Liu, Y., Emery, R., Chakrabarti, D., Burgard, W., Thrun, S.: Using EM to learn 3D models with mobile robots. In: Proceedings of the International Conference on Machine Learning (ICML) (2001)
11. Lorensen, W.E., Cline, H.E.: Marching cubes: A high resolution 3D surface construction algorithm. Computer Graphics (SIGGRAPH) 21(4), 163–169 (1987)
12. Martin, C., Thrun, S.: Online acquisition of compact volumetric maps with mobile robots. In: IEEE International Conference on Robotics and Automation (ICRA), Washington, DC (2002)
13. Nüchter, A., Wulf, O., Lingemann, K., Hertzberg, J., Wagner, B., Surmann, H.: 3d mapping with semantic knowledge. In: RoboCup International Symposium (2005)

14. Osada, R., Funkhouser, T., Chazelle, B., Dobkin, D.: Matching 3d models with shape distributions. In: Shape Modeling International, Genova, Italy (2001)
15. Taskar, B., Chatalbashev, V., Koller, D.: Learning Associative Markov Networks. In: Twenty First International Conference on Machine Learning (2004)
16. Taskar, B., Guestrin, C., Koller, D.: Max-Margin Markov Networks. In: Neural Information Processing Systems Conference (2003)
17. Thrun, S., Burgard, W., Fox, D.: A real-time algorithm for mobile robot mapping with applications to multi-robot and 3D mapping. In: Proc. of the IEEE International Conference on Robotics & Automation (ICRA) (2000)
18. Triebel, R., Burgard, W.: Improving simultaneous localization and mapping in 3d using global constraints. In: Proc. of the Twentieth National Conference on Artificial Intelligence (AAAI) (2005)
19. Triebel, R., Burgard, W.: Using hierarchical EM to extract planes from 3d range scans. In: Proc. of the IEEE International Conference on Robotics & Automation (ICRA) (2005)
20. Vandapel, N., Huber, D., Kapuria, A., Hebert, M.: Natural terrain classification using 3-d ladar data. In: IEEE International Conference on Robotics and Automation (2004)
21. Wulf, O., Arras, K., Christensen, H., Wagner, B.: 2d mapping of cluttered indoor environments by means of 3d perception. In: Proc. of the IEEE International Conference on Robotics & Automation (ICRA) (2004)

# Update Policy of Dense Maps: Efficient Algorithms and Sparse Representation

Manuel Yguel, Olivier Aycard, and Christian Laugier

$\varepsilon$-motion, GRAVIR-UJF-INRIA-INP Grenoble, France
firstname.lastname@inrialpes.fr

## 1 Introduction and Previous Work

Providing a robot with a fully detailed map is one appealing key for the Simultaneous Localisation and Mapping (SLAM) problem. It gives the robot a lot of hints to solve either the data association or the localisation problem itself. The more details are in the map, the more chances are that different places may appear differently, solving ambiguities. The more landmarks are used, the more accurate are the algorithms that solve the localisation problem since in a least square sense an approximation of the solution is more precise. Last, it helps a lot in the presence of a few dynamic objects because these moving parts of the environment remain marginal in the amount of data used to model the map and can thus be filtered out. For instance, the moving objects can be detected or cancelled in the localisation procedure by robust techniques using Monte-Carlo algorithms [6] or RANSAC [4].

The environment models that provide such possibilities are named *dense maps*. To build such a representation from range-finders there exist in the literature a lot of propositions: using the point clouds generated by the sensors [2], using grid-based representation [3], [5] or using more geometrical modelling involving feature fitting such as lines or planes [14], [13]. From our point of view the second kind of representation using occupancy grid (OG) is of particular interest since it offers a strong mathematical framework for updates or merges of maps even if the measurements or the map disagree. For example this property is useful if a part of the map, that was not observed, had changed a lot like the streets of a city where the cars park and go.

However as their name suggests it dense maps comes with a memory burden. For instance raw scan records of 361 points of a laser range-finder operating at $50Hz$ produce in 3 hours almost 195 millions of points which make $780Mb$ of data and a grid-based representation that use cubic cells with a side of $5cm$ needs 3.6 billions of cells to cover an area of $3km^2$ which make $28.8Gb$ for an (OG). Naturally this is a hard limitation for such maps and that is why the design of sparse representations is very important.

In this paper we present a review of update methods for grid-based maps like OGs under the requirement of a sparse representation emphasizing the pros and the cons then we propose a new update policy that is particularly appropriate to the sparse condition. In a first part grid based representations and update are presented. In a second part

C. Laugier and R. Siegwart (Eds.): Field and Service Robotics, STAR 42, pp. 23–33, 2008.
springerlink.com 

update policies are studied. Then in the third section is discussed how the new update policy modified the hierarchical algorithms for sparse OGs presented in our previous works. Finally results are presented in the last section.

## 2 Grid-Based Maps

In this section the process of information fusion in grid-based map is described.

### 2.1 Occupancy Grids

Grid-based representations were first introduced by Elfes and Moravec with the occupancy grids (OGs): in each cell of the grid a probabilistic estimation is maintained for the presence of an obstacle. This representation is closely related to range sensors for which sensor models exists that take into account measurement uncertainty to update the map.

*Sensor model*

- $Z$ a random variable[1] for the sensor range measurements in the set $\mathscr{Z}$.
- $O_{x,y} \in \mathscr{O} \equiv \{\text{occ}, \text{emp}\}$. $O_{x,y}$ is the state of the cell $(x,y)$, where $(x,y) \in \mathbb{Z}^2$. $\mathbb{Z}^2$ is the set of indexes of all the cells in the monitored area.
- the OGs needs three probability distribution to be defined: the *a priori* map occupancy $P_0(O_{x,y})$ and the two sensor models: $P(Z|[O_{x,y} = \text{occ}])$ and $P(Z|[O_{x,y} = \text{emp}])$. The two last modelling how uncertain a measurement is with respect to the obstacle location (fig. 1(a)).

*Log-ratio form of occupancy update*

As the occupancy is a binary variable, a quotient between the likelihoods of the two states of the variable is sufficient to describe the binary distribution. The new representation used is:

$$\text{odd}(O_{x,y}) = \log \frac{p([O_{x,y} = \text{occ}])}{p([O_{x,y} = \text{emp}])} \tag{1}$$

In the Bayesian update of the occupancy, the quotient makes the marginalization term disappear and thanks to a logarithm transformation, sums are sufficient for the inference:

$$\log \frac{p(\text{occ}|z)}{p(\text{emp}|z)} = \log \frac{p(\text{occ})}{p(\text{emp})} + \log \frac{p(z|\text{occ})}{p(z|\text{emp})} = odd_0 + odd(z) \tag{2}$$

where $odd_0$ named the initial estimation of the cell occupancy is log-ratio.

[1] For a certain variable $V$ we will note in upper case the variable, in lower case $v$ its realization, and we will note $p(v)$ for $P([V = v])$ the probability of a realization of the variable.

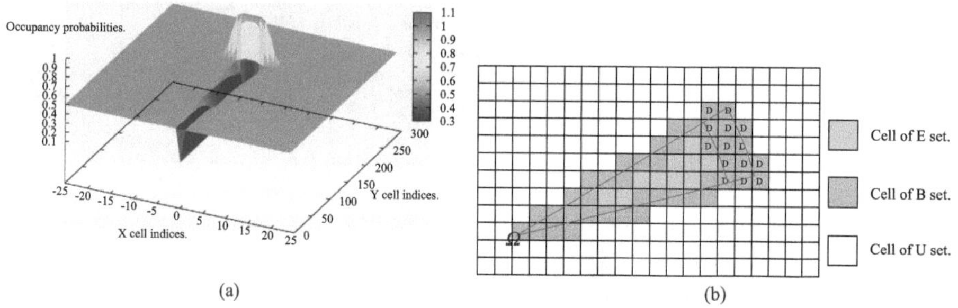


**Fig. 1.** (a) Update of a 2D OG after a sensor reading, initially each cell occupancy was unknown, *i.e.* 0.5 probability. The sensor beam has an aperture of 7 degrees. The sensor is positioned in (0,0). It thus is the just adding in each cell the value of the log-ratio of the two sensor models. (b) A range-finder beam. The range finder is located at $\Omega$ and its field of view is surrounded by red boundaries. It defines the three kind of cell types, unknown cells (U set) in white, empty cells (E set) in light blue and occupied cells: the one close to an obstacle marked with a "D" (B set).

### 2.2 Reflection Probability Maps

Reflection probability maps were described by Burgard and instead of the occupancy probability the map measures the probability of a reflection per grid cell and have the advantage of not requiring an explicit sensor model. For each cell two numbers are stored: the number of times a beam has traversed the cell (♯misses) and number of times a beam has ended in the cell (♯hits). Then the probability of reflection is estimated as:

$$p(O_{x,y}) = \frac{\sharp\text{hits}}{\sharp\text{hits} + \sharp\text{misses}} \tag{3}$$

### 2.3 Normal Transforms

The normal transforms store in addition of one of the previous kind of occupancy information the mean and the standard deviation of all the end-points of the beams that fall into the cell[2]. This representation was proposed by Biber in [1] and as a nice feature it reduces the problem of space discretization induced by the grid which greatly improved the localisation accuracy for instance.

## 3 Update Policies

In this section is studied the effect of long term sensor fusion on grid-based maps. In the first subsection the effect of the simple sensor fusion updates of the previous section are described highlighting the drawbacks. Then the usual max and clamping policies used to counter these drawbacks are reviewed. Last we present the policy we used along with the occupancy *a priory* that gives the best memory saving.

[2] Note that it supposed to store the number of time the cell was a beam end-point.

**Fig. 2.** Log-ratio occupancy maps from the Freiburg data set generated with various policies. For all the figure the color code is the following: the more bright is the color the more occupied is the cell. All color scales are defined from the min occupancy to the max occupancy of the maps. (a) No policy. (b) Max policy. (c) Clamping policy. (d) Exponential forgetting policy.

### 3.1 No Policy

When updates are performed using just the equation of the previous section, the map obtained suffers from two drawbacks: it is overconfident and it is highly irregular (fig. 2(a), fig. 3(a)). If carried in log-ratio space these two problems are evident through the unbounded sums along fusion updates. The number of observations required to change the state of a static cell is equal to the number of observations in the past that have defined this state. And if the robot have stayed a long time observing a part of the map, it will make a long time to change its state which is annoying since obstacles like car can be parked a long time before going quickly. This issue is called the map *overconfidence* problem. It can be understood as a memory problem since the map remember all agreeing observations. In particular as the vantage points of the sensor observations changes, all the minor changes in the pose of the sensor let a print in the map leading to very irregular representation. The term irregular means, here, that the function representing the map is far from a continuous function. And as there is no way to predict the patterns that appear in the map there is no way neither to compress the map efficiently.

However this update as a clear advantage: all the information are used to estimate the static parts of the map which made the location of the wall for instance very accurate (fig. 3(a)). It is worth notice that all these drawbacks and advantages are valid for all the grid-based representations.

### 3.2 Max Policy

A solution, usually use because it is conservative ( [8] or [12]) is to update using the maximum rule:

$$odd_t = \max(odd_{t-1}, odd(z)). \tag{4}$$

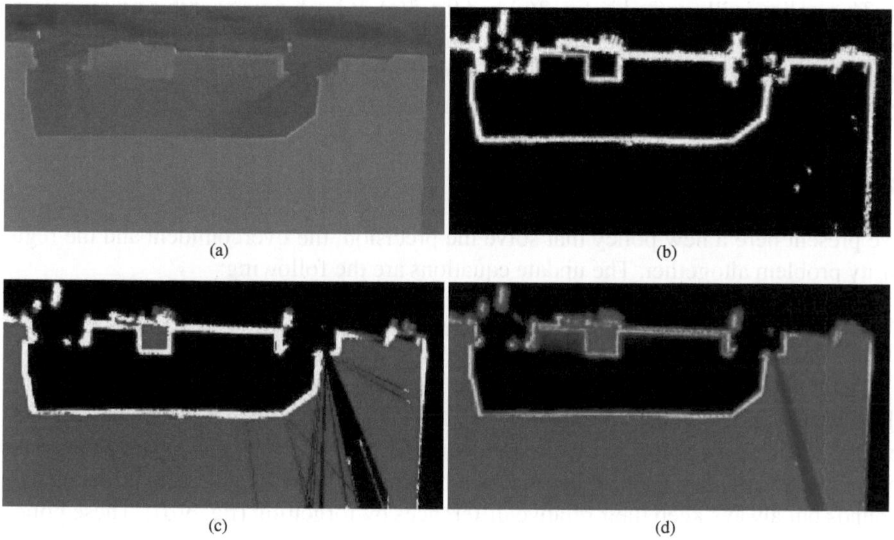

**Fig. 3.** Zoom on the map from the Freiburg data set generated with various policies. (a) No policy. (b) Max policy. (c) Clamping policy. (d) Exponential forgetting policy.

In this case, there is not any overconfidence problem for empty regions but an occupied area can never be empty anymore, the model cannot represent accurately the obstacles in case of false measurement or moving objects for instance. Furthermore the estimation of the obstacle position is very poor (thick wall in fig. 3(b)) which tends to give bad localisation. However the compression capabilities of such representation are interesting since most parts of the map stay at the original value when the space is empty (fig. 2(b)). For reflection probability maps the same effect can be obtained by just updating ♯hits.

### 3.3 Clamping Policy

A quantification policy discretized the range of occupancy values into a small set of possible values, like in [7]. This process is often done *a posteriori* since it is not obvious how to consistently update quantified values. For instance the minimum and maximum probabilities are not anymore respectively equivalent to 0 and 1 probabilities. The same policy is difficult to obtain with reflection probability maps since bounding ♯hits and ♯misses independently leads to wrong estimations. For log-ratio occupancy maps the update equation becomes:

$$odd_t = \max(\min(odd_{t-1} + odd(z), odd_{\max}), odd_{\min}). \quad (5)$$

where $\mathrm{odd}_{\max}$ and $\mathrm{odd}_{\min}$ are the predefined maximum and minimum log-ratio occupancy probabilities.

This policy is illustrated in fig. 2(c) and fig. 3(c). It is obvious that this policy suffers for the same problem of precision than the max policy. However, the state of the grid is reversible which make it more suitable for map updates. The regularity of the maps is good because most areas tend to converge towards the two extreme values.

### 3.4 Exponential Forgetting Policy

We present here a new policy that solve the precision, the overconfident and the regularity problem altogether. The update equations are the following:

$$odd_t = (1-\gamma)odd_{t-1} + \gamma odd(z). \tag{6}$$

The map is naturally bounded, since the $odd(z)$ is bounded and if $odd_{t-1} = odd(z)$ then $odd_t = odd_{t-1}$. Therefore the map is easy to compress since there are large coherent areas (fig. 2(d)) and is not overconfident. The map keeps precision since the peaks of occupancy do not disappear because neighboring cells do not all reach the occupancy bounds but always keep their relative differences by forgetting (fig. 3(d)). These policy is called exponential forgetting since a past observation is exponentially less important in the current occupancy estimation. It is worth to notice that ♯hits, ♯misses, mean and standard deviation can be updated the same way.

The choice of $\gamma$ determines the rate of forgetting and by trials we find that a low value is interesting in the case of the mapping with a laser-range finder at $10Hz$ ($\gamma = 0.3$ is taken in the experiments).

One drawback of this policy is that using naively in the unobserved area, it makes the cell occupancies converge toward a unknown state rapidly since in log-ratio the value of $odd(z)$ for this kind of area is zero. This problem can be avoided by only updating areas that are in the field of view of the sensor, in the occupied and empty zones only (see fig. 1(b))

### 3.5 Policy Mixing and Initial Conditions

All these different policies can be combined together to build new policies that solve some of the problems. For instance, since the empty space is often the most important part of a map, an interesting combination is to clamp the lower occupancy values while not bounding the upper ones. The compression abilities of such a map are good, and it was the policy we use in previous experiments [16]. In this case, an obstacle can easily appear but it disappears very slowly if it was observed a long time. It is conservative from an obstacle avoidance point of view while it could be really annoying when searching a free place in a car park.

An other important choice is the initialization of the occupancy value of the map. From a compression point of view the most interesting choice is to initialize the map with the empty bound[3] since it is the most probable value in an environment where a robot is supposed to evolve (see fig. 4(a) and fig. 4(b)).

[3] It is rarely done without any policy that bound the map since there is no really empty value, and the maps are often initialized with a uniform prior in those cases.

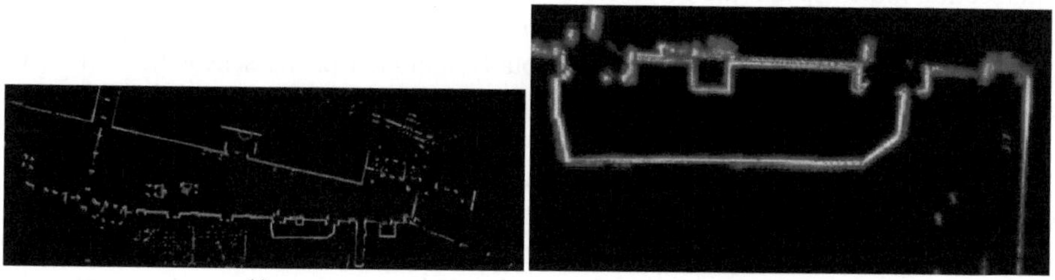

(a) (b)

**Fig. 4.** Map from the Freiburg data set generated with the exponential forgetting policy and a initialization to an empty map. (a) Entire map (b) Zoom on the map.

# 4 Sparse Representation through Hierarchical Map Organization

This section describe how the update policies change the hierarchical update algorithm developed in [16] for sparse representations.

## 4.1 Problem Statement

One popular method to compress a grid-based representation in the robotic community rely on building a tree based hierarchical representation of the map ( [7], [9], [10], [8], [16]). The idea is to represent large area with the same occupancy with a single node at an appropriate coarse scale. To be efficient these strategies suppose that the map have indeed large uniform occupancy area, therefore a policy that bound the map is always required at least for the main part of the map (which is the empty one). As stated in [15], it is often interesting to store differences between the scales instead of the plain values because small differences are the clue to notice uniform areas or a variant of this scheme like the Haar wavelet transform.

The problem in such representations is how to efficiently update the hierarchical representation taking into account sensor model uncertainties. In a previous work [16] we derive a hierarchical algorithm that allow to make efficient hierarchical updates with a clamping policy.and in particular updates that are hierarchical in the sens that large part of the map can be updated at large scale when a large part is discover as empty. That is very important in the car park situation when a car can occupied many cells at fine resolution for instance. The drawback of clamping-based policy is that they are either lacking in precision, the drawback when not bounding the occupancy is that the map is overconfident. In the following is exposed the modifications to bring to the hierarchical algorithm in order to use exponential forgetting policies.

## 4.2 Hierarchical Algorithm

The key idea in the hierarchical exploration of the grid space is to define a predicate: *existIntersection* that is true if a given set of grid cells intersect the volume defined by the field of view of the sensor beams (blue plus red cells in fig. 1(b)). The absence of intersection indicates that the given set of cells are outside the sensor field of view and don't need updating. For the case of *existIntersection* evaluating to true, a special sub case would be when the set of cells are totally included in the sensor field of view, then all the cells of the set belong to $E$ (blue cells in fig. 1(b)) and their occupancy are decreased by the same amount of $\text{odd}_{\text{emp}}$, eq. 2. As the algorithm is able to detect uniform regions recursively, the grid representation should allow the update of regions, and wavelets provide a natural mechanism for doing so. For each grid area, the *existIntersection* predicate guides the search. If there is intersection the traversal searches deeper into the grid hierarchy, *i.e.* exploring finer scales. Otherwise it stops at the current node.

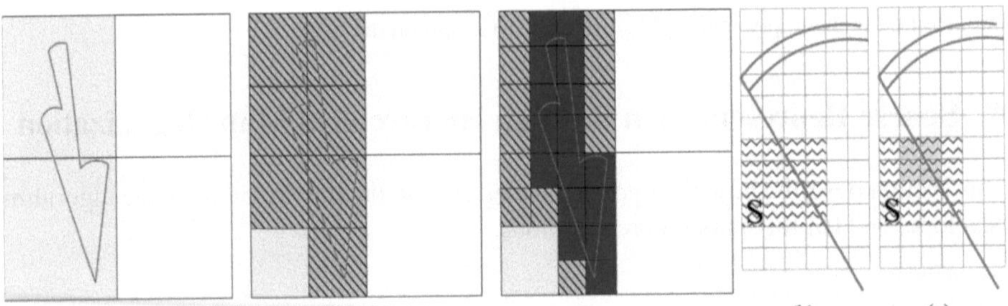


**Fig. 5.** (a) The hierarchical process of updating the grid: from the coarsest scale to the finest. To save computing time, area that are outside the polygon of view or totally included inside the area classified as empty are detected and processed early in the hierarchy. (b) and (c): two different cases for the iteration along a boundary of the field of view that separates $E$ set and $U$ set. Fig. 5(b) artificial separation, $S$ (with waves) was totally empty and the observation of a part of its interior (on the right of the red boundary) does not bring any information gain. Fig. 5(c) the separation brings information about the state of the yellow area that is inside the field of view (on the right of the red boundary). Just in this last case the algorithm 2 performs a search at a finer scale.

An update is almost always required for cells that are in the obstacle neighborhood (cells marked with 'D' in fig. 1(b) ) so iteration is always performed in area that contains such a cell. But for boundaries that separate cells that belongs to $U$ set and to $E$ set (white and blue cells in fig. 1(b)) iteration is required only if the $E$ set corrects the knowledge in the grid (fig. 5(c)) otherwise the iterations can stop early in the hierarchy (fig. 5(b)).

Further details and pseudo code can be found in [16].

### 4.3 Modification of the Update Algorithm under Exponential Forgetting Update Policy

The main contribution of the update algorithm with the wavelet representation is that for uniform area the update is hierarchical, however with the exponential forgetting policy all finer scales must forget too. This an important drawback of this policy for a hierarchical representation.

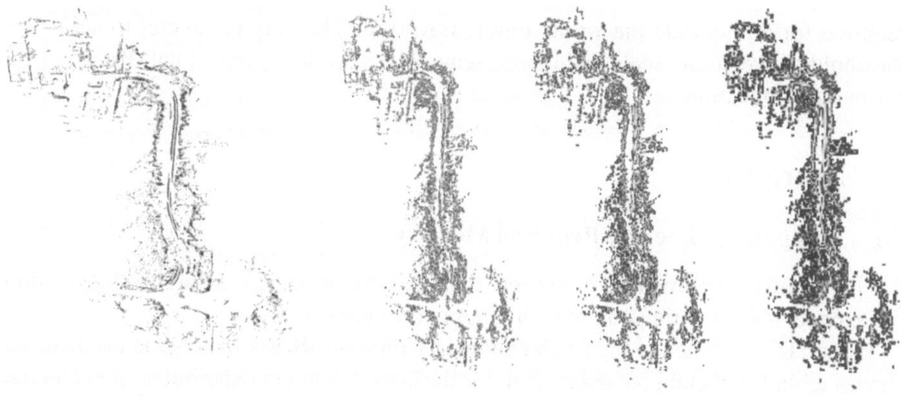

(a) (b) (c) (d)

**Fig. 6.** Results of algorithm 2 for 2D data: scaled views of the OG provided by different scales of WavOG representation (a) scale 0, cell size: $10cm \times 10cm$. (b) scale 2, cell size: $40cm \times 40cm$. (c) scale 3, cell size: $80cm \times 80cm$. (d) scale 4, cell size: $1.60m \times 1.60m$. The data where provided by CSIRO on a long-run experiment. The size of the map is approximately $450m \times 200m$. The number of cells vary from 9 Millions to 36000 from scale 0 to scale 4 and the ratio is one for 256 if the scale 4 is compared to the scale 0. It is obvious that the precision of the map decreases when the scale increases however the shape of the environment is sufficient enough for path planning since scale 3 in the big open area which is very interesting and promising for multiscale path planning algorithm for instance.

To solve this issue there is the possibility to always proceed to a complete traversing of the tree except for regions outside the sensor field-of-view (U set) or regions with occupancy that has reached one of the boundaries occupancy values. The solution proposed here improves this by noticing that a hierarchical update is worth when a big part of the environment being previously occupied becomes empty (like with parked car that leaves its place), in that case successive updates will lead to a complete empty area. Therefore updating the grid at fine scale is totally useless and time consuming, since big parts will be empty.

We propose to keep track of the empty updates performed at a coarse scale by recording the corresponding coarse nodes of the grid in a second tree $T_e$. In that tree is stored the corresponding nodes from which empty updates are required along with the number of time an empty update was performed. When updating the grid tree, $T_e$ is traversed too and if an update is required for child of a node in $T_e$ the empty updates are performed in a lazy fashion. The equations are the following:

$$\begin{aligned} odd_t &= (1-\gamma)^n odd_{t-1} + \gamma \mathrm{odd}_{\mathrm{emp}} \left(\sum_{i=0}^{n-1} (1-\gamma)^i\right), \\ &= (1-\gamma)^n (odd_{t-1} - \mathrm{odd}_{\mathrm{emp}}) + \mathrm{odd}_{\mathrm{emp}}. \end{aligned} \tag{7}$$

when $n$ is the number of cached empty update stored in the coarse node of $T_e$. $T_e$ is then updated by zeroing the cached updates at the coarse node. It the empty update is required for a fine scale the whole tree is traversed. The only parameter to set is the threshold between fine scale and coarse scale which we fix to match half the size of a car in our application.

# 5 Experiments

## 5.1 Computing Time and Required Memory

We performed experiments[4] on 2D real data with the modified hierarchical algorithm for exponential forgetting policy with moving obstacles.

In the 2D experiment a big truck equipped with four SICK LMS-291 at each corner carries a big hot metal container behind it during a 1.5 hours experiment, the data are noisy and the evolution property of the map is required by the presence of a lot of moving obstacles (in particular the hot metal container). The algorithm processes the 4 laser range-finder at real-time (40$Hz$ each). The processing time of the modified algorithm matches the processing time of the update algorithm with clamping policy but provides in addition a non overconfident map.

# 6 Conclusions and Future Works

## 6.1 Conclusions

The objective of this work is to present a review of the update policies in grid-based representation that makes this kind of environment modelling more suitable for compression and to present a new one based on exponential forgetting. This new update policy offers many advantages compare to the others because it present neither overconfidence nor loss of accuracy in the long term. However it remained to see how this new update pattern fit with our hierarchical update algorithm of a sparse grid representation. We have then presented a scheme that allows lazy updates in case of gross disappearance of objects like cars that leave a car park. In that case the algorithm delayed update as long as possible and in most of the case never needs to do them which makes it as efficient than the update algorithm with Haar wavelets and clamping update policy.

## 6.2 Future Works

In the future we will explore paralleling the computing of the hierarchical update algorithm since the tree structure and distributed memory organisation of the data structure seems be well designed for such a purpose.

# Acknowledgments

The authors would like to thank Christopher Tay Meng Keat, Christophe Braillon and Cedric Pradalier for their contributions and fruitful discussions. We express also our

[4] Every experiment was done with an Intel(R) Pentium(R) 4 CPU 3.00$GHz$.

gratitude to CSIRO for providing a consistent huge data set that allowed us to validate our method. We also thank Cyrill Stachniss for making corrected data sets available through his web page [11].

## References

1. Biber, P., Straßer, W.: The normal distributions transform: A new approach to laser scan matching. In: IEEE/RJS International Conference on Intelligent Robots and Systems (2003)
2. Cole, D., Newman, P.: Using laser range data for 3d slam in outdoor environments. In: Proc. of the IEEE International Conference on Robotics and Automation (ICRA), Florida (2006)
3. Elfes, A.: Occupancy grids: a probabilistic framework for robot perception and navigation. PhD thesis, Carnegie Mellon University (1989)
4. Fischler, M.A., Bolles, R.C.: Random sample consensus: a paradigm for model fitting with applications to image analysis and automated cartography. Commun. ACM 24(6), 381–395 (1981)
5. Grisetti, G., Stachniss, C., Burgard, W.: Improving grid-based slam with rao-blackwellized particle filters by adaptive proposals and selective resampling. In: Proc. of the IEEE International Conference on Robotics and Automation (ICRA), pp. 2443–2448 (2005)
6. Khan, Z., Balch, T., Dellaert, F.: Mcmc data association and sparse factorization updating for real time multitarget tracking with merged and multiple measurements. IEEE Transactions on Pattern Analysis and Machine Intelligence 28(12), 1960–1972 (2006)
7. Kraetzschmar, G.K., Gassull, G.P., Uhl, K.: Probabilistic quadtrees for variable-resolution mapping of large environments. In: Ribeiro, M.I., Victor, J.S. (eds.) Proceedings of the 5th IFAC/EURON Symposium on Intelligent Autonomous Vehicles (July 2004)
8. Payeur, P., Hébert, P., Laurendeau, D., Gosselin, C.: Probabilistic octree modeling of a 3-d dynamic environment. In: Proc. IEEE ICRA 1997, Albuquerque, NM, April 20-25, 1997, pp. 1289–1296 (1997)
9. Payeur, P., Laurendeau, D., Gosselin, C.: Merging uncertainty into probabilistic octree models of 3-d perturbed workspaces. In: Proc. VI 1998, Vancouver BC, June 18-20, 1998, pp. 439–446 (1998)
10. Payeur, P., Laurendeau, D., Gosselin, C.: Range data merging for probabilistic octree modeling of 3-d workspaces. In: Proc. ICRA 1998, Leuven, Belgium, May 16-21, 1998, pp. 3071–3078 (1998)
11. Stachniss, C.: Corrected robotic log-files, `http://www.informatik.uni-freiburg.de/stachnis/datasets.html`
12. Thrun, S., Burgard, W., Fox, D.: Probabilistic Robotics (Intelligent Robotics and Autonomous Agents), ch. 9, p. 231. The MIT Press, Cambridge (2005)
13. Triebel, R., Pfaff, P., Burgard, W.: Multi-level surface maps for outdoor terrain mapping and loop closing. In: Proc. of the International Conference on Intelligent Robots and Systems (IROS) (2006)
14. Vandorpe, J., Van Brussel, H., Xu, H.: Exact dynamic map building for a mobile robot using geometrical primitives produced by a 2d range finder. In: IEEE International Conference on Robotics and Automation, April 22-28, 1996, vol. 1, pp. 901–908 (1996)
15. Yguel, M., Aycard, O., Laugier, C.: Wavelet occupancy grids: a method for compact map building. In: Proc. of the Int. Conf. on Field and Service Robotics (2005)
16. Yguel, M., Keat, C.T.M., Braillon, C., Laugier, C., Aycard, O.: Dense mapping for range sensors: Efficient algorithms and sparse representations. In: Proceedings of Robotics: Science and Systems, Atlanta, GA, USA (June 2007)

# Assessing Map Quality Using Conditional Random Fields

Manjari Chandran-Ramesh and Paul Newman

Robotics Research Group, University of Oxford, Oxford - OX1 3PJ
{manjari,pnewman}@robots.ox.ac.uk

**Summary.** This paper is concerned with assessing the quality of work-space maps. While there has been much work in recent years on building maps of field settings, little attention has been given to endowing a machine with introspective competencies which would allow assessing the reliability/plausibility of the representation. We classify regions in 3D point-cloud maps into two binary classes — "plausible" or "suspicious". In this paper we concentrate on the classification of urban maps and use a Conditional Random Fields to model the intrinsic qualities of planar patches and crucially, their relationship to each other. A bipartite labelling of the map is acquired via application of the Graph Cut algorithm. We present results using data gathered by a mobile robot equipped with a 3D laser range sensor while operating in a typical urban setting.

## 1 Introduction and Background

Robust localisation, efficient navigation and exploration are essential competencies in mobile robotics. These capabilities crucially depend on an internal representation of an agent's workspace. Good theoretical progress in understanding the fundamentals of the mapping problem (including its parent problem - SLAM) has benefitted in particular the field robotics domain where *a-priori* maps are hard to come by and external navigation infrastructure is expensive or inconvenient to install. It comes as no surprise then, that the mapping of unknown environments has been and continues to be the focus of intense research. The past two decades have seen approaches to detecting loop closure while mapping (see, for example, [9]) as well as attempts to mapping dynamic environments [10],[14].

While recent mapping techniques are reasonably successful at building consistent representations of less complex workspaces, local map inconsistencies and errors are still commonplace in more complex environments. This is particularly problematic outdoors, where large distances lead to increased odometry error and the presence of dynamic objects such as pedestrians, cars or even leaves moving in the wind can lead to spurious sensor readings. Despite this known inadequacy of mapping algorithm, hardly any effort has been expended to determine the quality of a map once it is built. Clearly, the ability of automatically detecting possibly spurious workspace representations is of value, both for diagnostic purposes as well as an additional source of information about the affected area of the workspace itself.

C. Laugier and R. Siegwart (Eds.): Field and Service Robotics, STAR 42, pp. 35–48, 2008.
springerlink.com 

This work aims to addresses this lack of introspection by introducing a procedure to measure the quality of 3D laser maps of complex urban environments. The approach makes explicit use of the artificial nature of these work-spaces by focusing on geometric features which are commonly found in man-made environments. The intuition is that most parts of the world can be represented by planar structures where a correctly rendered point-cloud produces a well-defined, non-intersecting - almost 'crisp' - image of the environment. Thus, the approach first segments the given 3D point-cloud map into plane patches consisting of a subset of the original points. A context-sensitive classification framework is then employed to classify these plane patches according to the plausibility of their being rendered correctly. While numerous techniques are available in the area of classification, each with their merits and demerits [18], [7], capturing contextual information has been shown to work effectively in the framework of Conditional Random Fields [13], [15], [1], [17]. Hence this work uses conditional random fields for the purpose of classification.

This paper is structured as follows: Section 2 provides a brief description of Conditional Random Fields. Section 3 develops and describes the actual method of classification of the map and Section 4 describes the application of the approach to laser data from a large urban environment.

## 2 Conditional Random Fields

A Markov random field is an undirected graphical model which has a set of nodes each of which corresponds to a random variable or group of variables, as well as a set of links each of which connects a pair of nodes [4], [3], [8], [11]. A conditional random field (CRF) can be viewed as a Markov random field globally conditioned on the random variable representing the observation sequence [19].

Consider a labelling problem consisting of a network of $N$ nodes that represent the random variable, $X$, over the data sequence . Let these nodes be labelled by the random variable $Y$ and have $K$ possible labels: $y_i \in \{1, ..., K\}$. $X$ and $Y$ are jointly distributed. However, Lafferty *et al.* [13] developed a conditional model $p(y|x)$ that represented directly the conditional distribution of labels given the data sequence and did not require explicit modelling of the marginal $p(x)$. This conditional model became more suitable for classification tasks with overlapping attributes as it made no assumptions about the dependency structure [15].

Formally, a CRF can be defined as follows. Consider a graph $G = (V, E)$, such that $Y = (y_i)_{i \in V}$ or $Y$ is indexed by the vertices in $G$. When conditioned on $X$, if $Y$ obeys the Markov property with respect to $G$, then $(X, Y)$ is a CRF. This can be written as $p(y_v|x, y_w, w \neq v) = p(y_v|x, y_w, w \sim v)$, where $w \sim v$ implies that $w$ and $v$ are neighbors in $G$. This graphical structure can be used to factorize the distribution into positive potential functions that operate on a subset of nodes. To ensure that the potential functions fulfill the conditional independence criteria of undirected graphical models, the concept of fully connected sub-graph or clique is introduced. The potential functions operate on the set of nodes that form a clique and the conditional distribution is factorized into a product of clique potentials $\phi_c(x_c, y_c,)$ where $c \in C$ is a clique in the set of cliques $C$ and $x_c$ and $y_c$ are the nodes and its labels in that clique. This factorization is written as

$$p(y|x) = \frac{1}{Z(x)} \prod_{c \in C} \phi_c(x_c, y_c) \tag{1}$$

where $Z(x)$ is the partition function (normalization constant) given by $Z(x) = \sum_y \prod_{c \in C} \phi_c(x_c, y_c,)$. The partition function ensures that the conditional distribution $p(y|x)$ is correctly normalized.

The potential function represents the constraints on the configuration or the "compatibility" between the nodes in the clique. Typically, for ease of notation, pairwise CRFs are considered and hence the maximal clique is of size two. This implies that each clique has local or node potentials, $\phi_i(x_i, y_i)$ for each node $i$, and pairwise or edge potentials,$\phi_{ij}(x_{ij}, y_i, y_j)$ for an edge between nodes $i$ and $j$. The conditional distribution can now be written as

$$p(y|x) = \frac{1}{Z(x)} \prod_{i=1}^{N} \phi_i(x_i, y_i) \prod_{ij \in E} \phi_{ij}(x_{ij}, y_i, y_j) \tag{2}$$

where $Z(x)$ is the partition function given by $\sum_y \prod_{i=1}^{N} \phi_i(x_i, y_i) \prod_{ij \in E} \phi_{ij}(x_{ij}, y_i, y_j)$, $N$ is the number of nodes, $E$ is the set of edges $\{ij\}(i < j)$ in the graph, $\phi_i(x_i, y_i)$ is the node or unary potential, $\phi_{ij}(x_{ij}, y_i, y_j)$ is the edge or binary potential.

Using the Hammersley-Clifford fundamental theorem of random fields [13], the potential functions can now take the form

$$p_\Theta(y|x) = \frac{1}{Z(x)} exp\left( \sum_{i \in N, k} \mu_k f_k(x_i, y_i) + \sum_{ij \in E, k} \lambda_k g_k(x_{ij}, y_i, y_j) \right) \tag{3}$$

where $f_k$ and $g_k$ are vectors of local and pairwise features respectively and $\Theta = (\mu_1, \mu_2, ..., \mu_k; \lambda_1, \lambda_2, ..., \lambda_k)$ are the parameters or weights to be estimated from training data. Typically, the user designs the feature functions which are often real-valued or binary and the importance of each feature is represented in the weights. The above form of the potential functions guarantees non-negative values for the potentials.

There are two steps involved in using a CRF for the purpose of classification, namely, the parameter learning step and the inference step. The parameter learning step involves the estimation of the parameters $\Theta$ from training data. All cliques are made to share the same parameters to reduce the amount of training data required. Consider a training set $\{x^{(t)}, y^{(t)}\}$ that are independently and identically distributed. The distribution in equation (3) over all the training data as a function of the parameter set $\Theta$, is the likelihood given by $p(\{y^{(t)}\}|\{x^{(t)}\}, \Theta)$. The techniques maximum likelihood estimation (MLE) or maximum *a priori* estimation (MAP) can be used to estimate the parameter values from the training data. Typically, the logarithm of the likelihood is used for estimation. The log-likelihood for a CRF is given by

$$L(\Theta) = \sum_t \left[ \log \frac{1}{Z(x^t, \Theta)} + \sum_k \mu_k f_k(x^t, y^t) + \sum_k \lambda_k g_k(x^t, y^t) \right] \tag{4}$$

The above function is concave and hence there is a guaranteed convergence to the global maximum/minimum [19]. Parameter estimation in CRFs is an actively researched field and there exist various techniques. In this work, the maximum psuedo-likelihood method has been used to train the parameter set $\Theta$. For more details about this method, please refer [15], [2].

Once parameters have been estimated in the learning step, they can be used to infer the labels of an unlabelled data set. This is the inference step and is done by maximizing the conditional distribution of the labels given the feature vectors and the parameter set. This can be written as

$$Y = \arg\max_y p_\Theta(y|x) \tag{5}$$

where $Y$ is the array of labels for the nodes. Optimizations based on graph cuts [6], [12], [5], [16] are a popular method to do this kind of maximization as they are guaranteed to find the global maximum when binary labels are required.

## 3 Map Quality Assessment Using a CRF

The previous section outlined the general CRF framework as an undirected graphical model comprising of nodes and edges. In this section, the use of a pairwise CRF framework for the purpose of binary classification of portions of a given point-cloud map is described. As stated earlier, the given point-cloud is segmented into plane patches consisting of a subset of points. These plane patches constitute the nodes and their alignment with neighboring plane patches constitute the edges. When selecting the neighborhood for each plane, care must be taken that enough contextual information is provided while the computation costs of inference and learning are not compromised. In this work, the neighborhood system is defined as the *four* closest plane patches to the current plane patch with respect to euclidean distance within a radius of $s$ meters.

### 3.1 Plausible and Suspicious Portions of a Map

The binary labels considered for this work are "plausible" and "suspicious", that is, at the time of labelling the questions considered are – "is the subset of points acceptable as a plane patch?" and "is the alignment of the plane patch with respect to its neighbors reasonable?" Both these questions can be answered based on spatial geometry of the point-cloud. To check if a subset of points represent a plane patch, the metrics defined should measure how accurately the points fit a plane and whether the fitted plane is more two-dimensional than cubic in nature.

To check if the alignment between neighboring plane patches is reasonable, the various cases of alignments first needs to explored. These alignments can be broadly classified into plane patches that are parallel, those that are not parallel, but do not intersect and those that intersect. The alignments are shown in figure (1). Of these alignments, table (1) describes those that are considered acceptable and those that are not.

**Table 1.** Plausible and suspicious plane patch configurations

| Plausible | Suspicious |
|---|---|
| Parallel plane patches a reasonable distance apart. | Parallel plane patches separated only by a small distance. |
| Intersecting plane patches representing corners. | Intersecting plane patches that do not represent corners. |
| Plane patches that are not parallel but do not intersect. | |

These somewhat intuitive considerations based on geometry can be formalized within the CRF framework by way of the feature functions.

### 3.2 Feature Functions

This section describes the features that reconstruct the intuitive considerations into functions that can be used in the CRF as node and edge potentials. The feature vectors, $f_k$, $g_k$ along with the parameter set, $\Theta$ estimated in the learning step, is used to label the plane patches (refer equations (3),(4),(5) in the previous section). The features for the vector $f_k$ or node potential are defined by $x_{i1}$,..., $x_{ik}$ and the features for the vector $g_k$ or edge potential are defined as $x_{ij1}, ..., x_{ijk}$.

#### Node Features

Intuitively, these features describe the representation of the subset of points as a plane. Three geometric properties are considered:

1. the mean-squared error of fit, to evaluate how accurately the points fit a plane

$$x_{i1} = \frac{1}{\frac{1}{M_i}\sum_{j=1}^{M_i} d_{j \to i}^2} \tag{6}$$

2. the cosine-distance between the plane normal and the z-axis, to evaluate the direction of the plane as this work considers only vertical planes

$$x_{i2} = \frac{1}{\overrightarrow{Z}.\overrightarrow{n_i}} \tag{7}$$

3. and the aspect ratio of the plane patch as a measure of variation along the normal, to evaluate how two-dimensional the plane patch is

$$x_{i3} = \frac{V_i}{A_i} \tag{8}$$

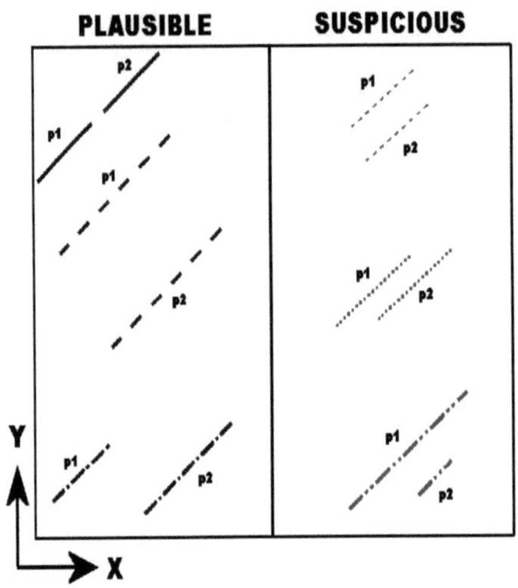


(a) Parallel Planes

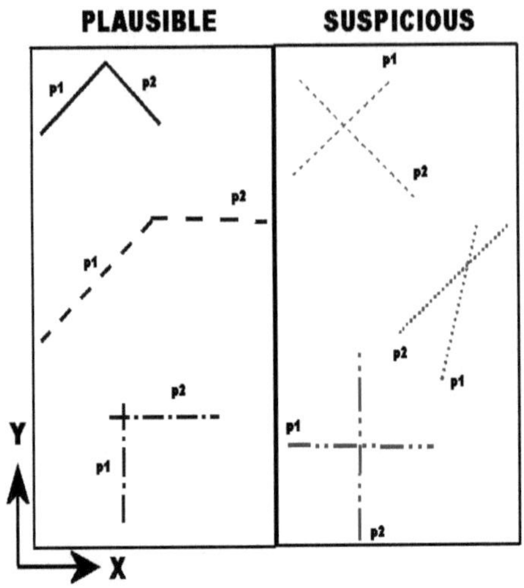


(b) Intersecting Planes

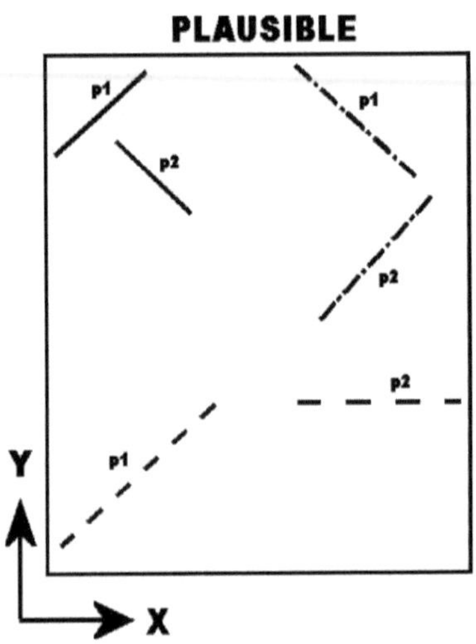


(c) Planes that are not parallel but do not intersect

**Fig. 1.** The figure shows the different alignments of neighboring plane patches in a plan view. The different hashing styles denote the neighboring pairs. The Z-direction is into the paper. All alignments considered as "plausible" are in blue and those considered as "suspicious" are in red.

In equations 6 - 8, $x_{i1}$, $x_{i2}$, $x_{i3}$ denote the features of the $i^{th}$ plane patch, $M_i$ refers to the number of points in the $i^{th}$ plane patch, $d_{j \to i}$ is the distance of each point $j$ in the $i^{th}$ plane patch to plane patch fitted, $Z$ is the Z axis vector, $n_i$ is the normal to the $i^{th}$ plane patch, $V_i$ is the volume of the $3D$ convex hull fitted to the points in the $i^{th}$ plane patch and $A_i$ is the area of the $2D$ convex hull fitted to the points of the $i^{th}$ plane patch, projected to the $XY$ plane.

**Edge Features**

The purpose of the edge features is to encode our intuitive understanding of how plausible or suspicious local plane configurations are (see Table 1).

Some measures that would give an idea of this alignment are the distance between the plane patches, the area of plane patches that overlap and if the plane patches intersect, the ratio of the distances between the patch edge and the line of intersection.

1. Plane separation: there are various methods of calculating the distance between adjacent plane patches. Here we consider only the shortest distance between points in both plane patches,

$$\rho_{ij} = min_{\forall_{m \in M_i} \forall_{n \in M_j}}(d_{mn}) \tag{9}$$

   where $\rho_{ij}$ denotes the closest distance between the plane patches $i$ and $j$, $M_i$ and $M_j$ represent the number of points in plane patches $i$ and $j$ respectively and $d_{mn}$ is the distance between points $m$ and $n$.

   Based on this distance, the a *plausibility measure* can be weighted such that plane patches close to each other are weighted 1.

$$W_{ij} = \exp(\frac{-\rho_{ij}}{\tau}) \tag{10}$$

   where $W_{ij}$ is the weight, $\rho_{ij}$ is the closest distance between the plane patches and $\tau$ is the distance at which plausibility of alignment between the plane patches becomes true.
2. Area of overlap: The maximum of the areas of the $i^{th}$ plane patch projected on the $j^{th}$ plane patch and of the $j^{th}$ plane patch projected on the $i^{th}$ plane patch is taken. This is then normalized with the minimum of areas of the $i^{th}$ and the $j^{th}$ plane patches projected onto the $XY$ plane. This is illustrated in figure (2(b)) and written as

$$A_{ij} = \frac{max(A_{i \to j}, A_{j \to i})}{min(A_i, A_j)} \tag{11}$$

   where $A_{ij}$ is the overlapping area, $A_{i \to j}$ is the area of $i^{th}$ plane patch projected on the $j^{th}$ plane patch, $A_{j \to i}$ is the area of the $j^{th}$ plane patch projected on the $i^{th}$ plane patch, $A_i$ and $A_j$ are the areas of the plane patches $i$ and $j$ respectively.
3. Location of line of intersection with respect to plane boundaries: this property concerns intersecting plane patches only and is defined as 0 when the plane patches are parallel. In this measure, the line of intersection between the patches is found and the distance between this line and the edges of the

patches are computed. The ratio between the minimum distance and maximum distance is taken. This is illustrated in figure (2(a)) and written as

$$R_{ij} = min(r_{ij}, r_{ji}) \tag{12}$$

where $R_{ij}$ is the minimum of ratio, $r_{ij}$ and $r_{ji}$, $r_{ij}$ is the ratio of the distances between the line of intersection and the boundaries of plane patch $p_1$ and $r_{ji}$ is the ratio of the distances between the line of intersection and the

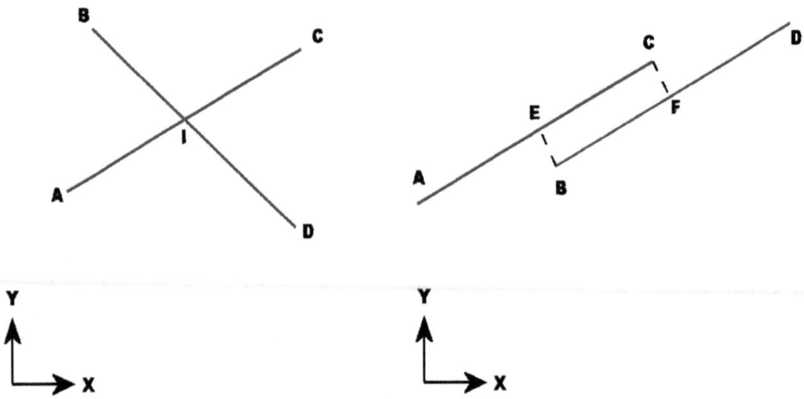


(a) Ratio Measure Calculation (b) Overlap Area Measure Calculation

**Fig. 2.** The figures illustrates the calculation of the ratio measure and overlapping area measure in the computations of $x_{ij}$, the measure of plausibility of geometry between the two plane patches. The planes patches, $AC$ and $BD$ are shown in plan view for better clarity with the Z-direction into the paper. In the ratio measure calculation, the line of intersection passes through point $I$. The ratio, $r_{ij}$ is $\frac{AI}{CI}$ and the ratio, $r_{ji}$ is $\frac{BI}{DI}$. The minimum of $r_{ij}$ and $r_{ji}$ gives the ratio $R_{ij}$. In the overlap area measure calculation, $A_{i\rightarrow j}$ is the area bounded by $EC$ of plane $AC$ and $A_{j\rightarrow i}$ is the area bounded by $BF$ of plane $BD.A_i$ is the area of plane $BD$ and $A_j$ is the area of plane $AC$. $A_{ij}$ is computed as given in equation (11).

The above properties can be combined into a quantitative *measure of plausibility* of the joint configuration of the $i^{th}$ and the $j^{th}$ plane patch, $x_{ij}$, as

$$x_{ij} = 1 - \{W_{ij}(\rho_{ij}) * (A_{ij} + R_{ij})\} \tag{13}$$

where the terms are as defined above. By definition, plausible configurations should attain high values and suspicious configurations should attain low values.

### 3.3 Model Learning and Inference

The set of CRF model parameters, $\Theta$, is learnt using a labelled training set of plane patches. For each patch in the set the unary features are evaluated using equ. 6 - 8 and the binary features (over a fixed size neighborhood) are evaluated using equ. 9 - 13. The so determined $x_i$ and $x_{ij}$ are substituted for $f_k$ and $g_k$ in equation 4 and the parameter set $\Theta$ is learnt using maximum psuedo-likelihood estimation [15].

Once the model parameters have been learnt the conditional distribution $p_\Theta(y|x)$ of new data can be classified by once again determining the relevant $x_i$ and $x_{ij}$ and substituting for $f_k$, $g_k$ and $\Theta$ in equation 3. This conditional probability is then maximized using Graph Cuts optimization [6] in order to determine the most likely set of labels $Y$ (equation 5).

## 4 Experimental Results

In this section, the effectiveness of the above algorithm is illustrated using data gathered in an urban environment by a mobile robot. A mobile platform equipped with a $3D$ SICK laser scanner was used to capture this data. The environment chosen was that of a typical urban environment consisting of office buildings, roadways, foliage, railings, people, moving cars and archways. The $3D$ point-cloud map built by the mapping algorithm was first segmented into plane patches using a region growing based approach [20]. Then these planes were manually labelled as "plausible" and "suspicious" based on how well the plane fitted the points and on the alignment of the planes with its neighbors. Approximately, $30,000$ points segmented into 150 planes that were representative of the data, was then used in the training step. Maximum psuedo-likelihood using the CRFtoolbox[1] was run until convergence and used to learn the parameter set, $\Theta$. On inspection of the parameter set, it was found that the weight for the mean-squared error of fit feature (equation (6)) was a factor of two less than the weights for the other node features. This is because of the relatively good plane segmentation performed. The weight for the edge feature was factor of ten higher than the the weights for the node features. This makes the labelling more sensitive to the alignments between the plane patches, which is desired.

Using this parameter set, $\Theta$, the plane patches from the test data were then labelled as "plausible" and "suspicious" in the inference step using Graph Cuts. Figure (3) shows a magnified section of a map that has been classified by the algorithm. The "plausible" portions are marked in blue and the "suspicious" portions in red. For clarity, the plan view of this portion is also shown above the $3D$ view. In figures (3(a), 3(c)), the "plausible" region of the map consists of a stretch of wall on either side of the robot's pathway. Since these two plane patches are at a reasonable distance apart, the classifier correctly labels this portion as "plausible". In figures (3(b), 3(d)), the two planes are intersecting. However, this does not represent a corner and it is not acceptable for two plane patches to intersect like this. These two plane patches are correctly labelled as "suspicious".

Figure (4) shows a map consisting of 1323 planes that has been classified by the algorithm. The same color code of blue for "plausible" and red for "suspicious"

[1] http://cs.ubc.ca/ murphyk/Software/CRF/crf.html

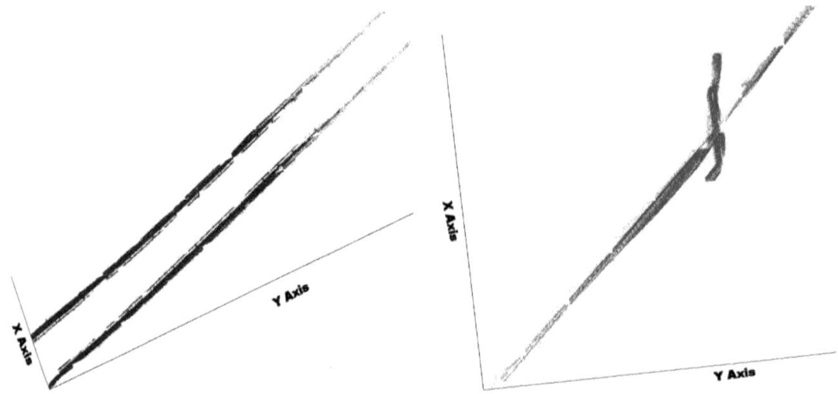


(a) Plan view of magnified section of map labelled “plausible”

(b) Plan view of magnified section of map labelled “suspicious”

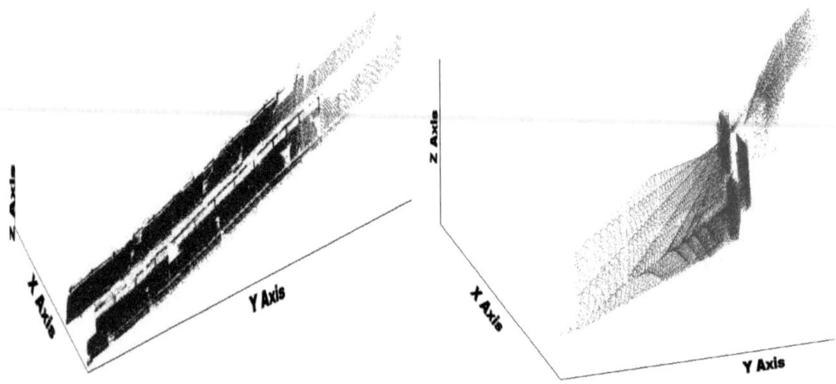


(c) $3D$ view of magnified section of map labelled “plausible”

(d) $3D$ view of magnified section of map labelled “suspicious”

**Fig. 3.** Figure shows the magnified section of a map in $3D$ that has been classified by the algorithm. The “plausible” portions are marked in blue and the “suspicious” portions in red. For clarity, the plan view of this portion is also shown, above the magnified portion. In (a, c), the two plane patches are at a reasonable distance apart. Hence the algorithm correctly labels this portion as “plausible”. In (b, d), there are two planes that intersect, however not representing a corner. This is correctly labelled as “suspicious”.

is used here as well. The pathway of the robot consists of an initial stretch of road lined with office buildings. The robot then goes around a building twice before continuing further down the path under an archway. The point-cloud is rendered from odometry.

Of particular interest are the regions marked $A$, $B$, $C$, $D$, $E$ and $F$. The region $A$ shows the region in the map that is marked as “plausible”. The environment is this region consists of a office buildings lined on either side of the road. It

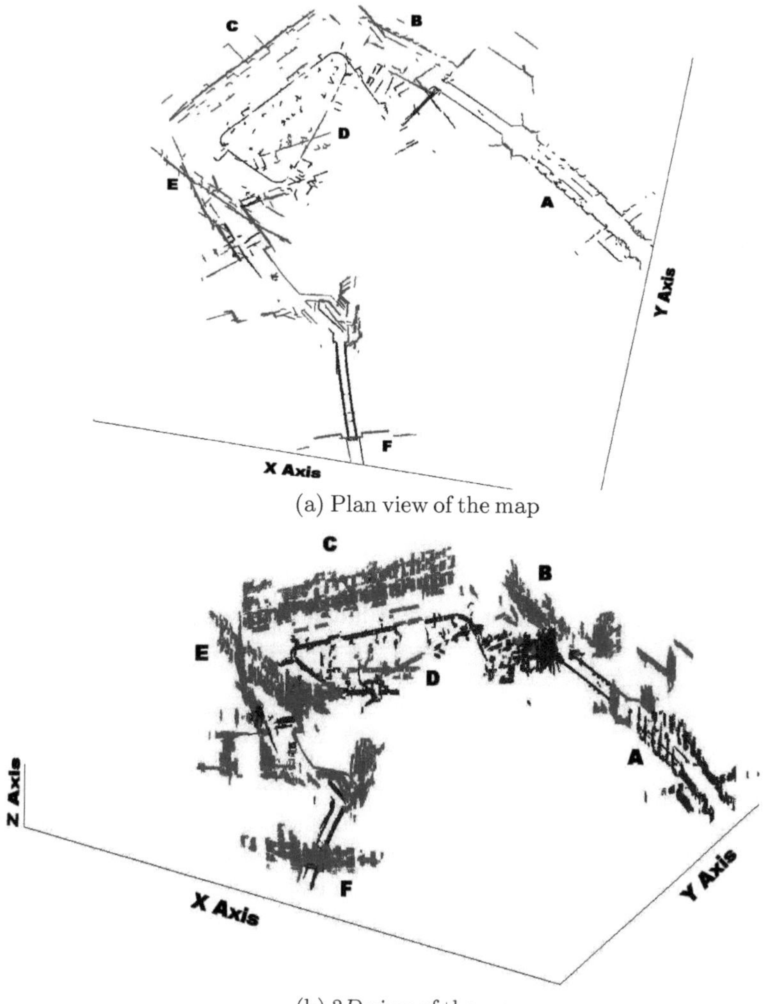


(a) Plan view of the map

(b) 3*D* view of the map

**Fig. 4.** Figure shows a 3*D* view of a map consisting of 1323 planes that has been classified by the algorithm. The plan view is shown above the 3*D* view for better clarity. The same color code of blue for "plausible" and red for "suspicious" is used here. Of particular interest are the regions marked *A*, *B*, *C*, *D*, *E* and *F*. In region *A*, the point-cloud alignment is in keeping with the environment and the algorithm has correctly labelled this region as "plausible". The regions *B*, *D* and *E* all have plane patches intersecting with each other due to the misalignment caused by the re-traversal of the robot. The three regions are correctly labelled "suspicious". When two plane patches are parallel and very close to each other, it is labelled as "suspicious". Hence region *C* is labelled "suspicious". Region *F* actually represents an archway in the environment. Since the map quality algorithm considers only simple plane geometry, this region is incorrectly labelled as "suspicious". Future work will include features to consider this case.

is seen that the point-cloud alignment is in keeping with the environment and the algorithm has correctly labelled this region as "plausible" (blue in color). The next portion of the map shows the re-traversal of the robot on that path causing misalignment of the plane patches. The regions $B$, $D$ and $E$ all have plane patches intersecting with each other which are considered as "suspicious" alignment in the algorithm. These the algorithm labels correctly as "suspicious" (red in color). The region $C$ is labelled as "suspicious" due to there being plane patches separated by a very small distance. The region $F$ is actually an archway in the environment. The segmentation algorithm returns planes that are intersecting in this region. However, since the map quality algorithm considers only simple plane geometry, this region is considered as two plane patches intersecting at the middle of the patches and hence labelled "suspicious". Future work will include features to consider cases such as archways as well. It is seen that the classifier labels the gross errors in mapping as "suspicious".

**Table 2.** Confusion Matrix for the map shown in Figure(4) consisting of 1323 planes. The percentage of correct classification was found to be 80.42%.

| Inferred \ Truth | Plausible | Suspicious |
|---|---|---|
| **Plausible** | 94.18% | 5.81% |
| **Suspicious** | 34.48% | 65.51% |

The percentage of correct classification was found to be 80.42%. Table (2) summarize the results achieved in terms of a confusion matrix. When the classifier returns a "plausible" label for a plane patch, it was found that 94.18% of the times the correct label was "plausible". The false positive rate was found to be 5.81%. Similarly, when the classifier returns a "suspicious" label for a plane patch, it was found that 65.51% of the times it was the correct label and the false negative rate was 34.48%. This is due to a conservative labelling of training data as only those planes that are definitely "plausible" should be labelled as that.

## 5 Conclusions

This work addresses the problem of measuring the quality of maps in retrospection. A measure of map quality is necessary for assessing how accurately the environment has been mapped, once the map is built. In this paper, a $3D$ point-cloud map is classified into binary classes of "plausible" and "suspicious". The problem of classifying the regions of the given map is modelled as a Conditional Random Field framework, by segmenting the map into plane patches and considering these plane patches as nodes. Since CRFs capture the neighborhood dependencies, the alignment between patches also contribute to the classification. A probability distribution of the array of labels is obtained which is maximized using graph cuts. The method is illustrated using data gathered by

$3D$ laser scanners from an urban environment. The algorithm labels the gross errors in mapping as "suspicious". The particular case of archways is erroneously classified as "suspicious". Future work will include features that considers this case as well.

## Acknowledgments

The authors would like to thank Dr.Ingmar Posner for his comments during this work as well as Mark Cummins for his help with the labelling of the laser data.

## References

1. Anguelov, D., Taskar, B., Chatalbashev, V., Koller, D., Gupta, D., Heitz, G., Ng, A.: Discriminative learning of markov random fields for segmentation of 3d scan data. In: CVPR 2005: Proceedings of the 2005 IEEE Computer Society Conference on Computer Vision and Pattern Recognition (CVPR 2005), vol. 2, pp. 169–176. IEEE Computer Society, Washington (2005)
2. Besag, J.: Statistical analysis of non-lattice data 24, 179–195 (1975)
3. Besag, J.: On the statistical analysis of dirty pictures. Journal of the Royal Statistical Society 48(3), 259–302 (1986)
4. Bishop, C.M.: Pattern Recognition and Machine Learning (Information Science and Statistics). Springer, New York (2006)
5. Boykov, Y., Kolmogorov, V.: An experimental comparison of min-cut/max-flow algorithms for energy minimization in vision. IEEE Trans. Pattern Anal. Mach. Intell. 26(9), 1124–1137 (2004)
6. Boykov, Y., Veksler, O., Zabih, R.: Fast approximate energy minimization via graph cuts. IEEE Trans. Pattern Anal. Mach. Intell. 23(11), 1222–1239 (2001)
7. Burges, C.J.C.: A tutorial on support vector machines for pattern recognition. Data Mining and Knowledge Discovery 2(2), 121–167 (1998)
8. Geman, S., Geman, D.: Stochastic relaxation, gibbs distribution and the bayesian restoration of images. IEEE Transactions on Pattern Analysis and Machine Intelligence, PAMI 6(6), 721–741 (1984)
9. Gutmann, J., Konolige, K.: Incremental mapping of large cyclic environments. In: Int. Symposium on Computational Intelligence in Robotics and Automation (1999)
10. Hähnel, D., Triebel, R., Burgard, W., Thrun, S.: Map building with mobile robots in dynamic environments. In: Proc. of the IEEE International Conference on Robotics and Automation (ICRA) (2003)
11. Kinderman, R., Snell, J.: Markov random fields and their applications. American Mathematical Society, Providence, RI (1980)
12. Kolmogorov, V., Zabin, R.: What energy functions can be minimized via graph cuts? Pattern Analysis and Machine Intelligence. IEEE Transactions on 26(2), 147–159 (2004)
13. Lafferty, J., McCallum, A., Pereira, F.: Conditional random fields: Probabilistic models for segmenting and labeling sequence data. In: Proc. 18th International Conf. on Machine Learning, pp. 282–289. Morgan Kaufmann, San Francisco (2001)
14. Leonard, J.J., Durrant-Whyte, H.F., Cox, I.J.: Dynamic map building for an autonomous mobile robot. Int. J. Rob. Res., 11(4), 286–298 (1992)

15. Liao, L., Fox, D., Kautz, H.: Extracting places and activities from gps traces using hierarchical conditional random fields. The International Journal of Robotics Research 26(1), 119–134 (2007)
16. Szeliski, R., Zabih, R., Scharstein, D., Veksler, O., Kolmogorov, V., Agarwala, A., Tappen, M., Rother, C.: A comparative study of energy minimization methods for markov random fields. In: European Conference on Computer Vision, vol. 2, pp. 16–29 (2006)
17. Triebel, R., Kersting, K., Burgard, W.: Robust 3D scan point classification using associative markov networks. In: Proc. of the International Conference on Robotics and Automation (ICRA) (2006)
18. Viola, P., Jones, M.: Robust real-time object detection. International Journal of Computer Vision (2002)
19. Wallach, H.M.: Conditional random fields: An introduction (2004)
20. Weingarten, J., Gruener, G., Siegwart, R.: A fast and robust 3D feature extraction algorithm for structured environment reconstruction. In: Proc. of the 11th International Conference on Advanced Robotics (ICAR) (2003)

# Part II

# Vision for Navigation

# Robust Vision-Based Underwater Target Identification and Homing Using Self-Similar Landmarks

Amaury Negre[1], Cédric Pradalier[2], and Matthew Dunbabin[2]

[1] INRIA Rhônes-Alpes, France
firstname.lastname@inrialpes.fr
[2] Autonomous Systems Laboratory, CSIRO ICT Centre, Australia
firstname.lastname@csiro.au

**Summary.** Next generation Autonomous Underwater Vehicles (AUVs) will be required to robustly identify underwater targets for tasks such as inspection, localisation and docking. Given their often unstructured operating environments, vision offers enormous potential in underwater navigation over more traditional methods, however, reliable target segmentation often plagues these systems. This paper addresses robust vision-based target recognition by presenting a novel scale and rotationally invariant target design and recognition routine based on Self-Similar Landmarks (SSL) that enables robust target pose estimation with respect to a single camera. These algorithms are applied to an AUV with controllers developed for vision-based docking with the target. Experimental results show that system performs exceptionally on limited processing power and demonstrates how the combined vision and controller systems enables robust target identification and docking in a variety of operating conditions.

## 1 Introduction

Target identification and homing are of particular interest in our research as we desire an Autonomous Underwater Vehicle (AUV) fitted with a camera and vision processing capabilities to accurately locate and home onto points of interest in relatively cluttered environments such as coral reefs, or in close proximity to subsea oil and gas structures. Due to issues such as multi-pathing and variable lighting, the performance of typical identification and homing methods such as acoustics or feature and color-based vision systems can degrade significantly. The ability to robustly identify objects in these environments can allow tasks such as reliable localisation, inspection, object collection and docking to complex structures.

Underwater targets for docking/homing can be either active or passive and are typically identified using either acoustics or vision systems. Active (energy emitting) targets are most common in literature due to their generally larger detection range, whereas passive targets can be further classified as artificial or natural and are typically detected using vision systems at closer distances.

Acoustics and vision are the most common ways of identifying and homing onto targets. Acoustics are advantageous in that they can operate over long ranges and in a variety of water conditions (visibility and lighting). However, for docking, acoustics degrade at very close ranges ($< 5$m) and in cluttered environments such reefs or close to the seafloor and subsea structures. Vision-based target identification has complimentary

C. Laugier and R. Siegwart (Eds.): Field and Service Robotics, STAR 42, pp. 51–60, 2008.
springerlink.com 

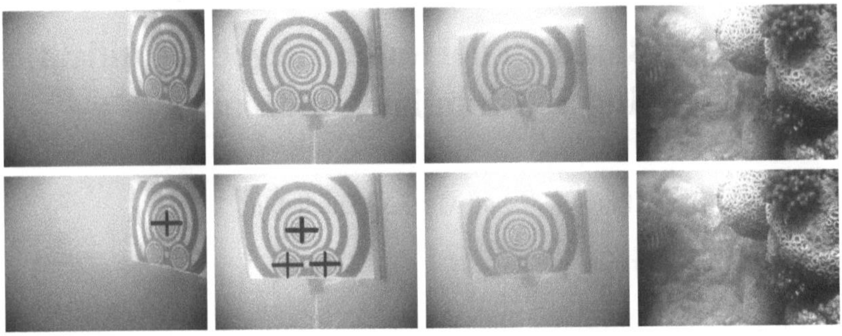

**Fig. 1.** SSL detection in marine environment. Upper row: original images, lower row: detection. From left to right: robustness to perspective and lighting discontinuity, minimum required contrast, insufficient contrast, absence of false positives in a natural environment.

properties. Vision is suitable for close range tracking and can accommodate changing and multiple targets, however, its performance degrades in turbid water and poor lighting conditions.

There are many examples of target identification and homing in the literature. The simplest approach is to home to an acoustic target using range and heading information from the target [8]. Although a reliable method at larger distances, at close ranges the performance becomes impractical for docking due to the high update rates required. Alternatively, Feezor [5] successfully demonstrated homing and docking using electromagnetic guidance in which an AUV was fitted with coils to sense field strength and orientation that enabled docking from a distance of 30m.

Improvements in the docking performance of AUVs have been considered by combining acoustics with vision [4]. Here, the acoustics/sonar provides the longer range homing direction with the vision providing guidance in the final stages of docking. A survey of vision-based target identification and tracking literature is provided by Dalgleish [2] describing the optic flow and feature based techniques commonly used for underwater tracking applications.

Many vision-based target identification schemes require active targets. Lee [7] uses a monocular vision system to identify a large circular target with an LED ring and 5 large lights, whereas Wang [9] describes a stereo vision system to locate and retrieve an underwater object which has a light emitting beacon on its surface.

Considering all these methods of target identification and homing, it was decided that our system would require passive targets due to energy requirements on behalf of the AUV and target itself for long duration operations. Additionally, when repeat visitations of a target over extended periods of time are required, natural landmarks can change significantly, requiring artificial landmarks. However, detection is required at relatively large distances and different orientations making color-based methods unreliable. Therefore, the target and identification routine must be robust, scale and rotationally invariant, as well as capable of running in real-time on an AUV's limited processing power. This paper considers the use of Self-Similar Landmarks (SSL) [1] as a robust, color, scale and rotational invariant means of target identification from which the information can be used to guide and AUV for homing and docking operations.

## 2 Self-Similar Landmarks

The notion of self-similar landmarks was first used in a robotic context by Scharstein and Briggs [1]. Their objective was to develop planar targets that would be detected easily with a standard perspective camera on a mobile indoor robot. To this end, the targets were designed to be invariant to change of scale. This is where the self-similarity is essential.

A p-similar function for $0 < p < 1$ is a function $f : \mathbb{R}^+ \to \mathbb{R}$ such that $\forall x > 0, f(x) = f(p \cdot x)$. It is essential to note that the p-similarity is invariant to change of scale ($\forall x > 0, \forall k > 0, f(k \cdot x) = f(p \cdot k \cdot x)$). In the context of a computer vision application, this property is interesting since, if a method to detect self-similarity is available, it will be possible to do so from whatever distance since, with a pin-hole camera, a change of range results in a change of scale in the observed image.

When designing a self-similarity detector, the fact that all constant functions are $p$-similar for all $p$ is problematic since any uniform region in an image will give a string response on the self-similarity detector. To solve this, [1] introduces the notion of anti-similarity: a function $f$ is $p$-antisimilar if $\forall x > 0, |f(x) - f(p \cdot x)| = 1$.

The $p$-similarity and $\sqrt{p}$-antisimilarity are then used by Scharstein and Briggs to define the "Self-similar square wave" function which is both p-similar and $\sqrt{p}$-antisimilar:

$$s_p(x) = \lfloor 2(\log_p x - \lfloor \log_p x \rfloor) \rfloor \tag{1}$$

This function maps $\mathbb{R}^+$ to $\{0, 1\}$. It is used to print black and white landmarks, as shown in fig. 2(a).

To recognise the self-similar landmarks, [1] defines a matching function that will respond strongly to a sequence of pixels looking like a $p$-similar and $\sqrt{p}$-antisimilar function. For a pixel located at $(x, y)$ in image $I$, this is achieved by integrating the $p$-similarity and $\sqrt{p}$-antisimilarity conditions on a window of length $w$ starting at $(x, y)$:

$$\begin{aligned} m^w_{+x}(x,y) = & \frac{1}{w}\int_0^w |I(x+\xi,y) - I(x+\sqrt{p}\cdot\xi,y)|\,d\xi \\ & - \frac{1}{w}\int_0^w |I(x+\xi,y) - I(x+p\cdot\xi,y)|\,d\xi \end{aligned} \tag{2}$$

In this equation, the window is made of the $w$ pixels after $(x, y)$, on an image line in the $x$-direction. We identify this function by the $+x$ subscript. The function can be obviously modified to use the $w$ pixels before, or vertically above, or below $(x, y)$. We name these functions $m^w_{-x}$, $m^w_{-y}$ and $m^w_{+y}$ respectively. In practical implementations, $w$ takes values in the $[20, 100]$ pixel range. As in [1], we use 40 as a good balance between robustness and computation cost.

### 2.1 A Modified SSL for Pose and Range Estimation

The two-dimensional landmark used by Briggs is self-similar on one dimension (horizontal for example) and constant on the other one. This landmark is not optimal for many applications, particularly underwater:

- This is not a point landmark and consequently it is hard to localise precisely in the image. In [1], the border of the landmark is detected, and a binary coding is used to identify the landmark. Our practical experience in outdoor experiments found this to be the weakest part of the algorithms.
- This brings us to our second point, this landmark is not robust to rotations. Since only the $m^w_{+x}$ function is used, the landmark can only be detected when the apparent rotation is below 45 degrees (from our experiments).
- Also, this landmark is not robust to motion blur: a blur in the direction of the $p$-similar function can make the landmark detection fail.

(a) Original SSL target (b) Circular 3-pattern SSL target. (c) SSL detector output

**Fig. 2.** The original(a) and circular SSL(b) target patterns used in this investigation and the output(c) of our circular SSL detector

In order to solve those problems, we designed a circular landmark where the intensity $I$ is self-similar and anti-similar in all directions:

$$I(\rho, \theta) = s_p(\rho) \tag{3}$$

where $\rho, \theta$ are the polar coordinate and $s_p$ the self-similar square wave function defined in Equation 1.

This landmark is interesting because it is rotationally invariant thanks to the circular geometry and the matching function exhibits a single maximum point in the center. Moreover, to improve robustness to noise and motion blur, we can apply the matching function in several directions (e.g. top, bottom, left and right).

In our application we also need to estimate the range and the pose of the target. Range could be estimated using a single circular landmark and stereo camera system. However, detection of distant objects requires large camera baselines and this also doubles the image processing requirements. In this research, a monocular vision was chosen to reduce computation burden and allow portability to other platforms. As we can detect only the center of the pattern we need at least 3 patterns to estimate the pose[1]. Our final

[1] It is well known that at most 8 poses will be consistent with the observation[6].

target (shown in Figure 2(b)) consists of 3 circular self-similar patterns on an equilateral triangle: one large pattern for detection at greater distances and two small patterns for the pose estimation at short range.

## 3 SSL Target Identification and Tracking

### 3.1 Target Detection

The first step in the detection algorithm consists of applying the matching function on every pixel of the image in four perpendicular directions:

$$M^w(x,y) = m^w_{+x}(x,y) + m^w_{-x}(x,y) + m^w_{+y}(x,y) + m^w_{-y}(x,y) \tag{4}$$

Next, the location of the local maxima of the matching function is determined with a threshold function applied to remove small outlier peaks.

The results and a typical response of the matching function are shown in Fig. 2(c). In our testing, this target appeared robust to occlusion (as long as the center point is visible), deformation resulting from bending or shaking the paper landmark, camera model (the same detection software worked unmodified on various focal length pin-hole cameras in air or water, and also on fish-eye and catadioptric cameras) and lighting change. For underwater detection, the only possible problematic visual effects are water turbidity and the reflection of the landmark by the water-air interface.

### 3.2 Pose and Range Estimation

Once the position of the landmarks in the image has been found, we then need to evaluate the pose of the target with respect to the camera. Three cases are possible according to the number of visible landmarks.

Case 1: only one visible landmark When the target is far from the camera, we typically detect only the large landmark. Therefore, when using a monocular camera, only the bearing can be determined.

Case 2: two visible landmarks In the case where only two landmarks are visible, there is no reliable way to detect which ones are seen. Consequently, we treat this case as if only the larger landmark was visible.

Case 3: three visible landmarks When the target is close enough, all landmarks are visible and it is possible (with some ambiguities) to compute the pose from three points.

### 3.3 Target Tracking

The underwater environment and the motion of the submarine make the target detection subject to noise and misdetection. In order to be less sensitive to these problems, we developed a target tracking system based on a particle filter (see [3] ).

Each particle represents a pose of the target in the robot reference, coded by a 7-dimensions vector: the first four coordinates represent the rotation as a quaternion and the last three coordinates represent the 3D position of the centroid of the target.

The particle filter is initialized when the system detects at least one landmark for the first time (or after a long period). We then initialise all particles around the position estimated as explained in Section 3.2. For the prediction phase of the particle filter, we use the odometric sensors (accelerometers, compass, and depthmeter) to update the particles with the estimated motion blurred with a gaussian noise. In the updating phase, the weight of the particles is estimated by projecting the virtual landmarks represented by the particle in the image and evaluating the distance to the observed landmark.

# 4 Practical Limitations of SSL in Field Robotics

Using SSL is computationally costly. The detection requires $O(n^2 w)$ non-sequential image accesses, where $n$ is the image size and $w$ the size of the integration window. The non-sequential pixel indexing aspect is critical for an optimized implementation but it prevents an efficient implementation using specialised processor instruction such as *MMX*, *SSE* or *SSE2*[2] and also prevents efficient caching of image data.

## 4.1 Improving Performance

As mentioned previously, the processor requirement is important to compute the self-similar matching function in real-time. Table 1 lists the measured performance of the proposed target identification technique on different computing platforms and image sizes.

To improve performance, a simple way is to only compute this function in a reduced Region Of Interest (ROI) in the image. Using the particle filter, we can predict the position of the 3 landmarks in the image and reduce the search region around these 3 positions. Typically, in our application the ROI is a 100x100 pixel rectangle around each predicted point, dividing the computation time by 3 for an image 640x480 (see Table 1). Nevertheless, to avoid the risk of losing the target, we maintain a search in the whole image at least every 10 frames.

**Table 1.** Computation time for different processor and different image sizes

| Processor | image size | FPS for whole image | FPS with tracking (ROI 100x100 pixels) |
|---|---|---|---|
| Pentium IV 3GHz | 640x480 | 2.8 | 8.7 |
| | 320x240 | 11.1 | 18.7 |
| Pentium M 1.4GHz | 640x480 | 2.1 | 6.8 |
| | 320x240 | 8.9 | 15.4 |

# 5 Experimental Results

The vision-based target identification algorithm presented above was evaluated on an AUV in semi-controlled (test tank) conditions. The purpose of these experiments was

[2] Our comparison of implementation using floating point operations, integer operations, MMX, SSE and SSE2 showed compiler optimised floating point implementation to be the most effective on our platform.

to evaluate the robustness and performance of the circular SSL tracking system, as well as the AUV's target identification and homing/docking capabilities under different operating scenarios.

The Starbug AUV was the platform used for these experiments. Starbug, shown in Fig. 3, uses vision as its primary sensory mode for navigation. The vision system consists of two stereo camera pairs, one looking forward along the $x$-axis for obstacle avoidance, and the other looking downwards along the $z$-axis for visual odometry and terrain following.

As the SSL algorithm requires only monocular vision, only one of the forward ($x$-axis) cameras was used in these experiments. Additionally, the roll, pitch and yaw measurements from the on-board IMU were made available to the target tracking algorithm. All visual and inertial measurements are shared amongst all the AUV's sub-systems using the DDX middleware.

The target consisted of the circular SSL geometry (Fig. 2(b)) printed on A3 paper which was laminated and glued to a rigid backing. Anchors and floats attached to the target enabled the distance from the sea-floor to be set, however, it could move with water currents.

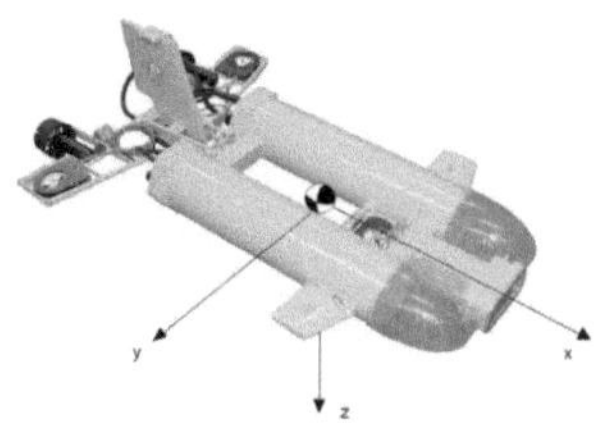


**Fig. 3.** The Starbug AUV used for docking experiments showing local coordinate system

## 5.1 Target Homing and Stand-Off Experiments

The experiments consisted of developing a target searching routine for the AUV and when detected, use the vision-based SSL tracking algorithm to guide (home) the AUV towards the target and maintain a stand-off distance from it.

The AUV target identification and docking behaviour consisted of three states; State(1) is the target search mode whereby the AUV performs a constant rate spin about its $z$-axis at varying depths. The AUV enters State(2) when the target has been identified and remains until the AUV's $x$-axis is pointing at the identified target's centre. Finally, State(3) is the homing/docking mode where the AUV moves towards the target and maintains a prespecified distance from the target.

Figure 4 shows results of a typical autonomous target identification and homing experiment in which it is desired to maintain a distance of 2m from the target. Here the target was placed at one end of the test tank and the AUV at the other. Figure 4 shows the target tracking state as well as demanded (dashed) and actual (solid) AUV yaw angle and depth. Additionally, the number of SSL patterns found (nbPoints) in the image is shown along with the estimated range from the target identification algorithm (see Section 3).

From Figure 4, it can be seen that the target is first identified at $t$=75s and as the AUV moves towards the target, the 2m standoff distance is maintained. At $t$=140s, the target was removed from the water and the AUV re-entered the search mode until $t$=225s where the target was replaced and the AUV reacquired and tracked it. This experiment was repeated many times with the system consistently able to identify and home onto the target.

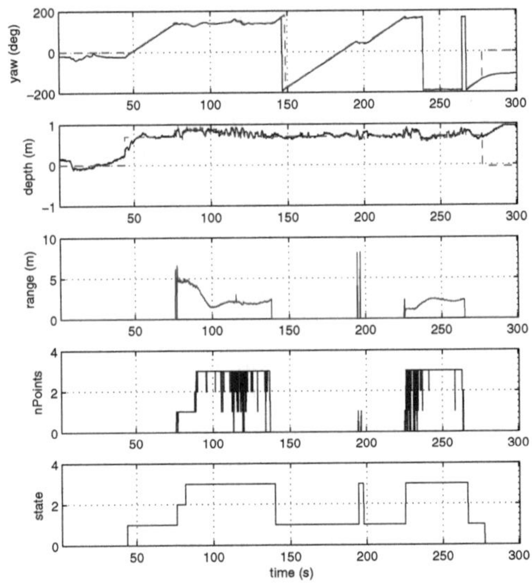


**Fig. 4.** AUV pose and SSL detection properties during a typical target identification and homing experiment. Here the AUV was required to identify the target and maintain a stand-off distance (range) of 2m.

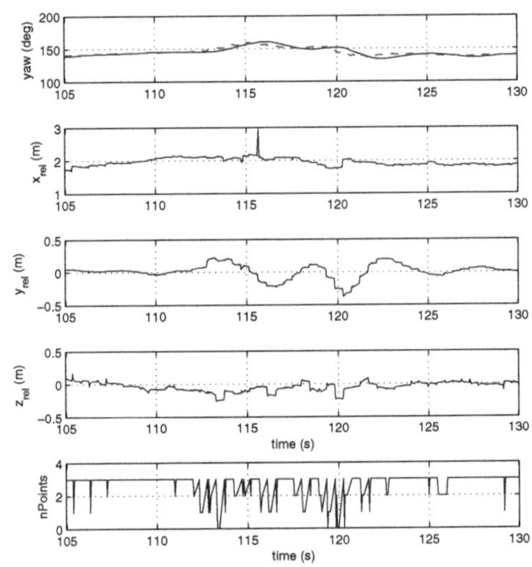


**Fig. 5.** Target tracking performance showing target position relative to AUV coordinate frame moving the target left and right and orienting it relative to the AUV

To demonstrate the robustness of the target tracking, during maintaining the 2m standoff distance shown in Fig. 4, the target was moved left then right before being returned to its original position. Figure 5 shows the AUV yaw angle as well as the estimated target position relative to the AUV coordinate frame. Here the lateral ($y_{rel}$) position of the target moves as the target moves left and right and the yaw angle varies to maintain the target directly in front of the AUV.

Finally, the ability of the system to dock with the target is demonstrated in Fig. 6 by setting the stand-off distance to zero. The figure shows the actual and demanded AUV yaw and depth, as well as the estimated range to the target and number of SSL circles tracked. Additionally, the ADC value of the frontal collision avoidance sensor is shown. A collision is detected when the ADC value exceeds 300. The AUV detects a collision with the target at $t$=511s. Again this experiment was repeated several times with consistent performance demonstrated.

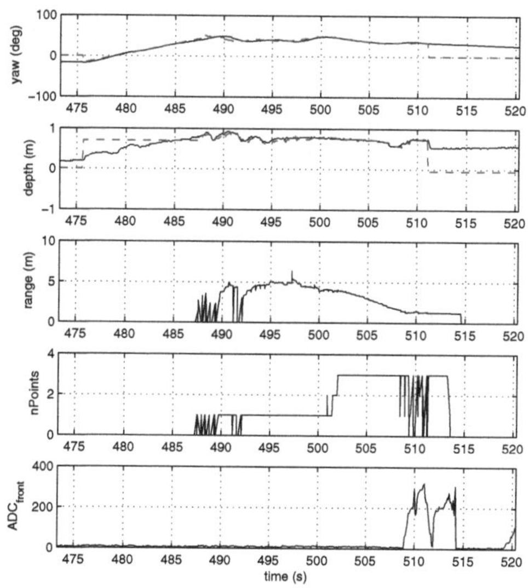


**Fig. 6.** AUV pose and SSL detection properties during a typical docking experiment where the AUV identifies and moves towards the target until the frontal collision sensor was triggered

# 6 Conclusions

Underwater vision-based tasks are typically complex to implement due to poor lighting conditions, refractions and moving objects such as fish. In this paper we developed a rotationally invariant circular self-similar landmark and demonstrated its use for target identification and in enabling vision-based docking using the Starbug AUV. The method provides an exceptionally robust landmark with very little sensitivity to camera model, distortion and observation range. The resulting docking task was proven effective through extensive pool trials. In May 2007, an experiment in uncontrolled reef environment has been conducted to demonstrate the applicability of our approach in the field. The SSL detection proved effective (see fig. 1), but our control was not reactive enough to perform the docking in presence of current. This problem will be approached in future research.

## References

1. Briggs, A., Scharstein, D., Braziunas, D., Dima, C., Wall, P.: Mobile robot navigation using self-similar landmarks. In: Proc. Int. Conf. on Robotics and Automation (2000)
2. Dalgleish, F., Tetlow, S., Allwood, R.: Vision-based navigation of unmanned underwater vehicles: A survey. part 2: Vision-based station-keeping and positioning. Journal of Marine Design and Operations (2005)
3. Doucet, A., De Freitas, N., Gordon, N. (eds.): Sequential Monte Carlo methods in practice (2001)
4. Evans, J., Redmond, P., Plakas, C., Hamilton, K., Lane, D.: Autonomous docking for Intervention-AUVs using sonar and vision-based real-time 3D pose estimation. In: Proc. OCEANS (2003)
5. Feezor, M., Sorrell, Y., Blankinship, P., Bellingham, J.: Autonomous underwater vehicle homing/docking via electromagnetic guidance. Journal of Oceanic Engineering (2001)
6. Hartley, R., Zisserman, A.: Multiple View Geometry in Computer Vision (2000)
7. Lee, P.-M., Jeon, B.-H., Kim, S.-M.: Visual servoing for underwater docking of an autonomous underwater vehicle with one camera. In: Proc. OCEANS 2003 (2003)
8. Stokey, R., Purcell, M., Forrester, N., Austin, T., R.G., Allen, B., van Alt, C.: A docking system for REMUS, an autonomous underwater vehicle. In: Proc. Oceans (1997)
9. Wang, H., Rock, S., Lee, M.: Experiments in automatic retrieval of underwater objects with an AUV. In: Proc. OCEANS 1995 MTS/IEEE (1995)

# Vision-Based Handling Tasks for an Autonomous Outdoor Forklift

Cédric Pradalier

CSIRO ICT Centre, Autonomous Systems Laboratory, Brisbane, Australia

**Summary.** In the aluminium industry, Hot Metal Carriers (HMCs) perform the task of transporting molten aluminium from the smelter (where the aluminium is made) to the casting shed where it is turned into block products. The vehicles weigh approximately 20 tonnes unloaded and resemble forklifts except they have a dedicated hook for manipulating the load, called the crucible. In the perspective of a full automation of this kind of vehicle, this paper main contribution is the implementation of robust outdoor crucible handling capacities, relying on the autonomous tracking of especially designed fiducials attached to the crucible handle.

## 1 Introduction

**Fig. 1.** Hot Metal Carrier picking up a crucible

Hot Metal Carriers (HMCs, see fig. 1) perform the task of transporting molten aluminium from the smelter (where the aluminium is made) to the casting house where it is turned into block products. The vehicles weigh approximately 20 tonnes unloaded and resemble forklifts except they have a dedicated hook for manipulating the load rather than fork tines. The molten aluminium is carried in large metal crucibles. The crucibles weigh approximately 2 tonnes and they can carry 8 tonnes of molten aluminium at 700 degrees Celsius. Therefore HMC operations are considered heavy, hot, and hazardous, with safety of operation a significant issue.

Our long term objective is the full automation of a HMC in order to fulfil tasks coherent with its industrial usage. The current state of our HMC automation has been described in [6]. This paper will address especially the implementation of robust outdoor crucible handling capacities based on computer vision. In an industrial context, it may be argued that laser-based implementation would provide better overall robustness. Nevertheless, we chose computer vision for its versatility and wider field of view.

C. Laugier and R. Siegwart (Eds.): Field and Service Robotics, STAR 42, pp. 61–70, 2008.
springerlink.com 

The vehicle is equipped with a mast that supports the pickup hook. The mast is controllable in tilt and the hook is also controllable in height. Mechanically, the hook is linked to the top of the mast and raising the top of the mast with an hydraulic cylinder raises the hook. These mechanical aspects of the vehicle will be important for the remaining of the paper since we equipped the vehicle with two pan/tilt/zoom cameras attached to the top of the mast. This is required in order to have a good vantage point toward the pickup scene, but as we will see later, this introduces complexity in the evaluation of the exact position of the camera in the vehicle frame of reference.

### 1.1 Related Work

As expected for such a widely used industrial task, forklift handling has attracted a lot of attraction in the robotics research community, as well as in the industry, and commercial products are already available. These products are, to our knowledge, mainly using laser range finders for most of their activities and limited to indoor environments.

In the research community, some work can be found on vision-based cargo handling operations. The list below is not exhaustive but rather an overview of what we have found to be typical approaches. Garibotto *et al.* [3] worked on an indoor forklift detecting and localising pallets using vision. The localisation is achieved by detecting a H-shaped landmark painted on the ground and the pallets are identified by the dark appearance of the tine engagement holes. This method is not generic enough to be adapted to our cargo and we estimated that then relying on shadows for outdoor vision based operations is unlikely to provide the required robustness.

Also, Seelinger *et al.* [5] presented a fiducial-based pallet localisation system. The fiducial they describe are circular discs used in the motion capture industry. They provide very good response to corner detectors such as Harris' detector, and may be good enough for an indoor controlled environment, but they are not distinctive enough for reliable outdoor detection.

Most of the control algorithms discussed in the above-mentioned research into load handling could be applied to our crucible handling task. However, the pallet sensing methods are not suitable due to their low saliency and reliability in an outdoor, industrial setting. Nevertheless, for an industrial application we decided that the most reliable approach was to use fiducials attached to the payload, and to design them so as to maximise the distinctiveness in the environment. The next sections will elaborate on this approach.

## 2 Crucible Operations

The key functionality of a Hot Metal Carrier is its ability to handle the crucible. Two main operational phases can be distinguished: crucible pickup and crucible drop off.

### 2.1 Crucible Drop Off

Drop off is an easy manoeuvre from an automation point-of-view. No sensing is required and a simple ballistic manoeuvre is sufficient (Figure 2).

### 2.2 Crucible Pickup

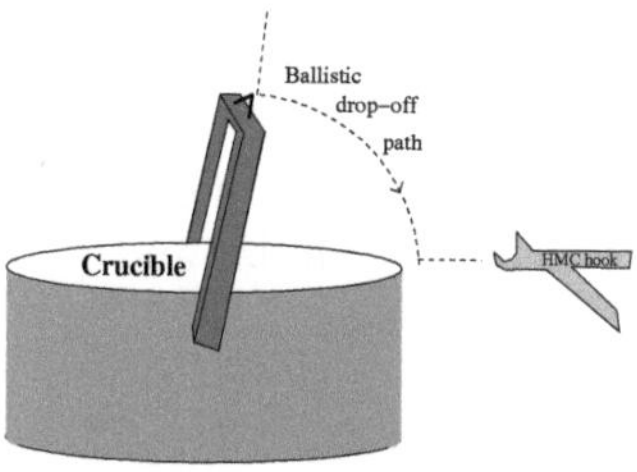


**Fig. 2.** Schematic of the hook movement during a drop off manoeuvre

The pickup manoeuvre is much harder than the drop off. It can be divided into two steps: first, an approach step where the hook is visually guided toward the pickup point in the middle of the crucible handle (figure 7), then the actual pickup. The latter is an easy manoeuvre, again a ballistic movement, similar to drop off (Figure 2).

The approach part is more complex. It is principally based on the on-board camera detecting the crucible from about 5 m. Using sensor-based servoing and not a memory of the position where the crucible was last dropped off is justified by two main argument: first, in a real site, several agents handle the crucible (cranes, lifts, and other HMCs). If some are not computer controlled or instrumented, there will be inaccuracies and uncertainties on the crucible drop off positions. Secondly, even during computer controlled drop off from our autonomous HMC, the crucible can swing around the hook and produce a different drop off position than previously. Most significantly, the crucible's orientation can vary in an unpredictable manner.

During the pickup manoeuvre the critical part is the crucible localisation in the vehicle frame. The description of our implementation of the crucible localisation forms the major part of this paper. Using the localisation, an approach trajectory can be implemented as described in section 5.

## 3 Landmark Based Localisation

### 3.1 Landmark Choice

Our aim is to implement a robust vision-based handling task for an outdoor forklift. To maximise the resilience of our system to external perturbation, we decided to use artificial fiducials that can be attached to our payload.

As mentioned before, some existing vision-based industrial applications use motion capture fiducials, i.e. small disks with black and white quadrants. These fiducials are not reliable enough in an outdoor uncontrolled environment, so we chose to use self-similar landmarks, as described by Scharstein and Briggs[1].

#### Self-similarity

A $p$-similar function for $0 < p < 1$ is a function $f : \mathbb{R}^+ \rightarrow \mathbb{R}$ such that $\forall x > 0, f(x) = f(p{\cdot}x)$. It is essential to note that the $p$-similarity is invariant to change of scale ($\forall x > 0, \forall k > 0, f(k \cdot x) = f(p \cdot k \cdot x)$). In the context of a computer vision application, this property is interesting since, if a method to detect self-similarity is available, it will be possible to do so from whatever distance[1] since, with a pin-hole camera, a change of range results in a change of scale in the observed image.

When designing a self-similarity detector, the fact that all constant functions are $p$-similar for all $p$ is problematic since any uniform region in an image will give a strong response on the self-similarity detector. To solve this, [1] introduces the notion of anti-similarity: a function $f$ is $p$-antisimilar if $\forall x > 0, |f(x) - f(p \cdot x)| = 1$.

The $p$-similarity and $\sqrt{p}$-antisimilarity are then used by Scharstein and Briggs to define the "Self-similar square wave" function which is both $p$-similar and $\sqrt{p}$-antisimilar (with $\lfloor \cdot \rfloor$ representing the floor function):

$$s_p(x) = \lfloor 2(\log_p x - \lfloor \log_p x \rfloor) \rfloor \tag{1}$$

This function maps $\mathbb{R}^+$ to $0, 1$. It is used to print black and white landmarks, as shown in fig. 3.

**Fig. 3.** Self-similar landmark as described in [1], with $p = \frac{2}{3}$

In order to recognise the self-similar landmarks, [1] defines a matching function that will respond strongly to a sequence of pixels looking like a $p$-similar and $\sqrt{p}$-antisimilar function. For a pixel located at $(x, y)$ in image $I$, this is achieved by integrating the $p$-similarity and $\sqrt{p}$-antisimilarity conditions on a window of length $w$ starting at $(x, y)$:

$$\begin{aligned} m^w_{+x}(x,y) = & \frac{1}{w}\int_0^w \left| I(x+\xi, y) - I(x + \sqrt(p)\cdot\xi, y) \right| d\xi \\ & - \frac{1}{w}\int_0^w \left| I(x+\xi, y) - I(x + p\cdot\xi, y) \right| d\xi \end{aligned} \tag{2}$$

In this equation, the window is made of the $w$ pixels after $(x, y)$, on an image row. We identify this function by the $+x$ subscript. The function can be obviously modified to use the $w$ pixels before, above, or below $(x, y)$. We name these functions $m^w_{-x}$, $m^w_{-y}$ and $m^w_{+y}$ respectively. In a practical implementation, $w$ would take values in the $[20, 100]$ pixel range. We use 40, as in [1] as a good balance between robustness and computation cost.

[1] As long as the resolution provides enough information on the observed object.

### Landmark Improvement

The single landmark used in [1] and shown in fig. 3 is very reliably detected, especially in an indoor setting. In a more variable setting such as outdoors, we found it suffers from two main defects. First, even if the vertical line of peaks in $m^w_{+x}(x, y)$ is always very salient and very well located horizontally, its vertical position is more uncertain: strong contrasts, JPEG encoding, sensor saturation or image slant can make the vertical position of the line very unreliable. Then, if we consider an application in an industrial environment, corrugated iron used to create shed walls can create a nearly self-similar pattern when observed with enough perspective. In this case, this would generate a substantial number of false positive.

To increase the robustness of our landmark, we have tested two alternatives: an orthogonal landmark made of two self-similar pattern patched together with a $90^o$ rotation, and a circular landmark where the self-similarity is radial instead of being axial. The circular landmark is less sensitive to rotation and motion blur, but more sensitive to occlusions than the orthogonal one. Consequently, the circular landmark was used for an underwater robot (see [4]) and the orthogonal one was used in this article. Both alternatives are depicted in fig. 4. Orthogonal landmarks can be made in two versions easily distinguishable: a *left* version, where the horizontal landmark goes from left to right, on the left of an up-to-down vertical landmark, and a *right* version, resulting of a flip of the other one over a vertical axis. The former is the one that appears in fig. 4.

Orthogonal landmark (left version) Circular landmark

**Fig. 4.** Improved self-similar landmark

To detect an orthogonal landmark, we use the self-similar landmark detector form [1]. For a given search orientation, this gives us a segment identified as laying on the right hand border of the landmark (see fig. 3). After the extraction of all the segments, for left-right, right-left and top-down orientation, we select only points where one segment is close enough to an other one, and roughly orthogonal. The image position of the landmark will then be the intersection of its two orthogonal segments.

## 3.2 Localisation

In order to localise the vehicle load, we need to detect and identify the fiducials attached to it, and from their location in the image, estimate their location with respect to the vehicle end-effector. In the context of our forklift handling task, our camera is mounted on a pan-tilt-zoom (PTZ) head, attached to the forklift mast. To map the image observations to world coordinates, we have two options: either calibrate the full kinematic chain linking the camera frame to the forklift reference frame, or calibrate directly the relation between pixel coordinates and world coordinates for a fixed mast pose. The remaining of this section will compare both approaches.

### Calibrating the Kinematic Chain

From a scientific point of view, this is the most attractive solution: it is more generic and closer to the physical system. On the other hand, the full kinematic chain, as described in fig. 5, is quite complex. On a large vehicle, most of these lengths or angles cannot be accurately measured, so they would have to be calibrated from indirect measurements. Even if this can be done, the accumulation of uncertainties along the kinematic chain will inevitably result in inaccuracies greater than what is acceptable for our application.

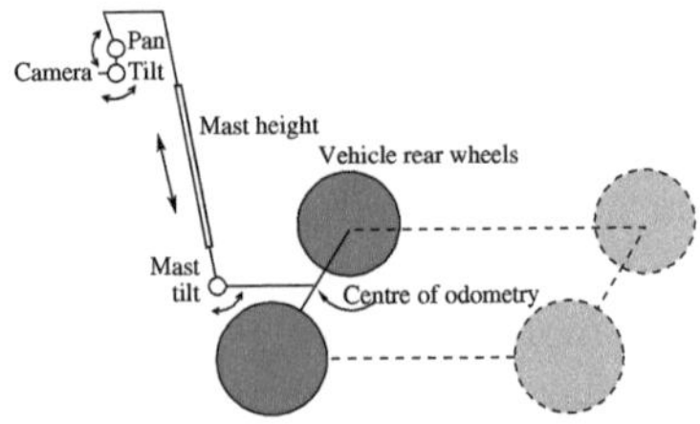


**Fig. 5.** Kinematic chain linking the camera lens to the vehicle centre of odometry

### Calibration of the Pixel-to-World Relation

As an alternative to the full calibration, we can assume that the payload localisation will only be required while the forklift mast is in a known constant position. To constrain the problem even more, we will assume that the vehicle moves in a planar environment, and that the payload fiducials are located at a known constant height above the ground. This simplifies the problem since the fiducials need only to be localised in a plane parallel to the ground plane. From now on, this plane will be named *target plane.*

Our objective now is to calibrate the relationship between pixel coordinates and target position in the target plane. With a fixed camera, this would only require to estimate an homography, which is easily estimated using the mapping of at least four points.

When considering a PTZ camera, an observation is not only the pixel coordinates $(i, j)$ but also the pan $(p)$, tilt $(t)$ and zoom $(z)$ of the camera. Nevertheless, a similar technique can be used. Let us first define two frames: the camera frame is aligned with the image plane and the optical axis, and the base frame is the frame of the PTZ head. The relationship between these frames is a rotation defined by the pan and tilt angles.

We can now define our mapping in two phases. First, for a given pan, tilt and zoom, an image pixel can be mapped to a unit vector going out of the origin, in the base frame. This mapping is bijective. Then, let us consider a plane $\mathcal{P}$ located between the camera and the target plane[2], at unit distance of the camera for instance. The mapping associating each unit vector $v$ to the intersection of the half-line supported by $v$ with $\mathcal{P}$, is invertible as long as the intersection exists. This mapping can be computed from purely geometric considerations and does not require calibration.

Finally, the image in $P$ of at least four points from the target plane is enough to estimate an homography mapping $P$ to the target plane.

It is important to notice that the resulting mapping, from $(i, j, p, t, z)$ to coordinates in the target plane, is the composition of two bijective mappings, and thus bijective too.

**Note:** It may be argued that this mapping model is less valuable than an estimation of the fixed $6DOF$ pose of the base frame in the vehicle reference frame. When developing this application, we effectively tried this method first, as well as, modelling the mapping using a $3^{rd}$-degree multivariate polynomial function. In our setting, the two-phase homography resulted in more accurate results.

## 4 Payload Localisation and Tracking

Our objective here is to describe how the vision system described in the previous sections can be used to localise and track our payload. Let us first recall the setup: our vehicle is equipped with a single hook that must be passed into the eye of the payload handle in order to realise a successful pickup. The pickup phase is the only time when the payload localisation is required.

The approach to the handle must be done on a trajectory perpendicular to the handle, consequently, we not only need to estimate the pose of the handle's eye, but also the orientation of the handle. Nevertheless, to simplify the problem, the payload handle is a massive rigid object and we can safely assume that it will always lay inside (or close enough to) the target plane.

**Fig. 6.** Fiducial and their placement on the crucible

In order to reliably and accurately obtain the orientation of the handle, we set up two orthogonal self-similar landmarks (one *left* and one *right*). From the camera located on the forklift mast, a typical tracking image is illustrated in fig. 6.

2 Theoretically only a plane parallel to the target plane would verify this, but in practice we only want a plane separating the camera and a region of interest in the target plane.

Using the landmark extractor described in section 3 and the mapping described in section 3.2, it will be nearly straight forward to localise the handle. Three aspects have to be taken into considerations:

*Conservation of field of view:* this is required to account for the vehicle movement while approaching, and to ensure that the landmark is big enough in the image frames. This is achieved by controlling the PTZ head to put the current position estimate in the centre of the image, with enough resolution.

*Long processing time:* The pickup of our handle requires a position accuracy better than 0.05 m. Consequently, even at very small speeds, the time spent processing the images must be accounted for when integrating each observation.

*Handle identification:* In our application, with only one payload available, we have not developed any specific identification software so far. Our landmarks are distinctive enough to avoid any false positives and to identify each side of the handle without ambiguity.

In order to be robust to multiple possible perturbations, the handle localisation is integrated into a particle filter[2] and fused with movement information coming from the vehicle odometry.

## 5 Crucible Approach

This section will describe the final piece of our application: the control of the approach. The payload localisation provides us with a continuous measurement of the payload position and orientation in the vehicle frame. Considering the non-holonomic properties of the vehicle, the best we can do is try to servo to the line orthogonal to the payload passing through the pickup point (see fig. 7).

The approach trajectory is done at constant speed $v$ while the steering of the vehicle is controlled using a standard control law:

$$\dot{\theta} = K_\theta \Delta\theta + K_y v \frac{\sin \Delta\theta}{\Delta\theta} \Delta y \qquad (3)$$

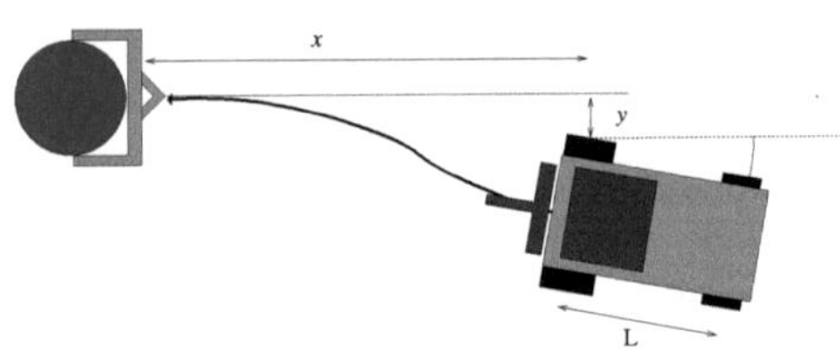


**Fig. 7.** Crucible approach trajectories and variables appearing in the control law

where $K_\theta$ and $K_y$ are tunable gains, and other variables are illustrated on fig. 7. According to the initial error, this controller may bring the vehicle to the pickup line. In this case the pickup can continue successfully. In other cases, we need to reliably detect this and generate a mitigation strategy. We currently detect this kind of failure by checking that the final position of the vehicle presents a value of $|\Delta y|$ smaller than 0.15 m and an orientation error smaller than 20 degrees. These values are arbitrary and adapted empirically to our system. Fig. 8 gives an example of an approach trajectory captured during 5 hours of continuous operations[6]. During this experiment, 100% of the pickups were successful.

**Fig. 8.** Example of a real approach trajectory

The reliability and predictability of the HMC autonomous manoeuvre is critical to its acceptance by the light metal industry. As an illustration of the repeatability of our current implementation, figure 9 shows the superimposition of half of the 58 pickups performed during our 5 hour run. It can be clearly seen that the paths are well contained in an envelope whose width is correlated with the variation of crucible position (dots and ellipse around crucible handle).

**Fig. 9.** Pickup approach trajectories, superimposition of 29 pickups. Dots and ellipse around the crucible handle show the variability of the crucible location across this set of manoeuvres. The left figure shows an overview of the stage, right figure shows a close-up on the paths themselves. For reference, the handle width is about 2 m and the pickup hole approximately 20 cm across.

## 6 Conclusions

The automation of Hot Metal Carriers, or general forklift, requires the implementation of robust handling capabilities. In this paper, we presented how we used especially designed artificial fiducials to implement a robust payload detection and localisation using pan-tilt-zoom cameras. In an ideal world a fiducial-less system will be more desirable, but within the current state of the art, this is unlikely to provide a robust enough solution for a real industrial deployment. Nevertheless, we are trying to develop such a system to evaluate how robust it can be, in comparison with our current one. Also, in this article, we did not address the long range detection of the payload, when a single view is not enough to detect the fiducials and a scanning strategy is required. We are working on this aspect of the problem and hope to present results soon.

## References

1. Briggs, A.J., Scharstein, D., Braziunas, D., Dima, C., Wall, P.: Mobile robot navigation using self-similar landmarks. In: Proc. of IEEE Conf. on Robotics and Automation (2000)
2. Doucet, A., De Freitas, N., Gordon, N. (eds.): Sequential Monte Carlo methods in practice (2001)
3. Garibotto, G., Masciangelo, S., Bassino, P., Coelho, C., Pavan, A., Marson, M., Bailey, G.E.: Industrial exploitation of computer vision in logistic automation:autonomous control of an intelligent forklift truck. In: Proc. of IEEE Conf. on Robotics and Automation (1998)
4. Negre, A., Pradalier, C., Dunbabin, M.: Robust vision-based underwater target identification & docking using self-similar landmarks. In: Conf. on Field and Service Robotics (submitted, 2007)
5. Seelinger, M., Yoder, J.-D.: Automatic pallet engagment by a vision guided forklift. In: Proc. of IEEE Conf. on Robotics and Automation (2005)
6. Tews, A., Pradalier, C., Roberts, J.: Autonomous hot metal carrier. In: Proc. of IEEE Conf. on Robotics and Automation (2007)

# Robust Feature Extraction and Matching for Omnidirectional Images

Davide Scaramuzza[1], Nicolas Criblez[2], Agostino Martinelli[3], and Roland Siegwart[4]

[1] Swiss Federal Institute of Technology Zurich
davide.scaramuzza@ieee.org
[2] Swiss Federal Institute of Technology Lausanne
nicolas.criblez@gmail.com
[3] INRIA
agostino.martinelli@ieee.org
[4] Swiss Federal Institute of Technology Zurich
r.siegwart@ieee.org

**Summary.** This paper presents a new and robust method for extracting and matching visual vertical features between images taken by an omnidirectional camera. Matching robustness is achieved by creating a descriptor which is unique and distinctive for each feature. Furthermore, the proposed descriptor is invariant to rotation. The robustness of the approach is validated through real experiments with a wheeled robot equipped with an omnidirectional camera. We show that vertical lines are very well extracted and tracked during the robot motion. At the end, we also present an application of our algorithm to the robot simultaneous localization and mapping in an unknown environment.

## 1 Introduction

One of the most important problems in vision based robot navigation systems is the search for correspondences in images taken from different viewpoints. In the last decades, the feature correspondence problem has been largely investigated for standard perspective cameras. Furthermore, some works have provided robust solutions for wide-baseline stereo matching, structure from motion, ego-motion estimation, and robot navigation (see [1], [2], [3], [4], [5], [6], [7], [8], and [9]). Some of these works normalize the region around each detected feature using a local affine transformation, which attempts to compensate for the distortion introduced by the perspective projection. However, such methods cannot be directly applied to images taken by omnidirectional imaging devices because of the non-linear distortions introduced by their large field of view. In order to apply those methods, one needs first to generate a perspective view out of the omnidirectional image, provided that the imaging model is known and that the omnidirectional camera possesses a single effective viewpoint [10]. An application of this approach can be found in [11]. There, the authors generate perspective views from each region of interest of the omnidirectional image.

C. Laugier and R. Siegwart (Eds.): Field and Service Robotics, STAR 42, pp. 71–81, 2008.
springerlink.com 

This image unwrapping removes the distortions of the omnidirectional imaging device and enables the use of state-of-the-art wide-baseline algorithms designed for perspective cameras. Nevertheless, other researchers have attempted to apply to omnidirectional images standard feature detectors and matching techniques, which have been traditionally employed for perspective images. In [15], for instance, the authors check the candidate correspondences between two views using RANSAC algorithm. Finally, other works have been developed, which extract one-dimensional features from new images called Epipolar plane images, under the assumption that the camera is moving on a flat surface [16]. These images are generated by converting each omnidirectional picture into a 1D circular image, which is obtained by averaging the scan lines of a cylindrical panorama. Then, 1D features are extracted directly from such kinds of images.

In this paper, the features we want to track from omnidirectional images are real world vertical features, which are predominant in structured environments. In our experiments, we used a wheeled robot equipped with an omnidirectional camera, which had the camera axis perpendicular to the direction of motion of the robot. Because of this settings and assuming the environment to be flat, all world vertical lines project into radial lines on the image plane.

The novelty of this paper consists of a robust method to match vertical lines between images taken by an omnidirectional camera during the motion of the robot. Matching robustness is achieved by creating a descriptor which is unique and distinctive for each feature. Furthermore, the proposed descriptor is invariant to rotation. This descriptor is based on the image gradients. The robustness of the approach was validated through real experiments by using a robot equipped with an omnidirectional camera. In this paper, we show that vertical lines are very well extracted and tracked during the robot motion. At the end, we also present an application of our algorithm to the robot Simultaneous Localization And Mapping (SLAM) in an unknown environment.

This paper is organized as follows. In section 2, we will discuss the vertical line extraction. In section 3, we will illustrate how to mark a feature by means of a descriptor, and, in section 4, we will describe the feature matching process. Finally, in section 5, we will present our experimental results and the application of our algorithm to SLAM.

## 2 Vertical Line Extraction

Our platform consists of a wheeled robot equipped with an omnidirectional camera looking upwards. The main advantage of such kind of camera is that it provides a 360° field of view of the scene, which gives a very rich and sparse information.

Figure 1 shows a sample picture taken by our omnidirectional camera. As the circular external boundary of the mirror is visible in the image, we use a circle detector to determine the location of the center.

In our arrangement, we set the camera-mirror system perpendicular to the floor where the robot moves. This guarantees that all vertical lines of the environment converge towards the image center. To extract the vertical lines, we first compute the image gradients (e.g. we used a Sobel filter), and then we keep only those gradients whose orientation looks towards the image center up to $\pm 5°$. This $10°$ tolerance allows us to deal with the effects of the floor irregularities in the projections of the 3D vertical lines. After this filtering, we apply non-maxima suppression and we end up with a binary edge map (Fig. 2).

The next step will be identifying the most reliable vertical lines. To this end, we divide the omnidirectional image into 720 predefined uniform sectors, which give us an angular resolution of $0.5°$. By summing up all binary pixels that vote for the same sector, we obtain the histogram shown in Fig. 3. As observed in Fig. 2, there are many potential vertical lines in a structured environment. To keep the most reliable and robust features, we choose only those lines whose length covers at least half of the angle of view (Fig. 3). Finally, we use non-maxima suppression to avoid the features to be too close (Fig. 4).

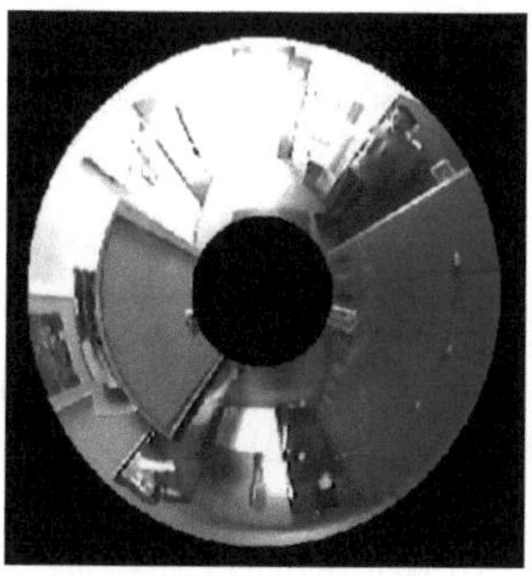

**Fig. 1.** An image taken by our omnidirectional camera

**Fig. 2.** The final binary edge map after non-maxima suppression

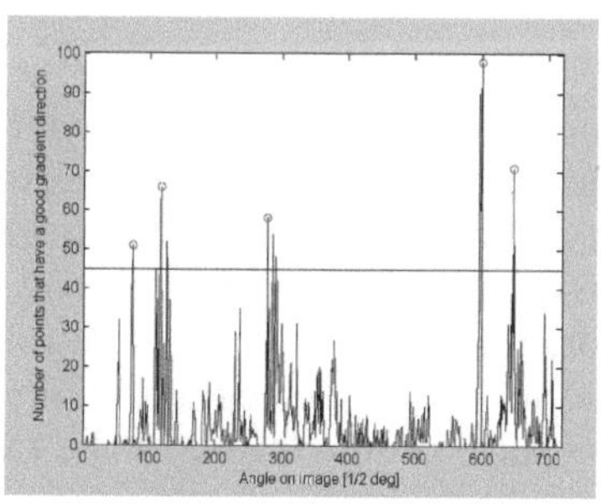


**Fig. 3.** The number of binary pixels voting for a given orientation angle. The orientation axis ranges from 0 to 720 half degrees.

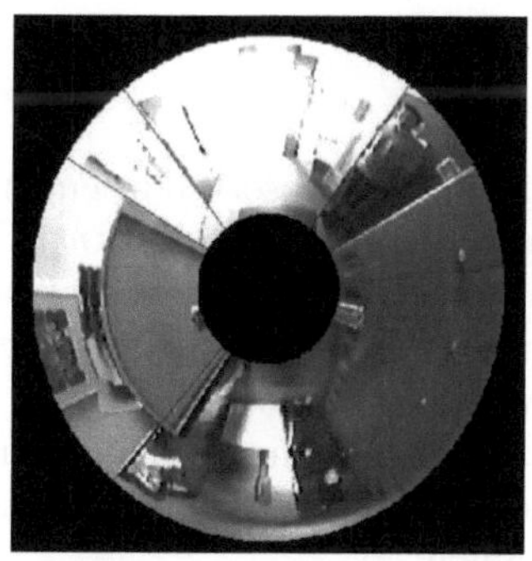

**Fig. 4.** Extraction of the most reliable vertical features from an omnidirectional image

# 3 Building the Descriptor

In section 4, we will describe our method for matching vertical lines between consecutive frames while the robot is moving in an unknown environment. To make the feature correspondence robust to false positives, each vertical line is given a descriptor, which is unique and distinctive for each feature. Furthermore, this descriptor is invariant to rotation. In this way, finding the correspondent of a vertical line can be done by looking for the line with the closest descriptor. In the next subsections, we will describe how to build this descriptor.

## 3.1 Rotation Invariance

To build the descriptor, we extract a predefined number of circular areas (namely 3 areas) in fixed position along a given radial line (Fig. 5). The centers of these circular areas are equally spaced and the radius is chosen such that the circles touch without overlapping. Then, each area is smoothed by a Gaussian window of variance $\sigma$ and the image gradients (magnitude and phase) are computed within each of these areas. The rotation invariance is achieved redefining the gradient phase relatively to the radial line's angle. (Fig. 5).

## 3.2 Orientation Histograms

To make the descriptor robust to false matches, we split each circular area into two parts and consider each one individually (Fig. 6). In this way, we preserve the information about what we have on the left and right sides of the feature. For each side of each circular area, we compute an orientation histogram (Fig. 7) of all gradient vectors. The whole orientation space (from -$\pi$ to $\pi$) is divided into $N_b$ equally spaced bins. In order to decide how much of a certain gradient magnitude $m$ belongs to the adjacent inferior bin $b$ and how much to the adjacent superior bin, each magnitude $m$ is weighted by the factor $(1 - w)$, where

$$w = N_b \cdot \frac{o - b}{2\pi} \tag{1}$$

with $o$ being the observed orientation in radians. Thus, $m(1-w)$ will vote for the adjacent inferior bin, while $mw$ will vote for the adjacent superior bin. According to what has been mentioned so far, each bin contains the sum of the weighted gradient magnitudes which belong to the correspondent orientation interval. We observed that this weighted sum made the orientation histogram more robust to image noise. The reader observe that the orientation histogram is already rotation invariant because the gradient angles have been referred to the radial line's angle.

## 3.3 Building the Feature Descriptor

The computed orientation histograms help to build the feature descriptor. Indeed, the descriptor is an $N$-element vector containing all histogram values of

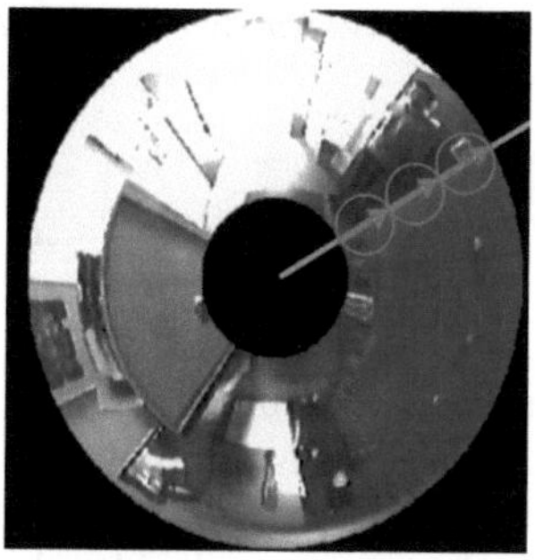

**Fig. 5.** Extraction of the circular areas. To have rotation invariance, the gradient phase is referred to the orientation of the vertical line.

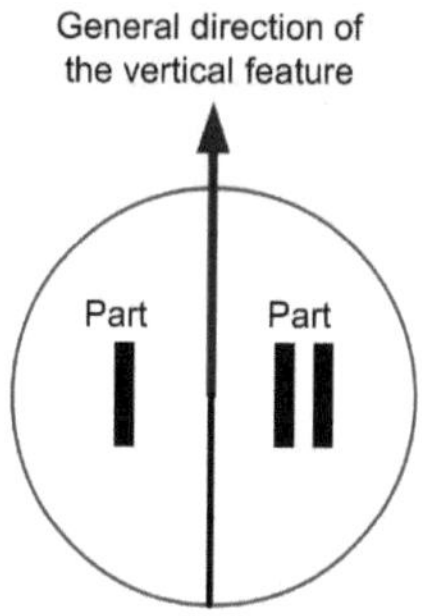


**Fig. 6.** Two parts of a circular area

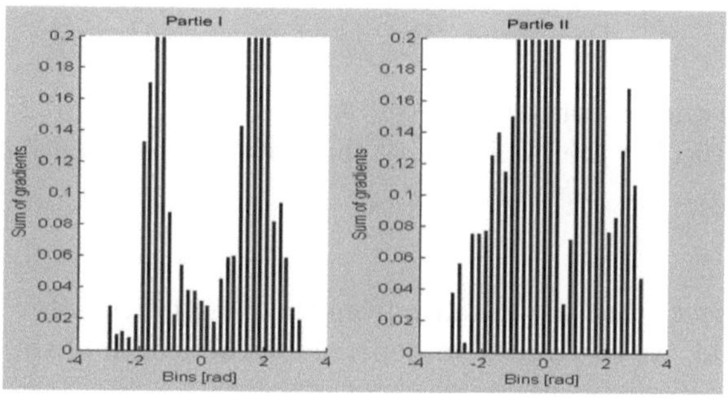


**Fig. 7.** An example of gradient orientation histograms for the left and right sides of a circular area

the circular areas. For instance, by extracting 3 circular areas for each vertical feature and choosing 30 bins for each histogram, the length of the feature descriptor will be

$$N = 3areas \cdot 2parts \cdot 30bins = 180 \quad (2)$$

Finally, it is important to know that all feature descriptors have the same length $N$. To achieve slight illumination invariance, every descriptor is normalized to 1. This choice relies on the hypothesis that the image intensity changes linearly with illumination. Although this is not true in nature, this approximation proved to work properly.

## 4 Feature Matching

As every vertical feature has its own descriptor, the correspondent of a vertical line in the consecutive images can be searched among the features with the closest descriptor. As a distance measure between two vector descriptors **A** and **B**, we use the Euclidean distance:

$$d(\mathbf{A},\mathbf{B}) = \sqrt{\sum_{k=1}^{N}(\mathbf{A}(k) - \mathbf{B}(k))^2} \quad (3)$$

As a consequence, the correspondent of a feature, in the current image, is expected to be the one, in the consecutive image, with the minimum distance. However, if a feature is no longer present in the next image, there will be a closest feature anyway. For this reason, we define three tests to decide whether a feature correspondent exists and which the correspondent is. Before describing the three criterions, let us introduce some definitions.
Say $\{\mathbf{A_1},\mathbf{A_2},...,\mathbf{A_{N_A}}\}$ and $\{\mathbf{B_1},\mathbf{B_2},...,\mathbf{B_{N_B}}\}$ two sets of feature descriptors extracted at time $t$ and $t-1$ respectively (where $N_A$ and $N_B$ are the number of features in the first and second image). Then, say

$$D_i = \{d(\mathbf{A_i},\mathbf{B_j}), j = 1,2,...,N_B)\} \quad (4)$$

the set of all distances between a given $\mathbf{A_i}$ and all $\mathbf{B_j}$ $(j = 1,2,,N_B)$. Finally, say $minD_i = min(D_i)$ the minimum of the distances between $\mathbf{A_i}$ and all $\mathbf{B_j}$.

*First Test*

The first test checks that the distance from the closest descriptor is smaller than a given threshold. As the threshold depends on the length of the descriptor, we set

$$minD_i = F_1 \cdot N \quad (5)$$

where $N$ is the descriptor length. By this criterion, we actually set a bound for the maximum acceptable distance to the closest descriptor.

*Second Test*

The second test checks that the distance from the closest descriptor is smaller enough than the mean of the distances from all other descriptors, that is:

$$minD_i = F_2 \cdot < D_i > \quad (6)$$

where $< D_i >$ is the mean value of $D_i$ and $F_2$ clearly ranges from 0 to 1. This criterion comes out of experimental results. In Table 1, we show a real comparison among the distances between the descriptor $\mathbf{A_1}$ at time $t$ and all the descriptors at time $t-1$. There, the descriptor $\mathbf{B_1}$ is the correct correspondent of $\mathbf{A_1}$. The reader might also observe that its distance is smaller than the mean of all other distances.

**Table 1.** The distances between the descriptor $\mathbf{A_1}$ at time $t$ and all descriptors $\mathbf{B_j}$ , $j = 1, 2, .., N_B$ at time $t - 1$

| B1 | B2 | B3 | B4 | B5 | B6 | B7 |
|---|---|---|---|---|---|---|
| 2.38 | 5.42 | 4.55 | 5.79 | 5.66 | 6.17 | 5.43 |

*Third Test*

Finally, the third test checks that the distance from the closest descriptor is smaller than the distance from the second closest descriptor:

$$minD_i = F_3 \cdot SecondSmallestDistance, \tag{7}$$

where $F_3$ clearly ranges from 0 to 1. As in the previous test, the third test raises from the observation that, if the correct correspondence exists, then there must be a big jump between the closest and the second closest descriptor.

Factors $F_1$, $F_2$, and $F_3$ are to be determined experimentally. The empirical values we used for these parameters are shown in Table 2.

**Table 2.** The parameters used by our algorithm with their empirical values

| | | |
|---|---|---|
| F1 = 0.0075 | F2 = 0.55 | F3 = 0.85 |

## 5 Experimental Results

In this section, we present some experimental results obtained by moving our robot in a real indoor environment. In the first subsection, we show the performance of our feature tracker during the motion of the robot, while in the second one, we present the results of our feature tracker applied to SLAM. In these experiments, the robot was moving at about 0.15 $m/s$. The image size was $640x480$ *pixels* and the frame rate was 3 $Hz$.

### 5.1 Performance of the Feature Tracking

In this experiment, we guided our robot through an office-like environment for about 70 meters. The results of the feature tracking are shown in Fig. 8. The plot refers to a short path of the whole trajectory while the robot was coping with an L-shaped trajectory. As the reader may observe, many features are detected and tracked over the time. Indeed, all lines of the plot appear to be smooth and homogeneous. Furthermore, the lines do not intersect, meaning that there was no false matching. We can also notice that the algorithm was able to match vertical elements even when the correspondent features were not observed in the previous image (e.g. observe the large gap between the dots pointed to by the arrow in Fig. 8). Indeed, when the correspondence is not found in the last frame, our algorithm starts looking into all previous frames (actually up to the twentieth frame), and stops when the correspondence is found. By zooming into the plot of Fig. 8, we found that some lines are given different numbers. For instance, feature number 24 is labeled as 31 after some frames. And the same happens with features 42 and 49. When this happens, it means that the algorithm found no correspondence for the current feature, and thus, the feature is labeled as a new entry, but in fact this is a false new entry. After having visually checked every single frame of the video sequence, we found 6 false matches and 22 false new entries. Comparing these errors to the 2631 corresponding pairs detected by the algorithm over the whole video sequence, we had 1.06% of mismatches. Furthermore, we found that the false matches occurred every time the camera was facing objects with repetitive texture. Thus, the ambiguity was caused by the presence of vertical elements which were almost identical. On the other hand, a few of false new entries occurred whenever the displacement of the robot between two successive images was big. However, the reader should observe that when a feature matches with no other previous feature of the last frames, it is better to believe this feature to be new rather than commit a false matching.

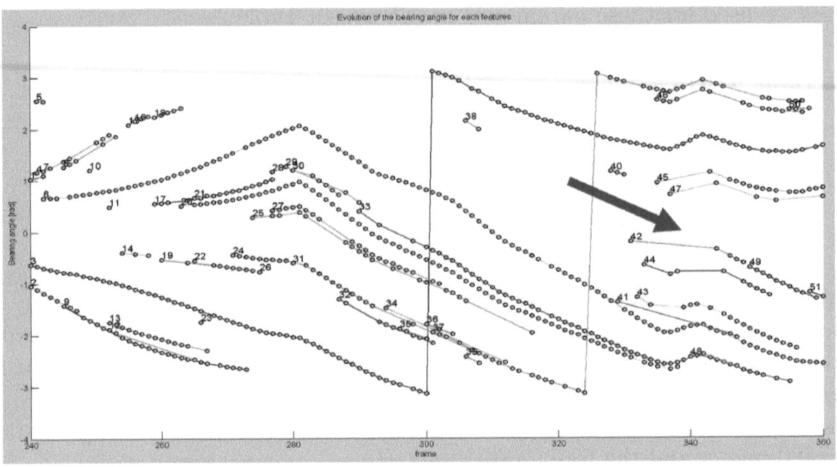


**Fig. 8.** Feature tracking during the robot motion. In $y$-axis, we have the angle of sight of each feature, and in the $x$-axis, the frame number. Each dot represents a feature detected in the current frame. The lines represent the tracked features. The number reported on some dot appears only when a new feature is detected.

## 5.2 Application to SLAM

We applied our feature tracker to two important problems in autonomous navigation, that is, sensor self-calibration and SLAM. Regarding the former, the results can be found in [12], [13], and [14]. In this section, we show only the results we obtained for SLAM. We implemented the standard EKF based SLAM. In particular, the EKF estimates the vector:

$$X = [x_r, y_r, \theta_r, X_1, Y_1, ..., X_{N_O}, Y_{N_O}]^T \tag{8}$$

where $[x_r, y_r, \theta_r]$ is the robot configuration, $X_i, Y_i$ are the Cartesian coordinates of the $i$-feature in the map, and $N_O$ is the number of observed features. Our mobile robot is equipped with wheel encoders and with the same omnidirectional camera adopted in the experiments described in section 5.1. The bearing observations provided by the omnidirectional camera consist of the vector $z$ whose components are:

$$\begin{aligned} z_1 &= arctan(\frac{Y_1 - y_r}{X_1 - x_r}) \\ &\vdots \\ z_{N_O} &= arctan(\frac{Y_{N_O} - y_r}{X_{N_O} - x_r}) \end{aligned} \tag{9}$$

To initialize a new feature in the map, consecutive bearing observations as in [14], which refer to the same feature, are integrated with the odometry. Then, the estimation is improved by integrating the information coming from all the bearing observations through the EKF. The result is shown in Fig. 9 where both the robot trajectory and the position of the features are shown.

**Fig. 9.** The results obtained by implementing a simple EKF-SLAM, which uses the proposed feature tracker. The black line is the trajectory estimated by using the odometry alone. The red line is the trajectory estimated by the EKF using both the odometry and vertical lines. The blue points represent the map ground truth provided by a laser range finder. The red circles are the detected verticals features.

## 6 Conclusion

In this paper we introduced and discussed a new and robust method to extract and match vertical lines between images taken by an omnidirectional camera. The basic idea to achieve robust feature matching consists of creating a descriptor which is unique and distinctive for each feature. Furthermore, this descriptor is invariant to rotation. To evaluate the performance of our approach, we performed real experiments where we evaluated the quality of the matching. We conclude that the proposed approach is very robust and precise. Finally, we adopted the proposed method to implement an EKF based SLAM.

## Acknowledgment

This work was conducted within the EU Integrated Projects COGNIRON (*The Cognitive Robot Companion*) and BACS (*Bayesian Approach to Cognitive Systems*). It was funded by the European Commission Division FP6-IST Future and Emerging Technologies under the contracts FP6-IST-002020 and FP6-IST-027140 respectively.

## References

1. Matas, J., Chum, O., Urban, M., Pajdla, T.: Robust wide baseline stereo from maximally stable extremal regions. In: Proc. 13th British Machine Vision Conference, Cardiff, UK, pp. 384–393 (2002)
2. Kadir, T., Zisserman, A., Brady, M.: An affine invariant salient region detector. In: Proc. 7th European Conference on Computer Vision, Prague, Czech Republic, vol. I, pp. 228–241 (2004)
3. Mikolajczyk, K., Schmid, C.: Indexing based on scale invariant interest points. In: Proceedingsof the 8th International Conference on Computer Vision, Vancouver, Canada, pp. 525–531 (2001)
4. Lowe, D.G.: Distinctive image features from scale-invariant keypoints. International Journal of Computer Vision 60(2), 91–110 (2004)
5. Lindeberg, T.: Feature detection with automatic scale selection. International Journal of Computer Vision 30(2), 79–116 (1998)
6. Mikolajczyk, K., Schmid, C.: An affine invariant interest point detector. In: Proc. 7th European Conference on Computer Vision, Copenhagen, Denmark, vol. I, p. 128 (2002)
7. Tuytelaars, T., Van Gool, L.: Matching widely separated views based on affine invariant regions. International Journal of Computer Vision 1(59), 61–85 (2004)
8. Baumberg, A.: Reliable feature matching across widely separated views. In: Proc. IEEE Conference on Computer Vision and Pattern Recognition, Hilton Head, South Carolina, pp. 774–781 (2000)
9. Tuytelaars, T., Van Gool, L.: Matching widely separated views based on affine invariant regions. International Journal of Computer Vision 1(59), 61–85 (2004)
10. Nayar, S.K.: Catadioptric omnidirectional camera. In: CVPR 1997: Proceedings of the 1997 Conference on Computer Vision and Pattern Recognition (CVPR 1997), p. 482. IEEE Computer Society, Los Alamitos (1997)
11. Mauthner, T., Fraundorfer, F., Bischof, H.: Region matching for omnidirectional images using virtual camera planes. In: Proc. of Computer Vision Winter Workshop 2006, Telc, Czech Republic (2006)
12. Scaramuzza, D., Martinelli, A., Siegwart, R.: A Flexible Technique for Accurate Omnidirectional Camera Calibration and Structure from Motion. In: Proc. of the IEEE International Conference on Vision Systems, New York (2006)
13. Scaramuzza, D., Martinelli, A., Siegwart, R.: A Toolbox for Easily Calibrating Omnidirectional Cameras. In: Proc. of the IEEE International Conference on Intelligent Systems, IROS 2006, Beijing, China (2006)

14. Martinelli, A., Scaramuzza, D., Siegwart, R.: Automatic Self-Calibration of a Vision System during Robot Motion. In: Proc. of the IEEE International Conference on Robotics and Automation, ICRA 2006, Orlando, Florida, USA (2006)
15. Micusik, B., Pajdla, T.: Structure from Motion with Wide Circular Field of View Cameras. IEEE Transactions on Pattern Analysis and Machine Intelligence 28(7), 1135–1149 (2006)
16. Briggs, A., Li, Y., Scharstein, D., Wilder, M.: Robot Navigation Using 1D Panoramic Images. In: Proc. of IEEE Intl. Conference on Robotics and Automation (ICRA 2006), Orlando, FL (May 2006)

# Dynamic Obstacle Detection Based on Probabilistic Moving Feature Recognition*

Roman Katz, Oliver Frank, Juan Nieto, and Eduardo Nebot

ARC Centre of Excellence for Autonomous Systems
Australian Centre for Field Robotics
The University of Sydney, Australia
{r.katz,o.frank,j.nieto,nebot}@cas.edu.au

**Summary.** This paper presents a framework to detect moving objects based on the recognition of moving features in the images. The classification scheme is based on a complete probabilistic representation of feature locations that relates the vehicle motion with the visual information. Experimental evaluation under different settings in an outdoor, urban environment shows the performance of the proposed architecture.

## 1 Introduction

The detection and tracking of moving obstacles from a moving vehicle is considered at the core of safe autonomous navigation research. Also, the development of driver assistance systems depends on robust and reliable moving obstacle detection techniques. Such advanced perception capabilities would provide these systems with the appropriate environmental information to plan actions or avoid collisions, considering risky agents as other vehicles, pedestrians and bikes.

Many of the approaches to tackle this problem based their efforts on single, either laser-based [6, 12] or camera-based [4, 7] schemes. These camera based approaches often use optical flow to infer the moving obstacles in the images, a calculation procedure that is usually computationally expensive. Efficient ways to compute the optical flow while providing a probabilistic representation have been recently presented [13]. On the other hand, some multisensor based approaches have been proposed lately [9]. However, substantial effort needs to be made in order to achieve higher levels of reliability [11]. In this regard, it seems that probabilistic based schemes performing multisensor integration might provide the appropriate basis for moving obstacles perception.

This paper proposes a framework to detect moving objects by means of discovering moving features in a probabilistic way. Features are tracked in the images, while a probabilistic representation of the predicted feature locations is calculated simultaneously. These predicted *epipolar areas* are then used to classify the features as static or dynamic. The paper is organized as follows: Section 2 presents an overview of the proposed architecture. Functional modules are later

* This work is supported by the ARC Centre of Excellence programme, funded by the Australian Research Council (ARC) and the New South Wales state government.

C. Laugier and R. Siegwart (Eds.): Field and Service Robotics, STAR 42, pp. 83–91, 2008.
springerlink.com 

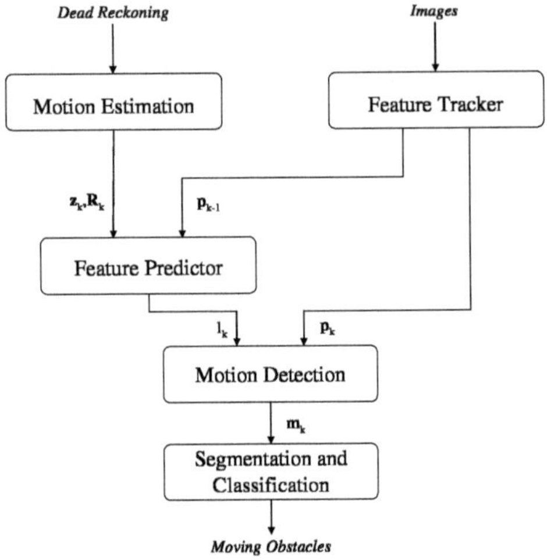


**Fig. 1.** High-level diagram of the proposed architecture showing each functional component

described in detail, including vehicle motion estimation in Section 3, and feature prediction and motion detection in Section 4. In Section 5 implementation issues are commented, and experimental results confirming the validity of the proposed model and its performance are exposed. Conclusions and further improvements are finally discussed in Section 6.

## 2 Detection of Moving Obstacles

In this section we describe a general framework aimed to discover moving objects from a moving vehicle equipped with a monocular camera and dead-reckoning sensors. A global, high-level diagram of the proposed architecture is depicted in Figure 1. The proposed scheme assumes that a camera is rigidly attached to a vehicle, and that features can be detected and tracked in the images provided by the calibrated camera, regardless of its pose. The scheme will provide sensible results as long as features can be detected and tracked on the moving obstacles. It is also assumed that an accurate estimation of the relative camera motion between poses can be obtained, based solely on dead reckoning sensors. We denote the camera pose on the vehicle at time $k$ as $\mathbf{x}_{v_k}$ with covariance matrix $\mathbf{P}_{v_k}$.

When an image is received, an initial set of features $\mathbf{p}_{k-1}$ is first detected. A *feature tracker* module tracks these feature points for a given number of successive images. We define the set of tracked features at time $t = k$ as the feature vector $\mathbf{p}_k$. Simultaneously, a *motion estimation* component estimates the relative pose (rotation $R$ and translation $T$) between the poses of the camera at the first and last frame of the tracked sequence. This procedure obtains a Gaussian representation of the relative pose – the displacement and attitude $\mathbf{z}_k$ between the poses together with its covariance matrix $\mathbf{R}_k$.

Once the relative pose is known, we can use this geometric displacement in the "world" coordinate frame to predict where the static features $\mathbf{p}_{k-1}$ should project onto the camera coordinate frame at time $k$. This is done by borrowing concepts from the epipolar geometry domain [5]. A relationship between the 3D coordinates of static points in the world coordinate and camera frame [8] at different moments can be found from the corresponding rigid-body transformation $(R,T)$ between the poses. Therefore, a *feature predictor* module predicts the location $l_k$ of the features $\mathbf{p}_{k-1}$ based on $\mathbf{z}_k$ as $l_k = \mathbf{j}(\mathbf{p}_{k-1}, \mathbf{z}_k, \mathbf{R}_k)$, with $\mathbf{j}$ the appropriate mapping function derived from the epipolar constraints. A *motion detection* module finally contrasts the tracked features $\mathbf{p}_k$ with their predicted locations $l_k$. The static features should fall in the predicted areas, since they are ruled by the epipolar geometry constraints. However, the features $\mathbf{m}_k$ from moving obstacles will have their own dynamics and will not comply with the epipolar constraints.

## 3 Motion Estimation and Feature Correspondence

A good estimation of the relative pose between camera frames is needed in order to relate geometric information contained in the images properly. This is done through dead-reckoning estimation of the vehicle position. Given that camera frames are available at high rates, dead-reckoning based position estimation becomes a reasonable approach because the drift is negligible.

### 3.1 State Estimation

In order to estimate the relative pose between consecutive frames, we keep track of consecutive poses and their corresponding covariance matrix. We define the state vector $\mathbf{x}$ at time $t = k$ as:

$$\mathbf{x}_k = \begin{bmatrix} \mathbf{x}_{v_{k-1}} \\ \mathbf{x}_{v_k} \end{bmatrix} = \begin{bmatrix} \mathbf{x}_{v_{k-1}} \\ \mathbf{f}(\mathbf{x}_{v_{k-1}}, \mathbf{u}) \end{bmatrix}, \tag{1}$$

where $\mathbf{x}_{v_{k-1}}$ and $\mathbf{x}_{v_k}$ are the vehicle poses at times $t = k-1$ and $t = k$, respectively. The function[1] $\mathbf{f}$ represents the relationship between the past estimated vehicle pose and the predicted one, according to the received control inputs $\mathbf{u}$. We propagate the covariance matrix $\mathbf{P}_k$ for the state vector at time $k$ as:

$$\mathbf{P}_k = \begin{bmatrix} \mathbf{P}_{v_{k-1}} & \mathbf{P}_{v_{k-1}} \nabla \mathbf{f}^T_{\mathbf{x}_{v_{k-1}}} \\ \nabla \mathbf{f}_{\mathbf{x}_{v_{k-1}}} \mathbf{P}^T_{v_{k-1}} & \nabla \mathbf{f}_{\mathbf{x}_{v_{k-1}}} \mathbf{P}_{v_{k-1}} \nabla \mathbf{f}^T_{\mathbf{x}_{v_{k-1}}} + \nabla \mathbf{f}_{\mathbf{u}} \mathbf{U} \nabla \mathbf{f}^T_{\mathbf{u}} \end{bmatrix}, \tag{2}$$

where the Jacobians [2] are:

$$\nabla \mathbf{f}_{\mathbf{x}_{v_{k-1}}} = \frac{\partial \mathbf{f}}{\partial \mathbf{x}_{v_{k-1}}} \text{ and } \nabla \mathbf{f}_{\mathbf{u}} = \frac{\partial \mathbf{f}}{\partial \mathbf{u}}$$

and $\mathbf{U}$ represents the covariance matrix of the control inputs noise $\mathbf{u}$.

[1] See [10] for details on the formulation of $\mathbf{f}$ based on wheel encoder and gyro inputs.
[2] For simplicity, we denote the Jacobian $\nabla \mathbf{a}_{\mathbf{b}} = \partial \mathbf{a} / \partial \mathbf{b}$ from now on.

### 3.2 Relative Motion

The relative pose at time $k$ between two vehicle poses can be estimated as:

$$\mathbf{z}_k = \mathbf{h}(\mathbf{x}_k) = \mathbf{h}(\mathbf{x}_{v_i}, \mathbf{x}_{v_j}), \tag{3}$$

where $\mathbf{x}_{v_i}$ and $\mathbf{x}_{v_j}$ refer to the vehicle poses at times $i = k-1$ and $j = k$, respectively. The covariance $\mathbf{R}_k$ of the relative pose can be obtained propagating the covariance $\mathbf{P}_k$ of the state by linearization of $\mathbf{h}$ as:

$$\mathbf{R}_k = \nabla\mathbf{h}_{\mathbf{x}_k}\mathbf{P}_k\nabla\mathbf{h}_{\mathbf{x}_k}^T. \tag{4}$$

### 3.3 Feature Correspondence

The correspondence of features[3] between consecutive images consists of a *feature selection* module that chooses candidate points, and a *feature tracking* module that matches them across different views. We denote the tracked set of feature points at $t = k$ as the set $\mathbf{p}_k$. The feature selection is based on the Harris' corner criterion, measuring the quality of local windows and applying thresholding to detect candidate points [8]. Once the candidate points are selected, the feature tracking is performed for the case of small baseline, considering that the sequence is taken from a moving camera triggered at high speed. A purely translation motion model is sufficient for our case given that consecutive images are taken from very close vantage points. This can be obtained by minimization of an error formulation based on the image brightness, as presented thoroughly in [8]. Results of this tracking scheme can be seen in Figure 2(a).

## 4 Feature Prediction and Motion Detection

The analysis of the dynamics of the features is based on the fact that their location can be predicted, given that features can be tracked between locations. The static features should fall in the predicted *epipolar areas*, since they are ruled by the epipolar geometry constraints [5]. However, the features belonging to moving obstacles will not comply with the epipolar constraints. Those features that do not fall within the expected areas can be considered as moving ones.

### 4.1 Feature Prediction

A prediction module first predicts the location $l_k$ of the features $\mathbf{p}_{k-1}$ based on the relative motion between the vehicle poses $\mathbf{z}_k$ and its covariance $\mathbf{R}_k$ as $l_k = \mathbf{j}(\mathbf{p}_{k-1}, \mathbf{z}_k, \mathbf{R}_k)$, where $\mathbf{j}$ is the composed mapping function derived from the epipolar constraints. We derive the mapping function as follows below.

[3] The proposed architecture does not rely on any particular type of feature detection or tracking mechanism.

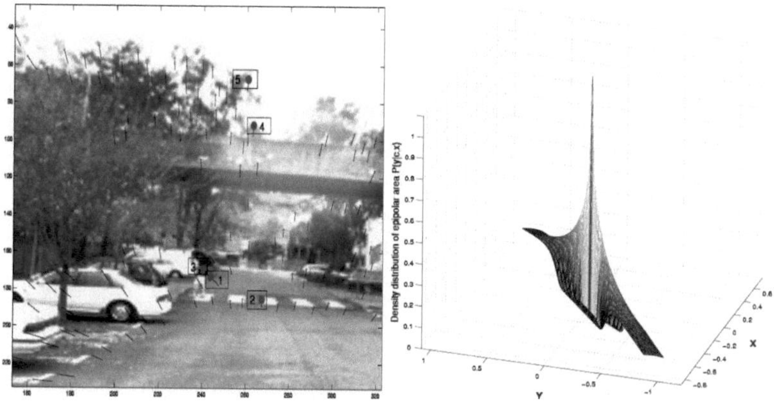


(a) (b)

**Fig. 2.** (a) shows tracked features at the last frame of an experimental sequence. The features recognized as dynamic (using a deterministic approach) are indicated with labeled dark boxes. (b) shows the density distribution for feature 3, as defined in (9) and (10).

We consider as the initial inputs the motion between poses $\mathbf{z}_k$ and its covariance $\mathbf{R}_k$. The essential matrix $\mathbf{E}$ [4] is defined as in [5] in terms of the rigid-body transformation $(R,T)$ between the poses:

$$\mathbf{E} = \mathbf{e}(\mathbf{z}) = T \times R = \begin{bmatrix} e_1 & e_2 & e_3 \\ e_4 & e_5 & e_6 \\ e_7 & e_8 & e_9 \end{bmatrix}, \tag{5}$$

where $\times$ is the cross-product, and $T$ and $R$ are the translation and rotation corresponding to the relative motion $\mathbf{z}$ between poses. If we rewrite this matrix as the vector $\mathbf{E} = \mathbf{e}(\mathbf{z}) = [e_1, \ldots, e_9]^T$, then its covariance $\mathbf{P_E}$ is:

$$\mathbf{P_E} = \nabla \mathbf{e_z} \mathbf{R}_k \nabla \mathbf{e}_\mathbf{z}^T. \tag{6}$$

We propagate now the essential matrix in the calculation of the epipolar areas – the predicted areas where the static features $\mathbf{p}_k$ should lie in. Following the epipolar constraints [5, 8], the epipolar lines in the image at time $k$ are determined by the coefficients of $\mathbf{c}$ as:

$$\mathbf{c} = [\mathbf{c}_1, \mathbf{c}_2, \mathbf{c}_3]^T = c(\mathbf{E}, \mathbf{p}_{k-1}) = \mathbf{E}\mathbf{p}_{k-1}, \tag{7}$$

where $\mathbf{E}$ is defined as in (5), and $\mathbf{p}_{k-1}$ are the feature points at $k-1$. Its covariance is:

$$\mathbf{P_c} = \nabla c_\mathbf{E} \mathbf{P_E} \nabla c_\mathbf{E}^T + \nabla c_{\mathbf{p}_{k-1}} \mathbf{P}_{\mathbf{p}_{k-1}} \nabla c_{\mathbf{p}_{k-1}}^T, \tag{8}$$

[4] The index $k$ is dropped for brevity from now on.

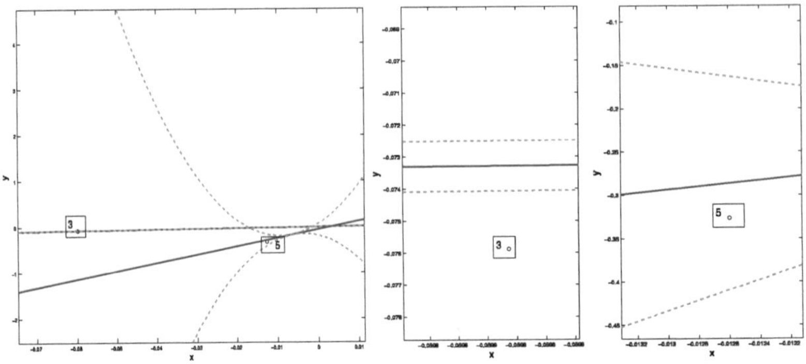

(a) (b) (c)

**Fig. 3.** (a) depicts the epipolar lines (red solid lines) for the feature points 3 and 5 (blue ∘) shown in Figure 2. Their 3-$\sigma$ boundaries (green dashed lines) are also displayed. Note that the 3-$\sigma$ boundaries and the epipolar line for feature 3 overlap. Closeup for features 3 and 5 shown in (b) and (c), respectively.

where $\mathbf{P_E}$ is defined as in (6), and $\mathbf{P}_{\mathbf{p}_{k-1}}$ is the uncertainty of the feature points at $k-1$. The epipolar lines are now defined using the coefficients from (7) as:

$$\mathbf{c}_1 x + \mathbf{c}_2 y + \mathbf{c}_3 = 0,$$

with $x$ and $y$ image coordinates at $k$. Hence:

$$y(x) = l(\mathbf{c}, x) = -\left(\frac{\mathbf{c}_3}{\mathbf{c}_2}\right) - \left(\frac{\mathbf{c}_1}{\mathbf{c}_2}\right) x, \tag{9}$$

with covariance:

$$\mathbf{P}_y(x) = \nabla l_{\mathbf{c}} \mathbf{P}_{\mathbf{c}} \nabla l_{\mathbf{c}}^T, \tag{10}$$

where $\mathbf{P_c}$ is defined as in (8). Therefore, the predicted location (epipolar area) on the image at $k$ for each of the features at $k-1$ is defined as a Gaussian distribution over $y$ for each value of $x$, that depends on the values of $\mathbf{c}$. As a generic example, the density function for the feature 3 in Figure 2(a) is presented in Figure 2(b).

### 4.2 Motion Detection

A motion detection module finally contrasts the tracked features $\mathbf{p}_k$ with their predicted locations $l_k$, to determine the moving features. In order to make the system robust, we consider the full uncertain representation of the epipolar areas. The decision criterion is probabilistic and based on gate validation techniques [2]. We first set up the gate based on the innovation and its predicted covariance, using the predicted epipolar areas and the tracked features (observation). The detection of a tracked feature within this gate classifies it as static, otherwise it is regarded as a moving feature. If we consider that there is no uncertainty in the position of the tracked features $\mathbf{p}_k$, this gating is equivalent to determining whether the tracked features are within a confidence interval, given by the epipolar area density distribution defined by (9) and (10).

To evaluate the performance of this probabilistic approach, we consider also a simple, deterministic criterion. Since, ideally, the tracked features should lie on (or very close to) the epipolar lines, we can then make decisions based on distance to these lines. Hence, we consider a threshold level $T_d$, such that a feature point $i$ is regarded as dynamic if:

$$\min\left(\|(x,y) - \mathbf{p}_k^i\| : (x,y) \in l^i\right) > T_d, \tag{11}$$

where $l^i$ is the epipolar line as defined in (9), and $\mathbf{p}_k^i$ is the tracked feature at $k$, both for the $i^{th}$ feature correspondence. The labeled features shown in Figure 2(a) were obtained using this simple approach. Features 1 and 3 correspond to a moving pedestrian, but features 2, 4 and 5 are indeed incorrectly detected, belonging to static objects in the scene.

The full probabilistic approach greatly increases the robustness of the classification procedure, eliminating many undesired outliers. For the same setting shown in Figure 2(a), only the features 1 and 3 were classified as dynamic. Figure 3(a) depicts the epipolar lines (red solid lines) for the feature points 3 and 5 (blue $\circ$) shown in Figure 2(a). Their 3-$\sigma$ boundaries (green dashed lines) are also displayed in Figure 3 (closeup for features 3 and 5 in (b) and (c), respectively). As can be seen, the moving features fall within the 3-$\sigma$ boundaries (feature 5 in Figure 3(c)), whereas the dynamic ones lie outside their boundaries (feature 3 in Figure 3(b)). These curves are also consistent with the conic envelopes for epipolar lines as described in [5].

## 5 Experimental Evaluation

In this section we present experimental results showing the performance of the proposed architecture. The datasets were obtained in an outdoor, urban environment near the ACFR building, using the system provided by [1]. The image resolution was $512 \times 384$, and we used the camera calibration toolbox [3] to obtain the camera parameters and to normalize the coordinates of the feature points. No uncertainty has been considered in these experiments for the feature points[5] such that $\mathbf{P}_{\mathbf{p}_{k-1}} = 0$ in (8). According to our experiments in urban environments, a 2D motion model of the vehicle ($\mathbf{x}_{v_k} = [x_{v_k}, y_{v_k}, \phi_{v_k}]^T$) has proven to be accurate enough since most of the time the vehicle is moving over a flat ground.

Figure 4 presents two sequences (top and bottom rows) of images where moving obstacles are detected from the moving vehicle. The general approach for moving obstacle detection is used successfully in this experimental setting. Several moving features are correctly detected, and the system is capable of tracking

[5] We assume that the camera is a “perfect” sensor after normalization, and that the tracking procedure does not introduce any additional uncertainty.

**Fig. 4.** Two sequences (top and bottom rows) showing the moving features detected from the moving vehicle

these over an extended period. No outliers are detected, thanks to the utilization of the more robust probabilistic scheme for feature prediction and motion detection described earlier. Other experiments under different settings have been completed[6].

## 6 Conclusions and Future Work

This paper presents a framework to detect moving objects based on the recognition of moving features in the images. The classification scheme is based on a complete probabilistic representation of feature locations that relates the vehicle motion with the visual information. Experimental evaluation under different settings in outdoor, urban environment shows the performance of the proposed architecture.

We are currently further improving our architecture in several directions. Firstly, we are considering the utilization of a full 3D motion model for the vehicle. This will be important to evaluate the robustness of the architecture in uneven and rough terrain. Secondly, we are extending the system in order to include the uncertainty in the tracking scheme and camera sensor. The proposed probabilistic scheme can readily support these uncertainty levels, however sensible measures need to be further estimated. And finally, we are integrating the

[6] More images and videos available at:
`http://www.acfr.usyd.edu.au/projects/research/ute/obstacledetection/`

detection and tracking scheme together with the segmentation and classification procedures. In this way, the low-level processes might take advantage of high-level information of detected moving objects through a feedback mechanism.

## References


1. ACFR, The University of Sydney and LCR, Universidad Nacional del Sur. PAATV/UTE Projects (2006)
2. Blackman, S., Popolif, R.: Design and Analysis of Modern Tracking Systems. Artech House Radar Library, New York (1999)
3. Bouguet, J.-Y.: Camera Calibration Toolbox for Matlab (2007)
4. Giachetti, A., Campani, M., Torre, V.: The Use of Optical Flow for Road Navigation. IEEE Transactions on Robotics and Automation 14(1), 34–48 (1998)
5. Hartley, R.I., Zisserman, A.: Multiple View Geometry in Computer Vision, 2nd edn. Cambridge University Press, Cambridge (2004)
6. Krishna, K.M., Kalra, P.K.: Detection, Tracking and Avoidance of Multiple Dynamic Objects. Journal of Intelligent and Robotic Systems 33, 371–408 (2002)
7. Kruger, W., Enkelmann, W., Rossle, S.: Real-time Estimation and Tracking of Optical Flow Vectors for Obstacle Detection. In: Intelligent Vehicles 1995 Symposium (1995)
8. Ma, Y., Soatto, S., Kosecka, J., Sastry, S.: An Invitation to 3-D Vision: From Images to Geometric Models. Springer, New York (2004)
9. Monteiro, G., Premebida, C., Peixoto, P., Nunes, U.: Tracking and Classification of Dynamic Obstacles Using Laser Range Finder and Vision. In: IEEE/RSJ International Workshop on Intelligent Robots and Systems, Beijing, China, IEEE, Los Alamitos (2006)
10. Nieto, J.: Detailed Environment Representation for the SLAM Problem. PhD thesis, University of Sydney, Australia (2005)
11. Sun, Z., Bebis, G., Miller, R.: On-Road Vehicle Detection: A Review. IEEE Transactions on Pattern Analysis and Machine Intelligence 28(5), 694–711 (2006)
12. Wang, B.: Simultaneous Localization, Mapping and Moving Object Tracking. PhD thesis, Robotics Institute, Carnegie Mellon University, USA (2004)
13. Woelk, F., Koch, R.: Fast Monocular Bayesian Detection of Independently Moving Objects by a Moving Observer. In: Rasmussen, C.E., Bülthoff, H.H., Schölkopf, B., Giese, M.A. (eds.) DAGM 2004. LNCS, vol. 3175, pp. 27–35. Springer, Heidelberg (2004)

# Mobile Robot Geometry Initialization from Single Camera

Daniel Pizarro[1], Manuel Mazo[1], Enrique Santiso[1], and Hideki Hashimoto[2]

[1] Department of Electronics, University of Alcala,
Esc. Politecnica, 28871 Alcala de Henares, Spain
{pizarro,mazo,santiso}@depeca.uah.es

[2] Hashimoto Lab, University of Tokyo
4-6-1 Komaba, Meguro-ku, Tokyo 153-8505, Japan
hashimoto@iis.u-tokyo.ac.jp

**Summary.** Using external cameras to achieve robot localization has been widely proposed in the area of Intelligent Spaces. Recently, an online approach that simultaneously obtains robot's pose and its 3D structure using a single external camera has been developed [8]. Such proposal relies on a proper initialization of pose and structure information of the robot. The present paper proposes a solution to initialization which consists of retrieving 3D structure and motion of a rigid object from a set of point matches measured by the camera. A batch Structure from Motion (SFM) approach is proposed along a short path. By incorporating odometry information available in the robot, the ambiguity generated by a single view in the solution is solved. We propose to describe robot's motion and image detection as statistical processes in which the uncertainty is properly modelled. Using a Gaussian equivalence of the processes involved, the SFM cost function is expressed as a Maximum Likelihood optimization. The paper shows the improvements of the approach in the presence of the usual odometry drift noise, compared with those using Euclidean distance as a likelihood. The proposed method is assessed on synthetic and real data.

## 1 Introduction

The usage of external cameras to obtain the pose of a robot is inspired in the concept of 'Intelligent Space' [6]. The aim is to provide the environment with intelligence and to reduce its necessity in the robots. As a consequence the space is composed by a set of sensors (*e.g.* Cameras ), a distributed intelligence, and a network layer, which allows communication with agents (*e.g. robots*). Under this framework, robot devices derive their localization and the high level navigation tasks from the space.

In this paper the usage of a single camera, external to the robot is proposed as a sensor device for robot localization. The ability of operating with a single sensor allows the system to work with few cameras, reducing heavily the cost and avoiding complex multiple camera configurations. There exist several approaches with the same aim that the present paper, by using artificial landmarks, either active [4] or passive ones [7]. In other works a geometrical model [10], is learnt previously to the tracking task in a supervised step. The common point of many

C. Laugier and R. Siegwart (Eds.): Field and Service Robotics, STAR 42, pp. 93–102, 2008.
springerlink.com 

of the proposals comes from the fact that high amount of prior knowledge is used for a posterior tracking.

In [8], a sequential algorithm that recovers the pose (with respect a global coordinate origin) and structure (set of 3D points from robot's structure) of a mobile robot is presented. The main novelty of this proposal is that it uses only as prior information the rigidity assumption in the geometry and the odometry readings present in the robot. The approach is based on retrieving the robot's pose by using a motion model and a sparse geometrical model which is learnt online. Such method depends heavily on a proper initialization of pose and structure before applying the online method.

In this work the initialization or self-calibration required in the online method is solved using one camera and odometry information. The initialization consists of a batch sequence of images of the robot taken during a short path which are synchronized with the odometry readings. In this paper a Structure from Motion (SFM) algorithm is proposed to solve the pose and structure by minimizing a likelihood function of the residuals in image frame. As the position estimation with odometry has a growing error the algorithm must tackle properly the uncertainty in the odometry metric reference.

In [2] the authors obtain the orientation and position of a camera with respect to a robot by using artificial markers and a set of predefined motions for calibration. The error in odometry position estimation and its influence in the solution is studied and the predefined motions are selected to minimize calibration error.

In the present paper we propose to describe robot's motion and image detection as statistical processes in which the uncertainty is properly modelled. Using a Gaussian equivalence of the processes involved, the SFM cost function is expressed as a Maximum Likelihood optimization. The main contribution of the paper is to show the improvements of this approach in the presence of the usual odometry drift noise, compared with those using Euclidean distance as a likelihood [9]. Besides, in this paper the natural structure of the robot is retrieved jointly with its pose.

We state the problem in §2. In §3 the likelihood distribution is analytically obtained from measurements and inputs. The final M.L cost function is presented in §4. Results on synthetic and real images are presented in §5. Finally, conclusions are presented in §6.

## 2 Problem Statement

As was commented before, the problem to solve is to achieve Structure from Motion by using one camera and the given odometry information which has growing error with robot's motion. To formalize the particularities of the problem, in this section it will be roughly commented on the notation and information involved.

Firstly, a set of discrete processes concerning pose, structure and measurements are presented. The number of time samples $k_N$ is considered to be finite and the proposed method runs as a batch process, so all the measurements and inputs are available from the beginning.

Robot's pose at time $k$ is described by a vector $x_k$. Usually for 3D motion (6 D.O.F) the vector is composed of 3 position components and 3 orientation angles $x_k = (r_k^x, r_k^y, r_k^z, \alpha, \beta, \gamma)$. For wheeled robots which its motion lie on a plane, the vector pose $x_k$ is reduced to 3 components $(r_k^x, r_k^y, \phi)$. Motion model $x_k = f(x_{k-1}, u_k)$ obtains actual position with respect to previous time and some input $u_k$ given by odometry (i.e. angular speed and linear speed of the robot).

The geometry of the robot is composed by a sparse set of $N$ 3D points $\mathcal{M} = \{m^1, \cdots, m^M\}$ referred from a local coordinate origin described by robot's pose $x_k$. The rigidity in geometry involves that the points $\mathcal{M}$ are static in time and no temporal subindex is necessary. Function $m_x^i = t(x_k, m^i)$ uses actual pose $x_k$ to express $m^i$ in the global coordinate origin.

The camera sensor follows a perfect "pin-hole" described by its 3x4 projection matrix $P$, which encodes intrinsic and extrinsic parameters. The camera projection model is expressed by the non-homogeneous transformation $y = h(m_x, P)$ which converts a 3D point $m_x$ expressed in a global coordinate origin into its 2D projection $y$ in the image plane using camera parameters $P$. The set of measurements which correspond to structure points is encoded in vector $Y_k = \{y_k^i \quad i = 1, \cdots, M\}$ where $y_k^i = h(t(x_k, m^i), P)$. Each measurement in image plane is obtained by using a fiducial detector which also is able to establish correspondences between images.

The objective proposed in this paper is to retrieve the initial pose $x_0$ and the set $\mathcal{M}$ given a set of image measurements $Y_{1:N_K}$ and odometry input information $u_{1:N_k}$.

In the next sections a statistical model of all processes is proposed. A random process is defined by its P.D.F and it is expressed in bold typography. (i.e. $\mathbf{x_k} = N(\hat{x_k}, \Sigma_k)$ is the Gaussian random process and $x_k$ is a realization of it).

## 3 Uncertainty Modeling

In this section, the objective is to obtain the statistical description of the processes stated in §2. By describing motion and image measurements as random processes, the uncertainty present in them is properly tackled and the odometry unbound error is directly modeled. In this paper it is shown how such representation improves the solution obtained in the reconstruction algorithm compared with a simple quadratic cost function.

The statistical models appear by including uncertainty noise variables into projection ($h$) and motion ($f$) models. We propose to approximate all statistical processes using Gaussian density distributions, which gives a meaningful description of the solution (mean) and uncertainty (covariance). The main difficulties come from the fact that the motion and projection models are in general non-linear equations, which transform the input Gaussian process and involves solving integral equations.

A first order approximation of the non linear function is used for propagating Gaussian processes. This approach has been widely used in Extended versions of Kalman Filtering [3]. It is known to suffer from bias [11], however it offers an analytical solution for the mean and covariance which will be necessary to minimize the likelihood cost function.

In the following sections the processes involved will be tackled in detail.

### 3.1 Motion Uncertainty

Initial solution for pose $x_0$ is represented by a Gaussian process $\mathbf{x_0} = N(\hat{x}_0, \Sigma_0)$ with a mean vector $\hat{x}_0$ and a covariance matrix $\Sigma_0$. Between time samples, $\mathbf{x_k}$ is consider a Markov Process with a transition kernel $p(x_k|x_{k-1}, u_k)$ ruled by the motion model $f$.

For simplicity, motion model used in this paper corresponds to a simple wheeled differential robot which moves over a ground plane. Pose parameters $x_k = (r_k, \phi_k)$ will be composed of position in the ground plane $r_k$ with respect a global coordinate origin $O$ and orientation $\phi_k$:

$$x_k = x_{k-1} + \begin{pmatrix} (Vl_k + W_k^1)\cos(\phi_{k-1} + \Omega_k + W_k^2) \\ (Vl_k + W_k^1)\sin(\phi_{k-1} + \Omega_k + W_k^2) \\ \Omega_k + W_k^2 \end{pmatrix}, \qquad (1)$$

Input from odometry is represented by the vector $u_k = (Vl_k, \Omega_k)$ with linear and angular speed components. Noise from odometry readings is modeled by a Gaussian process $\mathbf{W}_k = (\mathbf{W}_k^1, \mathbf{W}_k^2)$ which is statistically independent given $x_{k-1}$. Gaussian approximation of the distribution $p(x_k|x_0) = N(\hat{x_k}, \Sigma_k)$ is recovered by propagating the last pose $p(x_{k-1}|x_0)$ and the actual odometry $u_k$ through the transition kernel $p(x_k|x_{k-1}, u_k)$. The result is a Gaussian process with continuously growing variance usually presented in Dead-Reckoning [1] processes (See Figure 1).

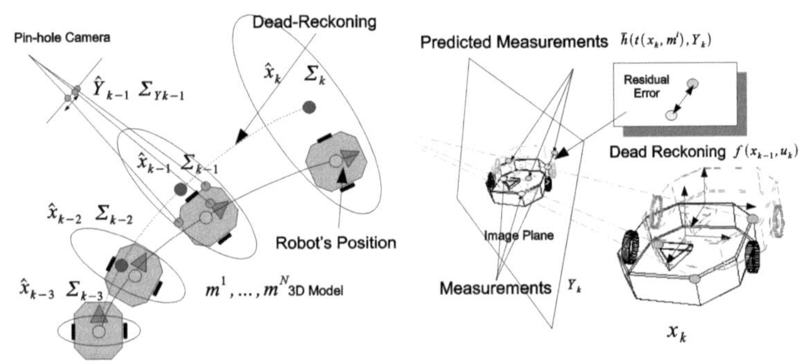


(a) Uncertainty Growing with odometry (b) Image residual information

**Fig. 1.**

### 3.2 Measurement Uncertainty

Measurement uncertainty depends on the method used to track fiducial points in the image. Measurements $y_k^i$ are jointly attached to form a vector $Y_k$ which will be described as a measurement process $\mathbf{Y_k}$. Each separate measurement $y_k^i$ is contaminated by additive Gaussian noise $\mathbf{v_k^i} = N(0, \Sigma_v)$. The noise components are independent between each other and are included in the noise process $\mathbf{V}_k$.

Likelihood distribution $p(Y_{1:k_N}|x_0, \mathcal{M})$ represents the joint distribution of the whole set of measurements in the sequence. It is obtained by combining projection model from every measurement $y_k^i$ into a global measurement equation $\bar{h}$:

$$Y_{1:k_N} = \bar{h}(x_{1:k_N}, Y_{1:k_N}, V_{1:k_N}) \quad V_k = \{V_k^i \quad i = 1, \cdots, M\}$$

$$\bar{h} = \left(h(t(x_1, m^1), P) + V_1^1, \cdots, h(t(x_{k_N}, m^M), P) + V_{k_N}^M\right) \in \mathbb{R}^{2 \cdot M \cdot k_N} \quad (2)$$

A Gaussian approximation of $p(Y_{1:k_N}|x_0, \mathcal{M}) = N(\hat{Y}_{1:k_N}, \Sigma_Y)$ is then obtained. It is used in this paper as the likelihood from which the solution to initial pose and structure is retrieved.

Usually, $\Sigma_Y \in \mathbb{R}^{n_Y \times n_Y}$ where $n_Y = 2Mk_N$, is a full correlated covariance matrix, in which all cross-correlation terms are present. The cross correlation terms will play an important role in the reconstruction step.

Assuming independence between each $\mathbf{x}_k$, the resulting covariance will be approximated by a box-diagonal one, where only terms from the same frame are correlated. In the same way, assuming that at time $k$ there exists independence between different measurements, the resulting covariance matrix will be box-diagonal where only coordinates of the same measurement can be correlated. (See (3)).

In the next section, the differences and the importance of correlated terms will be explained. In §5, some synthetic experiments are run to test the approaches selected for Gaussian propagation.

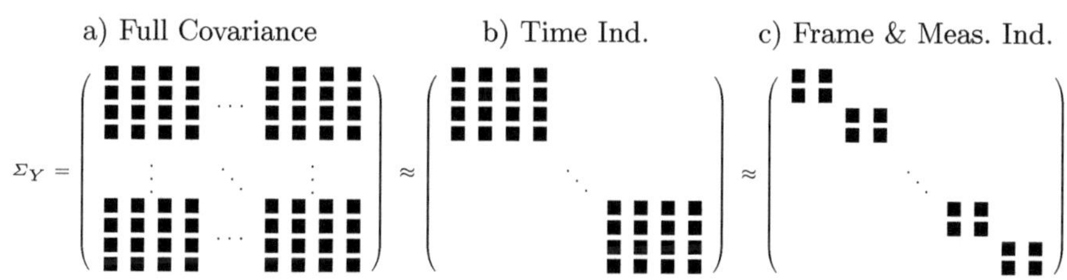


(3)

## 4 Maximum Likelihood Solution

Rigid structure from motion algorithms are at the moment in a mature state [5]. Thanks to the advances made in the last ten years, it can be considered a solved problem given that the correspondence between frames is correct. The usage of non-linear 'Bundle Adjustment' [9] with robust cost functions allows to obtain a solution for the SFM problem.

'Bundle Adjustment' makes use of reprojection errors to get robot initial pose $x_0$ and the set $\mathcal{M}$:

$$\min_{x_0, \mathcal{M}} \sum_{k}^{N_k} \sum_{i}^{M} \rho(y_k^i - h(t(x_k, m^i), P)), \quad (4)$$

where function $\rho$ strictly depends on the reprojection error statistical properties. In this paper the cost function $\rho$ is naturally obtained by maximizing the logarithm of the joint distribution $p(Y_{1:k_N}|x_0, \mathcal{M})$. By using Maximum Likelihood optimization, the cost function penalizes those residuals $y_k^i - h(t(x_k, m^i), P)$ which has more error due to odometry drift.

### 4.1 Maximum Likelihood Weighted Least Squares

By taking logarithm, the likelihood function proposed $p(Y_{1:k_N}|x_0, \mathcal{M})$, and approximated by a Gaussian function, is maximized by minimizing the following quadratic cost function:

$$\min_{x_0,\mathcal{M}} (Y_{1:k_N} - \hat{Y}_{1:k_N})^T (\Sigma_Y)^{-1} (Y_{1:k_N} - \hat{Y}_{1:k_N}) \tag{5}$$

It must be noticed that both $\Sigma_Y$ and $\hat{Y}_{1:k_N}$ are functions of the parameters to minimize $x_0$ and $\mathcal{M}$.

It is straightforward to see that when the covariance matrix $\Sigma_Y$ is the identity, the resulting cost function is directly the one presented in (4) with square of residuals as $\rho$.

Expression (5) can be optimized by using any non-linear iterative optimization method such as Gauss-Newton or Region-Trust algorithms as Levenberg Mardquardt.

### 4.2 Sparse Hessian

Each basic step of optimization involve the calculation of the following operations: We define as $\Phi_n = \{x_0, \mathcal{M}\}_n$ as the value of the parameters to minimize at iteration number $n$.

Given $\Phi_{n-1}$ the solution for the increment $\Phi_n = \Phi_{n-1} + \Delta\Phi$ at iteration $n$ is obtained by solving the following linear system:

$$J^T(\Sigma_Y)^{-1} J \Delta\Phi = J^T(\Sigma_Y)^{-1}(Y_{1:k_N} - \hat{Y}_{1:k_N}(\Phi_{n-1})), \tag{6}$$

where $J$ represents the first derivative of $\hat{Y}_{1:k_N}$ evaluated at $\Phi_{n-1}$.

Matrix $H = J^T(\Sigma_Y)^{-1} J$ which is a square matrix of the size of $\Phi$ represents the approximated Hessian.

At each iteration system (6) is solved by inverting H.

$$\Delta\Phi = (H)^{-1} b, \tag{7}$$

If $P_Y$ is a diagonal matrix, usually $H$ have a sparse structure. This allows a very important improvement in the computational complexity for its inversion. The situation is specially critic when $\mathcal{M}$ contains a large number of 3D points.

As we can see in Figure 2, the inclusion of a full covariance $\Sigma_Y$ or only a time correlated one, destroys the natural sparsity of the problem. The direct effect is that the systems become of higher complexity.

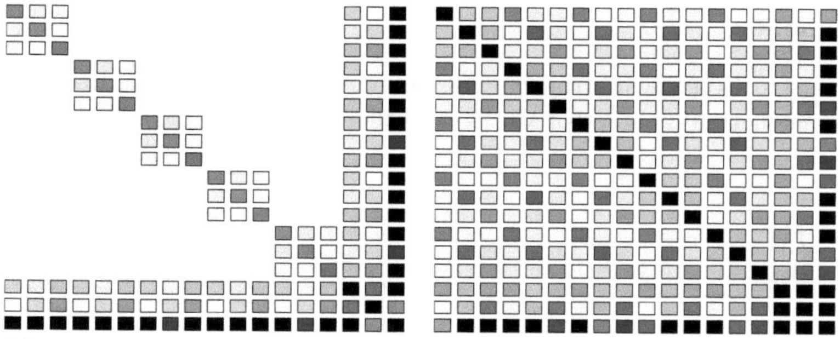

(a) Elements of H with diagonal or F.M.I Boxed diagonal $\Sigma_Y$

(b) Elements of H with full covariance or F.I Boxed diagonal $\Sigma_Y$

**Fig. 2.** Hessian Configurations

### 4.3 Initial Guess

As a non-linear iterative algorithm, the proposed optimization requires an initial guess for vector $\Phi$. The assumption is that if robot height is small compared to the distance to the camera the points from robot structure can be supposed to lie on the ground plane. Using the set of measurements from the first frame $Y_0$, and knowing the position of the ground plane, a set of 3D positions from global reference frame can be obtained, namely $X_0$. The set $\mathcal{M}$ is referred to the initial pose $x_0$. By using centroid of the set $X_0$ the position $r_0$ is easily obtained.

## 5 Results

In this section some results on synthetic generated data and real images are presented.

### 5.1 Synthetic Data

The experiment conditions are the following:

- A set of M=30 points $m_k^i$ initialized as a random Gaussian jointly with robot true initial position $r_0$. The geometry has a deviation around the center of rotation of $\sigma = 200(mm)$
- Camera calibration and ground position from a real configuration.
- Odometry noise variance $\Sigma_W$ is varied from $\sigma^2 I_{2x2}$ $\sigma = 0.1, \cdots,$ 20 (m/s,rad/s).
- Measurement noise $\Sigma_V$ with standard deviation of around 0.1 pixels.

The following algorithms will be tested:

1. Element Diagonal covariance estimation (E.D): By supposing that $\Sigma_Y$ is diagonal. No weighted optimization.
2. Frame Box Diagonal covariance matrix (F.D): See Figure 2(a).
3. Measurement Box Diagonal covariance (M.D): See Figure 2(b).

**Experiments:**

- Reconstruction mean error vs. Odometry noise variance. (Figure 3(a))
- Reconstruction mean error vs. % of outliers. (Figure 3 (b)).

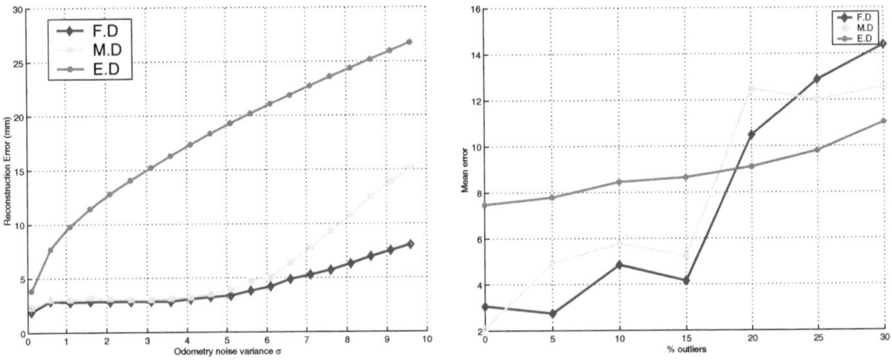


(a) Reconstruction error vs Odometry noise

(b) Real path and dead-reckon estimation

**Fig. 3.** Experiment 1

## 5.2 Real Data

A real system composed of a low cost camera and a real mobile robot is tested. The error initialization is the same used in the synthetic data.

The result of an initial calibration procedure is shown in Fig. 4 (b). As can be see five coplanar points are considered as initial state vector. The path required to obtain the estimation is shown in Fig. 4 (a).

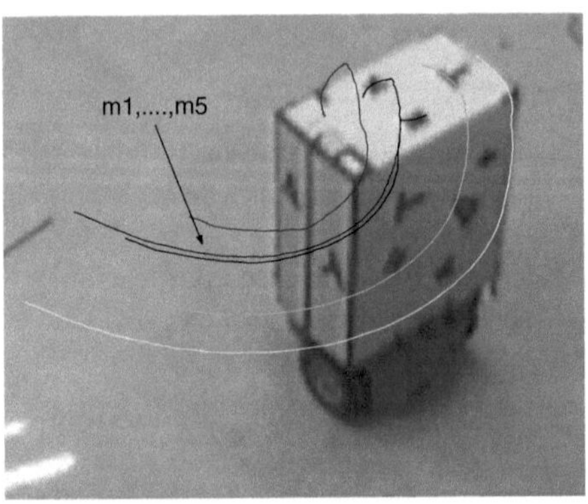


(a) Real robot initialization

| | | | |
|---|---|---|---|
| $m_x^1$ | 55.6421 | $m_x^2$ | 18.5460 |
| $m_y^1$ | −12.8574 | $m_y^2$ | 117.9783 |
| $m_z^1$ | 696.8408 | $m_z^2$ | 700.4888 |
| $m_x^3$ | −63.2563 | $m_x^4$ | 44.9386 |
| $m_y^3$ | −7.0530 | $m_y^4$ | 194.9733 |
| $m_z^3$ | 694.5169 | $m_z^4$ | 681.7358 |
| $m_x^5$ | −125.9381 | $x_0$ | 34.5282 |
| $m_y^5$ | −93.5649 | $y_0$ | 756.0392 |
| $m_z^5$ | 696.0449 | $\alpha_0$ | 111.2295 |

(b) Reconstructed table

**Fig. 4.** Real results

## 6 Conclusions

In this paper a method for computing robot geometry and pose from single camera is presented, with the aim of properly initializing a sequential approach. The inclusion of odometry allows metric reconstruction with single camera. Knowledge about the statistical properties of odometry readings and camera detection processes allows to propose a Maximum Likelihood approach which penalizes properly information which has more uncertainty as the last positions given by a dead-reckoning process. The equivalent approach is a weighted quadratic cost functions. The reconstruction accuracy is proved to be higher in the case of M.L versions. As can be seen in Figure 3 error is higher when no statistical information is used. Between the two approximations of the covariance matrix, the one which has more cross-correlated terms (F.D) achieves better performance when the error grows. The main drawback is that the more cross correlated terms the less sparse the Hessian becomes, which increases problem complexity. As a conclusion, for low error in odometry the Measurement Box Diagonal (M.D) approach will have the same behavior with the advantage that the sparsity of the problem remains equal to a Element Diagonal approach (E.D). Real results with coplanar points are also presented, achieving good accuracy with real odometry readings and image matching. This work was supported by the Ministry of Science and Technology under RESELAI project (reference TIN2006-14896-C02-01).

## References

1. Borenstein, J., Feng, L.: Measurement and correction of systematic odometry errors in mobile robots (1996)
2. Chen, H., Matsumoto, K., Ota, J., Arai, T.: Self-calibration of environmental camera for mobile robot navigation. Robotics and Autonomous Systems 55(3), 177–190 (2007)
3. Davison, A.J.: SLAM with a single camera. In: Workshop on Concurrent Mapping and Localization for Autonomous Mobile Robots, in conjunction with ICRA (2002)
4. Hada, Y., Hemeldan, E., Takase, K., Gakuhari, H.: Trajectory tracking control of a nonholonomic mobile robot using igps and odometry. In: Multisensor Fusion and Integration for Intelligent Systems, MFI 2003. Proceedings of IEEE International Conference on, July 30 - August 1, 2003, pp. 51–57 (2003)
5. Hartley, R., Zisserman, A.: Multiple View Geometry in Computer Vision, 2nd edn. Cambridge University Press, Cambridge (2003)
6. Lee, J.-H., Morioka, K., Hashimoto, H.: Intelligent space and mobile robots. In: The Industrial Information Technology Handbook, pp. 1–15 (2005)
7. Morioka, K., Lee, J.H., Hashimoto, H.: Human-following mobile robot in a distributed intelligent sensor network. Industrial Electronics, IEEE Transactions on 51(1), 229–237 (2004)
8. Pizarro, D., Santiso, E., Mazo, M.: Simultaneous localization and structure reconstruction of mobile robots with external cameras. In: ISIE 2005 (June 2005)
9. Triggs, B., McLauchlan, P., Hartley, R., Fitzgibbon, A.: Bundle adjustment – a modern synthesis. In: Triggs, B., Zisserman, A., Szeliski, R. (eds.) ICCV-WS 1999. LNCS, vol. 1883, pp. 298–372. Springer, Heidelberg (2000)

10. Sogo, T., Ishiguro, H., Ishida, T.: Acquisition of qualitative spatial representation by visual observation. In: IJCAI 1999 (1999)
11. van der Merwe, R., de Freitas, N., Doucet, A., Wan, E.: The unscented particle filter. Technical report, Cambridge University Engineering Department (August 2000)

# Part III

# Underwater Robots and Systems

# A Feature Based Navigation System for an Autonomous Underwater Robot

John Folkesson, Jacques Leederkerken, Rob Williams,
Andrew Patrikalakis, and John Leonard

Massachusetts Institute of Technology

**Summary.** We present a system for autonomous underwater navigation as implemented on a Nekton Ranger autonomous underwater vehicle, AUV. This is one of the first implementations of a practical application for simultaneous localization and mapping on an AUV. Besides being an application of real-time SLAM, the implemtation demonstrates a novel data fusion solution where data from 7 sources are fused at different time scales in 5 separate estimators. By modularizing the data fusion problem in this way each estimator can be tuned separately to provide output useful to the end goal of localizing the AUV, on an a priori map. The Ranger AUV is equipped with a BlueView blazed array sonar which is used to detect features in the underwater environment. Underwater testing results are presented. The features in these tests are deployed radar reflectors.

## 1 Introduction

Navigating underwater requires localization without GPS. Stationary features in the environment can be used to help the robot stay localized [1], [2], [3], [4]. Examples of sonar mapping can be seen in [5], [6] and [7]. This is commonly known as simultaneous localization and mapping, SLAM . Some underwater robots have the advantage of rather expensive sensors such as Doppler velocity logs for measuring the movement of the robot. In this application the robot must be expendable which limits the choice of sensors. Instead of a single high quality sensor of the motion we have a number of weaker sensors that must be fused to estimate the motion of the robot.

The problem of fusing data from dissimilar sensors is challenging for a number of reasons [8], [9], [10], [11], [12]. First there is a temporal issue in that some sensors need to be incorporated with a simple calculation at a high update rate while other information might need intensive calculation but at a much lower update rate. Another issue is observablity. Generally each sensor provides only a partial observation of the total state. The problem is complicated by the fact that some sensors provide measurements in the relative frame of the robot while others are in the absolute earth frame. Measurement errors that are correlated over time due to bias or environmental disturbances can lead to over confident estimates if the errors are treated as independent.

A more practical issue is that of getting the system to work robustly. This requires that it be possible to examine each sensor's contribution to the final

C. Laugier and R. Siegwart (Eds.): Field and Service Robotics, STAR 42, pp. 105–114, 2008.
springerlink.com 

estimate separately. Only then can one hope to understand the system well enough to tune the parameters.

## 2 Project Overview

The task of clearing a lane from sea to shore through a mine field has traditionally been carried out by human divers. This method is associated with very high casualty rates but in many cases there are no alternatives. It is therefore very desirable to come up with methods that do not cost lives.

The system presented here is part of a larger project. The larger mission scenario starts with a target field of underwater mines along a shoreline. The mine field will be scouted first by a REMUS AUV. The data collected from the REMUS will be used to create an a priori map of the mines and other strong sonar reflecting objects along the shore. A particular target mine will be selected. The Nekton Ranger will then be sent on a mission to locate and attach to the target. As the Ranger is a less expensive AUV than REMUS it is meant to represent an expendable robot that could be set to detonate an explosive charge after attaching itself to the target.

The a priori map will not be perfect. It will have an absolute accuracy of about 5 meters. The relative accuracy between nearby features will be considerably better. The Ranger will be be able to get to within 70 meters of the target on the surface and to dive to the target depth. The task then becomes one of distinguishing between several nearby targets to find the designated target. The AUV must then navigate to that target accurately.

The approach taken is to have the Ranger carry out simultaneous localization and mapping, SLAM, were the SLAM state is initialized with the a priori map including its uncertainty. By then matching the features detected by the Ranger to the a priori map the robot pose will become correlated to the a priori map. When this correlation is sufficiently accurate the target can be acquired with confidence.

The a priori map will initially have relatively tight correlations between features that go like the inverse of the distance between the features. The absolute uncertainty of the features will be much looser allowing the map to shift as a whole to agree with where the Ranger finds it. Thus upon matching a feature seen by the Ranger with an a priori map feature both the Ranger pose and the map will shift to reconcile any differences. The side scan sonar used by the REMUS to collect the data for the a priori map is different than the forward looking blazed array sonar on the Ranger. It is therefore likely that the maps will look different both in feature positions and numbers.

## 3 System Overview

Figure 1 is a schematic over the navigation system. Its inputs, raw motion data and sonar detections, are written to as data arrives. These writes then trigger the filters downstream to update as data is pushed and pulled through the system.

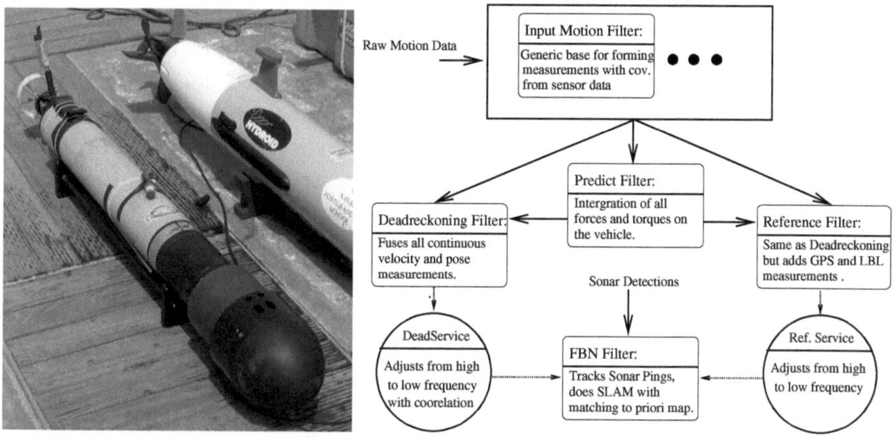


**Fig. 1.** The Nekton Ranger is shown to the left. The separate modules of the navigation system are shown on the right.

The data from the motion sensors is fed to sensor specific modules called input motion filters in figure 1. These model the individual sensors error characteristics. This sensor data is fused using five separately tuned estimators. Each estimator has a specific purpose and frequency as summarized in table 1 The outputs of the input motion filters are pushed to the predict filter and to two separate Extended Kalman Filters (EKF). The predict filter uses a model of the forces and torques acting on the vehicle to estimate the motion over short periods of time. Each of the EKF filters has a specific function. The dead-reckoning filter provides the best continuous estimate of the robot pose. This estimate is used as the frame for accumulating data from the sonar.

The reference EKF is identical to the dead-reckoning EKF aside from its inputs and outputs. GPS and long base line transponder data (LBL) are the additional inputs. These give measurements of the absolute x and y position of the robot. They also cause the estimated trajectory to have discontinuities in it.

**Table 1.** The five estimators that fuse the various sensor data are summarized

| Estimator | Output | Rate | Inputs |
|---|---|---|---|
| Predict Filter | Motion between measurements | 100 Hz | Propeller Actuation signals, Various parameters |
| Deadreck. EKF | Smooth pose trajectory for tracking | ≈ 30 Hz async | 3-Axis Compass, Depth, Altitude, Predict output |
| Reference EKF | Corrected xy and parameter estimates | ≈ 30 Hz async | 3-Axis Compass, Depth, Altitude, GPS, LBL, Predict output |
| Tracking Filter Squareroot SAM | Composite measurements | 1 to 2 Hz | Sonar Detections, Dead-reckoning output |
| Global EKF | Localization on an a priori map | .01 to .1 Hz | Composite Measurements and Reference output |

These 'jumps' are why the reference estimate is not used for feature tracking. The reference EKF also estimates more parameters than the dead-reckoning EKF. The additional parameter estimates such as water current velocity can only be estimated using the GPS or LBL. The corrected parameters and their covariance are then passed to the predict filter to be used in prediction for both EKF filters.

The feature based navigation, FBN, filter consists of two separate estimators. The tracking filter and the global EKF. The tracking filter accumulates information from the sonar detections over short snippets of the robot path, (SLAM). The local map and trajectory from these snippets are then used to form *composite measurements*. These are then used to update the global EKF. Matching to the prior map is done on the global EKF.

## 4 The Predict Filter

A propeller on the aft end of the robot is mounted on a universal joint that can be pan/tilt rotated relative to the robot. An estimate of the pan/tilt angle is used to decompose this propeller thrust into robot pitch/yaw torques and a force forward. The movements of the robot through the water are resisted by forces and torques proportional to the linear and angular velocities. These forces then produce a convergent velocity for a given RPM. The parameters are tuned to give the empirical .75 m/s terminal velocity at 1,200 RPM, Fig. 2. The prediction model includes damping, actuation transfer, water velocity, buoyancy, and wave surge parameters, all modeled as Gaussian.

These various forces and torques are integrated at a rate of 100 Hz using Newton's equations to provide the EKF filters with estimates of the motion between the sensor measurements. The EKF filter can be configured to estimate and pass to the predict filter values for any of the parameters. The predict filter will then used those along with its own defaults for the others when making predictions. The right plot of Fig. 2 shows the improvement in the dead-reckoning from using the reference EKF estimated parameters.

## 5 Dead-Reckoning EKF

The dead-reckoning EKF estimates the robot pose, its velocity and angular actuation parameters, a total of 15 dof. As input it has depth, altitude velocity, and orientation from a 3-axis compass. The output is a smooth and accurate estimate of the motion between two sonar pings.

The dead service module does the translation from the EKF outputs at around 30 Hz to two pose estimates at the ping rate (1-2 Hz). The two pose estimates are a 6 dof incremental change estimate $P_i$ and a 4 dof earth frame measurement, $P_e$. The two estimated poses are then able to account for the correlations between estimates taken at different times. The EKF provides snapshots of the uncertainty at the time of each sonar ping. We must break this uncertainty up into a part relative to the previous pose and a part relative to the earth. This is done by accumulating the predictions used in the EKF. Both poses' mean are

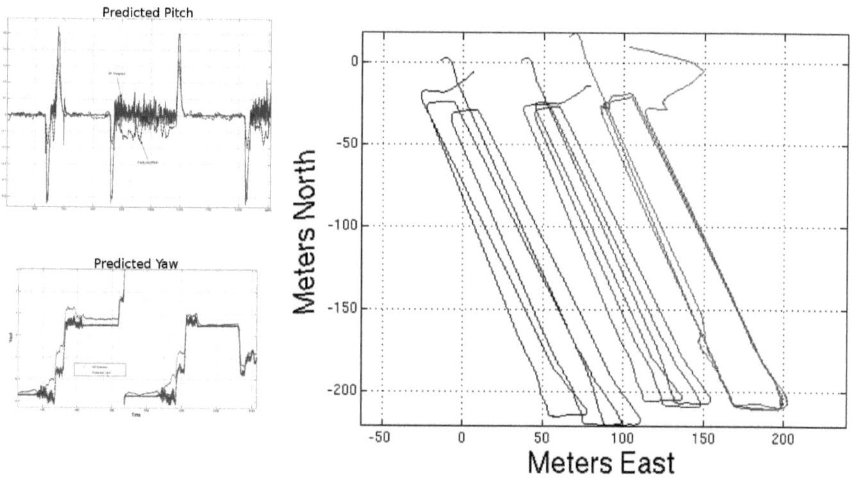


**Fig. 2.** The tuning of the prediction for the yaw and pitch angle is done by comparing to the 3D compass as shown on the bottom left (yaw) and. upper left (pitch). The figure to the right compares the dead-reckoning and reference estimates of the robot trajectory. The three paths have been offset to separate them for comparison. The path to the right is the reference EKF output. It uses the GPS during the turn arounds on the surface. The middle is the dead-reckoning EKF, sharing the predict filter with the reference EKF. This allows it to benefit from the parameter estimate made in the reference EKF. The left is the EKF, without any parameter sharing.

the EKF values. The covariance of $P_i$ is taken as the accumulated prediction value. This then gives a system of matrix equations that can be solved for the covariance of $P_e$. In this way we were able to have the tracking filter run at a much slower rate than the EKF filters. At the same time we had the exact pose and covariance estimate that the EKF provided.

## 6 Reference EKF

The reference EKF adds estimates of velocity damping, linear actuation, and water velocity parameters to the state vector estimated by the dead-reckoning EKF. A total of 27 dof. The purpose of the reference filter is quite different than that of the dead-reckoning EKF. The reference EKF is used to smooth the GPS and LBL measurements for use by the global EKF which operates at a much lower frequency. Only the xy parts are actually used. The global EKF will only use the reference xy estimate for an update if the covariance in xy has decreased (ie. when the Reference EKF has had a GPS update).

The reference EKF will only use a GPS measurement if its estimated covariance is less than that of the reference EKF. The reason for this seemingly odd update rule is that the GPS measurements have time correlated errors. Using consecutive GPS measurements and treating them as independent will cause the filter to become over confident very quickly. The rule softens this and essentially ensures that the covariance of the EKF will never get much lower than that of the single best GPS measurement.

## 7 The Tracking Filter

The tracking filter has the function of tracking features in the sonar data and using them to correct the motion of the robot over times of 10 to 100 seconds. These times correspond roughly to the time a feature is in the field of view of the sonar. The output of this filter is a composite measurement that combines the dead-reckoning and feature tracking information over this interval. So a composite measurement gives the positions of any features tracked, the estimated pose change for the robot and the absolute $(z,\theta,\phi,\psi)$ of the robot at the end of the interval.

The tracking is done in a small scale but highly detailed SLAM filter. This SLAM filter captures all the characteristics of the blazed array sonar. It estimates the entire path of the robot over the interval. Thus it adds 6 dimensions to its state for the robot pose on each new ping. The features are parameterized in the sensor frame of the first pose they were seen from. In that way the feature will move when this pose is re-estimated. The parameterization in that frame is the range and bearing to the feature. This ensures that the uncertainty in these parameters will be closer to Gaussian than simple Cartesian coordinates would be.

The tracking estimator is a square root filter [13]. This is a Gaussian estimator similar to the Extended Information Filter, EIF, which in turn is equivalent to the EKF. It has several important advantages. The most important is that as the square root of the information matrix is used the condition number of this matrix will be smaller. A smaller condition number means that the matrix equations that must be solved are numerically more well-posed. This will virtually always lead to an improvement over EKF or EIF for SLAM. Secondly, one can in our case do the updates in constant time. This is because the short snippets of path we use have no loops in them. Partial covariances can be computed in constant time as well. Estimating the full covariance would take a longer time but is actually never needed.

The details of this filter would take us beyond the scope of this paper and will appear in a later publication. Here we present the system as a whole and concentrate on how the parts work together in a particular application.

## 8 The Global EKF

The global EKF is updated with the composite measurements. This begins by augmenting the state with the pose and features of the composite measurement. The EKF state can then be updated using the composite measurement. The older previous pose can then be marginalized out. This implies simple deleting the rows and columns of the older pose. The new features can then be matched and merged with the older features including the a priori map features.

The matching of the features is done in three ways. First any features that were tracked in the tracking filter from one composite measurement to another are matched based on a key provided by the tracking filter. Second a threshold is placed on the mahanolobis distance between single features that are relatively well correlated to one another to determine matching features. Third multiple feature matches are search for. All of the pairs of possibly matched features are used a a basis over a hypothesis space. This space is search exhaustively if small enough or selectively if too large. The details of the selective search and match criteria are original and require an entire paper to describe. One part of our criteria is the joint compatibility test [14]. This forms a distance between sets of points which is essentially the Mahanolobis distance of a pseudo measurement equating the point in each set.

$$d = \eta S^{-1} \eta. \tag{1}$$

Where:

$$S = C_{\eta\eta} + J_{\eta,\mathbf{x}} C_{\mathbf{xx}} J_{\eta,\mathbf{x}}^T. \tag{2}$$

Here $C_{\eta\eta}$ is the covariance of this pseudo measurement (a parameter) and should reflect the finite size of the actual features, $C_{\mathbf{xx}}$ is the covariance of the EKF state and $J_{\eta,\mathbf{x}}$ is the Jacobian of the measurement. This is balanced against a threshold that is based on certain statistical measurements accumulated over the path of the robot.

The problem of finding candidate combinations to apply this test to is where the difficulty lies. The number of possibilities quickly becomes unmanageable as the number of features in the map increases. For this application we deal with a relatively small number of features and a very slow update rate for the maps. However we still need to limit this search at times.

We start by forming $matchdata_{ij}$ objects between each pair of features, labeled $i$ and $j$. This is a $N^2$ operation where $N$ is the number of features. The match data consists of $r_{ij}$, the Euclidean distance between the points,$c_{ij}$, the covariance of $rij$, and $a_{ij}$, the angle of the vector between them.

$$rij = |\mathbf{x}_i - \mathbf{x}_j|, \tag{3}$$

$$c_{ij} = J_{r,x} C_{\mathbf{xx}} J_{r,x}^T \tag{4}$$

$$aij = arctan(y_i - y_j, x_i - x_j) \tag{5}$$

$$matchdata_{ij} = (r_{ij}, c_{ij}, a_{ij}) \tag{6}$$

Remember that we are using a compass and will take full advantage of knowing the orientation of the robot and thus these angles in some cases. The cases that we feel reasonably confident about the angles between two features is when the two features are relatively tightly correlated. This means that they were either seen near each other in time as the robot moved or that they were matched to features that were in turn tightly coupled to one another.

We see that this concept of two features being tightly coupled is important. We define tightly coupled as $c_{ij}$ < some threshold. Having this we can form $N$ lists of $matchdata_{ij}$ objects where the $i^{th}$ list contains all the $j$ that are tightly coupled to $i$. These are the local patches that we will try to find multiple matches between.

Continuing to try and limit the number of possible matches we can note all pairs that are possibly the same point feature by either $r_{ij}$ or $rij^2/c_{ij}$ being relatively small. These will be stored on a list of possible matches. We can then iterate through this list and search for multiple matches between the local patches associated with each of the possibly matched features. We check if there are other possible matched features between the local patches by checking the list.

## 9 Experiments

The experiments consisted of a series of underwater tests. A target field was deployed consisting of 12 radar reflectors anchored 1 meter off the sea floor. These were meant to represent a mine field obstructing access to a part of the shoreline. The targets were deployed in three rows perpendicular to the shore. Two of the rows had five reflectors and a center row had 2 reflectors. This target field was then surveyed by the REMUS AUV (see Figure 1). This data was then used to form a hand made a priori map for the Ranger AUV to use, Figure 3 left.

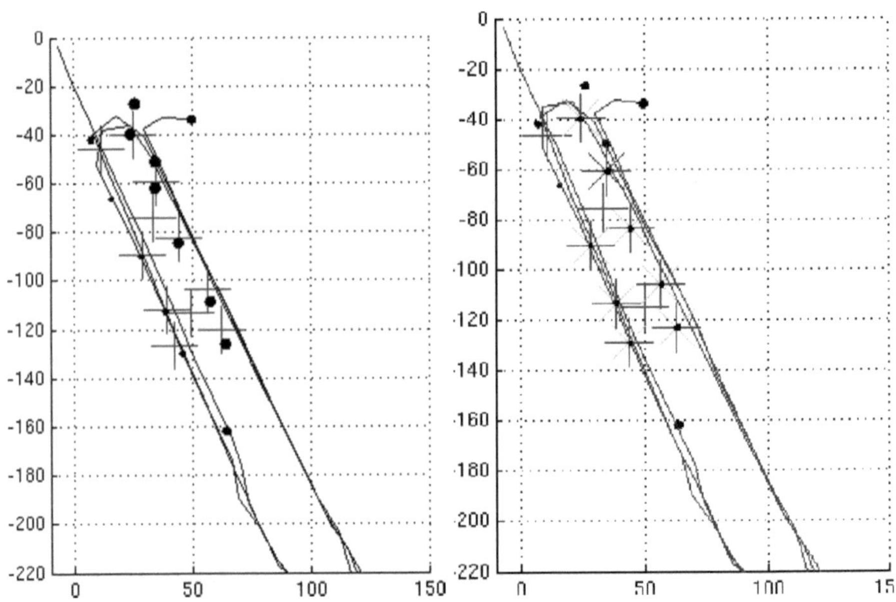


**Fig. 3.** On the left is the show the SLAM map as black ellipsoids and the a priori map as red crosses. This map made no attempt to match the two maps to one another. On the right is a map made with matching. The robot succeeded in matching on all 6 dives demonstrating the localization of the robot is working.

The final map made by the Ranger is seen to be very similar to the a priori map however it is shows some features displaced from the priori positions. The absolute xy of the map is of no significance here the important point is that the robot pose is correlated to the a priori map. This despite significant differences in the GPS location fix. The robot here dove six separate times and was able to correctly match features to the map each time. In at least one instance, (the first return to start location), the ranger FBN Filter gave better estimated positions then the Reference EKF had.

## 10 Conclusion

The system was able to localize the robot on the a priori map in real time. This will allow an autonomous AUV to home in on a pre-designated target. The approach of fusing the different sensor data using five separate estimators works well. This approach allows us to modularize the system which adds flexibility and presumably some robustness. The various estimates give us a more precise view of the overall estimation process than a single integrated estimator could give. We can then identify the source of any errors more quickly. These error then give us a valuable guide to adjustments and refinements of the estimators. The GPS was able to be used to improve the dead-reckoning estimate without introducing any discontinuities in the pose estimate. That thanks to a dual EKF implementation.

In the future we plan to work on the final control of the robot to the designated target. We would also like to add some active perception to the system by having the robot rotate slightly both in pitch and yaw when targets are in view. This will help the convergence of the tracked feature's covariances.

## References

1. Dissanayake, M.G., Newman, P., et al.: A solution to the simultaneous localization and map building (SLAM) problem. IEEE Trans. on Robotics and Automation 17(3), 229–241 (2001)
2. Newman, P., Leonard, J., Rikoski, R.: Towards constant-time slam on an autonomous underwater vehicle using synthetic aperture sonar. In: Proc. of the International Symposyum on Robotics Research (ISRR 2003) (2003)
3. Leonard, J., Rikoski, R., Newman, p., Bosse, M.: Mapping partially observable features from multiple uncertain vantage points. IJRR International Journal on Robotics Research 7(3), 943–975 (2002)
4. Leonard, J.J., Rikoski, R.J.: Incorporation of delayed decision making into stochastic mapping. In: Experimental Robotics VII. lecture Notes in Control and Information Sciences, vol. 271, pp. 533–542. Springer, Heidelberg (2001)
5. Ribas, D., Ridao, P., Neira, J., Tardos, J.: Slam using an imaging sonar for partially structured underwater environments. In: Proc. of the IEEE International Conference on Intelligent Robots and Systems(IROS 2006), IEEE, Los Alamitos (2006)
6. Leonard, J.J., Carpenter, R., Feder, H.J.S.: Stochastic mapping using forward look sonar. Robotica 19, 467–480 (2001)

7. Tena, I., de Raucourt, S., Petillot, Y., Lane, D.: Concurrent mapping and localization using sidescan sonar. IEEE Journal of Ocean Engineering 29(2), 442–456 (2004)
8. Castellanos, J.A., Neira, J., Tardós, J.D.: Multisensor fusion for simultaneous localization and map building. IEEE Transactions on Robotics and Automation 17(6), 908–914 (2001)
9. Durrant-Whyte, H.F., Rao, B.Y.S., Hu, H.: Toward a fully decentralized architecture for multisensor datafusion. In: Proc. of the IEEE International Conference on Robotics and Automation (ICRA 1990) (1990)
10. Moutarlier, P., Chatila, R.: Stochastic multisensory data fusion for mobile robot location and environmental modelling. In: Proc. of the International Symposium on Robotics Research, pp. 85–94 (1990)
11. Guivant, J.E., Masson, F.: Using absolute non-Gaussian non-white observations in Gaussian SLAM. In: Proc. of the IEEE International Conference on Robotics and Automation (ICRA 2005), vol. 1, pp. 338–343 (2005)
12. Sukkarieh, S., Nebot, E., Durrant-Whyte, H., Corba, M.: A high integrity IMU/GPS navigation loop for autonomous land vehicle applications. IEEE Trans. on Robotics and Automation 15(3), 572–578 (1999)
13. Dellaert, F.: Square root sam: Simultaneous location and mapping via square root information smoothing. In: Robotics: Science and Systems (2005)
14. Neira, J., Tardós, J.D.: Data association in stocastic mapping using the joint compatibility test. IEEE Transaction on Robotics and Automation 17(6), 890–897 (2001)

# Experiments in Navigation and Mapping with a Hovering AUV

George Kantor, Nathaniel Fairfield, Dominic Jonak, and David Wettergreen

The Robotics Institute
Carnegie Mellon University
Pittsburgh, Pennsylvania 15213 USA
{kantor,than,dom,dsw}@cmu.edu

**Summary.** This paper describes the basic control, navigation, and mapping methods and experiments a hovering autonomous underwater vehicle (AUV) designed to explore flooded cenotes in Mexico as part of the DEPTHX project. We describe the low level control system of the vehicle, and present a dead reckoning navigation filter that compensates for frequent Doppler velocity log (DVL) dropouts. Sonar data collected during autonomous excursions in a limestone quarry are used to generate a map of the quarry geometry.

## 1 Introduction

The DEPTHX (DEep Phreatic THermal eXplorer) project is a three-year NASA-funded effort whose primary objective is to use an autonomous vehicle to explore and characterize the unique biology of the Zacatón cenote. Zacatón, the world's deepest known limestone sinkhole, is a water-filled cavern that is at least 300 meters deep. The depths of Zacatón are geothermally heated with a high sulfur content and a lack of sunlight or dissolved oxygen, making this an ideal place to search for exotic microbial life [Gary, 2002]. The robotic exploration and search for microbial life in Zacatón is an analog mission for the search for life in the liquid water ocean beneath the frozen surface of Europa.

The DEPTHX robot (Figure 1) is a hovering autonomous underwater vehicle (AUV) designed to explore flooded caverns and tunnels while building 3D maps, collecting environmental data, and obtaining samples from the water column and cavern walls. To accomplish these tasks, the vehicle is equipped with a Doppler velocity logger (DVL), a ring laser gyro-based inertial navigation system (INS), a depth sensor, and an array of 54 narrow beam sonar transducers. In the Zacatón mission, the vehicle will use all of these sensors together to perform simultaneous localization and mapping (SLAM). This paper presents a preliminary field result where the INS, DVL, and depth sensor are used together to generate a dead reckoned position estimate and the sonar array is used to create a map based on that estimate. This technique was used to autonomously map a limestone quarry in Austin, Texas during field tests conducted in January of 2007.

The paper is organized as follows: Related research activities in using AUVs for science and mapping are briefly discussed in Section 2. Then we provide a description of the instrumentation of the DEPTHX vehicle in Section 3. This is followed by a description of our implementation of a Kalman filter to merge DVL and IMU measurements,

C. Laugier and R. Siegwart (Eds.): Field and Service Robotics, STAR 42, pp. 115–124, 2008.
springerlink.com 

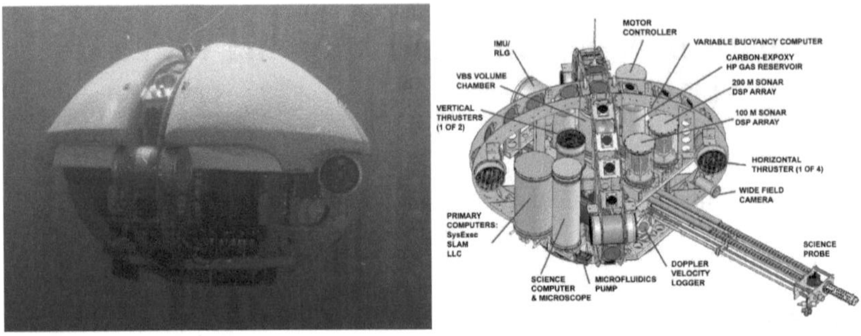


**Fig. 1.** Left: the DEPTHX AUV deployed in a test tank at Austin Research Laboratory. Right: A model of the DEPTHX vehicle structure and components. Eleven pressure vessels house computing, batteries, sensors, and science instruments. Diameter is approximately 2m. ©Stone Aerospace, 2006.

allowing for high performance dead reckoned position estimation even in cases where the DVL frequently drops out (Section 4). An overview of the vehicle control system is provided in Section 5. Experimental results are presented in Section 7, followed by some conclusions and a discussion of future work in Section 8.

## 2 Related Work

There are more than 1000 commercial underwater vehicles in operation today, performing tasks as diverse as archaeological excavation to drill-rig maintenance to oceanographic survey [Whitcomb, 2000].

There are a variety of techniques employed for determining the position of underwater vehicles – unfortunately GPS does not work underwater – and they can be divided into those that utilize emplaced infrastructure and those that do not. (See [Leonard et al., 1998] for a comprehensive survey.)

When accurate position is needed underwater, most AUVs and many human-driven remotely operated vehicles (ROVs) rely on a surveyed array of acoustic beacons, known as a long base-line (LBL) array. Acoustic beacons provide a fixed frame of reference for positioning the vehicle [Whitcomb, 2000]. [Yoerger et al., 2007] have shown detailed sea-floor mapping of subsea vents using an LBL array with the Autonomous Benthic Explorer.

Over large distances, or in underwater caves and tunnels, the performance of LBL systems is unknown due to signal attenuation, reverberation and multipath. Without a fixed LBL infrastructure, an AUV uses a combination of depth sensors, inertial sensors, and Doppler velocity sensors to compute a dead reckoned estimate of its position while at depth. With high accuracy attitude and depth sensors the uncertainty in the AUV's 3D pose (roll, pitch, yaw, $x$, $y$, $z$) is primarily in $x$ and $y$. Most underwater navigation systems are based on a Kalman Filter which combines Doppler velocity and inertial measurements [Larsen, 2000]. These systems report navigation errors as low as 0.015% of distance traveled, however error accrues drastically in situations where the

DVL is unable to make accurate velocity measurements. [Kirkwood et al., 2001] have developed an AUV for multi-day under-ice arctic survey.

Eventually the dead reckoned estimate will drift from true, and when the accumulated error exceeds what is required for the application, a correction must be made by (re)observing a known reference. A common approach is to surface and obtain position from a GPS. If LBL or surfacing is not an option, the position error can be bounded by simultaneous localization and mapping (SLAM) [Williams et al., 2000] [Dissanayake et al., 2001]. [Williams and Mahon, 2004] provide an example of near-bottom mapping with sonars and cameras of coral reefs using SLAM. [Roman, 2005] uses multibeam sonar maps to do SLAM over varied topography.

## 3 Vehicle Description

The DEPTHX vehicle is a hovering AUV that has been specifically designed for exploration of flooded caverns and tunnels. It has an ellipsoidal shape that measures approximately 1.5 meters in height and 1.9 meters in both length and width (Figure 1). Its dry mass is 1500kg. Four large pieces of syntactic foam are mounted on the top half of the vehicle, passively stabilizing the vehicle roll and pitch. The vehicle can move directly in the remaining four degrees of freedom (forward, starboard, down, and heading) using six thrusters driven by brushless DC motors. The cruising speed of the vehicle is about 0.2 meters per second. The vehicle is powered by two 56-volt Lithium-Ion battery stacks with a total capacity of 6.2 KWh, enough to supply the vehicle during an four-hour exploration mission.

The DEPTHX vehicle has a full suite of underwater navigation sensors, including a Honeywell HG2001AC INS, two Paroscientific Digiquartz depth sensors, and an RDI Navigator 600kHz DVL. The specifications for the INS are roll/pitch: $0.2^\circ\ 2\sigma$, yaw: $0.4^\circ 2\sigma$, for the DVL velocities 0.3 cm/s $1\sigma$, and for the depth sensors $0.01\%$ of full range (10 cm for our 1000m rated sensor). The two depth sensors are tared, or zeroed with respect to atmospheric pressure, at the start of each day. The DVL is mounted to the front of the vehicle facing forward and tilted down 30 degrees from horizontal, a nonstandard configuration for this instrument. The usual DVL configuration points straight down so that it can achieve lock on the ocean floor. In our application, it is difficult to predict the relative direction to surfaces useful for DVL lock. The top 280 meters of Zacatón is known to be a chimney with a diameter of approximately 80 meters [Fairfield et al., 2005], so the forward-looking configuration should allow the DVL to lock on to one of the vertical walls in most situations. This configuration can cause the DVL to lose bottom lock in more wide-open waters, such as the quarry that is the subject of this paper. Loss of bottom lock can also occur at extremely short ranges, or when passing over highly irregular terrain.

For our purposes, the raw roll, pitch, and yaw measurements provided by the IMU and the depth measurements provided by the depth sensors are accurate enough to be considered absolute measurements of those quantities. The task of determining the location of the vehicle is then reduced to the two dimensional problem of estimating its position in the lateral plane ($x$ and $y$).

For mapping, the vehicle has an array of 54 2° beam-width sonars that provide a constellation of range measurements around the vehicle. This array is in the shape of three great circles, a configuration that was arrived at after studying the suitability of various sonar geometries for the purposes of SLAM [Fairfield et al., 2005]. The sonars have long ranges (some 100m and others 200m) and the accuracy of the range measurements is fairly high (about 10cm), however the low resolution, update rate, and point density makes the mapping problem significantly more difficult than it is with ranging sensors like a laser scanner that provide fast, accurate, high-resolution range images.

## 4 Dead Reckoning Localization

The concept of dead reckoning from DVL velocities is straightforward: integrate the velocities to obtain a position estimate. Here we lay out the details of how this is done on the DEPTHX vehicle.

### 4.1 Dead Reckoning with DVL Velocities

The raw DVL measurement is a vector specifying the velocity of the DVL in the DVL's own coordinate frame. Using the notation of [Spong et al., 2006], we denote this vector by $\mathbf{v}_{dvl}^{dvl}$. We actually want to know the velocity of the origin of the vehicle frame (which is located at the centroid of the vehicle) represented in the world frame, which we denote by $\mathbf{v}_v^w$. To get this, we first compute the velocity of the vehicle in the vehicle frame:

$$\mathbf{v}_v^v = R_{dvl}^v \mathbf{v}_{dvl}^{dvl} - \omega_v^v \times \mathbf{o}^{dvl}.$$

Here, $R_{dvl}^v$ is the constant rotation matrix describing the orientation of the DVL relative to the vehicle frame; $\omega_v^v = [\omega_x \ \omega_y \ \omega_z]^T$ are the angular velocities in vehicle frame (as measured by the IMU); and $\mathbf{o}^{dvl}$ is the position of the DVL in the vehicle frame (or lever-arm.) It is then a simple matter to rotate the vehicle frame velocities into the world frame

$$\mathbf{v}_v^w = R_v^w \mathbf{v}_v^v,$$

where $R_v^w$ is the rotation matrix describing the orientation of the vehicle relative to the world frame, as measured by the IMU. The $x$ and $y$ components of $\mathbf{v}_v^w$ are then numerically integrated to get the new dead reckoned $x$ and $y$ position estimates using Euler's method.

### 4.2 Patching DVL Dropouts

As discussed above, the forward-looking configuration of the DVL together with uneven bottom conditions causes it to frequently lose bottom lock in open water situations. During these times, no DVL velocity measurements are available, making it impossible to update the dead reckoned position estimate. Another unfortunate symptom of losing bottom lock is that the last few DVL measurements before and the first few measurements after the loss of lock are likely to be noisy. In order to filter out these measurements, we used a simple first difference filter to discard DVL measurements which indicated a change in speed of more that 0.05 m/s – the vehicle does not have a very high acceleration.

Our approach to solving this dropout problem involves using the velocity estimates provided by the IMU to continue the dead reckoning solution when DVL measurements are unavailable. However, since the IMU velocity estimates result from integrating measurements from the IMU accelerometers, they are subject to significant drift. To address this problem, we use a Kalman filter to estimate the drift of the IMU velocity estimates during the times when the DVL measurements are good (i.e., when the DVL has a lock).

We implement this with two independent and identical Kalman filters. One filter is used to estimate the IMU velocity and drift in the world frame $x$ direction, the other is used to estimate the same parameters in the $y$ direction. The state of each filter is $q = [v,\ d]^T$, where $v$ is the world frame vehicle velocity in the relevant direction, and $d$ is the associated IMU drift term. If we define $q_k$ to be the value of the state at the time $t$ and $q_{k+1}$ to be the value of the state at the time $t + \Delta t$, then the state transition model that maps $(q_k, a_k)$ to $q_{k+1}$ is

$$q_{k+1} = \begin{bmatrix} 1 & \Delta t \\ 0 & 1 \end{bmatrix} q_k + \begin{bmatrix} 1 \\ 0 \end{bmatrix} a_k + w_k \triangleq F_k q_k + G a_k + w_k,$$

where the process noise $w_k$ is assumed to be white, zero-mean, and Gaussian with covariance matrix $Q$. The DVL provides a measurement of the velocity $v$, so the measurement model is

$$y_k = \begin{bmatrix} 1 & 0 \end{bmatrix} q_k + \nu_k \triangleq H_k q_k + \nu_k,$$

where the measurement noise $\nu_k$ is assumed to be white, zero-mean, and Gaussian with variance $R$. From here, the filter is implemented using the standard Kalman equations (see, e.g., [Maybeck, 1979]), with the small caveat that prediction steps are executed whenever an IMU measurement is received (about 50 Hz) and update steps are executed whenever a valid DVL measurement is received (about 4 Hz).

# 5 Vehicle Controller

The DEPTHX vehicle has a three-level control system that is used to guide the vehicle on its mission. The lowest level, aptly named the low level control system (LLCS), employs velocity feedback from the DVL and IMU in order to generate the thrust necessary to track a desired vehicle frame velocity command. The middle level, named the navigator, issues velocity commands to the LLCS in order to achieve the immediate goal. In this paper, the immediate goal is to drive the vehicle to a specified waypoint, however the navigator also is capable of executing more general behaviors such as wall following and obstacle avoidance. At the highest level is the system executive that, among other things, issues a series waypoint commands to the navigator in order to accomplish the overall mission.

## 5.1 Low Level Control System

The LLCS performs two basic functions: it uses velocity feedback to convert a vehicle frame velocity command into the vehicle frame thrust needed to track that velocity and it implements a mixing table in order to convert the vehicle frame thrust command into the necessary shaft torque[1] commands to each of the individual thrusters.

Velocity feedback is implemented in four independent loops, one for each of the vehicle's four degrees of freedom. Each loop contains an experimentally tuned PI controller. Note that this structure assumes that the components of vehicle frame velocity are not coupled by the dynamics of the vehicle, an assumption which not true. In particular, the forward and sideways velocity components of the vehicle will be highly coupled when the vehicle simultaneously rotates and moves in the lateral plane. Hence we can enforce the decoupled assumption by simply avoiding these type of motions, a restriction which is compatible with the slow, deliberate types of missions that the vehicle will undertake. In practice, however, the controller performs well even when such coupling motions are executed.

The thrust mixer maps the vehicle frame thrust command vector $F_v = [F_\omega, F_x, F_y, F_z]^T$ into a vector of $n$ individual thruster commands $T = [\tau_1, \tau_2, \tau_3, \ldots, \tau_n]^T$, where $n$ is the number of thrusters. It is implemented as a matrix multiplication, i.e., $T = MF_v$, where $M$ is computed as follows. First, let $A$ be the $6 \times n$ matrix whose $i$th column is given by

$$A_i = \begin{bmatrix} p_i \times D_i \\ D_i \end{bmatrix},$$

where $p_i$ is a vector describing the location of the $i$th thruster in the vehicle frame, and $D_i$ is a vector describing the positive thrust direction of the $i$th thruster in the vehicle frame. Now let $B$ be the $4 \times n$ matrix that is the bottom four rows of $A$. $B$ is the matrix that maps the individual thruster thrusts $T$ into the resulting vehicle frame thrust vector $F_v$. Assuming that the rows of $B$ are linearly independent, $M$ can then be found by taking the pseudoinverse of $B$:

$$M = B^T \left(BB^T\right)^{-1}.$$

Note that in the nominal case, the number of thrusters is $n = 6$. However, this formulation allows the mixing matrix to easily be recomputed in the event of thruster failure.

## 6 Mapping

In the DEPTHX mission to Zacatón, the vehicle will employ a sophisticated SLAM system that uses 3D evidence grids as maps and uses a Rao-Blackwellized particle filter to simultaneously estimate the most likely combination vehicle trajectory and world map [Fairfield et al., 2007]. In this paper, we present a simplistic method of generating a map of an explored area. Namely, we assume that the dead reckoned position estimate described above is correct and we use that estimate to determine the locations of points on the quarry walls sensed by the array of mapping sonars. By manually using GPS to measure the starting location of the vehicle, the resulting map can be registered into UTM coordinates. The details of our implementation of this process are described here.

[1] The relationship between shaft torque and thrust is very nearly linear, and we rely on the DriveBlok$^{TM}$ controller produced by MTS Systems Corp to implement the desired shaft torque on the brushless DC thruster motors.

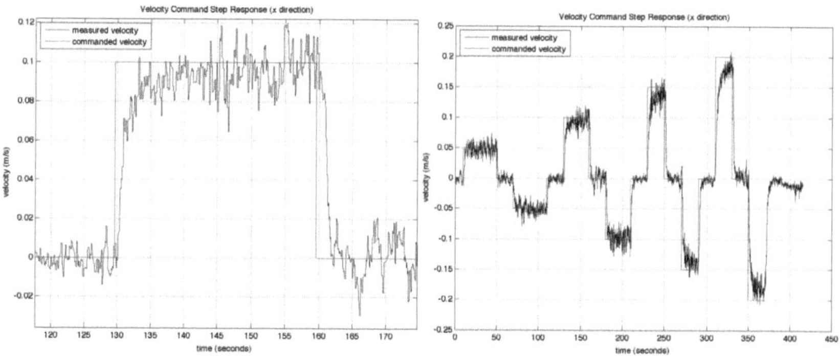


**Fig. 2.** Velocity tracking performance of the LLCS for various commanded velocity pulses. The duration of the leading and falling edge transients is on the order of 4 seconds.

First, the dead reckoned position estimate is combined with the IMU roll, pitch, and yaw measurements and the depth sensor measurement to generate a full 6-DOF pose estimate for the vehicle. This is represented by the homogeneous transformation matrix $H_v^w$. We also represent the pose of each mapping sonar relative to the vehicle frame as a homogeneous transformation matrix. Specifically, $H_{si}^v$ describes the 6-DOF pose of the $i$th sonar transducer relative to the vehicle frame. These matrices are constant and computed in advance. The sonar coordinate frames are defined such that the $x$ axis is aligned with the axis of the sonar beam. So given a sonar range measurement $r_i$, we can compute the world frame location of the sensed point as

$$\begin{bmatrix} p_i^w \\ 1 \end{bmatrix} = H_v^w H_{si}^v \begin{bmatrix} r_i \\ 0 \\ 0 \\ 1 \end{bmatrix} .$$

With this relationship, the mapping process is simply a matter of driving in a survey pattern while maintaining the vehicle pose estimate and computing the points $p_i^w$ as the sonar range measurements are received. These points are filtered to eliminate surface reflections and other noise. Note that we are treating the 2° sonar beam as a ray, which is a reasonable approximation at short ranges. Below, we describe building a simple map from the point cloud data.

We can then insert the sonar point cloud into a $k$d-tree [Bentley, 1975], and use the ANN library [Mount and Arya, 1997] to implement a simple weighted $k$-nearest neighbor algorithm to build a regularly sampled map.

## 7 Experimental Results

### 7.1 LLCS Results

The plots in Figure 2 demonstrate the typical performance we have observed for the low level control system. The vehicle tracks velocity commands well after a transient period

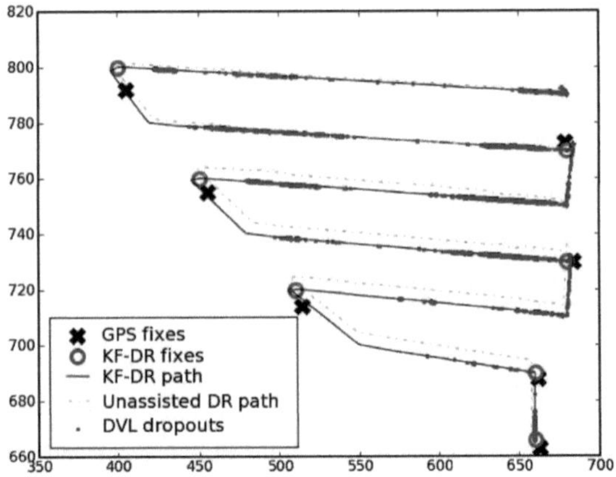


**Fig. 3.** A comparison of dead reckoning localization performance with and without the IMU velocity Kalman filter over a raster scan mission of the quarry. Coordinates are in truncated UTM northings and eastings.

of about 4 seconds. This performance should be sufficient to execute the anticipated DEPTHX missions.

### 7.2 Dead Reckoning Results

Figure 3 shows the performance of the dead reckoning system described in Section 4 during a long raster mission to map the quarry. This mission had a total path length of over a kilometer, at the end of which the positioning error was 3.16 meters. This gives a respectable dead reckoning accuracy of less than half a percent of distance traveled. It should be noted that GPS fixes used as ground truth in this experiment were taken using a hand held non-differential GPS receiver from a moving boat, so it is likely that the dead reckoning estimate is actually more accurate than the "ground truth".

The red dots in Figure 3 denote locations where the DVL failed to achieve bottom lock. During this 6300 second mission, the vehicle was without DVL measurements for a total of about 780 seconds. Some of these dropout periods were as long as sixty seconds. Given this DVL performance, it would have been impossible to achieve any sort of dead reckoning estimate without patching in the IMU velocities as estimated by the Kalman filter.

### 7.3 Mapping Results

In order to produce a map of the quarry, the vehicle performed several overlapping survey missions, designed to provide sonar coverage of all the navigable regions of the quarry. We then concatenated all the data points from these missions, and used the nearest-neighbor approach described above to build the regularly sampled map shown in Figure 4.

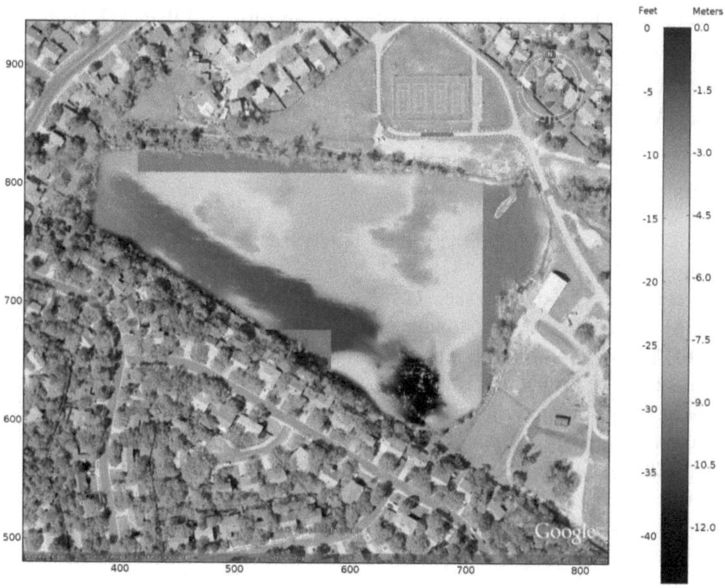


**Fig. 4.** A map of quarry geometry generated using weighted $k$-nearest neighbor interpolation. Coordinates are in truncated UTM northings and eastings. Background image courtesy Google, ©Europa Technologies, 2007.

# 8 Conclusions and Future Work

This paper demonstrates the most basic capabilities of the DEPTHX vehicle, a hovering AUV designed to explore and map flooded cenotes in Mexico as part of the DEPTHX project. We have demonstrated the ability to do basic motion control, achieve decent position estimates from dead reckoning with unreliable DVL data, and use the mapping sonars to generate a map. During the first half of 2007, more sophisticated capabilities will be developed and tested in a series of missions to the La Pilita and Zacatón cenotes in central Mexico. These abilities include more sophisticated motion control, including the ability to maneuver in close proximity along the cavern walls, scanning the walls for suitable locations to take a solid sample. In the area of localization and mapping, the vehicle will use its dead reckoned position and its sonars to build a 3D map – but it will also use this map to simultaneously localize itself, which allows the vehicle to correct for the gradual drift in the dead reckoned position. The dead reckoning capability demonstrated by our experiments in the quarry is a key element of the performance of the DEPTHX Simultaneous Localization and Mapping (SLAM) system, which is described in [Fairfield et al., 2007].


# Acknowledgments

We would like to thank the Hyde Park Baptist Church in Austin, Texas for allowing us to use their quarry for two weeks of robot testing. We would also like to thank the other members of the DEPTHX team for their roles in bringing the DEPTHX vehicle together. In particular, we thank John Kerr and Bill Stone at Stone Aerospace and Marcus Gary at the Department of Geology at the University of Texas, Austin. This work was funded by the NASA ASTEP program, grant NNG04GC09G.

## References

[Bentley, 1975] Bentley, J.: Multidimensional binary search trees for associative searching. Commun. ACM 18(9), 509–517 (1975)

[Dissanayake et al., 2001] Dissanayake, G., Newman, P., Clark, S., Durrant-Whyte, H.F., Csorba, M.: A solution to the simultaneous localisation and map building (SLAM) problem. IEEE Transactions on Robotics & Automation 17(3), 229–241 (2001)

[Fairfield et al., 2005] Fairfield, N., Kantor, G.A., Wettergreen, D.: Three dimensional evidence grids for SLAM in complex underwater environments. In: Proc. of the 14th Intl. Symposium of Unmanned Untethered Submersible Technology (2005)

[Fairfield et al., 2007] Fairfield, N., Kantor, G., Wettergreen, D.: Real-time slam with octree evidence grids for exploration in underwater tunnels. Journal of Field Robotics (to appear, 2007)

[Gary, 2002] Gary, M.O.: Understanding Zacatón: Exploration and initial interpretation of the world's deepest known phreatic sinkhole and related karst features, southern Tamaulipas, Mexico. In: Karst Frontiers, Karst Waters Institute Special Publication, vol. 7, pp. 141–145 (2002)

[Kirkwood et al., 2001] Kirkwood, W.J., Gashler, D., Thomas, H., O'Reilly, T.C., McEwen, R., Tervalon, N., Shane, F., Au, D., Sibenac, M., Konvalina, T., Bahlavouni, A., Bellingham, J.G.: Development of a long endurance autonomous underwater vehicle for ocean science exploration. In: IEEE Oceans 2001, pp. 1504–1512 (2001)

[Larsen, 2000] Larsen, M.B.: High performance doppler-inertial navigation experimental results. In: Proc. of IEEE/MTS OCEANS, pp. 1449–1456 (2000)

[Leonard et al., 1998] Leonard, J.J., Bennett, A.A., Smith, C.M., Feder, H.J.S.: Autonomous underwater vehicle navigation. Technical report, MIT Marine Robotics Laboratory (1998)

[Maybeck, 1979] Maybeck, P.S.: Stochastic models, estimation, and control. In: Mathematics in Science and Engineering, vol. 141, Academic Press, London (1979)

[Mount and Arya, 1997] Mount, D.M., Arya, S.: ANN: A library for approximate nearest neighbor searching. In: CGC 2nd Annual Fall Workshop on Computational Geometry (1997), `http://www.cs.umd.edu/mount/ANN`

[Roman, 2005] Roman, C.: Self Consistent Bathymetric Mapping from Robotic Vehicles in the Deep Ocean. PhD thesis, Massachusetts Institute of Technology & Woods Hole Oceanographic Institution (2005)

[Spong et al., 2006] Spong, M.W., Hutchinson, S., Vidyasagar, M.: Robot Modeling and Control. John Wiley and Sons, Inc., Chichester (2006)

[Whitcomb, 2000] Whitcomb, L.: Underwater robotics: Out of the research laboratory and into the field. In: IEEE 2000 International Conference on Robotics and Automation (2000)

[Williams et al., 2000] Williams, S.B., Newman, P., Dissanayake, G., Durrant-Whyte, H.: Autonomous underwater simultaneous localisation and map building. In: Proc of IEEE Conf. on Robotics and Automation (2000)

[Williams and Mahon, 2004] Williams, S.B., Mahon, I.: Simultaneous localisation and mapping on the great barrier reef. In: Proc. of IEEE Intl. Conf. on Robotics and Automation, vol. 2, pp. 1771–1776 (2004)

[Yoerger et al., 2007] Yoerger, D., Jakuba, M., Bradley, A., Bingham, B.: Techniques for deep sea near-bottom survey using an autonomous underwater vehicle. International Journal of Robotics Research 26(1), 41–54 (2007)

# Mobile Robot Sensing for Environmental Applications

Amarjeet Singh[1], Maxim A. Batalin[1], Michael Stealey[1], Victor Chen[1], Mark H Hansen[2], Thomas C. Harmon[3], Gaurav S. Sukhatme[4], and William J Kaiser[1]

[1] Department of Electrical Engineering, University of California, Los Angeles
[2] Department of Statistics, University of California, Los Angeles
[3] School of Engineering, University of California, Merced
[4] Computer Science Department, University of Southern California

**Summary.** This paper reports the first application of iterative experimental design methodology for high spatiotemporal resolution characterization of river and lake aquatic systems performed using mobile robot sensing systems. Both applications involve dynamic phenomena spread over large spatial domain: 1) Characterization of contaminant concentration and flow at the confluence of two major rivers displaying dynamics due to flow of the water; and 2) Characterization of rapidly evolving biological processes such as phytoplankton dynamics in a lake system. We describe the development and application of a new general purpose method for mobile robot sensing in such environments - Iterative experiment Design for Environmental Applications (IDEA). IDEA introduces in-field adaptation of mobile robotic sensing system. Analysis of the complex spatial and temporal structures associated with each observed environment is presented. Detailed characterization of the observed environment using IDEA methodology is used as an informed prior to improve the performance of the existing adaptive experimental design approaches for mobile robotic systems - stratified adaptive sampling and hierarchical non-stationary Gaussian Processes.

## 1 Introduction

A broad class of environmental sensing applications, both terrestrial and aquatic, require observing an environment that displays significant heterogeneity in both space and time. As an example, river observations are useful for answering the questions pertaining to the hydraulics and multi-dimensional river modeling [1], geomorphology, sediment transport and riparian habitat restoration [2]. These require high granularity measurements of coupled velocity and water quality parameters in river cross-sections. Flow of water in the rivers result in temporal heterogeneity while mixing of the two streams at a confluence zone results in spatial heterogeneity. Fig. 1 displays the temperature distribution observed in a lake and the contaminant distribution in terms of specific conductivity (in space and time) at a confluence of two rivers, displaying the high degree of heterogeneity in such environments.

Traditionally, aquatic environments were observed sparsely in both space and time because of the difficulties and the associated cost of continuously accessing these environments [3]. Over the last few years, with the development of various robotic and sensor network platforms, several prototype systems have been used for sensing the aquatic environments at different scales. Buoyed or moored platforms now exist which can provide vertical profiling capabilities with high temporal resolution over long time

C. Laugier and R. Siegwart (Eds.): Field and Service Robotics, STAR 42, pp. 125–135, 2008.
springerlink.com 

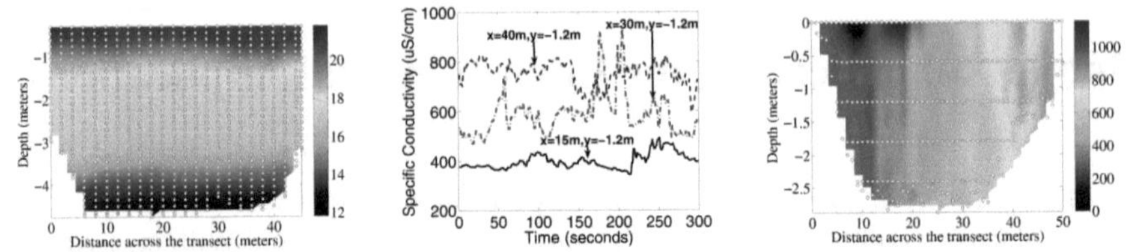


(a) Spatial variation of temperature ($^oC$) at Lake Fulmor (b) Temporal variation of specific conductivity at SJ (c) Spatial variation of specific conductivity ($\mu S/cm$) at SJ

**Fig. 1.** Spatio-temporal heterogeneity in environmental phenomena (Points represent observation locations)

periods at key locations [4]. Mobile robotic platforms such as boats [4], Autonomous Underwater Vehicles [5,6] and cabled robotic systems [7], each suited for a specific environment, have resulted in high resolution sampling in both space and time. However, literature survey reveals that this is the first time an actuated sensing system has been used for detailed characterization of applications such as phytoplankton growth in lake environment and coupled flow and water quality parameters in river environment.

In this paper, we present IDEA - Iterative experiment Design for Environmental Applications; an approach for characterizing an unknown environment under the constraints of rapidly evolving phenomena. This involves an in-field adaptation in the experiment design: to capture phenomena dynamics exploiting observations from prior models, iteratively executed experiments and the behavior of the underlying control processes (if known). We discuss how such an approach can optimize the limited campaign time and acquire data that characterizes the spatiotemporal distribution of the observed phenomena with high fidelity.

To illustrate IDEA and its validation in real world applications, we present case studies of two environmental sensing campaigns where such an approach was successfully followed and validated. The first campaign was executed at the confluence of two rivers, Merced river and San Joaquin river, in California from August 21-25, 2006 (hereafter referred to as SJ deployment). The second campaign was executed at a lake located in a sub alpine (1,600 m) coniferous forest within the San Jacinto mountains of Southern California [8] from August 28-31, 2006 (hereafter referred to as Fulmor deployment). Information collected using IDEA methodology is used to demonstrate the improvement in performance of two of the existing adaptive experiment design approaches: 1) Stratified adaptive sampling [9]; and 2) Hierarchical non-stationary Gaussian Process based modeling approach [10, 11].

## 2 System Description

An actuated mobile robotic system, Rapidly Deployable Networked Info Mechanical System [7] (hereafter called NIMS), was used during each of the two campaigns to navigate a sensor payload anywhere within the two dimensional cross-sections of the aquatic environments. Fig. 2a shows the schematic diagram of the system with the basic cable

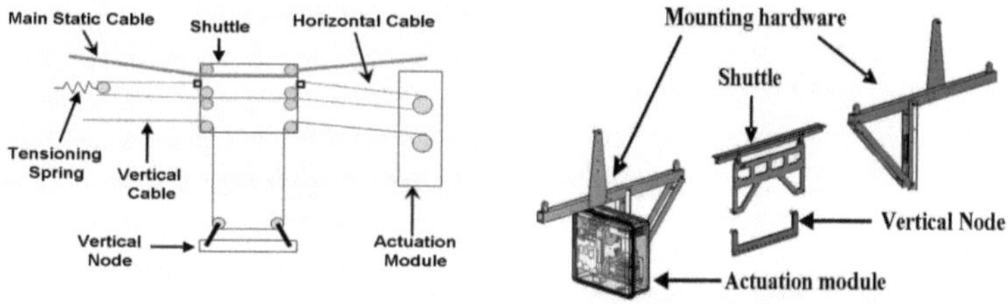


(a) Schematic view (b) Major system components

**Fig. 2.** NIMS system with basic cable setup

setup. It consists of three main cables: the static cable supporting the shuttle and the vertical node platform, a horizontal and a vertical cable used to control the corresponding motion of the sensor payload. Fig. 2b displays major system components that include the mounting hardware, actuation module and a shuttle. The static cable is used to support the mounting hardware which in turn is used to attach the actuation module at one end of the transect and horizontal and vertical cables at the other end.

The static cable is specified to support a maximum tension of 3700 pounds while the maximum tension on our system was close to 750 pounds. Tensioning spring at one end keeps the horizontal cable loop tight and prevents it from slipping on the spool on the other side. The vertical cable was tensioned manually. This proved to be sufficient, since the vertical cable was also kept under tension by the sensor payload. The shuttle and the vertical node platform are actuated through the horizontal and the vertical cables using two motors located inside the actuation module. Both of these motors are controlled simultaneously using a serial interface.

Several parameters of the phenomena were measured, using commercially available sensors [12, 13, 14], to validate and advance the understanding of interrelated physical, chemical and biological processes in the observed environments. These include temperature, pH, specific conductivity, Photosynthetically Active Radiation (PAR), depth, oxidation reduction potential, turbidity, Luminescent Dissolved Oxygen (LDO), nitrate, fluorometer and flow velocity among others. Additionally, a physical sampling device, designed and developed at UCLA specifically for these campaigns, was used to collect water samples for detailed lab analysis and measurement of other required parameters for which no commercial sensors are available. The device uses dual spring loaded syringes to collect water samples when actuated at any location within the two dimensional cross-section of the aquatic system.

This system architecture is an improvement from first generation system introduced in [15]. It allows for reduced node mass, increased horizontal and vertical node speed, constant and convenient access to the actuation control module and rapid yet flexible deployments. Use of cable based system enabled precise localization and ability to actuate heavy sensor payload. System design rationale and implementation details are provided in [7].

Software architecture included system for monitoring the real time data stream to provide immediate fault detection as well as to monitor the variability in the observed phenomena distribution in real time. Additionally, algorithms for adaptive experimental

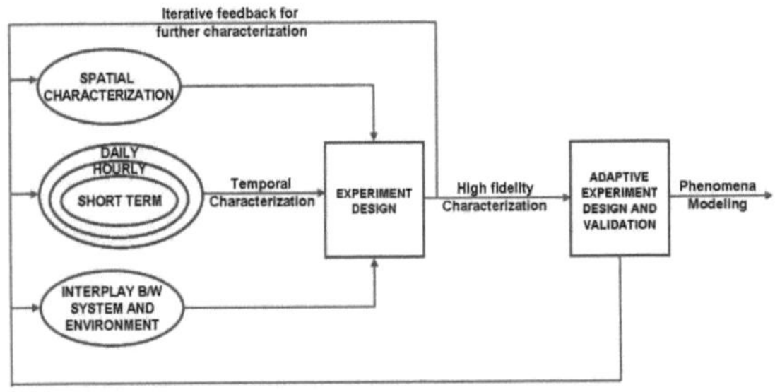


**Fig. 3.** Schematic view of IDEA methodology

design, adjusting to the observed phenomena distribution, were integrated in the field with the existing architecture to provide autonomous system operation.

## 3 IDEA (Iterative Experimental Design for Environmental Applications) Methodology

IDEA provides a methodology for in-field adaptation of experimental design to perform detailed characterization of the spatiotemporal distribution of the observed environment. The two successful campaigns discussed in this paper display the utility of IDEA methodology for actuating the mobile robotic system to perform such characterization. Fig. 3 displays the schematic view of IDEA methodology. Following is the detailed discussion of each module of IDEA methodology with an illustration of their general applicability to different sensing systems.

### 3.1 Characterizing the Spatial Variability

Dense spatial sampling is required to characterize the spatial distribution with high fidelity. In the case of statically distributed sensing system, this requires a dense, uniform distribution of sensor network. On the other hand, with an actuated sensing system such as NIMS, this requires performing deterministic, dense raster scans discounting the temporal variation in the phenomena distribution. In both cases, the initial spatial density can be decided based on the known characteristics of the observed phenomena. This can then be iteratively improved till it is sufficient to completely characterize the spatial variability in the observed phenomena with high fidelity. For each of our aquatic sensing applications, dense, deterministic scans (raster scans) were performed by actuating NIMS to navigate the sensor payload in the two-dimensional cross-section of the observed environment. The density of these raster scans was decided in consultation with the environmental scientists and aquatic biologists and was influenced by prior experiments in the same environment. This was found to be sufficient for high fidelity characterization based on the observed phenomena distribution.

### 3.2 Characterizing the Temporal Variability

Temporal variation in the phenomena distribution should be characterized at different scales. When using static sensor network, sampling frequency should be guided by observed temporal variation. These static sensors will provide both the short term and long term temporal variations. Though such measurements will be restricted to only a few locations in the observed environment (high resource cost makes it impractical to deploy highly dense static sensor network). In case of actuated systems, diurnal variation should be characterized by repeating the dense raster scans at different times of the day. Such scans will also provide periodic sampling checks for post deployment calibration, data integrity etc. and a baseline to characterize the performance of any adaptive sampling approach. At each of our two campaigns we performed repeated raster scans at different times of the day (and nights at Lake Fulmor), using NIMS, that enabled us to understand the mixing effects across different layers of water column.

Dwelling time at each sensing location, while performing raster scans, should be selected so as to satisfy both the requirements for observing short term temporal variations across the complete observed environment and the response time of the sensing system. For the San Joaquin deployment, dwelling time during each of the raster scans was selected to be 30 seconds to account for high variability due to constant flow in the river. On the other hand, for the physically static Lake Fulmor environment, the dwelling time for each of the raster scans was selected to be 10 seconds. Observed standard deviation at each of the sensing locations confirmed that these dwelling times for our campaign were sufficient to satisfy both the demands for sampling short term temporal distribution and the response time of the sensors.

### 3.3 Interplay between Sensing System and Observed Environment

In addition to characterizing the observed environment, experiments must be designed to characterize the constraints of the sampling system such as localization drift, interference in the observed environment and others, such that the integrity of the collected data can be established. For example, during Lake Fulmor campaign we designed experiments to characterize the mixing of water across different layers of water column due to the motion of NIMS inside the water. On the other hand, during San Joaquin deployment with significant flow of the river, we designed experiments to characterize the static deflection of the suspended, pendulous sensor system while making observations inside water.

### 3.4 Adaptive Experimental Design and Validation

Finally, interspersed with these experiments, adaptive experiments should be designed to enable sampling with larger spatial coverage without discounting the temporal dynamics in the phenomena. During the two campaigns, we designed several experiments that adapted to the specific observed distribution by varying the sampling density and dwelling time besides others. Post deployment, the collected data should be used to validate the known understanding of the observed environments and hypothesize the deviations. The detailed characterization performed using IDEA methodology can be used as informed prior for the design of effective adaptive sampling approaches.

## 4 Observations Following IDEA Methodology

We present here the analysis from several experiments performed by actuating NIMS in the two dimensional cross-section of the aquatic environments, following the IDEA methodology. Piecewise bilinear interpolation is performed between the observed locations to create the surface distributions. For each surface distribution, the mean for all the observations is considered as the observed value at that location. Points in the surface distributions, wherever applicable, represent the observation locations. As per the coordinate system, x-axis represent distance along the cross-section of the aquatic environment, while y-axis represent distance along the depth ($y = 0$ represent the water surface). For the San Joaquin deployment the water stream on the near side (lower x coordinate value) is coming from the Merced River while the water stream on the far side is coming from San Joaquin river. Due to the space constraints here we only present the analysis from SJ deployment. Detailed characterization of both SJ deployment and Fulmor deployment is discussed in [16]. Additionally, only a subset of the observed variables are shown for each campaign. However, it is important to note that these are representative of the similar trends shown by all other observed parameters in each of the two campaigns.

The scientific objective for the San Joaquin campaign was to characterize the transport and mixing phenomena at the confluence of two distinct rivers - Merced river (relatively low salinity) and the agricultural drainage-impacted San Joaquin River (relatively high salinity) by observing several parameters that may control the mixing behavior of the two streams. Multiple river cross-sections were observed during the campaign. These transects were chosen based on prior observations [17] in the same environment. Sampling density was selected in consultation with the scientists to be 1 meter across the river and 0.6 meters along the depth resulting in a total of 250 and 293 observation locations respectively at the two transects. Dwelling time at each of the observed location was selected to be 30 seconds.

### 4.1 Spatial Trends

Fig. 1c and 4a show the spatial distribution of specific conductivity and nitrate concentration respectively. Each of these observed distributions is uniform towards each end of the cross-section while displaying dynamic variability in the mixing zone in the middle, resulting in vertical stratification. Moreover, comparing Fig. 4d, that displays the distribution of specific conductivity at the cross-section of the second transect (closer to the confluence point of the two rivers), with Fig. 1c, one can observe increased mixing of water downstream at the first transect resulting in reduced gradient in specific conductivity between the two ends of the river.

### 4.2 Temporal Trends

Fig. 4c shows the distribution of specific conductivity at a depth of 1.2 meters during different times of the day. As the day progresses, gradient of specific conductivity between the two flows reduces. Fig. 4b shows the contour plot for the standard deviation observed at each of the sensing locations during the 30 seconds of dwelling time. As

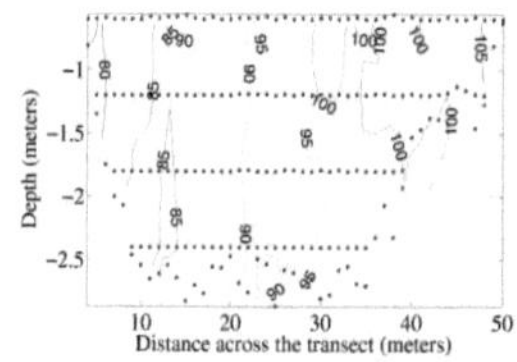


(a) Nitrate concentration ($\mu Mol/liter$) at transect-1

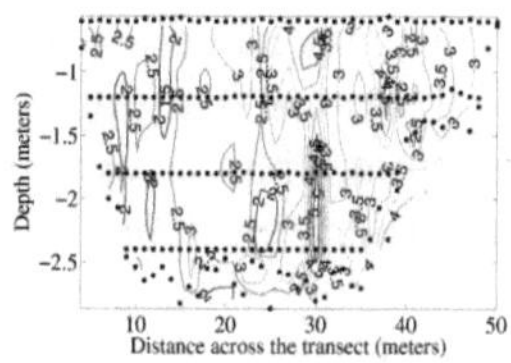


(b) Standard deviation for nitrate at transect-1

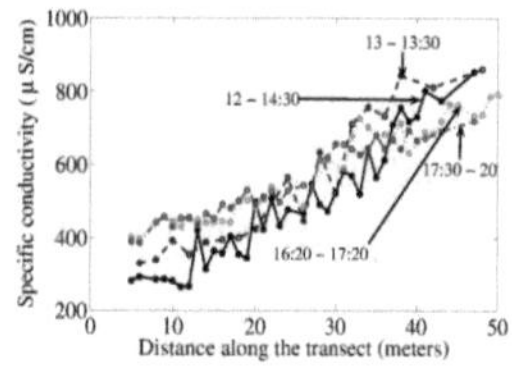


(c) Diurnal variations in specific conductivity at 1.2 m

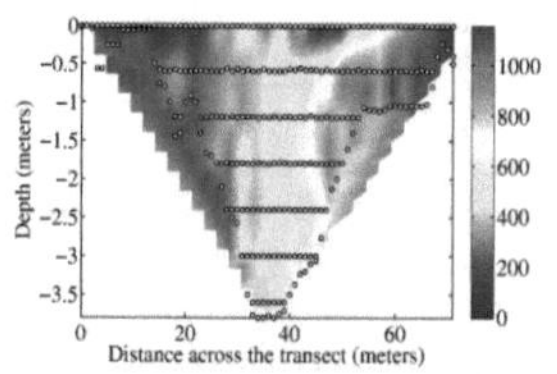


(d) Specific conductivity ($\mu S/cm$) at transect-2

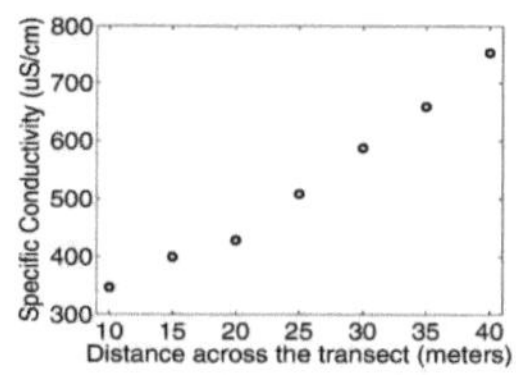


(e) Mean specific conductivity during 5 minute interval

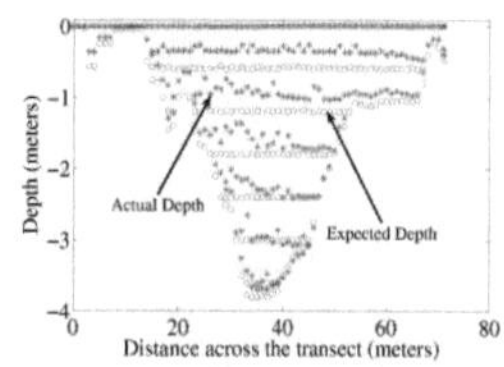


(f) Characterization of localization error along y axis

**Fig. 4.** Experimental results from San Joaquin deployment

expected, maximum standard deviation (which is not significant) is in the mixing zone in the middle. To characterize the temporal variations at point locations, as would have been captured by a static sensing device, the NIMS scanning pattern included a dwell for 5 minutes at several locations along the cross-section at a depth of 1.2 meters. Fig. 1b shows this distribution at three such locations while Fig. 4e displays the mean specific conductivity observed at each of these locations during the observation time of 5 minutes. As can be observed, there was higher variation during this short time towards the middle of the river (x=30) as compared to near end of the transect (x=15). This further confirms higher standard deviation and larger dynamic range in the middle mixing zone observed during raster scans.

### 4.3 Adaptive Experiments

Based on the observed uniformity in the spatial distribution and higher temporal variability in the mixing region, an adaptive experiment was designed where the total time of the experiment was fixed at 1 hour and the dwelling time at each location (only the locations at depths 1.2 meters and 2.4 meters were chosen for this experiment) was selected based on the observed standard deviation, instead of keeping it uniform. Such an approach reduced the dwelling time towards each shore to 10 seconds while increasing the dwelling time in the middle region to as high as 80 seconds. This experiment was performed from 16:20 to 17:20 and the results can be seen in Fig. 4c, labeled accordingly. Comparing it with a raster scan performed the previous day around the same time (17:30 - 20:00), one can observe that reducing the dwelling time towards each shore end provides similar results to a deterministic scan while requiring substantially lower total observation time.

Interplay between the robotic sensing system and observed environment, in this case, involves a possible static deflection of the suspended, pendulous sensor system due to the flow of water in the river. Since this may induce a localization error along the vertical (y) axis, measures must be taken to ensure that this is precisely characterized to be accounted for in any actuation algorithm. Thus, a depth sensor was used, and integrated with the sensor payload, to directly determine depth. Fig. 4f presents a comparison of depth reported by the actuation system (containing possible deflection error), and the actual system localization determined by the integrated depth sensor.

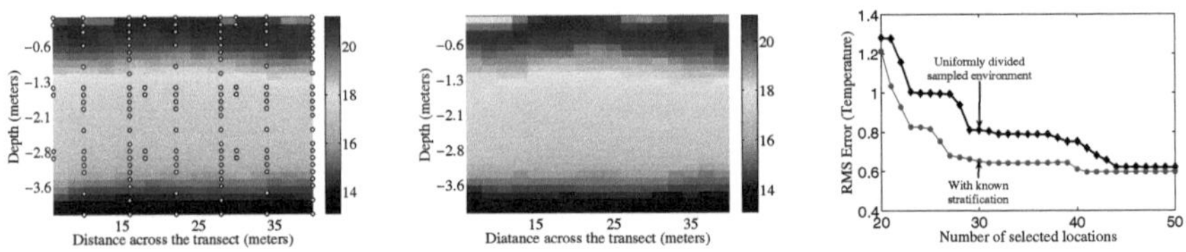


(a) Temperature field used for Adaptive Sampling.

(b) Estimation using stratified adaptive sampling

(c) Performance improvement for GP based approach

**Fig. 5.** Performance improvements in adaptive algorithms when using the information about phenomena distribution at Lake Fulmor as informed prior. Points in (a) represent the locations selected by stratified adaptive sampling algorithm.

# 5 Informed Priors for Improved Adaptive Approaches

Here, we discuss how two previously proposed adaptive approaches for environment sensing: (1) Stratified Adaptive Sampling [9]; and (2) Hierarchical Gaussian Process Based Modeling [10]; displayed improved performance when prior information regarding the spatial distribution of phenomena, as characterized using IDEA methodology, is known as an input.

## 5.1 Stratified Adaptive Sampling

Several adaptive sampling approaches have been proposed in literature that hierarchically stratify the observed environment, thus performing dense sampling only in the regions of interest (regions of high phenomena variability) [9]. For this purpose, a sparse scan of the environment is required in the first stage to extract the regions of high variability. In the second stage, dense observations are performed only in these regions. If the phenomena shows considerable variability then this approach will result in poor performance. However, if some information about the spatial structure of the phenomena distribution is available as a prior, then it can be used to bias the first stage sampling accordingly.

Fig. 5a displays a typical example of temperature distribution, observed using NIMS at Lake Fulmor. This field was used as an input distribution to be observed by the stratified adaptive algorithm, discussed in [9]. When no information was known about the phenomena distribution, the algorithm made observations at 270 locations (out of a total of 522 locations) and estimated the field with root mean square error of 0.04. However, when the vertical stratification information was given as prior, the algorithm observed at only 135 locations (a significant reduction in sampling time) and estimated the distribution with root mean square error of 0.11 (a modest increase in sampling error). Points in Fig. 5a represent the observation locations selected by the algorithm when vertical stratification information was known as an input. Fig. 5b represent the field distribution estimated by making observations at this subset of locations using local polynomial fitting of second order over the observed locations.

### 5.2 Gaussian Process Based Modeling

A common approach in statistical methods for addressing a spatially distributed phenomena is to use a rich class of probabilistic models called Gaussian Processes (GPs) [10]. Using such models, one can quantify the informativeness of a particular location, in terms of the uncertainty about our prediction of the phenomena, given the measurements made at already visited locations. To quantify this uncertainty, we used the mutual information (MI) criterion [11]. If the phenomenon is discretized into finitely many sensing locations $\mathcal{V}$, then for a set of locations $\mathcal{P}$, visited by the mobile robot, the MI criterion is defined as:

$$\mathrm{MI}(\mathcal{P}) \equiv H(\mathcal{X}_{\mathcal{V}\setminus\mathcal{P}}) - H(\mathcal{X}_{\mathcal{V}\setminus\mathcal{P}} \mid \mathcal{X}_{\mathcal{P}}), \tag{1}$$

where $H(\mathcal{X}_{\mathcal{V}\setminus\mathcal{P}})$ is the entropy of the unobserved locations, and $H(\mathcal{X}_{\mathcal{V}\setminus\mathcal{P}} \mid \mathcal{X}_{\mathcal{P}})$ is the conditional entropy after observing at locations $\mathcal{P}$. Hence mutual information measures the reduction in uncertainty at the unobserved locations.

In particular, we first learned a non-stationary GP model, (using an extension of [11]), by maximizing the marginal likelihood [10] using a subset of temperature data at Lake Fulmor. This non-stationary process was learned by dividing the complete region into smaller sub-regions and combining the locally-stationary GPs from each of these sub regions. The two algorithms compared are when: (1) The observed environment was divided into smaller sub-spaces uniformly; and (2) The sub-spaces were selected based on horizontal stratification. For both approaches, starting from initial greedy set of 20 locations, additional 30 locations are selected, again greedily, based on the mutual information criterion given in Equation (1). Greedy selection is performed as it had been proved to provide near optimal value of information [11]. Fig. 5c compares the root mean square error in prediction after making observations at each of the selected locations for the two approaches. GP based approach with known input information regarding the horizontal stratification performs better than the approach with uniformly divided observed environment.

## 6 Conclusions and Future Work

Traditional approaches for environmental sampling are not capable of capturing the spatiotemporal dynamics of a complex phenomena, spread over large spatial extent, with high resolution. In this paper we described the Iterative experiment Design for Environmental Applications (IDEA) approach that we developed and used in two recent field campaigns for observing the aquatic system. IDEA is based on an iterative execution of experiments that reveal phenomena behavior with an objective to perform detailed characterization of the dynamics in the observed phenomena distribution. We demonstrated the effectiveness of the IDEA methodology by providing a detailed characterization of two markedly different aquatic phenomena. Literature survey reveals that this is the first time that an iterative experiment design approach is used an a mobile robot sensing system for characterizing environmental phenomena. We demonstrated that with the use of actuated mobile robotic system, it is imperative to characterize the interplay between the sensing system and the observed environment such that the integrity of the collected data can be relied upon. We illustrated this characterization through experiments based on in-field observations such as characterizing the static deflection of the suspended pendulous sensor system in an environment with significant flow Observations from such experiments, when included in creating new models for experiment design, result in increased fidelity and reliability in the collected data.

Finally, we also illustrated how accurate and detailed characterization, performed by actuating NIMS using IDEA methodology, can be used to improve the performance of existing model-based or adaptive experimental design approaches. This improved performance is enabled by using the observed spatial structure as prior information for the observed environment. In the future we plan to investigate model based adaptive approaches, that can use the specific spatial and temporal structure existing in the real environment as prior information in real time. We also plan to further advance the theoretical treatment of several environmental phenomena and incorporate these into a model based experimental design approach. These steps will then extend the IDEA system to enable it to autonomously determine the set of experiments to perform based on previously collected data and the known understanding of the observed phenomena.

## Acknowledgment

This material is based upon work supported in part by the US National Science Foundation (NSF) under Grants ANI-00331481 and CCR-0120778. We also acknowledge the support of the team involved in the development of NIMS and the two campaigns discussed in this paper and Andreas Krause for the code for generating non-stationary GPs.

## References

1. Hardy, R., Bates, P., Anderson, M.: The importance of spatial resolution in hydraulic models for floodplain environments. Journal of Hydrology 216, 124–136 (1999)
2. Wheaton, J., Pasternack, G., Merz, J.: Spawning habitat rehabilitation i. conceptual approach and methods. Intl. Journal River Basis Management 2, 3–20 (2004)
3. Gelda, R., Effler, S.: Metabolic rate estimates for a eutrophic lake from diel dissolved oxygen signals. Hydrobiologia 485(16), 51–66 (2002)
4. Dhariwal, A., Zhang, B., Stauffer, B., Oberg, C., Sukhatme, G.S., Caron, D.A., Requicha, A.A.: Networked aquatic microbial observing system. In: IEEE ICRA (2006)
5. Blidberg, D.R.: The development of autonomous underwater vehicles (auvs); a brief summary. In: IEEE ICRA (2001)
6. Kumagai, M., Ura, T., Kuroda, Y., Walker, R.: A new autonomous underwater vehicle designed for lake environment monitoring. Advanced Robotics 16, 17–26 (2002)

7. Jordan, B., Batalin, M., Kaiser, W.: Nims rd: A rapidly deployable cable based robot. In: IEEE ICRA, Rome, Italy (May 2007)
8. (2006), http://www.jamesreserve.edu
9. Rahimi, M., Pon, R., Kaiser, W.J., Sukhatme, G.S., Estrin, D., Srivastavan, M.: Adaptive sampling for environmental robotics. In: IEEE IROS, pp. 3537–3544 (April 2004)
10. Seeger, M.: Gaussian processes for machine learning. Int. Jour. of Neural Systems 14(2), 69–106 (2004)
11. Krause, A., Singh, A., Guestrin, C.: Near-optimal sensor placements in gaussian processes: Theory, efficient algorithms and empirical studies. Journal of Machine Learning Research (submitted)
12. (2006), http://www.hachenvironmental.com
13. (2006), http://www.satlantic.com
14. (2006), http://www.sontek.com
15. Pon, R., Batalin, M., Gordon, J., Rahimi, M., Kaiser, W., Sukhatme, G., Srivastava, M., Estrin, D.: Networked infomechanical systems: A mobile wireless sensor network platform. In: IPSN 2005, pp. 376–381 (2005)
16. Singh, A., Batalin, M., Stealey, M., Chen, V., Hansen, M.H., Harmon, T.C., Sukhatme, G.S., Kaiser, W.J.: Idea: Iterative experiment design for environmental applications. CENS Technical Report 67 (November 2006)
17. Harmon, T.C., Ambrose, R.F., Gilbert, R.M., Fisher, J.C., Stealey, M., Kaiser, W.J.: High resolution river hydraulic and water quality characterization using rapidly deployable networked infomechanical systems (nims rd). Environmental Eng. Science (2006)

# Tracking Environmental Isoclines Using Polygonal Formations of Submersible Autonomous Vehicles

Shahab Kalantar and Uwe Zimmer

Australian National University,
Department of Information Engineering, Canberra, ACT 0200, Australia
shahab.kalantar@rsise.anu.edu.au

**Summary.** Knowledge of iso-contours of the underwater terrain can be used to reconstruct it using interpolation. Identifying a set of isoclines can be more efficient and less time-intensive than sweeping a large area. In this paper, we propose a system where a small number of agile underwater vehicles cooperatively maintain a polygonal formation on a plane above the terrain and use field values measured by the individual robots to locally reconstruct the field using interpolation schemes. The formation then tracks a desired iso-contour of the field by tracking the corresponding curve on the reconstructed field.

## 1 Introduction

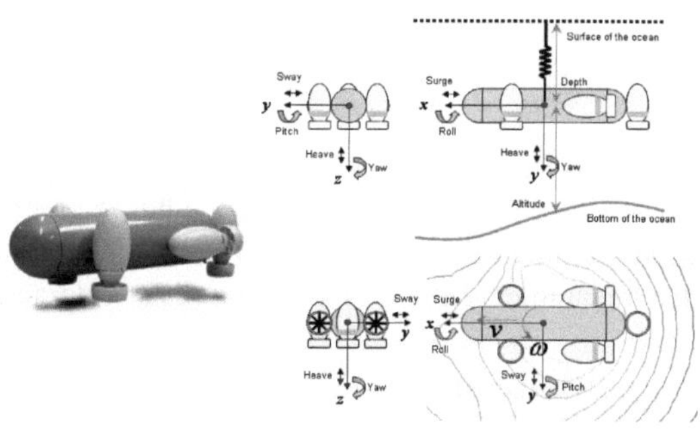


**Fig. 1.** Serafina underwater vehicle

Traditionally, to create a bathymetric map of the underwater terrain employing AUV's, a single bulky vehicle (moving at a certain depth) is towed to follow a path composed of a set of parallel transects (covering a large area) and record altitude values. The recorded information is then used to approximate the field in that region (see, e.g., [3]). An alternative approach is that the vehicle autonomously guide itself along lines of constant altitude. A set of these recorded iso-contours can then be used to reconstruct the field using Hermite interpolation [14]. An obvious benefit of such an approach is that only regions with high *variability* need to be covered. Given the local gradient and Hessian, this is an easy task to do. One only needs to climb the gradient path and follow the direction orthogonal to field gradient once at the desired isocline. In reality, though, these information are not readily available (have to be estimated) and are sometimes even meaningless. In [5], a single vehicle uses a history of past field values. In [6], UUV-gas theory is used where we only need to determine whether we are inside or outside the region surrounded by the iso-contour. A similar approach is described in [7]. In [4], the trajectory of a vehicle is modeled as a damped sinosoid whose parameters are

C. Laugier and R. Siegwart (Eds.): Field and Service Robotics, STAR 42, pp. 137–145, 2008.
springerlink.com 

adjusted to stabilize the vehicle on the contour. Rather than using a single vehicle, we can employ a formation of communicating vehicles. This way, local characteristics of the field can be computed more easily [9]. In [8], a four-vehicle system is described which estimates the gradient and the curvature to track a contour and adjust its size to minimize estimation error.

In this paper, we propose a polygonal formation and interpolation for local reconstruction based on only static instantaneous measurements of the field at the positions of the robots. The interesting aspect of this method is that the representation of the desired isocline in the reconstructed field can be used to convert the problem of tracking tangent to the true isocline to that of tracking a well-defined path. This is because, at each particular time instant, a reconstructed *picture* of the field is made available which contains a smooth curve representing the isoclne. Accordingly, robust methods for tracking such curves can be employed.

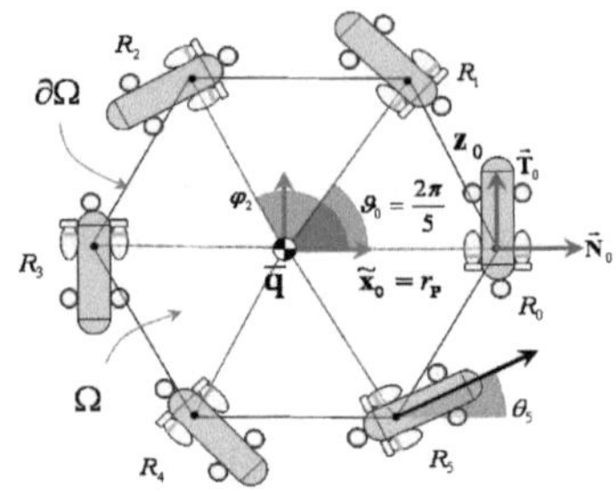

**Fig. 2.** Polygonal formation

As models for our physical platforms, we use Serafina [2], a small fast five thruster vehicle equipped with a suite of sensors (figure 1). Ongoing research is aiming at low-bandwidth short-range long-wave radio communication [12], optical communication [10], scheduling strategies [11], and acoustic range and bearing measurement [13]. In this paper, we will not be concerned with these issues. Furthermore, we will consider a very simple motion model for the vehicles, i.e., that of a non-holonomic unicycle moving on the plane:

$$\dot{x}(t) = v(t)\cos\theta(t)\ ,\quad \dot{y}(t) = v(t)\sin\theta(t)\ ,\quad \dot{\theta}(t) = \omega(t)\ , \tag{1}$$

where $v(t)$ and $\omega(t)$ are, respectively, desired linear and angular velocity inputs of the yaw-surge controller. Additionally, a heave controller keeps the vehicle at the desired depth and the roll and pitch controllers maintain these angles at zero. We also assume that the ambient flow is negligible or is counteracted by the controller.

## 2 Initializing and Maintaining Polygonal Formations

In this section, we explain how a polygonal formation (figure 2) of co-planar vehicles can be created and maintained. Given a desired radius $r_{\mathbf{P}}$, a deformation of the corresponding regular polygonal formation $\mathbf{P}$, composed of $N$ vehicles $R_i$, $i = \mathbf{0}, \ldots, N-\mathbf{1}$, with planar positions, denoted by $\mathbf{q}_i(t) = (x_i(t), y_i(t))^T \in \Re^{\mathbf{2}}$, and collective state $\mathbf{q}_{\mathbf{P}}(t) = [\mathbf{q}_0(t), \ldots, \mathbf{q}_{N-1}(t)]^T$, can be defined by the formation function [15]

$$\mathbf{G}_{\mathbf{P}}(\mathbf{q}_{\mathbf{P}}(t)) = (\tilde{\mathbf{x}}(t) - r_{\mathbf{P}}\vec{\mathbf{1}}_N)^T(\tilde{\mathbf{x}}(t) - r_{\mathbf{P}}\vec{\mathbf{1}}_N) + \tag{2}$$

$$\left(\vartheta(t) - \frac{2\pi}{N}\vec{\mathbf{1}}_N\right)^T\left(\vartheta(t) - \frac{2\pi}{N}\vec{\mathbf{1}}_N\right)$$

where $\mathbf{x}_i(t) = \mathbf{q}_i(t) - \bar{\mathbf{q}}(t)$, $\tilde{\mathbf{x}}_i(t) = \|\mathbf{x}_i(t)\|$,

$$\vartheta_i(t) = \cos^{-1}(\mathbf{n}(\mathbf{x}_i(t)) \cdot \mathbf{n}(\mathbf{x}_{i+1}(t))),\ \mathbf{n}(\hat{v}) = \hat{v}/\|\hat{v}\|\ , \tag{3}$$

$$\tilde{\mathbf{x}}(t) = [\tilde{\mathbf{x}}_0(t), \ldots, \tilde{\mathbf{x}}_{N-1}(t)]^T,\ \vartheta(t) = [\vartheta_0(t), \ldots, \vartheta_{N-1}(t)]^T . \tag{4}$$

$\bar{\mathbf{q}}(t)$ is the centre of mass and $\vec{\mathbf{1}}_N$ is a vector of size $N$ with all the entries equal to $\mathbf{1}$. The operations on the indexes are modulus $N$, so that $\mathbf{q}_N = \mathbf{q}_0$. Any root of $\mathbf{G}_\mathbf{P}(\mathbf{q}(t)) = \mathbf{0}$ gives a polygonal formation which is unique up to translation and rotation. More technically, $\mathbf{G}_\mathbf{P}^{-1}(0)/\ SO(2) \otimes \Re^2$ is a singleton. The position of the centre of mass $\bar{\mathbf{q}}(t)$ breaks the translational ($\Re^2$) symmetry and the angle $\theta_0(t)$ (which can be defined as the orientation of the formation) breaks the rotational ($\mathbf{SO}(2)$) symmetry. Letting $B_\varepsilon(u)$ denote an $\varepsilon$*-ball* around $u \in \Re^N$, an $\varepsilon$*-neighborhood* of a formation is defined by $B_\varepsilon(\mathbf{G}_\mathbf{P}^{-1}(0))$. A number of control strategies have been proposed in the literature for creating a polygonal formation from an arbitrary aggregate of robots. In [17], *cyclic pursuit* strategy is proposed for obtaining a regular polygon at equillibrium. In this scheme, each robot $R_i$ pursues $R_{i+1}$ with appropriate control inputs $v_i$ amd $\omega_i$. Once the vehicles have converged to a sufficiently small neighborhood of a polygonal formation, they enter a formation keeping phase. The control rules for this phase have to make sure that the formation does not deviate from an ideal polygon too much, or, in other words, it stays within $B_\varepsilon\ \mathbf{G}_\mathbf{P}^{-1}(0)$. We will use the concept of *virtual leaders* proposed in [15]. We design the desired position for the non-holonomic vehicle $R_i$ as $_{i_d}(t) = p_i(s_i)$ where $p_i(s_i)$ denotes the trajectory of $\mathbf{q}_{i_d}(t)$, parametrized by $s_i$, and

$$\dot{\mathbf{q}}_{i_d}(t) = \frac{\partial}{\partial s_i} p_i(s_i)\dot{s}_i\ , \tag{5}$$

$$\frac{\partial}{\partial s_i} p_i(s_i) = -\nabla_{\mathbf{q}_i(t)}\mathbf{G}_\mathbf{P}(\mathbf{q}_\mathbf{P}(t)) + \beta(\bar{\mathbf{q}}_d(t) - \bar{\mathbf{q}}(t))\ . \tag{6}$$

$\partial p_i(s_i)/\partial s_i$ is the tangent to the trajectory. $\dot{s}_i$ is the speed of traversal of the trajectory. $\mathbf{q}_{i_d}(t)$ can be regarded as the virtual leader of $R_i$. $\bar{\mathbf{q}}_d(t)$ is the desired position of the centre of mass and should be designed according to application. It plays the role of the virtual leader for the whole formation. Note that this control rule effectively decouples formation keeping and maneuvering which is very desirable. Control rules for non-holonomic vehicles ([15]) can be used for tracking. Initially, $\bar{\mathbf{q}}_d(t_0) = \bar{\mathbf{q}}(t_0)$. The trajectory for $\bar{\mathbf{q}}_d(t)$ can be represented as the parametrized curve $p_0(s_0)$ such that

$$\dot{\bar{\mathbf{q}}}_d(t) = \frac{\partial}{\partial s_0} p_0(s_0)\dot{s}_0\ . \tag{7}$$

$\partial p_0(s_0)/\partial s_0$ will be designed in later sections. The integrity of the formation is maintained by regulating the speed of traversal of the paths $p_i(s_i)$, i.e. $\dot{s}_i$. In [15.], expressions for $\dot{s}_i$ and $\dot{s}_0$ are given.

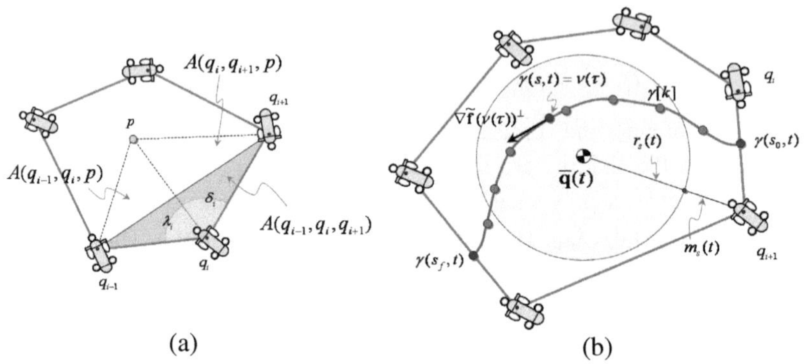


**Fig. 3.** (a) Interpolation on a convex polygon. (b) Construction of the interpolated isocline.

## 3 Field Interpolation by Polygonal Formations

Interpolation on polygons can be done by barycentric coordinates ([18],[19]). Let the domain $\Omega_{\mathbf{P}}(t) \subset \Re^2$ denote the inside of the (assumed convex) polygonal formation, with boundary $\partial\Omega$. Let $p \in \Omega_{\mathbf{P}}$ be a given point. Any vector of real numbers $\alpha(p, \mathbf{q_P}) = [\alpha_0, \alpha_1, \ldots, \alpha_{N-1}]^T$ is called the *generalized barycentric coordinates* of $p$ if **(1)** the coordinates have *linear precision*, i.e., they can reproduce a linear function exactly, $p = \alpha(p, \mathbf{q_P}) \cdot \mathbf{q_P}$, **(2)** the coordinates are non-negative and bounded to guarantee no under- or over-shooting in the coordinates, $0 \le \alpha_i \le 1$, **(3)** the coordinates form a *partition of unity* to assure constant precision and to make the formulation invariant to both translation and rotation, $\Sigma_i \alpha_i = \mathbf{1}$, **(4)** the coordinates are infinitely differentiable with respect to their arguments to ensure smoothness when a node $\mathbf{q}_i$ is moved, $\alpha_i \in C^\infty$, (5) we have $\alpha_i(\mathbf{q}_j, \mathbf{q_P}) = \delta_{ij}$, where $\delta_{ij}$ is the Kronecker delta function. Conditions 1 and 2 are called *affine combination*. Positivity of the coordinates (condition 2) is called *convex combination*. The functions $\alpha_i : \Re^2 \otimes \Re^{2N} \to \Re^+$, $i = 0, \ldots, N-1$, are also called $C^0$ *shape functions* associated with $p$.

Let $\mathbf{f} : \Re^2 \to \Re$ represent a field defined on the plane (altitude measurements). We assume that the vehicles can measure the field at their respective positions, giving the values $\mathbf{f}(q_i(t)) = \mathbf{f}_i(t)$. An *interpolant* (or *interpolation scheme*) of $\mathbf{f}$, based on polygon $\mathbf{P}$, is a function $\tilde{\mathbf{f}} : \Omega \to \Re$ defined by $\tilde{\mathbf{f}}(p) = \alpha_p(p, \mathbf{q_P})^T \mathbf{f_P}(t)$, where $\mathbf{f_P}(t) = [\mathbf{f}_0(t), \ldots, \mathbf{f}_{N-1}(t)]^T$. $\tilde{\mathbf{f}}(p)$ is the interpolated value of $\mathbf{f}(p)$. $\tilde{\mathbf{f}}$ generates a vector of barycentric coordinates and takes the inner product of this vector and the vector composed of field values at the vertices. The coordinates generated by $\tilde{\mathbf{f}}$ should of course satisfy the above conditions. Condition 5 above makes sure that the interpolated value at a vertex is equal to node data: $\tilde{\mathbf{f}}(\mathbf{q}_i(t)) = \mathbf{f}_i(t)$. Conditions 2 and 3 ensure that the interpolated values are bounded between the minimum and maximum of the nodal values: $\min_i \{\mathbf{f}_i\} \le \tilde{\mathbf{f}}(p) \le \max_i \{\mathbf{f}_i\}$. Along the edges of the polygon, the interpolant must be piece-wise linear (i.e., $C^0$). This can be stated as

$$\tilde{\mathbf{f}}(\tau) = \tau\mathbf{f}_i + (1-\tau)\mathbf{f}_{i+1}, \; q = \tau\mathbf{q}_i + (1-\tau)\mathbf{q}_{i+1}, \tag{8}$$

where $q \in \partial\Omega$ and $\tau \in [0, 1]$. If $w(p, \mathbf{q}_\mathbf{P}) = [w_0(p, \mathbf{q}_\mathbf{P}), \ldots, w_{N-1}(p, \mathbf{q}_\mathbf{P})]^T$ is a vector of real numbers such that $w(p, \mathbf{q}_\mathbf{P}) \cdot (\mathbf{q}_\mathbf{P} - p\mathbf{1}_N) = 0$, then partition of unity coordinates can be found by the formula

$$\alpha_i(p, \mathbf{q}_\mathbf{P}) = w_i(p, \mathbf{q}_\mathbf{P})/(\Sigma_k w_k(p, \mathbf{q}_\mathbf{P})) . \tag{9}$$

$w_i(p, \mathbf{q}_\mathbf{P})$'s are called (non-normalized) *weight functions*. There are a plethora of methods for defining shape functions. We will use a method based on *Wachspress* construction which can be used for irregular but convex polygons (thus providing for some robustness to deviations from a perfect polygon). In [20], a simple local formula is given for when $p$ is strictly inside the polygon:

$$w_i(p, \mathbf{q}_\mathbf{P}) = \frac{A(\mathbf{q}_{i-1}, \mathbf{q}_i, \mathbf{q}_{i+1})}{A(\mathbf{q}_{i-1}, \mathbf{q}_i, p)A(\mathbf{q}_i, \mathbf{q}_{i+1}, p)} = \frac{\cot(\delta_i(t)) + \cot(\lambda_i(t))}{\|p - \mathbf{q}_i(t)\|^2} . \tag{10}$$

Figure 3(a) shows the various terms used in these formulas. If $p$ is very close to the boundary, a simple linear interpolation should be used. To reconstruct a particular isoline of the interpolated field, we need to compute the gradient of $\tilde{\mathbf{f}}(p)$ :

$$\nabla_p\tilde{\mathbf{f}}(p) = \frac{\partial}{\partial p}\tilde{\mathbf{f}}(p) = \sum_{i=0}^{N-1}\frac{\partial}{\partial p}\alpha_i(p, \mathbf{q}_\mathbf{P})\mathbf{f}_i(t) , \tag{11}$$

To track a desired isoline value $\mathbf{c}_d$ of the underlying field, we construct the corresponding isoline of the interpolated field and track it instead. This *interpolated isoline* is our observation of the real iso-contour. Tracking this curve and making sure that it is accurate enough (i.e., close enough to the real one) are decoupled in our method. The desired curve is the solution to the equation $\tilde{\mathbf{f}}(p) = \mathbf{c}_d$, where $p \in \Omega$. We are assuming that this solution is a unique curve. To reconstruct the curve $\gamma(s, t)$, we first find its end-points ($\gamma(s_0, t)$ and $\gamma(s_f, t)$), which lie on edges. Next, we traverse the curve sampling it along the way. We can also define a *safety circle* sufficientlt away from the edges. See figure 3(b).

## 4 Isocline Tracking Control

Here, we present three simple tracking strategies.

**1.** The simplest way to track an isoline is to move the virtual leader (the desired goal for the centre of mass) towards the interpolated isoline and, at the same time, move along it. This behaviour is captured in the control law (figure 4(a))

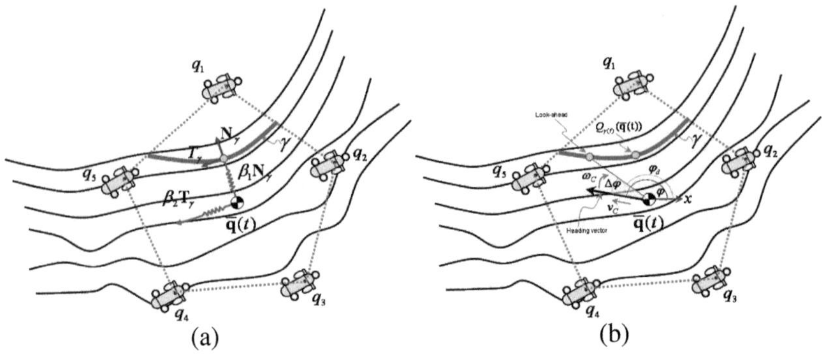


**Fig. 4.** (a) Tracking the reconstructed isocline. (b) Lateral and longitudinal control of the formation centre.

$$\frac{\partial}{\partial s_0}p_0(s_0) = a_\gamma(\beta_1\vec{\mathbf{N}}_\gamma(\tau_C, t) + \beta_2\vec{\mathbf{T}}_\gamma(\tau_C, t)) + (1 - a_\gamma)\beta_3(\mathbf{c}_d - \tilde{\mathbf{f}}(\bar{\mathbf{q}}(t)))\nabla_{\bar{\mathbf{q}}(t)}\tilde{\mathbf{f}}(\bar{\mathbf{q}}(t)) \qquad (12)$$

where $a_\gamma = 1$ if $\gamma$ exists and is zero otherwise, $\beta_3$ is a gain, $\tau_C$ denotes the value of the parametrization of $\gamma$ for which $\gamma(t, \tau_C)$ is the closest point to $\bar{\mathbf{q}}(t)$. This closest point is also denoted $Q_{\gamma(t)}(\bar{\mathbf{q}}(t))$. Also $\vec{\mathbf{N}}_\gamma(\tau_C, t) = \mathbf{n}(Q_{\gamma(t)}(\bar{\mathbf{q}}(t)))$ is normal to the curve at $\tau_C$, and $\vec{\mathbf{T}}_\gamma(\tau_C, t)$ is the local tangent at $\tau_C$. The gain $\beta_1$ determines how fast the formation should react to *normal* variations in the tracked position $\gamma(t, \tau_C)$ and $\beta_2$ determines the speed of traversal.

**2.** To reduce chattering when close to the isocline and dicourage moving along the tangent when far from it, we can make $\beta_1$ and $\beta_1$ functions of the normal distance. Let distance. Let us define $\beta_1 = v_N\sigma_{\varepsilon,\beta}(\mathbf{Q}(t))$, $\beta_2 = v_T(1 - \sigma_{\varepsilon,\beta}(\mathbf{Q}(t)))$, $\mathbf{Q}(t) = \| Q_{\gamma(t)}(\bar{\mathbf{q}}(t)) \|$, where $\sigma_{p,q}(u)$ is the sigmoidal function defined in 1. to implement a smooth switching mechanism between the two tangential and normal behaviours . $v_N$ and $v_T$ denote the nominal normal and tangential speeds. For particular values chosen for $\tilde{\varepsilon} > \mathbf{0}$ and small $\delta > \mathbf{0}$, if we set $\beta = |\ln(\delta/(1-\delta))|/4\tilde{\varepsilon}$, then we will have $\beta_1 \geq v_N(1-\delta)$ ($\beta_1 \leq v_N\delta$) and $\beta_2 \leq v_T\delta$ ($\beta_2 \geq v_T(1-\delta)$) when $\mathbf{Q}(t) \geq \varepsilon + \tilde{\varepsilon}$ ($\mathbf{Q}(t) \leq \varepsilon - \tilde{\varepsilon}$).

**3.** An alternative way of moving towards the reconstructed isocline is to model the motion of the centre of the formation as a non-holonomic device. To do this, we can associate a *heading vector* with the centre. In this case, instead of determining normal and tangential speeds, we should design *lateral* and *longitudinal* speeds. Also, instead of moving towards the closest point and sliding along the tangent to the curve, we should now designate a *lookahead point* on the curve which would serve as the desired goal for the centre of mass of the formation. Using a lookahead makes tracking more robust to fluctuations of the path (whose observation changes over time). figure 4(b) illustrates

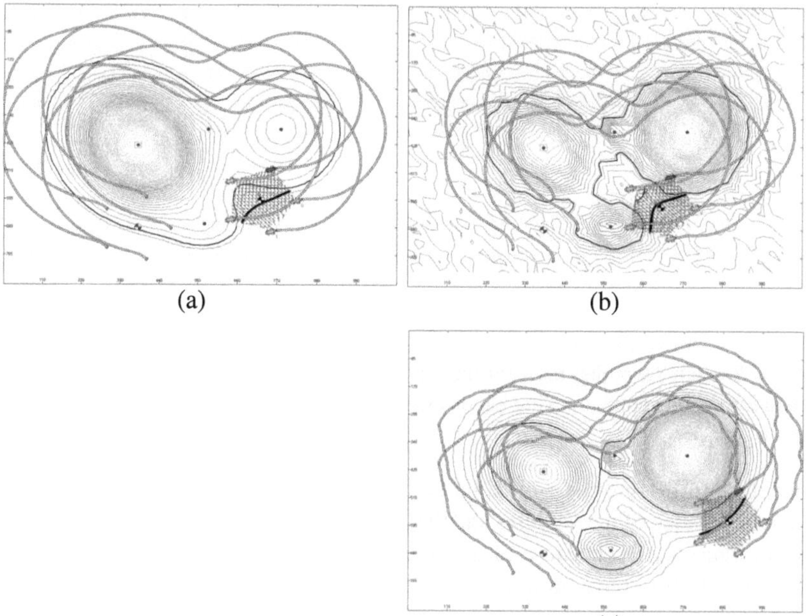

(a) (b) (c)

**Fig. 5.** (a) Small normal forcing. (b) Rugged terrain. (c) High normal gain.

this. Denoting by $\mathbf{L}$ this lookahead point, the virtual centre of mass is moved according to the following control laws ([16]) for longitudinal and lateral movements:

$$\begin{aligned} v_C(t) &= a_\gamma \frac{v_n}{\rho_T} \|\bar{\mathbf{q}}(t) - \mathbf{L}(\gamma, \bar{\mathbf{q}}(t))\| \cos(\Delta\varphi)\dot{s}_0 + \\ &(1 - a_\gamma)\beta_3(\mathbf{c}_d - \tilde{\mathbf{f}}(\bar{\mathbf{q}}(t)))\nabla_{\bar{\mathbf{q}}(t)}\tilde{\mathbf{f}}(\bar{\mathbf{q}}(t)) \end{aligned} \tag{13}$$

and $\omega_C(t) = a_\gamma(k\Delta\varphi + \dot{\varphi}_d)$, where $v_n$ is a nominal speed, $\rho_T$ is a desired distance between the centre of mass the lookahead and $k$ is a positive gain. $\dot{s}_0$ denotes the formation feedback. Note that with this control scheme, longitudinal and lateral *responsive-ness* play the same role as tangential and normal responsive-ness in the previous scheme. If they are designed carefully, it is possible to produce smooth motion even against very rugged terrain or the presence of large noises.

## 5 Simulations

Figure 5 (a) shows a simulation run for a formation with $N = 5$ vehicles and $r_{\mathbf{P}} = 100$. We have set $\beta_1 = 40$ and $\beta_2 = 400$. Also, $k_v = 5.5$ and $k_\omega = 50$. The simulations are synchronous and the vehicles carry on identical

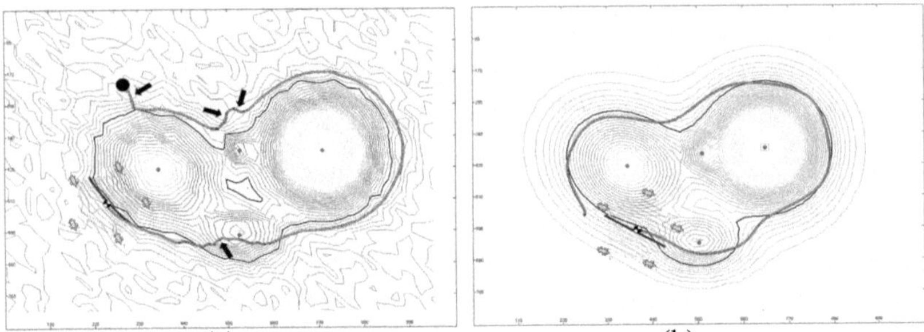

(a) (b)

**Fig. 6.** (a) Smooth switching between normal and tangential motions. (b) Non-holonomically modelled formation centre, tracking a lookahead on the reconstructed curve.

motions. It is furthermore assumed that the designed linear and angular velocities are within the tolerance of the vehicles and there is no error in measurement, although there is error in interpolation because of the size of the formation. As is seen, the produced path for the robots is quite smooth. The reason is that the formation is always moving *forward* on the line on which the tangent to the curve lies and the spring for normal motion is not very stiff. Figure 5(b) shows the same formation in a very rugged terrain. The maximum field value is 6000. Underlying smooth field values have been contaminated with a uniform noise in the range $[-60, 60]$. It is seen that the behaviour has not changed and the paths are still smooth. Figure 5(c) shows the paths for a formation on a smooth terrain with $\beta = 200$. Figure 6(a) shows a simulation run of a formation employing adaptive gain strategy. The initial position is marked by a black disk and the arrows show parts of the path where motion is mainly translation, along the normal, towards the curve. In this simulation $\tilde{\varepsilon} = 5$, $\varepsilon = 10$, $\delta = 0.00001$, and $v_N = v_T = 200$. Figure 6(b) shows a simulation run for a formation employing the third control strategy.

## 6 Conclusions and Future Research

In this paper, we presented a gradient-free method for tracking iso-contours using a robotic network. The basic idea was that of transforming the problem of following the gradient (and its conjugate) into that of tracking a curve which is an isocline of a locally reconstructed distribution. The particular polygonal shape of the formation allows the use of interpolation schemes. With this robust method, smooth paths can be produced even when the underlying field is very rugged or measurements and/or actuation are noisy. There are several issues which merit further consideration. First of all, to minimize interpolation error, the size of the formation has to be adaptively changed. Relevant initial results will be presented elsewhere. Secondly, the interpolation method used allows for some deviation from the rigid shape as long as it remains convex. We will, in the future, consider methods which can handle non-convex polygons as well. Future research should also aim at explicitly dealing with noise in various sensor/actuator modalities. Finally, autonomous selective sampling of isoclines should be addressed.

## References

1. Jaeger, H., Christaller, T.: Dual Dynamics: Designing Behaviour Systems for Autono-mous Robots. Artificial Life and Robotics 2 (1998)
2. http://users.rsise.anu.edu.au/~serafina/
3. Yoerger, D., et al.: High Resolution Mapping of a Fast Spreading Mid Ocean Ridge with the Autonomous Benthic Explorer. In: Proc. of the 11th Symp. on Unmanned Sub-mersible Technology, Durham, NH, USA (1999)
4. Rendas, M.J., Folcher, J.-P., Lourtie, I.M.: Contour Tracking with Video and Altmeter. Project SUMARE Deliverable 4.1 (2002)
5. Reet, A.V.: Contour Tracking for the REMUS Autonomous Underwater Vehicle. Mas-ter's thesis, Dept. of Mech. Eng., US Naval Postgraduate School (2005)
6. Kemp, M., Bertozzi, A.L., Marthaler, D.: Multi-UUV Perimeter Surveillance. IEEE/OES Conf. on Autonomous Underwater Vehicles (2004)
7. Bennett, A.B., Leonard, J.J., Bellingham, J.G.: Bottom Following for Survey Class Autonomous Underwater Vehicles. In: Proc. Int. Symp. on Unmanned Untethered Submersible Technology, New Hampshire (1995)
8. Zhang, F., Leonard, N.E.: Generating Contour Plots using Multiple Sensor Platforms. In: Proc. IEEE Swarm Intelligence Symposium (2005)
9. Ogren, P., Fiorelli, E., Leonard, N.E.: Formations with a Mission: Stable Coordination of Vehicle Group Maneuvres. In: Proc. 15th Int. Symp. on Math. Theory of Networks and Systems (2002)
10. Schill, F., Zimmer, U.R., Trumpf, J.: Visible Spectrum Optical Communication and Distance Sensing for Underwater Applications. In: Proc. of ACRA, Canberra, Australia (2004)
11. Schill, F., Zimmer, U.R., Trumpf, J.: Towards Optimal TDMA Scheduling for Robotic Swarm Communication. In: Proc. of TAROS, London (2005)
12. Schill, F., Zimmer, U.R., Trumpf.: Effective Communication in Schools of Submersibles. In: Proc. of OCEANS, Singapore (2006)
13. Kottege, N., Zimmer, U.R.: MLS-based, Distributed Bearing, Range, and Posture Estimation for Schools of Submersibles. In: Proc. of 10th Int. Symp. on Experimental Robot-ics, Rio de Janeiro (2006)
14. Hormann, K., Spinello, S., Schroder, P.: -Continuous Terrain Reconstruction from Sparse Contours, Munich, Germany. 8th Int. Fall Workshop on Vision and Visualiza-tion (2003)
15. Egerstedt, M., Hu, X.: Formation Constrained Multi-Agent Control. IEEE Trans. on Ro-botics and Automation 17(6) (2001)
16. Egerstedt, M., Hu, X., Stosky, A.: Control of Mobile Platforms Using a Virtual Vehicle Approach. IEEE Trans. on Automatic Control 46(11) (2001)
17. Marshall, J.A., Lin, Z., Brouke, M.E., Francis, B.A.: A Pursuit Strategy for Wheeled-Vehicle Formations. In: Proc. of 42nd IEEE Conf. on Decision and Control, Hawaii (2003)
18. Sukumar, N., Malsch, E.A.: Recent Advances in the Construction of Polygonal Finite Element Interpolants. Archives of Computational Methods in Engineering 13(1) (2006)
19. Floater, M.S., Hormann, K., Kos, G.: A General Construction of Barycentric Coordinates over Convex Polygons. Advances in Computational Mathematics 24(1) (2006)
20. Meyer, M., Lee, H., Barr, A., Desbrun, M.: Generalized Barycentric Coordinates on Ir-regular Polygons. Journal of Graphics Tools 7(1) (2002)

# A Miniature Biomimetic Robotic Fish and Its Realtime Path Planning*

Chao Zhou, Zhiqiang Cao, Shuo Wang, Xiang Dong, and Min Tan

Laboratory of Complex Systems and Intelligence Science, Institute of Automation,Chinese Academy of Sciences, Beijing, China
zhouchao@compsys.ia.ac.cn

**Summary.** This paper proposes a novel miniature biomimetic robotic fish based on single link with compact structure, high maneuverability and multiple sensors. The robotic fish mimics the motion of Thunniform mode, and the methods of propelling and turning are given. The information provided by infrared sensors and photosensitive sensors are analyzed and applied to the algorithms of obstacle avoidance, the light source searching and the dynamic light source tracking. Experimental results are given to show the motion ability and the validity of the proposed algorithm.

## 1 Introduction

There are more and more productive underwater activities and research work, and the Autonomous Underwater Vehicle (AUV) is paid more attentions on. Fish exhibit the advantages of the high efficiency, high maneuverability and low noise, which has the advantages over conventional marine vehicles powered by rotary propellers with the same power consumption. Roboticists explore robotic fish to develop small-size, efficient, maneuverable, and low noise AUV.

Investigation on biomimetic robotic fish systems has provided significant insights into both theory and application of robotics in recent years, whose appealing nature involves higher efficiency, more remarkable maneuverability and quieter actuation[1][2]. These advantages are of great benefit to applications in marine and military fields. Many theories are proposed to explore secrets of fish swimming mechanisms and summarize driving modes of fish motions [3]-[5]. Based on these theories, many prototypes of biomimetic robotic fish have been developed [1] [6]-[9].

Considering that there are some special occasions under the water, such as pipelines, gaps, the development of miniature robotic fish with high maneuverability and adaptability is of great advantages. On the other hand, the miniature robotic fish is cheap enough to be mass-produced, which may enhance the quality of the solution by cooperation among them. The miniature robotic fish should

* This work is funded by the National Natural Science Foundation of China (No. 60635010, No.60605026, No.50475179), 863 Project (No.2006AA11Z225), and CASIA Innovation Fund for Young Scientists.

C. Laugier and R. Siegwart (Eds.): Field and Service Robotics, STAR 42, pp. 147–156, 2008.
springerlink.com 

be elaborately designed because all the mechanical structures, control units, sensors, communication module and power supply are integrated into a very limited space, which brings new challenge.

There are some miniature robotic fish prototypes. Some of them utilize new materials as the actuators, such as ICPF, which can reduce the volume largely. However, the power dissipation is so much that the power source should be outside [7] [10], which will restrict the robotic fish's motion greatly because of the power line. Some miniature swimming robots are designed for special purposes, and the motion ability is suitable for special environment, for example, for the medical purposes [11], but it may not adapt the general task.

In this paper, the design of a miniature biomimetic robotic fish is given based on single link undulating, which is only 15 centimeters long. The method of propelling and turning are proposed based on the motion law of Thunniform mode. The robotic fish may execute tasks independently and autonomously. The algorithms of obstacle avoidance and light source tracking based on the infrared sensors and photosensitive sensors are introduced.

## 2 The Design of the Robotic Fish and Motion Control

### 2.1 The Structure of the Robotic Fish

According to the wavelength and the amplitude envelope of the propulsive wave underlying fish's behavior, the fish is divided into five different modes: *Anguilliform mode, Subcarangiform mode, Carangiform mode, Thunniform mode, Ostraciiform mode* [12].

*Thunniform mode* is by far the most efficient locomotion mode evolved in the aquatic environment, where thrust is generated with a lift-based method, allowing high cruising speeds to be maintained for long periods. Significant lateral movements occur only at the caudal fin (producing more than 90% of the thrust) and at the area near the narrow peduncle. In addition, the reduction of the length of the undulating part will simplify the structure, and reduce the volume on the whole. A novel Thunniform mode-based miniature robotic fish is designed, as shown in Fig.1. A servo motor is adopted to drive a lunate caudal fin as the propeller of the robotic fish, according to fish outline and its motion characteristics. The servo is agile and controllable in movement without additional driver, and more energy is saved than that of the driving unit adopting some new materials such as SMA.

Because of the volume restriction, all assembly units are highly cost-effective ones. The forebody is fish-shape thin hard plastic and a servo is fixed on it. The controller integrates the functions of information acquisition and processing, communication, control and motion decision to improve the integration level and reduce the volume and power dissipation. The tail is cover by thin flexible rubber film as waterproofer, which has little effect on the undulating. The wireless signal is sent by an antenna out of the water. The Li-Polymer Batteries is chosen for the high energy density. In the system of sensors, the infrared switch sensors are selected instead of the sonar sensors because of the volume. The average density

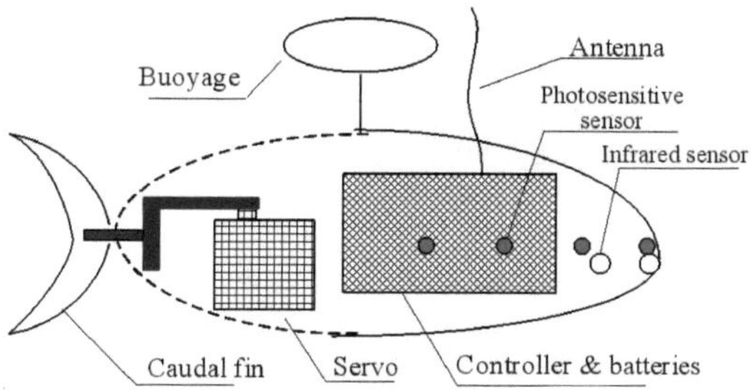


**Fig. 1.** The schematic structure of the robotic fish

of the robotic fish except the buoyage is made a little higher than water by fixed weight on bottom, and the robotic fish will float on water with the buoyage so that all the body is under water, which is favourable to the propelling and stability.

### 2.2 The Structure of the Robotic Fish

*Thunniform mode* motion involves the undulation of the entire body, whose large amplitude undulation is mainly confined to the last 1/3 part of the body. A Thunniform propulsive model is given to describe the motion in the frame of polar coordinate for caudal motion as shown in Fig.2.

$$A_{body}(t) = Amp \cdot \sin(\omega t) \tag{1}$$

where the $A_{body}$is the realtime angle of the tail, $Amp$ is the undulation amplitude, $\omega$ is the frequency of the wave($\omega$=$2\pi f$=$2\pi/T$).

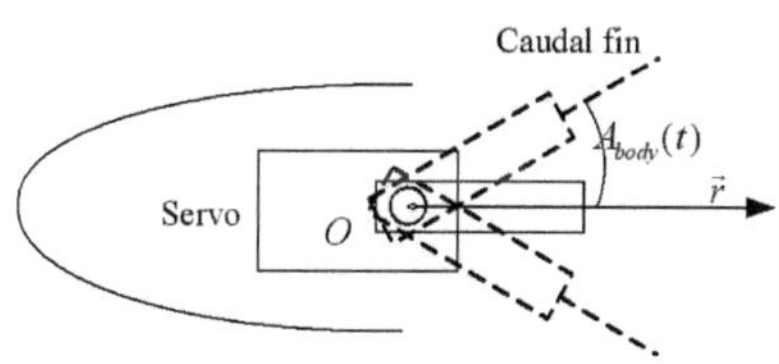


**Fig. 2.** The polar coordinate for caudal motion

(1) provides a method to design the robotic fish: choose the parameters including *Amp,* $\omega$, to determine the proper body wave and then develop a mechanical structure to fit the curve so that the fish's locomotion may be emulated. Discrete (1), then we have:

$$A_{body}(i) = Amp \cdot \sin(\frac{2\pi}{M}i) \tag{2}$$

where $i$ is the variable of spline curve sequence, $M$ is the body-wave resolution that represents the discrete degree of the overall traveling wave.

The interval of time and the movement distance between every two points is restricted by the maximum angle velocity and amplitude of the servo, which is described as follows.

$$\begin{cases} \max\{|A_{body}(i) - A_{body}(i-1)|\} \cdot f \cdot M < \Omega_{\max} \\ A_{body}(i) \leq Amp_{\max} \end{cases} \quad 1 \leq i \leq M \tag{3}$$

where$\Omega_{\max}$,$Amp_{\max}$are the maximum angular velocity and amplitude respectively. $\max\{A_{body}(i)\} \leq Amp$, so $Amp$ is proportional to $\max\{|A_{body}(i) - A_{body}(i-1)|\}$ with $M$ being constant. Let:

$$Amp = c \cdot \max\{|A_{body}(i) - A_{body}(i-1)|\} \tag{4}$$

Then the constraint equations between the $Amp$ and $f$ are obtained:

$$\begin{cases} Amp \leq Amp_{\max} \\ f \leq c\omega_{\max}/\left(M \cdot Amp\right) \end{cases} \tag{5}$$

The turn of the fish, according to the observation, is implemented by changing the axis of the body undulation. Based on it, the direction of the robotic fish may be controlled. A certain angle is superposed on the caudal fin. It is necessary to insure that the undulating rule marches (2) when turning, Rewriting (2), we have:

$$A_{body}(i) = Amp \cdot \sin(\frac{2\pi}{M}i) + A_{turn} \tag{6}$$

where $A_{turn}$ is the deflection angle of the caudal fin. This turning method is used in continuing, stable turning in the movement. The motion of servo should also fit the constraint of (3)

## 3 The Realtime Path Planning Based on Sensors Information

The path planning is necessary for the robotic fish to complete the task. Considering that the limitation of the volume, infrared sensors are chosen to detect the obstacle, and the photosensitive sensors to detect the light source. Firstly,

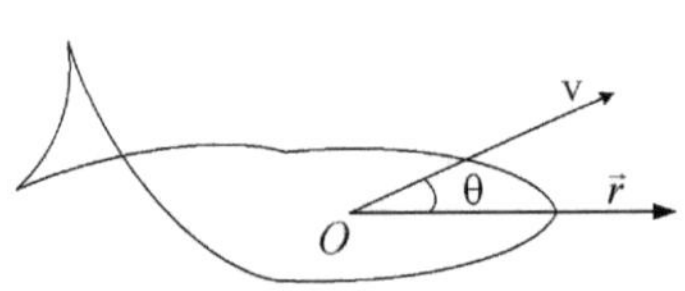


**Fig. 3.** The inertial frame of reference

the inertial frame of reference for the motion is defined as shown in Fig.3. $O$ is the center of inertia of the robotic fish, and is the direction of the rigid forebody.

The motion state of the robotic fish $S_j$at$t = j$ includes the velocity and direction, which is shown in (7).

$$S_j = [v_j, \theta_j]^T \tag{7}$$

We denote with F the motion function and we have:

$$S_j = F(S_{j-1}, \cdots, S_{j-n}, Act_j) \tag{8}$$

where $Act_j$is the action of the robotic fish at$t = j$,

$$Act_j = [Amp, f, A_{turn}]_j^T \tag{9}$$

It means that the action of robotic fish includes the frequency, amplitude, and deflection angle. The C-turn is not involved because it is not suitable for the path planning. In the implementation, all the three parameters are discrete and numbered, and the sets of them are$\{Amp^n\}, \{f^m\}, \{A_{turn}^l\}$, so the $Act$ is numbered, therefore, all the action of the robotic fish is:

$$\Phi_{act} = \{Act \,|\, Amp \in \{Amp^n\}, f \in \{f^m\}, A_{turn} \in \{A_{turn}^l\}\} \tag{10}$$

### 3.1 The Algorithm of Obstacle Avoidance

Three infrared sensors are installed on the robotic fish, the output of them are Boolean variables and labeled as $Ir_l$ (left), $Ir_m$ (middle), $Ir_r$ (right). '1'means there is obstacle. The distribution of the sensors is shown in Fig.4. $\theta$ is the angle between the velocity $v$ and **r**. $\alpha_r$, $\alpha_m$ and $\alpha_l$ are the detecting directions of three sensors, respectively.

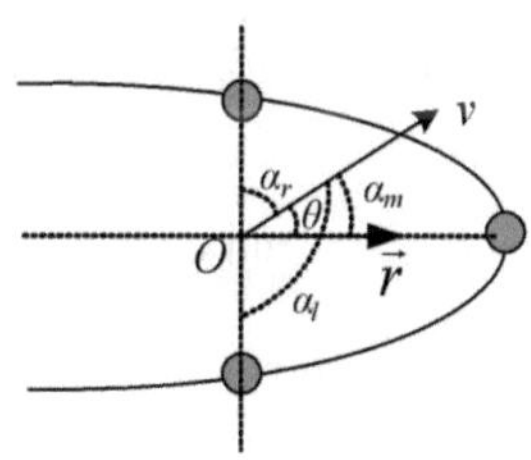


**Fig. 4.** The distribution of infrared sensors

The purpose of avoiding obstacle algorithm is to select the safest direction for robotic fish. The risk factor function at$t = j$ is defined as follows:

$$\begin{aligned} Risk_j &= -(Ir_{l,j} \cdot \alpha_{l,j} + Ir_{m,j} \cdot \alpha_{m,j} + Ir_{r,j} \cdot \alpha_{r,j}) \\ &= -[Ir_{l,j} \cdot (90^\circ + \theta_j) + Ir_{m,j} \cdot |\theta_j| + Ir_{r,j} \cdot (90^\circ - \theta_j)] \end{aligned} \tag{11}$$

The smaller $Risk_j$ is, the smaller the collision probability of robotic fish is. So the algorithm is to calculate the robotic fish action at $t = j+1$, $Act_{j+1}^* = [Amp^*, f^*, d_{turn}^*]^T$, which satisfy:

$$\begin{aligned}&\min_{Act_{j+1}\in\Phi_{act}}\{Risk_{j+1}\}\\&=\min_{Act_{j+1}\in\Phi_{act}}\{-[Ir_{l,j+1}\cdot(90^\circ+\theta_{j+1})+Ir_{m,j+1}\cdot|\theta_{j+1}|+Ir_{r,j+1}\cdot(90^\circ-\theta_{j+1})]\}\end{aligned} \tag{12}$$

The information of sensors, $Ir_{l,j+1}$ , $Ir_{m,j+1}$ and $Ir_{r,j+1}$ are unknown at $t = j$, so the values at $t = j$, $Ir_{l,j}$, $Ir_{m,j}$ and $Ir_{r,j}$ are accepted for estimation. Thus,

$$Act_{j+1}^* = \arg\min_{Act_{j+1}\in\Phi_{act}}\{-[Ir_{l,j}\cdot(90^\circ+\theta_{j+1})+Ir_{m,j}\cdot|\theta_{j+1}|+Ir_{r,j}\cdot(90^\circ-\theta_{j+1})]\} \tag{13}$$

### 3.2 The Algorithm of Light Source Searching

There are seven mini photosensitive sensors installed on the rigid forebody, and the distribution of the sensors is shown in Fig.5

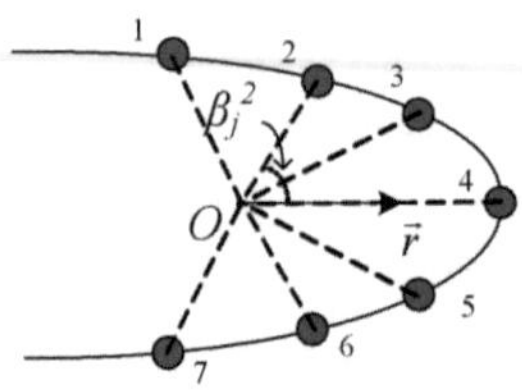


**Fig. 5.** The distribution of the photosensitive sensors

Label $\beta_j^k$ as the detecting directions of the $k^{th}$ sensor at $t = j$, as $\beta_j^2$ shown in Fig.5. The light intensity detected by the $k^{th}$ sensor is labeled as $I_k$ and the direction of the light source is calculated as follows:

Firstly, obtain maximum value of $I_k$:

$$I_{\max} = \max_{1\le k\le 7}\{I_k\} \tag{14}$$

If $I_{\max} < I_{env}$, it means that that are no light source, where $I_{env}$ is the light intensity of environment.

else

$$K = \arg\max_{1\le k\le 7}\{I_k\} \tag{15}$$

Secondly, determine the direction of the light source at $t = j$:

$$\beta_j^* = \frac{\sum\limits_{k=K-1}^{k=K+1}(I_k - I_{env})\cdot\beta_j^k}{\sum\limits_{k=K-1}^{k=K+1}(I_k - I_{env})} \tag{16}$$

From (16), it is seen that only $K^{th}$, $(K-1)^{th}$, $(K+1)^{th}$ sensors are considered because the robotic fish body rocks about when it swims, and the detecting for the light intensity is disturbed. Define the photokinesis function:

$$Light_j = |\theta_j - \beta_j^*| \tag{17}$$

Thus, the photokinesis action for the robotic fish at$t = j + 1$ is:

$$\begin{aligned} Act_{j+1}^* &= \arg \min_{\forall Act_{j+1} \in \Phi_{act}} \{Light_{j+1}\} \\ &= \arg \min_{\forall Act_{j+1} \in \Phi_{act}} \{|\theta_{j+1} - \beta_{j+1}^*|\} \end{aligned} \tag{18}$$

As well, $\beta_{j+1}^*$ are not known at$t = j$, and $\beta_j^*$ is introduced as the estimation. We have:

$$Act_{j+1}^* = \arg \min_{\forall Act_{j+1} \in \Phi_{act}} \{|\theta_{j+1} - \beta_j^*|\} \tag{19}$$

### 3.3 The Algorithm of Dynamic Light Source Tracking

In actual task, the goal (light source) is usually dynamic, which means the robotic fish has to track light source while avoiding obstacles. When the robotic fish is close to the light source, which is both the goal and obstacle, the corresponding conflict should be considered. Because of the limited information, the robotic fish can not distinguish the obstacle and moving light source. Therefore, a multiple objectives optimization approach with self adaptive weight is proposed to solve this problem. The action of the robotic fish is decided by both the infrared sensors and photosensitive sensors:

$$\begin{aligned} Act_{j+1}^* &= \arg \min_{\forall Act_{j+1} \in \Phi_{act}} \{M_1 \cdot Light_{j+1} + M_2 \cdot Risk_{j+1}\} \\ &= \arg \min_{\forall Act_{j+1} \in \Phi_{act}} \{M_1 \cdot |\theta_{j+1} - \beta_{j+1}^*| \\ &\quad - M_1 \cdot [Ir_{l,j+1} \cdot (90^\circ + \theta_{j+1}) + Ir_{m,j+1} \cdot |\theta_{j+1}| + Ir_{r,j+1} \cdot (90^\circ - \theta_{j+1})]\} \end{aligned} \tag{20}$$

where $M_1$ and $M_2$ are the weights of the photosensitive sensors and infrared sensors, respectively. As well, $\beta_{j+1}^*$and$Ir_{l,j+1}$are replaced by the estimated values, and we have:

$$\begin{aligned} Act_{j+1}^* &= \arg \min_{\forall Act_{j+1} \in \Phi_{act}} \{M_1 \cdot |\theta_{j+1} - \beta_j^*| \\ &\quad - M_1 \cdot [Ir_{l,j} \cdot (90^\circ + \theta_{j+1}) + Ir_{m,j} \cdot |\theta_{j+1}| + Ir_{r,j} \cdot (90^\circ - \theta_{j+1})]\} \end{aligned} \tag{21}$$

$M_1$ and $M_2$ are decided based on actual application. In the light tracking, it can be self adapted according to light intensity. When the light is weak, the robotic fish is far from the light source. If obstacle is detected, the algorithm of obstacle avoidance should be validated and $M_1 > M_2$ in this case. When the light intensity is high, the obstacle detected may be the light source, thus $M_1 < M_2$. When the light is very high, it means that the robot is very close to the goal, and the obstacle avoiding should be validated to avoid the collision with light source, therefore we have $M_1 > M_2$. In the experiment, we choose a piecewise function:

$$M_1 = \begin{cases} 2 & I_{\max} < I_a \\ 0.5 & I_a < I_{\max} < I_b \\ 2 & I_{\max} > I_b \end{cases} \qquad (22)$$
$$M_2 = 1$$

where $I_a$ and $I_b$ are the thresholds of light intensity, which are decided by the light source.

## 4 Experiments

The developed robotic fish prototype is shown in Fig.6.

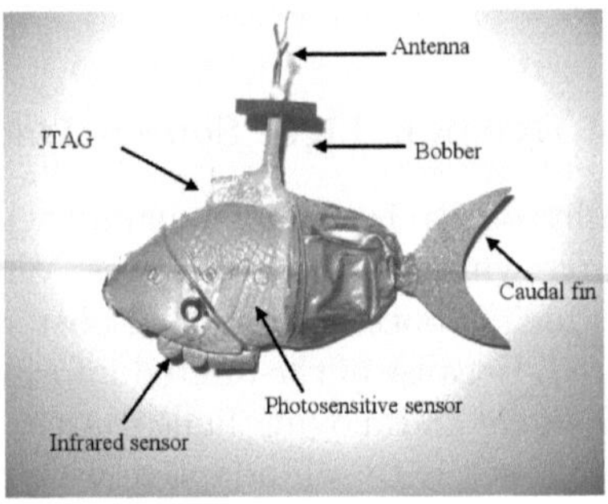


**Fig. 6.** The prototype of the robotic fish

The experiments are conducted in an experiment pool of 1.80m*1m to testify the miniature biomimetic robotic fish. An overhead camera is introduced to capture the video of the robotic fish motion and the video information is sent to the computer to recognize and record the realtime position of the robotic fish by the help of a platform for the motion control of robotic fish, MRFS (Multiple Robot fishes coordinate System). The top right part shows the result

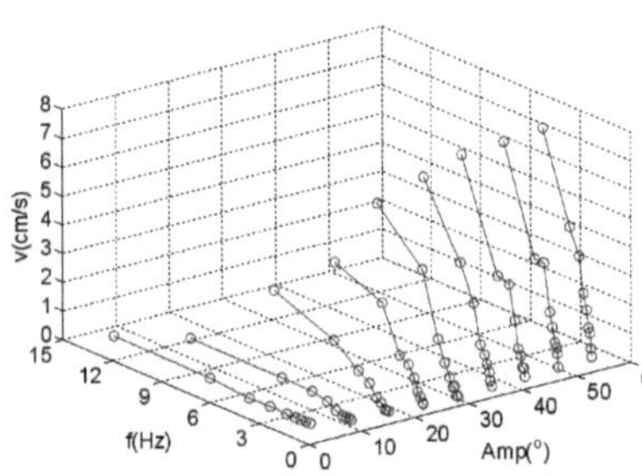


**Fig. 7.** The velocity at different frequencies and amplitudes

of the recognition. The position data is filtered, and the movement track is got by Least-squares spline approximation. The relative parameters are then calculated.

**Experiment 1:** Testing the velocity.

According to the law in (2) and the constraint of (5), the velocity at different frequencies and amplitudes are measured in detail. The results are shown in Fig.7. The maximal velocity is v=7.63cm/s.

**Experiment 2:** Dynamic light source tracking.

Firstly, the motion function F(*) is built as a look-up table. The results of (21) are calculated offline and stored in the MCU. A light source moves randomly in the pool, and the robotic fish is put away from the light source. The trajectories are shown in the following:

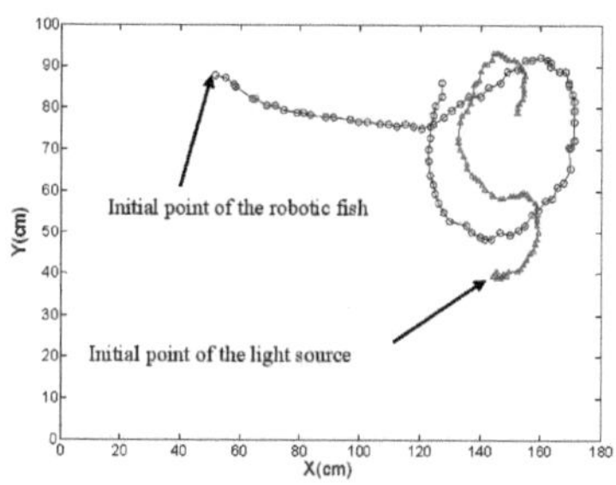


**Fig. 8.** The series of dynamic light source tracking

where '$\Delta$' is the trace of the light source, and 'O' is the trace of robotic fish. X and Y in the figure are the edges of the pool which is regarded as obstacles in this experiment. From the figure, the robotic fish traces the light source without collision with the light source or pool edges.

## 5 Conclusion

In this paper, a miniature biomimetic robotic fish is given, which reduces the volume of the robotic fish largely. The method of propelling, turning, and C-turning for the robotic fish is proposed based on the motion law of Thunniform mode. The realtime path planning is given based on the sensors information. Experiments validate the design and algorithms.

## References

1. Triantafyllou, M.S., Triantafyllou, G.S.: An efficient swimming machine. Scientific American 272(3), 64–70 (1995)
2. Sfakiotakis, M., Lane, D.M., Davies, J.B.C.: Review of fish swimming modes for aquatic locomotion. IEEE J. Oceanic Eng. 24(2), 237–252 (1999)

3. Lighthill, M.J.: Note on the swimming of slender fish. J. Fluid Mech 9, 305–317 (1960)
4. Wu, T.Y.: Swimming of a waving plate. J. Fluid Mech. 10, 321–344 (1961)
5. Tong, B.G.: Propulsive mechanism of fishs undulatory motion. Mechanics In Engineering 22, 69–74 (2000)
6. Kato, N.: Control Performance in the Horizontal Plane of a Robot fish with Mechanical Pectoral Fins. IEEE J. Ocean. Eng. 25, 121–129 (2000)
7. Fukuda, T., Kawamoto, A., Arai, F., Matsuura, H.: Mechnism and swimming experiment of micro mobile robot in water. In: Proceedings, IEEE Workshop on Micro Electro Mechanical Systems, pp. 273–278 (1994)
8. Liu, J., Hu, H.S.: Building a 3D simulator for autonomous navigation of robot fishes. IEEE IROS, 613–618 (2004)
9. Zhou, C., Cao, Z., Wang, S., Tan, M.: The Posture Control and 3-D Locomotion Implementation of Biomimetic Robot Fish. IEEE IROS, 2583–2588 (2006)
10. Zhang, W., Guo, S., Asaka, K.: A New Type of Hybrid Fish-like Microrobot. IJAC 4, 358–365 (2006)
11. Edd, J., Payen, S., Rubinsky, B., Stoller, M.L.: Biomimetic Propulsion for a Swimming Surgical Micro-Robot. IEEE IROS, 2583–2588 (2003)
12. Body/Caudal Fin Propulsion, `http://www.ece.eps.hw.ac.uk/Research/oceans/projects/flaps/bcfmodes.htm`
13. Domenfci, P., Blake, R.W.: Kinematics and Performance of Fish Fast-start Swimming. J. Exp. BinI. 200, 1165–1178 (1997)

# Development of a 5 Degree-of-Freedom Towfish and Its Control Strategy

Mohammad Aamir Khan, Aazir Khan, Matteo Zoppi, and Rezia Molfino

PMAR Robot Design Research Group - University of Genova, Genova, Italy
{Aamirkhan,khan,zoppi,molfino}@dimec.unige.it

**Summary.** This paper describes the development of a five degree of freedom towed underwater vehicle designed for condition monitoring of undersea pipelines. Its innovative design, with 6 fins actuated by 5 motors adding heave and sway capabilities to the yaw, roll and pitch movements possible in traditional designs, is explained. The parameters influencing the stability and the design in general are analyzed. A Control strategy for this design of highly coupled movements and control modes encompassing all the operational scenarios, are presented.

## 1 Introduction

In spite of significant costs due to the harsh conditions and specialized equipments required, undersea pipelines are indispensable due to geographical complexities and socio-political reasons. Because of the trawling equipments, anchors, heavy dropped objects and even ocean currents, it is necessary to protect the pipelines; and a prevalent way of doing this is by trenching and then burying the pipelines with seabed material [1]. Where the terrain is rough and rocky, levelling would also be required but pipeline would still pass through rocks along steep slopes.

Since the pipelines traverse all types of difficult trajectories and that over the period of time they get buried even deeper than the original installation due to the silt transfer or can even get uncovered, it is very challenging to track and then detect any leakage or fault along their length. Currently, gradiometers are more prevalently used to track pipelines while an ultrasonic detection system is employed for condition monitoring [2]. These gradiometers and leakage detection systems together with various auxiliary devices need to be carried by a vehicle which could be of any of these types: Remote Operated Vehicles (ROVs), Autonomous Underwater Vehicles (AUVs) and towed underwater vehicles (towfish) [3].

A towed undersea vehicle was selected to be developed for the gradiometer and leakage detection system, based on the following reasons. Firstly, it is much cheaper than both ROVs and AUVs because of the less complex control system and absence of the independent propulsion system. Secondly, although ROVs allow larger data transfer with mission boat due to "wired" connection with it, serving for power supply and tele-robotic control, they are best used for speeds

C. Laugier and R. Siegwart (Eds.): Field and Service Robotics, STAR 42, pp. 157–166, 2008.
springerlink.com 

of 1 m/sec or less [4]. This speed falls at the lower end of the operating range in which the gradiometer is required to operate. Thirdly, although AUVs have come a long way over the years, they still have a low data transfer rate [5].

To find undersea pipeline or cable buried in the ocean floor, previously, tone generators were used. This method traditionally can only detect cable and pipeline buried at shallow ocean floor depths [6]. Magnetometer can over come this shortcoming in case of ferrous or magnetized pipelines. Conventional total-field marine magnetometers are not suitable either because they are prone to masking by background magnetic signals [7].

A gradiometer is a specialized type of magnetometer that measures a gradient of the first spatial derivative of the total magnetic field. They usually consist of two sensors (simultaneously measuring the total field strength) separated by a fixed distance [3]. The linear estimate of the gradient of the ambient field is obtained by dividing the difference in measured intensity by the distance between the sensors (baseline distance). The orientation of the gradiometer will determine which component of the three-dimensional field is measured (usually the x, y or vertical component) [8]. Because of the ability to detect at greater depths and insensitivity to background magnetic signals they are frequently used to track pipelines.

For the detection of the leakage itself, various techniques are used in marine applications: high resolution color imaging sonars, side scan sonar, hydrophone, cathodic protection survey and ultrasonic testing [2]. According to the requirement of the end user, ultrasonic detection system was selected.

This paper describes the development of a towed underwater vehicle for tracking undersea pipeline and monitoring their state.

## 2 The Operational Requirements

The requirement was to develop an underwater vehicle with the capability to operate at a maximum depth of 300m, and to automatically track and monitor the condition of submerged oil pipelines. The required forward speed range is from 1.03m/s to 2.06m/s (2 to 4 knots) and the vehicle is towed from a ship by an umbilical. It should have the capability to track pipelines of diameter greater than 0.6m. The mobility requirement is that the vehicle should be able to move 16m vertically and 140m horizontally. This vertical figure gives enough freedom to avoid any obstacle and to follow any abrupt change. The horizontal mobility figure is based on the expected discrepancy in pipeline maps used to steer the GPS equipped towing boat roughly above the pipeline location. The vehicle should also have the translational degree of freedom in vertical and transverse directions together with all three rotational dofs (5dofs in total).

The stringent requirement of 5 degrees of freedom mobility was set on the bases of the several considerations. For the proper functioning of the gradiometer setup, the vehicle should move parallel to the pipeline at constant vertical and horizontal distance. The distance of the towfish from the pipes must not be less than 1.5m, a limit set by the condition monitoring system. The upper limit depends on the diameter of the pipe and is in the range of 5-8m. Underwater currents and energetic state of the sea has significant effect on the vehicle dynamics [9] and generate substantial hydrodynamic forces. As explained later, the umbilical also transmits strong disturbances to the towed vehicle [10, 11, 12] through the ship. Due to the versatile topography of the seabed, pipelines can traverse through rocky areas with sudden unpredictable rocky obstacles for the towfish. Submerged pumping stations, concrete anchorage blocks and other infrastructure forms another obstacle. Then relatively higher forward velocity constitutes another set of constraints calling for swifter manoeuvring. Also outlined in [13, 14] sufficient maneuvering authority to independently control heave, and sway in addition to the roll, yaw and pitch is a desirable feature for surveillance and inspection vehicles. So, it was considered more appropriate that the vehicle has 5 degrees of freedom.

## 3 State of Art

Generally underwater towed vehicles are capable of only three rotational movements [4, 5, 10, 15]. As, they approach any given point through rotational dofs., this deteriorates the quality of tracking when terrain is rocky and also when external disturbances are high. A more capable 4dof (3dof rotational and 1dof translational) design is discussed in [16]. It has 4 tail planes and 1 hydrodynamic depressor fin for depth control. Our design based on 5dof, allows the vehicle to quickly navigate laterally and vertically when required.

## 4 The Vehicle's Concept and Layout

This requirement of 5 degree of freedom makes the basic design of the vehicle substantially different from the existing vehicles and it is achieved by 6 fins controlled by 5 actuators. Labelled in Fig. 1a, fins 3 and 4 are coupled together and actuated by a single servo motor. Each of the remaining fins is actuated by a separate servo motor. The fins can be actuated to have both positive and negative angle of attack and can generate force in both directions (Lift force and down force) at the centre of pressure of the fin. These 5-dofs which are realized by the movements of the fins in various combinations are Heave, Sway, Roll, Pitch and Yaw (T2, T3, R4, R5 and R6). To illustrate the origin of the movements, a heave force F by fin 4 is shown to be applied. Due to offset from centre of gravity, roll moment and pitch moment also produce due to moment arms x and y, respectively. Similarly looking into the effect of each fins force we can summarize, Fig. 2, the pair of fin movements required to generate each of the 5dof movement alone. From these any form of combined movement can be generated.

The vehicle is "propelled" by a towline attached to the towing ship which pulls the vehicle forward. The towline transmits the disturbances coming from the ship, disturbances induced by the movement of the umbilical in the water

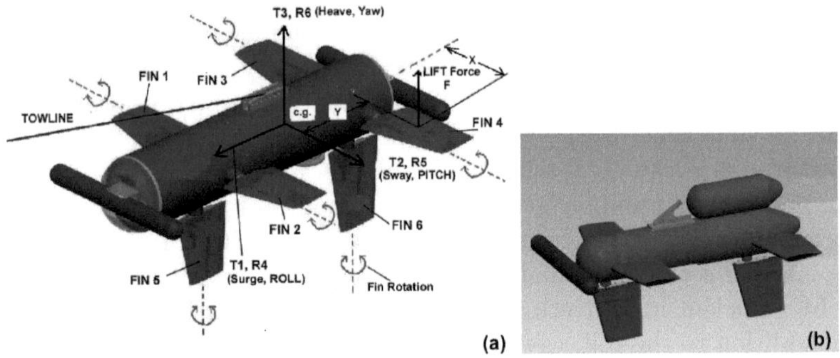


**Fig. 1.** (a) 3D CAD model of the towfish showing 5dof and force on fin 4 (b) An initial concept of the towfish

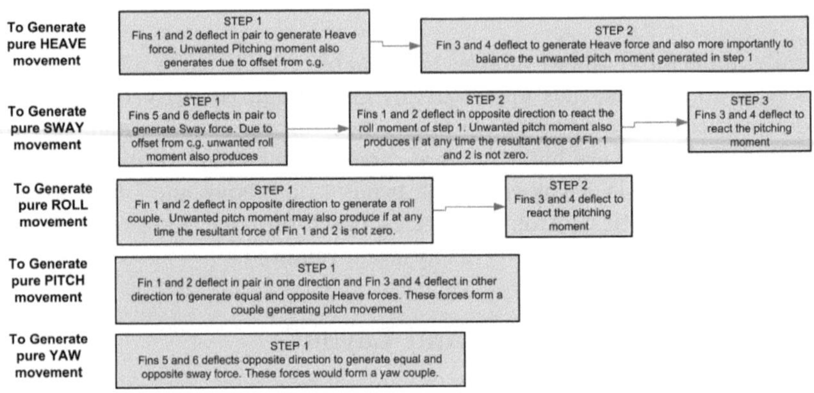


**Fig. 2.** Chart showing the generation of each of the 5 dof movement

and shedding vortices causing it to strum [17, 11]. On the bases of a fast control system and fin size, all of these except for axial disturbances can be compensated.

Vertical fins on the top of the vehicle could not be adopted because of hindrance to the towline, hence fins 1 and 2 remain to react the unbalance roll moment due to unsymmetrical vertical fins. An earlier concept design of shorter length with over-the-top control compartment is shown in Fig. 1b. The concept was dropped later due to the use of smaller control unit than the one perceived earlier, extra cost and hydrodynamically unbalanced shape.

## 5 Main Modules of the System

The vehicle is divided in three cascaded compartments joined using O-rings. The intermediate compartment is open and flooded with water. It has backbone structure, Fig. 3a, with reinforcement pipes and housing of the depth sonar mounted to it through a vertical pipe.

Both front and rear compartments are fixed to intermediate compartment through screws and normal access to the both is provided through front and rear covers. The cabling passes from front to rear compartment through the "backbone" pipe. Fig. 3b shows an exploded view of the vehicle. As the drive shafts attached to the fins is rotating, special type of lip sealing with dust wipers in series, are used.

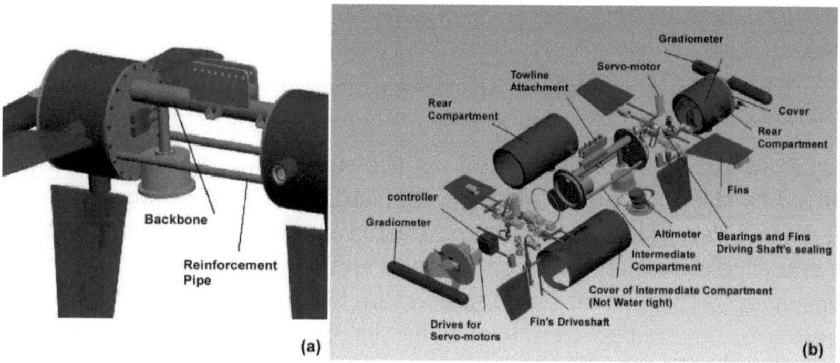


**Fig. 3.** (a) Intermediate compartment (b) Exploded 3D CAD view of the towfish

The 5 servo motors with reduction gears and optical encoders are identical and capable of providing peak torque of 25Nm. As mentioned by [18], for a well performing control system most of the fin deflections are within 8 degrees. Because of the higher manoeuvrability standards the fins deflection for our vehicle was calculated to be 15 degrees. So, computing [5] on the bases of response time, the rated speed of the selected motor (12060rpm) is much higher than the required 5000rpm.

To couple the motors to driveshaft fibreglass reinforced Neoprene timing belts with customized belt tensioning mechanism are used. A sonic altimeter monitors the distance of towfish from seabed in real time and a depth pressure transducer measures the depth of the vehicle under sea surface. Two gradiometers mounted in the transverse position both at the front and rear end of the towfish, are used to control yaw and let the ends accurately track the pipeline. Roll and pitch sensors are also used.

## 6 Mobility of the Towfish

Within water, the depth of the towline pulling the towfish is dependent on the length of the line, the type of material used and the weight of the towfish in water. With 21kg as the weight of the towfish in water, obtained from buoyancy calculations, we get the Fig. 4a for 7 different cases calculated through computer coding as tabulated in the Fig. 4b.

It can be seen that at 2 and 4 knots, the total vertical movement of the towfish is 16m and 50m, respectively and is generated by front wing with the help of

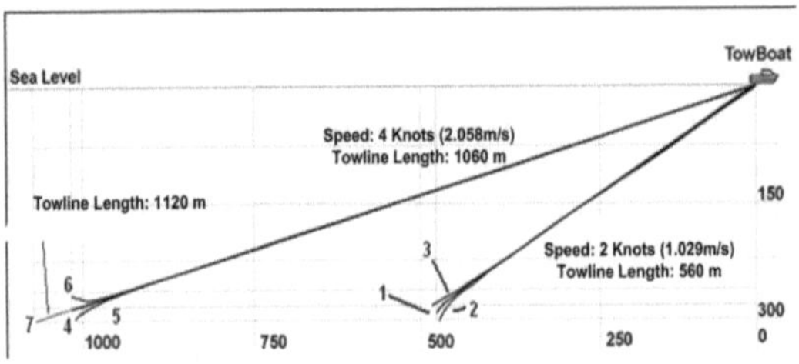


(a)

| | | CASE 1 | CASE 2 | CASE 3 | CASE 4 | CASE 5 | CASE 6 | CASE 7 |
|---|---|---|---|---|---|---|---|---|
| Speed | Knots | 2 | 2 | 2 | 4 | 4 | 4 | 4 |
| Lift/ Down Force | | 0 | 100 | -100 | 0 | 400 | -400 | 0 |
| Tension | N | 1236 | 1356 | 1116 | 1023 | 2246 | 1752 | 1783 |
| layback | m | 478 | 471 | 471 | 1717 | 1014 | 1024 | 1082 |
| Depth | m | 291 | 300 | 284 | 286 | 301 | 253 | 300 |
| Length of Towline | m | 560 | 560 | 560 | 1060 | 1060 | 1060 | 1120 |

(b)

**Fig. 4.** (a) Illustration of Towline for different cases detailed in (b)

rear wing to keep towfish horizontal. In the similar way, the lateral movement permitted by the tension in towline is 140m and 500m for the 2 knots and 4 knots speed respectively. These minimum permitted movements of 16m vertically and 140m transversely satisfy the requirements laid earlier.

These movements can be varied more at higher speeds. For example, if towfish is required to rise immediately increasing the speed from 2 knots to 6 knots, at the same towline length of 560m, will allow the towfish to rise by 73m. Of course, increasing the angle of attack of the fins allows even more and faster movement.

### 6.1 Fins Properties

A NACA 0012-63 profile is used for both front and rear wings. The maximum lift at front wing predicted by the polar diagram is 71 N at 2 knots and 286 N at 4 knots which counteracting the vertical and transverse component of towline force, satisfy the mobility requirements. Because of the downwash from the front wing, the behaviour of the rear wing (used for elevation as well as lift/down force) is more complex. This phenomenon has been accounted for within the stability analysis.

### 6.2 Disturbances through Towline

The disturbances generated by the movement of the umbilical in the water depend on the umbilical used, on the mass and drag of the depressor, if used, and on the mass and drag of the towfish. A large part of the strumming can be removed with the use of fairing.

As far as ship disturbances are concerned, disturbances on single tows are basically characterized by two modes: pendulum mode, similar to a mass swinging on a string; bowing mode, involving back and forth bowing of the umbilical. If the connection point of the umbilical is at or slightly forward from the centre of the gravity (CG), surge and heave of the ship tow point are transmitted and result in heave disturbance of the towfish, eventually with an additional pitch disturbance due to the rear fins [10]. Because in our design the tow point is slightly forward of the CG, surge and heave in the towing cause sway in the towfish, and towfish-induced kiting, cause sway and roll, [11].

Presence of the depressor helps reduce pitch disturbances but have no effect on yaw and roll disturbances [10] and on sway or surge disturbances [12]. As evident from following section, our vehicle has more robustness in pitching mode. So, it was decided not to use depressor since it could make yaw control more difficult.

### 6.3 Handling Characteristics

Based on a mathematical model of the dynamics of the system, Fig. 5 tabulates the handling characteristics of the towfish. When angle of attack is zero, at 4 knots, the natural frequency and damping ratio are within a satisfactory range with minimum oscillations when steadying the system for normal operation; where as at 2 knots, we have small over controlling oscillations that could be overcome at control systems. In both cases, it is a stable system requiring less help from the control system (unless required to control faster).

| Angle of Attack | Vehicle speed (Knots) | Natural frequency (rad/s) | Damping ratio | Time to half for pitch oscillation (s) |
|---|---|---|---|---|
| 0 | 2 | 0.8361 | 0.4508 | 3.3056 |
| 0 | 4 | 2.0304 | 0.4508 | 1.361 |
| Max | 2 | 3.2354 | 0.1556 | 1.3764 |
| Max | 4 | 0.82354 | 0.1556 | 3.3447 |

**Fig. 5.** Handling characteristics at different vehicle speeds

In the case of max. angle of attack, the towfish is required to vertically climb and fall without any pitching moment, and therefore the rear wing is driven to control that moment as well as provide some extra lift (or down force). System is still very stable. The handling characteristics are much more prone to over controlling oscillations, but those can be adjusted by precision controlling by the control system.

## 7 Control System Overview

Careful engineering of the hardware and software systems, supervising the robotic system mission, has been performed, particularly considering the equipment installed, the data exchange and the performance in different operative modes. Several aspects have been carefully investigated at the level of sensory system selection and control modules design: pipe engagement, pipe tracking, towfish deployment and recovery, pipe monitoring and inspecting.

The Fig. 6(a) shows the control and monitoring system layout. The control system comprises of a surface control unit and the monitoring unit running diagnostic applications, a surface-to-underwater interface and analogical/digital control unit; the towfish equipment, includes actuators, sensors, alarms besides the monitoring device module and various other equipment. The surface control

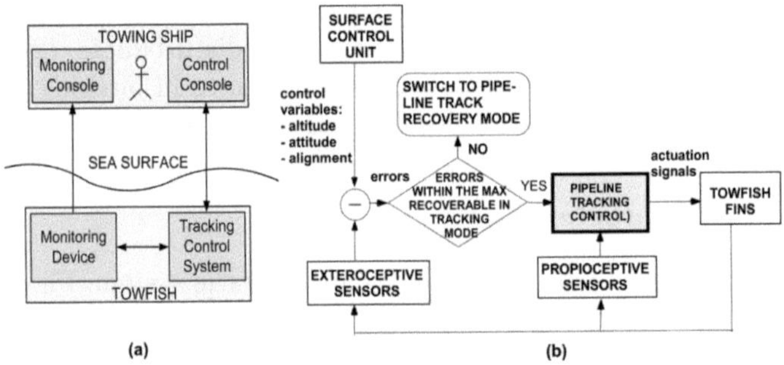


**Fig. 6.** Control system schemes

unit reads all sensor signals from the sub-sea unit and presents the scaled values on the monitor.

To enhance the system reliability and maintainability, a modular approach has been adopted in the control design. Operative missions are performed as sequence of different tasks, accomplished by pertinent control modules. Every module is designed to achieve a particular goal. A human supervisor can check, in real time, the progress of the missions. As far as possible and acceptable, standard equipment, software, and transmission protocols have been chosen.

Fig. 6(b) shows the main control loop of the system. Propioceptive sensors include the yaw, pitch and roll sensors, while altimeter, gradiometer, depth pressure transducer and pipeline monitoring system are grouped as exteroceptive sensors. They jointly measure the instantaneous state of the towfish in term of underwater depth, distance from seabed, roll, yaw and pitch angles and gradient of the magnetic flux. If the errors in position are beyond the range of normal tracking mode then track recovery mode is actuated and human intervention will be required to return the towfish on normal course through tele-control of fins and/or towline manipulation.

For illustration purpose, one towfish manoeuvring situation is described here in which movement of the towing ship suddenly shifts the towfish away from the pipeline causing the impairment in condition monitoring task. Data from the two gradiometers show that the vehicle needs to move in lateral direction swiftly and that yaw angle also needs to be corrected. The motion vector is determined by the 3 dimensional field of the gradiometer as shown in the Fig. 7a.

To get it back again on the normal course, a series of the fins movements would be required. Fin 5 and 6 would deflect to generate the lateral force and their differential forces would compensate of the yaw change as well. In doing so, a roll moment would be generated due to the offset of the centre of forces from towfish centreline. Fins 1 and 2 would deflect in opposite directions to react this roll moment. Any transient pitch moment or yaw moment (and movements) would be reacted or corrected by the fins pairs (3,4) and (5,6), respectively. The elevation can be corrected by the pairs (1,2) and (3,4).

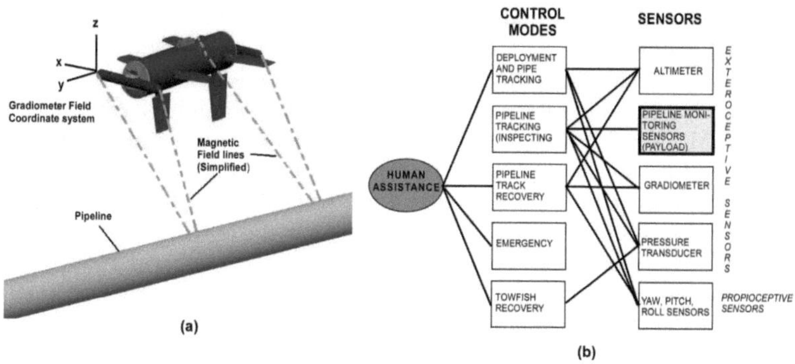


**Fig. 7.** (a) Illustration of how gradiometer helps to find and align with pipe (b) Control Modes

Various control modes and the sensors employed are shown in Fig. 7b. Three emergency modes are activated when sea-bed is very close, sea-surface is close and fins stall. Fins actuation to obtain maximum upwards and downwards forces are emergency measures for first and second mode. For the stall emergency mode, effective angle of attack of fins will be extracted from the yaw, roll and pitch sensors and maximum corrective measure will be taken at fins.

The amount of information displayed and the control functions available to the operator are set individually for each control mode. In emergency mode, the value of all sensors are displayed, and all controls are available for the operator. Each mode will have direct access to the equipment necessary for its limited operation. If other equipment shall be operated, the manual override is necessary. The emergency mode has access to all functions and can automatically override current tasks by shut down procedures. The alarms are arranged for a number of parameters; the shut down is enabled only for a limited number of catastrophical events.

## 8 Conclusion

This paper presented the design of a towfish which is intended to have high mobility and maneuverability than traditional designs to track the undersea pipelines more effectively. It also highlighted general design requirements for the application and presented a model of interaction between environment and towfish. Preliminary tests has been successfully conducted (high pressure tests and tracking of a metallic wire in a 500m pool).

## References

1. Naxans: White Paper Oil and Gas (2006), http://www.nexans.com/Corporate/2007/Nexans_White_Paper_Oil_Gas_April
2. Mastandrea, J.R., Miller, W.J., Clare, D.M.: Rapid Leak Detection for Sea Floor Pipelines: Development of Practical New Methods. NDE Technical Report No. NDE-90R0081601. NDE Environmental Corporation, California (1990)

3. Pozza, M., Hrvoic, D., Priddis, K.: Mapping Marine Ferrous Targets Using the SeaQuest Gradiometer System. Technical Report Marine Magnetics Corp (2003)
4. Linklater, A.: Design and simulation of a towed underwater vehicle. MS Thesis, Virginia Polytechnic Institute and State University, Virginia (2005)
5. Schuch, E.M.: Towfish Design, Simulation and Control. MS Thesis, Virginia Polytechnic Institute and State University, Virginia (2004)
6. Innovatum: Magnetization Technology Locates Undersea Pipes. Pipeline and Gas Journal 8, 227 (2000)
7. MacLeod, I.N., Vierra, S., Chavew, A.C.: Analytic signal and reduction-to-the-pole in the interpretation of total magnetic field data at low magnetic latitudes. In: Proc. of the third international congress of the Brazilian society of geophysics (1993)
8. Roest, W.R., Verhoef, J., Pilkington, M.: Geophysics 57, 116–125 (1992)
9. Ananthakrishnan, P., Decron, S.: Dynamic Stability of Small and Mini-Autonomous Underwater Vehicles: Part I. Analysis and Simulation for Mid water Applications, Technical Report, Florida Atlantic University, Florida (2000)
10. Preston, J.M.: Stability of Towfish used as Sonar Platforms. In: Oceans 1992 Proceedings of Mastering The Oceans Through Technology, Newport (1992)
11. Crane, J.W., Humphreys, D.E.: Modeling and simulation of underwater vehicles. In: SCS Summer Simulation Conference, Baltimore (1992)
12. Jiaming, W., Chwang, A.T.: Ocean Engineering 27, 455–472 (2000)
13. Licht, S., Polidoro, V., Flores, M., Hover, F.S., Triantafyllou, M.S.: IEEE Journal of Oceanic Engineering 29(3), 786–794 (2004)
14. Anan, C.S.: Autonomous Underwater Robot: Vision and Control. MS Thesis, The Australian National University, Canberra (2001)
15. Fossen, T.I.: Guidance and Control of Ocean Vehicles. John Wiley and Sons, New York (1994)
16. Buckham, B., Nahon, M., Seto, M., Zhao, X., Lambert, C.: Journal of Ocean Engineering 30, 453–470 (2003)
17. Chapman, D.A.: Ocean Engineering 9(3), 189–220 (1982)
18. Rodriguez, R.R.: Performance Evaluation of the Control Surface Actuators for the Towed Body for Mine Countermeasures Sensor Testing. Technical Report CSS/TR-97/20, Naval Surface Warfare Center, Panama City (1997)

# Part IV

# Wheeled Robots

# Autonomous Robotic Inspection for Lunar Surface Operations

Maria Bualat[1], Laurence Edwards[1], Terrence Fong[1], Michael Broxton[2], Lorenzo Flueckiger[2], Susan Y. Lee[3], Eric Park[2], Vinh To[3], Hans Utz[4], Vandi Verma[3], Clayton Kunz[2], and Matt MacMahon[5]

[1] NASA Ames Research Center, MS269-3, Moffett Field, CA 94035
Maria.G.Bualat@nasa.gov, Laurence.J.Edwards@nasa.gov, Terry.Fong@nasa.gov
[2] Carnegie Mellon University West
Michael.J.Broxton@nasa.gov, Lorenzo.J.Flueckiger@nasa.gov, ckunz@mail.arc.nasa.gov, epark@mail.arc.nasa.gov
[3] Research Institute for Advanced Computer Science
hutz@mail.arc.nasa.gov
[4] Perot Systems Corporation
sylee@mail.arc.nasa.gov, vinh2@email.arc.nasa.gov, vandi@email.arc.nasa.gov
[5] University of Texas, Austin, 1 University Station A8000, Austin, TX 78712
Matt@MacMahon.org

**Summary.** In this paper, we describe NASA Ames Research Center's K10 rover as used in the 2006 Coordinated Field Demonstration at Meteor Crater, Arizona. We briefly discuss the control software architecture and describe a high dynamic range imaging system and panoramic display system used for the remote inspection of an EVA crew vehicle.

## 1 Introduction

In preparation for returning humans to the Moon by 2020, NASA is dedicating significant effort to building exploration systems that are affordable, reliable, and effective. A key part of this effort is to develop tools and techniques that will allow humans to operate efficiently and safely on the lunar surface over the long-term. Given that cost pressures will keep astronaut teams small (e.g., four-person crews), robots will be needed to extend the effectiveness of these teams well beyond their individual human capacity. Robots, whether teleoperated or supervisory-controlled, can be used for "dull, dirty, or dangerous" tasks and other tasks that are not sensible or necessary for humans to perform unaided during exploration missions. Robots can also augment humans by differing from the human scale of performance in one or more parameters: force, size, speed, duration, precision, temperature, environmental exposure, cleanliness, etc. In particular, robots can be used for a wide range of surface operation tasks including (but not limited to) site survey, construction, routine maintenance, visual inspection, instrument placement, and transport. Involving robots in these tasks

C. Laugier and R. Siegwart (Eds.): Field and Service Robotics, STAR 42, pp. 169–178, 2008.
springerlink.com 

will provide three key benefits: (1) a significant reduction in the number and duration of extra-vehicular activity (EVA) sorties (and thus reduced cost and risk to crew), (2) more effective use of surface crew time (e.g., tasks can be off-loaded to ground control), and (3) enhanced mission capability (ability to reposition large payloads, greater sortie range, emergency resupply, etc.).

*Coordinated Field Demonstration at Meteor Crater*

During 2006, multiple NASA centers collaborated to develop human-robot systems for a variety of lunar surface activities. To provide a usage example of this collaboration a "Coordinated Field Demonstration" (CFD) was conducted involving robots from the NASA Ames Research Center (ARC), the Jet Propulsion Laboratory (JPL), and the NASA Johnson Space Center (JSC). The CFD took place during a two-week period in September 2006 near Meteor Crater, Arizona. The scenario for the CFD focused on operations that occur after the crew returns from a "typical" lunar EVA sortie on a manned vehicle (the JSC SCOUT rover) to a Lunar base site. At the base site, three robots supported post-sortie operations: the JPL ATHLETE robot equipped with a modular Pressurized Rover Compartment provided crew shelter, the JSC Robonaut off-loaded and transfered a payload from SCOUT, and the ARC K10 rover performed a "walk-around" visual inspection of SCOUT. The visual inspection ensured that SCOUT was safe for the next EVA.

*Visual Inspection Architecture*

The ARC K10 rover performed a remote walk-around visual inspection of the SCOUT rover. This was in lieu of having an astronaut perform the same task. Our goal was to create a system that would provide a remote human operator with imagery of sufficient resolution and dynamic range to determine EVA vehicle integrity after a sortie. The system must autonomously collect the inspection data and present it to the remote user in an easily navigated form.

The inspection task architecture is as follows:

- The K10 rover, described in Section 2, automatically collects high-resolution panoramic imagery at multiple inspection points. We describe the on-board software and executive in Sections 3 and 4, respectively.
- Inspection panoramas use high-dynamic range imaging, described in Section 5, in order to capture detail in deep shadow and strong highlight regions.
- K10 images from the inspection of SCOUT are transmitted to the control station, where they are stitched into seamless panoramas and displayed in the Panorama Viewer, described in Section 6.
- Ground control operators perform inspection by "exploring" the high-resolution panoramas.

In Section 7, we describe the CFD results and conclude in Section 8.

## 2 Inspection Rover

**Fig. 1.** K10 "Blue" Rover

The K10 rovers (Figure 1) are a series of robots designed to be a cost-efficient, easily maintainable robotic test-bed using as many commercial off-the-shelf components as possible. Each robot features a 4-wheel steer, 4-wheel drive rocker chassis and a top speed of 1 m/sec, comparable to human walking speed. Hard points on all sides allow attachment of additional components including antennas, masts, arms, and other equipment. K10's avionics are off-the-shelf with the exception of a few custom power electronics components. [1] The rover runs off of twenty 14.4V, 6.6 AH Li-Ion smart battery packs. The brain of K10 is an IBM Thinkpad X31 laptop that enables eased debugging and servicing while maintaining the reliability and computational power of a PC104 stack. K10's sensor suite includes a Novatel differential GPS system, a Honeywell digital compass, three sets of firewire stereo cameras, a Control Chief wireless e-stop, encoders and potentiometers for ground truth and wheel calibration, as well as temperature and power monitoring sensors. K10's inspection camera rig consists of a stereo pair of Point Grey Scorpion cameras mounted atop a Directed Perception pan-tilt unit. The cameras have a resolution of 1600x1200 and are configured with synchronized shutters. The vergence angle of the stereo pair is 5 degrees with a baseline of 36.2 cm making the ideal target-to-rover distance about 1 meter. A third camera is mounted in the center of the camera rig for taking HDR panoramas. In order to resolve millimeter scale features at distances up to 5m, we selected a 35mm Schneider Optics Xenoplan lens for the HDR panorama camera.

## 3 Rover Software

The K10 rover is one of a number of hardware platforms supported at ARC. We use our rovers for diverse experiments calling for various sensor and actuator configurations and we conduct multiple field tests requiring integration of various scientific instruments. In addition, the robot controller uses complex software components in applied computer vision and human-robot interaction. Finally the robot controller needs to smoothly integrate with interactive 3D visualization and monitoring systems for ground control.

To manage this complexity and create a scalable robotic system, we developed a Service-Oriented Architecture (SOA). Each component in this architecture provides specific functionality and exposes a clear interface to the others. The interfaces are defined using the Interface Definition Language (IDL) [2] and

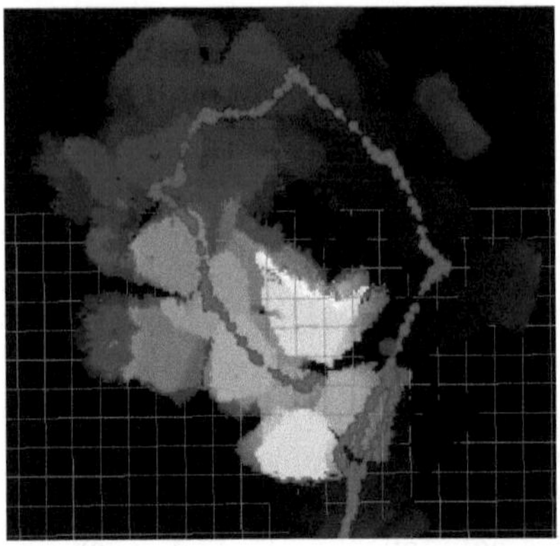

**Fig. 2.** Goodness map during an inspection task. The image shows safe areas (green) and dangerous obstacles (red). The confidence level (depicted by pixel brightness) decreases with time. SCOUTs rectangular profile is clearly seen in the center of the map. The other large red area is the ATHLETE robot sitting close to SCOUT.

communication between components relies on CORBA [3]. The main components used for this particular field demonstration are: locomotion, localization, navigation and panorama acquisition. Most of these components are built using facilities provided by the CLARAty (Coupled Layer Architecture for Robot Autonomy) framework [4].

*Locomotion*

The locomotion service encapsulates the locomotion framework provided by CLARAty [5]. This framework provides a common locomotion interface for any wheeled vehicle. A generic algorithm handles steering and drive actuation accounting for the vehicles kinematics and its capabilities. Communication with the motors is delegated to adaptations of the generic CLARAty motors for the K10 specific hardware.

*Localization*

The localization service computes rover pose estimates required by the navigation system to reach absolute GPS waypoints. K10 uses an Extended Kalman Filter with inputs from rover odometry, compass, inclinometer and differential GPS.

*Navigation*

The navigation service computes a safe path to a given goal. The navigator evaluates the environment and generates trajectories for the locomotor to execute using the sense-think-act navigation scheme realized by the CLARAty navigation framework [6]. Terrain reconstruction is performed with stereo vision using front and rear hazard detection camera pairs. The ARC stereo correlator [7] generates high fidelity 3D point clouds in 0.4 seconds on a 1.2GHz Pentium M. Rover path generation relies on the Morphin/D* algorithms implemented in CLARAty with specific input parameters for the K10 rover. An example of a goodness map and path generation is given in Figure 2.

*Panorama and HDR Enabled Service*

To obtain full coverage of the SCOUT at a nominal inspection distance of 2m, panoramas were acquired with approximately 90 degrees coverage in the horizontal direction and 45 degrees coverage vertically. The panoramic camera lens has a 12 degree horizontal and 9 degree vertical field of view. With 15% image overlap, achieving the desired SCOUT coverage requires acquisition of a 9x6 mosaic of images. The panorama service requires a specific start up sequence since it requires the camera and the pan-tilt services to be running. Panorama requests are handled by first sending the pan-tilt unit to the minimum pan angle and minimum tilt angle. Images are then taken using the inspection camera. The pan-tilt unit moves in a raster pattern with small pan-tilt increment angles to insure image overlap. If the service is started with HDR enabled, several images with multiple exposure settings are taken at each step increment.

*Benefits of the Service-Oriented Architecture*

Each of the above-described subsystems is a service component which only relies on abstract (IDL-defined) interfaces to other components, they are resolved at service startup time. The SOA allows us to group the services into dynamic libraries that can be loaded and configured at run time. This reduces the time needed to recompile and relink software. The memory footprint of the controller is also reduced since unused sensors and algorithms are not loaded into memory. The highly decoupled nature of the system helps avoid over-bloated code because deprecated services can be removed with minimal effort and risk.

## 4 PLEXIL Exec

The K10 rover operates in an uncertain environment. For example, the execution time of a traverse may vary depending on whether or not the rover encounters obstacles, components fail, or subtasks take longer to complete than anticipated. A robust control strategy must therefore implement desired behavior flexibly while ensuring safety in all conditions. An executive is a general software system that implements a control plan that specifies what to do in given situations. K10 uses the Universal Executive [8], a lightweight, efficient executive capable of executing PLEXIL plans. PLEXIL [8] is a language for representing control plans that is expressive and deterministic, and has formal semantics [9] that enable formal verification and validation. In addition, it allows for easy encoding of safety monitoring in the controller that runs parallel with task execution, thus ensuring safety constraints independent of individual tasks. K10's task was to perform a visual inspection of the SCOUT vehicle. A number of primary and secondary inspection points were defined. At each of these locations, K10 was to take a HDR panoramic image of SCOUT and send it to ground control for analysis. HDR panoramas from primary inspection points were mandatory, while inspections from secondary locations were to be performed only if the rover was operating well within the given time limit. During tests, the rover correctly skipped inspections at secondary tests to get back on schedule when obstacles slowed traversal between inspection points and when the rover was delayed by pose estimation inaccuracies. If no such delays were encountered, the robot acquired the whole set. During the final field demonstration runs, the time limit was very short, so the rover trimmed all secondary inspection points to accomplish the tight schedule.

## 5 HDR Imaging

The range of brightness values in most outdoor scenes far exceeds the dynamic range of commodity CCD and CMOS sensors. The luminance of a sunlit scene may vary by five or more orders of magnitude, whereas most solid-state sensors are 12-bit devices with a limited dynamic range of three orders of magnitude. Hence, even when the optimal exposure is measured and set for the scene, most images captured in a sunlit environment contain some regions that are either over- or under-exposed. Due to the absence of atmospheric dispersion and the anisotropic nature of sunlight on the moon, severe shadows result in very dark areas that will be underexposed in digital imagery, and correspondingly, when exposure is adjusted for shadowed areas, the sunlit areas will be harshly over-exposed. We have addressed this issue by capturing a bracket of several images

**Fig. 3.** Original low dynamic range mosaic (top) and high dynamic range mosaic (bottom)

with varying exposures. A bracket consists of images with evenly spaced exposures that range from underexposed (to capture bright regions) to overexposed (to capture dark regions). The number of images in the bracket can be increased until the full dynamic range of the scene has been captured. In our experiments, a bracket of five images with an exposure ratio of 1 photographic stop (i.e. a factor of two in shutter speed) between images proved sufficient. Once captured, the bracketed images are combined into a single high dynamic range (HDR) image using the weighted averaging technique of [10][11]. Combined in this manner, a HDR image contains well-exposed pixel information for every pixel in the image.

The resulting HDR image contains all of the information needed by the ground inspection team, however it is not ready for display; print media and most display technologies have inherently low dynamic range, thus the dynamic range of a HDR image must be reduced before it can be presented to the user. Simply scaling the pixel values linearly to fit in the dynamic range of the display yields poor results because the human visual system's response to luminance is approximately logarithmic rather than linear. Instead, we use the logarithmic brightness mapping function described in [12], which was designed to closely mimic the physiological response to brightness in the human visual system. This operation, generically called tone mapping, produces natural looking images that are well exposed across a very wide range of real-world brightnesses. Figure 3 shows a comparison of the inspection images before and after high dynamic range image processing. The low dynamic range mosaic is composed from a single exposure. The HDR images are composed from a bracket of five exposures, and then tone mapped for display using the algorithm in [12].

## 6 Panorama Stitching and Display

In order to provide a coherent context for visual inspection, individual images were projected onto a common spherical surface utilizing pointing information from rover telemetry, and merged into a single mosaiced panorama. As described above, the multiple exposures are combined into single images prior to panorama mosaic merging. To reduce discontinuities in the merged panorama due to differences in brightness between individual images, a weighted average of pixel values was used where images overlapped. Pixel weighting is based on the distance of a given pixel from its containing images edge. Given the high resolution of the images and the number of images acquired, a straightforward distance calculation is computationally prohibitive. A fast distance transform technique [13] was implemented which enabled generation of blended merged 12000x6000 pixel mosaics in less than five minutes. The merged panoramas were then converted into a multi-resolution image structure for interactive viewing in the Panorama Viewer described below. The Panorama Viewer was developed to allow inspection of the blended and merged panoramas. The multi-resolution structure of the final panorama image product enabled smooth, interactive navigation (zooming and panning) of the large images providing continuous contextual cues during the inspection process. The Panorama Viewer user interface also provided means to quickly switch between panoramas acquired at adjacent inspection stations, again allowing the user to maintain a sense of continuity while navigating the data. The Panorama Viewer was developed as part of the Ames VizExplorer component of the Ensemble software framework. Ensemble is being developed in collaboration with JPL as a framework for mission operations software development. The Panorama Viewer utilizes the OpenSceneGraph C++ library [14] wrapped in a Java Native Interface (JNI) for high performance rendering.

## 7 Field Demonstration Results

Overall the Coordinated Field Demonstration was successful, with the robots from each NASA center reliably completing their tasks. At the end of each EVA sortie, the suited astronauts parked the SCOUT rover within the designated inspection area and K10 was commanded to begin its walk-around. During the morning and early afternoon K10 had little problem properly navigating around SCOUT and taking inspection panoramas; however, after about 4pm the lighting conditions were poor enough to cause false obstacle detections leading to extremely long navigation times. Due to time constraints we did not acquire full HDR panoramas during every run. When running without acquiring panoramas at inspection points, K10's SCOUT inspection took an average of fifteen minutes. A total of 18 full and partial inspection runs were performed; however, only two HDR panoramas were acquired. Once the circumnavigation of SCOUT was complete, the inspection images were downlinked over the wireless network to the command trailer where the ground station automatically generated the stitched HDR panoramas. The HDR inspection panoramas contained sufficient resolution to enable a very detailed inspection of the SCOUT rover. Small details that would normally be obscured by deep shadow, such as a pebble stuck in the tread of the tire under the wheel well, were clearly visible. The rover control software's Service-Oriented Architecture provided us with the flexibility to adapt inspection runs to the conditions and time constraints encountered in the field. For example, a single flag in the configuration file allowed switching between HDR and non-HDR panoramas. The availability of the rover component interfaces across the network (using the CORBA middleware) provides seamless multi-mode control. Autonomous execution could be suspended to allow direct teleoperation using a dedicated GUI, and then resumed.

## 8 Conclusion

Using robots to perform remote equipment inspections is one way of reducing risk to humans during exploration missions by reducing the number and duration of EVA. In order to enable remote visual inspection by a habitat- or an earth-based user, the inspection system must be robust and autonomous enough to keep mission operations simple and efficient. The imagery returned by the inspection robot must be of sufficient resolution and dynamic range that significant damage can easily be spotted. We have created an inspection task architecture that addresses these requirements and have demonstrated the system in the field. Further use studies would help to better characterize and improve the efficiency of the system.

## Acknowledgments

The authors of this paper would like to acknowledge the support and sponsorship of the Exploration Systems Technology Development Program. We thank the interns and other members of the Intelligent Robotics Group at NASA Ames Research Center for their continuing support and inspiration. We would also like to thank our colleagues from NASA Johnson Space Center who provided us with invaluable support and made the Coordinated Field Demonstration possible, especially Ron Diftler and Darby Magruder.

## References

1. Bualat, M., Kobayashi, L., Lee, S.Y., Park, E.: Flexible Rover Architecture for Science Instrument Integration and Testing. In: Proceedings of AIAA Space 2006, San Jose, California (September 2006)
2. Object Management Group. OMG IDL (2006) [Online], `http://www.omg.org/gettingstarted/omgidl.htm`
3. Object Management Group. CORBA FAQ (2006) [Online], `http://www.omg.org/gettingstarted/corbafaq.htm`
4. Nesnas, I., Wright, A., Bajracharya, M., Simmons, R., Estlin, T.: CLARAty and challenges of developing interoperable robotic software. In: Proceedings of the 2003 IEEE/RSJ International Conference on Intelligent Robots and Systems, Las Vegas, Nevada (October 2003)
5. Bualat, M.G., Kunz, C.G., Wright, A.R., Nesnas, I.A.D.: Developing an autonomy infusion infrastructure for robotic exploration. In: Proceedings of the 2004 IEEE Aerospace Conference, March 2004, vol. 2, pp. 849–860 (2004)
6. Urmson, C., Simmons, R., Nesnas, I.: A generic framework for robotic navigation. In: Proceedings of the 2003 Aerospace Conference, March 2003, vol. 5, pp. 2463–2470 (2003)
7. Edwards, L., Sims, M., Kunz, C., Lees, D., Bowman, J.: Photo-realistic terrain modeling and visualization for Mars Exploration Rover science operations. In: Proceedings of the 2005 IEEE International Conference on Systems. Man and Cybernetics, vol. 2, pp. 1389–1395 (2005)
8. Verma, V., Jnsson, A., Pasareanu, C., Iatauro, M.: Universal Executive and PLEXIL: Engine and Language for Robust Spacecraft Control and Operations. In: Proceedings of AIAA Space 2006, San Jose, California (September 2006)
9. Munoz, C., Pasareanu, C., Dowek, G.: A Small-Step Semantics of PLEXIL, NASA Langley Research Center Technical Report
10. Reinhard, E., Ward, G., Pattanaik, S., Debevec, P.: High Dynamic Range Imaging: Acquisition, Display, and Image-Based Lighting. Morgan Kaufmann, San Francisco (2006)
11. Mitsunaga, T., Nayar, S.K.: Radiometric Self Calibration. In: 1999 IEEE Computer Society Conference on Computer Vision and Pattern Recognition (CVPR 1999), vol. 1, p. 1374 (1999)

12. Drago, F., Myszkowski, K., Annen, T., Chiba, N.: Adaptive Logarithmic Mapping For Displaying High Contrast Scenes. Computer Graphics Forum 22(3) (2003)
13. Felzenszwalb, P.F., Huttenlocher, D.P.: Distance Transforms of Sampled Functions, Technical Report TR2004-1963, Cornell Computing and Information Science (2004)
14. The OpenSceneGraph high performance 3D graphics toolkit (2006) [Online], http://www.openscenegraph.org

# Roving Faster Farther Cheaper

Matt Roman, David P. Miller, and Zac White

University of Oklahoma, Norman Oklahoma, USA
mattjroman@gmail.com

**Summary.** Current Mars rovers travel a few kilometers per year. Future rovers will need to travel a couple orders of magnitude faster than that in order to move from safe landing zones to areas of scientific interest – which may be many 10s of kilometers away from a suitable landing location. This paper describes a recent field test of the SR-II rover. SR-II is a 22kg solar powered autonomous rover, which during a week long traverse through rough Mars-like terrain was able to cover almost a kilometer per day.

## 1 Introduction

Rovers are important for conducting in-situ scientific analysis of objectives that are separated by many meters. In the search for answers to the history of Mars these science outcrops could be separated by tens of kilometers. The twin Mars Exploration Rovers have traversed over 15km between them, validating that long distance traverses are possible. However, this was only possible after multiple mission extensions. Under the current procedures for operating a rover on Mars this type of mission will take many years to complete and cost billions of dollars. More rovers could be sent in order to speed this process up, but the cost of the mission would only increase more.

The current state of planetary landing technology has a landing ellipse many kilometers wide. Missions to Mars are therefore limited in landing sites that have a high probability of a safe landing. However, the more interesting scientific outcrops may be located in an area where the odds of a safe landing are too risky. Fast long traverses could be used to get the instrument payload from a safe landing site to where it is needed.

In this paper we will describe the latest field test of the Solar Rover-II (SR-II) robot. SR-II uses a simplified suspension system but exhibits similar mobility capabilities to the rocker bogie suspension used on the 1997 Pathfinder [1] and 2003 MER missions [3]. It uses only two motors, which are located inside of the chassis to maintain a nominal temperature, instead of ten exposed to the atmosphere. SR-II is a combination of simplified mechanics and control algorithms using non-computationally intensive sensors under the influence of a simple operational strategy. It has been shown under realistic terrain conditions that traverses of 1km per day through rugged terrain are possible under autonomous control.

C. Laugier and R. Siegwart (Eds.): Field and Service Robotics, STAR 42, pp. 179–188, 2008.
springerlink.com 

**Fig. 1.** SR-II in Anza Borrego

## 2 Rover Hardware

SR-II is a four-wheeled rover with a ground clearance greater than a wheel diameter. It has 4-wheel drive and is skid steered using only two motors. The motors and drive train run through a hollow tubular suspension. At the center of the chassis is a geared differential that increases stability for the sensors and assures that all four wheels maintain ground contact with equal force. The rover mechanics are described in detail in [7] and [8].

### 2.1 Mast Cameras

A stereo camera pair with an integrated fiber optic lens for taking spectrometry readings can be seen on the mast at the top of figure 1. The mast is capable of rotating more than $360^o$ and tilted approximately $30^o$ above and below horizontal at a height of 1m from the ground. The pan and tilt drive mechanisms are integrated with the solar panel to save space inside the chassis.

### 2.2 Electronics

The rover electronics (see Figure 2) include a PowerPC based Mac mini and an XBC processor [2] real-time controller. The XBC provides a low-level interface to many of the sensors that keep track of the health of the rover.

**Fig. 2.** Left: Inside the SR2 Chassis: Mac-mini, XBC, Power management electronics and stereo navigation cameras (near top) are not currently used. Right: Close-up of the SR2 wheels used in 2006 field test.

A battery charging protection system was implemented using thermal sensors on each of the batteries to prevent over charging. Current meters signal how much power is flowing from the batteries and solar panel to the motors and electronics, the XBC uses these to detect a stalled motor. A PID control loop on the XBC uses back-EMF to position the mast and control the speed of the drive motors. However, a separate motor driver H-bridge circuit was created to handle the more powerful drive motors. During normal driving operations the XBC would simply relay the sensory information and drive motor positions to the Mac mini over a serial connection.

The primary sensors used for navigation and hazard avoidance are connected to the Mac mini through a serial port hub. Three Hokuyo URG-04LX scanning lasers are mounted to the chassis and underside of the solar panel on the forward end of the rover. The lasers prove to be effective in detecting obstacles 2 to 3 meters away even in the extreme desert sunlight.

A Honeywell HMR300 magnetic compass is used to simulate an IMU system. The compass provides pitch and roll angles up to $\pm 45^o$ as well as heading information to the Mac Mini. A hand held GPS was also attached to the rover but only to supply ground truth location data for logging. The rover did not use onboard GPS data for navigation.

### 2.3 Solar Power

Two 12 volt 4.5Ah Ni-MH batteries are used to store energy from the solar panel. The $0.72m^2$ solar panel has 6 strings with 31 silicon cells per string potentially giving a maximum of 120 watts. The avionics used between 17 and 60 watts depending on the computational loads. On level ground the mobility system requires only about 10 to 11 watts to cruise at 15cm/second. On much more difficult terrain the drive motors can pull up to 50 watts to climb over rocks greater than a wheel diameter. The power system has not yet been optimized but in the transition to a flight system for Mars there would be a decrease of about 50% in solar flux but also a decrease in rover weight by about two-thirds. Thus the present ratio of power generation to use is approximately correct.

### 2.4 The Wheels

The wheel design uses a three piece tire made of sheet aluminum less than 1mm thick. The center band is 200mm in diameter by 30mm wide and is completely smooth. It is backed by the main supporting structure of the wheel, figure 2, a 5 spoked ring that connects to the wheel hub and axle. The middle portion of the center band is effectively rigid, the inner and outer edge bands are fastened to the rim of the center band. The edge bands are conically shaped inward at a $20^o$ angle from the centerline making the wheel 110mm wide. These edge bands have 18 raised grousers fastened at a $45^o$ angle to the centerline of the axle. The concept is to have as little drag as possible when driving straight on hard surfaces while maintaining the ability to climb over various obstacles. The combination of the thin sheet metal and the 5 spoked support ring allow the outer edges of the tire to deform, making more surface contact with the terrain which allows the rover to climb more effectively.

## 3 Remote User Interface

CommandCenter is a GUI program that allows the remote user to communicate with SR-II [9]. All mission plans are posted on CommandCenter and uploaded to the rover. Its primary purpose is to view the health of the rover and show updated position information. The map on the GUI displays where the rover believes it is and where other way-points are in relation to it. CommandCenter can also be used to tele-operate the rover if it needs help or needs to be put to sleep. New way-points can also be added to the plan that include the option of taking panoramic images once the rover reaches them.

## 4 Onboard Autonomy

The rover receives a list of way-points from the CommandCenter application described above. The way-points are in rectilinear coordinates where the origin was the rover's starting position at the beginning of the traverse and the Y-axis is aligned with North.

The onboard system checks to see if there are any outstanding way-points. If so, it takes the next point, calculates the vector between the current position and the way-point and starts toward the new point. If there are no points, it waits for a point or another command. Each way-point may contain an optional science operation (e.g., record a stereo panorama, take a spectrometer reading, etc). The rover will pause at the way-point and do the attached operation, if one exists, before proceeding to the next way-point.

If the rover is moving toward a way-point, it may encounter an obstacle or hazard. If an obstacle is detected by the forward laser in the rover's path (a 70cm wide swath up to 2m in front of the rover or the way-point, whichever is closer), then the full sweep of laser data is searched to check if the rover should go around the obstacle to the left or right. The forward laser sweeps the semi-circle forward of the rover along the plane defined by the skid plate of the chassis (approximately 1 wheel diameter above the surface). This is the maximum height of an obstacle or step that the rover can reliably surmount. If the entire forward path is blocked, the rover will continue forward at reduced speed.

A laser mounted at a $45^o$ downward angle detects the distance to the ground along a line approximately 60cm in front of the wheels as the rover moves. Both positive and negative obstacles can be detected by this sensor. If a positive or negative obstacle is detected in front of the wheels or a positive obstacle is detected in front of the body, the rover will turn to avoid the obstacle; stopping and performing a point turn if needed. The definition of an obstacle is determined in part by the roll and pitch of the rover at that time. E.g., if the rover is going down hill then a positive obstacle is less of an obstacle, and may simply be the bottom of the slope.

Obstacles detected by the angled laser that are in front of a wheel are avoided by the rover initiating a turn during its forward progress. Obstacles in front of the body are avoided by a turn in place until a clear path is found.

When avoiding an obstacle, the rover must pass the obstacle by two vehicle lengths before the rover will try and resume its path toward the way-point. If another obstacle is detected prior to reaching this safe path distance, the rover will avoid and proceed 2 vehicle lengths past the most recently detected obstacle before resuming its course. If the rover turns left at the first obstacle it detects, it will turn left at all future obstacles until it has been able to resume its path to the way-point. A similar strategy is pursued if the rover had turned right.

When the obstacle reaches within a specified distance of the goal way-point (currently three rover lengths), it will move toward the next way-point. If the rover fails to make progress toward a way-point for a specified amount of time (currently set at 15 minutes) meaning that the rover is no closer to the way-point now then it was fifteen minutes ago, it will stop and call for help. This will usually be due to a way-point being placed in an area that cannot be reached by the rover. This scenario has not occurred in a real field test, though it has been demonstrated in a test situation.

At any time, if the rover detects that it only has a few minutes of battery life left (determined by battery voltage and load), then it will initiate a timed sleep. If this is before 3pm local time, then the rover will power down all of the sensors and put the Mac to sleep, allowing the batteries to charge for twenty minutes. If it is after 3pm, it is assumed that insufficient sunlight is available for a quick charge, and the rover will sleep until the following morning.

The operator can interrupt the rover at any time (except during a sleep, whose interruption currently requires physical contact) to put the rover in a manual mode. When the rover gets into a situation where it is calling for help, going into manual mode is the only way to get the rover to move. In manual mode the rover can be teleoperated, and/or the way-point list may be edited.

The on board autonomy is written in Common Lisp running on LispWorks on the Mac-Mini. LispWorks provides support for multi-threading (which is used to decouple sensing from mobility operations) and external function calls (to C and objective C functions which control the sensors and communications to command center respectively).

## 5 Field Test Results

The Anza Borrego field test site was selected because it reflects similar geological features that are analogous to the most easily accessible layered or sedimentary materials on Mars [4] (e.g., Figure 3). The topography is relatively low, with elements of non-traversable terrain such as steep slopes, escarpments, deep channels and large rocks relative to the scale of the rover. This configuration of terrain elements is seen on Mars where layered rock is created by wind and lake-water sediments. The test site is located near the western edge of the Salton Sea Lake where similar processes are taking place [6].

**Fig. 3.** Left: SR-II successfully navigated through 100s of meters of sedimentary materials. Right: a wheel rim caught on an edge of one of the rocks requiring a short tele-operation to extricate it.

The field test was completed in 8 days where SR-II was able to traverse over 5km along a predetermined path of way-points. On several days the rover almost traversed a kilometer. On average fewer than three hours (and never more than four and one half) of driving time were spent each day. This limitation was due to the elevated air temperatures (in excess of $46^oC$) during prime solar power time (noon). The high heat reduced the solar panel's efficiency limiting the recharging of the rover's batteries, but more severely it limited the endurance of the field test personal.

Google Earth images were used to select the way-points the rover would follow. Prior to the start of the test a total of 148 way-points were imported by CommandCenter. The software converted the Lat/Long coordinates from the KML file given by Google Earth into relative positions in meters from the start. The operator selected various target way-points to conduct science, then sent the final compiled way point list to the rover as a plan. During the test, the rover continuously recorded sensor values and GPS positions into a log file.

Each day of driving was completed by letting SR-II autonomously navigate from way point to way point. During the traverse the rover operator intervention occurred only 6 times. This situation was resolved by tele-operating the rover a few meters away from the incident spot. Three of these events were caused by rocks catching the inner edge of the rovers wheel as it made a skid turn (Figure 3). Similar situations may be solved in the future by adding hub caps to decrease the chances that a protruding rock will snag a wheel. In the current cases rover would initially detect a stalled drive motor and begin a series of maneuvers to free itself. When this failed to free the rover a message was sent to CommandCenter indicating what caused the rover to stop and ask how to proceed. On one occasion the rover slid on a slope and was angled at what it considered a severe roll angle. It called for help and was tele-operated about a meter whereupon it could continue on its own. The other two occasions of operator intervention were due to operator impatience – the rover looked as if the rover might have difficulty making its way around a feature and a short manual tele-operation was done instead. In most cases, cooler heads prevailed and the rover was allowed to find its own way.

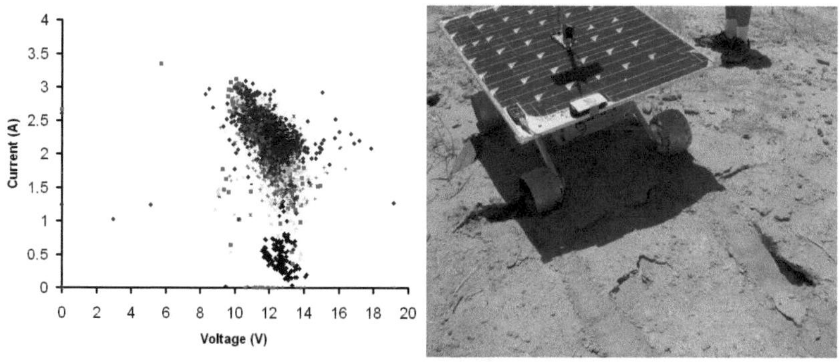


**Fig. 4.** Left: Solar panel power characteristics. Right: SR-II autonomously driving out of a sink hole.

The power system on the rover was also being tested while out in the field. The current system allowed the the rover to operate for about three hours. Late in the day the rover would become starved for power because the batteries had been drained making it difficult for the rover to climb over rocks. It is apparent that the batteries were being used as the primary power source. While in the field, the surface of the solar panel was approximately $30^oC$ higher than originally planned for. The average power output of the panel was less than 35 watts (figure 4). The lack of power from the solar panel is due to the decrease in voltage as the solar cells heat up. In the event that the voltage dropped too low the rover put itself into sleep mode without any complications.

Analysis of the motor current data from the total 5km traverse showed that on average only about 17 watts of power is needed for the mobility system. During the experiment the rover drove across various surface textures, from loose sand to hard packed soil that occasionally collapsed into a sinkhole, figure 4.

The log files from all 8 days of traverse were converted into Google Maps KML files and uploaded to Google Earth. Figure 5 shows location data of the path the rover took overlayed on the desired path. This allows us to view the accuracy of the rover's computed position which is based on odometry and compass heading. On day 6 the rovers path deviated significantly when a previous version of the control software was loaded which did not take into account magnetic declination for the compass. Aside

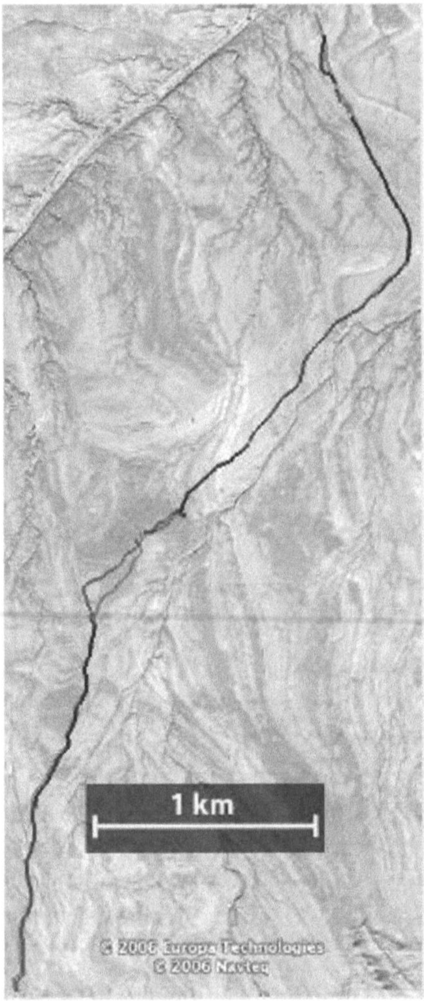


**Fig. 5.** Each section of the path is a single days drive

from this instance, dead-reckoning errors caused by wheel slippage was less than 2%, and usually less than 1% of the distance traveled. After a kilometer long traverse the rover would actually be within ten meters or so of where it thought it was.

The dead reckoning errors seem low, but have been consistent over a number of trials with various types of terrain (though not sand). Our current theory for these results is that slippage errors, over long enough traverses, tend to cancel each other out. This phenomenon has also been observed in Pacific canoe voyages[1].

[1] Personal communication from Ben Finney, University of Hawaii, April 2007.

## 6 Conclusion

The simplified mobility system of SR-II is quite capable of traversing Mars-like terrain while being very power efficient in most cases. More power is required when the rover changes direction compared to Rocky style rovers. However, only a small portion of the traverse is spent turning, even on complex terrain. As a result of the simpler mobility and despite the comparatively large amount of power used on SR-II for the computation and communication (when compared to similar class vehicles such as Sojourner) it has a much larger range per sol [5].

The hazard avoidance system was acceptable in finding a safe path between way-points. This is significant because the hazard avoidance system is much simpler than that used by MER, and SR-II has, theoretically, lower overall mobility than MER [3]. These two factors might lead one to expect that the rover would be more prone to get lost or get stuck or have other problems related to ground hazards. Yet this was not the case.

Traverses of a kilometer or more per day are possible with a single communication cycle, rather than the few meters currently being done on Mars. We believe that future robotic missions to Mars should accomplish more scientific goals by relying on the autonomous capabilities of the rover. We believe that the operational philosophy of providing numerous, widely spaced way-points per communication cycle was largely responsible for the significant distances traveled per day. This difference in operations is more significant for improved performance than the hardware and software differences between SR-II and MER. The rover covered as much ground as it could – there was always more for it to do.

Additional field tests are planned for 2007. We will use an improved solar panel/battery system and wheel design. We expect increased endurance from the new power system. The new wheels will hopefully further reduce the already rare occurrences where operator intervention is required. Experiments comparing stereo navigation with the current laser scanner system are also planned.

## Acknowledgments

This project has been funded on part by NASA's Mars Technology Program and MSSS Inc.

## References

1. Bickler, D.: The new family of jpl planetary surface vehicles. In: Moura, D. (ed.) Missions, Technologies and Design of Planetary Mobile Vehicles, Toulouse France, Cepadues-Editions Publisher (1993) (IAF-97 - Q.2.07 1997)
2. LeGrand, R., Machulis, K., Miller, D.P., Sargent, R., Wright, A.: The XBC: a modern low-cost mobile robot controller. In: Proceedings of IROS 2005, IEEE Press, Los Alamitos (August 2005)
3. Lindemann, R.A., Bickler, D.B., Harrington, B.D., Oritz, G.M., Voorhees, C.J.: Mars exploration rover mobility development - mechanical mobility design, development, and testing. Robotics & Automation Magazine, IEEE 13(2), 19–26 (2006)

4. Malin, M.C., Edgett, K.S.: Mars global surveyor mars orbiter camera: Interplanetary cruise through primary mission. J. Geophys. Res. 106 (E10), 23429–23570 (2001)
5. Miller, D.P., Hunt, T.S., Roman, M.J.: Experiments & analysis of the role of solar power in limiting mars rover range. In: Proceedings of the IROS 2003 Conference (October 2003)
6. Miller, D.P., Ravine, M.: Semi-autonomous rover operations: An integrated hardware and software approach for more capable mars rover missions. In: Proceedings of the NASA Science Technology Conference (June 2007)
7. Matthew, J.: Roman. Design and analysis of a four wheeled planetary rover. Master's thesis, University of Oklahoma (August 2005)
8. Roman, M.J., Miller, D.P.: Four is enough. In: Proceedings of Space 2006, p, September 2006, p. AIAA–2006–7399. AIAA (2006)
9. White, Z., Miller, D.P., Roman, M., Ravine, M., Flippo, D., Mills, B., Nixon, B., Noe, E., Malin, M.: Control and operations for a long duration solar powered mars rover. In: WWDC 2006 Scientific Development Poster Session, Apple Computer Corp. (2006)

# Characterization and Comparison of Rover Locomotion Performance Based on Kinematic Aspects

Thomas Thueer and Roland Siegwart

Autonomous Systems Lab, ETH Zurich, 8092 Zurich, Switzerland
{thomas.thueer,roland.siegwart}@mavt.ethz.ch

**Summary.** Evaluation and comparison of locomotion performance of rovers is a difficult, though very important issue. The performance is influenced by a large number of parameters. In this work, three different rovers were analyzed from a kinematic point of view. Based on a kinematic model, the optimal velocities at the actual position were calculated for all wheels and used for characterization of the suspension of the different rovers. Simulation results show significant differences between the rovers and thus, the utility of the chosen metric. It is shown that a substantial reduction of slip can be achieved by integrating kinematics in a model-based velocity controller.

## 1 Introduction

Many rovers have been presented in the past by different people and institutions, however, comparison of performance was widely neglected and the advantages of new systems were highlighted without an absolute reference. It is difficult to compare systems that differ in all dimensions, e.g. wheel diameter, footprint, centre of gravity (CoG), and which were developed with respect to different requirements but comparison can yield important information about the systems' real capabilities.

An effort in this direction was made at the Autonomous Systems Lab (ASL) in order to compare a selection of wheeled, passive systems based on precisely defined metrics. The first steps towards a complete evaluation of these systems included the development of a quasi-static simulation tool [1], an analysis of a wide selection of locomotion systems [2] and a validation of the simulation results on real breadboards of selected rovers [3].

This paper describes the continuation of the rover comparison based on another aspect: kinematics. Several people have presented work that focuses on kinematics for different purposes. Forward kinematics was used in simulation for the estimation of rover position and heading in [4]. Wheel actuation commands can be derived for a desired rover motion by means of an inverse kinematics model [5]. [6] and [7] included rover kinematics in the estimation of the wheel-ground contact angle and [8] developed a kinematic observer for articulated rovers.

C. Laugier and R. Siegwart (Eds.): Field and Service Robotics, STAR 42, pp. 189–198, 2008.
springerlink.com

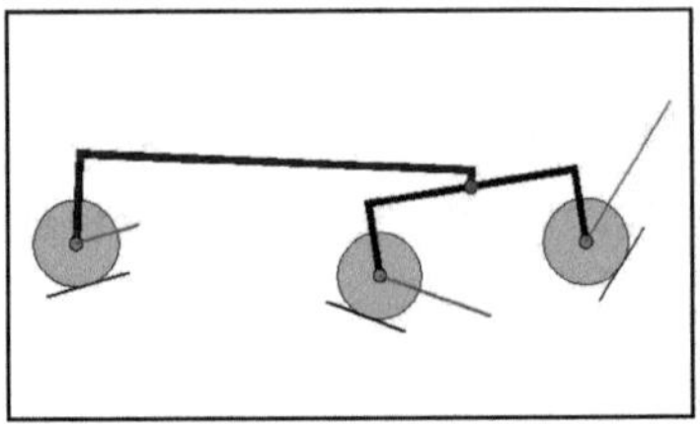

**Fig. 1.** Example of optimal velocity distribution depending on rover state

A kinematic model can provide valuable information because the ability of articulated rovers to adapt to uneven terrain makes it difficult to relate rover motion to wheel motion and requires the wheels to move at different speeds. In this context rover control plays a central role. Several control strategies have been presented which aim at increasing the rover's performance by synchronizing wheel velocities [9], setting optimal torques depending on the rover's state [10] or actively adapting the rover's configuration based on kinematic information [11].

This work's main focus lies on the development of a metric that helps characterizing the kinematic potential of different suspension types. Kinematics are used to calculate the optimal velocity distribution depending on the rover's state in rough terrain. Fig. 1 shows that optimal velocities on the wheels differ significantly if kinematic constraints of the suspension have to be respected. The bigger the difference between optimal velocities, the more difficult it is to satisfy the constraints. Violation of such constraints induces slip. Therefore, the different rovers will be analyzed with regards to the velocity distribution that is caused by their suspension mechanism and the resulting slip. Additionally, the kinematic model is integrated into a control loop which adapts the velocity commands for each wheel separately in order to reduce slip.

The next section describes how the kinematic models were derived for the different rovers. Then the metrics are introduced which allow to properly compare locomotion performance. Further, the simulation setup is presented in which the tests were performed before the results are discussed. Finally, the presented work is summarized.

## 2 Kinematic Models of Rovers

In this section the analyzed rovers are briefly presented and the kinematic modeling is explained. For reasons of consistency, the same rovers were selected for the kinematic analysis as in [3] where more information can be found about the systems. The selected rovers are shown in Table 1. The kinematic models were simplified such that they still respect the geometrical constraints imposed by the suspension system, thus maintaining identical behavior as the real rovers. Since the original rovers differ with respect to many parameters, they were all

**Table 1.** Selected rovers: MER, CRAB and RCL-E

| Breadboard | Schematic view | Kinematic model |
|---|---|---|
| **MER** by NASA (rocker-bogie type) | | |
|  | 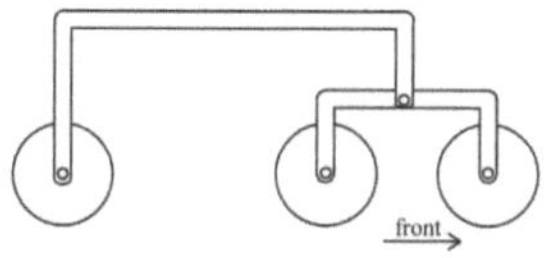  | 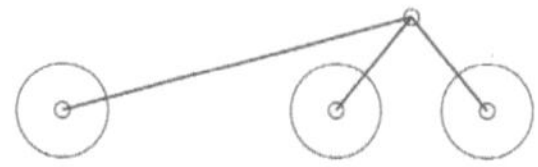 |
| **CRAB** by ASL (symmetric structure based on four parallel bogies) | | |
|  | 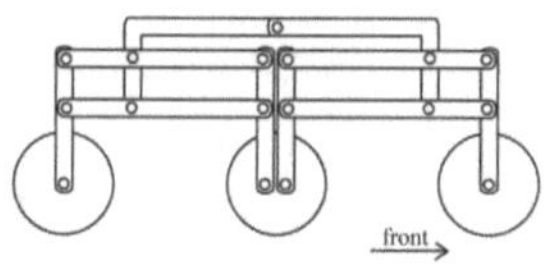  | 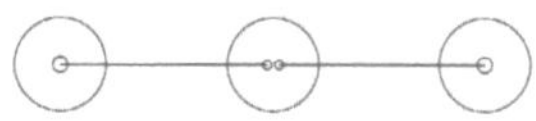 |
| **RCL-E** by RCL (three parallel bogies, no differential mechanism) | | |
|  | 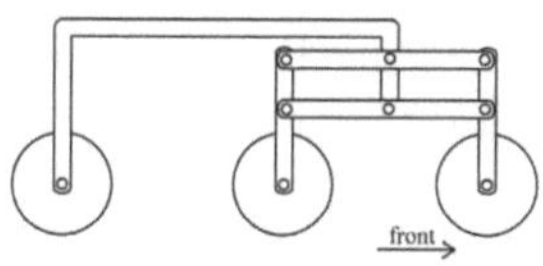  | 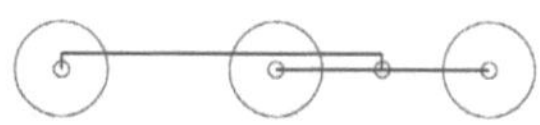 |

normalized to the approximate dimensions of the Mars Exploration Rovers[1] (MER) [12] with the same total mass (~180kg), wheels (⌀0.25m), foot print (~1.5m) and CoG. This normalization allows for a proper comparison of the suspension types regardless of terrain characteristics.

Rough-terrain robots usually consist of several rigid elements connected through joints of a certain number of degrees of freedom (DoF) resulting in a structure that has one system DoF. This allows the rovers to move along an uneven terrain without loosing contact (in most cases). Rigid body kinematics for closed loop systems was used to represent these characteristics and establish the rover models. The modeling process is explained on the example of the MER as it is depicted in Fig. 2.

The system is defined through the vectors of constant length AD, DB and DC, wheel-ground contact angles ($\gamma_i$) as well as orientation of rocker ($\alpha$) and bogie ($\beta$) with respect to the inertial system. In reality $\alpha$ could be measured by

[1] Nominal driving direction of MER (trailing bogie) in opposite direction to definition in Table 1.

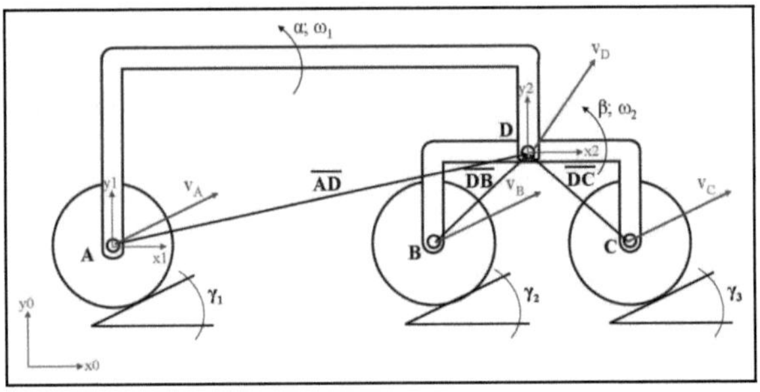


**Fig. 2.** Kinematic model of MER with parameter definition

means of an inertial measurement unit (IMU) and $\beta$ with an angular sensor in D that measures the relative movement between rocker and bogie. Measuring the wheel-ground contact angles requires tactile wheels as in [13]. The development of tactile wheels is an ongoing project at ASL.

The unknown parameters are the magnitudes of the velocities in A, B, C and D (given that the wheels always touch the ground), as well as the rotational velocities of rocker ($\omega_1$) and bogie ($\omega_2$). The velocity in D is not of interest; therefore it is expressed by means of the other velocities yielding an equation system with four equations (2D) and five unknowns. As it was mentioned before rovers have one DoF, which means that one velocity can be chosen as input. Thus, the equation system has exactly one solution. $\vartheta_D$ is expressed in the inertial system through A, B and C.

$$\vartheta_{D_A} := \vartheta_D = \vartheta_A + \omega_1 \times {}^0_1R(\alpha)\ {}^1\overline{AD} \tag{1}$$

$$\vartheta_{D_B} := \vartheta_D = \vartheta_B + \omega_2 \times {}^0_2R(\beta)\ {}^2\overline{DB} \tag{2}$$

$$\vartheta_{D_C} := \vartheta_D = \vartheta_C + \omega_2 \times {}^0_2R(\beta)\ {}^2\overline{DC} \tag{3}$$

where $\vartheta_i$ = velocity in $i$ w.r.t. inertial system,
$\omega_i$ = rotational velocity of system $i$ w.r.t. inertial system,
${}^i_jR(m)$ = transformation from system $j$ to $i$ by rotation of angle $m$,
${}^i\overline{XY}$ = vector from $X$ to $Y$ expressed in coordinate system $i$.

The equation system becomes

$$\vartheta_{D_A} = \vartheta_{D_B}; \qquad \vartheta_{D_A} = \vartheta_{D_C} \tag{4}$$

and can be written as

$$Ax = b \quad with \quad x = (\vartheta_B, \vartheta_C, \omega_1, \omega_2)^T \tag{5}$$

Equation 5 can be easily solved for $x$ with $\vartheta_A$ as input. If the velocity input is given at B or C the system has to be modified such that $\vartheta_B$ or $\vartheta_C$ appear in $b$. Models of the same form were also generated for CRAB and RCL-E.

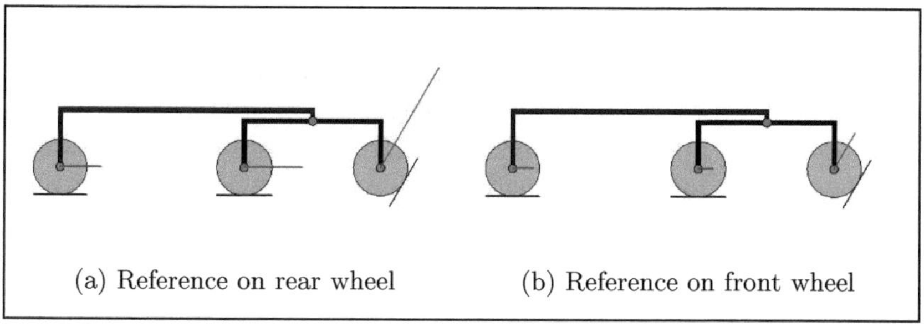


**Fig. 3.** Velocity distribution with different reference wheels

Since the kinematic model takes only one input velocity special attention has to be paid to the selection of the reference where this input is set. This problem of reference selection needs extended explanation.

In Fig. 3 the MER is depicted twice in the same position but with different reference wheels. The velocity of the reference wheel is the same for both cases (red line). The green lines indicate the corresponding optimal velocities on the other wheels. Given, the rover is on a flat surface and encounters a step-like obstacle, either the front wheel has to speed up the obstacle (Fig. 3(a)) or the rear wheel has to slow down significantly (Fig. 3(b)) in order to satisfy the kinematic constraints. In the first case, the front wheel is very likely to slip because it cannot accelerate sufficiently fast. In the second case, slip may be avoided on the front wheel, but the global velocity of the rover is considerably reduced. This kind of problem is inherent to the articulated structure of the suspension system. It is very important to treat this issue of reference selection carefully because it can influence the performance of the system.

## 3 Metrics

It is important to have metrics that precisely define what is considered a good or bad performance. The metrics used in this study are described below.

### 3.1 Difference between Input and Optimal Velocities: $\Delta vel_{opt}$

This metric is a measure for the risk of violation of kinematic constraints through deviation from optimal velocity. This is the case, e.g., if a simple velocity control is used which sets equal speed on all wheels.

$$\Delta vel_{opt} = \sum_{i=1}^{n} |\vartheta_{ref} - \vartheta_{opt_i}| \quad with \quad i \neq ref \tag{6}$$

where $\vartheta_{ref}$ = velocity of reference wheel,
$\vartheta_{opt_i}$ = optimal velocity of wheel $i$,
$n$ = number of wheels.

As it was mentioned above, uneven terrain leads to different optimal velocities on the wheels. If the controller is not able to set the velocities accordingly, slip occurs because kinematic constraints are broken. $\Delta vel_{opt}$ is an indicator how good a suspension system adapts to the terrain while respecting kinematic constraints and not needing a sophisticated controller. If this value is small it is easier to control the rover because the kinematics impose similar wheel velocities, and the risk of slip is smaller.

### 3.2 Slip

Slip is defined as the difference between the displacement of a wheel measured at the wheel center point and the displacement derived from wheel rotation measurement with encoders.

$$slip = \sum_{i=1}^{n} |\Delta pos_{wheelcenter_i} - \Delta pos_{encoder_i}| \tag{7}$$

Slip is bad for the odometry and it is a waste of energy because it does not contribute to the movement of the rover. Therefore this value must be small for good performance.

## 4 Simulation Setup

All simulations were performed using the software Working Model 2D (WM2D) by Design-Simulation [14]. It provides a nice graphical user interface (GUI) for easy definition of simulation runs. Fig. 4 depicts a normalized rover model on an randomly generated, uneven terrain. WM2D has an integrated interface to exchange data with Matlab in which the kinematic model is implemented (Fig. 5). The motors are velocity controlled, i.e., constant velocity input.

## 5 Simulation Results

The same simulations were performed for all three rovers under equal conditions. The results are presented in this section. First, the parameters that define a simulation are described.

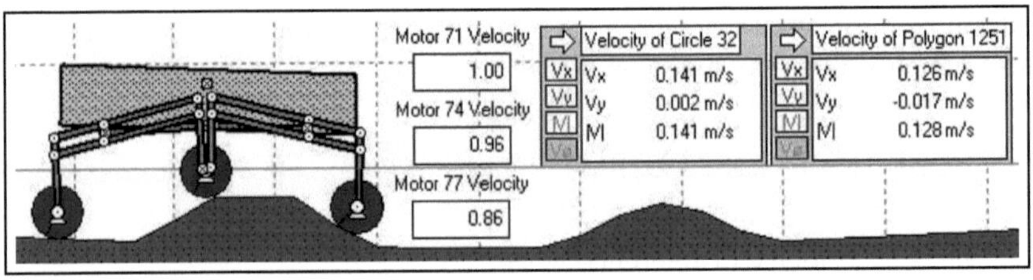


**Fig. 4.** Graphical user interface WM2D with CRAB

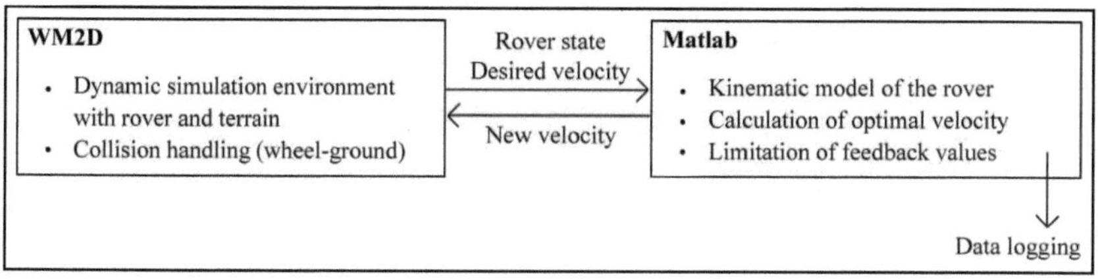


**Fig. 5.** Interaction of WM2D and Matlab

- **Terrain**
  The selection of the terrains was motivated by the idea to have two distinct types: uneven terrain to represent a real environment (24m long, sinusoidal obstacles with defined max. height); artificial benchmark obstacle that allows analyzing what happens during obstacle negotiation due to decoupled climbing phases of the wheels (truncated pyramid, height equal to wheel diameter).
- **Control**
  First, the rovers were assessed with constant speed control, then the optimal velocity information was added to the controller (model-based).
- **Reference wheel**
  Indicates the wheel used as reference (1: rear; 2: middle; 3: front).

Table 2 lists the results for the metric $\Delta vel_{opt}$. The values correspond to the sum of $\Delta vel_{opt}$ over the whole simulation run. For these tests constant speed control was used in combination with the artificial obstacle (tests 1-3) and uneven terrain (tests 7-9).

Tests 1-3 show that there is a significant difference between the performance of MER (~15-27 m/s) and the other rovers (~10-13 m/s), CRAB and RCL-E. This means that, if a constant speed control was used on the rovers, the error would be much bigger on MER or, in other words, MER has a higher need to adapt the wheel velocities in order to satisfy the kinematic constraints and reduce slip. The performance is influenced by the selection of the reference wheel for all rovers, whereas the results for MER show the biggest influence, i.e., the performance varies by almost 50%. It is interesting that the performance of CRAB reflects the symmetrical suspension with same results for front and rear wheel. The results from tests 1-3 are confirmed by tests 7-9 on uneven terrain.

Table 3 shows the performances regarding the slip metric on two different terrains, as well as with constant speed and model-based control. Note that the selection of the reference wheel does not influence the results in constant speed mode because all wheels have the same input velocity. CRAB and RCL-E show similarly good performance while slip for MER is substantially bigger. This

**Table 2.** Simulation results for metric $\Delta vel_{opt}$

| Rover | Ref. wheel | Test | $\sum(\Delta vel_{opt})$ [m/s] | Test | $\sum(\Delta vel_{opt})$ [m/s] |
|---|---|---|---|---|---|
| MER | 1 | 1 | 27.87 | 7 | 93.71 |
| CRAB | | | 12.17 | | 37.53 |
| RCL-E | | | 12.72 | | 35.69 |
| MER | 2 | 2 | 15.87 | 8 | 55.46 |
| CRAB | | | 10.00 | | 28.00 |
| RCL-E | | | 11.12 | | 29.96 |
| MER | 3 | 3 | 17.25 | 9 | 55.20 |
| CRAB | | | 12.02 | | 37.87 |
| RCL-E | | | 11.69 | | 33.70 |
| Terrain type | | Truncated pyramid | | Uneven terrain 24m | |

**Table 3.** Simulation results for metric slip

| Rover | Ref. wheel | Test | Slip (control type) | | Test | Slip (control type) | |
|---|---|---|---|---|---|---|---|
| | | | const.speed | model-based | | const.speed | model-based |
| MER | 1 | 1/4 | 1.12 | 0.61 | 7/10 | 4.19 | 2.46 |
| CRAB | | | 0.64 | 0.24 | | 1.89 | 1.00 |
| RCL-E | | | 0.69 | 0.21 | | 1.97 | 0.94 |
| MER | 2 | 2/5 | 1.12 | 0.49 | 8/11 | 4.19 | 1.99 |
| CRAB | | | 0.64 | 0.20 | | 1.89 | 0.85 |
| RCL-E | | | 0.69 | 0.22 | | 1.97 | 1.23 |
| MER | 3 | 3/6 | 1.12 | 0.51 | 9/12 | 4.19 | 1.94 |
| CRAB | | | 0.64 | 0.22 | | 1.89 | 1.02 |
| RCL-E | | | 0.69 | 0.22 | | 1.97 | 0.99 |
| Terrain type | | Truncated pyramid | | | Uneven terrain 24m | | |

ranking is the same for all simulation types whether the rovers move on uneven terrain or have to cross the truncated pyramid.

As it was to be expected, a significant improvement can be observed for all rovers if the model-based control is activated, i.e., the input velocities are adapted according to the optimal velocities from the kinematic model. The performance can be improved by up to 70%. However, considerable differences in performance, and thus the ranking, remain.

Slip and optimal velocity of all rovers for test 7 are depicted in Fig. 6. The graphs illustrate the results discussed above: similar slip values of CRAB and RCL-E; more than twice the final value of total slip of MER. The deviation of optimal velocity from input velocity (0.13 [m/s]) is the biggest for MER.

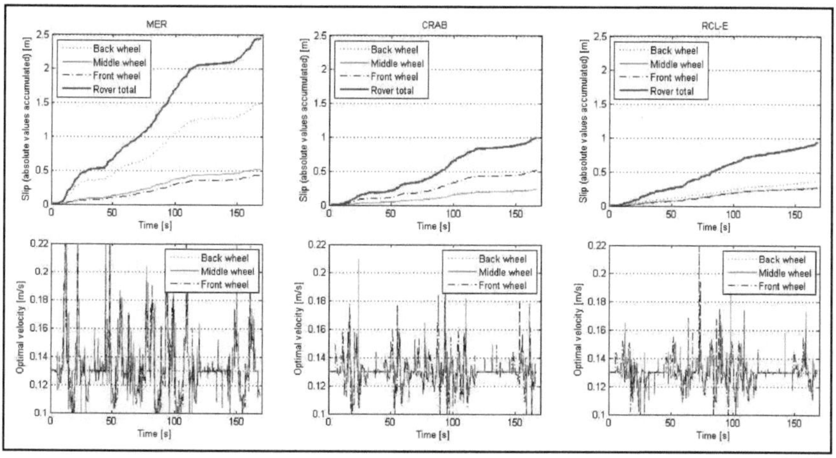


**Fig. 6.** Slip and optimal velocity of all rovers with rear wheel as reference

# 6 Conclusion and Future Work

Rover locomotion performance was analyzed from a kinematic point of view. The main focus was laid on characterization of suspension systems. Additionally to a simple analysis three different systems were compared based on two different metrics. The new metric $\Delta vel_{opt}$ was introduced with the aim to characterize the suspension systems in terms of compliance with kinematic constraints while moving on uneven terrain. For this, the optimal wheel velocities depending on the rover's state were calculated by means of a kinematic model. Slip was used as a second metric because it occurs if kinematic constraints are violated. A large number of simulations was performed with different settings. The rovers CRAB and RCL-E performed well with respect to both presented metrics while MER's performance was significantly inferior. It was also shown how integration of kinematics in a model-based controller improves performance.

In future work motor dynamics will be added to the simulation environment and kinematic investigations will be extended.

# References


1. Krebs, A., Thueer, T., Michaud, S., Siegwart, R.: Performance optimization for passive rover structure: A 2d quasi-static tool. In: IEEE International Conference on Intelligent Robots and Systems (IROS 2006), Beijing, China (2006)
2. Thueer, T., Krebs, A., Siegwart, R.: Comprehensive locomotion performance evaluation of all-terrain robots. In: IEEE International Conference on Intelligent Robots and Systems (IROS 2006), Beijing, China (2006)
3. Thueer, T., Krebs, A., Lamon, P., Siegwart, R.: Performance comparison of rough-terrain robots - simulation and hardware. International Journal of Field Robotics 24(3)(Special Issue on Space Robotics Part I) (2007)

4. Hacot, H., Dubowsky, S., Bidaud, P.: Analysis and simulation of a rocker-bogie exploration rover. In: CISM-IFToMM Symposium, Paris, France (1998)
5. Tarokh, M., McDermott, G., Hayati, S., Hung, J.: Kinematic modeling of a high mobility mars rover. In: IEEE International Conference on Robotics and Automation (ICRA 1999), Detroit, Michigan, USA, vol. 2, pp. 992–998 (1999)
6. Iagnemma, K., Dubowsky, S.: Vehicle wheel-ground contact angle estimation: with application to mobile robot traction control. In: International Symposium on Advances in Robot Kinematics (ARK 2000) (2000)
7. Peynot, T., Lacroix, S.: Enhanced locomotion control for a planetary rover. In: IEEE Internation Conference on Intelligent Robots and Systems (IROS 2003), Las Vegas, USA, vol. 1, pp. 311–316 (2003)
8. Balaram, J.: Kinematic state estimation for a mars rover. Robotica 18(12) (2000)
9. Baumgartner, E.T., Aghazarian, H., Trebi-Ollennu, A.: Rover localization results for the fido rover. In: McKee, G.T., Schenker, P.S. (eds.) SPIE Proc. Vol. 4571, Sensor Fusion and Decentralized Control in Autonomous Robotic Systems IV, Newton, MA, USA (2001) (wheel synchronisation control algorithm)
10. Lamon, P., Siegwart, R.: Wheel torque control in rough terrain - modeling and simulation. In: IEEE International Conference on Robotics and Automation (ICRA 2005), Barcelona, Spain (2005)
11. Grand, C., Amar, F.B., Bidaud, P.: Kinematic analysis and stability optimization of a reconfigurable legged-wheeled mini-rover. In: Unmanned Ground Vehicle Technology IV, Orlando, FL, USA, vol. 4715, pp. 295–302. SPIE (2002)
12. Lindemann, R.A., Voorhees, C.J.: Mars exploration rover mobility assembly design, test and performance. In: IEEE International Conference on Systems, Man and Cybernetics, Waikoloa, Hawaii, USA, vol. 1, pp. 450–455 (2005)
13. Lauria, M., Piguet, Y., Siegwart, R.: Octopus - an autonomous wheeled climbing robot. In: Fifth International Conference on Climbing and Walking Robots, Bury St Edmunds and London, UK, Professional Engineering Publishing Limited (2002)
14. Design-Simulation. Working model 2d (2006), `http://www.design-simulation.com/wm2d/`

# Part V

# Localization and Sensor Fusion

# Robot Position Estimation on a RFID-Tagged Smart Floor

Thomas Kämpke[1], Boris Kluge[1], Erwin Prassler[2], and Matthias Strobel[1]

[1] InMach Intelligente Maschinen GmbH, 89081 Ulm, Germany
{kaempke,kluge,strobel}@inmach.de
[2] Bonn-Rhein-Sieg Univ. of Applied Science, 53757 Sankt Augustin, Germany
erwin.prassler@fh-brs.de

**Summary.** We describe a navigation and position estimation system for mobile service robots which is based on RFID technology. This system avoids deficiencies of existing solutions and offers high flexibility and accuracy at moderate cost. The core idea of the system is to structure the work space by means of a *Smart Floor* in a manner which enables and supports reliable navigation and positioning with absolute accuracy over large distances. The "intelligence" of the floor is in a dense, area-wide network of thousands of RFID transponder, which are mounted underneath the regular floor covering and quasi serve as radio beacons. The system has been presented at CeBIT 2006 in Hannover by invitation of the German Ministry of Education and Research. InMach Intelligente Maschinen has further received the Walter Reis Innovation Award for Service Robotics for this innovative solution. It is important to note that although the primary focus of the work described in this paper is on the guidance of service robots in large-scale indoor environments, the Smart Floor described below provides the infrastructure for all kinds of location-based services which might be of relevance in a large facility. The economic value of these services may even exceed the economic value of automated cleaning.

## 1 Introduction

In this paper we describe a navigation and position estimation system for mobile service robots which is based on RFID technology. The design of this system avoids deficiencies of existing solutions and offers high flexibility and accuracy at moderate cost. The core idea of this system is, like for vehicles guided by wires or magnetic beacons buried in the floor, to structure the work space in a manner which enables and supports reliable navigation and positioning with absolute accuracy over large distances.

Structuring of the work space, however, does not involve the application of any visible markers or wires. Instead the workspace is equipped with a so-called smart floor – a key component of our approach – which neither in its basic physical consistency nor in its optical appearance differs much from a regular floor covering. The "intelligence" of the floor is in a dense, area-wide network of thousands of RFID transponder, which are mounted underneath the regular floor covering and quasi serve as radio beacons. A commercial version of such a

C. Laugier and R. Siegwart (Eds.): Field and Service Robotics, STAR 42, pp. 201–211, 2008.
springerlink.com 

smart floor has been developed by the German company Vorwerk Teppichwerke, which is a partner of InMach Intelligent Maschinen GmbH.

The second key component in our approach is the actual RFID based vehicle navigation system. While the vehicle is in motion, this navigation system senses the identifiers of RFID transponders in its proximity and computes an estimate of the absolute position and orientation of the vehicle and uses these data to guide the vehicles locomotion along a planned path.

Through their synergy these two key components *smart floor* and *RFID based navigation* provide a low-cost, flexible guidance system with absolute position accuracy for mobile service robots and AGVs. This solution has the potential not only to re-animate but also to boost the market for mobile robots for professional services which has ben almost disappearing in the past years.

The navigation system described in this paper, has been presented at CeBIT 2006 in Hannover by invitation of the German Ministry of Education and Research. InMach Intelligente Maschinen has further received the Walter Reis Innovation Award for Service Robotics for this innovative solution.

It is important to note that although the primary focus of the work described in this paper is on the guidance of service robots in large-scale indoor environments, the Smart Floor described below provides the infrastructure for all kinds of location-based services which might be of relevance in a large facility. The economic value of these services may exceed the economic value of automated cleaning by far.

## 2 State of the Art

The RFID based navigation system described here is not so much the result of a basic research activity or a prove of concept. It is an industrial prototype which is currently under field testing and which is supposed to reach series-production readiness by the end of 2007. It will be used in two professional applications first: in an *automatically guided transportation vehicle* (AGV) for hospital logistics and in a robot for professional cleaning. This short review of related work will therefore focus on professional guidance and navigation systems for mobile (service) robots and only briefly discuss basic research on RFID based navigation and smart floors. The state of the art in professional navigation systems will be shown in a few selected, highly representative systems.

### 2.1 State of the Art in Commercial Mobile Service Robot Guidance

Using mobile robots for professional services for example in the area of facility management is not a very original idea. Since the beginning of the nineties several attempts have been undertaken, to automatize burdensome, dirty, low-qualified tasks by means of mobile service robots. The following systems are representative for the state of the art in this field.

Helpmate was the first mobile service robot, which was specifically designed for logistic and transportation tasks in hospitals (1995). HelpMate had a rather extensive sensor configuration, which included amongst others a laser range finder, a set of sonar sensors, and a gyroscope. It used a floor plan and permanently installed markers for navigation. Helpmate was preferably used for transportation of food, medication, and laundry. In spite of the public and media interest, the commercial use of HelpMate has not become a success story. In total only 150 units were manufactured and installed [1]. Helpmate cost approximately US$ 110.000 per unit. Typically the system were not sold but leased for an hourly rate of US$ 4 to 6. The yearly lease was US$ 110.000. It is not completely unlikely that the low number of installed systems is related to the rather high initial cost and the insufficient price-performance ratio.

A representative of the more recent developments in the area of service robots for professional cleaning is the robotic floor scrubber ST82 R Variotech of Hefter Cleantech, Germany. Its navigation system SINAS (Siemens Navigation System) included two laser range finder, an array of sonar sensors, a gyroscope and an odometry system. Using its sensors the system was capable of generating a map of its work space. ST82 R Variotech was distributed for approximately 50.000 Euro. Amongst others it was installed in a Dutch supermarket chain. One unit which was installed at the airport of Manchester, UK, was give to the Queen Elizabeth II Hospital in Welwyn Garden City, Hertfordshire in 2001. The manufacturing and sale of ST82 R Variotech has been discontinued [2].

### 2.2 Recent Developments in the Area of AGVs

For many years automatic guidance of vehicles by induction wires or magnetic beacons or optical markers were state of the art ([3, 5, 4]). Not until recently a sensor, which has significantly influenced and fostered basic research in mobile robotics, namely the SICK PLS and LMS sensor, has also expanded into the development of automatically guided vehicles (AGVs). A representative development is the system Transcar LTC-2 by Swisslog Telelift GmbH ([4]). Transcar LTC-2 uses two laser range finder (one in the front and one in the rear area) to sense and measure the environment. To estimate the position of the vehicle the perceived range images are compared and aligned with a give floor plan. One would expect that the system by using laser range finders has become significantly more flexible than traditional, wire-guided systems. This flexibility has its limits, however. For reliable position estimation and navigation a complete metric model (map) of the environment is required. In a second topological map way points for navigation are defined. The approach quasi implements a virtual automatic guidance. Cost for pre-structuring the work space can be avoided this way. However every change of the environment for consistency reasons requires an update of the metric map of the work space. Such an update may require a significant effort and cause significant cost. Automatically mapping the workspace, which has its own deficiencies, is not foreseen in this approach. Navigation in an unknown environment seems to be impossible.

### 2.3 RFID Navigation and Smart Floors in Basic Research

There is not much basic research in RFID navigation and smart floor. The literature is accordingly sparse. Based on the patent situation it seems that industry has discovered this topic before the academic community. There is actually an significant number of patents or patent applications out there which address similar or almost identical problems: given we can distribute a sufficient number of RFID transponders over the workspace, how can a robot use such an infrastructure and the information encoded in the transponders to estimate its position and execute its mission. The approaches described in [10, 9, 6, 7, 8] address the same problem and provide very similar solutions. Typically the position estimation is solved by aligning and/or fitting the observations of one or more tags and their approximate position sensed by odometry to an established frame with known tag positions.

A frequently cited academic publication is the work by Hähnel et al. [11]. They generate maps of RFID tags with mobile robots and compute a posterior about the position of a tag after the trajectory and the map has been generated with a highly accurate FastSLAM algorithm for laser range scans. A concept of a Smart Floor equipped with RFID transponders was presented in [12]. This smart floor is primarily used for identifying and tracking people.

## 3 RFID Navigation on Smart Floor

The above described systems are representative for the state of the art in the area of commercial mobile service robots and automatically guided vehicles and their guidance systems. It is noticeable that even the "free navigating" systems lack a sufficient flexibility, which would allow a customer to deploy the vehicles in an partially unknown environment without costly system adaptations or modifications of the environment.

A feature, which is shared by the systems described in the previous section, is the significant cost for the navigation and guidance systems. These guidance systems use at least one, often several expensive laser range finder. The price for this component by itself (ca. 3500 Euro per sensor) exclude, that the vehicles can be brought to the market for less than 20.000 Euros. For many service robot applications, amongst them professional cleaning this is definite knock out criterion. All cleaning robot developments for professional cleaning in the past – approximately 15 in number – have died because of their cost regardless of the tremendous economic potential of automation of cleaning.

The innovative approach presented in the section below is a navigation system which avoids the deficiencies (lack of flexibility, hight cost) identified above. The core idea of this approach is to pre-structure the work space like in earlier approaches using induction loops and magnetic beacons in order to enable reliable navigation and absolute positioning with high accuracy over long distances. Our navigation system operates on a so called smart floor, a dense, area-wide network of thousands of RFID transponder, which are mounted underneath the regular floor covering and quasi serve as radio beacons.

The guidance system gets by without expensive range sensors like laser range finder and uses regular sonar sensors as they are used in parking assistance systems in the automotive industry. An extra "sensor" which is used for position estimation is an RFID reader. These devices are commercially available for prices starting at 100 Euros.

The additional cost for the smart floor will amount to 13 to 15 Euro per square meter. If one compares these additional cost to the cost for expensive range sensors, then 1000 square meters of smart floor cost approximately as much as four laser range finders, which allows to equip two vehicle with range sensors yielding for example a 360 degree planar view or some non-planar view configuration.

The two key components of RFID based Robot Navigation on a Smart Floor outlined in the preceding section will be described in somewhat more detail below. Furthermore two specific service robot systems, using RFID based robot navigation will be described.

### 3.1 RFID Makes the Floor Smart

The core idea of the Smart Floor is to distribute and integrate RFID transponders in the floor in an area-wide manner and use the transponders for various purposes. Since the integration in the floor covering itself would complicate the manufacturing and increase the price of the floor covering and would have to be implemented separately and specifically for every type of floor covering, our partners Vorwerk Teppichwerke and FutureShape have been developing a so-called *Smart Underlay* (see Fig. 1). The smart underlay is separate meadow-like textile, in which the RFID transponders are integrated typically in a regular grid. This underlay can be placed nearly below every non-metallic floor covering. The material has furthermore an acoustic insulation effect. At the IT fair CeBIT 2006 in Hannover, for example, smart underlay was placed below wood (laminate), plastic, tiles, and carpet. Together with the floor covering on top of it the smart underlay forms the so-called *Smart Floor* (see Fig. 4).

The operation principle of RFID transponder is as follows. A reading device (reader) creates an electro-magnetic field over an antenna. RFID transponders located in this field will be charged by the emitted energy and enables to respond with an electro-magnetic echo. This echo typically carries an identifier plus various information such as co-ordinates, traffic information, environmental information, service information and alike, which is received by the reader. Transponders cannot only be read but also be written with information.

The piece costs for transponders basically depend on the storage capacity. The price per tag is currently around 0,30 to 0,50 Euro. Due to very tight competition a decay of this price to approx. 0,10 Euro per transponder is not unlikely.

The RFID tags integrated in a smart underlay or a smart floor can serve various purposes:

- They can be used for position estimation and tracking of robots as much as of humans moving in large facilities such as warehouses or public buildings.
- They can store location specific information regarding the execution and performance of a service. For example, one could store the date of the last cleaning service and the code of a specific cleaning liquid or procedure.

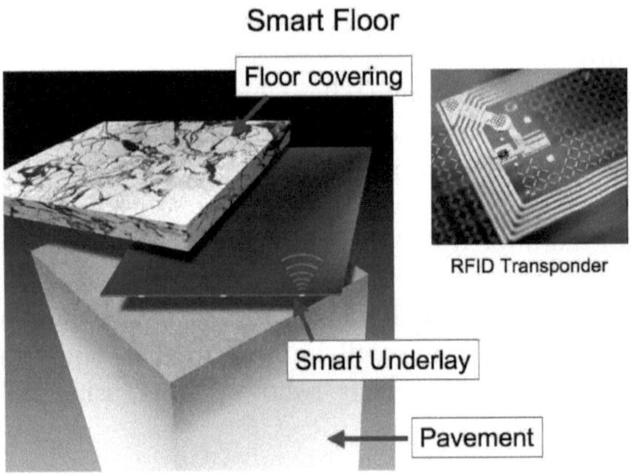


**Fig. 1.** Design Smart Underlay and Smart Floor

- They could store information concerning the guidance of robot vehicles such as speed limits, temporary or permanent restricted areas.
- They can provide the localization infrastructure for any arbitrary location-based service in a large facility.

It is worth noting, that location-based information regarding the environment and the objects therein comes with absolute position accuracy. This property allows one to fuse and integrate the perceptions of many robots with many different sensor suits over time in a central map. Through this integration of time and the perceptions of many robots an increasingly accurate and complete image of the environment arises. All robots can access this central map "for free" and at any time and use it for safe locomotion and execution of their mission. The RFID readers, which are used to read the information stored in the tags, quasi serve as additional, low-cost but highly accurate sensor.

In the extreme case, a robot could navigation almost blindly only by using its RFID readers, the RFID transponders, and the central map. This, however, requires the information in the central map to be up to date and to represent a correct and complete image of the environment.

These features do not only contribute to an increased reliability and accuracy but also have a significant effect on the implementation of a low-cost service robot system. The generation of a sufficiently accurate central map, requires, if at all, only a very small number of robots with precise and expensive sensors such as laser radars or stereo cameras. All other robots can be equipped with less accurate, inexpensive sensors such as sonars, infrared, or tactile sensors, since they can complete their environmental perception by the information in the central map. Over time and over many observations the integration and fusion even of inaccurate sensor data yields an increasingly accurate model of the environment.

### 3.2 Self-localization on a Smart Foor

A robot that uses RFID transponders for planar self-localization may be equipped with one or several RFID readers each having one or several antennas. Operating several antennas from one reading device can be facilitated by time-multiplexing. Any antenna is directed towards the floor. Obviously, a large antenna lobe allows for reading all transponders from a large area but each at a small spatial resolution and a small antenna lobe allows for the opposite.

#### Single Point Observations

Let the spatial resolution of transponder observation be high enough to allow for estimates as single points in robot coordinates. Also, it is assumed that at most one transponder can be seen at any moment and transponder observation does not provide any directional information. These assumptions correspond to a "point-shaped" antenna.

When the robot moves for a certain time or distance, it records all transponder readings in one local map. Whenever self-localization is invoked, the local map is matched to a global map of the smart floor. The global map contains the positions of all transponders in world coordinates.

Self-localization under the present conditions admits the computation of the position only. The orientation is inferred from the robot odometry.

The present map matching is nothing but point cloud matching with the correspondence problem being trivially solved since transponder identifiers are unique. Translation and rotation of the local map to the global map adhere to the least squares approach

$$\min_{\varphi\in[0,360^o),\,\vartheta\in I\!R^2} \sum_{i=1}^{n} ||T_i^{world} - (Rot_\varphi T_i^{robot} + \vartheta)||^2. \tag{1}$$

The summation index $n$ equals the number of transponder observations, with multiple observations of the same transponder being handled as observations of different transponders with possibly the same coordinates. $T_i^{robot}$ and $T_i^{world}$ are the coordinates of transponder $i$ in the robot and the world frame respectively. The translation vector $\vartheta$ may vary unboundedly in the plane and the rotation is denoted by a standard rotation matrix $Rot_\varphi$.

The optimal rotation angle is computed from two candidates values. The first value is $\varphi_1 \in [0, 180^o)$ with

$$\tan\varphi_1 = \frac{\sum_{i=1}^{n} det(T_i^{robot} - \bar{T}^{robot}\; T_i^{world} - \bar{T}^{world})}{\sum_{i=1}^{n} (T_i^{robot} - \bar{T}^{robot})^t (T_i^{world} - \bar{T}^{world})} \tag{2}$$

if the denominator is different from zero, else $\varphi_1 = 90^o$ and the second candidate angle is $\varphi_2 = \varphi_1 + 180^o$. The superscript $t$ denotes the transpose of vectors and the superscript $^-$ denotes average vectors.

The optimal rotation angle is obtained from value insertion as

$$\varphi_0 = \arg\min_{\varphi_j} \sum_{i=1}^{n} ||T_i^{world} - \bar{T}^{world} - (Rot_{\varphi_j} T_i^{robot} - \bar{T}^{robot})||^2 \tag{3}$$

and the optimal translation vector is $\vartheta_0 = \bar{T}^{world} - Rot_{\varphi_0} \bar{T}^{robot}$.

Position and orientation of the robot frame are eventually transformed into the world frame by

$$P^{world} = Rot_{\varphi_0} P^{robot} + \vartheta_0 \tag{4}$$

$$\varphi^{world} = \varphi^{robot} + \varphi_0 \bmod 360^o. \tag{5}$$

### Single Line Observations

A suitable antenna shape for mobile robots is that of a rod antenna or a slim rectangle that comes close to a rod. The antenna is mounted crossways to the machine direction and it should extend over the whole robot width. The latter results in a high likelihood to observe all transponders that are overpassed.

A straight robot motion is considered first. Whenever a transponder is traversed, it can be assumed to lie on a line segment that indicates the rod position at the observation moment. All the segments are parallel since the motion is straight. The distances between successive observations of transponders $i-1$ and $i$ are recorded as $D_{i-1,i}$, see Fig. 2.

The observed distance between transponder traversals being directional rather than Euclidean distances between transponder positions allows to compute the robot orientation in world coordinates. Let the robot make $n$ transponder traversals along one straight line. Its direction in world coordinates is a solution of the minimization problem

$$\min_{\varphi \in [0,360^o)} \sum_{i=2}^{n} \Big( (\cos\varphi, \sin\varphi)\,(T_i^{world} - T_{i-1}^{world}) - D_{i-1,i} \Big)^2 . \tag{6}$$

This problem can be solved numerically by iterated bisection. The robot position is not unique with respect to the observations but it can be narrowed down to a segment such that all projected transponder positions fall in that segment. From this segment, the center of gravity of all projections can be chosen as position estimate of the robot while it observed the first transponder. The current position in world coordinates is then inferred as in the foregoing section. The approach can be extended from motion along a single line to other paths, in particular to parallel lines as they are typically used by cleaning robots.

## 4 Field Evaluation

The results of a first quantitative evaluation of the position estimation performance of our approach is shown in Fig. 3a). For this evaluation a test arena of

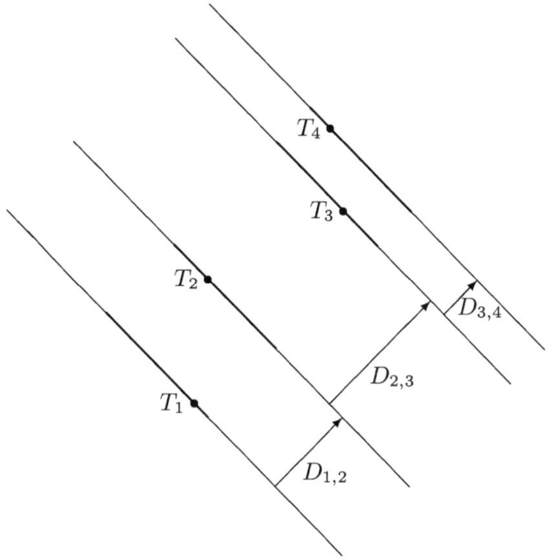


**Fig. 2.** Straight robot motion with transponder overpasses in the order $1, 2, 3, 4$ with successive distances $D_{1,2}, D_{2,3}, D_{3,4}$. The rod positions (bold segments) on the parallels are unknown and the exact positions of the transponders relative to the rod positions are unknown.

$4\,\text{m} \times 3\,\text{m}$ was set up, with RFID tags laid out in intervals of 50 cm. Within this test arena the robot was sent on a rectangular course for about 10 minutes at a speed of approximately 30 cm/sec. During the experiment the robot traveled approximately 170 meters. The position was estimated approximately every 200 millisecond or every 5 to 6 cm. As can be seen in Fig. 3a), the trajectory fitted to the estimated position forms a smooth path.

In Fig. 3b), the error between ground truth position and estimated position is shown. Ground truth was obtained by means of a specifically designed measuring system involving a laser range finder, which of course also contributes some error to the measurement. In the diagram in Fig. 3b) we can see that there is some significant position errors in the beginning of the experiment. Such significant errors occur only during the initialization phase of the algorithm. Once the system has the first fix, the errors drop drastically. It is also worth noting that the error in the described experiment stayed within bounds defined by the tag distance and the reading distance of the RFID reader. This, however, is not a guaranteed bound. If the reader does not discover a tag for a longer period of time, the position error might as well grow beyond this bound.

The RFID based navigation system described in this paper has matured to a level which allows and suggests excessive field testing. Three cleaning robots equipped with prototypes of the navigation system have been demonstrated during the international IT fair CeBIT 2006 in Hannover (see Fig. 4). During this exhibit the robots were in operation for approximately 6 to 8 hours per day,

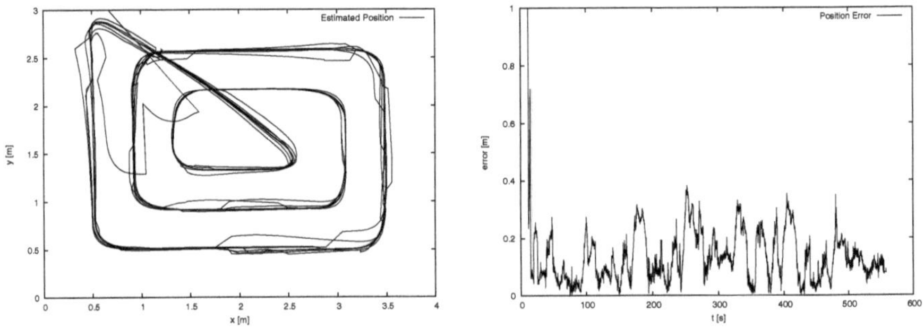


**Fig. 3.** a) Positions estimated along a rectangular course over a traveled distance of approx. 170 meters; b) Position error between ground truth position and estimated position.

seven days long. Occasional mulfunctions of the robots were due to mechanical problems and problem with the collision detection sensors. The RFID navigation component itself worked without any problems. Currently an extended field test and a permanent installation of a smart floor in a gym and a cleaning robot navigating with RFID is in preparation.

**Fig. 4.** CeBIT 2006: first field tests for RFID based navigation of robo40 cleaning robots

## 5 Conclusion

We have presented a novel position estimation system for mobile service robots using RFID transponders in a so-called smart floor. The approach has several advantages over existing solutions. It provides absolute position accuracy over arbitrary distances at low-cost. The sensor(s) needed for absolute position estimation are one or more RFID reader devices at price of 50 to 200 Euro each (depending on range and frequency). Additional features and advantages are easy installability and moderate installation cost, flexibility, and a spectrum of functions provided by the programmable smart floor.

Altogether the described RFID based robot navigation system has the potential not only to revitalize but to boost the technology and the stagnating market in the field of automatically guided transportation vehicles for industrial applications as much as the development of mobile service robots for professional applications.

## Acknowledgment

This work was partially supported by the German Ministry of Education under grant no. 01IME01J in the context of the German Service Robotics Initiative DESIRE.

## References

1. Robot navigation technology, `http://www.atp.nist.gov/eao/sp950-1/helpmate.htm`
2. BBC News: Hospital recruits robot cleaner, `http://news.bbc.co.uk/1/hi/health/1396433.stm`
3. Fraunhofer IPA Technologieforum. Fahrerlose Transportsysteme (FTS) (2005)
4. Fottner, J.: FTS-Einsatz in der Krankenhauslogistik. In: Fraunhofer IPA Technologieforum. Fahrerlose Transportsysteme (FTS) (2005)
5. Osterhoff, W.: Technologische Innovationen und Trendentwicklungen bei FTS. In: Fraunhofer IPA Technologieforum. Fahrerlose Transportsysteme (FTS) (2005)
6. Song, J.-G., Lee, S.-Y.: Mobile Robot Using RF Module. US 6.597,143 B2. Samsung Kwangju Electronics Co. (2001)
7. Jung, S.-I.: Automatic Guided System and Control Method Thereof. US 6.493.614 B1. Samsung Electronics Co. (2002)
8. Lee, Y.-K., Choe, W.: Apparatus and Method for Measuring Position of Mobile Robot. US 2003/0236590 A1 Samsung Electronics Co. (2003)
9. Piotrowski, T.: RFID Navigation System. US 2003/0080901 A1. Koninkljike Philips Electronics N.V. (2003)
10. Douglas, B., Bencel, J.: System for Automatic Guided Vehicle Navigation. US 6.049.745. FMC Corporation (1997)
11. Hähnel, D., Burgard, W., Fox, D., Fishkin, K., Philipose, M.: Mapping and Localization with RFID Technology. In: Proc. IEEE Int. Conf. on Robotics and Automation ICRA, New Orleans (2004)
12. Orr, R.J., Abowd, G.D.: The Smart Floor: A Mechanism for Natural User Identification and Tracking. In: Proc. Conf. on Human Factors in Computing Systems (2000)

# Monte Carlo Localization in Outdoor Terrains Using Multi-Level Surface Maps

Rainer Kümmerle[1], Rudolph Triebel[1,2], Patrick Pfaff[1], and Wolfram Burgard[1]

[1] Department of Computer Science, University of Freiburg, Germany
{kuemmerl,triebel,pfaff,burgard}@informatik.uni-freiburg.de
[2] Autonomous Systems Lab, Swiss Federal Institute of Technology, Zurich, Switzerland
rudolph.triebel@mavt.ethz.ch

**Summary.** In this paper we consider the problem of mobile robot localization with range sensors in outdoor environments. Our approach applies a particle filter to estimate the full six-dimensional state of the robot. To represent the environment we utilize multi-level surface maps which allow the robot to represent vertical structures and multiple levels in the environment. We describe probabilistic motion and sensor models to calculate the proposal distribution and to evaluate the likelihood of observations. Experimental results obtained with a mobile robot in an outdoor environment indicate that our approach can be used to robustly and accurately localize an outdoor vehicle. The experiments also demonstrate that multi-level surface maps lead to a significantly better localization performance than standard elevation maps.

## 1 Introduction

The problem of mobile robot localization with range sensors in outdoor environments arises whenever GPS signals are missing because of occlusions caused by buildings, bridges, or trees. In such situations, a mobile robot typically has to estimate its position in the environment using its other sensors and a map of the environment. In this paper, we consider the problem of localizing a mobile robot in outdoor environments by matching laser range measurements to a given map of the environment. One of the most popular representations for outdoor environments are elevation maps [3, 9, 10, 13]. The key idea underlying elevation maps is to store the $2\frac{1}{2}$-dimensional height information of the terrain in a two-dimensional grid, which corresponds to a representation of the horizontal surfaces of the environment. Whereas the knowledge about the horizontal surfaces is well suited to support traversability analysis and path planning, it provides only weak support for the localization of the vehicle. Modeling only the horizontal surfaces, namely, means that vertical structures, which are frequently perceived by ground based vehicles, cannot be used to support localization. To avoid this problem, multi-level surface (MLS) maps [16] have been introduced. They can be regarded as an extension of the classical elevation maps as they additionally represent intervals corresponding to vertical objects in the environment. A further disadvantage of elevation maps is that they cannot represent multiple levels. This, for example, is important when mobile robots are deployed in environments with bridges or underpasses.

Fig. 1 depicts examples of an elevation map (left) and the corresponding MLS map (right) of the campus at the University of Freiburg. As can be seen from the images,

C. Laugier and R. Siegwart (Eds.): Field and Service Robotics, STAR 42, pp. 213–222, 2008.
springerlink.com 

**Fig. 1.** Elevation Map (left) and multi-level surface (MLS) map (right) of the Freiburg campus. The MLS map represents vertical structures more accurately and can deal with multiple surfaces that can be traversed by the robot.

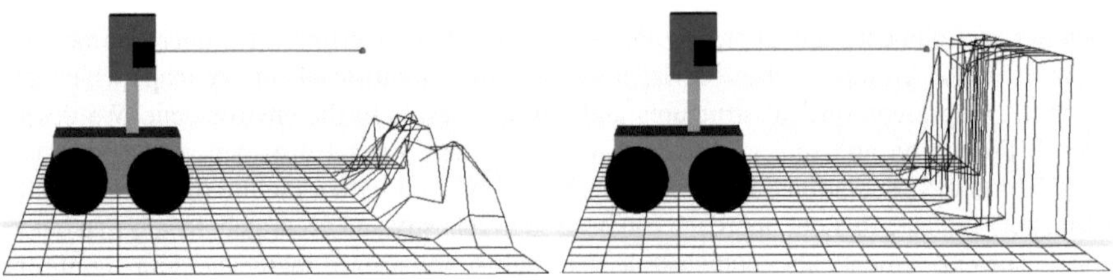

**Fig. 2.** Advantage of the MLS map approach in comparison to the standard elevation maps. In contrast to the MLS map (right) the elevation map (left) lacks the ability to model vertical structures, because it averages over all measured height values. Since the distance of the endpoint of a laser beam to the closest point in the elevation map can have substantial deviations from the true distance, localization becomes harder.

the MLS map is able to represent the environment more accurately than the elevation map. In the MLS map, objects like trees and walls are represented properly. Another example is shown in Fig. 2. Here the robot is located in front of a wall. In the elevation map, the wall is not represented correctly, because the height values obtained from beams reflected by the wall are averaged which is the typical approach in elevation maps. This can lead to a poor estimate of the measurement likelihood at the particular robot position. In contrast, when the MLS map is used, one obtains a better value of the likelihood, because the wall is modeled correctly.

In this paper we present an approach to use the multi-level surface maps for localization. We present probabilistic motion and observation models and describe how these models can be utilized in a probabilistic localization scheme. We furthermore evaluate how the localization performance changes when standard elevation maps are used instead of MLS maps.

This paper is organized as follows. After discussing related work in the next section, we briefly describe Monte Carlo localization in Section 3. Our approach to apply the MLS maps is presented in Sections 4 and 5. Finally, in Section 6, we present experimental results illustrating the advantages of applying MLS maps for localization in outdoor environments.

## 2 Related Work

The problem of localization of mobile robots in outdoor environments with range sensors or cameras has been studied intensively in the past. For example, Adams *et al.* [1] extract predefined features from range scanners and apply a particle filter for localization. Davison and Kita [4] utilize a Kalman filter for vision-based localization with point features on non-flat surfaces. Recently, Agrawal and Konolige [2] presented an approach to robot localization in outdoor terrains based on feature points that are tracked across frames in stereo images. Lingemann *et al.* [11] recently described a method for fast localization in in- and outdoor environments. Their system operates on raw data sets, which results in huge memory requirements. Additionally, they apply a scan-matching routine for localization, which does not facilitate global localization. To reduce the memory requirements of outdoor terrain representations, several researchers applied elevation maps [3, 9, 10, 13]. A probabilistic approach to localize a planetary rover in such elevation maps has been described by Olson [12]. In this system, elevation maps were sufficient to robustly localize the vehicle, mainly because the number of vertical and overhanging objects is negligible in environments like on Mars. However, environments on earth contain many objects like buildings or trees which have vertical or even overhanging surfaces. To address this issue, Pfaff *et al.* [14] extended the elevation map approach by a cell classification. In this approach the individual cells are divided into three classes: cells which have been observed from above, cells with vertical objects, and cells with overhanging objects. This extension has been utilized to improve the data association during the scan matching process and enabled the robot to traverse cells with overhanging objects. In contrast to MLS maps, however, this approach still lacks the ability to store multiple surfaces at one position, which prevents the robot from dealing with situations, in which it has to traverse a bridge and move through the corresponding underpass. The goal of this paper is to develop a probabilistic localization method based on MLS maps and to demonstrate that the more accurate representation of the environment results in improved localization capabilities.

## 3 Monte Carlo Localization

To estimate the pose $\mathbf{x} = (x, y, z, \varphi, \vartheta, \psi)$ of the robot in its environment, we consider probabilistic localization, which follows the recursive Bayesian filtering scheme. The key idea of this approach is to maintain a probability density $p(\mathbf{x}_t \mid \mathbf{z}_{1:t}, \mathbf{u}_{0:t-1})$ of the robot's location $\mathbf{x}_t$ at time $t$ given all observations $\mathbf{z}_{1:t}$ up to time $t$ and all control inputs $\mathbf{u}_{0:t-1}$ up to time $t-1$. This posterior is updated as follows:

$$p(\mathbf{x}_t \mid \mathbf{z}_{1:t}, \mathbf{u}_{0:t-1}) = \alpha \cdot p(\mathbf{z}_t \mid \mathbf{x}_t) \cdot \int p(\mathbf{x}_t \mid \mathbf{u}_{t-1}, \mathbf{x}_{t-1}) \cdot p(\mathbf{x}_{t-1})\, d\mathbf{x}_{t-1}. \tag{1}$$

Here, $\alpha$ is a normalization constant ensuring that $p(\mathbf{x}_t \mid \mathbf{z}_{1:t}, \mathbf{u}_{0:t-1})$ sums up to one over all $\mathbf{x}_t$. The terms to be described in Eqn. (1) are the *prediction model* $p(\mathbf{x}_t \mid \mathbf{u}_{t-1}, \mathbf{x}_{t-1})$ and the *sensor model* $p(\mathbf{z}_t \mid \mathbf{x}_t)$. One major contribution of this paper is an appropriate computation of these models in the case that an MLS map is given.

For the implementation of the described filtering scheme, we use a sample-based approach which is commonly known as *Monte Carlo localization* [5]. Monte-Carlo localization is a variant of particle filtering [6] where each particle corresponds to a possible robot pose and has an assigned weight $w_i$. The *belief update* from Eqn. (1) is performed by the following two alternating steps:

1. In the **prediction step**, we draw for each particle with weight $w_i$ a new particle according to $w_i$ and to the prediction model $p(\mathbf{x}_t \mid \mathbf{u}_{t-1}, \mathbf{x}_{t-1})$.
2. In the **correction step**, a new observation $\mathbf{z}_t$ is integrated. This is done by assigning a new weight $w_i$ to each particle according to the sensor model $p(\mathbf{z}_t \mid \mathbf{x}_t)$.

Furthermore, the particle set needs to be re-sampled according to the assigned weights to obtain a good approximation of the pose distribution with a finite number of particles. However, the resampling step can remove good samples from the filter which can lead to particle impoverishment. Thus, we need to find a criterion for deciding when to perform the resampling step. In this paper, we calculate the number $N_{eff}$ of effective particles according to the formula proposed by Doucet *et al.* [6]

$$N_{eff} = \frac{1}{\sum_{i=1}^{N} \left(\tilde{w}_i^{\,2}\right)}, \tag{2}$$

where $\tilde{w}_i$ refers to the normalized weight of sample $i$. In particular, we only resample if $N_{eff}$ drops below the threshold of $\frac{N}{2}$ where $N$ is the number of samples. In the past, this approach has already successfully been applied in the context of the simultaneous mapping and localization (SLAM) problem [8].

## 4 Prediction Model for MLS Maps

The prediction model $p(\mathbf{x}_t \mid \mathbf{u}_{t-1}, \mathbf{x}_{t-1})$ we use is based on an approach introduced by Eliazar *et al.* [7]. It reflects systematic errors such as drift, as well as the uncertainty in the execution of an action $\mathbf{u} = (x_u, y_u, \theta_u)$, where $(x_u, y_u)$ is the translation and $\theta_u$ the rotation angle. To incorporate this 2D motion into our 3D map we proceed as follows. First, we obtain a possible outcome $(x_v, y_v, \theta_v)$ of the action by applying the probabilistic model. Then, we adapt the motion vector $\mathbf{v} = (x_v, y_v)$ to the shape of the 3D surface traversed by the robot. This surface is obtained from the given MLS map and consists of planar square patches. To adapt the motion vector, we discretize it into segments of length $c$, which is the cell size of the MLS map, in our case 0.1 m. For each segment, we determine the corresponding surface patch $\mathcal{S}$ and rotate the segment according to the orientation $(\varphi_{\mathcal{S}}, \vartheta_{\mathcal{S}})$ of the patch, where $\varphi_{\mathcal{S}}$ is the rotation about the $x$-axis and $\vartheta_{\mathcal{S}}$ the rotation about the $y$-axis. The patch orientation is computed from the normal vector $\mathbf{n}_{\mathcal{S}}$ of the patch $\mathcal{S}$, which in turn is obtained by fitting a plane into the local vicinity of $\mathcal{S}$. The normal vector computation is done beforehand and constitutes an extension to the framework of MLS maps. In general, it is not robust against noise and small errors in the MLS map, which results in an uncertainty of the patch orientation. In our approach, we model this uncertainty by adding Gaussian noise to the orientation parameters $\varphi_{\mathcal{S}}$ and $\vartheta_{\mathcal{S}}$. Thus, our prediction model expresses the uncertainty in 5 out of 6 position parameters – $x$, $y$ and $\psi$ by the 2D motion model and $\varphi$ and $\vartheta$ by our 3D extension. For the last one – the height value $z$ – we have the constraint that the robot must stay on the ground. Therefore, we adjust the $z$-value manually whenever it is too high or too low.

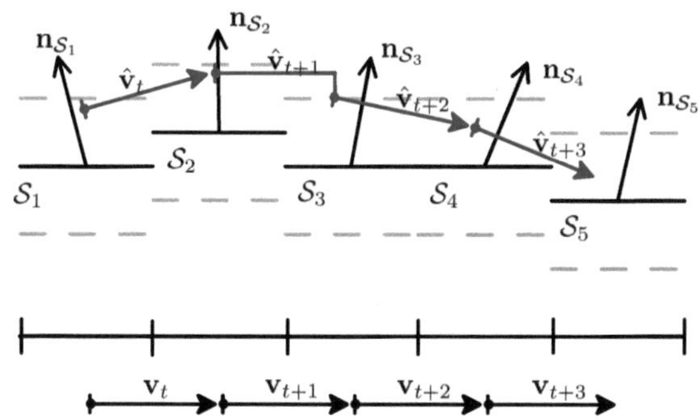


**Fig. 3.** Application of our prediction model to a series of 2D motion vectors (black). They are rotated to estimate the 3D motion vectors (red). The dashed line indicates the tolerance interval for the $z$-coordinate.

This is illustrated in Fig. 3. Finally, after concatenating all transformed motion vector segments, we obtain a new 3D motion vector $\hat{\mathbf{v}}$ which is added to the current estimate of the robot position $\mathbf{x}_{t-1}$ to obtain a new position estimate $\mathbf{x}_t$.

## 5 Endpoint Sensor Model for MLS Maps

In our sensor model, we treat each beam independently and determine the likelihood of a whole laser scan by factorizing over all beams. Thus, we have

$$p(\mathbf{z} \mid \mathbf{x}) = \prod_{k=1}^{K} p(z^k \mid \mathbf{x}) \tag{3}$$

where $K$ is the number of beams in each laser measurement $\mathbf{z}$. In Eqn. (3) and in the following, we drop the index $t$ for convenience. Our sensor model $p(z^k \mid \mathbf{x})$ is based on an approach that has been introduced by Thrun [15] as *likelihood fields* (LF) or *end point model*. In particular, we formulate the sensor model $p(z^k \mid \mathbf{x})$ for each particular beam as a mixture of three different distributions:

$$p(z^k \mid \mathbf{x}) = \alpha_{hit} p_{hit}(z^k \mid \mathbf{x}) + \alpha_{rand} p_{rand}(z^k \mid \mathbf{x}) + \alpha_{max} p_{max}(z^k \mid \mathbf{x}), \tag{4}$$

where $p_{hit}$ is a normal distribution $\mathcal{N}(0, \sigma^2)$ that models situations in which the sensor detects an obstacle. Random measurements are modeled using a uniform distribution $p_{rand}(z^k \mid \mathbf{x})$. Maximum range measurements are covered by a point mass distribution $p_{max}(z^k \mid \mathbf{x})$. These three distributions are weighted by the non-negative parameters $\alpha_{hit}$, $\alpha_{rand}$, and $\alpha_{max}$, which sum up to one. The values for $\alpha_{hit}$, $\alpha_{rand}$, $\alpha_{max}$, and $\sigma^2$ used in our current implementation have been determined empirically.

In the end point model, the probability $p_{hit}(z_k \mid \mathbf{x})$ only depends on the distance $d^k$ between the end point of the $k$-th laser beam and the closest obstacle in the map. Thus, the physical property of the laser beam is ignored, because the model just uses the end point and does not consider the beam characteristic of the laser. Therefore, we need to

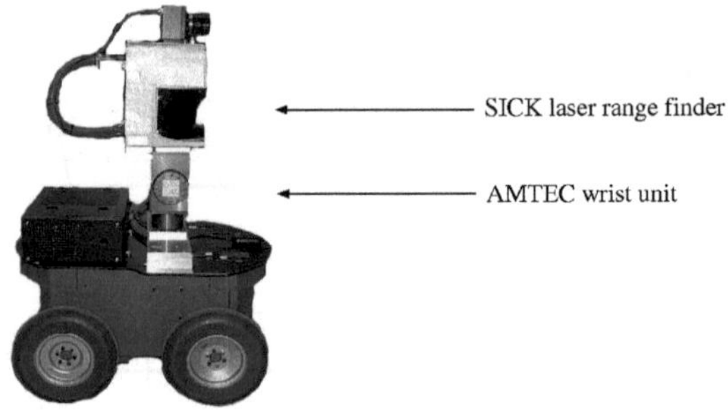


**Fig. 4.** Robot Herbert used for the experiments

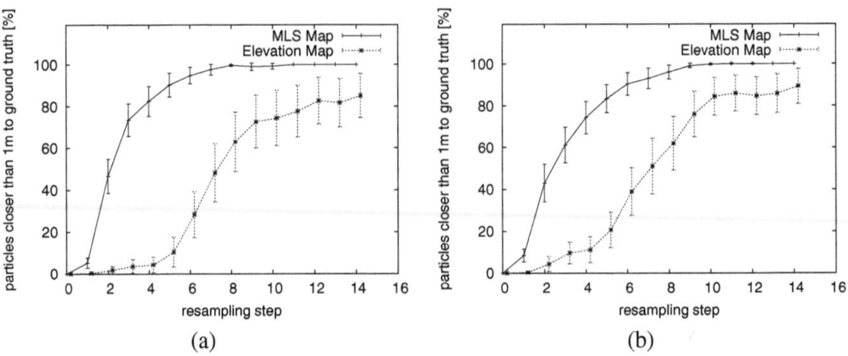


**Fig. 5.** Convergence of the particles to the true position of the robot with (a) 500,000 and (b) 1,000,000 particles. The $x$-axes depict the number of resampling steps, while the $y$-axes show the percentage of particles that are closer than $1m$ to the true position.

calculate the global coordinates for a beam end point. If we denote the angle of the $k$-th beam relative to the zero angle with $\zeta^k$, then the end point $\tilde{\mathbf{p}}^k = (\tilde{x}^k, \tilde{y}^k, \tilde{z}^k)^T$ of that beam in the robot's own coordinate frame is calculated as

$$\begin{pmatrix} \tilde{x}^k \\ \tilde{y}^k \\ \tilde{z}^k \end{pmatrix} = \begin{pmatrix} \hat{x} \\ \hat{y} \\ \hat{z} \end{pmatrix} + R z^k \begin{pmatrix} \cos(\zeta^k) \\ \sin(\zeta^k) \\ 0 \end{pmatrix}, \tag{5}$$

where $(\hat{x}, \hat{y}, \hat{z})^T$ denotes the position of the sensor at time $t$ and $R$ is a rotation matrix that expresses the 3D sensor orientation in the robot's coordinate frame. For a given robot pose $\mathbf{x} = (x, y, z, \varphi, \vartheta, \psi)$ at time $t$ we can compute the global coordinates $\mathbf{p}^k = (x^k, y^k, z^k)^T$ of the $k$-th beam end point $\mathbf{p}^k$ as follows

$$\begin{pmatrix} x^k \\ y^k \\ z^k \end{pmatrix} = R(\varphi, \vartheta, \psi) \begin{pmatrix} \tilde{x}^k \\ \tilde{y}^k \\ \tilde{z}^k \end{pmatrix} + \begin{pmatrix} x \\ y \\ z \end{pmatrix}, \tag{6}$$

where $R(\varphi, \vartheta, \psi)$ denotes the rotation matrix for the given Euler angles $\varphi$, $\vartheta$, and $\psi$. In MLS maps, obstacles are represented as *vertical surface patches*, which can be seen as vertical segments of occupied space. Unfortunately, there is no efficient way to find the closest of all vertical segments to a given beam end point. Therefore, we use an approximation by uniformly sampling a set $\mathcal{P}$ of 3D points from all vertical patches. The distance $d^k$ of the $k$-th beam end point $\mathbf{p}^k$ to the closest obstacle is then approximated as the Euclidean distance $d(\mathbf{p}^k, \mathcal{P})$ between $\mathbf{p}^k$ and $\mathcal{P}$. This distance can be efficiently calculated by storing all points from $\mathcal{P}$ in a $k$D-tree.

Equations. (5) and (6) describe a 3D transform $T(z^k; \mathbf{x})$ of the measurement $z^k$ at position $\mathbf{x}$. Using this and the fact that $p_{hit}$ is Gaussian, we can compute $p_{hit}$ as

$$p_{hit}(z^k \mid \mathbf{x}) \approx \frac{1}{\sqrt{2\pi\sigma^2}} \exp\left(-\frac{1}{2}\left(\frac{d(\mathbf{p}^k, \mathcal{P})}{\sigma}\right)^2\right), \tag{7}$$

where $\mathbf{p}^k = T(z^k; \mathbf{x})$. Plugging this into Eqn. (4) and the result into Eqn. (3), we obtain the entire sensor model.

## 6 Experimental Results

The sensor and prediction models have been implemented in a particle filter algorithm and evaluated on real data acquired with a mobile robot. The robot is a Pioneer II AT system equipped with a SICK LMS laser range scanner and an AMTEC wrist unit, which is used as a pan/tilt device for the laser (see Figure 4). During the experiments described in this section, the laser was directed horizontally. The experiments are designed to investigate if the MLS map approach facilitates mobile robot localization and whether it yields better localization performance than the elevation maps.

### 6.1 Global Localization

The first set of experiments is designed to evaluate the performance of the MLS map approach in the context of a global localization task. Figure 5 depicts the convergence of the particles to the true position of the robot with 500,000 and 1,000,000 particles. Whereas the x-axis corresponds to the resampling step, the y-axis shows the number of particles in percent that are closer than 1 m to the true position, which has been computed by a tracking experiment with 100,000 particles. Shown are the evolutions of these numbers when the MCL is applied on standard elevation maps and on MLS maps. Note that the elevation map does not reach 100%. This is due to the fact that the sensor model for the standard elevation map relies on a highly smoothed likelihood function, which is good for global localization but does not achieve maximal accuracy during tracking. The application of a more peaked sensor model in the case of the standard elevation map would lead to much higher divergence rates. In both cases, a t-test showed that it is significantly better to apply the MLS maps than the standard elevation maps for the global localization task. Experiments with 250,000 and 750,000 particles showed the same behavior. Figure 6 shows the number of successful localizations for the two different map representations and for different numbers of particles. Here, we assumed that the localization was achieved when every particle differed by at most 1 m from the true location of the robot. We can see that the global localization performs more robust on the MLS map than on the standard elevation map.

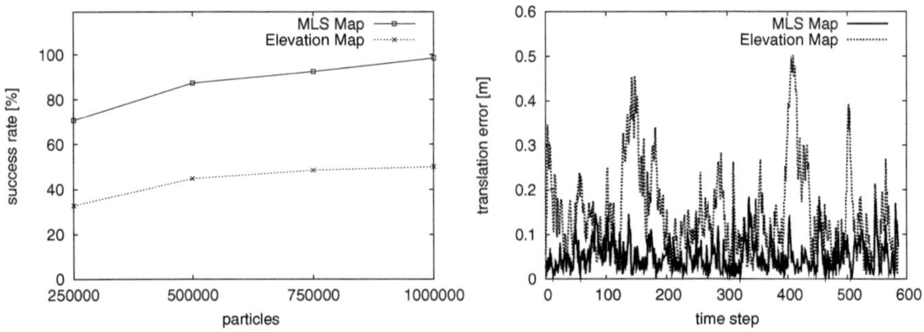


**Fig. 6.** The left image depicts the number of successful localizations after 15 resampling steps for the two different map representations for particle numbers from 250,000 up to 1,000,000. The right image shows the average localization error over all particles for a tracking experiment with 1,000 particles. In average the use of the MLS maps leads to smaller errors.

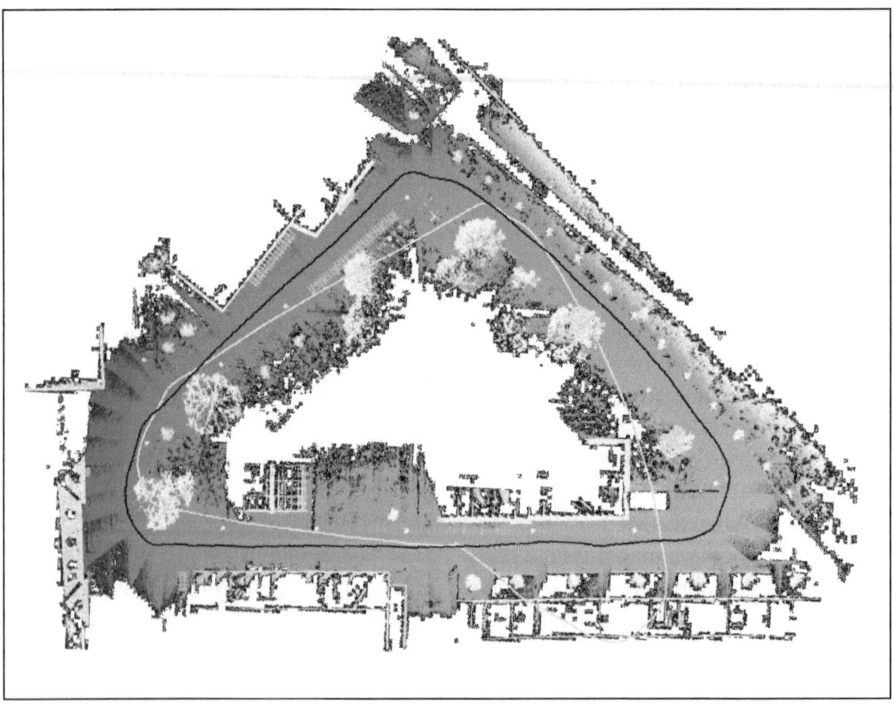

**Fig. 7.** MLS map used for the localization experiments. The area represented by this map spans approximately 195 by 146 meters. The blue / dark grey line shows the localized robot poses. The yellow / light grey line shows the pure odometry. The traversed trajectory has a length of 284 meters.

### 6.2 Tracking

We also carried out experiments, in which we analyzed the accuracy of the MLS map approach in the context of a position tracking task. To obtain the corresponding data set, we steered along a loop in our campus environment. The traversed trajectory has a length of 284 meters. Figure 7 depicts a top view of the MLS map of our test environment. The blue / dark grey line shows the localized robot poses. The yellow / light grey line shows the pure odometry. Figure 6 depicts the average localization error for a tracking experiment with 1,000 particles. As can be seen from the figure, the MLS map approach outperforms the standard elevation map approach. The tracking experiments have been computed online on a standard PC with an AMD Athlon 64 3200+ processor. In the practical experiments we found that the use of the MLS maps results in a computational overhead of no more than 10% compared to elevation maps.

## 7 Conclusions

In this paper, we presented an approach to Monte-Carlo localization (MCL) using multi-level surface (MLS) maps. We applied the main concepts of MCL to this new 3D map representation by adapting the prediction model and the sensor model to MLS maps. Furthermore, we showed in experiments that MLS maps are better suited for the task of local and global localization than standard elevation maps. The slightly increased runtime due to the higher accuracy of MLS maps is compensated with a significantly faster convergence of the particle filter. This makes MLS maps useful for outdoor localization in cases in which no GPS signal is available.

## Acknowledgment

This work has partly been supported by the German Research Foundation (DFG) within the Research Training Group 1103 and under contract number SFB/TR-8 and by the EC under contract number FP6-IST-027140-BACS and FP6-IST-034120, Action line: Micro/Nano Based Sub-Systems.

## References

1. Adams, M., Zhang, S., Xie, L.: Particle filter based outdoor robot localization using natural features extracted from laser scanners. In: Proc. of the IEEE Int. Conf. on Robotics & Automation (ICRA) (2004)
2. Agrawal, M., Konolige, K.: Real-time localization in outdoor environments using stereo vision and inexpensive gps. In: International Conference on Pattern Recognition (ICPR) (2006)
3. Bares, J., Hebert, M., Kanade, T., Krotkov, E., Mitchell, T., Simmons, R., Whittaker, W.R.L.: Ambler: An autonomous rover for planetary exploration. IEEE Computer Society Press 22(6), 18–22 (1989)
4. Davison, A., Kita, N.: 3d simultaneous localisation and map-building using active vision for a robot moving on undulating terrain. In: Proc. IEEE Conference on Computer Vision and Pattern Recognition, Kauai, December 2001, IEEE Computer Society Press, Los Alamitos (2001)

5. Dellaert, F., Fox, D., Burgard, W., Thrun, S.: Monte Carlo localization for mobile robots. In: Proc. of the IEEE Int. Conf. on Robotics & Automation (ICRA) (1999)
6. Doucet, A., de Freitas, N., Gordan, N. (eds.): Sequential Monte-Carlo Methods in Practice. Springer, Heidelberg (2001)
7. Eliazar, A., Parr, R.: Learning probabilistic motion models for mobile robots. In: Proc. of the International Conference on Machine Learning (ICML) (2004)
8. Grisetti, G., Stachniss, C., Burgard, W.: Improving grid-based slam with rao-blackwellized particle filters by adaptive proposals and selective resampling. In: Proc. of the IEEE Int. Conf. on Robotics & Automation (ICRA), Barcelona, Spain, pp. 2443–2448 (2005)
9. Hebert, M., Caillas, C., Krotkov, E., Kweon, I.S., Kanade, T.: Terrain mapping for a roving planetary explorer. In: Proc. of the IEEE Int. Conf. on Robotics & Automation (ICRA), pp. 997–1002 (1989)
10. Lacroix, S., Mallet, A., Bonnafous, D., Bauzil, G., Fleury, S., Herrb, M., Chatila, R.: Autonomous rover navigation on unknown terrains: Functions and integration. International Journal of Robotics Research 21(10-11), 917–942 (2002)
11. Lingemann, K., Surmann, H., Nüchter, A., Hertzberg, J.: High-speed laser localization for mobile robots. Journal of Robotics & Autonomous Systems 51(4), 275–296 (2005)
12. Olson, C.F.: Probabilistic self-localization for mobile robots. IEEE Transactions on Robotics and Automation 16(1), 55–66 (2000)
13. Parra, C., Murrieta-Cid, R., Devy, M., Briot, M.: 3-d modelling and robot localization from visual and range data in natural scenes. In: Christensen, H.I. (ed.) ICVS 1999. LNCS, vol. 1542, pp. 450–468. Springer, Heidelberg (1998)
14. Pfaff, P., Triebel, R., Burgard, W.: An efficient extension to elevation maps for outdoor terrain mapping and loop closing. International Journal of Robotics Research (2007)
15. Thrun, S.: A probabilistic online mapping algorithm for teams of mobile robots. International Journal of Robotics Research 20(5), 335–363 (2001)
16. Triebel, R., Pfaff, P., Burgard, W.: Multi level surface maps for outdoor terrain mapping and loop closing. In: Proc. of the IEEE/RSJ Int. Conf. on Intelligent Robots and Systems (IROS) (2006)

# Estimating Vehicle Position Using Few Range Measurements to a Known Curved Surface

Michael A. Kokko[1] and Franz S. Hover[2]

Massachusetts Institute of Technology, Cambridge, MA
[1]kokkom@alum.mit.edu, [2]hover@mit.edu

**Summary.** This paper presents a method by which data from two or more rangefinders can be combined to estimate vehicle position when operating in proximity to a surface of known curvature. Estimates are combined in an Extended Kalman Filter which accounts for the expected transformation of sensor noise. Although developed for use with Autonomous Underwater Vehicles, this curvature-based localization technique can be applied to any mobile agent capable of measuring its range to a previously-mapped surface.

## 1 Introduction

Many autonomous vehicles rely on navigation data obtained from absolute position and orientation sensors including GPS units and compasses; however, it is often necessary to operate vehicles in environments that preclude the use of such sensors. Our present application, for example, involves inspecting ship hulls with an Autonomous Underwater Vehicle (AUV) in harbor environments where neither GPS nor compass readings are available. Various sensor-based navigation strategies are possible for the ship hull inspection task. One approach is to deploy and calibrate a network of transponders to provide localization information [10]. Another approach is to use acoustic or optical images to estimate the 3D structure of the ship hull and concurrently estimate vehicle pose [1, 2, 8]. Our prototype inspection vehicle currently obtains odometry information by means of a four-beam Teledyne RDI Doppler Velocity Log (DVL) sensor which tracks relative motion between the AUV and its target hull.

In this paper[1], we propose an alternative method of navigation based on a small number of range measurements taken to a known surface. We envision this capability to provide on-line, real-time servoing in close proximity to curved sections of target ship hulls. Similar work has been done by Madhavan and Durrant-Whyte [6] in which they navigate based on points of maximum curvature extracted from observed surfaces. Several other authors including Sekmen

[1] This work and the development of the Hovering Autonomous Underwater Vehicle have been supported by the Office of Naval Research under grant N00014-06-1-0043 which is monitored by Dr. T.F. Swean.

C. Laugier and R. Siegwart (Eds.): Field and Service Robotics, STAR 42, pp. 223–232, 2008.
springerlink.com 

[7] have considered the problem of identifying the location and curvature of cylindrical targets from a mobile agent. Kondo and Ura also use curvature in [5] to track features with their AUV remaining at a desired relative heading.

The present approach differs from these in that we produce actual vehicle position estimates by exploiting the known geometric properties of an observed surface. Details of the observation geometry are considered and the derived position estimates are integrated into an Extended Kalman Filter (EKF) framework. Both simulated navigation results as well as data obtained from a real-time surface craft implementation are presented.

## 2 Vehicle Platform

The Hovering Autonomous Underwater Vehicle (HAUV) in this work measures roughly 1.0m x 0.7m x 0.4m and weighs approximately 80kg. More detailed descriptions of the physical system are presented in [9] (see also Fig. 1). Equipped with both a DVL and a Sound Metrics DIDSON imaging sonar, the HAUV has been designed to servo laterally along a ship hull, inspecting its surface for the presence of mines. Since both compass and GPS signals are useless in these conditions, the HAUV integrates velocity estimates derived from DVL beam measurements to track its position along the hull. Meanwhile, the DIDSON captures high-resolution imagery which can later be reconstructed in the form of a mosaic to view the entire inspection area.

**Fig. 1.** The HAUV with its sensor payload comprised of a DIDSON imaging sonar (left sensor) and a DVL (right sensor)

## 3 Dynamic Model

A state-space model of a simplified HAUV-like system can be developed for use in simulation. The vehicle states are taken to include vehicle-relative velocities (two planar and one rotational) as well as global Cartesian locations and vehicle yaw. That is:

$$\mathbf{x}(t) = \left[ u \; v \; \dot{\theta}_v \; x_v \; y_v \; \theta_v \right]^T \tag{1}$$

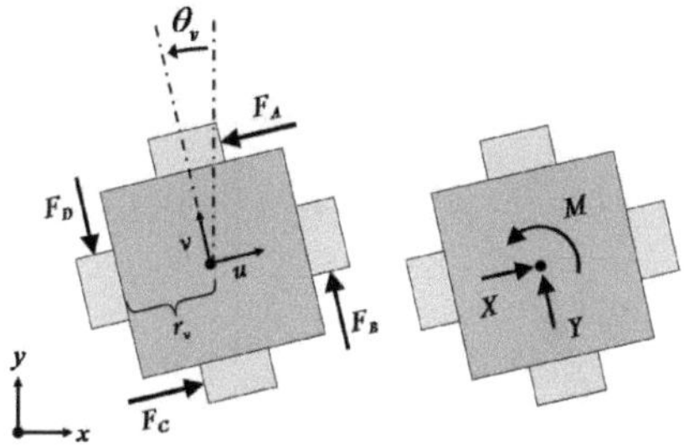


**Fig. 2.** Free-body diagram of the simplified HAUV dynamic model indicating chosen thrust and yaw conventions

At low velocities, when the effects of hydrodynamic drag on the system are negligible, the system dynamics propagate as:

$$\dot{\mathbf{x}}(t) = \mathbf{f}(\mathbf{x}(t), \mathbf{u_v}(t), t) + \mathbf{w}(t) \tag{2}$$

$$\mathbf{z_k} = \mathbf{h}(\mathbf{x}(t_k)) + \mathbf{v_k} \tag{3}$$

In this representation, $\mathbf{x}(t)$ are the vehicle states, $\mathbf{u_v}(t)$ are control inputs $\{X,Y,M\}$, $\mathbf{z_k}$ are measurements, and $\mathbf{w}(t)$ and $\mathbf{v_k}$ add Gaussian noise with respective variances of $\mathbf{Q}$ and $\mathbf{R_k}$. Notice that the vehicle state vector propagates continuously while measurements are made in discrete time. The nonlinear state propagation function $\mathbf{f}(\mathbf{x}(t), \mathbf{u_v}(t), t)$ can be defined with respect to vehicle mass ($m$) and inertia ($J$) as:

$$\mathbf{f}(\mathbf{x}(t), \mathbf{u_v}(t), t) = \left[X/m \; Y/m \; M/J \; c_1 \; c_2 \; \dot{\theta}_v\right]^T \tag{4}$$

where:

$$c_1 = u \cos\theta_v - v \sin\theta_v$$
$$c_2 = u \sin\theta_v + v \cos\theta_v$$

A simple PD control scheme drives the estimated state $\hat{\mathbf{x}}(t)$ to a reference state $\mathbf{x_r}(t)$ by means of an error vector $\tilde{\mathbf{x}}(t) = \mathbf{x_r}(t) - \hat{\mathbf{x}}(t)$. The controller was developed with dynamics linearized about the hovering condition which provides a means for selecting an adequate compensator of the form $\mathbf{u_v}(t) = -\mathbf{K_c}\tilde{\mathbf{x}}(t)$. Proportional and derivative control gains were chosen to place the compensated system poles such that the $\theta_v$ step response settles in roughly 0.5 seconds while the $x_v$ and $y_v$ responses settle about four times slower (each with minimal overshoot). These selections are consistent with the dynamic performance of the actual vehicle in operation.

## 4 Observation Model

Our approach to curvature-based navigation requires two or more range-measurements taken at fixed agent-relative angles. On the HAUV, these range

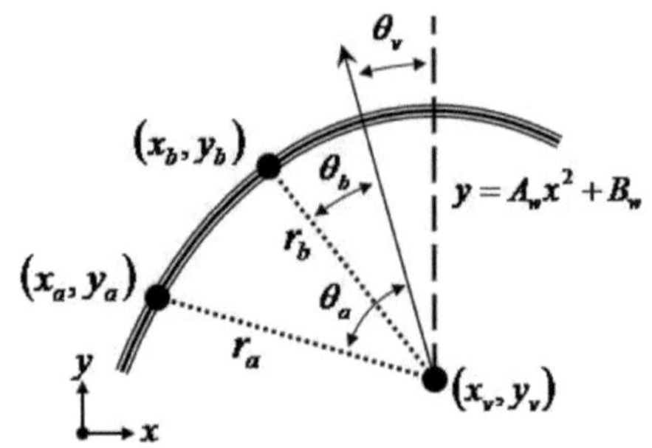


**Fig. 3.** Observation geometry modeled to simulate range data received from the DVL sensor on the HAUV. For the DVL case, $\theta_a = +20.7°$ and $\theta_b = -20.7°$.

measurements can be made by pitching the DVL down 22° to emulate a pair of range-finding beams which lie in-plane with the vehicle and are separated by 41.4°. We assume that an accurate yaw position $\theta_v$ is available, at least in the short term, from an inertial measurement unit[2], and that range measurements are taken between the agent and a target of known parabolic ($y = A_w x^2 + B_w$) curvature. With these assumptions, two range measurements $\{r_a, r_b\}$ taken at bearings $\{\theta_a, \theta_b\}$ are sufficient to produce an estimate of vehicle planar position $\{x_v, y_v\}$. Note that although $A_w$ is negative in Fig. 3 to model a parabolic pocket, there is no *algorithmic* difference when using a positive value corresponding to a protrusion as might be found on a ship hull.

$$x_v = \frac{r_a c_a / A_w - r_b c_b / A_w - r_a^2 s_a^2 + r_b^2 s_b^2}{-2 r_a s_a + 2 r_b s_b} \tag{5}$$

$$y_v = \tfrac{1}{2}\left\{A_w\left[(x_v - r_a s_a)^2 + (x_v - r_b s_b)^2\right] + 2B_w - r_a c_a - r_b c_b\right\} \tag{6}$$

where:

$$s_a = \sin(\theta_v + \theta_a)\ \ s_b = \sin(\theta_v + \theta_b)$$
$$c_a = \cos(\theta_v + \theta_a)\ \ c_b = \cos(\theta_v + \theta_b)$$

While only two range measurements are needed to develop a position estimate, it is clear that this procedure can be applied to the general $n$-beam scenario where $n \geq 2$. When redundant range information ($n > 2$) is available, $m = \frac{n!}{2(n-2)!}$ distinct estimates of vehicle position can be combined by filtering. In this case, the observation vector at timestep $k$ becomes:

$$\mathbf{z_k} = \left[x_{v1}\ y_{v1}\ \cdots\ x_{vm}\ y_{vm}\ \theta_v\right]^T \tag{7}$$

The multi-beam sensor scenario is, in fact, a simplified version of scanning sonar observation. The algorithm presented here is directly applicable to the scanning case, allowing vehicle position estimates to be developed when these more

[2] If only rate gyros are available, vehicle yaw position must be calibrated with respect to a known feature prior to generating position estimates.

advanced sensors are employed in environments of known curvature. It is also worth noting that this localization technique may still be applicable when features cannot be adequately described by a single parabolic curve. Estimation equations similar to (5) and (6) have been developed for sinusoidal pockets for example, and the same procedure can be applied to other types of curvature. Additionally, multiple curves can be combined in a piecewise manner to facilitate vehicle navigation along complex targets like ship hulls. In some cases the correct estimate my need to be selected from a set of candidate positions based on recent navigation history or an explicit switching metric.

In addition to the geometric constraints derived in (5) and (6), it is also important to understand the noise properties of the resulting position estimate. By calculating the partial derivatives of these equations with respect to each range measurement we observe how sensor noise translates into estimate uncer-

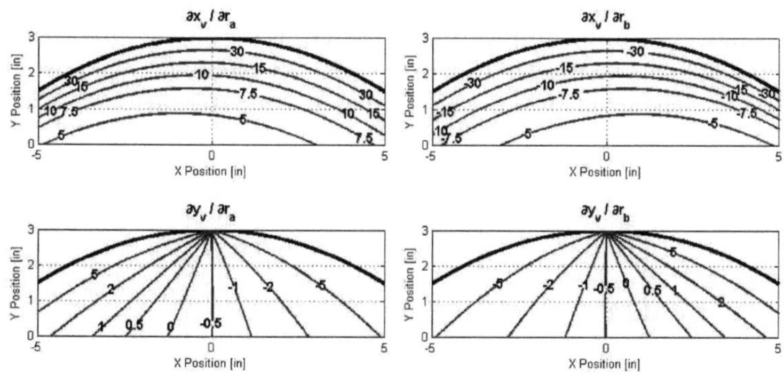


**Fig. 4.** Partial derivatives of estimated vehicle position shown for an agent with no yaw and a DVL-like sensor configuration (two symmetric range-finder beams separated by 41.4°) observing a surface conforming to the equation $y = -0.06x^2 + 3$

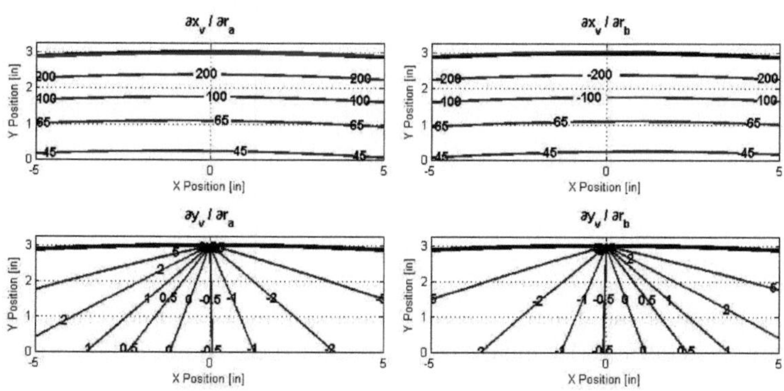


**Fig. 5.** Partial derivatives as in Fig. 4 with a target fitting $y = -0.005x^2+3$. Notice that the horizontal position estimate noise greatly increases as the target surface flattens.

tainty. Figs. 4 and 5 show graphical representations of these derivatives which indicate that the noise on $x_v$ depends strongly upon the distance from the wall while the noise gradient on $y_v$ is much more gentle throughout. Notice that the noise on $x_v$ becomes substantially higher when the surface flattens. In Fig. 5, for example, the vehicle would have to move roughly 26m away from the wall to achieve $\partial x_v / \partial r_a$ and $\partial x_v / \partial r_b$ values of $\pm 5$ which occur at a range of only 2m in Fig. 4. Thus, this navigation scheme is most useful in situations where highly curved structures can be observed at an appropriate distance.

## 5 Nonlinear Estimator

An Extended Kalman Filter can be employed to combine estimates developed based on the observations in $\mathbf{z_k}$. At each timestep, the filtering process begins by calculating the Jacobian of the nonlinear state propagation function $\mathbf{F}(\mathbf{x}(t), t)$ as well as the Jacobian of the observation model $\mathbf{H_k}$ using (8) and (9), respectively. Notice that the first two rows of the observation Jacobian are repeated $m$ times, where $m$ reflects the number of range-pair-based estimates included in $\mathbf{z_k}$ (which itself has $2m + 1$ rows as seen in (7)).

$$\mathbf{F}(\mathbf{x}(t), t) = \begin{bmatrix} 0 & 0 & 0\,0\,0\,0 \\ 0 & 0 & 0\,0\,0\,0 \\ 0 & 0 & 0\,0\,0\,0 \\ \cos\theta_v & -\sin\theta_v & 0\,0\,0\,c_3 \\ \sin\theta_v & \cos\theta_v & 0\,0\,0\,c_4 \\ 0 & 0 & 1\,0\,0\,0 \end{bmatrix} \tag{8}$$

where:

$$c_3 = -u\dot{\theta}_v \sin\theta_v - v\dot{\theta}_v \cos\theta_v$$
$$c_4 = u\dot{\theta}_v \cos\theta_v - v\dot{\theta}_v \sin\theta_v$$

$$\mathbf{H_k} = \begin{bmatrix} 0\,0\,0\,1\,0\,0 \\ 0\,0\,0\,0\,1\,0 \\ 0\,0\,0\,1\,0\,0 \\ 0\,0\,0\,0\,1\,0 \\ \vdots\,\vdots\,\vdots\,\vdots\,\vdots\,\vdots \\ 0\,0\,0\,0\,0\,1 \end{bmatrix} \tag{9}$$

With the Jacobians calculated, *a priori* state $\hat{\mathbf{x}}_\mathbf{k}^-$ and error covariance $\mathbf{P_k}^-$ estimates can be generated by means of the standard continuous-discrete EKF formulation found in [3]. When new curvature-based position estimates are available in $\mathbf{z_k}$, a Kalman gain can be calculated and the *a priori* estimates updated. The predicted observation vector $\mathbf{h_k}(\hat{\mathbf{x}}_\mathbf{k}^-)$ and the observation noise covariance $\mathbf{R_k}$ are dependent upon vehicle position and must be recomputed at each timestep.

$$\mathbf{h_k}(\hat{\mathbf{x}}_\mathbf{k}^-) = \left[\hat{x}_v\ \hat{y}_v \cdots \hat{x}_v\ \hat{y}_v\ \hat{\theta}_v\right]^T \tag{10}$$

$$\mathbf{R_k} = \begin{bmatrix} R_{1,1} & \cdots & R_{1,2m} & 0 \\ \vdots & \ddots & \vdots & \vdots \\ R_{2m,1} & \cdots & R_{2m,2m} & 0 \\ 0 & \cdots & 0 & \sigma^2_{\theta_v} \end{bmatrix} \tag{11}$$

where:

$$R_{i,j} = \frac{\partial z_i}{\partial r_a}\frac{\partial z_j}{\partial r_a}\sigma^2_{r_a} + \frac{\partial z_i}{\partial r_b}\frac{\partial z_j}{\partial r_b}\sigma^2_{r_b}$$

In summary, the curvature-based estimation algorithm follows a simple procedure at each timestep (k) where new range data is available:

1. Calculate Jacobians with (8) and (9) and perform the EKF prediction step propagating the state and covariance estimates forward.
2. Capture a set of $n$ range measurements taken between the mobile agent and a surface of known curvature.
3. Calculate a set of $m$ agent position estimates $\{x_{vi}, y_{vi}\}$ using (5) and (6) where each estimate is based on a unique pair of range measurements. These positions are a subset of $\mathbf{z}_k$ as defined in (7).
4. Construct a predicted observation vector $\mathbf{h_k}(\hat{\mathbf{x}}_\mathbf{k}^-)$ as in (10) using the predicted state and compute the observation covariance matrix $\mathbf{R_k}$ with (11).
5. Calculate the Kalman gain and perform the EKF correction step to generate new state and error covariance estimates.

## 6 Simulation Results

The performance of the simplified dynamic system developed in Section 3 with curvature-based observations provided to an EKF has been simulated for a variety of target and beam configurations. Sensor readings were taken in these simulations discretely at 10Hz as is typical for a DVL, and both the actual vehicle state and the EKF were propagated continuously via the Runge-Kutta method between timesteps. Range sensor noise standard deviations of 6mm were assumed, as was a nominal yaw sensor deviation of 0.005rad. Diagonal elements of the $\mathbf{Q}$ matrix representing $X$, $Y$, and $M$ variances of $(5\text{N})^2$, $(4\text{N})^2$ and $(0.5\text{Nm})^2$ were chosen to ensure a faster response from the estimator than from the controller.

Fig. 6 presents simulation results which indicate that relatively accurate trajectory-following is feasible using curvature-based navigation. As expected, position errors in the $x_v$ direction are larger in magnitude than those in the $y_v$ direction ($\sigma_{x,err}$ = 2.9cm vs $\sigma_{y,err}$ = 1.1cm) due to the increased sensitivity predicted by Figs. 4 and 5. Since position estimate noise rapidly increases close to the observed surface, better results can be obtained by operating farther from the wall.

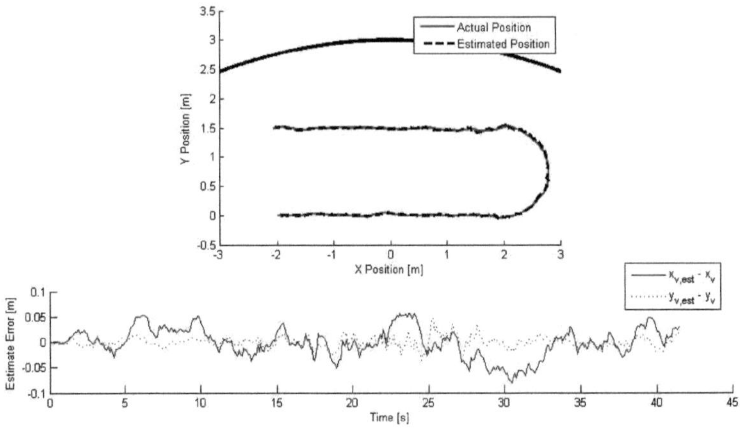


**Fig. 6.** Results of a typical simulation with no vehicle yaw, two sensor beams spaced 41.4° apart, and range error deviations of 1mm. The vehicle followed a 'U'-shaped path from the lower-left at approximately 0.25m/s.

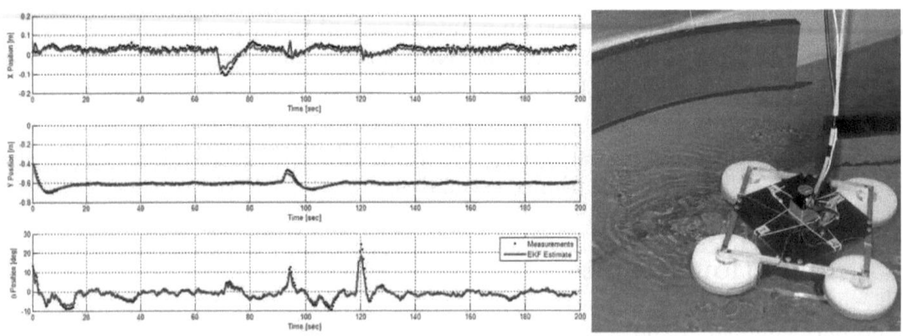

**Fig. 7.** The raft robot servoing relative to a parabolic target. Data from this 200-second test demonstrate the system's ability to hold station and recover from $x$, $y$, and $\theta$ disturbances.

# 7 Real-Time Implementation

A small raft robot equipped with a Hokuyo URG miniature laser scanner was constructed to validate the range-based navigation procedure prior to implementation on the HAUV. Two range measurements separated by 41.4° were taken to a parabolic wall fitting the equation[3] $y = -0.015x^2$ and all data from the laser were used to determine vehicle yaw. A fully discrete EKF was implemented in C++ on a Mac Mini PC and integrated into a control loop running at 2.5Hz (limited by sensor communication latencies). This relatively slow sampling rate

[3] The units of $A_w$ are $[\text{in}^{-1}]$ for the target employed in this experiment.

bounded the bandwidth of both the PD controller and the estimator, leading to long response and settling times.

The raft ran a number of missions during which it transited to and held station at (0m, −0.635m) relative to the target vertex. Results from a typical test are shown in Fig. 7. During the final 50 seconds of this mission the $x$ and $y$ estimate standard deviations are observed to be 8.1mm and 4.1mm (respectively) which are reasonably close to the predicted deviations of 8.9mm and 2.4mm calculated using (11). This result confirms that localization accuracy is indeed a function target-relative agent pose as shown in Figs. 4 and 5. The magnitude of the experimental estimate noise is significantly smaller than the simulated results shown in Fig. 6 due in part to more accurate laser scanner range measurements ($\sigma_r = 3.3$mm).

## 8 Conclusions and Future Work

Extending curvature-based localization to underwater applications using sonar range measurements will necessarily involve an examination of acoustic reflection from curved structures. It is expected that high-frequency, narrow-beam sonar systems similar to the DVL will provide the most accurate measurements. The navigation scheme itself is expected to prove useful in situations where sections of the target hull can be modeled with reasonable accuracy prior to inspection (e.g. from ship's lines); however, its effectiveness will diminish as the hull curvature flattens as viewed from the sensing position unless we move away a commeasurate amount. Potential extensions of this work including multi-curve and three-dimensional navigation are discussed in [4].

Although more advanced image and scanning sonar based navigation techniques have been widely adopted, the present curvature-based solution is attractive because of its simplicity, minimal computational requirements, and application to navigation in truly hostile environments. The algorithm presented in this paper relies on a minimum of two range measurements which need not always be provided by a DVL as we consider here; rather, this method can be implemented on nearly any planar land, air, or undersea craft equipped with at least two range-finders and a yaw sensor. The curvature-based localization technique can either serve as a redundant pose estimator to more complex scanning techniques or it can assume full localization responsibility in situations where more sophisticated sensors are inoperable, e.g., replacing a vision-based system in dark or murky environments.

## References

1. Eustice, R., Pizarro, O., Singh, H.: Visually Augmented Navigation in an Unstructured Environment Using a Delayed State History. In: Proceedings of the 2004 IEEE International Conference on Robotics and Automation, New Orleans, USA, vol. 1, pp. 25–32 (2004)
2. Feder, H.J.S., Leonard, J.J., Smith, C.M.: Adaptive mobile robot navigation and mapping. The International Journal of Robotics Research 18(7) (July 1999)

3. Gelb, A. (ed.): Applied Optimal Estimation. The MIT Press, Cambridge (1974)
4. Kokko, M.: Range-based navigation of AUVs operating near ship hulls. Master's thesis, Massachusetts Institute of Technology (2007)
5. Kondo, H., Ura, T.: Navigation of an AUV for investigation of underwater structures. Control Engineering Practice 12(12), 1551–1559 (2004)
6. Madhavan, R., Durrant-Whyte, H.F.: Natural landmark-based autonomous vehicle navigation. Robotics and Autonomous Systems 46(2), 79–95 (2004)
7. Sekmen, A.S., Barshan, B.: Estimation of object location and radius of curvature using ultrasonic sonar. Applied Acoustics 62(7), 841–865 (2001)
8. Tardos, J.D., Neira, J., Newman, P.M., Leonard, J.J.: Robust Mapping and Localization in Indoor Environments Using Sonar Data. The International Journal of Robotics Research 21(4), 311–330 (2002), `http://ijr.sagepub.com/cgi/content/abstract/21/4/311`
9. Vaganay, J., Elkins, M., Esposito, D., O'Halloran, W., Hover, F., Kokko, M.: Ship Hull Inspection with the HAUV: US Navy and NATO Demonstration Results. In: Proceedings of IEEE / MTS Oceans 2006 (2006)
10. Whitcomb, L., Yoerger, D., Singh, H.: Towards Precision Robotic Manuvering, Survey and Manipulation in Unstructured Undersea Environments. In: Robotics Research - The Eighth International Symposium, pp. 45–54. Springer, London (1998)

# Sensor Registration for Robotic Applications

Alen Alempijevic, Sarath Kodagoda, and Gamini Dissanayake

ARC Centre of Excellence for Autonomous Systems (CAS),
Faculty of Engineering,
University of Technology Sydney, NSW 2007, Australia
{a.alempijevic,s.kodagoda,g.dissanayake}@cas.edu.au

**Summary.** Multi-sensor data fusion plays an essential role in most robotic applications. Appropriate registration of information from different sensors is a fundamental requirement in multi-sensor data fusion. Registration requires significant effort particularly when sensor signals do not have direct geometric interpretations, observer dynamics are unknown and occlusions are present. In this paper, we propose Mutual Information (MI) based sensor registration which exploits the effect of a *common cause* in the observed space on the sensor outputs that does not require any prior knowledge of relative poses of the observers. Simulation results are presented to substantiate the claim that the algorithm is capable of registering the sensors in the presence of substantial observer dynamics.

## 1 Introduction

With the recent advancements in sensing technologies, it is not unreasonable to assume that sensors will be omnipresent in the near future. Multi-sensor integration and data fusion are essential in order to exploit complementary information from different sensor modalities for achieving robust perception. However, when the relationships between different sensors or sensory signals are unknown, data fusion is not tractable. For example, sensor registration is essential for the fusion of information from a camera mounted in the environment with a mobile robot mounted laser data or fusion of video signals with audio signals. Once the sensor registration is established, multi-sensory data can be fused to perform a variety of tasks in robotics, sensor networks and biomedical engineering.

Mutual information (MI) analysis is well established in the field of medical imaging [1] as a method for image registration and feature selection from a diverse range of sensor modalities. Furthermore, MI has been used as a method to detect statistical relationship or as a measure of coupling between signals [2]. However, traditional MI approaches have a substantial computational cost, in order of a factorial in the number of samples. Fisher et al [3] proposed a methodology for substantially reducing the computational cost of MI. This method is based on unsupervised learning of a nonparametric sensing model by maximizing mutual information between signal streams in a lower dimensional space. Experimental results were presented to pin point the mouth of a speaker on a video

C. Laugier and R. Siegwart (Eds.): Field and Service Robotics, STAR 42, pp. 233–242, 2008.
springerlink.com 

stream analyzing video and audio signals [5]. This method, however, assumes that there is no relative motion of the speaker in successive image frames. It can not deal with the situations where multiple common sources influence the two sensory signals. These limitations are substantial in most of the robotics applications, which we are embarking on.

In our previous work [6],we have extended the Fisher's idea to incorporate dynamic objects in the environment to establish the sensor registration. Moving targets provide very little instant information at the signal level. However, tracked dynamic targets provide substantial amount of information in the feature level. The target attributes were carefully chosen giving due regards to sensor modality and rate of change of the attribute. The MI based approach was utilized to register the common feature attributes in two sensory signals providing sensor registration. The algorithm was further extended to incorporate the maneuvering targets with possible occlusions as in [10]. However, only stationary observers were considered so far, which is a significant limitation, particularly for robotic applications. In this paper, we have further developed this algorithm to register and calibrate the sensors with moving observers even when target occlusions are present. We do not assume any prior knowledge of relative localization between the observers. The target tracking problem is relatively complex due to the motion of the observer, maneuvering dynamics of the target and occlusions. Multiple model approach, in particular the Interacting Multiple Model (IMM) [7], provides one of the most effective frameworks for tracking maneuvering targets. Sequential Probability Ratio Test (SPRT) is used to eliminate false tracks that get initiated due to anomalies in the background and spurious data [8].

The paper is organized as follows. Section 2 discusses the Fisher's MI based approach for sensor registration. In Section 3, the problem of tracking maneuvering targets with moving observers is formulated. IMM with SPRT is discussed in Section 4. Simulation results are presented in Section 5. Section 6 concludes the paper providing future direction of the research.

## 2 Mutual Information Based Sensor Registration

In this section we describe an information theoretic approach for sensor registration, which is based on [3]-[5]. One important aspect of this method is it does not assume any prior information about relationships among either sensory signals or sensors.

**Fisher's Mutual Information Algorithm:** Let $X^a \in \Re^{N_a}$ and $X^b \in \Re^{N_b}$ are two high dimensional random variables. Those are passed through functions $Y^a = g(X^a, H_a)$ and $Y^b = g(X^b, H_b)$ where $Y^a \in \Re^{M_a}$, $Y^b \in \Re^{M_b}$ with $M_a << N_a$ and $M_b << N_b$. Although the mapping can be done through any differentiable function, here, $H_a$ and $H_b$ are treated as coefficients of linear projections. The idea is now to choose the projection coefficients to optimize the MI, which is,

$$I(Y^a; Y^b) = h(Y^a) + h(Y^b) - h(Y^a, Y^b) \tag{1}$$

where, $h(.)$ is the differential entropy, which can be defined for a random variable $Y$ with density $p_Y$ as,

$$h(Y) = -\int_{\Omega} p_Y(y) log(p_Y(y))dy \quad (2)$$

Our goal is to maximize the mutual information between two random variables $Y^a$ and $Y^b$. This can be achieved by maximizing the entropies of $Y^a$ and $Y^b$ and minimizing the joint entropy, $h(Y^a, Y^b)$. Entropy is maximum in a uniform distribution, which is therefore chosen as the desired distribution in the transformed space. The density of the samples are estimated using Parzen density estimator [11]. (2) is linearized with a second order Taylor series expanded around the uniform distribution and the difference between entropies of desired and estimated densities are calculated. The updating term now becomes [4],

$$\Delta y_i^{(k)} = b_r(y_i^{(k-1)}) - \frac{1}{N}\sum_{j \neq i} \kappa_a(y_i^{(k-1)} - y_j^{(k-1)}, \sum) \quad (3)$$

$$b_r(y_i)_j \approx \frac{1}{d}\left(\kappa(y_i + \frac{d}{2}, \sum)_j - \kappa(y_i - \frac{d}{2}, \sum)_j\right)$$
$$\kappa_a(y, \sum) = \kappa(y, \sum) * \kappa'(y, \sum)$$
$$\kappa_a(y, \sum) = -(2^{M+1}\pi^{M/2}\sigma^{M+2})^{-1} exp(-\frac{y^T y}{4\sigma^2})y$$

where, $\kappa_a(.)$ is a kernel: a Gaussian pdf is assumed here. $y_i$ symbolize a sample of either $Y^a$ or $Y^b$, $M = M_a, M_b$ or $M_a + M_b$ based on the term of (1) that is considered. The $j^{th}$ element of $b_r(y_i)$ is defined as $b_r(y_i)_j$. $d$ is the support of the output space. $N$ is the number of samples. The adaptation procedure includes the update rule (3) followed by a least squares solution for $h_a$ and $h_b$ until a maximum is reached. The stopping criteria used is,

$$\delta = \frac{max(\Delta_{NN}) - min(\Delta_{NN})}{max(\Delta)} \quad (4)$$

where, the term $\Delta_{NN}$ is the nearest neighbour distance in the resulting output distribution, max(.) and min(.) are the maximum distance and minimum distance between any two points in the output space. The numerator is a measure of uniformity of the output space and the denominator is a measure of how well the output space is filled.

## 3 Process and Observation Models for Tracking with Moving Observers

In this section, we formulate the tracking problem of maneuvering targets with a moving observer. When the observer is stationery, the targets tracked in the respective sensory spaces can be used to calculate necessary attributes to be used with MI. However, if the observers are moving, the attributes calculated on the tracked objects in the sensory spaces is correlated to the observer movement.

This may cause the same target's attributes calculated from different moving observers to be non informative resulting in the MI based sensor registration to fail. This problem can be solved, if the targets are being tracked in their respective locally defined world frames. Here, we assume the observer localization to a known uncertainty using some external means (eg. SLAM) in these local coordinate frames. However, we do not assume any knowledge of the relative location of the observers.

### 3.1 Process Model

The target dynamics in the world coordinates, W, (Fig.1(a)) are modelled by a combination of constant speed and turn rate models.

$$\dot{\mathbf{x}}_w(t) = \mathbf{f}(\mathbf{x}_w(t), \omega, \boldsymbol{\nu}(t), t) \tag{5}$$

where, $\dot{\mathbf{x}}_w(t) = [x,\ \dot{x},\ y,\ \dot{y}]$, $\{x,\ y\}$ are the position coordinates of the target in $x$ and $y$ directions and $\{\dot{x},\ \dot{y}\}$ are the respective velocities in the world coordinates. $\omega$ is the turn rate of the target.

### 3.2 Observation Model

Here we assume range and bearing observations. Therefore the observation model in the sensory space is,

$$\mathbf{z}(t) = \mathbf{h}(\mathbf{x}_w(t), \mathbf{x}_s(t), \varpi(t), t) \tag{6}$$

where, $\mathbf{z}(t) = [r\ \theta]^T$ are range and bearing to the target in the sensory space. $\mathbf{x}_s(t)$ is the pose of the observer in the world reference frame. $\varpi(t)$ is the measurement noise (zero mean Gaussian).

## 4 Tracker Based on Interacting Multiple Model (IMM)

One of the best algorithms to track maneuvering targets with stationary/moving observers is the IMM algorithm. In the following, the IMM is used to track maneuvering targets whilst SPRT is used for track initiation, maintenance and deletion. Once the targets are detected and tracked, the attributes of the targets are calculated and used in MI based sensor registration. The IMM algorithm [7] is shown in the Fig.1(b).

False tracks may be initiated due to spurious target detections. The true track confirmation and false track deletion are handled using the ideas from integrated probabilistic data association (IPDA) [9] with SPRT [8]. Using the Markov relationship, the probability of existence of the true target,$P_T\ (k+1|k)$ before the receipt of data in scan $k+1$ is [8],

$$P_T(k+1|k) = P_{22}P_T(k|k) + P_{12}[1 - P_T(k|k)] \tag{7}$$

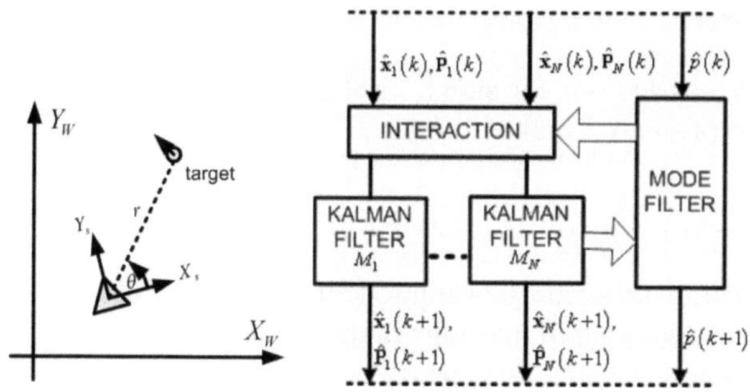


**Fig. 1.** a) Coordinate frames: $W$- world coordinates, $S$ - sensor coordinates and b) The IMM algorithm [10]: $\hat{\mathbf{x}}_i$, $\hat{\mathbf{P}}_i$ define the states and their covariance matrices for $i = 1, \ldots.., N$ hypotheses. $\hat{p}(t)$ is the mode probability.

where $P_{22}$ is the probability of transition from observable state to observable state, whilst $P_{12}$ is the probability of transition from unobservable state to observable state. Then, the update of the probability of target existence is [9],

$$P_T(k+1|k+1) = \frac{1-\delta_{k+1}}{1-\delta_{k+1}P_T(k+1|k)} P_T(k+1|k) \tag{8}$$

where $\delta_{k+1}$ is defined as,

$$\delta_{k+1} = \begin{cases} P_D P_G & N_{k+1} = 0 \\ P_D P_G \left[1 - \overline{V} \sum_{i=1}^{N_{k+1}} \frac{1}{P_G (2\pi)^{M/2} \sqrt{|S(k+1|k)|}} e^{-d_i^2/2}\right] & Otherwise \end{cases} \tag{9}$$

$P_D$ is the probability of detection, $P_G$ is the gate probability, $V_G$ is the gate volume, $N_{k+1}$ is the number of measurements inside the validation gate, $S$ is the innovation covariance, and $d_i^2$ is the normalized innovation squared of the $i^{th}$ measurement. The Log Likelihood Ratio, LLR, can now be defined as [8],

$$LLR_{k+1} = ln\left(\frac{P_T}{1-P_T}\right) \tag{10}$$

Once the LLR is obtained, track confirmation and termination thresholds are determined using the SPRT [8] as,

$$LLR_{k+1} \geq ln\left(\tfrac{1-\beta_T}{\alpha_T}\right), \text{ declare track confirmation}$$
$$ln\left(\tfrac{\beta_T}{1-\alpha_T}\right) < LLR_{k+1} < ln\left(\tfrac{1-\beta_T}{\alpha_T}\right), \text{ continue test}$$
$$LLR_{k+1} \leq ln\left(\tfrac{\beta_T}{1-\alpha_T}\right), \text{ delete track}$$

where,$\alpha_T$ is the probability of false track confirmation, and $\beta_T$ is the probability of true track termination.

## 5 Simulation Results

Simulations are carried out for signal level and feature level sensor registration. The results are presented bellow.

### 5.1 Signal Level Sensor Registration

The MI algorithm was initially evaluated through simulations. Twenty two different signals were generated and four random signals from the set were supplied to sensor 1 and another four to sensor 2 with only one common signal (signal 1 in sensor 1 and signal 3 in sensor 2 in Fig.2. Calculation of Mutual Information was performed on 80 samples imposing the following criteria on the variation of $\delta$. Firstly the convergence limit must be less than 0.085, convergence variation contained in a 0.001 limit and a minimal convergence length of 10 iterations. It is to be noted that the convergence criterion is dependant on the number of samples obtained from the signal and (4). With the imposed criteria, a match was found in 28 iterations (Fig.2(b)).

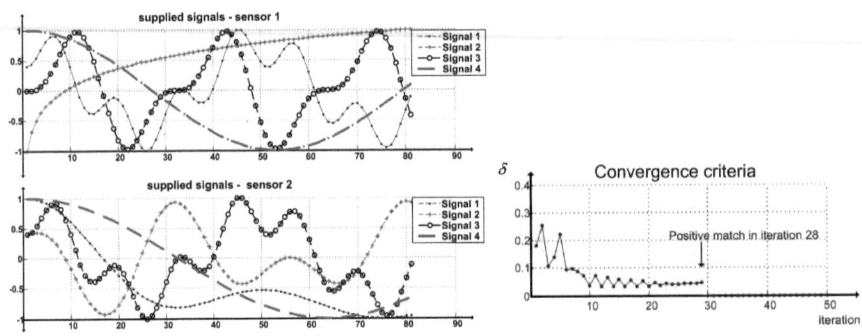


(a) (b)

**Fig. 2.** Signal level sensor registration a) Simulated signal streams and b) MI learning curve

Table 1 shows the calculated MIs with the highest MIs are denoted by symbol *. The algorithm correctly identifies common signals between sensors as signal 1 of sensor one and signal 3 of sensor two, which agrees with the simulation.

**Table 1.** MI values

| Sensor/Signal | 1 | 2 | 3 | 4 |
|---|---|---|---|---|
| 1 | 0.7744* | 0.1050 | 0.0085 | 0.0490 |
| 2 | 0.1596 | 0.1124 | 0.7372* | 0.1758 |

### 5.2 Sensor Registration Based on Moving Observers and Moving Targets

The targets can be tracked in the sensory space and attributes can be estimated even with moving observers. However, this introduces correlations of the observer motion in the estimated target attributes. Since the observers have independent dynamics, the attributes are not mutually informative. Therefore, the targets have to be tracked in a locally defined world reference frames. Fig.3(a) shows a simulation study carried out with two moving observers (O1 and O2) and three maneuvering targets (T1, T2 and T3). Fig.3(b) shows the extracted target

**Table 2.** MI values: a-angle, s-speed, r-range, T-target

| Sensor/Signal | a.T1 | a.T2 | a.T3 | s.T1 | s.T2 | s.T3 | r.T1 | r.T2 | r.T3 |
|---|---|---|---|---|---|---|---|---|---|
| 1 | 0.9 | 50 | 15 | 0.06 | 78* | 0.2 | 0.1 | 0.2 | 0.1 |
| 2 | 19.6 | 22 | 3.8 | 1.9 | 100* | 0.5 | 1.5 | 1 | 0.9 |

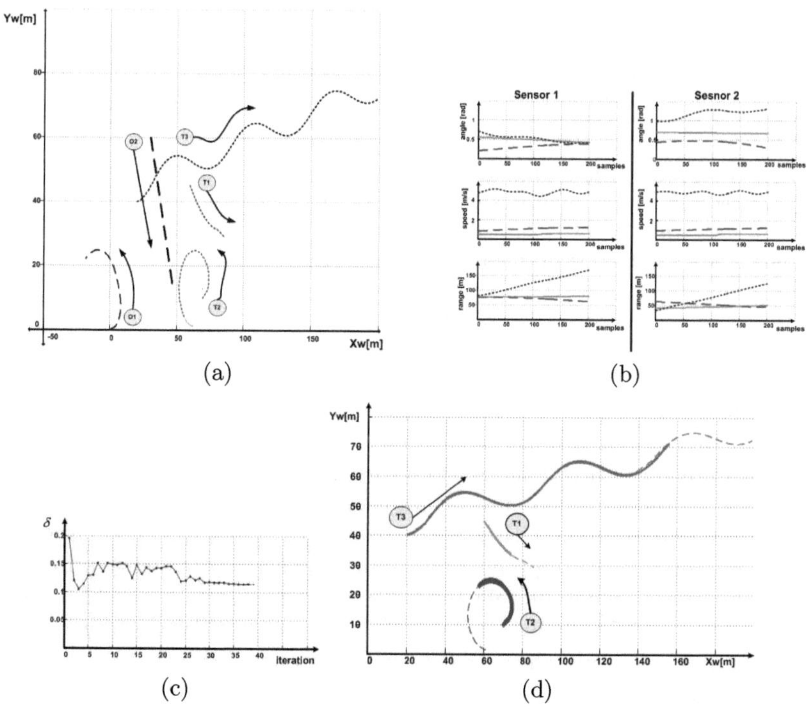


**Fig. 3.** Sensor registration with moving observers and moving targets (a) Simulation environment: O - observer, T - target, (b) Extracted target attributes, *solid line*- target 1, *dashed line* - target 2, *dotted line* - target 3, (c) Learning curve, and (d) All data transformed into sensor 1 reference frame

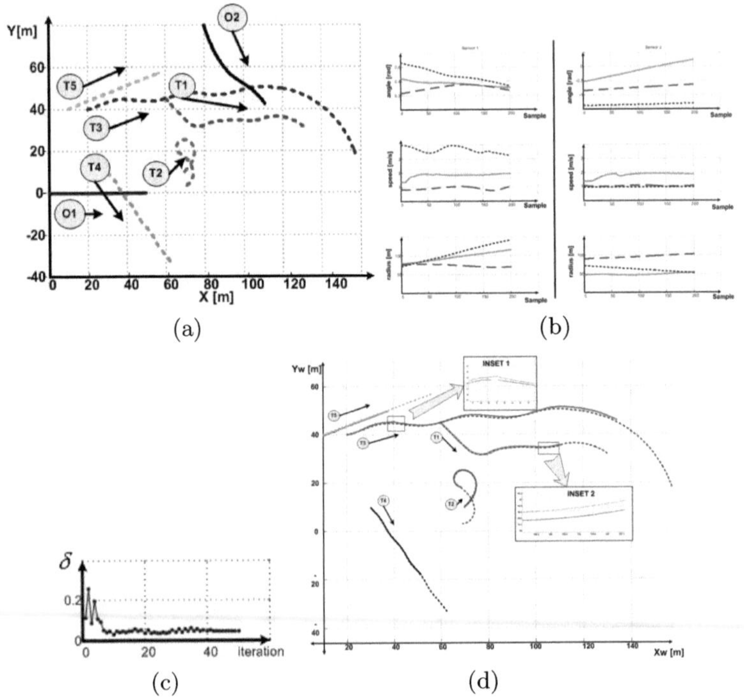


**Fig. 4.** Sensor registration with moving observers and moving targets experiencing occlusions (a) Simulation environment: O - observer, T - target, (b) Extracted target attributes, *solid line*- target 1, *dashed line* - target 2, *dotted line* - target 3, (c) Learning curve, and (d) All data transformed into sensor 1 reference frame

attributes in respective observers' locally defined world coordinates. The MI based algorithm converges at 39th iteration as depicted in Fig.3(c), detecting the most informative attribute as the speed of Target 2 (Table 2). The path of Target 2 in the two local world co-ordinate frames is then used to compute the relative calibration between the sensors through a least square fit. The tracked target poses in observer 2 can now be transformed onto the locally defined world reference frame of observer 1 (Fig.3(d)). It is to be noted that the target tracks of both observers are overlapping, confirming correct sensor registration.

The main requirement of this sensor registration strategy is that at least one *common cause* should be observed by the both observers. Any number of targets that influence only one sensor may be present. Simulation in Fig.4 is used to demonstrate this scenario and the effect of target occlusion. Two moving observers (O1 and O2) and five maneuvering targets (T1 to T5) are present. Observer O1 can see targets T1, T2 and T3 while observer O2 can see targets T1, T4 and T5. Occlusions were introduced at inset 1 and 2 in Fig.4(d). Fig.4(b) shows the extracted target attributes in a locally defined world coordinates. Two hundred samples of the target attributes were used in the MI based algorithm,

**Table 3.** MI values: a-angle, s-speed, r-range, T-target

| Sensor/Signal | a.T1 | a.T2/T4 | a.T3/T5 | s.T1 | s.T2/T4 | s.T3/T5 | r.T1 | r.T2/T4 | r.T3/T5 |
|---|---|---|---|---|---|---|---|---|---|
| 1 | 18.2* | 12 | 8.2 | 2.5 | 0.9 | 0.2 | 0.9 | 0.8 | 0.9 |
| 2 | 87* | 21 | 36.9 | 1.2 | 1.1 | 32.5 | 1.2 | 1.1 | 2.4 |

which converged at iteration 48 as depicted in Fig.4(c), resulting in the angle to target 1 being the most informative (Table 3). Fig.4(d) depicts that the observer 2 data transformed on to the observer 1 are aligned with each other, which shows that the sensor registration is correct. This confirms that MI based algorithm correctly uses *common causes* rather than distracted by other targets. While the targets are occluded, the IMM estimates their pose. Therefore, the algorithm is also immune to target occlusions.

## 6 Conclusions and Future Work

One of the main requirements of multi-sensor fusion is sensor registration. In mobile robotics applications, sensor registration is complicated due to the presence of moving observers and occlusions, and the dynamic nature of the environment. Under these conditions signal level information provided by a common cause can not be easily interpreted. Therefore, we proposed to track the dynamic targets in the environment and use their attributes as the informative signals in a MI based approach for sensor registration. The target tracking problem is nontrivial due to the unknown dynamics of the targets, false detections, and occlusions. These problems were successfully handled by IMM, and SPRT. Then the MI based algorithm was successfully used for sensor registration. However, one of the deficiencies of the current algorithm is that it requires high computational effort as it requires the accumulation of a large sample set to perform the sensor registration. Improving computational efficiency to achieve real time operation is the subject of our future work.

**Acknowledgements.** This work is supported by the ARC Centre of Excellence programme, funded by the Australian Research Council (ARC) and the New South Wales State Government.

## References

1. Pluim, J.P.W., Maintz, J.B.A., Viergever, M.A.: Mutual-information-based registration of medical images,a survey. IEEE Transactions on Medical Imaging 22, 986–1004 (2003)
2. Pompe, B., Blidh, P., Hoyer, D., Eiselt, M.: Using mutual information to measure coupling in the cardiorespiratory system. IEEE Engineering in Medicine and Biology Magazine 17, 32–39 (1998)

3. Fisher III, J.W., Principe, J.C.: Entropy manipulation of arbitrary nonlinear mappings. In: Proceedings of the IEEE Workshop on Neural Networks for Signal Processing, pp. 14–23 (1997)
4. Fisher, J., Wainwright, M.J., Sudderth, E.B., Willsky, A.S.: Statistical and Information-Theoretic Methods for Self-Organization and Fusion of Multimodal, Networked Sensors. International Journal of High Performance Computing Applications 16, 337–353 (2002)
5. Fisher III, J.W., Darrell, T.: Speaker association with signal-level audiovisual fusion. IEEE Transactions on Multimedia 6, 406–413 (2004)
6. Alempijevic, A., Kodagoda, S., Underwood, J., Kumar, S., Dissanayake, G.: Mutual Information based Sensor Registration and Calibration. In: Proceedings of the IEEE/RSJ International Conference on Intelligent Robots and Systems (IROS 2006), Beijing, pp. 25–30 (2006)
7. Blom, H.A.P., Bar-Shalom, Y.: The interacting multiple model algorithm for systems with Markovian switching coefficients. The IEEE Transactions on Automatic Control 33(8), 780–783 (1988)
8. Blackman, S., Popoli, R.: Design and Analysis of Modern Tracking Systems. Artech House, Boston (1999)
9. Musicki, D., Evans, R., Stankovic, S.: Integrated probabilistic data association. IEEE Transactions on Automatic Control 39(6), 1237–1241 (1994)
10. Kodagoda, K.R.S., Alempijevic, A., Underwood, J., Kumar, S., Dissanayake, G.: Sensor Registration and Calibration using Moving Targets. In: Proceedings of nineth International Conference on Control, Automation, Robotics and Vision (ICARCV 2006), pp. 830–835 (2006)
11. Parzen, E.: On estimation of probability density function and mode. Annals of Mathematical Statistics 33, 1065–1076 (1962)

# Part VI

# Aerial Navigation and Visual Tracking

# Reactive Collision Avoidance for Unmanned Aerial Vehicles Using Doppler Radar

Andrew Viquerat, Lachlan Blackhall, Alistair Reid, Salah Sukkarieh, and Graham Brooker

ARC Centre of Excellence for Autonomous Systems,
School of Aerospace, Mechanical and Mechatronic Engineering,
The University of Sydney, NSW, 2006, Australia
viquerat@gmail.com, lachlanblackhall@gmail.com, alistaireid@hotmail.com, s.sukkarieh@cas.edu.au, g.brooker@cas.edu.au

**Summary.** Research into reactive collision avoidance for unmanned aerial vehicles has been conducted on unmanned terrestrial and mini aerial vehicles utilising active Doppler radar obstacle detection sensors. Flight tests conducted by flying a mini UAV at an obstacle have confirmed that a simple reactive collision avoidance algorithm enables aerial vehicles to autonomously avoid obstacles. This builds upon simulation work and results obtained using a terrestrial vehicle that had already confirmed that active sensors and a reactive collision avoidance algorithm are able to successfully find a collision free path through an obstacle field.

## 1 Introduction

The ability to deploy UAVs across a multitude of operating environments and mission scenarios has rendered them exceedingly useful for both defence and commercial applications and this has fuelled their rapid development in the last decade. Collision avoidance is a critical autonomous behaviour that is required to achieve a suitably robust platform for mission deployment.

Collision avoidance systems require both obstacle detection sensors and a collision avoidance algorithm that utilises the information obtained from the sensor(s) to determine a path through the obstacle field. Sensors used in existing collision avoidance systems can be broken into two types; active and passive. Examples of passive sensors include optical flow sensors or CCD devices with embedded optical flow algorithms [1] [2], monocular or stereo vision systems using object detection and/or extraction [3] [4] and infrared cameras, which can be used to identify obstacles by the thermal wavelengths they emit. Active sensors include ultrasonic devices, SOund Navigation And Ranging (SONAR), active infrared devices, Laser [5], Doppler measurement radar, conventional radar and Ultra Wideband (UWB) radar [6] [7]. With the exception of IR sensors which rely on radiation emitted by the object due to its temperature, any form of passive system requires an external lighting source, which may limit operation in low lighting or low contrast environments. Both Optical flow and Doppler radar sensors require relative motion between the vehicle and target to detect obstacles.

C. Laugier and R. Siegwart (Eds.): Field and Service Robotics, STAR 42, pp. 245–254, 2008.
springerlink.com 

As the UAV is continually moving this is not considered a limitation. The angular resolution of the sensors mentioned above varies and is a key consideration in the design process. In selecting a suitable sensor for obstacle detection it is imperative that the size, speed, onboard power resources and payload capabilities of the vehicle are considered. Environmental factors such as the available light and obstacle density also impact upon the choice of a suitable sensor system.

Current collision avoidance methods have utilised a variety of algorithms. Sampling based path planners, which include probabilistic roadmap algorithms and rapidly random trees [8] [9], reactive collision avoidance based on optical flow [1] [2] [5], occupancy grids [10] and receding horizon [11] collision avoidance algorithms have all been used with some success.

## 2 Test Vehicles for Collision Avoidance

A terrestrial test vehicle allowed preliminary testing and validation of the obstacle detection sensors and the collision avoidance algorithm before implementation on a UAV. A remote control car was augmented to allow the collision avoidance algorithm to control the steering mechanism while the throttle was controlled via the remote. The collision avoidance algorithm did not have control of the throttle because on the equivalent fixed wing aerial vehicle it is not possible to slow down in the presence of obstacles without changing the flight conditions and risking the initiation of a stall.

The second vehicle to be tested was a mini UAV. This UAV has large control surfaces to facilitate rapid collision avoidance and is shown in Fig. 1.

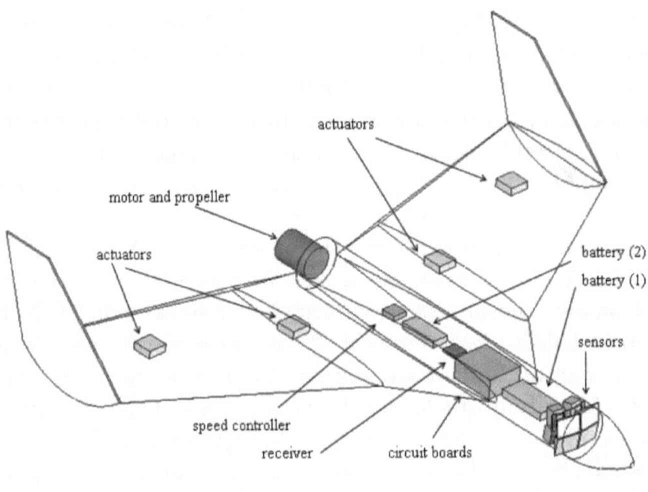


**Fig. 1.** The UAV designed and built for this project has a mass of 1.503kg (Airframe, wingtips and control surfaces 566.3g, Actuators & actuator arms 4 x 23g, Motor and propeller 213.7g, Speed controller 42g, Receiver 20g, Sensors 4 x 35.8g, Circuit Boards 161g, Battery (1) 180g, Battery (2) 85g). (Image courtesy of Alex Davidson).

## 3 Obstacle Detection Sensors

Continuous wave microwave Doppler radars utilising the synthetic aperture technique of Doppler beam-sharpening are ideal for this application. For the aircraft above, the radar unit was developed to have a mass of 304g, use less than 5W and be able to detect the presence of an obstacle at 10-15m (1-2s of flight time) with an operation rate of 10Hz. It was also desired that the angular resolution be less than 20 degrees.

Pulsed Doppler radar can be used to give target velocity, range and in some cases angular location with respect to the observer. Existing work has focussed on the use of this technology for improved automobile safety. Arguably the most successful vehicular collision warning system to date is VORAD (Vehicle Onboard RADar) based on the work of Woll [12]. The VORAD system operates at 24.125GHz to overcome atmospheric attenuation and performs a 4096 point Fast Fourier Transform (FFT), giving an obstacle detection update rate of between 15 and 20Hz. A large horn antenna (approx 20x14cm) with a gain of 30dB and a -3dB beamwidth of 4° allows a highly directional radar beam that consumes only 0.5mW as compared to the 5mW transmitters used in this project. To the authors knowledge, no UAV collision avoidance system to date has made use of a Doppler radar obstacle detection sensor system.

## 4 Operation of the Doppler Radar Units

The approximate formula for the Doppler shift which occurs when a target moves much slower than the speed of the reflected wave is given by:

$$f_d = \frac{2v_t}{\lambda} \cos\theta \tag{1}$$

This can be rearranged to find the angle which the target forms to the radar's velocity vector:

$$\theta = \cos^{-1}\left(\frac{f_d \lambda}{2v_t}\right) \tag{2}$$

Note that this equation requires the difference on speed between a target and a vehicle to be known. Here, we assume that all targets are stationary, with a moving vehicle-mounted radar. By measuring the airspeed of the UAV it is possible to ascertain the relative speed between the UAV and a stationary target.

The system implemented for this project consists of four microwave Doppler transducers mounted to illuminate the four quadrants of the forward field of view of the UAV. The reason for this will become apparent following the discussion of the reactive collision avoidance algorithm in Section 6. The output from these sensors is digitally sampled and a 64 point FFT with fixed sampling rate is run over each waveform allowing the peak frequency in each of the radar outputs to be determined [13]. A Pitot-static tube linked to a Micro-Electro-Mechanical

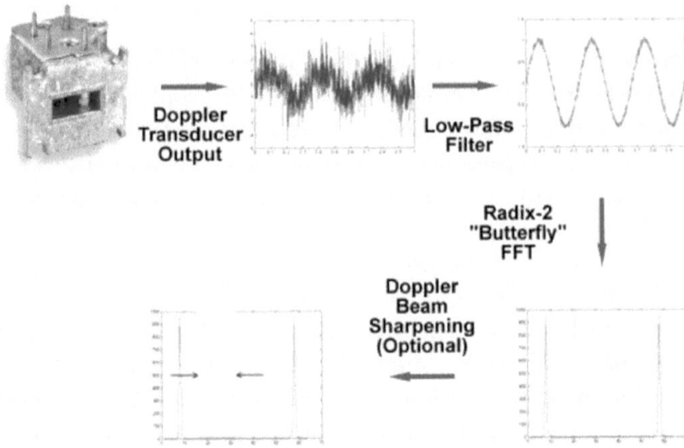


**Fig. 2.** Operation of the Doppler radar obstacle detection sensors

Systems (MEMS) pressure sensor allowed the airspeed to be determined, providing the relative speed of the UAV to a stationary obstacle to be determined. Knowledge of the radar parameters and the vehicle speed allow the presence of obstacles in each of the four quadrants to be determined. The use of Doppler beam sharpening allows readings that correspond to obstacles which do not pose a threat to the vehicle's current trajectory to be discarded.

The system can be dynamically configured to operate with or without Doppler beam sharpening allowing the benefits of the synthetic aperture technique to be evaluated.

**Fig. 3.** The antenna configuration used within the UAV allows four individual radar units to monitor the flight path of the UAV for obstacles

## 5 Results and Discussion of the Doppler Radar Units

Initial tests on the Doppler transducers have demonstrated that the digital sampling and FFT were able to detect the presence of obstacles in the path of the radar units. The features of these units are summarised in Table 1. The update rate and accuracy of the Doppler radar are competing requirements for a given processing power and thus 10Hz and a 64bit FFT provide a suitable trade off between the demands.

**Table 1.** Final Doppler radar sensor suite specifications

| Parameter | Specification |
|---|---|
| Power | 3.7W |
| Weight | 304g (382g with sensor mount for UAV nose) |
| Radar Dome Diameter | 13cm |
| Range | 10-15m for a large tree trunk or wall |
| Target Angular Resolution | 15° |
| Update Rate | 10Hz (7Hz in beam-sharpening mode) |
| Cost (No Labour) | $500 (AUD) |

Testing of the radar units progressed in two phases. The radar units were first tested in isolation with relative motion between the unit and a reflecting target. These tests ascertained that the size of the FFT was adequate for this purpose. In addition, airflow of known velocity was used to verify the operation of the Doppler beam-sharpening.

The second phase involved the use of the system (with only two radars operating) in an obstacle detection application mounted on a small remote controlled terrestrial vehicle. The tests were performed without Doppler beam-sharpening and full discussion and analysis can be found in Section 7.

An obvious assertion that must be made about Doppler radar is that the magnitude of the return signal is dependant on both the size and material type of the obstacle[14] [15]. For this reason the range will be dependant on the environment in which the radar system is used, and the type of obstacles the UAV will encounter. However, it must be remembered that the obstacles we expect to encounter (trees, buildings & the ground) will all provide a significant magnitude of radar return and thus the use of Doppler radar for obstacle detection is appropriate.

## 6 A Reactive Collision Avoidance Algorithm

The purpose of the collision avoidance system is to find a continuous, unimpeded trajectory autonomously. At each point in time, the attainable flight trajectories of the UAV are determined by manoeuvrability and flight speed constraints. The region of space monitored by the radar units is divided into a series of

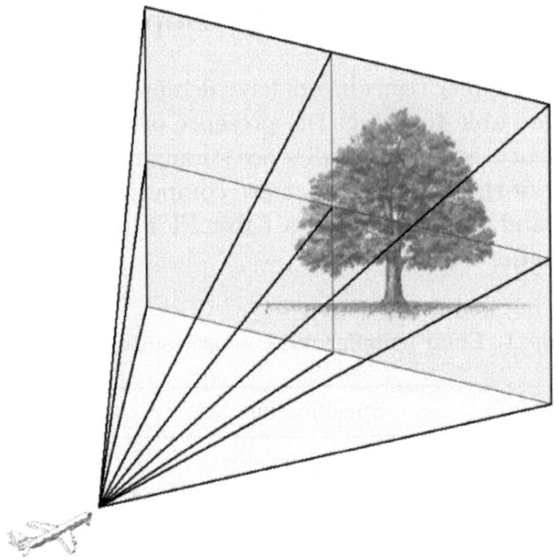

**Fig. 4.** The forward image plane decomposed into quadrants. Each quadrant is monitored for the presence of obstacles by a dedicated Doppler radar unit.

sub-regions (Fig. 4), each corresponding to a different trajectory direction. The authors propose the use of the term activation to describe the danger of flying into a given sub-region, where a higher activation corresponds to more or larger obstacles. For this purpose, the obstacle detection sensors must return higher magnitude values for closer and denser obstacles.

The measured radar returns are attributed to the sub-region from which they are taken using statistical estimation theory and the trajectory for the next time step is determined by choosing the sub-region with the lowest activation. In this application, the algorithm is used every 0.1s, which corresponds to the update rate of the Doppler radar units.

In this application, four dedicated Doppler radar units monitor the region in front of the UAV. The reactive algorithm will guide the UAV into the region monitored by the Doppler radar unit with the lowest return radar signal. By choosing the field of view of the sensors sufficiently wide it is considered unlikely that the collision avoidance algorithm will choose a trajectory towards a previously avoided obstacle. As the algorithm is calculating the trajectory at each point in time based only on the current sensor readings, it is considered to be making stateless decisions.

Existing collision avoidance systems have typically focused on stateful collision avoidance. The authors propose a new approach where the collision avoidance system is stateless, because stateless machines are capable of rapid analysis of time variant data that does not exhibit strong correlation.

For a UAV in a dense obstacle field, it is probable that the ability of the collision avoidance system to predict the obstacle field of a given trajectory will be poor, due to the limited computing power available coupled with the highly dynamic motion of the platform. This dynamic motion is likely to be even more apparent on mini and micro air vehicles, because aerodynamic effects and manoeuvre behaviours are amplified by the low wing loading and the small inertial resistance provided by the airframe to motion. This would suggest that a purely reactive collision avoidance system could provide computational benefits over existing collision avoidance algorithms and implementations, without significant performance compromises.

It must be noted that the reactive nature of the collision avoidance system means that if a continuous unimpeded path does not exist then the algorithm will result in the loss of the UAV as the algorithm uses only the sensor readings at each time step in choosing the best trajectory for the next period of flight time.

## 7 Experimental Results and Discussion

Preliminary results were obtained using a remote control car fitted with the collision avoidance system (utilising only two obstacle detection radar units). The vehicle was driven directly towards an obstacle approximately 60cm in diameter. In these tests, the vehicle was traveling about 1m/s and was successfully able to avoid the obstacle. Figure 5 shows the vehicle steering right to avoid the obstacle, as the right radar unit registered a lower activation than the left, and hence a reduced danger of collision.

Further tests on the ground vehicle have confirmed that this algorithm is able to navigate through a cluttered obstacle field in real-time despite its limited computational resources. Similar trials are yet to be shown on the air vehicle.

To validate the functionality of the collision avoidance algorithm and sensor suite, the UAV discussed previously was flown towards an obstacle and the resulting trajectory observed. The obstacle utilised was a tethered radar reflector. The reflector was designed to minimize risk to the UAV, giving a strong radar reflection and with a low mass in case a collision did occur. In this test the UAV was flown towards the target by a human pilot and then the autonomous mode was activated. Once in autonomous operation the collision avoidance system had control of the ailerons allowing a lateral avoidance manoeuvre to be performed. Photographs of an avoidance manoeuvre are shown in Fig. 6. In this sequence of images the UAV approaches the target, from the left, at a speed of approximately 8m/s and then initiates a collision avoidance manoeuvre to the left to avoid the obstacle. Given the range of the radar units is approximately 10m, this approach speed gives the UAV a window of approximately one second to complete its avoidance manoeuvre.

The preliminary tests above were designed to validate the methodology of this approach and have demonstrated that reactive collision avoidance has the potential to allow air vehicles, especially those with limited computational resources, to navigate through obstacle fields using Doppler radar units.

While early tests have been successful, additional flight tests need to be conducted to further validate the approach used. A significant limitation observed

(a) (b) (c)

**Fig. 5.** The terrestrial test vehicle avoiding an obstacle in its path using Doppler radar obstacle detection sensors and a reactive collision avoidance algorithm

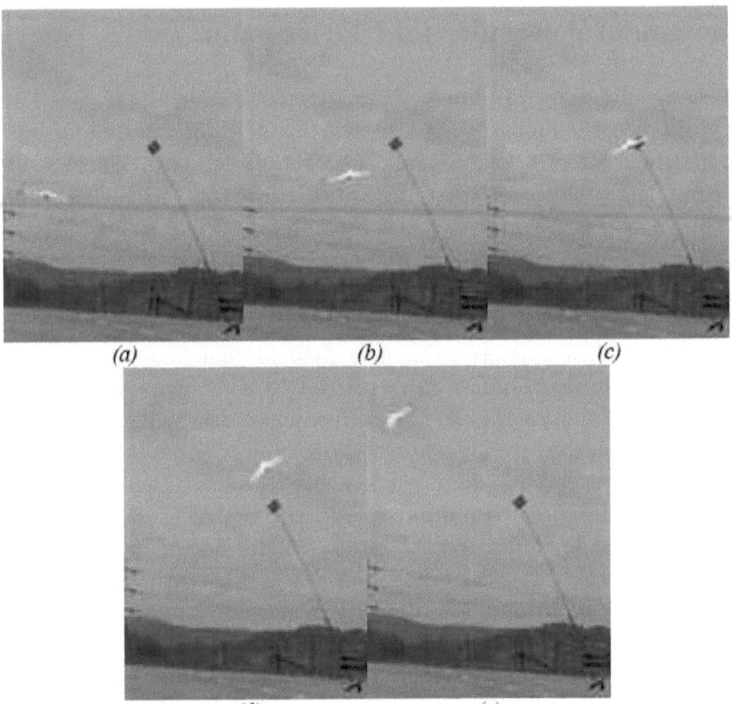

(a) (b) (c) (d) (e)

**Fig. 6.** The UAV autonomously avoiding an obstacle by rapidly turning away

thus far is the accurate range of the Doppler radar units. An increase in range would increase the manoeuvre horizon (the time allowed for manoeuvres prior to a collision) for a given obstacle allowing the UAV to avoid significantly larger obstacles in the flight path. In the current implementation, noticeable noise is induced in the radar signal by the signal-processing electronics mounted within the UAV. This noise is significantly decreasing the usable range of the units.

Increased shielding between the radar units and the other electronics should increase the accurate range of the Dopler radar units allowing this system to be utilised by UAVs with higher operational speeds.

## 8 Conclusion

Existing approaches to collision avoidance have favoured short term dynamic trajectory planning to solve the collision avoidance problem. Unfortunately, such methods become computationally expensive in dense obstacle fields when the collision avoidance system must rapidly process large amounts of data to prevent potential collisions. The fully reactive approach is advantageous because its computational costs are independent of the number of obstacles present in the obstacle field. For this reason, it becomes feasible for aerial platforms of any size. This design shares its philosophy with optical flow sensors, providing effective performance with a low computational complexity.

In addition to improving the radar unit shielding, future work may involve applying the synthetic aperture radar technique of Doppler beam-sharpening. This technique does not rely on any kind of antenna directionality. Rather, it uses velocity components to 'spotlight' obstacles directly in the path of a moving vehicle. A continuous wave microwave Doppler radar system which makes use of Doppler beam-sharpening has potential as the basis for a collision avoidance system in aerial vehicles and further work on the concept is certainly justified.

## Acknowledgements

This work is supported by the ARC Centres of Excellence programme funded by the Australian Research Council (ARC) and the New South Wales State Government. The authors must thank Mr Jeremy Randle and Mr Alan Trinder who worked tirelessly to ready the UAV for flight trials and also to Dr Hugh Stone and Mr Alex Davidson who undertook the computer modelling and flight simulation of the UAV.

## References

1. Barrows, G.: Mixed-Mode VLSI Optic Flow Sensors For Micro Air Vehicles. PhD thesis, University of Maryland, College Park (1999)
2. Zufferey, J.: Bio-Inspired Vision-Based Flying Robots. PhD thesis, Ecole Polytechnique Federale De Lausanne (2005)
3. McGee, T., Sengupta, R., Hedrick, K.: Obstacle detection for small autonomous aircraft using sky segmentation. In: Proceedings of the IEEE International Conference on Robotics and Automation (2005)
4. Lee, D., Beard, R., Merrel, P., Zhan, P.: See and avoidance behaviors for autonomous navigation. In: SPIE Opics East, Robotics Technologies and Architectures, Mobile Robot XVI, vol. 5609-05 (2004)

5. Griffiths, S., Saunders, J., Curtis, A., McLain, T., Beard, R.: Obstacle and terrain, avoidance for miniature aerial vehicles. In: IEEE Robotics and Automation Magazine (to appear)
6. Gresham, I., et al.: Ultra-wideband radar sensors for short-range vehicular applications. Transactions on Microwave Theory and Techniques 52, 2105–2122 (2004)
7. Fontana, R., Richley, E., Marzullo, A., Beard, L., Mulloy, R., Knight, E.: An ultra wideband radar for micro air vehicles air vehicle applications. In: IEEE Conference on Ultra Wideband Systems and Technologies (2002)
8. Lingelbach, F.: Path Planning using Probabilistic Cell Decomposition. PhD thesis (2005)
9. Pettersson, P.: Sampling-based Path Planning for an Autonomous Helicopter. PhD thesis, Linkoping Institute of Technology at Linkoping University (2006)
10. Sinopoli, B., Micheli, M., Donato, G., Koo, J.: Vision based navigation for an unmanned air vehicle. In: Proceedings of the IEEE International Conference on Robotics and Automation, pp. 1757–1765 (2001)
11. Frew, E., Langelaan, J., Joo, S.: Adaptive receding horizon control for vision based navigation of small unmanned aircraft. In: Proceedings of the 2006 American Control Conference (2006)
12. Woll, J.: Vorad collision warning radar, pp. 369–372 (1995)
13. Brigham, E.: The Fast Fourier Transform. Prentice-Hall Inc., Englewood Cliffs (1974)
14. Sarabandi, K., Park, M.: A radar cross-section model for power lines at millimeter-wave frequencies. IEEE Transactions on Antennas and Propagation 51, 2353–2360 (2003)
15. Yonemoto, N., Yamamotoa, K., Yamadaa, K., Yasuib, H., Tanakab, N., Migliaccioc, C., Dauvignacc, J., Pichotc, C.: Performance of obstacle detection and collision warning system for civil helicopters. Proc. SPIE 6226 (2006)

# Combined Visual and Inertial Navigation for an Unmanned Aerial Vehicle

Jonathan Kelly, Srikanth Saripalli, and Gaurav S. Sukhatme

Robotic Embedded Systems Laboratory
University of Southern California
Los Angeles, California, USA 90089-0781
{jonathsk,srik,gaurav}@usc.edu

**Summary.** We describe an UAV navigation system which combines stereo visual odometry with inertial measurements from an IMU. Our approach fuses the motion estimates from both sensors in an extended Kalman filter to determine vehicle position and attitude. We present results using data from a robotic helicopter, in which the visual and inertial system produced a final position estimate within 1% of the measured GPS position, over a flight distance of more than 400 meters. Our results show that the combination of visual and inertial sensing reduced overall positioning error by nearly an order of magnitude compared to visual odometry alone.

## 1 Introduction

Unmanned aerial vehicles (UAVs) typically depend on GPS to provide absolute positioning information for navigation. There are situations, however, in which GPS may be unavailable, for example when flying through urban canyons. In this paper, we describe a passive system that allows an UAV to navigate reliably without GPS or other wide-area positioning information. The system combines stereo *visual odometry* with measurements from an inertial measurement unit (IMU) to continuously estimate the pose (position and attitude) of the vehicle.

For the work presented here, we use a robotic helicopter as our UAV platform. Helicopters are able to rapidly change all of their 6-DOF pose parameters simultaneously, making this a particularly challenging estimation problem. Our solution is designed to take advantage of complimentary IMU and camera sensor characteristics. The IMU is able to accurately measure rapid changes in angular rotation rates and linear accelerations, but is subject to unbounded low-frequency drift. Visual motion estimates, in contrast, are generally more accurate when the cameras' field of view changes relatively slowly. By fusing their output, each sensor is able to compensate for the weaknesses inherent in the other.

We formulate visual odometry as a maximum-likelihood estimation (MLE) problem, where point landmarks are tracked across sequential stereo image pairs; the changes in triangulated landmark positions are used to determine robot ego-motion. These visual measurements are then optimally combined with data from the on-board IMU in an extended Kalman filter (EKF). Although the resulting trajectory estimates are produced by integration, we demonstrate that

C. Laugier and R. Siegwart (Eds.): Field and Service Robotics, STAR 42, pp. 255–264, 2008.
springerlink.com 

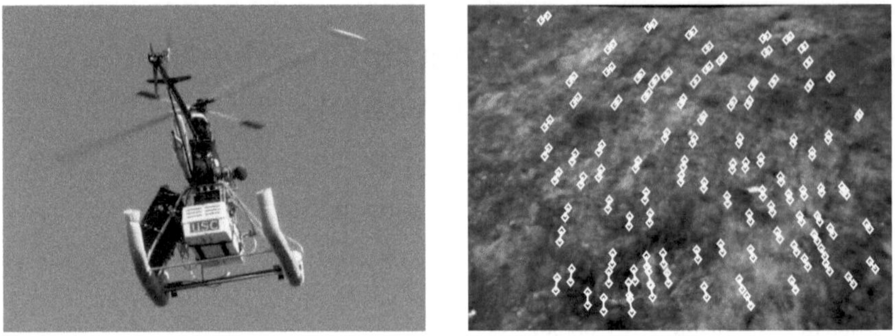

(a) (b)

**Fig. 1.** (a) Helicopter during a data acquisition flight. The avionics box (white) is located beneath the main helicopter chassis; the vision computer (black) is mounted on left-hand side of the vehicle. Both stereo cameras are also visible near the front of the landing skids. (b) Two-frame feature tracks overlaid on a left stereo image.

an absolute positioning accuracy to within 1% of the measured GPS position can be achieved, over flight distances of more than 400 meters. As a percentage, this difference is almost an order of magnitude lower than the result using visual odometry alone.

## 2 Related Work

Visual odometry (VO) describes the process of incrementally estimating changes in robot pose by identifying and tracking visual landmarks in the environment. Many VO implementations use stereo cameras, as stereo allows the depth of landmarks to be calculated directly from known camera geometry. However, monocular, omnidirectional [1] and spherical cameras [2] have been employed as well. In particular, Nistér et al. have developed a VO system that operates in real-time with either monocular or stereo imagery [3]. VO is also used on-board the NASA Mars Exploration Rover (MER) robots [4] – the rovers rely on visual estimates during periods when wheel odometry is unreliable, such as when driving over high-slip terrain.

Examples of combined visual and inertial sensing include a system described by Roumeliotis et al. in [5], designed to safely and precisely land a spacecraft on the surface of a planetary body. An EKF is used to fuse monocular camera, laser altimeter, and IMU data. The laser altimeter measurements provide an absolute scale factor for the single-camera motion estimates. Unlike [5], we determine scale and depth directly from stereo correspondences.

In [6], Amidi et al. present a visual odometer designed specifically for an autonomous helicopter. They estimate vehicle attitude using gyroscopes and vehicle position by tracking ground targets with stereo cameras. To reduce computational demands, the targets are assumed to be planar and the camera view is assumed to change slowly with time. Our approach allows for operation over varied terrain, as long as the captured images have sufficient texture.

Finally, the work discussed here is closely related to aerial simultaneous localization and mapping (see, for example, [7]), although we are primarily concerned with point-to-point navigation, and so do not maintain a map of landmark positions.

## 3 Helicopter Platform and Vision System

AVATAR (Autonomous Vehicle Aerial Tracking and Reconnaissance), our experimental testbed, is a robotic helicopter built on a modified Bergen RC chassis. The helicopter is equipped with a suite of avionics hardware including a Novatel RT-2 GPS receiver, Inertial Science ISIS inertial measurement unit, PNI TCM2-50 electronic compass, and a PC-104 Linux computer for autonomous control. Our ISIS IMU has roll, pitch and yaw rate gyros and three independent single-axis accelerometers. A full description of the control system is given in [8].

For our visual navigation studies, the helicopter carries an additional Mini-ITX Linux computer and two color FireWire cameras from Videre Design. Each camera has a resolution of 640×480 pixels and a field of view of 50 degrees horizontally. The cameras are mounted on a stereo bench with a 50 cm baseline.

## 4 Visual Odometry

Our stereo visual odometry algorithm is based on the approach described by Matthies and Shafer in [9] and refined by Matthies in [10]. We track point landmarks across sequential stereo image pairs, and find the incremental change in camera pose by aligning corresponding sets of triangulated landmark positions. The steps in the algorithm are described in more detail below.

### 4.1 Feature Tracking and Stereo Triangulation

Given an initial stereo pair, the first step in the motion estimation process is to identify landmarks that can be tracked reliably. We use the KLT algorithm [11] to find salient feature points in the left stereo image. For each selected left image point, we then search for a corresponding point in the right image using normalized cross-correlation with a 15-pixel square correlation window. Our stereo cameras are accurately calibrated before each flight, allowing us to limit this correspondence search to a narrow region centered on the right epipolar line. If the correlation score for the best match in the right image is below a fixed threshold (0.75 in our implementation), the point pair is not included in further processing. For matches above the threshold, we determine the sub-pixel disparity by fitting a biquadratic polynomial to the correlation values in a 3×3 region around the integer correlation peak.

The 3D positions of the landmarks are found by stereo triangulation. To model error in the position estimates, we consider the image coordinates as point vectors corrupted by zero-mean, white Gaussian noise. The noise covariance is approximated directly from the curvature of the biquadratic polynomial used for the sub-pixel disparity calculation. A 3D covariance ellipsoid is computed for each landmark using standard Gaussian error propagation [4].

As each new stereo pair is acquired, the KLT tracker updates the 2D positions of all tracked landmarks in the most recent left camera image. The 3D positions of the landmarks are then re-triangulated with respect to the current stereo bench coordinate frame. When a landmark is no longer visible in either the left or right camera image, it is replaced by a newly-initialized landmark, in order to track an approximately constant number of landmarks over time. In our present implementation, we attempt to track 200 landmarks from frame to frame.

### 4.2 Robust Visual Motion Estimation

At each time step, the triangulation procedure above yields two sets of corresponding 3D landmark observations, before and after the helicopter has undergone an unknown rotation $R$ and translation $T$. The relationship between each pair of observations is:

$$P_{i,c} = RP_{i,p} + T + e_i \tag{1}$$

where $P = [p_x, p_y, p_z]^T$ is a 3D position vector, the first subscript indexes the specific landmark, and the second subscript denotes whether the observation was made at the current or previous time step. The vector $e_i$ is a zero-mean, white Gaussian noise term that models the combined errors in both the current and previous position estimates. There is one such equation for each observed landmark.

To solve (1), we linearize using the first-order Taylor expansion of the rotation matrix $R$ with respect to roll, pitch and yaw angles, $\Theta = [\alpha, \beta, \gamma]^T$:

$$P_{i,c} \approx R_0 P_{i,p} + J_{i,p}(\Theta - \Theta_0) + T + e_i \tag{2}$$

Here, $J_{i,p}$ is the Jacobian for landmark $i$ with respect to $\Theta$, evaluated at an initial rotation $\Theta_0$. The noise vector $e_i$ has covariance $\Sigma_i = \Sigma_{i,c} + R_0 \Sigma_{i,p} R_0$, where the point covariance matrices $\Sigma_{i,c}$, $\Sigma_{i,p}$ are determined during the stereo triangulation step. Using our Gaussian error model, the MLE values for $\Theta$ and $T$ are found by minimizing the objective function:

$$M(\Theta, T) = \sum_i r_i W_i r_i^T \tag{3}$$

$$r_i = P_{i,c} - R_0 P_{i,p} - J_{i,p}(\Theta - \Theta_0) - T \tag{4}$$

where $W_i$ is the inverse covariance matrix for $e_i$. After differentiating (3) with respect to $\Theta$ and $T$ and setting the result to zero, we obtain:

$$\left[\sum_{i=0}^{n} H_i^T W_i H_i\right] \begin{bmatrix} \Theta \\ T \end{bmatrix} = \left[\sum_{i=0}^{n} H_i^T W_i L_i\right] \tag{5}$$

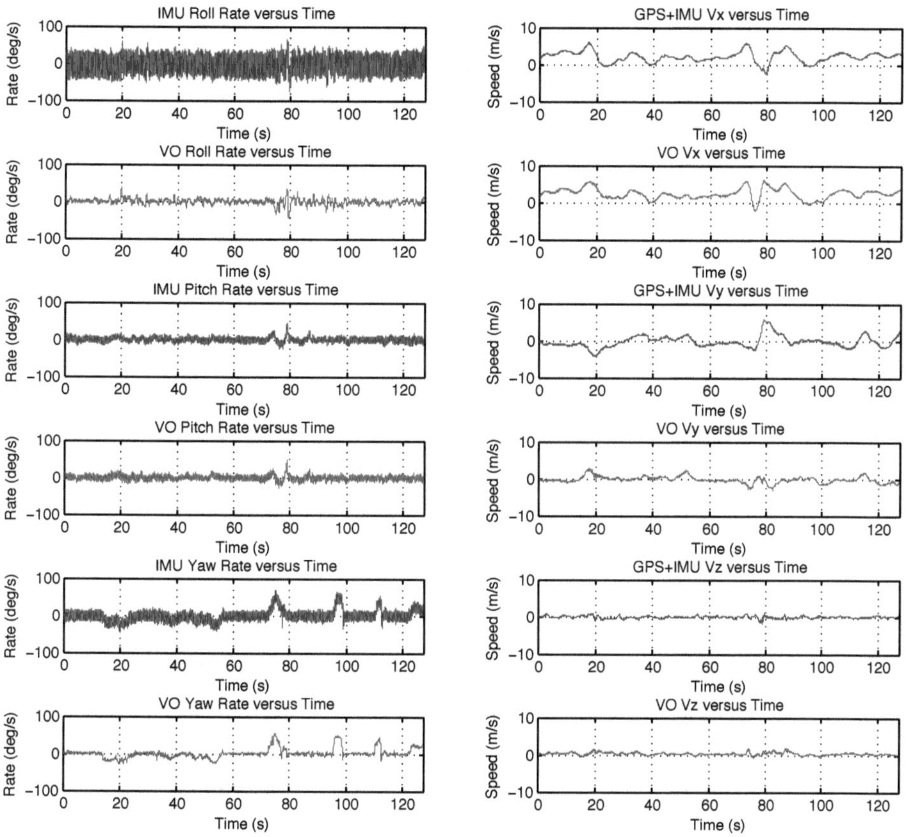


**Fig. 2.** Comparison of angular rates measured by the IMU and by VO (left column), and linear velocities measured by GPS+IMU and by VO (right column). All values are expressed in the helicopter body coordinate frame.

with $H = [J_{i,p}\ I]$ and $L_i = P_{i,c} - R_0 P_{i,p} + J_{i,p}\Theta_0$; a detailed derivation of this result can be found in [10].

The final motion estimate is obtained by iteratively computing (5), using the previous estimate of $R$ the as the value of $R_0$ for the following iteration. The covariance matrix for the result is:

$$\Sigma_{\Theta,T} = \left[ \sum_i H_i^T W_i H_i \right]^{-1} \tag{6}$$

To ensure that the estimation algorithm is robust against incorrect data associations and tracking errors, we embed the computation of (5) in a RANSAC procedure [12].

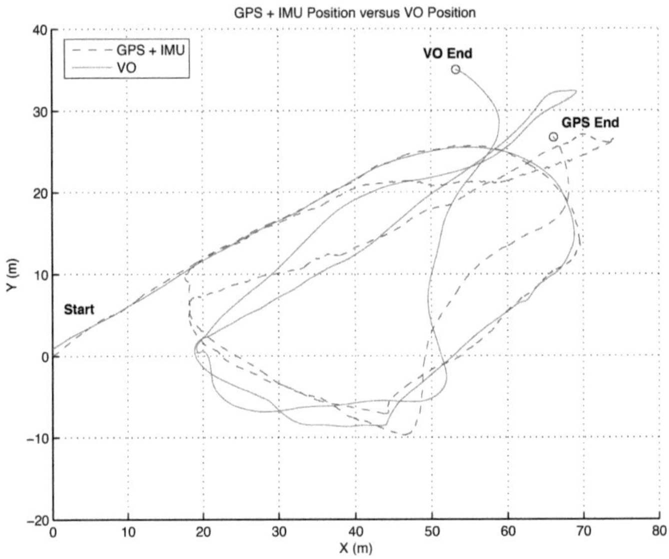


**Fig. 3.** Filtered GPS+IMU position estimate versus VO position estimate. The VO trajectory exhibits a characteristic misalignment due to the integration of small orientation errors over time. The difference between the final GPS+IMU position and the final VO position is 14.7 meters (after a 405.5 meter flight).

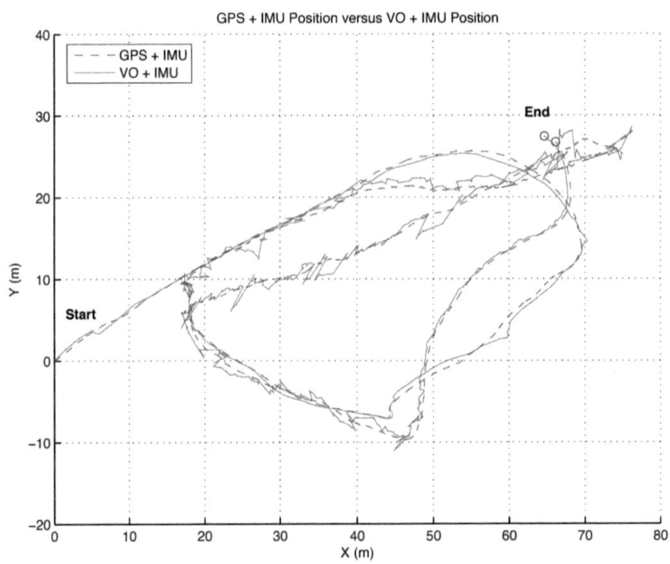


**Fig. 4.** Filtered GPS+IMU position estimate versus filtered VO+IMU position estimate. A single end label is shown, as the difference between the final GPS+IMU position and the final VO+IMU position is only 1.6 meters (after a 405.5 meter flight).

## 5 Extended Kalman Filter-Based State Estimation

Visual and inertial measurements are fused in an extended Kalman filter (EKF) to produce an estimate of the vehicle state. We use the continuous-discrete formulation of the EKF, in which the state estimate is propagated according to the underlying continuous-time non-linear system dynamics, while measurement updates are made at discrete time steps. Our state vector includes the position of the helicopter in the global NED frame, the velocity of the helicopter in the body frame, the attitude of the helicopter, gyroscope biases, accelerometer biases and the magnitude of the gravity vector. Further details on the EKF implementation are available in [13].

The filter process model is driven by the IMU linear accelerations and angular velocities, which substitute for control inputs in the system dynamics equations. Our use of the continuous-discrete form of the EKF is particularly important in this case, since measurements from the IMU arrive at approximately 100 Hz, at least three times faster than measurements from the vision system. We incorporate the relative (frame-to-frame) visual motion estimates into the filter as linear and angular velocities, after transforming from the stereo bench coordinate frame to the helicopter body coordinate frame.

## 6 Experiments

To determine the performance of the proposed navigation system, we ran several experiments with the AVATAR helicopter at our test site in Downey, California. During each experiment, the helicopter was flown manually while we logged data from the on-board GPS receiver, IMU and the stereo cameras. The data were analyzed offline.

For the flight trials, our stereo cameras were mounted above the landing skids at the front of the vehicle, with the camera optical axes pointed at an angle of 60 degrees down from horizontal. We selected this oblique pointing angle, instead of a nadir pointing angle, so that we could perform several additional tracking and obstacle detection experiments using the same data. Images were captured from both cameras at 30 frames per second.

In the remainder of the paper we present results from our longest experiment, with a duration of 127.4 seconds and a ground track of 405.5 meters, as measured by the GPS receiver. The helicopter had a maximum forward velocity of 6.0 meters per second and a maximum yaw rate of 68 degrees per second. Vehicle altitude varied from approximately three meters to nine meters above ground level. Our shorter experiments followed similar flight profiles and produced comparable results (in terms of accuracy) to those described below.

Over the course of the longest flight a total of 3822 stereo image pairs were acquired. Figure 1b shows an image from the left stereo camera taken at $t = 74$ seconds, when the helicopter was five meters above the terrain surface. The KLT tracker was able to track an average of 163.6 features (out of 200 selected features) across sequential left image frames. After stereo triangulation and outlier removal, between 27 and 102 landmarks were used to compute the incremental visual motion estimates.

**Table 1.** Comparison of position estimates produced by GPS+IMU, VO alone and VO+IMU. Total flight distance, as measured by GPS, was 405.5 meters in $x$, $y$.

| Sensors Used | Final Position $(x, y)$ [m] | Final Position Error [m] | Average Position Error [m] | Maximum Position Error [m] |
|---|---|---|---|---|
| GPS+IMU | (66.13, 26.72) | – | – | – |
| VO alone | (53.39, 34.12) | 15. 14 | 4.66 | 15. 14 |
| VO+IMU | (64.63, 27.42) | 1. 65 | 1.27 | 5. 58 |

## 7 Discussion

We evaluated the accuracy of the system by comparing the output from the Kalman filter using GPS and IMU measurements with the integrated output from VO alone, and with the output from the Kalman filter using both VO and IMU measurements. This comparison is summarized in Table 1. In the text below, we use the terms 'GPS+IMU' to denote estimates produced using filtered GPS and IMU data, 'VO' to denote estimates produced using visual odometry alone, and 'VO+IMU' to denote estimates produced using filtered visual odometry and IMU data. Since we expect the GPS position estimate to be the most accurate over longer time intervals, we used the GPS+IMU flight path as our reference trajectory. Before plotting the ground tracks in Figures 3 and 4, we performed a least squares alignment of the first 30 frames (one second) of VO and VO+IMU data with the filtered GPS+IMU output.

Plots of the three-axis VO and IMU angular rate measurements are shown in the left column of Figure 2. There is strong agreement between the IMU pitch and yaw rates and the VO pitch and yaw rates, confirming that both sensors can correctly measure changes in vehicle orientation[1]. The measured GPS+IMU and VO linear velocities are plotted in the right column of Figure 2. In this case, the $x$-axis and $z$-axis velocities measured by the GPS+IMU coincide approximately with the velocities measured by VO. However, there are portions of $y$-axis velocity curves that do not correlate well. A preliminary analysis indicates that the discrepancies may be due to poor VO output; over the regions in which the GPS+IMU and VO values differ, the number of landmarks used to compute the visual motion estimate dropped below the average for the flight. We are currently investigating this issue.

The final position estimate computed using VO has an error of 3.7% relative to the GPS+IMU position. This is comparable to previous visual odometry results for ground robots [3], although ground robots usually move more slowly than the helicopter. The final position estimate computed using filtered VO+IMU data differs by only 0.4% from the GPS+IMU position – this difference is almost an order of magnitude less, as a percentage. More importantly, although the

[1] The roll gyro on our IMU is particularly noisy, and so we did not compare the roll rates directly.

VO+IMU position estimate is noisy, it tracks the GPS+IMU estimate over the entire flight. The deviation in the VO position increases with time.

An important aspect of the VO algorithm is that the quality of the output depends upon accurate stereo triangulation. Our imagery was acquired at an average height of approximately 6 meters above the terrain surface, producing consistent triangulated landmark positions. However, we have found that above a height of 12 to 15 meters, it becomes very difficult to select a large set of inlying landmark points, although the majority of the image features are tracked reliably. The reason is that, as vehicle altitude increases and the left-right stereo disparity decreases, small errors in the disparity estimates lead to large changes in the triangulated landmark depths relative to the stereo bench. This, in turn, means that corresponding landmarks points often cannot be aligned along the depth axis without significantly increasing the error tolerance in the RANSAC procedure, which degrades the quality of the overall motion estimate.

## 8 Conclusions and Future Work

This paper presented a visual and inertial system that allows an UAV to navigate over significant distances without GPS or other wide-area positioning information. Fusing inertial measurements with visual odometry was shown to reduce positioning error by almost an order of magnitude compared to using visual odometry alone. We expect that with further analysis of the visual odometry output and tuning of the Kalman filter, we should be able to improve the system's overall performance for longer trajectories. Additionally, we are working to implement a version of the software which operates in real time on-board the helicopter.

## Acknowledgments

This work was funded in part by the National Science Foundation under grants CNS-0325875, CNS-0540420, CNS-0520305 and CCF-0120778 and by a gift from the Okawa Foundation. Jonathan Kelly was supported by an NSERC postgraduate scholarship from the Government of Canada. We would like to thank our safety pilot Alan Butler for his assistance with the AVATAR flight trials.

## References

1. Corke, P.I., Strelow, D., Singh, S.: Omnidirectional visual odometry for a planetary rover. In: Proc. IEEE/RSJ Int'l Conf. Intelligent Robots and Systems (IROS 2004), vol. 4, pp. 4007–4012 (September 2004)
2. Levin, A., Szeliski, R.: Visual odometry and map correlation. In: Proc. IEEE Computer Soc. Conf. Computer Vision and Pattern Recognition (CVPR 2004), vol. 1, pp. 611–618 (2004)
3. Nistér, D., Naroditsky, O., Bergen, J.: Visual odometry. In: Proc. IEEE Computer Soc. Conf. Computer Vision and Pattern Recognition (CVPR 2004), vol. 1, pp. 652–659 (June 2004)

4. Cheng, Y., Maimone, M.W., Matthies, L.: Visual odometry on the Mars exploration rovers. In: Proc. IEEE Int'l Conf. Systems, Man and Cybernetics, Big Island, USA, vol. 1, pp. 903–910 (October 2005)
5. Roumeliotis, S.I., Johnson, A.E., Montgomery, J.F.: Augmenting inertial navigation with image-based motion estimation. In: Proc. IEEE Int'l Conf. Robotics and Automation (ICRA 2002), Washington D.C., USA, vol. 4, pp. 4326–4333 (May 2002)
6. Amidi, O., Kanade, T., Fujita, K.: A visual odometer for autonomous helicopter flight. J. Robotics and Autonomous Systems 28, 185–193 (1999)
7. Langelaan, J., Rock, S.: Passive GPS-free navigation for small UAVs. In: Proc. IEEE Aerospace Conf., Big Sky, USA, pp. 1–9 (March 2005)
8. Saripalli, S., Montgomery, J.F., Sukhatme, G.S.: Visually-guided landing of an unmanned aerial vehicle. IEEE Trans. Robotics and Automation 19(3), 371–381 (2003)
9. Matthies, L., Shafer, S.: Error modeling in stereo navigation. IEEE J. Robotics and Automation RA-3(3), 239–250 (1987)
10. Matthies, L.: Dyamic stereo vision. Ph.D. dissertation, Carnegie Mellon University (October 1989)
11. Shi, J., Tomasi, C.: Good features to track. In: Proc. IEEE Conf. Computer Vision and Pattern Recognition (CVPR 1994), Seattle, USA, pp. 593–600 (June 1994)
12. Fischler, M.A., Bolles, R.C.: Random sample consensus: a paradigm for model fitting with applications to image analysis and automated cartography. Comm. ACM 24(6), 381–395 (1981)
13. Saripalli, S., Roberts, J.M., Corke, P.I., Buskey, G., Sukhatme, G.S.: A tale of two helicopters. In: Proc. IEEE/RSJ Int'l Conf. Intelligent Robots and Systems (IROS 2003), Las Vegas, USA, vol. 1, pp. 805–810 (October 2003)

# Pedestrian Shape Extraction by Means of Active Contours

Massimo Bertozzi[1], Alberto Broggi[1], Stefano Ghidoni[1], and Michael Del Rose[2]

[1] Dip. Ing. Informazione, Parma, Italy
{bertozzi,broggi,ghidoni}@ce.unipr.it
[2] U.S.Army TARDEC, Warren, MI, U.S.A.
mike.delrose@us.army.mil

**Summary.** This article presents a shape extraction and results of a preliminary validation stage for a pedestrian detection system based on the use of active contours. The complete system is based on the use of both far infrared and visible cameras to detect areas that potentially contain pedestrians; in order to validate and filter such result a refinement of the human shape by means of active contours is performed followed by a neural network based filtering.

## 1 Introduction

During the last years, pedestrian detection has been a key topic of the research on intelligent vehicles. This is due to the many applications of this functionality, like driver assistance, surveillance, or automatic driving systems; moreover, the heavy investments made by almost all car manufacturers on this kind of research proves that particular attention is now focused on improving road safety, especially for reducing the high number of pedestrians being injured every year. Also the U.S. Army is actively developing systems for obstacle detection, path following and anti-tamper surveillance, for its robotic fleet [9, 11].

Finding pedestrians from a moving vehicle is, however, one of the most challenging tasks in the artificial vision field, since a pedestrian is one of the most *deformable* object that can appear in a scene. Moreover, the automotive environment is often barely unstructured, incredibly variable, and apparently moving, due to the fact that the camera itself is in motion; therefore, really few assumptions can be made on the scene.

This paper describes a module for human shape filtering developed for integration into a vision-based obstacle detection system to be installed on an autonomous military vehicle. This system is able to detect all obstacles appearing in the scene and is based on the simultaneous use of two stereo camera systems: two far infrared cameras and two daylight cameras [4]. The first stages of this system provide a reliable detection of image areas that potentially contain pedestrians; following stages are devoted to refine and filter this rough results to validate the pedestrians presence. The filtering is based on a multi-vote system; several approaches are independently used to analyze areas of attention, and

C. Laugier and R. Siegwart (Eds.): Field and Service Robotics, STAR 42, pp. 265–274, 2008.
springerlink.com 

each subsystem outputs a vote describing how much the obstacle is likely to be a pedestrian. Then, a final validation is done, based on all votes.

This paper describes the subsystem developed for extracting the object shape, in order to provide a vote based on it. However, the evaluation of the object shape is currently under development, therefore only results on shape extraction are detailed.

This paper is organized as follows: section 2 describes related work in pedestrian detection systems based on artificial vision. The pedestrian detection system is discussed in 3 and the module for active contours-based shape detection algorithm is detailed in section 4. Finally, section 5 ends the paper presenting few results and remarks on the system.

## 2 Related Work

For the U.S. Army the use of vision as a primary sensor for the detection of human shapes is a natural choice since cameras are non-invasive sensors and therefore do not emit signals.

Vision-based systems for pedestrian detection have been developed exploiting different approaches; like the use of monocular [16, 20] or stereo [17, 6] vision. Many systems based on the use of a stationary camera employ simple segmentation techniques to obtain foreground region; but this approach fails when the pedestrians has to be detected from moving platforms. Most of the current approaches for pedestrian detection using moving cameras treats the problem as a recognition task: a foreground detection is followed by a recognition step to verify the presence of a pedestrian. Some systems use motion detection [15,6] or stereo analysis [21] as a means of segmentation.

Other systems, substitute the segmentation step with a focus-of-attention approach, where salient regions in feature maps are considered as candidates for pedestrians. In the GOLD system [1], vertical symmetries are associated with potential pedestrians. In [5] the local image entropy directs the focus-of-attention followed by a model-matching module.

For what concerns the recognition phase, recent research are often motion-based, shape-based or multi-cue based. Motion based approaches use the periodicity of human gait or gait patterns for pedestrian detection [6, 5]. These approaches seems to be more reliable than shape based ones, but they require temporal information and are unable to correctly classify pedestrians that are still or have an unusual gait pattern.

Shape-base approaches are based on pedestrians' appearance, so can detect both moving and stationary people [1,3]. In these cases, the challenge is to model the several variations in the shapes, pose, size and appearance of humans and their backgrounds. Basic shape analysis methods consist in matching a template onto candidate foreground regions. In [10] a tree-based hierarchy of human silhouettes is constructed and the matching follows a coarse-to-fine approach. In [13, 18] probabilistic templates are used to take in account the possible variations in human shape. To accomplish the recognition step, other systems systems exploit pattern recognition based on the use of classifiers or in combination with a shape analysis with gait detection [10, 14].

## 3 System Description

The active contours-based algorithm described in this work has been developed as a part of a tetra-vision based pedestrian system [2,4]. The whole architecture is based on the simultaneous use of two far infrared and two daylight cameras. Thanks to this approach, the system is able to detect obstacles and pedestrians when the use of infrared devices is more appropriate (night, low illumination conditions...) or, conversely, in the case visible cameras are more suitable for the detection (hot, sunny environments...).

In fact, FIR images convey a type of information that is very different from those in the visible spectrum. In the infrared domain the image of an object depends on the amount of heat it emits, namely, it is generally related to its temperature (see fig. 1). Conversely, in the visible domain, objects appearance depends on how the surface of the object reflects the incident light as well as on the illumination conditions.

Since humans usually emit more heat than other objects like trees, background, or road artifacts, the thermal shape can be often successfully exploited for pedestrian detection. In such cases, pedestrians are in fact brighter than the background. Unfortunately, other road participants or artifacts emit heat as well (cars, heated buildings...). Moreover, infrared images are blurred, have a poor resolution and the contrast is low compared with rich and colorful visible images.

Consequently, both visible and far infrared images are used for reducing the search space.

Figure 2 depicts the overall algorithm flow for the complete pedestrian system. Different approaches have been developed for the initial detection in the two image domains: warm areas detection, vertical edges detection, and an approach based on the simultaneous computation of disparity space images in the two domains [2,4].

These first stages of detection output a list of areas of attention in which pedestrians can be potentially detected. Each area of attention is labelled using a bounding box. A symmetry-based approach is further used to refine this rough result in order to resize bounding boxes or to separate bounding boxes that can contain more pedestrians.

These two steps in the processing, barely take in account specific features of pedestrians; in fact, only symmetrical and size considerations are used to compute the list of bounding boxes. Therefore, independent validators are used to evaluate the presence of human shapes inside the bounding boxes. These stages exploits specific pedestrian characteristics to discard false positives from the list of bounding boxes.

A final decision step is used to balance the votes of validators for each bounding box.

**Fig. 1.** Examples of typical scenarios in FIR and visible images

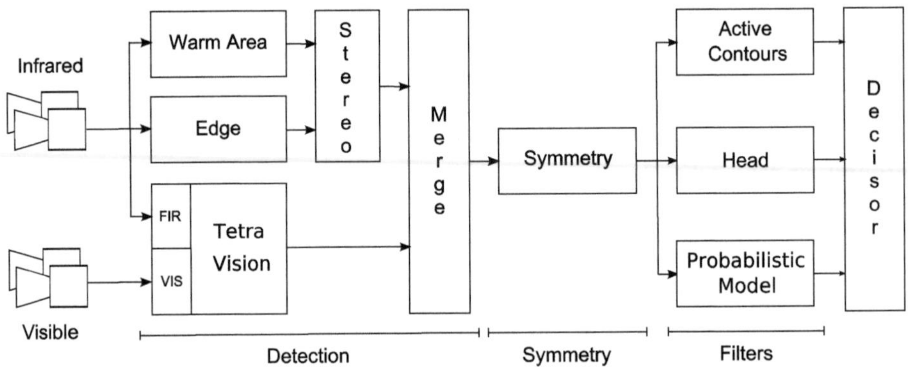


**Fig. 2.** Overall algorithm flow

# 4 Shape Extraction

As previously discussed, the pedestrian validation step is composed by several validators, each one supplying a vote, that is then provided to the final evaluation step. The validator detailed in this section is based on the analysis of a pedestrian shape, which can be extracted using the well-known active contour models, also known as snakes.

## 4.1 Active Contour Models

Active contour models are widely used in pattern recognition for extracting an object shape.

First introduced by Kass et al. [12], this topic has been extensively explored also in the last years. Basically, a snake is a curve described by the parametric equation $\mathbf{v}(s) = (x(s), y(s))$, where $s$ is the normalized length, assuming values in the range $[0, 1]$. This continuous curve becomes, in a discrete domain, a set of points, that are pushed by some energies that depend on the specific problem being addressed. Every point composing the snake reaches a local energy minimum; this means that the active contour does not find a global optimum position; rather, since it is based on local minimization, the final position strongly depends on the initial condition, i.e. the initial snake position.

Because initial stages of the pedestrian detection system provide a bounding box for each detected object, the snake initial position can be chosen as the bounding box contour; then, a contracting behavior should be impressed, to force the snake to move inside the bounding box. Other energies must also be introduced to make the snake stop when the object contour is reached.

Forces, and associated energies, can be divided into two different categories: internal and external. Internal energy only depends on the topology of the snake, and controls the continuity of the curve derivatives; it is evaluated by the equation:

$$E_{int} = \alpha(s)|\mathbf{v}_s(s)|^2 + \beta(s)|\mathbf{v}_{ss}(s)|^2 \tag{1}$$

where $\mathbf{v}_s(s)$ and $\mathbf{v}_{ss}(s)$ are, respectively, the first and second derivatives of $\mathbf{v}(s)$ with respect to $s$. The first contribution appearing in the sum represents the tension of the snake, that is responsible for the elastic behavior; the second one gives the snake resistance to bending; $\alpha(s)$ and $\beta(s)$ are weights.

Therefore, internal energy controls the snake mechanical properties, but is independent of the image; external energy, on the contrary, causes the snake to be attracted to the desired features, and should therefore be a function of the image.

Analytically, the snake will try to minimize the whole energy balance, given by the equation:

$$E_{snake} = \int_0^1 \left(E_{int}\left(\mathbf{v}(s)\right) + E_{ext}\left(\mathbf{v}(s)\right)\right) \mathrm{d}s \tag{2}$$

Because energies are the only way to control a snake, a proper choice of both internal and external energies should be made. In particular, the external energy depending on the image must decrease in the regions where the snake should be attracted. In the following, the energies adopted to obtain an object shape are described.

As previously said, the initial snake position was chosen to be along the bounding box contour. In this system both visible and far infrared images are available, but the latter seems much more convenient when dealing with pedestrians, due to the thermal difference between a human being and the background [4].

To extract a pedestrian shape, the Sobel filter output is a useful starting point; moreover, the edge image is needed also by previous steps of the recognition algorithm, so it is already available. A Gaussian smoothing filter is then applied to enlarge the edges, and therefore the area capable of influencing the snake behavior. The resulting image is then associated with an energy field that pushes the snake towards the edges: for this reason, the brighter a pixel in that image, the lowest the associated energy; in this way, *snaxels* (the points into which the snake is discretized) are attracted by the strongest edges, see fig. 3.

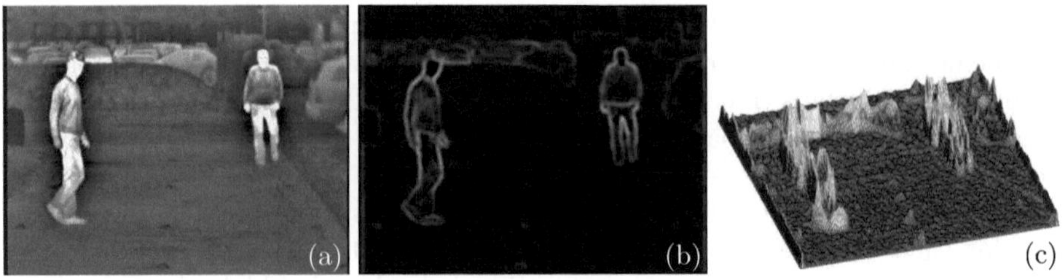

**Fig. 3.** Energy field due to edges: (a) original image, (b) edge image obtained using Sobel operator and gaussian smooth, (c) edge energy functional with inverted sign, to obtain a more effective graphical representation

Bright regions of the original FIR image are also considered. In fact, smoothed edges does not accurately define the object contour (mainly because they are smoothed), so snake contraction has to be arrested by bright regions in the FIR image, that can belong to a portion of a human body (see fig. 4). This method lets the snake correctly adapt to a body shape in a lot of situations, and it should also be noticed that this mechanism works only if there are hot regions inside the bounding box; a useful side effect, then, is an excessive snake contraction when there are not warm blobs inside a bounding box

**Fig. 4.** Energy field due to the image: (a) original image, (b) intensity energy functional with inverted sign, to obtain a more effective graphical representation

The minimum energy location is found by iteratively moving each snaxel, following an *energy minimization algorithm*. Many of them were proposed in the literature. For this application, the greedy snake algorithm [19], applied on $5 \times 5$ neighborhood, was adopted.

During the initial iterations, the snake tends to contract, due to the elastic energy; this tendency stops when some other energy counterweights it: for instance, the presence of edges, or a light image region. While adapting to the object shape, the snake length decreases, as well as the mean distance between two adjacent snaxels. Since this mean distance is a value that affects the internal energy, to keep almost constant the elastic property also during strong contraction, the snake is periodically resampled using a fixed step; in this way some unwanted snaxels accumulation can be avoided.

## 5 Discussion

The developed system has been tested in different situations using an experimental vehicle equipped with the tetra-vision system. As previously said, the contour extraction technique has been developed for pedestrian validation; this means that the snake has to be further processed in order to obtain a vote, say, the probability that it contains a pedestrian. This part of the system is still under development; however, some work has already been done: the first idea was to exploit a shape-matching algorithm, like those described in [7,8], to compare the final shape of the snake with a pre-built database of typical pedestrian shapes. This technique turned out to be not very effective, because of the high number of the false positives it generates. In fact, pedestrians can be seen in a wide variety of poses and postures, so that it is difficult to describe what a *pedestrian shape* is; for this reason, a high number of different shapes have to be considered, therefore causing a lot of false positives.

The second approach chosen for pedestrian classification is based on a neural network, an approach that gives good results when the problem description turns out to be complex; some encouraging results have already been achieved, but a lot of testing is still needed in order to evaluate precise recognition statistics. Preliminary results, however, show that shape analysis based on a neural network has a particularly low false positive rate. This validator, then, will become one of the voting modules described in paragraph 3.

Some shape extraction results are now presented. In fig. 5 some example of the contraction mechanism are reported: the white lines are the snakes in the initial position, that is, on the bounding box contour, while the snakes after energy minimization are drawn in yellow. Some examples are presented for a close pedestrian, fig. 5.a, and for a distant pedestrian and a scooter, fig. 5.b. Some shape extraction issues are also presented: in fig. 5.c, pedestrians shapes are correctly found, but the contour of the pedestrian in the middle also includes a warm background object; in fig. 5.d the importance of the initial snake position is highlighted: the head is not detected because it is outside of the initial snake position (in white).

Some shape extraction results are presented when the FIR images are not optimal, like those acquired in summer, under heavy direct sunlight; in this condition, many object in the background become warm, and the assumption that a pedestrian has a higher temperature than the background is not satisfied. This causes some errors in the contraction process, so that the snake in the final position does not completely adhere to the pedestrian contour, but also include some background details (fig. 5.e); in fig. 5.f, some body parts (the legs in this case) are not included in the extracted shape, because they are colder than the background. In both cases, however, the pedestrian shape is acceptably extracted.

The above discussion focused on how snakes are able to adapt to pedestrians' shape. Looking at the images, a qualitative evaluation of the shape extraction algorithm can be performed: results seem reasonable, also in bad working conditions. Since the snake-based voting system is still under development, only

**Fig. 5.** Results: in (a) and (b), shape extraction of a close and distant pedestrian, respectively; the white snake represents the initial position, while the yellow one is the final configuration. In (c), a group of pedestrians is analyzed; in (d), a typical issue connected with a wrong initial snake disposition is shown: the head is outside the extracted shape because it was also outside the bounding box. In (e) and (f) some results in a difficult working condition are presented, i.e. during summer, when a lot of background objects appear bright, due to the high temperature.

preliminary quantitative results can be provided: an initial release of the neural network classificator has been tested on about 1000 frames: 73 % of all pedestrians were correctly classified, as well as 91 % of non pedestrians. These preliminary results seem promising, but the net still needs some tuning, especially to improve correct classification rate.

## Acknowledgment

This work has been supported by the European Research Office of the U. S. Army under contract number N62558-05-P-0380.

## References

1. Bertozzi, M., Broggi, A., Fascioli, A., Sechi, M.: Shape-based Pedestrian Detection. In: Procs. IEEE Intelligent Vehicles Symposium 2000, Detroit, USA, October. 2000, pp. 215–220 (2000), doi:10.1109/IVS.2000.898344.
2. Bertozzi, M., Broggi, A., Felisa, M., Vezzoni, G., Del Rose, M.: Low-level Pedestrian Detection by means of Visible and Far Infra-red Tetra-vision. In: Procs. IEEE Intelligent Vehicles Symposium 2006, Tokyo, Japan, June 2006, pp. 231–236 (2006), doi:10.1109/IVS.2006.1689633.
3. Beymer, D., Konolige, K.: Real-time Tracking of Multiple People using Continuous Detection. In: Procs. Intl. Conf. on Computer Vision, Kerkyra (1999)
4. Broggi, A., Bertozzi, M., Felisa, M., Grisleri, P., Ghidoni, S., Vezzoni, G., Hilario Gómez, C., Del Rose, M.: Pedestrian Detection by means of Far-infrared Stereo Vision. Computer Vision and Image Understanding 106(2), 194–204 (2007)
5. Curio, C., Edelbrunner, J., Kalinke, T., Tzomakas, C., von Seelen, W.: Walking Pedestrian Recognition. IEEE Trans. on Intelligent Transportation Systems 1(3), 155–163 (2000)
6. Cutler, R., Davis, L.S.: Robust Real-time Periodic Motion Detection, Analysis and Applications. IEEE Trans. on Pattern Analysis and Machine Intelligence 22(8), 781–796 (2000)
7. Dao, M.-S., G., F., Natale, B.D., Massa, A.: Edge potential functions and genetic algorithms for shape-based image retrieval. In: Procs. IEEE Intl. Conf. on Image Processing (ICIP 2003), Barcelona, Spain, September 2003, vol. 2, pp. 729–732 (2003)
8. Dao, M.-S., Natale, F.G.B.D., Massa, A.: Efficient Shape Matching Using Weighted Edge Potential Function. In: Procs. $13^{th}$ Intl. Conf. on Image Analysis and Processing (ICIAP 2005), Cagliari, Italy (September 2005)
9. Del Rose, M., Frederick, P.: Pedestrian Detection. In: Procs. Intelligent Vehicle Systems Symposium, Traverse City, USA (2005)
10. Gavrila, D.M.: Pedestrian Detection from a Moving Vehicle. In: Procs. of European Conference on Computer Vision, vol. 2, pp. 37–49 (2000)
11. Kania, R., Del Rose, M., Frederick, P.: Autonomous Robotic Following Using Vision Based Techniques. In: Procs. Ground Vehicle Survivability Symposium, Monterey, USA (2005)
12. Kass, M., Witkin, A., Terzopoulos, D.: Snakes: Active Contour Models. Intl. Journal of Computer Vision 1(4), 321–331 (1988)

13. Nanda, H., Davis, L.: Probabilistic Template Based Pedestrian Detection in Infrared Videos. In: Procs. IEEE Intelligent Vehicles Symposium 2002, Paris, France (June 2002)
14. Philomin, V., Duraiswami, R., Davis, L.: Pedestrian Tracking from a Moving Vehicle. In: Procs. IEEE Intelligent Vehicles Symposium 2000, Detroit, USA, October 2000, pp. 350–355 (2000)
15. Polana, R., Nelson, R.C.: Detection and Recognition of Periodic, Non-rigid Motion. Internation Journal of Computer Vision 23(3), 261–282 (1997)
16. Shashua, A., Gdalyahu, Y., Hayun, G.: Pedestrian Detection for Driving Assistance Systems: Single-frame Classification and System level Performance. In: Procs. IEEE Intelligent Vehicles Symposium 2004, Parma, Italy (June 2004)
17. Shimizu, H., Poggie, T.: Direction Estimation of Pedestrian from Multiple Still Images. In: Procs. IEEE Intelligent Vehicles Symposium 2004, Parma, Italy (June 2004)
18. Stauffer, C., Grimson, W.E.L.: Similarity Templates for Detection and Recognition. Procs. IEEE Intl. Conf. on Computer Vision and Pattern Recognition 1, 221–228 (2001)
19. Williams, D.J., Shah, M.: A Fast Algorithm for Active Contours and Curvature Estimation. CVGIP: Image Understanding 55(1), 14–26 (1992)
20. Zhao, L.: Dressed Human Modeling, Detection, and Parts Localization. Ph.D. dissertation, Carnegie Mellon University (2001)
21. Zhao, L., Thorpe, C.: Stereo and neural network-based pedestrian detection. IEEE Trans. on Intelligent Transportation Systems 1(3), 148–154 (2000)

# Part VII

# Teleoperation

# The Small Outdoor Rover MERLIN and Its Assistance System for Tele-operations

Daniel Eck[1], Manuel Stahl[1], and Klaus Schilling[1,2]

[1] Steinbeis Transferzentrum ARS, Würzburg
[2] University Würzburg, Informatik VII: Robotics and Telematics, Am Hubland, D-97074 Würzburg, Germany
schi@informatik.uni-wuerzburg.de

**Summary.** Tele-operated rovers offer huge application potential in the areas of emergency support, surveillance and security, in particular if they are small and easy to transport. In this context, the MERLIN outdoor rovers have been designed for robust operations in harsh outdoor environments at a total mass below 20 kg. The MERLIN rover system design aspects are described, including the rover on-board data processing and the sensor configuration. Special emphasis is on the remote operations assistance system, composed of the human-machine interface for appropriate presentation of relevant data at the tele-operator workplace and the autonomous reaction capabilities coordinated by the on-board data handling system.

## 1 Introduction

Emergencies and harsh environments often demand dangerous operations, where human personal is to be replaced as far as possible by robots [1, 2, 3]. While in the 90ies the emphasis was on large vehicles, carrying sophisticated sensors for autonomous operations, currently the emphasis is rather towards smaller tele-operated vehicles, enabling better access to narrow work environments. In a broad rover application range including space exploration [4], search and rescue [2] and surveillance, this tendency can be observed. Modern miniaturisation techniques in electronics support this trend without reduction in capabilities, in particular for on-board data processing and control systems, as well as for sensorics [5].

The MERLIN rovers (Mobile Experimental Rover for Locomotion and Intelligent Navigation) were developed in the 90ies, as sensor test vehicles for the European Mars rover design effort MIDD by the Steinbeis center ARS [6]. A new element in this rover family is the Outdoor MERLIN in the mass range between 15 and 20 kg. Therefore, this contribution addresses the system design of small agile mobile robots for tele-operations.

In the following chapter the rover system design aspects are addressed with emphasis on the on-board data processing and sensors. The third chapter deals with the implemented tele-operation approaches, including autonomous control functionalities. Experiences from tests to characterize the rover performance are summarized.

C. Laugier and R. Siegwart (Eds.): Field and Service Robotics, STAR 42, pp. 277–286, 2008.
springerlink.com 

## 2 System Design of the MERLIN Rover

The design requirements relate to a payload mass of about 5 kg at a total rover mass of max 20 kg, a maximum velocity of 40 km/h (on road), an operational period of 1 hour, and robust performance in harsh outdoor environments.

**Fig. 1.** The Outdoor MERLIN in drive configuration during the test at the European Land Robotics trial ELROB 2005

To comply with these requirements the MERLIN platform concept has been modified, whereby a new chassis was developed. Furthermore all subsystems of the MERLIN platform concept have been adapted to the new chassis and a new robot, the Outdoor MERLIN, was developed. The Outdoor MERLIN system is composed of four main parts, the mobile robot "'Outdoor MERLIN"', the control station, the radio link and the payload, described in more detail in this section.

### 2.1 Payload

The MERLIN vehicle provides a standardized payload interface system, enabling the quick installation of alternative payloads. The payload interface options include by example a grab arm, a surveillance mechanism or a mine search mechanism. Payloads can be changed fast, in order to adapt the robot to the execution of different tasks. The mounting slot providing the mechanical interface is placed in the central axis of the vehicle to keep the center of mass in a well defined area. The control of the different payloads is realized via a CAN bus. The modular software concept allows quick integration of new payloads. A regulated power supply of 12 V is implemented in two independent lines for electronics and for actuators.

**Fig. 2.** The Outdoor MERLIN with the deployed standard mast configuration, elevating the camera for monitoring purposes to 1 m height for better panoramic view

The standard payload configuration is a camera on top of a robotic arm used as mast. The arm can be moved up to approximately 100 cm elevation. The head is able to change yaw and pitch orientation. It carries two cameras: a surveillance camera with high zoom capabilities and a drive camera with a wide angle objective. So the operator is able to identify objects in the area around the robot, which can be zoomed in for detailed observation. When the operator drives the MERLIN, he uses the drive camera, looking at the environment in front of the MERLIN, including the front wheels, too. Thus the distances to the obstacles can be easily identified, enabling better and faster reactions. Alternative configurations payloads on the mast include also infrared cameras for use during night time.

### 2.2 Control Station

To tele-operate the MERLIN, a portable control station was developed. The station includes a laptop with joystick, radio devices to send and receive data, the antennas and a charger for the robot's battery packs. All components of the station are stored in a box with the size of 40 x 60 x 40 $cm^3$ and a total weight of 21.5 kg. The antenna system consists of a deployable mast and 3 antennas. One bar antenna is for the uplink to the robot. Another bar antenna and a beam antenna are used for the downlink. The Laptop computer hosts the MERLIN Control Software (MRCS) [7] to control the robot and its subsystems. In addition, software for the video and audio treatment is available, too. There is the possibility to operate the control station with a 12 V battery. Therefore it is very mobile and is not relying on the electricity network. An alternative 220 V access option is integrated, too.

### 2.3 Mobile Vehicle

For the energy-efficient use in a broad spectrum of environments, a robust vehicle with four wheels was designed. The four wheel drive system can be operated with wheels of different diameters. The front axle is steerable. The rover has enough ground clearance to safely pass obstacles of at least 10 cm. Maximum speed is up to 40 km/h, but for safety reasons the velocity is limited according to path curvature. The low weight below 15 kg (without payload) supports transportability and enables it to be carried by a single person. It is equipped by an electric motor with a high torque, enabling it to handle significant slopes. The motor provides a power of 1600 watt. The dimensions of the robot are 50 cm length, 40 cm width and 60 cm height (without antenna).

Sensors were chosen to be of maximum robustness and reliability for a broad spectrum of operational environments. Different sensors supply information about the vehicle's status, the position and orientation of the robot, as well as about the environment. Two cameras observe the range before and behind the vehicle. The vehicle is equipped with a basic configuration of

- 4 ultrasonic sensors for obstacle avoidance: 3 looking into drive direction, one placed in the rear.
- 2 infrared distance sensors for obstacle avoidance looking at the right and left side.
- 1 incremental sensor for motor propulsion.
- 1 current sensor for motor current.
- 2 b/w cameras on front and rear.
- 1 gyroscope for yaw rate.
- 2 inclinometers for pitch and roll.
- 1 GPS sensor for global position and absolute orientation.

The sensors were extensively tested to verify the suitability for a broad range of typical environments. A particular challenge are electromagnetic cleanliness requirements within the densely packed volume on-board the rover. Intensive shielding of motors to protect from the induced electromagnetic fields are mandatory.

### 2.4 Radio Link

The communication link consists of two components, the uplink from the control station to the vehicle, transports commands from operator, and the downlink in opposite direction, transports the audio and video signal as well as sensor data. Both directions are using a separate radio transmitter. The range of both radio transmitting sets is about 500-1000 meter. The maximum data rate of the uplink is 9600 baud. For transmitting video and audio signal, the downlink has an maximum data rate of 4.9 Mbps. The delay of both radio transmitting sets is about 0.5 seconds and depends on the link quality.

For the communication between control station and Outdoor MERLIN a flexible communication protocol is used. It is optimized to save bandwidth and to be open for the integration of new components into the communication protocol.

## 3 Assistance System for Tele-operations

A human operator is challenged by the remote control of mobile robots in outdoor environments, taking into account environmental characteristics as well as vehicle kinematics and dynamics [8, 9]. Moreover, the quality of his work achievable depends on the feedback received from the robot.

Hence, the operator has the need for support to increase mission performance [7]. This support can be provided in two ways:

**Passive support** the operator is consulted and advised by the system
**Active support** the system intervenes by improving the operator's instructions

Passive assistance is characterized by its consulting nature. In contrast to intervening commands generated by the operator, it provides advice in specific operation scenarios. The quality of the support varies significantly with the quality, quantity and priorities of the information available. Passive support takes place at the Remote Control and is applied through the user interface by visual and audio feedback, which allows the operator to control the robot remotely. Active support is implemented on the robot in the field mainly due to the need for short control loops, which could otherwise be unnecessarily delayed by data transmission via radio link. As defined above, the active support analyzes the operator's commands in relation to the current system status and characteristics of the environment (e.g. current vehicle velocity, threatening obstacles), and adapts his commands to them [7]. In particular, emphasis was placed on implementing autonomous reactions (for obstacle avoidance, "'back to base"') and drive assistance systems (for velocity control, maximum centrifugal force) on the robot. The following section focuses on the investigation of active support mechanisms including their classification, organization and realization.

### 3.1 User Interface

The control software is based on a modular concept and is implemented with JAVA as plug-ins for the operating system Eclipse. Therefore, it is easy to integrate control and visualization modules for new hardware components or to adapt the graphic user interface (GUI). Each visualization module has its own window within the application window, which can be moved in an operator defined integrated display area. The operator has the option to save different standard user interfaces for the typical missions. Figure 3 presents an exemplary GUI for the rover tele-operator.

The software supports commercial joysticks or gamepads as input device to control the mobile robot. Often needed functions can be linked with the buttons of these input devices.

### 3.2 Operation Modes

There exist several operation modes that can be seen as presets to adapt the robot configuration to the current task.

**Ready:** This is the initial state after power up. All subsystems are initialized, tested and started. The system waits for user commands. It also indicates that all systems were started successfully.

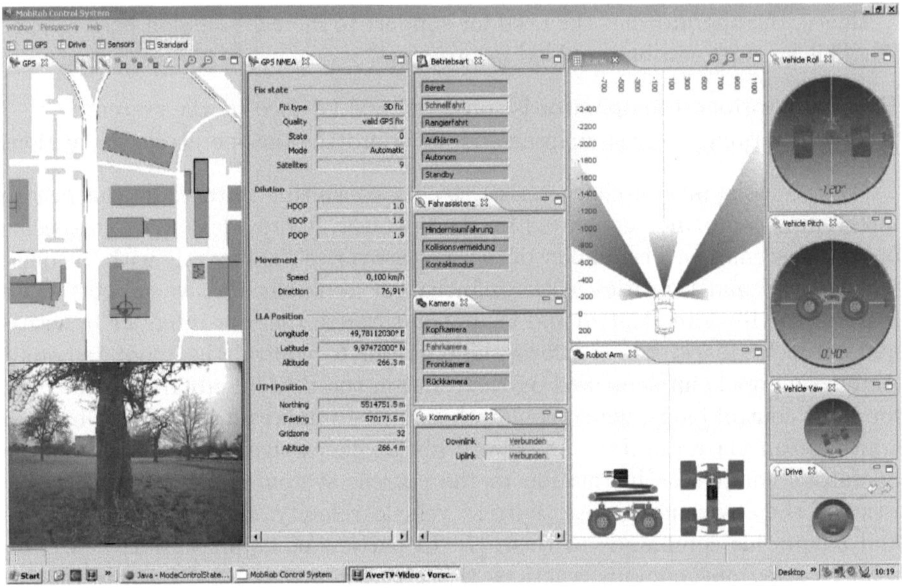


**Fig. 3.** Typical GUI with integrated windows for navigation, rover status

**Fast Driving:** The user can drive the vehicle with full speed. Assistance systems can be activated on demand. Payload operation is reduced because the center of mass must not be altered. (i.e. for the camera payload only the head can move but elevation is disabled).

**Slow Driving:** The maximum speed is limited to provide sensitive steering for maneuvers in narrow places. The payload is active but input is focused on driving.

**Surveillance:** The vehicle can not be moved. Input mapping is changed to provide better control of the payload. All sensor systems are still active and provide information.

**Standby:** Energy consumption is reduced to a minimum. Driving is not possible. All sensor systems are still active and provide information to the user.

**Autonomy:** In autonomy mode the vehicle generates driving commands itself. As soon as the user submits new commands and they are received on the vehicle, the autonomy mode gets disabled and the operation mode changes to fast driving. In autonomy mode different behaviors are available that can be activated by the user or automatically.

### 3.3 Basic Drive Assistance

The operator has no permission to disable the features of this level from the autonomous assistance. Nevertheless he can adjust the behavior of each feature by applying situation-dependent presets, which include the needed parameters

**Table 1.** Comparison of operation modes

| Operation Mode | Velocity | Sensors | Payload | Joystick |
|---|---|---|---|---|
| Ready | 0 km/h | Active | Inactive | Inactive |
| Fast Driving | 12 km/h | Active | Limited | Driving / Payload |
| Slow Driving | 4 km/h | Active | Limited | Driving / Payload |
| Surveillance | 0 km/h | Active | Active | Payload |
| Standby | 0 km/h | Active | Inactive | Inactive |
| Autonomy | 10 km/h | Active | Inactive | Inactive |

for the features. This allows for a problem-specific adaptation at this level, but always maintains the minimal support necessary.

It is quite difficult for the operator to assess the current velocity and acceleration of the robot due to delayed data transmission, limited field of vision or inadequate feedback quality. Furthermore, it is nearly impossible for the operator to adjust his commands in real time to variable environmental conditions. For example, it is not possible to maintain the velocity of the vehicle, when the externally applied forces change (downhill force, friction, etc.). Challenges of the implementation are:

- large range of possible velocities
- many different types of environments
- variable robot configuration due to payload changes

The system gets periodic speed commands as an input. Derived from these commands, trajectories are generated in order to modify the current speed towards the desired value, while limiting acceleration forces (throttle, break) avoid overturning of the vehicle. A PID controller calculates the resulting output value, which is delivered to the motor controller in order to achieve the desired velocity of the vehicle.

### 3.4 Extended Drive Assistance

To support the operator during a mission in the "'autonomy mode"', the robot must be able to react to obstacles, which are detected on his course. The operator notices obstacles very often retarded or too late, so that he is not able to react fast enough. To avoid this situation, an extended driving system was implemented, which processes data from the environment of the vehicle. A key contribution is from ultrasonic and infrared ranging sensors. If the vehicle approaches an obstacle, two autonomous behavior patterns can be chosen: to drive around the obstacle or to push the obstacle.

### 3.5 Autonomous Features

The implementation of partially autonomous functions is necessary. In case the radio link is disturbed, the vehicle must be able to return autonomously until contact to the radio is re-established. Additionally, the operator is able to plan a path for the mission in advance by defining a set of GPS coordinates. The vehicle will try to reach each point avoiding obstacles in its way, even without the communication link. The MERLIN vehicle is able to drive along these paths alone without instructions, while the operator can concentrate on monitoring the area. The autonomous abilities are based on the GPS sensor data, supported by inertial navigation sensors if not enough reference satellites are in the field of view. The accuracy of these GPS coordinates, as well as the quality of the autonomous functions depends on the accuracy of the GPS signals.

## 4 Performance Tests

According to the huge operations range, the robot was subjected to a large number of extensive tests. The robot test area of University Würzburg consists of steep and shallow slopes, many different ground surfaces (asphalt, gravel, grassland, dirt, sand) and various kinds of obstacles (buildings, cars, trees, lampposts, snow piles, tunnels).

Testing took place in a wide range of weather conditions like snowfalls, rains and intense sunshine.

Depending on conditions of the particular test area like roughness, traction and inclination, the configuration of the basic drive assistance and the composition of autonomous reactions were adapted. All tests were accomplished by several operators whose experiences influenced the parameterization and configuration of the drive assistance as well as the overall development process.

In reference tests, the robot was not equipped with any kind of autonomy, not even with the basic drive assistance. The operator could hardly control the acceleration generated by the powerful motor directly, causing frequent turnovers.

**Fig. 4.** The rover test area of University Würzburg

Testing the basic drive assistance was characterized by finding parameters for the PID controller in a way that the operator was able to drive safely without noticing the assistance system. This allowed controlling the robot reliably in the operator's field of view.

To emphasize the efficiency of the control algorithm an outdoor test was done driving on grassland and then climbing a hill. The results show that the velocity can be kept in a close region (standard deviation about 6of the desired speed. This is especially necessary when operating with full autonomy.

Controlling the robot outside this area using the small experimental onboard camera led to collisions in many cases due to the highly restrictive field of vision. The faster the robot was driven, the more insufficient was the tele-operations performance of the operator.

The next step was the addition of various sets of autonomous reactions which increased the quality of robot control significantly. Depending on the composition of autonomous features, it was possible to enhance the operator's ability to react in critical situations. For example, navigating the robot through a tree grove in the testing area was possible without colliding with any trees.

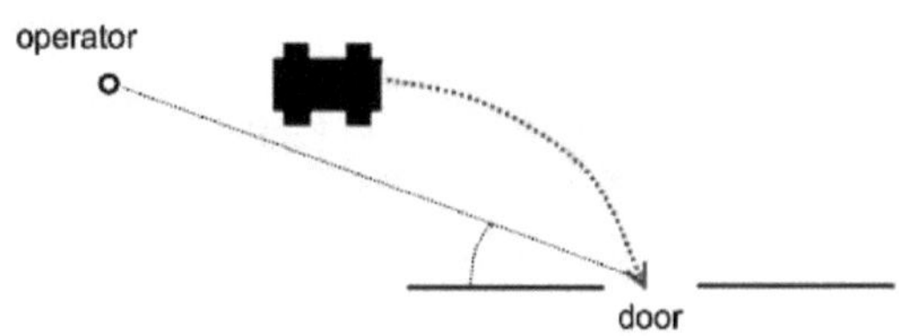


**Fig. 5.** Obstacle avoidance test setup

**Table 2.** Gain in mission time with obstacle avoidance

| **Angle** | 90° | 60° | 30° | 0° |
|---|---|---|---|---|
| **Improvement** | -20% | 0% | +10% | +20% |

Table 2 shows some test results, where the task was to drive through a door as quickly as possible. This was done without and with obstacle avoidance. The view angle of the operator relative to the door was shifted four times. The values give the average improvement of the time it took to drive through the door with the obstacle avoidance compared to the time it took without the assistance.

If the operator has a clear view of what the vehicle is doing, the driving assistance can lead to a slightly worse performance. In this case, the reason is the assistance system breaking the vehicle before passing through the door. But as soon as the operator has to guess, where the vehicle is exactly, the assistance system reduces the risk of collisions and also improves the time it takes to complete the mission.

## 5 Conclusions

The Outdoor MERLIN$^{TM}$ vehicles are characterized by compact dimensions at a mass below 20 kg and robust operations. In order to support save tele-operations, a drive assistant system has been developed based on autonomous functionalities implemented in the on-board microprocessor, as well as by a data interpretation and analysis system implemented in the remote control station. Preliminary tests exhibit a safer, improved performance, when tele-operations is supported by this drive assistance system.

## References

1. Sheridan, T.B.: Telerobotics, automation and human supervisory control. The MIT Press, Cambridge (1992)
2. Baier, H., Driewer, F., Schilling, K.: Robot/human rescue teams: A user requirement analysis. Advanced Robotic 19, 819–838 (2005)
3. Driewer, F., Schilling, K.: Remote control of mobile robots for emergencies. In: Proceedings of the 16th IFAC World Congress (2005) (Tu–E15–TO/3)
4. Schilling, K.: Control of planetary rovers. special issue of Control Engineering Practice 5, 823 (1997)
5. Everett, H.R.: Sensors for mobile robots. AK Peters (1995)
6. Schilling, K., Q.M.: The merlin vehicles for outdoor applications. In: SPIE conference proceedings Unmanned Ground Vehicle Technology IV, Orlando, pp. 43–49 (2002)
7. Frank, M., Busch, S., P.D.: Teleoperations of a mobile outdoor robot with adjustable autonomy. In: ACIDCA-ICMI Proceedings, pp. O–8 (2005)
8. Siegwart, R., Nourbakhsh, I.R.: Introduction to Autonomous Mobile Robots. Bradford Books (2004)
9. Kiencke, U., Nielsen, L.: Automotive Control Systems. Springer, Heidelberg (2000)

# Visually-Guided Robot Navigation: From Artificial to Natural Landmarks

Enric Celaya, Jose-Luis Albarral, Pablo Jiménez, and Carme Torras

Institut de Robòtica i Informàtica Industrial (CSIC-UPC), Llorens i Artigas 4-6, 08028 Barcelona, Spain
{celaya,albarral,jimenez,torras}@iri.upc.edu

**Summary.** Landmark-based navigation in unknown unstructured environments is far from solved. The bottleneck nowadays seems to be the fast detection of reliable visual references in the image stream as the robot moves. In our research, we have decoupled the navigation issues from this visual bottleneck, by first using artificial landmarks that could be easily detected and identified. Once we had a navigation system working, we developed a strategy to detect and track salient regions along image streams by just performing on-line pixel sampling. This strategy continuously updates the mean and covariances of the salient regions, as well as creates, deletes and merges regions according to the sample flow. Regions detected as salient can be considered as potential landmarks to be used in the navigation task.

## 1 Introduction

The robotics community has devoted many efforts to specific indoor navigation schemes for wheeled robots, often strongly dependent on structured visual features [1]. Nowadays, the demand for outdoor applications is increasing, and the research attention is shifting towards more general, though perhaps less accurate, navigation strategies that rely on unstructured visual cues.

Not only unstructured environments pose new challenges, but also the mobility capabilities of the robots themselves -outdoor robots usually use legs, tracks, or deformable frames- imply that concepts like what is actually an obstacle have to be reconsidered. Even the very goal of outdoor navigation is usually different from that of indoors. The typical goal in indoor navigation applications is to reach a desired place in a previously known environment, usually using a map, whereas in outdoor navigation, e.g., in a rescue task, the environment is often unknown and the goal must be specified on-line and redefined during the navigation task, based on the observations made from the positions already reached. Full teleoperation is not always possible, as shown by NASA's Mars Exploration Rovers project [2, 3] operating Spirit and Opportunity rovers in the surface of Mars, due to the time delay introduced by the travel of the control signal between the Earth and Mars. The possibility of fully autonomous navigation may also be severely limited by the processing capabilities available to the rover's CPU.

Thus, in applications such as planetary exploration or rescue in catastrophic areas, partly teleoperated robots, able to reach autonomously a destination

C. Laugier and R. Siegwart (Eds.): Field and Service Robotics, STAR 42, pp. 287–296, 2008.
springerlink.com 

marked by a human operator on the image as seen by the robot, are agreed to be very handy. Updating the target as the robot advances, permits long-range journeys even with low-bandwidth and long-delay communications. Additionally, it would be desirable that these robots were able to find their way back autonomously without relying on any a priori knowledge of the environment. This so-called homing problem has been addressed in some works [4], where visual homing with bearings-based control is experimented in natural environments by using correlation of SIFT features [5] in images taken from different robot poses.

In our visually-guided navigation approach we have addressed these issues by relying on landmarks. Only few previous works have tackled a similar visually-guided landmark-based approach. In [6] a landmark-based environmental model is incrementally built using color and stereo range information. Then, the landmarks in the model are used for the robot localization and for long-range robot navigation, through the selection of different landmarks that form a sequence of sub-goals that the robot must successively reach. The system differs from ours in that, instead of being user-guided, it performs an automatic goal selection by searching for the landmark with the highest elevation peak among all detected landmarks in the scene.

The paper is structured as follows: Section 2 describes our approach to visually-guided navigation. In Section 3, we show the viability of the approach by testing it for indoor navigation using artificial landmarks. Section 4 introduces our approach to natural landmark detection, and some results are shown for real image sequences taken by a mobile robot in an outdoor setting. Finally, some conclusions and future work are pointed out.

## 2 Our Approach to Visually-Guided Navigation

A problem we had to face while designing our visually-guided navigation system was how to specify the navigation target. Map-based approaches, and those based on absolute coordinate systems like GPS, were discarded since we required our system to be useful also in completely unknown environments. Ideally, the user should be able to select any object in sight and mark it as navigation target. As the navigation proceeds, the user could select new targets, thus guiding the robot towards the desired location in a way similar to what a human explorer would do himself.

The problem with this approach is that, what can be a clearly identifiable object for the user, may be difficult to discern for an artificial vision system, so that the robot can be lost very soon. Our strategy to solve this problem consists in limiting the selection of possible targets to those that the robot is able to identify with its vision system. Thus, it is the system that first shows to the user the available landmarks, and the user may select one of them as the target. To overcome the limitations that this approach could have in environments with sparse identifiable landmarks, it is possible for the user to directly drive the robot in the desired direction.

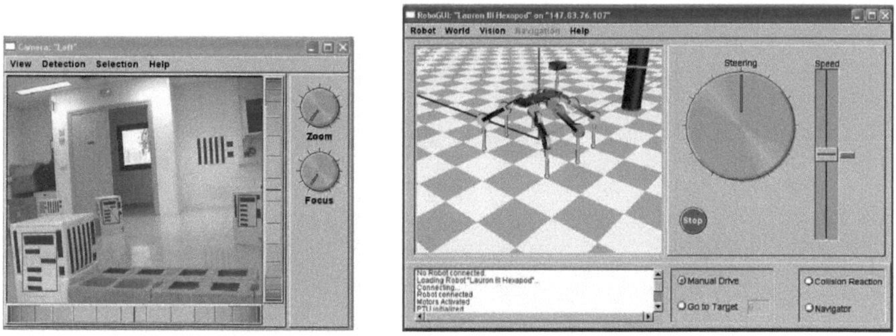


**Fig. 1.** Camera Window and Robot Control Interface

### 2.1 The Navigation Interface

Fig. 1 shows the two main windows of the navigation interface: The *Camera Window* and the *Robot Control Interface*. In the first, the user can see the images taken from the camera of the robot and control its gaze direction. Superimposed on the image, landmarks detected by the vision system are marked in red. The user may select one of these landmarks as the navigation target with a mouse click. Alternatively, the target can be selected using the Robot Control Interface, where landmarks are displayed in a simulated environment. In developing our platform for visually-guided navigation, our aim was to achieve a robot-independent control, so that the user could use it without being concerned about specific features of the robot that is performing the navigation task. The only assumptions we made about the robot are that it carries a camera, it is endowed with appropriate sensors to detect obstacles, and it is able of a short-term estimation of its own displacements (e.g. by odometry). For the tests of the system, two robots as different as a four wheeled Pioneer II and a six-legged Lauron III (Fig. 2) have been used.

The interface allows two control modes: *Manual Drive* and *Go to Target*. In the Manual Drive mode, the user directly controls the robot speed and direction with a slider and a driving wheel. Once a landmark has been selected as navigation target, the *Go to Target* mode can be triggered to let the robot reach the target in a completely autonomous way. Both modes can use the option *collision reaction*, which implements a high priority behavior that performs a pre-defined evading maneuver when an obstacle is detected. Obstacle detection is specific of each robot. In the case of the Pioneer, an activation of the bumpers or a null advance of the robot in response of a movement command are interpreted as obstacles. In the legged robot, the activation above a threshold of the front infrared sensor, the collision with one leg at a given height, or the impossibility to find a stable support, are all considered as obstacles. Both robots also consider very close landmarks as obstacles.

In the *Go to Target* mode, by default, the robot simply heads directly towards the direction of the target. This simple reactive form of goal attraction, together

**Fig. 2.** The six-legged robot Lauron III used in the experiments

with the collision reaction behavior, is enough to succeed in simple environments where path planning is not strictly necessary. For more complex environments, the *Navigator* option can be used to invoke a landmark-based navigation algorithm with map building and path planning capabilities. Our current implementation features the algorithm described in [7], but other landmark-based navigation algorithms could be used.

## 3 Navigating with Artificial Landmarks

A major problem in landmark-based navigation outdoors is a reliable identification of landmarks. In order to validate our visually-guided navigation approach, we isolated the landmark recognition problem and provisionally solved it by using artificial landmarks. For this, we designed a set of landmarks inspired by the work of Scharstein and Briggs in [8], who proposed the use of self-similar patterns, invariant to scaling, rotation, and viewing angle, whose detection indicates the presence of a landmark. A second pattern attached to the landmark composing a binary code served as identifier. This design has the drawback that most of the landmark area is used to indicate the presence of a landmark, but only a small area is devoted to the landmark code. This may prevent a correct identification of landmarks that are far away.

To improve on this, we modified the landmark design so that the code pattern is at the same time part of the landmark-presence indicator. The result is a set of landmarks like those marked in red in fig. 1, consisting in a header rectangle and five half-sized ones, whose relative positions, at the right or the left, allows the composition of $2^5 = 32$ different codes. Landmark detection tests performed with different cameras resulted in identification rates near 100% for non-occluded landmarks in the distance range from 60cm to almost 5m, provided the image is not too blurred due to fast camera movements. No false identifications of landmarks have been observed.

The navigation interface has been tested indoors using artificial landmarks with robots as different as a four wheeled Pioneer II and a six-legged Lauron III (Fig. 2). Users not familiar with the interface could easily drive the robot using

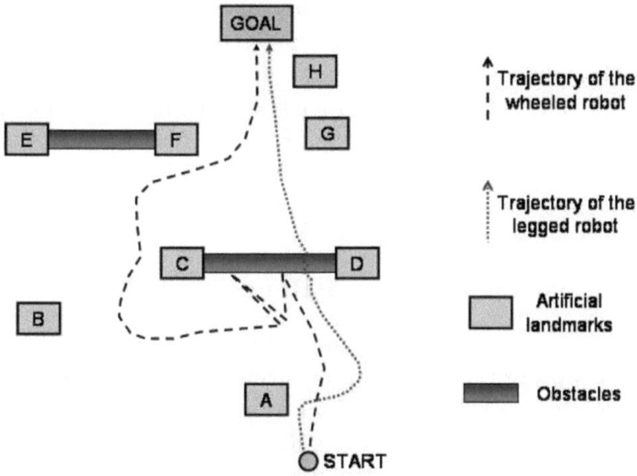


**Fig. 3.** Navigation experiment

the manual mode, no matter what robot was at work. On the other hand, the simple *Go to Target* mode, with the Navigator inactive, shows itself to be very intuitive. The collision reaction capability allows for a successful task completion even when the target is behind some obstacle. An unanticipated way to drive a robot was soon discovered: taking the selected landmark away causes the robot to track it, despite the fact that it is continuously moving or suddenly changes its position. With respect to the Navigator mode, it could be seen that the interaction between the navigation algorithm and the collision reaction behavior, naturally adapts the resulting trajectory to the actual capabilities of the robot performing the task, a feature that could be hard to implement in a more traditional map-based approach. To show this, we put the two robots in the same environment, and launched the navigator with the same target (Fig. 3). Since the legged robot has a much greater capability for obstacle crossing, what for the wheeled robot was an obstacle that caused the navigator to plan an alternative path, was considered just as an irregular patch of terrain by the legged robot, which followed its way walking over it with no further planning. Note that, for a map-based navigator to do the same, it should be provided with robot-specific maps, with different distributions of obstacles and free space.

## 4 Natural Landmarks

Having shown that our visually-guided navigation approach was satisfactory, our next focus was on solving the problem of natural landmarks detection. For this, we relied on the assumption that useful landmarks must be salient, i.e., they must constitute distinctive regions in the image, so that its repeated detection and identification is facilitated. The saliency of a region is not determined by the absolute value of any intrinsic magnitude, but rather, by the contrast or difference of this value with respect to the value of the same magnitude in the surroundings [9]. This is also known as *opponency* [10]. Many variables can be used to define the saliency of a region, like color components, intensity, or feature orientation.Works like [11] compute saliency as a combination of the opponency values of these three variables. In what follows, we present a very simple approach that just considers RGB color values, but the same idea could be used with more informative features.

### 4.1 Detecting Salient Regions: A Random Exploration Approach

Instead of processing each incoming individual frame captured by the camera, the sequence of snapshots is treated as a continuously varying image. To this end, individual pixels are picked up at random from the current image and processed independently of whether the frame has already changed or not. In this way, a frame-rate and resolution-independent processing system is obtained, and the power of the computational resources only affects the number of pixels that can be processed per time step.

In the process, a number of so-called *units* are competing for the most relevant parts of the image, which we call Regions of Interest: each time a pixel is examined, the unit that best *responds* to it is selected and *updated.* By *responding* we mean to have a similar color while being in its proximity, and by *updating*, the process of adjusting the location and dimensions of the unit so as to improve its response to this pixel. In the case that no unit responds to a pixel, it may be considered as the seed of a new unit.

Units have elliptic receptive fields in the image plane, which adapt through successive updates to cover regions whose pixels have similar colors. These ellipses arise naturally from considering normal distributions for pixels taken at random from each region: the center of the ellipse corresponds to the mean value of the pixels' positions, whereas its dimensions (the major and the minor axes) and orientation are given by the covariances. Each unit also has a spherical receptive field in the RGB color space centered at the mean color value and fixed radius, which is a parameter of the system. Units are implemented as data structures with the following fields:

- Center vector $\mathbf{U}_{xy}$ and covariance matrix $\mathbf{\Sigma}_{xy}$ in the image plane,
- Center vector $\mathbf{U}_{rgb}$ in the color space,
- Contrast $C_U$, a scalar that measures saliency, as explained below,
- Creation date, to record the time since which the unit exists,
- Counter $Updates_U$ for the number of times a unit has been updated,
- Counter $Inside_U$ for the number of times an input pixel has fallen inside the receptive field of the unit in the image plane, and
- Strength $S_U$, a scalar that estimates the current proportion between the number of pixels to which the unit responds and those lying into the unit's receptive field in the image plane.

In the main loop of the algorithm, a random input pixel $I$ is selected and, for each unit $U$, its Mahalanobis distance in the image coordinates and Euclidean distance in color space are computed as:

$$Mdist_{xy}(U,I) = \sqrt{(\mathbf{I}_{xy} - \mathbf{U}_{xy})^{\top}\mathbf{\Sigma}_{xy}^{-1}(\mathbf{I}_{xy} - \mathbf{U}_{xy})}, \tag{1}$$

where $\mathbf{\Sigma}_{xy}$ is the covariance matrix of unit $U$, and

$$Edist^2_{RGB}(U,I) = \sum_i (I_i - U_i)^2, \qquad \text{with } i \in \{r,g,b\}. \tag{2}$$

The Mahalanobis distance is used in the image space in order to take into account the shape of the spatial distribution. If both distances are below given thresholds (MAXDISTXY and MAXDISTRGB, respectively) then the unit is said to respond to the input pixel. Among the units that respond, the one which is closest in the image space, is considered as the *winner*.

**Updating the Winner Unit**

The center of the unit is approached to the coordinates of the pixel in the image plane as well as in color space:

$$U_i \leftarrow U_i + \gamma d_i, \tag{3}$$
$$\text{where } d_i = I_i - U_i, \text{whith } i \in \{x,y,r,g,b\} \text{ and } 0 < \gamma < 1$$

Covariances in image space determine the dimensions and orientation of the elliptic receptive field of the unit, and are updated according to:

$$\sigma_{ij} \leftarrow \sigma_{ij} + \gamma(d_i d_j - \sigma_{ij}), \qquad \text{where } i,j \in \{x,y\}. \tag{4}$$

Finally, the counter of updates is incremented by 1:

$$Updates_U \leftarrow Updates_U + 1 \tag{5}$$

**Updating Other Units**

If the Mahalanobis distance from the input pixel to a non-winner unit is below three times MAXDISTXY, the pixel is considered to lay in the unit's neighborhood, and the pixel color is used to update the unit contrast:

$$C_U \leftarrow \alpha C_U + (1-\alpha)\sqrt{\sum_{i\in\{r,g,b\}} (I_i - U_i)^2}, \qquad 0 < \alpha < 1 \tag{6}$$

i.e., increasing or decreasing according to the Euclidean distance in the color space between the input pixel and the unit.

The updating of the strength $S_U$ and $Inside_U$ is done for all units for which the Mahalanobis distance of the input pixel is below MAXDISTXY, i.e., the pixel is in the receptive field of the unit. While $Inside_U$ is simply incremented by one, the strength is increased when the unit responds to the input (i.e., also chromatically) according to:

$$S_U \leftarrow \beta S_U + (1 - \beta), \qquad \text{with } 0 < \beta < 1. \tag{7}$$

If the unit does not respond to the input color, $S_U$ is decreased according to:

$$S_U \leftarrow \beta S_U \tag{8}$$

### Removing Irrelevant Units

To avoid an unlimited proliferation of units, their number is limited by a parameter of the system. To allow the creation of new units when the maximum number has been reached, the less useful ones must be removed. This is done in the following way: First, the unit with the lowest strength is sought. If its strength is below a given value, it is assumed that the region it represents is no longer there, and the unit is removed. Otherwise, a unit will only be replaced provided a contrast estimation of the unit to be created is above that of the lowest-contrast unit. The contrast estimation is given by the distance in color space of the input pixel to its spatially closest unit.

The new unit is initialized with a circular receptive field with center at the position of the input pixel and radius equal to its distance to the closest unit.

### Merging of Units

If two units respond to a given pixel, they are probably representing different parts of the same region and can be merged. The center and covariance values of the merged unit are computed as a weighted sum of those of the merging ones, with weights proportional to the respective area and strength, roughly corresponding to the "mass", or number of pixels each unit responds to:

$$Weight(U) = Area(U) S_U \tag{9}$$

The area is computed from the covariances as

$$Area(U) = \sqrt{((\sigma_{xx} + \sigma_{yy} + \Delta) \cdot (\sigma_{xx} + \sigma_{yy} - \Delta))} \tag{10}$$
$$\Delta = (\sigma_{xx} - \sigma_{yy})^2 + 4\sigma_{xy}^2$$

The contrast of the merged unit is set to the highest contrast value of the two original ones, and for the strength, the weighted sum of the strengths is computed, this time using the respective values of $Updates_U$ as weights.

### Output of the System

As for the output of the system, only the units that correspond to the most salient regions are returned as possible landmarks. To this end, units covering too large regions of the image are discarded, as they usually capture the background and are not useful for navigation. Regions that are too small are also discarded to remove isolated pixels or noise. Finally, units that have not been updated a minimum number of times are not considered, since they are still not reliable enough. From the remaining units, those with contrast values above a given threshold are selected and output as landmarks.

### 4.2 Navigation Using Natural Landmarks

Fig. 4 shows some snapshots of a preliminary test of the use of the salience detector for the guidance of a robot outdoors. The legged robot Lauron III is facing the situation shown in frame 1, in which the blue bag is chosen as initial navigation target. After some progress towards it (frame 2), two bright objects enter the field of view and are identified as salient landmarks by the system (frame 3). One of these objects is selected as a new target and the robot proceeds towards it letting the old target back (frames 4-6).

**Fig. 4.** Six snapshots of the navigation experiment with the robot Lauron III

## 5 Conclusions

We have presented a working system for visually-guided navigation, which differs from others in the following features: (1) The interface is robot-independent even if the notion of obstacle varies from one robot to another, depending on their locomotion capabilities. Experiments have shown that the same navigation algorithm generates different trajectories for a legged and a wheeled robot. (2) A partial teleoperation scheme has been adopted, in which the goal is selected by the user among a set of landmarks previously detected by the robot. (3) Navigation issues have been decoupled from vision ones by designing artificial landmarks that could be easily detected and identified. (4) Once we had a navigation system working, artificial landmarks were replaced by natural ones. A color-contrast saliency detection algorithm has been developed which relies on randomly sampling pixels from the continuous video stream acquired by an on-board camera. This algorithm has been tested on a legged robot in an outdoor scenario with promising results.

**Acknowledgments.** This work has been supported by the Spanish *Ministerio de Ciencia y Tecnología* and FEDER under the project SIRVENT (DPI2003-05193-C02-01), as well as the European Union under the project PACO-PLUS (FP6-2004-IST-4-27657). We thank the IIIA for making available the environment for the navigation experiments and performing the tests with the Pioneer 2, and also for providing the implementation of the navigator.

## References

1. DeSouza, G., Kak, A.: Vision for mobile robot navigation: a survey. IEEE Transactions on Pattern Analysis and Machine Intelligence 24(2), 237–267 (2002)
2. Leger, P., Deen, R., Bonitz, R.: Remote Image Analysis for Mars Exploration Rover Mobility and Manipulation Operations. In: IEEE Int. Conf. on Systems, Man and Cybernetics (2005)
3. Maxwell, S., Cooper, B., Hartman, F., Leger, C., Wright, J.: The Best of Both Worlds: Integrating Textual and Visual Command Interfaces for Mars Rover Operations. In: IEEE Int. Conf. on Systems, Man and Cybernetics (2005)
4. Kirigin, I., Singh, S.: Bearings based robot homing with robust landmark matching and limited horizon view. Technical Report CMU-RI-TR-05-02, Carnegie Mellon University (2005)
5. Lowe, D.: Distinctive Image Features from Scale-Invariant Keypoints. International Journal of Computer Vision 60, 91–110 (2004)
6. Murrieta-Cid, R., Parra, C., Devy, M.: Visual navigation in natural environments: from range and color data to a landmark-based model. Autonomous Robots 3(2), 143–168 (2002)
7. Busquets, D., Sierra, C., de Mántaras, R.L.: A multi-agent approach to qualitative landmark-based navigation. Autonomous Robots 15, 129–153 (2003)
8. Scharstein, D., Briggs, A.J.: Real-time recognition of self-similar landmarks. Image and Vision Computing 19, 763–772 (2000)
9. Nothdurft, H.C.: Saliency from Feature Contrast: Additivity Across Dimensions. Vision Research 40, 1183–1201 (2000)
10. Todt, E., Torras, C.: Detection of natural landmarks through multiscale opponent features. In: 15th Int. Conf. on Pattern Recognition, Barcelona, Spain, pp. 976–979 (2000)
11. Itti, L., Koch, C., Niebur, E.: A Model of Saliency-based visual Attention for Rapid Scene Analysis. IEEE Transactions on Pattern Analysis and Machine Intelligence 20(11), 1254–1259 (1998)

# An Advanced Teleoperation Testbed

Bill Ross[1], John Bares[1], David Stager[1], Larry Jackel[2], and Mike Perschbacher[3]

[1] The Robotics Institute, Carnegie Mellon University, 5000 Forbes Ave, Pittsburgh, PA 15213, USA
br@cs.cmu.edu
[2] DARPA TTO Project Manager
[3] RovnoTech

**Abstract.** Due to the technology available, most previous work in teleoperated robotics used relatively low-resolution video links and provided limited perceptual feedback to the teleoperator. In most cases, these projects reported only limited teleoperator success compared to vehicles with human drivers on-board. We set out to build a high-fidelity teleoperation system which takes advantage of recent technological advances. This system permits highly capable teleoperation and has allowed us to begin to investigate the minimum system requirements for effective teleoperation.

## 1 Introduction

While there have been a great number of teleoperated machines built, including some practical machines in daily commercial use, teleoperation success in highly unstructured environments has been comparatively limited. The reason for this may be simply that high resolution video cameras, video transmission systems, and displays have only recently become readily available.

The goal of this project was to develop a very high-fidelity teleoperation system for the CMU Crusher robot, to test this system in challenging, real-world terrain, and to produce estimates of the minimum parameters of an highly effective teleoperation system for use in unstructured environments.

Sponsored by Defense Advanced Research Projects Agency TTO Program under DARPA CMO Agreement MDA972-01-9-0005. The views and conclusions contained in this document are those of the authors and should not be interpreted as representing the official policies, either expressly or implied, of the Defense Advanced Research Projects Agency or the U.S. Government. Approved for Public Release, Distribution Unlimited.

## 2 Background

There have been a number of excellent studies done on the impact of system capabilities on a teleoperator's performance [1-10]. In most of these studies the conclusion was that teleoperators seldom perform as well as human drivers and often performed poorly, with collisions and vehicle rollovers being common [10]. For the

C. Laugier and R. Siegwart (Eds.): Field and Service Robotics, STAR 42, pp. 297–304, 2008.
springerlink.com 

most part, however, these studies tested only relatively low-quality video systems with limited resolution and field of view, and with high latencies. Most of these tests were also conducted either in simulation[1][5], or in real-world terrain with relatively structured paths providing only limited challenges for teleoperators.

## 3 System Overview

While the performance targets for our system were ambitious, we never contemplated building a permanent system for continuous use. In many cases, we used what was available quickly and affordably to get the system up and running.

**Fig. 1.** The Crusher robot

*Vehicle:* The system is based on the UGCV Crusher robot. Crusher is a 6,800 kg, six-wheeled, hybrid-powered robot that has shown to be extremely capable in the most severe terrain. Crusher's active suspension provides a smooth ride for the camera system over rough ground at up to 12 meters per second.

*Cameras and displays:* The system is based on five 1600 by 1200 pixel, single-CCD, Bayer-pattern color cameras which give an aggregate field of view of 200 by 30 degrees. The overall resolution is 40 pixels per degree (1.5 MOA per pixel) in the field of view. This resolution is about 4 times the linear pixel resolution provided by typical television systems. Images are displayed on standard LCD monitors with matching resolution which are sized and positioned so that the operator's field of view closely matches that of the cameras in the real world.

*Communications:* Primary communications between the vehicle and the control station is carried by a 1 km single-mode CWDM optical fiber which can handle 16 separate data channels. Currently we use five of these channels for the video feeds and a sixth for a gigabit Ethernet connection to the vehicle which carries vehicle IMU data, vehicle audio, and control information. The total data transmitted over the fiber is about 8.5 gigabits per second with almost no latency added by the fiber link. The fiber itself is a 1.6mm high-strength tactical cable with an aramid strength member and a polyurethane jacket which has proven to be surprisingly robust. The fiber is automatically reeled in and out using an iRobot spooler which is slaved to the vehicle odometry.

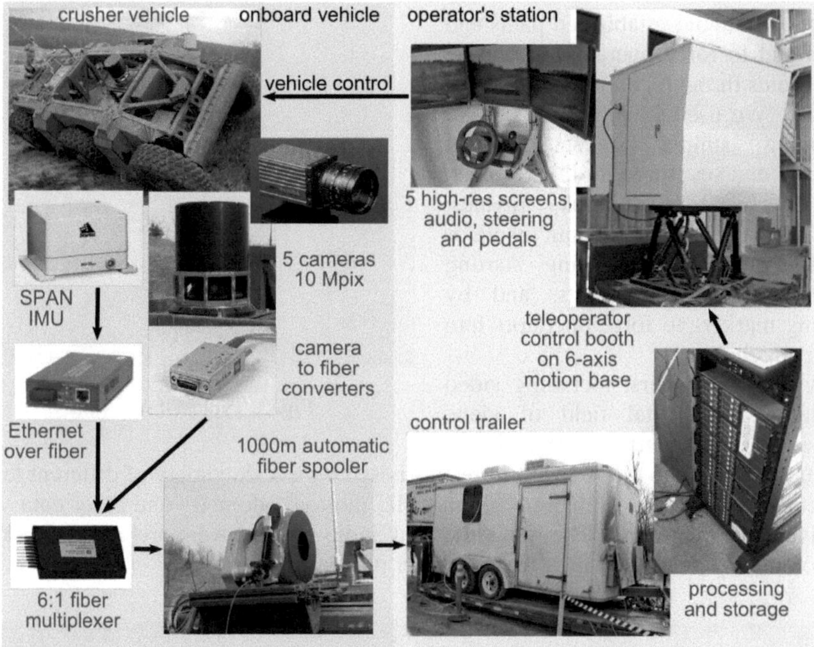


**Fig. 2.** Overall teleoperation system layout

*Control trailer:* The fiber connection from Crusher leads to a control and computing trailer. Five dual-Opteron machines with high-speed RAID arrays and high-end GPUs process the video feeds and display them in real-time while also saving the unprocessed data to disk. The video is processed using both CPU and GPU programs to perform Bayer filtering, to fine tune the image quality, and to intentionally degrade the image if desired. A high level control program ensures synchronization between the video and audio streams and allows the user to add artificial latency to the system if desired.

*Operator's booth:* The teleoperator rides in a booth mounted on a 6-axis motion base which is capable of greater than 1g of acceleration. The booth includes five video displays which are duplicates of the camera views in the control trailer, stereo speakers, and a standard steering wheel and pedals interface to drive the vehicle.

## 4 Trial Courses and Methods

We conducted 7 days of testing on the site of an large reclaimed strip mine. This area includes forest, hills, ravines, streams and ponds. During the week of testing we performed 56 trials with a total teleoperated distance of 36 km.

The operator's booth is equipped with a real-time position display on a low-resolution map of the test area. In each trial, the operator was given a goal point on his map and was instructed to reach the goal as quickly as possible. Goals were designed to avoid established paths and roads and to force operators to choose new routes through complex and varied terrain. We used 11 different courses, including simple courses around barrels, courses through thick forest, and courses through rocky ravines. Courses were varied to limit operator familiarization by changing starting positions and goal points, and by placing markers to force operators into new path choices.

Fig. 3. Typical test terrain

System parameters, including video resolution, horizontal field of view, frame-rate, and latency, were varied at the start of each run to produce test results based on a wide variety of different levels of perceptual fidelity. Vehicle position, IMU motion, odometry and other data were archived for every trial. The raw video and audio streams were archived to disk for approximately one-third of the trials for later playback and review.

**Fig. 4.** Eight example trial paths plotted from GPS logs

## 5 Results

Due to the limited number of runs which we were able to perform, the complex unstructured terrain, and the large number of variables involved, the statistics we will present must represent a honest best effort rather than a highly rigorous analysis of carefully controlled variables. In our analysis, however, we have relied as much as possible on quantitative results and in this paper we will only present statistics which are supported by at least 10 trials. Where significant, we will also present subjective impressions reported by the teleoperator and by observers.

**Table 1.** Latency

| Total Latency (ms) | Avg Speed (m/s) | Avg Time Stopped (sec) |
|---|---|---|
| 360 | 2.47 | 64 |
| 480 | 2.20 | 96 |
| 960 | 1.57 | 168 |
| 1320 | 1.15 | 246 |

*Latency:* Total latency (control lag + perception delay) proved to have the greatest impact on teleoperator performance in our testing. Doubling the total latency from 480ms to 960ms resulted in a 29% reduction in average vehicle speed, and a 75% increase in time the vehicle was stopped by the operator for planning or other reasons. Teleoperator performance declined noticeably, even for experienced operators, with any increase in latency. Operators agreed with NASA Ames results which show [2] that higher latencies force the adoption of a "move and wait" control strategy above a certain latency threshold. While the Ames results put this threshold at 300ms, we felt that drivers of our vehicle were able to adapt to lower levels of latency and were only forced to alter their driving style at levels of 480ms or above. Subjectively, high latency was the factor which teleoperators found most annoying and challenging. Comments such as "latency is horrible" and "latency is killing me" were commonplace.

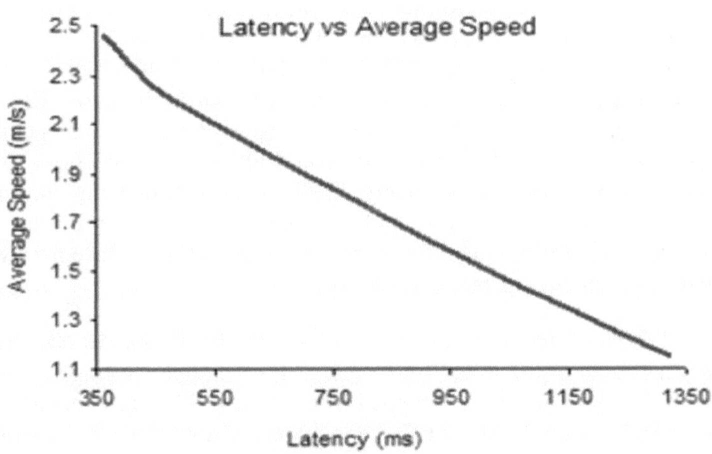


**Fig. 5.** Overall latency vs average vehicle speed Graph based on 29 trials

*Resolution:* The impact of display resolution on teleoperator performance was less noticeable than latency, but was still significant. Compared to full resolution trials (1600x1200 pixels per screen), trials with screen resolutions of 320x240 pixels saw a

**Table 2.** Resolution

| Pixels Per Screen | Avg Speed (m/s) | Avg Time Stopped (sec) |
|---|---|---|
| 320x240 | 2.3 | 138 |
| 800x600 | 2.6 | 72 |
| 1600x1200 | 3.0 | 2 |

23% reduction in average speed and a factor of 69 increase in time stopped. Low resolution video was also found to contribute to poor path planning decisions. In one example, a teleoperator aimed for a "gap" in the forest which turned out to be blocked by trees not visible in the low-resolution view. In another example, a teleoperator reported that he was driving slowly because he could not clearly distinguish obstacles ahead of the vehicle. When the teleoperator was given highest resolution only on the center monitor and low resolution on the remaining four monitors, he reported that he was intentionally yawing the vehicle back and forth to use the higher resolution of the center camera. Subjectively, the higher resolution views seemed to contribute greatly to the sense of realism and presence felt by the teleoperator and to increased operator confidence.

**Fig. 6.** Field of view options

*Field of view:* Trials were performed with the horizontal field of view (HFOV) ranging from 40 to 200 degrees and the impact of a limited field of view was found to be significant. Average speed with a 200 degree HFOV was 40% higher than that for trials with a 40 degree HFOV. The time stopped with a 40 degree HFOV was 2 times greater than with a 120 degree HFOV. Paths chosen by teleoperators with narrow fields of view were often poor and several times it was necessary to back the vehicle out of dead ends. Subjectively, field of view had a big impact on the teleoperators' situational awareness and on their understanding of the vehicle's position and motion. Several operators expressed a special preference for a wide field of view when the vehicle was moving quickly.

*Frame rate:* Display frame rate was tested at 30, 25, 20, 15 and 10 Hz. Significant effects only appeared at the lower frame rates. For example, average vehicle speed dropped by 37% when frame rate dropped from 20 to 10 Hz. Teleoperators reported feeling comfortable with 25 Hz but flickering became noticeable and uncomfortable at 20 Hz. Lower frame rates were more challenging and teleoperators reported that it was more difficult to judge vehicle speed when the frame rate was 15 Hz or lower.

*Audio:* Our system included stereo microphones on the Crusher robot and stereo speakers for playback in the teleoperator's booth. Crusher is a hybrid diesel/electric vehicle and is very quiet when running only on batteries, which allowed the teleoperator to clearly hear what was going on around the vehicle. Teleoperators commented that the audio was useful in providing cues such as tree branch and rock strikes, tire slippage, drive motor load and vehicle speed. Subjectively, audio seemed to add considerably to situational awareness and the teleoperators' feeling of telepresence, and teleoperators complained when the audio was turned off.

*Motion:* While the 1g motion base could not faithfully replicate every motion of the vehicle, it could move quickly enough to be uncomfortable when Crusher hit large bumps. Most people who teleoperated the vehicle requested that the gain be turned down on the motion to reduce discomfort over rough terrain. The most experienced teleoperator came to prefer a 30% gain on the motion feedback. Nevertheless, teleoperators reported that the motion was very useful on slopes, over rough terrain, and any time a clear horizon was not visible. Motion sickness was never severe, but was noticeable for some operators.

*Operator's choice trials:* We conducted several trials in which we limited only the teleoperator's total video bandwidth, but allowed him to choose how to spend that bandwidth and to change his allocation at any time during the trial. For example, the teleoperator might choose a single screen at full resolution and full frame-rate, or all five screens at low resolution. This approach proved to be very successful and was well liked by teleoperators. For general driving, teleoperators preferred high frame-rate with medium resolution ahead and low resolution on the sides with a full 200 degree field of view. In complex environments, looking far ahead, and planning routes, teleoperators often traded off field of view or frame rate for higher resolution in the area of interest. This may be a highly effective method for reducing bandwidth requirements in future teleoperated systems.

## 6 Conclusions

While this paper represents results from only one week of testing, a number of compelling results have emerged. The first observation is that teleoperation with a high-fidelity system is very effective. With the highest quality settings, the feeling of "being in the vehicle" was very strong. Several teleoperators emerged from the control booth believing that they were at the location of the vehicle a kilometer away. Using this system, teleoperators were able to drive effectively, confidently, and safely at the maximum speed of the Crusher vehicle through highly complex and unstructured off-road terrain to the point that safety vehicles, including a HMMWV, often had trouble keeping up.

In many instances the quality of the teleoperation was so good that it surpassed close-proximity visual radio-controlled operation by a skilled operator. We conjecture that the teleoperator benefited from better sightlines provided by the cameras placed directly on the vehicle as well as the physical feedback from the motion base. The subtly, speed, and precision of teleoperation observed in our experiments has convinced the authors that major research advances are required before a fully autonomous vehicle can achieve comparable operation.

This work has also led us to draw some conclusions which may be useful for future teleoperated systems:

- Latency should be minimized and is severe above 400-500ms
- A wide FOV is a great help for situational awareness
- A 20-25 Hz frame rate is acceptable to many operators
- High resolution is essential at long distances in complex terrain
- Audio and motion feedback provide realism and useful cues
- Bandwidth can be reduced by giving the operator choices

## Acknowledgements

Many people went out of their way to help us with this work. In particular, we're especially grateful to: Dave LaRose, Brian Beyer, Jon Chu, Megan Grimm, Mark Waldbaum, Roger Boulet, Mike Sergi, Don Salvadori, David Rice, Chuck Amurgis, Chris Fromme, and Bill Curtis who all supported this work.

## References

1. Ellis, S.R., Mania, K., Adelstein, B.D., Hill, M.: Generalizeability of Latency Detection in a variety of Virtual Environments. In: Human Factors and Ergonomics Society 48th Annual meeting, New Orleans, USA (2004)
2. Ellis, S.R., Young, M.J., Adelstein, B.D., Ehrlich, S.M.: Discrimination of changes in latency during head movement. In: Proceedings of the HCI International 1999 (1999)
3. Holloway, R.L.: Registration error analysis for augmented reality. Presence, 6, 4, 413–432 (1997)
4. Ellis, S.R., Young, M.J., Adelstein, B.D., Ehrlich, S.M.: Discrimination of changes of latency during voluntary hand movement of virtual objects. In: Proceedings of the Human Factors and Ergonomics Society, Houston, Texas, September 27-October 2 (1999)
5. Vries, S., Padmos, P.: Steering a Simulated Unmanned Aerial Vehicle Using a Head-Slaved Camera and HMD. In: ECCV/HCI 2004, vol. 3058, pp. 24–33 (1997)
6. Kasper, E., Haworth, L., Szoboszlay, Z., King, R., Halmos, Z.: Effects of in-flight field of view restriction on rotorcraft pilot head movement. In: SPIE, vol. 3058, pp. 34–45 (1997)
7. McLean, G.F., Prescott, B., Podhorodeski, R.: Teleoperated system performance evaluation. Systems, Man and Cybernetics, IEEE Transactions on 24(5), 796–804 (1994)
8. Volbracht, S., Shahrbabaki, K., Domik, G., Fels, G.: Perspective viewing, anaglyph stereo or shutter glass stereo? Visual Languages, 1996. In: Proceedings, IEEE Symposium, pp. 192–193. September 3-6 (1996)
9. Adlin, T.: Human Factors Issues Associated with Image-Compression for Low Data Rate Remote Driving. In: Proceedings of SPIE, vol. 1700, pp. 489–502 (1992)
10. Douglas, E.M.: Experiences and Results in Teleoperation of Land Vehicles. SANDIA REPORT SAND90–0299 UC–515 (printed, April 1990)

# Teleoperation of a Surgical Robot Via Airborne Wireless Radio and Transatlantic Internet Links

Mitchell J.H. Lum[1], Diana C.W. Friedman[1], Hawkeye H.I. King[1], Regina Donlin[1], Ganesh Sankaranarayanan[1], Timoty J. Broderick[2], Mika N. Sinanan[1], Jacob Rosen[1], and Blake Hannaford[1]

[1] University of Washington, BioRobotics Lab, Seattle, WA, USA
{mitchlum,dwarden,hawkeye1,rdonlin,ganeshs,mssurg,rosen, blake}@u.washington.edu

[2] University of Cincinnati, Center for Surgical Innovation, Cincinnati, OH, USA
brodertj@ucmail.uc.edu

**Summary.** Robotic assisted surgery generates the possibility of remote operation between surgeon and patient. We need better understanding of the engineering issues involved in operating a surgical robot in remote locations and through novel communication links between surgeon and surgery site. This paper describes two recent experiments in which we tested the RAVEN, a new prototype surgical robot manipulation system, in field and laboratory conditions. In the first experiment, the RAVEN was deployed in a pasture and ran on generator power. Telecommunication with the surgical control station was provided by a novel airborne radio link supported by an unmanned aerial vehicle. In the second experiment, the RAVEN was teleoperated via Internet between Imperial College in London and the BioRobotics Lab at the University of Washington in Seattle. Data are reported on surgeon completion times for basic tasks and on network latency experience. The results are a small step towards teleoperated surgical robots which can be rapidly deployed in emergency situations in the field.

## 1 Introduction

Remote environments limit access to power and telecommunication resources needed by telesurgery systems. Mobile Robotic Telesurgery (MRT) systems that allow a remote surgeon to operate on a patient, regardless of their location or environment, need to overcome these resource limitations. In June 2006, we conducted experiments in pasture land outside Simi Valley, California using a novel MRT system comprised of the RAVEN surgical robot and an Unmanned Aerial Vehicle (UAV)-based communication system. The primary objective of this project was to demonstrate in the field a MRT system that would eventually allow a remote surgeon to operate on a patient regardless of their location or environment.

Network limitations including latency, bandwidth, jitter, packet loss and loss of signal, make telesurgical intervention in extreme environments difficult. While other existing topologies such as geosynchronous satellites can provide wireless communication, long latency precludes robust use of satellite communications in

C. Laugier and R. Siegwart (Eds.): Field and Service Robotics, STAR 42, pp. 305–314, 2008.
springerlink.com 

telesurgery. Geosynchronus orbit is 35,900 km above the earth's surface, and a complete loop from surgeon to robot and back requires two trips up and down: a 0.48 second round trip at the speed of light.

Various types of UAVs can operate at altitudes from 100 to 20,000 meters and carry equipment to provide robust, low-latency, high quality communications. In these experiments, the communication link was provided by a wireless digital datalink onboard a Puma UAV by AeroVironment Inc. (Simi Valley, CA). The Puma is a 2-meter wingspan, Small Unmanned Aerial Vehicle (SUAV) that flies at altitudes below 5000 meters above sea level and can provide line-of-sight communication up to a distance of 12 km with low gain antennas and 20 km with higher gain antennas. Approximately 4,000 Pumas are currently deployed across the world in military applications.

**Literature Review.** The earliest teleoperators, invented by Ray Goertz of Argonne National Labs in the late 1940's and early 1950's [1] are mechanical hands, coupled through a system of cables and pulleys to a remote handle which the operator can control at a safe distance. These mechanisms were very effective and are still in use, but are limited by their mechanical nature to short distances and similar size scales.

After continuous development at many labs [2, 3], teleoperation technology has matured enough in the last 15 years to be applied to surgery. By connecting the surgeon to the tools through teleoperators, the tools can be made smaller than the human hand, more dexterous in small body cavities, and, perhaps most revolutionary, can connect surgeon and patient across large distances.

In the early 1990's SRI International developed a two-handed teleoperated surgery unit for DARPA [4]. This highly influential project encouraged the startup of two companies to address the civilian surgery market, Computer Motion Inc. (Goleta, CA), and Intuitive Surgical Inc. (ISI) of Silicon Valley, CA. Both of these companies developed FDA approved surgical robot systems, the Zeus from Computer Motion and the da Vinci from ISI [5]. In 2003 they merged under the name of ISI. Over 300 da Vinci systems are in use around the world today. Both of these systems are teleoperators, but neither had the capability to separate the surgeon and patient by more than a few feet.

In 2001, a team led by Dr. Jacques Marescaux worked with Computer Motion to develop a specially modified Zeus system to perform the first remote surgical procedure (laparoscopic cholecystectomy) on a human patient [6, 7]. Dr. Marescaux controlled the robot from New York, and the patient was in Strasbourg, France. Using a similarly modified Zeus, Dr. Mehran Anvari of McMaster University, has treated about 25 patients in Northern Canada from his offices in southern Ontario [8]. Recently, Dr. Timothy Broderick has evaluated remote surgical technology in extreme environments. With support from the US Army and NASA, in 2005 he completed the NEEMO 9 mission — a two week stay in the Aquarius undersea habitat, maintained by the US National Oceanographic and Atmospheric Administration (NOAA), 19 meters underwater off the Florida Keys. An updated version of SRI's manipulator was deployed in this cramped habitat and Dr. Anvari successfully operated the robot from Canada with up to

2 seconds of time delay [9], performing simulated surgical procedures and the handling of simulated lunar rocks.

A major user of remote surgical technologies could be the military. Can combat casualties be reduced or ameliorated by getting surgical care to the soldier faster than getting the soldier to the care? Surgeons are a scarce resource in the military; with remote surgical technology they could be rapidly deployed right to where they are needed, switching electronically from one battlefield to another in seconds, without the need to "scrub in".

The world has seen a tremendous loss of life due to natural disasters. The scope of MRT applications will reach the civillian market as well, providing emergency medical care to disaster zones. Rather than a limited number of surgeons and emergency medical personel working non-stop in the disaster area, remote surgery relief stations could be deployed to the region and surgeons from around the world could connect in, providing trauma and other critical healthcare.

The need for small and deployable surgical telerobotic systems led to recent development of a new surgical robot, the RAVEN, by the authors at the University of Washington [10]. This manuscript presents experiences from the first field trials. The extreme environment provided difficulties that resulted in little time for rigorous evaluation of the quality of surgical performance but did lay the initial ground work for future experiments.

## 2 Methods

**RAVEN Surgical Robot System.** A detailed discussion of the RAVEN's design was reported in Feruary 2006 [10]. The system consists of a patient site (slave), a surgeon site (master) and a communication link between them. The patient site (Figure 1) consists of two surgical manipulators that are positioned over the patient by passive macro-positioning arms. The surgeon site consists of a surgeon console featuring two PHANToM Omni devices to control each of the surgical manipulators, a foot pedal, a surgeon's interface screen and a video feed from the patient site. The communication link between patient and surgeon sites is a standard Internet connection using UDP connectionless protocol for minimum time delay.

**Field readiness preparations.** Many research systems are conceived, developed and tested in a lab environment. To fully demonstrate the applicability of a system it must be deployed into the environment for which it was designed. The ability to provide surgical care in rural, remote or extreme environments will allow surgeons to provide immediate care in battlefield or disaster emergencies. Thus, the RAVEN must not only operate in field conditions but must also be easy to transport and quick to set up.

In the lab, the RAVEN is set up on a fully articulated operating table that is large, heavy, and cumbersome. For field testing the RAVEN manipulators are mounted on a small portable operating table, fabricated from modular aluminum extrusions. It was rigid and heavy enough to be very stable, but light enough that

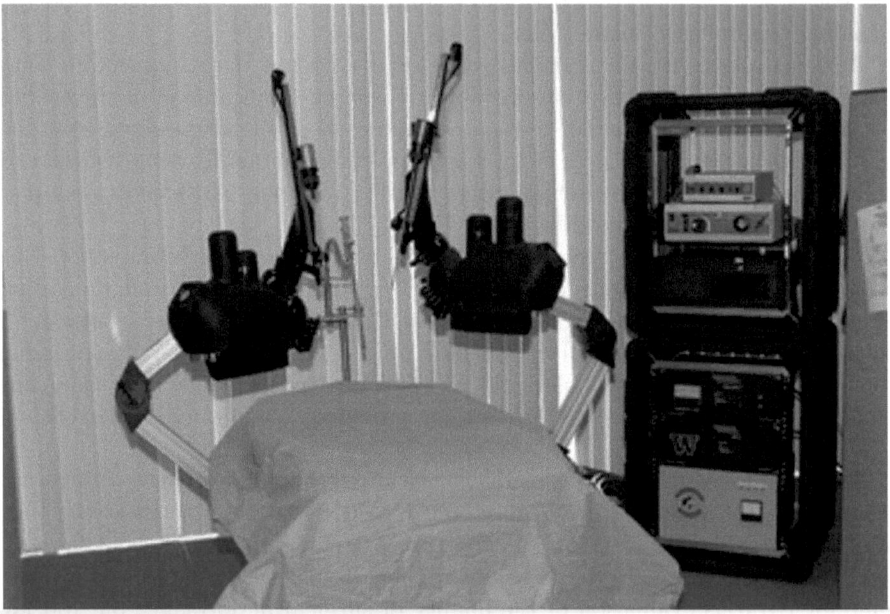

**Fig. 1.** Photograph of the surgical robot system being checked and tested at AeroVironment's headquarters prior to field deployment.

two people could easily move it. The table has adjustable mounting locations for the positioning arms, creating a quickly adjustable platform upon which the surgical manipulators and the "patient" (Figure 2) were mounted.

Power-off brakes on the surgical manipulators are a key piece of the safety system. If the system loses power the brakes will automatically engage, keeping the manipulators' tools from falling due to gravity. However, during packing/unpacking and setup it is necessary to reorient the arms of the surgical manipulators. A battery override provides a simple and inexpensive solution. On each manipulator a 1/8" headphone jack was wired in parallel to existing brake circuitry. When the manipulators are disconnected, a 9V battery connected to an 1/8" headphone plug can be used to power the brake circuitry. This is particularly useful when removing the manipulators from the positioning arms and placing them in their transport cases.

The major evironmental concerns we expected to face in the desert included dust, heat and a lack of clean power. To protect the motor packs against dust and debris, 3-piece covers were rapid prototyped out of ABS plastic. Vents allow motor heat to dissipate and fan mounts were included so that active cooling from standard 2.5" computer fans could be added. In the field, the whole system was powered by construction grade internal combustion 110V electric generators. To protect the computers and sensitive electronics from generator power spikes, a 1200W line regulator from APC was used at the patient site.

For safe transport of the system, custom foam lined cases for the surgical manipulators, surgical tools, and the master console devices were obtained from H.H. Bauer Cases (Vancouver, WA). The major electronic components, including computers, power supplies, Maxon brushless motor amplifier and our USB2.0 interface device, were mounted inside two SKB Industrial Roto-Shock Rack cases. The Roto-Shock cases provide shock isolation between the components, mounted to its internal frame, and the exterior hard-shell. The entire system, including the two surgical manpulators, portable OR table, SKB cases, surgeon console, tools and back-up equipment, fit easily in a Chevy Express Cargo van. The system could be deployed and set up in under an hour.

**Experimental Protocol.** An experimental protocol was developed to demonstrate system dexterity including touching a series of landmarks on a latex glove stretched over a box (Figure 2). The gloved box was marked with a circle, and a grid of 10 landmarks spaced 1 cm apart left to right and 0.5 cm apart toward and away from the camera. The landmarks were numbered 1-10 row by row starting with 1 at the upper left, and finishing with 10 at the lower right. The following five tasks were part of the experimental protocol.

1. Right hand touch each landmark in numeric order.
2. Left hand touch each landmark in numeric order.
3. Touch each landmark in numeric order using alternating left and right hands. Right hand touches the odd numbered landmarks, left hand touches the even numbered ones.
4. Right hand trace inner edge of circle clockwise.
5. Left hand trace inner edge of circle clockwise.

The tasks on the gloved box tested both the surgeon and manipulators ability to perform bi-manual gross positioning tasks. Beyond the experimental protocol, a surgeon performed intracorporeal knot tying (suturing) on the gloved box to demonstrate dexterous manipulation.

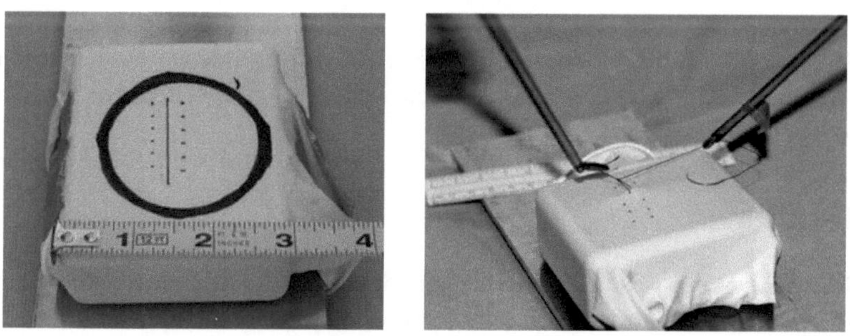


**Fig. 2.** (left) Experimental protocol was performed on a rubber glove stretched over a small box (right) Suture tied sucessfully on a latex glove

## 3 Experiments

**Telesurgery Field Experiment: Simi Valley, CA.** In early June of 2006, the RAVEN was driven from Seattle, WA and deployed and tested a few kilometers north of Simi Valley, CA for telesurgery experiments on an inanimate model.

The fielded system consisted of two stations connected by a wireless datalink relayed by UAV (Figure 3). The system was powered by generators and was set up under portable tents in an isolated field. Separated by a distance of 100 meters, the patient site (surgical manipulators) and the surgeon site (master control console) were connected via a wireless digital datalink relayed by AeroVironment's PUMA unmanned aircraft up to 1 km away at an altitude of about 100 meters. The datalink provided by AeroVironment utilized Internet Protocol communication at a rate of 1MB per second between the two sites allowing the master/slave communication architecture to remain unmodified for this experiment.

A single NTSC color video signal was provided using a digital video camera which acquired a closeup picture of the patient site. A second video channel was sometimes used to display the overall operating environment. HaiVision Inc. (Montreal, Canada) provided a hardware codec that transmitted the video signal at 800kbps in MPEG-2 format.

During three days of field deployment, kinematic data of the surgeons' commands and the surgical manipulator motions were collected along with network characterization data as two surgeons performed the experimental protocol described above. Control information from the master side was transmitted at 100 packets per second in the field trial and 1000 packets per second in a later Transatlantic trial (below).

**Transatlantic Experiment: Seattle, WA, USA to London, England.** In July 2006, the system was tested in a long distance teleoperation experiment in collaboration with Imperial College (London, England). At the surgeon site two PHANToM 6-DOF Premium haptic devices were equipped with our surgeon console software. iChat software (Apple Computer, Inc) was used for close-up video feedback of the surgical site. Skype video transmitted an overall view of the operating room. Audio, video, and RAVEN control signals were all sent through standard Internet. Two surgeons performed the same experimental protocol described above.

## 4 Results

The RAVEN system was originally designed so that the kinematic control signals would be sent from the surgeon site to the patient site at a 1kHz rate. During the field experiment packet loss through the UAV's wireless data link became problematic. When the kinematic control signals were transmitted at full rate the packet loss was approximately 80%. The bandwidth of the system was scaled back so that kinematic control signals were sent at 100Hz and video bandwidth was 800kB/s, resulting in reduced packet loss of between 3%-15%. The surgeons

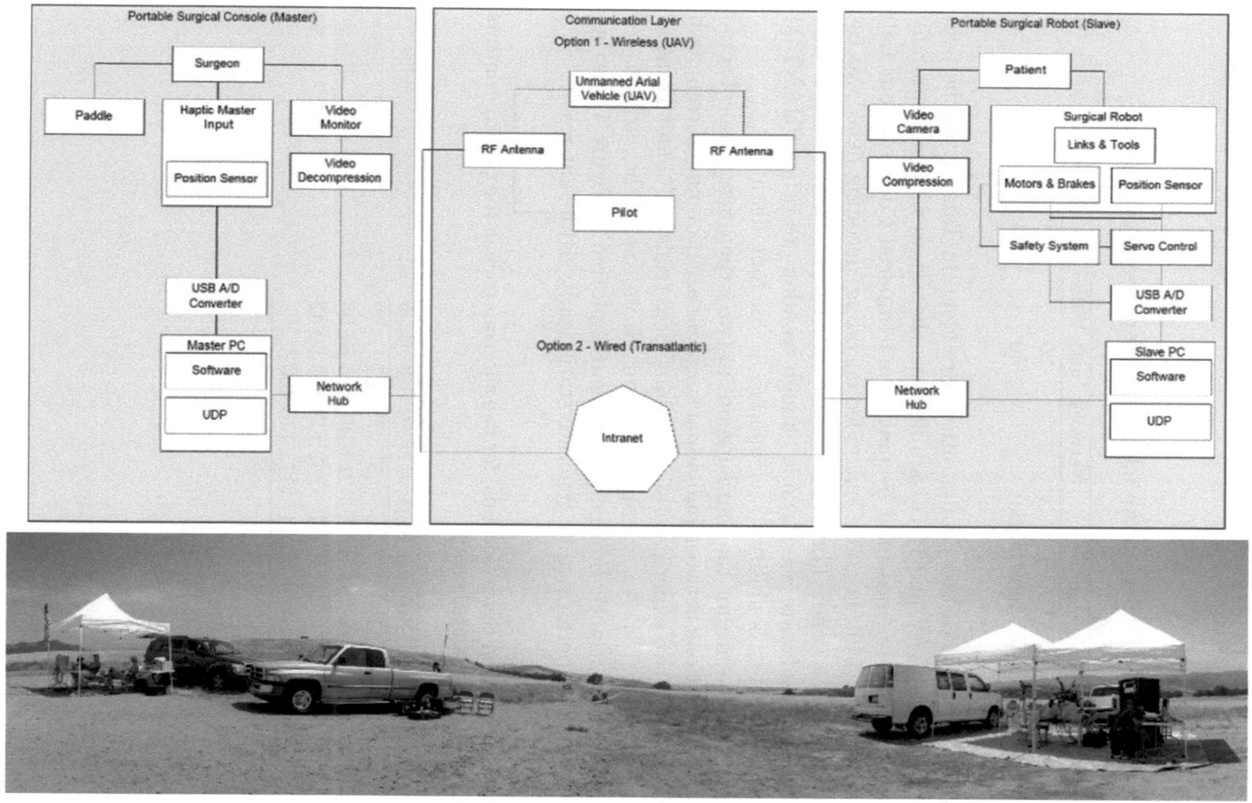


**Fig. 3.** Telesurgery experimental setup. (top) Block diagram of the system with UAV wireless link and wired Internet connection options. (bottom) Panoramic photo of the field site outside Simi Valley, CA. The surgeon site at left, patient site at right. UAV is visible in center of photo on ground.

did not notice any decrease in manipulator performance but did comment on increased pixelation in the video feed. While decreased video quality was noticable they did not feel it affected their task performance. During the transatlantic experiments no network bottlenecks were encoutered and the system was run at full bandwidth.

**Table 1.** Time delays in two telerobotic surgery configurations

| Configuration | Network (ms) | Video (ms) |
|---|---|---|
| Wireless (UAV) | 20 | 200 |
| Wired (Transatlantic) | 140 | 1000 |

Time delays for the two test configurations are given in Table 1. In the field experiment, with the UAV and HaiVision devices, the delays were substantially shorter. Internet latency (measured by ping) from Imperial College to our Lab in Seattle was about 140 ms and iChat video encoding/decoding delay was estimated at about 1 second. The transatlantic experiments showed both that the master console software was flexible enough to adapt to other PHANToM devices, and the RAVENs ability to teleoperate across long distances.

Task performance data for the four surgeons in two experiments are collected in Table 2. The variations between individual surgeons are pronounced compared to those between the two experimental conditions: field trials ("F") and Transatlantic Internet link ("T"). An example trajectory from one of the surgeons for Task 4 from the transatlantic Internet experiment is plotted in Figure 4. Further experiments must be performed to generate statistically significant results.

**Table 2.** Task completion times (seconds) for two surgeons each in the two experiments (F for field, and T for transatlantic)

| Subject | Exp. | 1 | 2 | 3 | 4 | 5 |
|---|---|---|---|---|---|---|
| 1 | F | 57 | 75 | 63 | | |
| 2 | F | 213 | 161 | 323 | 64 | 98 |
| 3 | T | 52 | 118 | 110 | 32 | 42 |
| 4 | T | 173 | 146 | 430 | | |

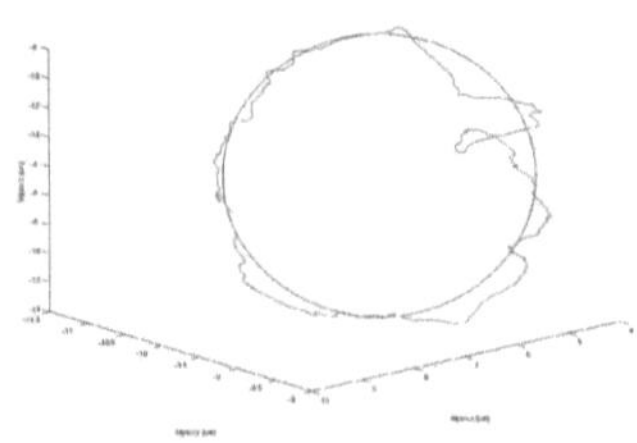

**Fig. 4.** Tool tip trajectory of surgeon performing Task 4 (see text) in the transatlantic Internet link experimment. Circle overlayed for perspective.

## 5 Conclusions

We have reported initial tests of a surgical robot in the field and when controlled over a global Internet link. Preliminary data indicate that under both conditions the robot can be effectively controlled in basic positioning tasks. The rigors of the field test exposed some weaknesses such as low generator voltage output (which automatically shut down our power conditioners) and poor control of tool orientation due to mechanical interferences inside the mechanism (which have now been fixed).

For some applications, the 1-10 km range of the low altitude UAV will limit applicability. Future MRT experiments using a High Altitude Long Endurance (HALE) UAV (Aerovironment Inc.), to provide ranges on the order of 10-50 km, will allow further reaching emergency care to areas where conventional telecommunication connection is unavailable.

A novel feature of our current system is a low cost highly portable surgeon user interface built entirely from commercially available hardware [11]. In our field trial, we obtained very high video quality using dedicated codec hardware. In our intercontinental trial, we substituted freely available video chatting software with surprisingly good results. Although it is unlikely that basic technology like iChat video will statisfy clinical surgeons, it provides a very convenient platform for experimentation. Our low cost surgeon console facilitated the ease with which we were able to establish our intercontinental teleoperation collaboration. It will allow for cost effective collaboration with many more labs in the future.

Futher development of the RAVEN will focus on increasing robustness of the system and a more rigorous scientific method for evaluating telerobotic surgical tasks. Deploying the system into a field environment and executing these experiments demonstrated the feasibility of Mobile Robotic Telesurgery (MRT) in remote environments and with long communication links. Development in this area will allow patients in rural, remote or extreme environments to receive lifesaving surgical interventions.

**Acknowledgements.** The authors would like to acknowledge support from the US Army TATRC program grant number W81XWH-05-2-0080, our colleagues, Brett Harnett, Charles Doarn and Lynn Huffman at the University of Cincinnati, Julian Leung, George Mylonas, Sir Ara Darzi and Ghuang-Zhong Yang at Imperial College in London, those at AeroVironment, and HaiVision Inc., and contributions from Clint Bland, Hiroki Mizumi and Levi Miller in the BioRobotics Lab.

## References

1. Goertz, R.C., Thompson, W.M.: Electronically controlled manipulator. Nucleonics (November 1954)
2. Bejczy, A.K.: Sensors, controls, and man-machine interface for advanced teleoperation. Science 208, 1327–1335 (1980)
3. A.K., Szakaly, Z.: Universal computer control systems (uccs) for space telerobots. Proceedings, IEEE ICRA 4 (March 1987)

4. Hill, J.W., Green, P.S., Jensen, J.F., Gorfu, Y., Shah, A.S.: Telepresence surgery demonstration system. Proceedings, IEEE ICRA 3, May 8-13 (1994)
5. Guthart, G.S., Salisbury, J.K.: The intuitive telesurgery system: Overview and application. Proceedings IEEE ICRA (April 2000)
6. Marescaux, J.: Transatlantic robot-assisted telesurgery. Nature 413 (September 2001)
7. Marescaux, J., Leroy, J., Rubino, F., Smith, M., Vix, M., Simone, M., Mutter, D.: Transcontinental robot-assisted remote telesurgery: Feasibility and potential applications advances in surgical technique. Annals of Surgery 235(4) (April 2002)
8. Pirisi, A.: Telerobotics brings surgical skills to remote communities. The Lancet 361(9371)
9. Anvari, M., Broderick, T., Stein, H., Chapman, T., Ghodoussi, M., Birch, D.W., Mckinley, C., Trudeau, P., Dutta, S., Goldsmith, C.H.: The impact of latency on surgical precision and task completion during robotic-assisted remote telepresence surgery. Computer Aided Surgery 10(2) (March 2005)
10. Lum, M.J.H., Trimble, D., Rosen, J., Fodero, K., King, H., Sankaranarayanan, G., Dosher, J., Leushke, R., Martin-Anderson, B., Sinanan, M.N., Hannaford, B.: Multidisciplinary approach for developing a new minimally invasive surgical robot system. In: BioRob Conference, Pisa, Italy (February 2006)
11. Sankaranarayanan, G., King, H.I., Ko, S.Y., Lum, M.J.H., Friedman, D.C.W., Rosen, J., Hannaford, B.: Portable surgery master station for mobile robotic telesurgery. In: ROBOCOMM, Athens, Greece (2007)

# Part VIII

# SLAM

# SIFT Based Graphical SLAM on a Packbot

John Folkesson and Henrik Christensen

[1] Dept. of Mech. Engineering
Massachusetts Institute of Technology
Cambridge, MA 02139
johnfolk@mit.edu

[2] Center for Robotics and Intelligent Machines
Georgia Institute of Technology,
Atlanta, GA 30332
hic@cc.gatech.edu

**Summary.** We present an implementation of Simultaneous Localization and Mapping ($SLAM$) that uses infrared ($IR$) camera images collected at 10 Hz from a Packbot robot. The Packbot has a number of challenging characteristics with regard to vision based SLAM. The robot travels on tracks which causes the odometry to be poor especially while turning. The IMU is of relatively low quality as well making the drift in the motion prediction greater than on conventional robots. In addition, the very low placement of the camera and its fixed orientation looking forward is not ideal for estimating motion from the images. Several novel ideas are tested here. Harris corners are extracted from every $5^{th}$ frame and used as image features for our SLAM. Scale Invariant Feature Transform, $SIFT$, descriptors are formed from each of these. These are used to match image features over these 5 frame intervals. Lucas-Kanade tracking is done to find corresponding pixels in the frames between the SIFT frames. This allows a substantial computational savings over doing SIFT matching every frame. The epipolar constraints between all these matches that are implied by the dead-reckoning are used to further test the matches and eliminate poor features. Finally, the features are initialized on the map at once using an inverse depth parameterization which eliminates the delay in initialization of the 3D point features.

## 1 Introduction

The goal of much research over the last decade has been to give robots a sense of location and direction comparable to that of humans. Humans have very limited sense of their ego motion with eyes closed. With eyes opened one can move with much more confidence. Judgements of distance and orientation are greatly improved. Comparable automatic robot vision capabilities have not yet been demonstrated. Much progress has been made [1], [2], [3], [4], [5],[6].

The problem has been given the name of vision based simultaneous localization and mapping SLAM. An issue in making maps from features seen in images is the lack of depth information in a single image [7]. Most of the SLAM methodologies estimate a maximum likelihood solution given the motion and feature measurements. This estimate typically assumes that the map and robot pose can be described by Gaussian distributions. Features seen in a single camera image correspond to real 3D objects. The probability distribution of such an object's location given the image has the shape of an infinite cone

C. Laugier and R. Siegwart (Eds.): Field and Service Robotics, STAR 42, pp. 317–328, 2008.
springerlink.com 

in Cartesian 3 space. This causes considerable difficulty for SLAM methods that use xyz as the feature representation. This problem has been cleverly solved [8] by utilizing a representation in terms of the bearing from the pose of the camera at the first observation and the inverse depth in that frame. The 6 parameters of the camera pose for this frame also become part of the parameterization. The result is a 9 parameter representation of the 3D point whose distribution can be well approximated by a Gaussian.

Another issue in vision based SLAM is the extraction of pixel level features from images [9], [10] and matching these with images of the same objects from different vantage points. Scale Invariant Feature Transform, SIFT, features have been use for SLAM [11], [12]. SIFT features have relatively weak descriptors associated with them. These descriptors have the advantage of invariance with respect to scale. They are also fairly similar for changes in viewing angle up to about 15-20 degrees. The descriptors alone can not be used to match features however and must be combined with other evidence such as epipolar constraints or the relative locations of other matching points.

## 2 Packbot Robot

Figure 1 shows the Packbot robot. This robot is designed to reach inaccessible places. It can travel up and down stairs, on its back and even withstand drops of several meters. It communicates over a wireless Ethernet connection. To help it stay localized, it has GPS, an inexpensive IMU and encoders on its two tracks. The GPS is of no value indoors and the robot is meant to be used in GPS deprived areas. The encoders on the tracks give reasonably good information on distance traveled while the robot moves straight on level ground with good traction. Obviously there is much error when slippage occurs on turns and poor surfaces. The IMU helps reduce this somewhat but can not correct all of the errors.

The robot is also equipped with two cameras, one a wide angle camera and the other an IR camera. We have used the IR camera to improve the motion estimate by doing

**Fig. 1.** The Packbot robot has a rectangular window in the front. Behind this sits an IR camera and a wide angle camera. Left is an example of one of the IR images from the Packbot. The green + signs are detected Harris corners. The size of theses + signs indicates the scale of the corresponding SIFT descriptor.

SLAM on Harris corners extracted from the images matched using SIFT descriptors. Figure 1 shows a typical image from the IR camera. As can be seen in this image Harris corners in the image do not always correspond to well defined objects in 3 dimensional space. Notice the reflection of the window in the glass door of the bookcase on the left of the image. Also notice the intersection of objects inside the far room with the edge of the doorway that give rise to Harris corner image features. Distinguishing these type of 2D image features from ones that are better defined in the real 3D world is the challenge.

## 3 Graphical SLAM Method

The Graphical SLAM method has been presented in [13] and [14]. We refer to those papers for the mathematical details. The measurements of the motion over some interval of time from the encoders and the IMU give us some probability distribution over the robot poses at the beginning and end of the interval. This is the probability of the measurements given the state (times a normalization factor that does not depend on the state). The negative log of this is the energy contribution of the dead reckoning over the interval. Thus the tradeoff between different measurements of the motion can be made based on a minimization principle on a graph of robot pose nodes.

The camera images give us addition information on the motion and additional states corresponding to features seen in the images. These additional feature states become feature nodes in our graph. The measurements of them by the camera create energy terms relating them to the robot poses. So that the graph consists of state nodes for the 6 dof[1] robot poses at the instant in time of the camera frames and the feature nodes of the 3D points that gave rise to image features in those frames. The connections between these state nodes is through energy nodes that correspond to measurements. The energy nodes can be connected to one or more state node[2].

The main advantage of working with this graph construction is that information can be gathered more flexibly and the linearization errors are greatly reduced. The difference is that we do not marginalize out pose states from the system. Having the pose states allows one to add and remove measurements from the solution. One can change data associations after testing how the energy would be changed. As a general principle one can say that it is easier to find representations that fit well to Gaussian distributions by increasing the dimension of the state space. This is true for the robot pose and, as we will see, for the feature part of the state as well.

The down side of increasing the dimensions is higher computation cost. In fact the full graphical solution gets unmanageable very quickly. However two escapes can be found. First the growth of the graph by adding more measurements does not change the solution for states far removed from the added measurements. That is unless the measurements include some new global constraints such as closing of a loop. So as long as the new measurements are not matched to older features one can simple relax the end of the graph, a constant time update. This leaves out the global localization problem completely. One is concerned with local accuracy only and the global solution is not important.

[1] Degrees of freedom: $(x, y, z, \theta, \phi, \psi)$.

[2] Gravity or compass measurements give rise to energy nodes attached to only one state node.

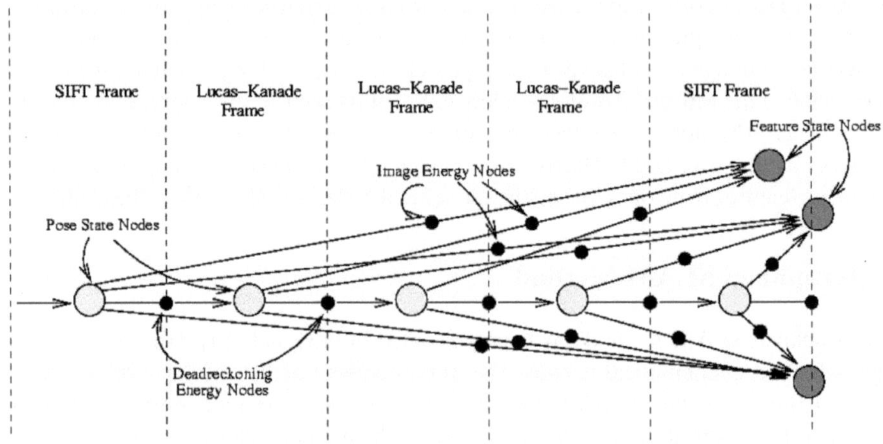


**Fig. 2.** Here we show a 5 frame segment of the graph. There are 3 Lucas-Kanade frames in this illustration. We used 4 such frames in our experiments.

The other escape is to marginalize out part of the state from mature sections of the graph. This will allow global matching with reasonable complexity while maintaining local accuracy of the high dimensional representation. This is done by forming what we call star nodes. Star nodes are simple an energy node that replaces a set of energy nodes with a linear approximation of their energies. Having linearized the energy, one can then marginalize out the pose nodes that only connect to this one star node. By doing this in a local frame one obtains invariance with respect to large transformations of sections of the map. Thus once can still rotate the solution by say 20 degrees in order to close a loop without invalidating the linearizations.

One thus maintains the full state for the part of the graph connected to currently observed features. Then for the older section one forms a star node that marginalizes out a section of the robot path. The star node is allowed to grow until it has eliminated a predetermined number of pose nodes or has a maximum number of features. Then a new star node is begun.

## 4 The Feature Tracking

The SLAM algorithm is only one part of an implementation. The results will also depend on the tracking/matching of features in the sequence of image frames. Harris corners have an advantage over other image features[3] in that they have well defined pixel locations in the image. Those locations may or may not correspond to well defined 3D locations in the real world.

The matching of the SIFT descriptors normally gives a good indication of pixel level correspondences between frames. The difficulty in this implementation was that it took

[3] Such as for instance, difference of Gaussians.

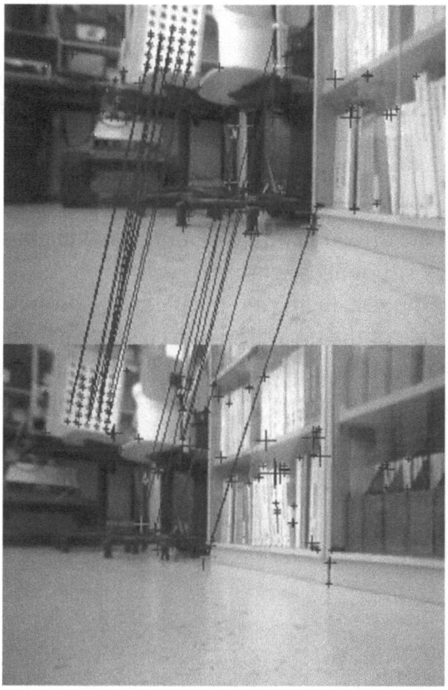

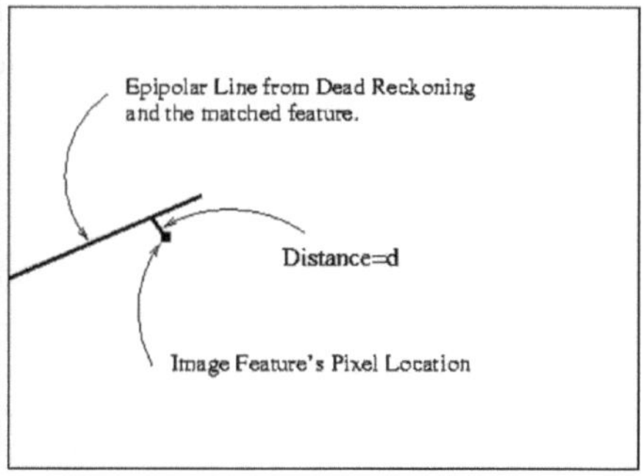


**Fig. 3.** The left is one example of poor SIFT matching between frames. Here the robot motion between the images leads to significant change in the images. The nature of the scene is leading to similar SIFT descriptors for different parts of the image. Still many of the points match correctly. Notice however that the calibration grid that happens to be in the frame is in fact confusing the SIFT descriptor matching. The repetitive pattern is the problem. On the right we show the geometry of the constraint energy nodes. We have two bearings, $\mathbf{b}_1$ and $\mathbf{b}_2$ to the point feature and an estimate of $\mathbf{Dx}$ the transformation between the camera frames. The energy of this constraint would be $\theta^2/(2C)$, where C, the measurement variance in $\theta$, is found from the errors in the bearing angles. The two ranges $r_1$ and $r_2$ are calculated to minimize $d$. In the case of poorly defined ranges as when the bearings are in the $\mathbf{Dx}$ direction we simply use a default distance of 5 m for $r_1$.

about 100 msec per image to extract the Harris corners and calculate the SIFT descriptors for them. We would like to use a 10 Hz frame rate as the images can change rapidly as the robot moves about. We can only hope to achieve cpu utilizations of around 50-70% in the real time application. Add to this the need to do SLAM calculations and one realizes that SIFT descriptors can not be extracted from every frame.

The solution we chose was to do Harris/SIFT extraction on every $5^{th}$ frame. Then for the intervening frames we interpolate the SIFT matches on each side using Lucas-Kanade tracking [15]. This tracking maximizes the correlation to a templet of the pixels around our feature to achieve sub pixel tracking accuracy. When the tracking failed to join the two bordering SIFT points smoothly the tracked pixel matches were not used. In this way we could reduce the utilization by a factor of 5 giving us a margin for the other computations that needed to be done.

Having tracked an image feature over 6 frames we apply the epipolar constraint to the motion in the image sequence. This we do in by utilizing our graph of pose nodes. We form temporary constraint energy nodes connecting two poses. The energy of these constraint nodes is defined in figure 3. There we have shown the two camera frames as having no relative rotation. The actual deadreckoning estimate is for the full 6 dof transform between frames including any rotation. This is then used to find the constraint. One advantage of this definition of the constraint energy is that the epipolar constraint can be applied for features seen in the direction of the motion. Such features have no epipolar line but the constraint energy still makes sense.

These constraints are one dimensional and three of them can specify the rotation beween two frames given the translation. The scale of the the translation can not be found but the energy is a function of the translation direction as well. We can differentiate the constraint energy with respect the pose states. This will give us a gradient and hessian to relax the graph with.

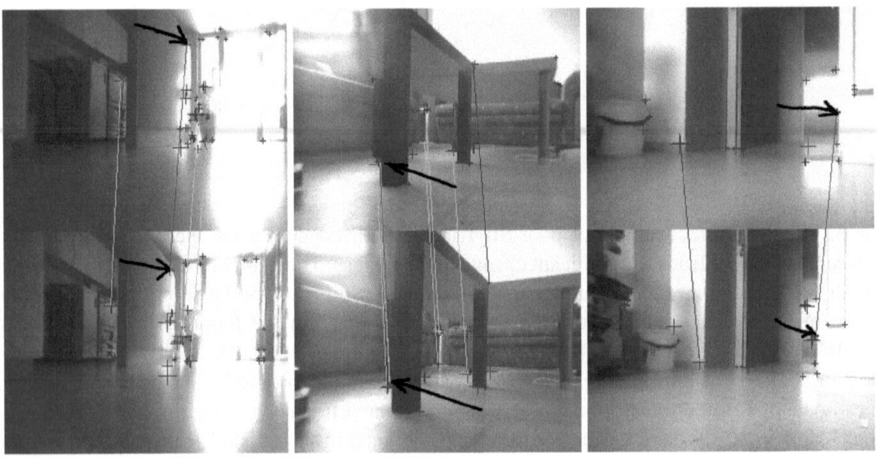

**Fig. 4.** Here the dark (red) lines connect the matches that became fully initialized Cartesian points. The light gray lines connect matches that were used with the inverse depth parameterization to create constraints between frames. The purple lines show features that have been identified as 'bad' (indicted also by the arrows). These image features were formed by two objects at differing depths or reflected light on a curved wall (furthest left).

We add such constraint nodes for all the new correspondences that we found by SIFT matching and Lucas-Kanade tracking. Then we relax the pose states attached to these constraint nodes. The resulting energy of each constraint is then tested. If any are above some threshold we mark the image feature as bad and remove all its energy nodes. Image features marked as bad will be kept to be matched to again and any future image features that match to bad features will not be used. It is hoped that this will eliminate some of the occlusion type features, see figure 4. We then relax the graph again until no constraint energy is over the threshold.

Having tested the epipolar constraints we remove the temporary constraint nodes. By modeling the 3D points that gave rise to the image features we can form two dimensional constraints between each pose and feature node. These are then stronger constraints than the epipolar one dimensional constraints between two pose nodes. They require forming 3D point feature state nodes.

We do this and then add energy nodes to the 3D point feature nodes. These energy nodes have an energy based on the innovation in the predicted pixel locations in the images. We do this without any initialization delay by using a special inverse depth representation as explained in the next section. We accumulate the information on each feature in this way. If the standard deviation in the estimated depth becomes less than 30% of the depth we initialize a full Cartesian represented point feature. Only such features will be added to a global matching database. This global database entry will include the surrounding SIFT points as part of the context used in matching. This global database was formed but not matched to here. Global matching is the next stage of this research.

Figure 4 shows some examples of the constraints used to improve the odometry. In these example there were some initialized Cartesian points as well as a some inverse depth type constraints. There were also features labled as bad. These were image features that did not correspond to well defined 3D points.

The benefit of features to the motion estimate decreases rapidly after about 3 or 4 matches are found between frames. On the other hand, matching to poor features or miss matching can cause the estimate to worsen. We therefore only keep the best matches (those that have the lowest constraint energy) plus any matches from any initialized Cartesian points. When features were scarce in the images we relaxed our threshold to allow some matches.

## 5 The Feature Representation

We did not begin by implementing the inverse depth parameterized features. Instead we began by implementing a conventional Cartesian coordinated representation for the features. This meant that we had to wait to initialize the point features until we had enough lateral motion to triangulate the feature location into good Gaussian ellipsiods in the Cartesian feature space. For features that did not resolve the depth sufficiently we simply left the epipolar constraint nodes on the graph. These pairwise constraints between poses are weaker than the constraints of forming a 3D feature but they do not require any depth. The results were an improvement over dead-reckoning alone but not nearly as good as that which we obtained by switching to the inverse depth parameterization.

The idea of the inverse depth parameterization is to use the pose the feature was first seen from as part of its parameterization. Then one takes the bearing angles or pixel locations in that frame as two other parameters. This eliminates the problem of the ever widening cone of possible feature location that a Cartesian xyz representation would have. By using the bearing angles as parameters the covariance is a well defined ellipse in the bearing space.

For the ninth and last parameter of the features one chooses the inverse of the depth of the feature. The reason for this is that the depth itself could initially be anything from a few cm to infinity. This range will be transformed to a finite range for the inverse depth.

The feature can then be initialized at once without needing to accumulate information on it. The distribution in this parameter space is well approximated by a Gaussian from the start.

This parameterization was rather easy to implement as we already had the pose as part of our state.

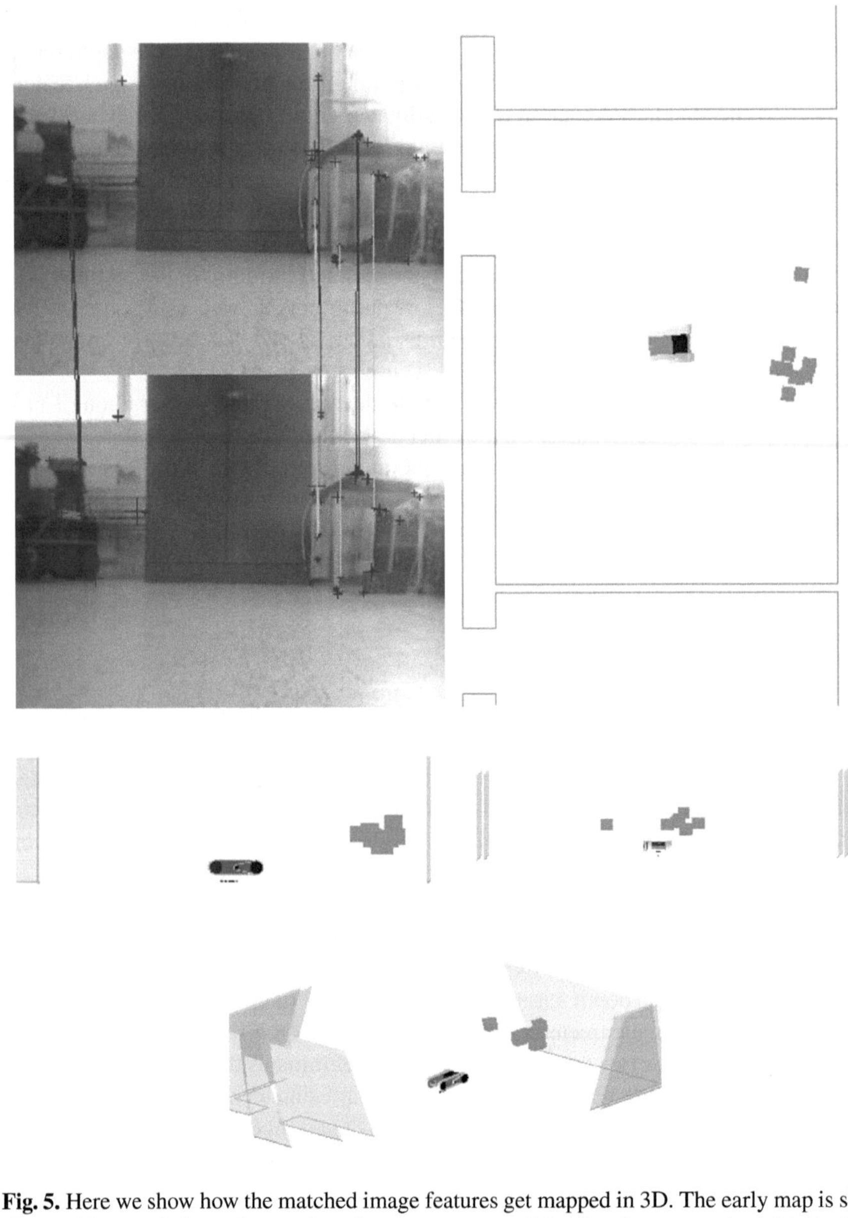

**Fig. 5.** Here we show how the matched image features get mapped in 3D. The early map is shown along with one of the early matches that shows some of the first features to be initialized. The map is shown from 4 angles top, side back and from a downward angle.

**Fig. 6.** We show on the left the dead-reckoning estimate of the robot path around a small loop. The corrected path and the map of point features (shown as squares) is shown on the right. The field of view is forward the whole time so the robot does not see the whole room as it moves. Also there is no attempt to match features seen at the start with those seen at the end.

**Table 1.** This shows the estimated ending pose for the robot around the small loop. The ending position for the dead-reckoning is off by about 18 cm and 0.13 rads. By using the IR camera and the inverse depth parameterized features we get about 4 cm error and .02 rads error. The error in positioning the robot to line up with the start point was less than 1 cm and .01 rads. 117 entries were made in the global database, (initialized Cartesian points) and there were 1230 frames.

| Test | x (m) | y (m) | z (m) | $\theta$ (rads) | $\phi$ (rads) | $\psi$ (rads) |
|---|---|---|---|---|---|---|
| True | 0.000 | 0.000 | 0.085 | -1.4300 | 0.0000 | 0.0000 |
| Dead-reckoning | 0.104 | -0.148 | 0.085 | -1.3088 | -0.0012 | -0.0012 |
| Inverse Depth Features | 0.042 | 0.001 | 0.087 | -1.4531 | -0.0006 | -0.0008 |

## 6 Experiments

The experiments were carried out in out lab. We collected the images and telemetry data transmitted from the robot on a laptop computer connected via the wireless Ethernet. The images were directly saved in the 320X240 grey-scale jpg format. The first stage of the feature extraction was an undistortion of the images using the openCV library.

The first path was a simple circle around a table in the middle of the room. By carefully marking the floor we were able to return to exactly the same pose at the end of the test. This gave us a relatively easy data set to work with and the ground truth of returning to the same pose. Note that global matching of the SIFT features was not

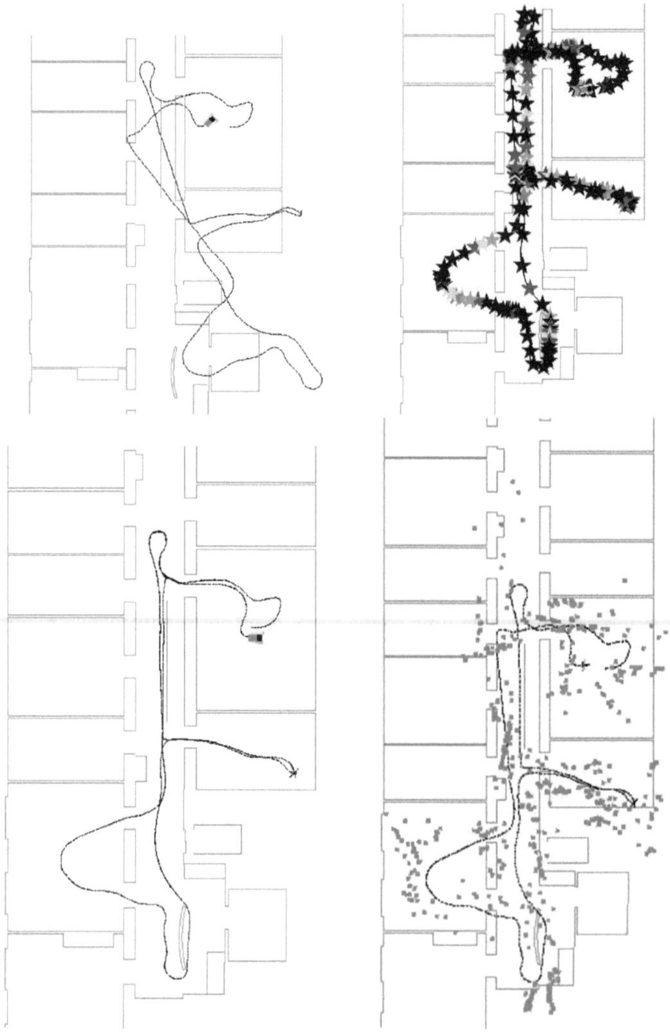

**Fig. 7.** On the top left we show the dead-reckoning estimate of the path for our second test. As one can see there are larger errors each time the robot turns. On the bottom left we show the best corrected path we were able to obtain using the image data. One can see the motion estimate can be rather good. This map was the result of excessive parameter tuning but represents the ideal result we would like to get more robustly. On the bottom right we have a more typical good result of the corrected path that one obtains from using the image data. There we show the features as well. Generally one always gets some improvement regardless of the tuning but the amount varies from a little better to very good depending on the tuning and the images. The top right shows another typical result where we have displayed the star nodes that form the local map patches in our graph. The two maps on the right have been rotated about the start position, (3 and 5 degrees), in order to line the average path up with the building. better. These maps had about 650 fully initialized Cartesian features and 8,600 frames.

done so the program would not recognize the features from the end as being the same as those at the beginning. This global loop closing is future work and not done here. Our goal here is to try and minimize the drift of the estimate, a sort of visual odometry.

Figure 5 shows how the map looks in 3D just after the first few points have been added. The results are shown in figure 6. As you can see there were no major problems in this small test and the estimate is nearly prefect. In table 1 we summarize the error correction around the loop.

The second test was a much longer path which lead out into the corridor and into two other rooms before returning to the exact starting location. The results are shown in figure 7. Here you see that the tuning of the filter had a significant impact on the results. A perfect loop closing without any global matching being done was not possible although we were able to get close on some occasions. The problem here as compared to the first case was not so much the length of the path but rather the variety of environments encountered. In some parts of the path there were simple no features found in the images. This a weakness that can not be overcome. In other cases occlusion type features could not be entirely filtered out of the data by the epipolar constraints.

## 7 Conclusion

Despite some unavoidable short comings we saw a consistent improvement over dead-reckoning and for areas that had many good features the results were near perfect corrections. This should give us a good base on which to build global matching and constraints on. The SIFT descriptors are after all designed specifically for global matching and should facilitate the next phase of this research.

The results indicated clearly that the inverse depth parameterization does solve the problem of initialization features seen in a single camera frame. This allowed us to enforce two dimensional constraints on the motion between two frames as compared to the one dimensional epipolar constraint. That in turn lead to better visual odometry estimates.

The use of Harris corners in combination with SIFT descriptors gave mixed results. We would like to have better criteria for selecting features in the images that would give features that were not only salient in the images but also gave bearings to true well defined 3D points. Our tracking method based on the epipolar constraint energy nodes did eliminate most of the bad image features.

The fundamental problem we encountered was that of lack of any features at all in many of the images. This was a problem when the field of view was a blank wall or contained a bright window that blinded the camera. The only way to recover from such a situation is to recognize some part of the environment later on so that loop closing can be done to correct the path.

## Acknowledgment

This work was supported by the Swedish defense agency FMV. This support is gratefully acknowledged.

## References


1. Davison, A.: Real-time simultaneous localization and mapping with a single camera. In: Proc. of the IEEE International Conference on Computer Vision (2003)
2. Davison, A., Cid, Y.G., Kita, N.: Real-time 3D SLAM with awide-angle vision. In: Ribeiro, M.I., Santos-Victor, J. (eds.) Intelligent Autonomous Vehicles (IAV-2004). IFAC/EURON, IFAC/Elsevier (2004)
3. Burschka, D., Hager, G.D.: V-GPS(SLAM): Vision-based inertial system for mobile robots. In: Proc. of the IEEE International Conference on Robotics and Automation (ICRA 2004), vol. 1, pp. 409–415 (2004)
4. Newman, P., Cole, D., Ho, K.: Outdoor SLAM using visual appearance and laser ranging. In: Proc. of the IEEE International Conference on Robotics and Automation (ICRA 2006), vol. 1, pp. 1180–1188 (2006)
5. Folkesson, J., Jensfelt, P., Christensen, H.I.: Vision SLAM in the measurement subspace. In: Proc. of the IEEE International Conference on Robotics and Automation (ICRA 2005) (2005)
6. Jensfelt, P., Kragic, D., Folkesson, J., Björkman, M.: A framework for vision based bearing only 3D SLAM. In: Proc. of the IEEE International Conference on Robotics and Automation (ICRA 2006), Orlando, FL (May 2006)
7. Sola, J., Monin, A., Devy, M., Lemaire, T.: Undelayed initialization in bearing only slam. In: Proc. of the IEEE/RJS International Conference on Intelligent Robots and Systems (IROS 2005), vol. 1, pp. 2499–2504 (2005)
8. Montiel, J., Civera, J., Davison, A.J.: Unified inverse depth parametrization for monocular SLAM. In: Proc. of the Robotics Science and Systems Conference (RSS 2006), vol. 1 (2006)
9. Harris, C.J., Stephens, M.: A combined corner and edge detector. In: Proc. 4th Alvey Vision Conference, Manchester, pp. 147–151 (1988)
10. Lowe, D.G.: Object recognition from local scale-invariant features. In: Proc. of the International Conference on Computer Vision (ICCV 1999), Corfu, Greece, pp. 1150–57 (1999)
11. Se, S., Lowe, D.G., Little, J.: Mobile robot localization and mapping with uncertainty using scale-invariant visual landmarks. International Journal of Robotics Research 21(8), 735–758 (2002)
12. Barfoot, T.D.: Online visual motion estimation using fastslam with sift features. In: Proc. of the IEEE International Conference on Intelligent Robots and Systems (IROS 2005). IEEE/RSJ, vol. 1, pp. 579–585 (2005)
13. Folkesson, J., Jensfelt, P., Christensen, H.I.: Graphical SLAM using vision and the measurement subspace. In: Proc. of the IEEE/RSJ International Conference on Intellegent Robots and Systems (IROS 2005) (2005)
14. Folkesson, J., Christensen, H.I.: Graphical SLAM for outdoor applications. Journal of Field Service Robotics (to appear, 2006)
15. Lucas, B.D., Kanande, T.: An iterative image registration technique with and application to stereo vision. In: Proc. of the International Joint Conference on Artificial intelligence (1981)

# Look-Ahead Proposals for Robust Grid-Based SLAM

Slawomir Grzonka, Christian Plagemann, Giorgio Grisetti, and Wolfram Burgard

University of Freiburg
Department of Computer Science
Georges-Koehler-Allee, 79110 Freiburg, Germany
{grzonka,plagem,grisetti,burgard}@informatik.uni-freiburg.de

**Summary.** Simultaneous Localization and Mapping (SLAM) is one of the classical problems in mobile robotics. The task is to build a map of the environment using on-board sensors while at the same time localizing the robot relative to this map. Rao-Blackwellized particle filters have emerged as a powerful technique for solving the SLAM problem in a wide variety of environments. It is a well-known fact for sampling-based approaches that the choice of the proposal distribution greatly influences the robustness and efficiency achievable by the algorithm. In this paper, we present a significantly improved proposal distribution for grid-based SLAM, which utilizes whole sequences of sensor measurements rather than only the most recent one. We have implemented our system on a real robot and evaluated its performance on standard data sets as well as in hard outdoor settings with few and ambiguous features. Our approach improves the localization accuracy and the map quality. At the same time, it substantially reduces the risk of mapping failures.

## 1 Introduction

The ability to construct models of natural and human-build environments is widely regarded as a precondition for autonomous service robots. Such models are required for basic tasks such as localization and motion planning. The combination of a Rao-Blackwellized particle filter (RBPF) with occupancy grid maps represents an effective and flexible solution to the SLAM problem as it only makes mild assumptions about the structure of the environment. Theoretically, given infinitely many particles, RBPFs always converge to the correct map. In practice however, only a finite number of particles can be used. This number inherently limits the uncertainty which can be represented by the filter and in this way can lead to divergence. Consider, for example, the environment depicted in the upper picture of Figure 1. This environment consists of two box-like landmarks, which cannot be perceived at the same time due to a limited sensor range. The geometric shape of these boxes needs to be represented accurately in the map as otherwise the estimate about the orientation of the robot relative to the landmark quickly gets lost. When the robot drives around a box several times the standard RBPF mapping approach typically turns the squared shape into a circle, since slight pose errors are introduced by the limited number of particles and by the suboptimal proposal distribution used. As a consequence, the particle filter loses track of the heading of the

C. Laugier and R. Siegwart (Eds.): Field and Service Robotics, STAR 42, pp. 329–338, 2008.
springerlink.com 

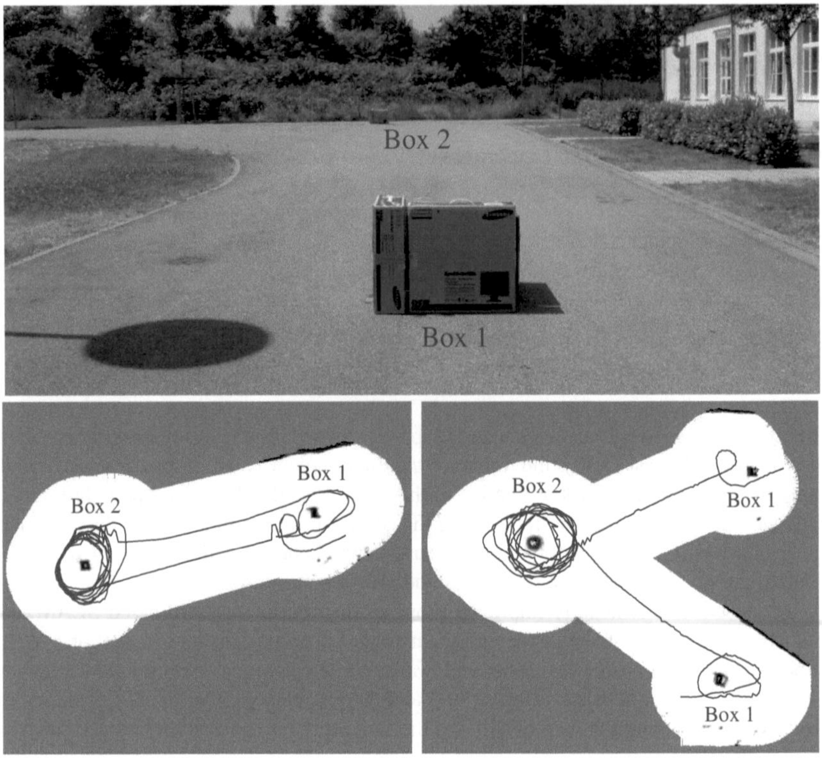


**Fig. 1.** A mobile robot with a limited sensor range navigates through the low structured environment depicted in the upper image, which consists of two cardboard boxes placed on flat, open terrain. The robot starts at *Box 1*, moves to *Box 2*, and revolves around it several times before returning to its origin. Standard mapping approaches cannot deal with the large amount of pose uncertainty build up when circling around the second box, which results in a seriously diverged map (lower right). In comparison, our approach using $k$-step look-ahead proposals retains the squared shape of the boxes and yields an accurate map (lower left). Note, that due to the limited laser range the hedge was visible only at Box 1.

robot and yields seriously wrong maps like the one depicted in the lower right diagram of Figure 1.

In this paper, we present the so-called look-ahead mapping approach which uses improved proposals for the robot pose to increase the accuracy of the particles after the proposal step. This is achieved by performing independent, $k$-step localization runs for the individual particles to better align them to their local maps. In this way, the filter can deal with higher levels of noise and operate in less structured environments. Experiments with real robots demonstrate that our approach can handle situations in which state-of-the-art RBPF-based approaches, like the one using a motion model based on the odometry error [10] or laser scan-matching error [5] or the one using dynamically adapted proposal distributions based on local laser scan-matching [4] fail.

## 2 Related Work

Originally, Murphy *et al.* [1, 10] have introduced Rao-Blackwellized particle filters as an effective means to solve the SLAM problem. In a RBPF, each particle represents a possible robot trajectory and a map. The framework has been subsequently extended by Montemerlo *et al.* [8, 9] for SLAM with discrete sets of landmarks. To efficiently learn accurate grid maps, Eliazar and Parr [3] described an efficient map representation. Additionally, Hähnel *et al.* [5] proposed an improved motion model that substantially reduces the number of particles required. Recently, Howard presented an extension towards multi-robot systems [6] in which he describes how to effectively merge the information obtained by different robots. Furthermore, Grisetti *et al.* [4] proposed an extension of the approach by Hähnel *et al.* [5]. Instead of using a fixed proposal distribution, their algorithm computes an improved proposal distribution on a per-particle basis, in a way similar to FastSLAM-2, as presented by Montemerlo *et al.* [7] for the case of landmark-based mapping.

All of the above mentioned approaches perform mapping and localization as the data is available, that is, the pose and the map uncertainty result from the very same set of observations processed. In the work presented here, we use $k$ "future" measurements to build up a look-ahead proposal distribution for the robot pose. In this way, the uncertainty in the robot pose is reduced and consequently, less map uncertainty has to be represented by the filter.

## 3 Mapping with Rao-Blackwellized Particle Filters

The key idea of the Rao-Blackwellized particle filter for SLAM is to reason about possible robot trajectories and the corresponding maps using a sample-based representation [5]. Formally, the task is to estimate the joint posterior $p(\mathbf{x}_{1:t}, \mathbf{m} \mid \mathbf{z}_{1:t}, \mathbf{u}_{1:t})$ of the map $\mathbf{m}$ and the trajectory $\mathbf{x}_{1:t} = \mathbf{x}_1, \ldots, \mathbf{x}_t$ of the robot, given observations $\mathbf{z}_{1:t} = \mathbf{z}_1, \ldots, \mathbf{z}_t$ and odometry measurements $\mathbf{u}_{1:t} = \mathbf{u}_1, \ldots, \mathbf{u}_t$. In the particle filter framework, the posterior after each time step is represented by a set of weighted trajectories $\mathbf{x}_{1:t}^{[j]}$ and the corresponding maps $\mathbf{m}_t^{[j]}$ generated from these trajectories. Using the factorization

$$p(\mathbf{x}_{1:t}, \mathbf{m}_t \mid \mathbf{z}_{1:t}, \mathbf{u}_{1:t}) = p(\mathbf{m}_t \mid \mathbf{x}_{1:t}, \mathbf{z}_{1:t}) \cdot p(\mathbf{x}_{1:t} \mid \mathbf{z}_{1:t}, \mathbf{u}_{1:t}) \,, \tag{1}$$

we can derive a recursive filter that in each iteration updates the trajectory samples $\mathbf{x}_{1:t}^{i}$ and then analytically builds the corresponding map $\mathbf{m}_t^{i}$. Each filter iteration consists of the following steps:

1. *Sampling*: The next generation of particles $\{\mathbf{x}_t^{[j]}\}$ is obtained from the generation $\{\mathbf{x}_{t-1}^{[j]}\}$ by sampling from a proposal distribution $\pi$. Often, a probabilistic odometry-based motion model is used for $\pi$.
2. *Importance Weighting*: Importance weights $w_t^{[j]}$ are assigned to the individual particles according to

$$w_t^{[j]} = \frac{p(\mathbf{x}_{1:t}^{[j]} \mid \mathbf{z}_{1:t}, \mathbf{u}_{1:t})}{\pi(\mathbf{x}_{1:t}^{[j]} \mid \mathbf{z}_{1:t}, \mathbf{u}_{1:t})} \propto \frac{p(\mathbf{z}_t \mid \mathbf{m}_{t-1}^{[j]}, \mathbf{x}_t^{[j]}) p(\mathbf{x}_t^{[j]} \mid \mathbf{x}_{t-1}^{[j]}, \mathbf{u}_t)}{\pi(\mathbf{x}_t \mid \mathbf{x}_{1:t-1}^{[j]}, \mathbf{z}_{1:t}, \mathbf{u}_{1:t})} \cdot w_{t-1}^{[j]} \,. \tag{2}$$

The weights account for the fact that the proposal distribution $\pi$ is in general not equal to the target distribution of successor states [2].

3. *Resampling*: Particles are drawn with replacement proportional to their importance weight.
4. *Map Estimation*: For each particle, the corresponding map estimate $p(\mathbf{m}^{[j]} \mid \mathbf{x}^{[j]}_{1:t}, \mathbf{z}_{1:t})$ is computed based on the trajectory $\mathbf{x}^{[j]}_{1:t}$ of that sample and the history of observations $\mathbf{z}_{1:t}$.

The robustness and efficiency of this procedure strongly depends on the proposal distribution $\pi$ that is used to sample the new state hypotheses in the selection step. If the proposal distribution differs too much from the true posterior, there is a high risk of filter divergence. In the following section, we introduce a concrete proposal distribution $\pi$ that utilizes a set of "future" sensor measurements to yield better pose estimates. The resulting weight update equation is straightforward to implement, while the new approach is more robust in standard and hard, poorly-structured environments.

## 4 Look-ahead Proposals

The standard RBPF mapping approach fails in situations, in which the particle distribution significantly differs from the true posterior. This can happen when the proposal distribution $\pi$ provides a bad approximation to the true one, or when the environment does not provide enough structure to allow proper particle weighting. The latter situation is illustrated in the lower right diagram of Figure 1. Due to the limited structure of the environment, the robot is unable to localize itself properly and looses the fine structure of Box 2. Before returning to its starting location, the robot is clearly de-localized, such that Box 1 appears twice in the resulting map. Such a divergence can either be avoided by using an extremely large number of particles or by directing the given number of particles to more accurate locations. We follow the latter strategy by computing the pose prediction in each iteration based on the $k$ next sensory inputs instead of just one. These $k$ measurements are used to better localize each particle within its own map. Concretely, for each mapping particle at time $t-1$, we draw $l$ localization particles and localize them $k$ steps ahead within their map. The resulting pose posterior at time $t+k$ is then used to sample the successor pose of the mapping particle at time $t$. This process is visualized in Figure 2.

By using the additional sensory input, a more informed and thus more accurate proposal for the robot pose can be computed. Formally, the idea is to compute a better estimate for the pose $\mathbf{x}_t$, using the previous position $\mathbf{x}_{t-1}$, the commands (odometry) $\mathbf{u}_{t:t+k}$, and the measurements $\mathbf{z}_{t:t+k}$ up to time-index $t+k$. As stated in Equation (1), the Rao-Blackwellized particle filter is an approach for sequentially estimating the distribution

$$p(\mathbf{m}_t|\mathbf{x}_{1:t}, \mathbf{z}_{1:t})\, p(\mathbf{x}_{1:t}|\mathbf{z}_{1:t}, \mathbf{u}_{1:t}) \,. \tag{3}$$

Here, $p(\mathbf{x}_{1:t}|\mathbf{z}_{1:t}, \mathbf{u}_{1:t})$ is approximated using a particle filter, whose samples we will call *SLAM particles*. In the standard approach, the SLAM particles are drawn from a proposal distribution based on the motion model $p(\mathbf{x}_t|\mathbf{x}_{t-1}, \mathbf{u}_t)$. In our approach, we

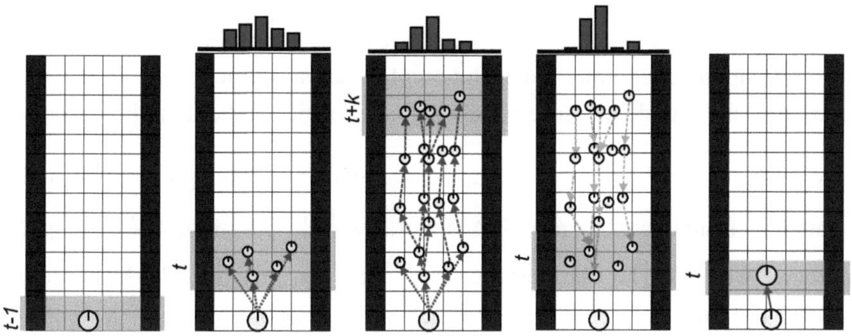


**Fig. 2.** The algorithm for look-ahead pose sampling using $k$ steps: For each mapping particle at time $t-1$ (first diagram), draw $l$ localizing particles (second). Move them according to $u_{t+i}$ and weight them according to $z_{t+i}$ until $t+k$ (third diagram). Propagate back the weights of the particles to the initial situation at time $t$ (forth) and draw the successor state according to this distribution (right diagram). The charts above the diagrams visualize the current weights of the individual localization particles at the highlighted timeindex.

use the more informed proposal $p(\mathbf{x}_t|\mathbf{x}_{t-1}, \mathbf{u}_t, \mathbf{z}_{t:t+k}, \mathbf{u}_{t+1:t+k}, \mathbf{m}_{t-1})$. Such a proposal results from performing an independent $k$-step localization for each SLAM particle. The localization algorithm is initialized with the map and the robot pose of the corresponding SLAM particle at time $t-1$. Our proposal can be rewritten more compactly, omitting $\mathbf{m}_{t-1}$, as $p(\mathbf{x}_t|\mathbf{x}_{t-1}, \mathbf{u}_{t:t+k}, \mathbf{z}_{t:t+k})$, which we can rewrite as

$$p(\mathbf{x}_t|\mathbf{x}_{t-1}, \mathbf{u}_{t:t+k}, \mathbf{z}_{t:t+k}) = \int p(\mathbf{x}_{t:t+k}|\mathbf{x}_{t-1}, \mathbf{u}_{t:t+k}, \mathbf{z}_{t:t+k})\, d\mathbf{x}_{t+1:t+k}. \tag{4}$$

Given the poses $\mathbf{x}_t \ldots \mathbf{x}_{t+k}$ and the map $\mathbf{m}_{t-1}$ the obsersavtions $\mathbf{z}_t \ldots \mathbf{z}_{t+k}$ are independant. Therefore, we can rewrite the term inside the integral as

$$p(\mathbf{x}_{t:t+k}|\mathbf{x}_{t-1}, \mathbf{u}_{t:t+k}, \mathbf{z}_{t:t+k}) = \eta \prod_{j=t}^{t+k} p(\mathbf{z}_j|\mathbf{x}_j) \cdot \prod_{j=t}^{t+k} p(\mathbf{x}_j|\mathbf{x}_{j-1}, \mathbf{u}_j). \tag{5}$$

Here $\eta = p(\mathbf{z}_{t:t+k}|\mathbf{x}_{t-1}\mathbf{u}_{t:t+k})^{-1}$ is the Bayes normalizer. A particle approximation of Equation 5 (the localizer particles) can be obtained by sampling a sequence of poses according to the sequence of motion commands. Each sample $i$ has a weight $v_{t:t+k}^{(i)}$ proportional to the likelihood of the chain of observations starting at $t-1$.

$$\mathbf{x}_{t:t+k}^{(i)} \sim \prod_{j=t}^{t+k} p(\mathbf{x}_j^{(i)}|\mathbf{x}_{j-1}^{(i)}, \mathbf{u}_j) \qquad v_{t:t+k}^{(i)} = \prod_{j=t}^{t+k} p(\mathbf{z}_j^{(i)}|\mathbf{x}_j^{(i)}). \tag{6}$$

A sampled approximation of the integral in Eq. 4

$$p(\mathbf{x}_t|\mathbf{x}_{t-1}, \mathbf{u}_{t:t+k}, \mathbf{z}_{t:t+k}) \sim \left\langle \mathbf{x}_t^{(i)}, v_t^{(i)} \right\rangle \tag{7}$$

is recovered from the samples in Eq. 6. This is done by truncating each trajectory $\mathbf{x}_{t:t+k}^{(i)}$ at time $t$ and by back-propagating the weights of the trajectory, according to Fig. 2. Due to resampling operations, the evolution of these trajectories can be described as a tree [3], thus the sample $\mathbf{x}_t^{(i)}$ receives a weight $v_t^{(i)}$ which is the sum of the weights of its successors.

The set $\mathcal{S} = \left\{\left\langle \mathbf{x}_t^{(i)}, v_t^{(i)} \right\rangle\right\}$ is a sampled approximation of our proposal distribution. We can draw from this set $N$ new SLAM particles $\left\{\mathbf{x}_t^{(j)}\right\}$ for time $t$, according to the importance weights. The true posterior $p(\mathbf{x}_t^{(j)}|\mathbf{x}_{t-1}^{(j)}, \mathbf{z}_t, \mathbf{u}_t)$ (the SLAM particles) is recovered from this set by assigning to each newly drawn sample $\mathbf{x}_t^{(j)}$ a weight $w_t^{[j]}$ according to the importance sampling principle (see Eq. 2):

$$w_t^{[j]} \propto w_{t-1}^{[j]} \frac{p(\mathbf{x}_t^{(j)}|\mathbf{x}_{t-1}^{(j)}, \mathbf{z}_t, \mathbf{u}_t)}{p(\mathbf{x}_t^{(j)}|\mathbf{x}_{t-1}^{(j)}, \mathbf{u}_{t:t+k}, \mathbf{z}_{t:t+k})} \tag{8}$$

$$\propto w_{t-1}^{[j]} \frac{p(\mathbf{z}_t|\mathbf{x}_t^{(j)})p(\mathbf{x}_t^{(j)}|\mathbf{x}_{t-1}^{(j)}, \mathbf{u}_t)}{p(\mathbf{x}_t^{(j)}|\mathbf{x}_{t-1}^{(j)}, \mathbf{u}_{t:t+k}, \mathbf{z}_{t:t+k})} \tag{9}$$

$$\propto w_{t-1}^{[j]} \frac{p(\mathbf{z}_t|\mathbf{x}_t^{(j)})}{v_t^{(j)}} \tag{10}$$

The last step follows directly from Equation 7 and the fact that the trajectories $\mathbf{x}_{t:t+k}^{(i)}$ had been drawn including the motion $\mathbf{u}_t$ between $t-1$ and $t$. Note that for updating the weight of a SLAM particle, we just need the weight of the drawn successor state at time $t$ (first iteration of the nested localization run) and the corresponding weight at time $t+k(=v_t)$. Both values are readily available from the localization run. Equation (10) assures, that no information (odometry, measurements) is integrated more than once. Thus, the presented approach does not change the estimated posterior distribution, but only the way it is represented by the limited particle set of the filter.

## 5 Experiments

In this section, we present a set of experiments demonstrating that our mapping approach can be applied in both indoor and outdoor scenarios. Furthermore, we compared our approach using localization proposals (termed *LP* in this section) to a state-of-the-art technique [4], which uses a dynamically adapted proposal based on scan matching (*SMP*) and to a standard RBPF using odometry-based proposals (*OP*). The results show that our approach is more robust in poorly structured situations, while also providing high quality maps in more structured environments.

### *Measuring Map Quality*

To assess the quality of the resulting maps we evaluate a measure denoted as *revisiting accuracy* (RA), which reflects the error in the robot pose estimate at revisited places relative to previous visits. This measure captures the internal consistency of a map. It

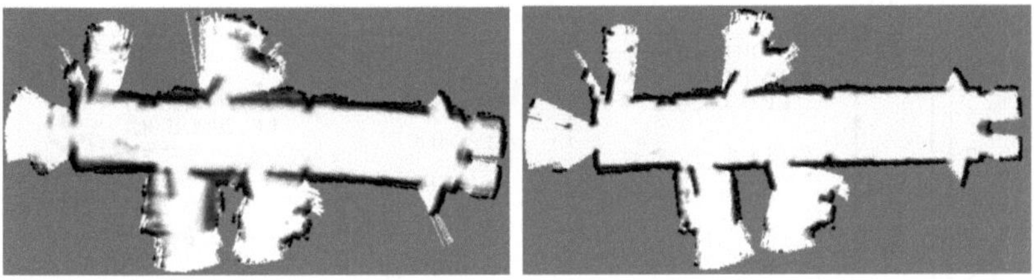

**Fig. 3.** Maps obtained from the indoor environment using *OP* (left) and *LP* (right)

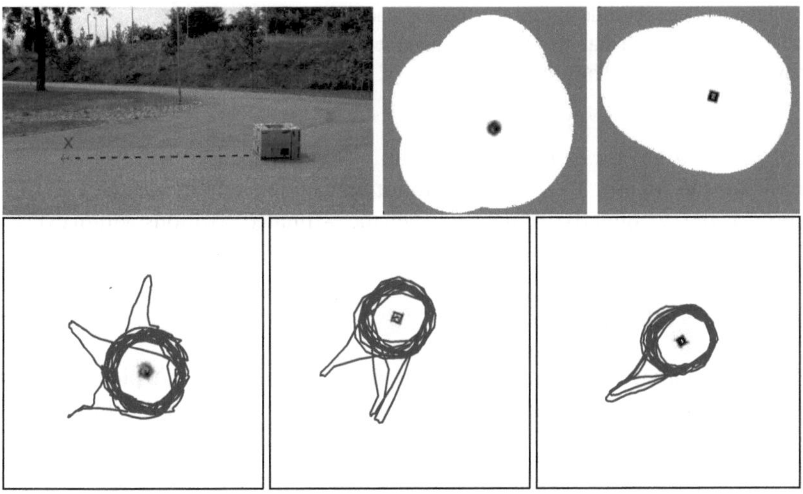

**Fig. 4.** Mapping a low-structured environment (upper left): Maps obtained by *OP* (upper middle) and our approach (upper right). Traveled paths of the robot using *OP* (lower left), *SMP* (lower middle) and our *LP* approach (lower right). All approaches used 50 mapping particles. Our approach used additionally 100 localization particles per map and a look-ahead of $k = 5$.

therefore better reflects the practical usefulness of the map than for example the absolute mean squared error of the corrected trajectory relative to an assumed true one. The latter measure is overly sensitive to distortions on a global level, which does in general not lead to practical problems, and it requires a ground truth trajectory.

For calculating the *revisiting accuracy*, we add color markers to the ground at places that are to be traversed several times during the experiment. We then recorded the timestamps at which the robot passed over these checkpoints. Let $p$ be a checkpoint visited at times $t_1$ and $t_2$. The revisiting accuracy for this location is defined as

$$\epsilon^{p}_{t_1,t_2} = \sum_{j} w^{[j]} \sqrt{(1-\lambda)((x^{[j]}_{t_1} - x^{[j]}_{t_2})^2 + (y^{[j]}_{t_1} - y^{[j]}_{t_2})^2) + \lambda(\theta^{[j]}_{t_1} - \theta^{[j]}_{t_2})^2} \,.$$

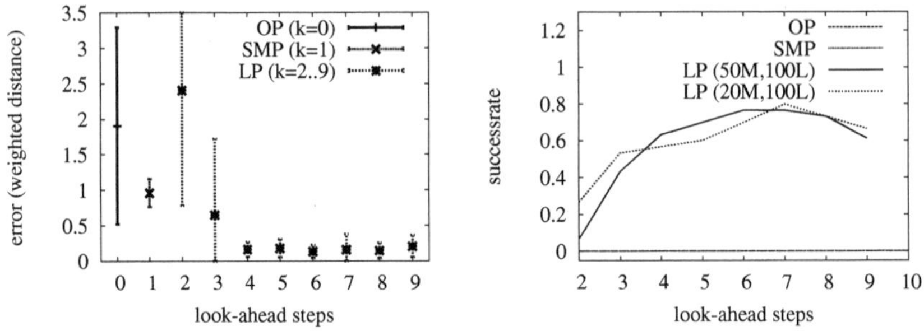


**Fig. 5.** Mean and standard deviation for the approaches (left) and the ratio of runs with an error less than 0.2 (right). Note that the *OP*-RBPF and the *SMP*-RBPF approach never had an revisiting accuracy error below 0.2. Thus the success rates of both approaches is 0.

Here, $x, y$, and $\theta$ are the components of the pose vector and $\lambda \in [0, 1]$ is a weighing factor for the rotational component. Intuitively, the revisiting accuracy for a given checkpoint is the (weighted) distance between the estimated positions within every map, weighted by the corresponding map-probability.

### *Mapping an Office Environment*

We tested our approach in the highly structured office environment depicted in Figure 3. Given this dataset, we compared our approach (*LP*) to *SMP* and *OP* in terms of the revisiting accuracy measure defined above. Figure 3 depicts the maps obtained using *OP* (left diagram) and our approach (right diagram) for 20 mapping particles. In our approach, we used 50 localization particles per map particle and a look-ahead of $k = 3$. The map generated using the *SMP* is similar to ours. The localization accuracy is between 5 cm and 10 cm for *OP* and less than 5 cm for both, our approach (*LP*) and *SMP*. Overall, a localization error of less than 10 cm is considerably small with the given grid resolution of 5 cm per cell.

### *Mapping a Low-Structured Outdoor Environment*

We compared our approach to those proposed by Hähnel *et al.* [5] and Grisetti *et al.* [4] in the outdoor environment depicted in the upper left picture of Figure 4. The robot started at the checkpoint "X", moved around the cardboard box 15 times. Thereby it returned to the checkpoint every 5th run. The laser range was limited to 4 meters, so that the box was the only visible map element. Figure 4 depicts the mapping results of the *OP* and our *LP* approach (second and third upper diagram) using 50 SLAM particles. For generating the map, we used a look-ahead of $k = 5$ and 100 localization particles per SLAM particle. The three lower diagrams in Figure 4 depict typical paths estimated by the three mapping approaches. The distance between the checkpoint "X" and the box is around 4 m. As can be seen from these results, *OP* fails to map this environment and yields a highly inconsistent map. Although, when using *SMP*, the map quality is higher and the box has retained its squared shape, the resulting map is still inconsistent. Our approach, in contrast, is able to deal with this hard situation and generates accurate maps. To quantitatively evaluate the performance of the approaches, we analyzed the mapping results for different numbers of particles. Each approach was executed 30 times for each setting. We measured $\epsilon^X_{t_{\text{first}},t_{\text{last}}}$, with $t_{\text{first}}$ and $t_{\text{last}}$ being the timestamps, when the robot moved over the checkpoint "X" for the first and the last time respectively. Figure 5 (left) gives the results of this experiment. For reasons of better readability we depicted the results obtained by 50 mapping particles only. Furthermore, Figure 5 (right) depicts the success rate of the runs, defined as the ratio of runs where the revisiting error $\epsilon$ was less than 0.2.

It can be seen from these experiments, that in situations with hardly identifiable features, the performance obtained using our proposal is higher than the one obtained by the other grid based RBPF mappers [4, 5]. Although the RBPF based on *SMP* outperforms the standard RBPF, it achieves significantly less accurate maps than the technique proposed in this paper. Our approach is able to "look through gaps" of low structure until enough structure is provided for proper localization. In this way, it can handle such a hard environment even with less mapping particles (see figure 5 (right), curve obtained with 20 mapping particles and 100 localization particles for each map).

## 6 Conclusions

In this paper, we presented an extension to the Rao-Blackwellized particle filter for simultaneous localization and mapping (SLAM), which significantly improves the mapping in environments which are poorly structured. Our approach uses a novel proposal distribution that looks $k$ steps "ahead in time" to yield a more informed pose estimate for the mapping decision. We provided a mathematical derivation of the approach and showed, that the weight update for our improved proposal takes a simple form. Our method has been implemented and tested on real robot data sets. We compared our technique to two state of the art mapping techniques. Experimental results suggest that our approach yields better maps, especially in environments with sparse features that facilitate accurate localization of particles.

## Acknowledgment

This work has been supported by the EC under contract number FP6-IST-034120, Action line: Micro/Nano Based Sub-Systems.

## References

1. Doucet, A., de Freitas, J.F.G., Murphy, K., Russel, S.: Rao-Blackwellized partcile filtering for dynamic bayesian networks. In: Proc. of the Conf. on Uncertainty in Artificial Intelligence (UAI), Stanford, CA, USA, pp. 176–183 (2000)
2. Doucet, A., de Freitas, N., Gordan, N.: Sequential Monte-Carlo Methods in Practice. Springer, Heidelberg (2001)
3. Eliazar, A., Parr, R.: DP-SLAM: Fast, robust simultainous localization and mapping without predetermined landmarks. In: Proc. of the Int. Conf. on Artificial Intelligence (IJCAI), Acapulco, Mexico, pp. 1135–1142 (2003)

4. Grisetti, G., Stachniss, C., Burgard, W.: Improving grid-based slam with Rao-Blackwellized particle filters by adaptive proposals and selective resampling. In: Proc. of the IEEE Int. Conf. on Robotics & Automation (ICRA), Barcelona, Spain, pp. 2443–2448 (2005)
5. Hähnel, D., Burgard, W., Fox, D., Thrun, S.: An efficient FastSLAM algorithm for generating maps of large-scale cyclic environments from raw laser range measurements. In: Proc. of the IEEE/RSJ Int. Conf. on Intelligent Robots and Systems (IROS), Las Vegas, NV, USA, pp. 206–211 (2003)
6. Howard, A.: Multi-robot simultaneous localization and mapping using particle filters. In: Robotics: Science and Systems, Cambridge, MA, USA (2005)
7. Montemerlo, M., Thrun, S., Koller, D., Wegbreit, B.: FastSLAM 2.0: An improved particle filtering algorithm for simultaneous localization and mapping that provably converges. In: Proc. of the Int. Conf. on Artificial Intelligence (IJCAI), Acapulco, Mexico, pp. 1151–1156 (2003)
8. Montemerlo, M., Thrun, S.: Simultaneous localization and mapping with unknown data association using FastSLAM. In: Proc. of the IEEE Int. Conf. on Robotics & Automation (ICRA), Taipei, Taiwan (2003)
9. Montemerlo, M., Thrun, S., Koller, D., Wegbreit, B.: FastSLAM: A factored solution to simultaneous localization and mapping. In: Proc. of the National Conference on Artificial Intelligence (AAAI), Edmonton, Canada, pp. 593–598 (2002)
10. Murphy, K.: Bayesian map learning in dynamic environments. In: Proc. of the Conf. on Neural Information Processing Systems (NIPS), Denver, CO, USA, pp. 1015–1021 (1999)

# Tradeoffs in SLAM with Sparse Information Filters

Zhan Wang, Shoudong Huang, and Gamini Dissanayake

ARC Centre of Excellence for Autonomous Systems (CAS),
Faculty of Engineering, University of Technology, Sydney, Australia
{zwang,sdhuang,gdissa}@eng.uts.edu.au

**Summary.** Designing filters exploiting the sparseness of the information matrix for efficiently solving the simultaneous localization and mapping (SLAM) problem has attracted significant attention during the recent past. The main contribution of this paper is a review of the various sparse information filters proposed in the literature to date, in particular, the compromises used to achieve sparseness. Two of the most recent algorithms that the authors have implemented, Exactly Sparse Extended Information Filter (ESEIF) by Walter et al. [5] and the D-SLAM by Wang et al. [6] are discussed and analyzed in detail. It is proposed that this analysis can stimulate developing a framework suitable for evaluating the relative merits of SLAM algorithms.

## 1 Introduction

Use of an Extended Information Filter (EIF) to solve SLAM to gain computational savings has been demonstrated by a number of researchers. A notable result has been that from Thrun et al. [2] where a sparsification process is used to reduce the number of non-zero elements in the information matrix. Another way to ensure that the information matrix is sparse is to include both features and a sequence of robot poses in the state vector [13] [3]. Alternatively SLAM can also be formulated with a state vector containing only robot poses, as shown in the Exactly Sparse Delayed State Filter (ESDSF) [4]. Very recently, two SLAM algorithms, which can achieve exactly sparse information matrix without including previous robot poses in the state vector, were developed independently. One is the Exactly Sparse Extended Information Filter (ESEIF) by Walter et al. [5] and the other is the D-SLAM by Wang et al. [6].

All above algorithms rely on the sparse structure of the information matrix to achieve computational benefits. Sparseness is achieved through various tradeoffs. Although the estimation process becomes efficient, probabilistic data association requires the recovery of the relevant elements of the state as well as the associated covariance matrix, and may have a significant impact on the computational cost of a practical implementation. Although each algorithm claims different advantages, no comprehensive examination is so far available to make clear the relative merits and the compromises used. This paper addresses this important issue by providing a classification and a qualitative analysis of SLAM algorithms

C. Laugier and R. Siegwart (Eds.): Field and Service Robotics, STAR 42, pp. 339–348, 2008.
springerlink.com 

**Table 1.** A summary of sparse information filters used in SLAM – here $N$ is the total number of 2D features and $M$ is the number of selected 3D robot poses

| SLAM Algorithm | Dimension of state vector | Consistency | Information loss | Covariance recovery for data association |
|---|---|---|---|---|
| SEIF [2] | 2N+3 | No[5] | – | approximate |
| SAM [3][7] | 3M+2N | Yes | No | exact [7] |
| ESDSF [4] | $3M$ | Yes | See Section 2.3 | approximate |
| ESEIF [5] | 2N+3 | Yes | Yes | approximate |
| D-SLAM [6] | 2N | Yes | Yes | exact |

based on sparse information filters published to date. ESEIF and D-SLAM are analyzed in detail and a quantitative comparison is provided. It is hoped that this paper will stimulate further research towards the development of a comprehensive strategy and benchmark datasets for evaluating SLAM algorithms under a variety of different conditions.

The paper is organized as follows. Section 2 summarizes different sparse information filters for SLAM. Simulation and experiment comparisons of ESEIF and D-SLAM are provided in Section 3. Section 4 concludes the paper.

## 2 Summary of Sparse Information Filters in SLAM

This section classifies the sparse information filters published so far based on the strategy used to achieve sparseness. Only the algorithms based on a filtering process are discussed. Other valuable contributions that use sparse representations such as [10], [1], [11] and [9] are not examined.

### 2.1 Approximate the Information Matrix in EIF SLAM

An empirical finding that normalized information matrix obtained when the SLAM problem is formulated in the information form is approximately sparse, motivates the work by Thrun et al. [2], Sparse Extended Information Filter (SEIF). Theoretical explanation for this observation was later presented by Frese [13]. Sparsification essentially removes the weak links in the information matrix by setting elements that are smaller than a given threshold to zero, while strengthening other links to make up for the effect of this change.

State estimates for robot pose and a subset of features that are needed for computing Jacobians are recovered by solving a set of linear equations with a sparse coefficient matrix using relaxation. Data association is solved by approximating the data association probability that an observation under consideration originates from a feature in the map and using the standard maximum likelihood method. However, Walter et al. [5] demonstrates that the sparsification process leads to inconsistent estimates.

### 2.2 Including Previous Robot Poses and Features in the State Vector

When all the previous robot poses and all the features are included in the state vector, the SLAM problem becomes a static estimation problem. This situation is discussed in Frese [13]. A considerable amount of off-diagonal elements in the information matrix are exactly zero and the information matrix is exactly sparse. This state vector is used in Square Root SAM [3].

In [3], SLAM is formulated as a linear least squares problem and is solved by factorizing the smoothing information matrix using Choleskey or QR factorization in a batch or incremental manner. Exact state and covariance recovery can be achieved by exploiting the special sparse structure of the factorization matrix resulting from the QR factorization [7]. The advantage of the SAM algorithm is that the quality of the estimate can be better than traditional EKF SLAM since in each step the estimates of all previous robot poses are updated together with the feature location estimates through the smoothing process. Thus the linearization error is reduced. Therefore, the SAM algorithm is less prone to estimator inconsistency that can arise due to linearization errors than all other EIF algorithms discussed in this paper.

However, the sparseness of the information matrix is achieved through increasing the state dimension, which keeps increasing even when robot is revisiting previously explored regions. Therefore, the computational cost increases over time and is not bounded by the number of features in the environment.

### 2.3 Including Only Robot Poses in the State Vector

By computing the relationship between two consecutive robot poses using the observation made at each pose, SLAM can be solved using a state vector containing only robot poses. The resulting information matrix is exactly sparse, as shown in the Exactly Sparse Delayed State Filter (ESDSF) [4].

A suboptimal partial state recovery is achieved by keeping the irrelevant states fixed at their current estimates and solving a set of sparse equations relating the state, the information matrix and the information vector. Data association is solved by estimating a bound for the covariance matrix. The key advantage of ESDSF is that it is suitable for the scenarios where features are difficult to extract or the number of features is too large as compared with robot poses, as demonstrated by the excellent maps shown in [4].

However, the resulting "map" is only an alignment of a sequence of observations (such as images or laser scans). There are no statistical map updates, thus improvements to the state estimates achieved through feature location updates in traditional SLAM is not present. Extent of the information loss due to this has not yet been analyzed.

### 2.4 Including Only Features in the State Vector

D-SLAM, recently proposed by Wang et al. [6] uses a state vector that only contains the feature locations to generate maps of an environment. Robot location estimate is obtained through a concurrent yet separate process.

In D-SLAM mapping, the original measurements relating the robot and features are first transformed into relative distances and angles among features. Then these transformed measurements are fused into the map using EIF. It is shown that only the features that are observed at the same time instant have links in the information matrix making it exactly sparse. The extent of sparseness is governed by the sensor range and feature density. Localization is performed by combining two estimates: one is obtained by solving a "kidnapped robot problem"; the other is obtained by a local EKF SLAM where only the features currently observed are retained in the state vector. The two correlated estimates are fused by Covariance Intersection (CI) [6].

Exact state and covariance recovery is achieved by preconditioned Conjugated Gradient (PCG). A good preconditioner produced by an iterative Cholesky factorization method by exploiting the similarity between the information matrices of successive steps is used to make the PCG efficient. Data association is solved by a combination of the standard maximum likelihood approach and a chi-square test.

There is some information loss in D-SLAM, and this is further addressed in Section 3.

### 2.5 Periodic Marginalization of Robot Pose

The ESEIF algorithm by Walter et al. [5] achieves sparse information matrix by periodically marginalizing out and relocating robot.

Similar to SEIF, ESEIF exploits the fact that when the robot location is marginalized out from the state vector, new links will only be built up among the features that were previously linked with the robot in the information matrix. The set of features that are linked with the robot is called "active features". Thus the information matrix will be sparse if the number of "active features" is bounded. In contrast to the sparsification process in SEIF, ESEIF controls the number of "active features" by "kidnapping" the robot when the number of "active features" is about to become larger than a predefined threshold $\Gamma_a$. This is followed by "relocating" the robot using a set of selected measurements. Thus, the EIF information matrix is kept sparse without any approximations that can lead to inconsistency. The extent of sparseness is controlled by the active feature bound $\Gamma_a$, the sensor range and feature density.

There is some information loss in ESEIF due to "kidnapping" and "relocating" the robot. This is further addressed in Section 3.

# 3 Comparison of Information Loss in ESEIF and D-SLAM

ESEIF [5] and D-SLAM [6] have been selected for a detailed comparison as both these algorithms have a state vector of the same character and do not use any approximations to achieve sparseness. As the authors do not have access to an efficient implementation of ESEIF, a quantitative analysis of the computational cost can not be presented. This section, therefore, focuses on the extent of information loss.

## 3.1 Comparison Using 1D Simulations

An analysis of the mapping performance of ESEIF and D-SLAM based on a linear one-dimensional form of the SLAM problem is presented below. The 1D simulation is used so that the effects due to linearization errors are avoided.

The scenario consists of a set of uniformly distributed features arranged on a straight line with a interval of $1m$. The robot, moving forward and backward along this line, can measure the distances to features within its sensor range ($9m$). At each time step, robot observes 9 features. The robot only makes one observation before moving.

Four parameters are considered: the active feature bound $\Gamma_a$, the process noise, the robot speed and the time steps. Extensive simulations were conducted by varying these parameters over a large range of values. The trace of the submatrix of the covariance matrix corresponding to all features from ESEIF, $P_{mm}$, and the trace of the covariance matrix from D-SLAM mapping, $P$, are used to indicate the accumulated information. The results obtained are summarized in Table 2, from which the following conclusions can be drawn.

(1) When the active feature bound $\Gamma_a$ is smaller than the number of feature in each observation, ESEIF performs worse than D-SLAM, because in this case robot is marginalized out and relocated at each step in ESEIF. When the active feature bound $\Gamma_a$ is gradually increased, number of instances where the robot needs to be relocated reduces and at some point ESEIF map becomes better than that obtained using D-SLAM.

(2) When all other parameters are fixed, the larger the robot speed, the larger the range of $\Gamma_a$ in which D-SLAM map is better (compare rows $4-6$ with rows $1-3$ in Table 2). This is because when robot runs faster, robot relocation step becomes more frequent resulting in more information loss.

(3) When all other parameters are fixed and the process noise is small enough, ESEIF map is always better than D-SLAM as long as $\Gamma_a$ is at least equal to the the number of features in one observation (see rows $10-12$ in Table 2). When all other parameters are fixed, the larger the process noise, the larger the range of $\Gamma_a$ in which D-SLAM map is better (compare rows $7-9$ with rows $1-3$ in Table 2). With larger process noise, process model contains less information, thus the information loss in D-SLAM is less.

(4) When other parameters are fixed, the smaller the time steps, the larger the range of $\Gamma_a$ in which ESEIF map is better (compare rows $13-15$ with rows

**Table 2.** Evaluation of mapping in ESEIF and D-SLAM using 1D simulations

| Test No. | Active feature bound $\Gamma_a$ | Process noise (m/s) | Speed (m/s) | Time steps | D-SLAM trace of $P$ | ESEIF trace of $P_{mm}$ | Trace ratio $\frac{D-SLAM}{ESEIF}$ |
|---|---|---|---|---|---|---|---|
| 1 | 8 | 0.5 | 1 | 280 | 0.087099 | 0.099094 | 0.87895 |
| 2 | 10 | 0.5 | 1 | 280 | 0.087099 | 0.087564 | 0.99468 |
| 3 | 11 | 0.5 | 1 | 280 | 0.087099 | 0.087074 | 1.00029 |
| 4 | 8 | 0.5 | *2* | 280 | 0.423266 | 0.518746 | 0.87895 |
| 5 | *23* | 0.5 | *2* | 280 | 0.423266 | 0.423358 | 0.99978 |
| 6 | *24* | 0.5 | *2* | 280 | 0.423266 | 0.423229 | 1.00009 |
| 7 | 8 | *2* | 1 | 280 | 0.087099 | 0.104539 | 0.83317 |
| 8 | *33* | *2* | 1 | 280 | 0.087099 | 0.087105 | 0.99994 |
| 9 | *34* | *2* | 1 | 280 | 0.087099 | 0.087097 | 1.00002 |
| 10 | 8 | *0.2* | 1 | 280 | 0.087099 | 0.089111 | 0.97743 |
| 11 | 9 | *0.2* | 1 | 280 | 0.087099 | 0.083668 | 1.04101 |
| 12 | 10 | *0.2* | 1 | 280 | 0.087099 | 0.08311 | 1.04801 |
| 13 | 8 | 0.5 | 1 | *600* | 0.236643 | 0.281696 | 0.84006 |
| 14 | *13* | 0.5 | 1 | *600* | 0.236643 | 0.23673 | 0.99963 |
| 15 | *14* | 0.5 | 1 | *600* | 0.236643 | 0.236378 | 1.00112 |

$1-3$ in Table 2). This is due to the different accumulation speed of information loss in the two algorithms.

### 3.2 Comparison Using 2D Simulations

A 2D simulation using larger number of features was conducted to further compare the two algorithms. The environment used is a 35 meter square with 144 features arranged in uniformly spaced rows and columns. The robot trajectory are shown in Figures 1(a) and 1(d). A sensor with a field of view of 180 degrees and a range of 6 meters is simulated to generate relative range and bearing measurements between the robot and the features.

In this simulation, all the parameters are set to values for an indoor scenario. The standard deviations of the noises in robot speed and turnrate are 0.05 m/s and 4 degree/s respectively, and those of the noises in sensor range and bearing are 0.1 m and 1 degree respectively. Robot explores in the environment at the speed of 0.4 m/s for 533 time steps. At each time step, robot can observe $5-6$ features. $\Gamma_a$ is set to be 10.

As expected, with small process noise and robot speed, ESEIF provides more accurate map than D-SLAM as shown in Figures 1(a) and 1(d). This can be seen more clearly in Figures 1(c) and 1(f). Both algorithms provide consistent

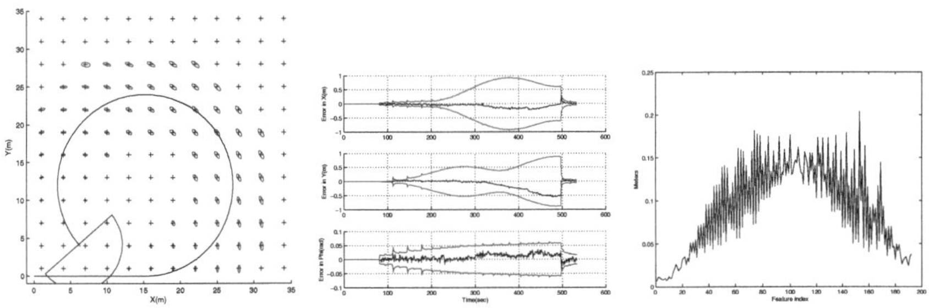

(a) Map from ESIEF (crosses indicate feature true locations; ellipses indicate $2\sigma$ error bounds)

(b) Robot location estimation errors and $2\sigma$ error bounds from ESEIF

(c) Final $1\sigma$ error bounds for all feature location estimates from ESEIF

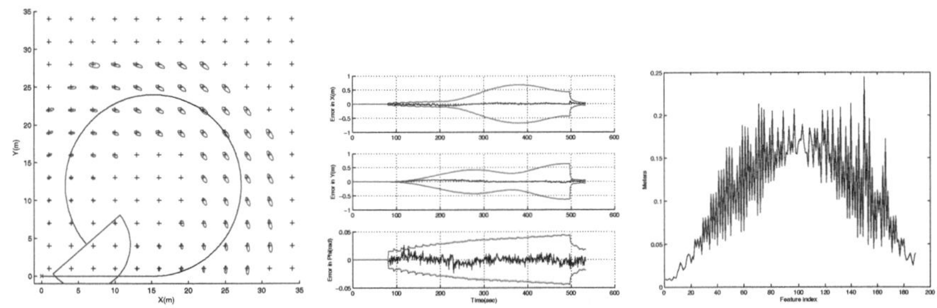

(d) Map from D-SLAM (crosses indicate feature true locations; ellipses indicate $2\sigma$ error bounds)

(e) Robot location estimation errors and $2\sigma$ error bounds from D-SLAM

(f) Final $1\sigma$ error bounds for all feature location estimates from D-SLAM

**Fig. 1.** Simulation results

robot location estimation. However, robot location estimation in D-SLAM is more accurate as shown in Figures 1(b) and 1(e).

### 3.3 Experimental Comparison

The large-scale outdoor Victoria Park data set [14] is also used for the comparison. As the ground truth is not available, it is impossible to use this data set for evaluating whether a given algorithm is consistent. Furthermore, some sparse information filter based SLAM papers use this data set with different pre-defined data associations, and different noise parameters. It is, therefore, unreasonable to compare the outcomes from this data set except as a proof that the algorithms concerned are able to deal with some of the practical issues such as non-stationery objects and spurious measurements. Interestingly, the gross character of the vehicle trajectories reported in some publications differs from the results presented by the original authors who collected this data. Figures extracted from the original publications of D-SLAM and ESEIF are presented to highlight these issues.

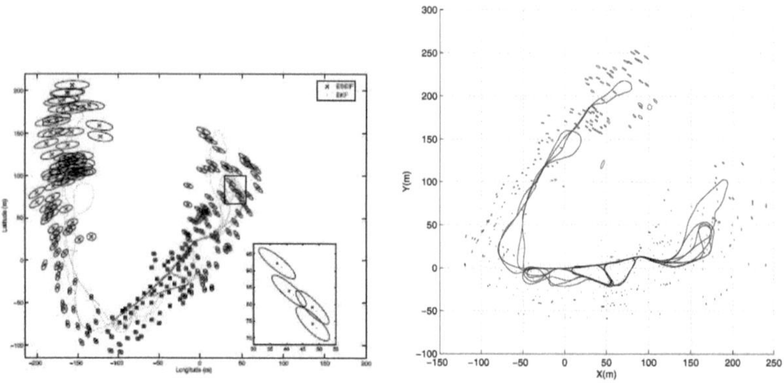


(a) Map and robot trajectory from ESIEF (ellipses indicate $3\sigma$ error bounds; the figure is from [5])

(b) Map and robot trajectory from D-SLAM (ellipses indicate $3\sigma$ error bounds)

**Fig. 2.** Outdoor, large-scale implementation using Victoria Park data set

### 3.4 Analysis of Comparison Results between ESEIF and D-SLAM

Since D-SLAM transforms the original measurements to relative distances and angles among features, the information from process model is lost in mapping, as compared with traditional EKF SLAM. In localization, the information loss comes from using CI, which is a conservative way to fuse two pieces of information with unknown correlations. The main factors that affect the extent of information loss include: process and sensor noises, feature density, and time steps before robot starts moving. This is analyzed in detail in [6].

In ESEIF, when marginalizing out and relocating robot, the information from process model is lost. This process results in poor robot location estimation for the following step and the loss of the correlations between robot and features, as compared with traditional EKF SLAM. These correlations together with the robot uncertainty govern the improvement to the feature location estimates. Information is also lost due to the fact that one part of measurements ($z_\alpha$ in [5]) is used in normal update and another part ($z_\beta$ in [5]) is used in relocating robot. During this process, the implicit information that these two parts of measurements are made from the same robot location is not exploited. The main factor governing the extent of information loss is how often robot is kidnapped and relocated. This is mainly determined by the active feature bound $\Gamma_a$, sensor range and the robot speed.

## 4 Conclusions and Discussions

Following the pioneering work by Thrun et al. [2], many different sparse information filters have been developed for SLAM. This paper summarized the relative merits of different sparse information filters. Especially, the extent of information loss of D-SLAM and ESEIF is compared in detail.

When the number of robot poses (at which observations are made) is not significant as compared with the total number of features, including all the robot poses in the state vector will produce better estimation than traditional EKF SLAM due to the smoothing of previous robot poses [3]. In the case when features are difficult to extract or the number of features is too large as compared with robot poses. The view-based ESDSF [4] is a good option since only robot poses are included in the state vector. When the number of robot poses is large as compared with the total number of features, ESEIF [5] and D-SLAM [6] are two efficient exactly sparse information filters.

Overall, sparse information filters provide promising solutions for the SLAM problems. The computational saving is significant although there are some tradeoffs. More work is necessary to further investigate the recovery of state and covariance matrix such that real time data association can be achieved. Furthermore, it is perhaps a critical requirement that SLAM researchers collaborate to generate a set of benchmark data sets, both simulated and real-life, as well as performance metrics so that plethora of emerging SLAM algorithms can be compared in a consistent setting. The recently proposed sparse information filters based on submaps, Tectonic SAM [17], Treemap [18], D-SLAM local map joining [15], Sparse local submap joining filter (SLSJF) [16], which are not discussed in this paper due to the space limitation, introduce significant flexibilities in achieving the sparseness and the comparison of these algorithms remain the future work.

## Acknowledgment

This work is supported by the ARC Centre of Excellence programme, funded by the Australian Research Council (ARC) and the New South Wales State Government.

## References

1. Frese, U., Larsson, P., Duckett, T.: A multilevel relaxation algorithm for simultaneous localization and mapping. IEEE Transactions on Robotics 21(2), 196–207 (2005)
2. Thrun, S., Liu, Y., Koller, D., Ng, A.Y., Ghahramani, Z., Durrant-Whyte, H.: Simultaneous localization and mapping with sparse extended information filters. International Journal of Robotics Research 23(7-8), 693–716 (2004)
3. Dellaert, F., Kaess, M.: Square Root SAM: simultaneous location and mapping via square root information smoothing. International Journal of Robotics Research 25(12), 1181–1203 (2006)
4. Eustice, R.M., Singh, H., Leonard, J.: Exactly sparse delayed-state filters for view-based SLAM. IEEE Transactions on Robotics 22(6), 1100–1114 (2006)
5. Walter, M., Eustice, R., Leonard, J.: Exactly sparse Extended Information filters for feature-based SLAM. International Journal of Robotics Research 26(4), 335–359 (2007)

6. Wang, Z., Huang, S., Dissanayake, G.: D-SLAM: a decoupled solution to simultaneous localization and mapping. International Journal of Robotics Research 26(2), 187–204 (2007)
7. Kaess, M., Ranganathan, A., Dellaert, F.: Fast incremental square root information smoothing. In: Proc.International Joint Conferences on Artificial Intelligence (IJCAI), pp. 2129–2134 (2007)
8. Krauthausen, P., Kipp, A., Dellaert, F.: Exploiting locality in SLAM by Nested Dissection. In: Proc. Robotics: Science and Systems (2006)
9. Paskin, M.: Thin junction tree filters for simultaneous localization and mapping. In: Proc.International Joint Conferences on Artificial Intelligence (IJCAI), pp. 1157–1164 (2003)
10. Folkesson, J., Christensen, H.I.: Graphical SLAM - a self-correcting map. In: Proc. IEEE International Conference on Robotics and Automation, pp. 383–390 (2004)
11. Thrun, S., Montemerlo, M.: The GraphSLAM algorithm with applications to large-scale mapping of urban structures. International Journal of Robotics Research 25(5-6), 403–429 (2004)
12. Eustice, R., Singh, H., Leonard, J., Walter, M., Ballard, R.: Visually navigating the RMS Titanic with SLAM information filters. In: Proc. Robotics: Science and Systems (2005)
13. Frese, U.: A proof for the approximate sparsity of SLAM information matrices. In: Proc. IEEE International Conference on Robotics and Automation, pp. 331–337 (2005)
14. Guivant, J.E., Nebot, E.M.: Optimization of the simultaneous localization and map building (SLAM) algorithm for real time implementation. IEEE Transactions on Robotics and Automation 17(3), 242–257 (2001)
15. Huang, S., Wang, Z., Dissanayake, G.: Mapping large scale environments using relative position information among landmarks. In: Proc. International Conference on Robotics and Automation, pp. 2297–2302 (2006)
16. Wang, Z.: Exactly sparse information filters for simultaneous localization and mapping. Ph.D. Thesis. Centre of Excellence for Autonomous Systems, University of Techology, Sydney (2007)
17. Ni, K., Steedly, D., Dellaert, F.: Tectonic SAM: exact, out-of-core, submap-based SLAM. In: Proc. IEEE International Conference on Robotics and Automation, pp. 1678–1685 (2007)
18. Frese, U.: Treemap: An O(log n) algorithm for indoor simultaneous localization and mapping. Autonomus Robots 21, 103–122 (2006)

# Robust Real-Time Local Laser Scanner Registration with Uncertainty Estimation

Justin Carlson, Charles Thorpe, and David L. Duke

Carnegie Mellon University - Qatar Campus
{justinca,cet,dlduke}@ri.cmu.edu

**Summary.** We present a fast, robust method for registering successive laser rangefinder scans. Correspondences between the current scan and previous scans are determined. Gaussian uncertainties of the correspondences are generated from the data, and are used to fuse the data together into a unified egomotion estimate using a Kalman process. Robustness is increased by using a RANSAC variant to avoid invalid point correspondences. The algorithm is very fast; computational and memory requirements are $O(nlogn)$ where $n$ is the number of points in a scan. Additionally, a covariance suitable for use in SLAM and filter techniques is cogenerated with the egomotion estimate. Results in large indoor environments are presented.

## 1 Introduction and Motivation

Maps are extremely useful in mobile robotics; they are of immense value to a robot tasked with localization, planning, and navigation. At present, it is common to use dead reckoning with range sensing in simultaneous localization and mapping (SLAM) algorithms to bound global error in unknown environments. The effectiveness of such algorithms depends in large part on the quality of dead reckoning. As dead reckoning degrades, the search for landmark correspondences grows, and becomes the dominating factor in most SLAM approaches. Although high-quality dead reckoning solutions exist, their components, such as fiber-optic gyroscopes, tend to be expensive.

At the same time, laser rangefinder registration is capable of providing significant improvements to dead reckoning estimates; however, to be useful in a larger context, such registrations must provide useful error metrics to higher level systems. This work incorporates data-derived uncertainties in scanner registrations to both improve registration results and generate accurate uncertainty estimates to higher level systems.

## 2 Related Work

Most of the popular methods for laser scanner registration find their roots in the Iterative Closest Point algorithm presented in [3]. [10] and [11] present a modified ICP-based approach to estimate robot egomotion using different

C. Laugier and R. Siegwart (Eds.): Field and Service Robotics, STAR 42, pp. 349–357, 2008.
springerlink.com 

correspondence algorithms to estimate rotation and translation, but the work doesn't take into account the quality of individual point matches nor does it generate an estimate of the quality of the registration. In [14] and [13], a Bayesian framework is added to separate moving objects from stationary landmarks; this work is orthogonal to and compatible with that framework. [5] and [4] use a grid to divide previous scans into localized bins. Covariances are calculated for these bins, and new scans are registered into old ones by numerical maximization of a pseudo probability density function. However, the artificial division of the world and required selection of representation is not straightforward.

[2] presents two methods for generating covariances from the Iterative Dual Correspondence algorithm; a real-time approach is based on examining the minimized error function, and an offline approach shifts registration around a minimum and uses a correlation metric to estimate registration uncertainty. The offline approach generates very good results, but is not suitable for a real-time system.

[12] uses Gaussian noise models to improve estimation of linear surfaces in laser scans. [16] uses line segments to generate localization matches. However, these methods require specific structure in the environment; this method does not.

## 3 Problem Definition

Consider a mobile robot which is capable of translation and rotation on a flat surface. This robot is equipped with a rigidly fixed range sensor which generates sets of range returns from a plane parallel with the surface of motion. At periodic intervals while moving, the robot takes a set of readings. The sensor and environment are assumed to provide a significant number of returns, but the environment is specifically not constrained to be linear or smooth.

In order to estimate the robot's path, we wish to generate an estimate of the rigid transformation needed to most accurately register successive sets of measurements. To enable incorporation into a higher level framework, we wish to simultaneously generate a corresponding metric of the uncertainty of the estimate.

## 4 Algorithm

We deal with two sets of returns: $\boldsymbol{A}$ is collected, the robot moves, then $\boldsymbol{B}$ is collected. The motion of the robot between scans is $m = (m_x, m_y, m_\theta)^T$. We seek an estimate $\hat{m}$ of this motion as well as a covariance $Q$ for that estimate.

Each return in a set is parameterized as a Cartesian coordinate in a frame with its origin at the robot's center of rotation at the time $\boldsymbol{A}$ was gathered. Generating this parameterization for $\boldsymbol{B}$ requires some initial value for $\hat{m}$. If no information is available to seed the registation process, the initial estimate can be 0. Alternatively, the initial estimate can come from odometry or a motion model.

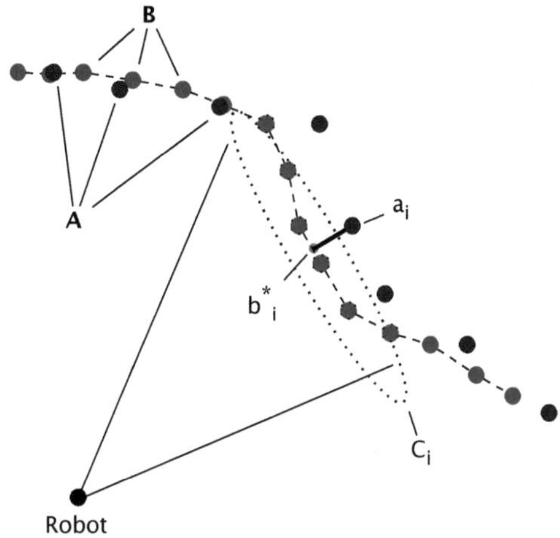


**Fig. 1.** Sample Match and Covariance Ellipse

For each point $a_i$ in $\boldsymbol{A}$, we find $b_i^*$, the closest point on the polyline defined by the bearing-sorted points of $\boldsymbol{B}$. All returns in $\boldsymbol{B}$ within a fixed bearing window of $b_i^*$ are used to calculate a 2×2 covariance matrix, $R_i$.

Figure 1 illustrates the various aspects of a single generated point match.

The next step is best discussed in the framework of an update step of a Kalman Filter. One popular formulation of the filter, from [15], can be summarized as follows:

Given a state estimate $\hat{x}_k$ at time $k$ with associated covariance matrix $P$, if we have an observation of the form:

$$z_k = H_k x_k + v_k$$

where

$$p(v_k) \sim N(0, R_k)$$

$\hat{x}$ and $P$ are updated as follows:

$$K_k = P_k H_k^T (H_k P_k H_k^T + R)^{-1}$$

$$\hat{x}_{k+1} = \hat{x}_k + K_k(z_k - H\hat{x}_k)$$

$$P_{k+1} = (I - K_k H_k) P_k$$

We initialize the filter with a $\hat{x}$ of 0 and we let $P_0$ be an arbitrarily large diagonal matrix. We consider each point match $z_i = b_i^* - a_i$ to be an independent observation of the state to be estimated, with associated covariance $R_i$. The mapping from state to observation is nonlinear, so we substitute the Jacobian:

$$H = \begin{pmatrix} 1 & 0 & -d_i sin(\phi_i) \\ 0 & 1 & d_i cos(\phi_i) \end{pmatrix}$$

where $d_i = ||a_i||$ and $\phi_i = atan2(a_{iy}, a_{ix})$.

After processing all the point correspondences, the filter contains an incremental refinement to the overall transformation:

$$\hat{m}_{k+1} = \hat{m}_k + \hat{x}$$

As with ICP, as the estimate is refined, better correspondences are found between scans. Thus, we generate new matches with the refined estimate and iterate until convergence. When converged, $Q$ is simply the calculated $P$ from the final iteration.

## 5 Robustness

This algorithm is a modified least squares optimization. As such, it inherits the sensitivity of such methods to outliers. The most common cause of such outliers is moving objects in the scene.

The resulting scan registrations are greatly improved by adding a RANSAC step to the point selection [6]. The process for doing so is straightforward. $n$ point matches are drawn from the registration, and used to generate an estimate and covariance. Inliers from the rest of the set are determined from the Mahalanobis distance of the generated estimate. If a sufficent proportion of the points are inliers, the overall estimate is recalculated using the inliers, otherwise a new set of $n$ points is drawn. Emprically, generating a sample with 4 single point motion estimates does an excellent job of rejecting outliers while not being computationally burdensome.

## 6 Test Platform

Data is presented from two different robotic platforms.

Indoor results are presented using data gathered from a heavily modified Segway Robotic Mobility Platform (RMP). The RMP carries a SICK LMS mounted 45 cm from the ground. Due to the inverted pendulum nature of the RMP motion, the SICK is mounted on a servo which allows the robot to keep it nominally parallel to the ground. The RMP provides state updates, including odometry and kinematic estimates, at 100 Hz. The SICK provides scans of 181 returns spaced 1 degree apart at 75 Hz.

The RMP data sets presented here were gathered at speeds of up to 2.5 meters per second, and angular rates of up to 2 radians per second.

Additional testing used data sets collected at Stanford and graciously provided online at [8]. The Stanford data sets were collected in the Gates building using a SICK LMS on a Pioneer XT.

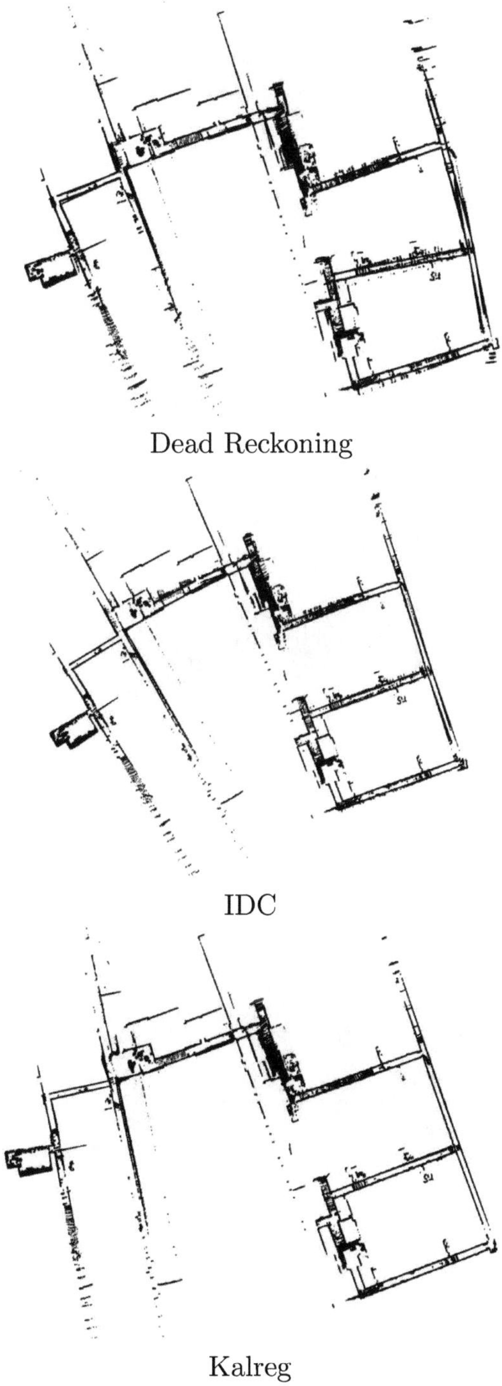


**Fig. 2.** Result: CMU-Qatar building, 31000 scans

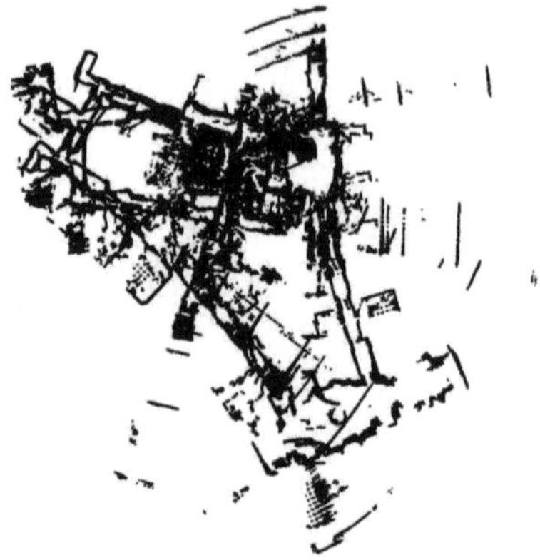

Dead Reckoning

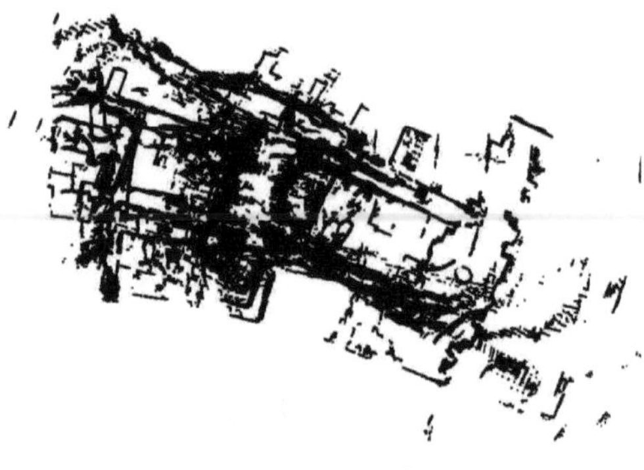

IDC

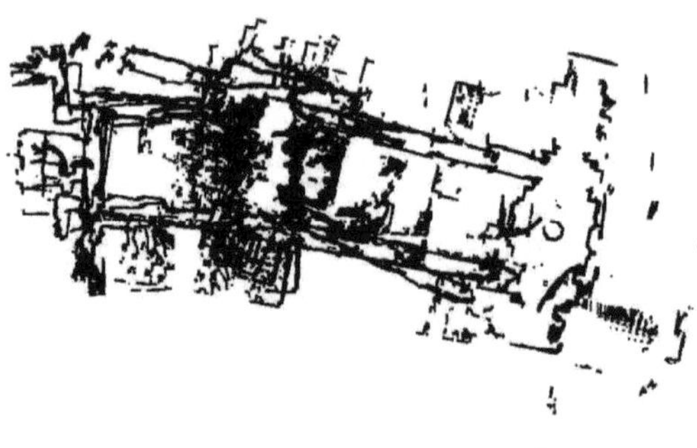

Kalreg

**Fig. 3.** Result: Stanford Gates building, 50,000 scans

## 7 Accuracy

For this paper, we call our algorithm "Kalreg", and compare our results against both dead reckoning and an implementation of the Iterative Dual Correspondence algorithm [10], a widely used scan-matching system.

To make comparisons more valid, results shown use odometry to seed the registration process in all cases. The "Kalreg" results do not fuse the odometry measurements into the final registration.

Figure 2 shows a run consisting of about 31,000 registered scans over a path of approximate 750 meters. Both registration algorithms encounter some difficulties with an open area in the middle of the run; the heading estimate in both cases is noticeably skewed.

Figure 3 uses the first 50,000 scans of a traversal of the Gates building at Stanford. This data set is intended for SLAM-type applications, and contains many loops and backtracks. Despite the lack of loop closure in these results, our algorithm generates an overall registration which is visibly more consistent than the IDC result.

## 8 Complexity and Performance

The algorithm is designed to be very efficient and fast. With the exception of finding correspondences, the entire algorithm is linear in the number of points in a scan.

Given a bearing-ordered sensor, several shortcuts can be used to find reasonable candidates for matches in constant time. In the general case, points can be sorted into bearing order, making approximate correspondence search an $O(nlog(n))$ proposition.

Other computational factors that are significant are the number of RANSAC trials needed to find a consistent estimate and the number of iterations needed to converge. In the environments tested, where most of the scene is static, neither factor appears to be onerous.

Interestingly, the uncertainty estimates for matches reduce the iterations required for convergence, often significantly. One problem in many ICP approaches is the tendency of linear scene segments correspondences with near-zero offsets to cause a "drag" effect on the convergence. The original ICP paper proposes a form of gradient descent to bypass the problem. In this method, such segments are associated with appropriate covariances, so the "drag" effect is dramatically lessened, obviating such a requirement.

The authors' most recent C++ implementation runs comfortably in real-time on a 2.4 Ghz Pentium 4 workstation. It is possible to perform the bulk of necessary computations as highly pipelinable parallelized vector operations using SIMD instructions. The authors did write one mildly optimized implementation which was capable of registering scans at 320 Hz. With more effort, this likely could be improved by an order of magnitude.

## 9 Conclusion

This paper put forth a new method for estimating robot egomotion with explicit uncertainty reasoning. The algorithm provides good registration results comfortably in real-time, and generates uncertainty estimates concurrently.

## 10 Future Work

One of the goals in development of this algorithm was to generate useful but pessimistic error estimates for registration. Ongoing work is being carried out on validation of the generated error estimates using loop closures. Additionally, there are no obvious obstacles to extending the developed framework into registering points in three dimensions.

## References


[1] Bar-Shalom, Y., Li, X.-R.: Estimation and Tracking: Principles, Techniques and Software. Artech House, Inc. (1993)

[2] Bengtsson, O., Baerveldt, A.-J.: Robot localization based on scan-matching - estimating the covariance matrix for the idc algorithm. Robotics and Autonomous Systems 44, 29–40 (2003)

[3] Besl, P.J., McKay, N.D.: A method for registration of 3-d shapes. IEEE Trans. Pattern Anal. Mach. Intell. 14(2), 239–256 (1992), doi:10.1109/34.121791

[4] Biber, P., Fleck, S., Straßer, W.: A probabilistic framework for robust and accurate matching of point clouds. In: Rasmussen, C.E., Bülthoff, H.H., Schölkopf, B., Giese, M.A. (eds.) DAGM 2004. LNCS, vol. 3175, Springer, Heidelberg (2004)

[5] Biber, P., Straßer, W.: The normal distributions transform: A new approach to laser scan matching. In: IEEE/RJS International Conference on Intelligent Robots and Systems (2003)

[6] Fischler, M.A., Bolles, R.C.: Random sample consensus: a paradigm for model fitting with applications to image analysis and automated cartography. Commun. ACM 24(6), 381–395 (1981), doi:10.1145/358669.358692

[7] Guivant, J., Nebot, E.: Optimization of the simultaneous localization and map building algorithm for real time implementation. IEEE Transactions on Robotics and Automation (May 2001)

[8] Howard, A., Roy, N.: The robotics data set repository (radish) (2007)

[9] Lindsey, J.K.: Introduction to Applied Statistics: A Modelling Approach, 2nd edn. Oxford University Press, Oxford (2004)

[10] Lu, F., Milios, E.: Robot pose estimation in unknown environments by matching 2d range scans. In: CVPR 1994, pp. 935–938 (1994)

[11] Lu, F., Milios, E.: Globally consistent range scan alignment for environment mapping. Auton. Robots 4(4), 333–349 (1997)

[12] Pfister, S.T., Roumeliotis, S.I., Burdick, J.W.: Weighted line fitting algorithms for mobile robot map building and efficient data representation. In: Proceedings of the 2003 IEEE International Conference on Robotics and Automation, pp. 1304–1311 (2003)

[13] Wang, C.-C.: Simultaneous Localization, Mapping and Moving Object Tracking. PhD thesis, Robotics Institute, Carnegie Mellon University, Pittsburgh, PA (April 2004)

[14] Wang, C.-C., Thorpe, C., Thrun, S.: Online simultaneous localization and mapping with detection and tracking of moving objects: Theory and results from a ground vehicle in crowded urban areas. In: Proceedings of the IEEE International Conference on Robotics and Automation (ICRA), Taipei, Taiwan (September 2003)

[15] Welch, G., Bishop, G.: An introduction to the Kalman filter. Technical report, University of North Carolina at Chapel Hill (2006)

[16] Zezhong, X., Jilin, L., Zhiyu, X.: Scan matching based on cls relationships. In: IEEE International Conference on Robotics, Intelligent Systems, and Signal Processing, October 2003, pp. 99–104 (2003)

# A FastSLAM Algorithm Based on the Unscented Filtering with Adaptive Selective Resampling*

Manuel Cugliari and Francesco Martinelli

DISP, Università di Roma Tor Vergata, via del Politecnico, I-00133 Roma, Italy
mcugliari@tiscali.it, martinelli@disp.uniroma2.it

**Summary.** A FastSLAM approach to the SLAM problem is considered in this paper. An improvement to the classical FastSLAM algorithm has been obtained by replacing the Extended Kalman Filters used in the prediction step and in the feature update with Unscented Kalman Filters and by introducing an adaptive selective resampling. The simulations confirm the effectiveness of the proposed modifications.

## 1 Introduction

The Simultaneous Localization and Map Building (SLAM) problem has recently received a large attention (see, e.g., [4, 13]). The SLAM problem consists in a mobile robot moving in an unknown environment, which attempts to estimate its own position and to realize a spatial map. The environment is described through a set of natural landmarks extracted by the robot from the surrounding.

The SLAM can be faced as a state estimation problem, where the state includes the robot and the landmark coordinates, and can be approached with the well-known Extended Kalman Filter (EKF) [1]. EKF-based SLAM algorithms have two limitations: they present a high computational complexity and are prone to the data association problem when, like in this paper, the correspondence between measurements and landmarks is assumed unknown. An Unscented Kalman Filter (UKF) based solution to the SLAM problem [10] has been recently proposed. The UKF has been introduced in the context of non-linear system filtering by [7, 8]. Even if more performant, the resulting approach is similar to the EKF-SLAM since the conditional independence among the features position estimates given the robot path (introduced by the FastSLAM approach) is not exploited. The FastSLAM approach [11, 12] exploits this independence: it uses a modified particle filter to estimate the robot pose and each particle is equipped with $N$ EKF to estimate the position of the $N$ landmarks. The resulting algorithm is an application of the Rao-Blackwellized particle filter [5]. The FastSLAM has a lower complexity w.r.t. the EKF-SLAM and is more robust regarding the data association problem since each particle performs its own data association. Nevertheless, the robot pose estimated by the classical FastSLAM is based on the EKF and may incur into divergence due to

* This work has been partially supported by MIUR under grant PRIN 2005092439. The authors would like to thank F. Romanelli and L. Spinello for their valuable initial support.

C. Laugier and R. Siegwart (Eds.): Field and Service Robotics, STAR 42, pp. 359–368, 2008.
springerlink.com 

the approximations introduced by linearization. The idea of replacing EKF with UKF to improve the proposal distribution of a particle filter has been introduced in [15] in a general framework. In addition, the use of UKF to cope with nonlinearities associated with a stereo camera observation model was exploited in [2], where a FastSLAM algorithm was developed in an experimental context. In our paper, the idea of using the Unscented Transformation (UT) both in the observation and in the motion model has been combined with an adaptive selective resampling, based on an adaptive threshold, which significantly improves the selective resampling described in [5, 6], where a fixed threshold was considered. With these modifications, the robot pose error is drastically reduced (around one order of magnitude in our simulation results) and a more consistent map is obtained with respect to the classical FastSLAM 2.0 algorithm described in [12].

## 2 SLAM and FastSLAM Background

The SLAM problem can be correctly described through a Markov chain or, more in general, through a dynamic Bayesian network. The robot pose $s_t$ changes according to a probabilistic law $p(s_t|u_t, s_{t-1})$ referred to as *motion model*, where $u_t \in \mathbb{R}^{n_u}$ is the control at time $t$. Following a common notation, the superscript $^t$ denotes a set of variables from time 1 up to time t, the subscript $_t$ indicates the value of the variable at time $t$. In planar SLAM $s_t = [x_t, y_t, \alpha_t]^T$, where $x$, $y$ are the coordinates of the robot and $\alpha$ is its orientation. The environment is represented through $N$ fixed landmarks $\vartheta_n$, $n = 1, ..., N$, described by their coordinates relative to the global coordinate frame. Proprioceptive data (like encoders) and exteroceptive data (like, e.g., lasers) are available to the robot which can sense landmarks and know their distance and bearing, relative to its local coordinate frame. $z_t$ denotes a landmark measurement at time $t$ and for mathematical convenience it is assumed that a single feature at a time is observed. These landmark measurements are noisy and a *measurement model* $p(z_t|s_t, \vartheta_{n_t}, n_t)$ is considered for them, with $\vartheta_{n_t}$, $n_t \in 1, ...N$ being the particular landmark observed at time $t$. We assume in our formulation that data association is unknown. Both the *motion* and the *measurement models* are in general governed by non-linear deterministic functions, respectively $h$ and $g$, affected by noise [14]:

$$p(s_t|u_t, s_{t-1}) = h(s_{t-1}, u_t + \delta_t) \tag{1}$$

$$p(z_t|s_t, \vartheta_{n_t}, n_t) = g(\vartheta_{n_t}, s_t) + \varepsilon_t \tag{2}$$

The noises $\delta_t$ and $\varepsilon_t$ are assumed normally distributed with zero mean and covariance given, respectively, by $P_t$ and $R_t$. With respect to [14], a direct effect of $\delta_t$ on $u_t$ has been assumed. This is the case in several kinematic models and in particular for the one considered in Section 5. Formally stated, the SLAM problem consists in the determination of the location of all the landmarks $\vartheta_n$ and of the robot pose $s_t$ only using the information of the controls $u^t = \{u_1, ..., u_t\}$ and the measurements $z^t = \{z_1, ..., z_t\}$. That is, it consists in the determination of the posterior

$$p(s_t, \Theta|z^t, u^t, n^t) \tag{3}$$

where $\Theta = \{\vartheta_1, \ldots, \vartheta_N\}$ is the set of all the landmarks. The FastSLAM approach heavily relies on the conditional independence present in the SLAM problem which allows to factorize the posterior over the robot path as follows [14]:

$$p(s^t, \Theta | z^t, u^t, n^t) = p(s^t | z^t, u^t, n^t) \prod_n p(\vartheta_n | s^t, z^t, n^t) \tag{4}$$

A particle filter, similar to the *Monte Carlo Localization* (MCL) algorithm [3], estimates the robot path, maintaining a set $S_t$ of $M$ particles $s^{t,[m]}$, $m = 1, \ldots, M$, which approximates the posterior $p(s^t | z^t, u^t, n^t)$. A temporary set of particles is generated at time $t$ according to $S_{t-1}$, using the measurements $z_t$ and the controls $u_t$. Since the set $S_{t-1}$ corresponds to $p(s^{t-1} | z^{t-1}, u^{t-1}, n^{t-1})$, the new set is distributed according to $p(s^t | s^{t-1}, z^t, u^t, n^t)$. This distribution is often referred to as the *proposal distribution* of the particle filter. The new set is generated by sampling (with replacement) $M$ particles proportionally to an *importance factor* $w_t^{[m]}$:

$$w_t^{[m]} = \frac{\text{target distribution}}{\text{proposal distribution}} = \frac{p(s^t | z^t, u^t, n^t)}{p(s^t | s^{t-1}, z^t, u^t, n^t)} \tag{5}$$

This step, usually known as resampling process, accounts for the difference between the target and the proposal distribution. Next, the FastSLAM updates the posterior $p(\vartheta_n | s^t, z^t, n^t)$ conditioning it on the robot path. So, the full posterior over the robot path and the landmark position is approximated through the set:

$$S_t = \{s^{t,[m]}, \mu_{1,t}^{[m]}, \Sigma_{1,t}^{[m]}, \ldots, \mu_{N,t}^{[m]}, \Sigma_{N,t}^{[m]}\}_{m=1,\ldots,M} \tag{6}$$

where $\mu_{n,t}^{[m]}$ and $\Sigma_{n,t}^{[m]}$ denote the mean and the covariance of the normal distribution representing the estimate provided by the $m$-th particle about the $n$-th landmark position. The FastSLAM performs the position estimate update of landmark $n_t$ through the EKF while the position estimate of the other landmarks remains unchanged. Therefore there are $N \cdot M$ low dimensional EKF (one for each landmark relative to each particle). The FastSLAM samples the pose $s_t^{[m]}$ from the posterior:

$$s_t^{[m]} \sim p(s_t | s^{t-1,[m]}, z^t, u^t, n^t) \tag{7}$$

The posterior in (7) is approximated through the following expression [14], which allows to perform the sampling procedure:

$$s_t^{[m]} \sim p(s_t | s^{t-1,[m]}, z^t, u^t, n^t) \overset{Bayes, Markov}{=} \eta^{[m]} \int \left[ p(z_t | \vartheta_{n_t}, s_t, n_t) p(\vartheta_{n_t} | s^{t-1,[m]}, z^{t-1}, n^{t-1}) \right] d\vartheta_{n_t} \; p(s_t | s_{t-1}^{[m]}, u_t) \tag{8}$$

In the remaining of the paper only the modified steps of the FastSLAM 2.0 algorithm will be discussed. The other steps can be found in [12] and [14].

## 3 The Proposed Unscented Based Improvements

We focus our attention on the rightmost term in (8) because an accurate approximation of this term drastically reduces the ambiguity in the data association.

Let $\Xi_t$ be the set representing the evolution of the particles belonging to the set $S_{t-1}$ at time $t-1$, according to noisy $u_t$. $\Xi_t$ has the following structure:

$$\Xi_t = \{\langle \xi_t^{[1]}, Q_t^{[1]} \rangle, ..., \langle \xi_t^{[M]}, Q_t^{[M]} \rangle\} \tag{9}$$

where $\xi_t^{[m]}$ is the new $m$-th particle pose and $Q_t^{[m]}$ is the covariance matrix relative to the $m$-th particle. FastSLAM 2.0 computes $\Xi_t$ by putting each particle $s_{t-1}^{[m]}$ into the motion model and taking the obtained value as the mean of a Gaussian distribution approximating the particle pose. That is[1]:

$$^s\xi_t^{[m]} = h(s_{t-1}^{[m]}, u_t), \quad ^sQ_t^{[m]} = G_{s_{t-1}} Q_{t-1}^{[m]} G_{s_{t-1}}^T + G_{u_t} P_t G_{u_t}^T \tag{10}$$

where $G_{s_{t-1}}$ and $G_{u_t}$ are the Jacobians of $h$ with respect to $s_{t-1}$ and $u_t$. This linearization-based approach, which neglects the second and higher order terms of the Taylor expansion of $s_{t-1}^{[m]}$, sometimes produces uncorrect *clouds* of particles for the motion-only evolution. We attempt to provide a more accurate set $\Xi_t$ exploiting the UT. First, the pose vector is augmented with the control noise vector $\delta_t \in \mathbb{R}^{n_u}$ to give an $n_a = 3+n_u$ dimensional random vector $s_{t-1}^{a,[m]} := [s_{t-1}^{[m]}; \delta_t] \in \mathbb{R}^{n_a}$, with augmented covariance matrix:

$$Q_{t-1}^{[m]'} = \begin{bmatrix} Q_{t-1}^{[m]} & 0 \\ 0 & P_t \end{bmatrix} \tag{11}$$

The cross terms in $Q_{t-1}^{[m]'}$ are 0 since the pose $s_{t-1}^{[m]}$ and the noise $\delta_t$ are independent. Then, a set of (*sigma points*) $\{\mathcal{X}^{i,[m]}\}$, each $\mathcal{X}^{i,[m]} \in \mathbb{R}^{n_a}$, and corresponding weights $\{W^{i,[m]}\}$, $i = 0, 1, \ldots, p$, are generated according to [7]: the weights $W^{i,[m]}$ can be positive or negative but, to provide an unbiased estimate, must sum to one. In addition, to match the mean and the covariance, it must be:

$$\sum_{i=0}^{p} W^{i,[m]} \mathcal{X}^{i,[m]} = \bar{s}_{t-1}^{a,[m]}, \quad \sum_{i=0}^{p} W^{i,[m]} (\mathcal{X}^{i,[m]} - \bar{s}_{t-1}^{a,[m]})(\mathcal{X}^{i,[m]} - \bar{s}_{t-1}^{a,[m]})^T = Q_{t-1}^{[m]'} \tag{12}$$

where $\bar{s}_{t-1}^{a,[m]} = \begin{bmatrix} s_{t-1}^{[m]} \\ 0 \end{bmatrix}$.

Next, the non-linear motion model is applied, in turn, to each point ($i = 0, ..., p$) for the $m$-th particle to yield a cloud of transformed points $\mathcal{Z}^{i,[m]} \in \mathbb{R}^3$:

$$\mathcal{Z}^{i,[m]} = h(\mathcal{X}_{1:3}^{i,[m]}, u_t + \mathcal{X}_{4:n_a}^{i,[m]}) \tag{13}$$

[1] We denote the standard computation of $\xi_t^{[m]}$ by $^s\xi_t^{[m]}$, while we denote by $^u\xi_t^{[m]}$ the computation performed through the unscented approach. A similar notation applies to the covariance matrix.

where $\mathcal{X}_{r:j}^{i,[m]}$ denotes a vector comprising the components from $r$ to $j$ of $\mathcal{X}^{i,[m]}$. Now, according to [7], the predicted particle pose $\xi_t^{[m]}$ is given by the weighted average of the transformed points:

$$^{u}\xi_t^{[m]} = \sum_{i=0}^{p} W^{i,[m]} \mathcal{Z}^{i,[m]} \tag{14}$$

and the covariance is the weighted outer product of the transformed points:

$$^{u}Q_t = \sum_{i=0}^{p} W^{i,[m]} (\mathcal{Z}^{i,[m]} - {^{u}\xi_t^{[m]}})(\mathcal{Z}^{i,[m]} - {^{u}\xi_t^{[m]}})^T \tag{15}$$

After estimating the particles pose, FastSLAM 2.0 carries out the landmark location estimation by using a set of EKF. We rather provide an unscented-based feature update (UKF). The UT enables one to avoid the linearization of the perceptual model and the common problems that this solution involves. Details are omitted for space reasons. The update of the landmark location is capital for the correct data association and consequently for the estimated pose corrections.

## 4 The Proposed Adaptive Selective Resampling

Resampling is the step that has a major influence on the performance of the particle filter and it consists in replacing particles with low importance factor $w_t^{[m]}$ by particles with a higher weight. This process may erroneously eliminate good particles, causing the phenomenon known as particles impoverishment or *depletion*. To overcome this problem, it is important to determine *when* and *how* the resampling must be performed. Liu [9] introduced the effective number of particles

$$N_{eff} = \frac{1}{\sum_{m=1}^{M} (w^{[m]})^2} \tag{16}$$

which reflects the particles dispersion and consequently is also a measure of the approximation quality of the true posterior: it would be equal to the real number $M$ of particles if the samples were drawn from the true posterior. Our approach extends the one reported in [5, 6] where the resampling is performed when the effective number $N_{eff}$ of particles goes below a fixed threshold. A fixed threshold however does not take into account of the past evolution of $N_{eff}$, evolution which contains important information on the quality of the posterior approximation. For example, if the effective number of particles is not very high but it slightly oscillates in the time around a constant value, this probably corresponds to a good posterior approximation. In this case a resampling performed when one of the small oscillations of $N_{eff}$ slightly goes below the threshold would be undermining. It would be better to decrease the threshold, avoiding the resampling, at least until this effective number remains above a minimum acceptable value. Also the average number of effective particles in a time interval plays an important role in the characterization of the quality of the approximation over that interval. Therefore, instead of a static threshold, in this paper a dynamic threshold has

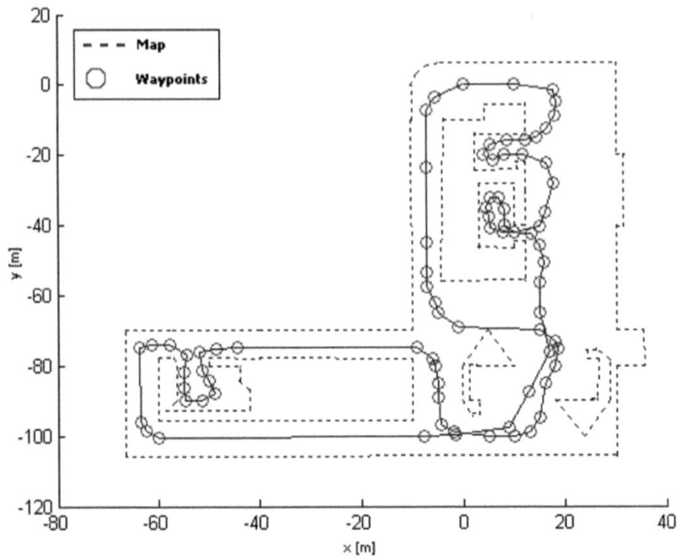


**Fig. 1.** Map (dashed black line) and waypoints (red circle). The robot controls are computed according to the waypoints. The blue solid line simply links the waypoints.

been introduced, whose value depends on the past history of $N_{eff}$. The idea above can be formally stated as follows. We take into account of the recent past of $N_{eff}$ considering a window of observation comprising $k$ time steps. First, denote with $\upsilon_{\tau-1}$ the value of the threshold in the window $\tau - 1$ and with $N_{min}$ and $N_{max}$, respectively, the minimum and the maximum value that the threshold may assume. Our heuristic is to set $N_{min} = 0.6M$ and $N_{max} = 0.9M$. Observe that $N_{max}$ must be taken below $M$ since a threshold too near to $M$ would imply a too large resampling frequency. The new $\upsilon_\tau$ is taken according to how rapidly $N_{eff}$ changes and to how many effective particles are present in window $\tau - 1$. If $k$ denotes the number of time steps in the observation window and $N_{eff}(t)$ is the effective number of particles at the $t$-th time step in the window, a measure $\alpha$ of the mean (normalized) number of particles in window $\tau - 1$ and a measure $\beta$ of the (normalized) variation of $N_{eff}$ can be expressed by:

$$\alpha = \frac{1}{Mk}\sum_{t=1}^{k-1}\frac{N_{eff}(t+1)+N_{eff}(t)}{2}, \qquad \beta = \frac{N_{eff}(k)-N_{eff}(1)}{(k-1)(M-N_{min})} \tag{17}$$

Then, the following two quantities associated with $\alpha$ and $\beta$ are computed:

$$N_\alpha = \alpha(N_{max} - N_{min}) + N_{min} \tag{18}$$

$$N_\beta = (N_{max} - N_{min})|\beta|^{\frac{1}{N_{min}}} + N_{min} \tag{19}$$

The new threshold $\upsilon_\tau$ can be obtained as a proper weighted average of $N_\alpha$ and $N_\beta$:

$$\upsilon_\tau = aN_\alpha + bN_\beta \tag{20}$$

with $a, b > 0$, $a + b = 1$, two suitable parameters. Notice that $\alpha$ near to one denotes a large average value of $N_{eff}$ and the threshold can be increased. This is in fact obtained, since, when $\alpha$ approaches one, $N_\alpha$ moves linearly to $N_{max}$ and the threshold is increased, according to (20). The non-linear function $N_\beta$ takes into account of the variation of $N_{eff}$. Now, when $\beta$ is near to zero, the variation of $N_{eff}$ in the window is small. A low variation of $N_{eff}$, even if $N_{eff}$ is not large, means an agreement among particles with respect to the approximation of the true posterior and the resampling should be avoided. This is in fact obtained since the threshold, according to (20), with a rate dependent on the exponent of $|\beta|$ in (19), is decreased. We choose a function like the one in (19) for its high variability near to zero. On the other hand, a $|\beta|$ quite greater than 0 corresponds to a large variation of $N_{eff}$ in the window and the threshold is increased, making a resampling more likely to occur.

## 5 Simulation Results

The robot considered in the simulations is characterized by a car-like kinematic model (e.g. [4]) and is equipped with a range-bearing laser with a maximum range of 20 meters and a 180 degrees frontal field-of-view. The indoor environment and the way-points defining its target path are graphically depicted in Fig. 1. The robot controls (velocity and steering inputs) are generated in order to follow, from the initial pose $s_0 = \langle 0, 0, 0 \rangle$, the blue solid line in Fig. 1. The total traveled distance is about 700 m. Range and bearing readings, as well as the controls, are perturbed by Gaussian noises with zero mean and diagonal covariance matrices, respectively, $R_t$ and $P_t$. If $u_t = [u_{t,v};\, u_{t,s}]$ is the control vector, with $u_{t,v}$ and $u_{t,s}$ the velocity and the steering inputs, and $\delta_t = [\delta_{t,v};\, \delta_{t,s}]$ the corresponding control noise vector, then $P_t = E[\delta_t \delta_t^T] = [\sigma_v^2,\, 0; 0,\, \sigma_s^2]$. In the simulations reported below, $\sigma_v = 0.01m/s$ and $\sigma_s = 0.017rad$. We have assumed, for the noise on the range and the bearing of each laser reading, a standard deviation respectively of $0.01m$ and of $0.017rad$.

The FastSLAM algorithm, with the modifications proposed in this paper (with parameters $a = 0.7$, $b = 0.3$, $k = 150$ and $p = 2\,n_a + 1 = 11$, where $a$ and $b$ are needed in (20), $k$ is the number of steps in each observation window and $p$ the number of Sigma points) was tested extensively in various conditions (high and low noises for the odometry and for the laser measurements, different parameters for the data association and, also, different environments). Our methods reduce the error between the real path and the estimated path (at each time step the particle with the greater weight is considered as the best estimate of the robot pose).

In Fig. 2 a comparison between the FastSLAM with and without modifications is illustrated for the robot coordinates $\langle x, y, \alpha \rangle$ error (Frames (a) and (b)) and for the absolute robot position error (Frames (c) and (d)). The robot, running the FastSLAM algorithm with the proposed modifications and with a number of particles $M = 60$, incurs in an average mean absolute error (MAE) (average computed over 50 simulation runs, with different noise realizations) of 0.03 m with standard deviation 0.007 m. With the standard FastSLAM (with the same number of particles and simulation runs) the average mean absolute error obtained is 0.62 m with standard deviation 0.02 m. The Frames (e) and (f) of Fig. 2 show the effectiveness of the adaptive selective resampling

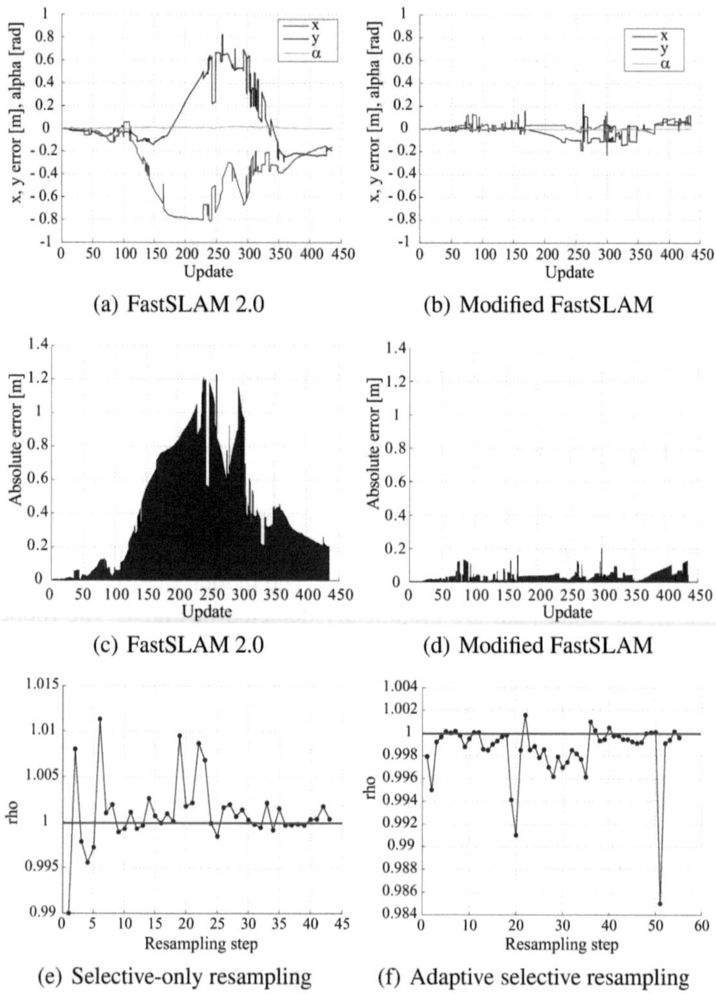


**Fig. 2.** Comparison of the standard FastSLAM and the FastSLAM with our modifications for the $\langle x, y, \alpha \rangle$ error ((a) and (b)) and for the pose error ((c) and (d); (e) and (f) show the ratio $\rho$ after the selective-only resampling step (e) and the adaptive selective resampling step (f).

(Frame (f)) w.r.t. the selective resampling (Frame (e)) by reporting the ratio $\rho$ between the absolute position error when a resampling occurs and the absolute position error in the previous time step. That is:

$$\rho = \frac{e_{tr}}{e_{tr-1}} \tag{21}$$

where $e_{tr}$ is the pose error after the resampling step and $e_{tr-1}$ is the same quantity in the previous time step. With the adaptive selective resampling this ratio is almost always less than one, i.e. the resampling almost always reduces the pose error. Fig. 3

shows the maps generated using the FastSLAM modified and the classical FastSLAM algorithms. In the first case the real path (green) and the estimated path (black) are almost always overlapping and the robot closes the loop with a position error of only 0.0414 m; the classical FastSLAM closes the loop with a position error of 0.2040 m and there are pronounced gaps between the real path and the estimated path; both the algorithms generate a consistent map, but with the proposed modifications the quality of the map is so high that it can be magnified up to 3 cm of resolution without an observable significant error; on the other hand, with the classical FastSLAM algorithm, a map reconstruction error can be highlighted with a smaller magnification level (1 m is enough).

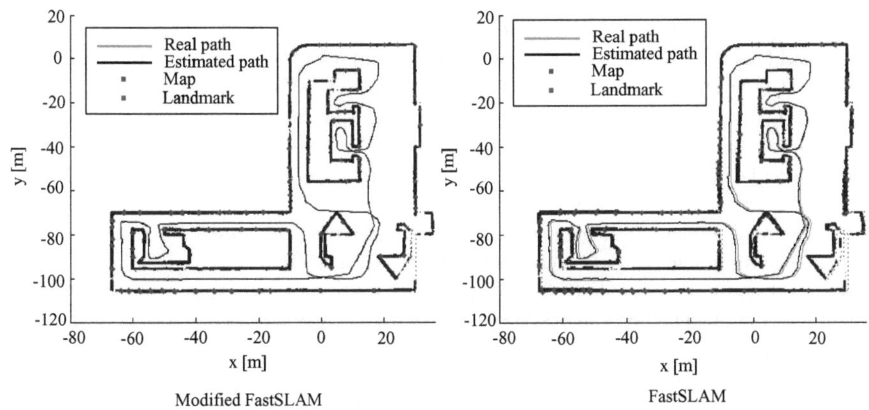


**Fig. 3.** Final 2-D map generated with FastSLAM and our modification of the FastSLAM

# 6 Conclusions

This work provides some modifications to the FastSLAM 2.0 algorithm. Unscented Transformations replace the EKF (avoiding linearizations) generating more accurate sets of particles $S_t$ and improving the proposal distribution from which the set of particles is drawn. The paper has emphasized the importance of providing accurate *clouds* of particles for motion-only evolution. This allows to generate a more consistent proposal and consequently a better set $S_t$. In addition an adaptive selective resampling has been introduced in the resampling step of the particle filter to significantly reduce w.r.t. the classical FastSLAM the risk of ill-timed resampling operations. Finally the set of EKF used in the classical FastSLAM approach for the landmark position estimate has been replaced by a set of UKF. The proposed modifications substantially improve the performance of the classical FastSLAM algorithm decreasing of about one order of magnitude the robot pose estimation error, as shown through an extensive campaign of simulations. The effectiveness of the adaptive selective resampling has been also illustrated in the paper.

## References


1. Andrade-Cetto, J., Sanfeliu, A.: Environment Learning for Indoor Mobile Robots. Springer, Heidelberg (2006)
2. Barfoot, T.D.: Online Visual Motion Estimation using FastSLAM with SIFT Features. In: Proceedings of Intelligent Robots and Systems (IROS 2005) (2005)
3. Dellaert, F., Fox, D., Burgard, W., Thrun, S.: Monte Carlo localization for mobile robots. In: Proc. of the IEEE Int. Conf. on Robotics & Automation (ICRA) (1998)
4. Dissanayake, M.W.M.G., Newman, P., Clark, S., Durrant-Whyte, H.F., Csorba, M.: A solution to the simultaneous localization and map building (SLAM) problem. IEEE Trans. on Robotics and Automation 17(3), 229–241 (2001)
5. Doucet, A., de Freitas, N., Murphy, K., Russell, S.: Rao-Blackwellised particle filtering for dynamic Bayesian networks. In: UAI 2000 (Uncertainty in AI) (2000)
6. Grisetti, G., Stachniss, C., Burgard, W.: Improving grid-based SLAM with Rao-Blackwellized particle filters by adaptive proposals and selective resampling. In: IEEE International Conference on Robotics & Automation (ICRA) (2005)
7. Julier, S.J., Uhlmann, J.K.: Unscented filtering and non-linear estimation. Proceedings of the IEEE 92(3), 401–422 (2004)
8. Julier, S.J.: The spherical simplex Unscented Transformation. In: Proc. of the American Control Conference (ACC), Denver (2003)
9. Liu, J.S.: Metropolized independent sampling with comparisons to rejection sampling and importance sampling. Statistics and Computing 6, 113–119 (1996)
10. Martinez-Cantin, R., Castellanos, J.A.: Unscented SLAM for Large-Scale Outdoor Environments. In: Intelligent Robots and Systems (IROS 2005) (2005)
11. Montemerlo, M., Thrun, S., Koller, D., Wegbreit, B.: FastSLAM: A Factored Solution to the Simultaneous Localization and Mapping Problem. In: AAAI-2002 (2002)
12. Montemerlo, M., Thrun, S., Koller, D., Wegbreit, B.: FastSLAM 2.0: An improved particle filtering algorithm for simultaneous localization and mapping that provably converges. In: Proc. of Int. Conf. on Artificial Intelligence (IJCAI) (2003)
13. Newman, P.: On the Structure and Solution of the Simultaneous Localisation and Map Building Problem. PhD thesis, Univ. of Sydney (2000)
14. Thrun, S., Montemerlo, M., Koller, D., Wegbreit, B., Nieto, J., Nebot, E.: FastSLAM: An efficient solution to the simultaneous localization and mapping problem with unknown data association. Journal of Machine Learning Research (2004)
15. van der Merwe, R., Doucet, A., de Freitas, N., Wan, E.: The Unscented Particle Filter. In: Leen, T.K., Dietterich, T.G., Tresp, V. (eds.) Advances in Neural Information Processing Systems (NIPS13) (2000)

# Tree Measurement and Simultaneous Localization and Mapping System for Forest Harvesters

Matti Öhman[1], Mikko Miettinen[2], Kosti Kannas[3], Jaakko Jutila[4], Arto Visala[5], and Pekka Forsman[6]

Helsinki University of Technology (TKK)
[1]Matti.Ohman@tkk.fi, [2]Mikko.Miettinen@tkk.fi, [3]Kosti.Kannas@tkk.fi, [4]Jaakko.Jutila@tkk.fi, [5]Arto.Visala@tkk.fi, [6]Pekka.Forsman@tkk.fi

**Summary.** For the last decades, measurement and automation systems in Nordic cut-to-length forestry machines have evolved gradually. These heavy duty machines are lighter, faster and more accurate than ever before but the basic technologies and operation have remained the same. In many respects, their current automation systems have reached their limits. The Forestrix project studies how advances in mobile robotics could be applied in the field of forestry machine automation. Machine vision systems and scanning laser range finders have established themselves as standard equipment in mobile robotics. With the new sensor and computing technologies it is possible to get information about the surrounding forest, such as tree diameters, positions and stand density. This information can be used on-line in operator's decision support system, or off-line in a forest asset management system. This paper describes the prototype measurement platform and the software algorithms developed in the Forestrix project. Results from tests with an all terrain vehicle are also presented.

## 1 Introduction

The Forestrix project studies advanced sensor systems for forest harvesters. One of the sensors used is a scanning laser range finder (henceforth a laser scanner). As the machine is moving in the forest the laser scanner can take multiple scans of the trees it passes from different directions. This allows the measurement system to use stochastic models to greatly improve the quality of the collected information. However, combining the different measurements requires that the system is able to reliably relate the measurements to correct trees. In the literature this is known as the data association problem. It is easy to solve if the path of the moving sensor is known, in which case we have a simple mapping problem. Unfortunately, the path cannot be directly measured without a very accurate and expensive navigation sensor, such as RTK-DGPS[1]. However, if

C. Laugier and R. Siegwart (Eds.): Field and Service Robotics, STAR 42, pp. 369–378, 2008.
springerlink.com 

the association problem can be solved without the exact sensor path, the sensor path can be calculated relative to the trees being measured. In the literature, this is known as the simultaneous localization and mapping (henceforth SLAM) problem.

In the Forestrix project the laser scanner is used for SLAM[2]. A machine vision system is used to augment the map generated by the SLAM algorithm. With the help of the laser scanner, the machine vision system divides camera images into sub images. Each sub image contains one dominant tree trunk. Different edge detection algorithms are used to extract the vertical edges of the dominant tree trunk. Using a calibrated camera and range information from the laser scanner, it should be possible to measure the actual diameter of the tree trunk with greater accuracy that is attainable with a laser scanner alone. The machine vision system should also be able to identify the tree type i.e. whether it is pine, spruce, birch, or other deciduous tree. This additional information should be stored in the tree map that is incrementally built by the SLAM algorithm.

### 1.1 Data Collection System

The data collection system consists of 2D and 3D laser scanners, machine vision camera, differential GPS receiver, MEMS inertial measurement unit and PC computer. The 2D laser scanner is used continuously while the 3D laser scanner is used only for reference measurements while the platform is stationary. The 3D scanner is unsuitable for continuous measurements because it takes several minutes to create a single 3D scan. The data collection system is build on an all terrain vehicle which is shown in Figure 1.

**Fig. 1.** ATV test platform

The machine vision camera is synchronized to the 2D scanner. A special hardware frequency divider is used to transform the synchronization signal from the laser scanner so that it can be used to trigger the camera which operates at a much lower frequency. The differential GPS is the same model that is commonly used in forest harvesters. It has higher sensitivity than most consumer level receivers so that it can maintain better satellite fix even in demanding forest environments. MEMS inertial measurement unit is used to provide information of the pose of the platform.

## 2 Feature Based SLAM

The overall structure of the feature based SLAM algorithm is shown in Figure 2. The SLAM is based on the raw 2D laser scanner data. The first step is the feature extraction. In this application, the features are the surrounding trees. The tree features identified from a single scan are called echoes due to their uncertain nature. In a clean forest with little or no underbrush it is a relatively simple task to find the tree trunks from the raw laser scans. An example scan and its segmentation are shown in Figure 3. However, in more dense forests, finding the trunks can be extremely difficult if not impossible. The dense vegetation also reduces the effective measurement range drastically. Even in relatively clean forests the blind areas behind the nearest trunks can be substantial.

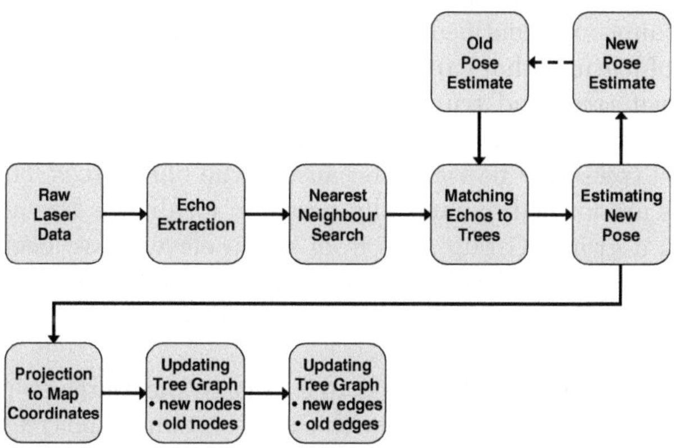


**Fig. 2.** Information flow in feature based SLAM

While it is straightforward to extract the echoes from the raw scans, it is very difficult to identify individual trees. This problem of matching features is usually referred as the association problem in the SLAM literature. Many SLAM applications get around of this problem by using beacons that can be unambiguously identified. However, this approach is not feasible in the proposed

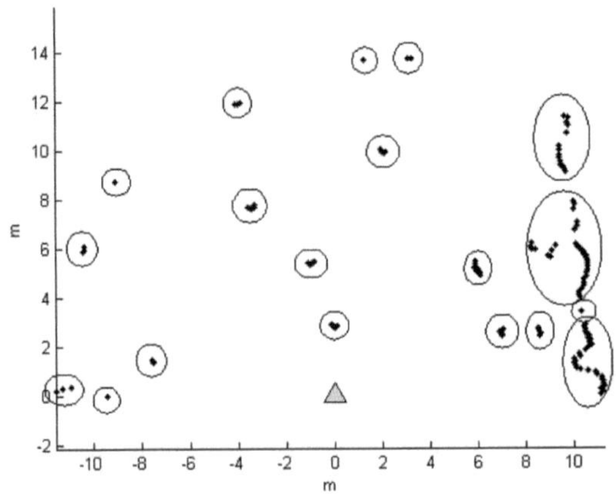


**Fig. 3.** The segmentation of a single scan from a well-managed pine forest

forest mapping application. A single tree trunk is difficult to identify because they all look the same for the laser scanner. Even the variation of trunk diameter can be very small in a well managed forest.

### 2.1 Tree Group Matching

Instead of trying to identify individual trees, it is better to identify groups of trees. These groups are identified in the third step in Figure 2. The tree groups offer a variety of features that can be used for identification: distances and angles between adjacent trees and trunk diameters. One of the central challenges in this approach is that the tree groups are not constant. Depending on current position, not all trees may be visible because of the blind areas. For this reason, the tree groups are formed dynamically based on locality: a tree and its nearest neighbors form a group. Group sizes from six to eleven have been tested. The angular resolution is of the laser scanner is 1 , 0.5 , or 0.25 degrees depending on its operating mode. The measurement uncertainties make it harder to identify trees at greater distances.

The matching of echoes to the tree map is done in the fourth step. The matching is done a group at a time starting with the closest one. The tree map is stored as a graph structure which makes this a graph matching problem. However, there are some issues that will have to be taken into account. First, the echo graphs are usually incomplete because of the blind areas. Second, also the tree graph is usually incomplete, because the map is incrementally built. The matching algorithm uses distances between adjacent tree trunks to match tree groups. Distances between trees are easy to work with because they are translation and rotation invariant. Relative angles are more difficult to work with because they need an additional reference point. If three trees are matched together in a triangular configuration, as shown in Figure 4, the distance information alone describes the full geometry making the relative angles completely redundant. For this reason, the relative angles are not used for matching tree groups. Absolute angles (e.g. relative to magnetic North) could be useful for matching purposes, but the current sensor platform does not have a reliable and accurate instrument for measuring absolute angles.

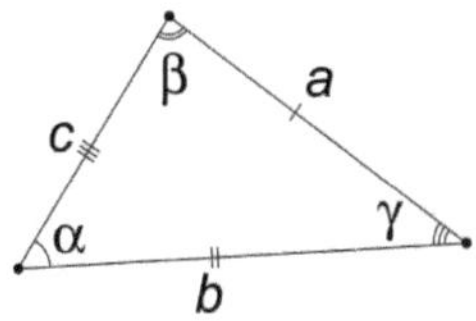


**Fig. 4.** Distances and relative angles contain the same information

After the echoes have been matched to trees, a simple algorithm is used to estimate the new pose. First, the "centers of gravities" of both set of points are calculated and the echo set is then translated so that the CoGs coincide. Second, the average angle between all echo-tree pairs is calculated and the echo set is then rotated by this angle. Care should be taken that all the angles are calculated on the correct cycle. This algorithm is based on the realistic assumption that the subsequent scans differ only by a translation and a rotation.

### 2.2 Tree Graph Updates

In the final phases the echoes are projected to map coordinates. At this point they are classified either as new or old trees. New trees are added to the tree graph and diameter information is accumulated to old trees. The closer the scan is taken, the better the diameter information usually is. New edges are also inserted into the tree graph and old edges are updated to reflect the new information. Currently only the distances between the trees are recorded but later angles will be used as well.

One of the problems of the presented algorithm is that it makes the optimistic assumption that all echoes really are trees. This helps the algorithm to acquire new trees and to keep going, but if echoes of branches are falsely interpreted as trees they will be added permanently to the tree map. To solve this problem another algorithm was added to remove false positives from the tree map. The principle of this algorithm is shown in Figure 5.

The algorithm works by projecting laser scans to map coordinates and then using a collision detection algorithm to find intersections between the "rays" and the trees in the map. If there are more "rays" passing through a tree than there are valid measurements of that tree, the tree is removed from the map. This second algorithm was added as an afterthought and it reflects the inability of the first algorithm to record what is sometimes referred as negative information. Future improvements of the algorithm may use local occupancy grids to keep track of areas that are known to be free of trees.

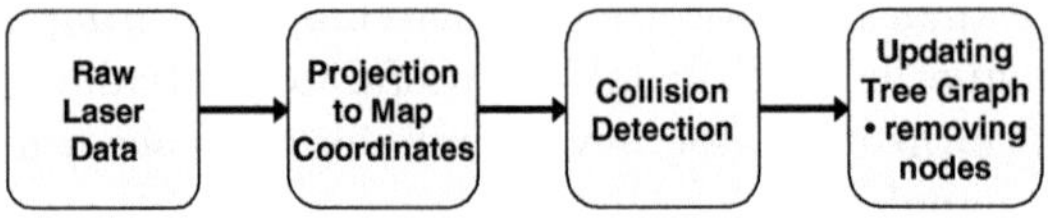


**Fig. 5.** Removal of false trees in feature based SLAM

The tree graph is a hybrid data structure. The graph structure contains the nodes and the edges between them. Each edge has length and in future versions also direction. However, each node has also position in Cartesian coordinates. The possible mismatch between Cartesian positions and edge lengths can be used by a third algorithm to distribute the accumulated error evenly e.g. when a large loop is being closed. Currently this is done with a simple iterative algorithm that tries to minimize and distribute the tensions in the tree graph.

### 2.3 SLAM Results

The current version of the software is written in the Java programming language. The software can read input data either from data files or from the actual sensors. The data files are essential for testing the system in a laboratory environment. The software can connect to the 2D laser scanner and the DGPS receiver. The SLAM path can be matched to the DGPS path. As a result the algorithm also gives the tree positions in absolute map coordinates. The resulting absolute accuracy is mostly limited by the DGPS receiver.

A tree map produced by Forestrix SLAM is shown in Figure 6. The DGPS path is marked with a narrow line and the SLAM path is drawn with a bold line. Measured tree positions are also shown. These positions and diameters can be compared to hand measured tree information provided by METLA (Finnish Forest Research Institute). At a measurement distance of 1m, the maximum observed diameter error is 1cm. The error increases linearly with distance up to 10m, where it is 6cm. After that the error increases more rapidly. These tests were done using 0.5 laser scanner resolution. A higher 0.25 resolution is expected to improve performance. At this time the described SLAM algorithm works in well defined forests.

## 3 Machine Vision System

Forest is one of the hardest environments for computer vision as it is extremely organic and unpredictable. It is structured in the sense that most trees grow in nearly vertical direction but otherwise the vegetation is very diverse in shape and density. The ground is not flat and there are arbitrarily shaped objects such as big rocks. The shooting is done in varying lighting and weather conditions and there are a lot of occlusion and shadows that complicate the situation. Taking images while moving introduces its own problems as well.

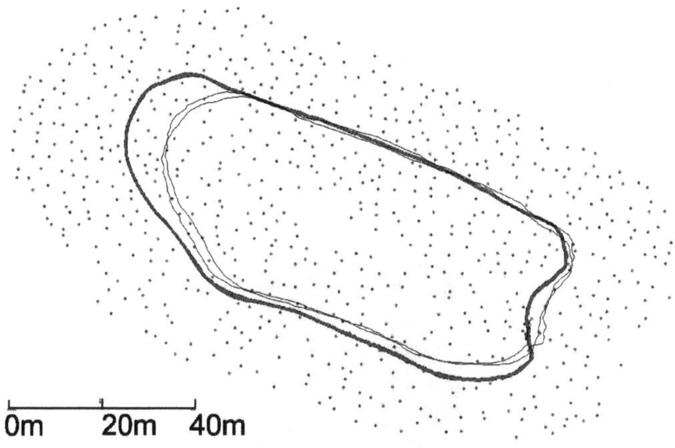


**Fig. 6.** Tree map produced by Forestrix SLAM (the path is traversed twice)

### 3.1 Tree Edge Extraction

To measure the tree diameter from an image, the feature extraction algorithm has to find the edges that correspond to the borders of the tree. The problem is, despite of all the filtering, that there are usually many edges to choose from, and the correct edges are not necessarily the strongest ones. The contrast between occluded trees is usually very poor while the contrast between the lower part of the trees and the ground is generally quite good. Seasonal variation, weather, time of the day and many other aspects make sure fixed parameters cannot be used for feature extraction.

There are many approaches to edge detection from traditional gradient based methods to neural networks and other more complex methods[3]. They are based on different assumptions and use different edge models or different optimality criteria. Assessing edge detector performance is difficult because there is no available ground truth of which pixels should really be edges and which not. Movement during exposure, uneven lighting, quantization errors and noise make the edges more difficult to detect.

The feature extraction part of the implemented algorithm starts with optional homomorphic filtering followed by vertical Fourier filtering. After the filtering stage the algorithm has a choice of using either Canny[4] or phase congruency based edge detector[5], [6]. In the default setting the algorithm will first use Canny with no prior homomorphism. If the algorithm doesn't seem to find enough features, it makes another pass with homomorphism enabled and Canny replaced by phase congruency edge detector. The Canny edge detector is chosen as the main edge detector here because it is well known and tested, it is fast and it performs well. The more advanced phase congruency based detector is chosen as an alternative approach which produces considerably different results than Canny. The idea is to tackle the easy situations quickly with Canny and try the phase congruency detector with the hard low contrast cases. The MATLAB implementations of both edge-detectors are made by Kovesi[7]. The Canny algorithm used here implements the enhancements proposed by Fleck[8]. The raw output of both edge detectors were iterated with varying hysteresis threshold values until enough edge features were detected. Figure 7 illustrates how phase congruency can manage low contrast situations better than Canny.

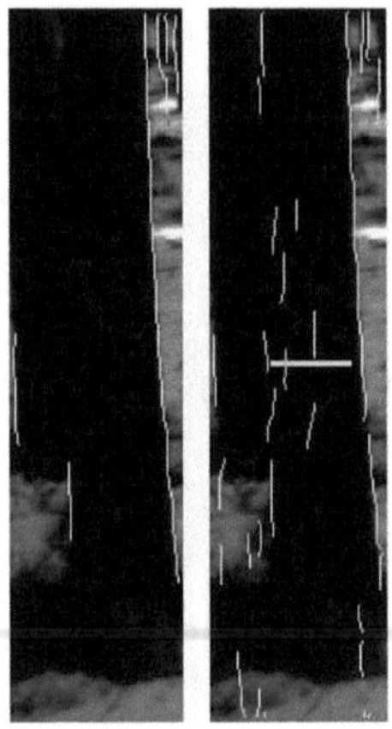

**Fig. 7.** A comparison between Canny edge detector algorithm (left) and phase congruency (right)

### 3.2 Tree Edge Selection

For selection, the edges are projected to form a 1D signal. Choosing the right edges, when the data can be quite arbitrary, is the hardest part of the algorithm. In this approach, the strongest edge is selected first and it is matched with as strong edge as possible when constrained by the minimum and maximum estimates of the tree width provided by the laser scanner. The chosen edges are required to be more than 5 pixels strong. The final measuring point is chosen by taking into account the neighborhood of the chosen edge in a form of a weighted average filter. The size of the filter depends on the distance of the tree in question and can vary between 3 for trees far away to 9 bins for trees very near. Figure 8 shows how the left side edge is shifted more to the left because of its strong neighborhood (chosen points are marked with diamonds). This edge selection method totally loses the vertical dimension and assumes that the trees are positioned straight along the window. That is the price of a simplified approach. Lens distortions are corrected[9] for the chosen two points and the measurement is transformed to metric units with pinhole camera model and calibration information. Processing one image takes usually a few seconds with a mid-level PC.

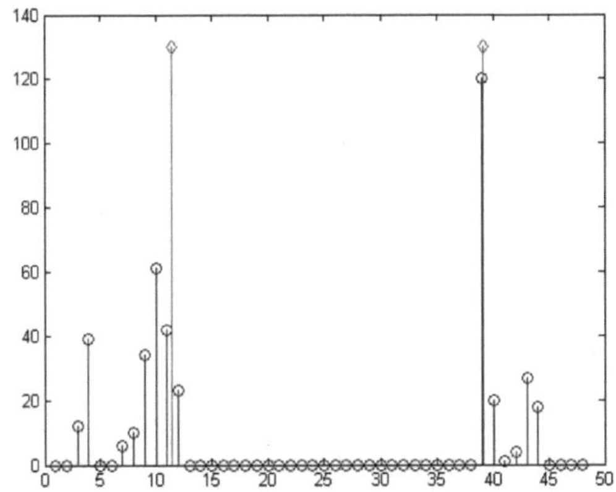


**Fig. 8.** Edge selection from 1D signal (vertical pixel sum vs. x-axis pixel)

### 3.3 Machine Vision Results

A data set of 191 images was analyzed with automatic batch processing and also with manual classification where unreliable cases were not included. In the hand picked case the mean diameter error is 20mm with 20mm standard deviation, while for the automatic processing the figures are 0.4mm and 34mm respectively. The automatic results are not as good as they seem as the maximum errors for diameters can be as large as 100mm, because the algorithm often measures something completely different than the desired tree. The reliability of the system does not meet any useable levels. However, the system has demonstrated that a laser scanner and a camera can work together in an intelligent and usable way. This research has also uncovered some fundamental problems of digital forest imaging and paved the way for further studies.

## 4 Conclusions and Future Work

The current algorithms implemented in Java programming language run somewhat slower in modern laptop computers than what is required for real-time applications. Some parts of the presented SLAM algorithm can be easily tuned for better performance. Java is a safe language which makes it easy to program with but it is not the best programming language performance-wise. Also interfacing it with exotic hardware such as 2D laser scanners can be difficult. C++ implementation may have to be considered in later phases of the project.

A forest is very tough environment for precision instruments. Luckily there are 2D laser scanners and DGPS receivers that are designed for outdoor use. The forest terrain is often quite rough, which adds additional challenge for the sensor system. It may be necessary to tilt the sensor package when the harvester is working in an inclined position or traversing slopes. More development work is needed to find better solutions to the association problem. The current system works well for small loops (50m x 100m). However, the accumulation of errors may be a problem for larger loops. Identifying trees may also be a problem in more dense and cluttered forest environments. Digital forest imaging has turned out to be very challenging research area, but it is essential for future forest mapping systems which will require features such as tree species recognition. Digital imaging systems have the ability to see textures and other tree qualities that laser scanners cannot perceive.

## References


1. Hellström, T., Johansson, T., Ringdahl, O., Georgsson, F., Prorok, K., Sandström, U.: Development of an Autonomous Path Tracking Forest Machine. In: The 5th International Conference on Field and Service Robotics (FSR 2005), Port Douglas, Australia, July 29–31 (2005)
2. Miettinen, M., Öhman, M., Visala, A., Forsman, P.: Simultaneous Localization and Mapping for Forest Harvesters. In: The 2007 IEEE International Conference on Robotics and Automation (ICRA 2007), Rome, Italy, April 10–14 (2007)
3. Ziou, D., Tabbone, S.: Edge detection techniques – an overview. International Journal of Pattern Recognition and Image Analysis 8, 537–559 (1998)
4. Canny, J.: A computational approach to edge detection. IEEE Trans. Pattern Analysis and Machine Intelligence 8, 679–698 (1986)
5. Kovesi, P.: Image features from phase congruency. Videre: Journal of Computer Vision Research 1(3) (1999)
6. Kovesi, P.: Edges are not just steps. In: The 5th Asian Conference on Computer Vision (ACCV 2002) (January 2002)
7. Kovesi, P.: MATLAB and Octave functions for computer vision and image processing. School of Computer Science & Software Engineering, The University of Western Australia (Referenced 4.12.2006) (2006), http://www.csse.uwa.edu.au/~pk/research/matlabfns/
8. Fleck, M.: Some defects in finite-difference edge finders. IEEE Trans. Pattern Analysis and Machine Intelligence 14(3) (March 1992)
9. Mallon, J., Whelan, P.: Precise radial un-distortion of images. In: Proc. the 17th Int. Conf. on Pattern Recognition (ICPR 2004) (2004)

# Part IX

# Learning and Scene Estimation

# Modelling Smooth Paths Using Gaussian Processes

Meng Keat Christopher Tay[1] and Christian Laugier[2]

INRIA Rhône-Alpes, Laboratory LIG, team e-Motion
[1]tay@inrialpes.fr, [2]christian.laugier@inrialpes.fr

**Summary.** A generative model based on the gaussian mixture model and gaussian processes is presented in this paper. Typical motion paths are learnt and then used for motion prediction using this model. The principal novel aspect of this approach is the modelling of paths using gaussian processes. It allows the representation of smooth trajectories and avoids discretization problems found in most existing methods. Gaussian processes not only provides a comprehensive and formal theoretical framework to work with, it also lends itself naturally to path clustering using gaussian mixture models. Learning is performed using expectation maximization where the E-Step uses variational methods to maximize its lower bound before optimization over parameters are performed in the M-Step.

## 1 Introduction

The analysis of motion patterns is important for performing motion prediction in robotic applications. For most environments, objects do not move around randomly. The objects often move according to a set of typical motion pattern. One such example is the traffic cross junction where vehicles follow a certain pattern of movements according to the traffic rules. It is hence possible to learn the set of typical motions in a certain scene. Prediction can then be performed based on the knowledge of the typical motions learnt from past observations. Such predictions can be used in applications such as anomaly detection or improved navigation capabilities [5] [3] [4].

Most existing models for modelling motion patterns requires the discretization of the state space [8] [5] [3]. These approaches mostly model motion trajectories as transitions between discretized states. The main disadvantage of discretization is the need to determine the discretization of the state spaces and the association of observations to these discretized state spaces.

Most existing models are based on the "learn then predict" approach where a database of observations are available to perform batch learning before prediction can then be performed. In these approaches, the discretization problem again makes the problem of path clustering unnatural and difficult. However, [8] described a "learn and predict" approach where learning and prediction can be performed on line while continuously making a stream of observations. The

C. Laugier and R. Siegwart (Eds.): Field and Service Robotics, STAR 42, pp. 381–390, 2008.
springerlink.com 

model described in [8] does not require the clustering of paths because the model is able to adapt the topology of the state space within the hidden markov model [6] using growing neural gas [2].

In this paper, we propose a novel path modelling approach based on gaussian processes [7]. Gaussian processes makes it possible to represent paths as continuous functions in a probabilistic manner. The problem of discretization is conveniently side stepped. Prediction on the future path taken can be performed in a theoretically proper probabilistic framework.

The proposed model enables one to pose questions in a probabilistic manner such as the distribution of predicted trajectories (see fig. 1). Similarly, it is also possible to obtain probability distribution of trajectories in unobserved portions (see fig. 2). Since we model a single motion trajectory as a gaussian process, the probability distribution over trajectories are analytic and easily evaluated.

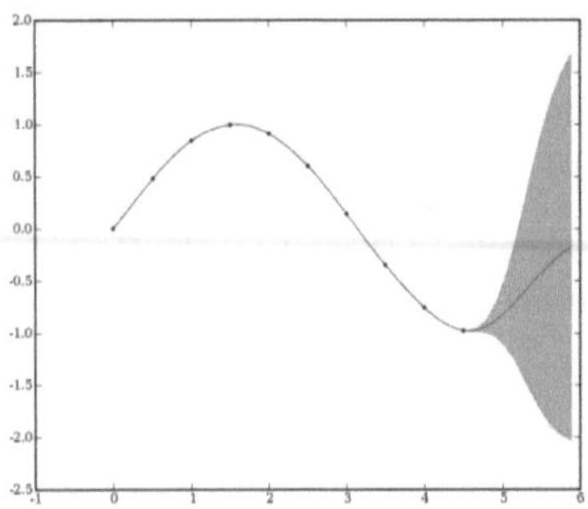

**Fig. 1.** Variance and mean of motion trajectory prediction. Dots represents observations.

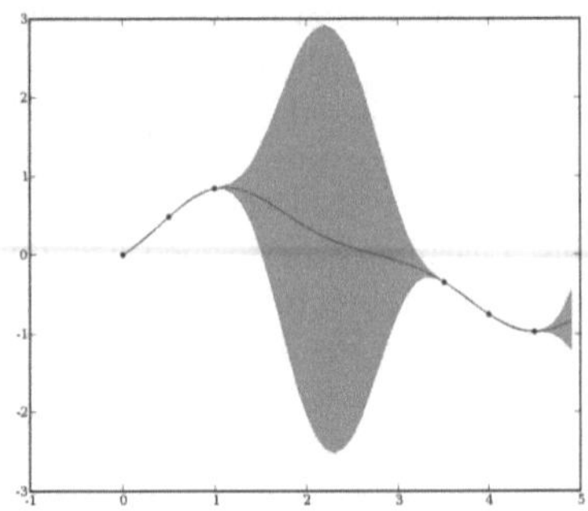

**Fig. 2.** Variance and mean of unobserved motion trajectory portion

The approach described in this paper falls into the "learn then predict" category. Thanks to the gaussian process model for paths, the well known mixture of gaussians model is naturally used to perform clustering. In this paper, the expectation maximization approach is adopted. During the E-step, variational approximation is used to estimate the probability distribution over latent variables. The M-step maximizes the lower bound with respect to gaussian process parameters. This approach enables the automatic determination of the number of clusters while obtaining a point estimate for gaussian process parameters.

In section 2, we present the generative model based on the gaussian mixture model to model the set of paths in a scene. Learning of the model and its parameters are described in section 3 using expectation maximization. Path predictions based on the gaussian mixture model and gaussian process are described in section 4. The paper finishes with some results in section 5 and conclusion in section 6.

## 2 Probabilistic Model

We consider the problem of learning a set of typical paths in a certain scene. Inputs into the system consists of the path of objects in the scene. The path taken by each object is a sequence of $(x, y)_t$ cartesian coordinates representing the position of the object at various instances. Each typical path is represented by a gaussian process and the different typical paths in a scene is modelled using the mixture of gaussians.

### 2.1 Representing Typical Paths Using Gaussian Processes

A short description of gaussian processes is presented here. For more details, please refer to [7]. A gaussian process is a generalization of the gaussian probability distribution in function space. Given a collection of random variables, any finite number of which are gaussian distributed [7]. In this case, the set of random variables are $\{f(x_1), f(x_2), ..., f(x_N)\}$. It can be represented mathematically using the mean function and covariance function $k(x, x')$ as:

$$f(x) \sim G(m(x), k(x, x')) \tag{1}$$

$$m(x) = E[f(x)] \tag{2}$$

$$k(x, x') = E[(f(x) - m(x))(f(x') - m(x'))] \tag{3}$$

Where $G(\mu, \Sigma)$ represents a gaussian distribution with mean $\mu$ and covariance $\Sigma$. $k(x, x')$ is expressed as a covariance function which are functions with domain from the input space. In this paper, the squared exponential covariance function is adopted:

$$cov(f(x), f(x')) = k(x, x') = \theta_0^2 exp - \frac{(x - x')^2}{\theta_1^2} + \theta_2^2 \delta_{ij} \tag{4}$$

In our model, each path corresponding to a typical path that an object takes in a given environment is represented as 2 gaussian processes, one each to represent the path in $x$ and $y$ axes as we assume the movements in the $x$ and $y$ axes to be independent. The mean of these gaussian processes is the mean function (in the $x$ and $y$ axes each) representing the path. A gaussian distribution over the different paths for a typical path can thus be represented using gaussian processes. As the gaussian distribution takes place within a function space, it is infinite dimensional and no discretization is required.

A single observation of a path is represented as two $D$ dimensional vectors. One vector represents the sequence of positions along the cartesian $x$ axis and the other for the $y$ axis. The likelihood for a set of $N$ path observations for a single typical path can then be written as:

$$L_x = \prod_{n=1}^{N} G(x_n | \mu_x, \Sigma_x) \tag{5}$$

$$L_y = \prod_{n=1}^{N} G(y_n | \mu_y, \Sigma_y) \tag{6}$$

Where $x_n$ and $y_n$ are vectors of $x$ and $y$ positions for the $n^{th}$ observation. $\mu_x$, $\mu_y$, $\Sigma_x$ and $\Sigma_y$ are the mean vectors and covariances of the $x$ and $y$ positions for the typical path. In almost all cases, the sequence of observations of position for path observations are all of different length. In this case, a fixed dimension $D$ can be chosen and the $D$ positions can be obtained from $D$ samples equally distributed along the length of the path.

## 2.2 Multiple Typical Trajectories Using Gaussian Mixture Model

Since each observed path corresponds to two $D$ dimensional vectors and that a typical motion path is gaussian distributed, the observed paths are gaussian distributed from the generative point of view. A single typical motion path is a $D$ dimensional gaussian distribution and this view can be easily extended to the case of representing several typical motion paths using mixture of gaussians. Each path observation $n$ is probabilistically associated with one of the mixture components $k$ via the variable $Z_{nk} = \{0, 1\}$ where $\sum_k Z_{nk} = 1$. The prior distribution for all $Z_{nk}$, all path observations and component means are:

$$P(Z|\pi) = \prod_{n=1}^{N} \prod_{k=1}^{K} \pi_k^{Z_{nk}} \tag{7}$$

$$P(X|Z, \mu_x, \Theta) = \prod_{n=1}^{N} \prod_{k=1}^{K} G(x_n|\mu_{xk}, C(\Theta))^{Z_{nk}} \tag{8}$$

$$P(Y|Z, \mu_x, \Theta) = \prod_{n=1}^{N} \prod_{k=1}^{K} G(y_n|\mu_{yk}, C(\Theta))^{Z_{nk}} \tag{9}$$

$$P(\mu_x|\Theta) = \prod_{k=1}^{K} G(\mu_{xk}|m_{xk}, b_{xk}C(\Theta)) \tag{10}$$

$$P(\mu_y|\Theta) = \prod_{k=1}^{K} G(\mu_{yk}|m_{yk}, b_{yk}C(\Theta)) \tag{11}$$

where $x_n$ and $y_n$ are the $x$ and $y$ vectors of a path observation. $\mu_{xk}$ and $\mu_{yk}$ are $x$ and $y$ means of cluster component $k$, with each component weight being $\pi_k$. $m_{xk}$ and $m_{yk}$ are means of the prior distribution over the mean components. $b_{xk}$ and $b_{yk}$ are scaling constants to $C(\Theta_k)$, the covariance matrix constructed using the squared exponential function (eq. 5) parameterized by $\Theta_k$.

The decomposition of the joint distribution over the variables can be expressed as:

$$\begin{aligned} P(X, Y, \mu_x, \mu_y, Z|\pi, \Theta) = P(X|Z, \mu_x, \Theta)P(Y|Z, \mu_y, \Theta) \\ P(\mu_x|\Theta)P(\mu_y|\Theta)P(Z|\pi) \end{aligned} \tag{12}$$

# 3 Model Learning

The parameters for the probabilistic model in section 2 are learnt from the data using the expectation maximization (EM) algorithm. Consider the general model of data $X$, latent variables $Z$ and model parametrized by $\Theta$. The goal in EM is to maximize the log likelihood:

$$\begin{aligned}\ln P(X|\Theta) &= \ln \int_Z P(X, Z|\Theta) dZ \\ &\geq L(Q, \Theta) + KL(Q \parallel P)\end{aligned} \tag{13}$$

where $Q(Z)$ is the joint distribution over latent parameters. $KL(Q \parallel P)$ is the kullback-leibler divergence and lower bound $L$ defined as:

$$L(Q, \Theta) = \int_Z Q(Z) \ln \frac{P(X, Z|\Theta)}{Q(Z)} dZ \tag{14}$$

Maximization of eqn. 13 is equivalent to minimizing $KL(Q \parallel P)$. This occurs when $Q(Z) = P(X|\Theta)$. This minimization can be performed by approximating $Q(Z)$ with a constrained family of variational approximations:

$$Q(Z) = \prod_i Q_i(Z_i) \tag{15}$$

In this case, the minimization of $KL(Q \parallel P)$ can be obtained by iteratively computing for each $Q_i(Z_i)$:

$$Q_i(Z_i) = \frac{exp\langle \ln P(X, Z|\Theta)\rangle_{i \neq j}}{\int exp\langle \ln P(X, Z|\Theta)\rangle_{i \neq j} dz_i} \tag{16}$$

This variational maximization of the lowerbound $L(Q, \Theta)$ constitutes the E-Step. During the M-Step, the log likelihood is maximized with respect to parameter $\Theta$:

$$\Theta_{max} = \arg\max_{\Theta} \ln P(X|\Theta) \tag{17}$$

The E-Step and M-Step are iteratively computed till convergence.

We shall next present in more detail the expectation and maximization step especially within the context of trajectory modelling.

## 3.1 Expectation Step

This technique can be applied to the mixture of gaussians model. A similar model can be found in [1]. The variational approximation and posteriors can thus be obtained as follows:

$$Q(\mu_x, \mu_y, Z) = Q(\mu_x) Q(\mu_y) Q(Z) \tag{18}$$

$$Q(Z) = \prod_{n=1}^{N} \prod_{k=1}^{K} r_{nk}^{z_{nk}} \tag{19}$$

$$Q(\mu_x) = \prod_{k=1}^{K} G(\mu_{xk}|M_{xk}, B_{xk}C(\Theta)) \tag{20}$$

$$Q(\mu_y) = \prod_{k=1}^{K} G(\mu_{yk}|M_{yk}, B_{yk}C(\Theta)) \tag{21}$$

(22)

The variational update equations are as follows:

$$\begin{aligned} \ln P_{nk} = & - \ln |C(\Theta)| - D \ln 2\pi \\ & - \frac{1}{2} \left\langle (x_n - \mu_{xk})^T C(\Theta)^{-1} (x_n - \mu_{xk}) \right\rangle_{\mu_x k} \\ & - \frac{1}{2} \left\langle (y_n - \mu_{yk})^T C(\Theta)^{-1} (x_n - \mu_{yk}) \right\rangle_{\mu_y k} + \ln \pi_k \end{aligned} \tag{23}$$

$$r_{nk} = \frac{P_{nk}}{\sum_j P_{nj}} \tag{24}$$

$$N_k = \sum_{n=1}^{N} r_{nk} \tag{25}$$

$$\bar{x}_k = \frac{1}{N_k} \sum_{n=1}^{N} r_{nk} x_n \tag{26}$$

$$B_{xk}^{-1} = N_K + b_{xk}^{-1} \tag{27}$$

$$M_{xk} = (N_K \bar{x}_k + m_{xk} b_{xk}^{-1}) B_{xk} \tag{28}$$

Equations 26, 27 and 28 are similar for $\bar{y}_k, B_{yk}^{-1}$ and $M_{yk}$.

### 3.2 Maximization

The maximization involves the optimization of the marginal log likelihood given by:

$$\ln P(X, Y|\Theta) = \ln \sum_Z \int_{\mu_x} \int_{\mu_y} P(X, Y, \mu_x, \mu_y, Z|\pi, \Theta) d\mu_x d\mu_y \tag{29}$$

Since eqn 29 is not tractable, we can approximate it with the lower bound (refer also to eqn. 14):

$$\begin{aligned} L = & \langle lnP(X|Z, \mu_x, \Theta) \rangle + \langle lnP(Y|Z, \mu_y, \Theta) \rangle \\ & \langle lnP(\mu_x|\Theta) \rangle + \langle lnP(\mu_y|\Theta) \rangle + \langle lnP(Z|\pi) \rangle \\ & - \langle lnQ(\mu_x) \rangle - \langle Q(\mu_y) \rangle - \langle Q(Z) \rangle \end{aligned} \tag{30}$$

Optimization of the lower bound in equation 30 with respect to the gaussian process parameters $\Theta$. Optimization algorithms such as conjugate gradient or grid based search can be used.

By setting the first derivate of the lower bound with respect to $\pi$ to zero and imposing the constraint of $\sum_i \pi_i = 1$ with lagrange multipliers, $\pi$ can be updated by:

$$\pi_i = \frac{N_i}{N} \tag{31}$$

## 4 Path Prediction

When performing path prediction, the input is a partially observed path of dimension $M < D$. For the case of a $D$ dimensional gaussian with $x_1$ of dimension $M$ and $x_2$ of dimension $D - M$:

$$P'(x_1, x_2) \sim G\left(\begin{bmatrix}\mu_1 \\ \mu_2\end{bmatrix}, \begin{bmatrix}\Sigma_{11} & \Sigma_{12} \\ \Sigma_{21} & \Sigma_{22}\end{bmatrix}\right) \tag{32}$$

The probability of a partial path observation of dimension $M$ belonging to a gaussian of dimension $D$ is evaluated by integrating over the $D - M$ dimensions of the gaussian distribution to yield the marginal gaussian distribution:

$$P'(x_1) \sim G(\mu_1, \Sigma_{11}) \tag{33}$$

The prediction of a path $x_2$ given observation $x_1$ can be obtained by the gaussian conditional distribution for each cluster $k$:

$$P'_k(x_2|x_1) \sim G\left(\mu_2 + \Sigma_{21}\Sigma_{11}(x_1 - \mu_1), \Sigma_{22} - \Sigma_{21}\Sigma_{11}^{-1}\Sigma_{21}^T\right) \tag{34}$$

In order to choose the suitable clusters that corresponds to the observations made so far, the mahanalobis distance can be calculated and then gated based on the appropriate chi-square values.

Figure 3 shows an example of the prediction where the predicted path mean and variance are represented by the 'bars'. Clusters for prediction were selected according to the chi-square statistic corresponding to the 95% confidence interval. For each cluster, the gaussian distribution of the predicted path can be obtained using eq. 34.

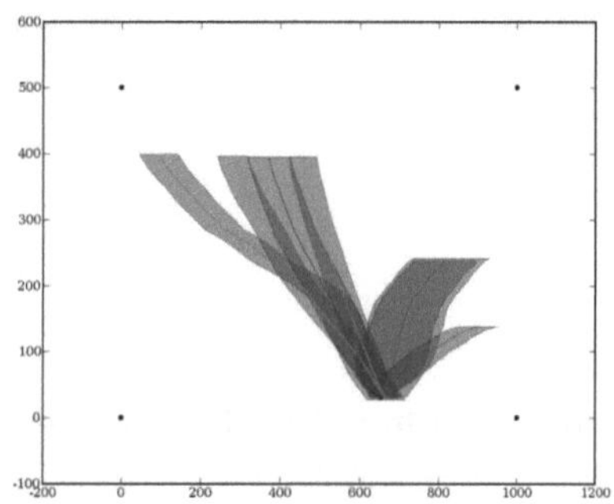


**Fig. 3.** Prediction paths and path variance to 2 standard deviations

## 5 Results

Experiments were conducted based on simulated data. 36 trajectories typical motion were manually defined for the generation of the different motions. Figure 4 shows the ensemble of motion in the training data set.

It has been noted in [1] that the variational updates were initialized using the K-Means clustering algorithm. We assumed a maximal number of 100 clusters during the K-Means stage. And the expectation maximization algorithms were performed with 100 clusters with equal cluster component weights. Result of the k-means cluster algorithm is show in figure 5.

The number of cluster components can be determined from the weight of the cluster components. The learning algorithm on the training data set allowed

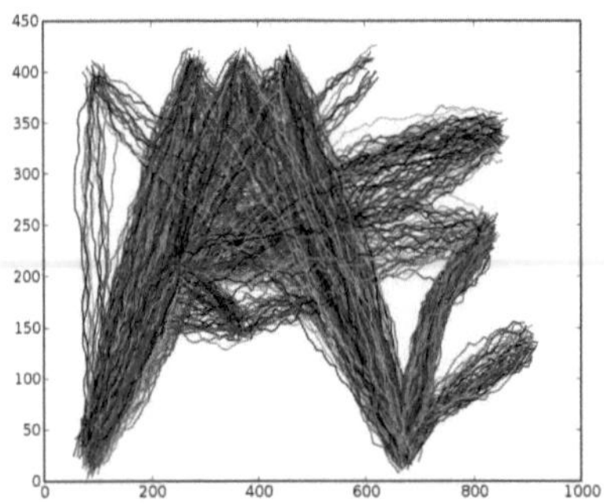


**Fig. 4.** Training Data Motions

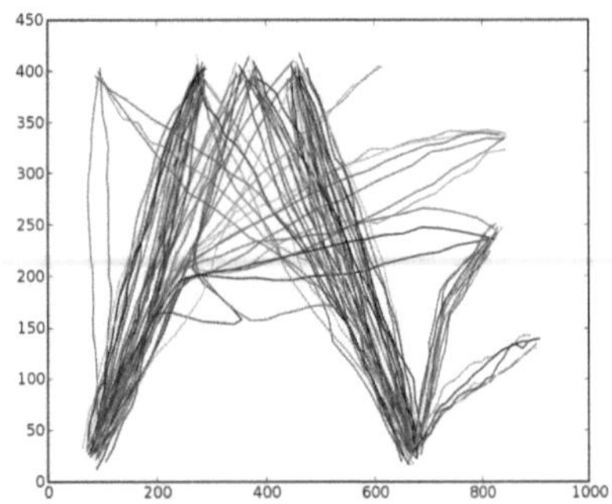


**Fig. 5.** The output from the K-Means algorithm with 100 clusters

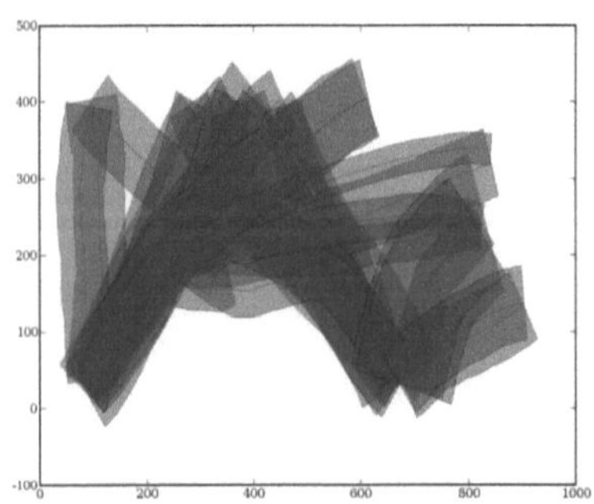


**Fig. 6.** 26 typical motion recovered from the training data (different gaussian process hyperparameters for each cluster)

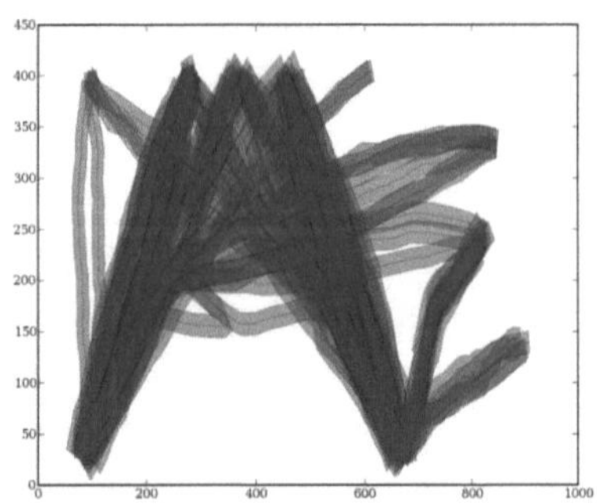


**Fig. 7.** Typical motion recovered from the training data set (same gaussian process hyperparameters across all clusters)

the suppression of clusters with component weights of nearly zero ($< 10^{-5}$) very quickly early in the EM iterations. The learning algorithm allows the domination of gaussian components over the others when clustering in the same regions. 26 gaussian components were recovered after the clustering as shown in figure 6.

The suppression of unrepresentative cluster components with negligible weights works well only when gaussian process hyperparameters are different for each cluster. However, when the gaussian process hyperparameters are shared across clusters, the variance of the different clusters are not well expressed and as a result, it is difficult to obtain cluster components with negligible component weights (see figure 7).

## 6 Conclusion and Perspectives

We described a generative approach to model motion paths using gaussian processes. The typical motion paths are smooth and avoids the problems associated with discretization. The representation of motion paths with gaussian process lends itself naturally to clustering using the gaussian mixture model. The "learn then predict" approach is performed using the EM algorithm. The lower bound of the EM algorithm is maximized using variational methods in the E-Step. The M-Step optimizes the model parameters.

We are currently working on extending this method such that it can work in real time in an incremental fashion. It is also interesting to explore the modelling of dependencies between the $x$ and $y$ axes. Applications to path planning are also envisaged.

Finally the authors would like to acknowledge the financial support from the French Embassy of Singapore. Thanks goes as well to Dizan Vasquez and Pierre Dangauthier for their interesting discussions.

## References

1. Corduneanu, A., Bishop, C.M.: Variational bayesian model selection for mixture distributions. In: Proceedings Eighth International Conference on Artificial Intelligence and Statistics, pp. 27–34 (2001)
2. Fritzke, B.: A growing neural gas network learns topologies. Advances in Neural Information Processing Systems (1995)
3. Hu, W., Xiao, X., Fu, Z., Xie, D., Tan, T., Maybank, S.: A system for learning statistical motion patterns. IEEE PAMI 28(9) (September 2006)
4. Kruse, E., Gutsche, R., Wahl, F.M.: Acquisition of statistical motion patterns in dynamic environments and their application to mobile robot motion planning. In: Proc. of the IEE Int. Conf. on Robotics and Systems, pp. 712–717 (1997)
5. Bennewitz, M., Burgard, W., Cielniak, G., Thrun, S.: Learning motion patterns of people for compliant robot motion. The International Journal of Robotics Research (IJRR) 24(1) (2005)
6. Rabiner, L.R.: A tutorial on hidden Markov models and selected applications in speech recognition. Proceedings of the IEEE 77(2), 257–286 (1989)

7. Rasmussen, C.E., Williams, C.K.I.: Gaussian Processes for Machine Learning. MIT Press, Cambridge (2006)
8. Vasquez, D., Fraichard, T., Aycard, O., Laugier, C.: Intentional motion on-line learning and prediction. In: Proc. of the Int. Conf. on Field and Service Robotics, Port Douglas (AU) (July 2005)

# Data-Driven Identification of Group Dynamics for Motion Prediction and Control

Mac Schwager[1], Dean Anderson[2], and Daniela Rus[1]

[1] Computer Science and Artificial Intelligence Lab
MIT, Cambridge, MA 02139, USA
schwager@mit.edu, rus@csail.mit.edu
[2] USDA-ARS, Jornada Experimental Range
Las Cruces, NM 88003, USA

**Summary.** A decentralized model structure for representing groups of coupled dynamic agents is proposed, and the Least Squares method is used for fitting model parameters based on observed position data. The physically motivated, difference equation model combines effects from agent dynamics, interactions between agents, and interactions between each agent and its environment. The technique is implemented to identify a model for a group of three cows using GPS tracking data. The model is shown to capture overall characteristics of the group as well as attributes of individual group members. Applications to surveillance, prediction, and control of various kinds of groups of dynamical agents are suggested.

## 1 Introduction

We wish to model groups of coupled dynamic agents, such as flocks, swarms, and herds, using measured data from those agents. For example, we would like to be able to use the trajectories of people in a crowd to develop dynamical models that capture the behaviors of the crowd as a whole. This is a prohibitively complicated problem in general, however, we provide a practical solution by restricting our attention to a special model structure. We propose a difference equation model that is decentralized and nonlinear, though it is designed to be linear-in-parameters. We use the Least Squares method to fit model parameters to position data from a group of agents. Such a model may then be used, for example, to predict future states of the group, to determine individual roles of agents within the group (e.g. leaders vs. followers), or, ultimately, to control the group. This could help ranchers to manage overgrazing by livestock, or allow animal behavioral scientists to determine social roles within animal groups. In the case of people, the ability to model group behavior has numerous applications in surveillance, urban planning, and crowd control. Also, this technique can be used to design controllers for groups of engineered agents to mimic the behaviors of natural agents.

The large body of work on designing and analyzing models of flocks, swarms, and similar decentralized dynamical systems ( [7, 3] and references therein) does not incorporate learning from data to tune such models. However, other related learning problems have been investigated. For example, in [1] a system identification technique is used to model global properties of a swarm of robots over time using observed data from the robots. There also has been considerable activity in learning behavioral

C. Laugier and R. Siegwart (Eds.): Field and Service Robotics, STAR 42, pp. 391–400, 2008.
springerlink.com 

models of individual natural agents. In [5] and [6], system identification is carried out on switching linear systems to learn models of the honey bee waggle dance and human hand motion, respectively, and in [2] a technique is used to find Motion Description Language (MDL) codes from observed ants.

Our work introduces two specific innovations. The first is in applying system identification to learn the parameters of a decentralized dynamical system from data. The second is in the model structure itself, which uses a novel, non-gradient field to express the interaction between each agent and its environment. We demonstrate the technique by fitting a model to GPS data from a group of three free-ranging cows. The model is validated by testing the whiteness of the residual error, and by comparing global statistics of simulations verse the actual data. We also show how the resulting model can be used to control simple robots. Recently, the technique has been validated with data from ten cows, though we do not present those results here.

## 2 Model Description

We consider a linear-in-parameters model structure with three naturally distinct parts to describe the motion of coupled physical agents moving over a plane surface. Firstly, each agent is given internal dynamics to enforce the constrains of Newtons laws. Secondly, a force[1] is applied to each agent from its interaction with each of the other agents in the group. Thirdly, a force is applied to each agent as a function of its position in the environment. All remaining effects are modeled as a white noise process.

### 2.1 Individual Agent Dynamics

Supposing a group of $m$ agents, the proposed model structure for an individual agent $i \in \{1,\dots,m\}$ can be written in state-space, difference equation form as

$$x_i^{\tau+1} = \begin{bmatrix} 1 & 0 & \Delta t & 0 \\ 0 & 1 & 0 & \Delta t \\ 0 & 0 & a_i & 0 \\ 0 & 0 & 0 & a_i \end{bmatrix} x_i^\tau + \begin{bmatrix} 0 & 0 \\ 0 & 0 \\ 1 & 0 \\ 0 & 1 \end{bmatrix} \left( \sum_{j=1, j\neq i}^{m} f_{ij}(p_i^\tau, p_j^\tau) + g_i(p_i^\tau) + w_i^\tau \right). \quad (1)$$

Agent $i$'s state $x_i^\tau = [e_i^\tau \; n_i^\tau \; u_i^\tau \; v_i^\tau]^T$ consists of its East position, North position, Eastern component of velocity, and Northern component of velocity after the $\tau$th iteration, and its position is given by, $p_i^\tau = [e_i^\tau \; n_i^\tau]^T$. The time step $\Delta t$ is given by $t^{\tau+1} - t^\tau$, and we assume it is constant for all $\tau$. The term $a_i$ represents damping, $a_i = 1$ for zero damping, and $|a_i| < 1$ for stable systems. The function $f_{ij}(p_i^\tau, p_j^\tau)$ determines the coupling force applied by agent $j$ to agent $i$. The function $g_i(p_i^\tau)$ represents the force applied by the environment to the agent at point $p_i^\tau$. Finally, $w_i^\tau$ is a zero-mean, stationary, Gaussian white noise process uncorrelated with $p_j$ $\forall j$ used to model the unpredictable decision-motive processes of agent $i$. Nonholonomic constraints which are often present

[1] In this work the term "force" is used in a metaphoric sense. When we talk of a "force" we are referring to the intention of the agent to accelerate in a particular way using it's own motive mechanisms.

in mobile agents, such as people, cattle, or automobiles, are neglected in this treatment, though they could be incorporated with an increase in the complexity of the model structure. Note that the force terms are only applied to affect changes in velocity in accordance with Newton's second law.

### 2.2 Agent-to-Agent Interaction Force

Dropping the $\tau$ superscripts for clarity, the form of the agent coupling force $f_{ij}(p_i, p_j)$ is given by

$$f_{ij}(p_i, p_j) = \left(\theta_{1_{ij}} - \frac{\theta_{2ij}}{\|p_j - p_i\|}\right) n_{ij}, \tag{2}$$

where $n_{ij} = (p_j - p_i)/\|p_j - p_i\|$ is the unit vector along the line from $p_i$ to $p_j$ (henceforth, $\|\cdot\|$ will denote the $\ell^2$ norm). This is the simplest of a family of force laws commonly used in computational models of physical, multi-body systems (e.g. kinetic gas models). The important feature of this family is that an agent is repulsed from its neighbor at close distances and attracted to its neighbor at far distances. To see this property clearly, examine the magnitude of (2) given by $\|f_{ij}\| = \theta_{1_{ij}} - \theta_{2_{ij}}/\|p_j - p_i\|$, and shown in the left of Figure 1.

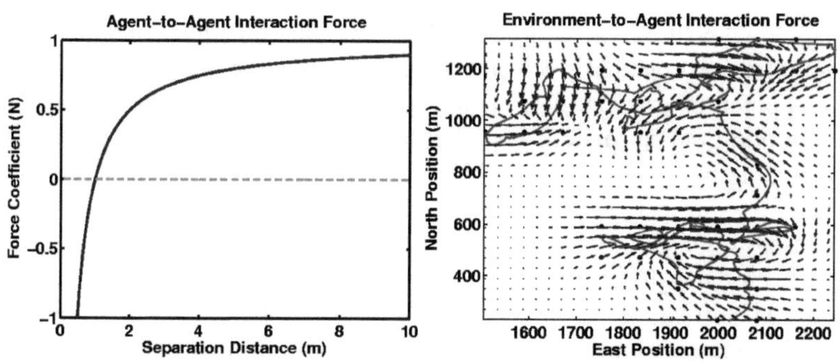


**Fig. 1.** The magnitude of the agent-to-agent interaction force is shown on the left for $\theta_1 = \theta_2 = 1$. On the right, the vector field representing the force felt by an agent at each point on the plane is shown for an example agent trajectory. The swirling patterns evident in the field are made possible by a novel parameterization.

After some manipulation, the sum of $f_{ij}$ over all neighbors $j$ can be expressed as

$$\sum_{j \neq i} f_{ij} = \begin{bmatrix} \phi_{fu_i} \\ \phi_{fv_i} \end{bmatrix} \theta_{f_i}, \tag{3}$$

where

$$\begin{aligned}
\phi_{fu_i} &= \left[ \frac{(e_1 - e_i)}{\|p_1 - p_i\|} \cdots \frac{(e_m - e_i)}{\|p_m - p_i\|} \; \frac{-(e_1 - e_i)}{\|p_1 - p_i\|^2} \cdots \frac{-(e_m - e_i)}{\|p_m - p_i\|^2} \right] \\
\phi_{fv_i} &= \left[ \frac{(n_1 - n_i)}{\|p_1 - p_i\|} \cdots \frac{(n_m - n_i)}{\|p_m - p_i\|} \; \frac{-(n_1 - n_i)}{\|p_1 - p_i\|^2} \cdots \frac{-(n_m - n_i)}{\|p_m - p_i\|^2} \right], \text{ and} \\
\theta_{f_i} &= \left[ \theta_{1_{i1}} \cdots \theta_{1_{im}} \; \theta_{2_{i1}} \cdots \theta_{2_{im}} \right]^T,
\end{aligned}$$

permitting a slight abuse of notation since the indices $j = i$ are excluded from the above vectors. This notation will be useful in what follows.

## 2.3 Environment-to-Agent Interaction Force

The agent's preference for certain paths in the environment is modeled as a nonlinear mapping from each point on the plane to a force vector felt by the agent. To this end, two networks of Gaussian basis functions are used, one for each of two perpendicular force components.

In particular, the function $g_i(p_i)$ can be written

$$g_i(p_i) = \begin{bmatrix} \theta_{u_{i1}} & \cdots & \theta_{u_{in}} \\ \theta_{v_{i1}} & \cdots & \theta_{v_{in}} \end{bmatrix} \begin{bmatrix} \phi_{g_1}(p_i) \\ \vdots \\ \phi_{g_n}(p_i) \end{bmatrix}, \tag{4}$$

where $\phi_{g_{ik}}(p_i) = \frac{1}{\sigma_{ik}\sqrt{2\pi}} \exp(-\frac{\|p_i - \gamma_{ik}\|^2}{2\sigma_{ik}^2})$, and $k \in \{1, \ldots, n\}$. Each Gaussian is centered at $\gamma_{ik}$, with standard deviation $\sigma_{ik}$, and its strength is represented by the unknown parameters $\theta_{u_{ik}}$ for the Eastern component, and $\theta_{v_{ik}}$ for the Northern component. Gaussian basis functions were chosen for their familiarity; the objective being to demonstrate the modeling approach with a minimum of complications. A number of other basis functions types could be used, including wavelets, sigmoidal functions, or splines.

It is important to note that a vector-field parameterized in this way is *not* a potential gradient. A potential gradient field cannot admit circulation around closed paths.[2] We introduce a non-gradient parameterization to enable circulation, as one can imagine agents intending to traverse closed orbits on the plane. For example, a cow may have a routine of passing between a water source, a shaded tree, and a grassy patch in a periodic fashion.

Figure 1, on the right, shows a plot of an example force-field parameterized in the above way. The arrows show the forces induced by the field, the black dots show the centers of the Gaussian functions, $\gamma_{ik}$, and the red curve shows the path of an agent over the vector field. The swirling patterns evident in the vector field would be impossible if it were a gradient field.

The expression in (4) can be put into a different form to match that of (3). In particular

$$g_i(p_i) = \begin{bmatrix} \phi_{gu_i} \\ \phi_{gv_i} \end{bmatrix} \theta_{g_i}, \tag{5}$$

where

$$\phi_{gu_i} = [\, \phi_{g_1} \ \cdots \ \phi_{g_n} \ \cdots \ 0 \cdots \,],$$
$$\phi_{gv_i} = [\, \cdots 0 \cdots \ \phi_{g_1} \ \cdots \ \phi_{g_n} \,], \text{ and}$$
$$\theta_{g_i} = [\, \theta_{u_{i1}} \ \cdots \ \theta_{u_{in}} \ \theta_{v_{i1}} \ \cdots \ \theta_{v_{in}} \,]^T.$$

Although somewhat awkward, this form will become useful in what follows.

---

[2] Proof: Let $\Psi(p)$ be a potential function and $V(p) = -\text{grad}(\Psi)$ its gradient field. Then $\text{curl}(V) = \text{curl}(-\text{grad}(\Psi)) = 0$, thus by Green's Theorem, $\oint_s V ds = \int_{A_s} \text{curl}(V) dA = 0$, where $s$ is any closed curve on the plane, and $A_s$ is the area enclosed by $s$.

## 3 System Identification with Least Squares Fitting

The model that was described in Section 2 can be manipulated into a form so that its parameters can be fitted using the Least Squares method for system identification [4]. We assume that only position measurements, $p_i^\tau$, $\tau = 1,...,N$, are available to perform the fitting. We can eliminate $u_i$ and $v_i$ from the dynamics in (1) to provide a second order equation in the position only. Notice that from (1) we can write

$$p_i^{\tau+1} = p_i^\tau + \Delta t \begin{bmatrix} u_i^\tau \\ v_i^\tau \end{bmatrix}, \tag{6}$$

and

$$\begin{bmatrix} u_i^{\tau+1} \\ v_i^{\tau+1} \end{bmatrix} = a_i \begin{bmatrix} u_i^\tau \\ v_i^\tau \end{bmatrix} + \sum_{j=1, j\neq i}^{m} f_{ij}^\tau + g_i^\tau + w_i^\tau. \tag{7}$$

We can solve (6) for $[u_i^\tau \ \ v_i^\tau]^T$ and substitute into the right hand side of (7). We then substitute the result back into the right hand side of (6), shifting time indices appropriately, to obtain the desired expression

$$p_i^{\tau+2} = p_i^{\tau+1} + (p_i^{\tau+1} - p_i^\tau)a_i + \Delta t \left( \sum_{j=1, j\neq i}^{m} f_{ij}^\tau + g_i^\tau + w_i^\tau \right).$$

We can use the above expression to formulate a one-step-ahead predictor. First, define the combined regressor vectors $\phi_{u_i}^\tau = [\,(e_i^{\tau+1} - e_i^\tau)/\Delta t \ \phi_{fu_i}^\tau \ \phi_{gu_i}^\tau\,]$, and $\phi_{v_i}^\tau = [\,(n_i^{\tau+1} - n_i^\tau)/\Delta t \ \phi_{fv_i}^\tau \ \phi_{gv_i}^\tau\,]$, and a combined parameter vector $\theta_i = [\,a_i \ \theta_{f_i}^T \ \theta_{g_i}^T\,]^T$. By taking the expectation conditioned on the positions, substituting (3) and (5) for $\sum_{j\neq i} f_{ij}$ and $g_i$, respectively, then making use of the combined regressor and parameter vectors we get

$$\hat{p}_i^{\tau+2} = p_i^{\tau+1} + \Delta t \begin{bmatrix} \phi_{u_i}^\tau \\ \phi_{v_i}^\tau \end{bmatrix} \theta_i, \tag{8}$$

where $\hat{p}_i^{\tau+2}$ is the expected value of $p_i$ after $\tau+2$ time steps, given positions up to $\tau+1$, and $w_i^\tau$ drops out in the conditional expectation.

The Least Squares method is now implemented to find the optimal model parameters. Specifically, we wish to find the parameters, $\theta_i$, to minimize the mean squared prediction error over all available time steps. The mean squared prediction error can be written $J_i = 1/(N-2)\sum_{\tau=1}^{N-2}(p_i^{\tau+2} - \hat{p}_i^{\tau+2})^T(p_i^{\tau+2} - \hat{p}_i^{\tau+2})$. Substituting into $J_i$ with (8) and (6) yields

$$J_i = \frac{\Delta t^2}{N-2}(Y_i - \Phi_i\theta_i)^T(Y_i - \Phi_i\theta_i), \tag{9}$$

where

$$Y_i = [\,u_i^2 \ \dots \ u_i^{N-1} \ v_i^2 \ \dots \ v_i^{N-1}\,]^T, \text{ and } \Phi_i = [\,\phi_{u_i}^{1\,T} \ \dots \ \phi_{u_i}^{N-2\,T} \ \phi_{v_i}^{1\,T} \ \dots \ \phi_{v_i}^{N-2\,T}\,]^T,$$

and $u_i^\tau$ and $v_i^\tau$ are obtain from (6). The least squares problem is then formulated as $\theta_i^* = \arg\min_{\theta_i} J_i(\theta_i)$. Following the typical procedure for solving the least squares problem we find that

$$\theta_i^* = [\Phi_i^T\Phi_i]^{-1}\Phi_i^T Y_i. \tag{10}$$

The right hand side of (10) consists entirely of measured data while the left hand side is the vector which represents the optimal parameters of the model. We assume that the data are rich enough that the matrix inversion in (10) is possible. The deep implications of this invertibility are discussed in [4]. The myriad merits and deficiencies of Least Squares fitting compared with other learning methods will not be discussed in this work.

The resulting residual error in the fitting process is now used to define $\hat{w}_i$, so that

$$w_i^\tau = \begin{bmatrix} u_i^{\tau+1} \\ v_i^{\tau+1} \end{bmatrix} - \begin{bmatrix} \phi_{u_i}^\tau \theta_i^* \\ \phi_{v_i}^\tau \theta_i^* \end{bmatrix}.$$

If the "true" system dynamics are represented by the fitted model, we expect to find that $w_i$ is zero-mean, stationary, Gaussian white noise. Specifically, for perfect fitting, $E[w_i(t)w_i^T(t+\tau)] = \delta(\tau)Q_i$, where $\delta(\tau)$ is the Kronecker delta function. Therefore, the "whiteness" of $w_i$ can be used as an indicator of the goodness of fit that has been achieved. For simulation purposes, we would assume $w_i$ is white noise and design $Q_i$ to be equal to the empirical covariance of $w_i$.

## 4 Modeling a Group of Cows

The method presented above was used to model the dynamics of a herd of cows. Data collected from actual cows were used for fitting the model parameters and for evaluating the resulting model.

Data were collected in a single trial from three cows wearing GPS devices. For each cow, data consisted of approximately 3100 GPS position entries collected at 1Hz. The data for all animals were artificially synchronized to a common clock using a standard linear interpolation. The characteristic time scale of cow dynamics is considerably longer than 1 second, thus such an interpolation is expected to have a negligible effect on modeling results.

### 4.1 Least Square Results

The data were used to find model parameters as described in section 3. The panels in Figure 2 show the force coefficients $F_{ij}(p_i, p_j)$ for the three cows. For each cow the two plots show the interaction force with each of the two other cows in the herd. Note that the forces are not necessarily pair-wise symmetric, that is, $F_{ij} \neq F_{ji}$ in general.

The environment-to-agent vector fields are shown in the panels of Figure 3 for the three cows. The black dots show the centers of the Gaussian basis functions, $\gamma_{ki}$, the blue arrows show the direction and magnitude of the force felt by a cow at each point, and the red dots show the position data used for regression. The Gaussian centers were spaced over an even grid containing the trajectory of the cow. If the trajectory did not enter the grid region of a Gaussian function, it was dropped from the network. This primitive pruning algorithm was used for simplicity; more complex algorithms could be employed. The Gaussian widths were chosen to be 2/3 the length of the grid space occupied by the Gaussian.

The ability of the model to capture the relevant qualities of the data was investigated by examining the covariance function of the residual error, $w_i$, in comparison with the

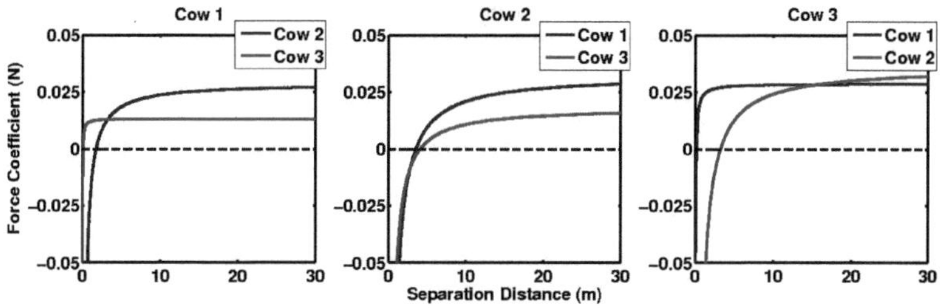


**Fig. 2.** The agent-to-agent interaction forces are shown for the three cows. Each curve represents the size of the force imposed by one cow on another as a function of the distance between the cows. A positive value is attractive while a negative value is repulsive.

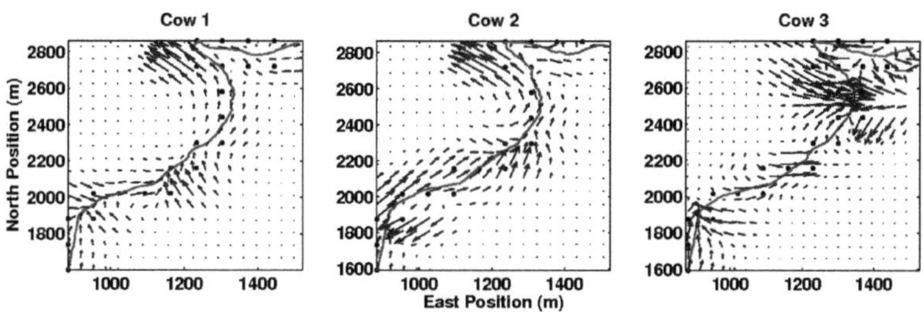


**Fig. 3.** The environment-to-agent force field is shown for the three cows. The black dots are the centers of the Gaussian functions and the red curve shows the data used for regression, marking the cow's path over the region.

covariance function of the velocity as calculated by (6). The top of Figure 4 shows the empirically calculated auto-covariance of the Eastern component of the velocity for Cow 1, and the bottom figure shows the empirically calculated auto-covariance of the corresponding residual error. The dotted lines indicate a 90% whiteness confidence interval. This means that, for each point on the auto-covariance function, a stationary Gaussian white noise process would have produced an empirical auto-covariance within the dotted lines with probability .9. Notice that the strong temporal correlation in the velocity has been greatly reduced in the residual, indicating that the model has captured a good deal of the predictable qualities of the data. Furthermore, a large majority of the residual auto-covariance points lie inside the confidence interval, giving good reason to believe that it is a "nearly white" signal. The plots for the other cows are excluded in the interests of space, though the results are similar.

### 4.2 Herd Simulation

Simulation experiments were carried out with the models fitted in Section 4.1. We simulated a group of three simple mobile robots controlled to have the dynamics in (1) with the parameters found in section 4.1. The difference equations were numerically solved in a Matlab environment.

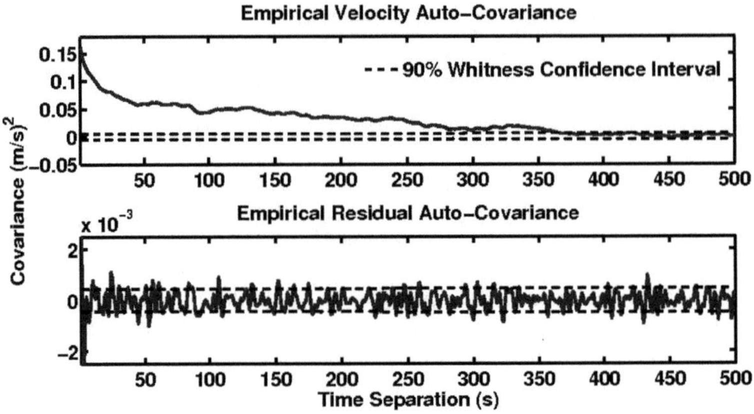


**Fig. 4.** The empirical covariance function for the Eastern component of the velocity is shown in the top plot, and for the error residual in the bottom plot. The dotted lines indicate a 90% whiteness confidence interval, meaning that a Gaussian white noise process would have generated an empirical auto-covariance inside the interval with probability .9 at each point. Clearly the residual error is "more white" than the original velocity.

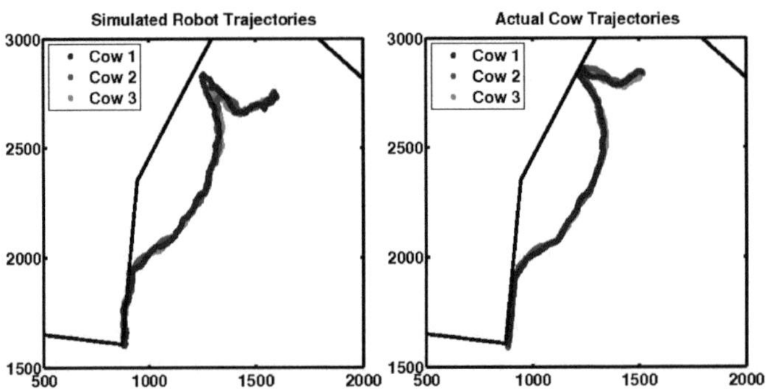


**Fig. 5.** The left picture shows trajectories of a team of simulated robots controlled to behave like a herd of cows. The robots use dynamical laws generated from the procedure described in this work. Their positions are laid over the fence lines of the paddock where the original data were collected, though they have no direct knowledge of fence positions. The right picture shows the actual cow data over the same time window.

The trajectories of the robots from a typical simulation are shown in the left side of Figure 5 laid over a schematic showing the fences of the paddock where the actual cow data were recorded. The trajectories of the simulation are similar to those of the real cows. What is perhaps more surprising is that the simulated robots track the fence lines, as did the real cows. This tendency is captured solely through the agent-to-environment force field (described in section 2.3), as the model has no direct knowledge of where fence lines may lie. Furthermore, statistics were gathered for the simulated robots and

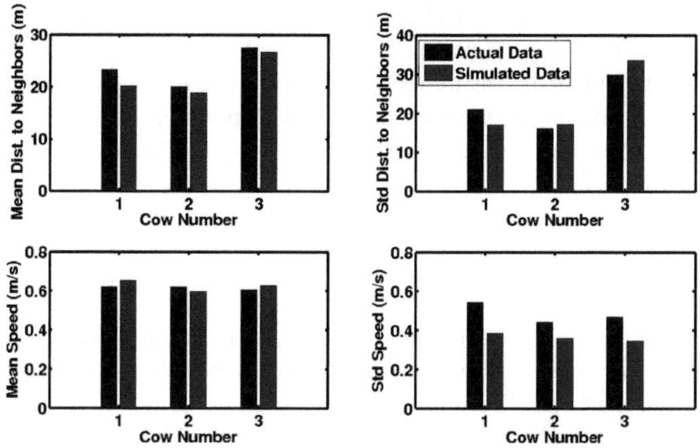


**Fig. 6.** The bar charts compare various statistics for the actual cow data and the simulated robots. The top charts show the mean and standard deviation of the distance from one cow to the other two cows in the group. The bottom charts show the mean and standard deviation of the speed of each cow.

compared with those from the cow data. Figure 6 shows a comparison of the two sets of statistics. Notice that global properties as measured by these statistics are preserved by the model.

## 5 Conclusions

In this paper we presented a method to generate models of groups of dynamical agents (e.g. flocks, herds, and swarms) using observations of the agents' positions over time. We formulated a physically motivated difference equation model, and used the Least Squares system identification method to fit the model to data from a herd of three cows. Many avenues remain open for the extension of this work. Most importantly, the method should be tested on data from much larger groups, though such data is difficult to collect. Recently, we have validated the technique with data from ten cows with results very similar to those presented here, however data from an order of magnitude more animals would likely reveal collective behaviors not seen in smaller herds. It would also be interesting to compare this model structure with others, to use different learning algorithms, to explore recursive learning algorithms for on-line learning, or to examine analytical properties of the learned models.

## Acknowledgments

This work was done in the Distributed Robotics Laboratory (DRL) at CSAIL-MIT and was supported in part by NSF grant IIS-0513628 and the MURI SWARMS project. We are grateful for this support. We thank the DRL lab members for stimulating feedback and discussions. We thank Iuliu Vasilescu and to Carrick Detweiler for designing and building the hardware used to collect the animal data. The cattle data was collected in collaboration with Dr. Dean Anderson on the Jornada Experimental Range under the New Mexico State University Institutional Animal Care and Use Committee (IACUC), Number 2007-001. We thank the staff at the US Department of Agriculture - Agricultural Research Service, Jornada Experimental Range (USDA-ARS-JER) for their support in using their facilities and animals.

## References

1. Correll, N., Martinoli, A.: System identification of self-organizing robotic swarms. In: Proceedings of $8^{th}$ Int. Symp. on Distributed Autonomous Robotic Systems, Minneapolis/St. Paul, Minnesota, USA (July 12-14, 2006)
2. Delmotte, F., Egerstedt, M., Austin, A.: Data-driven generation of low-complexity control programs. International Journal of Hybrid Systems 4(1&2), 53–72 (2004)
3. Jadbabaie, A., Lin, J., Morse, A.S.: Coordination of groups of mobile autonomous agents using nearest neighbor rules. IEEE Transactions on Automatic Control 48(6), 988–1001 (2003)
4. Ljung, L.: System Identification: Theory for the User. Prentice-Hall, Upper Saddle River, New Jersey (1999)
5. Oh, S.M., Rehg, J.M., Balch, T., Dellaert, F.: Data-driven MCMC for learning and inference in switching linear dynamic systems. In: Proceedings of $20^{th}$ National Conference on Artificial Intelligence, Pittsburgh, USA (2005)
6. Pavlovic, V., Rehg, J.M., MacCormick, J.: Advances in Neural Information Processing Systems 13 (NIPS 2000). In: Learning Switching Linear Models of Human Motion, MIT Press, Cambridge (2001)
7. Reynolds, C.: Flocks, herds, and schools: A distributed behavioral model. Computer Graphics 21, 25–34 (1987)

# Unsupervised Detection of Artificial Objects in Outdoor Environments

Luciano Spinello[1] and Roland Siegwart[2]

ASL, Swiss Federal Institute of Technology Zurich
[1]luciano.spinello@mavt.ethz.ch, [2]roland.siegwart@mavt.ethz.ch

**Summary.** This paper presents a novel unsupervised sensor fusion method to detect artificial objects in outdoor environments. We define artificial objects present in outdoor environments using structure and appearance information: an artificial object is composed of several smooth surfaces with sufficiently extended area and distinctive colors with respect to the environment. Structure information is obtained extracting smooth linked surfaces from 3D range data. Appearance information is computed on image data through a color processing. The problem of fusing these two different kinds of information is addressed through the use of a Bayesian sensor fusion approach. A probability map is built and then clustered with a 2.5D unsupervised self organizing network. This method defines objects according to the distribution of the probability values in the fusion map. The resulting clusters are labeled with their mean probability value representing the confidence the detected regions have of being artificial objects. In order to show the validity of the proposed algorithm some experiments have been performed in real outdoor environments showing promising results.

## 1 Introduction

Scene understanding and object detection in outdoor environments play an important role in a number of mobile robotics problems, including outdoor robot navigation, object tracking and prediction.

Several methods are used in computer vision to achieve object detection. Classifier based methods are often successfully used: cascade of boosted classifiers [1] (or SVMs) as well as template based methods. Due to its insensitivity to illumination, range data is well suited for object detection but a dense range measurement is required for a reliable object detection [2]. In this paper, we propose an approach to address the problem of artificial object detection in outdoor environments using a method that relies on range and image data.

We define artificial objects present in outdoor environments using structure and appearance information: an artificial object is composed of several smooth surfaces with sufficiently extended area and distinctive colors with respect to the environment. Smooth surfaces and distinctive colors are key characteristics of mostly every kind of human-made object. Thus, these features can be representative of the presence of artificial objects due to unstructured nature of

C. Laugier and R. Siegwart (Eds.): Field and Service Robotics, STAR 42, pp. 401–410, 2008.
springerlink.com 

outdoor environments. Range data processing is achieved to obtain structure information and image data processing is executed to obtain appearance information. Nevertheless it's possible to have cases in which structural information is more relevant than appearance information (or viceversa). It's therefore necessary to manage the fusion of the information in a probabilistic manner using a Bayesian modeling approach.

The technique presented here aims to segment the space in labeled regions of interest and it's conceived to be fast and deployable in field robots. This technique can be a valid hypothesis generator for further reasoning (i.e. object classification) in the salient 3D space.

The novelty of this article is to present a multiple cues approach to unsupervisely detect artificial objects in outdoor environment based on 3D smooth surface extraction and salient color blobs. Moreover, object segmentation is achieved through a fast self organizing network in 2.5D.

## 2 Structure Information Processing

Salient structure information for artificial objects in outdoor environments is defined by a metric on the smoothness and flatness of sampled 3D points. An outdoor scenario is an highly unstructured environment: natural objects are generally constituted by irregular 3D shapes with complex surface profiles (i.e. trees, leaves, rocks etc). Artificial human-made objects are instead usually symmetric and smooth surfaced because of common design/object production phases. It is therefore acceptable, for the aim of this paper, to consider surfaces with peculiar properties of smoothness and area as interesting geometrical cues of an artificial object.

In order to define regions of common smoothness a process to compute local tangent planes is needed. The fitted plane of the point $p_i$, in the $k$-neighborhood subset, is defined by the centroid $o_i$ and the normal $n_i$ computed using a PCA plane fitting approach.

The normals of the planes can be orientated only in the opposite direction of the laser ray, thus the sign of the resulting normal is therefore adjusted.

A further geometrical processing is needed to define common smoothness regions. The normal orientation of each centroid is compared with the others in the neighborhood. When the angle between two consecutive normals is smaller than the threshold $\alpha_{max}$ then the region is grown with that vector. A new region is created when it's not more possible to find unvisited neighbors.

The point cloud is segmented in smooth regions when every normal has been analyzed . A value $h_i^l$ is stored for each region related to its inverse area measurement. $H^l = h_1^l \dots h_w^l$ defines the set of $w$ smooth patches. A penalty weight is assigned to each horizontal surface found proportional to its position in the $Z$ axis. This procedure aims to give a low weight value to smooth ground surfaces.

Furthermore, an adjacency graph is built among the regions to explain the geometric topology. A graph exploration algorithm is thus applied to link contiguous smooth surfaces.

# 3 Appearance Information Processing

The purpose of camera image processing is to obtain the appearance information of artificial objects. In order to define salient color blobs in the picture a two-steps technique is applied: two axis color clustering and a color quantization.

Firstly, an image enhancement process is applied. Two morphological image operations are used to simplify color diversity: erosion and dilation. The color space selected for the image processing is the HSB (Hue, Saturation and Brightness) due the similarities of this color model to the way humans tend to perceive colors.

## 3.1 Unsupervised Color Clustering and Quantization

The purpose of any clustering method is to group entities on the basis of similarity of features. In our case we want to cluster in the color space the main color blob to detect the principal color palette present in the image. Furthermore we want to achieve an unsupervised color clustering.

The Self Organizing Network (SON or Self Organizing Map) is a neural network often used for unsupervised learning, but it can be also applied as an unsupervised clustering method [3]. It has the advantage of adaptively computing the number of clusters and it has also a low computational complexity. Due the color perception importance, only the data present on the plane Hue-Saturation is fed to the SON. Therefore the topology of the self organizing network is a 2D squared grid and the computational clustering performance is assured.

The main color cluster is detected calculating the biggest cluster area in the color plane. The remaining image color information, without the principal mean colors, is then HSB color quantized to discard the most common outdoor natural palette.

The coarse filtering of discarding *greens*, *browns*, and too dark colors is achieved using several not-admissible 3D color windows in the image color space. This process introduces some a-priori knowledge in the algorithm, but it is a reasonable assumption for natural environments. In order to take in account the mean luminosity present in the picture, $B_{env}$ is computed as the mean environment brightness. Thus, it's possible to change the minimum value of the brightness for the coarse color filtering windows to $B_{min} = \frac{B_{env}}{2}$. Therefore the palette quantization process is adaptive to the environment brightness changes.

The Sobel operator is applied on each channel of each color blob to compute an approximation of the gradient of the image intensity. Therefore, using the information theory, it's possible to compute the blob entropy value, considering each channel as an independent intensity image:

$$N_i = card(blob_i) \quad p = \frac{hist(blob_i)}{N_i}$$

$$h_i^c = [-\sum_{N_i} p\ log_2(p)]_H + [-\sum_{N_i} p\ log_2(p)]_S + [-\sum_{N_i} p\ log_2(p)]_B \tag{1}$$

$$h_z = \arg\max_{[1,w]}(h_i^c) \tag{2}$$

The set $H^c = (1 - \frac{h_1^c}{h_z}) \ldots (1 - \frac{h_w^c}{h_z})$ defines the weighted set of $w$ extracted color blobs. Image entropy is a quantity which is used to describe the amount of information richness present in the image blob. Following the principle of the artificial object appearance given, less entropy a color patch has, more its confidence value is high. A low energy patch is a more reasonable evidence of an artificial object as it encodes an almost uniform color patch.

## 4 Information Fusion

Let's resume the information retrieved using range and intensity data processing before explaining the information fusion. Structure and appearance characteristics of artificial objects are retrieved. Therefore a careful merging of these incomplete information it's necessary to handle the partial and noisy information.

Structure information is defined in $\Re^3$ space, appearance information lays in $\Re^2$ space. A dimension reduction is defined for the structure information space $\Re^3 \rightarrow \Re^2$ projecting the selected smooth surfaces onto the image plane[1]. Furthermore, the structure information that falls out of the boundaries of the viewing area is discarded.

3D range laser to omnidirectional camera rigid transformation is calibrated using a method based on [4].

Structure and appearance information are considered cues of same importance, thus the same confidence level of detection should be given in the fusion information process. The information fusion is addressed using a Bayesian modeling approach. The variables used to formalize the problem are:

- $\phi$ : it describes the existence of an artificial object
- $\theta_l$ : it encodes the structure information
- $\theta_c$ : it encodes the appearance information

Starting from the joint distribution and applying recursively the conjunction rule we obtain the decomposition:

$$P(\phi \wedge \theta_l \wedge \theta_c) = P(\phi)\ P(\theta_l|\phi)\ P(\theta_c|\phi) \tag{3}$$

In equation 3 the phenomenon $\phi$ is considered to be the main reason for the contingency of the structure and appearance information, thus knowing the cause $\phi$ of the readings the variables $\theta_l$ and $\theta_c$ are independent. In general, this hypothesis is not always satisfied, but it is often used in literature and it has the main advantage of considerably reducing the complexity of the computation.

The conditional probability that defines the information fusion is:

$$P(\phi|\theta_l \wedge \theta_c) = \frac{P(\phi)\ P(\theta_l|\phi)\ P(\theta_c|\phi)}{\sum_\phi (P(\phi)\ P(\theta_l|\phi)\ P(\theta_c|\phi))} \tag{4}$$

$$P(\phi = true) = p_s \tag{5}$$

[1] A previously calibrated camera-laser system is supposed.

If the method of artificial object detection is, for example, used as a part of a warning system, then false positive rates can be tolerated and $p_s$ in equation 5 can be set to an high value. In contrast, if it's included as part of active vehicle control a more conservative choice is needed.

Equation 4 is evaluated for each point of structure and appearance information in the fusion plane. Thus $P(\theta_l|\phi)$ is defined for each point by its value $h_i^l$ of the $i$-surface patch. $P(\theta_c|\phi)$ is defined for each point by its value $h_i^c$ of the $i$-color blob.

The resulting probability map has to be clustered to segment objects. Local maxima in the fusion map are thus segmented with a 2.5D self organizing network.

The weights of the SON, during the learning phase, are updated according to the probability value of the fusion map. The proposed cluster algorithm, described in section 4.1, has the advantage to cluster in 2D space, maintaining a fast execution time, and it has also the advantage of shaping the clusters according to the amount of confidence present.

### 4.1 2.5D Self Organizing Network for Clustering

The network is composed of $T = M \times N$ nodes connected each other with undirected edges [5] arranged in a circular or squared grid, with $M$ rows and $N$ columns. Every node is connected with four other nodes (excluding the nodes located on the network border) and it has two associated variables: its mean value $\mu_i(x_i, y_i)$ and the counter $c_i \in [0, D]$, where $D$ represents the size of input data elements. The weight $e_{i,j}$ is stored for every arc that connects the node $i$ with the node $j$. Without loss of generality, a circular topology SON is described here that will be used in the experiments. The Cartesian coordinates of a node constituting the grid are computed iteratively:

$$n_{i_x} = (R + R_\Delta)cos(\alpha + \alpha_\Delta) \tag{6}$$

$$n_{i_y} = (R + R_\Delta)sin(\alpha + \alpha_\Delta) \tag{7}$$

where $R$ is the current radius of the SON grid, $R_\Delta$ is the radius increment, $\alpha$ is the current angle, $\alpha_\Delta$ is the angle increment. $R_\Delta$ and $\alpha_\Delta$ are chosen beforehand.

During the initialization of the algorithm, every node is placed to obtain a regular circular grid and every $c_i$ and $e_{i,j}$ are set to zero. The learning phase is processed every time an input data of the fusion plane is fed to the SON. Let's define

$$p(x, y, v)$$

as data point expressed by its Cartesian coordinates $x, y$ in the fusion plane and its value $v_i$. The first step of the learning phase consists in the selection of the two network nodes closest to the input data. Only the Cartesian coordinates of the data input $p_{x,y}$ are considered in this process:

$$w_1 = \arg \min_{i \in [0,T]} \| p_{x,y} - \mu_i \| \tag{8}$$

$$w_2 = \arg \min_{i \in [0,T]/w_1} \| p_{x,y} - \mu_i \| \tag{9}$$

The update values for $c_{w_1}$ and $e_{w_1,w_2}$ are set:

$$e_{w_1,w_2} = e_{w_1,w_2} + 1 \tag{10}$$

$$c_{w_i} = 1 + c_{w_i} - (1 - p_v) \tag{11}$$

In equation 11 the counter dynamic $c_{w_i}$ is changed according to the value present in the input data $p_v$. The mean of the closest node and its four neighbors is then modified:

$$\mu_{w_1} = \mu_{w_1} + \frac{e_w}{c_{w_1}}(p_{x,y} - \mu_{w_1}) \cdot p_v \tag{12}$$

$$\mu_i = \mu_i + \frac{e_n}{c_i}(p_{x,y} - \mu_i) \cdot p_v \quad \forall_i \in neigh(w_i) \tag{13}$$

The update equations 12, 13 and 11 are modified with respect to the theory proposed in [5] to keep in account the third coordinate that define the 2.5D space.

It's important to notice that the mean $\mu_i$ changes more if an high value $v_i$ is present in the input data set. Viceversa, if $p_v$ is a small value, $\mu_i$ reflects small changes. After the learning process phase, the cluster representation is then accomplished using a graph cutting approach: arcs having low weight values are cut, leaving connected components with high edge values[2].

The Self Organizing Network here proposed has several advantages with respect to classic clustering techniques:

- The maximum cardinality of detectable clusters is not defined by the user.
- The algorithm complexity of the algorithm is $T\ N_t$, where $N_t$ is the number of input points
- The clusters shapes are influenced by the third value present in each point of the map and thus the resulting clusters will more easily segment local maxima.

### 4.2 Region Labeling

A label and a mean probability value are finally assigned to each detected cluster. The fusion plane is now completely segmented in regions where, with certain confidence, artificial objects can be found.

## 5 Experimental Results

We demonstrate the performance of the proposed algorithm using a data set acquired in a real outdoor environment to test the detection of artificial objects.

The platform used to acquire data is composed of a rotating laser rangefinder and an omnidirectional camera mounted on the rooftop of a Daimler-Chrysler Smart vehicle. The SICK laser rangefinder is mounted with its scan line in vertical position; the rotation arounbd the $Z$ axis is given by a stepper motor.

[2] Full explaination of the cluster representation is given in [5].

The revolution frequency of the laser is $0.5Hz$. In a complete rotation the laser spans $\phi = [-50°, 50°]$ and $\theta = [0°, 360°]$ with a vertical resolution of 1°. The omnidirectional camera is composed by a standard firewire camera with an hyperbolic lens. The images captured have the resolution of $640x480px$. The data is retrieved in a static environment.

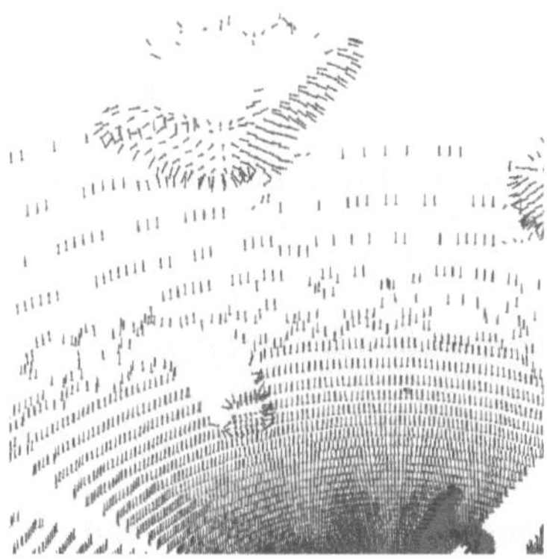

**Fig. 1.** Normals (in violet) and centroids (in blue) in the 3D laser point cloud scene

The data retrieved from the 3D rotating laser is processed to compute local normals as presented in Fig. 1. Smooth linked surfaces are identified and are shown in Fig. 2. Note that the radial sectors missing in Fig. 1 and Fig. 2 are not caused by the algorithm of range data processing but they are caused by lack of data retrieved from the 3D laser. The data is then retrieved from the omnidirectional camera. In order to reduce computation time, a circular area of interest is set; the minimum bounding circle is set beforehand and represents the minimum radius of useful field of view. The most far away point from laser range data set is taken, then its 2D correspondence in the image plane is computed and the maximum radius for the image area of interest is set. The removal of the main color region is presented in Fig. 3.

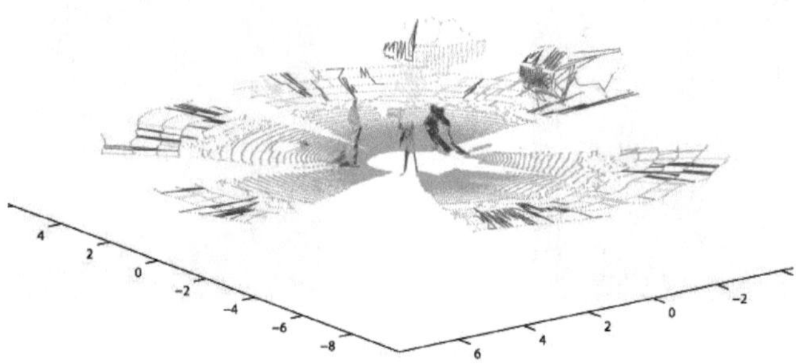

**Fig. 2.** The surfaces extracted from the 3D range data set are depicted in color

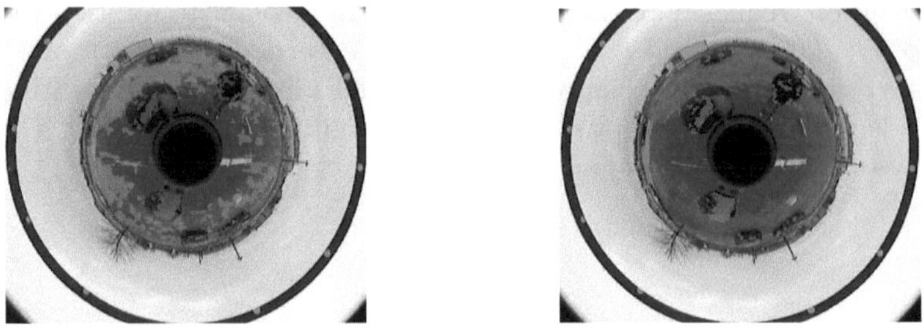

**Fig. 3.** Omnidirectional camera ROI is defined by the two red circles. Results of color clustering are shown in the left figure, final color blobs in the right figure.

Therefore the color quantization is executed and the final image processing results are shown in Fig. 3.

To optimize the performance, the 3D laser and image processing are computed in parallel threads. The 3D smooth surfaces from the laser data processing are projected onto the omnidirectional camera plane and the information fusion takes place Fig. 4. The 2.5D self organizing network used is a neural network with circular topology in order to obtain a better data coverage in the fusion plane, which is the omnidirectional camera image plane. Therefore, the segmentation is performed. The labeled clusters are presented in Fig. 5. Three men, some cars, and some colored boxes are detected in the scene, every cluster with probability $> 0.25$ is shown in the 3D model figure presented in Fig. 6.

Even though the image processing involves a dynamic brightness technique, the algorithm still suffers from poor light conditions. In this case the resulting object detection has less probability and less spatial precision because the algorithm relies principally on range data processing. Similar results are obtained if reflective surfaces with low SNR are present in the laser data; in that case the algorithm relies more on appearance information. The distance also plays an important role in the detection phase. Distant surfaces have a low $h_i^l$ due the big distance among far

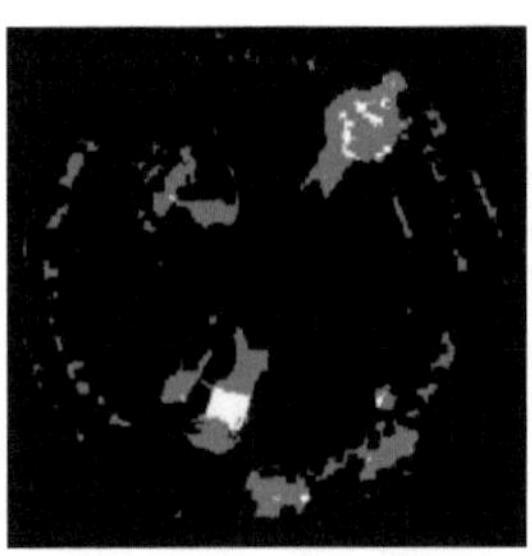

**Fig. 4.** Fusion plane is populated with structure information and appearance information probability

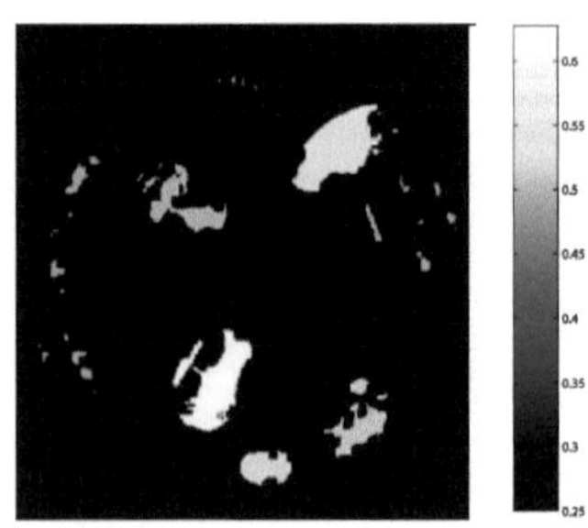


**Fig. 5.** Culsters expressing artificial objects are segmented from the fusion plane. The colorbar describes the clusters confidence.

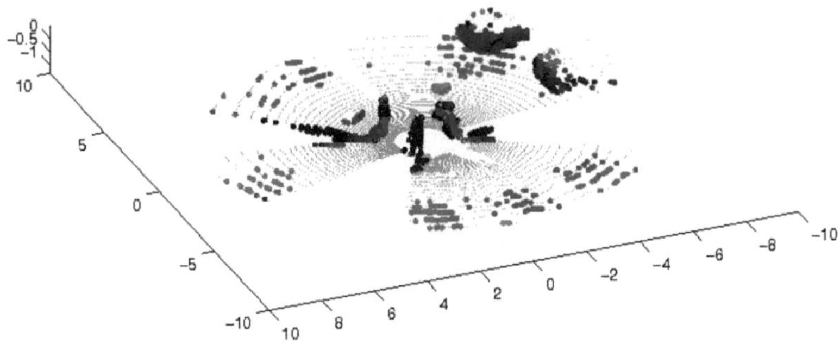


**Fig. 6.** 3D point cloud in which colored points represent detected artificial objects

points with respect to the range sensor position. In the experimental platform used the hyperbolic mirror of the omnidirectional camera gives a very small resolution of peripheral image, resulting a poor detection of far away objects.

The algorithm is completely implemented in C code, using ProBT lib [6] for Bayesian fusion, and it takes 1.2$s$ to generate the results on a Intel Centrino 1.60 Ghz based system.

## 6 Conclusions and Future Works

This paper presented a novel unsupervised sensor fusion method to detect artificial objects in outdoor environments. We define artificial objects present in outdoor environments using structure and appearance information. The problem of fusing different kinds of information is addressed through the use of a Bayesian sensor fusion approach.

Further extensions of this work can be achieved maintaining the same probabilistic fusion method and adding other structure information (i.e. shape factors) and appearance information (i.e. texture analysis). Moreover the value $p_s$ of equation 5 could be learned and tuned according to the type of outdoor environment in which the mobile platform is moving. Tracking of detected objects could also be used to improve the detection performance.

## Acknowledgment

This work was conducted and funded within the EU Integrated Projects BACS (Bayesian Approach to Cognitive Systems) - FP6-IST-027140

## References

1. Viola, P., Jones: Robust real-time object detection. In: IEEE Workshop on Statistical and Theories of Computer Vision (2001)
2. Rodgers, J., Anguelov, D., Pang, H.-C., Koller, D.: Object pose detection in range scan data. In: Int. Conference on Computer Vision and Pattern Recognition (CVPR) (2006)

3. Vesanto, J., Alhoniemi, E.: Clustering of the self-organizing map. IEEE Transactions on Neural Networks (May 2000)
4. Scaramuzza, D., Martinelli, A.: Flexible technique for accurate omnidirectional camera calibration and structure from motion. In: Fourth IEEE International Conference on Computer Vision Systems (ICVS 2006), Oxford, England (2006)
5. Vasquez, A.D., Fraichard, T.: A novel self organizing network to perform fast moving object extraction from video streams. In: Proceedings of the IEEE/RSJ Intenational Conference on Intelligent Robots and Systems (IROS), Beijing, China (2006)
6. Probayes, Probt library, `http://www.probayes.com/`

# Online Learning of Multiple Perceptual Models for Navigation in Unknown Terrain

Greg Grudic, Jane Mulligan, Michael Otte, and Adam Bates

University of Colorado at Boulder, Boulder CO, USA
Jane.Mulligan@Colorado.edu

**Summary.** Autonomous robots in unknown and unstructured environments must be able to distinguish safe and unsafe terrain in order to navigate effectively. Stereo depth data is effective in the near field, but agents should also be able to observe and learn perceptual models for identifying traversable surfaces and obstacles in the far field. As the robot passes through the environment however, the appearance of ground plane and obstacles may vary, for example in open fields versus tree cover or paved versus gravel or dirt tracks. In this paper we describe a working robot navigation system based primarily on colour imaging, which learns sets of fast, efficient density-based models online. As the robot moves through the environment the system chooses whether to apply current models, discard inappropriate models or acquire new ones. These models operate on complex natural images and are acquired and used in real time as the robot navigates.

## 1 Introduction

The system described in this paper operates on an autonomous wheeled vehicle designed for the DARPA Learning Applied to Ground Robots (LAGR) program (Fig. 1). The goal of the program is to apply Machine Learning techniques to autonomous navigation in unknown, unstructured terrains [8].

The LAGR platform is equipped with Stereo camera systems, but camera resolutions and geometry constrain the resolution and discrimination of stereo depth data making it useful to about 10m. As a result considerable effort by program participants has been devoted to using image colour and texture for classification of "traversable terrain" and obstacles in the far field. Frequently a standard Learning approach such as Neural Nets, SVM or Decision Trees [5, 7, 1] is applied to build a single model for the terrain at a particular test site. This approach has several drawbacks. First the models are large and difficult to adapt online when terrain appearance changes: a single model cannot capture the diversity of terrain necessary for far field navigation. The models require examples of traversable and non-traversable terrain, where often one may only be able to confidently identify traversable regions. Finally, models are blindly applied across the image, even in regions sufficiently dissimilar to the training set to make their results meaningless. The models offer no means to identify where their classifications can be applied with high confidence.

C. Laugier and R. Siegwart (Eds.): Field and Service Robotics, STAR 42, pp. 411–420, 2008.
springerlink.com 

**Fig. 1.** DARPA LAGR platform

Extensive work has been done on autonomous driving in structured environments such as highway settings [4]. The work described here instead deals with autonomous navigation in natural outdoor environments where paths and traversable surfaces are not distinctively marked. In contrast with systems which depend on detailed reconstruction of a rover's environment [2], our system exploits reliable Stereo depth readings in the near field to identify the image properties associated with obstacles and traversable surfaces. These image samples are used to learn single class density models which respond to obstacle or traversable regions in the far field. Our approach builds a set of models as it traverses a new environment. At each frame the current set of models is applied to the image, and if none explains the data a new model is constructed and added to the set. This allows the robot to pass across varied surfaces avoiding obstacles which differ from one area to the next, as when moving from a rocky open field to a path with tree cover. Because we construct separate single class density-based models for traversable and non-traversable terrain, models respond only to terrain that is similar to the training data from which they were constructed.

Online learning of multiple models has not yet been used extensively in Robotics. Bredeche et al. [3] describe an object identification system which learns to label 3 kinds of objects in images based on a human operator's labeling of an initial image set. Rule-based classifiers are built for HSV image features over a range of scales and structures. As the robot runs the model outcomes are combined based on weights determined by their performance on the validation set. After a fixed number of frames, new models are constructed and the old models evaluated, and possibly replaced, based on their performance over the sequence. Our system uses only Stereo readings to provide learning and validation data for modeling the appearance of obstacles and traversable terrain. New models are constructed whenever existing models fail to agree with near-field Stereo data.

This notion of *concept drift*, where a target (such as the appearance of traversable surfaces) varies over time has been more extensively examined in the Machine Learning community [9, 6, 11]. Most of this work addresses theoretical issues of bounds on performance based on assumptions about issues such as the rate of drift. They do not however offer a structured approach to questions such as which models to apply to the current sample, when to add new models or discard old ones.

The main questions when maintaining a set of models for a changing environment are: how to determine when the current models are no longer effective, and thus when to construct new models; and for large numbers of models, how to choose which to evaluate at a particular instant.

## 2 System Overview

The LAGR platform has two "eye" computers each devoted to a separate Stereo rig and associated image processing. A third computer handles planning and control functions. The main issues for our work are how to construct the image costmaps which classify near and far field regions as traversable and non-traversable, and how these are used to plan appropriate behaviour.

### 2.1 Features

An important question in appearance-based terrain classification is what image features should our model construction be based on? We have experimented with a number of colour encodings and local feature descriptors for classifying traversable and non-traversable terrain in natural outdoor scenes. Currently Normalized RGB $(R/(R+G+B), G/(R+G+B), B/(R+G+B))$ has proven to be the best performer in terms of speed of calculation and classification accuracy. We have also found 1D colour histograms (based on non-normalized RGB) effective for disambiguating terrain classes, but these are much slower to compute. We continue to pursue a separate project to propose and evaluate features for our classification task.

### 2.2 Combining Cost Maps

The current system generates cost masks based on Stereo and learned appearance-based terrain classification. Initially these masks are the same size as the image, but are mapped to the robot's local groundplane grid before they are passed to the planner.

For each new frame a feature image is constructed and appearance based classifiers are applied. Two cost masks are constructed, one for obstacles and the other for traversable surfaces. When classifying regions of ground plane using a single class classifier we can only be sure that areas that are classified as in the class are in the ground plane. This doesn't mean that the non classified regions are not in the ground plane. A second mask that classifies obstacle regions is needed. This creates a check for the non-classified regions in each mask. If a point in the ground plane mask has a low probability of being in the ground plane then the decision as to whether that pixel is an obstacle is left to the obstacle classifier. This is also true for the obstacle mask. Therefore the assumption that non-classified pixels are obstacles is only made when the ground plane classification is low and the obstacle classification is high.

The two cost masks must be combined in a meaningful way. We take the difference between the obstacle mask $O$ and the groundplane mask $G$, $O - G$ (scaled from -1 to 1). When the two cost masks disagree, i.e. ground plane mask equals 1 and obstacle mask equals 1 then a 0 is returned. When the cost masks agree, i.e. ground plane mask equals 1 and the obstacle mask equals 0 then a -1 is returned. This means that when sent to the planner the cost in that location will be reduced. When the planner is sent a positive value, the cost in that location is increased.

### 2.3 Planning System

The planning system is responsible for accumulating data provided by the vision system, mapping it to the world frame, and then computing a route toward some predetermined goal. In our case, the goal is a GPS coordinate and the route is chosen in order to minimize a specialized cost function.

Our planning system is a modified version of a state-based planner provided by Daniel Lee at the University of Pennsylvania. Behaviour in each state $S_t$ is dictated by events that have occurred during the previous state $S_{t-1}$, as well as information provided by a global state $S_{Global}$. Information from $S_{Global}$ is usually related to outside sources (bumper hits, human input, messages from the eyes, etc.), while information from $S_{t-1}$ pertains to tasks that the planner has scheduled for itself (i.e. the calculation of a new path or movement of a certain distance). $S_{Global}$ also assumes responsibility for placing cost information into a persistent global cost map $MAP_{Global}$.

$MAP_{Global}$ is a 2D bird's eye view of the area that the robot has previously sensed and/or can currently perceive. In practice, the map is stored in a 2-dimensional occupancy grid that maintains the relative spatial location of real-world cost information. The occupancy grid has a granularity of 0.5 meters, and contains cost information from the Stereo and appearance-based cost maps. These two forms of information can be weighted differently depending on the navigation task defined at runtime (i.e. learned models will be more useful than stereo data if the task involves water avoidance or navigation through tall grass). The costs in the occupancy grid always fall on a spectrum of belief about the navigability of their associated regions in the real-world. High cost means that an area is less navigable than low cost.

Most of the planner's time is spent finding a path through $MAP_{Global}$ from the current location to the goal. It searches for a path using the $A*$ search algorithm, where a path is composed of a set of adjacent grid elements of $MAP_{Global}$. The cost function should reflect some combination of navigability and distance. A principled way to optimize over these two variables is to represent low navigability as a force that the robot must overcome; and then minimize the amount of work that the robot must do in order to reach the goal. Thus, the cost function to minimize is given by:

$$\textit{Navigable Work} = C_1 D_1 + C_2 D_2 + ... + C_{\text{goal}} D_{\text{goal}}$$

Where $C_i$ and $D_i$ represent the cost and length of path segment i, respectively. In practice $C_i$ values are normalized between 1 and 10 to accommodate A*. Once a path to the goal has been found, the planner interacts with the robot's low level motors and servos to navigate through the real-world until there is a change in state or until a predetermined amount of time has passed. In the latter case, the planner reenters StatePlan and the process repeats itself.

# 3 The Learning Framework

## 3.1 Fast Density Models

The approach taken to far field terrain identification is to build two types of density models - one for identifying traversable terrain and the second for non-traversable terrain. The traversable terrain density models are constructed by sampling image regions that are associated with traversable terrain, and similarly for non-traversable terrain density models.

A number of powerful techniques have been proposed in the Machine Learning and Statistics communities for density estimation [10]. Although these techniques can be effective in high dimensional spaces (such as those addressed in this paper), invariably they are too computationally intensive to be applied to real-time robot navigation. In this paper, we present an efficient framework for density estimation that is suitable for real-time terrain classification. Our framework builds many small, fast density models, which are combined to produce a final density model for the entire image.

Assume a set of $N$ examples, each of dimension $d$, extracted from an image region that is known to be traversable. We symbolize these as $\{\mathbf{x}_1, ..., \mathbf{x}_N\}$, where each $\mathbf{x}_i \in \Re^d$ for all $i \in \{1, ..., N\}$ is a column vector representing a set of features extracted from an image. Next, the NULL space and the Basis space (i.e. the Principle Components ordered according to relevance) of this data are computed using Singular Value Decomposition. We symbolize the NULL space and Basis space matrices as $N_{tr}$ and $B_{tr}$ respectively (the subscript $tr$ indicates that a traversable model is being constructed). Note that $N_{tr}$ has $d$ rows and $m$ columns, $B_{tr}$ has $d$ rows and $k$ columns, where $k + m = d$ (note that $k$ and $d$ are determined in a standard way based on the floating point precision of the CPU). The NULL space is used to identify when new sensor readings do NOT fall within the scope of the model (i.e. if any new feature vector falls in the null space of the model, the model outputs 0). Therefore, given some new feature vector $\mathbf{x}$, the model outputs 0 if $\|N_{tr}^{'}\mathbf{x}\| < \varepsilon$, where $\varepsilon$ is determined by the floating point of the CPU.

Next the remaining data is projected into the Basis space $B_{tr}$, giving $\mathbf{b}_i = B_{tr}^{'}\mathbf{x}_i$, for all $i = 1, ..., N$, where each $\mathbf{b}_i \in \Re^k$. This data is divided into two approximately equal parts - one set is the training set $\{\mathbf{b}_1^t, ..., \mathbf{b}_{N_t}^t\}$, and the other is the validation set $\{\mathbf{b}_1^v, ..., \mathbf{b}_{N_v}^v\}$, where $t$ refers to training data, $v$ refers to the validation data, $N_t$ refers to the number in the training set, and $N_v$ refers to the number in the validation set (note that $N_v + N_t = N$). A linear function is then defined $g_{\text{tr}}(\mathbf{x}) = \mathbf{p}^{'}\mathbf{x}$, where $\mathbf{p}$ is the first column of the Basis space $B_{tr}$, corresponding to the largest eigenvalue of the data PCA space, and thus the coordinate which contains the greatest data range (ie. $g_{\text{tr}}$ is the first component of the projected data $b_i$). Next $g_{\text{tr}}(\mathbf{x})$ is converted to a density model using a one dimensional histogram model called $h_{\text{tr}}(g_{\text{tr}}(\mathbf{x}))$. This histogram density model is constructed by choosing the bin size that maximizes the negative log likelihood of the first dimensions of the validation set $\{\mathbf{b}_1^v, ..., \mathbf{b}_{N_v}^v\}$. By maximizing the negative log likelihood on the validation data, we ensure that the resulting density model is most representative of the training data. A typical histogram density model is depicted in Figure 2(a). The remaining $k-1$ dimensions are used to find the minimum and maximum values that $\{\mathbf{b}_1^t, ..., \mathbf{b}_{N_t}^t\}$, and $\{\mathbf{b}_1^v, ..., \mathbf{b}_{N_v}^v\}$ have along these coordinates - any new image samples falling outside these ranges result in the model outputting 0. Finally, if training data is available from the non-traversable class, it is used to adjust the density model $h_{\text{tr}}(g_{\text{tr}}(\mathbf{x}))$ such that the traversable training data is always more probable than the non-traversable data in each bin of the histogram. This ensures that the models are as selective as possible in predicting traversable regions. This algorithm constitutes an efficient and effective framework for bounded density estimation for robotics applications.

The same procedure is used to construct density models for non-traversable terrain, giving $g_{\text{ntr}}(\mathbf{x})$ and $h_{\text{ntr}}(g_{\text{ntr}})$.

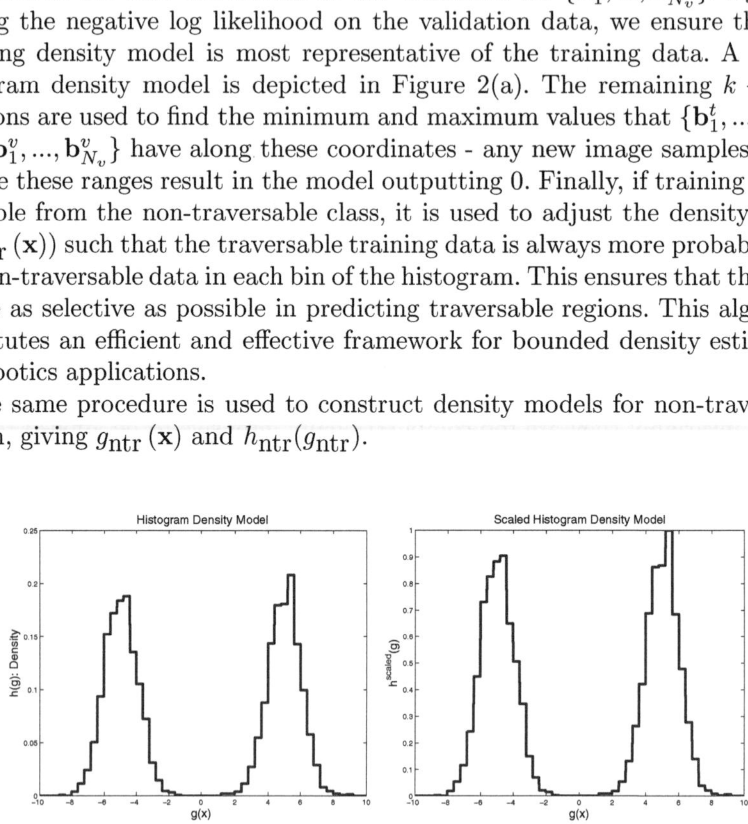


(a) Histogram Density Model

(b) Scaled Histogram Density Model for Image Segmentation

**Fig. 2.** Producing Density Models For Image Classification

### 3.2 Converting a Density Model to a Segmented Image

Every model $h_{\text{ntr}}(g_{\text{ntr}})$ is used to identify nontraversable terrain in images, and similarly, every model $h_{\text{tr}}(g_{\text{tr}})$ is used to identify traversable terrain in images. This is done by scaling the density models to range between 0 and 1 as in Figure 2(b), which is obtained by taking the density model in Figure 2(a) and dividing by its maximum value. Once this is done, every segment of the image can be classified as shown in Figure 3. Figure 3(a) shows the original image, Figure 3(b) shows the resulting segmentation after passing the feature image through the traversable terrain scaled histogram model $h_{\mathrm{tr}}(g_{\mathrm{tr}})$, and Figure 3(c) shows the resulting segmentation after passing the image through the non-traversable terrain scaled histogram model $h_{\mathrm{ntr}}(g_{\mathrm{ntr}})$. A key property of these scaled density models is that, in every region where the image is dark, the density models have no opinion on whether the region is traversable or non-traversable. This implies that the model is able to predict when image regions are too dissimilar from the data used to construct it, and therefore no predictions can be made using it. As discussed below, this property gives a formal framework for deciding when new models must be added, which models are most appropriate for a given image, and for combining multiple models.

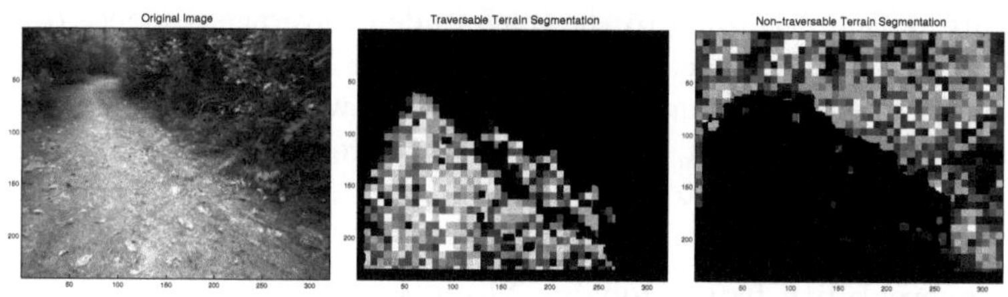


(a) Original Image (b) Traversable Segmentation (c) Non-traversable Segmentation

**Fig. 3.** Using the Scaled Density Models to segment traversable and non-traversable image regions. Image brightness is proportional to confidence in segmentation. Training data was obtained using Stereo in the near range.

### 3.3 Learning Multiple Models

In the proposed framework, we construct a set of traversable terrain models $\{h^1_{\mathrm{tr}}(g^1_{\mathrm{tr}}), ..., h^{N_{tr}}_{\mathrm{tr}})(g^{N_{tr}}_{\mathrm{tr}})\}$, as well as a set of non-traversable models $\{h^1_{\mathrm{ntr}}(g^1_{\mathrm{ntr}}), ..., h^{N_{ntr}}_{\mathrm{ntr}}(g^{N_{ntr}}_{\mathrm{ntr}})\}$. Each model represents a different type of terrain that the robot encounters as it navigates, and all are constructed in real time. The decision on how many models are needed is made automatically: whenever the current set of models do not explain the observations made by Stereo, new models are added. Thus, if the current set of traversable models do not label a traversable portion of the image as being traversable, a new traversable model is constructed. Similarly, if a region in the image is identified as non-traversable, and the current non-traversable models do not label this region as such, then a new non-traversable model is constructed to account for this new terrain type. As the robot traverses a variety of different terrains, more models are needed, resulting in potentially thousands of models.

### 3.4 Choosing Model Subsets

Each density model is representative of a particular terrain type, and as the terrain changes over time, models may not be appropriate for every image. In particular, if a model outputs zero everywhere in an image, then it was constructed from a terrain type that does not appear in the image, and therefore the model should not be used. Similarly if a traversable model has a non-zero response in an image region that Stereo designates NOT traversable, then it should not be used on the current image. An identical argument applies to non-traversable models. Thus the models that are applied to any image are sampled from models that are consistent with near field observation in the image. Even after this selection process, there still may be many hundreds of models that are consistent with an image. In order to ensure real-time operation, not all consistent models can be applied to every image. As a result, the consistent models are ranked according to the magnitude of average response over small random regions of the image. The models with the highest average response are most consistent with the image, while the models with the lowest are least consistent. The first $N$ highest ranked models, where $N$ is dictated by real time considerations, are applied to the image.

### 3.5 Traversability from Density Models

Once the subset of models is chosen for application to the image, the final traversable image classification is done by taking a maximum over all models, over every region of the image. A similar operation is done for generating the non-traversable classification. The justification for this max operation is that the models that are applied to an image are all consistent with its near field observations. When there is disagreement between the traversable and non-traversable classifications, then the robot errs on the side of exploration and labels the inconsistent parts as traversable, leaving Stereo to correct this possible mislabeling when the robot gets close enough to the region in question. This type of inconsistency is rare because of the way the models are chosen, but can nonetheless occur.

## 4 Experimental Results

The current set of density models are constructed based on image regions identified as traversable or non-traversable by near-field Stereo. The feature vectors used are formed from 7 by 7 patches of normalized RGB pixels. Thus, the dimension of each density model constructed is $7 \times 7 \times 3 = 147$.

This system has been applied to a wide variety of real tests (under the DARPA LAGR program) in unstructured outdoor environments. The density models were able to segment difficult terrain in the far field, well beyond stereo range, as demonstrated in Figure 4, where the colour difference between the non traversable vegetation and the ground is difficult to distinguish. In all the environments tested to date, never more that 47 traversable terrain models, and 43 non-traversable terrain models were required to model the terrain. This demonstrates the flexibility of the individual models. In addition, at each frame up to 20 models where applied (chosen according to the procedure defined Section 3.4) at a rate of about 5 frames per second. The current implementation of this system is written in MATLAB, and we anticipate significant speedups in frame rates when these models are implemented in C. It is also worth noting that systems which classify terrain beyond stereo range have consistently outperformed robots that only use stereo. The system described here also has the benefit of fast online learning and quick adaptation to new environments.

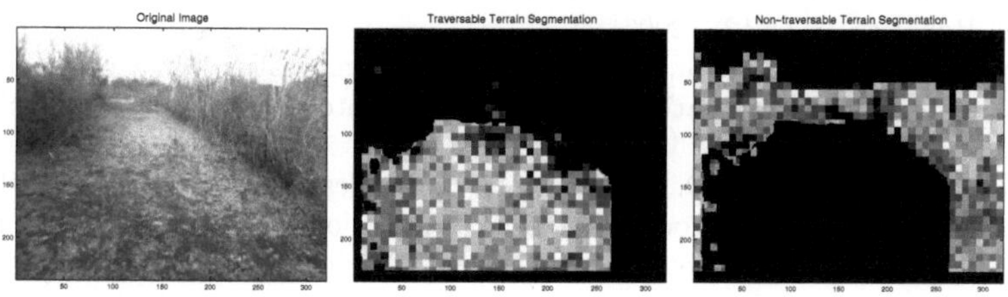


(a) Original Image (b) Traversable Segmentation (c) Non-traversable Segmentation

**Fig. 4.** Using Scaled Density Models to identify traversable and non-traversable image regions. Image brightness is proportional to confidence in classification. Training data was obtained using Stereo in the near range.

## 5 Conclusion

For autonomous robots traversing unknown terrain, Stereo depth data alone is too "short sighted". Considerable effort has been devoted to learning models of the appearance of traversable terrain to extend near field Stereo observations of ground plane and obstacles to the far field.

In this paper we describe a successful robot system which learns multiple models of obstacle and ground plane online as the terrain varies over the robot's trajectory. We describe an efficient way of choosing the best subset of models for each image, out of a set of models, possibly learned over the robot's lifetime.

The models constructed are density-based and thus respond only to inputs which are sufficiently similar to the data from which they were built. This allows the system to identify novel terrain when its models stop responding, thus triggering the construction of new models. Model construction, selection and evaluation are real-time online operations.

Future efforts include finding optimal image features for terrain classification and allowing each model to identify a preferred feature set for the terrain it encodes.

## Acknowledgment

We gratefully acknowledge the equipment and financial support provided by the DARPA LAGR Program (DOD AFRL award no. FA8650-07-C-7702) and NSF CNS-0430593.

## References

1. Albus, J., Bostelman, R., Chang, T., Hong, T., Shackleford, W., Shneier, M.: Learning in a hierarchical control system: 4d/rcs in the darpa lagr program. Journal of Field Robotics 23(11-12) (2006)
2. Bonnafous, D., Lacroix, S., Simeon, T.: Motion generation for a rover on rough terrains. In: Proc. International Conference on Intelligent Robotics and Systems (2001)
3. Bredeche, N., Zucker, J.-D., Zhongzhi, S.: Online learning for object identification by a mobile robot. In: Proc. 2003 IEEE International Symposium on Computational Intelligence in Robotics and Automation, Kobe, Japan, pp. 630–635 (July 2003)
4. Dickmanns, E.D.: Vehicles capable of dynamic vision. In: Proc. 15th Intl. Joint Conf. on Artificial Intelligence (IJCAI 1997), vol. 2, pp. 1577–1592 (1997)
5. Happold, M., Ollis, M., Johnson, N.: Enhancing supervised terrain classification with predictive unsupervised learning. In: Proceedings of Robotics: Science and Systems, Cambridge, USA (June 2006)
6. Helmbold, D., Long, P.: Tracking drifting concepts by minimizing disagreements. Technical Report UCSC-CRL-91-26, University of California, Santa Cruz, CA, 950–964 (1992)
7. Howard, A., Turmon, M., Matthies, L., Tang, B., Angelova, A., Mjolsness, E.: Towards learned traversability for robot navigation: From underfoot to the far field. Journal of Field Robotics 23(11-12) (2006)
8. Jackel, L.D., Krotkov, E., Perschbacher, M., Pippine, J., Sullivan, C.: The darpa lagr program: Goals, challenges, methodologies, and initial results. Journal of Field Robotics 23(11-12) (2006)
9. Kolter, J.Z., Maloof, M.A.: Using additive expert ensembles to cope with concept drift. In: Proc. Intl. Conference on Machine Learning (2005)
10. Vincent, P., Bengio, Y.: Manifold parzen windows. In: Thrun, S., Becker, S., Obermayer, K. (eds.) Advances in Neural Information Processing Systems 15, pp. 825–832. MIT Press, Cambridge, MA (2003)
11. Widmer, G., Kubat, M.: Learning in the presence of concept drift and hiddent contexts. Machine Learning 23, 69–101 (1996)

# Part X

# Multirobot Systems

# Implementation of Formation Transition System Using Synchronization in a Mobile Robot Group

Takaaki Shimone[1], Daisuke Kurabayashi[2], Kunio Okita[3], and Tetsuro Funato[4]

[1] Kansai Electric Power Inc.
shimone@irs.ctrl.titech.ac.jp
[2] Tokyo Institute of Technology
{dkura,funato}@irs.ctrl.titech.ac.jp
[3] RICOH co., ltd.

**Summary.** This study proposes an algorithm for the passage of a mobile robot group through a narrow space. We deal with a robot group composed of one leader and multiple unlabeled followers. A steady formation is realized by generating a Delaunay diagram using the profiles of surrounding robots. Further, we resolved a deadlock situation by using the phase gap of nonlinear oscillators to which the profiles of the surrounding obstacles are provided as inputs.We implement the proposed system in a real multi-robot system in order to demonstrate its effectiveness.

## 1 Introduction

In a system comprising several robots, it is effective to arrange them in a formation in order to facilitate their navigation. In such systems, it is sufficient to control (or assign a desired path to) a leader robot. Other followers maintain their appropriate relative positions and follow the leader to reach their destination. There exist several proposals regarding grouping and maintaining formation control.

Das et al. have proposed an elegant control algorithm of formation control among multiple vehicles by using omni-direction camera [1]. They have discussed design of possible controllers under description of a directed graph among labeled robots. They also have proposed formation switching algorithm under the order of labeled robot.

In this paper, we are going to deal with a group of anonymous, unlabeled robots. For the fault-tolerance of a robot group, considering breakdowns of some members in a group, we need to consider such unlabeled robots. Especially, we focus on solving deadlock caused by geometrical features of environment. One of the most common system to navigate unlabeled robot group is potential method [2] [3]. However, when a robot group encounters obstacles, the two objectives of maintaining a formation and avoiding obstacles often conflict with each other. A deadlock situation is often caused by the symmetric features of the environment around a robot group (Fig. 1).

Here, we propose and implement an algorithm that is compatible with both maintaining a formation and resolving deadlocks. For the former, we employ a

C. Laugier and R. Siegwart (Eds.): Field and Service Robotics, STAR 42, pp. 423–432, 2008.
springerlink.com 

Delaunay diagram generated using the profiles of surrounding robots. A geometrically steady formation is maintained by acquiring the best adjacent relation from this diagram. For the latter, we employ nonlinear oscillators that input the profiles of surrounding obstacles. Deadlocks are resolved by using the phase gaps of nonlinear oscillators that transit in different states. We have implemented simple synchronization system, which is also utilized to distinguish robots from obstacles.

# 2 Model of a Robot Group

## 2.1 Problem Settlement

In this research, a robot group is composed of a single high-performance (or manually operated) leader and multiple followers that follow the leader.

We assume that all the robots have the following common characteristics:

- The shape of a robot is a circle of radius $R$.
- The robots can move in omni direction.
- The robots can measure the relative position of an object within their constant range (sight range).
- The robots can acquire the internal state of the nonlinear oscillator of other robot.

The objective of this research is that the leader should guide all the followers to the destination in an environment with obstacles, as shown in Fig. 2.

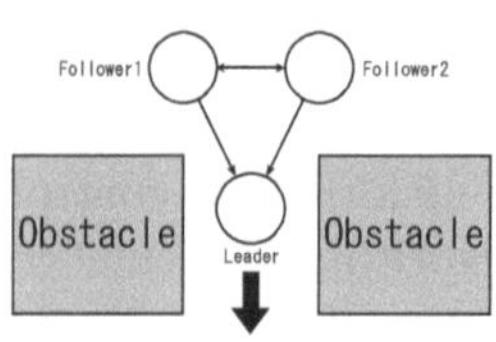


**Fig. 1.** Example of a deadlock situation for a robot group

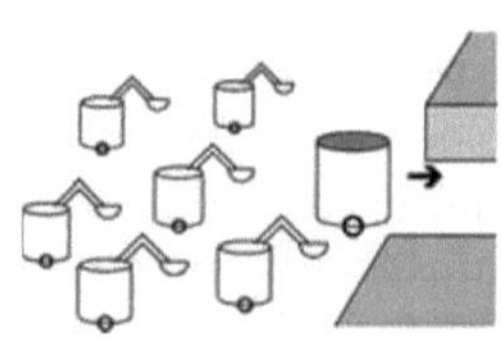

**Fig. 2.** Sketch of a robot group that is entering a narrow corridor guided by a leader

## 2.2 Formulation of Followers

We assume that a robot group is led by a leader. The robots in this group move according to a simple system (1) [4].

$$ {}^i\dot{x} = \sum_{j\in \mathrm{NBR_i}} \rho_{ij}\tau \frac{{}^ix_j}{|{}^ix_j|} + \sum_{j\in \mathrm{OBJECTS_i}} \delta \frac{{}^i\tilde{x}_j}{|{}^i\tilde{x}_j|} \tag{1} $$

where ${}^i\dot{x}$ and ${}^ix_j$ represent the relative velocity of robot $i$ and its distance from robot $j$, respectively, in terms of the coordinates of robot $i$; ${}^i\tilde{x}_j$ represents the relative distance between the nearest point of object $j$ and the center of $i$; $\tau$ and $\delta$ are constants; $\rho_{ij}$ represents the strength of the robot linkage between $i$ and $j$; $\mathrm{NBR_i}$ represents a set of the neighboring robots; and $\mathrm{OBJECTS_i}$ represents a set of visible objects.

In this research, we implement a robot formation by obtaining $\mathrm{NBR_i}$ autonomously. Furthermore, a deadlock situation is resolved by adjusting the balance of $\rho_{ij}$.

### 2.3 Finding Appropriate Neighbors

We employ Delaunay diagram so that each robot can determine its neighboring robots autonomously. The Delaunay diagram is a graph in which a natural adjacent relation is derived from a geometrical one. We can generate this diagram from the profiles of the surrounding robots. The neighboring robots of a given robot are the ones adjacent to it, as shown in Fig. 3.

**Fig. 3.** Determination of neighboring robots

## 3 Resolving Deadlock by Formation Switching

### 3.1 Conceptual Design by Using Oscillators

We employ van der Pol (VDP) oscillator [5], which is a well-known nonlinear oscillator. The dynamics of this oscillator can be expressed as (2),

$$\ddot{p}_i - \varepsilon(1 - p_i^2)\dot{p}_i + \omega^2 p_i = 0 \tag{2}$$

where $p_i$, $\varepsilon$, and $\omega$ indicate the state variable of oscillator $i$, the nonlinearity, and the eigen frequency, respectively. Typically, connected VDP oscillators can synchronize with each other. We formulate the connected oscillators by using (3):

$$p_i(t+1) \leftarrow p_i(t+1) + \frac{1}{|N_i|} \sum_{j \in N_i} \gamma_j (p_j(t) - p_i(t)) \tag{3}$$

where $N(i)$ and $\gamma_j$ represent the assembly of connected oscillators, and the gain of interactions among other oscillators, respectively.

The VDP oscillator can synchronize with a periodic external input similar to the synchronization with other VDP oscillators. Let us introduce an external input $F_i$ as follows:

$$\ddot{p}_i - \varepsilon(1 - p_i^2)\dot{p}_i + \omega^2 p_i = F_i \tag{4}$$

Here, let us suppose quite simple expression: $F_i = A_i \cos(\Omega t + \phi_i)$. Assuming that $\omega \approx \Omega$, VDP oscillator synchronizes with an external input. This VDP oscillator exhibits a response dissimilar to that of other oscillators that have not synchronized with an external input.

We use these characteristics of VDP oscillator. We implement VDP oscillator in each robot. By communicating each other, VDP oscillators make interactions, which may synchronize them. Without external input, the VDP oscillator of each robot is synchronized with others. In this case, phase gaps among oscillators becomes quite small. By arranging linkage among robots according to the phase gap, we can make a formation of robot tight. On the other hand, by generating external signals according to detected obstacles, they receive a forcing oscillation and correspondingly converge into different states. Then, the phase gaps becomes larger and unequal. Thus, we can expect resolving a deadlock situation by adjusting the balance of the strength of robot linkage corresponding to the phase gaps. By the concept, when passing through a narrow space, a robot group undergoes formation transition by acquiring parameter $\gamma_j$ corresponding to the environment.

### 3.2 Sensor Signals: Periodic External Input

We propose a method for generating an external input from the profiles of the surrounding obstacles. We assume that the robots are equipped with a LASER distance scanner that can measure the relative position of the object within a certain sight range. In reality, this sensor rotates a LASER beam and produces a profile of the area around the robot, as shown in Fig.4. Because the sensor makes scanning in certain period, which is very much faster than the speed of robot motion, we regard that the obtained series of data is a periodic signal. If we extract characteristics of the sensor signal in a specific range of frequency, the VDP may synchronize with the external input, so that synchronization among VDPs may be disturbed. Because objects far from a robot does not disturb it, we translate measured distance by robot $i$, $\hat{d}_i$, into $d_i$ as follows:

$$d_i(\theta_i) \propto \frac{1}{\hat{d}_i(\theta_i)} \tag{5}$$

Fig. 4 illustrates schematic view of sensing. Robot i and j obtain profiles of surroundings as left and right graph, respectively. Even if they are in symmetric positions, the generated graphs are different. Thus, even in a deadlock caused by symmetric features, followers can generate an asymmetric signal. However, the output of the scanning sensor may contain high frequency signals that cannot be modulated by the oscillator.We apply the discrete Fourier transformation [6] to extract the dominant property. Then, we simplify the sensor output as (6):

$$F_i = |c_1| \cos(\omega t + \phi_1) \tag{6}$$

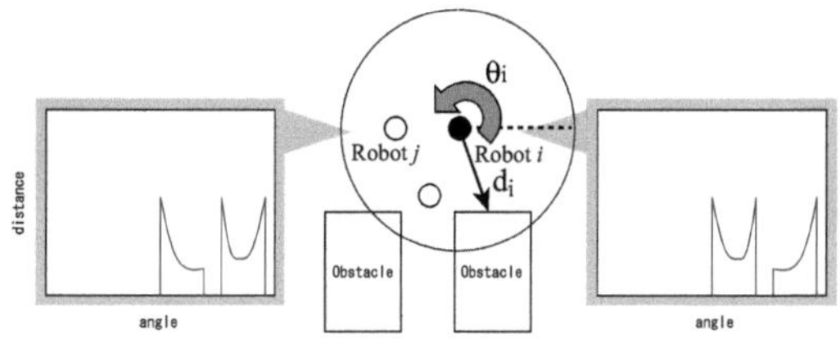


**Fig. 4.** Profile scanning of surroundings

where $|c_n|$ and $\phi_n$ represent the size and phase of the Fourier coefficient, respectively.

### 3.3 Mapping of $\rho_{ij}$

We change the strength of the robot linkage $\rho_{ij}$ according to the phase gap $\theta_{ij}$ between $i$ and $j$. Thus, mapping from the phase gaps to $\rho_{ij}$ is required. Let us consider the ideal formation transition, shown in Fig. 5, when passing through a narrow space. This phenomenon is caused by the decrease in the strength of the robot linkage with the increase in distance. To realize the above mentioned phenomenon, we calculate $\gamma_j$ using (7). This calculated parameter can reduce the phase gap when the distance of the robot linkage increases. Finally, the phase gap is mapped to the strength of robot linkage by increasing monotonically.

$$\gamma_j(D_{ij}) = \gamma_0 + \frac{\gamma_{up}}{1+e^{\alpha(D_{ij}-\theta_1)}} + \frac{\gamma_{up}}{1+e^{\alpha(D_{ij}-\theta_2)}} \tag{7}$$

where $\gamma_0$ and $\gamma_{up}$ are certain statics. Then, we set the mapping of $\rho_{ij}$ as (8).

$$\rho_{ij}(|\theta_{ij}|) = K|\theta_{ij}| + 1 \tag{8}$$

where $K$ is a fixed parameter. To determine $K$, we have carried out a heuristic optimization considering the time required to resolve a default deadlock situation. In this study, we define two default deadlock situations, as shown in Fig. 6(a) and (b). We search for a valid range of $K$ that can be used for mapping by adjusting its value. Accordingly, the following results are obtained: $11.5 \leq K \leq 28.5$ and $12.0 \leq K \leq 30.0$. We adopt a common region of the abovementioned ranges $*12.0 \leq K \leq 28.5$) as the meaning solution as shown in Fig. 6(c).

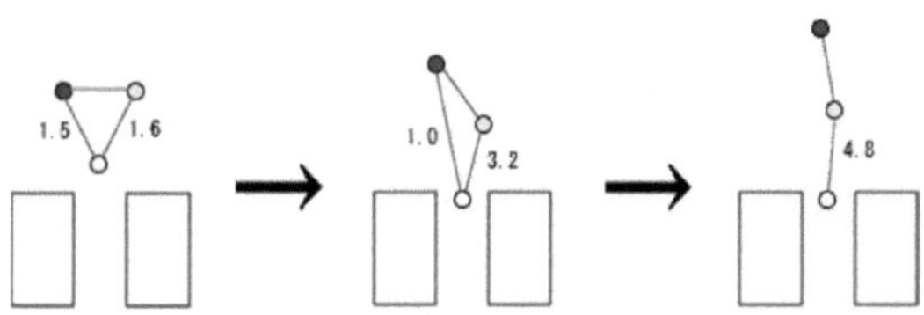


**Fig. 5.** Schematic view of formation transition

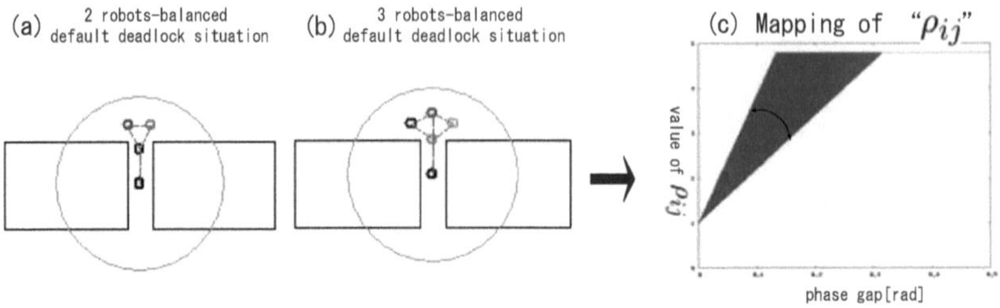


**Fig. 6.** Valid range of parameter K

### 3.4 Simulation of a Robot Group

We conducted a computer simulation to verify the transitions of the phase gaps and the movements of the robot group. Table 1 lists the simulation parameters.

**Table 1.** Simulation parameters

| Radius | View Radius | $\tau$ | $\delta$ | $K$ | Range of narrow space |
|---|---|---|---|---|---|
| 0.3 | 5.0 | 50 | 300 | 12.0 | 1.5 |

Fig. 7 illustrates the transitions of the group formation comprising five robots. Based on the transitions of $\rho_{ij}$ shown in Fig. 8, we can state that the proposed algorithm changes the gain successfully in order to maintain the formation. The robots modified their formation according to the environment.

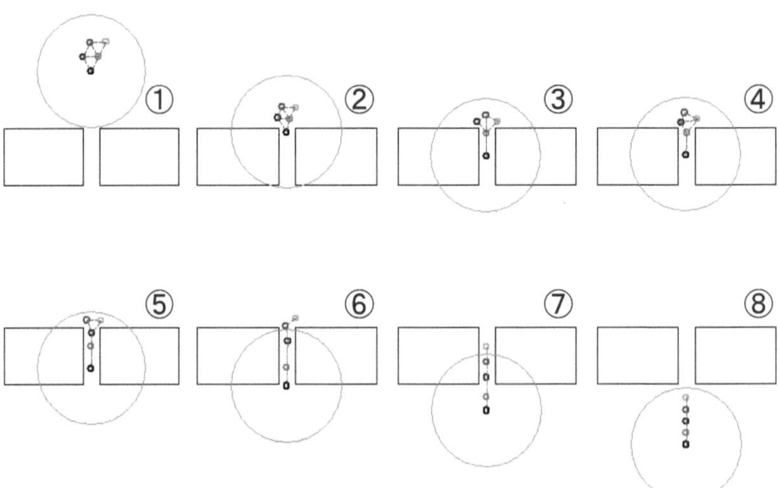


**Fig. 7.** Formation transition in simulation

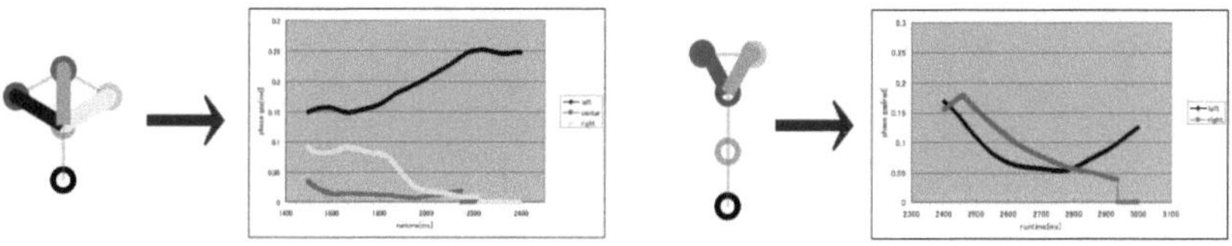

(a) 3 robots-balanced default deadlock situation (as shown in ③~⑤)

(b) 2 robots-balanced default deadlock situation (as shown in ⑤~⑥)

**Fig. 8.** Transitions of $\rho_{ij}$ (gain) to maintain a formation

# 4 Implementation

## 4.1 Robot Platform

The robot used as both leader and follower is shown in Fig. 9. This two-wheeled robot is based on **e-puck**, small autonomous mobile robot base [7]. The shape of this robot is a cylinder of 8.5[cm] in the diameter. The robot is equipped with a small micro computer, dsPIC. This unit can calculate the amount of robotic movement and state of an oscillator in real time.

A laser-scanning distance sensor URG-04LX [8] is attached on the top of the robot. This sensor composes a laser source and rotating mirror, and it can measure a distance of approximately 240 [deg] using the time-of-flight of the laser ray. Using this sensor, the robot acquires the relative position of other robots for formation maintenance and cyclic signal for external input to the oscillator.

## 4.2 Synchronization of the Oscillators

The robot is equipped with light-emitting elements composed of LED units and light-sensitive elements of phototransistors in eight directions. All light-emitting elements flash when the phase of the oscillator becomes 0[deg]. Using its light-sensitive elements, the robot can acquire the timing of phase 0 of another robot

**Fig. 9.** The two wheeled robot

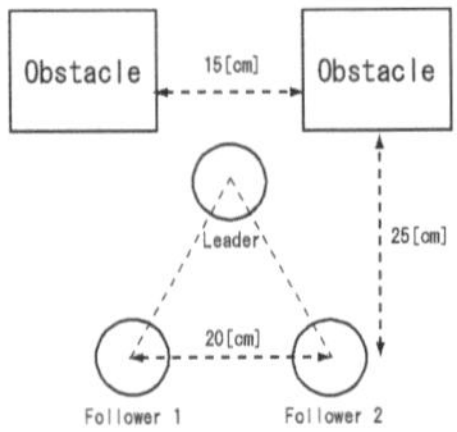


**Fig. 10.** The experimental setup

by detecting flash of another robot. As a result, the robot can estimate phase $\theta_j$ of other robots at any instance because the cycle of all the oscillators is constant in this study. The limit cycle of the oscillator can be approximated to an elliptic orbit as follows:

$$\frac{p^2}{4} + \frac{\dot{p}^2}{4\omega^2} = 1 \tag{9}$$

By introducing $\theta_j$ in (9), we obtain the internal state of another robot $p_j$ as follows:

$$p_j = 2\cos(\arctan(\frac{\tan\theta_j}{\omega})) \tag{10}$$

By introducing $p_j$ in (3), we implement the synchronization of the oscillators.

By using this system, we verified the synchronization and transition to an asynchronous state of oscillators in the real robots. The experimental setup is shown in Fig. 10. We position three robots with differing initial phases, and started synchronization procedure. Once the synchronization confirmed, we place

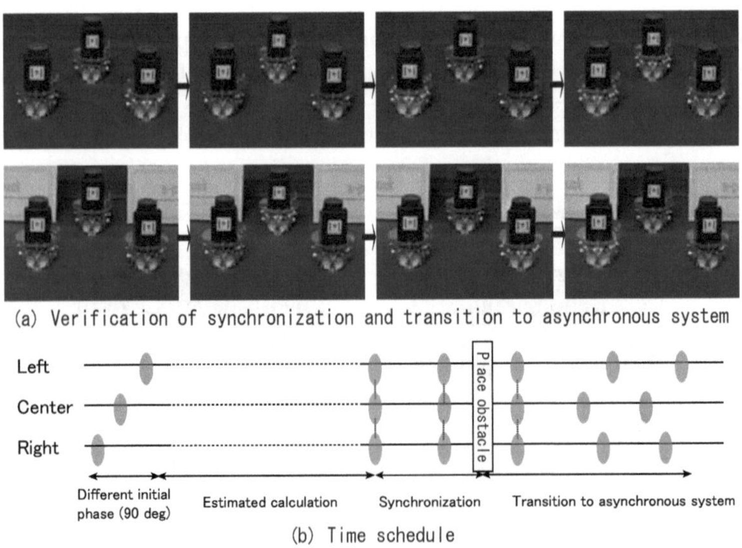


**Fig. 11.** The experiment of synchronization

a narrow space (obstacle) in front of the leader in order to observe the phase gaps between the followers' oscillators.

Fig. 11(a) and (b) shows the snapshots and the time schedule of the experiment, respectively. Based on these results, we can see that the differing initial phases are synchronized by the estimated calculation and that the three oscillators generate a phase gap when the obstacle is introduced.

### 4.3 Formation Transition

Finally, we conducted an experiment using three robots. One of them is a leader that goes directly to the destination, while the other are followers . Fig. 12 shows the snapshots of the experiment. By using the proposed algorithm and system, the two followers successfully modified their triangular formation into a linear one in order to pass through a narrow space.

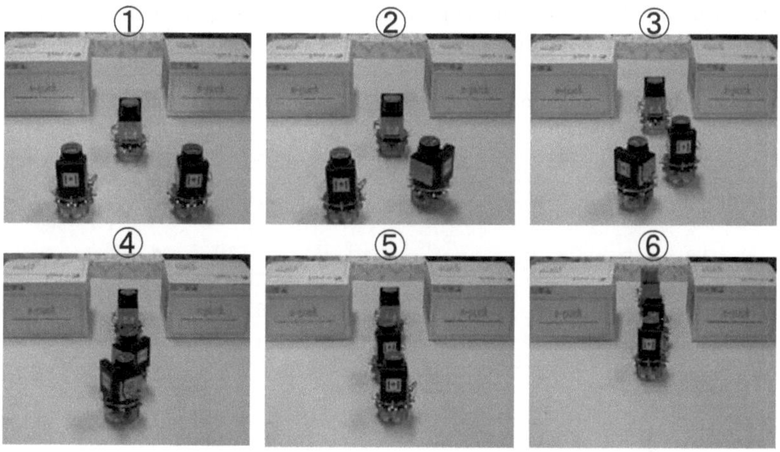


**Fig. 12.** Result of the experiment involving three robots

## 5 Conclusions

In this paper, we proposed a formation transition algorithm to adapt to the geometrical features of an environment. We consider a robot group composed of one leader and multiple unlabeled followers. We employed Delaunay diagram to maintain a geometrically steady formation and nonlinear oscillators to avoid deadlocks in a situation involving symmetric features. We verified the effectiveness of the algorithm by simulations and experiments.

## Acknowledgement

This research is partially supported by Grants-in-Aid for Scientific Research, MEXT, Japan, under contract, 17075007, and 17680016.

## References

1. Das, A.K., Fierro, R., Kumar, V., Ostrowski, J.P., Spletzer, J., Taylor, C.J.: A Vision-Based Formation Control Framework. IEEE Trans. Robotics and Automat. 18, 813–825 (2002)
2. Ota, J., Arai, T.: Motion Planning of Multiple Mobile Robots Using Dynamic Groups. In: Proc.1993 IEEE Int. Conf. Robotics and Automat., pp. 28–33 (1993)
3. Ota, J., Arai, T., Yokogawa, Y.: Distributed Strategy-Making Method in Multiple Mobile Robot System. Distributed Autonomous Robotics Systems, 103–106 (1994)
4. Yamaguchi, H.: A Cooperative Hunting Behavior by Mobile Robot Troops. In: Proc.1998 IEEE, Robotics and Automat, pp. 3204–3209 (1998)
5. Shimizu, M., Ishiguro, A., Kawakatsu, T., Masubuchi, Y., Doi, M.: Coherent Swarming from Local Interaction by Exploiting Molecular Dynamics and Stokesian Dynamics Methods. In: Proc. Int. Conf. Intelligent Robots and Systems, pp. 1604–1619 (2003)
6. Tolimieri, R., An, M., Lu, C.: Algorithms for Discrete Fourier Transform and Convolution (Signal Processing and Digital Filtering). Springer, Heidelberg (1997)
7. EPFL education robot, `http://www.e-puck.org/`
8. Hokuyo Automatic, `http://www.hokuyo-aut.co.jp/02sensor/07scanner/urg.html`

# Coordinated Search in Cluttered Environments Using Range from Multiple Robots

Geoffrey Hollinger, Joseph Djugash, and Sanjiv Singh

Carnegie Mellon University, Pittsburgh, PA USA
gholling@andrew.cmu.edu, robojoe@cmu.edu, ssingh@ri.cmu.edu

**Summary.** In this paper, we describe real-time methods for incorporating non-line-of-sight range measurements into a framework for finding a non-adversarial target in cluttered environments using multiple robotic searchers. We extend previous coordinated search strategies to utilize information from noisy non-line-of-sight range measurements. Sensors using ultra-wideband radio are becoming available that provide range measurements to targets even when they are occluded. We present two Bayesian methods for updating the expected location of a mobile target and integrating these updates into planning. We present simulated results in a complex museum environment as well as on mobile robots. Our results show the success of our algorithms at utilizing information from measurements in a coordinated search framework.

## 1 Introduction

The problem of coordinating multiple robots to search for a non-adversarial target is relevant to many applications throughout robotics. Urban search and rescue teams in military and first response scenarios often need to search for both moving and stationary targets in cluttered environments. The major application that has inspired our work is the assistance of first responders in disaster scenarios. Previous work by Liao et al. has examined the potential for range-only measurements to track a mobile first responder in disaster scenarios [8]. As in this previous work, we assume that non-line-of-sight range measurements to a target are available. Ultra-wideband radios are currently in development that can provide range measurements through walls using a synchronized time-of-arrival method [9]. The availability of these sensors motivates the development of algorithms to utilize their potential.

This paper describes real-time methods for incorporating noisy non-line-of-sight sensor measurements of range to a target into a probabilistic multi-robot coordinated search framework. Applications to first response assistance have led us to make the assumption that the target is *non-adversarial.* This assumption is consistent with the task of locating a moving first responder or survivor in a disaster scenario. We discuss our underlying framework and then outline two Bayesian methods for utilizing information from sensors. This paper is an extension of previous work by Hollinger, Kehagias, and Singh on locating a non-adversarial target in the minimum expected time [6]. The novelties of this

C. Laugier and R. Siegwart (Eds.): Field and Service Robotics, STAR 42, pp. 433–442, 2008.
springerlink.com 

paper include the formulation and testing of the measurement update methods, coordinated search results on the museum environment, and the implementation of this work on mobile robots.

This paper is organized as follows. Section 2 discusses related work in the areas of tracking, search, and pursuit-evasion. Section 3 describes the problem of finding a non-adversarial target with multiple searchers. Section 4 presents our algorithms for coordinated search and incorporating measurements. Section 5 gives both simulated results and results on mobile robots. Finally, Section 6 draws conclusions and discusses future directions for this work.

## 2 Related Work

Research in the areas of tracking, pursuit-evasion, and multi-robot coordination are all relevant to the coordinated search problem. The potential for multi-robot assistance in first response scenarios is examined at a high level by Kumar, Rus, and Singh [7]. They propose a general system for autonomous robotic assistance for first responders. More specific solutions to the first response assistance problem have been formulated by Djugash el al. [2] and later by Liao et al. [8]. These works utilize an ad hoc network of range-only sensors to track first responders, map the locations of new nodes in the network, and coordinate mobile nodes to minimize uncertainty.

Recent work in tracking mobile targets is also relevant to the coordinated search problem. Ferris, Hahnel, and Fox develop an algorithm for tracking a mobile target using wireless signal strength [3]. Their algorithm discertizes the environment in a similar manner to ours and then uses a particle filter to estimate position. We provide an alternative method for position estimation that closely couples with planning and remains computationally tractable. Our method also does not require prior training data, and it is completely online. As a tradeoff, our method is less appropriate than theirs for sensors with very complex or unknown models.

Coordinated search is closely related to work in the field of pursuit-evasion. One of the first algorithms for capturing an adversarial evader in an indoor environment was developed by Guibas et al. [5]. Their algorithm discretizes a polygonal environment based on visibility boundaries and searches the resulting information space for a path that guarantees capture. The problem studied by Guibas and LaValle differs from the one we discuss in this paper in that it assumes an adversarial evader. Sarmiento et al. also directly studied the coordinated search problem by developing strategies for searching indoor environments for a stationary target using a cost heuristic [10]. In previous work, we developed a heuristic graph search solution to the coordinated search problem for a mobile, non-adversarial evader [6].

## 3 Problem Definition

To describe the state of the world in a coordinated search scenario, we must develop representations for the environment and the locations of the searchers and targets. Assume that the state of a searcher at time $t$ is known to be $X^P(t)$, and the state of a target at any time $t$ is known with a certain probability to be $X^E(t)$. In a polygonal environment, this problem is continuous. To collapse the problem into a discrete domain, discretize the environment into a number of cells, $1 \ldots N$. The state of the $i^{th}$ searcher is now fully defined by $X_i^P(t) = n$ where $n$ is the cell in which the searcher is currently located. Now, define a capture event at time $t$ as the occurrence of $X_i^P(t) = X^E(t)$ for any searcher $i$. To define the state of the target, let $p$ be a row vector such that $p = [p_0, \ldots, p_N]$ where each value represents the probability that the target is in the corresponding cell. Let the value $p_0$ represent the probability that the target has already been found by the searchers. Refer to this as the "capture state." The vector $p$ now defines a probability distribution over the target's position in the environment (with the addition of a capture state).

The searchers' goal is to minimize the expected time of reaching a capture event. Thus, the searchers seek to maximize the probability that the target is in the capture state at any given time $t$. The coordination problem is then defined as the determination of paths for the searchers such that the probability of capture is maximized at any given time. The problem of incorporating measurements becomes one of modifying the probability vector $p$ to take into account new information from sensors.

## 4 Algorithm Description

### 4.1 Planning and Coordination

To reduce the task of finding a mobile target in a polygonal environment from a continuous to a discrete planning problem, we discretize a polygonal map into convex cells to form an undirected graph. Figure 1 shows an example discretization for a museum environment. The museum environment is challenging for coordinated search because it contains many cycles, which give a moving target many paths to avoid detection.

To integrate a motion model of the target into our pursuit-evasion framework and better define capture events, we use "capture" and "dispersion" matrices to modify the target's state vector. We can represent a capture event by defining a matrix that moves probability from the searcher's current cell to the capture state. Similarly, we can define dispersion matrices to represent the expected motion of the target in the environment. Concatenating the capture and dispersion matrices yields a forward model for the target's state vector at each time step.

Since the expected time of capture of the target cannot be known exactly during planning, we use a heuristic cost function to guide coordination of robotic searchers. The entropy of the target's estimated state vector provides a heuristic that directly relates to expected time of capture. Adding a summation to the cost function over a path allows the searchers to look several cells ahead to determine the optimal search path. When planning using cost functions, locations in the environment must be assigned as replanning points. These locations represent the points in the environment between which the robots move. The most obvious

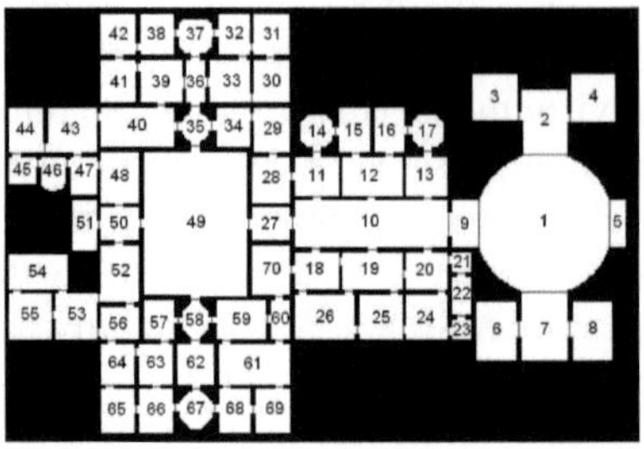


**Fig. 1.** Museum map used for simulated coordinated search trials. Gray lines show discretization boundaries. The museum map has many cycles by which moving targets can escape detection.

replanning points to use are the centroids of cells because they provide quick access to adjacent cells.

Once a cost function has been defined, coordination strategies for robotic searchers can be found by enumerating possible paths on the discretized floor plan. To develop a decoupled planning algorithm, allow each searcher to enumerate all possible paths on the graph to a given depth while assuming that the states of the other searchers are fixed. This prevents the search space from growing in complexity with the number of searchers. The searchers can passively coordinate with each other by communicating their current paths. Each searcher can then plan for its own path while assuming that the other searchers will adhere to theirs. The coordination algorithm presented in this section is given in much greater detail in previous work [6].

### 4.2 Incorporation of Measurements

To incorporate noisy, sporadic measurements into our coordinated search framework, we propose two methods for modifying the target's state probability vector given new sensor data. Both methods can be seen as estimating a posterior distribution given a prior distribution on the target's state. Thus, they both fit into a Bayesian probabilistic framework. For all cells $x_i$, we seek to estimate the posterior distribution $P(x_i^t) = P(x_i^t|z, x_i^{t-1})$ given a measurement $z$ and a prior $P(x_i^{t^-})$. The prior is an estimate of the evader's position after the application of the dispersion matrix, $P(x_i^{t^-}) = P(x_i^t|x_i^{t-1})P(x_i^{t-1})$. Once we calculate a posterior, we can renormalize and define the target's state probability vector as $p_t = [P(x_c^t), P(x_1^t), \ldots, P(x_n^t)]$, where $P(x_c^t)$ is the probability that evader has already been captured (the capture state). The value of $P(x_c^t)$ remains constant since we assume that noisy measurements cannot capture the target.

#### Sampling Method

Our first method for incorporating sensor data draws samples from a measurement distribution and uses these samples to directly estimate the posterior distribution on the target's state. Assume that we receive a measurement $z$ with an arbitrary noise model. We can draw $n$ samples from this distribution to represent the "true" measurement.[1] In Euclidean space, each sample represents an $(x, y)$ position of the target, which corresponds to a cell on the map discretization. We then weight each sample according to the prior on the cell in which it falls. This allows for the calculation of the posterior distribution on each cell as in Equation 1.

$$P(x_i^t|z, x_i^{t-1}) = \eta P(x_i^{t^-})n_i, \tag{1}$$

where $n_i$ is the number of samples that fall in a cell $x_i$, and $\eta$ is a normalizing constant.

In this paper, we examine the incorporation of range measurements into the coordinated search framework. A measurement $z$ is a single range value, and we assume that its noise is Gaussian with variance $\sigma^2$. We can draw sample range values from this Gaussian, and these values map to circles in Euclidean space. To map to $(x, y)$ coordinates, we use Bresenham's circle drawing algorithm [1] on a finely discretized grid and count each hit as a sample. This takes full advantage of the sampled range values and greatly decreases the number of samples necessary to properly estimate the posterior.

## Centroid Method

Our second method seeks to improve computational efficiency by calculating the likelihood of a measurement given the target is in each cell. Using Bayes's rule, the posterior can be rewritten as in Equation 2.

$$P(x_i^t|z, x_i^{t-1}) = \eta P(z|x_i^t)P(x_i^{t^-}), \tag{2}$$

where $\eta$ is a normalizing constant.

This reduces the problem of calculating the posterior to that of calculating a likelihood term $P(z|x_i^t)$. Since each cell is represented as a continuous set of points in the map plane, this calculation is difficult. To reduce the complexity of the problem, we make the assumption that the target will remain close to the centroid of each cell.[2] Thus, each cell can be reduced to its centroid. Since cells are convex, the centroid is guaranteed to be inside the cell. The calculation of $P(z|x_i^t)$ is now one of calculating a likelihood at a point. This is a common problem in state estimation, and is prevalent throughout Monte Carlo methods and particle filter literature [3].

For range measurements, the problem of calculating $P(z|x_i^t)$ is simply that of determining the expected range value for each cell centroid. Let $r_i^c$ be the range

[1] There are other methods for estimating the posterior through sampling, i.e. maintaining particle estimates of the target's position. Our method is particularly appealing for range measurements because it allows us to sample a 1D space.

[2] Another technique for generating likelihoods is to finely discretize the environment and calculate a likelihood at each point. This is equivalent to the histogram filter in the literature. The centroid method can be seen as a special case of the histogram filter.

from a ranging sensor to the centroid of cell $x_i$ and let $r$ be the received range measurement with assumed Gaussian noise variance $\sigma^2$. The likelihood is then calculated as in Equation 3.

$$P(z|x_i^t) = N(r; r_i^c, \sigma^2) \tag{3}$$

**Comparison of Methods**

The sampling and centroid methods each have strengths and weaknesses in different environments. The sampling method biases towards larger cells since those cells will contain more sampled points. In the case of a moving target, the target will spend more time in larger cells, thus warranting this bias. A drawback to the sampling method is that it is computationally expensive if the number of samples grows large. Thus, to maintain real-time measurement incorporation, the number of samples must be kept small. The sampling method is particularly appropriate for small environments (requiring few samples) with highly irregular cells. Such cells are difficult to approximate simply by their centroid.

The major disadvantage of the centroid method is that it makes the strong assumption that the target is near the centroid of the cell. To account for this assumption, a value for $\sigma$ must be used that is greater than the underlying noise model of the sensor. This degrades the accuracy of the sensor model. Unlike the sampling method, the centroid method does not bias towards larger cells. The major advantage of the centroid method is that it is extremely fast, only requiring one calculation for each cell in the environment. Thus, the centroid method is appropriate for large environments with regularly shaped cells.

## 5 Results

### 5.1 Simulated Results

We implemented our algorithms on a multi-agent coordinated search simulation in C++ on a 3.2 GHz Pentium 4 processor [6]. This simulation allows for multiple searchers, realistic modeling of ranging measurements, and both stationary and moving targets. For moving trials, we assumed that the average speed of the target is 1 $m/s$, and that it moves holonomically between cell boundaries. We tested our algorithms with varying numbers of searchers on a museum map (shown in Fig. 1). The searchers start in the same location for all trials, and the location of the target is initialized at random on the map. Each searcher holds a mobile ranging sensor, and a stationary ranging sensor is placed at the top left corner of the map. The stationary sensor corresponds to one placed outside of a building and helps to disambiguate target position uncertainty caused by symmetry. The simulated ranging sensors were given a variance of 2 $m^2$ and were given a 10% chance of providing a measurement each second.

Fig. 2 shows average capture times for moving and stationary targets with varying numbers of searchers. The results show significant improvement using measurements with both the centroid and the sampling methods. Additionally,

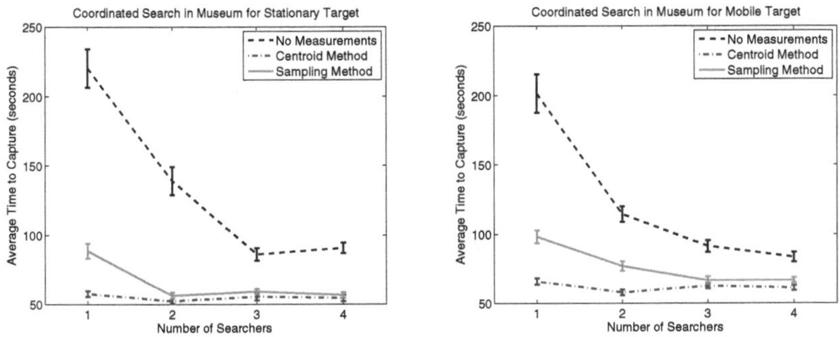


**Fig. 2.** Simulated coordinated search results in a museum comparing two methods for incorporating range measurements. Graphs show average capture times versus number of searchers for a stationary target (left) and a moving target (right). Error bars are one SEM, and averages are over 200 trials. Graphs are zoomed in for comparison.

the centroid method outperformed the sampling method. This is likely due to the limitations on the number of samples necessary to maintain real-time operation (500 samples were used for each measurement). Furthermore, the sampling method's bias towards large cells is detrimental when the target is stationary. Conversely, the centroid method's assumption that the target remains close to the centroid of the cell is valid for moving targets in small cells, which is most often the case on this map.

### 5.2 Robot Results

To further confirm our algorithms, we implemented them on a mobile robot. We used a single Pioneer One robot from ActivMedia, equipped with a laser range scanner, as a searcher and a Nomad Scout as a mobile target. We utilized the Player/Stage software to control the robots [4]. The Pioneer searcher used the Adaptive Monte Carlo Localization (AMCL) algorithm packaged with Player/Stage to localize within the known environment. The Nomad Scout utilized a pseudo-random walk algorithm with sonar sensors.

For experiments that required the incorporation of measurements to the target, a *Parrot* ranging sensor was used to acquire the measurements. The *Parrot* sensors form a ranging sensor network that uses radio to communicate data and ultrasonic signals to determine range between nodes [11]. A total of three Parrots were used in our experiments. The first was mounted on top of the searcher, the second was placed on top of the target, and the third was placed at a surveyed location near a corner of the map.

We conducted experiments within a 20 $ft \times 17$ $ft$ room, which we partitioned as in Fig. 3 using 1 $ft$ cardboard walls. Since the test environment contains a cycle, it is impossible to guarantee capture of a quickly moving target with a single searcher [5]. We placed both mobile and stationary Parrots at a height of approximately 1.5 $ft$ above the floor, making them taller than the cardboard walls.

By placing the Parrots above the walls, we were able to convert the otherwise line-of-sight restricted Parrot sensors to "non-line-of-sight" ranging sensors. This simulates the range measurements that one could receive with ultra-wideband radio ranging sensors.

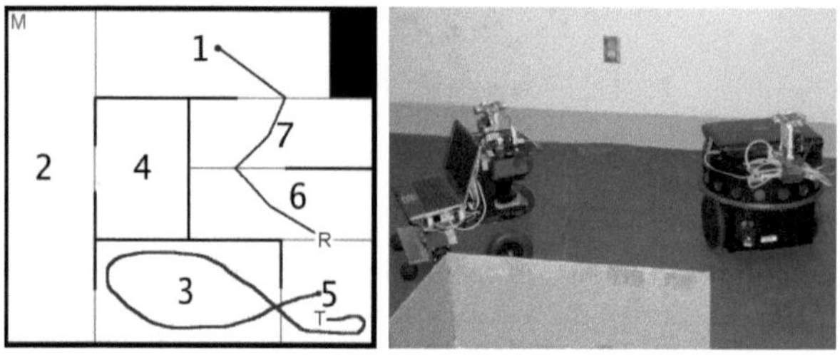


**Fig. 3.** Simple map used for coordinated search trials on Pioneer robot platforms (left) and corresponding photograph of robots in test environment (right). Gray lines show discretization boundaries, 'R' shows position of robotic searcher, 'T' shows position of target, and 'M' shows position of stationary measurement beacon. Blue and red traces show paths of searcher and target respectively (reconstructed from AMCL output). The robots are in the same locations on the map as they are in the photograph.

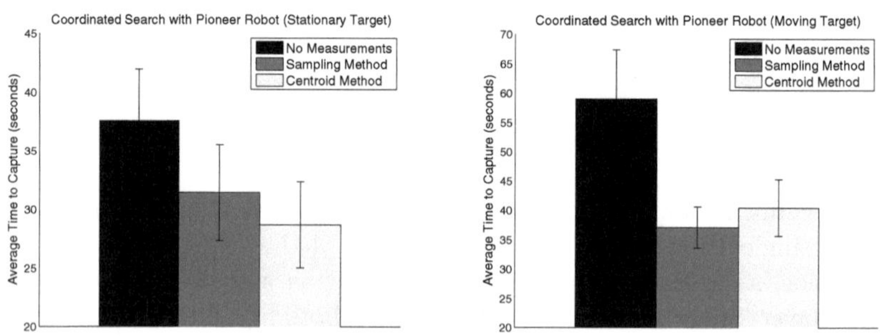


**Fig. 4.** Coordinated search results on a Pioneer robot comparing two methods for incorporating range measurements. Graphs show average capture times for a single searcher finding a stationary target (left) and a moving target (right). Error bars are one SEM, and averages are over 24 trials. Graphs are zoomed in for comparison.

In ultrasonic-based ranging systems, there are two types of noises that affect range estimation, multipath reflections and random noise. For the Parrots, the latter is easily modeled as a zero-mean gaussian with a variance of 0.5 $m^2$. In order to detect and remove the multipath measurements from processing, we added a chi-squared test to our algorithm. This helps to filter out erroneous measurements caused by multipath.

Fig. 4 gives average times to capture for trials using the Pioneer and Nomad over a total of 144 runs. The robot results confirm the simulated results by showing that the centroid and sampling methods outperform the case without sensor data. They also show an improvement of the sampling method over the centroid method for moving targets and the reverse for stationary targets. This is likely due to the tendency for moving targets to remain in large cells and the sampling method's bias towards larger cells. Qualitatively, the searcher moved more quickly towards the target's cell while incorporating measurements. This demonstrates the success of our methods on a mobile robot platform.

## 6 Conclusions and Future Work

We have presented two techniques for incorporating non-line-of-sight range measurements into a coordinated search framework. Both techniques utilize a Bayesian framework for estimating the posterior distribution of a target's position given a prior estimate and a new measurement. Our techniques generalize to most sensors with a known noise model, and they scale well with increasing searchers. We make the assumption that a target is non-adversarial and present a heuristic graph search approach to minimizing the expected time to capture. We have presented simulated results showing that our methods remain tractable in a large environment with cycles. Our simulated results also show that both of our techniques greatly outperform the case without measurements. Additionally, we have implemented our algorithm on a Pioneer robot utilizing sonar ranging sensors. The results on the Pioneer confirm the simulated results and demonstrate the real-time performance of our methods.

A major extension of this work is the implementation of our coordinated search algorithms on multiple robots in large-scale environments. We plan to move out of the controlled laboratory environment and into an office building similar to map used in simulation. We will also use ultra-wideband ranging radios to provide non-line-of-sight sensing in this environment. Our algorithms provide an important step towards fully utilizing the new technology of ultra-wideband ranging radios.

## Acknowledgements

We thank Athanasios Kehagias for his insightful comments. This work is supported by the National Science Foundation under Grant No. IIS-0426945.

## References

1. Bresenham, J.E.: Algorithm for computer control of a digital plotter. IBM Systems Journal 4(1) (1965)
2. Djugash, J., Singh, S., Corke, P.: Further results with localization and mapping using range from radio. In: Proc. 5th Int'l Conf. on Field and Service Robotics (2005)
3. Ferris, B., Hahnel, D., Fox, D.: Gaussian processes for signal strength-based location estimation. In: Proc. of Robotics Science and Systems (2006)

4. Gerkey, B., Vaughan, R., Howard, A.: The Player/Stage project: Tools for multi-robot and distributed sensor systems. In: Proc. Int'l Conf. on Advanced Robotics (2003)
5. Guibas, L., Latombe, J., LaValle, S., Lin, D., Motwani, R.: Visibility-based pursuit-evasion in a polygonal environment. Int'l Journal of Computational Geometry and Applications 9(5), 471–494 (1999)
6. Hollinger, G., Kehagias, A., Singh, S.: Probabilistic strategies for pursuit in cluttered environments with multiple robots. In: Proc. Int'l Conf. on Robotics and Automation (2007)
7. Kumar, V., Rus, D., Singh, S.: Robot and sensor networks for first responders. Pervasive Computing, 24–33 (2004)
8. Liao, E., Hollinger, G., Djugash, J., Singh, S.: Preliminary results in tracking mobile targets using range sensors from multiple robots. In: Distributed Autonomous Robotic Systems (2006)
9. Saeed, R., Khatun, S., Ali, B.M., Khazani, M.: Performance of ultra-wideband time-of-arrival estimation enhanced with synchronization scheme. ECTI Trans. on Electrical Eng., Electronics and Communication 4(1) (2006)
10. Sarmiento, A., Murrieta-Cid, R., Hutchinson, S.: A multi-robot strategy for rapidly searching a polygonal environment. In: Proc. 9th Ibero-American Conf. on AI (2004)
11. Zhang, W., Djugash, J., Singh, S.: Parrots: A range measuring sensor network. Technical Report CMU-RI-TR-06-05, Robotics Institute, Carnegie Mellon University, Pittsburgh, PA

# A Plan Manager for Multi-robot Systems

Sylvain Joyeux, Rachid Alami, and Simon Lacroix

LAAS-CNRS, University of Toulouse, Toulouse, France
firstname.lastname@laas.fr

**Summary.** This paper presents a software component, the plan database, which provides the services needed to build and execute plans in a multi-robot context. This plan database handles fully dynamic plans (insertion and removal of tasks), provides tools for safe concurrent execution and modification of plans, and handles distributed plan supervision without permanent robot-to-robot communication. The proposed concept is illustrated by a simulated example that involves a rover and an UAV in an initially unmapped environment.

## 1 Introduction

One can distinguish two main schemes in today's distributed robotics architectures. On the one hand, some approaches rely on a team-management layer which sends orders to a mono-robot supervision system (for instance Teamwork [9] and TraderBots [1]). In these approaches, interactions between robots are reactively managed, and are not pro-actively managed within a plan structure. Systems based on such a scheme cannot predict that, for instance, interactions will be needed in the future. Also, the team-management layer is limited to managing high-level tasks, which makes difficult to handle tighter interaction modalities, like opportunistic cooperation, which take place at a lower abstraction level.

On the other hand, some approaches rely in mono-robot supervision systems that can send orders transparently to other robots: in FIRE [8], a robot can change the state of a remote robot, but not pro-actively, and it assumes that communication is available at all times, which can not be taken for granted in multi-robot systems. Coupling such supervision layers with a team-management component gives better result (like FIRE, which is a modified Task Desciption Language coupled with a Contract-Net Protocol task allocation scheme), but the resulting systems still have the problem of the above architectures: the two layers do not share all the information they have, and so the interaction managed by the upper layer can only loosely take into account the interaction in the lowest one. They lack a common place in which to describe all the interactions in place.

To our knowledge, three architecture have implemented plan-based cooperation: GPGP and its associated task representation TAEMS [6], the COMETS [3] architecture and Machinetta [7]. In COMETS, a plan which can contain joint

C. Laugier and R. Siegwart (Eds.): Field and Service Robotics, STAR 42, pp. 443–452, 2008.
springerlink.com 

tasks[1] is produced by a Shop2 planner. The joint tasks are then negotiated among the team by an "Interaction Manager". The resulting plan (developed joint tasks and mono-robot tasks) is then sent as an unique plan to an execution component. In GPGP, the joint plan is modified through a whiteboard mechanism by all involved robots. When they all agree on the result, each robot sends it to its own scheduler which starts its execution. Machinetta defines a plan-based interaction mechanism and an architecture for the interaction. The main focus of COMETS and Machinetta is to define a generic team management system and the one of GPGP is multi-robot planning. All three are tied to their particular interaction scheme. None has been designed to be flexible enough to integrate different plan production mechanisms and different team management schemes.

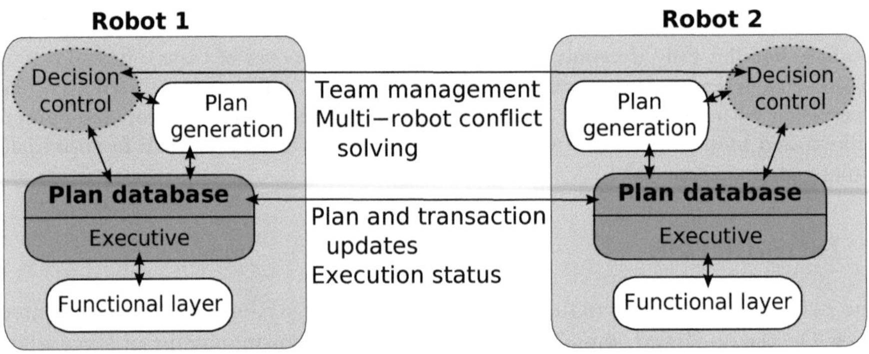


**Fig. 1.** In this architecture, each robot plan is managed by its own plan database. Parts of these plans are shared among databases. Decision control manages plan generation, team management and solves conflicts between execution and planning.

In this paper, we present a plan database component, whose role is to provide the services required to manage and execute multi-robot plans. The component is generically designed, and is not associated to specific plan generation systems or specific team management schemes. This plan database acts as the middleman between the functional layer and plan generation components (Fig. 1). In this architecture, the planners are responsible for producing coherent plans either for the robot alone or for the team, and the functional layer is a service layer which provides the perception and action algorithms that interfaces the robot and the real world. Between these two, we introduce three components: the *plan database* maintains a plan, which is a graph of tasks and events defining what the robot may do in the future and how it will do it. This plan is continuously interpreted by the *executive* to produce actual actions, using an event-based model. The *decision control* component has two roles: first, it may call the planners for instance to adapt the plan for new missions, for contingency planning

[1] A joint task is a task which is executed by more than one robot at a time.

and because of cooperation possibilities with other robots. Second, it handles the choices that have to be made during execution: since our framework allows simultaneous planning and execution, conflicts will arise between the main plan being executed and the partial plans being built. In a multi-robot context, it also solves conflicts between the needs of the robot and the needs of the team.

The plan database (pDB) has been designed with the following goals:

- provide the tools required to build multi-robot plans, and use plans as a basis of negotiation. In a pDB, one can build plans that involve multiple robots, make these robots negotiate about the new plan (distributed plan modification) and, if they agree, commit them to the result.
- manage plan modifications atomically: the executive must not execute partial plans that are being generated by planners, as they are not sound yet.
- express multi-robot execution contexts, and handle them during execution. In particular, plan execution is able to handle the fact that robots cannot communicate at all times.
- allow integration of external tools to handle multi-robot decision processes like task allocation and multi-robot planning.

## 2 Plan-Based Cooperation in a Rover/UAV Scenario

As a supporting example to present the concepts associated to the pDB, we outline here a scenario which has been partially realized in simulation[2]. In this scenario, the mission to achieve is a rover `GoTo` task in an initially unknown terrain. For that purpose, the rover is endowed with algorithms which instance a perception/decision/action loop, summarized by the three tasks `Bmap::TravPercep`, `Nav::Path` and `P3d::TrackPath`. These tasks respectively build a traversability map from perceived data, plan a path in the traversability map and execute the planned path. The rover is assisted by an UAV, which is able to update the traversability information of the rover thanks to a vision process (`Plid::TravPercep`). We assume the UAV flies at an obstacle free elevation, so that its movement can be handled by a simple waypoint navigation scheme.

On the basis of this simple scenario, we will show in this section how the rover-UAV cooperation can be expressed as relations between individual robot plans, and how this joint plan is managed in our plan database.

Fig. 2 presents the initial rover plan, in which `P3d::TrackPath` follows the waypoint list, `Nav::Path`, established by `Nav::PathPlanning` on the basis of the results of `Bmap::TravPercep` [3]. The arrows between tasks express relationships between tasks, which are described in section 3.1.

We selected for this scenario an opportunistic method to initialize the interaction: the rover does not know beforehand that an UAV will help to realize its mission. When the UAV detects the rover presence by using an automatic network discovery mechanism, it will add *triggers* on the rover's plan. These triggers are used to match specific patterns in plans, and a list of known patterns

[2] Its realization with real robots is currently under way.

[3] `POM::Localization` summarizes all the localization processes on-board the rover.

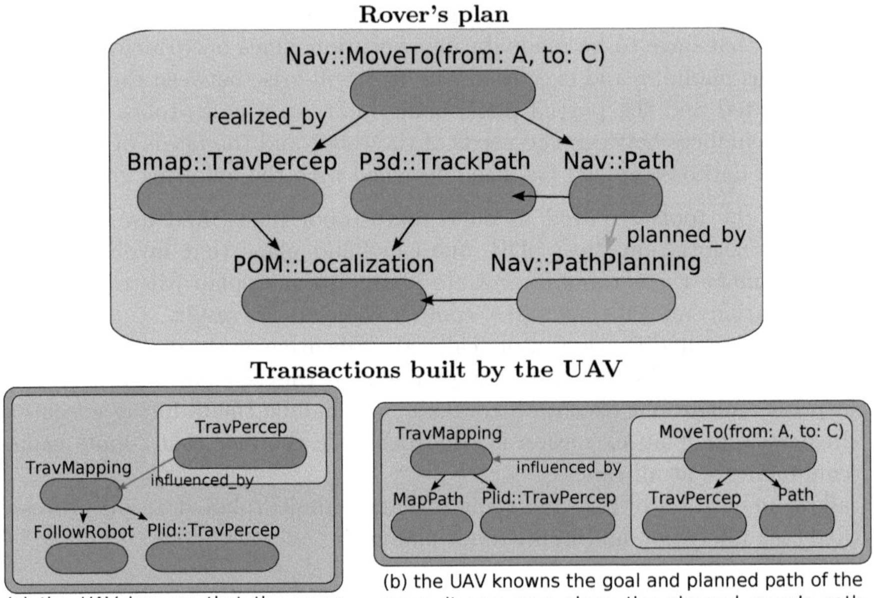


(a) the UAV knowns that the rover depends on traversability perception. It acts as a remote sensor, following the rover (FollowRobot)

(b) the UAV knowns the goal and planned path of the rover. It can map along the planned rover's path (MapAlongPath). This way, the UAV takes the rover projected needs into account.

**Fig. 2.** *(top)* partial view of the initial rover plan. *(bottom)* two alternative cooperation scheme, based on two different patterns found in the rover's plan, in transactions built by the UAV.

are defined in each robot controller. The UAV will therefore be notified of a pattern it is able to interpret.

The two alternatives presented in Fig. 2 are based on two different triggers. When multiple triggers match the same task pattern, the choice of the modality is not a decision made by the pDB, but by the decision control. The pDB only notifies decision control what task set triggered what pattern. To achieve this choice, decision control can for instance build multiple transactions with the multiple possible outcomes and compare the resulting plans.

Let's assume that the (*b*) modality is chosen by decision control. The UAV then uses a planner to produce the plan, based on its partial knowledge of the rover's plan. This planner cannot directly change the current rover plan: first, partial plans are not directly executable, and second, if the UAV had such an ability, it would mean that the rover is not able to fully control its activities. We therefore need a way to change plans outside the plan being executed, so that *(i)* the executive only sees complete executable plans and *(ii)* one robot can propose a plan modification to another. This is achieved by *transactions*, that represent a plan changeset, *i.e.* set of modifications required to get from the current main plan to the desired final plan. Transactions can be shared and synchronized across pDBs.

So, while the UAV generates its proposal, the rover is not aware of the transaction. Once the transaction is complete from the UAV point of view, the transaction is sent to the rover. The rover can modify it, in which case the changes are sent back to the UAV: both robots use the transaction as a whiteboard to build their joint plan. Once both agree on the new plan, they change their main plan accordingly at the same time, and can start its execution. If a joint plan cannot be found, the transaction is simply discarded.

# 3 A Distributed Plan Database

The plan database has been designed as a tool for plan management, in both mono-robot and multi-robot systems. In a multi-robot context, a single pDB is able to express and manipulate plans where tasks executed by the local robot (local tasks) are interacting with tasks executed by other robots (remote tasks), or even the joint tasks that imply more than a single robot. Note that in order to reduce plan complexity, there is no need for one robot to know *everything* about another's plan: a pDB is informed only about the remote tasks it is interacting with (section 3.3). This allows to keep each pDB plan at a tractable size regardless of the number of pDB currently interacting.

The section first describes the pDB model of tasks and events, and how these objects interact which each other. It then presents how multi-robot plans are managed and built, and finally the way to handle situations where two interacting pDBs do not have communication, which is a key feature in a multi-robot system.

## 3.1 Plan Model

We summarize here the plan model used in the pDB – more details on this model, and the general plan execution mechanisms and on error recovery are detailed in [4].

In the pDB, plans are graphs of two kind of objects: events and tasks. Events describe *what* occurs during task execution (for instance a `stopped` event is emitted when a task has stopped), and *the way* they occur: an interruptible task has a *controllable* `stopped` event, which means that the controller can call the event command and make the task stop. The main event relation is the *signal* relation: if an event $e_1$ signals a controlable event $e_2$, then $e_2$'s command will be called when $e_1$ is emitted. The signal graph therefore describes the action(s) to take when an event is emitted.

Tasks describe processes, and the task relations describes the interactions between these processes. Tasks models are defined as a set of events, the basic ones being `started`, `failed`, `success` and `stopped`, which define milestones in the task execution. These task models are put into a hierarchy in which the more abstract task models are refined in less abstract ones. In the plan, these models are used to define task *instances*, which are the *active* objects: they determine when the events should be emitted, handle event commands, and can locally repair the plan in case of exceptions. Defining a task model hierarchy has the advantage that, in multi-robot contexts, a robot does not necessarily have to know all the models known by all the pDBs. Instead, it can use an intermediate level of abstraction to represent another's specific task implementation (compare the specific models of the rover and the more generic ones on Fig. 2).

Three task relations are currently defined:

- The `realized_by` relation is used to express dependency: the successful execution of `Nav::MoveTo` requires its two child tasks to be successfully achieved.
- The `planned_by` relation expresses that finding a way to execute an action is handled by a specific task.
- The `influenced_by` relation: the rover motion and the UAV traversability mapping do not have a strong dependency, but the execution time and efficiency of the motion can be greatly improved thanks to the UAV's `Plid::TravPercep` task.

### 3.2 Multi-robot Plans

In order to represent what tasks are executed by whom, a task instance has an *ownership* attribute, which is set to the list of pDBs which are executing this task. For local tasks, ownership is naturally set to the local pDB only, for remote tasks to the remote pDB which is handling the task, and for joint tasks to all the pDBs involved in the joint task.

During execution, a pDB is transparently notified of all changes that are done to remote tasks that are present in its own main plan, in a manner robust to communication failures (section 3.4). This includes execution status (signalling, event emission), and changes in the event and task graphs, which allows the distributed execution and adaptation of multi-robot plans. However, we can easily see that if it were possible to freely create relations between local and remote objects, then one pDB would be able to take decisions for another, which should only be possible through negotiation: the golden rule of multi-robot plan management in our system is that a remote pDB cannot change a robot plan without its consent. To ensure that, the following rules are in effect:

- to *remove a relation*, it is sufficient to only own one of the two objects involved. A pDB can for instance freely remove a `realized_by` relation between one of its own task and a task owned by another pDB (or a joint task). This is needed because one robot should be able to remove itself from a joint plan without negotiation, in cases of emergency for instance.
- any robot can remove itself at any time from the list of owners of a joint task. Ownership removal is then notified to the other pDBs.
- as for the removal of task relations, to *add or remove an event relation*, the local pDB must be the owner of the child in the relation: it allows one pDB to synchronize itself on events of another pDB, which does not really affect the other pDB.

Any other modification involving an object not exclusively owned by the local pDB is not allowed in the main plan. To handle the negotiation process needed for these modifications, we added *distributed transactions*, which act as distributed whiteboard to change plans.

### 3.3 Building Multi-robot Plans

In a mono-robot pDB, a transaction is a whiteboard used to build a set of plan modifications without modifying the plan being executed. The pDB can then synchronously apply all the changes to the main plan (or discard them). It ensures that the main plan is always sound, provided that the planners produce sound plans.

In a multi-robot context, transactions are shared among pDBs, and therefore they can change multiple plans at the same time. They are a central tool for negotiation: one robot builds a partial multi-robot transaction, which can then be modified by others, until an agreement has been found on the new joint plan, in which case the transaction is *committed* into all the involved plans. The pDB ensures that the changeset contained in the transaction is applied at the same time on all involved plans or not at all. A robot can therefore assume that, if new multi-robot relations have been added to its plan, the same relations have been added on the other pDBs as well.

The update mechanism mentioned earlier uses subscription: a pDB subscribes to remote tasks to get all updates about them. Obviously, it must be automatically subscribed to remote tasks that are related to local tasks by the object graphs. One of the role of plan building through distributed transactions is to determine what tasks, outside of the automatically subscribed ones, are relevant to the remote pDBs.

### 3.4 Connection Management

Multi-robot systems cannot take communication for granted: interactions between plan databases can therefore not rely on a permanent network link. In our system, two interacting pDBs are *connected*. Connections are either *alive* if a communication link exists, or *dead* if there is no communication link.

In the plan, connections are represented by `ConnectionTask` instances, with remote tasks being dependent of this connection task. Therefore, if a separate component determines that we cannot rely on the remote robot, for instance because the connection has been dead for too long or the robot is late at a rendezvous, it simply stops the corresponding `ConnectionTask` and lets the usual pDB recovery mechanisms apply (not described here, see [4]). These monitoring situations can also be expressed in the plan, for instance by making the `ConnectionTask` `realized_by` monitoring tasks. That way, `ConnectionTask` will be marked as `failed` if the monitoring fails.

Moreover, remote tasks are represented locally by task proxies, which are passive objects in general. Specific proxies can be defined for specific task models to predict the remote task state in a disconnected scenario, like the proxies of

Machinetta [7]. This allows to make the plan continue its execution without having a communication link. Recovering after a bad prediction is however an issue with which our system is not dealing.

# 4 Implementation and Simulation Results

The pDB is currently implemented[4] in the Ruby language. We use the object-oriented capabilities of Ruby as a way to define task models and task instances as classes and objects. In distributed contexts, Ruby allows to create classes on-the-fly, which allows to map unknown remote models to anonymous local classes. Moreover, developing the system in a general-purpose language promotes code reuse in supervisors. The development of our controllers shows that a great level of modularity can be achieved by sharing code between the task models on the one hand, and defining libraries of often-used plan modification operators on the other hand.

Two controllers, one for the rover and one for the UAV have been implemented, using an already existing set of Genom [2] modules that implement the robots' functional layer. Pocosim [5], our simulation system, allowed us to use most of the modules in simulation without modification. We only replaced the image acquisition chain by modules (`plid` on the UAV and `bitmap` on the rover) which read a pre-computed traversability map (Fig. 3). These modules offer the same interface as the real ones: we keep the two layers as close as possible to use the same controller in simulated and real environments. Since no noise is added to perception, the functional layer behaviour is of course much better than in real conditions.

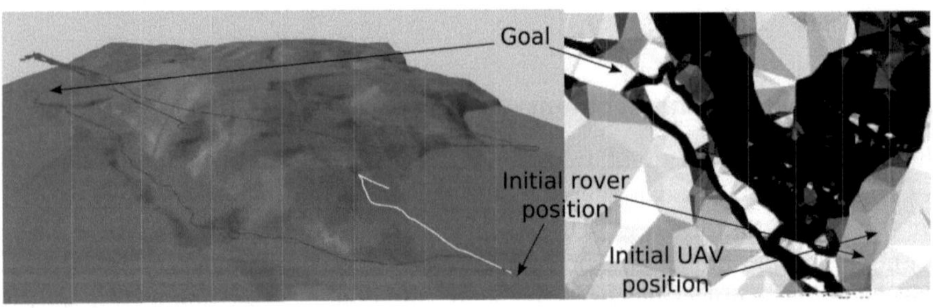


**Fig. 3.** (*left*) rover trajectories with (red) and without (white) the UAV's help, and the UAV trajectory in map-along-path mode. (*right*) Traversability map, white being perfectly traversable and black not traversable.

## 4.1 Plan Execution

The joint plans built during negotiation include all the information required to manage its execution. The pDB offers two error recovery mechanisms: either

[4] Available at `http://www.laas.fr/~sjoyeux/research.php`

errors are handled in the plan (conditional plans), or an exception mechanism (not described here) is used. In multi-robot contexts, exception handling is not distributed: it is supposed to be a synchronous operation, and thus cannot be done in the asynchronous communication scheme we use. If multi-robot error recovery is needed, it shall be either directly expressed in the plan or done reactively by negociation between the involved decision control components.

During execution, the UAV is therefore kept informed of any update of the rover's path (`Path`). It can use this information to build and update its own mapping path and send the traversability updates to the rover. The rover then replans its path, sends the updated path to the UAV, and so on.

The interaction finishes either when (*i*) the influenced tasks (parent tasks in the `influenced_by` relation) have successfully finished, which would announce the success of the joint plan, or (*ii*) when the plan structure from which the systems initiated the interaction has ceased to exist. In that case, the rover's pDB will notify the UAV of this change, and the UAV can then decide to change its plan accordingly. Note that it is not an automatic process of the pDB plan management, but a decision to be taken by the UAV.

### 4.2 Simulation Results

We have run three different kinds of simulations: the rover alone, with the UAV as a remote sensor, and with the UAV which maps along the rover path (Fig. 3 displays the resulting trajectories). In these simulations, the rover alone don't always reach its goal. The two interaction schemes are consistently successful. The rover reaches its goal (at 200 metres) in 10 minutes. The two setups are equivalent in this simulation because the rover quickly gets trapped in the corridor we see in the terrain.

## 5 Conclusion

The contributions of this paper are the multi-robot extension to a plan model, and the set of tools embedded in the plan database to build and execute multi-robot plans. The pDB transactions allow to cooperatively build multi-robot plans, negotiate on them and commit them. Its distribution mechanisms allow to manage the plan during its distributed execution, including situations where communication is not permanent. Finally, by not relying on particular interaction schemes, the pDB is generic enough to allow integration of most of the existing schemes.

The framework described in this paper has shown promising results for the integration of a modular functional layer like Genom in a plan management system, and for the development of interaction schemes. In the future, it should act in various projects as an integration tool for interaction schemes, planning tools and functional layers. Our main objective is now to integrate temporal information in the plans, which is required to properly handle communication loss for instance by handling the setup of rendezvous points.

## References


1. Dias, B.: TraderBots: A New Paradigm For Robust and Efficient Multirobot Coordination in Dynamic Environments. PhD thesis (2004)
2. Fleury, S., Herrb, M., Chatila, R.: Genom: A tool for the specification and the implementation of operating modules in a distributed robot architecture. In: Proceedings of IROS (1997)
3. Gancet, J., Hattenberger, G., Alami, R., Lacroix, S.: Task planning and control for a multi-UAV system: architecture and algorithms. In: Proceedings of IEEE IROS (2005)
4. Joyeux, S., Lacroix, S., Alami, R.: A software component for simultaneous plan execution and adaptation. In: Proceedings of the IEEE IROS (2007) (to be published)
5. Joyeux, S., Lampe, A., Lacroix, S., Alami, R.: Simulation in the LAAS architecture. In: Workshop in Software development in robotics, ICRA (2005)
6. Lesser, V., Decker, K., Wagner, T., Carver, N., Garvey, A., Horling, B., Neiman, D., Podorozhny, R., NagendraPrasad, M., Raja, A., Vincent, R., Xuan, P., Zhang, X.: Evolution of the GPGP/TAEMS Domain-Independent Coordination Framework. Autonomous Agents and Multi-Agent Systems 9(1), 87–143 (2004)
7. Schurr, N., Okamoto, S., Maheswaran, R.T., Scerri, P., Tambe, M.: Evolution of a teamwork model. Cognitive Modeling and Multi-Agent Interactions (2005)
8. Simmons, R., Smith, T., Dias, M.B., Goldberg, D., Hershberger, D., Stentz, A., Zlot, R.M.: A layered architecture for coordination of mobile robots. In: Proceedings From the NRL Workshop On Multi-Robot Systems (2002)
9. Tambe, M.: Agent architectures for flexible, practical teamwork. In: Proceedings of the 14th National Conference On AI and the 9th Innovative Applications of AI Conference, AAAI, Menlo Park (1996)

# Development of a Networked Robotic System for Disaster Mitigation

## -System Description of Multi-robot System and Performance Tests of the Robots-

Keiji Nagatani[1], Kazuya Yoshida[1], Kiyoshi Kiyokawa[2], Yasushi Yagi[2], Tadashi Adachi[3], Hiroaki Saitoh[3], Toshiya Suzuki[4], and Osamu Takizawa[5]

[1] Tohoku University
[2] Osaka University
[3] IHI Aerospace Co., Ltd.
[4] Eizoh Co., Ltd.
[5] National Institute of Information and Communications Technology

**Summary.** In this paper, the research project named "networked robotic system for disaster mitigation" is introduced. The project aims to develop key technologies for multiple robots to be teleoperated through a wireless communication network, which includes a satellite communication link, for surveillance tasks at a disaster site. The robotic system consists of a large-scale outdoor robot and a group of smaller indoor robots. The large-scale robot will serve as a carrier for the smaller robots which are deployed inside a partly-collapsed building. A three-dimensional range sensor and an omnidirectional camera are used as tools to ease the teleoperation for the human operator. This paper presents the mission scenario and the development status of the proposed networked multi-robotic system. The results of the performance tests are also reported.

**Keywords:** Disaster mitigation, Surveillance robots, Ad-hoc network, Teleoperation, Omnidirectional camera, Laser range sensor, Augmented reality.

## 1 Introduction

Development of the robotic systems for search and rescue operations receives increasing attention and national priority after the Hanshin-Awaji earthquake in 1995 (Japan) and the World Trade Center attack in 2001 (U.S.A.). In such disasters or incidents, priority tasks are grasping a whole picture of the damages and gathering the information on the victims' locations as quick as possible. To reduce the risk of human rescue crews and to help their operation, the introduction of remotely operated robots for immediate surveillance is strongly expected to the robotics community [1].

Since 2003, we have been working on a newly initiated project based on a networked robotic system to help mitigating the damages of disasters [2], under the support of the Japanese Ministry of Internal Affairs and Communications (MIC). The project is organized by Tohoku University, Osaka University, IHI Aerospace, Eizoh Co., Ltd and NICT (National Institute of Information and

C. Laugier and R. Siegwart (Eds.): Field and Service Robotics, STAR 42, pp. 453–462, 2008.
springerlink.com 

Communications Technology), aiming at the development of a robotic system for surveillance of a remote disaster site.

The target environment is assumed as a disaster area under a large-scale urban earthquake. There are concrete buildings half collapsed or almost destroyed. Road networks are damaged so that firefighters have a difficult access. Conventional communication networks may be also damaged. In such conditions, the first priority in any rescue procedure is to gather the information for the immediate assessment of the damages and locate the victims in each specific site. To be able to tackle such tasks with the robotic technology, we have developed a robotic system which consists of a large-scale outdoor robot (hereafter termed as a "parent robot") and small-scale indoor robots (hereafter termed as "child robots"). There have been several robotic research projects related to marsupial type robots to deal with uneven and hazardous environments (e.g.[3]). One of the features in the proposed system is that each child robot has a relatively higher locomotion capability (without cables).

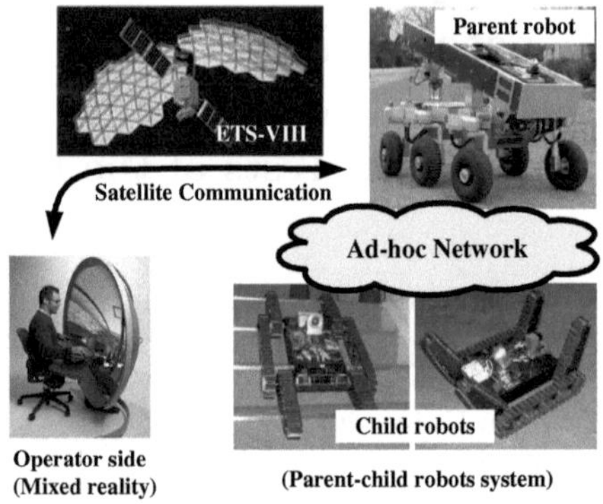


**Fig. 1.** Project overview

The proposed mission scenario is described as follows:

1. A set of the "parent-child robotic system" is carried and deployed as a cargo of a large-scale helicopter onto a disaster site. (This part is out of this project.)
2. The parent robot navigates the outdoor and acquires 3-dimensional range data of the environment to construct a map of the site.
3. If a half collapsed building in which some victims may exist is identified, the parent robot approaches and deploys child robots from the opening of the building. Using an extendable ramp, the child robots can be deployed from a window on a higher floor (see Fig 2).
4. The child robots explore inside the building to map the environment and to find victims.
5. Both of the control commands and the obtained data of the site are transmitted via wireless network from and to the operators located in a safe, remote site.

Three key issues to realize the above scenario are summarized as, (1) **development of a robotic system** that can be deployed in the disaster site and can perform surveillance tasks remotely, (2) **development of the augmented reality technology** to display 3-dimensional geometry of the disaster site in a fashion that remote human operators can grasp the images of the site with reality and, (3) **development of a combined satellite-based IP communication and local ad-hoc networking system** to secure the control and data transmission of the robots. Generally, satellite-based communication links have the

**Fig. 2.** An overview of the parent robot "MURROS-Grande" and its ramp

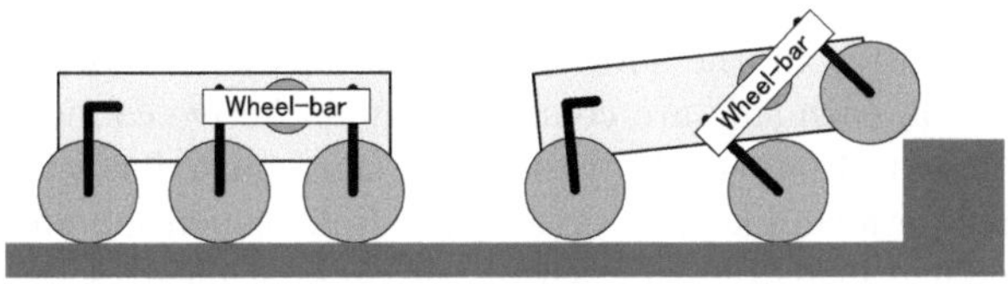


**Fig. 3.** Description of the wheel-bar of the parent robot

disadvantage of a lower transmission bandwidth and time-delay, however they have greater advantage given by the fact that the satellites are not damaged by the disasters on the ground. A concept figure is depicted in Fig.1.

Each research institute in this project is assigned to one of the above issues as sub-goals. Finally, the project will perform the demonstrations of the integrated technologies.

In this paper, interim progress reports on each sub-goal are made from Section 2 to Section 4. Preliminary integrated performance tests of the robots are summarized in Section 5.

## 2 Parent-Child Robotic System (Sub-goal #1)

### 2.1 System Description of the Parent Robot

The parent robot named MURROS (Multi-Purpose Rescue Robot System)-Grande is required to negotiate uneven terrain such as a boulder field or rubble strewn surface, and to locate itself to deploy the child robots into higher floors of the collapsed buildings. Therefore, the Grande is designed to have six wheels with a linkage suspension and carry an extendable ramp, then developed by IHI Aerospace Co., Ltd. An overview of the robot is shown in Fig.2.

The locomotion is based on an electrical 6-wheel-drive and a 6-wheel-steering, with the maximum traveling velocity of 4.0 [km/h]. The dimensions of the robot are 2.3 [m] in length, 1.3 [m] in width, and 0.8 [m] in height. It mounts the three-segmented extendable ramp which can deploy child-robots up to the height of 5 [m]. The maximum length of the ramp is about 7 [m]. The Grande weighs about 300 [kg] without the ramp.

**Fig. 4.** The child robots “MURROS-Piccolo” (left) and mounted on parent robot (right)

One of the features in the locomotion mechanism of the Grande is that front two wheels on each side are connected by an active connecting rod called wheel-bar. If the wheel-bars on both sides are harmonized as with a differential mechanism, they perform like a passive rocker-link suspension and contribute to equalize the vertical loads on each wheel. Bt making most of its active motion capability, the Grande can go over 0.5 [m] steps as shown in Fig.3.

As for the mapping and localization, the robot equips a 3-dimensional image and range sensor unit (shown in Section 3.4). It also mounts GPS receivers, inclination sensors and rate gyros.

## 2.2 System Description of the Child Robots

The child robot named MURROS-Piccolo is required to be capable of climbing up/down stairs and overcoming obstacles to search victims inside the buildings or houses. Therefore, the locomotion mechanism with segmented multi-tracks designed by Technocraft was used as a mobile base (6 tracks in total). An overview of the Piccolo is shown in Fig.4.

Also as shown in Fig.5, by changing the up/down angles of the front and rear pair of sub-tracks, the Piccolo can overcome an obstacle. The dimensions of the Piccolo are 1.0 [m] in length, 0.5 [m] in width, and 0.5 [m] in height. It weighs about 30 [kg]. The maximum speed of locomotion is about 1.8 [km/h], skid-steer is used for steering maneuvers.

As for the sensors, the robot equips a 3-dimensional range sensor unit, a conventional pan-tilt web-camera, inclination sensors and rate gyros. Particularly, the 3-dimensional range sensor is useful in the teleoperation system discussed in Section 2.3.

**Fig. 5.** The child robot overcoming an obstacle

## 2.3 Teleoperation Using 3-Dimensional Maps

The time delay and narrow bandwidth are major drawbacks in the satellite-based communication. In such situation, transmitting the video stream is not a

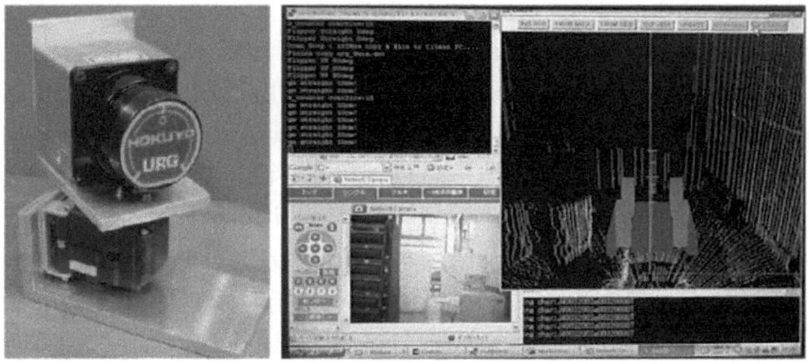

**Fig. 6.** The 3-dimensional range sensor (left) and the operator's GUI (right)

smart idea since it requires a huge bandwidth though its content information is very dilute. On the other hand, 3D ranging date is much more informative and useful for the display of the robot working site to remote operators.

The left picture in the Fig. 6 depicts the range sensor unit mounted on the MURROS-Piccolo. The sensor unit is composed of a small laser range finder (URG-04LX, Hokuyo Co. Ltd.) and a Smart Motor (DX-117, Dynamixel) that rotates the laser range finder around the vertical axis in 180 [deg]. It can obtain range data in every 0.35 [deg] in both vertical and horizontal directions, and the maximum measuring range is about 4.0 [m]. The scanning speed is about 1 [min / 180 deg], but it can be increased by rotating the the Smart Motor faster although the horizontal resolution will be decreased accordingly.

For the teleoperation of the robot, "Stop-and-Go" strategy was taken to cope with the bandwidth limitation and time delay issue. The procedure is as follows:

S1: [robot] Stop and obtain a local 3-dimensional map by the laser sensor. (hereafter called "3D-map").
S2: [robot → operator] Send the 3D-map data to the operator.
S3: [operator] Display the 3D-map from an arbitrary view point and superimpose a virtual robot model on it.
S4: [operator] Drive the virtual robot using an input device such as a joystick in the 3D-map.
S5: [operator → robot] The motion trace of the virtual robot is sent to the real robot as its reference trajectory.
S6: [robot] Move along the given trajectory based on the on-board odomeory and rate gyro measurment.
S7: Go back to S1 to obtain a new 3D-map.

The right side of the Fig. 6 depicts the operator's GUI for the teleoperation of the child robot. Based on the above procedure, the operator controls the real robot as if controling the virtual robot in the virtual environment.

Advantages of this strategy are as follows: (1) The operator can choose any view direction and any zoom-size of the virtual environment. In general, the images of a bird-eye view is much more useful and comfortable than those of an on-board camera. (2) The proposed procedure requires far small amount of data to transmit, comparing to the video stream. (3) The proposed procedure can cope with any amount of time delay. For example, some 10 to 30 minutes are needed for the remote driving of a Mars rover, but the proposed procedure is still effective.

There are some points to be improved: (1) The robot motion control in the step S6 relies on the onboard odometory, but when the wheels or tracks slip, the odometory becomes unreliable. Particularly, the skid-steering of the tracked robot generates large odometry errors. The problem of track's slippage is well described in [4], and we already proposed how to improve the reliability of the odometry for the tracked robots. Details are in [5]. (2) The mapped area can be extended by combining multiple 3d-maps. For this purpose, an improved scan-matching technique should be applied. (3) The virtual environment represented by a 3D-map looks artificial, not realistic. To improve the reality, the technique to superimpose real picture images onto the map segments as texture is proposed in the following section.

## 3 Omnivision and 3-Dimensional Range Sensor Unit for Effective Telepresence (Sub-goal #2)

### 3.1 Development of a Hanging Mechanism

When a robot runs on bumpy terrains, the acquired map data are tilted if the sensor is fixed to the robot's body, then the transformation is necessary to fit in a vertical-horizontal coordinate system. In the proposed prototype model, a "hanging mechanism" was developed using a passive gimbal mechanism (shown in Fig. 7) which consists of two rotary-disk-dampers (FDT-47A, Fuji-Latex Co., Ltd). By hanging the sensor unit, it passively keeps the sensor's pose vertically and horizontally by its own weight.

**Fig. 7.** Hanging mechanism

### 3.2 Non-isotropic Omnidirectional Imaging System

Considering both high-speed locomotion and surveillance tasks, high-resolution image and wide-angle-view are effective, but both are under the trade-off usually. To achieve these two features simultaneously, non-isotropic omnidirectional imaging system has been developed in Osaka university [6]. The sensor has the Horizontally fixed viewpoints Biconical Paraboloidal (HBP) mirror which generates an inhomogeneous angular resolution. It enables the operator obtain not only an omni-directional scene but high-resolution views in a particular direction (in our case, in front). Figure 8 shows an overview of the sensor and an obtained image. We can see that the resolution at the front view is larger than at the side view (by comparison between the characters "HBP").

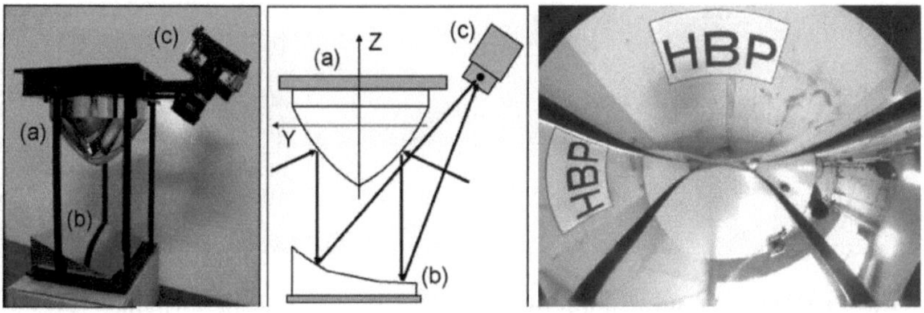


**Fig. 8.** Non-isotropic omnidirectional imaging system and obtained image

**Fig. 9.** Proposed sensor unit and obtained 3-dimensional information

### 3.3 The 3-Dimensional Range Sensor for Outdoor

To obtain 3D map data in outdoor environment, a scanning system is developed, which has almost the same design as indoor-type shown in Section 2.3, but SICK LMS291 is used as the laser range finder. The scanning speed is about 1 [min / 180 deg]. In this case, the sensor acquires range data in every 0.5 [deg] in both vertical and horizontal direction, and the maximum range is about 30 [m].

### 3.4 Integration of the Sensor Unit

Integrating the above sub-components, "Omni-Vision and 3-dimensional Range Sensor Unit" was completed [7]. In Fig. 9, the sensor unit mounted on the parent robot is shown on the left, and an example of the textured 3-dimensional map is shown on the right.

## 4 Combined Satellite and Local Ad-Hoc Network System (Sub-goal #3)

### 4.1 Engineering Test Satellite VIII

The Japanese Engineering Test Satellite VIII (ETS-VIII), developed by Japan Aerospace Exploration Agency (JAXA), is a technology demonstration satellite

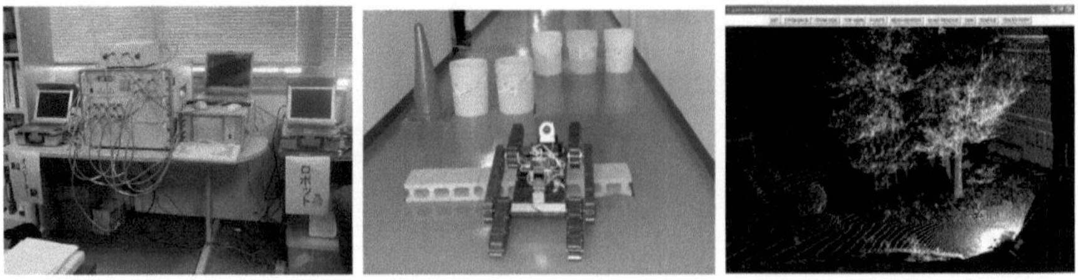

**Fig. 10.** The satellite simulator and the teleoperation performance tests

for providing IP-based wireless network connection all over Japan and nearby countries. This satellite works as a repeater station between two portable terminals. One is located at the operators' side, and the other is located in disaster site. The satellite has the capability of up to 1.5 [Mbps] communication link, but in the experiments, we anticipate much lower bandwidth.

### 4.2 Wireless Ad-Hoc Network

In this research, commercial ad-hoc network systems (produced by Decentra and Thinktube) were tested and evaluated for the operation of the child robots using multi-hop capabilities.

### 4.3 Performance Test of Teleoperation Using Satellite Simulator

As the ETS-VIII is not opened for the user experiments, the system performances were tested using the satellite communication simulator provided by JAXA. The appearance of the simulator is shown in the left of Fig. 10. It simulate the transmission throughput under the bandwidth limitation, time delay, and packet loss in a realistic situation.

Using the simulator, three basic performance tests were carried out on December, 2006. The first test was to measure the effective time delay in transmitting data files. Packet-collision problems were observed in the case of a communication rate of 768 [Kbps] or lower, when sending 2 [MB] data of the map data with 5.0 [sec] intervals. Depending on the collision, the time delay varied between 1 and 2 seconds.

The second test was to remotely operate the MURROS-Piccolo in the indoor environment. The setup for the test is shown at the center of Fig. 10. To operate the robot, the 3D-map-based teleoperation system described in Section 2.3) was used. Ten people (that includes three skilled operators) performed a short course teleoperation based only the virtual model in the 3D-map without any other clues.

During the performance test, we observed that a conventional velocity-based control (the traveling speed of the robot is given by the tilt-angle of the joystick-bar) was very hard even for experienced operators in the case of unstable 1-2 [sec] time delay. Therefore, in the performance test, we updated the control scheme from the velocity-based one to an "inching" position control. In this new control interface, if the operator clicks a button once the robot is commanded to move a fixed distance (10 [cm] in our case) or certain rotational angle (10 [deg] in our case). After that, total navigation time and safety were improved drastically. We did not perform a quantitative test yet, but obviously, the result suggests that a position-based-control has a big advantage under time-delay conditions.

The third test was to remotely operate a robot in the outdoor environment. The robot carries the 3-dimensional range sensor described in Section 3.3 and a pair of RTK-GPS for the localization. The right side picture of Fig. 10 is one example of the 3D-map obtained in the outdoor environment, and the position and pose of the robot is superimposed based on the RTK-GPS data. The teleoperation by skilled operator was succeeded in navigating the robot slowly. However, sometimes, the GPS denied cases were caused by the occlusion and multi-pass of the GPS satellite signals due to the surrounding buildings, then the accuracy was disturbed in the localization.

## 5 Performance Test of the Proposed Parent-Child Robotic System

To confirm the proposed parent-child robotic system for disaster mitigation tasks, an open-demonstration was organized on March 10th, 2006. It was performed in a field and a building of IHI Aerospace Co., Ltd. in Kawagoe, Saitama, Japan. In this performance test, the following scenario was demonstrated.

1. The parent robot, MURROS-Grande is controlled through a visual feedback.
2. The Grande performs the acquisition of 3-dimensional range data and image to construct a textured 3D-map.
3. The Grande approaches a building and extends the extendable ramp to the second floor's opening in the height of 4.2[m].
4. Two child robots, MURROS-Piccolos climb the ramp to enter the building.
5. Two Piccolos are operated for remote searching mission with constructing 3D-maps. One tries to climb up the staircases and the other goes down the stairs.

At the time of this performance test, as the ad-hoc network system was not implemented yet, a conventional wireless LAN (IEEE 802.11 b/g) was preliminary set inside the building.

The parent robot, Grande, succeeded in negotiating with bumps and gaps in the outdoor environment, and deploying two child robots, Piccolos, onto the 2nd floor of the building. (A picture is shown in Fig. 2). One of the two child robots climbed up to 3rd floor through the stairs and found a dummy victim successfully. The other child robot stepped down to the 1st floor smoothly.

The whole scenario was performed as expected, then we concluded that the installation of the basic functions for the proposed robotic system was successfully completed, except the ad-hoc network system.

## 6 Conclusion

In this paper, the research project named "networked robotic system for disaster mitigation" was introduced. The project has three sub-goals, which are (1) the development of parent-child robotic system, (2) the development of omni-vision and 3-dimensional range sensor unit, and (3) the development of a combined satellite-based IP communication and local ad-hoc networking system. This paper described the system concepts and state-of-the-art of the project. Finally, the performance tests at the open demonstration on March 10th, 2006 was summarized along with the proposed mission scenario. The performance tests validate the advanced technologies for rough terrain mobility in both outdoor and indoor environments, 3D ranging data acquisition and 3D-map-based teleoperation, which were developed in this projects and applicable to many other mobile robots for the search and rescue missions.

## References

1. Special issue of advanced robotics: Advanced research and development of robotics for search and rescue. Journal of Advanced Robotics 19(3), 219–347 (2005)
2. Vargas, A.E.M., Mizuuchi, K., Endo, D., Rohmer, E., Nagatani, K., Yoshida, K.: Development of a networked robotic system for disaster mitigation –navigation system based on 3d geometry acquisition–. In: Proceedings of the 2006 IEEE International Conference on Intelligent Robots and Systems (IROS 2006), Beijing, China, pp. 4821–4826 (2006)
3. Murphy, R., Ausmus, M., Bugajska, M., Ellis, T., Johnson, T., Kelley, N., Kiefer, J., Pollock, L.: Marsupial-like mobile robot societies. IEEE Intelligent Systems 15(3), 364–365 (1999)
4. Wong, J.Y.: Theory of Ground Vehicles. John Wiley & Sons, Chichester (1978)
5. Nagatani, K., Endo, D., Yoshida, K.: Improvement of odometry accuracy of crawler vehicle with consideration of slippage. In: IEEE International Conference on Robotics and Automation (2007) (page to be appeared)
6. Kondo, K., Yagi, Y., Yachida, M.: Non-isotropic omnidirectional imaging system for an autonomous mobile robot. In: IEEE International Conference on Robotics and Automation, pp. 18–22 (2005)
7. Saitoh, K., Machida, T., Kiyokawa, K., Takemura, H.: A 2d-3d integrated interface for mobile robot control using omnidirectional images and 3d geometric models. In: IEEE and ACM International Sympoisium on Mixed Augmented Reality (ISMAR), pp. 173–176 (2006)

# Cooperative Localization for Multi-robot Incorporating Proprioceptive/Exteroceptive Position Sensors*

Jihong Lee[1], Kyounghwan Jo[2], and ChoulSoo Jang[3]

[1,2] BK21 Mechatronics Group at Chungnam National University, Daejeon, Republic of Korea
jihong@cnu.ac.kr, neoview@cnu.ac.kr
[3] Intelligent Robot Research Division at Electronics and Telecommunications Research Institute, Daejeon, Republic of Korea
jangcs@etri.re.kr

**Summary.** This paper presents a new method of cooperative localization for multiple robots utilizing correlation between GPS errors of common mode in shared workspace. Assuming that GPS data of individual robot are correlated strongly as the distance between robots are close, we utilize the differential position data between the robots to refine robot's position data. Under artificial environment for simulation with imposed model error to robot motion and GPS sensor data error, it is confirmed that the proposed method provides improved localization accuracy [9]. In addition, we present a practical solution to accumulated position error in traveling long distance.

## 1 Introduction

Mobile robots require capability to estimate their position in order to navigate autonomously in their work space. Consequently, localization by sensor-based method has been researched as one of the most essential problems in mobile robotics. In previous researches, a number of works on localization of single robot have been reported [1, 2]. However recently, many robotic applications require that robots work in collaboration in common workspace to perform a task. In such tasks as multiple robots operate in close, we need more precise absolute localization or relative localization of multiple robots in order to avoid collisions with each other.

The multiple robot system, in comparison with the single robot system, has the advantages of collecting and integrating multiple sensor data from different robots. Accordingly, the system can obtain better localization performance and increase the robustness of the localization accuracy for each robot by fusing collected multiple sensor data. In addition, if each robot is equipped with heterogeneous sensor, the system can improve the overall localization accuracy.

Despite of above-mentioned advantages, most existing localization researches for multiple robots have not utilized such advantages [3,4]. Even in multiple robot systems each robot estimates its position by its own sensor data.

* This work was partly supported by the IT R&D program of MIC/IITA [2005-S-033-03, Embedded Component Technology and Standardization for URC].

C. Laugier and R. Siegwart (Eds.): Field and Service Robotics, STAR 42, pp. 463–471, 2008.
springerlink.com 

Recently, some research works start to focus on integrating the sensor data from multiple sources in order to remove the uncertainty of the absolute or relative position of mobile robot. Most of research commonly used methods include triangulation [7], Kalman filter [5], and MCL (Monte Carlo Localization) [8]. These approaches have improved position accuracy as contrast with the methods utilizing information from single robot. On the other hand, these approaches are relatively complex because they deal with statistic parameters. In addition, they are usually applicable to robots in indoor.

In our previous research, we presented an approach for cooperative multiple robot localization utilizing correlation between GPS errors of robots [9]. The proposed method was relatively simple compared with previously methods (include triangulation, Kalman filter, and MCL) and provided better localization accuracy compared with relying on the resource of each robot.

In this paper, we propose a practical solution to accumulated position error of cooperative multiple robot localization in traveling long distance. We define two operational parameters. One is the ratio of standard deviation between proprioceptive and exteroceptive position sensors. The other is a refresh interval to replace accumulated DR data with GPS data providing fixed standard deviation of position error. Determining the two operational parameters, we can improve the cooperative localization accuracy of multiple robots.

## 2 Cooperative Localization

### 2.1 Assumptions

In order to deal with localization problem for multiple robots, we assume the followings:

1) Each robot has GPS receiver, odometer, and gyro sensor in order to localize absolute position and relative position by model-based dead-reckoning.
2) The closer robots have the stronger correlation between GPS data errors.
3) All robots are equipped with communication devices that allow other robots or remote control station to receive position data measured from each robot.

### 2.2 Cooperative Localization Algorithm

Based on above-mentioned assumptions, we describe mathematical details of the proposed cooperative localization algorithm with the nomenclature shown in Table 1.

If $m$ robots work in common space, the position of $k$th robot estimated by robot model at time $i$+1 as:

$$\hat{P}_{i+1}^{k} = \overline{P}_{i}^{k} + (V_{i+1} + e_{i+1})\delta t \qquad (1)$$

for $k = 1,\ldots,m$. And GPS data $\widetilde{P}_{i+1}^{k}$ are sampled at time $i$+1.

**Table 1.** Nomenclature for Cooperative Localization Algorithm

| Symbol | Description |
|---|---|
| $\overline{P}_i^k$ | Position of *k*th robot estimated by cooperative localization algorithm at time *i* |
| $\hat{P}_i^k$ | Position of *k*th robot estimated by robot model at time *i* |
| $\widetilde{P}_i^k$ | Position of *k*th robot estimated by GPS data at time *i* |
| $V_i$ | True velocity of robot between of time *i*-1 and time *i* |
| $e_i$ | Velocity error caused by slip between robot wheels and ground between of time *i*-1 and time *i* |
| $\delta t$ | Sampling period |
| $W_i^{kj}$ | Weight factor for compensation between *k*th robot and *j*th robots at time *i* |
| $\Delta P_i^k$ | GPS data error of *k*th robot estimated by difference between GPS data and robot model at time *i* |
| $l_i^{kj}$ | Differential position data between *k*th robot and *j*th robot at time *i* |
| $L_i^k$ | Parameter determined from total distance between robots at time *i* |

Then, the key idea of proposed cooperative localization algorithm is described as follows:

$$\overline{P}_{i+1}^k = \widetilde{P}_{i+1}^k - \sum_{j=1}^{m} (W_{i+1}^{kj} \Delta P_{i+1}^k) \tag{2}$$

$$W_{i+1}^{kj} = \frac{1/(l_{i+1}^{kj} + 1)}{L_{i+1}^k} \tag{3}$$

$$L_{i+1}^k = \sum_{j=1}^{m} 1/(l_{i+1}^{kj} + 1) \tag{4}$$

$$l_{i+1}^{kj} = | \hat{P}_{i+1}^k - \hat{P}_{i+1}^j | \tag{5}$$

$$\Delta P_{i+1}^k = \widetilde{P}_{i+1}^k - \hat{P}_{i+1}^k \tag{6}$$

for $k = 1,\ldots,m$.

The main idea is motivated from the concept of DGPS which compensates error by comparing true data and noisy sensed data. In this approach we adopt the quasi true value as $\hat{P}_i^k$ and noisy data as $\tilde{P}_i^k$ . And we find the cases where our algorithm can be applied in view point of statistic parameters of the sensors.

Both the weight term, $W_{i+1}^{kj}$, and the reciprocal term of total distance between robots, $L_{i+1}^k$, include $1/(l_{i+1}^{kj}+1)$ term. When compensate GPS errors, we fuse the compensation quantity according to the distance between the robots, $1/(l_{i+1}^{kj}+1)$. Note that the closer robots contribute more than the farther robots in the fusion.

### 2.3 Refresh Interval

Most mobile robot applications employ two basic position estimation methods: absolute/relative positioning or proprioceptive/exteroceptive position. Absolute positioning methods usually rely on landmark, map matching, satellite-based GPS data, or navigation beacons. In this paper, we assume that robots can receive the GPS data for absolute positioning, but it doesn't care the other proprioceptive sensors.

Relative positioning method is usually Dead Reckoning (DR) based on odometer and gyro. Odometer and gyro are inexpensive, simple, and easy to accomplish in real time. On the other hand, odometer has its unbounded accumulation of errors, and also gyro has relatively large drift rates which cause unbounded growth in orientation errors. Therefore, it is necessary to update absolute position periodically in order to reduce potential for unbounded growth of errors.

In proposed method, we have found the proper period called refresh interval to update GPS position data. The localization performance of the proposed method can be improved by replacing $\hat{P}_i^k$ with $\tilde{P}_i^k$ , when errors of the $\hat{P}_i^k$ are bigger than errors of the $\tilde{P}_i^k$ .

## 3 Simulation Results

In this section, we simulate the proposed cooperative localization algorithm according to the virtual position. The traveling space for multiple robots is a square 40 meter on a side. We limited the maximum gap between robots to 1 meter in order to avoid collisions between the robots. The robots for simulation were assumed to cylindrical shape with a radius of 50 centimeter. The maximum velocity of robots limits 1 meter per second. There are 10 mobile robots traveling randomly on a 2D-flat platform.

We compare the proposed cooperative localization (CL) method for multiple robots with the single robot localization (SL) method by fusing dead reckoning data and the GPS data relying on the resource of each robot. In addition, we present examples to prevent position data estimated by proposed method from accumulating errors.

### 3.1 Virtual Position Error

In order to simulate the proposed cooperative localization method, it is necessary to construct simulation environment of a piece assumptions mentioned in section 2. We have to make two kinds of errors. One is the GPS errors of all robots respectively according to probability distribution of proper standard deviation. The other is the model errors of all robots generated by cause such as slip between the wheels of robot and the ground.

First, we assume that the GPS data of individual robot are correlated strongly as the distance between robots are close. In other words, we assume that the GPS data error of individual is correlated linearly as the distance between robots. It is described how to make GPS data error in previous research [9].

Seconds, we assume that robot model error is generated as maximum 0.1m and minimum 0m per 1m in traveling distance. It means that the model errors between 0 and 0.1m are the same probability. It is possible to calculate standard deviation from continuous probability distribution. The calculated standard deviation is about 0.02887m per 1 second. In this paper, we simulate the cooperative localization method using this standard deviation value for model error.

### 3.2 The Ratio of Standard Deviation between Sensors

In previous case studies [9], we have found that the proposed method is valid in the proper ratio. The valid ratio ranges of standard deviation between the DR error and GPS error per unit time are about 1:2 ~ 1:5 in Fig.1.

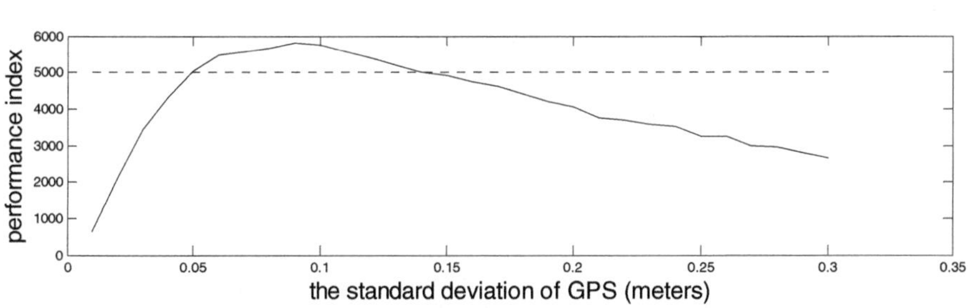


**Fig. 1.** The localization performance of proposed method according to the standard deviation of GPS when the standard deviation of DR is 0.02887m. The range below dot is valid.

### 3.3 Case 1 : $\sigma_{DR} = 0.02887m$, $\sigma_{GPS} = 0.08661m$, Operating Time : 300s, and Refresh Interval : 100s

In this case, there are robots' operating time of 300 seconds and compensating period of 100 second. Fig. 2 and Fig. 4 show the real trajectory and position errors for robot 6 during 300s respectively. Fig. 3 shows the comparison of the accumulated distance

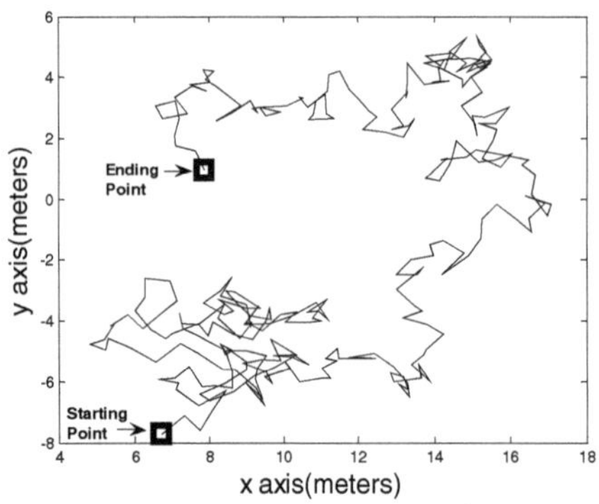


Fig. 2. Trajectory of a robot 10 in case 1

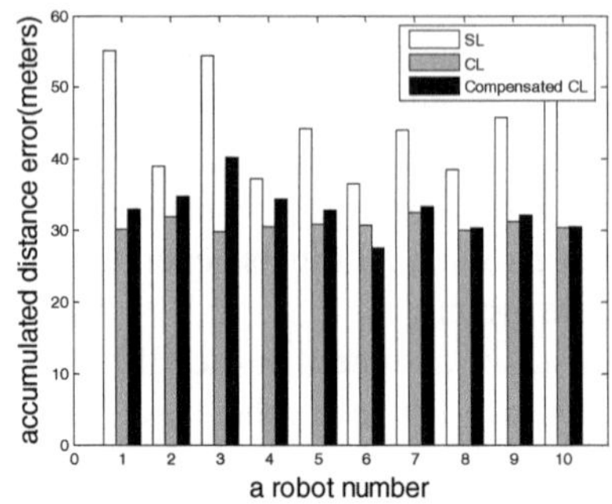


Fig. 3. Accumulated position errors of 10 robots in case 1

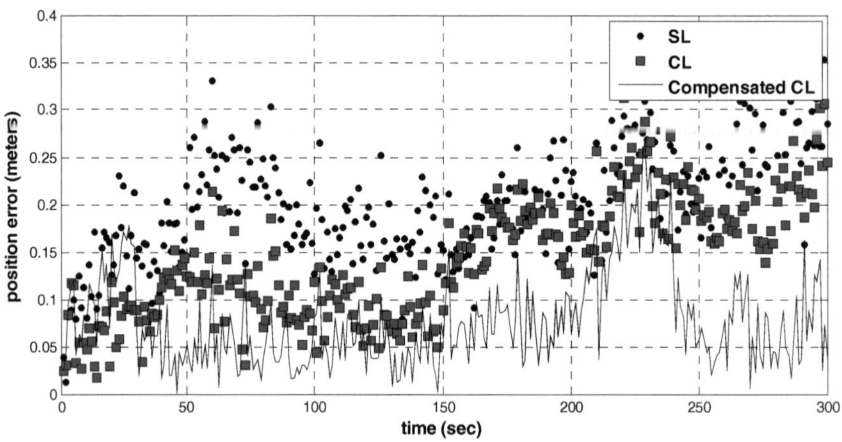


Fig. 4. Position errors of robot 6 in case 1

errors for 10 robots obtained by three different localization methods. From the results of this case, we know that it doesn't suitable for most robots to replace accumulated DR data with simultaneous GPS data per 100 Seconds.

### 3.4 Case 2 : $\sigma_{DR} = 0.02887m$, $\sigma_{GPS} = 0.08661m$, Operating Time : 300s, and Refresh Interval : 50s

In this case, there are robots' operating time of 300 seconds and refresh interval of 50 second. Fig. 5 and Fig. 7 show the real trajectory and position errors for robot 9 respectively. Fig. 6 shows the comparison of the accumulated distance errors for 10 robots obtained by three different localization methods. From the results of this case, we present the effect of replacing DR data accumulated with simultaneous GPS data per 50 Seconds.

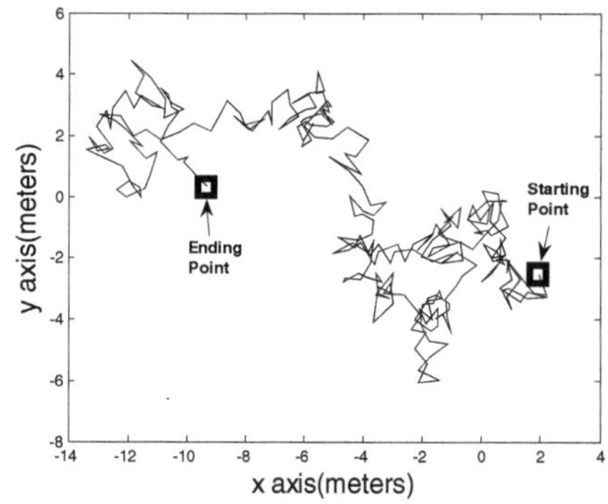


**Fig. 5.** Trajectory of a robot 5 in case 2

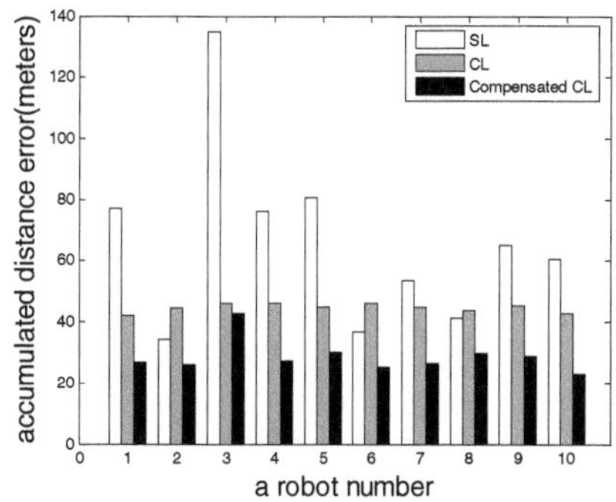


**Fig. 6.** Accumulated position errors of 10 robots in case 2

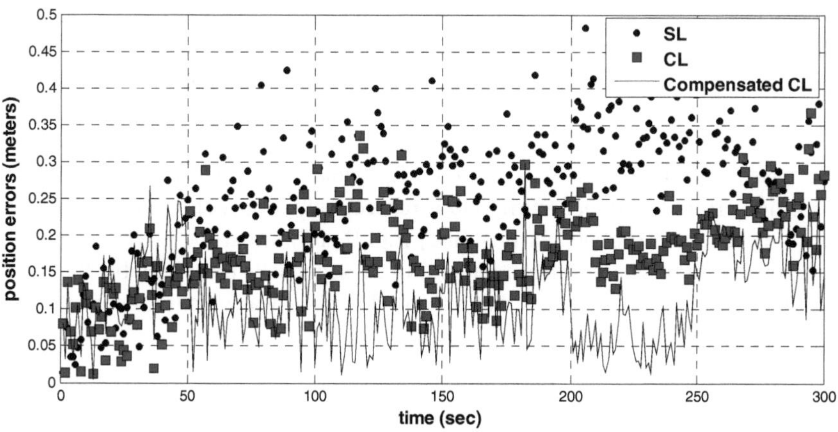


**Fig. 7.** Position errors of robot 9 in case 2

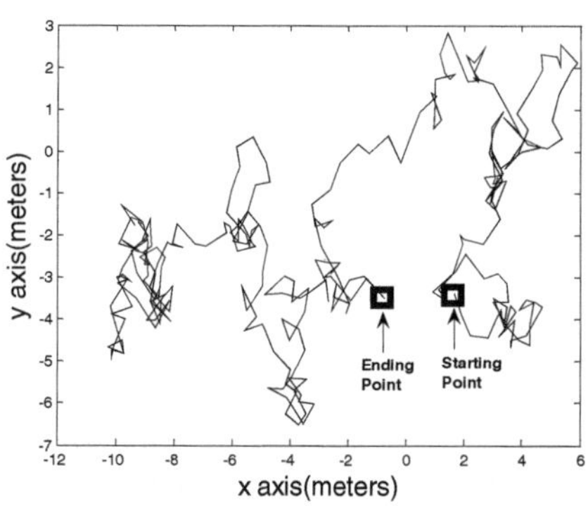


**Fig. 8.** Trajectory of a robot 5 in case 3

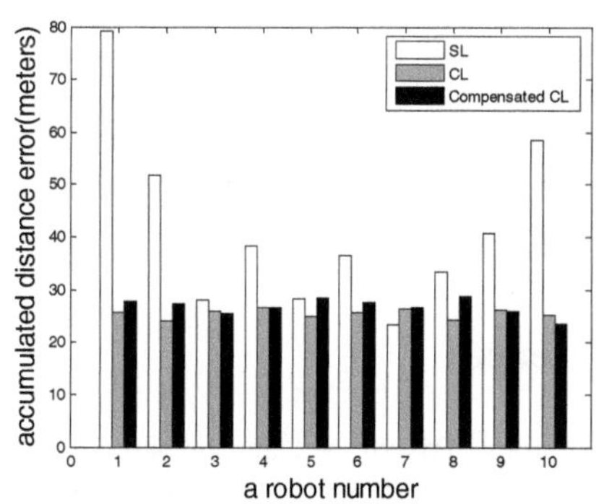


**Fig. 9.** Accumulated position errors of 10 robots in case 3

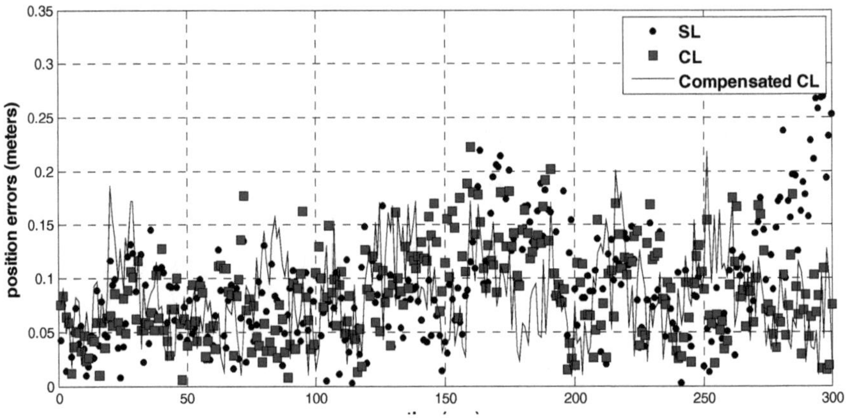


**Fig. 10.** Position errors of robot 3 in case 3

### 3.5 Case 3 : $\sigma_{DR} = 0.02887m$, $\sigma_{GPS} = 0.08661m$, Operating Time : 300s, and Refresh Interval : 10s

In this case, refresh interval is 10 second. Fig. 8 and Fig. 10 show the real trajectory and position errors for robot 3 during 300s respectively. Fig. 9 shows the accumulated distance errors for 10 robots. In this case, Fig. 9 shows that the errors of all robots aren't reduced. In this case, refresh interval of 50 seconds is more suitable than that of 10 seconds. As a result, we know that proposed method was able to improve localization accuracy in proper refresh interval (about 30s ~ 80s).

## 4 Conclusion and the Future Study

In this paper, the proposed method is motivated from the concept of DGPS which utilizes correlation between errors in common mode, and summarized in simple mathematical formula compared with existing methods. Simulation results show that the proposed method can achieve better localization performance in such cases as replacing the accumulated DR data with the GPS data providing fixed standard deviation for position error by proper refresh interval. In the future study, we will apply proposed cooperative localization algorithm to real multiple robot system.

## References

[1] Gourley, Trivedi, M.: Sensor based obstacle avoidance and mapping for fast mobile robots. In: Proc. I994 IEEE Int. Conf. Robotics and Automation, May 8-13, pp. 1306–1311 (1994)

[2] Borenstein, J., Feng, L.: Measurement and correction of systematic odometry errors in mobile robots. IEEE Trans. Robot. Automat. 12, 869–880 (1996)

[3] Ferrari, C., Pagello, E., Ota, J., Arai, T.: Multirobot motion coordination in space and time. Robot. Auton. Syst. 25(3/4), 219–229 (1998)
[4] Borenstein, J., Feng, L.: Gyrodometer: A New Method for Combining Data from Gyros and Odometry in Mobile Robots. In: Proc. 1996 IEEE Int. Conf. Robotics and Automation, April 1996, pp. 423–428 (1996)
[5] Roumeliotis, S.I., Bekey, G.A.: Collective localization: a distributed Kalman filter approach to localization of groups of mobile robots. In: Proc.2000 IEEE Int. Conf. Robotics and Automation, April 24–28, pp. 2958–2965 (2000)
[6] Roumeliotis, S.I., Bekey, G.A.: Distributed multirobot localization. IEEE Trans. Robot. Automat. 18, 780–795 (2002)
[7] Rekleitis, I., Dudek, G., Milios, E.: Experiments in free space triangulation using cooperative localization. In: Proc. 2003 IEEE/RSJ Int. Conf. Intelligent Robots Systems, October 2003, pp. 1777–1782 (2003)
[8] Liu, J., Yuan, K., Zou, W., Yang, Q.: Monte calro multi-robot localization based on grid cells and characteristic particles. In: Proc. 2005 IEEE/ASME Int. Conf. Advanced Intelligent Mechatronics, July 24-28, 2005, pp. 510–515 (2005)
[9] Lee, J., Jo, K.: Cooperative Multi-Robot Localization using Differential Position Data. In: Proc. 2007 IEEE/ASME Int. Conf. Advanced Intelligent Mechatronics (September 4-7, 2007)

# Part XI

# Field Robotics - Systems and Applications

# Experiments in Autonomous Reversing of a Tractor-Trailer System

Cédric Pradalier and Kane Usher

CSIRO ICT Centre, Autonomous Systems Laboratory, Brisbane, Australia
firstname.lastname@csiro.au

**Summary.** Trailer reversing is a problem frequently considered in the literature, usually with fairly complex non-linear control theory based approaches. In this paper, we overview our simple method for stabilizing a tractor-trailer system to a trajectory based on the notion of controlling the hitch-angle of the trailer rather than the steering angle of the tractor. The performance of this control method, as implemented on the CSIRO Autonomous Tractor, is then experimentally compared against a number of human drivers, showing that this method performs as well as a moderately skilled human driver, even though the system is significantly handicapped in terms of steering actuation speed and by errors in localization.

## 1 Introduction

Reversing of tractor-trailer systems is a problem frequently addressed in the literature due to the common, real-world occurrences of this problem, along with the interesting mathematical properties of this type of system. Industrial occurrences for reversing tractor-trailer systems include transportation, cargo handling and agriculture. In the main, professional drivers have very high skill levels and experience little problem with this task. However, there are a number of examples of recreational activities which require a driver to position a trailer, boat and caravan owners to name a few. In Australia, the 'grey-nomad' phenomenom sees a growing number of retirees, a large portion of whom are inexperienced in towing a caravan, embarking on a journey that takes them around the country. It is hard to imagine that throughout this journey that they would not be required to reverse this tractor-trailer system, a task which can be quite arduous for drivers of limited experience and ability. This is but one example in which a driver-aid or automation system could be of significant use.

Most approaches address the problem of reversing a tractor-trailer system through the development of relatively complex, non-linear controllers. Our approach is based on closing an inner-loop on the angle between the tractor and trailer using a simple Proportional Integral (PI) controller, and then closing an outer loop on the trajectory using a path tracker developed for articulated vehicles. This paper presents results from experiments which evaluate the

C. Laugier and R. Siegwart (Eds.): Field and Service Robotics, STAR 42, pp. 475–484, 2008.
springerlink.com 

performance of this approach in long duration reversing tasks, and compares its performance to that of human drivers with a range of skill levels.

### 1.1 Background

In the scientific literature, this problem has been approached in three ways: through a *change of variables* which brings the system into a standard form, *optimal control* methods, and *learning-based* approaches.

The *change of variables* approach relies on a transformation which brings the system into a chained (or similar) form, from which standard non-linear control theory can be applied, see for example [2]. These techniques have the advantage of being theoretically generalizable to any number of trailers. Their main disadvantage, with respect to the work presented in this paper, is their complexity and high sensitivity to numerical approximation. This complexity makes it difficult to implement these methods reliably, and also poses difficulties in the tuning of these systems — after the change of variables, many of the control parameters no longer have a physical meaning making the design of a tuning strategy challenging and somewhat non-intuitive. As for numerical stability, the chained-form-based conversion for instance requires several trigonometric operators and inversions, and the resulting control often requires inversion of matrices which can become singular at some point of the navigation.

The *optimal control* based methods use optimization schemes to derive a sequence of demands which will control the tractor-trailer system onto the desired trajectory or path, see for example [1]. In simple terms, these methods use a vehicle model and a simulation process to compute the control commands that will lead to the best tracking of the trajectory. This requires accurate models of the vehicle's behaviour and a heavy reliance on computational resources. Deviations from the model, which in real-world implementations are inevitable, lead to errors, and these methods are also open-loop meaning that re-planning is necessary to deal with errors in localization.

Finally, the *learning-based* controllers seek to ease the computational burdens of the previous methods by providing a mapping between the current vehicle state, the desired state and the required inputs to reach the desired state, see for example [7]. Essentially, a simulated model of the tractor-trailer system is used to try many different possible methods and parameters. 'Learning' occurs by searching the parameter space for the best set of methods/parameters which are then encoded into, for example, a neural network or look-up table, which maps the current to desired configuration. The main drawback of these techniques is the learning itself: if learning occurs from a model of the system, then errors in the model are clearly problematic; if learning occurs on the real vehicle, then there are clearly safety issues since it is not possible to predict the behaviour of the vehicle in the learning phase. Also, even if the best set of parameters performs well in practice, it is hard to guarantee its performance, which can be problematic in an application where reliability is an issue.

Several patents have been proposed for automatically reversing a tractor-trailer system [10, 5, 6, 3]. However, these patents are focused on reversing on straight trajectories.

### 1.2 Originality

The control scheme we propose and evaluate in this paper has several advantages over the schemes proposed in the scientific and patent literature. It is able to control the tractor-trailer along straight and curved trajectories and it does not require any mechanical modification of the tractor-trailer system other than the ability to sense the hitch-angle. Our control scheme is simple to implement and tune, numerically stable, and easy to predict and understand. Its main drawback is that, in its current state, it is not generalizable to any number of trailers.

## 2 Experimental Platform

The platform used in these experiments is the CSIRO Autonomous Tractor (AT), as shown in Fig. 1(a). It is an Ackerman steered, ride-on mower which has been retro-fitted with an array of actuators, sensors, and a computer system enabling the implementation and testing of control and navigation algorithms. For full details of the vehicle's design, refer to [11].

The trailer hitch-angle is sensed using a set of string-pot encoders. Much of the existing research presumes that the trailer hitch-point lies over the centre of the rear-axle, as seen on a prime-mover and trailer combination. Here we consider the case where the hitch-point is offset from the rear-axle, an example being a standard automobile towing a trailer, which represents a further distinction of this work from the existing literature.

## 3 Hitch-Angle Stabilization

Briefly, the method used to control the tractor-trailer hitch-angle when reversing is simply a PI controller:

$$\phi = K_p(\psi^\star - \psi) + K_I \int_0^t (\psi^\star - \psi) du \tag{1}$$

where $\psi^\star$ is the demanded hitch-angle, $\phi$ is the vehicle steering angle and $K_p$ and $K_I$ are the proportional and integral gains. Due to the non-linear nature of the system, the demand to the system has to be modified in order to reduce the steady state error when stabilizing to non-zero set-points. The demanded hitch-angle $\psi^d$ that will make the system converge to a desired angle $\psi^\star$ cab be approximated by:

$$\psi^d(\psi^\star) = \frac{K_P L_1 - L + K_P L_2}{K_P(L_1 + L_2)} \psi^\star \tag{2}$$

(a) The Autonomous Tractor with its trailer.

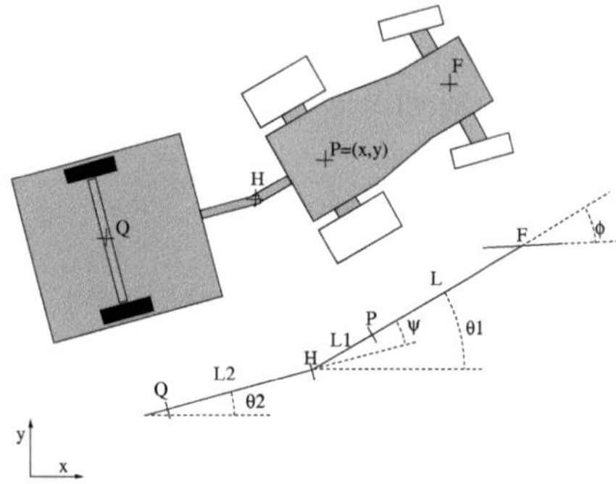


(b) Kinematic model of the tractor-trailer system.

**Fig. 1.** The Autonomous Tractor-Trailer system used in these experiments

and the controller becomes:

$$\phi = K_p(\psi^d(\psi^\star) - \psi) + K_I \int_0^t (\psi^\star - \psi)du \tag{3}$$

In this system, the integral term is not essential for the stability of the control. It is added to account for minor inaccuracies remaining after the proportional control. Such inaccuracies can result from the linearization leading to $\psi^d(\psi^\star)$, or from errors in the vehicle model (e.g $L$, $L_1$, and $L_2$ in fig. 1(b)). The hitch-angle controller is described more completely in [8].

## 4 Stabilization of an Articulated Vehicle to a Trajectory

To stabilize the vehicle to a trajectory – *i.e.* a sequence of vehicle states indexed by time – we use a path tracking controller derived for an articulated vehicle.

Our control law is inspired from [9], which aims at stabilizing the vehicle on a path, i.e. a 2D curve in the plane. It relies on three error measurements, as depicted in Fig. 2(a): $\xi_y$ the lateral error, $\xi_\theta$ the heading error, and $\xi_\kappa$ the curvature error. In the original work, the authors controlled the hitch-angle rate of the articulated vehicle using these measurements. Here we control the hitch-angle and the control law is defined as:

$$\psi = K_y\xi_y + K_\theta\xi_\theta + K_\kappa\xi_\kappa \tag{4}$$

where $K_y$, $K_\theta$ and $K_\kappa$ are tuning parameters. The resulting hitch-angle is then fed to the hitch-angle stabilization law described in Section 3. Speed is controlled using a standard pursuit approach to maintain the vehicle's progression along the trajectory and switching logic is used to deal with any changes of direction, as occurs when the onset of jack-knifing is detected or there is a cusp in the trajectory (see [8] for details). The trajectory-tracking system is described in block diagram form in Fig. 2(b).

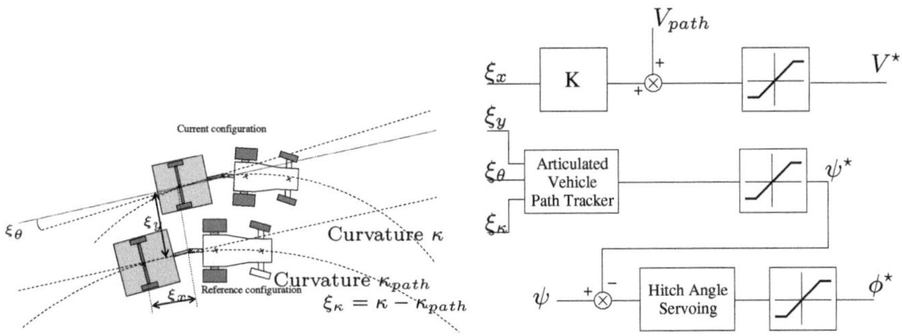


(a) Trajectory error definitions.

(b) Block diagram for the reversing control law.

**Fig. 2.** The trajectory controller and the definition of path errors

## 5 Experimental Results

This section presents results obtained when tracking various reversing trajectories, first using odometry based localization, then several tests using an external localization estimate. The external localization estimates are provided by a particle-filter based method using the vehicle odometry and sparse reflective beacons sensed with a front-mounted laser range-finder (details on the localization system can be found in [4]). This localization estimate is 'drift-free' but comes at the cost of localization discontinuities when corrections are applied on spotting a beacon. Such discontinuities are especially challenging for the trajectory tracking system.

### 5.1 Reversing on a Circle with Fixed Radius

In this first set of experiments, we define reference trajectories as arcs of circles of various radii. The pose of the vehicle is computed using odometry information. The advantage of this setting is that the vehicle pose is a very clean and smooth signal. The disadvantage is that odometry localization is known to drift over time, and consequently, is not suitable for real-life, long range, robotic operations.

Fig. 3 shows tracking results for a circular path, with radii of 7.5 m and 15 m. As can be observed, the tracking is very accurate for radii of 15 m. For a radius of 7.5 m, the maximum achievable curvature is reached and the system cannot converge to the required trajectory. However, it should be noted that the system converges to a stable orbit, which is the best it can do to follow the required curvature.

### 5.2 Reversing on a Badly Planned Trajectory

In this second set of experiments, we stretch the system by introducing a more complex path as depicted in Fig. 4(a). We consider this trajectory "badly

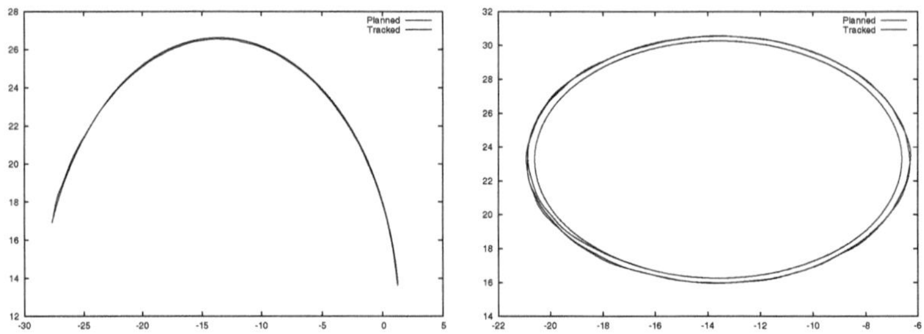


(a) radius = 15 m (b) radius = 7.5 m

**Fig. 3.** Trajectory tracking with odometry based localization

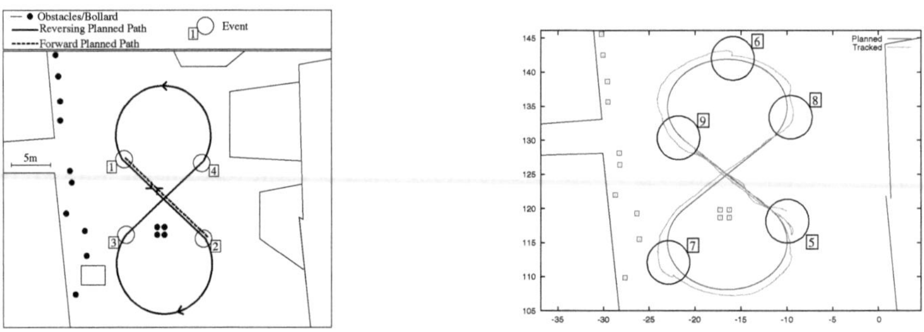


(a) Planned trajectory. (b) The tracked trajectory.

**Fig. 4.** Performance for a poorly planned trajectory which does not account for curvature limitations or path discontinuities. *events* are denoted by the boxed numbers.

planned" since it does not take into account the limitations of the vehicle, especially the limited turning radius and the very strong limitation on the hitch-angle rate. In addition, the trajectory contains path discontinuities in which straight lines join circular segments — such discontinuities are challenging for the control law since they require a very fast change of hitch-angle. Our objective here is to show the performance of our system at the limits of its nominal specification. To further challenge the controller, we also use the external localization system rather than the vehicle odometry.

Fig. 4(b) illustrates how the planned path is tracked by our tractor-trailer systems. From this experiment, we note that on straight line segments the system converges reasonably fast. On the curved segments, the curvature is beyond the vehicle capabilities and the tracked trajectory is offset from the planned one. Transitions between the curved and straight segments of the path (events 5, 8 and 9) introduce a discontinuity which the system struggles to deal with but nevertheless recovers from. Localization discontinuities (events 6 and 7) are handled much more gracefully than the path discontinuities. Finally the effects of the switching logic are illustrated at event 5.

From these experiments, it is clear that in order to obtain very accurate tracking, the path planner needs to take into account the strong constraints of this system, in particular, the small maximum trackable curvature and the need for smooth curvature profiles. However, the trajectory control performed extremely well within these constraints.

### 5.3 Reversing on a Pre-planned Trajectory

In this set of experiments, our objective was to demonstrate the performance of our system in a real-world situation. To this end, we designed a path across our research centre. This path is approximately 170 m long and was planned manually using hand-selected way points and interpolating splines. This approach guarantees a smooth and continuous trajectory. For these experiments, the system relied on the external localization estimates.

The resulting path is depicted in Fig. 5(a). It is composed of two segments: a short forward motion to help align the system with the trajectory, and a long reversing segment of approximately 160 m. The main constraints when designing this trajectory were to keep the vehicle close to the middle of the road, and to minimize the path curvature.

Fig. 5(b) and 5(c) show the trajectories realized by our controller on two seperate runs. More accurately, the trajectories represent the path of the centre of the AT's rear axle. Examples of localization jumps (corrections) can be seen at the top end of all trajectories, and around $(55, 0)$ in Run $B$ and $C$. It is important to note that our system stays stable around the reference trajectory, even when the localization estimation is very unstable. In some situations, e.g. close to $(-10, 10)$ in run $A$ and $C$, or close to $(75, 0)$ in run $B$, the localization estimate oscillates between both sides of the reference trajectory. If this occurs at the wrong frequency, it can bring the system to a jack-knife situation. In all cases, the preliminary signs of this event where detected by the system, and a short forward motion was initiated to realign the vehicle with the reference trajectory.

### 5.4 Evaluation: Comparison with Human Drivers

As a final experimental evaluation of our reversing system, we tested a number of human drivers in performing a trajectory similar to that performed by the autonomous system outlined in Section 5.3. Skill levels of these drivers ranged from novice to professional. For practical reasons, the drivers were required to keep the vehicle in the middle of the road, rather than following the same trajectory as the automation system.

We evaluated their ability to stabilize the trailer, to avoid jack-knife situations, and the statistical properties of their steering input. Figure 5.d shows an example of human controlled trajectory, in comparison with autonomous system ones.

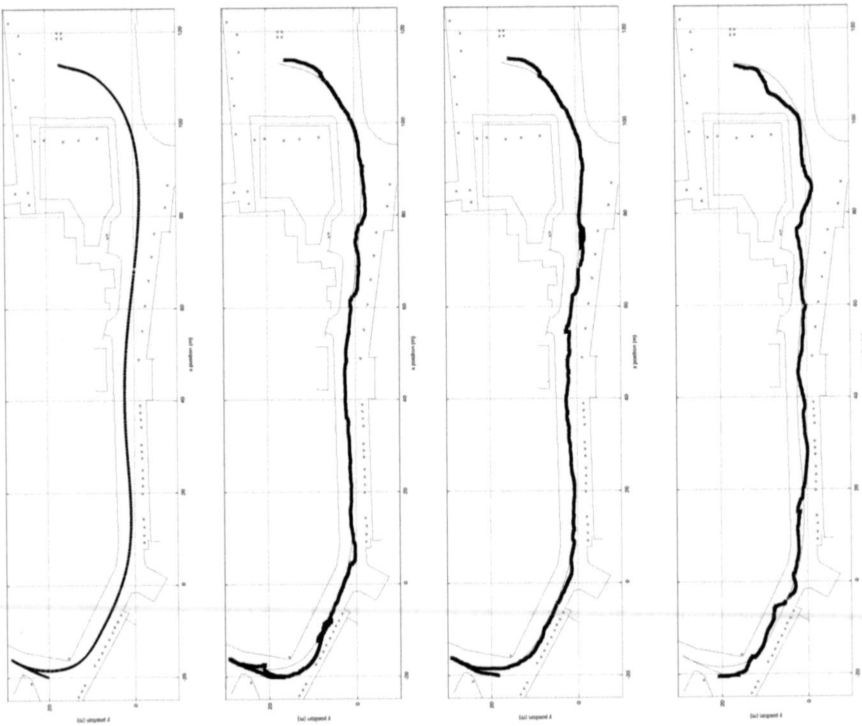

(a) Reference (b) Run A (c) Run B (d) Subject 4.

**Fig. 5.** Reversing trajectories of the autonomous system on two independent runs, compared with a human driver on the same trajectory (Subject 4)

## Hitch-angle Stabilization

Fig. 6(a) depicts the statistical properties of the measured hitch-angle, for each run, as box diagrams. Each line represents one run, with numbers referring to the different subjects. Lines 10 to 12 refer to the autonomous runs. On these box diagrams, the red circles are points considered outliers, in the context of a normal distribution. They are particularly important since they represent measurements resulting from unusual events, such as the correction of a jack-knife situation.

For the autonomous runs we note that the mean hitch angle is slightly shifted toward positive values. This is expected, since the trajectory is close to a long right-hand turn. When comparing the hitch-angle range of the autonomous system with the human drivers, it is clear that the autonomous system keeps much tighter control of the angle than most of the human drivers and is comparable with the professional driver (driver 1). Nevertheless, it must be noted that this performance is achieved at the cost of a much lower ground speed ($0.3m/s$ instead of $1.0m/s$).

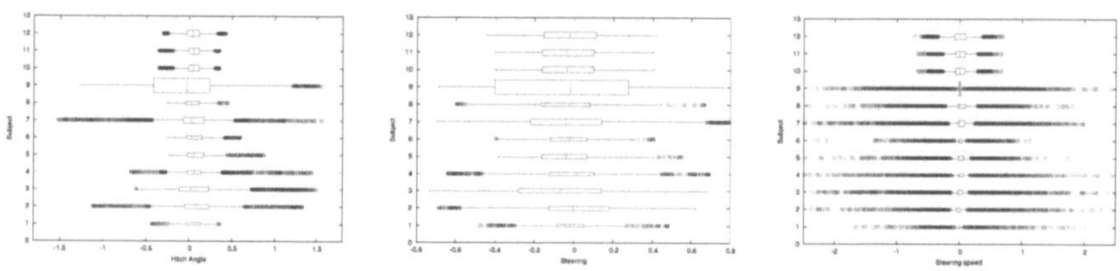

(a) Hitch-angle stabilization. (b) Steering angle input. (c) Steering angle rate.

**Fig. 6.** Comparisons of the capabilities of human drivers and the autonomous system

Fig. 6(b) and 6(c) illustrate the properties of the steering angle and its first derivative, recorded for the human drivers and the autonomous system. These graphs demonstrate the reduced control range and capabilities available to the autonomous system in comparison with the capabilities of human drivers — the accessible steering range available to the automation system is reduced by 40% in comparison to the human drivers and the accessible steering speed and steering acceleration are reduced by about 70%. From these observations, it is clear that human drivers use their additional range of control to drive faster while demonstrating similar hitch-angle stabilization performance.

Finally, it is interesting to note that with driving capabilities reduced by 70%, the performance of our system is similar to a professional driver, albeit with a velocity reduced by about 70%. We are yet to determine if this is a coincidence.

## 6 Conclusion

This article evaluated a new control scheme for a tractor-trailer system. This scheme is based on a two layer control loop: first a hitch-angle stabilization loop controls the angle between tractor and trailer, then a path tracking control loop, initially designed for an articulated mining vehicle, is adapted to our tractor-trailer system.

The main advantage of this approach over traditional methods is its simplicity of implementation. Only a few parameters need to be tuned and they all have a clear physical meaning. Although simple, this control scheme relies on a mathematically sound background.

Finally, this control law has been implemented on a real vehicle and experiments were conducted on challenging trajectories. Given the limited dynamic performance of our platform (slow response time, loose components, low speed actuation), the control law exhibited excellent convergence and stability properties. Furthermore, our controller also compared well with a range of human drivers on a similar trajectory, even though the control system is significantly handicapped in terms of its actuation capabilities.

**Acknowledgments.** This work was funded by the CSIRO ICT Centre under the ROVER and Dependable Field Robotics projects. The authors would like to thank the ASL team for their support. Special thanks go to Polly Alexander, Stephen Brosnan, Peter Corke, Elliot Duff, Paul Flick, Leslie Overs, Jonathan Roberts, Ashley Tews, John Whitham and Graeme Winstanley who all contributed to the development of the AT.

## References

1. Altafini, C., Speranzon, A., Wahlberg, B.: A feedback control scheme for reversing a truck and trailer vehicle. IEEE Transactions on Robotics and Automation (2001)
2. Bolzern, P., DeSantis, R.M., Locatelli, A.: An input-output linearization approach to the control of an n-body articulated vehicle. Journal of Dynamic Systems, Measurement, and Control (2001)
3. Carter, D.N., Lormor, J.R.: Vehicle steering aid system, GB patent GB2398050 (2004)
4. Duff, E., Usher, K., Ridley, P.: Swing loader traffic control. Technical Report ICT 06/142, CSIRO ICT Centre, Brisbane, Australia (January 2006)
5. Kandall, R.G.: Vehicular steering system for reverse paths, US patent US5247442 (1993)
6. Karl, S.: Automatic reverse steering system for passenger vehicles with trailer, DE patent DE3538338 (1987)
7. Koza, J.R.: A genetic approach to finding a controller to back up a tractor-trailer truck. In: Proceedings of the 1992 American Control Conference (1992)
8. Pradalier, C., Usher, K.: A simple and efficient control scheme to reverse a tractor-trailer system on a trajectory. In: Proc. of IEEE Int. Conf. on Robotics and Automation, Roma, Italy (2007)
9. Ridley, P., Corke, P.: Load haul dump vehicle kinematics and control. Journal of Dyn. Sys, Measurement and Control (2003)
10. Robert, S.D.: Trailer backing up device and method, US patent US2004215374 (2004)
11. Usher, K.: Visual homing for a car-like vehicle. PhD thesis, Queensland University of Technology, Brisbane, Australia (June 2005)

# ATRS - A Technology-Based Solution to Automobility for Wheelchair Users

Chao Gao[1], Ira Hoffman[2], Thomas Panzarella[2], and John Spletzer[1]

[1] Lehigh University, Bethlehem, PA, USA
spletzer@cse.lehigh.edu
[2] Freedom Sciences, LLC, Philadelphia, PA, USA
tpanzarella@freedomsciences.com

**Summary.** In this paper, we present the Automated Transport and Retrieval System (ATRS). ATRS represents an alternative to van conversions for automobile drivers with lower body disabilities. It employs robotics and automation technologies that integrate into a standard mini-van or sport utility vehicle (SUV). At the core of ATRS is a "smart" wheelchair that navigates between the driver's position and a powered lift at the rear of the vehicle - eliminating the need for an attendant. From an automation perspective, autonomously docking the wheelchair onto the lift platform presented the most significant technical challenge during system development. This was driven by geometry constraints, which limited clearance between the chair wheels and the lift platform rails. To solve this problem, we employed an LMS291 LIDAR in conjunction with an Extended Kalman Filter for reliable and accurate wheelchair localization. Coupled with a hybrid controller design, the system has proven to be exceptionally robust. This was validated through extensive simulation and experimental results, culminating in a three-day demonstration at the 2006 World Congress and Exposition on Disabilities where the system completed over 300 consecutive cycles without a failure.

## 1 Introduction and Motivation

According to the U.S. Bureau of Transportation, over six million people with disabilities have difficulties in obtaining the transportation they need [1]. This is a major contributor to the unemployment rate of the disabled population nationally, estimated at over 65% by the U.S. Census Bureau [2].

A van conversion currently represents the sole personal transportation solution for an individual in a wheelchair. Van conversions start with a standard van produced by a major automotive manufacturer. The van is subsequently modified or "converted" by another company, usually a specialized mobility equipment manufacturing company or mobility dealer. The modifications are permanent, and include extensive changes to the chassis, frame, and interior. Typical modifications include removing and lowering the vehicle floor, and relocating/replacing major subsystems such as the gas tank, fuel system, and heating/cooling systems of the vehicle [3]. While enabling independent mobility, van conversions represent a costly and unsafe transportation solution for wheelchair users.

To eliminate these shortcomings, we have developed a technology-based alternative to van conversions for wheelchair users: the Automated Transport and Retrieval

C. Laugier and R. Siegwart (Eds.): Field and Service Robotics, STAR 42, pp. 485–494, 2008.
springerlink.com 

System (ATRS). ATRS employs robotics and automation technologies, and can be integrated into a standard mini-van or sport utility vehicle (SUV). At the core of ATRS is a "smart" wheelchair system that autonomously navigates between the driver's position and a powered lift at the rear of the vehicle. A primary benefit of this paradigm is the operator and chair are separated during vehicle operations as well as entry/exit. This eliminates the potential for injuries or deaths caused by both improper securement (as the operator is seated in a crash tested seat system) as well as lift malfunctions. Furthermore, by eliminating the *drastic* and *permanent* vehicle modifications associated with van conversions, ATRS will cost significantly less.

## 2 Related Work

Extensive work has been done in order to increase the safety levels of power wheelchairs while minimizing the level of human intervention. In these systems, the human operator is responsible for high-level decisions while the low-level control of the wheelchair is autonomous.

The Tin Man system [4], developed by the KISS Institute, automates some of the navigation and steering operations for indoor environments. The Wheelesley project [5], based on a Tin Man wheelchair, is designed for both indoor and outdoor environments. The chair is controlled through a graphical user interface that has successfully been integrated with an eye tracking system and with single switch scanning as input methods. The TAO Project [6] provides basic collision avoidance, navigation in an indoor corridor, and passage through narrow doorways. The system also provides landmark-based navigation that requires a topological map of the environment. The NavChair assistive wheelchair navigation system [7] uses feedback from ultrasonic sensors and offers obstacle avoidance, door passage, and wall following modes. More recently, the SmartChair [8] uses a virtual interface displayed by an on-board projection system to implement a shared control framework that enables the user to interact with the wheelchair while it is performing an autonomous task.

A common theme in the above research is the robotics technology has been applied to assist or augment the skills of the chair operator. In contrast, the ATRS wheelchair is in fact capable of autonomous vehicle navigation in outdoor environments. This can be realized because the operator is never seated in the chair during autonomous operations, and the chair always operates in the vicinity of the automobile. The former constraint mitigates operator safety issues, while the latter provides significant, invariant landmarks/features in an otherwise unstructured environment.

## 3 System Overview

In describing the ATRS operational procedures, we refer to Figure 1. When the operator returns to his/her automobile, a keyless entry is used to both unlock the vehicle and to deploy the traversing driver's seat. The operator then positions the wheelchair, and performs a seat-to-seat transfer (pose A). After this, the wheelchair is deployed to the rear of the vehicle (pose B). In our proof-of-concept system, this side traversal was

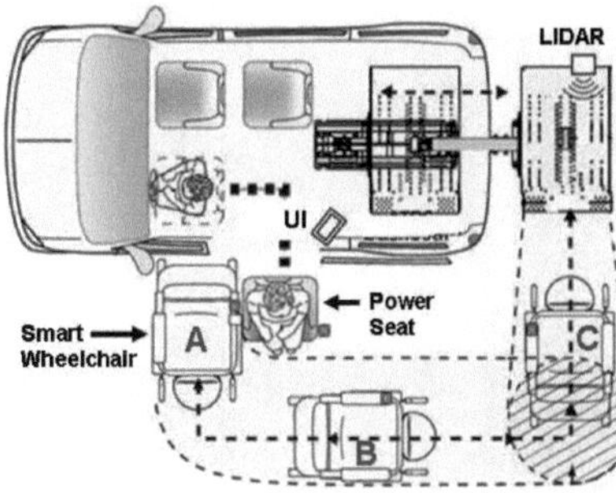


**Fig. 1.** ATRS concept diagram illustrating the primary system components

completely autonomous [9]. In the current system – referred to colloquially as "ATRS-Lite" – the wheelchair is remotely controlled by the vehicle operator via a joystick located at the UI. Once the chair enters the LIDAR's field-of-view at the rear of the vehicle (pose C), it is automatically tracked. The UI then cues the operator to place the wheelchair into "docking" mode. This enables the van-side computer to transmit real-time control inputs to the chair over a dedicated RF link for reliable docking (locking in place) onto the lift platform (pose D). With the chair docked, the operator actuates the lift via the UI – stowing the platform and chair into the vehicle cargo area. The process is repeated in reverse when disembarking from the automobile. We should emphasize that when not operating autonomously, the ATRS wheelchair is placed in "manual mode," and operates no differently than any other powered wheelchair

The primary focus of this paper is the development of a reliable, autonomous means for docking (and undocking) the ATRS wheelchair onto (and off of) the vehicle's lift platform. Our current approach employs a SICK LMS291-S14 LIDAR system in conjunction with an Extended Kalman Filter (EKF) for estimating the chair pose. EKF techniques for feature-based mobile robot localization were pioneered by Durrant-Whyte and others (*e.g.*, [10]), and such an approach has also proven to be well suited for our application.

## 4 Wheelchair Localization

To reliably execute docking under a broad range of environmental conditions, wheelchair localization requires both robust feature segmentation as well as accurate pose estimation with respect to the lift platform. The platform mounted LMS291 LIDAR system provides bearing, range, and reflectivity measurements that are leveraged for robust feature segmentation. The positions of these features - in conjunction with the control inputs to the chair - are then used as input to an Extended Kalman Filter that estimates the wheelchair pose over time. Details of the localization process are as follows.

### 4.1 Feature Segmentation

The primary sensor used for estimating the wheelchair pose with respect to the lift platform is a SICK LMS291 LIDAR. Figure 2 (left) illustrates a typical integration of

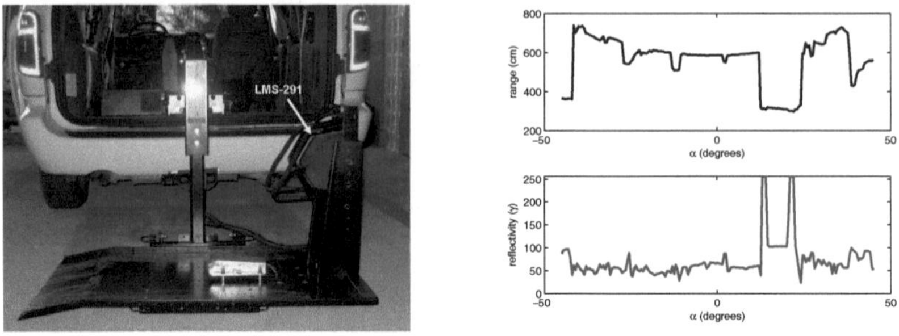


**Fig. 2.** (left) The $Tracker^{TM}$ lift platform with integrated LMS291. The LIDAR housing provides both environmental protection, as well as improves the system SNR. (right) Range (top) and reflectivity (bottom) data from a single LIDAR scan. As was typical, reflectivity measurements alone were adequate for segmenting the two target features.

the LIDAR into the vehicle lift platform. The LMS291 measures the line-of-sight range to objects in the environment over a 90° degree field of view with a discretization of $0.5°$. Each of these measurements can be written as a tuple of the form $\mathbf{z}_m = [r, \alpha, \gamma]^T_m$, $m = 0...180$, where $r_m$ and $\gamma_m$ denote the measured range to and reflectivity of the $m^{th}$ feature at a bearing of $\alpha_m = \frac{m}{2} - 45°$ with respect to the LIDAR sensor frame $\mathcal{L}$. We exploit the reflectivity measurements to greatly simplify the segmentation problem. A pair of target features $\{t_l, t_r\}$ made from retro-reflective material are permanently affixed to the wheelchair chassis. When imaged by the LIDAR, a significant portion of the incident beam is reflected directly back to the detector, saturating the photo-diode (Figure 2 (left)). This allows a simple threshold on reflectivity $\gamma_{min}$ to be used as the primary filter for segmenting the target features. An additional level of filtering is based upon a range constraint $r_{max}$. As the wheelchair is presented in the immediate vicinity of the lift platform, targets at excessive ranges (*e.g.* $>$ 4 meters away) can immediately be disqualified from potential features of interest. From these two filters and assuming a ground plane constraint, we construct a valid feature set

$$F = \begin{bmatrix} r_n \cos \alpha_n \\ r_n \sin \alpha_n \end{bmatrix}, \; s.t. \; r_n < r_{max}, \; \gamma_n > \gamma_{min} \tag{1}$$

A final level of filtering exploits a priori knowledge of the relative geometry of the wheelchair targets. Candidate targets $T \supseteq \{t_l, t_r\}$ are identified via clustering the candidate feature set $F$ in Euclidean space using the actual target size (plus a tolerance) as a constraint. Pairs of targets $t_i, t_j \in T$ are then examined using the actual target distance $||t_l - t_r||$ as a binary filter to identify the correct target pair in the LIDAR scan. If a valid target pair cannot be identified, the operator is alerted to take corrective action (*i.e.*, reposition the chair) and the process repeated.

### 4.2 Pose Estimation

An Extended Kalman Filter (EKF) is used to estimate the chair pose. The prediction step employs feedback from the on-board encoders, while the correction step leverages the target positions recovered from the segmentation process. The process model for the EKF is based upon the standard kinematic motion model for the differential-drive wheelchair, and is not covered here in the interest of brevity. Instead, we focus upon the measurement update phase of the EKF where position estimates of the two retro-reflective targets are used to correct the pose estimate. A straight-forward approach to modeling the LIDAR measurements would be to derive the measurement Jacobian $H$ based upon the range and bearing measurements. There are potential shortcomings with such an approach. The corresponding measurement equations (for bearings in particular) are highly non-linear, and are not well modeled by a first-order approximation. Ultimately, this could lead to poor filter performance. Instead, with each LIDAR scan we directly estimate the position of the features in the world frame. This is straightforward from the available LIDAR measurements

$$\mathbf{z}(t_k) = \begin{bmatrix} x_l \\ y_l \end{bmatrix}_{t_k} + r_i \begin{bmatrix} \cos(\alpha_i + \theta_l) \\ \sin(\alpha_i + \theta_l) \end{bmatrix}_{t_k} \qquad i \in \{1, 2\} \tag{2}$$

where $[z_{x_i}, z_{y_i}]^T$ corresponds to the position of the $i^{th}$ feature in our world frame $\mathcal{W}$, $[x_l, y_l, \theta_l]^T$ reflect the coordinate transformation from the LIDAR sensor frame $\mathcal{L}$ to $\mathcal{W}$, and $(r_i, \alpha_i)$ denote the range and bearing measurements to the $i^{th}$ target with respect to $\mathcal{L}$. These feature positions are then are used directly by the filter. This is akin to using GPS measurements in the filter rather than the raw range measurements from which they are formed - a common practice in mobile robotics. The corresponding measurement equations for the filter can then be written as

$$\mathbf{h}_i(t_k) = \begin{bmatrix} x \\ y \end{bmatrix}_{t_k} + \begin{bmatrix} \cos\theta & -\sin\theta \\ \sin\theta & \cos\theta \end{bmatrix}_{t_k} \begin{bmatrix} a_i \\ b_i \end{bmatrix} \tag{3}$$

where $[a_i, b_i]^T$ denotes the fixed position of the $i^{th}$ target in the robot frame $\mathcal{R}$, and $\theta$ corresponds to the *predicted* wheelchair orientation at the $k^{th}$ time-step. The resulting measurement Jacobian is

$$H_i(t_k) = \begin{bmatrix} 1 & 0 & -a_i \sin\theta - b_i \cos\theta \\ 0 & 1 & a_i \cos\theta - b_i \sin\theta \end{bmatrix} \tag{4}$$

One issue remaining is properly modeling the covariance of the resulting sensor measurements. Each range and bearing measurement defines a new coordinate frame with basis vectors $u_1 = [\cos\alpha_i, \sin\alpha_i]^T$, $u_2 = [\sin\alpha_i, -\cos\alpha_i]^T$. The uncertainty in the $u_1$ direction corresponds directly to the variance of the range measurement $\sigma_{u_1}^2 = \sigma_r^2$. Uncertainty in the $u_2$ direction is a function of the uncertainty of the bearing angle measurement where $\sigma_{u_2}^2 = r_i^2 \sin^2 \sigma_\alpha$. Noting that for the LMS291, $\sigma_\alpha$ is in fact quite small ($< 1°$), this can be very well approximated by $\sigma_{u_2}^2 = r_i^2 \sigma_\alpha^2$. Since this transformation is linear in terms of $\sigma_\alpha$, we expect the characteristics of the (assumed Gaussian) noise of the bearing measurements to be preserved.

The last step is to transform the corresponding covariance matrix to $\mathcal{W}$. This can readily be accomplished through the similarity transform

$$R_i(t_k) = T_i(t_k) \begin{bmatrix} \sigma_r^2 & 0 \\ 0 & r_i^2 \sigma_\alpha^2 \end{bmatrix} T_i(t_k)^T \tag{5}$$

where

$$T_i(t_k) = \begin{bmatrix} \cos(\alpha_i + \theta_l) & -\sin(\alpha_i + \theta_l) \\ \sin(\alpha_i + \theta_l) & \cos(\alpha_i + \theta_l) \end{bmatrix}_{t_k} \tag{6}$$

and where $R_i$ is the measurement covariance for the $i^{th}$ target. The correction step for the EKF can then be written in standard form.

## 5 Wheelchair Control

From an automation perspective, two aspects to the control problem must be considered: *motor* control and *motion* control. The motion controller generates higher level velocity commands vanside based upon the current chair pose as estimated via the localization system presented in Section 4. These are in turn transmitted to the powerchair over a dedicated RF link, which regulates the wheel velocities to achieve the objective linear and angular velocities for docking.

### 5.1 Motor Control

The wheelchair employs a differential drive system where the motion controller transmits objective linear and angular velocities, which are in turn mapped to wheel velocities. These are then regulated via a PID controller implemented in software on the chair's embedded PC. Feedback to the PID is provided via high-resolution quadrature encoders that measure right and left wheel travel $(\Delta\phi_L, \Delta\phi_R)$ at 100Hz. To properly model motor controller performance, experiments were conducted to characterize the latency between controller input and chair actuation. These trials indicated typical latencies of 500-600ms between motor controller input and wheelchair actuation. When considering the nominal linear velocity of the chair was 40 cm/s, and the clearances associated with docking were on the order of 4 cm, such a delay was significant. This influenced the design of the motion controller, as discussed in the sequel.

### 5.2 Motion Control

Motion controller design was influenced by real-world constraints associated with system use. These included not only the controller latency, but also docking clearances and the constrained ground area adjacent to the vehicle for navigation. As such, our motion planner employed a hybrid control design consisting of two primary controller modes: course-correction, and path-following. In this paradigm, gross y-position errors were first corrected (when necessary) in the course-correction phase before proceeding to path-following for docking. We now describe each mode in greater detail.

**Path-Following Phase**

Path-following was the primary controller mode, and the one used for docking onto the lift platform. It employs a traditional PD controller derived using I/O feedback linearization techniques

$$\omega = -k_v \tan\theta - \frac{k_p y}{v \cos\theta} \tag{7}$$

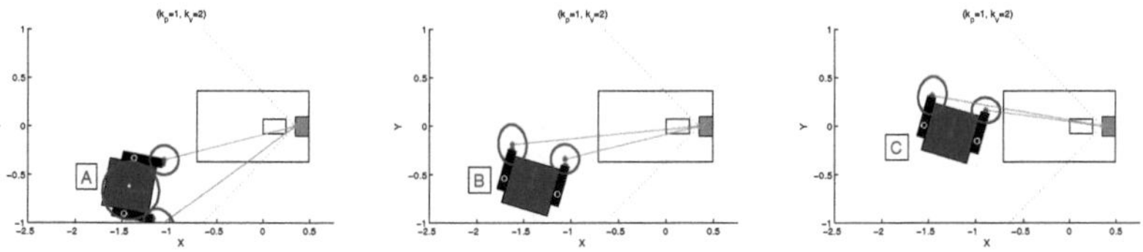


**Fig. 3.** Docking trial simulation illustrating controller course-correction to eliminate gross y-position errors. At position C, the wheelchair switches to path-following mode for final docking. In this example, $\theta_{max} = 75°$.

where $\omega$ and $v$ are the desired linear and angular velocities transmitted to the chair, $v$ is assumed piecewise constant, and $k_v, k_p$ are positive controller gains. For safety considerations, we specify maximum linear and angular velocities $[v_{max}, \omega_{max}]^T$ for the chair. Typical values were 0.4 m/s and 0.8 rad/s in practice (significantly less than what can be achieved by the actual hardware). To accommodate these limits, we borrow from [11] and constrain the actual controller inputs to

$$\omega_{act} = \mathcal{S}(\omega) \arg\min\{|\omega|, \omega_{max}\} \ , \ v_{act} = \tfrac{\omega_{act}}{\omega} v_{max} \tag{8}$$

where $\mathcal{S}()$ corresponds to the *sign* function. These constraints ensure that while the wheelchair will no longer follow the same trajectory specified by (7), it will follow the same path while protecting against actuator saturation.

One further refinement was made immediately preceding path-follwing mode by an orientation correction. The intent was to find an initial orientation $\theta^*$ such that the magnitude of $\omega_0$ is minimized - and ideally zero. From (7), we obtain two possible solutions:

$$\theta^* = \left\{ -\arcsin\left(\frac{k_p y}{k_v v}\right), -\arcsin\left(\frac{k_v v}{k_p y}\right) \right\} \tag{9}$$

So, for the case where $|(k_p y)/(k_v v)| \leq 1$ there is an initial orientation for our path follower that requires zero initial angular velocity. Fortunately, our configuration parameters allow such an orientation to be readily achieved. Thus, all initial orientation error can be removed prior to initializing the path-follower controller.

### Course-Correction Phase

To enhance ATRS docking reliability, a course-correction mode is also incorporated to address gross y-position errors. This controller phase is activated after initial localization in autonomous mode *only if* it is determined that the path-following mode would be at risk for failing to dock the chair at the handoff location provided by the operator (*e.g.*, for large y-position errors with the chair left too near the lift platform). In this event, we again exploit the chair's two degrees of mobility to align the chair along the x-axis in our world frame. This is accomplished through the following set of control inputs.

$$d\theta_R = -\theta_{max}\mathcal{S}(y_0) - \theta_0 \ , \ dx_R = \frac{abs(y_0)}{\sin\theta_{max}} \tag{10}$$

where $y_0, \theta_0$ denote the initial y-position and orientation of the robot, respectively, and $\theta_{max}$ corresponds to a maximum allowable orientation angle for the wheelchair that ensures both retro-reflective targets will be visible to the LIDAR (somewhat less than 90°). The effect of these inputs is illustrated in simulation at Figure 3. The first reorients the wheelchair from its initial pose (A) to one more normal to the x-axis (B). After this, the chair translates to approximately $y = 0$ (C). At this point, the controller switches to path-following, where the initial orientation correction will reorient the wheelchair for docking. The net result is a dramatic reduction in the settling distance, which also reduces the impact of latency on controller robustness.

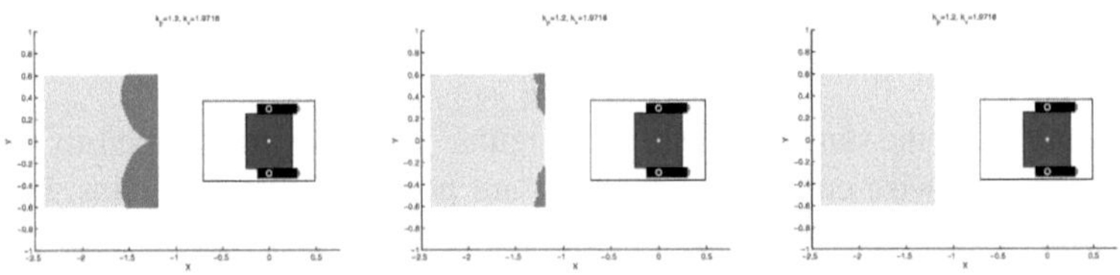

**Fig. 4.** Controller performance with the PD control law (left), orientation correction integrated (center), and course-correction added (right). The latter eliminated residual poses associated with docking failure. Simulation resolution was 1 $cm^2$.

## 6 Simulation Results

Docking performance was first extensively evaluated in simulation. Our simulation model echoed the real-world system characteristics to the extent possible. It integrated the EKF for localization, and included (incompletely) modeled estimates for controller latency, process noise in the odometry system, and measurement noise for the LIDAR. Monte-Carlo simulations were then run over the range of feasible poses to identify potential failure conditions for docking. These were also useful in demonstrating the evolution of the controller strategy.

Figure 4 illustrates the results of a Monte-Carlo simulation used to assess controller performance. Green and red dots reflected whether the corresponding initial position resulted in a successful or failed dock attempt, respectively. For purposes of this simulation, success corresponded to orientation and y-position errors less than 10° and 5 cm, respectively, before the chair reached the ramp of the lift platform. Figure 4 (left) reflects the performance of the path-follower component from (7) alone. Controller performance is improved when path-following is preceded by an initial orientation correction (center), while all failure modes were eliminated with the integration of the course-correction component (right).

As a testament to the fidelity of the simulation, the gains used for the path-follower controller on the actual vehicle were nearly identical to those obtained through the simulation process.

## 7 Experimental Results and Future Work

Over the past six months, the beta ATRS has been tested across a range of conditions. This included three days of continuous demonstrations at the World Congress Exposition on Disabilities (WCD 2006) in November 2006. Conference participants were also

**Fig. 5.** ATRS demonstration at WCD2006. The operator performs a seat-to-seat transfer (left), and then remotely controls the wheelchair to the vicinity of the lift platform (center). At this point, the LIDAR tracks the chair while sending real-time control inputs over a dedicated RF link. Autonomous operations conclude with the chair successfully docked (right).

given the opportunity to test the system. Over 300 cycles of docking and undocking were conducted during this time without a single failure. A sample trial is illustrated at Figure 5. A video of this same trial can be viewed at *http://vader.cse.lehigh.edu/*

While we are optimistic that this framework is sufficiently robust under real-world conditions, additional work remains. In the shorter term, this includes integrating a gyroscope with the odometry system to detect wheel slippage, as well as actuating the LIDAR pitch to relax our ground-plane assumption and reduce the fiducial size. In the longer term, we are also investigating an active vision system to be used in conjunction with (or as an alternative to) the LMS291.

## Acknowledgments

Special thanks to the entire ATRS Team, especially Mike Martin and Tom Panzarella, Sr. (Cook Technologies). Our research results are based upon work supported by the National Science Foundation Partnerships for Innovation (PFI) Program under Grant No. 0650115. Any opinions, findings, and conclusions or recommendations expressed in this material are those of the author(s) and do not necessarily reflect the views of the National Science Foundation. This project was also funded in part by a grant from the Commonwealth of Pennsylvania, Department of Community and Economic Development.

## References

1. National Council on Disabilities. In: The Current State of Transportation for People with Disabilities in the United States, 1331 F Street, NW, Suite 850, Washington, D.C. 20004 (June 2005)
2. Stern, S., Brault, M.: Disability Data From the American Community Survey: A Brief Examination of the Effects of Question Redesign in 2003 (January 2005)
3. Wood, T.: When your plan is a van: Decisions, Decisions A Guide to Asking the Right Questions and Choosing the Best Vehicle for You. Quest - Journal of the Muscular Dystrophy Association 11(2) (March-April 2004)
4. Miller, D., Slack, M.: Design and testing of a low-cost robotic wheelchair prototype. Autonomous Robots 1(3) (1995)

5. Yanco, H.: Wheelesley, a robotic wheelchair system: indoor navigation and user interface. In: Assistive Technology and Artificial Intelligence. LNCS (LNAI), pp. 256–268 (1998)
6. Gomi, T., Griffith, A.: Developing intelligent wheelchairs for the handicapped. In: Assitive Technology and Artificial Intelligence. LNCS (LNAI), pp. 150–178 (1998)
7. Simpson, R., Levine, S.: Automatic adaptation in the NavChair assistive wheelchair navigation system. IEEE Transactions on Rehabilitation Engineering 7(4), 452–463 (1999)
8. Parikh, S.P., Grassi, V., Kumar, V., Okamoto, J.: Incorporating user inputs in motion planning for a smart wheelchair. In: Proc. of the IEEE International Conference on Robotics and Automation (ICRA), New Orleans, LA, pp. 2043–2048 (2004)
9. Sermeno-Villalta, H., Spletzer, J.: Vision-based Control of a Smart Wheelchair for the Automated Transport and Retrieval System. In: Proceedings of the 2006 IEEE International Conference on Robotics and Automation (ICRA 2006), Orlando, FL, USA (May 2006)
10. Leonard, J., Durrant-Whyte, H.F.: Mobile robot localization by tracking geometric beacons. IEEE Trans. on Robotics and Automation 7(3), 376–382 (1991)
11. Oriolo, G., Luca, A., Vendittelli, M.: Wmr control via dynamic feedback linearization: Design (2002)

# Design and Validation of a Mobile Robot for Power Line Inspection and Maintenance

Serge Montambault[1] and Nicolas Pouliot[2]

Hydro-Québec Research Institute (IREQ) 1740 Lionel-Boulet Blvd,
Varennes Québec, Canada, J3X 1S1
[1] montambault.serge@ireq.ca, [2] pouliot.nicolas@ireq.ca

**Summary.** This paper presents the LineScout Technology, a mobile teleoperated robot for power line inspection and maintenance. Optimizing several geometric parameters achieved a compact design that was successfully tested over many line configurations and obstacle sequences. An overview of the technology is presented, including a description of the control strategy, followed by a section focusing on key aspects of the prototype thorough validation. Working on live lines, up to 735 kV and 1,000 A, means that the technology must be robust to electromagnetic interference. The third generation prototype, tested in laboratory and in field conditions, is now ready to undertake inspection pilot projects.

## 1 Introduction

Power line inspection and maintenance activities have lagged behind in the use of robotic technologies even though power line maintenance work must be carried out routinely in hostile environments and hard-to-reach locations. Working on energized lines is now a necessity for most maintenance jobs. Transmission grid owners are looking towards innovation as transmission systems are under pressure from increasing loads and stringent reliability requirements [1-2].

Although some tasks have been successfully automated [3] and more research teams have been active in mobile robotics in recent years [4-5], most research results published in the past are based on concepts or prototype technologies that never made it onto transmission networks [6-7].

Following an extensive review of technologies applied to transmission line inspection and maintenance [8], the authors undertook the development of LineScout Technology as part of a Hydro-Québec research program that aims at providing robotic technologies to perform inspection and maintenance on its transmission lines [9]. From the very beginning, the key objective was to design a robust, reliable and field oriented mobile platform that would be applicable to real inspection and maintenance tasks, aligned with power line maintenance teams' considerations.

## 2 System Requirements

A complete review of transmission line components and circuit configurations was conducted in order to identify the wide variety of obstacles that must be cleared by the robot. Among these, warning spheres (0.76-m diameter, Fig. 1A) and corona rings

C. Laugier and R. Siegwart (Eds.): Field and Service Robotics, STAR 42, pp. 495–504, 2008.
springerlink.com 

(Fig. 1B) were chosen as the longest obstacle to cross. Other examples of critical components identified are double insulator string configurations and vibration dampers potentially damaged by excessive vibration (Fig. 1C, 1D).

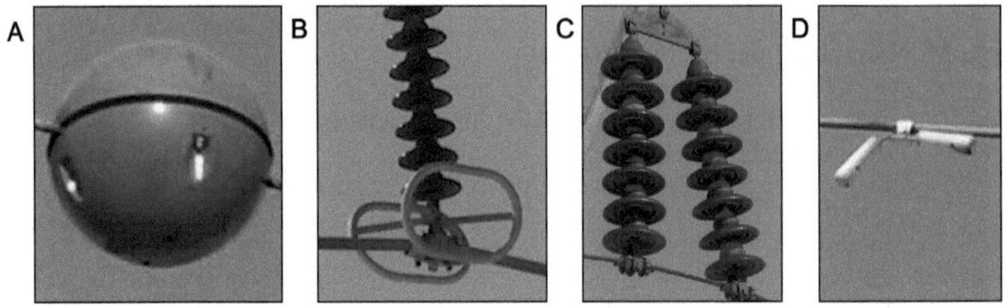


**Fig. 1.** Identified obstacles

Since inspection must be performed on energized lines, one of the most important requirements, besides being robust to electromagnetic interference (EMI), is to keep the size and weight of the platform to a strict minimum in order to maintain a safe dielectric distance from ground and other circuits. Table 1 shows other constraints and target values.

**Table 1.** Design requirements

| Line components | Value |
|---|---|
| Conductor or splice diameter | 12 – 85 mm |
| Maximum obstacle length | 0.76 m |
| Maximum conductor temperature | 95.0°C |
| Number of conductors | 1 – 4 |
| Maximum slope in span | 30° |
| Ambient operating temperature | 0°C – 40°C |
| Platform | |
| Weight | 100 kg |
| Length x Height | 1.37m x 0.75m |
| Traction force | 500 N |
| Linear speed | 1.0 m/s |
| Battery life | 5.0 hours |
| Communication signal range | 5.0 km |
| EMI robustness | 735 kV, 1000 |

# 3 Design of the Technology

## 3.1 Mechanical Concept

Using one conductor as a support, rolling with wheels for locomotion not only allows moving quickly and efficiently along the power line, but also makes it possible to roll over some obstacles (compression splices, vibration dampers). To clear other types of obstacles, the approach chosen uses the sequence schematized in Fig. 2.

LineScout is built around three independent frames (Fig. 2): the wheel frame (dark frame), which includes two motorized rubber wheels called “traction wheels”, the arm frame (light frame), with two arms and two grippers, and the center frame (white circle), which links together the first two frames (called “extremity frames”) and allows them to slide and pivot. As an obstacle is reached, the arm frame is deployed so that the two arms and grippers can temporarily support the robot while the wheel frame is transferred on the other side of the obstacle. To do so, the wheels themselves are flipped down under the obstacle.

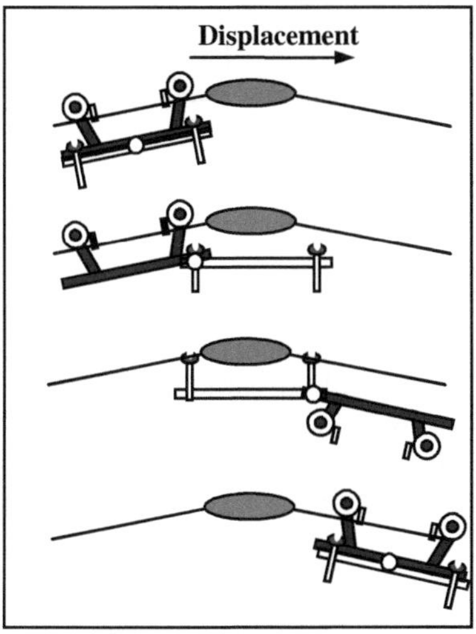


**Fig. 2.** Obstacle-clearing sequence

**Fig. 3.** The LineScout mobile platform

The development of a solution based on the approach in Fig. 2 resulted in the prototype shown in Fig. 3. Besides the traction wheels, the wheel frame includes a camera mounted on a pan-and-tilt unit as well as a pair of safety rollers for platform stability when crossing obstacles. The center frame supports the electronics cabinet and battery pack. Finally, it is the arm frame, which includes two other cameras on pan-and-tilt units, that supports most of the application modules mounted on the mobile platform.

The center frame is a strategic system of the mobile platform since it generates the movement of the two other frames by sliding them in opposite directions, while supporting about 40% of the platform’s weight. By doing so, a gradual weight transfer is obtained during the obstacle clearance sequence and this greatly minimizes the cantilevered load applied when extremity frames are in total extension, as in Fig. 4. Therefore, the center frame has to be stiff for predictable platform behavior yet flexible enough to adapt to particular situations, like a change in direction of the conductor at a suspension clamp. Some amount of passive flexibility between the frames was therefore maintained.

**Fig. 4.** LineScout about to clear a spacer damper

### 3.2 Geometrical Optimization

One of the main advantages of this concept in comparison to other existing prototype is that it is neither limited to a specific size of obstacle nor to a specific distance between adjacent obstacles, making it versatile and adaptable to various sequence of obstacles. In fact, six different clearing sequences were identified as strategies for crossing consecutive obstacles, depending on their sizes and the distance between them. Three of these variants are presented in Fig. 5. Also on this figure, $Lo_{MAX}$ is

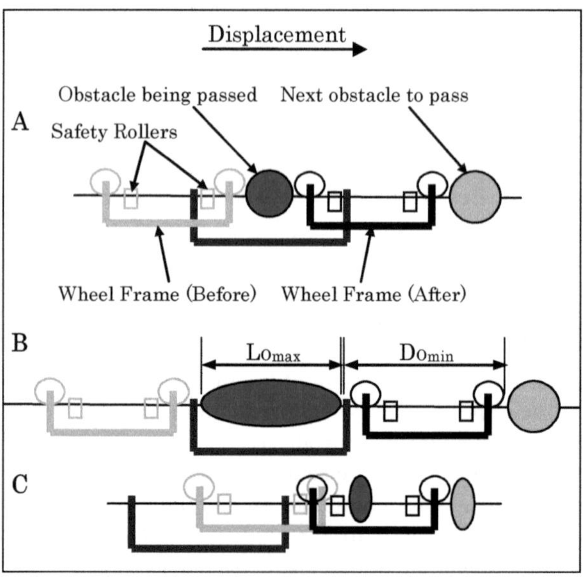


**Fig. 5.** Examples of clearing sequences

**Table 2.** Critical geometric parameters

| Parameter description | Symbol |
|---|---|
| External diameter of the traction wheels | Dt |
| Wheelbase of the traction wheels | Et |
| Wheelbase of the safety rollers | Er |
| Wheelbase of the grippers | Ea |
| Distance from conductor to center of rotation | Lz |
| Translation range of movement | Tmax |
| Rotation range of movement | Rmax |
| Grippers' range of movement | Amax |

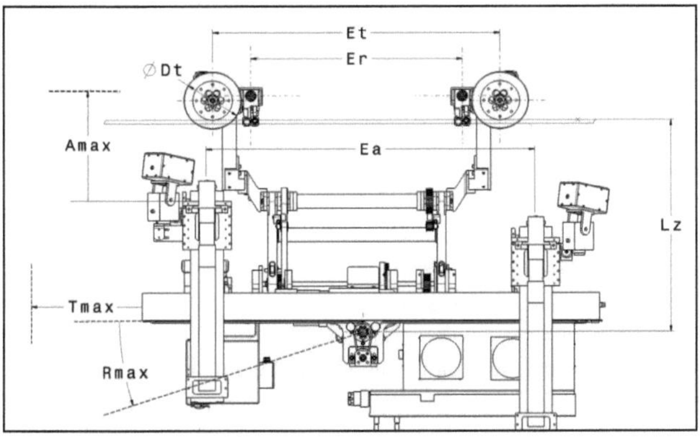


**Fig. 6.** Critical parameters of the mobile platform

defined as the maximum length of a crossable obstacle, and $Do_{MIN}$ is defined as the minimum distance between two consecutive obstacles.

In order to optimize the LineScout geometry according to the range of expected possibilities, eight key parameters were defined, as summarized by Table 2 and Fig. 6. Values of the critical geometric parameters of LineScout have then been optimized based the six possible sequences for clearing line component configurations and based on the expected relation between $Lo_{MAX}$ and $Do_{MIN}$.

This optimization process assures that at least one clearing sequence is usable for any given situation. Another key advantage of this optimization is the resulting relative compactness: a vehicle with an overall length of only 1.37 m is able to clear obstacles 0.76 m in diameter. This design still allows a relatively long (0.79 m) wheelbase making the vehicle very stable and able to carry a larger payload.

### 3.3 Electronics and Control

An overview of the onboard electronics is presented in Fig. 7A. The electronics cabinet holds the onboard computing system which includes 19 custom boards using PIC microcontroller technology. These boards are dedicated to motion control (wheel motors, motion motors, and pan-and-tilt motors), DC-DC converters, communication systems (data and video), protection circuits, video image management and data acquisition. Also, 12-V, 24-V and 48-V buses carry power to the boards and components. An RS-485 bus is used for communication between the boards.

The data communication system operates at 900 MHz (spread spectrum technology) while the video transmission system operates at 2.4 GHz. Even though harvesting power from the powerline is technically feasible, Lithium-Ions battery pack that allows for full day autonomy was a simpler, more reliable, and much lighter choice.

All seventeen motors (counting the pan-and-tilt units) and their encoders and brakes, plus all cameras and sensors are linked to the central electronics cabinet by a total of 30 m of double-shielded cables that are securely coupled by military-type connectors.

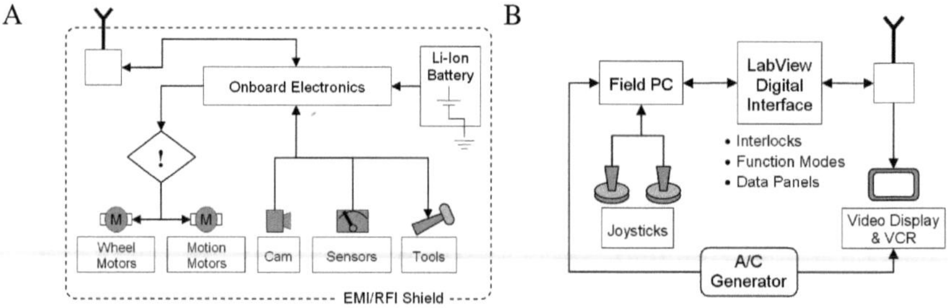


**Fig. 7.** Onboard electronics and Ground Control Unit schematics

The ground control unit (Fig. 7B) consists of a portable computer (military standards) with touch screen interface and two industrial three-axis joysticks equipped with five push buttons. Control software was developed on LabView software. A video display unit allows images from the mobile platform to be viewed for control of the platform and archiving of inspection images.

One of the determinant decisions taken at the beginning of the project was to make the LineScout a teleoperated technology, instead of an autonomous vehicle. Of course it had to be as intuitive to operate as possible, but its initial purpose is more or less to extend the reach of experienced linemen, allowing them to safely inspect and maintain the network that they know well. This is another key difference between this technology and previous ones and this is likely to facilitate its introduction into service.

The strategy adopted to control the mobile platform and allow a human operator to handle so many axis of movement with two joysticks is based on the notion of modes. Modes are based on a series of actions relevant to a particular sequence of movement. In displacement mode, for instance, only the traction wheel motors and safety roller motors can be actuated by the operator by mean of the actuation joystick, the other joystick being always dedicated to the control of cameras and pan-and-tilt units.

When a mode change is necessary, interlocks manage the transition to make sure it follows a safe and reliable sequence of operations. Most motor movements are speed-controlled and limited through a mixed strategy based on position feedback and a peak current limit reached as the actuated axis approaches a mechanical limit.

A watchdog (represented by an exclamation mark on Fig. 7A) reduces power to the motors automatically if unexpected commands are received or loss of transmission suddenly occurs. A hardware emergency stop, located near the joysticks, can also activate the watchdog. What has been defined as "sleep mode" allows minimal energy consumption, all systems being down except for data communications.

The operator receives feedback mainly from the three cameras. Two-axis inclinometers that return the platform pitch and roll in absolute values along with each motor position feedback are also used to animate a 3D model of the plateform in a separate window of the graphical interface.

Even though the current control strategy only requires so few sensors, plans are laid out to progressively increase the level of automation and the onboard electronics allow for additions as they would seem necessary. Obstacle detection and identification will eventually allow autonomous selection of the appropriate clearing sequence. Instantaneous weight-on-wheel measurement, Global Positioning System (GPS), and thermal measurement are also examples of potential short-term additions.

### 3.4 Implemented Application Modules

A total of four controllable cameras can be used to give very flexible and complementary points of view for visual inspection of line components, such as cylindrical warning markers, suspension clamps and insulator strings as they are being passed. Splice condition evaluation based on measuring the electrical resistance of the splice is an application readily implemented on LineScout Technology. A motorized system was developed for positioning an off-the-shelf sensor used to collect information on the compression splice's electrical resistance. Development was recently completed of a module to temporarily repair broken conductor strands by installing a copper clamp close to the damaged area. In addition to collecting visual information for better repair job planning, this module eliminates the risk of further deterioration of the conductor as broken strands unravel under windy conditions.

## 4 Validation

A systematic approach was adopted from day one to ensure the technology's reliability and safe operation. After three years of development and testing, a validated, field-oriented prototype is ready to undertake inspection pilot projects onto strategic power utility equipments.

Validation included lab testing (software simulation, test beds for subsystems, testing on a power line mockup, live-line testing, demos and training) and field testing (IREQ's experimental power line, de-energized lines, energized lines and pilot projects).

### 4.1 Failure Mode and Effect Analysis

Literature [10] suggests approaches to assess the reliability of systems being designed. The starting point of our analysis was to identify undesirable events and their consequences on the powerline system (component integrity, service continuity, worker safety and public safety) and on the robot itself. Most probable causes of these events were then systematically identified as all subsystems of the mobile platform were broken down into their simplest parts.

Failures of previous generations of the mobile platform were classified into three sources: mechanical, electronics and control/software. With the first-generation prototype, most failures (80%) came from their mechanics and electronics: harness and connectors, belt transmission systems, motor shaft couplings, motor amplifier controllers, communication system and electric brakes. Solutions have since been applied for each weak point identified.

The latest generation of the platform is used to implement fault detection and recovery, addressing electronics and control failures or the most probable cause of potential failures. Live-line testing, which is a completely different context for experiments, pointed to failures of communication system components, harness grounds and shields as well as electronic protection circuitry.

Another key benefit of this exercise was to provide a detailed maintenance procedure and a pre-installation verification list.

### 4.2 Endurance Testing

For over two years, the mobile platform has been tested under indoor and outdoor simulations of maintenance operations. Throughout these tests, failures, operator mistakes and surprising behaviors were logged and subsequently solved. These tests were conducted mainly to observe any mechanical system premature wear, battery performance, board and component temperature rise, overall design performance, electronic board and component behavior under peak power draw, cable routing, actuator choices, structure strength and flexibility, additional feedback required and PID control loop gain adjustment.

The mobile platform rolled over 75 km of conductors and crossed 700 obstacles. The experience gained from this endurance testing was introduced into the design of second and third generation prototypes.

Traction force measurements on various slope conditions and stability evaluation were conducted. Cold-temperature operation of LineScout was also validated after the robot spent the entire night outside at -13°C and functioned normally.

### 4.3 EMI Robustness

The electromagnetic environment surrounding a live 735-kV line with an average load of 1,000 A is quite a challenge for electronics that operate at voltage between 12 V and 48 V. Boards were designed carefully respecting basic principles of EMI robustness. Since the entire structure must conduct electricity to minimize any difference in potential developed across components, particular care was taken in designing high-quality shields (connectors, harness, electronics cabinet, battery and peripheral systems). The traction wheels are made out of a conductive natural rubber specially developed and molded for this application. All other components in contact with the energized conductor, such as the safety rollers and grippers, are covered by a protective sleeve made of soft conductive nylon that does not scar or dent conductors.

An incremental testing strategy was adopted regarding electromagnetic immunity of the systems. Hydro-Quebec Research Institute (IREQ) expertise in this field made it possible to quickly focus on key issues and design robust onboard electronics. A high-voltage test chamber was then used to generate the electromagnetic equivalent of a 500-kV transmission line. On Fig. 8, the live line installation of the LineScout is simulated as an electric arc is applied for 20 seconds on different parts of the robot (mechanical systems, harness and cameras). Areas with strong corona emissions were identified using a UV camera and mechanical parts were modified accordingly. LineScout was then tested on a conductor with a 1,000 A load, submitting its electronics to a strong magnetic field.

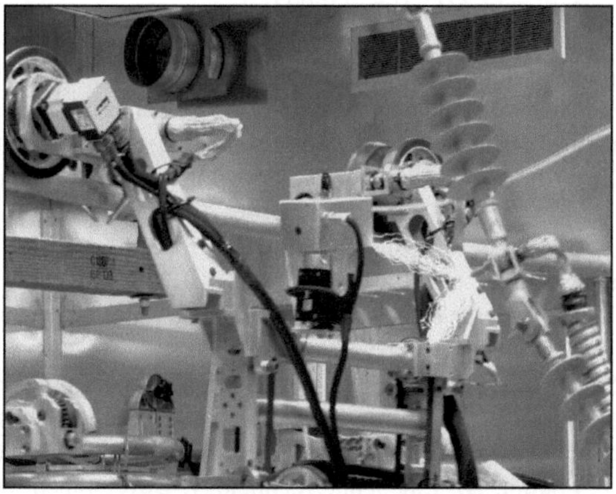

**Fig. 8.** LineScout being tested at 500 kV in IREQ's facility

The field tests for in situ validation on energized power lines were conducted on a live 315-kV line. After successful testing, the mobile platform was finally tested on a live 735-kV line.

## 5 Conclusion and Future Work

A mobile robotic platform able to cross most of the obstacles found on transmission lines has been presented, for which a patent application was filed in 2004 [11]. The third-generation prototype has been tested and validated under field conditions. Several features of the technology make it distinct with respect to previous attempts to incorporate robotic into the field of power utilities. Among them is the reasonable payload it can carry, its versatility to adapt to unforeseen sequence of obstacle, the intuitive teleoperation type of control, its reliability, and the capacity to add sensors and tools in a modular way.

Obstacle detection and identification will eventually allow autonomous selection of the appropriate strategy for crossing obstacles. Another high-value application for LineScout Technology is visual inspection of line components that are found on conductor bundles. The robotic arm needed for this task, under development, will also lend itself to splice condition evaluation and other live-line work.

## References


1. Montambault, S., Pouliot, N.: On the Economic and Strategic Impact of Robotics Applied to Transmission Line Maintenance. In: 7th International Conference on Live Maintenance (ICOLIM 2004), Bucharest, Romania (2004)
2. Jiang, B., Mamishev, V.: Robotic monitoring of power systems. IEEE Transactions on Power Delivery 19(3) (2004)
3. Campos, M., et al.: A Mobile Manipulator for Installation and Removal of Aircraft Warning Spheres on Aerial Power Transmission Lines, Universidade Federal de Minas Gerais, 31270-901, Belo Horizonte, MG, Brazil (2002)
4. Zhou, F.Y., Wang, J.D., Li, Y.B., Wang, J., Xiao, H.R.: Control of an inspection robot for 110 kV power transmission lines based on expert system design methods. In: Proceedings of the 2005 IEEE Conference on Control Applications, Toronto, Canada (2005)
5. Rochia, J., Sequeira, J.: The development of a robotic system for maintenance and inspection of power lines. In: Proceedings of the 35th International Symposium on Robotics, Paris (2004)
6. Earp, G.: EA Technology's Robotics Program. In: Proceedings of Robotics for use in the Electricity Industry (1996)
7. Sawada, J., et al.: A Mobile Robot for Inspection of Power Transmission Lines. IEEE Transactions on Power Delivery 6(1), 309–315 (1991)
8. Montambault, S., Pouliot, N.: The HQ LineROVer: Contributing to Innovation in Transmission Line Maintenance. In: 10th International Conference on Transmission and Distribution Construction. Operation and Live-Line Maintenance (ESMO 2003) (2003)
9. Montambault, S., Pouliot, N.: LineScout Technology: Development of an Inspection Robot Capable of Clearing Obstacles While Operating on a Live Line. In: Operation and Live-Line Maintenance -ESMO 2006. 11th International Conference on Transmission and Distribution Construction (2006)
10. Villemeur, A.: Sûreté de fonctionnement des systèmes industriels, Editions Eyrolles (1988), ISBN# (2212)016158
11. Montambault, S., et al.: Remote-Controlled Vehicle Designed to be Mounted on a Support and Capable of Clearing an Obstacle, Patent Application, WO2005/101600 (2005)

# Kinematics and Sensor and Control Systems of the Fully Automated Facade Cleaning Robot SIRIUSc for Fraunhofer Headquarters in Munich

Norbert Elkmann, Mario Lucke, Tino Krüger,
Dietmar Kunst, and Thomas Stürze

Robotic Systems, Fraunhofer Institute for Factory Operation and Automation,
Sandtorstrasse 22, 39106 Magdeburg, Germany

**Summary.** The Fraunhofer Institute for Factory Operation and Automation IFF has developed the automatic facade cleaning robot SIRIUSc for use on the Fraunhofer-Gesellschaft's headquarters, a high-rise building in Munich, Germany. The building has a height of $80m$, its facade an area of $4000m^2$. Apart from the robot that moves along and cleans the facade, the complete, fully automated system consists of a fully automated gantry that secures, supplies energy to and above all positions the robot. Part of the project involved completely automating a standard gantry, which is an integral part of the complete facade cleaning robot system. This paper presents an overview of the significant basic functions of the robot and the gantry, emphasizing the kinematics and control and sensor systems for navigation and the cleaning sequence that employs the robot and gantry's extensive fully automatic functions.

## 1 Introduction

The complete facade cleaning robot system consists of more than the robot alone (Figure. 1). The rooftop gantry positions the robot at the top of the facade on every pane path to be cleaned.

**Fig. 1.** Rooftop gantry and facade cleaning robot atop the building

The robot then descends the facade vertically and cleans as it ascends. Four cables connected to the gantry atop the building safeguard the robot against

C. Laugier and R. Siegwart (Eds.): Field and Service Robotics, STAR 42, pp. 505–512, 2008.
springerlink.com 

falling. Since the cables must be taut to ensure the robot is secure, the cables are also used to position the robot as the winding and unwinding winch on the rooftop gantry moves it vertically along the building and to bear the load of the robot. Cables transmit data and supply power too. The robot is one of the first fully automatic systems of this kind used in a public setting. The robot weighs $450kg$, the gantry $5000kg$.

## 2 Rooftop Gantry Crane

The gantry (Figure. 2) possesses three degrees of freedom relevant for positioning the robot on the facade: Movement along the rails and two rotary cantilever arms.

Measuring systems help position the gantry on the roof to start on the pane path to be cleaned. The gantry has two asynchronous drives to move along the rails. Two encoders determine the gantry's position. The gantry is positioned by a continuous controller that reads the pane path positions taught out of a data module of the control system. Another redundant measuring system has been installed to counteract the slip.

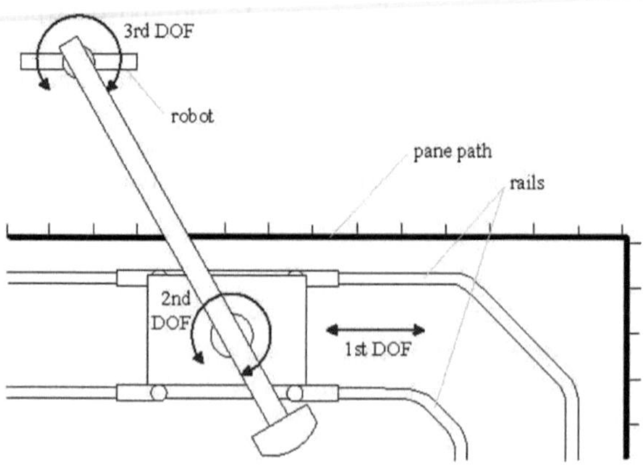


**Fig. 2.** Gantry's degrees of freedom to position and deposit the robot on the facade

When the encoder target position is reached, the transponders mounted on each of the pane path target positions must be matched with the transponder system. If the transponder assigned to each pane path can be read out successfully, the target position has been reached and the positioning of the cantilever arm can begin. Otherwise, an error code is generated and the start on the pane path must be repeated.

A synchronous movement of the two rotary cantilever arm drives sets the robot on the facade. To achieve a high level of security when positioning the robot, its orientation parallel to the facade during motion must be guaranteed. A cascaded controller in the gantry control system ensures this.

Since the rails execute a curve and do not run parallel to the facade, a particular challenge is positioning the robot and thus the gantry on the corners of the facade. The gantry's traveling mechanism (translation) and the cantilever arms (rotation) must run sequentially synchronized to prevent any collision of the robot with the facade. The gantry must proceed along the rails to reduce the robot's distance to the facade and to deposit the robot on it. Simultaneously, this negatively influences the robot's orientation plane-parallel to the facade. To simplify the positioning process, the traveling mechanism is guided in a second step to the window pane target position and then the cantilever arms are moved again so that the robot reaches its ideal position (a distance of approximately 100mm and parallel to the facade). All operations are fully automated. The gantry is a standard piece of equipment that was completely automated in the project.

## 3 SIRIUSc Kinematics

Above all, the modular kinematics (Figure. 3) ensures the robot remains in constant contact with the facade and can navigate a multitude of typical obstacles and move quickly along a facade. The kinematics is based on a structure of two pairs of linear modules, the so-called "'advanced sliding module mechanism"' [8,9]. Two linear modules constitute one pair that performs the same linear movement, thus ensuring secure and stable contact with the facade.

Servo drives move each of the outer or inner sucker units to the next position on a pane within a frame. The linear modules with their suckers are positioned in such a way that the suckers are located above or below the horizontal pane frames when docking onto the facade. Once a sucker unit has suctioned on, the drive's brake is deactivated and the other unit's suckers are released and retracted. The winding and unwinding of the securing cables on the gantry lift winch produces the robot's upward and downward motion on the facade. The alternation of the outer and inner sucker units with $2 \times 3$ suckers apiece produces the robot's walking motion. As the robot is lifted or lowered to the desired position, the other activated linear module pair is moved to the next free position.

Eddy current sensors mounted on the top and bottom of the robot's body detect and store the positions of horizontal pane frames.

One of the requirements for outdoor operation in high winds is that the robot be able to correct its direction of motion, should it drift a bit off course. To this end, the inner pair of linear modules is tilt adjustable to enable small steering movements to keep the robot on a straight path. Two drift scanners have been mounted on the robot to control drift compensation. The scanners' job is to detect the robot's position in relation to pane frames (vertical pane jambs) in order to systematically control the inner linear modules. A drift scanner consist of $2 \times 4$ eddy current sensors (spaced $50mm$), which a controller assembly switches off and on in succession. This is necessary since the eddy current sensors would otherwise interfere with one another. Depending on the robot's direction of movement and which eddy current sensors detect the metallic pane frames,

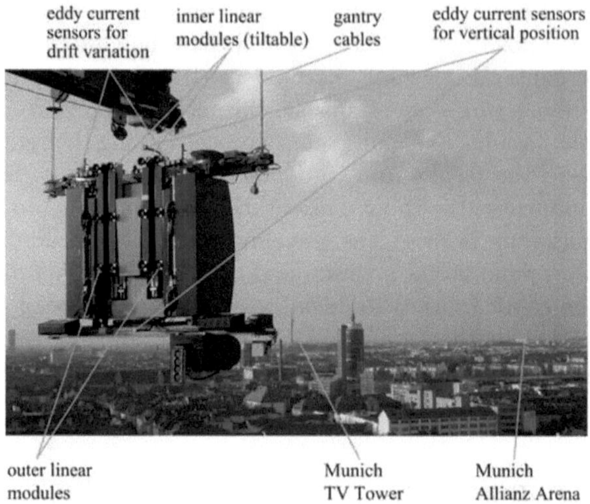


**Fig. 3.** SIRIUSc kinematics and navigation sensors

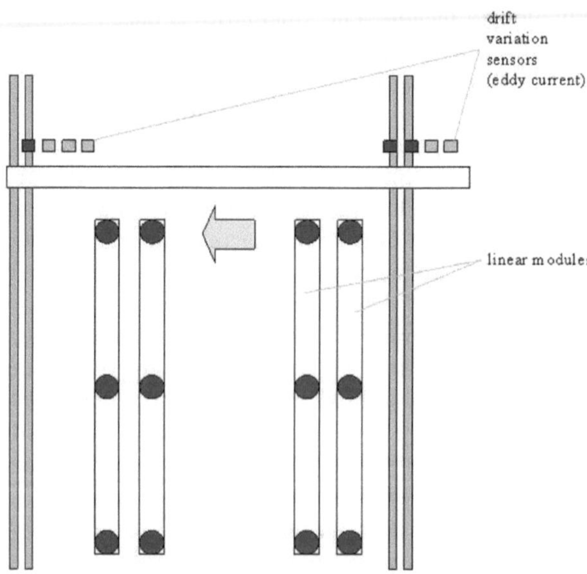


**Fig. 4.** Schematic diagram of the detection of drift variation

the pneumatic drift compensation drive turns the inner linear guide right, to the middle or left.

The pane frames are made of various materials and profiles. This represents a special challenge. In extensive testing with various types of sensors, only novel eddy current sensors with a detection distance of 0–50$mm$ were sufficiently reliable. None of the optical or other sensors were sufficiently reliable under the given conditions (rain, reflected sunlight on the panes).

## 4 Cleaning Sequence

First, the gantry automatically positions the robot before the facade. Since the rails on the roof and the panes on the facade are not exactly parallel to one another, the robot must be set down supported by two tactile distance sensors mounted on it. In the process, all three of the gantry's degrees of freedom are controlled. Once the robot is located only $100mm$ in front of the uppermost pane of the facade, four vacuum suckers with a stroke of $250mm$ located on the robot are retracted. After the four suckers have developed a vacuum, the robot is pulled onto the pane and the robot's vacuum suckers used for moving along the facade establish the robot's contact to the facade.

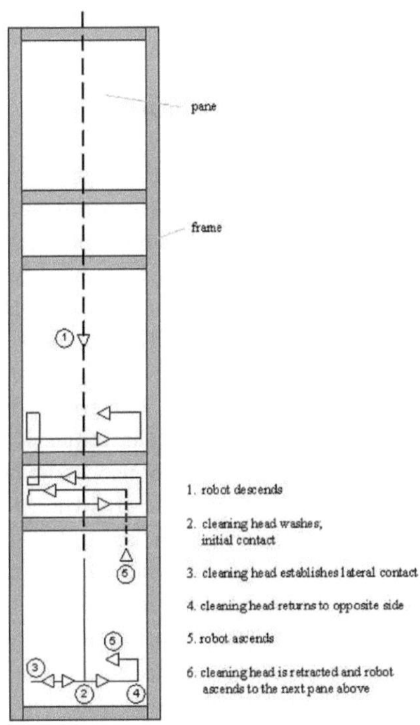


**Fig. 5.** Robot and cleaning head motion (operation/sequence) on a pane path

The robot descends to the bottommost pane (see Section 3 on page 4) and is moved vertically by the cable winch on the gantry, the robot's suckers always being in contact with the facade. Sensors continuously register the pane frames (horizontal and vertical). Consequently, the direction of the robot's vertical movement is corrected and the walking algorithm is controlled. The walking algorithm positions the robot's linear modules in such a way that the maximum number of suckers is always on the facade and the suckers are always positioned on the panes and not on the frames. Once the bottom pane has been reached, the cleaning head is positioned on the pane to be cleaned (Figure. 5). The water cycle and the rotating cleaning brushes are activated. The cleaning head runs sideways until it meets the frame. Tactile sensors on all four sides of the cleaning head detect the pane frames. If the cleaning head has contact with the frame, the robot continues moving downward until the lower pane jamb has also been detected. On the lower pane jamb, the head is moved to the opposite side of the pane, cleaning the pane. Once the head has traveled back and forth one time, the robot moves upward the height of the cleaning head. When the upper pane frame is detected, the cleaning head is run once more laterally, the cleaning water suctioned up and the cleaning head retracted from the pane.

When the uppermost pane has been reached and the complete pane path cleaned, the robot disengages from the facade. Once the cleaning of a vertical pane path has concluded, the gantry moves the cleaning robot one pane width laterally.

## 5 Control System and Navigation

The heart of the system is its control system, which receives and combines sensor data and operator instructions to generate robot actions. Selected for it is stability and modularity, a programmable logic controller (PLC) is used for the robot and the gantry control system. The PLC is on board and controls the entire system. It synchronizes the walking mechanism with the trolley and cleaning head. All robot motions and actions are fully automated. The robot was programmed modularly so that it can be transferred to a large number of different facades with minimal reprogramming work. The robot does not start out with information on all the obstacles it will face in its path. Rather, it keeps track of the surface and obstacles currently under it. Sensors identify and measure obstacles and window frames. The PLC then generates the appropriate step lengths for the robot to successfully walk over obstacles and frames. The PLC ensures that vacuum suckers directly over an obstacle are not engaged while simultaneously maximizing the number of vacuum suckers in contact with the facade at any given moment. Like SIRIUS, the cleaning head does not require detailed information on the surface of the facade. Rather, the cleaning system has its own sensors that detect obstacles and end positions. The sensor signals are also incorporated in the onboard PLC program, which in turn uses the information to generate the necessary cleaning head movements.

The only manual input information the robot requires before starting off on a given surface is end point data such as the height of the structure or building, off limit zones and the maximum length of obstacles. The robot automatically detects any other necessary surface information during operation. An operator master display is also located in the building so that an operator can monitor the robot's progress. A remote maintenance module allows downloading the robot's

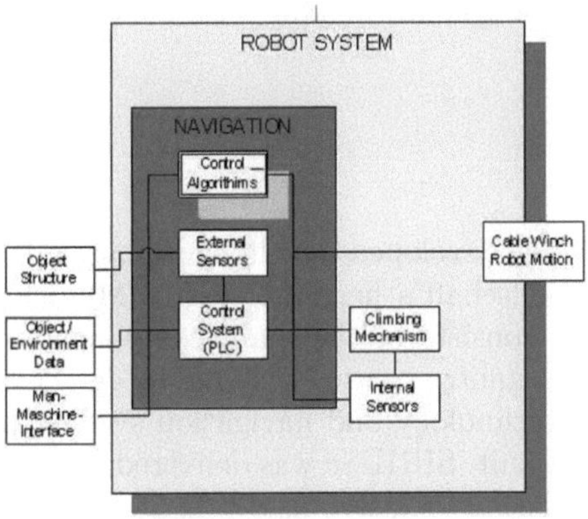


**Fig. 6.** Control system concept

**Fig. 7.** SIRIUSc on the high-rise building of the Fraunhofer-Gesellschaft's headquarters in Munich

status and sensor data, uploading new program modules and executing simple operator commands such as the motion commands for a pair of linear modules. All this can be done over the Internet.

Little knowledge about the general structure of a building's surface is needed before robot movement can be generated. The input data includes end positions, moving distances and path characteristics. This a priori data is supplemented by online sensors that detect the facade surface and search for possible obstacles. In addition to identifying obstacles, the external sensor technology also corrects the direction of motion. Sensors detect where the robot must deviate from a path, e.g. girders or window and panel seals. The robot control system communicates with the building control system, making sure all windows are closed in areas being cleaned.

## 6 Conclusion

The Fraunhofer IFF has developed a fully automated facade cleaning robot for the Fraunhofer-Gesellschaft's headquarters in Munich. Developed for vertical facades, SIRIUSc consists of the main components of robots mechanics and kinematics, rooftop gantry, sensor systems to detect facade shape, frames and obstacles, control technology and navigation system, power supply system and integrated cleaning unit. SIRIUSc was delivered to the facility management at the Fraunhofer-Gesellschaft's headquarters in 2006. The facility management staff is now able to clean the facade without any technical support from the researchers at the Fraunhofer IFF.

## References

1. Hirose, S., Kawabe, K.: Ceiling Walk Climbing Robot Ninja-II. In: Proceedings of CLAWAR 1998, First International Symposium on Mobile, Climbing and Walking Robots, Brussels, October 26-28, 1998, pp. 143–147 (1998)
2. Cusack, M.M., Thomas, J.G.: Robotics for the Inspection of Vertical Surfaces of Buildings and Structures. In: 25th ISIR, pp. 287–295.
3. Elkmann, N., Felsch, T., Sack, M., Boehme, T.: Modular Climbing Robot for Outdoor Operations. In: Proceedings of CLAWAR 1999. Second International Conference on Climbing and Walking Robots, Portsmouth, September 13-15, 1999, pp. 413–419 (1999)
4. Elkmann, N., Felsch, T., Sack, M., Boehme, T., Hortig, J., Saenz, J.: Modular Climbing Robot for Service Sector Applications. Industrial Robot 26(6) (1999)
5. Elkmann, N., Felsch, T., Sack, M., Boehme, T., Saenz, J.: SIRIUS: Modular Climbing Robot for Facade Cleaning and Other Service Jobs. In: International Conference on Field and Service Robotics FSR 2001, Helsinki (2001)
6. Felsch, T., Elkmann, N., Sack, M., Saenz, J.: Concepts of Service Robots for Facade Cleaning. In: International Symposium on Robotics ISR 2001, Seoul (2001)
7. Sack, M., Elkmann, N., Felsch, T., Boehme, T.: Intelligent Control of Modular Kinematics: The Robot Platform SIRIUS. In: International Symposium on Intelligent Control ISIC 2002, Vancouver, Canada (2002)
8. Elkmann, N., Felsch, T., Sack, M., Saenz, J., Hortig, J.: Innovative Service Robot Systems for Facade Cleaning of Difficult-to-Access Areas. In: International Conference on Intelligent Robots and Systems IROS 2002, Zurich (2002)
9. Elkmann, N., Lucke, M., Krüger, T., Kunst, D., Stürze, T.: SIRIUSc: Fully Automatic Facade Cleaning Robot for a High-rise Building in Munich. In: ISR/ROBOTIK 2006, Munich, Germany (2006)

# Automated Steam Turbine Straddle Root Disc Head Inspection

Roland Moser[1], Chris Udell[2], and Alistair Montgomery[2]

[1] ALSTOM Power Service Switzerland, 5401 Baden, Switzerland
[2] ALSTOM Power Service United Kingdom, Rugby, UK

**Abstract.** The non-destructive condition assessment of power generation steam turbines is key to ensure the power plants availability. As fossil and nuclear power plants come of age, the inspection intervals get shorter yet the inspections have to be performed with ever increasing speed and accuracy. The power industry's answer to that challenge is the automation of these non-destructive tests using inspection robots. This paper presents a versatile inspection system that is designed to assess the condition of steam turbine blade roots, an area that is under particularly high mechanical load due to the high centrifugal forces under operating conditions. Using advanced ultrasonic sensors the system performs autonomously the whole inspection, transmitting the sensor data to a human expert who has to decide if the turbine is still fit for purpose. Reported are the inspection delivery concepts and also the basics of its ultrasonic sensors are introduced. The operational industrial robot, its limits and performances are discussed presented and discussed.

## 1 Introduction

The steam turbines that entrain the generators of fossil or nuclear power plants have their turbine blades mounted on discs. Different attachment concepts ensure that the blades are tightly fitted on these rotor discs yet allow removal of the blades for replacement, repair or refurbishment. The most common blade attachment concepts are sketched in figure 1:

Figure 1 sketches a typical high pressure steam turbine rotor with its discs mounted on discs that are solid with the rotor. The blades, typically about 50 per row, are slid on a kind of dovetail groove. Three different types of these grooves are of interest for the presented project: The so called "T-Root", the "Straddle –Root", and the "Firtree-Root" grooves. The blades feature a matching shape, providing a tight fit.

During operation, the centrifugal forces lead to a load on the blades that can be extremely high e.g. for a typical last stage blade 660MW turbine the load is around 250 tons per blade at 3000rpm. The entire centrifugal load is reacted by the blade-root attachment, the grooves depicted in figure 1.

The high centrifugal stresses combined with steam chemistry can cause stress corrosion cracking. Such cracks can grow over time and, if not detected, can result in catastrophic failure. Figure 2 shows a photograph of such a straddle root disc with an indication of its crack locations and an example of catastrophic failure of a T-root blade attachment system.

Regular inspections are required to check for such cracks in the blade and rotor disc head root. As the access to the rotor discs is difficult and restricted to the distance

C. Laugier and R. Siegwart (Eds.): Field and Service Robotics, STAR 42, pp. 513–520, 2008.
springerlink.com 

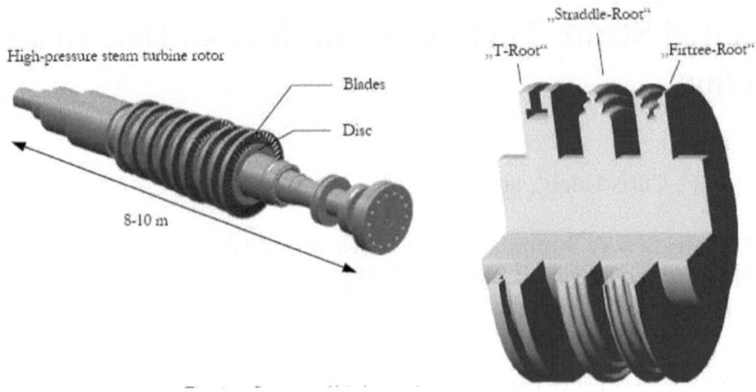


**Fig. 1.** Concept of blade attachment

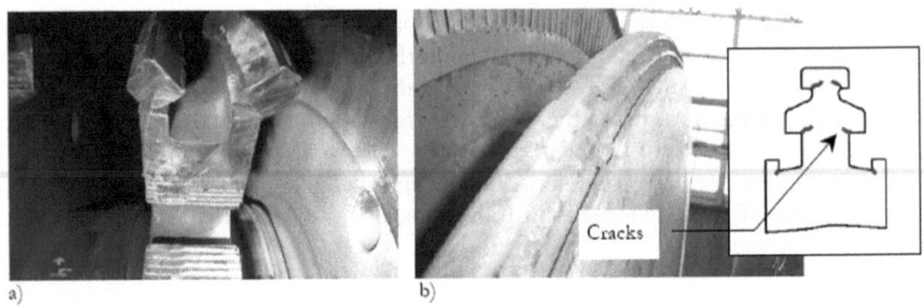


**Fig. 2.** a) Failure of T-Root due to non-detected cracking b) Steam turbine Straddle-Root discs with and without blades. Note the notch in the unbladed disc where the blades are slipped onto the root.

between two adjacent discs, the manually performed ultrasonic inspections are time consuming. To manually inspect, one radii is inspected at a time, resulting in an overall inspection time of 1 day per disc.

In order to speed up this inspection, thus increasing the availability of the powerplant, an automated robotic solution is proposed.

## 2 Concept

### A. Specifications

The proposed automated steam turbine root inspection should be able to be performed in the following situations:

- On site (at the power plant), with the rotor still in the stator. The upper casing would be removed and one has free access to the upper half of the rotor. With its hydraulic oil bearings operating, the rotor can be rotated manually.
- On site, with the rotor taken out of the stator. This requires heavy crane operation but is common practice during medium overhauls. The rotor is put

on a stand on the turbine floor, generally besides the empty stator. The rotor cannot be turned manually in this situation but is accessible from all sides.

- On site, with the rotor out, but this time also the blade have been taken out. Only the bare rotor is on its stand, also accessible from all sides.
- In the workshop, the rotor with or without blades, sitting on a stand.

Also freshly refurbished discs have to be checked for structural integrity. These refurbishments are generally done by successive weld-overlays on a stump of the old disc. Prior to a machining of a new blade root (see figure 1), the built-up metal must be scanned.

## B. Inspection Vehicle

The inspection vehicle must, in order to fulfill the aforementioned specifications, be capable to fit in between two adjacent blade rows and to run around a non-rotateable rotor.

As the rotors of steam turbines are made out of forged ferromagnetic alloys, magnetism as a means of keeping the vehicle stuck on the rotor is the solution of choice.

Mechanical serration or pneumatic suction as a fixing concept would cause additional actuators, additional supply leads to the vehicle (air conducts or power supply for the mechanical serration actuators) and present an additional system failure risk – with power down, the robot gets either stuck or falls off the rotor.

With the intention to mitigate the power-down failure risk to a maximum and to minimize the number of supply leads to the vehicle, permanent magnetic wheels were chosen.

Using magnetic wheels equipped with permanent magnets, the attraction force density is significantly higher than in the case of active magnetic wheels using electric coil magnets. Also, no additional power leads are needed and as long as the motor and its gearbox that drive said magnetic wheels have a reversible reduction, the vehicle can be pulled out in the case of power-down failure. Figure 3 shows the chosen vehicle concept:

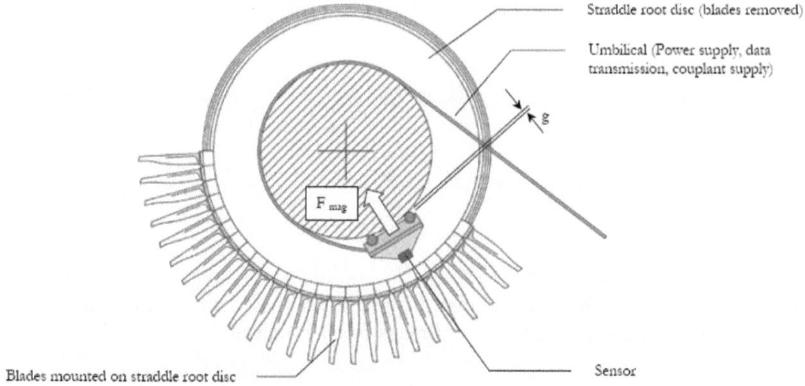


**Fig. 3.** Vehicle concept using permanent magnetic wheels. Fmag designated the magnetic force created by the permanent magnets in the wheels.

The reason why the magnets are placed in the wheels and not in the vehicle body as one might suggest as a simplification is, that with changing diameters, the clearance g between ferromagnetic rotor and the vehicle body change. As the magnetic attraction force would decrease following an inverse square law in function of g, the vehicle could carry a very reduced payload while running on large diameter rotors.

## C. Sensor

Figure 4 shows, for the case of a straddle-root disc, the zones that are prone to stress-corrosion cracking. These cracks are surface breaking. The MPI and conventional ultrasonic methods are widely accepted as reliable and are best practice in NDT workshops today. MPI is only usable if the disc is de-bladed. MPI is of limited use for automated inspections as they both require applying a liquid, doing an additional operation (applying an external magnetic field) – and to take a visual picture of the scene.

As for conventional ultrasonic inspection, a separate sensor for each disc groove would be required, and also these sets of sensors would differ between the different disc geometries (figure 1).

The method of choice is phased-array ultrasonic testing. Figure 4 shows the principle of this NDT technology as well as the typical inspection environment of a straddle root steam turbine:

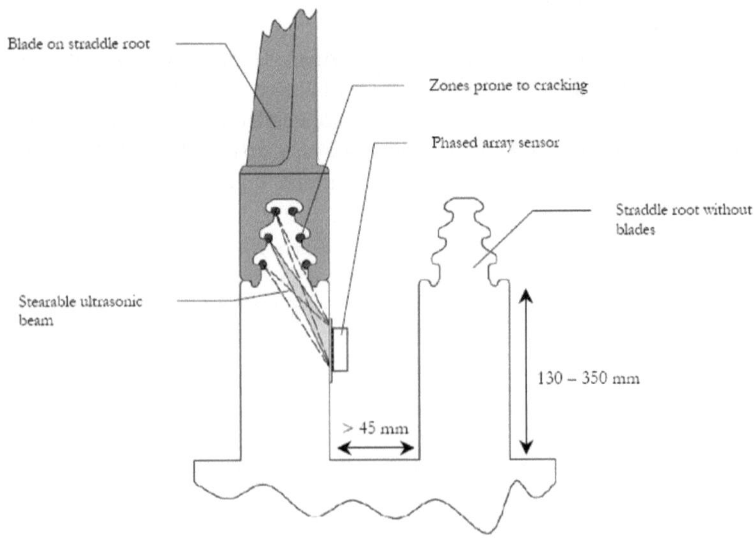


**Fig. 4.** Example of a typical straddle root disc analysed by a phased array ultrasonic transducer: Cracking-prone zones and overall geometric dimensions are indicated

In the last decade phased array ultrasonics has become widely accepted as a reliable NDT technology for the condition assessment of machinery and structures.

The principle is that, instead of having one single ultrasonic transducer, a one or two-dimensional array of such transducers is realized. If all of these elements would be actuated in the very same moment, the system would behave like a conventional monolithic ultrasonic actuator of similar dimensions. But, if the various elements are actuated with a slight mutual time difference, the individual ultrasonic waves issued from each element form constructive or destructive interference with their neighboring waves, resulting in a directed beam. This beam can be steered in direction and in its focal point by adequately actuating the different ultrasonic elements.

It becomes clear that, using phased array ultrasonics, a universal sensor can be used that can be programmed for all concerned geometries shown in figure 1.

## 3 Implementation

Figure 5 shows the disc inspection robot with its components. Not shown are the umbilical, the data acquisition unit for the ultrasonic phased array data as well as the motor drive unit.

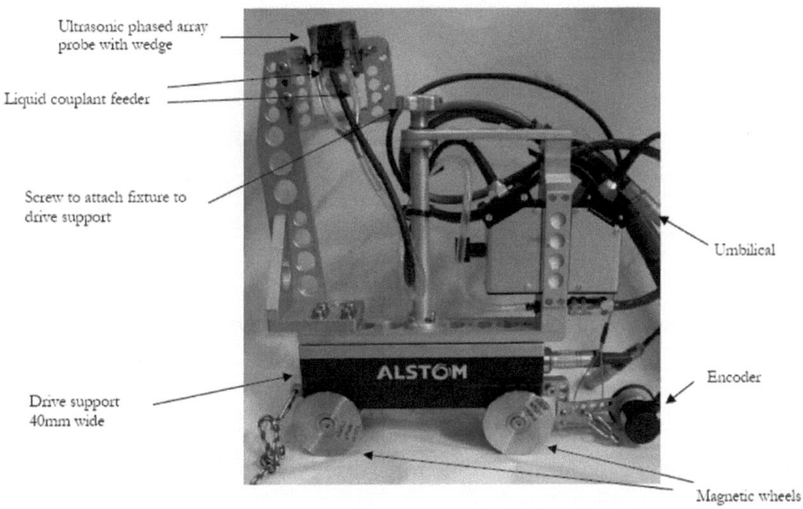


**Fig. 5.** Automated rotor disc inspection system

Atop a drive support with four direct driven magnetic wheels is mounted a sensor fixture, allowing for a swift regulation of the sensor position and height. The versatile phased array sensor is coupled to the inspected disc by means of a water-based couplant, fed through a hydraulic hose in the small air-gap between sensor and disc. The stepper servo motor inside the drive support features an encoder, an additional encoding device is fixed on the system. This improves the positioning accuracy of the system in the event of slipping of the magnetic wheels.

Figure 6 shows the system in operation. Depicted is the test-rig with a straddle-root disc without blades and a restored disc freshly welded up to its final diameter, prior to machining. Artificial cracks (Notches machined using electric discharge machining) as well as real cracks are used for validation of the method.

The system is capable of carrying a payload of up to 5 kg and is designed to run around a disc with a tangential velocity of 3 m/min top-speed. This top-speed is too fast for most ultrasonic sensors, they limit the speed by their data acquisition rate to 0.5 m/min. But, if one desires to inspect the rotor or the blades with Eddy-current testing technology (searching for surface-opening cracks) this top speed is adequate.

The inspection vehicle is secured via a tether cable. This is usually imposed by the owner of the power plant, since the worst case scenario is that a inspection robot gets stuck inside a generator, turbine or boiler, causing additional outage-time and labour.

**Fig. 6.** Automated rotor disc inspection system in operation on a test rotor

# 4 Results

The developed automated steam turbine rotor disc inspection proved to work satisfactorily.

All crack-prone zones of a straddle-root disc could be analyzed using one single ultrasonic transducer – a 32 element linear phased array device. The different ultrasonic beam angles can be set by an adequate actuation of the various ultrasonic elements and is done via the control software.

All real cracks as well as the artificial notches that were introduced on the test rotor were detected without any false calls. The resolution of the sensor proved to be small enough to detect all cracks that can compromise the structural integrity of the rotor.

Once set up, an operation that takes about 10 minutes, the automated inspection device is analyzing one side of the rotor in less than 10 minutes (depending on the rotor diameter). Compared to the conventional manual method with up to 4 hours of labor per disc side, this is a massive time gain. As the system is scaleable, the total inspection time for an entire rotor is also in the range of 10 minutes. Adding set-up and dismantle time, a total inspection time of 30 minutes seems achievable. This

**Fig. 7.** Illustration of two inspection systems simultaneously inspection a straddle root disc

extremely short time allows to fit the inspection - that previously took up to several days - between two other outage tasks. Figure 7 shows the concept of scalability: Two systems per disc allow inspecting the whole rotor at once, if necessary.

The system proved to provide repeatable results unmatched by any manual inspections. The human error is engineered out, the only source of a false diagnosis is now a misinterpretation of the ultrasonic data. This high degree of repeatability is the basis for reliable follow-up inspection during the life of a steam turbine rotor. The growth of the cracks can be monitored and the end of the rotor lifetime can now be predicted with a high degree of confidence.

The system's versatile modularity provides a very short set-up time for the inspection of a new rotor geometry. Figure 8 depicts the inspection of rotor discs with different diameters:

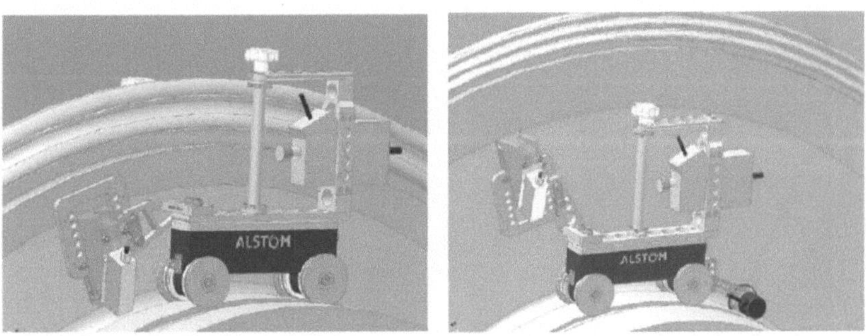


**Fig. 8.** Illustration of the inspection of discs with different diameters

## 5 Conclusion

This paper presents an inspection robot capable of inspecting the narrow space between adjacent steam turbine blade discs. Using magnetic wheels, it is capable of running around the rotor, covering its full circumference. The mechanical integrity of the disc is analyzed using ultrasonic phased array non-destructive testing technology. It was shown that the realized solution is highly compact and versatile. It can be adapted to all the common types of steam turbine blade discs by a simple readjustment of the sensor suspension mechanism and a reprogramming of the phased array sensor software.

The novel inspection robot showed performances unmatched by conventional manual inspection, in terms of accuracy and reliability as well as in terms of speed.

## References

[1] Slocum, A.H.: Magnebots – A Magnetic Wheels Based Overhead Transportation Concept. In: Proceedings of the 1st International Workshop on Advances in Service Robotics, Italy (March 2003)

[2] Guy, W.: Magnetic Wheel, U.S. Patent 3,690,393 (September 1972)

[3] Friedrich, M., Gatzoulis, L., Hayward, G., Galbraith, W.: Small Inspection Vehicles for Non-destructive Testing Applications. In: Proceeding of the 8th Int. Conf. on Climbing and Walking Robots and the Support Technologies for Mobile Machines (CLAWAR 2005), pp. 927–934 (2005)

[4] Lafontaine, G., Cancre, F.: Potential of Ultrasonic Phased Arrays for Faster, Better and Cheaper Inspections. NDT.net 5(10) (October 2000), http://www.ndt.net/article/v05n10/lafont2/lafont2.htm

[5] R/D Tech, Introduction to Phased Array Ultrasonic Technology Applications: R/D Tech Guideline, Quebec City, Canada, R/D Tech (2004)

# Design and Field Testing of an Autonomous Underground Tramming System

Joshua A. Marshall[1] and Timothy D. Barfoot[2]

[1] MDA Space Missions, 9445 Airport Road, Brampton, ON L7E 2K9 Canada
joshua.marshall@mdacorporation.com
[2] University of Toronto Institute for Aerospace Studies, 4925 Dufferin Street, Toronto, ON M3H 5T6 Canada
tim.barfoot@utoronto.ca

**Summary.** This paper describes the design, implementation, and field testing of an infrastructureless system for autonomous tramming (or hauling) of a center-articulated underground mining vehicle. Described is the development of a fast, reliable, and robust "autotramming" technology that does not require the installation of fixed infrastructure throughout the mine.

## 1 Introduction

Mining companies and equipment manufacturers are pursuing improved efficiency, productivity, and safety in underground mining operations by robotizing some of the functions of underground vehicles. For example, the repetitive "load-haul-dump" cycle is well suited to automation. In this case, a vehicle called a load-haul-dump (LHD) machine is often used to excavate fragmented rock, haul it to an assigned location, and then dump the material before returning for another load (see Figure 1). These vehicles are traditionally driven manually by an operator on the vehicle, and more recently by radio remote and teleoperation [1]. However, the hazardous nature of underground environments, driver safety and fatigue, labor costs, and the cyclic nature of this task are all motivations to seek an autonomous tramming solution.

Early "autotramming" attempts worked by outfitting the mine with signal-emitting cables [2], light-emitting ropes [1], or retroreflective tape [3], which allowed a vehicle to find its way through tunnels by tracking the installed infrastructure. Other systems have used fixed beacons, placed strategically throughout the mine [4].

Existing infrastructureless systems have typically employed a suite of sensors for estimating the vehicle's state with respect to its environment and with reference to some form of topological map. Unfortunately, systems that rely entirely on classifying tunnel topology for navigation have been shown to not operate robustly due to inevitable misclassifications that occur in practice [5, 6, 7, 8]. Inspired by advances in mobile robotics, some have considered the application of alternative poly-line map-based localization techniques [9, 10, 11]. Our approach is similar in philosophy to this work, but is novel in a number of ways.

C. Laugier and R. Siegwart (Eds.): Field and Service Robotics, STAR 42, pp. 521–530, 2008.
springerlink.com 

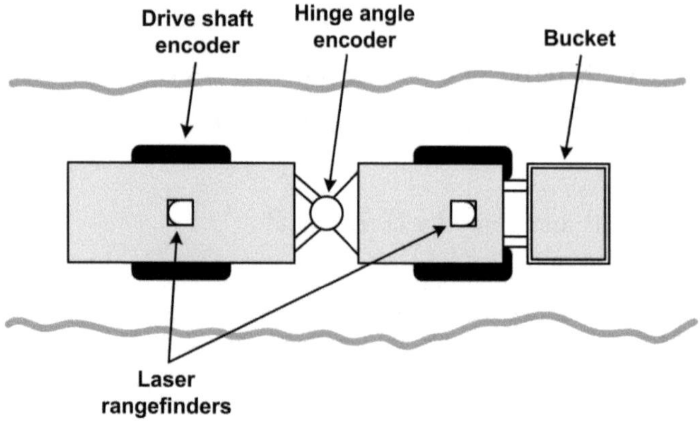


**Fig. 1.** LHD test vehicle schematic

Most notably, we employ a sequence of local self-generated grid-type maps [12] rather than use a monolithic map. The use of grid-type maps makes our system robust in the presence of environmental variations (e.g., non-standard intersections and/or passageways, which occur often in underground mining) and the use of local maps avoids the need to create a globally consistent map (i.e., we avoid solving a simultaneous localization and mapping problem).

## 2 System Architecture and Design

Our autotramming system's operation consists of three steps:

**Teaching.** The vehicle is driven along a desired route, either by an operator on the vehicle, or by remote/teleoperation, while sensor data is logged (front and rear SICK laser rangefinders, a hinge angle encoder, and a drive shaft encoder for measuring displacement of the vehicle).

**Route Profiling.** Data logged during the teaching step is processed (offline) and converted into a format suitable for use by the estimation and control algorithms during playback. The output is referred to as a *route profile*, which contains information about the traveled path, a sequence of overlapping metric maps along the path, a record of any pause points (e.g., for dumping material), as well as a vehicle speed profile to be tracked during playback. Loading and dumping are beyond the scope this paper but our system is compatible with manual or automated loading/dumping.

**Playback.** In this step, the system autonomously plays back a route profile generated by the teaching and route profiling steps. During playback, data from the specified route profile is used by navigation and guidance algorithms to estimate longitudinal, lateral, heading, and vehicle speed errors at discrete instants. The control system then stabilizes these errors so that the vehicle follows the profiled path at the desired speed.

Once profiled, a route can be played back many times. It is expected that re-profiling would only be necessary if significant changes to the environment were made; e.g., due to excavation. Since the vehicle is driven along a collision-free path during teaching, complex path planning is not required. However, the system does include short-range guidance algorithms designed to stop the vehicle should the profiled path be subsequently obstructed during playback.

### 2.1 Route Profiling and Atlas Maps

A *route profile* consists of four components: a path profile, a pause profile, a sequence of locally-consistent metric maps, and a speed profile. Firstly, a sequence of locations that are equally spaced (typically 0.5 m) along the path are created, called *path points.* We then associate with each path point the configuration of the vehicle at that point during the teaching step by interpolating the preprocessed logged data. Thus, the sequence of path points and associated data comprise the *path profile.*

Locally-consistent *metric* maps of the mine environment along the path profile are generated using both odometry and rangefinder data. Each map is an *occupancy grid* [13]. For localization in underground mines, this approach is much more flexible than a system that must classify tunnel topology in that it will work regardless of the shape of the walls, so long as the maps are of sufficient resolution. However, the use of a single monolithic map to represent the mine environment suffers from two key problems. Firstly, in some situations high memory usage is required. Secondly, map inconsistencies can result on longer traverses when a vehicle closes a loop or crosses its own path (e.g., see Figure 2). To address these problems, we employ a sequence or *atlas* of metric maps attached along the path to form an overall route profile; which is to say, the system does not rely on one monolithic map and an absolute frame of reference. The underlying idea is to create a situation in which the vehicle's path exists in a high-dimensional space wherein it never intersects itself [12].

Figure 3 illustrates this notion. A locally-consistent metric map is associated with each position $(x, y, d)$ along the vehicle path, where $(x, y)$ is the vehicle's location in the mine and $d$ its distance along the path. Figures 2 and 4 show example monolithic and atlas maps generated from real data acquired during one of our field tests. The monolithic map has a resolution of 0.3 m, while the atlas maps each have a resolution of 0.1 m.

### 2.2 Playback

Our objective was to create a system that permits a large articulated vehicle to robustly track the path specified by a route profile. During playback, this has been achieved through the design of a two-timescale control system. At the slower timescale, or *outer loop*, are localization and path-tracking algorithms that work to reject lateral and heading path errors. At the faster timescale, or *inner loop*, are rate estimators and two controllers that track reference steering rates and vehicle speeds. The underlying justification for this two-timescale design is

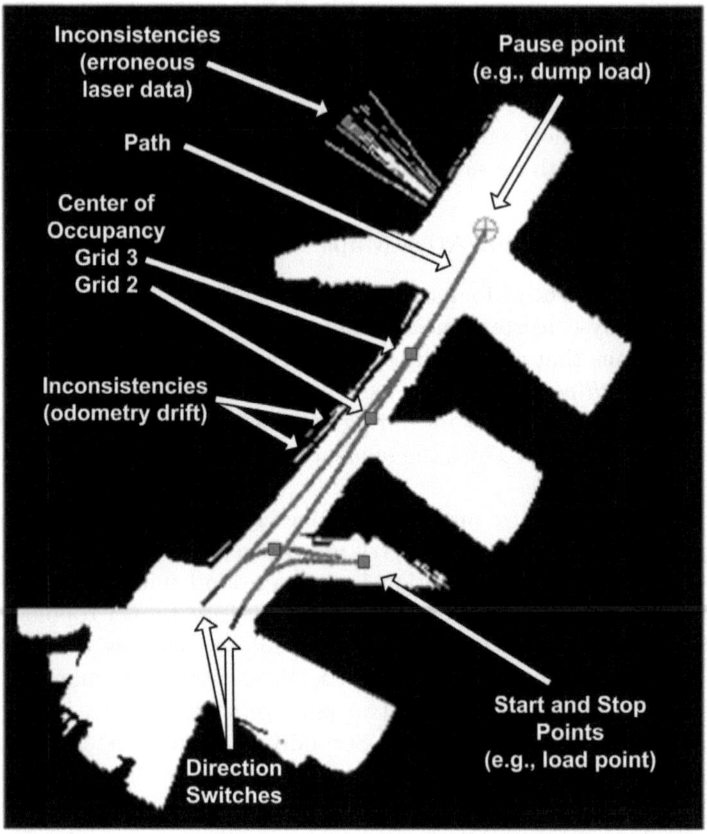


**Fig. 2.** Example monolithic occupancy grid created from real data, with vehicle path shown; the small squares indicate the centers of the atlas maps in Figure 4; the circle with cross hairs indicates a pause point

founded on the assumption that we can specify sufficient bandwidth separation between the nested control loops.

### Metric Map Localization

The localization problem solved here is one of estimating the vehicle's pose as it travels through a (locally) known environment. Recently, a number of techniques have been developed in the mobile robotics community that *globally* localize a robot in a known environment. Many of these techniques use a *particle filter* representation of the vehicle's pose [14]. An initial design using a particle filter was shown to work in simulation, but required the use of too many particles (e.g., $> 100$) for convergence from a reasonable initial pose estimate. Variations requiring fewer particles and computational resources exist, but are complex to implement. Moreover, the task at hand does not actually require a solution to the global localization problem. Instead, we chose to implement a variation of

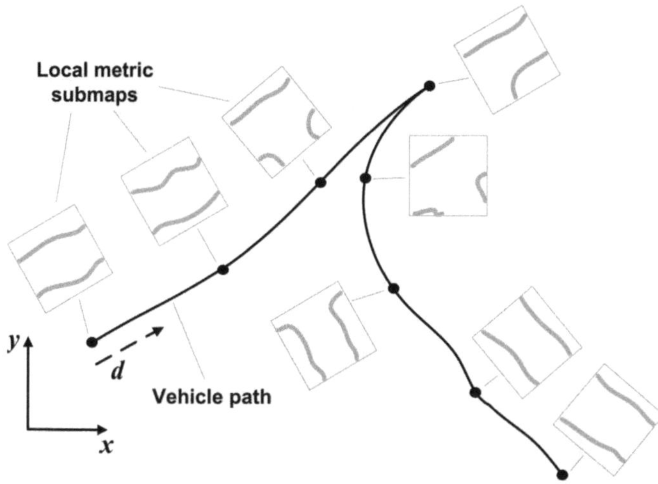


**Fig. 3.** Schematic of sequence of atlas maps along the vehicle path

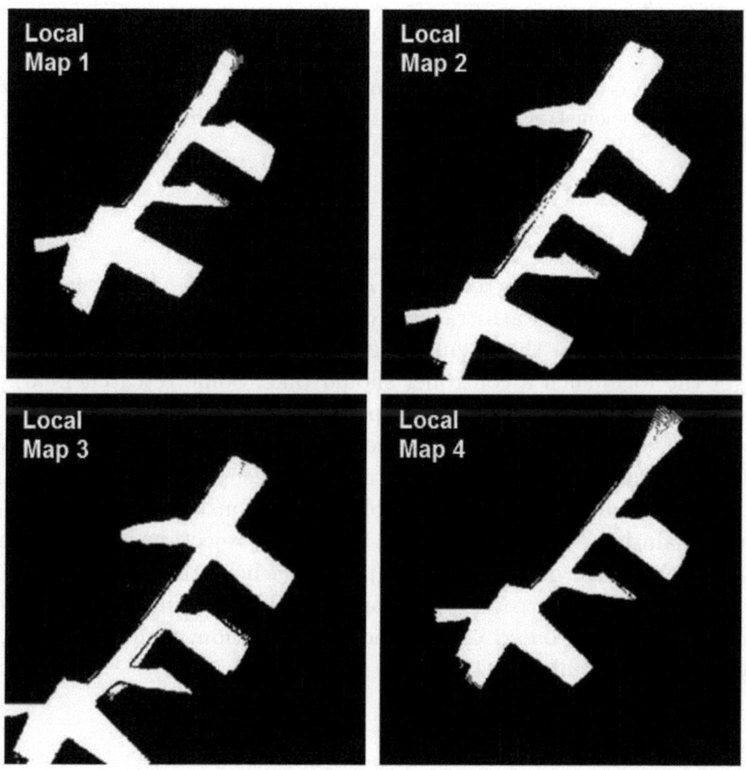


**Fig. 4.** Example atlas of maps created from real data; the centers of these maps correspond to the small squares in Figure 2

the Unscented Kalman Filter (UKF) [15, 16] to position with respect to the locally consistent submaps defined during route profiling. The inputs to the UKF algorithm are the laser rangefinder data as well as wheel odometry (i.e., hinge angle and wheel speed); the final outputs are heading and lateral errors of the vehicle with respect to the profiled path. Figure 5 shows how the laser rangefinder data is used to determine these path- tracking errors.

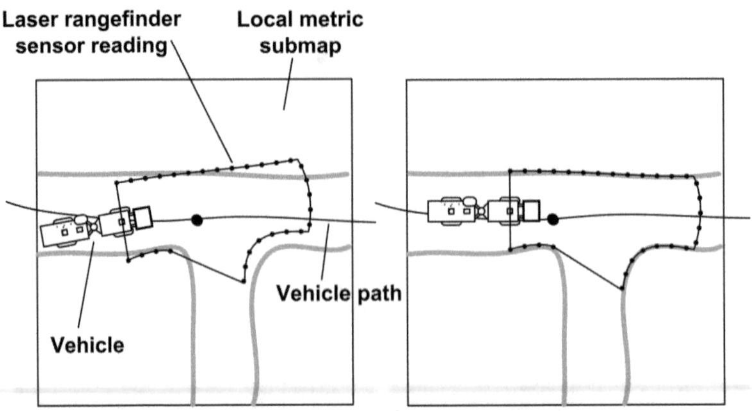


**Fig. 5.** Laser data is used to position within the local submap (which is attached to the vehicle path). It is then straightforward to determine the vehicle errors with respect to the path.

### Path Tracking

A path-tracking controller is required to guide the vehicle along the path specified by the route profile. Path-tracking control for articulated vehicles has been extensively discussed in the engineering literature, yet there is some disagreement over the form such a controller should take. Some researchers argue that the position of both front and rear components of the vehicle should be tracked; others suggest that wheel slip should be explicitly accounted for [17]. We have found that, in practise, neither of these tasks are necessary under a two-timescale control architecture if inner-loop controllers are robust enough to handle these model uncertainties. Moreover, we have found that rejection of only the heading error and lateral error of the front component (when traveling forward) is necessary. We obtain these error signals from the UKF algorithm and then employ a nonlinear controller, based on feedback linearization, to drive them to zero, thereby forcing the vehicle to track the profiled path.

## 3 Underground Field Trials

Field testing of the autotramming system was carried out during April 2006. Realistic test scenarios were devised and manual baseline times for traverses were obtained by timing the performance of a human operator driving the same route as the autotramming system. More than 100 tests were performed during integration and testing.

### 3.1 System Integration

The autotramming system design was initially created through modeling and simulation and then field tested on a 10-tonne capacity LHD, shown in Figure 6. The autotramming code was run on a real-time operating system, with control and localization algorithms updated at 25 Hz. SICK rangefinders with 180° field of view were used.

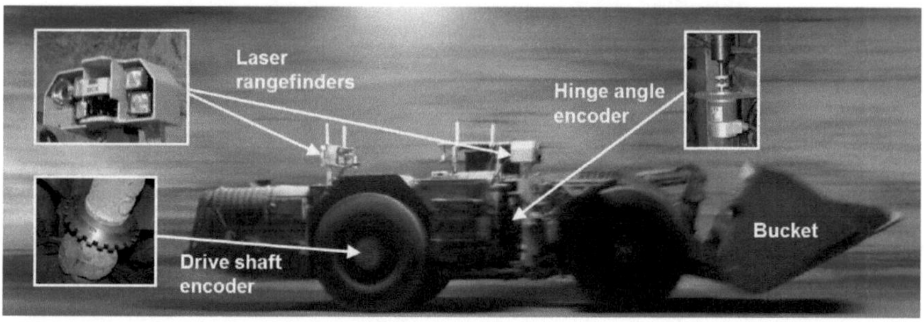


**Fig. 6.** 10-tonne capacity test LHD, with sensor layout

### 3.2 Field Testing Summary

A total of eight test scenarios were conducted: T1, T2, T3 T4E, T4L, T5E, T5L, T7. Figure 2 shows the layout of the T5 scenarios, which involved backing out of a crosscut containing a load point, switching directions and proceeding at high speed down the main drift (with $< 1$ m side clearance), parking at dump point, and then reversing this sequence to return to the load point. Each test was several hundred meters in total tramming distance. Some scenarios were conducted with the LHD bucket both empty, denoted 'E', and loaded, denoted 'L', to evaluate the performance of the system in both cases.

Average operator times for seven test scenarios are shown in Table 1 together with measured average times taken by the autotramming system to successfully complete the same route. Note that, in some cases the autotramming system actually performed better than the human operator. By examining the speed data logs, it was found that a manual operator is generally faster at performing direction switches, while the autotramming system can often perform faster on straight routes, especially when the drift width is small.

Markings made on the ground were used to estimate the positioning accuracy and repeatability of the autotramming system. An example photograph depicting this procedure is provided in Figure 7. The markings on the measurement ruler are spaced at 10 cm intervals. The measured errors for all tests are provided

**Table 1.** Average manual baseline times (in seconds)

| Test Scenario | T1 | T2 | T3 | T4E | T4L | T5E | T5L | Total |
|---|---|---|---|---|---|---|---|---|
| Human Operator | 41.8 | 27.0 | 57.0 | 85.0 | 90.0 | 81.0 | 83.0 | 464.8 |
| Autotramming | 41.3 | 25.8 | 59.5 | 97.3 | 89.8 | 81.8 | 85.0 | 480.5 |

**Fig. 7.** Example positioning measurement from a test run, where the orthogonal lines marked 'T2' show the desired pose of the wheel's outer edge

**Table 2.** Longitudinal and lateral positioning errors (cm) covering all measured stops for the relevant test runs

| Error / Test | T1 | T2 | T3 | T4E | T4L | T5Ea | T5Eb | T5La | T5Lb | T7 |
|---|---|---|---|---|---|---|---|---|---|---|
| Long. | 0 | 5 | -10 | -22 | -20 | 60 | -20 | 40 | -25 | 0 |
| | 0 | 0 | -10 | -20 | -30 | 60 | -25 | 30 | -25 | 5 |
| | 0 | 5 | -10 | -25 | -15 | 80 | -20 | 35 | -25 | 5 |
| | -5 | 0 | -10 | -30 | -20 | 70 | -25 | 30 | -20 | 5 |
| Lat. | 45 | 10 | 0 | -15 | -10 | 0 | 20 | 10 | -10 | 15 |
| | 40 | 10 | 0 | -15 | -15 | 0 | 25 | 0 | -15 | 5 |
| | 40 | 10 | 0 | -15 | -10 | 20 | 20 | 10 | -10 | 5 |
| | 45 | 15 | 0 | -10 | -10 | 20 | 20 | 0 | -10 | 0 |

in Table 2, which can also be used to infer repeatability. For the T5 cases, the position error was measured at the dump point, denoted 'a', and the end of the route, denoted 'b'.

In all but one test case the vehicle positioned within the ±50 cm tolerance set as a project goal. In this particular case, poor ground conditions contributed to a larger than normal longitudinal error. The target finishing point happened to be located on just the other side of a significant pothole, which forced the vehicle to increase throttle in order to get out of the hole, causing the vehicle to overshoot the target. It should be noted that the errors were generally very consistent within a given test scenario, most likely owing to the fact that the same route profile was used for each test run. It should be possible to remove a large portion of these systematic errors through route-specific tuning of the route profile.

## 4 Conclusion

The autotramming system described in this paper was designed with the aim of overcoming the speed and robustness deficiencies of existing technologies by applying map-based localization using an atlas of metric maps together with a suitable control architecture. Given the dynamic and unstructured nature of underground mines and the difficulties in controlling large hydraulically actuated vehicles, the presented results are very encouraging. The system has now been successfully operating in an underground mine test environment for more than 12 months (at the time of writing).

This is not to say that limitations do not exist and that challenges do not remain. The present design requires that individual routes be taught manually; this is because a *global* map of the mine environment is never generated. It would be a useful feature to have some form of globally consistent map from which route profiles could be derived without requiring that the route being physically taught.

## Acknowledgments

We would like to acknowledge R. Corcoran, D. Parry, R. Jakola, R. Ward, L. Bloomquist, and the many others who contributed to this project's success.

## References

1. Baiden, G.R.: Combining teleoperation with vehicle guidance for improving LHD productivity at Inco Ltd. CIM Bulletin 87(981), 36–39 (1994)
2. Amdahl, K., Lundström, M.: Automatic truck saves money underground. World Mining, 40–44 (November 1972)
3. Wiklund, U., Andersson, U., Hyypä, K.: AGV navigation by angle measurements. In: Proceedings of the 6th International Conference on Automated Guided Vehicle Systems, Brussels, Belgium, October 1988, pp. 199–212 (1988)
4. Scheding, S., Dissanayake, G., Nebot, E.M., Durrant-Whyte, H.: An experiment in autonomous navigation of an underground mining vehicle. IEEE Transactions on Robotics and Automation 15(1), 85–95 (1999)
5. Béranger, V., Hervé, J.-Y.: Recognition of intersections in corridor environments. In: Proceedings of the 13th International Conference on Pattern Recognition, Vienna, Austria, August 1996, pp. 133–137 (1996)
6. Steele, J., Ganesh, C., Kleve, A.: Control and scale model simulation of sensor-guided LHD mining machines. IEEE Transactions on Industry Applications 29(6), 1232–1238 (1993)
7. Roberts, J.M., Duff, E.S., Corke, P.I., Sikka, P., Winstanley, G.J., Cunningham, J.B.: Autonomous control of underground mining vehicles using reactive navigation. In: Proceedings of the 2000 IEEE Conference on Robotics and Automation, San Francisco, CA, April 2000, pp. 3790–3795 (2000)

8. Silver, D., Ferguson, D., Morris, A., Thayer, S.: Feature extraction for topological mine maps. In: Proc. of the 2004 IEEE/RSJ International Conference on Intelligent Robots and Systems, Sendai, Japan, September 2004, pp. 773–779 (2004)
9. Schaffer, G.K., Stentz, A., Whittaker, W.L., Fitzpatrick, K.W.: Position estimator for underground mine equipment. IEEE Transactions on Industry Applications 28(5), 1131–1140 (1992)
10. Madhavan, R., Dissanayake, M.W.M.G., Durrant-Whyte, H.F.: Autonomous underground navigation of an LHD using a combined ICP-EKF approach. In: Proceedings of the 1998 IEEE Conference on Robotics and Automation, Leuven, Belgium, May 1998, pp. 3703–3708 (1998)
11. Mäkelä, H.: Overview of LHD navigation without artificial beacons. Robotics and Autonomous Systems 36, 21–35 (2001)
12. Howard, A.: Multi-robot mapping using manifold representations. In: Proceedings of the IEEE International Conference on Robotics and Automation, New Orleans, Louisiana, pp. 4198–4203 (2004)
13. Elfes, A., Moravec, H.: High resolution maps from wide angle sonar. In: Proc. of the IEEE Int. Conf. on Rob. and Aut., St. Louis, MO, pp. 116–121 (1985)
14. Thrun, S., Fox, D., Burgard, W., Dellaert, F.: Robust Monte Carlo localization for mobile robots. Artificial Intelligence 128(1–2), 99–141 (2001)
15. Julier, S., Uhlmann, J.: A general method for approximating nonlinear transformations of probability distributions. Technical report, Department of Engineering Science, University of Oxford, Oxford, UK (November 1996)
16. Wan, E.A., van der Merwe, R.: The unscented kalman filter for nonlinear estimation. In: Proc. of the IEEE AS-SPCC, Lake Louise, AB (October 2000)
17. Ridley, P., Corke, P.: Autonomous control of an underground mining vehicle. In: Proceedings of the 2001 Australian Conference on Robotics and Automation, Sydney, Australia, November 2001, pp. 26–31 (2001)

# Autonomous Fruit Picking Machine: A Robotic Apple Harvester

Johan Baeten[1], Kevin Donné[2], Sven Boedrij[2], Wim Beckers[2], and Eric Claesen[1,2]

[1] Fac. of Industrial Sciences and Technology, Katholieke Hogeschool Limburg, Belgium
Johan.Baeten@iwt.khlim.be

[2] ACRO: Automation Centre for Research and Education, Katholieke Hogeschool Limburg, Belgium
Eric.Claesen@iwt.khlim.be

**Summary.** This paper describes the construction and functionality of an Autonomous Fruit Picking Machine (AFPM) for robotic apple harvesting. The key element for the success of the AFPM is the integrated approach which combines state of the art industrial components with the newly designed flexible gripper. The gripper consist of a silicone funnel with a camera mounted inside. The proposed concepts guarantee adequate control of the autonomous fruit harvesting operation globally and of the fruit picking cycle particularly. Extensive experiments in the field validate the functionality of the AFPM.

## 1 Introduction

The use of robots is no longer strictly limited to industrial environments. Also for outdoor activities, robotic systems are increasingly combined with new technologies to automate labour intensive work, such as e.g. apple harvesting [2, 10]. This paper describes the feasibility study for and the development of an Autonomous Fruit Picking Machine (AFPM)[1].

There are two main approaches in robotic apple harvesting being *bulk* [9, 10] or *apple by apple* harvesting [3, 11].

Peterson et al. [9] developed a mechanical bulk robotic harvester for apples grown on narrow, inclined trellises. This type of bulk harvesting requires, in addition to the canopy-like growth habit, uniform fruit ripeness at harvest, firm fruit, resistant to damage, and short/stiff limbs [10].

The use of an apple by apple picking system, although inherently slower, does not suffer from any of the above restrictions. Moreover, only apples of satisfactory size and maturity are selected for harvesting and can be sorted out immediately. An apple by apple picking system does, however, require an adequate *fruit gripper*. The gripper is the key element in the success of automated apple by apple harvesting. A good gripper ought to preserve the quality of the

[1] Funded by IWT-Vlaanderen under TETRA 40196.

C. Laugier and R. Siegwart (Eds.): Field and Service Robotics, STAR 42, pp. 531–539, 2008.
springerlink.com 

apple and should not damage the tree (nor the apple) during the picking cycle. Setiawan et al. [11] propose a low cost gripper with flexible inflatable parts. Our gripper consists of a flexible silicone funnel and uses suction to pick the apple.

An indispensable part of any autonomous apple harvesting machine is the vision system used to locate the apple [5, 10, 11, 12, 13]. Bulanon et al. [3] use a (RT) machine vision system to recognize the location of the fruit centre and the abscission layer of the peduncle. In contrast to most researchers, in our approach the camera is positioned in the centre of the gripper. This simplifies the calibration of the set-up and ensures adequate control.

According to the classification given by Hutchinson et al. [4], our image based control is closer to a look and move strategy than to visual servoing. There is, however, margin for improvement.

Despite previous developments towards a harvesting platform navigating autonomously between orchard rows [1], we chose not (yet) to implement an automated navigation, such as e.g. in [6] or [7], for reasons of both development time as well as safety approval.

The aim of this project was to prove the feasibility and demonstrate the functionality of an Autonomous Fruit Picking Machine (AFPM), by using existing state of the art (industrial) components. This resulted in the prototype AFPM, described in the following sections. First, section 2 describes the overall construction of the AFPM. Section 3 presents our new, patented gripper designed specifically for the apple harvesting task. Section 4 goes into the control details for one picking cycle. Finally, sections 5 and 6 summarize the results of the field experiments and conclude this paper.

## 2 Overall Construction of AFPM

The AFPM is built on a platform mounted behind an agriculture tractor. Figure 1 shows the first version of the AFPM. Figure 2 illustrates schematically the functional layout and data flow. In order to reduce the development period of the AFPM, although overkill, an industrial robot (Panasonic VR006L) is chosen as manipulator. The AFPM further consists of a tractor-driven generator for power supply, a (2D) horizontal stabilization unit, a 7th external vertical axis to enlarge the operation range, a safety scanning device, a central control unit, a touch panel PC with Human Machine Interface, a canopy and all around curtain to even out light conditions and, finally, a fruit gripper designed specifically for this task, with a camera mounted in the centre of the gripper. The flexible gripper (described in section 3) guarantees a firm grip without damaging the fruit and serves in fact as the mouth of a vacuum cleaner.

The 2D horizontal stabilization consists of two hydraulic feet and one turn over cylinder, configured as a 3-point suspension. Controlled by two level sensors, the system ensures a stable positioning of the robot platform during the picking cycle.

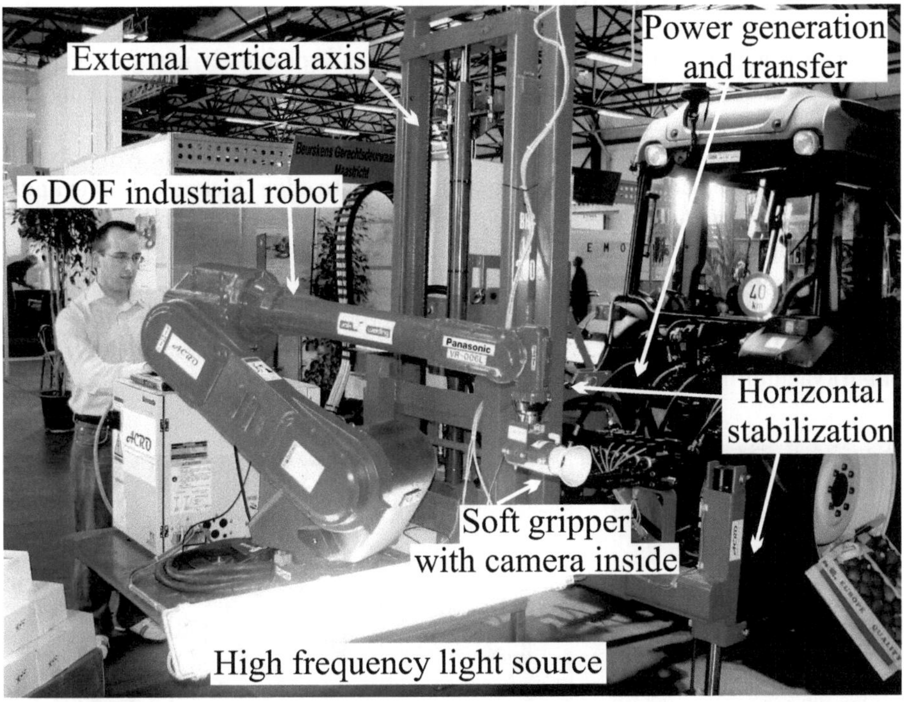


**Fig. 1.** Construction of the AFPM

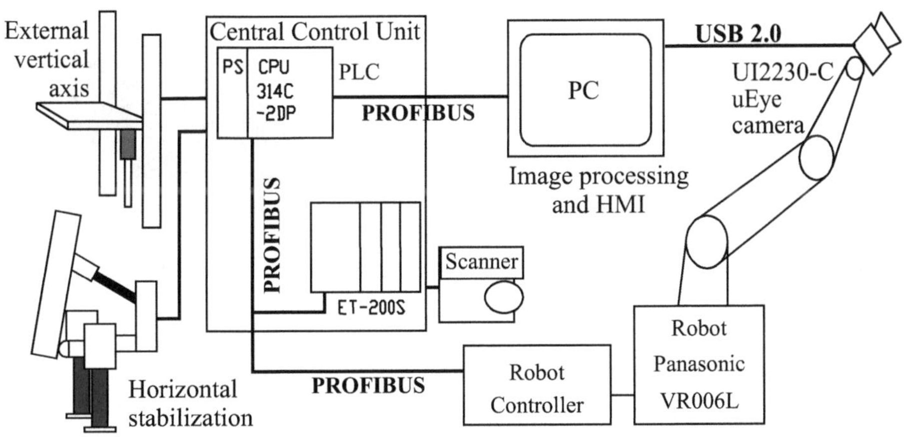


**Fig. 2.** Functional scheme of the AFPM

## 3 New Flexible Gripper

To pick the apple with as little energy as possible, a completely new fruit gripper had to be designed. Its patented design has two main functions: catching the apple and enclosing the camera. Figure 3 gives some examples.

The gripper assumes the shape of the apple and encloses it firmly. After several prototypes, the optimal funnel shape with respect to edge thickness, funnel angle and size with a trade off between flexibility and firmness was designed and tested. The current version has a maximum diameter of 10.5 cm. The gripping function is activated by vacuum suction. It is e.g. possible the pick up an apple by just pushing the gripper onto the apple and closing the vacuum port with your finger. Experiments show that even an elongated exposure of the apple to the (rather small) vacuum levels[2] used does not damage the apple in any way.

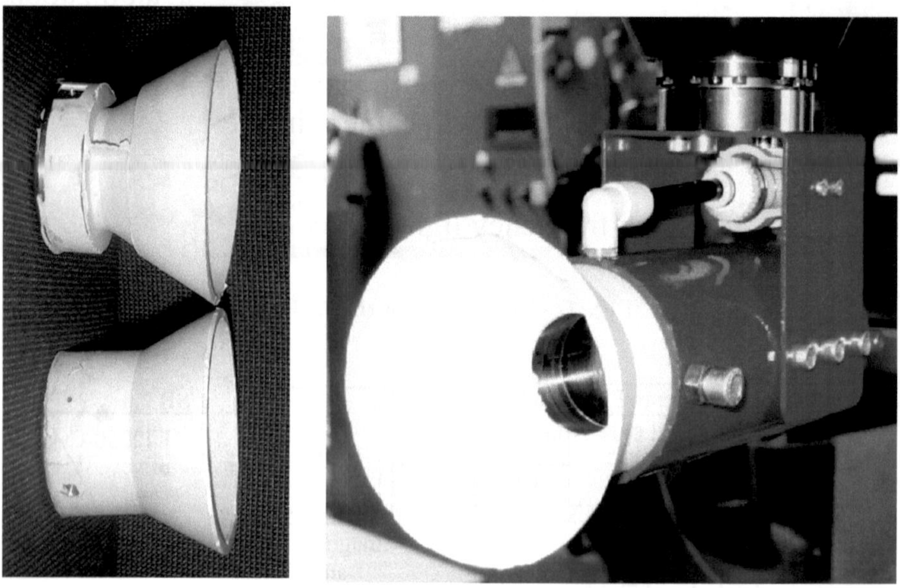

**Fig. 3.** Left: two examples of silicone gripper; right: gripper mounted on robot with camera inside

Placing the camera in the centre of the gripper offers numerous advantages. First of all, the gripper is always in line with the camera and thus with the image, which simplifies the (necessary) coordinate transformation from image to robot. Furthermore, the position of the camera is fully controllable. The camera can always point its optical axis to the apple (see section 4 and simulation figure 7). This reduces image distortion and eliminates the necessity for thorough calibration.

[2] The magnitude of underpressure ranges from 150 mbar to 230 mbar with a flow rate of 200 $m^3/h$.

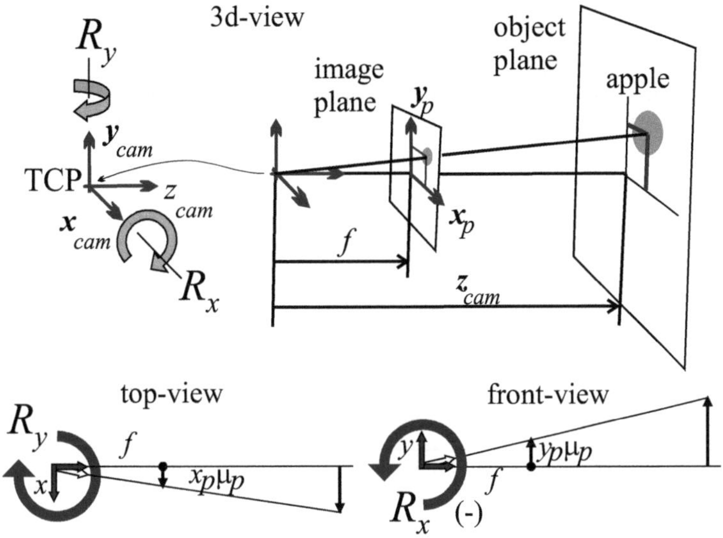


**Fig. 4.** 3D, top and front view of the camera model defining the camera frame and illustrating the computation of the rotation angles $\theta_x$ and $\theta_y$, needed to centre the apple in the image

# 4 Approach for Fruit Picking Cycle

The autonomous harvesting operation is hierarchically structured in three levels. Once the AFPM is stationed in front of the tree with active stabilization (first level), it scans the tree from 40 look-out positions or sectors (second level). For each sector, all ripe apples are listed and picked one by one in a looped task (third level). A picking operation consists of following steps:

1. The position of the apple in the image, possibly after declustering, is determined. Only ripe apples with qualified size are selected.
2. The camera rotates around $x$- and $y$-axes, by $\theta_x$ and $\theta_y$ respectively, in order to point the optical axis straight to the apple. Positioning the camera by only rotating the wrist results in a small and therefore fast robot movement. Figure 4 defines the used set-up. The rotation angles yield:

$$\theta_x = -\arctan(y_p \mu_p / f) \tag{1}$$

for the rotation around the $x$-axis and

$$\theta_y = \arctan(x_p \mu_p / f) \tag{2}$$

for the (simultaneous) rotation around the $y$-axis, with $f$ the focal length [mm], $\mu_p$ the pixel-size [mm/pix] and $x_p, y_p$ the measured centre of the apple in the image plane [pix]. As equations 1 and 2 show, the rotation angles do not depend on the distance to the apple $z_{cam}$. Only the focal length $f$ needs to be calibrated.

Even if the focal length is not exactly known, the final offset of the centre of the gripper with respect to the centre of the apple will lie within the margin of the gripper: e.g. a (rather large) error of 10% on the focal length causes an error of $1.5^o$ on $\theta_x$ (or $\theta_y$) resulting in an offset of 1.9 cm for an apple at 1 m distance. Due to the funnel-like design of our gripper, this offset will not cause a faulty picking cycle.

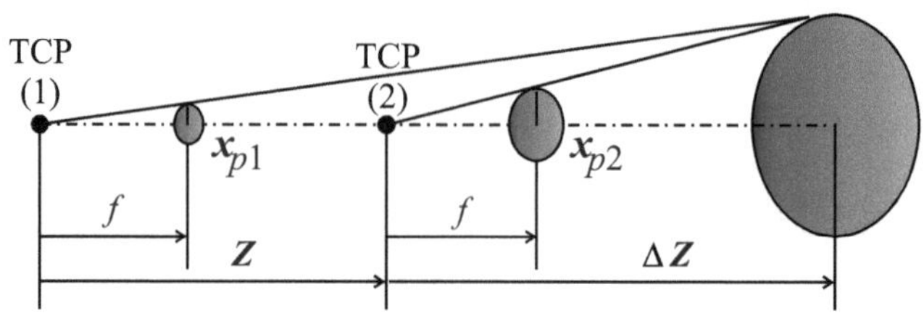


**Fig. 5.** Illustration of the remaining distance calculation

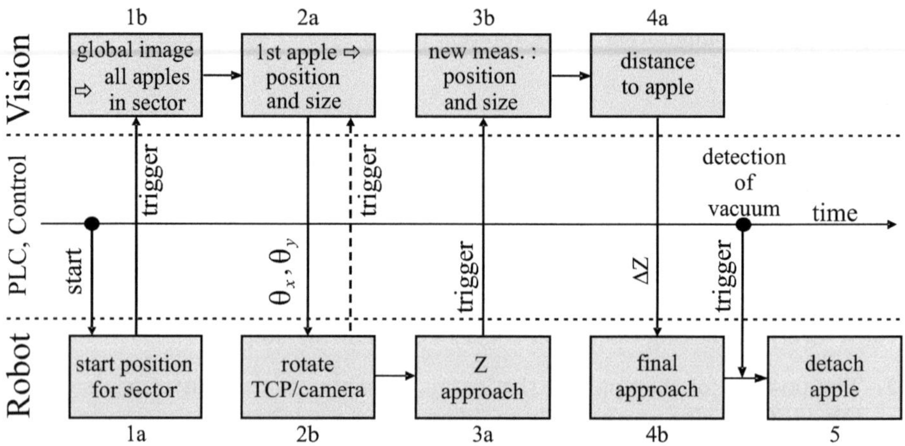


**Fig. 6.** Schematic overview of the control flow between vision system, PLC and robot during one apple-picking cycle

3. While approaching the apple, several images are processed to calculate by triangulation the remaining distance to the apple. Given the distance $Z$, travelled between two images, and $x_{p1}$, $x_{p2}$ the measured radii of the (corresponding) apple from those two images, the remaining distance $\Delta Z$ to the apple yields

$$\Delta Z = \frac{x_{p1} Z}{x_{p2} - x_{p1}}. \tag{3}$$

Figure 5 illustrates equation 3. Since the apple remains centred in the image, the correlation of a given apple in subsequent images is trivial.

4. Once the apple is within a given range of the gripper, the vacuum suction is activated. The detection of (a small level of) vacuum triggers the next step.
5. The apple is picked by rotating it and tilting it softly and is then put aside. The actual wrist/robot movement differs from section to section. It is programmed and optimized in advance with a minimum tree intrusion in mind. A standard apple transporting and collecting system still needs to be added.

Figure 6 gives a schematic overview of these 5 steps. The indicated numbers correspond to the step list for the picking cycle. The image processing is conducted on an industrial pentium IV 2 GHz PC with 1 GB RAM under Windows. The actual image processing[3] is programmed with Halcon software [8] and takes about 0.6 seconds (for step 1). Note that the complete image processing is only needed at the start of a picking cycle. Subsequent image processing should be limited to a small portion of the image and scaled down to less functional steps.

## 5 Field Experiments

Field experiments conducted over a period of several weeks in a Jonagold orchard show that the overall performance of the AFPM is more than satisfactory. The stabilization experiences no difficulties with possible rapid robot motion. Thanks to the roof and all-around cover, as shown in figure 7, the AFPM works even under poor weather conditions, although also here improvements are still possible.

The experiments demonstrate that about 80% of the apples (with diameter range from 6 cm to 11 cm) are detected and harvested. However, stem pulls of 30% are still too high. However, on should take into account the fact that the apples were treated to prolong the harvest period, causing a more firm connection of the stems to the limbs. Nevertheless, fine tuning step 5 in section 4 may lower this fault margin.

Apart from stem pulls, the apples show no bruises or damage at all. Apples not harvested are either not detectable by the vision system or not reachable for the robot manipulator. This problem can be partially solved by trimming the orchard rows into smaller hedges.

In the current setup the overall cycle time to pick one apple is 8 to 10 sec. Eliminating the current communication bottleneck between the vision system and the robot through the central controller ((trigger) signals in figure 6) will improve the overall picking cycle time.

Extra care should be taken in selecting the first apple in a cluster in order not to push off other apples. Additional vision information about nearby sharp limbs could also help to protect the gripper itself from damage.

[3] A full description of the image processing, however, falls outside the scope of this article but will be the subject of future publications.

**Fig. 7.** Left: simulation model; right: field experiments

## 6 Conclusion

The first results of the AFPM are very promising. All the necessary components are fitted together and operate as planned, hereby proving the feasibility and functionality of the AFPM. There is, however, still margin for improvement. The bottleneck in communication lies with the connection/communication between the vision-PC and the central control unit. Future work will focus on improving the bandwidth of this connection and on optimizing the image processing. The aim is to reduce the picking cycle period from an average of 9 to about 5 seconds (or less). In that case, the productivity of the AFPM will be close to the work load of about 6 workers, which makes the machine economically viable.

Special attention will go to improving the actual apple detach movement by taking into account the relative pose of the apple with respect to the branch. This should lower the stem pull percentage. Further research is also planned to automate the navigation trough the orchard and to investigate the harvesting of pears.

## References

1. Appeltants, W., Ilsbroux, W.: Automatische sturing van een plukwagen (Autonomous guidance for a harvester). MA Thesis, Katholieke Hogeschool Limburg, Belgium (1995)
2. Bulanon, D.M., Kataoka, T., Okamoto, H., Hata, S.: Feedback control of manipulator using machine vision for robotic apple harvesting. In: Proceedings of ASAE, Paper No. 053114, Tampa, USA (2005)

3. Bulanon, D.M., Kataoka, T., Ukamoto, H., Hata, S.: Development of a real-time machine vision system for the apple harvesting robot. In: SICE Annual Conference in Sapporo, Hokkaido Inst. of Techn., Japan, pp. 595–598 (2004)
4. Hutchinson, S., Hager, G., Corke, P.: A tutorial on visual servo control. IEEE Trans. on Robotics and Automation 12(5), 651–670 (1996)
5. Jimenez, A., Ceres, R., Pons, J.: A survey of computer vision methods for locating fruits on trees. Transactions of the ASAE 43(6), 1911–1920 (2000)
6. Katypitiya, J., Eaton, R., Cole, A., Meyer, C., Rodnay, G.: Automation of an agricultural tractor for fruit picking. In: IEEE Int. Conf. on Robotics and Automation, Barcelona, Spain, pp. 3212–3217 (2005)
7. Mateus, D., Aviña, G., Devy, M.: Robot visual navigation in semi-structured outdoor environments. In: IEEE Int. Conf. on Robotics and Automation, Barcelona, Spain, pp. 4702–4707 (2005)
8. MVTec Software GmbH, München, Germany, Halcon 7.1, Quick Guide, 2b edn. (2006)
9. Peterson, D.L., Bennedsen, B.S., Anger, W.C., Wolford, S.D.: A systems approach to robotic bulk harvesting of apples. Transactions of the ASAE 42(4), 871–876 (1999)
10. Peterson, D.L.: Harvest mechanization progress and prospects for fresh market quality deciduous tree fruits. HortTechnology 15(1), 72–75 (2005)
11. Setiawan, A.I., Furukawa, T., Preston, A.: A low-cost gripper for an apple picking robot. In: IEEE Int. Conf. on Robotics and Automation, New Orleans, LA, pp. 4448–4453 (2004)
12. Stajnko, D., Èmelik, Z.: Modelling of apple fruit growth by application of image analysis. Agriculturae Conspectus Scientificus 70(2), 59–64 (2005)
13. Zhao, J., Tow, J., Katupitiya, J.: On-tree fruit recognition using texture properties and color data. In: Int. Conf. on Intelligent Robots and Systems, Alberta, Canada, pp. 3993–3998 (2005)

# Standing Assistance System for Rehabilitation Walker

Daisuke Chugo[1], Wataru Matsuoka[1], Songmin Jia[1], Kunikatsu Takase[1], and Hajime Asama[2]

[1] The University of Electro-Communications, 1-5-1, Chofugaoka, Chofu, Tokyo, Japan
{chugo,matsuoka,jia,takase}@is.uec.ac.jp
[2] The University of Tokyo, 5-1-5, Kashiwanoha, Kashiwa-shi, Chiba, Japan
asama@race.u-tokyo.ac.jp

**Summary.** This paper proposes a walker system with power assistance device for standing up motion. Our system focuses on domestic uses for aged person who needs nursing in their daily life. Our key ideas are two topics. The first topic is new assistance mechanism with four parallel linkages. Our proposed manipulator mechanism requires only smaller actuators and realizes rigid structure with lighter linkages comparing with general manipulator. Thus, we can design our assistance system compactly with low-cost using our mechanism. The second topic is the combination of force and position control. According to the patient's posture during standing up, our control system selects more appropriate control method from them. We use the reference of standing-up motion which is based on the typical standing up motion by nursing specialist for realizing the natural assistance. The performance of our proposed assistance system is verified by computer simulations and experiments using our prototype.

**Keywords:** Walker; Standing up motion; Force control; Position control.

## 1 Introduction

In Japan, the population ratio of senior citizen who is 65 years old or more exceeds 20[%] at January 2004 and rapid aging in Japanese society will advance in the future. [1] In aging society, many elderly people cannot perform normal daily household, work related and recreational activities because of decrease in force generating capacity of their body. Today, the 23.5[%] of elderly person who does not stay at the hospital cannot perform daily life without nursing by other people. [2] For their independent life, they need assistance system which enable them to perform daily life easily in their home even if their physical strength reduces.

Standing up motion is the most serious and important operation in daily life for elderly person who doesn't have enough physical strength. [3, 4] In typical bad case, elderly person who doesn't have enough physical strength will cannot operate standing up motion and will falls into the wheelchair life or bedridden life. Furthermore, if once elderly person falls into such life, the decrease of physical strength will be promoted because he will not use his own physical strength.

C. Laugier and R. Siegwart (Eds.): Field and Service Robotics, STAR 42, pp. 541–550, 2008.
springerlink.com 

[5] Therefore, force assistance system is required to use part of the remaining strength of the patient for standing-up motion in order not to reduce their muscular strength.

In previous works, many researchers developed power assistance devices for standing up motion. However, these devices are large scale and they are not suitable for family use. [6, 7] Furthermore, these devices assist all necessary power for standing up and they do not discuss the using the remaining physical strength of patients. Therefore, there is a risk of promoting the decrease of their physical strength.

In this paper, we develop a walker system with power assistance device for standing up motion. Our system is based on a walker which is popular device for aged person in their daily life and realizes the standing up motion using the support pad which is actuated by our developed assistance mechanism with three degrees of freedom. For using the remaining physical strength, our system uses the motion pattern which is based on the typical standing up motion by nursing specialist as control reference. [9]

Our key ideas are two topics. The first is new assistance mechanism with four parallel linkages which enables the system to be rigid and compact. The second is the combination of force and position control. Using our control scheme, the patients can stand up with fewer loads and can use their own remaining physical strength during the motion. We verify the performance of our proposed assistance system through simulations and experiments using our prototype.

## 2 System Configuration

### 2.1 Assistance Mechanism

Fig.1 shows our proposed assistance system. The system consists of support pad with three degrees of freedom and the walker system. (Fig.1(a)) The support pad is actuated by our new assistance mechanism with four parallel linkages.

Fig.1(b) shows the frame-kinematic model of our assistance mechanism. The position of the support pad (2DOF, red lines in Fig.1(a)) is coordinated by Actuator 1 and Actuator 2 which are equipped on O point. Actuator 1 drives Link1 ($\alpha$) and Actuator 2 drives Link 2 ($\beta$). Using four parallel linkages mechanism, two actuators can generate the position of support pad. The inclination of the support pad (1DOF, the blue line in Fig.1(a)) is coordinated by Actuator 3 which is equipped on P point.

The advantages of our proposed mechanism are two topics. The first is that two main actuators (Actuator 1 and 2) keep only weight of linkages. In general manipulator, the actuator of lower part supports not only weight of linkages but also actuators of upper part. Therefore, the actuators of lower part are required high output traction and tend to be heavy. On the other hand, using our mechanism, main actuators are mounted on the walker body (O point) and they are required to keep only weight of linkages. As the result, we can use smaller actuators for our developed assistance mechanism.

The second is parallel linkage mechanism. The parallel linkage mechanism is rigid for a twist load, therefore, system realizes same strength with lighter linkages comparing with the general manipulator. Using our proposed mechanism, we can use smaller actuators and lighter linkages, and our system realizes compact design with low cost.

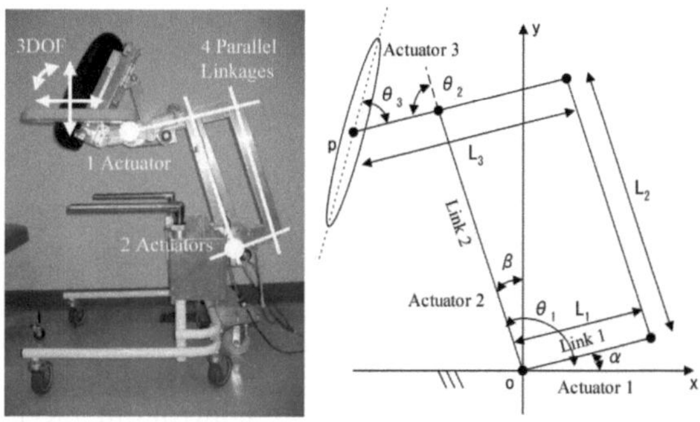


(a) Prototype system (b) Parallel Linkage Mechanism

**Fig. 1.** Our assistance system

### 2.2 Kinematics

In this section, we derive the inverse kinematics of our proposed linkage mechanism. In our proposed mechanism, we can derive the position of P point (We define its coordinate as $(x_p, y_p)$) using $\theta_1$ and $\theta_2$. However, it is $\alpha$ and $\beta$ that is used to control the actuators. Therefore, we set angular values and length of linkages as in Fig.1(b).

$$\theta_1 = \frac{\pi}{2} + \beta, \qquad \theta_2 = \frac{\pi}{2} + \alpha - \beta \tag{1}$$

$$l_1 = L_2, \qquad l_2 = L_3 - L_1 \tag{2}$$

Now, we consider the geometric relationships among the position of P point and these angular values, we can derive (3) and (4). From (3) and (4), $\theta_2$ is (5)

$$x_p = l_1 \cos\theta_1 + l_2 \cos(\theta_1 + \theta_2) \tag{3}$$

$$y_p = l_1 \sin\theta_1 + l_2 \sin(\theta_1 + \theta_2) \tag{4}$$

$$\theta_2 = \arccos\left(\frac{x_p^2 + y_p^2 - l_1^2 - l_2^2}{2 l_1 l_2}\right) \tag{5}$$

We set $k_1$ and $k_2$ as (6), (3) and (4) are expressed as (7) and (8).

$$k_1 = l_1 + l_2 \cos\theta_2, \qquad k_2 = l_2 \sin\theta_2 \tag{6}$$

$$x_p = k_1 \cos\theta_1 - k_2 \sin\theta_1 \tag{7}$$

$$y_p = k_1 \sin\theta_1 + k_2 \cos\theta_1 \tag{8}$$

Furthermore, we set $r$ and $\gamma$ as (9), $k_1$ and $k_2$ are expressed as (10). Using (10), (7) and (8) are expressed as (11) and (12).

$$r = \sqrt{x_p^2 + y_p^2}, \qquad \tan\gamma = \frac{k_2}{k_1} \tag{9}$$

$$k_1 = r\cos\gamma, \qquad k_2 = r\sin\gamma \tag{10}$$

$$x_p = r\cos(\gamma + \theta_1) \tag{11}$$

$$y_p = r\sin(\gamma + \theta_1) \tag{12}$$

From (11) and (12), we can derive (13).

$$\tan(\gamma + \theta_1) = \frac{y_p}{x_p} \tag{13}$$

Thus, $\theta_1$ is (14). Using (5) and (14), we can control our mechanism.

$$\theta_1 = \arctan\left(\frac{y_p}{x_p}\right) - \arctan\left(\frac{l_2\sin\theta_2}{l_1 + l_2\cos\theta_2}\right) \tag{14}$$

# 3 Motion Control

## 3.1 Motion by Nursing Specialist

In previous study, a lot of standing up motions for assistance are proposed. Kamiya [10] proposed the standing up motion which uses remaining physical strength of the patients maximum based on her experience as nursing specialist. In our previous work, we analyze this standing up motion and find that Kamiya scheme is effective to enable standing up motion with smaller load. [9]

Thus, we measure the angular values among the linkages, which reflect the relationship of body segments during standing up motion with Kamiya scheme. The angular value is derived using the body landmark as shown in Fig.2(a). We assume the standing up motion is symmetrical and we discuss the motion as movement of the linkages model on 2D plane. [8]

From measuring results, the trunk needs to incline to forward direction during lifting up from chair as shown in Fig.2(b). Y-axis shows the angular value (Pelvis and trunk, knee, ankle) and X-axis shows the movement pattern ($\hat{s}$) [14] which means the ratio of standing up operation as (15).

$$\hat{s} = \frac{t}{t_s} \tag{15}$$

where $t_s$ is required time to the standing up operation and $t$ is present time.

In order to realize the natural standing up motion by Kamiya scheme, we use these results as track references.

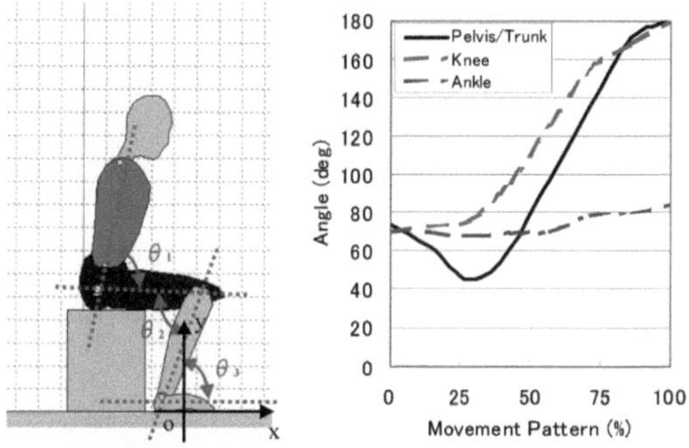


(a) Coordination (b) Angular Value

$\theta_1$, $\theta_2$ and $\theta_3$ show the angles of pelvis/trunk, knee and ankle, respectively.

**Fig. 2.** Standing-up motion with Kamiya scheme

### 3.2 Derivation of Control Reference

In this section, we derive the control reference of our assistance system which can realize the standing up motion proposed by Kamiya using a computer simulation. Fig.3(a) shows the simulation setup. The parameters are chosen from a standard body data of adult male as shown in Table 1. [11][12] We use the Working Model 2D as a physical simulator and MATLAB as a controller. Both applications are linked by Dynamic Data Exchange function on Windows OS.

From these simulation results, Fig.3(b) shows position references of the support pad and Fig.3(c) shows reference angle. The coordination is defined as shown in Fig.3(a). In Fig.3(b), the start point is lower and the end point is upper. Using these tracks as the position control reference, our assistance system can realize the standing up motion which Kamiya proposes.

### 3.3 Force Control

For realizing the motion of Kamiya scheme using the remaining physical strength of patients, the required condition for standing-up motion is different at each phase during standing up motion. [9]

- When the patient lifts up his body, it is required to reduce the load of knee joint. Therefore, it is required to force assistance.
- In other phases, it is required to maintain the standing up motion. Therefore, it is required to position maintaining assistance.

In order to fulfill both conditions, we propose new control scheme. Proposed control scheme combines dumping control and position control. The dumping control is suitable for the control of the objects with contact [13]. We use a force sensor attached on the support pad (P point in Fig.3(a)) for switching condition between the position control and the dumping control.

$$v_i = v_i^{ref} - B\left(F - F_0\right) - K\left(x_i - x_i^{ref}\right) \tag{16}$$

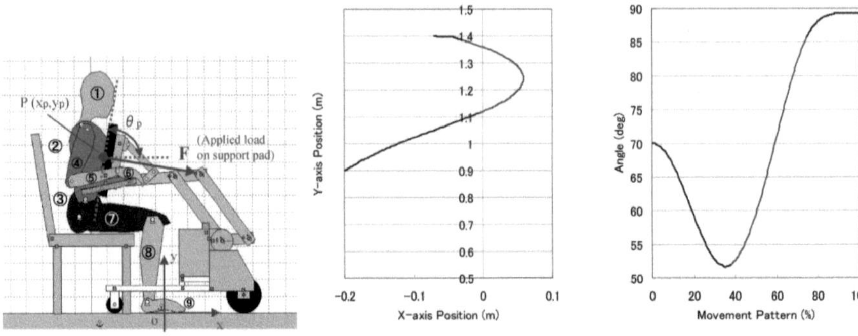


(a) Simulation Setup (b) Position Reference (c) Angle Reference
Red line shows the force control mode. (We discuss in next section.)

**Fig. 3.** Derived control references

**Table 1.** Human Parameters

| Number | Link Name | Mass [kg] | Length [m] | Width [m] |
|---|---|---|---|---|
| 1 | Head | 5.9 | 0.28 | 0.21 |
| 2 | Trunk | 27.2 | 0.48 | 0.23 |
| 3 | Hip | 18.1 | 0.23 | 0.23 |
| 4 | Humerus | 4.5 | 0.39 | 0.12 |
| 5 | Arm | 2.7 | 0.35 | 0.08 |
| 6 | Hand | 0.5 | 0.2 | 0.07 |
| 7 | Femur | 9.1 | 0.61 | 0.17 |
| 8 | Leg | 4.5 | 0.56 | 0.16 |
| 9 | Foot | 0.8 | 0.26 | 0.11 |

where $F$ is the applied force on the support pad as shown in Fig.3(a). $F_0$ is the threshold which selects force or position control. $v_i^{ref}$ is the velocity reference and $x_i^{ref}$ is the position reference derived by reference tracks in Fig.3(b) and (c). $v_i$ is actucal control velocity reference. $B$ and $K$ are constants.

Using this control scheme, when the patient lifts up his body and the load is large, he will lean on the support pad aggressively and the applied force on the support pad will be large. Therefore, our system can detect suitable condition between the position control and the dumping control.

### 3.4 Computer Simulation

We verify the performance of our control scheme by the computer simulation. In this experiment, the human model stands up with Kamiya motion and our assistance system assists him using our proposed control scheme. Furthermore, we compare the result by our proposed scheme with the result using only the position control reference.

We use the control references as shown in Fig.3(b) and (c) which are derived from standing up motion with Kamiya scheme in previous section. The simulation parameters are chosen from Table 1. The coordination is defined in Fig.3(a).

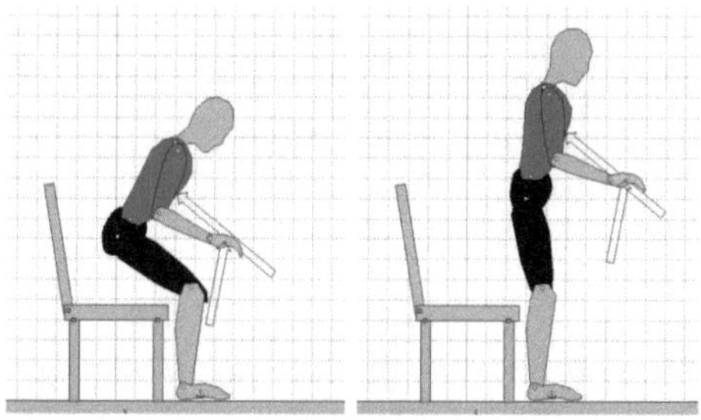

**Fig. 4.** Simulation Results

**Table 2.** Simulation Results

| | | Pelvis/Trunk | Knee | Ankle |
|---|---|---|---|---|
| Only Position Control | Peak(Nm/kg) | 0.48 | 0.59 | 0.38 |
| | Workload(Ws) | 29.0 | 39.3 | 31.0 |
| Proposed Control | Peak(Nm/kg) | 0.49 | 0.50 | 0.38 |
| | Workload(Ws) | 26.9 | 36.3 | 28.8 |

Furthermore, we set $F_0$=150[N] as a threshold which is derived experimentally. Red lines in Fig.3(b) and (c) show that the system selects the force control.

Fig.4 shows the simulation results. Fig.4 is standing up motion using our proposed assistance control. Allows in Fig.4 show the applied assistance force to the patient. The support pad applies the force vertically to the chest and the hand. Using our proposed control scheme, we verified that our assistance system realizes the natural standing up motion.

Table 2 shows the maximum output and output power of each joint. From these results, our proposed control scheme reduces maximum output into 0.5[Nm/kg] comparing with the result by the position control. In general, if the applied load to each joint is heavier than 0.5[Nm/kg], it is difficult to stand up for the elderly person. [15] Therefore, the patient can adapt this motion using his remaining physical strength.

Furthermore, with our proposed control scheme, the workload of each joint is maintained comparing with the result by the position control. During standing up motion, the patient is required to use his 92[%] of physical strength comparing with physical strength without the force assistance control. This means our assistance system can use part of his remaining strength in order not to reduce muscular strength. Therefore, we can verify that the system selects more appropriate control method and our proposed control scheme is effective.

## 4 Experiment

Here, we verify the performance of our prototype system by the experiment. In this experiment, our prototype system assists the patient with control references shown in Fig.3(b) and (c). As the result of the experiment, our system can assist

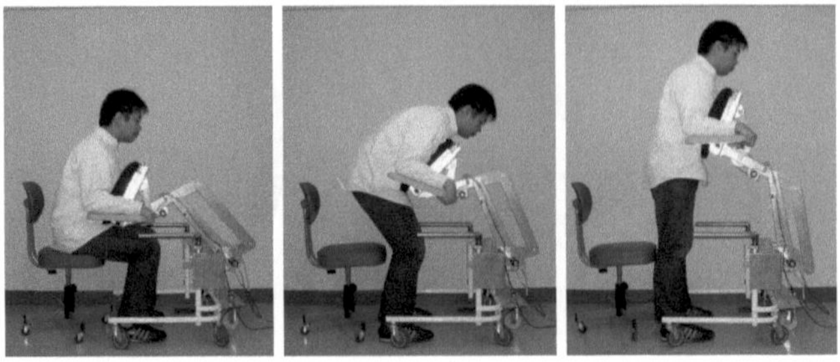

**Fig. 5.** Experimental results

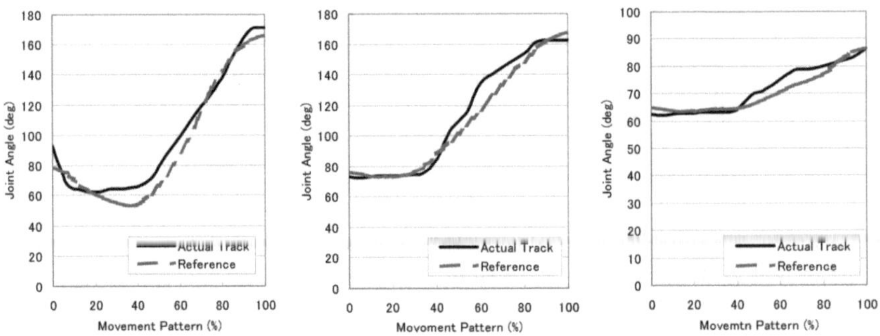


(a) Pelvis / Trunk (b) Knee (c) Ankle

**Fig. 6.** Angular value of each joint

the patient as shown in Fig.5. The height of the patient is 170[cm] and the system lifts him at 30[sec].

Fig.6 shows the tracks of the position of the patient's waist, knee and ankle joint and their control references. From Fig.6, both tracks are almost same line and this means our assistance system realizes the natural standing-up motion by nursing specialist.

## 5 Conclusion

In this paper, we develop the novel assistance system for the standing up motion. Our system focuses on family use and system can assist the elderly person using part of their remaining strength, in order not to reduce muscular strength.

In order to fulfill this condition, we propose new assistance manipulator mechanism with parallel linkages. Our developed mechanism enables the assistance system to be rigid and compact with low costs.

Furthermore, we design the novel control scheme which combines the dumping and the position control. According to the posture of the patient during standing up motion, our control system can select more appropriate control method from them. Using our assistance system, the load of knee joint reduces to 0.5[Nm/kg], which the elderly person can adapt with their own physical strength, generally. Our system requires the patient to use his 92[%] of physical strength comparing with physical strength without the force assistance control. This means our assistance system can use part of his remaining strength in order not to reduce muscular strength.

In our future work, we will develop the active walker function, which enables to assist the patients during standing, walking and seating motion continuously.

## Acknowledgement

This study is supported in part by *France Bed Medical Home Care Research Subsidy Foundation.*

## References

1. Statistics Bureau, Ministry of Internal Affairs and Communications. Current Population Estimates as of October 1, Japan (2004) (February 2007), `http://www.stat.go.jp/english/data/jinsui/2004np/index.htm`
2. Ministry of Health, Labour and Welfare. Annual Reports on Health and Welfare 2001 Social Security and National Life, Japan (2001) (February 2007), `http://www.mhlw.go.jp/toukei/saikin/hw/k-tyosa/k-tyosa01/4-3.html`
3. Alexander, N.B., Schultz, A.B., Warwick, D.N.: Rising From a Chair: Effects of Age and Functional Ability on Performance Biomechanics. J. of Geometry: MEDICAL SCIENCES 46(3), M91–98 (1991)
4. Hughes, M.A., Schenkman, M.L.: Chair rise strategy in the functionally impaired elderly. J. of Rehabilitation Research and Development 33(4), 409–412 (1996)
5. Hirvensalo, M., Rantanen, T., Heikkinen, E.: Mobility difficulties and physical activity as predictors of morality and loss of independence in the community-living older population. J. of the American Geriatric Society 48, 493–498 (2000)
6. Nagai, K., Nakanishi, I., Hanabusa, H.: Assistance of self-transfer of patients using a power-assisting device. In: Proc. of the IEEE Int. Conf. on Robotics and Automation, pp. 4008–4015 (2003)
7. Funakubo, A., Tanishiro, H., Fukui, Y.: Power Assist System for Transfer Aid. J. of the Society of Instrument and Control Engineers 40(5), 391–395 (2001)
8. Nuzik, S., Lamb, R., Vansant, A., Hirt, S.: Sit-to-Stand Movement Pattern, A kinematic Study. Physical Therapy 66(11), 1708–1713 (1986)
9. Chugo, D., Okada, E., Kawabata, K., Kaetsu, H., Asama, H., Miyake, N., Kosuge, K.: Force Assistance Control for Standing-Up Motion. In: Proc. of IEEE/RAS-EMBS Int. Conf. on Biomedical Robotics and Biomechatoronics, p. F132 (2006)
10. Kamiya, K.: Development and evaluation of life support technology in nursing. In: Proc. of 7th RACE Symp. Research into Intelligent Artifacts for the Generalization of Engineering, The University of Tokyo, pp. 116–121 (2005)
11. Digital Human Research Center, AIST. Human Body Properties Database (March 2006), `http://www.dh.aist.go.jp/bodyDB/index-e.html`
12. Omori, K., Yamazaki, Y., Yokoyama, H., Aoki, U., Kasahara, M., Hiraki, K.: The relationship between strength in the lower extremity and the ability to stand up from a chair in elderly inpatients. Sogo Rehabilitation 30(2), 167–171 (2001)

13. Sugihara, T., Kawabata, K., Kaetsu, H., Asama, H., Kosuge, K., Mishima, T.: Development of a Reasonable Force Sensor for a Standing-up and Sitting Motion Support System. In: Proc. of Robotics and Mechatronics Conf., pp. 1P1–H-11 (2004)
14. Schenkman, M., Berger, R.A., Riley, P.O., Mann, R.W., Hodge, W.A.: Whole-Body Movements During Rising to Standing from Sitting. Physical Therapy 70(10), 638–648 (1990)
15. Fisher, N.M., Pendergast, D.R., Calkins, E.C.: Maximal Isometric Torque of Knee Extension as a Function of Muscle Length in Subjects of Advancing Age. Arch. Phys. Med. Rehabil. 71(10), 729–734 (1990)

# Magnetic Wall Climbing Robot for Thin Surfaces with Specific Obstacles

Wolfgang Fischer[1], Fabien Tâche[2], and Roland Siegwart[3]

[1] Swiss Federal Institute of Technology Zürich
wfischer@ethz.ch
[2] Swiss Federal Institute of Technology Zürich
ftache@ethz.ch
[3] Swiss Federal Institute of Technology Zürich
rsiegwart@ethz.ch

**Summary.** This paper describes a novel solution to a mobile climbing robot on magnetic wheels, designed for inspecting the interior surfaces in gas tanks made out of thin metal sheets. These surfaces were inaccessible by previous climbing robots due to the following restrictions:

1. Ridges, where the magnetic force decreases to almost zero
2. Angular transitions between the surfaces (135°)
3. Thin metal sheets that cannot provide high magnetic forces

The main optimization criterion for this robot was to design it as light as possible, as the surface was also considered to be very fragile. As the here described type of application is very special and was not examined much in previous publications, this work also stresses on the early analysis phase. This phase mainly consists of tests to optimize magnetic wheels for thin surfaces and mechanical calculations for robots on magnetic wheels.

The chosen concept is described in detail, explaining how the robot moves around and passes the obstacles. The analysis of the most critical cases is presented, as well as some details about magnetic wheels and actuators.

## 1 Introduction

The robot described here was designed for inspecting gas tanks that are made out of thin metal sheets and are installed in huge ships. From time to time, they have to be analyzed for leaks, especially at the welds. For doing this, helium is injected in the surrounding structure from outside. A sensor that can detect this helium then is moved to all places inside the tank to find the position of the leak. Until now, this sensor was carried by a balloon that was operated manually, using some ropes. As this method was very slow and imprecise, a better inspection system, preferably consisting of a climbing robot on magnetic wheels, had to be developed. As the environment cannot support much force, the main goal was to make this robot as light as possible. To simplify the control and increase the reliability, another method was using only few actuators. To ensure a correct functionality, the most critical risks were analyzed. This analysis does not only incorporate the possibility of some components within the robot breaking down. It also accounts for the risks of plastically deforming the environment, falling or slipping.

C. Laugier and R. Siegwart (Eds.): Field and Service Robotics, STAR 42, pp. 551–561, 2008.
springerlink.com 

The work is structured into seven sections: Section 2 provides a short overview on the state of the art considering mainly climbing robots on magnetic wheels and their mechanisms to pass obstacles. Section 3 explains the application and deduces the most critical requirements. In Section 4 the first tests and analyses are presented that were necessary to obtain the best concept in an effective way. Sections 5 and 6 describe the chosen concept and explain how it works in the most critical cases. The last section provides a short summary, together with an outlook on future versions of the robot.

## 2 State of the Art

The robot analyzed here is a mobile climbing robot that uses magnetic wheels for also moving on vertical walls and on the ceiling. In order to pass difficult obstacles, it uses active elements within the structure. For better structuring, this section is divided into the following 3 fields of technology.

### 2.1 Mobile Climbing Robots

Mobile climbing robots are used in various applications - not only for non-destructive testing, but also for cleaning, repairing and other tasks. As the applications are very different, there exist many types of robots. The two main criteria to distinguish these types are the adhesion mechanism and the locomotion principle. According to [1], the most common adhesion mechanisms are magnetism [2]-[12], vacuum suction [13]-[15], specific attachment devices such as rails [17] or pegs and grippers/clamps [16]. The locomotion can be based either on wheels [2]-[8], tracks [11], legs [9] or arms. For both profiting from the simple control of wheeled locomotion and the high obstacle passing capabilities of legged locomotion, some robots use active elements within the structure ([8], [15], and [19]). As the here described robot uses magnetic wheels and a structure with active elements, the following descriptions mainly stress on this type.

### 2.2 Climbing Robots on Magnetic Wheels

The simplest structure for a climbing robot is a vehicle on two (like the "Magnebots" [2]) or more magnetic wheels. These wheels increase the normal force to the ground and thus allow for being more independent from the direction of gravity. The most common structure of such wheels is described in U.S. Patent 3,690,393 [3] and drawn in Fig. 3: A cylindrical magnet in the middle; two plates out of magnetic steel and of slightly bigger diameter on both sides (to better conduct the magnetic force into the surface); and a thin layer of rubber around the steel plates (to increase the friction to the ground from $\mu$=0.3-0.5 to $\mu$=0.5-0.8). Sometimes the wheels are also equipped with some special mechanisms that facilitate passing angular transitions. These mechanisms can either be passive [6] or active [7]. The most common applications for these robots are non-destructive testing on the outside surface of storage tanks [4], [5] and general transport tasks [2].

### 2.3 Active Elements in the Structure

Active elements in the structure are often used in wheeled robots to profit from both the advantages of wheeled and legged locomotion – high speed and simple control (wheels) combined with an extended mobility (legs). Examples for this type are the "Alicia³" climbing robot with suction pumps [15] or the "Octopus" rough-terrain robot [19]. In [8], the "Pipe Inspection Robot (PIR)" is described. This robot uses 3 pairs of magnetic wheels that can be lifted with linear actuators in order to pass obstacles on the outside surfaces of pipes. Its structure has some similarities to our robot, but needs more actuators (26 in the latest version that is also able to turn) and is not able to move on very thin surfaces where the magnetic force gets saturated.

## 3 Application and Requirements

As already pointed out in the introduction, the goal is to develop a climbing robot that can bring a sensor module (approximately 1kg) to every point in these specific gas tanks (Fig. 1) – an extremely difficult environment where it was impossible to go with previous climbing robots due to the following difficulties:

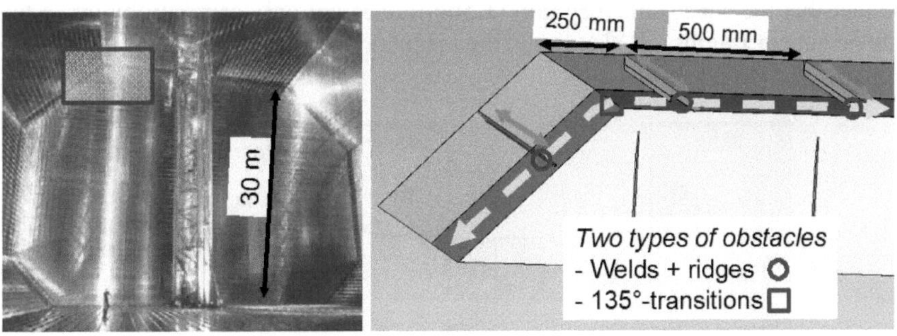


**Fig. 1.** The environment to inspect

### 3.1 Obstacles: Ridges and Angular 135°-Transitions

The most difficult obstacles in this environment are the ridges. Their height ranges up to 40mm. Whereas just rolling onto them would not be difficult with big enough wheels, the most critical problem is that the magnetic force on top of these ridges decreases to less than 15% of the original value. Thus, when rolling on a vertical wall or on the ceiling, the robot would slip or fall down if the attraction force is not compensated by another wheel that stays on the ground.

The other type of obstacle is the angular 135°-transition between two surfaces. As the surface is not rounded, the wheels get in contact at two points. When moving away from these points whilst entering the new surface, the robot needs a lot more traction than for just climbing a vertical wall.

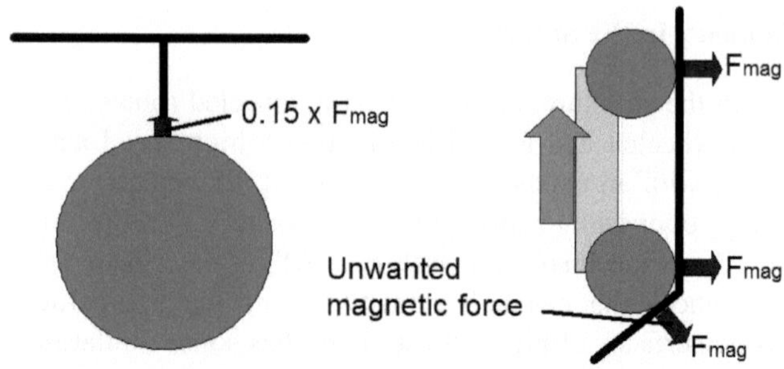


**Fig. 2.** Obstacles: force decrease on ridges (left); unwanted forces at angular 135° transitions (right)

Both obstacles only are encountered on paths perpendicular to the ridges (Fig. 1, dashed yellow line). On horizontal paths (Fig. 1, green line) there are no obstacles. The distance between the ridges is always 500mm, between the last ridges and the 135° transition it is only 250mm. This distance limits the maneuvering space of the robot.

## 3.2 Reduced Magnetic Force on Thin Sheet Material

The fact that the sheets are very thin (1.5mm) does not only cause problems in terms of environment fragility. It also limits the magnetic forces:

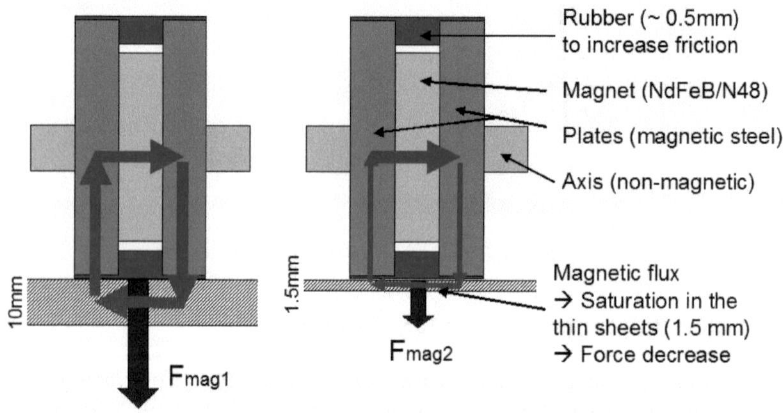


**Fig. 3.** Simple magnetic wheel [3] and the force decrease on thin ground

In order to result in high attraction forces, the magnetic flux has to go mainly through the metallic surface of the environment, as the magnetic resistance there is lower than in air. If this metallic surface is very thin, the magnetic flux becomes saturated there and the force is then limited. In our case we measured approximately 90N for an Ø60mm-wheel that could hold 200N on 10mm thick metal plates. For this reason, preventing the robot from slipping or falling down in the most difficult situations was crucial.

### 3.3 General Goals

Apart from just moving around in this difficult environment, the goals were to achieve a horizontal speed of at least 5m/min, to keep the mass below 10 kg and to keep the whole system as simple and reliable as possible.

## 4 First Analysis and Wheel Optimization

At first, the prior idea – the fact that a robot on magnetic wheels really is the best concept for this application – had to be verified. When analyzing alternatives, all of them did not seem as promising: A robot with suction cups and a pump would be heavy, complex and not safe in case of power interruption. A robot on clamps or on magnetic legs would not be fast enough to achieve the required minimum speed (5m/min). Installing rails on the whole surface would be too much effort. Robots on magnetic tracks did not seem promising either; as such tracks would be difficult to turn and be even worse to pass the ridges. Finally, a locomotion based on magnetic wheels seemed to be the best possibility. For better estimating the necessary properties of such a robot, some more analysis in this field of technology was necessary, consisting of some tests, calculations and simulations.

### 4.1 Tests and Improvement of the Magnetic Wheels

For testing the magnetic force, we used a tension spring balance and pulled the wheels off from different types of surfaces. Whereas on thick plates (10mm), 200N were measured, on the thin sheets (1.5mm) only 90N were possible.

For obtaining enough force on this thin surface also, two magnetic wheels were set in parallel. For making the whole structure lighter, the wheel plates were made in a conic shape. This structure saves about 30% of mass without decreasing the magnetic force. For increasing the friction and protecting the environment, all wheel plates are covered with a thin rubber.

### 4.2 Analysis of the Ridges

Together with the above tests, we did also analyze the behavior when rolling over the ridges. As already pointed out in the previous section (Fig. 2) the magnetic force thereby decreases to only 15%. Thus, in order to pass these ridges, an extra mechanism is necessary that can compensate for this force decrease. Among various possibilities for obtaining this goal, the one with two non-motorized extra wheels outside the central body and driven by linear actuators was considered as the best. How this mechanism passes the ridges is described in section VI.

### 4.3 Mechanical Calculation 135°-transitions, Roofs and Vertical Walls

The 135°-transitions were calculated for all 8 cases if inclination and both directions each. As the speed is supposed to be very slow, these calculations could be done with a static model. Estimating a mass of 10kg and realistic values for the geometry, we estimated a required friction coefficient of $\mu_{required}$=0.37 and a necessary torque of 3.5Nm in each wheel pair.

Also, the robot on the roof and when passing the ridges on vertical walls (with traction on only one motorized wheel pair, Fig. 5) was analyzed. These two cases resulted to be less critical than the 135°-transitions: The necessary motor torque for passing ridges on a vertical wall is only about 50% of the maximum value, and also the required friction coefficient $\mu_{required}$ (risk against slipping) was calculated to be less critical.

The risk against falling off the roof was not seen as critical either, as for passing ridges and 135°-transitions, the magnetic force has to be strong enough to prevent slipping.

## 5 The Chosen Concept

Among many promising concepts, the following one (Fig. 4) was selected to be the best, considering simplicity, mass, security against slipping and maneuvering space. It consists of 4 independently motorized wheel units, each of them being equipped with a steering motor, such that it can turn on spot. The wheels are designed as described in the last section – with conic shaped plates and always two in parallel.

For passing the ridges, two lifter units are situated outside the central body. These lifter units use linear actuators consisting of a spindle inside a stepper motor. This type of actuator has a much better force/mass ratio than other actuators of the same size (300N at only 300g) and thus was preferred. To guide the movement, linear slide bearings are used. The total mass was estimated to 10kg in the CAD software. This value will perhaps increase a little bit, when all sensors and motor controllers will be included. The magnetic force was measured to 170N for each wheel unit and to 2 x 90N for each small wheel unit on the lifter. The total size amounts to 200x200x300mm³, approximately.

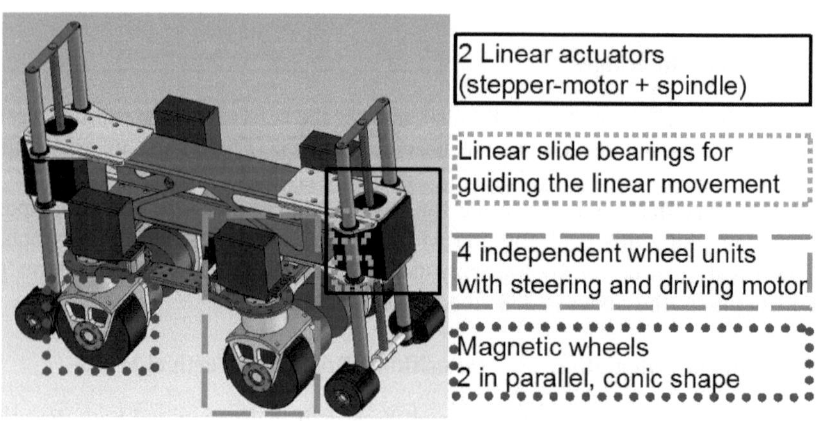


**Fig. 4.** Overview on the chosen concept

# 6 Behavior in the Most Critical Cases

## 6.1 Ridges

For passing the ridges, the main wheels can be lifted off the ground. During this lift-off, the missing attraction force is taken over by the non-motorized extra wheels (Fig. 5). For saving mass, space and reducing the force in the linear actuators, these extra wheels are smaller and less strong than the main wheels ($F_{magE}$ = 2 x 90N instead of 2 x 170N). As they only have to keep the robot on the wall but do not have to transmit traction forces, this force was calculated to be enough.

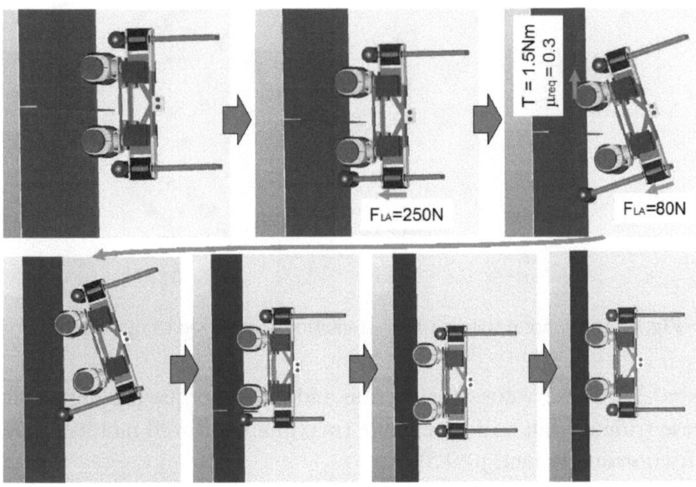


**Fig. 5.** Sequence of the robot passing a ridge

With realistic values for geometry, mass and magnetic force (m=10kg and $F_{mag}$=170N) the maximum required force in the linear actuator was calculated to 250N (allowed value: 300N). This force occurs, when the main wheel units are detached from the ground. The required maximal force during the rest of the movement (originating from the mass and the friction losses in the linear slide bearings) was also calculated and is much smaller (80N).

The required torque in each drive motor was calculated to 1.5Nm. The required friction coefficient was calculated to $\mu_{req}$=0.3 (measured value: $\mu \geq 0.5$). Thus the robot is safe against slipping.

## 6.2 Angular 135°-transitions

For passing the 135°-transitions, the robot has to roll over the obstacle with a torque distribution that is optimized to have the same required friction coefficient in all contact points (Fig. 6). In the worst case the required motor torque has to be 1.5Nm in the front wheels and 3.5Nm in the rear ones. The maximum required friction coefficient

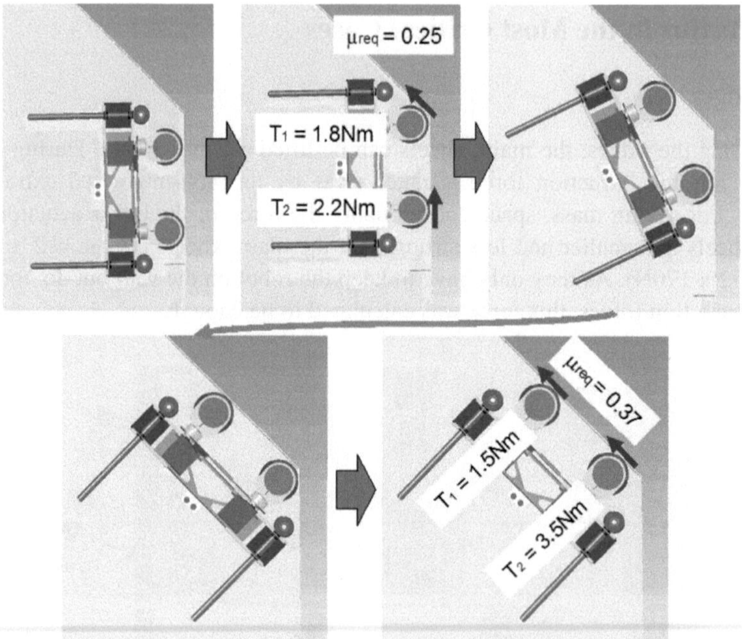


**Fig. 6.** The robot passing a 135°-transition (optimized torque distribution)

is then $\mu_{req}$=0.37. These values for friction and motor torque are higher than for the previous case (ridges), but still acceptable (maximum allowed motor torque: 4.5 Nm; measured friction coefficient: $\mu \geq 0.5$).

### 6.3 Turning

For switching from the vertical path to the horizontal one, the wheels turn on spot (Fig. 7). Each wheel is actuated by one steering motor. Here, assuming a friction

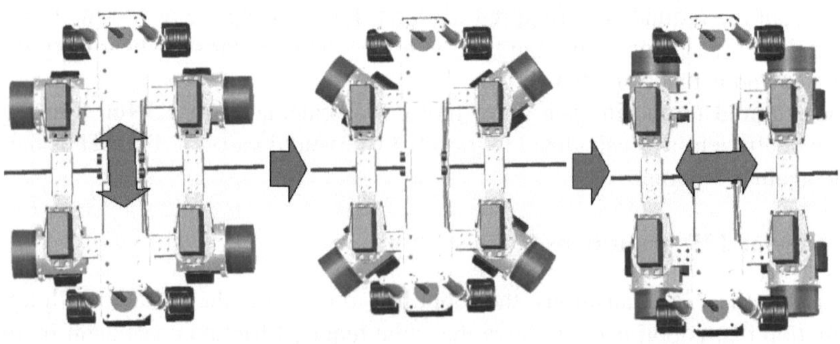

**Fig. 7.** The robot switching from vertical to horizontal paths

coefficient of $\mu$=0.8 (estimated maximum value) and a distance between the two wheels in each pair of d=20mm, the necessary steering torque resolves to 2.7Nm in each steering axis. As the planned actuators (servo motors) can provide 3.5Nm, turning the wheels is not a problem either.

The possibility to turn the whole robot instead of only the wheels was analyzed as well. In this configuration, fewer actuators would have been needed. However, it could not be used, as the allowed maneuvering space between the ridges and the 135°-transitions was not enough to perform this movement (Fig. 1).

## 7 Conclusion and Outlook

In this work, a novel solution to a magnetic wheeled climbing robot for thin and fragile surfaces was developed, with the capability to work in all inclinations of the surface, to pass two types of difficult obstacles and to turn on spot. Due to optimizations in structure and components, its mass will be around 10kg, which also allows for going in fragile environments without destroying them. The size will be around 200x200x300mm³ – small enough to fit well in the environment, but big enough to pass ridges of 40mm height. As only 10 actuators are used (instead of 26 in the most similar robot that was found in the state of the art [7]), cost, complexity and reliability will be limited to a reasonable range.

Due to measurements on already built parts (magnetic wheels and actuators), calculations and simulations, we could prove that all actuators will be strong enough and that the robot will not slip or lose contact at any time. Even for the most critical case – the risk of slipping in the 135°-transitions – the calculated required friction coefficient ($\mu_{required}$=0.37) lies significantly below the minimum value that was measured with a real wheel ($\mu_{measured} \geq 0.5$). Thus, it is assured that the concept will work well in the specified application – an environment that was impossible to be accessed by previous climbing robots.

However, when our industrial partners analyzed again the environment in these gas tanks, it resulted that the specifications that formed the base for developing this concept were not complete: On the largest part of the surface – except the last 500mm up to the neighboring walls (Fig. 1 grey area) – the sheet thickness is only 0.7mm instead of 1.5mm, resulting in a significant loss of magnetic force there. After doing more detailed stress tests, it also resulted that on these sheets the allowed mass should be only 5kg – instead of 10kg as it was specified before. However, on the thick surface even 20kg could be allowed. With these changed specifications, a safe operation could not be assured any more with the design described above.

Thus, for dealing with these changed specifications, the whole concept had to be adapted. The advanced design has already been started and will be realized soon. It is based on the same obstacle passing mechanism, but uses the additional idea to separate the whole system into two robots instead of going everywhere with only one big and complex one:

The bigger robot passes the obstacles and always stays on the thick surface where the magnetic attraction is strong and 20kg of robot mass is allowed. For the horizontal paths (no obstacles, but only 5 kg allowed), it sends out a smaller and simpler robot with only one degree of freedom. For the bigger one, the same obstacle-passing mechanism will be reused, on a structure that is optimized to the now changed requirements: No turning, but carrying the smaller robot.

## References


1. Balaguer, C., Virk, G., Armada, M.: Robot applications against gravity. In: Robotics & Automation Magazine, March 2006, vol. 13(1), pp. 5–6. IEEE, Los Alamitos (2006)
2. Slocum, H., Awtar, S., Hart, J.: Magnebots: Magnetic Wheels Based Overhead Transportation Concept. In: Conference paper from IFAC Mechatronics 2002, Berkeley (December 2002)
3. Guy, W.: Magnetic Wheel. U.S. Patent 3,690,393 (1972)
4. Sogi, T., Kawaguchi, Y., Morisaki, H., Ohkawa, K., Kai, N., Hayakawa, H.: Inspection robot for spherical storage tanks. In: Industrial Electronics Society. IECON 2000. 26th Annual Confjerence of the IEEE, 22-28 October, 2000, vol. 1, pp. 393–398 (2000)
5. Jireh Industries Ltd: Tripod – a Revolution in Remote Transportation. Industrial product, http://www.jireh-industries.com/
6. Kawaguchi, Y., Yoshida, I., Kurumatani, H., Kikuta, T., Yamada, Y.: Internal pipe inspection robot. In: Robotics and Automation. 1995 IEEE International Conference, vol. 1, 21-27 May, 1995, pp. 857–862 (1995)
7. Tâche, F., Fischer, W., Siegwart, R.: Adapted Magnetic Wheels for Robots Inspecting Narrow Pipes with Complex Shaped Structures. submitted for publication at IEEE/ASME International Conference on Advanced Intelligent Mechatronics (September 4-7, 2007)
8. Yukawa, T., Okano, H., Komatsubara, S.: Mechanisms for the movement of piping inspection robot with magnetic elements. In: Proceedings of the IASTED International Conference ROBOTICS AND APPLICATIONS, Cambridge, USA525390 (2005)
9. Grieco, J.C., Prieto, M., Armada, M., Gonzalez de Santos, P.: A six-legged climbing robot for high payloads. In: Proceedings of the 1998 IEEE International Conference on Control Applications, September 1-4, 1998, vol. 1, pp. 446–450 (1998)
10. Hirose, S., Tsutsumitake, H.: Disk Rover: A Wall-climbing Robot Using Permanent Magnetic Discs. In: Proceedings of the 1992 IEEE/RSJ International Conference on Intelligent Robots and Systems, July 7-10, 1992, vol. 3, pp. 2074–2079 (1992)
11. Shen, W.: Proposed wall climbing robot with permanent magnetic tracks for inspecting oil tanks. In: Mechatronics and Automation. 2005 IEEE International Conference, July 29-August 1, vol. 4, pp. 2072–2077 (2005)
12. Kitai, S., Tsuru, K., Hirose, S.: The proposal of swarm type wall climbing robot system "Anchor Climber" - design and examination of the adhering mobile unit. In: 2005 IEEE/RSJ International Conference on Intelligent Robots and Systems, August 2-6, 2005, pp. 475–480 (2005)
13. Xiao, J., Sadegh, A., Elliott, M., Calle, A., Persad, A., Chiu, H.M.: Design of Mobile Robots with Wall Climbing Capability (overlook on pneumatic robots). In: Advanced Intelligent Mechatronics. 2005 IEEE/ASME International Conference, pp. 438–443 (2005)
14. Pack, R.T., Christopher Jr., J.L., Kawamura, K.: A Rubbertuator-based structure-climbing inspection robot. In: Robotics and Automation. 1997 IEEE International Conference, April 20-25, 1997, vol. 3, pp. 1869–1874.
15. Longo, D., Muscato, G.: Alicia/sup 3/ climbing robot: a three-module robot for automatic wall inspection. In: Robotics & Automation Magazine, March 2006, vol. 13(1), pp. 42–50. IEEE, Los Alamitos (2006)

16. Aracil, R., Saltaren, R.J., Reinoso, O.: A climbing parallel robot: a robot to climb along tubular and metallic structures. Robotics & Automation Magazine 13(1), 16–22 (2006)
17. Yamamoto, S.: Development of inspection robot for nuclear power plant. In: 1992 IEEE International Conference on Robotics and Automation, May 12-14, 1992, vol. 2, pp. 1559–1566 (1992)
18. Balaguer, C., Gimenez, A., Huete, A.J., Sabatini, A.M., Topping, M., Bolmsjo, G.: The MATS robot: service climbing robot for personal assistance. In: Robotics & Automation Magazine, March 2006, vol. 13(1), pp. 51–58. IEEE, Los Alamitos (2006)
19. Lauria, M., Piguet, Y., Siegwart, R.: Octopus - An Autonomous Wheeled Climbing Robot. In: Proceedings of the Fifth International Conference on Climbing and Walking Robots (2002)

# Picking up an Unknown Object through Autonomous Modeling and Grasp Planning by a Mobile Manipulator

Kimitoshi Yamazaki[1], Masahiro Tomono[2], and Takashi Tsubouchi[3]

[1] The Univ. of Tokyo, 7-3-1 Hongo, Bunkyo-ku, Tokyo, Japan
yamazaki@jsk.t.u-tokyo.ac.jp
[2] Toyo Univ., 2100 Kujirai, Kawagoe, Saitama, Japan
tomono@eng.toyo.ac.jp
[3] Univ. of Tsukuba, 1-1-1 Tennodai, Tsukuba, Ibaraki, Japan
tsubo@roboken.esys.tsukuba.ac.jp

**Summary.** This paper describes a novel framework for object picking and carrying task by a mobile manipulator. Conventionally, researches on mobile manipulator cope well with object manipulation task with utilizing predefined knowledge or specific tools. So these researches have an essential problem that a new target object cannot be added without relatively many preparation. On the other hand, in our framework of a robot system, because the robot can create the 3D shape model of the object and can plan a grasp pose for the object autonomously under the condition that only the position of the object is given to the robot. Experimental results show the effectiveness of our robot system to be implemented our proposed method.

## 1 Introduction

Issues on mobile manipulators are being actively studied. such robots can be utilized in a large variety of tasks such as carrying objects from one place to another. Conventionally, there are several researches to cope well with such challenging tasks with utilizing predefined object models [4] [5], ID tags [1] or QR codes [3]. In these issues, how to give or what kind of information of the object for grasp management is focused on. So these researches have an essential problem that a new target object cannot be added without a heavy help by manually or a special tools.

We proposed approaches which differs from conventional researches. In our approach, there are two special policies for autonomous working. That is, an image based modeling method for creating object dense model [7] was developed. Moreover, a planning method for detecting grasp pose by inputting the model [8] was developed. Utilizing these functions related to object manipulation, it is expected that the robot will be capable of picking up and carrying task to unknown objects with giving only a little information by manual.

The measure purpose of this paper is to propose a solution for the picking up and carrying task based on automatic object modeling. We aim to develop

C. Laugier and R. Siegwart (Eds.): Field and Service Robotics, STAR 42, pp. 563–571, 2008.
springerlink.com 

a framework of a robot system which can handle small objects which generally exist in home or office environments with following conditions:

- No additional information on the object and environment are needed.
- No such information as the object shape or how to grasp it are needed.
- 3D grasping is performed through searching the best solution from many other feasible solutions.
- The object models and planning results can be easy reused if these have already acquired at once.

That is, the robot can add its target objects without constraint of shapes or types except one constraint that the object has some texture on its surface for object modeling in our framework.

## 2 Assumed Task and Research Issues

### 2.1 Assumed Task

Object carrying task by a mobile manipulator is performed according to the following procedure:

1. The object is put on a desk and the position of the object is given to the robot manually.
2. The robot moves to the desk, and
3. picks up the object by the manipulator, and
4. conveys the object and hands it to a person.

The item 3. are a main topic in this paper.

In this task, we assume that there are no constraint on object shape and no tags or marks on the object. In the meantime, we assume that relatively much natural texture can be found on the object surface, and the object has equivalent size that human can grasp it by one hand.

Because Jaw Gripper hand which is a compact and light weight hand has an advantage for a mobile manipulator, such hand is utilized for grasping the object. We assume that the finger tips contact with the object with some area and not with at a point.

### 2.2 Issues and Approach

There are two important challenges in our policy. The one is that the robot creates an object model autonomously. Utilizing multi-viewpoint images which are captured from a single camera mounted on a manipulator, the robot creates an object model as dense 3D points. The other is that the robot detects a grasp pose autonomously. Utilizing the automatic created model, grasp poses of the manipulator are planned automatically. That is, only the position information of the object is needed to the robot at manual instruction.

It is important that between modeling and grasp planning should be connected by a proper model representation. For achieving it, we apply a model representation named "oriented points". The object model is represented as 3D dense points that each point has normal information against object surface. This representation has an advantage to autonomous modeling because of its simple data structure. On the other hand, another advantage is found in grasp planning because the normal information enables to plan grasp poses effectively. In the meantime, sufficient countermeasures are needed against the shape error of the object model which is obtained from a series of images. As one of the countermeasure, the area contact in the grasp planning is performed to cancel the difference of the model shape and real shape of the object. The object modeling method is described in section 3, and the grasp planning method is described in section 4.

# 3 Object Modeling

## 3.1 Approach to Modeling

To acquire 3D model, SFM (structure from motion) method [7] is utilized. The 3D model means a reconstructed object surface which is represented by dense 3D points. Such model has simple data structure so that the model can be acquired by camera or LRF sensor autonomously without complex calculation or knowledge based data transform. Moreover, it can represent the various shapes of the objects.

We take an approach to create the object model by utilizing image streams which capture the object from multiple views. Because the object model created by our method can have both 3D shape information and photometric information, it is expected that the model will be highly reusable for the purpose of object recognition by the robot.

## 3.2 Modeling Procedure

Figure 1 shows modeling outline. The model is acquired according to following procedure: first, feature points in images are extracted and tracked from a small area which has strong intensity by using KLT-tracker [6]. From these points, object sparse model and camera poses are acquired by means of SFM. Next, the object dense shape is reconstructed from a close pair of images (Figure 1, (1)). Image pixels in each triangle patches (Figure 1, (2)) whose three vertices are adjoining and are common feature points in the pair of the images. At the same time, normal information is added to each reconstructed pixels (Figure 1, (3)) This process is applied to all the pairs of images and all the results for 3D locations of the triangular patches are integrated to obtain the 3D dense object model.

As a result, quite a number of points are reconstructed in online. Unfortunately, these are caused of consuming abundant memory because of the data redundancy. To solve the problem, we apply to translate the model to voxel representation by means of a method described in section 5.

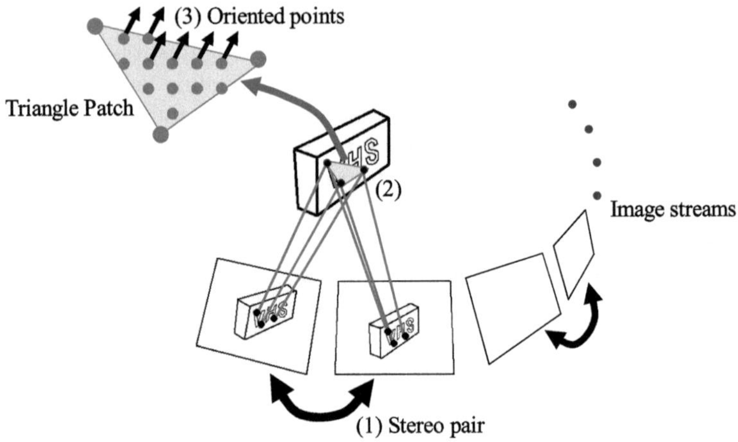


**Fig. 1.** Surface model reconstruction

## 4 Grasp Planning

The purpose of grasp planning is to find reasonable grasp pose with making use of the reconstructed dense 3D model on the spot.

### 4.1 Approach to Grasp Planning

Grasp planning in this study has two major issues: (1) how to plan a grasp pose efficiently from the 3D model, and (2) how to find a stable grasp from the 3D model, even though the 3D model has redundant data and errors in shape. We propose grasp planning to answer the above two issues.

For detecting the best grasp pose to pick up the object, we take an approach to evaluate by utilizing two criteria. The one is that contact area between the hand and the object model. The other is that a gravity balance depending on grasp position on the object.

### 4.2 Planning Method

In our planning, "stable grasp" is evaluated by the lowest sum total of three functions as follows [8]:

$$F = w_1 F_1(P_1, \mathbf{x}_h^o) + w_2 F_2(P_1, \mathbf{x}_h^o) + w_3 F_3(P_1, \mathbf{x}_h^o, \mathbf{x}_G, map) \quad (1)$$

where $P_1$ is a center point of finger plane on the hand. $\mathbf{x}_h^o$ is a hand pose in object coordinates, $\mathbf{x}_G$ is the pose of a mobile platform in world coordinates, and $map$ is environmental map which is given in advance. $w_i$ is a weight.

$F_1(\cdot)$ represents the function of contact area between the hand and the object. The evaluation value become smaller if the hand pose has more contact

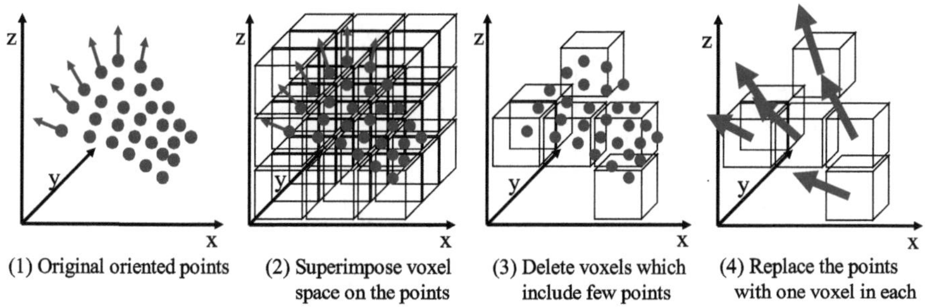


**Fig. 2.** Surface model reconstruction

area. $F_2(\cdot)$ represents the function of a gravity balance of the object. The evaluation value become small if a moment of the object is small. $F_3(\cdot)$ represents the function of the grasping pose. The evaluation value becomes small if manipulability related to grasp by the mobile manipulator is large [9]. The policy of grasp planning is to find $P_1$, $\mathbf{x}_h^o$ and $\mathbf{x}_G$ which minimize the function of $F$.

Oriented points touched to $P_1$ are selected in order, grasp poses are evaluated utilizing eq.(1). In this planning, it is important to reduce vain contact between finger and the object model and to select oriented points which can have good evaluation. The former can be achieved to restrict the direction of the contact by utilizing normal information of each point because the normal indicates the shape of object surface. The latter is described in next Section.

## 5 Model Representation for Efficient Implementation

To improve the performance of our framework, we take an approach to translate an object model into voxelized representation. A voxel in this paper means a small cubic bin, so an object model is represented by a group of voxels.

### 5.1 Voxel Representation

Oriented points described in section 3 has redundant data for grasp planning. By transforming from these points to voxelized model, redundant data can be reduced. Moreover, this process has effects of reducing the shape error of the object model. The size of voxel is set with 2mm or 5mm based on an allowable shape error in the grasp planning.

A new voxelized model is acquired as "thin" model according to the following procedure. At first, 3D-space is squared up fine cubic voxels (Figure 2, (2)). Voxels are deleted except special voxels which include several oriented points in it (Figure 2, (3)). Next, orientation component is averaged in each voxel (Figure 2, (4)). Finally, oriented points are acquired as each position is the center of each voxel and the orientation is an average value.

As it is necessary to yield the moment of inertia of the object for calculation of $F_2(\cdot)$, the model must be volumetric. For this purpose, once a voxel space

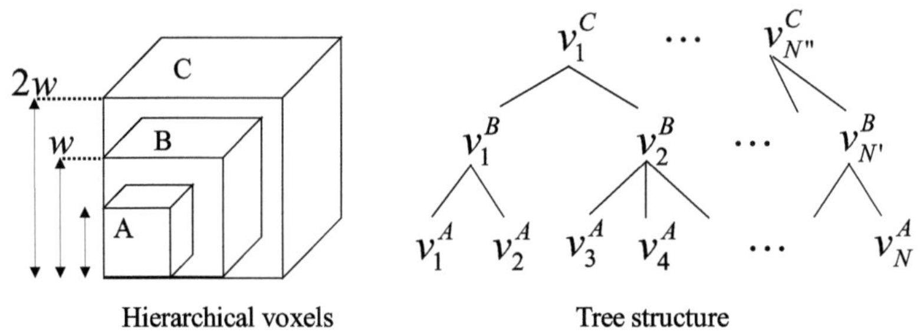


**Fig. 3.** Hierarchy model representation

including all the part of the model is defined. Then, the voxels of outside of the object are pruned away. The reminder of the voxels is the volumetric model of the object. This process consumes few time. So this is one of the advantage of voxelized model.

### 5.2 Hierarchical Data Representation

The method as mentioned in 5.1 can reduce the number of grasp pose searching. However, the searching has potential to be still capable of improving. We notice that there are somewhat points which obviously need not to be checked by eq.(1). From this reason, we propose hierarchical data representation which can exclude points before judging the quality of grasp pose. Using the new formed model, the searching can be performed at some parts of object model where will have rich contact area with fingers.

The hierarchical representation is similar to octree. Octree is often used for judging collision in the field of computer graphics. The transformation procedure is as follows: at first, initial voxels which construct original voxelized model are set hierarchical A. Next, other voxel space which is constructed $w$ times larger voxels than hierarchical A is superimposed on the voxels of hierarchical A. A new model is represented by the larger voxels which are set hierarchical B. In this processing, only voxels of hierarchical B are adopted if these voxels include over predefined threshold about the number of voxels of hierarchical A which have similar orientation. The same hierarchy construction is performed from hierarchical B to hierarchical C, too. As a result, one voxel of hierarchical C includes several voxels of hierarchical A as shown in Figure 3. Because these voxels of hierarchcal A exist in neighbor and has similar orientation, the area can be expected that it supplies rich contact area with finger.

In the grasp pose searching, voxels of hierarchical C are selected in order and evaluate by eq.(1) about inner voxels which belong to hierarchical A. This can achieve efficient searching with selecting only voxels which are guaranteed good evaluation result about contact area.

| Types of objects / Results | Rectangular box | Cup | Toy | Stapler | Ornament |
|---|---|---|---|---|---|
| Voxels | 807 | 986 | 966 | 934 | 1531 |
| Grasp pose Candidates | 64 | 37 | 49 | 48 | 58 |
| Processing time (A) [sec] | 0.96 | 1.01 | 0.94 | 0.86 | 1.36 |
| Processing time (B) [sec] | 9.3 | 6.3 | 6.7 | 7.2 | 16.3 |

**Fig. 4.** Trials to several shapes of objects

# 6 Experiments

## 6.1 Trials to Several Shapes of Objects

Object modeling and grasp planning are tried to several shapes and types of objects. Figure 4 shows the results. There are five objects which have relatively rich texture on its surface. The processing time of these object modeling took about 100 sec through 134th images in offline.

Notice that the planning results are not related to object shape. In these results, dozens of grasp poses can be found from each created models about 1 second (Pentium M, 2.0 GHz) as shown in processing times (A). Processing times (B) as shown in Figure 4 indicate the time of no utilizing hierarchy data representation describe in section 5.1. The more performance of the planning is achieved by utilizing our proposed method.

On the other hand, some problems are cleared up through this experiments. For instance, because an area where has no texture cannot reconstructed by our modeling method, grasp poses which touch to inner of the cup are not selected in grasp planning.

## 6.2 Implementation to Assumed Task

Figure 5 shows a mobile manipulator system to be used and an experimental environment. A small PET bottle is set on a desk as a target object. The goal of this experiment is to hand the object to a person who stands another place from the object position.

Environment map which includes the position of the object was given in advance. Moreover, the initial position of the robot and the position of the person were given in advance, too. In this condition, the robot planned its motion trajectory automatically by using artificial potential method [2].

Figure 6 shows experimental result. The robot (1)moved to the front of the desk, (2) created object model over capturing images, (3) grasped the object by applying the result of grasp planning, (4)moved to the front of a person, and

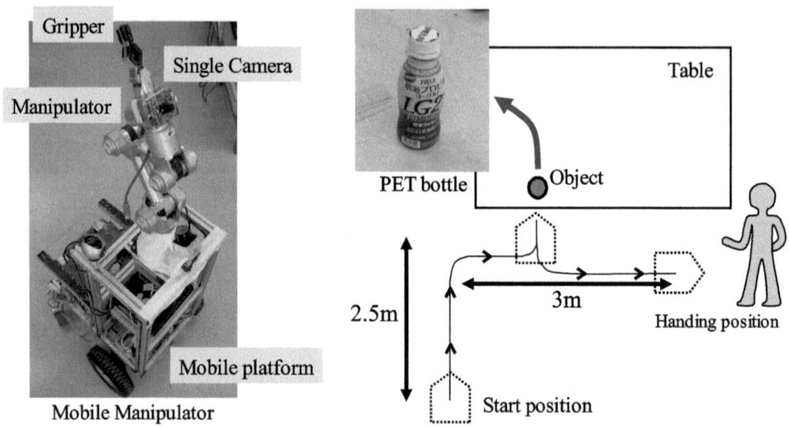


**Fig. 5.** Settings

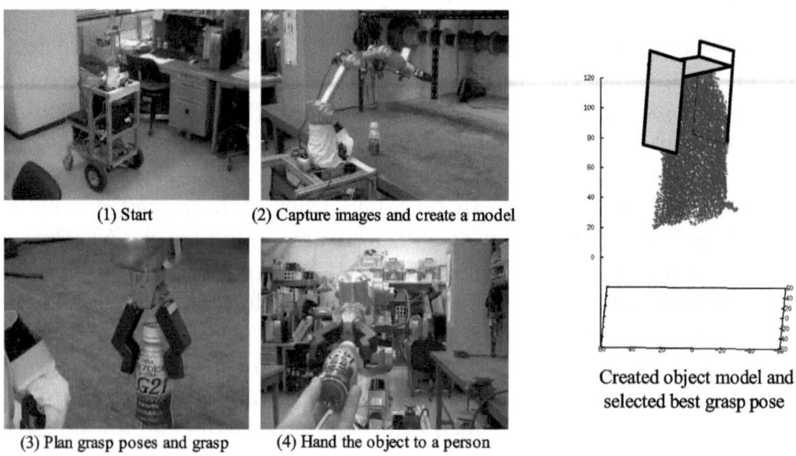


**Fig. 6.** Experimental result

handed it. All processing was performed by two laptop computers (Pentium M, 2.0GHz, Celeron, 1.0GHz) on the robot. As a result, image capturing and object modeling was needed about 360 sec and grasp planning was needed 0.65 sec in this experiment. In this experiment, final grasp pose is detected from only the sum of $F_1(\cdot)$ and $F_2(\cdot)$, and the inverse kinematics is calculated instead of $F_3(\cdot)$.

## 7 Conclusion

In this paper, a mobile manipulator system which can pick up an object in real environment autonomously was introduced. In this system, because automatic object modeling and grasp planning are applied with having proper data relation, object picking and carrying task can be achieved under the condition that only the position of the object is given to the robot.

## References


1. Chong, N.Y., Tanie, K.: Object Directive Manipulation Through RFID. In: Proc. of Int. Conf. on Control, Automation, and Systems, pp. 22–25 (2003)
2. Connolly, C., Burns, J., Weiss, R.: Path Planning using Laplace's Equation. In: Proc. of the IEEE Intl. Conf. on Robotics and Automation, pp. 2102–2106 (1990)
3. Katsuki, R., Ohta, J., Mizuta, T., Kito, T., Arai, T., Ueyama, T., Nishiyama, T.: Design of Artificial Marks to Determine 3D Pose By Monocular Vision. In: Proc. 2003 IEEE Int. Conf. Robotics and Automation, pp. 995–1000 (2003)
4. Miura, J., Shirai, Y., Shimada, N.: Development of a Personal Service Robot with User-Friendly Interfaces. In: 4th Int. Conf. on Field and Service Robotics, pp. 293–298 (2003)
5. Nagatani, K., Yuta, S.: Autonomous Mobile Robot Navigation Including Door Opening Behabior-System Integration of Mobile Manipulator to Adapt Real Environment-. In: Intl. Conf. on FSR, pp. 208–215 (1997)
6. Shi, J., Tomasi, C.: Good Features to Track. In: IEEE Computer Society Conf. on Computer Vision and Pattern Recognition, pp. 593–600 (1994)
7. Yamazaki, K., Tomono, M., Tsubouchi, T., Yuta, S.: 3D Object Modeling by a Camera mounted on a Mobile Robot. In: Proc. of the 2004 IEEE Int. Conf. Robotics and Automation (2004)
8. Yamazaki, K., Tomono, M., Tsubouchi, T., Yuta, S.: A Grasp Planning for Picking up an Unknown Object for a Mobile Manipulator. In: Proc. of the 2006 IEEE Int. Conf. Robotics and Automation (2006)
9. Yamazaki, K., Tomono, M., Tsubouchi, T., Yuta, S.: Motion Planning for a Mobile Manipulator Based on Joint Motions for Error Recovery. In: Proc. of the 2006 IEEE Int. Conf. Intelligent Robots and Systems, pp. 7–12 (2006)

# Part XII

# Path Planning

# An Architecture for Automated Driving in Urban Environments

Gang Chen and Thierry Fraichard

Inria Rhône-Alpes & LIG-CNRS Lab., Grenoble (FR)
firstname.lastname@inrialpes.fr

**Summary.** This paper presents a novel navigation architecture for automated car-like vehicles in urban environments. Motion safety is a critical issue in such environments given that they are partially known and highly dynamic with moving objects (other vehicles, pedestrians...). The main feature of the navigation architecture proposed is its ability to make *safe* motion decisions *in real-time*, thus taking into account the harsh constraints imposed by the type of environments considered. The architecture is based upon an efficient publish/subscribe middleware system that allows modularity in design and the easy integration of the key functional components required for autonomous navigation, namely perception, localisation, mapping, real-time motion planning and motion tracking. After an overall presentation of the architecture and its main modules, the paper focuses on the "motion" components of the architecture. Experimental results carried out on a simulated Cycab vehicle are presented.

## 1 Introduction

Autonomous navigation of intelligent vehicles in urban environments requires to solve a number of challenging problems in domains as different as perception, localization, environment modelling, reasoning and decision-making, control, *etc.* The problem of designing and integrating these functionalities within a single navigation architecture is of fundamental importance. Since Shakey's pioneering attempts at navigating around autonomously in the late sixties [1], the number and variety of autonomous navigation architectures that have been proposed is large (see [2]). From the motion determination perspective, these navigation architectures can be broadly classified into *deliberative* (*aka motion planning-based*) versus *reactive* approaches: deliberative approaches aim at computing a complete motion all the way to the goal using motion planning techniques, whereas reactive approaches determine the motion to be executed during the next time-step only. Deliberative approaches have to solve a motion planning problem [3]. They require a model of the environment as complete as possible and their intrinsic complexity is such that it may preclude their application in dynamic environments: indeed, the vehicle has only a limited time to determine its future course of action (by standing still for too long, it might be collided by one

C. Laugier and R. Siegwart (Eds.): Field and Service Robotics, STAR 42, pp. 575–584, 2008.
springerlink.com 

of the moving objects). Reactive approaches on the other hand can operate on-line using local sensor information: they can be used in any kind of environment whether unknown, changing or dynamic. This accounts for the large number of reactive approaches that have been developed over the years, *eg* [4, 5, 6, 7], *etc.* Most of today's reactive approaches however face a major challenge: as shown in [8], motion safety in dynamic environments is not guaranteed (in the sense that the vehicle may end up in a situation where a collision inevitably occurs at some point in the future).

The primary contribution of this paper is a motion planning module that takes into account these two constraints, namely the *real-time* and *safety* constraints. It is achieved thanks to the two concepts of Partial Motion Planning (PMP) [9] and Inevitable Collision States (ICS) [10]. PMP is a planning scheme that takes into account the real-time constraint explicitly by generating partial motions iteratively at each time-step. Like reactive decision schemes, PMP faces the safety issue. ICS are called upon to address this issue. An ICS is a state for which, no matter what the future trajectory followed by the vehicle is, a collision with an object eventually occurs. For obvious safety reasons, a vehicle should never end up in an ICS. By computing *ICS-free partial motion* at each time-step, the vehicle safety *can be guaranteed* in real-time.

The secondary contribution of this paper is a presentation of the navigation architecture hosting the PMP-ICS motion planner. It is based upon an efficient publish/subscribe middleware system named DDX [11] that allows modularity in design and the easy integration of the key functional components required for autonomous navigation, namely perception, localization, world modelling, motion planning and motion tracking.

The paper is organised as follows: an overall description of the navigation architecture is given first in §2. The three layers of this architecture are respectively detailed in §3–5. Experimental results are finally presented in §6.

## 2 Navigation Architecture Overview

The architecture presented in this paper is a DDX-based real-time modular architecture. DDX is a publish/subscribe middleware [11] that is used to provide the navigation modules with an abstract view of the Cycab, its sensors and its environment. Fig. 1 depicts the overall architecture. Below the DDX layer is the Cycab layer: it features the Cycab, its sensors and the environment either in simulation or for real. Above the DDX layer is the Navigation layer: it features the different key modules required for autonomous navigation: localization, world modelling, motion planning and motion tracking (these modules are detailed in §5). To complete the architecture, a Geographic Information System (GIS) is used to provide static information about the environment (road geometry and topology, traffic signs and traffic rules...). The dynamic information about the environment (other vehicles, pedestrians...) is computed by the Cycab layer

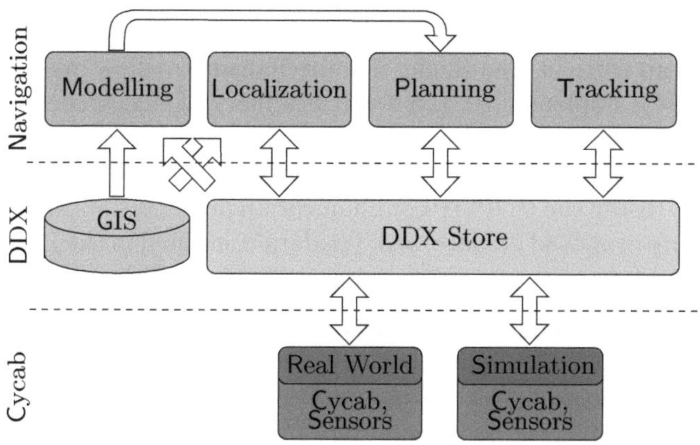


**Fig. 1.** Functional view of the navigation architecture

through sensor-data processing (or directly by the simulator). The following sections describe the Cycab, the DDX and the Navigation layers respectively.

## 3 Cycab Layer

The Cycab vehicle is a lightweight urban electric vehicle which is specifically designed for downtown areas (Fig.2). It integrates four motor wheels and a motorized mechanical jack for steering. Micro-controllers are used to control the motor-wheels and the steering mechanism. An embedded PC under Linux RTAI is used for the overall control of the vehicle. Two CAN (Controller Area Network) buses are used for communication between the different hardware components of the vehicle. It can be equipped with various sensors such as GPS, IMU, video cameras and range sensors (see http://www-lara.inria.fr/cycaba). A Cycab Simulator has been designed to facilitate the design and test of the automated driving algorithms that will be implemented on the real Cycab. It is based upon the MGengine simulation engine (http://mgengine.sourceforge.net) and permits the simulation of the Cycab vehicle, its sensors and its environment (see http://cycabtk.gforge.inria.fr).

**Fig. 2.** The Cycab vehicle (left) and the Cycab simulator (right)

## 4 DDX Layer

DDX provides an efficient communication mechanism to allow multiple processes to share data. It is implemented as a *Store*, *ie* a block of shared memory (possibly distributed over several computers), that is used to store shared information. The *Catalog* function is used to ensure the coherence of the information contained in the different stores (using the UDP/IP communication protocol). As far as the navigation architecture proposed is concerned, the data contained in the DDX store comprises four main data structures concerning either the Cycab or its environment:

- **Cycab:** information concerning the Cycab:
  - *CycabState*: encoder values, wheel velocities.
  - *CycabCommand*: actuator commands (speed, steering angle).
  - *CycabPose*: position and orientation of the Cycab.
- **Landmarks:** position of the observed landmarks, *ie* the salient features of the environment used for absolute localization.
- **Trajectory:** nominal trajectory that is to be executed by the vehicle (see §5.2). It is a sequence of (state, time) couples.
- **Moving objects:** list of the moving objects observed in the environment. Each moving object is characterized by it shape, position, orientation and velocity.

**Table 1.** DDX Store: who is using what?

| Module | Input/Output Data |
|---|---|
| Localization | Input: CycabPose, CycabState, GIS, Landmarks<br>Output: CycabPose |
| World Modelling | Input: GIS, Moving Objects<br>Output: Future Model |
| Motion Planning | Input: Future Model<br>Output: Trajectory |
| Motion Tracking | Input: CycabPose, Trajectory<br>Output: CycabCommand |

Table 1 summarizes how these data structures are used by the different navigation modules. The GIS data used by Localization is the list of the landmarks' position. World Modelling on the other hand gets the road geometry from GIS. Future Model is a description of the current state of the environment (fixed and moving objects) plus a prediction of the future motion of the moving objects (see §5.1).

## 5 Navigation Layer

This section presents the main modules used for autonomous navigation. There are four of them: *World Modelling* and *Localization* that deals with building a model of the vehicle's environment and localizing the vehicle inside this model. *Motion Planning* and *Motion Tracking* respectively deals with computing and executing a trajectory. *World Modelling* and *Motion Planning* are described in the next two sections. Due to lack of space, *Localization* and *Motion Tracking* are not described here, the reader is referred to [12] and [13] instead.

## 5.1 World Modelling

The primary purpose of the World Modelling module is to build a model of the environment of the vehicle that can be used for autonomous navigation purposes. Road-like environments feature both *fixed objects* (such as the road limits ) and *moving objects* (such as other vehicles and pedestrians) and the World Model must represent them both.

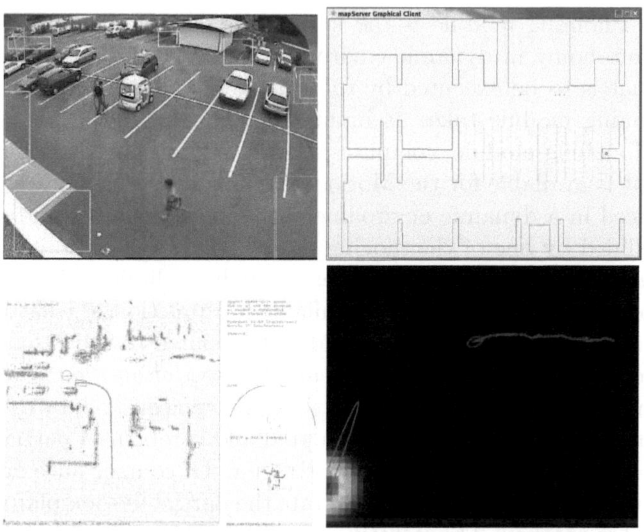

**Fig. 3.** View of the parking lot (top left), of the corresponding map (top right), of the moving objects tracking process (bottom left), and of the prediction of the future motion of one moving object (bottom right)

Static information about the environment, *eg* limits of the roadway, geometry of the obstacles, are assumed to be known a priori and made available in a Geographic Information System (GIS). For the purpose of navigation in urban environments, a 2D map of the environment (*ie* a set of polygonal obstacles) suffices (Fig. 3-top right). The structure of the roadway is also included in the GIS: it is represented as oriented lanes connected together in a network. This structure is exploited by the Motion Planning module to determine the route that is to be followed in order to reach a given goal.

In the architecture proposed, it is assumed that the information about the moving objects (linear and angular velocity/acceleration, *etc.*) is a direct output of a sensor-processing step corresponding to the different sensors used. In other words, the detection and tracking of the moving objects is performed in the Cycab layer, not the Navigation layer. Experiments have been carried out using video cameras and laser range sensor, and the tracking techniques presented in [14] and [9] (Fig. 3-bottom left).

Finally, noting that motion planning involves a certain degree of reasoning about the future, motion prediction regarding the future behaviour of the moving objects is required. In the architecture proposed, motion prediction relies upon the assumption that pedestrians and vehicles do not move randomly but follow typical "motion patterns" which may be learned and then used in a prediction phase (Fig. 3-bottom right; see [15] for details on the prediction scheme).

### 5.2 Motion Planning

The Motion Planning module is the key component of the solution proposed for motion autonomy in dynamic environments. Its purpose is to compute the trajectory that is to be executed by the vehicle in order to reach its goal. The Motion Planning module takes as input the model of the future provided by the World Modelling module, computes a trajectory and places it into the DDX store where it is available for the Motion Tracking module.

When placed in a dynamic environment, a vehicle cannot stand still since it might be collided by one of the moving objects. In a situation like this, a *real-time constraint* is imposed to the vehicle: it only has a limited time to determine its future course of action. The time available is a function of what is called the *dynamicity* of the environment which is directly related to the dynamics of both the moving objects and the robotic system. *Partial Motion Planning* (PMP) is a planning scheme that takes into account the real-time constraint explicitly: when the time available is over, PMP is interrupted and it returns a partial motion, *ie* a motion that may not necessarily reach the goal. Of course, since only a partial motion is computed, it is necessary to iterate the partial motion planning process until the goal is reached. The iterative nature of PMP is doubly required since the model of the future is based upon predictions whose validity duration is limited in most cases. An iterative planning scheme permits to take into account the unexpected changes of the environment by updating the predictions at a given frequency (which is also determined by the environment dynamicity). Let us consider the PMP cycle starting at time $t_i$, it comprises three steps:

(1) An updated model of the future is acquired (provided by the World Modelling module).
(2) The state-time space of the vehicle is searched using an incremental exploration method that builds a tree rooted at the state $s(t_{i+1})$ with $t_{i+1} = t_i + \delta_p$ where $\delta_p$ is the planning time available.
(3) At time $t_{i+1}$, the current cycle is over, the best partial trajectory $\Pi(i)$ of the tree is selected according to a given criterion (safety, length, *etc.*). It is discretized and placed into the DDX store.

PMP cycles until the last state of the planned trajectory reaches a neighbourhood of the goal state. An incremental search method is used to explore the state-space. It is based upon the Rapidly-Exploring Random Tree (RRT) technique [3] that incrementally expands a tree rooted at the start state. This method being incremental in nature, it can be interrupted at any time. Classically, RRT computes collision-free trajectories. In the approach proposed, the usual geometric collision-checker is replaced by an Inevitable Collision State-checker [16] that ensures that the vehicle will never end up in a situation eventually yielding a collision later in the future.

## 6 Experiments

The different modules of the navigation architecture are implemented in C++ under Linux. The DDX framework allows the different navigation functionalities/modules to be distributed over different computers. When the real Cycab is used, its embedded core software communicates with the rest of the application through wireless Ethernet. Experiments on autonomous navigation has

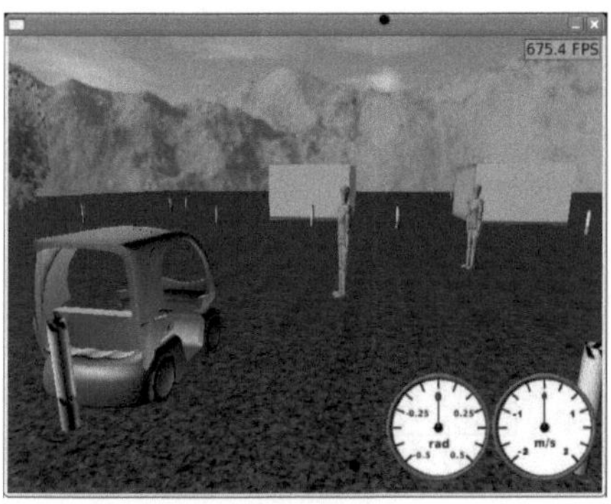


(a) Simple dynamic environment

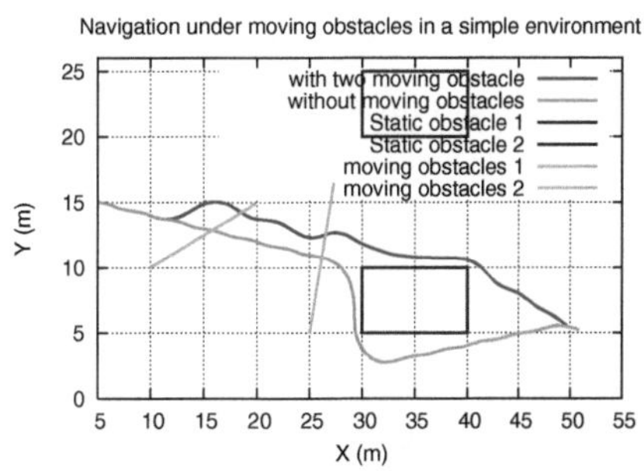


(b) Motion

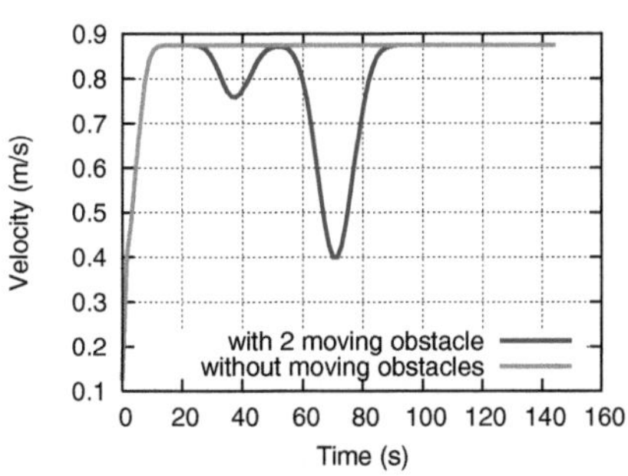


(c) Velocity

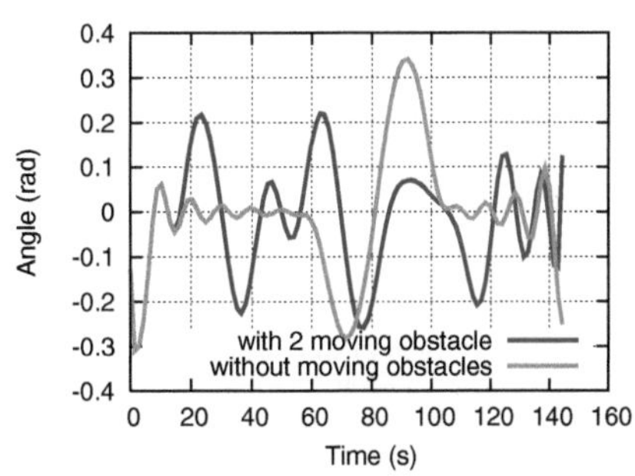


(d) Steering angle

**Fig. 4.** Experiment in a test environment

been carried out in simulation. So far, only the localization and tracking modules have been tested on the real Cycab. Autonomous navigation experiments with the real Cycab are underway. Simulations for two different scenarios have been studied to test the real-time planning performance of PMP. The PMP cycle time has been set to $1s$ and the motion tracking cycle time is $50ms$.

### 6.1 Test Environment

The first experiment is done in a dynamic 2D environment (size $60 \times 30m$), featuring two static rectangular objects and two moving disk objects (Fig. 4(a)). The starting and the goal pose of the Cycab are $(5, 15, 0)$ and $(50, 5, 0)$ respectively. The moving objects are programmed to move with a constant velocity (moving upwards). Fig. 4(b) shows the setup of this experiment and the output of the motion planning process. The safe motion planned for the Cycab is the red line passing between the two rectangular objects. In comparison, the green line passing below the two rectangular objects is the trajectory obtained when the moving objects are not present. Figs 4(c) and 4(d) depicts the velocity and steering angle profile along both trajectories. The difference between the two

(a) Parking environment

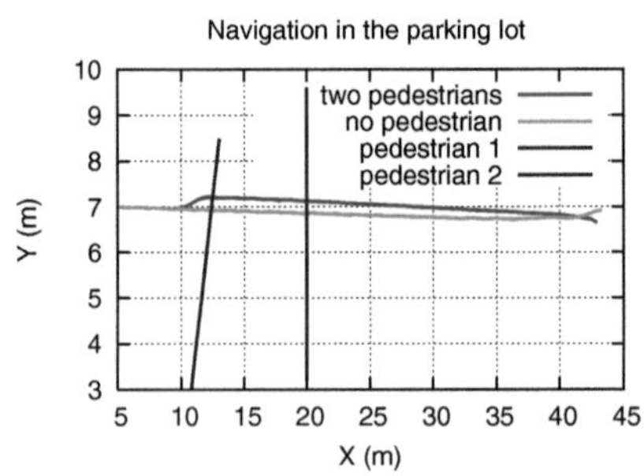


(b) Motion

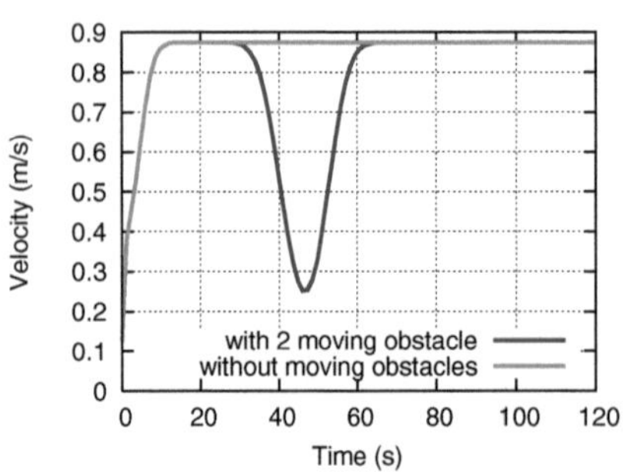


(c) Velocity

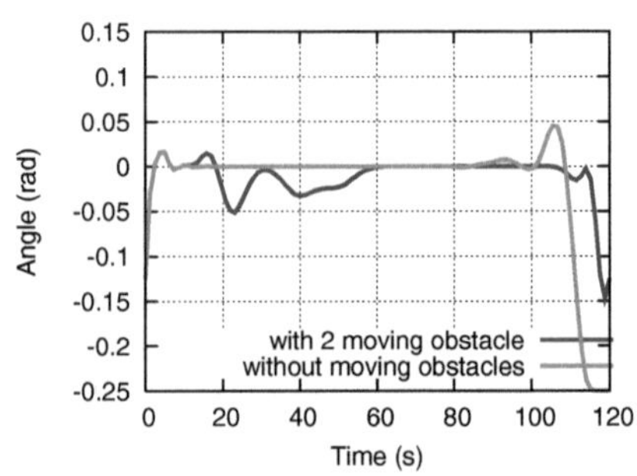


(d) Steering angle

**Fig. 5.** Experiment in the parking lot of Inria Rhône-Alpes trajectories is clearly due to the presence of the moving objects. Notice how the vehicle modify its course and slows down twice in order to give way to the moving objects.

### 6.2 Parking Lot of Inria Rhône-Alpes

The second experiment is done in a 2D model of the parking lot of Inria Rhône-Alpes (Fig. 5(a)). This environment is cluttered with twenty-six fixed objects and two pedestrians. From the motion planning point of view, this environment imposes more collision-avoidance constraints than the first scenario. The starting and the goal pose of the Cycab are $(5, 7, 0)$ and $(43, 7, 0.1)$ respectively. The pedestrians move upwards on the roadway. Fig. 5(b) shows the setup of this experiment and the output of the motion planning process. It also features the trajectory obtained when the moving objects are not present. Figs 5(c) and 5(d) depicts the velocity and steering angle profile along both trajectories. In this scenario, because of the extra constraint imposed by the fixed objects, the two trajectories are geometrically close (there is little room for manoeuvring). Most of the differences occur in the velocity profile.

## 7 Conclusions and Future Works

This paper has presented a novel navigation architecture for automated car-like vehicles in urban environments. The main feature of this navigation architecture is its ability to make *safe* motion decisions *in real-time*, thus taking into account the harsh constraints imposed by the type of environments considered (partially known with highly dynamic moving objects). Experimental results carried out on a simulation platform in a parking environment has demonstrated the ability to navigate safely in dynamic environments. Future works will include further experiments with a real vehicle.

## References

1. Nilsson, N.J.: Shakey the robot, AI Center, SRI International, Menlo Park, CA (US), Technical note 323 (April 1984)
2. Nourbakhsh, I.R., Siegwart, R.: Introduction to Autonomous Mobile Robots. MIT Press, Cambridge (2004)
3. Lavalle, S.: Planning Algorithms. Cambridge University Press, Cambridge (2006)
4. Khatib, O.: Real-time obstacle avoidance for manipulators and mobile robots. International Journal of Robotics Research 5(1) (1986)
5. Fox, D., Burgard, W., Thrun, S.: The dynamic window approach to collision avoidance. IEEE Robotics and Automation Magazine 4(1), 23–33 (1997)
6. Brock, O., Khatib, O.: High-speed navigation using the global dynamic window approach. In: IEEE Trans. on Robotics and Automation, Detroit, MI (US), pp. 341–346 (May 1999)
7. Minguez, J., Montano, L.: Nearness diagram (ND) navigation: collision avoidance in troublesome scenarios. IEEE Trans. on Robotics and Automation 20(1), 45–59 (2004)

8. Fraichard, T.: A short paper about safety. In: Proc. of the IEEE Int. Conf. on Robotics and Automation, Rome (IT) (April 2007)
9. Benenson, R., Petti, S., Fraichard, T., Parent, M.: Toward urban driverless vehicles. Int. Journal of Vehicle Autonomous Systems (August 2006) (accepted for publication)
10. Fraichard, T., Asama, H.: Inevitable collision states - a step towards safer robots? Advanced Robotics 18(10), 1001–1024 (2004)
11. Corke, P., Sikka, P., Roberts, J., Duff, E.: DDX: A distributed software architecture for robotic systems. In: Proc. of the Australian Conf. on Robotics and Automation, Canberra (AU) (December 2004)
12. Pradalier, C., Sekhavat, S.: Simultaneous localization and mapping using the geometric projection filter and correspondence graph matching. Advanced Robotics (2004)
13. Chen, G., Fraichard, T.: A real-time navigation architecture for automated vehicles in urban environments. In: Proc. of the IEEE Intelligent Vehicles Symposium, Istanbul (TR) (June 2007)
14. Vasquez, D., Fraichard, T.: A novel self organizing network to perform fast moving object extraction from video streams. In: Proc. of the IEEE-RSJ Int. Conf. on Intelligent Robots and Systems, Beijing (CN) (October 2006)
15. Vasquez, D., Fraichard, T., Aycard, O., Laugier, C.: Intentional motion on-line learning and prediction. Machine Vision and Applications (November 2006) (accepted for publication)
16. Parthasarathi, R., Fraichard, T.: An inevitable collision state-checker for a car-like vehicle. In: Proc. of the IEEE Int. Conf. on Robotics and Automation, Roma (IT) (April 2007)

# State Space Sampling of Feasible Motions for High Performance Mobile Robot Navigation in Highly Constrained Environments

Thomas M. Howard, Colin J. Green, and Alonzo Kelly

The Robotics Institute
Carnegie Mellon University
5000 Forbes Avenue, Pittsburgh, PA 15213, USA
{thoward,cjgreen,alonzo}@ri.cmu.edu

**Abstract.** Sampling in the space of controls or actions is a well-established method for ensuring feasible local motion plans. However, as mobile robots advance in performance and competence in complex outdoor environments, this classical motion planning technique ceases to be effective. When environmental constraints severely limit the space of acceptable motions or when global motion planning expresses strong preferences, a state space sampling strategy is more effective. While this has been clear for some time, the practical question is how to achieve it while also satisfying the severe constraints of vehicle dynamic feasibility. This paper presents an effective algorithm for state space sampling based on a model-based trajectory generation approach. This method enables high-speed navigation in highly constrained and/or partially known environments such as trails, roadways, and dense off-road obstacle fields.

## 1 Introduction

Outdoor mobile robot navigation is a challenging problem because environments are often complex and only partially known, dynamics can be difficult to predict accurately, and both planning time and computational resources are limited.

The dynamics of a vehicle can be modeled by a nonlinear differential equation of the form:

$$\dot{\mathbf{x}} = \mathbf{f}(\mathbf{x}, \mathbf{u}, t) \tag{1}$$

where $\mathbf{u}$ is called the input or control vector and $\mathbf{x}$ is the state vector and both are time-varying points in input and state spaces respectively. The complexity of such accurate models of mobility combined with the scale of outdoor mobile robot navigation leads to a difficult tradeoff between the computational demands of perceptive intelligence at the local level and deliberative intelligence at the global level. It is difficult to be both smart and fast when computation is limited.

A common approach to this problem is to use a multi-level motion planning strategy to generate behavior that is both intelligent and responsive. A large-scale motion plan is generated infrequently based on simplified dynamic models and coarse representations of the environment. This global planner is deliberative and it understands the gross topology of the environment. A finer-scale plan is generated more frequently by a local planner that utilizes higher fidelity dynamic models and finer resolution representations of the environment. The local planner provides obstacle

C. Laugier and R. Siegwart (Eds.): Field and Service Robotics, STAR 42, pp. 585–593, 2008.
springerlink.com 

avoidance and ensures dynamic feasibility in the near-term while the global planner provides high-level guidance (e.g. waypoint navigation). This paper deals with generating local motion planning search spaces which satisfy feasibility and environmental constraints while exploiting global guidance.

## 1.1 Motivation

This paper will address a difficult issue that arises in the context of the above two-tiered architecture, and indeed in differentially constrained motion planning in general. The formulation of the local planning problem involves constraints and utilities that are most conveniently expressed in two spaces:

- State Space: Those arising from the environment
- Control Space: Those arising from vehicle mobility

Traditional approaches to local motion planning involve searching alternatives expressed in control space because such alternatives are inherently feasible. Feasibility matters because commanding infeasible actions will lead to collisions or inability to execute other critical maneuvers. When most feasible motions are likely to satisfy environmental constraints, control space sampling is an effective approach. However, if the environment imposes severe limits on acceptable motions, this traditional approach does not work well.

Global guidance is expressed fundamentally in terms of a utility or constraint field over state space. For example, a global path is specified and minimum deviation is desirable, or a road lane is specified and its edges cannot be breached except in lane change maneuvers. Alternatively, a navigation function might associate with every point in state space its expected cost to the goal and its gradient specifies the preferred vehicle orientation at that point. Such guidance is highly valuable because the global path is less likely to contain obstacles, it is often an optimal global solution based on a coarse but informative map, and it may pass through narrow safe regions whose traversal is critical to reaching the goal.

Figure 1 illustrates how sampling in control space is a very poor approach under such conditions whereas state space sampling is ideal.

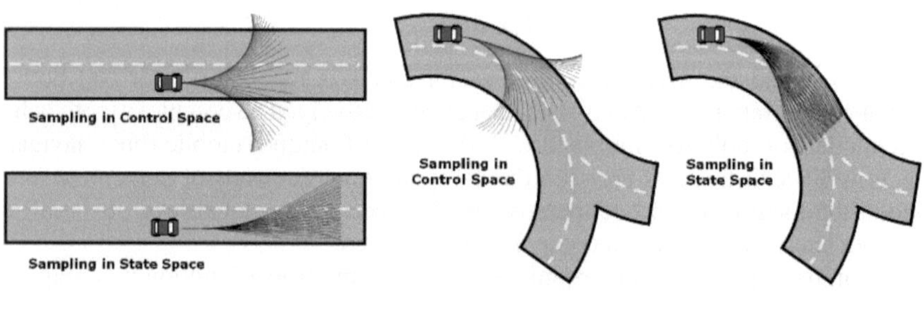


**Fig. 1.** Motion Planning in Constrained Environments. Search spaces generated by sampling in control space vs. state space are shown in environments that are highly constrained (e.g. road networks). Virtually all options generated by sampling in control space leave the lane or are oriented to do so eventually, whereas those generated by sampling in state space remain within the road network.

A second consideration is path sampling efficiency. All approaches to planning fundamentally consider a finite number of alternatives and pick the best one, but some sets of alternatives are better than others are. Separation matters because nearly identical paths will likely intersect the same obstacles and waste computation. We define a well-separated set of trajectories to be one that covers a majority of the state space with a minimum of overlap. When a global path is specified, it would be useful to control the distribution of trajectory end states to ensure that they are both near and oriented along the global path.

For example, consider the search space illustrated in Figure 2. It was generated by uniformly sampling in control space for different initial conditions. Notice that the trajectories are denser in the direction opposite to the initial curvature ($\kappa_0$) because the vehicle's maximum turning rate ($d\kappa/dt$) leads to the same output for several distinct inputs. Sampling in state space permits direct control over the spacing of the endpoints of these trajectories.

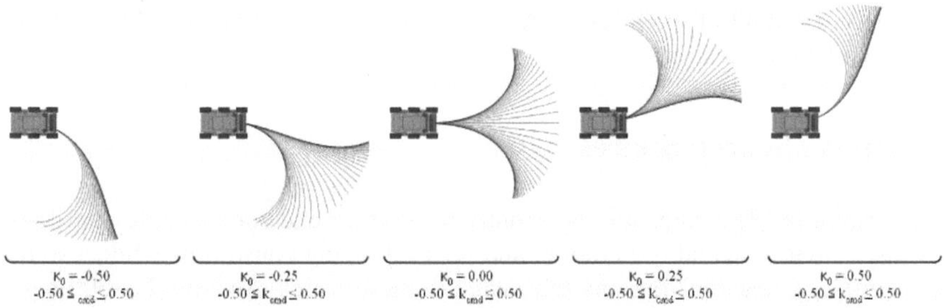


**Fig. 2.** Irregular Mapping from Control Space to State Space. Accurate dynamic simulations of a set of vehicle controls (constant curvature arcs uniformly sampled between ±0.50 radians/meter) are shown for different initial vehicle curvatures. The responses to the controls are not uniformly separated despite the uniform separation of the controls.

## 1.2 Related Work

Some of the earliest work in outdoor model-based mobile robot motion planning appears in [1], where the local motion planning search space is generated by sampling in the control space of curvature. Each control is passed through a vehicle dynamics model to estimate the response of the vehicle to the control. The shape of the response is highly dependent on the vehicle model and the initial vehicle state (curvatures and velocities).

Similar approaches have been adapted in a variety of other mobile robots [5]. The method presented in [2] samples trajectories around the arc that reacquires a target path, where "nudges" and "swerves" represent small and large lateral offsets (adjustments of the solution trajectory) respectively. Another method for generating local motion planning search space is egographs [7]. This approach generates a well-separated dynamically feasible search space for a limited set of initial vehicle states offline.

### 1.3 Technical Approach

An effective local planning search space would ideally be optimal, efficient, and robust. The search space would be optimal if it could maximally exploit global guidance, efficient if it could control path separation, and robust if it searched only feasible motions. Recent advances in real-time model-based trajectory generation have provided the capability to make progress towards achieving these goals. In [3] a general method is presented that computes control inputs that satisfy a pair of boundary states subject to the vehicle dynamics model, and in [4] we apply it to the problem of path following in the absence of obstacles.

This paper improves on [4] by generating a set of feasible actions by sampling in the surrounding state space. By using the model-based trajectory generation algorithm to generate motions between the current vehicle state and a set of terminal states on the boundary of the local motion planning search space, the state space sampling technique is superior in its efficiency (mean separation of trajectories) and robustness to initial conditions than its control space sampling counterparts. The approach in this paper differs from all of the prior work in its capacity to generate more expressive local motion planning search spaces.

## 2 Adaptive Search Spaces

The algorithm in [3] creates an opportunity to produce end state sample distributions that satisfy environmental, separation, and path following constraints while also being able to produce feasible motions that achieve these states. Section 2.1 outlines our general approach to structuring the search space adaptation algorithm while Section 2.2 discusses tuning the feasible set shape parameters for real vehicle applications.

### 2.1 Adaptive Search Space Design

A general description of state space sampling techniques is to generate a set of actions by solving for trajectories between n boundary state pairs ($\mathbf{x_p}$). The first state in each pair is the initial or current state of the vehicle ($\mathbf{x_i}$) while the second state is the terminal state ($\mathbf{x_f}$) which is reached at the end of each trajectory.

$$\mathbf{x_p} = \begin{bmatrix} \mathbf{x}_{i,0} & \mathbf{x}_{i,1} & \cdots & \mathbf{x}_{i,0} \\ \mathbf{x}_{f,0} & \mathbf{x}_{f,1} & \cdots & \mathbf{x}_{f,n} \end{bmatrix} \text{ where } \mathbf{x} = \begin{bmatrix} x & y & \psi & k & v & \ldots \end{bmatrix}^T \quad (2)$$

Note that this definition is independent of the method of generating the search space. We will utilize the model-based trajectory generation algorithm presented in [3] because of its ability to generate feasible actions between boundary state pairs in real-time, although other techniques could be substituted. Our approach is based on designing rules and parameters that define the shape of the outline of the trajetories and adjust the terminal states based on global guidance and initial vehicle state information.

For example, consider the example search spaces generated using this approach shown in Figure 3. Here, terminal state positions are selected uniformly at a horizon (constant radius) of 5.0 meters from the center of the robot between ±45 degrees from

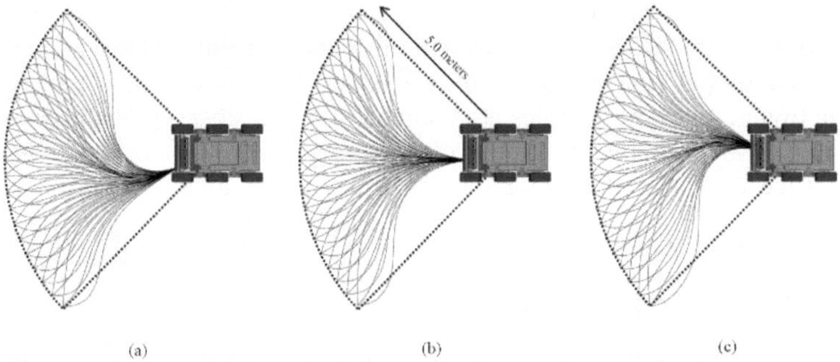


**Fig. 3.** Uniform Terminal State Sampling for Adaptive Search Space Generation with Varied Initial States. The ability to control the shape of the search space for three initial states: (a) turning left, (b) straight, and (c) turning right. Notice the sampling is more uniformly distributed than in Figure 2.

the forward central axis of the vehicle. Three headings at each terminal state position are considered, one aligned with the ray cast from the vehicle to the terminal state position and two others offset at a terminal heading adjustment angle (±45 degrees).

The result of this approach is a set of sophisticated maneuvers that are roughly equidistant, making this an efficient (non-redundant) search space. Notice that even in the face of varied initial vehicle states, the search space shape adapts to the dynamic constraints of the vehicle. The trajectories reach the same set of well-separated terminal states, in contrast to the results of control space sampling techniques (Figure 2).

## 2.2 Utilizing Global Guidance Information

On the assumption than deviation from global guidance is less likely to lead the goal, a local motion planning search space will improved if it biases its search to be most consistent with global guidance. We typically use a global planner which continu

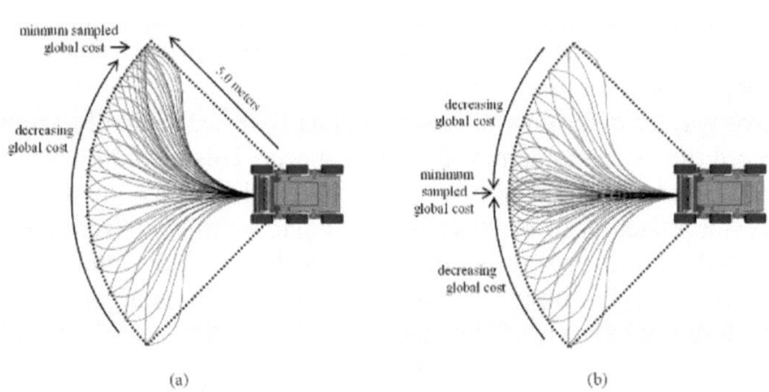


**Fig. 4.** Focused Terminal State Sampling for Adaptive Search Space Generation. The ability to exploit global guidance via state space sampling generates local motion planning search spaces that are denser in the direction of minimum global cost (and therefore more likely to be obstacle-free). Examples (a) and (b) shows the same setup from Figure 3 but focused in the direction of minimum global cost.

ously provides a navigation function (cost from any point to the goal) to the local planner. Given such information, it is better to sample terminal states at a higher density in lower global cost regions and at a lower density in higher cost regions, as shown in Figure 4. Some samples are retained in higher cost regions because the low cost regions produced by the global planner may not reflect actual dynamic constraints of the vehicle, and the global planner may not be able to react quickly to perceived obstacles.

### 2.3 Model-Based Trajectory Generation

To generate each motion, the real-time model-based trajectory generation algorithm in [3] is called for each boundary state pair in $\mathbf{x_p}$. A model-based approach is used to determine the actual command required to complete the necessary motion. The vehicle model is important because the mobile robot will execute these commands directly. The model is itself a tradeoff between speed (dynamics are computed tens of thousands of times per second) and fidelity (how well it can accurately predict the response to the inputs). We have found first-order models of linear and angular velocity response to be an effective tradeoff that encodes the major constraints of motion feasibility.

In general, model-based trajectory generation techniques can be used to solve for motions for arbitrary vehicle state constraints (positions, headings, curvatures, rates of curvatures, velocities, etc…). Here, the trajectory generator is only required to meet terminal position and heading constraints since the vehicle will likely never execute the entire motions between replanning cycles. In contrast to terminal state, the full initial vehicle state ($\mathbf{x_i}$) is necessary to initialize the vehicle dynamic model used by the algorithm.

### 2.4 Adaptive Search Envelope Determination

The envelope of the adaptive search space depends heavily of the dynamic limitations of the mobile robot. The horizon of the search space shape should adapt to the current speed because of braking distance and obstacle avoidance limitations. Likewise, the range of feasible terminal heading angles can vary heavily for vehicles with small curvature or angular acceleration limits.

An effective way to determine envelope is simply to exhaustively sample control space and record the extremes achieved in state space. This method would have to be employed off-line and it could not be used for vehicles whose models adjust over time. Another approach is to evaluate several aggressive maneuvers (such as max-turn right and left) to form rough bounds on reachable positions and headings at the horizon. This approach is simple, fast, and it can be used for adaptive vehicle models so we have preferred it over the off-line method.

## 3 Experiments and Experimental Results

In order to evaluate the performance of our approach, the adaptive search space algorithm presented in Section 2 has been tested in a series of simulation and field experiments. The simulation experiments are comparisons against arc-based (uniform sampling in control space) local motion planner search spaces in a series of randomized worlds. The field experiments consisted of a series of long distance missions. Sections 3.1, 3.2, 3.3, and 3.4 discuss the vehicle test platform, simulation setup, simulation results, and field experiment results and observations respectively.

### 3.1 Vehicle Test Platform

The vehicle used for this series of simulation and field experiments was Crusher (Figure 5), a six-wheel skid-steered mobile robot that is the current platform for the DARPA UGVC-Perceptor Integrated (UPI) project. The simulation uses the same simplified dynamic model of the vehicle (which represents the system delays and the linear and angular acceleration limits) as the one used by the planning system, so execution of commands is nearly ideal. The fidelity of the model is reduced in the field experiments, as effects such as wheel slip and sliding are not predicted as well on the vehicle as they are in simulation.

**Fig. 5.** Crusher. The current platform of the DARPA UGCV-PerceptOR Integrated (UPI) program.

### 3.2 Simulation Setup

To test the effectiveness of this adaptive search space on a statistically significant number of cases, a series of simulations were performed. The simulation consists of a rectangular vehicle driving through a randomly generated world while trying to reach the goal waypoint on the other side of the world. As the vehicle traverses the simulated world the global planner, Field D* [8], constantly provides the global path to goal and a field (navigation function) of path costs surrounding the vehicle. The cost computed from the convolution of the vehicle body along each trajectory, combined with the D* path cost at the end of that trajectory, comprises the score used to select the trajectory to follow. That trajectory is then followed during the next planning cycle, and a new trajectory selected at the end of that cycle. For simplicity, the planner is not allowed to perform any stopping or backup maneuvers and the vehicle speed is pre-computed based on obstacle cost and density.

The simulated worlds used are a set of three, connected real-valued obstacle fields filled with randomly positioned circular obstacles of different densities (a single example is shown in Figure 6). The target speeds in the high-, medium-, and low-density obstacle fields were 1.5 meters/second, 6.0 meters/second, and 9.0 meters/second respectively. The vehicle model in the simulator is the same vehicle model used by the trajectory generator (for the adaptive search space) and the vehicle motion simulator (for the arc-based search space). To ensure a fair comparison of search spaces in the test, both the adaptive and arc-based sets are comprised of the same number of trajectories (99). The arc-based set forward simulates 99 different constant curvatures whereas the adaptive set plans motions to three different headings at each of 33 different positions biased by the minimum D* path cost.

### 3.3 Simulation Results

One hundred simulated runs were performed for each of the local planner search spaces. For each pair of runs one simulated world is generated and both the arcs-based and adaptive search spaces were tested on that world. One of the representative simulation runs comparing the two approaches is shown in Figure 6. In all but one of the simulated worlds the adaptive search space outperformed the arc-based search space. On average, the overall path cost from the start to the end states was 24.8% lower for the adaptive search space, demonstrating that the improved flexibility and efficiency of the adaptive search space provides a performance advantage over constant curvature-control versions. In the best and worst cases, the adaptive search space had 79.5% lower and 20.0% higher overall path cost respectively. Lower overall path cost directly relates to probability of mission success as higher cost regions pose threats to vehicle survival.

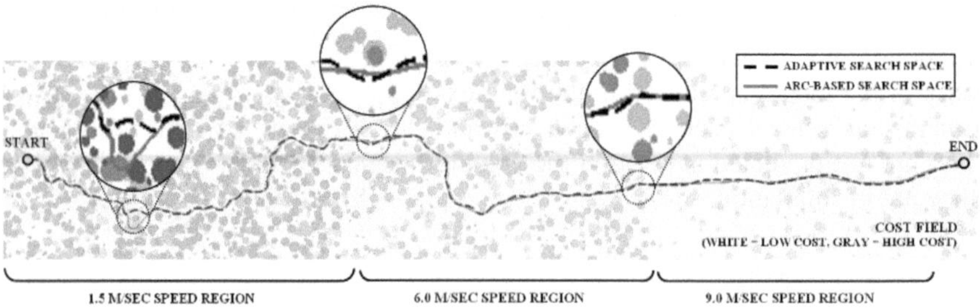


**Fig. 6.** Simulation of arc-based vs. adaptive search spaces. Each type of motion planning scheme (arc-based and adaptive search spaces) was evaluated in a representative simulated environment. In this particular example, the adaptive search space has a 43.0% lower cost along the path.

### 3.4 Field Experiment Results

Variants of the adaptive search space have been integrated and field tested on Crusher mobile robot (Figure 5). The algorithm demonstrated the capability to dodge obstacles and reacquire paths at speeds up to 12 meters/second in off-road environments. Figure 7 shows an example of the local planner utilizing the adaptive search space from field data. The minimum cost trajectory selected is a swerve maneuver that avoids the high cost regions in the front right and left of the robot. The new search space so consistently outperforms the older arc-based one that it has replaced it in all field tests.

**Fig. 7.** Field Experiments of Dynamically Adaptive Search Spaces

## 4 Conclusions and Future Work

We have leveraged our own recent work in model-based trajectory generation to create a capacity to navigate effectively in difficult environments while preserving inherent feasibility of local motion plans. Key aspects of the technique include explicit computation of the shape of the feasible set, and sampling strategies that adjust the distribution of samples to exploit global guidance. The benefits of the adaptive search space have been demonstrated in simulation and confirmed in off-road navigation.

Current and future work of this algorithm include more extensive field testing and refinement of the search space design to more optimally represent the set of all feasible motions of the vehicle at different speeds. Also, a more gradual decay of the fidelity of the dynamic model (as opposed to the drop-off between the local and global motion planner) will lead to robots that make better local decisions based on future dynamic limitations.

## Acknowledgement

This work was sponsored by the Defense Advanced Research Projects Agency (DARPA) under contract "Unmanned Ground Combat Vehicle – PerceptOR Integration" (contract number MDA972-01-9-0005). The views and conclusions contained in this document are those of the authors and should not be interpreted as representing the official policies, either expressed or implied, of the U.S. Government.

## References

[1] Kelly, A., Stentz, A.: Rough Terrain Autonomous Mobility - Part 2: An Active Vision, Predictive Control Approach. Autonomous Robots 5, 163–198 (1998)

[2] Thrun, S., Montemerlo, M., Dahlkamp, H., Stavens, D., Aron, A., Diebel, J., Fong, P., Gale, J., Halpenny, M., Goffmann, G., Lau, K., Oakley, C., Palatucci, M., Pratt, V., Stang, P.: Stanley: The Robot that Won the DARPA Grand Challenge. Journal of Field Robotics 23(9), 661–692 (2005)

[3] Howard, T.M., Kelly, A.: Optimal Rough Terrain Trajectory Generation for Wheeled Mobile Robots. International Journal of Robotics Research 26(2), 141–166 (2007)

[4] Howard, T.M., Knepper, R.A., Kelly, A.: Constrained Optimization Path Following of Wheeled Mobile Robots in Natural Terrain. In: International Symposium on Experimental Robotics 2006 (July 2006)

[5] Wettergreen, D., Tompkins, P., Urmson, C., Wagner, M., Whittaker, W.: Sun-Synchronous Robotics Exploration: Technical Description and Field Experimentation. International Journal of Robotics Research 24(1), 3–30 (2005)

[6] Green, C., Kelly, A.: Optimal Sampling in Space of Paths: Preliminary Results. Technical Report CMU-RI-TR-06-51. Robotics Institute, Carnegie Mellon University (November 2006)

[7] Lacze, A., Moscovitz, Y., DeClaris, N., Murphy, K.: Path planning for autonomous vehicles driving over rough terrain. In: Proceedings of the 1998 IEEE ISIC/CIRA/ISAS Joint Conference (1998)

[8] Ferguson, D., Stentz, A.: Field D*: An Interpolation-Based Path Planner and Replanner. In: Proceedings of the International Symposium on Robotics Research (ISRR) (October 2005)

# Author Index

# Springer Tracts in Advanced Robotics

**Edited by B. Siciliano, O. Khatib and F. Groen**

**Vol. 42:** Laugier, C.; Siegwart, R. (Eds.)
Field and Service Robotics
597 p. 2008 [978-3-540-75403-9]

**Vol. 41:** Milford, M.J.
Robot Navigation from Nature
194 p. 2008 [978-3-540-77519-5]

**Vol. 40:** Birglen, L.; Laliberté, T.; Gosselin, C.
Underactuated Robotic Hands
241 p. 2008 [978-3-540-77458-7]

**Vol. 39:** Khatib, O.; Kumar, V.; Rus, D. (Eds.)
Experimental Robotics
563 p. 2008 [978-3-540-77456-3]

**Vol. 38:** Jefferies, M.E.; Yeap, W.-K. (Eds.)
Robotics and Cognitive Approaches to Spatial Mapping
328 p. 2008 [978-3-540-75386-5]

**Vol. 37:** Ollero, A.; Maza, I. (Eds.)
Multiple Heterogeneous Unmanned Aerial Vehicles
233 p. 2007 [978-3-540-73957-9]

**Vol. 36:** Buehler, M.; Iagnemma, K.; Singh, S. (Eds.)
The 2005 DARPA Grand Challenge – The Great Robot Race
520 p. 2007 [978-3-540-73428-4]

**Vol. 35:** Laugier, C.; Chatila, R. (Eds.)
Autonomous Navigation in Dynamic Environments
169 p. 2007 [978-3-540-73421-5]

**Vol. 34:** Wisse, M.; van der Linde, R.Q.
Delft Pneumatic Bipeds
136 p. 2007 [978-3-540-72807-8]

**Vol. 33:** Kong, X.; Gosselin, C.
Type Synthesis of Parallel Mechanisms
272 p. 2007 [978-3-540-71989-2]

**Vol. 32:** Milutinović, D.; Lima, P.
Cells and Robots – Modeling and Control of Large-Size Agent Populations
130 p. 2007 [978-3-540-71981-6]

**Vol. 31:** Ferre, M.; Buss, M.; Aracil, R.; Melchiorri, C.; Balaguer C. (Eds.)
Advances in Telerobotics
500 p. 2007 [978-3-540-71363-0]

**Vol. 30:** Brugali, D. (Ed.)
Software Engineering for Experimental Robotics
490 p. 2007 [978-3-540-68949-2]

**Vol. 29:** Secchi, C.; Stramigioli, S.; Fantuzzi, C.
Control of Interactive Robotic Interfaces – A Port-Hamiltonian Approach
225 p. 2007 [978-3-540-49712-7]

**Vol. 28:** Thrun, S.; Brooks, R.; Durrant-Whyte, H. (Eds.)
Robotics Research – Results of the 12th International Symposium ISRR
602 p. 2007 [978-3-540-48110-2]

**Vol. 27:** Montemerlo, M.; Thrun, S.
FastSLAM – A Scalable Method for the Simultaneous Localization and Mapping Problem in Robotics
120 p. 2007 [978-3-540-46399-3]

**Vol. 26:** Taylor, G.; Kleeman, L.
Visual Perception and Robotic Manipulation – 3D Object Recognition, Tracking and Hand-Eye Coordination
218 p. 2007 [978-3-540-33454-5]

**Vol. 25:** Corke, P.; Sukkarieh, S. (Eds.)
Field and Service Robotics – Results of the 5th International Conference
580 p. 2006 [978-3-540-33452-1]

**Vol. 24:** Yuta, S.; Asama, H.; Thrun, S.; Prassler, E.; Tsubouchi, T. (Eds.)
Field and Service Robotics – Recent Advances in Research and Applications
550 p. 2006 [978-3-540-32801-8]

**Vol. 23:** Andrade-Cetto, J,; Sanfeliu, A.
Environment Learning for Indoor Mobile Robots – A Stochastic State Estimation Approach to Simultaneous Localization and Map Building
130 p. 2006 [978-3-540-32795-0]

**Vol. 22:** Christensen, H.I. (Ed.)
European Robotics Symposium 2006
209 p. 2006 [978-3-540-32688-5]

**Vol. 21:** Ang Jr., H.; Khatib, O. (Eds.)
Experimental Robotics IX – The 9th International Symposium on Experimental Robotics
618 p. 2006 [978-3-540-28816-9]

**Vol. 20:** Xu, Y.; Ou, Y.
Control of Single Wheel Robots
188 p. 2005 [978-3-540-28184-9]

**Vol. 19:** Lefebvre, T.; Bruyninckx, H.;
De Schutter, J. Nonlinear Kalman Filtering
for Force-Controlled Robot Tasks
280 p. 2005 [978-3-540-28023-1]

**Vol. 18:** Barbagli, F.; Prattichizzo, D.;
Salisbury, K. (Eds.)
Multi-point Interaction with Real
and Virtual Objects
281 p. 2005 [978-3-540-26036-3]

**Vol. 17:** Erdmann, M.; Hsu, D.; Overmars, M.;
van der Stappen, F.A (Eds.)
Algorithmic Foundations of Robotics VI
472 p. 2005 [978-3-540-25728-8]

**Vol. 16:** Cuesta, F.; Ollero, A.
Intelligent Mobile Robot Navigation
224 p. 2005 [978-3-540-23956-7]

**Vol. 15:** Dario, P.; Chatila R. (Eds.)
Robotics Research – The Eleventh
International Symposium
595 p. 2005 [978-3-540-23214-8]

**Vol. 14:** Prassler, E.; Lawitzky, G.; Stopp, A.;
Grunwald, G.; Hägele, M.; Dillmann, R.;
Iossifidis. I. (Eds.)
Advances in Human-Robot Interaction
414 p. 2005 [978-3-540-23211-7]

**Vol. 13:** Chung, W.
Nonholonomic Manipulators
115 p. 2004 [978-3-540-22108-1]

**Vol. 12:** Iagnemma K.; Dubowsky, S.
Mobile Robots in Rough Terrain –
Estimation, Motion Planning, and Control
with Application to Planetary Rovers
123 p. 2004 [978-3-540-21968-2]

**Vol. 11:** Kim, J.-H.; Kim, D.-H.; Kim, Y.-J.;
Seow, K.-T.
Soccer Robotics
353 p. 2004 [978-3-540-21859-3]

**Vol. 10:** Siciliano, B.; De Luca, A.; Melchiorri, C.;
Casalino, G. (Eds.)
Advances in Control of Articulated and
Mobile Robots
259 p. 2004 [978-3-540-20783-2]

**Vol. 9:** Yamane, K.
Simulating and Generating Motions
of Human Figures
176 p. 2004 [978-3-540-20317-9]

**Vol. 8:** Baeten, J.; De Schutter, J.
Integrated Visual Servoing and Force
Control – The Task Frame Approach
198 p. 2004 [978-3-540-40475-0]

**Vol. 7:** Boissonnat, J.-D.; Burdick, J.;
Goldberg, K.; Hutchinson, S. (Eds.)
Algorithmic Foundations of Robotics V
577 p. 2004 [978-3-540-40476-7]

**Vol. 6:** Jarvis, R.A.; Zelinsky, A. (Eds.)
Robotics Research – The Tenth
International Symposium
580 p. 2003 [978-3-540-00550-6]

**Vol. 5:** Siciliano, B.; Dario, P. (Eds.)
Experimental Robotics VIII – Proceedings
of the 8th International Symposium ISER02
685 p. 2003 [978-3-540-00305-2]

**Vol. 4:** Bicchi, A.; Christensen, H.I.;
Prattichizzo, D. (Eds.)
Control Problems in Robotics
296 p. 2003 [978-3-540-00251-2]

**Vol. 3:** Natale, C.
Interaction Control of Robot Manipulators –
Six-degrees-of-freedom Tasks
120 p. 2003 [978-3-540-00159-1]

**Vol. 2:** Antonelli, G.
Underwater Robots – Motion and Force Control
of Vehicle-Manipulator Systems
268 p. 2006 [978-3-540-31752-4]

**Vol. 1:** Caccavale, F.; Villani, L. (Eds.)
Fault Diagnosis and Fault Tolerance for
Mechatronic Systems – Recent Advances
191 p. 2003 [978-3-540-44159-5]

Springer Nature Customer Service Center GmbH
Europaplatz 3, 69115 Heidelberg, Germany

MIX
Papier aus verantwortungsvollen Quellen
Paper from responsible sources
FSC® C105338

If you have any concerns about our products,
you can contact us on
**ProductSafety@springernature.com**

In case Publisher is established outside the EU,
the EU authorized representative is:
**Springer Nature Customer Service Center GmbH**
**Europaplatz 3, 69115 Heidelberg, Germany**

Printed by Libri Plureos GmbH
in Hamburg, Germany